Theory of Preliminary Test and Stein-Type Estimation with Applications

WILEY SERIES IN PROBABILITY AND STATISTICS

Established by WALTER A. SHEWHART and SAMUEL S. WILKS

A complete list of the titles in this series appears at the end of this volume.

Theory of Preliminary Test and Stein-Type Estimation with Applications

A. K. Md. Ehsanes Saleh
Carleton University
Ottawa, Canada

WILEY-INTERSCIENCE
A JOHN WILEY & SONS, INC., PUBLICATION

Published by John Wiley & Sons, Inc., Hoboken, New Jersey.
Published simultaneously in Canada.

For general information on our other products and services or for technical support, please contact our Customer Care Department within the United States at (800) 762-2974, outside the United States at (317) 572-3993 or fax (317) 572-4002.

Wiley also publishes its books in a variety of electronic formats. Some content that appears in print may not be available in electronic formats. For more information about Wiley products, visit our web site at www.wiley.com.

Library of Congress Cataloging-in-Publication Data:

Saleh, A. K. Md. Ehsanes.
Theory of preliminary test and Stein-type estimation with applications. / A.K. Md. Ehsanes Saleh.
p. cm.
Includes bibliographical references and index.
ISBN: 978-0-471-56375-4

1. Parameter estimation. 2. Regression analysis. 3. Bayesian statistical decision theory. I. Title.

QA276.8.S257 2006

2005050196

10 9 8 7 6 5 4 3 2 1

To SHAHIDARA, my wife

Contents

List of Figures

List of Tables

Preface

The estimation of parameters of a model with "uncertain prior information" on the parameters of interest began with Bancroft (1944, **Annals of Mathematical Statistics** 15: 190–204). Bancroft introduced the "preliminary test estimation" on the classical front, although Bayesian methods already existed. But the real breakthrough came when Stein (1956, **Proceedings of the Third Berkeley Symposium** 1, pp. 197–206) and James and Stein (1961, **Proceedings of the Fourth Berkeley Symposium on Mathematical Statistics and Probability** 2, pp. 361–379) proved that the **sample mean** in a multivariate normal model is not admissible, under a quadratic loss, for dimensions more than two. This discovery, known as the Stein paradox, basically undermined the criteria, like unbiasedness, equivariance, and the MLE properties, that were the backbone of statistical estimation theory. Stein's theory does not only improve the point estimators of location parameters (mean, regression parameters, etc.) but also improves the traditional confidence sets by sets recentered at the Stein estimators. Preliminary test estimation of locations generally performs poorly and is not uniformly better than the MLEs/LSEs, but Stein (1964, **Annals of the Institute of Statistical Mathematics** 16, pp. 155–160) showed that, in the case of variance estimation, preliminary test estimators are uniformly better than the usual variance estimators. In addition, the preliminary test approach improves upon the standard James-Stein estimators and leads to admissible (generalized Bayes) estimators.

Due to the immense impact of Stein's approach on estimation, scores of technical papers appeared in the literature in various areas of applications. The book **The Statistical Implications of Pre-test and Stein-Rule Estimations in Econometrics** by Judge and Bock (1978), was the first attempt directed toward the econometrics area of applications. Recently, another book, **Improving Efficiency by Shrinkage: The James-Stein and Ridge Regression Estimators** by Marvin H. J. Gruber (1998) has directed the estimation toward student/teacher as descriptive applied research material. So far, there has been no book of statistics on the topic of preliminary test and Stein-type estimation. This led me to develop such a book that would be beneficial to graduate students, researchers, and teachers alike.

The mathematics and statistics prerequisites for this book are modest. It is assumed that the reader has had a basic course in mathematical statistics, preferably having used textbooks such as **An Introduction to Probability Theory and Mathematical Statistics** by Rohatgi and Saleh (2001, Wiley), **Statistical Inference** by Casella and Berger (1990, Brooks/Cole), **Mathematical Statistics: basic ideas and selected topics** by Bickel and Doksum (2001, Prentice Hall) or **Introduction to Mathematical Statistics** by Hogg, McKean and Craig (2005, Prentice Hall).

The aim of the book is to provide a clear and balanced introduction to preliminary test and Stein-type estimation theory for students/teachers of statistics programs. We start with the two-sample problem of pooling means in a general setup, in order to help the readers understand the results and calculations that are being used in every step of the development for the solution to the problem. Then, from chapter to chapter, we raise the level of discussion.

The book consists of twelve chapters. Chapter 1 gives the introduction to the preliminary test and Stein-type estimation, with details on each model, including simple linear regression, the ANOVA model, the parallelism model, the multiple regression model, and the multivariate simple linear model. Chapter 2 contains some basic results used in the book. Chapter 3 covers the introduction to the preliminary test estimation, with applications to the one-sample problems, two-sample problems, and simple linear models. Also included is Stein's variance estimation. Chapter 4 gives an introduction to Stein-type estimation. In addition, the chapter shows that the Stein-type estimation of the mean vector is an empirical Bayes estimator, and the Sclove, Morris, and Radhkrishnan (1972, **Annals of Mathematical Statistics** 43: 1481–1490) modification is carried out to obtain the positive-rule Stein estimator. Some asymptotic results are given for nonnormal situations. To complete the chapter, recentered confidence sets are studied for the mean vector.

Chapter 5 contains similar details of the ANOVA model, and Chapter 6, deals with the parallelism models. Chapter 7 contains the discussion of the multiple regression model with subset restrictions, while Chapter 8 is concerned with the multiple regression model with stochastic subset restrictions. The topic of ridge regression introduced by Horel and Kennard (1970, **Technometrics** 12: 55–67) is presented in Chapter 9. Chapter 10 contains the multiple regression models with autocorrelated errors. Finally, in Chapters 11 and 12 we discuss the one-sample to two-sample problems and simple linear models in a multivariate setup and some discrete data models.

The contents of the book may be covered in one semester. Various problems are included to enhance the knowledge of application of the theory.

I am grateful to my wife, Shahidara Saleh, for her support over our 54 years of marriage. While I remained a student all my life, she bore more than her share of chores and provided me with words of encouragement. My granddaughters, Jasmine and Sarah, watched me work on the manuscript, prepared tea at frequent intervals, and as a result, Sarah became interested in mathematics and statistics.

Thanks are due to Professor E.O. Kreyszig, who read the manuscript thoroughly, with great interest, and edited the early versions page by page to bring it to this final form. I am also grateful to Professor Vijay K. Rohatgi for checking some portions of the book for clarity.

I wish to thank Gillian Murray for diligently TEXing the manuscript; without her help, the book could not have been completed. Also, special thanks are due to H. M. Kim, who read the typed version and produced all the graphs and tables in the book, as well as to Ann Woodside, Drs. Bashir Ullah, M. Oulde Haye, Patrick Farrell, Shahjahan Khan, and B. M. Golam Kibria, who read various chapters and made many good suggestions to improve readability. My thanks are due to Wiley & Sons, Inc. and to authors for the copyright permissions on several books which helped me to finish the project.

Finally, I am grateful to NSERC for supporting my research for the last three decades, the outcome of which is, besides extensive publications, twelve Ph.D. theses, several postdoctorals and M.Sc. students, and this book.

A. K. Md. Ehsanes Saleh

Chapter 1

Introduction

Outline

In problems of statistical inference, the use of prior information on some or all of the parameters in a statistical model usually leads to improved inference procedures. For some or all parameters of the model, the *prior* information may be known or may be *uncertain*. The known prior information is generally incorporated in the model in the form of a *constraint*, giving rise to a *restricted* model. The analysis of such restricted models leads to an improved statistical procedure when such constraints hold compared to the *unrestricted* model. The estimators (or tests) resulting from a restricted (unrestricted) model are the restricted (unrestricted) estimators (or tests) of the parameters of the model. The validity and efficiency of restricted model analysis retains its properties over the restricted parameter space induced by the constraints, while the same holds for the unrestricted model analysis over the entire parameter space. Therefore, the results of an analysis of the restricted/unrestricted model need to weigh the loss of efficiency against the validity of the constraints in order to choose between two extreme inference techniques. Choosing restricted estimation/tests may be justified when full confidence may be placed in the prior information.

When we encounter statistical models with "uncertain prior information" in the form of a constraint on the parameters, we may consider the "uncertain constraints" as a "nuisance parameter" in the statistical analysis of the full model. The preferred procedure to eliminate this uncertainty of the prior information in a model is to use the Fisher's recipe. This procedure calls for elimination of the nuisance parameter by a so-called preliminary test. The **restricted** or **unrestricted** model is chosen based on the validity of the uncertain prior information. The result is a compromise inference procedure between the two extremes.

Bancroft (1944, 1964, 1965) was first to implement the idea of *preliminary test estimation* (PTE) in an ANOVA setup to analyze the effect of *preliminary test* (PT) on the estimation of variance. The idea was borrowed from a suggestion in Snedecor's (1938) book on testing differences between two means after testing the equality of variances that are unknown. If the test accepts the equality of two variances, then the usual t-test is to be used with pooled estimate of variance; otherwise, it falls in the category of Behrens'-Fisher problem. As it became clear, the preliminary test estimation procedure depends heavily on the level of significance and yields only two extreme choices as an estimator. The natural question is what would be the optimum size of the level of significance in a preliminary test estimation procedure. Later Han and Bancroft (1968) provided a maximin procedure to determine the size of a preliminary test for the estimation of a mean in a two-sample problem to obtain an optimum "preliminary test estimator" of the mean.

Mosteller (1948) looked at the problem from a Bayesian point of view and almost gave the idea of empirical Bayes estimation. Kitagawa (1963) followed the idea of preliminary test on successive occasions and obtained the distribution and moments of the preliminary test estimators in the two-sample problem. Bozivich, Bancroft, and Hartley (1956) studied the properties of the size of the preliminary test as well as the power of the test followed by a preliminary test. Bennett (1952, 1956), Huntsberger (1955), and Asano and Sato (1962) followed the application of preliminary test approach in various directions. Cohen (1965) proved that preliminary test estimators are *inadmissible* with no alternative suggestion for a superior estimator.

Charles Stein (1955, 1956; see also 1981) followed by James and Stein (1961) discovered a paradoxical statistic which undermined a century and a half of work on estimation theory, going back to Gauss, Legendre, Fisher, and Rao. The result was that the sample mean vector is inadmissible under a quadratic loss for the mean vector of a p-dimensional multivariate normal distribution for $p \geq 3$. After 25 years of resistance to Stein's ideas, the paradox has diminished and Stein's ideas are being incorporated into many applied and theoretical statistics. Not only that, Efron (1995) in his article "The Statistical Century" lists "empirical Bayes and James-Stein estimations" as important future topics among a dozen of his favorite topics in applied statistics.

In traditional statistical methodology, usually the *sample mean* is appropriate to estimate the *population mean*, as it can be proved that no other estimation rule is uniformly better than the sample mean. The paradoxical element of Stein's work is that it contradicts this basic law of statistical theory. This assertion is shown by the baseball data analysis of Efron and Morris (1973, 1975). For example, if we have *three or more* baseball players and we are interested in predicting future batting averages for each of the players, then the statistician who uses Stein's method can expect to predict the future batting more accurately than the predicted values based on the separate batting average of the players. Stein's method can be illustrated based on the famous baseball data of Efron and Morris (1975). Table 1.1 shows the first 45 batting averages for 18 players.

Table 1.1.1 Batting Averages of 18 Players

Player	**1**	**2**	**3**	**4**	**5**	**6**	**7**	**8**	**9**
Average after 45	0.400	0.478	0.356	0.333	0.311	0.311	0.289	0.267	0.244
Predicted average	0.294	0.289	0.245	0.280	0.275	0.275	0.270	0.266	0.261
Player	**10**	**11**	**12**	**13**	**14**	**15**	**16**	**17**	**18**
Average after 45	0.244	0.222	0.222	0.222	0.222	0.222	0.200	0.178	0.156
Predicted average	0.261	0.256	0.256	0.256	0.256	0.256	0.252	0.241	0.243

Source: Adapted from table 1 of Efron and Morris (1975)

The first step in applying Stein's method is to compute the **grand mean** of all these averages, which is $\hat{\theta} = 0.2654$ in this case. The idea is to **shrink** the 18 averages toward the **grand mean**, which is a reasonable quantity under consideration. The next step is to determine the weighted squared distance between the 18 averages in the table and 18 grand means values, which becomes 19.045, and calculate the adjusting factor $c = [1 - 15/19.045] = 0.212$, which is called the shrinkage constant. Here 15 equals the degrees of freedom (d.f.) of the weighted distance minus two. Then the formula for the ith player's predicted average is obtained by the *James-Stein's formula* for the ith predicted average $\hat{\theta}_i^{\mathrm{JS}}$ of the ith player, namely, $\hat{\theta}_i^{\mathrm{JS}} = \hat{\theta} + c(\tilde{\theta}_i - \hat{\theta})$, $i = 1, 2, \ldots, 18$. This means that the first player has the predicted average

$$\hat{\theta}_1^{\mathrm{S}} = 0.265 + 0.212(0.400 - 0.265) = 0.294,$$

and the 18th player has the average

$$\hat{\theta}_{18}^{\mathrm{S}} = 0.265 + 0.212(0.156 - 0.265) = 0.243\,.$$

Actual batting averages were 0.346 and 0.200, respectively. The preliminary test estimator in this case is 0.265 at the 0.15 level of significance. The usual sample mean vector has the following properties: (1) best unbiased estimator, (2) best equivariant estimator, (3) maximum likelihood estimator, and (4) minimax estimator. Yet the performance is inferior in terms of the expected squared loss relative to the James-Stein estimator.

Observe from Figure 1.1 that the predicted batting averages are more concentrated around the grand mean compared to the original batting averages. Stein's method pulls the original estimates toward the grand mean.

Berger (1980a, 1985) has a detailed account of some of these developments, mostly related to the classical multinormal and specific types of exponential

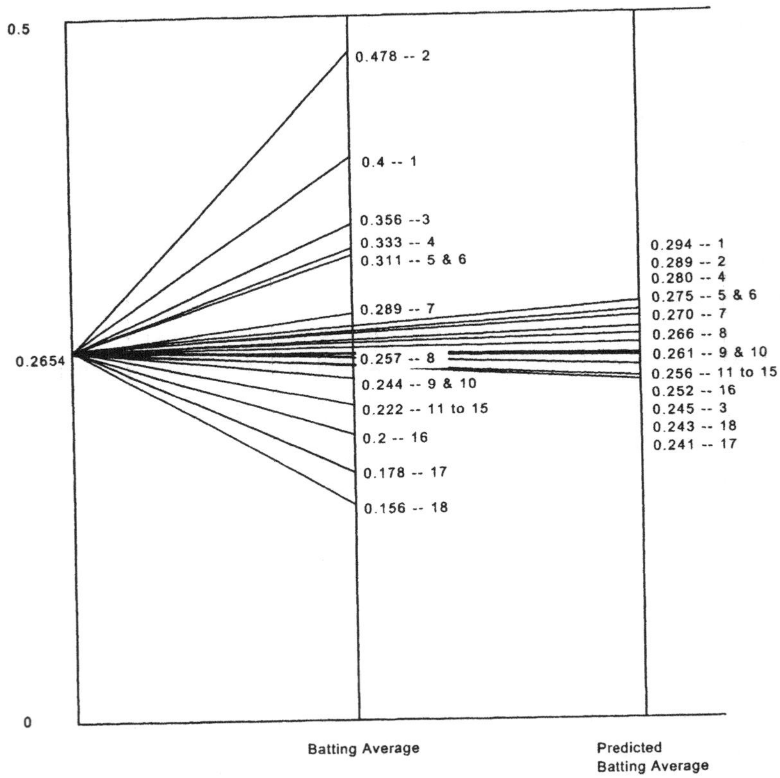

Figure 1.1 Display of predicted batting averages based on Stein's formula

families of distributions. For discrete multivariate Stein-type estimators we refer the reader to Ghosh, Hwang, and Tsui (1983).

Estimators that uniformly improve over standard (least squares or maximum likelihood estimators among others) estimators are usually termed Stein-type estimators in honour of Charles Stein. Basically, Stein-type estimators may be obtained via a decision-theoretic approach (risk-difference approach) due to Stein or the empirical Bayes approach (EBE) due to Robbins (1955) and Efron and Morris (1972, 1973, 1977) (see also Casella, 1985). "Empirical Bayes" is a term that has many meanings, reflecting different approaches to solving problems. Efron and Morris (1973) applied this method to justify the Stein-type estimation. The general approach of empirical Bayes estimation may be depicted as

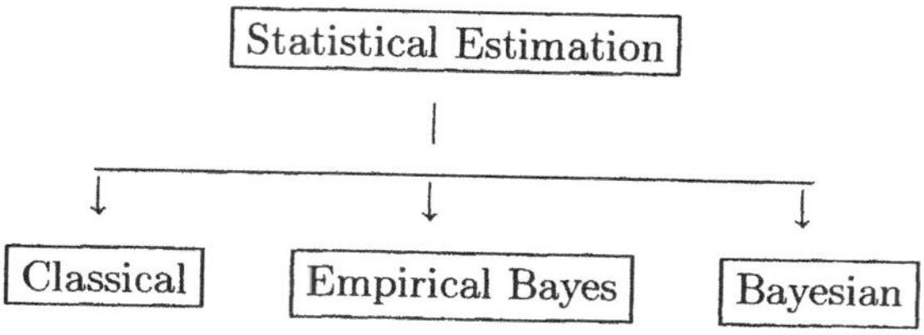

The empirical Bayes method sits in between the classical (Neyman-Pearson) and the Bayesian approach borrowing pieces from each.

However, Saleh and Sen (1978–1986), Sen and Saleh (1979–1987), and Casella (1985) pointed out that the Stein-type estimators involve an appropriate test-statistics for testing the adequacy of an uncertain prior information on the parameter space, which is incorporated in the actual formulation of the estimator. Stein-type estimators adjust the *unrestricted* estimator (for the full model) by an amount of the difference between unrestricted and restricted estimators scaled by the adjusted test-statistics for the uncertain prior information. Generally, these test-statistics are the normalized distance between the unrestricted and restricted estimators and follow a noncentral chi-square or F-distribution with appropriate degrees of freedom. The risk or MSE of Stein-type estimators depends on the noncentrality parameter, which reflects the distance between the full model and the restricted model.

The preliminary test estimators (PTE) are the precursors of the Stein-type estimators, and a careful look at the PTE reveals that a simple replacement of the indicator function by a multiple of the reciprocal of the test-statistics defines the Stein-type estimators. This procedure will be known in this book as the preliminary test approach (or alternatively, we may call it a **quasi-empirical Bayes approach**) to shrinkage estimators. This procedure has the far-fetched implication that it combines the two diverse areas of robust estimation and shrinkage estimation. On this ground Saleh and Sen (1978–1986) took this path to broaden the improvement of standard rank estimators to shrinkage estimators.

Among many benefits of Stein-type estimators is that they can be used as a predictive tool for events of interest. For example, they could be used to predict batting averages of individual baseball players from a collection of many batting averages of several baseball players in a season (see Efron, 1975), or meta-analysis of several case control studies, or forecasting, based on time-series data-fitting autoregressive models among others. A broad area known as "small area estimation" has evolved based on the Stein-type estimation for predicting in a small area where the sample size is very small. The methods described in this book provide the most useful techniques for combining data from various sources, such as in meta-analysis and data mining and in many other modern topics.

1.1 Objective of This Book

The main object of the book is to provide the readers with the mathematical and statistical foundations of the preliminary test and shrinkage estimation. This will enlarge the scope for use in applied areas. Basically we will present analytical properties of the five estimators: unrestricted estimator (UE), restricted estimator (RE), preliminary test estimator (PTE), Stein-type shrinkage estimator (SE), and positive-rule shrinkage estimator (PRSE) to show how they can be applied to the standard models, such as linear models, parallelism models, ANOVA models, multiple regression model, and their variations, multivariate models, and discrete data models. We usually assume the error distribution in these models to be normal. We also include nonparametric methods and models with *nonnormal* errors and provide asymptotic results.

1.2 Statistical Decision Principle

The basic elements of a statistical decision theory are the following:
(1) A random experiment represented by the triplet $(\mathfrak{X}, \mathcal{B}, \mathcal{P}_\theta)$ where $\mathfrak{X}$ is the sample space of the experiment, $\mathcal{B}$ is the σ-algebra generated by $\mathfrak{X}$, and $\mathcal{P}_\theta$ is the class of probability distributions over $\mathfrak{X}$ indexed by the parameter $\theta \in \Omega$, where Ω is the *parameter space.*
(2) A nonempty set $\mathcal{A}$ of actions at the disposal of a statistician.
(3) A loss function, $L(\theta, a)$, that is a real-valued function defined on $\Omega \times \mathcal{A}$.

These elements are related as shown:

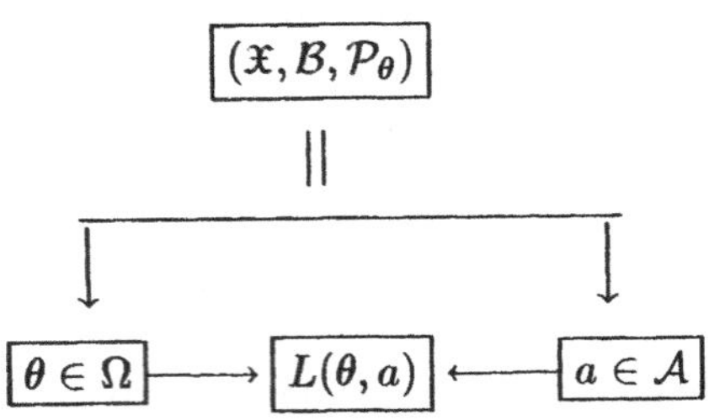

If $\mathbf{X}_n = \mathbf{x}_n = (x_1, \ldots, x_n)'$ is observed, the statistician takes the action $a = \delta(\mathbf{x}_n) \in \mathcal{A}$, where δ is a decision rule belonging to a nonempty set $\mathcal{D}$.

If $\mathcal{A} = \Omega$, then the problem is one of estimation. If $\mathcal{A} = \{a_1, a_2\}$, then any decision rule δ partitions the sample space $\mathfrak{X}$ into two disjoint sets C and $\overline{C}$. If $\delta(\mathbf{x}) = a_1 \in C$ or $\delta(\mathbf{x}) = a_2 \in \overline{C}$, then action a_1 or a_2 is taken, respectively. This problem relates to testing of hypotheses, where a_1 relates to rejecting a null–hypothesis and a_2 relates to accepting the null hypothesis.

The loss incurred during the process of choosing elements of $a \in \mathcal{A}$ is defined by $L(\theta, a)$. The action a is the result based on a function $\delta \in \mathcal{D}$ such that $\delta(\mathbf{x}) = a$. Then the loss $L(\theta, \delta(\mathbf{x}))$ is indicative of the action $a \in \mathcal{A}$. It is then a random variable. The average of the loss $L(\theta, \delta(\mathbf{x}))$ is called the risk $R(\theta, \delta) = E_\theta\{L(\theta, \delta(\mathbf{x}))\} = \int L(\theta, \delta(\mathbf{x}))dP_\theta(\mathbf{x})$ associated with the use of the decision rule δ from the class $\mathcal{D}$.

The fundamental problem of statistical decision theory is the selection of $\delta \in \mathcal{D}$ such that the risk $R(\theta, \delta)$ is minimum for all $\delta \in \mathcal{D}$. We thus need to specify some criterion to compare various decision rules.

Definition 1.1. A decision rule δ^* is said to dominate another decision rule δ^0 if for all $\theta \in \Omega$, $R(\theta, \delta^*) \leq R(\theta, \delta^0)$. If, in addition, for some $\theta \in \Omega$, $R(\theta, \delta^*) < R(\theta, \delta^0)$, then δ^* strictly dominates δ^0.

Definition 1.2. A decision rule δ^* is said to be admissible if for any other decision rule, $\delta^0 \in \mathcal{D}$, $R(\theta, \delta^*) \leq R(\theta, \delta^0)$ for all $\theta \in \Omega$.

The criteria used to limit the class of $\mathcal{D}$ of decision rules are "unbiasedness" "invariance" and "sufficiency". Another possible criterion is the "minimax" principle.

Definition 1.3. A decision rule $\delta^*(x)$ is minimax within the class $\mathcal{D}$, if $\delta^* \in \mathcal{D}$ and

$$\sup_{\theta \in \Omega} R(\theta, \delta^*) \leq \sup_{\theta \in \Omega} R(\theta, \delta) \quad \text{for all } \delta \in \mathcal{D}.$$

The Bayes principle with respect to a prior distribution $\pi(\theta)$ is frequently used to choose a minimax decision rule. Then the Bayes risk of the decision rule $\delta(\mathbf{x})$ is given by

$$\rho(\pi, \delta) = E_\pi[R(\theta, \delta)].$$

Definition 1.4. A decision rule δ^B is called a Bayes decision rule if

$$\rho(\pi, \delta^B) = \inf_{\delta \in \mathcal{D}} \rho(\pi, \delta).$$

For the estimation of $\theta \in \Omega \subset R^1$ based on the loss function $(\theta - \delta)^2$, a Bayes decision rule δ^B is given by

$$\delta^B(x) = E[\theta | X = x] = \int \theta h(\theta | x) d\theta;$$

this is called the Bayes estimator of θ, where $h(\theta|x)$ is the posterior distribution of $\boldsymbol{\theta}$ given $X = x$. For details the readers are referred to Rohatgi and Saleh (2001).

One of many methods of determining minimax decision rules is to find a Bayes rule based on the prior $\pi(\theta)$, say, $\delta^B(x)$ such that the risk, $R(\theta, \delta^B) =$ constant for all $\theta \in \Omega$; then δ^B is a minimax decision rule. See Rohatgi and Saleh (2001, thm 8.8.2) in connection with the estimation problem.

1.3 Quadratic Loss Function

In this book we evaluate the estimators in a decision-theoretic setup based on quadratic loss functions among other loss functions such as (1) the absolute deviation function $|\theta - \delta(x)|$ and (2) the Linex function $[\exp(a\delta(x)) - b\delta(x) - 1]$ for analytical simplicity.

Let $\boldsymbol{\delta} = (\delta_1, \dots, \delta_p)'$ be a vector of decision rules based on a sample of size n and let $\boldsymbol{\theta} = (\theta_1, \dots, \theta_p)' \in R^p$. Then we have a weighted square loss function defined by

$$\begin{aligned} L(\boldsymbol{\theta}, \boldsymbol{\delta}) &= n(\boldsymbol{\delta}(\mathbf{x}) - \boldsymbol{\theta})'\mathbf{W}(\boldsymbol{\delta}(\mathbf{x}) - \boldsymbol{\theta}) \\ &= n||\boldsymbol{\delta}(\mathbf{x}) - \boldsymbol{\theta}||^2_{\mathbf{W}}, \end{aligned}$$

where $\mathbf{W}$ is a positive semi-definite (p.s.d.) matrix of weights. As a special case we also discuss unweighted quadratic loss

$$L(\boldsymbol{\theta}, \boldsymbol{\delta}) = n||\boldsymbol{\delta}(\mathbf{x}) - \boldsymbol{\theta}||^2,$$

where $\mathbf{W} = \mathbf{I}_p$ is the identity matrix.

Then the corresponding risk function is given by

$$\begin{aligned} R(\boldsymbol{\theta}, \boldsymbol{\delta}) &= nE_{\boldsymbol{\theta}}||\boldsymbol{\delta}(\mathbf{x}) - \boldsymbol{\theta}||^2_{\mathbf{W}} \\ &= nE_{\boldsymbol{\theta}}||\boldsymbol{\delta}(\mathbf{x}) - \boldsymbol{\theta}||^2, \quad \text{if } \mathbf{W} = \mathbf{I}_p. \end{aligned}$$

Under the weighted quadratic loss function, the Bayes estimator is the mean of the posteriori distribution given $X = x$ corresponding to the prior distribution $\pi(\theta)$. If $\mathbf{W} = \mathbf{I}_p$, then we obtain an unbiased estimator with minimum variance. If we formulate the estimator as unbiased and linear, then the results coincide with the Gauss-Markoff theory having minimum risk or best linear unbiased estimator (BLUE).

Another quantity we may consider is the mean square error (MSE) matrix defined by

$$\mathbf{M} = nE_{\boldsymbol{\theta}}[(\boldsymbol{\delta}(\mathbf{x}) - \boldsymbol{\theta})(\boldsymbol{\delta}(\mathbf{x}) - \boldsymbol{\theta})'].$$

Note that the weighted quadratic risk function is $R(\boldsymbol{\theta}, \boldsymbol{\delta}) = \text{tr}(\mathbf{MW})$. The idea of using the criterion based on $\mathbf{M}$ is to find estimators whose mean square error is small, while the $R(\boldsymbol{\theta}, \boldsymbol{\delta})$ minimizes the weighted sum of the mean squares.

An estimator $\boldsymbol{\delta}$ is as good as $\boldsymbol{\delta}^0$ if

$$R(\boldsymbol{\ell}'\boldsymbol{\theta}, \boldsymbol{\ell}'\boldsymbol{\delta}) \leq R(\boldsymbol{\ell}'\boldsymbol{\theta}, \boldsymbol{\ell}'\boldsymbol{\delta}^0) \quad \text{for all } \boldsymbol{\theta} \in \mathcal{R}^p$$

for a given nonzero vector $\boldsymbol{\ell} = (\ell_1, \ldots, \ell_p)'$. In other words, we must have

$$\boldsymbol{\ell}' \left\{ nE_{\boldsymbol{\theta}}[(\boldsymbol{\delta}^0(x) - \boldsymbol{\theta})(\boldsymbol{\delta}^0(\mathbf{x}) - \boldsymbol{\theta})'] - nE_{\boldsymbol{\theta}}[(\boldsymbol{\delta}(\mathbf{x}) - \boldsymbol{\theta})(\boldsymbol{\delta}(\mathbf{x}) - \boldsymbol{\theta})'] \right\} \boldsymbol{\ell} \geq 0$$

for all $\boldsymbol{\ell} \neq \mathbf{0}$ and for all $\boldsymbol{\theta} \in \mathcal{R}^p$ for the mean square error.

1.4 Some Statistical Models with Preliminaries

In this section we consider some basic statistical models that are frequently used in applied statistical methodology, with preliminary information with regard to estimation and testing of hypotheses. These will be used to discuss the preliminary test and shrinkage estimation in later chapters. Some nonparametric methods are also given with preliminaries in Chapter 2.

1.4.1 Mean and Simple Linear Models

Consider the simple model

$$\mathbf{Y} = \theta \mathbf{1}_n + \boldsymbol{\varepsilon}, \tag{1.4.1}$$

where $\mathbf{Y} = (Y_1, \ldots, Y_n)'$ is the observation vector of n components, $\mathbf{1}_n = (1, \ldots, 1)'$ is a vector of an n-tuple of one's, θ is a constant, and $\varepsilon = (\varepsilon_1, \ldots, \varepsilon_n)'$ is a vector of n independent components distributed as $\mathcal{N}_n(\mathbf{0}, \sigma^2\mathbf{I}_n)$, $\mathbf{I}_n$ is the identity matrix of order n. Using the sample information and the specification (1.4.1), we obtain the *unrestricted estimator* (UE) of θ by minimizing the quadratic form

$$(\mathbf{Y} - \theta\mathbf{1}_n)'(\mathbf{Y} - \theta\mathbf{1}_n) \tag{1.4.2}$$

as

$$\tilde{\theta}_n = (\mathbf{1}_n'\mathbf{1}_n)^{-1}\mathbf{1}_n'\mathbf{Y} = \overline{Y}, \tag{1.4.3}$$

where $\overline{Y}$ is the sample mean, distributed as $\mathcal{N}(\theta, \sigma^2/n)$. Further, the unbiased estimator of σ^2 is given by

$$s_y^2 = \frac{1}{n-1}(\mathbf{Y} - \tilde{\theta}_n\mathbf{1}_n)'(\mathbf{Y} - \tilde{\theta}_n\mathbf{1}_n). \tag{1.4.4}$$

In order to test the null hypothesis $H_0 : \theta = \theta_0$ against $H_A : \theta \neq \theta_0$, we use the test-statistic

$$\mathcal{L}_n = \frac{n|\tilde{\theta}_n - \theta_0|^2}{\sigma^2} \quad \text{if } \sigma^2 \text{ is known}$$
$$= \frac{n|\tilde{\theta}_n - \theta_0|^2}{s_y^2} \quad \text{if } \sigma^2 \text{ is unknown.} \tag{1.4.5}$$

The exact distribution of $\mathcal{L}_n$ under H_A (σ^2 known) follows a noncentral chi-square distribution with one degree of freedom (d.f.) and noncentrality parameter $\Delta^2/2$, where

$$\Delta^2 = \frac{n|\theta - \theta_0|^2}{\sigma^2}, \tag{1.4.6}$$

while $\mathcal{L}_n$ (σ^2 unknown) follows a noncentral F-distribution with $(1, n-1)$ d.f. having the same noncentrality parameter (1.4.6). If $\theta = \theta_0$, the distributions above reduce to the central chi-square and the F-distribution, respectively.

Next, consider the simple linear model

$$\mathbf{Y} = \theta \mathbf{1}_n + \beta \mathbf{x} + \boldsymbol{\varepsilon}, \tag{1.4.7}$$

where $\mathbf{Y} = (Y_1, \ldots, Y_n)'$ and $\mathbf{x} = (x_1, \ldots, x_n)'$ are the observation vector and fixed vector of known constants, while $\boldsymbol{\varepsilon} = (\varepsilon_1, \ldots, \varepsilon_n)'$ is the error vector of the model distributed as $\mathcal{N}_n(\mathbf{0}, \sigma^2 \mathbf{I}_n)$. Based on the LS/ML principle, the estimators of $(\theta, \beta)'$ is given by

$$\begin{pmatrix} \tilde{\theta}_n \\ \tilde{\beta}_n \end{pmatrix} = \begin{pmatrix} \bar{y} - \tilde{\beta}_n \bar{x} \\ \frac{1}{Q}\{\mathbf{x}'\mathbf{y} - \frac{1}{n}(\mathbf{1}_n'\mathbf{x})(\mathbf{1}_n'\mathbf{y})\} \end{pmatrix}, \tag{1.4.8}$$

where

$$Q = \mathbf{x}'\mathbf{x} - \frac{1}{n}(\mathbf{1}_n'\mathbf{x})^2. \tag{1.4.9}$$

The exact distribution of $(\tilde{\theta}_n, \tilde{\beta}_n)'$ is a bivariate normal distribution with mean vector $(\theta, \beta)'$ and covariance matrix

$$\sigma^2 \begin{pmatrix} \frac{1}{n} + \frac{\bar{x}^2}{Q} & -\frac{\bar{x}}{Q} \\ -\frac{\bar{x}}{Q} & \frac{1}{Q} \end{pmatrix}. \tag{1.4.10}$$

In order to test the null hypothesis $H_0 : \beta = \beta_0$ against $H_A : \beta \neq \beta_0$, we use the test-statistic

$$\mathcal{L}_n = \frac{|\tilde{\beta}_n - \beta_0|^2 Q}{s_e^2}, \tag{1.4.11}$$

where

$$s_e^2 = \frac{1}{n-2}[\{\mathbf{y}'\mathbf{y} - \frac{1}{n}(\mathbf{1}_n'\mathbf{y})^2\} - \tilde{\beta}_n^2 Q]. \tag{1.4.12}$$

The exact distribution of $\mathcal{L}_n$ under H_A follows a noncentral F-distribution with $(1, n-2)$ d.f. and noncentrality parameter $\Delta^2/2$,

$$\Delta^2 = \frac{|\beta - \beta_0|^2}{\sigma^2} Q. \tag{1.4.13}$$

Under H_0, it follows a central F-distribution. Similarly, in order to test the null hypothesis $H_0 : \theta = \theta_0$ against $H_A : \theta \neq \theta_0$, we use the test-statistic

$$\mathcal{L}_n = \frac{|\tilde{\theta}_n - \theta_0|^2}{s_e^2(\frac{1}{n} + \frac{\overline{x}^2}{Q})}, \tag{1.4.14}$$

which follows a noncentral F-distribution with $(1, n-2)$ d.f. and noncentrality parameter $\Delta^2/2$, where

$$\Delta^2 = \frac{|\theta - \theta_0|^2}{\sigma^2(\frac{1}{n} + \frac{\overline{x}^2}{Q})}. \tag{1.4.15}$$

Under H_0, it follows a central F-distribution.

Further, in order to test the null hypothesis $H_0 : \theta = \theta_0$, $\beta = \beta_0$ against $H_A : \theta \neq \theta_0$, $\beta \neq \beta_0$, we use the test-statistic

$$\mathcal{L}_n = s_e^{-2}(\tilde{\theta}_n - \theta_0, \tilde{\beta}_n - \beta_0)' \begin{pmatrix} n & n\overline{x} \\ n\overline{x} & Q + n\overline{x}^2 \end{pmatrix} \begin{pmatrix} \tilde{\theta}_n - \theta_0 \\ \tilde{\beta}_n - \beta_0 \end{pmatrix} \tag{1.4.16}$$

whose exact distribution is a noncentral F-distribution with $(2, n-2)$ d.f. and noncentrality parameter $\Delta^2/2$, where

$$\Delta^2 = \frac{1}{\sigma^2}(\theta - \theta_0, \beta - \beta_0)' \begin{pmatrix} n & n\overline{x} \\ n\overline{x} & Q + n\overline{x}^2 \end{pmatrix} \begin{pmatrix} \theta - \theta_0 \\ \beta - \beta_0 \end{pmatrix}. \tag{1.4.17}$$

Under H_0, $\mathcal{L}_n$ follows a central F-distribution.

Finally, let $\mathbf{x} = (0, \dots, 0, 1, \dots, 1)'$ in the linear model where there are n_1 zero's and n_2 1's in the vector. Then we obtain the two-sample problem of estimating means and testing difference between two means. Here

$$\mu_1 = \theta, \quad \mu_2 = \theta + \beta, \quad \text{and} \quad \mu_2 - \mu_1 = \beta. \tag{1.4.18}$$

Also $\overline{x} = \frac{n_2}{n_1+n_2}$, $Q = \frac{n_1 n_2}{n_1+n_2}$, and $\frac{\overline{x}^2}{Q} = \frac{n_2}{n_1(n_1+n_2)}$. Thus, the mle of μ_1 is $\overline{y}_1$ and that of μ_2 is $\overline{y}_2$. For testing $H_0 : \mu_2 = \mu_1$ against $H_A : \mu_2 \neq \mu_1$, we use the likelihood test

$$\mathcal{L}_n = \frac{n_1 n_2}{n_1 + n_2} \frac{(\overline{y}_2 - \overline{y}_1)^2}{s_y^2}, \tag{1.4.19}$$

$$s_y^2 = \frac{1}{n_1 + n_2 - 2}\left\{ \sum_{i=1}^{n_1} (y_i - \overline{y}_1)^2 + \sum_{j=1}^{n_2} (y_j - \overline{y}_2)^2 \right\}, \tag{1.4.20}$$

where $\mathcal{L}_n$ follows a noncentral F-distribution with $(1, n_1 + n_2 - 2)$ d.f. and noncentrality parameter $\Delta^2/2$

$$\Delta^2 = \frac{n_1 n_2}{n_1 + n_2}\left(\frac{\mu_2 - \mu_1}{\sigma}\right)^2. \tag{1.4.21}$$

Under H_0, it follows a central F-distribution. These models will be considered in Chapter 3 for further estimators of θ or $(\theta, \beta)'$.

1.4.2 One-Sample Multivariate Model

Let $\mathbf{Y}_1, \mathbf{Y}_2, \ldots, \mathbf{Y}_N$ be N observation vectors satisfying the model

$$\mathbf{Y}_\alpha = \boldsymbol{\theta} + \boldsymbol{\varepsilon}_\alpha \quad \alpha = 1, 2, \ldots, N. \tag{1.4.22}$$

Here $\mathbf{Y}_\alpha = (Y_{\alpha 1}, \ldots, Y_{\alpha p})'$, $\boldsymbol{\theta} = (\theta_1, \ldots, \theta_p)'$, and $\boldsymbol{\varepsilon}_\alpha = (\varepsilon_{\alpha 1}, \ldots, \varepsilon_{\alpha p})'$ for $\boldsymbol{\varepsilon}_\alpha \sim \mathcal{N}_p(\mathbf{0}, \boldsymbol{\Sigma})$; $\boldsymbol{\Sigma}$ is the covariance matrix of the error vectors $\boldsymbol{\varepsilon}_\alpha$, and the error vectors $\boldsymbol{\varepsilon}_1, \boldsymbol{\varepsilon}_2, \ldots, \boldsymbol{\varepsilon}_N$ are mutually independent.

The LSE/MLE of $\boldsymbol{\theta}$ is given by the sample mean vector $\overline{\mathbf{Y}} = (\overline{Y}_1, \ldots, \overline{Y}_p)'$ such that

$$\overline{\mathbf{Y}} \sim \mathcal{N}_p\Big(\boldsymbol{\theta}, \frac{1}{N}\boldsymbol{\Sigma}\Big), \quad \overline{\mathbf{Y}} = \frac{1}{N}\sum_{\alpha=1}^{N} \mathbf{Y}_\alpha. \tag{1.4.23}$$

In order to test the null hypothesis $H_0 : \boldsymbol{\theta} = \boldsymbol{\theta}_0$ against $H_A : \boldsymbol{\theta} \neq \boldsymbol{\theta}_0$, we use the likelihood ratio test-statistic

$$\begin{aligned}
\mathcal{L}_N &= N(\overline{\mathbf{Y}} - \boldsymbol{\theta}_0)'\boldsymbol{\Sigma}^{-1}(\overline{\mathbf{Y}} - \boldsymbol{\theta}_0) \quad \text{if } \boldsymbol{\Sigma} \text{ is known} \\
&= N\|\overline{\mathbf{Y}} - \boldsymbol{\theta}_0\|^2_{\boldsymbol{\Sigma}^{-1}}, \\
\mathcal{L}_N &= N(\overline{\mathbf{Y}} - \boldsymbol{\theta}_0)'\mathbf{S}^{-1}(\overline{\mathbf{Y}} - \boldsymbol{\theta}_0) \quad \text{if } \boldsymbol{\Sigma} \text{ is unknown} \\
&= N\|\overline{\mathbf{Y}} - \boldsymbol{\theta}_0\|^2_{\mathbf{S}^{-1}} = T^2 \quad \text{(Hotelling's } T^2\text{)},
\end{aligned} \tag{1.4.24}$$

where

$$\mathbf{S} = \sum_{\alpha=1}^{N}(\mathbf{Y}_\alpha - \overline{\mathbf{Y}})(\mathbf{Y}_\alpha - \overline{\mathbf{Y}})'.$$

We then write

$$F = \frac{(N-p)}{(N-1)p}T^2.$$

For detailed information, see Anderson (1984), Srivastava and Khatri (1979), and Rao (1973),

The exact distribution of $\mathcal{L}_n$ (when $\mathbf{\Sigma}$ is known) is a noncentral chi-square distribution with p d.f. and (when $\mathbf{\Sigma}$ is unknown) it follows a noncentral F-distribution with $(p, N-p)$ d.f., having the noncentrality parameter $\Delta^2/2$ in both cases, where

$$\begin{aligned}\Delta^2 &= N(\boldsymbol{\theta}-\boldsymbol{\theta}_0)'\mathbf{\Sigma}^{-1}(\boldsymbol{\theta}-\boldsymbol{\theta}_0)\\ &= N\|\boldsymbol{\theta}-\boldsymbol{\theta}_0\|^2_{\mathbf{\Sigma}^{-1}}.\end{aligned} \tag{1.4.25}$$

If $\boldsymbol{\theta}=\boldsymbol{\theta}_0$, then it follows a central chi-squared and F-distribution, respectively.

Now, if we assume $\mathbf{\Sigma}=\sigma^2 I_p$, we have to estimate only one parameter in the covariance matrix. Estimation of the mean vector $\boldsymbol{\theta}=(\theta_1,\dots,\theta_p)'$ is considered in Chapter 4 when $\mathbf{\Sigma}=\mathbf{I}_p$, and the general estimation problem when $\mathbf{\Sigma}$ is unknown is considered in Chapter 11.

1.4.3 ANOVA Models

Let the observation vector $\mathbf{Y}$ be modeled as

$$\mathbf{Y}=\mathbf{B}\boldsymbol{\theta}+\boldsymbol{\varepsilon}, \tag{1.4.26}$$

where

$$\begin{aligned}
\mathbf{Y} &= (y_{11},\dots,y_{1n_1};y_{21};\dots y_{2n_2};\dots;y_{p1},\dots,y_{pn_p})' && (1.4.27a)\\
\boldsymbol{\theta} &= (\theta_1,\dots,\theta_p)' && (1.4.27b)\\
\boldsymbol{\varepsilon} &= (\varepsilon_{11},\dots,\varepsilon_{1n_1};\dots;\varepsilon_{p1},\dots,\varepsilon_{pn_p})' && (1.4.27c)\\
\mathbf{B} &= \text{Block diagonal matrix} = \text{Diag}(\mathbf{1}_{n_1},\dots,\mathbf{1}_{n_p}) && (1.4.27d)\\
\mathbf{1}_{n_i} &= (1,\dots,1)' \text{ a } n_i\text{-tuple of ones} && (1.4.27e)\\
n &= n_1+\dots+n_p. && (1.4.27f)
\end{aligned}$$

The joint distribution of the components of $\boldsymbol{\varepsilon}$ is $\mathcal{N}_n(\mathbf{0},\sigma^2\mathbf{B})$, and σ^2 is the variance of errors.

Using the LS principle/MLE method, we obtain the estimator of $\boldsymbol{\theta}$ as

$$\tilde{\boldsymbol{\theta}}_n=(\mathbf{B}'\mathbf{B})^{-1}\mathbf{B}'Y=\mathbf{N}^{-1}\mathbf{B}'\mathbf{Y}, \tag{1.4.28}$$

where $\mathbf{N}=\mathbf{B}'\mathbf{B}=\text{Diag}(n_1,\dots,n_p)$. The corresponding unbiased estimator of σ^2 is then given by

$$s_e^2=\frac{1}{n-p}(\mathbf{Y}-\mathbf{B}\tilde{\boldsymbol{\theta}}_n)'(\mathbf{Y}-\mathbf{B}\tilde{\boldsymbol{\theta}}_n). \tag{1.4.29}$$

Moreover, $\tilde{\boldsymbol{\theta}}_n\sim\mathcal{N}_p(\boldsymbol{\theta},\sigma^2\mathbf{N}^{-1})$ and $(n-p)s_e^2/\sigma^2$ follows a central chi-square distribution with $(n-p)$ d.f. independent of $\tilde{\boldsymbol{\theta}}_n$.

In order to test the null hypothesis $H_0 : \boldsymbol{\theta} = \theta_0 \mathbf{1}_p$ against $H_A = \boldsymbol{\theta} \neq \theta_0 \mathbf{1}_p$, where θ_0 is a constant, we use the likelihood ratio test-statistic

$$\mathcal{L}_n = \frac{(\tilde{\boldsymbol{\theta}}_n' \mathbf{H}'\mathbf{N}\mathbf{H}\tilde{\boldsymbol{\theta}}_n)}{(p-1)s_e^2}, \tag{1.4.30}$$

where $\mathbf{H} = (I_p - \frac{1}{n}\mathbf{1}_p\mathbf{1}_p'\mathbf{N})$ and $\hat{\theta}_0 \mathbf{1}_p = \frac{1}{n}\mathbf{1}_p\mathbf{1}_p'\mathbf{N}\tilde{\boldsymbol{\theta}}_n$ is the pooled estimator of $\boldsymbol{\theta}$. The exact distribution of $\mathcal{L}_n$ is a noncentral F-distribution with $(p-1, n-p)$ d.f. and noncentrality parameter $\Delta^2/2$ where

$$\Delta^2 = \frac{(\boldsymbol{\theta}'\mathbf{H}'\mathbf{N}\mathbf{H}\boldsymbol{\theta})}{\sigma^2}. \tag{1.4.31}$$

If $\boldsymbol{\theta} = \theta_0 \mathbf{1}_p$, it follows the central F-distribution. This ANOVA model will be the subject of Chapter 5.

1.4.4 Parallelism Models

Let

$$\mathbf{Y}_{n_i} = \theta_i \mathbf{1}_{n_i} + \beta_i \mathbf{x}_i + \boldsymbol{\varepsilon}_i, \quad i = 1, 2, \ldots, p, \tag{1.4.32}$$

where $\mathbf{Y}_{n_i} = (y_{11}, \ldots, y_{1n_i})'$, $\mathbf{x}_i = (x_{i1}, \ldots, x_{in_i})'$ and $\boldsymbol{\varepsilon}_i = (\varepsilon_{i1}, \ldots, \varepsilon_{in_i})'$. In addition, $\boldsymbol{\varepsilon}_i \sim N(\mathbf{0}, \sigma^2 \mathbf{I}_{n_i})$, where $\mathbf{I}_{n_i}$ is the identity matrix of order n_i. Thus, we have p linear models with different intercepts and slope parameters. If $\beta_1 = \cdots = \beta_p = \beta$ (same), then we have p parallel simple linear models used in many problems of bioassay and shelf-life determinations in pharmaceutical problems. Now, consider the LSE/MLE of the parameters $\{(\theta_i, \beta_i)'; i = 1, \ldots, p\}$. Let us define $\overline{\mathbf{Y}} = (\overline{Y}_1, \ldots, \overline{Y}_p)'$ as the vector of means of the observations in p models, $\overline{\mathbf{x}} = (\overline{x}_1, \ldots, \overline{x}_p)'$ being the mean vector of the known x-vectors, and let $\tilde{\boldsymbol{\beta}}_n = (\tilde{\beta}, \ldots, \tilde{\beta}_p)'$ be the LSE/MLE of the slopes in p-models. Then the intercept vector $\boldsymbol{\theta} = (\theta_1, \ldots, \theta_p)'$ may be estimated as $\tilde{\boldsymbol{\theta}}_n = (\tilde{\theta}_1, \ldots, \tilde{\theta}_p)'$, where

$$\tilde{\theta}_i = \overline{Y}_i - \tilde{\beta}_i \overline{x}_i, \quad i = 1, \ldots, p. \tag{1.4.33}$$

Let

$$\mathbf{T}_n = \text{Diag}(\overline{x}_1, \ldots, \overline{x}_p). \tag{1.4.34}$$

Then we can write the vector $\tilde{\boldsymbol{\theta}}_n = (\tilde{\theta}_1, \ldots, \tilde{\theta}_p)'$ compactly as

$$\tilde{\boldsymbol{\theta}}_n = \overline{\mathbf{Y}} - \mathbf{T}_n \tilde{\boldsymbol{\beta}}_n. \tag{1.4.35}$$

If we want to test the null hypothesis $H_0 : \boldsymbol{\beta} = \beta_0 \mathbf{1}_p$, where β_0 is a specified constant, against $H_A : \boldsymbol{\beta} \neq \beta_0 \mathbf{1}_p$ (parallel vs. not parallel), we can use the LR test-statistic

$$\mathcal{L}_n = \frac{(\tilde{\boldsymbol{\beta}}_n' \mathbf{H}'\mathbf{D}_{22}^{-1}\mathbf{H}\tilde{\boldsymbol{\beta}}_n)}{s_e^2} \tag{1.4.36}$$

where

$$\mathbf{H} = \mathbf{I}_p - \frac{1}{nQ}\mathbf{1}_p\mathbf{1}_p'\mathbf{D}_{22}^{-1}, \quad nQ = \sum_{i=1}^{p} n_iQ_i, \tag{1.4.37a}$$

$$\mathbf{D}_{22}^{-1} = \text{Diag}(n_1Q_1, \dots, n_pQ_p), \tag{1.4.37b}$$

$$n_iQ_i = \mathbf{x}_i'\mathbf{x}_i - \frac{1}{n_i}(\mathbf{1}_{n_i}'\mathbf{x}_i), \tag{1.4.37c}$$

$$s_e^2 = (n-2p)^{-1}\sum_{i=1}^{p}(\mathbf{Y}_i - \tilde{\theta}_i\mathbf{1}_{n_i} - \tilde{\beta}_i\mathbf{x}_i)'(\mathbf{Y}_i - \tilde{\theta}_i\mathbf{1}_{n_i} - \tilde{\beta}_i\mathbf{x}_i). \tag{1.4.37d}$$

The exact distribution of $\mathcal{L}_n$ follows a noncentral F-distribution with $(p, n-2p)$ d.f. and noncentrality parameter $\Delta^2/2$, where

$$\Delta^2 = \frac{1}{\sigma^2}\boldsymbol{\beta}'\mathbf{H}'\mathbf{D}_{22}^{-1}\mathbf{H}\boldsymbol{\beta} = \frac{1}{\sigma^2}(\boldsymbol{\beta} - \beta_0\mathbf{1}_p)'\mathbf{D}_{22}(\boldsymbol{\beta} - \beta_0\mathbf{1}_p). \tag{1.4.38}$$

Under H_0, $\mathcal{L}_n$ follows a central F-distribution with $(p-1, n-2p)$ d.f.
We will discuss this model in Chapter 6.

1.4.5 Multiple Regression Model and General Linear Hypothesis

Consider the multiple regression model

$$\mathbf{Y} = \mathbf{X}\boldsymbol{\beta} + \boldsymbol{\varepsilon}, \tag{1.4.39}$$

where Y is $n \times 1$ vector, $\mathbf{X}$ is a $n \times p$ matrix of known constants, $\boldsymbol{\beta} = (\beta_1, \dots, \beta_p)'$ is a $p \times 1$ vector and $\boldsymbol{\varepsilon}$ is a $n \times 1$ vector of errors distributed as $N_n(\mathbf{0}, \sigma^2\mathbf{I}_n)$. In many applied problems one formulates the response variable $\mathbf{Y}$ as above and tries to assess the linear hypotheses of the form

$$H_0 : \underset{q\times p}{\mathbf{H}} \ \underset{p\times 1}{\boldsymbol{\beta}} = \underset{q\times 1}{\mathbf{h}}\ . \tag{1.4.40}$$

In this case, we are interested in estimating $\boldsymbol{\beta}$ under the constraint $\mathbf{H}\boldsymbol{\beta} = \mathbf{h}$ and to test a hypothesis concerning the constraint. Thus, the unrestricted LSE of $\boldsymbol{\beta}$ is given by

$$\tilde{\boldsymbol{\beta}}_n = (\mathbf{X}'\mathbf{X})^{-1}\mathbf{X}'Y = \mathbf{C}^{-1}\mathbf{X}'\mathbf{Y}, \quad \mathbf{C} = \mathbf{X}'\mathbf{X}. \tag{1.4.41}$$

When the restriction is applied, we obtain the estimator $\hat{\boldsymbol{\beta}}_n$,

$$\hat{\boldsymbol{\beta}}_n = \tilde{\boldsymbol{\beta}}_n - \mathbf{C}^{-1}\mathbf{H}^{-1}(\mathbf{H}\mathbf{C}^{-1}\mathbf{H}')^{-1}(\mathbf{H}\tilde{\boldsymbol{\beta}}_n - \mathbf{h}). \tag{1.4.42}$$

In order to test the null hypothesis $H_0 : \mathbf{H}\boldsymbol{\beta} = \mathbf{h}$ against $H_A : \mathbf{H}\boldsymbol{\beta} \neq \mathbf{h}$, we use the test-statistic

$$\mathcal{L}_n = q^{-1}s_e^{-2}\left\{(\mathbf{H}\tilde{\boldsymbol{\beta}}_n - \mathbf{h})'(\mathbf{H}\mathbf{C}^{-1}\mathbf{H}')^{-1}(\mathbf{H}\tilde{\boldsymbol{\beta}}_n - \mathbf{h})\right\}, \tag{1.4.43}$$

where

$$s_e^2 = (n-p)^{-1}(\mathbf{Y} - \mathbf{X}\tilde{\boldsymbol{\beta}}_n)'(\mathbf{Y} - \mathbf{X}\tilde{\boldsymbol{\beta}}_n). \tag{1.4.44}$$

The exact distribution of $\mathcal{L}_n$ is a noncentral F-distribution with $(q, n-p)$ d.f. and noncentrality parameter $\Delta^2/2$, where

$$\Delta^2 = (\mathbf{H}\boldsymbol{\beta} - \mathbf{h})'(\mathbf{H}\mathbf{C}^{-1}\mathbf{H})^{-1}(\mathbf{H}\boldsymbol{\beta} - \mathbf{h})/\sigma^2. \tag{1.4.45}$$

Under H_0, $\mathcal{L}_n$ follows a central F-distribution. Further, the estimation of $\boldsymbol{\beta}$ will be discussed in Chapter 7.

There are three important variations of the basic model discussed above:

(1) $(\mathbf{X}'\mathbf{X})$ is a singular/ill-conditioned matrix that prevents a reasonable inverse of $(\mathbf{X}'\mathbf{X})$. This means that there is multicollinearity among the elements of $\mathbf{X}$, and $\mathbf{X}$ has rank $(\leq p)$ or one of the characteristic roots of $\mathbf{X}'\mathbf{X}$ is very small.

(2) The restriction $\mathbf{h} = \mathbf{H}\boldsymbol{\beta}$ is not exact; rather, it is of the form

$$\mathbf{h} = \mathbf{H}\boldsymbol{\beta} + \mathbf{v}, \quad \mathbf{v} \sim \mathcal{N}_q(\boldsymbol{\delta}, \sigma^2\boldsymbol{\Omega}), \tag{1.4.46}$$

where $\boldsymbol{\delta} \neq \mathbf{0}$ and $\boldsymbol{\Omega}$ is a $q \times q$ matrix of known constants that reflects the precision of an earlier sample study.

(3) The error vector in the model (1.4.39) has components related by a first-order stationery autoregressive process.

In case (1), the LSE of $\boldsymbol{\beta}$ is obtained by solving the normal equations

$$(\mathbf{X}'\mathbf{X})\boldsymbol{\beta} = \mathbf{X}'\mathbf{Y}. \tag{1.4.47}$$

The general solution for $\boldsymbol{\beta}$ is $\boldsymbol{\beta} = G\mathbf{X}'\mathbf{Y}$ where G is the generalized inverse of $(\mathbf{X}'\mathbf{X})$ and the solution for $\boldsymbol{\beta}$ is non unique, and for the non–full rank, there exists an infinite number of solutions. Hoerl and Kennard (1970) devised a method to overcome the difficulties above. Thus, they proposed the solution called the "ridge estimator" of $\boldsymbol{\beta}$, which is defined by

$$\hat{\boldsymbol{\beta}}_{HK} = (\mathbf{X}'\mathbf{X} + k\mathbf{I}_p)^{-1}\mathbf{X}'\mathbf{Y} = (\mathbf{C} + k\mathbf{I}_p)^{-1}\mathbf{X}'\mathbf{Y}, \tag{1.4.48}$$

where $k > 0$ is a positive real number known as "ridge constant" for the estimator of $\boldsymbol{\beta}$. The computation of the trace of

$$(\mathbf{C})^{-1} \quad \text{and} \quad (\mathbf{C} + k\mathbf{I}_p)^{-1} \tag{1.4.49}$$

shows that

$$\text{tr}(\mathbf{C}^{-1}) = \sum_{i=1}^{p} \frac{1}{\lambda_i}, \tag{1.4.50}$$

where $\lambda_1, \ldots, \lambda_p$ are the characteristic roots of $\mathbf{X}'\mathbf{X} = \mathbf{C}$ and

$$\text{tr}[C + k\mathbf{I}_p]^{-1} = \sum_{i=1}^{p} \frac{\lambda_i}{(\lambda_i + k)^2}. \tag{1.4.51}$$

Hence,

$$\text{tr}[\mathbf{C} + k\mathbf{I}_p]^{-1} \leq \text{tr}(\mathbf{C}^{-1}), \tag{1.4.52}$$

and the ridge estimator $\hat{\beta}_{HK}$ has smaller variability than $\tilde{\beta}_n$.

There are various ways of deriving the ridge estimator of β and the basic problem becomes that of the estimation of k.

The ridge estimator will be discussed in Chapter 9.

As for the case (2), consider the model

$$\mathbf{Y} = \mathbf{X}\beta + \varepsilon \tag{1.4.53}$$

subject to $\mathbf{h} = \mathbf{H}\beta + \mathbf{v}$, where $\mathbf{v} \sim \mathcal{N}_q(\mathbf{0}, \sigma^2\mathbf{\Omega})$, instead of assuming $\mathbf{v}$ to be non stochastic equaling $\mathbf{0}$.

Here, $\mathbf{\Omega}$ is a known $q \times q$ invertible matrix. The covariance matrix may reflect the information from previous samples or some prior information regarding the sizes of the elements of $\mathbf{\Omega}$. Thus, we can write

$$\begin{bmatrix} \mathbf{Y} \\ \mathbf{h} \end{bmatrix} = \begin{bmatrix} \mathbf{X} \\ \mathbf{H} \end{bmatrix} \beta + \begin{bmatrix} \varepsilon \\ \mathbf{v} \end{bmatrix}, \tag{1.4.54}$$

where

$$\begin{bmatrix} \varepsilon \\ \mathbf{v} \end{bmatrix} \sim \mathcal{N}_{n+q}\left\{ \begin{pmatrix} \mathbf{0} \\ \delta \end{pmatrix}; \sigma^2 \begin{pmatrix} \mathbf{I}_n & \mathbf{0} \\ \mathbf{0} & \mathbf{\Omega} \end{pmatrix} \right\}. \tag{1.4.55}$$

After suitable transformations we can write the model above as

$$\begin{bmatrix} \mathbf{Y} \\ \mathbf{\Omega}^{-1/2}\mathbf{h} \end{bmatrix} = \begin{bmatrix} \mathbf{X} & \mathbf{0} \\ \mathbf{0} & \mathbf{\Omega}^{-1/2} \end{bmatrix} \begin{bmatrix} \beta \\ \mathbf{H}\beta + \delta \end{bmatrix} + \begin{bmatrix} \varepsilon \\ \Omega^{-1/2}(\mathbf{v} - \delta) \end{bmatrix} \tag{1.4.56}$$

subject to

$$[-\mathbf{H}, \mathbf{I}_p] \begin{bmatrix} \beta \\ \mathbf{H}\beta + \delta \end{bmatrix} = \mathbf{0} \quad \text{or} \quad \delta = \mathbf{0}. \tag{1.4.57}$$

It is clearly

$$\begin{bmatrix} \varepsilon \\ \Omega^{-1/2}(\mathbf{v} - \delta) \end{bmatrix} \sim \mathcal{N}_{n+q}\left\{ \begin{pmatrix} \mathbf{0} \\ \mathbf{0} \end{pmatrix}; \sigma^2 \begin{pmatrix} \mathbf{I}_n & \mathbf{0} \\ \mathbf{0} & \mathbf{I}_q \end{pmatrix} \right\}. \tag{1.4.58}$$

Now, the *unrestricted estimator* (i.e. without model restriction) of $\begin{bmatrix} \beta \\ \mathbf{H}\beta + \delta \end{bmatrix}$ is $\begin{bmatrix} \tilde{\beta}_n \\ \mathbf{h} \end{bmatrix}$ by the LSE principle, with the unbiased estimator of σ^2 as

$$s_e^2 = \frac{1}{n-p}(\mathbf{Y}'\mathbf{Y} - \mathbf{Y}'\mathbf{X}\mathbf{C}^{-1}\mathbf{X}'\mathbf{Y}). \tag{1.4.59}$$

The restricted estimator of the parameters are

$$\begin{bmatrix} \hat{\beta}_n \\ \widehat{R\beta+\delta} \end{bmatrix} = \begin{bmatrix} \tilde{\beta}_n - \mathbf{C}^{-1}\mathbf{H}'[(\mathbf{H}\mathbf{C}^{-1}\mathbf{H}') + \boldsymbol{\Omega}]^{-1}(\mathbf{H}\tilde{\beta}_n - \mathbf{h}) \\ \mathbf{h} - \boldsymbol{\Omega}[(\mathbf{H}\mathbf{C}^{-1}\mathbf{H}') + \boldsymbol{\Omega}]^{-1}(\mathbf{H}\tilde{\beta}_n - \mathbf{h}) \end{bmatrix}. \tag{1.4.60}$$

The test-statistic for testing the null hypothesis $H_0 : \mathbf{h} = \mathbf{H}\boldsymbol{\beta}$ is given by

$$\mathcal{L}_n = \frac{(\mathbf{H}\tilde{\beta}_n - \mathbf{h})'[\mathbf{H}\mathbf{C}^{-1}\mathbf{H}' + \boldsymbol{\Omega}]^{-1}\mathbf{H}\tilde{\beta}_n - \mathbf{h})}{qs_e^2}, \tag{1.4.61}$$

which under $H_A : \mathbf{h} = \mathbf{H}\boldsymbol{\beta} + \boldsymbol{\delta},\ \ \boldsymbol{\delta} \neq \mathbf{0}$ follows a noncentral F-distribution with $(q, n-p)$ d.f. and noncentrality parameter $\Delta^2/2$, where

$$\Delta^2 = \frac{\boldsymbol{\delta}'[\mathbf{H}\mathbf{C}^{-1}\mathbf{H}' + \boldsymbol{\Omega}]^{-1}\boldsymbol{\delta}}{\sigma^2}. \tag{1.4.62}$$

This model will be discussed in Chapter 10. Some background material may be obtained from Graybill (1976), Gruber (1998), Judge and Bock (1978) among others.

Finally, the regression model that arises particularly in econometrics may be stated as

$$\mathbf{Y} = \mathbf{X}\boldsymbol{\beta} + \boldsymbol{\varepsilon},$$

where $\boldsymbol{\varepsilon} = (\varepsilon_1, \dots, \varepsilon_n)'$ and the components of $\boldsymbol{\varepsilon}$ are related by a first-order stationary autoregressive process

$$e_t = \rho e_{t-1} + v_t, \quad E(v_t) = 0, \quad \mathrm{Var}(v_t^2) = \sigma_v^2, \quad \text{for all } t, \tag{1.4.63}$$

and $|\rho| < 1$. This autoregressive specification implies that

$$E(\mathbf{ee}') = \sigma^2\Omega = \sigma^2 \begin{pmatrix} 1 & \rho & \dots & \rho^{n-1} \\ \rho & 1 & \dots & \rho^{n-2} \\ \vdots & \vdots & & \vdots \\ \rho^{n-1} & \rho^{n-2} & \dots & 1 \end{pmatrix}, \tag{1.4.64}$$

where

$$\Omega^{-1} = \begin{pmatrix} 1 & -\rho & 0 & \dots & 0 \\ -\rho & 1+\rho^2 & 0 & \dots & 0 \\ \vdots & \vdots & \vdots & & \vdots \\ & & & 1+\rho^2 & -\rho \\ 0 & & \dots & -\rho & 1 \end{pmatrix} \tag{1.4.65}$$

and has $n-1$ characteristic roots equaling $(1-\rho^2)^{-1}$ with one root equal to one. Using the generalized LSE principle, we obtain the LSE defined by

$$\mathbf{b}(\Omega) = (X\Omega^{-1}\mathbf{X})^{-1}\mathbf{X}'\Omega^{-1}\mathbf{Y}, \tag{1.4.66}$$

where $E[\mathbf{b}(\Omega)] = \boldsymbol{\beta}$ and $\mathrm{Cov}[\mathbf{b}(\Omega)] = \sigma^2(\mathbf{X}'\Omega^{-1}\mathbf{X})^{-1}$ whenever Ω is known. If Ω is unknown, then we estimate ρ, say, by $\hat{\rho}$ and plug in $\hat{\Omega}(\hat{\rho})$ in the expression above to obtain $\mathbf{b}(\hat{\Omega})$. Various other estimates are desired and will be obtained in Chapter 10.

1.4.6 Simple Multivariate Linear Model

Consider the simple multivariate linear model

$$\mathbf{Y}_\alpha = \boldsymbol{\theta} + \boldsymbol{\beta} x_\alpha + \boldsymbol{\varepsilon}_\alpha, \quad \boldsymbol{\varepsilon}_\alpha \sim \mathcal{N}_p(\mathbf{0}, \boldsymbol{\Sigma}), \quad \alpha = 1, \dots, N, \tag{1.4.67}$$

where $\mathbf{Y}_\alpha = (Y_{\alpha 1}, \dots, Y_{\alpha p})'$ is the observation vector, x_α is a fixed known constant and $\boldsymbol{\varepsilon}_\alpha = (\varepsilon_{\alpha 1}, \dots, \varepsilon_{\alpha p})' \sim \mathcal{N}_p(\mathbf{0}, \boldsymbol{\Sigma})$. The LSE/MLE of $\boldsymbol{\theta}$ and $\boldsymbol{\beta}$ are given by

$$\tilde{\boldsymbol{\theta}}_N = \overline{\mathbf{Y}} - \tilde{\boldsymbol{\beta}}_N \overline{x}, \tag{1.4.68a}$$

$$\tilde{\boldsymbol{\beta}}_N = \sum_{\alpha=1}^{N} \frac{(\mathbf{Y}_\alpha - \overline{\mathbf{Y}})(x_\alpha - \overline{x})}{Q} \tag{1.4.68b}$$

with

$$Q = \sum_{\alpha=1}^{N} (x_\alpha - \overline{x})^2. \tag{1.4.68c}$$

In order to test the null hypothesis $H_0 : \boldsymbol{\beta} = \mathbf{0}$ against $H_A : \boldsymbol{\beta} \neq \mathbf{0}$, we use the test-statistic

$$\mathcal{L}_N = Q(\tilde{\boldsymbol{\beta}}_N' \mathbf{S}^{-1} \tilde{\boldsymbol{\beta}}_N) = T^2 \quad \text{(Hotelling's } T^2\text{)}, \tag{1.4.69}$$

where

$$(N-2)\mathbf{S} = \sum_{\alpha=1}^{N} \left[(\mathbf{Y}_\alpha - \overline{\mathbf{Y}}) - \tilde{\boldsymbol{\beta}}_N(x_\alpha - \overline{x})\right]\left[(\overline{\mathbf{Y}}_\alpha - \overline{\mathbf{Y}}) - \tilde{\boldsymbol{\beta}}_N(x_\alpha - \overline{x})\right]' \tag{1.4.70}$$

is the unbiased estimator of $\boldsymbol{\Sigma}$.

The exact distribution of $\mathcal{L}_N$ is the Hotelling's T^2 distribution. Equivalently, $\mathcal{L}_n$ has a noncentral F-distribution with $(p, N-p)$ d.f. and noncentrality parameter $\Delta^2/2$, where

$$\Delta^2 = Q(\boldsymbol{\beta}' \boldsymbol{\Sigma}^{-1} \boldsymbol{\beta}). \tag{1.4.71}$$

As a special case, let us assume $x_\alpha = 0$ if $\alpha = 1, \dots, N_1$, and 1 if $\alpha = N_1 + 1, \dots, N$. Then we have the two-sample multivariate mean problem where

$$\boldsymbol{\mu}_1 = \boldsymbol{\theta}, \quad \boldsymbol{\mu}_2 = \boldsymbol{\theta} + \boldsymbol{\beta}, \quad \text{and} \quad \boldsymbol{\beta} = \boldsymbol{\mu}_2 - \boldsymbol{\mu}_1. \tag{1.4.72}$$

Also,

$$\overline{x} = \frac{N_1}{N_1 + N_2} \quad \text{and} \quad Q = \frac{N_1 N_2}{N_1 + N_2}. \tag{1.4.73}$$

The mle of $\boldsymbol{\mu}_1$ is $\overline{Y}_1$ and that of $\boldsymbol{\mu}_2$ is $\overline{\mathbf{Y}}_2$ based on the samples of sizes N_1 and N_2, respectively. Further, the mle of $\boldsymbol{\Sigma}$ is the pooled sample covariance

$$\mathbf{S}_p = \frac{1}{(N_1+N_2-2)}\left\{\sum_{\alpha=1}^{N_1}(\mathbf{Y}_\alpha - \overline{\mathbf{Y}}_1)(\mathbf{Y}_\alpha - \overline{\mathbf{Y}}_1)' + \sum_{\alpha=N_1+1}^{N}(\mathbf{Y}_\alpha - \overline{\mathbf{Y}}_2)(\mathbf{Y}_\alpha - \overline{\mathbf{Y}}_2)'\right\}. \quad (1.4.74)$$

In order to test the null hypothesis $H_0 : \boldsymbol{\mu}_2 = \boldsymbol{\mu}_1$ against $H_A : \boldsymbol{\mu}_2 \neq \boldsymbol{\mu}_1$, we use the LR test-statistic as

$$\begin{aligned}\mathcal{L}_N &= \frac{N_1N_2}{N_1+N_2}(\overline{\mathbf{Y}}_2 - \overline{\mathbf{Y}}_1)'\mathbf{S}_p^{-1}(\overline{\mathbf{Y}}_2 - \overline{\mathbf{Y}}_1)\\ &= T^2 \quad \text{(Hotelling's } T^2\text{)}\end{aligned} \quad (1.4.75)$$

The exact distribution is a noncentral F distribution with $(p, N_1 + N_2 - p)$ d.f. and noncentrality parameter $\Delta^2/2$, where

$$\Delta^2 = \frac{N_1N_2}{N_1+N_2}(\boldsymbol{\mu}_2 - \boldsymbol{\mu}_1)'\boldsymbol{\Sigma}^{-1}(\boldsymbol{\mu}_2 - \boldsymbol{\mu}_1). \quad (1.4.76)$$

Under H_0, it follows a central F-distribution.

More estimation will be discussed in Chapter 11 for the regression and the two-sample problems.

1.4.7 Discrete Data Models

Here we consider three selected discrete data models.

Product Bernoulli Distributions. Let $\{(x_{i1}, \dots, x_{in_i}|i = 1, \dots, p\}$ be a set of mutually independent Bernoulli random variables (r.v.'s) with joint distribution

$$\prod_{i=1}^{b}\prod_{j=1}^{n_i}\theta_i^{x_{ij}}(1-\theta_i)^{1-x_{ij}}. \quad (1.4.77)$$

Let $\mathbf{Y} = (y_1, \dots, y_p)$, $y_i = \sum_{j=1}^{n_i} x_{ij}$. Then the MLE of $\boldsymbol{\theta} = (\theta_1, \dots, \theta_p)'$ is given by $\tilde{\theta}_n = (\tilde{\theta}_1, \dots, \tilde{\theta}_p)'$, where $\tilde{\theta}_i = n_i^{-1}y_i$, $i = 1, \dots, p$. If $\theta_1 = \dots = \theta_p = \theta_0$ (unknown), then the MLE of θ_0 is $\hat{\theta}_{0n} = n^{-1}(y_1 + \dots + y_p)$. In vector notation, we then write

$$\begin{aligned}\tilde{\boldsymbol{\theta}}_n &= (\tilde{\theta}_1, \dots, \tilde{\theta}_p)',\\ \hat{\boldsymbol{\theta}}_n &= \hat{\theta}_{0n}\mathbf{1}_p, \quad \mathbf{1}_p = (1, \dots, 1)',\\ \hat{\boldsymbol{\theta}}_n &= \mathbf{1}_p\mathbf{1}_p'\mathbf{N}\tilde{\boldsymbol{\theta}}_n, \quad \mathbf{N} = \text{Diag}(n_1, \dots, n_p).\end{aligned} \quad (1.4.78)$$

In order to test the null hypothesis $H_0 : \boldsymbol{\theta} = \theta_0 \mathbf{1}_p$, we use the test-statistic

$$D_n = n(\tilde{\boldsymbol{\theta}}_n - \hat{\boldsymbol{\theta}}_n)' \hat{\Sigma}_n^{-1} (\tilde{\boldsymbol{\theta}}_n - \hat{\boldsymbol{\theta}}_n), \tag{1.4.79}$$

where

$$\hat{\Sigma}_n^{-1} = [\hat{\theta}_{0n}(1 - \hat{\theta}_{0n})]^{-1} \operatorname{Diag}\left(\frac{n_1}{n}, \dots \frac{n_p}{n}\right).$$

As $n \to \infty$, D_n approximately follows a chi-square distribution with $(p-1)$ d.f. under H_0.

Product Binomial Distributions. Consider k mutually independent 2×2 tables where each row is the outcome

$$\begin{array}{|c|c|}\hline x_{11} & n_{11} - x_{11} \\ \hline x_{21} & n_{21} - x_{21} \\ \hline \end{array} \cdots \begin{array}{|c|c|}\hline x_{ij} & n_{ij} - x_{ij} \\ \hline x_{2j} & n_{2j} - x_{2j} \\ \hline \end{array} \cdots \begin{array}{|c|c|}\hline x_{1k} & n_{1k} - x_{1k} \\ \hline x_{2k} & n_{2k} - x_{2k} \\ \hline \end{array} \tag{1.4.80}$$

of a binomial experiment. Then, the joint distribution of $\{(x_{1j}, x_{2j}) | j = 1, \dots, k\}$ is the product binomial given by

$$\prod_{j=1}^{k} \prod_{i=1}^{2} \binom{n_{ij}}{x_{ij}} \theta_{ij}^{x_{ij}} (1 - \theta_{ij})^{n_{ij} - x_{ij}}. \tag{1.4.81}$$

We are interested mostly in the estimation of the "odds-ratios", $\boldsymbol{\psi} = (\psi_1, \dots, \psi_k)'$ and homogeneity of the odds-ratios, meaning $\boldsymbol{\psi} = \psi_0 \mathbf{1}_k$ (ψ_0 is a scalar), where

$$\psi_j = \frac{\theta_{1j}(1 - \theta_{2j})}{\theta_{2j}(1 - \theta_{1j})}, \quad j = 1, \dots, k. \tag{1.4.82}$$

By invariance properties of MLE, we have the mle of $\boldsymbol{\psi}$ as

$$\tilde{\boldsymbol{\psi}}_n = (\tilde{\psi}_1, \dots, \tilde{\psi}_n)', \quad \tilde{\psi}_j = \frac{\tilde{\theta}_{1j}(1 - \tilde{\theta}_{2j})}{\tilde{\theta}_{2j}(1 - \tilde{\theta}_{1j})}, \tag{1.4.83}$$

and $\tilde{\theta}_{1j}$ and $\tilde{\theta}_{2j}$ are the MLE of θ_{1j} and θ_{2j}, respectively. If $\boldsymbol{\psi} = \psi_0 \mathbf{1}_k$ holds, then the common estimator of ψ_0 due to Gart (1992) is given by

$$\hat{\psi}_{0n} = \hat{\omega}^{-1} \sum_{j=1}^{k} \hat{\omega}_j \tilde{\psi}_j, \quad \hat{\omega} = \sum_{j=1}^{k} \hat{\omega}_j, \tag{1.4.84}$$

where

$$\hat{\omega}_j^{-1} = \frac{1}{n_{1j}\tilde{\theta}_{1j}(1 - \tilde{\theta}_{1j})} + \frac{1}{n_{2j}\tilde{\theta}_{2j}(1 - \tilde{\theta}_{2j})}, \quad j = 1, \dots, k. \tag{1.4.85}$$

In order to test the homogeneity of the odds-ratios, we consider the Wald-type statistic

$$D_n = \hat{\psi}_{0n}^{-2} (\tilde{\boldsymbol{\psi}}_n - \hat{\omega}_{0n} \mathbf{1}_k)' \hat{\mathbf{W}} (\tilde{\boldsymbol{\psi}} - \hat{\psi}_{0n} \mathbf{1}_k), \tag{1.4.86}$$

where $\hat{\mathbf{W}} = \text{Diag}(\hat{\omega}_1, \ldots, \hat{\omega}_k)$.

The asymptotic distribution of D_n follows a central chi-square distribution with $(k-1)$ d.f. under $H_0 : \boldsymbol{\psi} = \psi_0 \mathbf{1}_k$.

Product Multinomial Distributions. Consider a $(r \times c)$-contingency tables for two traits. Let x_{ij} stand for the cell frequency of the (i,j)-cell for $i = 1, \ldots, r\ (\geq 2)$ and $j = 1, \ldots, c\ (\geq 2)$, and let $x_{i+} = \sum_{j=1}^{c} x_{ij}$, $x_{+j} = \sum_{i=1}^{r} x_{ij}$, and $n = \sum_{i=1}^{r}\sum_{j=1}^{c} x_{ij}$. The probability distribution of the vector frequency $\mathbf{n} = (x_{11} \ldots x_{1c}, x_{21}, \ldots, x_{2c}, \ldots, x_{r1}, \ldots, x_{rc})'$ is given by the product multinomial distribution

$$n! \left\{ \prod^{r} \prod^{c} (x_{ij}!) \right\}^{-1} \prod_{i=1}^{r} \prod_{j=1}^{c} (\theta_{ij})^{x_{ij}}, \tag{1.4.87}$$

where $\boldsymbol{\theta} = \{\theta_{ij} | i = 1, \ldots, r,\ j = 1, \ldots, c\}$ with $\boldsymbol{\theta}' \mathbf{1}_k = 1$. Define

$$\theta_{i+} = \sum_{j=1}^{c} \theta_{ij} \quad \text{and} \quad \theta_{+j} = \sum_{i=1}^{r} \theta_{ij}. \tag{1.4.88}$$

If there is independence structure in the table, then we have

$$\theta_{ij} = \theta_{i+} \cdot \theta_{+j} \quad \forall\ (i,j). \tag{1.4.89}$$

Note that in (1.4.87) we have $rc - 1$ free parameters, while in (1.4.89) we deal with $r + c - 2$ free parameters.

In order to test for the null hypothesis

$$H_0 : \theta_{ij} = \theta_{i+} \cdot \theta_{+j} \quad \text{for all} \quad (i,j),$$

we consider the following chi-square test:

$$D_n = n(\tilde{\boldsymbol{\theta}}_n - \hat{\boldsymbol{\theta}}_n)' \Sigma_n^{-1} (\tilde{\boldsymbol{\theta}}_n - \hat{\boldsymbol{\theta}}_n), \tag{1.4.90}$$

where

$$\tilde{\boldsymbol{\theta}}_n = (\tilde{\theta}_{1+}, \ldots, \tilde{\theta}_{1c}, \ldots, \tilde{\theta}_{r1}, \ldots, \tilde{\theta}_{rc})', \quad \tilde{\theta}_{ij} = \frac{x_{ij}}{m}$$

and

$$\hat{\boldsymbol{\theta}}_n = (\hat{\theta}_{1+}, \ldots, \hat{\theta}_{1c}, \ldots, \hat{\theta}_{r1}, \ldots, \hat{\theta}_{rc})',$$

with $\hat{\theta}_{ij} = \hat{\theta}_{i+} \cdot \hat{\theta}_{+j}$, $\hat{\theta}_{i+} = x_{i+}/n$ and $\hat{\theta}_{+j} = x_{+j}/n$. Also,

$$\hat{\boldsymbol{\Sigma}} = \text{Diag}(\hat{\theta}_{11}, \ldots, \hat{\theta}_{1c}, \ldots, \hat{\theta}_{r1}, \ldots, \hat{\theta}_{rc}). \tag{1.4.91}$$

It may be shown, following Bishop et al. (1975) and Agresti (1990), that D_n closely follows the chi-square distribution with $(r-1)(c-1)$ d.f. as $n \to \infty$.

The improved estimators of the parameters of the three models will be the subject of Chapter 12.

1.5 Organization of the Book

The book is divided into 12 chapters that cover most of the useful models in applied statistics. In Chapter 2 we present properties of the normal and chi-square distributions together with results involving the multivariate normal and discrete distributions. Matrix results and formulas for the calculations of risks of the estimators, together with the Stein formulas, are provided for ready references. Some preliminaries of nonparametric methods are also included in this chapter.

Preliminary test estimation is introduced in Chapter 3 with the simple linear model. Chapter 4 involves the introduction of Stein-type shrinkage estimation in a simple multivariate model. A general development of the estimation of the mean vector when the covariance matrix is unknown is deferred until Chapter 11. The ANOVA model is discussed in Chapter 5, and Chapter 6 extends the results of the simple linear model to several linear models that may be parallel.

Chapter 7 deals with the general linear model with linear hypothesis and estimation regression parameters. Chapters 8 and 9 contain extensions of Chapter 7 in two ways: (1) parameters are restricted by stochastic constraints, and (2) the design matrix may be ill-conditioned, leading to "ridge regression." In Chapter 10 we consider the linear regression model where the errors are generated by a first-order stationary autoregressive process. Chapter 11 deals with the general problem of estimation of parameters with one-sample, two-sample, and simple multivariate linear models when the error distribution is a p-dimensional normal. Chapter 12 deals with three basic discrete data models, namely, product Bernoulli distributions, product binomial distributions, and the product multinomial distributions and the related estimation of the parameters and application to meta-analysis.

1.6 Conclusions

In this chapter we presented a historical perspective on the preliminary test and Stein-type estimators, illustrated by an example of the "baseball data" and the objective of the book. We also discussed the decision-theoretic approach to estimation and testing. Finally, we have presented the models that will be covered in Chapters 2 through 12 of this book.

1.7 Problems

1. (Refer to Section 1.4.1) Consider the model $\mathbf{Y} = \theta \mathbf{1}_n + \boldsymbol{\varepsilon}$.
   Show that

   (a) $\tilde{\theta}_n = \overline{Y}$, (b) $s_y^2 = \frac{1}{n-1}(\mathbf{Y} - \tilde{\theta}_n \mathbf{1}_n)'(\mathbf{Y} - \tilde{\theta}_n \mathbf{1}_n)$

(c) Show that the likelihood ratio test for $H_0 : \theta = \theta_0$ versus $H_A : \theta \neq \theta_0$ is given by

$$\mathcal{L}_n = \frac{n|\tilde{\theta}_n - \theta_0|^2}{\sigma^2} \quad \text{if } \sigma^2 \text{ is known}$$
$$= \frac{n|\tilde{\theta}_n - \theta_0|^2}{s_y^2} \quad \text{if } \sigma^2 \text{ is unknown.}$$

(d) What is the distribution of $\mathcal{L}_n$ under H_0 and under H_A?

2. Consider the model (Equation 1.4.7)

$$\mathbf{Y} = \theta \mathbf{1}_n + \beta \mathbf{x} + \varepsilon, \quad \varepsilon \sim N_n(0, \sigma^2 I_n)$$

(a) Show that the LSE/MLE of (θ, β) is given by

$$\begin{pmatrix} \overline{Y} - \tilde{\beta}_n \overline{x} \\ \frac{1}{Q}[\mathbf{x}'\mathbf{y} - \frac{1}{n}(\mathbf{1}_n'\mathbf{x})(\mathbf{1}_n'\mathbf{y})] \end{pmatrix}$$

with covariance matrix

$$\sigma^2 \begin{pmatrix} \frac{1}{n} + \frac{\overline{x}^2}{Q} & -\frac{\overline{x}}{Q} \\ -\frac{\overline{x}}{Q} & \frac{1}{Q} \end{pmatrix}.$$

(b) Show that the likelihood ratio test for $H_0 : \beta = \beta_0$ against $H_A : \beta \neq \beta_0$ is given by

$$\mathcal{L}_n = \frac{n|\tilde{\beta}_n - \beta_0|^2}{\sigma^2} Q \quad \text{if } \sigma^2 \text{ is known}$$
$$= \frac{n|\tilde{\beta}_n - \beta_0|^2}{s_y^2} Q \quad \text{if } \sigma^2 \text{ is unknown.}$$

(c) Determine the distribution of $\mathcal{L}_n$.

3. Consider the two-sample problem by taking $\mathbf{x} = (0, \dots, 0, 1, \dots, 1)'$ in the linear model, then verify (1.4.19)-(1.4.21) and distribution of $\mathcal{L}_n$.

4. Let $\mathbf{Y}_1, \mathbf{Y}_2, \dots, \mathbf{Y}_N$ be N p-vector observations

$$\mathbf{Y}_\alpha = \boldsymbol{\theta} + \boldsymbol{\varepsilon}_\alpha, \quad \alpha = 1, \dots, N, \quad \boldsymbol{\varepsilon}_\alpha \sim N_p(\mathbf{0}, \Sigma).$$

Let $\overline{\mathbf{Y}} = N^{-1} \sum_{\alpha=1}^N \mathbf{Y}_\alpha$ and $S = \sum_{\alpha=1}^N (\mathbf{Y}_\alpha - \overline{\mathbf{Y}})(\mathbf{Y}_\alpha - \overline{\mathbf{Y}})'$.

(a) Show that

$$\mathcal{L}_n = N(\overline{\mathbf{Y}} - \theta_0)'\boldsymbol{\Sigma}^{-1}(\overline{Y} - \theta_0) \quad \text{if } \Sigma \text{ is known}$$
$$= N(\overline{\mathbf{Y}} - \theta_0)'S^{-1}(\overline{Y} - \theta_0) \quad \text{if } \Sigma \text{ is unknown}$$

is the likelihood ratio statistic for testing $H_0 : \boldsymbol{\theta} = \boldsymbol{\theta}_0$ against $H_A : \boldsymbol{\theta} \neq \boldsymbol{\theta}_0$.

(b) Determine the distribution of $\mathcal{L}_n$ under H_0 and H_A.

5. Consider the linear model (ANOVA)

$$\mathbf{Y} = \mathbf{B}\boldsymbol{\theta} + \boldsymbol{\varepsilon}, \quad \boldsymbol{\varepsilon} \sim N_n(\mathbf{0}, \sigma^2 B)$$

as defined by (1.4.26)–(1.4.27), where $\boldsymbol{\theta} = (\theta_1, \ldots, \theta_p)'$.

(a) Show that the likelihood ratio statistic for testing $H_0 : \boldsymbol{\theta} = \theta_0 \mathbf{1}_p$ against $H_A : \boldsymbol{\theta} \neq \theta_0 \mathbf{1}_p$ is given by

$$\mathcal{L}_n = \frac{1}{\sigma^2} \tilde{\theta}_n' \mathbf{H}'\mathbf{N}\mathbf{H}\tilde{\boldsymbol{\theta}}_n \quad \text{if } \sigma^2 \text{ is known}$$

$$= \frac{1}{(p-1)s_e^2} (\tilde{\theta}_n' \mathbf{H}'\mathbf{N}\mathbf{H}\tilde{\boldsymbol{\theta}}_n) \quad \text{if } \sigma^2 \text{ is unknown.}$$

(b) What is the distribution of $\mathcal{L}_n$ under H_0 as well as under H_A?

6. Consider p simple linear models

$$\mathbf{Y}_{n_i} = \theta_i \mathbf{1}_{n_i} + \beta_i \mathbf{x}_i + \boldsymbol{\varepsilon}_i, \quad i = 1, \ldots, p,$$

where $\boldsymbol{\varepsilon}_i \sim N_{n_i}(0, \sigma^2 \mathbf{I}_{n_i})$. Show that LSE/MLE of $\boldsymbol{\theta} = (\theta_1, \ldots, \theta_p)'$ and that $\boldsymbol{\beta} = (\beta_1, \ldots, \beta_p)'$ are

$$\tilde{\boldsymbol{\theta}}_n = (\tilde{\theta}_1, \ldots, \tilde{\theta}_p), \quad \tilde{\theta}_i = \overline{Y}_i - \tilde{\beta}_i \overline{x}_i, \quad i = 1, \ldots, p,$$

and

$$\tilde{\boldsymbol{\beta}}_n = (\tilde{\beta}_1, \ldots, \tilde{\beta}_p)', \quad \tilde{\beta}_i = \frac{\left[\mathbf{x}_i - \frac{(\mathbf{1}_{n_i}' \mathbf{x}_i)}{n}\right]' \left[\mathbf{Y}_{n_i} - \frac{(\mathbf{1}_{n_i}' \mathbf{Y}_{n_i})}{n_i}\right]}{n_i Q_i}.$$

In order to test the null hypothesis $H_0 : \boldsymbol{\beta} = \beta_0 \mathbf{1}_p$ against $H_A : \boldsymbol{\beta} \neq \beta_0 \mathbf{1}_p$, the likelihood ratio statistic is given by

$$\mathcal{L}_n = \frac{(\tilde{\boldsymbol{\beta}}_n' \mathbf{H}'\mathbf{D}_{22}^{-1} \mathbf{H}\boldsymbol{\beta}_n)}{(p-1)s_e^2}$$

(by (1.4.36)), where $D_{22}^{-1} = \mathrm{Diag}(n_1 Q_1, \ldots, n_p Q_p)$.

7. Consider the multiple regression model

$$\mathbf{Y} = \mathbf{X}\boldsymbol{\beta} + \boldsymbol{\varepsilon}, \quad \boldsymbol{\varepsilon} \sim N_n(\mathbf{0}, \sigma^2 I_n).$$

Find the likelihood ratio test for the null hypothesis $H_0 : \mathbf{H}\boldsymbol{\beta} = \mathbf{h}$ against $H_A : H\boldsymbol{\beta} \neq \mathbf{h}$ given by (1.4.40), and determine its distribution.

8. Consider the multivariate simple linear model

$$\mathbf{Y}_\alpha = \boldsymbol{\theta} + \boldsymbol{\beta} x_\alpha + \boldsymbol{\varepsilon}_\alpha, \quad \boldsymbol{\varepsilon}_\alpha \sim N_p(\mathbf{0}, \Sigma).$$

Show that the likelihood ratio statistic for testing $H_0 : \boldsymbol{\beta} = \mathbf{0}$ against $H_A : \boldsymbol{\beta} \neq \mathbf{0}$ is given by

$$\mathcal{L}_N = Q(\tilde{\boldsymbol{\beta}}_N' \mathbf{S}^{-1} \tilde{\boldsymbol{\beta}}_N)$$

where $\mathbf{S}$ is defined by (1.4.70). What is the distribution of $\mathcal{L}_N$?

9. (Refer to Section 1.4.7a)

   (a) For the product Bernoulli distributions verify that the MLE of $\boldsymbol{\theta}$ is $\tilde{\boldsymbol{\theta}}_n = (\tilde{\theta}_1, \dots, \tilde{\theta}_p)'$.

   (b) Show that $\text{Cov}(\tilde{\boldsymbol{\theta}}_n) = \text{Diag}\left(\frac{\theta_1(1-\theta_1)}{n_1}, \dots, \frac{\theta_p(1-\theta_p)}{n_p}\right)$.

   (c) If $\boldsymbol{\theta} = \theta_0 \mathbf{1}_p$, show that the MLE of $\theta_0 \mathbf{1}_p$ is

   $$\hat{\boldsymbol{\theta}}_{0n} = \frac{\mathbf{1}_p \mathbf{1}_p' \mathbf{N}}{n} \tilde{\boldsymbol{\theta}}_n, \quad \mathbf{N} = \text{Diag}(n_1, \dots, n_p)$$

   with $\text{Cov}(\hat{\boldsymbol{\theta}}_{0n}) = \dfrac{\mathbf{1}_p \mathbf{1}_p'}{n}$.

   (d) Consider the r.v. $(\tilde{\boldsymbol{\theta}}_n - \hat{\boldsymbol{\theta}}_{0n}) = \left(I_p - \frac{\mathbf{1}_p \mathbf{1}_p' \mathbf{N}}{n}\right) \tilde{\boldsymbol{\theta}}_n$. Show that under $H_0 : \boldsymbol{\theta} = \theta_0 \mathbf{1}_p$

   $$\text{Cov}[(\tilde{\boldsymbol{\theta}}_n - \hat{\boldsymbol{\theta}}_{0n})] = \theta_0(1-\theta_0)\left(I_p - \frac{\mathbf{1}_p \mathbf{1}_p' \mathbf{N}}{n}\right) \mathbf{N}^{-1}.$$

   (e) Show that

   $$D_n = n(\tilde{\boldsymbol{\theta}}_n - \hat{\boldsymbol{\theta}}_{0n})' \Sigma_n^{-1} (\tilde{\boldsymbol{\theta}}_n - \hat{\boldsymbol{\theta}}_{0n}),$$

   where $\Sigma_n^{-1} = [\hat{\theta}_0(1-\hat{\theta}_0)]^{-1} \text{Diag}\left(n_1/n, \dots, n_p/n\right)$, approximately follows the chi-square distribution with $(p-1)$ d.f. under H_0.

10. (Refer to Section 1.4.7 and k (2×2) contingency tables) Define

$$\tilde{\psi}_j = \frac{x_{1j}(n_{2j} - x_{2j})}{x_{2j}(n_{1j} - x_{1j})}, \quad j = 1, \dots, k.$$

   (a) Show that the approximate variance of $\tilde{\psi}_j$ is

   $$\text{Var}(\tilde{\psi}_j) = \frac{1}{x_{1j}} + \frac{1}{n_{1j} - x_{1j}} + \frac{1}{x_{2j}} + \frac{1}{n_{2j} - x_{2j}} = w_j^{-1} \quad \text{(say)}$$

   $j = 1, 2, \dots, k$, as $n_{+j} = n_{1j} + n_{2j} \to \infty$ in such a way that $n_{ij}/n_{+j} \to \lambda_{ij}$ $(< \infty)$.

(b) Let $\hat{\psi}_{0n} = w^{-1}\sum_{j=1}^{k} w_j\tilde{\psi}_j$. Show that the asymptotic variance of $\hat{\psi}_{0n}$ is $\psi_0^2 w^{-1}$.

(c) Show that

$$D_n = \hat{\psi}_{0n}^{-2}(\tilde{\psi}_n - \hat{\psi}_{0n}\mathbf{1}_k)'\mathbf{W}(\tilde{\psi}_n - \hat{\psi}_{0n}\mathbf{1}_k), \quad \mathbf{W} = \text{Diag}(\hat{\omega}_1, \dots, \hat{\omega}_k)$$

follow approximately the chi-square distribution with $(k-1)$ d.f. as $n \to \infty$.

11. Let $(x_1, \dots, x_n)'$ follow the multinomial distribution

$$\frac{n!}{\prod_{j=1}^{k}(x_j!)}\prod_{j=1}^{k}\theta_j^{x_j}, \quad \sum_{j=1}^{k}\theta_j = 1.$$

(a) Show that the MLE of $\boldsymbol{\theta} = (\theta_1, \dots, \theta_k)'$ is $\tilde{\boldsymbol{\theta}}_n = (\tilde{\theta}_1, \dots, \tilde{\theta}_k)'$, $\tilde{\theta}_j = \frac{x_j}{n}$, $j = 1, \dots, k$.

(b) $n\,\text{Cov}(\tilde{\boldsymbol{\theta}}_n) = \text{Diag}(\theta_1, \dots, \theta_k) - \boldsymbol{\theta}\boldsymbol{\theta}' = \boldsymbol{\Sigma}(\theta)$.

12. (Continuation of Problem 11)

(a) Show that

$$\sqrt{n}(\tilde{\boldsymbol{\theta}}_n - \boldsymbol{\theta}) \overset{\mathcal{D}}{\simeq} N_k\{\mathbf{0}, \boldsymbol{\Sigma}(\theta)\}.$$

(b) $D_n = n(\tilde{\theta}_n - n\boldsymbol{\theta})'\boldsymbol{\Sigma}_n^{-1}(\tilde{\boldsymbol{\theta}}_n - n\boldsymbol{\theta})$ approximately follows the chi-square distribution with k d.f. as $n \to \infty$.

(c) (Refer to Section 1.4.7c) For the $r \times c$ contingency table, show that

$$D_n = n(\tilde{\boldsymbol{\theta}}_n - \hat{\boldsymbol{\theta}}_n)'\hat{\boldsymbol{\Sigma}}_n^{-1}(\tilde{\boldsymbol{\theta}}_n - \hat{\boldsymbol{\theta}}_n)$$

where $\tilde{\boldsymbol{\theta}}_n$, $\hat{\boldsymbol{\theta}}_n$ and $\hat{\boldsymbol{\Sigma}}_n$ are given by (1.4.90)–(1.4.91). Show that D_n approximately follows the chi-square distribution with $(r-1)(c-1)$ d.f. as $n \to \infty$ under $H_0 : \boldsymbol{\theta}_{ij} = \theta_{i+} \cdot \theta_{+j}$ for all (i, j).

Chapter 2

Preliminaries

Outline

In this chapter we provide summary of the properties of statistics related to normal and chi-square distributions. Particularly, the Stein Lemma and its extension is provided for application in subsequent chapters. In addition, we provide some core results on matrix theory, multivariate discrete and continuous distributions, beta distributions, large sample results, and nonparametric theory for application in subsequent chapters.

2.1 Normal Distribution

The most important distribution in statistical theory is the normal distribution. The probability density function (pdf) of the normal distribution may be denoted as $\mathcal{N}(\mu, \sigma^2)$, which is written as follows:

$$f(x; \mu, \sigma) = \frac{1}{\sigma\sqrt{2\pi}} \exp\left\{ -\frac{1}{2}\left(\frac{x-\mu}{\sigma}\right)^2 \right\}, \quad \begin{matrix} x, \mu \in R' \\ \sigma \in R^+ \end{matrix} \tag{2.1.1}$$

where μ is the mean and σ^2 is the variance of the random variable X. Some of the properties of a normal variable may be listed as follows:

- If X is $\mathcal{N}(\mu, \sigma^2)$, then $Z = X - \mu/\sigma$ is $\mathcal{N}(0,1)$, the standard normal variable.
- If X_1 and X_2 are independent, such that $X_i \sim \mathcal{N}(\mu_i, \sigma_i^2)$, $(i = 1, 2)$, then $X_1 \pm X_2$ is distributed as a normal variable with mean $\mu_1 \pm \mu_2$ and variance $\sigma_1^2 + \sigma_2^2$.
- If $X_1, \cdots, X_n$ are n independent $\mathcal{N}(\mu, \sigma^2)$ variables, then the sample mean $\bar{X} = \frac{1}{n}(X_1 + \cdots + X_n)$ is $\mathcal{N}(\mu, \frac{\sigma^2}{n})$.
- If $X_1, \cdots, X_n$ are independent random variables having any distribution with common mean μ and common variance σ^2, then $\sqrt{n}(\bar{X} - \mu)/\sigma$ is distributed as $\mathcal{N}(0,1)$ as $n \to \infty$. This is known as the central limit theorem (CLT). These properties are amply listed in Cramer (1946), Hogg, McKean and Craig (2005), Rohatgi and Saleh (2001), Judge and Bock (1978), and Bickel and Doksum (2001) among many others.
- **(Stein's Indentity)** If X is $\mathcal{N}(\theta, \sigma^2)$ and if $\varphi(X)$ is a differentiable function satisfying $E|\varphi'(X)| < \infty$, then

$$E[\varphi(X)(X - \theta)] = \sigma^2 E[\varphi'(X)]. \tag{2.1.2}$$

- **(Extended Stein's Indentity)** If $\mathbf{X} = (X_1, \cdots, X_p)'$ is $\mathcal{N}_p(\boldsymbol{\theta}, \sigma^2\mathbf{I}_p)$ and if a function $\varphi(\mathbf{X}) = (\varphi_1(\mathbf{X}), \cdots, \varphi_p(\mathbf{X}))'$ is partially differentiable, that is,

$$\left|\frac{\partial \varphi_i(\mathbf{X})}{\partial X_i}\right| < \infty, \qquad i = 1, \cdots, p, \tag{2.1.3a}$$

and $\varphi_i(\mathbf{X})$ is a continuous function of X_i for all vectors, $\mathbf{X}_{[i]} = (X_1, \cdots, X_{i-1}, X_{i+1}, \cdots, X_p)'$ then

$$E\Big\{\big[\varphi(\mathbf{X})\big]'(\mathbf{X} - \boldsymbol{\theta})\Big\} = \sigma^2 E\Big[\sum_{i=1}^{p} \frac{\partial \varphi_i(\mathbf{X})}{\partial X_i}\Big] \tag{2.1.3b}$$

provided that

$$E\|\varphi(\mathbf{X})\|^2 < \infty.$$

Note that for $p = 1$, (2.1.3b) reduces to (2.1.2). For the Stein identities, see Casella and Berger (1990) and Hoffmann (1992) among others.

2.2 Chi-square Distribution and Properties

If X is $\mathcal{N}(0,1)$, then X^2 is distributed as a central chi-square variable with one degree of freedom denoted as χ_1^2. However, if X is $\mathcal{N}(\mu, \sigma^2)$, then X^2 is distributed as a noncentral chi-square variable $\chi_1^2(\Delta^2)$ with one degrees of freedom and noncentrality parameter $\frac{\mu^2}{2\sigma^2} = \frac{1}{2}\Delta^2$. The pdf of a noncentral chi-square variable with ν degrees of freedom and noncentrality parameter $\frac{1}{2}\Delta^2$ may be written as

$$h_\nu(\chi^2 : \Delta^2) = \exp\left\{-\frac{\Delta^2}{2}\right\} \sum_{r \geq 0} \frac{(\Delta^2/2)^r}{r!} h_{\nu+2r}(\chi^2; 0), \qquad (2.2.1a)$$

where $h_{\nu+2r}(\chi^2; 0)$ is the pdf of the central chi-square distribution with $\nu + 2r$ degrees of freedom. The pdf of a central chi-square distribution with ν degrees of freedom may be written as

$$h_\nu(\chi^2 : 0) = \frac{(\chi^2)^{\nu/2-1} e^{-\chi^2/2}}{2^{\nu/2}\Gamma(\frac{\nu}{2})}. \qquad (2.2.1b)$$

The mean and variance of this distribution are ν and 2ν, respectively. The first four cumulants of the noncentral chi-square distribution at (2.2.1a) are given by

$$\kappa_1 = \nu + \Delta^2, \quad \kappa_2 = 2(\nu + 2\Delta^2), \quad \kappa_3 = 8(\nu + 3\Delta^2),$$

and

$$\kappa_4 = 48(\nu + 4\Delta^2). \qquad (2.2.2)$$

We denote by $H_\nu(x; \Delta^2)$ the cdf of a noncentral chi-square distribution with ν degrees of freedom, and noncentrality parameter $\Delta^2/2$. Some of the results involving noncentral chi-square distributions (see also Judge and Bock, 1978, Appendix B) are given below in theorem/corollary form.

Note that

$$H_{\nu+2}(x; 0) = -h_{\nu+2}(x; 0) + H_\nu(x; 0). \qquad (2.2.3)$$

Theorem 1. Let χ^2_ν be a central chi-square variable with ν degrees of freedom and let $\varphi(\chi^2_\nu)$ be a measurable function of χ^2_ν. Then

$$E\Big[\chi^2_\nu \varphi(\chi^2_\nu)\Big] = \nu E\Big[\varphi(\chi^2_{\nu+2})\Big]. \qquad (2.2.4)$$

Theorem 2. Let $\chi^2_\nu(\Delta^2)$ be a noncentral chi-square variable with ν degrees of freedom and noncentrality parameter $\Delta^2/2$, and let $\varphi(\chi^2_\nu(\Delta^2))$ be a measurable function of $\chi^2_\nu(\Delta^2)$. Then

$$E\Big[\chi^2_\nu(\Delta^2)\varphi(\chi^2_\nu(\Delta^2))\Big] = \nu E\Big[\varphi(\chi^2_{\nu+2}(\Delta^2))\Big] + \Delta^2 E\Big[\varphi(\chi^2_{\nu+4}(\Delta^2))\Big]. \qquad (2.2.5)$$

Theorem 3. Let $\chi^2_\nu(\Delta^2)$ be a noncentral chi-square variable with ν degrees of freedom and noncentrality parameter $\Delta^2/2$, and let $\varphi(\chi^2_\nu(\Delta^2))$ be a measurable function of $\chi^2_\nu(\Delta^2)$. Then

$$E\Big\{\Big[\chi^2_\nu(\Delta^2)\Big]^2 \varphi(\chi^2_\nu(\Delta^2))\Big\} = \nu(\nu+2)E\Big[\varphi(\chi^2_{\nu+4}(\Delta^2))\Big]$$
$$+2(\nu+2)\Delta^2 E\Big[\varphi(\chi^2_{\nu+6}(\Delta^2))\Big] + \Delta^4 E\Big[\varphi(\chi^2_{\nu+8}(\Delta^2))\Big]. \qquad (2.2.6)$$

Further to the above, we have the following results involving the expectation of a product of a normal variable and functions of a noncentral chi-square variable:

Theorem 4. If $\mathcal{Z} \sim \mathcal{N}(\Delta, 1)$ and $\varphi(\mathcal{Z}^2)$ is a Borel measurable function, then

$$E\Big[\mathcal{Z}\varphi(\mathcal{Z}^2)\Big] = \Delta E\Big[\varphi(\chi_3^2(\Delta^2))\Big]. \tag{2.2.7}$$

Theorem 5. If $\mathcal{Z} \sim \mathcal{N}(\Delta, 1)$ and $\varphi(\mathcal{Z}^2)$ is a Borel measurable function, then

$$E\Big[\mathcal{Z}^2\varphi(\mathcal{Z}^2)\Big] = E\Big[\varphi(\chi_3^2(\Delta^2))\Big] + \Delta^2 E\Big[\varphi(\chi_5^2(\Delta^2))\Big]. \tag{2.2.8}$$

Theorem 6. Let $\mathbf{Z} = (Z_1, \cdots, Z_p)'$ be p-dimensional normal vector distributed as $\mathcal{N}_p(\boldsymbol{\mu}, \mathbf{I}_p)$. Then for a measurable function φ we have

$$E\Big[\mathbf{Z}\varphi(\mathbf{Z}'\mathbf{Z})\Big] = \boldsymbol{\mu} E\Big[\varphi(\chi_{p+2}^2(\Delta^2))\Big]. \tag{2.2.9}$$

Theorem 7. Let $\mathbf{Z} = (Z_1, \cdots, Z_p)'$ be a p-dimensional normal vector distributed as $\mathcal{N}_p(\boldsymbol{\mu}, \mathbf{I}_p)$. Then for a measurable function φ we have

$$E\Big[\mathbf{Z}\mathbf{Z}'\varphi(\mathbf{Z}'\mathbf{Z})\Big] = \mathbf{I}_p E\Big[\varphi(\chi_{p+2}^2(\Delta^2))\Big] + \boldsymbol{\mu}\boldsymbol{\mu}' E\Big[\varphi(\chi_{p+4}^2(\Delta^2))\Big]. \tag{2.2.10}$$

Theorem 8. Let $\mathbf{Z} = (Z_1, \cdots, Z_p)'$ be a p-dimensional normal vector distributed as $\mathcal{N}_p(\boldsymbol{\mu}, \mathbf{I}_p)$. Then for a measurable function φ we have for a positive definite matrix $\mathbf{A}$

$$E\Big[\mathbf{Z}'\mathbf{A}\mathbf{Z}\varphi(\mathbf{Z}'\mathbf{Z})\Big] = \mathrm{tr}(\mathbf{A}) E\Big[\varphi(\chi_{p+2}^2(\Delta^2))\Big] + \boldsymbol{\mu}'\mathbf{A}\boldsymbol{\mu} E\Big[\varphi(\chi_{p+4}^2(\Delta^2))\Big]. \tag{2.2.11}$$

For proof of Theorems 1 through 8 see Judge and Bock (1978).

Finally, we may obtain the following useful identities:

$$E\Big[\chi_{p+2}^{-2}(\Delta^2)\Big] = e^{-\Delta^2/2} \sum_{r\geq 0} \frac{1}{r!}(\Delta^2/2)^r \frac{1}{p+2r} = E_r(p+2r)^{-1}, \tag{2.2.13a}$$

where E_r stands for the expectation with respect to a Poisson variable r with mean $\Delta^2/2$.

$$\begin{aligned} E\Big[\chi_{p+2}^{-4}(\Delta^2)\Big] &= e^{-\frac{\Delta^2}{2}} \textstyle\sum_{r\geq 0} \frac{1}{r!}\left(\frac{\Delta^2}{2}\right)^r \frac{1}{(p+2r)(p+2r-2)} \\ &= E_r[(p+2r)(p+2r-2)]^{-1}. \end{aligned} \tag{2.2.13b}$$

$$\Delta^2 E\Big[\chi_{p+2}^{-2}(\Delta^2)\Big] = 1 - (p-2)E\Big[\chi_p^{-2}(\Delta^2)\Big]. \tag{2.2.13c}$$

$$\Delta^2 E\Big[\chi_{p+4}^{-4}(\Delta^2)\Big] = E\Big[\chi_{p+2}^{-2}(\Delta^2)\Big] - (p-2)E\Big[\chi_{p+2}^{-4}(\Delta^2)\Big]. \tag{2.2.13d}$$

$$2E\left[\chi_{p+2}^{-4}(\Delta^2)\right] = E\left[\chi_p^{-2}(\Delta^2)\right] - E\left[\chi_{p+2}^{-2}(\Delta^2)\right]. \tag{2.2.13e}$$

$$E\left[\chi_p^{-2}(\Delta^2)\right] \leq E\left[\chi_p^{-2}\right] = \frac{1}{p-2} \quad (p > 2). \tag{2.2.13f}$$

$$E\left[\chi_{p+2}^{-2} I(\chi_{p+2}^2(\Delta^2) < c)\right] = E_r(p+2r)^{-1} H_{p+2r}(c;0), \tag{2.2.13g}$$

$$E\left[\chi_{p+2}^{-4}(\Delta^2) I(\chi_{p+2}^2(\Delta^2) < c)\right] = E_r(p+2r)^{-1}(p-2+2r)^{-1} H_{p-2+2r}(c;0). \tag{2.2.13h}$$

2.3 Some Results from Multivariate Normal Theory

Let $\mathbf{x}$ be a p-variate normal vector with mean vector $\boldsymbol{\mu}$ and covariate $\boldsymbol{\Sigma}$ be denoted by $\mathcal{N}_p(\boldsymbol{\mu}, \boldsymbol{\Sigma})$. If $\mathbf{x}_1, \cdots, \mathbf{x}_N$ be a random sample of size N from $\mathcal{N}_p(\boldsymbol{\mu}, \boldsymbol{\Sigma})$.

The joint pdf of $\mathbf{x}_1, \cdots, \mathbf{x}_N$ is given by

$$\left\{(2\pi)^{p/2}|\boldsymbol{\Sigma}|\right\}^{-N/2} \exp\left\{-\frac{1}{2}\sum_{\alpha=1}^{N}(\mathbf{x}_\alpha - \boldsymbol{\mu})'\boldsymbol{\Sigma}^{-1}(\mathbf{x}_\alpha - \boldsymbol{\mu})\right\} \tag{2.3.1}$$

Let $\bar{\mathbf{x}} = \frac{1}{N}\sum_{\alpha=1}^{N}\mathbf{x}_\alpha$, and let

$$\mathbf{V} = \sum_{\alpha=1}^{N}(\mathbf{x}_\alpha - \bar{\mathbf{x}})(\mathbf{x}_\alpha - \bar{\mathbf{x}})'. \tag{2.3.2}$$

Then we have the decomposition given by

$$\sum_{\alpha=1}^{N}(\mathbf{x}_\alpha - \boldsymbol{\mu})(\mathbf{x}_\alpha - \boldsymbol{\mu})' = \mathbf{V} + N(\bar{\mathbf{x}}_\alpha - \boldsymbol{\mu})(\bar{\mathbf{x}}_\alpha - \boldsymbol{\mu})'. \tag{2.3.3}$$

Note that $\bar{\mathbf{x}}$ and $\mathbf{V}/N-1$ are unbiased estimators of $\boldsymbol{\mu}$ and $\boldsymbol{\Sigma}$, respectively, while $\mathbf{V}/N$ is the mle of $\boldsymbol{\Sigma}$ and is a biased estimator of $\boldsymbol{\Sigma}$. With respect to $\mathcal{N}_p(\boldsymbol{\mu}, \boldsymbol{\Sigma})$ the following properties may be noted:

1. $\bar{\mathbf{X}} \sim \mathcal{N}_p(\mu, \frac{1}{N}\boldsymbol{\Sigma})$
2. $\mathbf{V} \sim W_p(\boldsymbol{\Sigma}, n)$, where $W_p(\boldsymbol{\Sigma}, n)$ stands for the Wishart distribution with $n = N-1$ degrees of freedom. Then the following results may be verified
   a. $E[\mathbf{V}] = n\boldsymbol{\Sigma}$,
   b. $E[\mathbf{VAV}] = n(n+1)\boldsymbol{\Sigma}\mathbf{A}\boldsymbol{\Sigma} + n\text{tr}(\mathbf{A}\boldsymbol{\Sigma})\boldsymbol{\Sigma}$, where $\mathbf{A}$ is symmetric, and

c. $E[\mathbf{V}^{-1}] = \frac{1}{n-p-1}\mathbf{\Sigma}^{-1}$.

3. Let $T^2 = Nn\bar{\mathbf{x}}'\mathbf{V}^{-1}\bar{\mathbf{x}}$. Then

$$\frac{n-p+1}{p}\frac{T^2}{n} = \frac{N-p}{p}\frac{T^2}{n} \stackrel{\mathcal{D}}{=} F_{p,N-p}(\Delta^2) \tag{2.3.4}$$

where $\mathcal{D}$ stands for "equal in distribution" and $F_{p,N-p}(\Delta^2)$ denotes a noncentral F-distribution with $(p, N-p)$ degrees of freedom and noncentrality parameter $\Delta^2/2$, where $\Delta^2 = N\boldsymbol{\mu}'\mathbf{\Sigma}^{-1}\boldsymbol{\mu}$. Further, we note that

$$\frac{T^2}{n} = \frac{N\bar{\mathbf{x}}'\mathbf{V}^{-1}\bar{\mathbf{x}}}{N\bar{\mathbf{x}}'\mathbf{\Sigma}^{-1}\bar{\mathbf{x}}}N\bar{\mathbf{x}}'\mathbf{\Sigma}^{-1}\bar{\mathbf{x}} \stackrel{\mathcal{D}}{=} \frac{\chi_p^2(\Delta^2)}{\chi_{N-p}^2} = \frac{p}{N-p}F_{p,N-p}(\Delta^2), \tag{2.3.5}$$

which proves the assertion (2.3.4). Here $F_{m,n}(\Delta^2)$ stands for the noncentral F-variable with (m, n) degrees of freedom and noncentral parameter $\Delta^2/2$.

Theorem 1. If $\mathbf{X} \sim \mathcal{N}_p(\boldsymbol{\theta}, \mathbf{\Sigma})$, then

$$E[\mathbf{X}'\mathbf{A}\mathbf{X}] = \text{tr}[\mathbf{A}\mathbf{\Sigma}] + \boldsymbol{\theta}'\mathbf{A}\boldsymbol{\theta}.$$

Theorem 2. Let $\mathbf{C} \sim W_p(n, \mathbf{\Sigma})$. Then for any vector $\boldsymbol{\beta}$ and measurable function g we have

$$E\Big[g(\boldsymbol{\beta}'\mathbf{C}\boldsymbol{\beta})\mathbf{C}^{-1}\Big] = \frac{n-2}{n-p-1}E\Big[\frac{g(\boldsymbol{\beta}'\mathbf{\Sigma}\boldsymbol{\beta}\chi_n^2)}{\chi_n^2}\Big]\mathbf{\Sigma}^{-1}.$$

Theorem 3. If $\mathbf{X}_1, \mathbf{X}_2, \ldots, \mathbf{X}_N$ are i.i.d.r.v. with distribution $\mathcal{N}_p(\boldsymbol{\theta}, \mathbf{\Sigma})$ and if $\boldsymbol{\theta}$ has the prior distribution $\mathcal{N}_p(\boldsymbol{\theta}_0, \mathbf{A})$, then the joint distribution of $(\boldsymbol{\theta}', \bar{\mathbf{X}}_n')'$ is

$$\mathcal{N}_{2p}\left\{\begin{pmatrix} \boldsymbol{\theta}_0 \\ \boldsymbol{\theta}_0 \end{pmatrix}, \begin{pmatrix} \mathbf{A} & \mathbf{A} \\ \mathbf{A} & (N^{-1}\mathbf{\Sigma} + \mathbf{A}) \end{pmatrix}\right\}.$$

For more properties of the multivariate normal distribution, see Anderson (1984) and Srivastava and Khatri (1979) among others.

2.4 Beta Distribution and Applications

Let $\chi_{\nu_1}^2$ and $\chi_{\nu_2}^2$ be two independent central chi-square variables with ν_1 and ν_2 degrees of freedom , respectively. Define

$$X = \frac{\chi_{\nu_1}^2}{\chi_{\nu_1}^2 + \chi_{\nu_2}^2}. \tag{2.4.1}$$

Then X is distributed as a beta-variable with the pdf given by

$$\frac{1}{B(\frac{\nu_1}{2}, \frac{\nu_2}{2})}x^{\nu_1/2-1}(1-x)^{\nu_2/2-1}, \qquad 0 < x < 1. \tag{2.4.2}$$

The mean and variance of X are $\nu_1/\nu_1+\nu_2$ and $\nu_1\nu_2(\nu_1+\nu_2)^{-1}(\nu_1+\nu_2+1)^{-1}$, respectively.

Application I. Compute $E[XI(X>c)]$, where $I(A)$ is the indicator function of the set A:

$$\begin{aligned}
E[XI(X>c)] &= \int_c^1 \frac{x^{\nu_1/2}(1-x)^{\nu_2/2-1}}{B(\frac{\nu_1}{2},\frac{\nu_2}{2})}dx \\
&= \frac{B(\frac{\nu_1}{2}+1,\frac{\nu_2}{2})}{B(\frac{\nu_1}{2},\frac{\nu_2}{2})}\int_c^1 \frac{x^{\nu_1+2/2-1}(1-x)^{\nu_2/2-1}}{B(\frac{\nu_1+1}{2},\frac{\nu_2}{2})}dx \\
&= \frac{\nu_1}{\nu_1+\nu_2}\int_c^1 \frac{x^{\nu_1+2/2-1}(1-x)^{\nu_2/2-1}}{B(\frac{\nu_1+2}{2},\frac{\nu_2}{2})}dx, \qquad (2.4.3)
\end{aligned}$$

since

$$\frac{B(\frac{\nu_1}{2}+1,\frac{\nu_2}{2})}{B(\frac{\nu_1}{2},\frac{\nu_2}{2})} = \frac{\Gamma(\frac{\nu_1+2}{2})\Gamma(\frac{\nu_1+\nu_2}{2})}{\Gamma(\frac{\nu_1}{2})\Gamma(\frac{\nu_1+\nu_2+2}{2})} = \frac{\nu_1}{\nu_1+\nu_2}.$$

Application II. If $F^*_{p,m} = \chi^2_p/\chi^2_m$ with χ^2_p and χ^2_m independent chi-square variables, then the statistic $G = \chi^2_p + \chi^2_m$ is **independent** of $F^*_{p,m}$. Thus, we write

$$\chi^2_p = G\frac{F^*_{p,m}}{1+F^*_{p,m}}. \qquad (2.4.4)$$

Now, find the $E[\chi^2_p I(F^*_{p,m}>c)]$. In this case, we write

$$\begin{aligned}
E[\chi^2_p I(F^*_{p,m}>c)] &= E\Big[G\Big(\frac{F^*_{p,m}}{1+F^*_{p,m}}\Big)I(F^*_{p,m}>c)\Big] \\
&= (p+m)E\Big[\Big(\frac{F^*_{p,m}}{1+F^*_{p,m}}\Big)I(F^*_{p,m}>c)\Big] \\
&= p\int_{(c/1+c)}^1 \frac{x^{\frac{p}{2}}(1-x)^{\frac{m}{2}-1}}{B\Big(\frac{p+2}{2},\frac{m}{2}\Big)}dx, \qquad (2.4.5)
\end{aligned}$$

where x is a $B\Big(\frac{p+2}{2},\frac{m}{2}\Big)$-variable.

Application III. Let $T = \chi^2_{p+2r} + \chi^2_m$ and $F^*_{p+2r} = \chi^2_{p+2r}/\chi^2_m$. Consider $E[\varphi(F^*_{p+2r,m})\chi^2_{p+2r}]$, where φ is a B-measurable function. Then

$$\begin{aligned}
E\Big[\varphi(F^*_{p+2r,m})\chi^2_{p+2r}\Big] &= E\Big[\varphi(F^*_{p+2r,m})\Big(\frac{TF^*_{p+2r,m}}{1+F^*_{p+2r,m}}\Big)\Big] \\
&= (p+2r+m)E\Big[\varphi(F^*_{p+2r,m})\Big(\frac{F^*_{p+2r,m}}{1+F^*_{p+2r,m}}\Big)\Big] \\
&= (p+2r)E\big[\varphi(F^*_{p+2+2r,m})\big]. \qquad (2.4.6)
\end{aligned}$$

Also,

$$E\left[\varphi(F^*_{p+2r,m})(1+F^*_{p+2r,m}){F^*}^{-1}_{p+2r,m}\right] = \frac{m+p-2+2r}{p-2+2r}E[\varphi(F^*_{p-2+2r,m})] \tag{2.4.7}$$

and

$$E\left[\varphi(F^*_{p+2r,m})F^{*-1}_{p+2r,m}\right] = \frac{m}{p-2+2r}E\left[\varphi(F^*_{p-2+2r,m+2})\right].$$

For more information on these properties, see Sclove et al. (1972).

2.5 Discrete Distributions

In this section we only discuss the binomial and the multinomial discrete distributions with their properties useful for our purposes.

2.5.1 Binomial Distribution

Let $x_1, \ldots, x_n$ be n independent Bernoulli variables with parameter θ. Then $y = \sum_{i=1}^n x_i$ follows the binomial distribution

$$\binom{n}{y}\theta^y(1-\theta)^{n-y}, \quad y = 0, 1, \ldots, n. \tag{2.5.1}$$

The first moment and the next three central moments of y are given by $n\theta$, $n\theta(1-\theta)$, $n\theta(1-\theta)(1-2\theta)$, and $n\theta(1-\theta)[1+3(n-2)\theta(1-\theta)]$, respectively. One may see that the skewness and kurtosis of the distribution tend to 0 and 3, as $n \to \infty$, supporting the fact that the binomial distribution may be approximated by the normal distribution and we have the De Moivre-Laplace theorem (see Renyi, 1970, page 204).

Theorem 1. The binomial distribution

$$\left\{\binom{n}{y}\theta^y(1-\theta)^{n-y}, \quad y = 0, 1, \ldots, n\right\}$$

may be approximated as follows:

$$\binom{n}{y}\theta^y(1-\theta)^{n-y} \simeq \frac{\exp\left\{-\frac{1}{2}\frac{(y-n\theta)^2}{n\theta(1-\theta)}\right\}}{\sqrt{2\pi n\theta(1-\theta)}} \quad \text{if } |y-n\theta| = 0(n^{2/3}), \tag{2.5.2}$$

and if $-\infty < a < b < +\infty$, then we have

$$\lim_{n\to\infty} \sum_{a<\frac{y-n\theta}{\sqrt{n\theta(1-\theta)}}<b} \binom{n}{y}\theta^y(1-\theta)^{n-y} = \frac{1}{\sqrt{2\pi}}\int_a^b e^{-z^2/2}dz. \tag{2.5.3}$$

Further, we may show

Theorem 2. The cumulative probability for the binomial distribution satisfies

$$\sum_{y<n\theta+z\sqrt{n\theta(1-\theta)}} \binom{n}{y}\theta^y(1-\theta)^{n-y} = \Phi(z) + O(n^{-1/2}) \tag{2.5.4}$$

uniformly in $z \in (-\infty, \infty)$, where $\Phi(z)$ is the cdf of the standard normal variable.

2.5.2 Multinomial Distribution

Let us consider a population which may be classified into k mutually exclusive categories, namely $A_1, \ldots, A_k$. Let $(x_1, \ldots, x_k)'$ be the vector of frequencies observed in the k categories $A_1 \ldots A_k$ in n trials. Then the probability distribution of $\mathbf{x} = (x_1, \ldots, x_k)'$ is

$$n!\left(\prod_{j=1}^{k} x_j!\right)^{-1} \theta_1^{x_1}\theta_2^{x_1}\ldots\theta_k^{x_k}, \tag{2.5.5}$$

where $n = x_1 + \cdots + x_k$, $\boldsymbol{\theta} = (\theta_1, \ldots, \theta_k)'$, $\boldsymbol{\theta}'\mathbf{1}_k = 1$, and $P(A_i) = \theta_i$ $(i = 1, \ldots, k)$. The mean and covariance matrix of $\mathbf{x}$ are

$$E(\mathbf{X}) = n\boldsymbol{\theta} \quad \text{and} \quad \text{Cov}(\mathbf{X}) = n(\mathbf{D}_{\boldsymbol{\theta}} - \boldsymbol{\theta}\boldsymbol{\theta}'\} = n\mathbf{V}_{\boldsymbol{\theta}}, \tag{2.5.6}$$

where $\mathbf{D}_{\boldsymbol{\theta}} = \text{Diag}(\theta_1, \ldots, \theta_k)$.

The following theorem (see Renyi, 1970, pp. 211–212) is an extension of Theorems 2.5.1 and 2.5.2 with respect to the binomial distribution:

Theorem 3.

$$n!\left\{\prod_{j=1}^{k} x_j!\right\}^{-1} \prod_{j=1}^{k}\theta_j^{x_j}$$

$$= \frac{1}{(\sqrt{2\pi n})|\mathbf{D}_{\boldsymbol{\theta}}|^{1/2}} \exp\left\{-\frac{1}{2}(\mathbf{x} - n\boldsymbol{\theta})'\mathbf{D}_{\boldsymbol{\theta}}^{-1}(\mathbf{x} - n\boldsymbol{\theta})\right\} + O(n^{-1/2}). \tag{2.5.7}$$

Now, setting $Z_i = (x_i - n\theta_i)/\sqrt{n\theta_i(1-\theta_i)}$, we obtain (2.5.7) as

$$\frac{1}{\sqrt{2\pi n}|\mathbf{D}_{\boldsymbol{\theta}}|^{1/2}} \exp\left\{-\frac{1}{2}\mathbf{Z}'\mathbf{B}\mathbf{Z}\right\} + O(n^{-1/2}), \tag{2.5.8}$$

where $\mathbf{Z} = (Z_1, \ldots, Z_n)'$ and

$$\mathbf{B} = ((b_{ij}))$$

with

$$b_{jj} = \theta_j\frac{(1-\theta_j)}{\theta_k} + (1-\theta_j), \quad j = 1, \ldots, k$$

and

$$b_{jk} = \frac{\theta_j(1-\theta_j)\theta_k(1-\theta_k)}{\theta_k^2}, \quad j \neq k. \tag{2.5.9}$$

Hence,

$$\lim \sum_{x_j \leq n\theta_j + z_j\sqrt{n\theta_j(1-\theta_j)}} n! \left\{ \prod_{i=1}^{k} (x_i!) \right\}^{-1} \prod_{j=1}^{k} \theta_j^{x_j}$$

$$= \int_{-\infty}^{Z_1} \cdots \int_{-\infty}^{Z_{k-1}} \frac{\exp\{-\frac{1}{2}\mathbf{Z}'\mathbf{B}\mathbf{Z}\}}{(\sqrt{2\pi n}\sqrt{\theta_1, \theta_2, \cdots, \theta_k}} dz_i \ldots dz_{k-1} + O(n^{-1/2}). \tag{2.5.10}$$

For more properties of the binomial and multinomial distributions see Cramér (1946), Kendall and Stuart (1963), Bishop, Fienberg, and Holland (1975), Agresti (1990), Renyi (1970), and Sen and Singer (1993) among others.

2.6 Matrix Results

The following results are presented as theorems or corollaries:

Let $\mathbf{A}$ be an $n \times n$ matrix. Then the roots of the determinantal (characteristic equation)

$$|\mathbf{A} - \lambda\mathbf{I}| = 0 \tag{2.6.1}$$

are called the characteristic roots of $\mathbf{A}$ or eigenvalues of $\mathbf{A}$.

Theorem 1. If $\mathbf{A}$ is an $n \times n$ matrix, then there exists an orthogonal matrix $\mathbf{\Gamma}$ such that $\mathbf{\Gamma}'\mathbf{A}\mathbf{\Gamma} = \mathbf{D} = \text{Diag}(d_1, \cdots, d_n)$. If $\mathbf{A}$ is positive semidefinite, then $d_i \geq 0, i = 1, \cdots, n$, and if $\mathbf{A}$ is positive definite, then $d_i > 0, i = 1, \cdots, n$.

Theorem 2. Consider the partition of the nonsingular matrix $\mathbf{A}$ given by

$$\mathbf{A} = \begin{pmatrix} \mathbf{A}_{11} & \mathbf{A}_{12} \\ \mathbf{A}_{21} & \mathbf{A}_{22} \end{pmatrix}, \tag{2.6.2}$$

where $\mathbf{A}_{22}$ is nonsingular. Let $\mathbf{A}_{11.2} = \mathbf{A}_{11} - \mathbf{A}_{12}\mathbf{A}_{22}^{-1}\mathbf{A}_{21}$. Then

$$\mathbf{A}^{-1} = \begin{pmatrix} \mathbf{A}_{11.2}^{-1} & -\mathbf{A}_{11.2}^{-1}\mathbf{A}_{12}\mathbf{A}_{22} \\ -\mathbf{A}_{22}^{-1}\mathbf{A}_{21}\mathbf{A}_{11.2} & \mathbf{A}_{22}^{-1}\mathbf{A}_{21}\mathbf{A}_{11.2}\mathbf{A}_{12}\mathbf{A}_{22}^{-1} + \mathbf{A}_{22}^{-1} \end{pmatrix}. \tag{2.6.3}$$

Further,

$$\mathbf{A}_{22.1}^{-1} = \mathbf{A}_{22}^{-1}\mathbf{A}_{21}\mathbf{A}_{11.2}^{-1}\mathbf{A}_{12}\mathbf{A}_{22}^{-1} + \mathbf{A}_{22}^{-1}. \tag{2.6.4}$$

Theorem 3. If $\mathbf{A}$ is an $n \times n$ positive definite matrix and $\mathbf{B}$ is an $n \times m$ $(m < n)$ matrix with rank m, then $\mathbf{B}'\mathbf{A}\mathbf{B}$ is positive definite.

If $\mathbf{A}$ is an $n \times n$ idempotent matrix, meaning $\mathbf{A} = \mathbf{A}'$ and $\mathbf{AA} = \mathbf{A}$, then as long as the rank of $\mathbf{A}$ is r, $\mathbf{A}$ has r positive roots and $n - r$ roots equal to 0, the rank of $\mathbf{A}$ is equal to $\text{tr}(\mathbf{A})$, and there exists an orthogonal matrix $\mathbf{\Gamma}$ such that

$$\mathbf{\Gamma}'\mathbf{A}\mathbf{\Gamma} = \begin{pmatrix} \mathbf{I}_r & \mathbf{0} \\ \mathbf{0} & \mathbf{0}_{n-r} \end{pmatrix}. \tag{2.6.5}$$

Theorem 4. If $\mathbf{A}$ and $\mathbf{B}$ are two symmetric matrices, a necessary and sufficient condition for an orthogonal matrix $\mathbf{\Gamma}$ to exist, such that $\mathbf{\Gamma}'\mathbf{A}\mathbf{\Gamma} = \mathbf{D}_1$ and $\mathbf{\Gamma}'\mathbf{B}\mathbf{\Gamma} = \mathbf{D}_2$ where $\mathbf{D}_1$ and $\mathbf{D}_2$ are both diagonal matrices, is that $\mathbf{A}$ and $\mathbf{B}$ satisfy $\mathbf{AB} = \mathbf{BA}$, meaning $\mathbf{A}$ and $\mathbf{B}$ commute.

Theorem 5 (Courant Theorem). Let $\lambda_1, \cdots, \lambda_n$ be the characteristic roots of an $n \times n$ matrix $\mathbf{A}$ such that $\min \lambda_i = \lambda_1, \max \lambda_i = \lambda_n$, and let $v_1, \cdots, v_n$ be the characteristic vectors. Then $\mathbf{A} = \lambda_1 v_1 v_1' + \cdots + \lambda_n v_n v_n', \mathbf{I} = v_1 v_1' + \cdots + v_n v_n' : \sup\left(\frac{\mathbf{x}'\mathbf{A}\mathbf{x}}{\mathbf{x}\mathbf{x}'}\right) = \lambda_n$ and $\inf\left(\frac{\mathbf{x}'\mathbf{A}\mathbf{x}}{\mathbf{x}\mathbf{x}'}\right) = \lambda_1$. Hence,

$$Ch_{\min}(\mathbf{A}) \leq \frac{\mathbf{x}'\mathbf{A}\mathbf{x}}{\mathbf{x}\mathbf{x}'} < Ch_{\max}(\mathbf{A}), \tag{2.6.6}$$

where

$$\min_i \lambda_i = Ch_{\min}(\mathbf{A}) \qquad \text{and} \qquad \max_i \lambda_i = Ch_{\max}(\mathbf{A}). \tag{2.6.7}$$

Theorem 6. If $\mathbf{A}$ is an $n \times n$ matrix and $\mathbf{b}$ is an $n \times 1$ vector. Then

$$\max_{\mathbf{x}} \frac{\mathbf{x}'\mathbf{b}\mathbf{b}'\mathbf{x}}{\mathbf{x}'\mathbf{A}\mathbf{x}} = \mathbf{b}'\mathbf{A}^{-1}\mathbf{b}. \tag{2.6.8}$$

Theorem 7. If $\mathbf{A}$ is an $n \times n$ positive definite matrix and $\mathbf{b}$ is an $n \times 1$ vector, a necessary and sufficient condition for $\left(\mathbf{A}^{-1} - \mathbf{b}\mathbf{b}'\right)$ to be positive definite is that $\mathbf{b}'\mathbf{A}\mathbf{b}$ is less than 1. Further, $\mathbf{b}\mathbf{b}' - \mathbf{A}^{-1}$ cannot be negative definite for n exceeding 1.

Theorem 8. Let $\mathbf{A}$ and $\mathbf{D}$ be nonsingular matrices of order p_1 and p_2 and $\mathbf{B}$ be a $p_1 \times p_2$ matrix. Then

$$\left(\mathbf{A} + \mathbf{BDB}'\right)^{-1} = \mathbf{A}^{-1} - \mathbf{A}^{-1}\mathbf{B}\left(\mathbf{B}'\mathbf{A}^{-1}\mathbf{B} + \mathbf{D}^{-1}\right)^{-1}\mathbf{B}'\mathbf{A}^{-1}. \tag{2.6.9}$$

Theorem 9. If $\mathbf{A}$ and $\mathbf{D}$ are square matrices such that $\mathbf{A}$ is nonsingular, then

$$\begin{vmatrix} \mathbf{A} & \mathbf{C} \\ \mathbf{B} & \mathbf{D} \end{vmatrix} = |\mathbf{A}||\mathbf{D} - \mathbf{B}\mathbf{A}^{-1}\mathbf{C}|. \tag{2.6.10}$$

Theorem 10. If $\mathbf{A}$ be a $p \times p$ matrix and $\mathbf{U}$ a column $p \times 1$ vector, then

$$|\mathbf{A} + \mathbf{U}\mathbf{U}'| = |\mathbf{A}|\left(\mathbf{I} + \mathbf{U}'\mathbf{A}^{-1}\mathbf{U}\right) \qquad \text{if} \qquad |\mathbf{A}| \neq 0 \tag{2.6.11}$$

Application I.

$$\left(\sigma^2\mathbf{I}_n + \tau^2\mathbf{P}_X\right)^{-1} = \sigma^{-2}\left(\mathbf{I}_n - (1-B)\mathbf{P}_X\right), \tag{2.6.12}$$

where $B = \sigma^2/(\sigma^2 + \tau^2)$ and $\mathbf{P}_X = \mathbf{X}(\mathbf{X}'\mathbf{X})^{-1}\mathbf{X}'$ and $\mathbf{X}$ is a $n \times p$ matrix. Here, $\mathbf{B} = \mathbf{X}$ and $\mathbf{D} = (\mathbf{X}'\mathbf{X})^{-1}$ of Theorem 8.

Application II.

$$(aI + b\mathbf{W}\mathbf{W}')^{-1} = \frac{1}{a}\left(I + \frac{b}{a + b\mathbf{W}'\mathbf{W}}\right) \tag{2.6.13}$$

and

$$\left|aI + b\mathbf{W}\mathbf{W}'\right| = a^p\left(1 + \frac{b}{a}\mathbf{W}'\mathbf{W}\right). \tag{2.6.14}$$

Theorem 11. Let $\mathbf{A}$ be a $p \times p$ positive definite matrix, $\mathbf{B}$ a $p \times m$ matrix, and $\mathbf{B}_1$ a $p \times n$ matrix such that the rank of $\mathbf{B}_1$ equals p minus the rank of $\mathbf{B}$ and $\mathbf{B}'\mathbf{B} = \mathbf{0}$. Then

$$\mathbf{A}^{-1} - \mathbf{A}^{-1}\mathbf{B}(\mathbf{B}'\mathbf{A}^{-1}\mathbf{B})^{-1}\mathbf{B}'\mathbf{A}^{-1} = \mathbf{B}_1(\mathbf{B}_1'\mathbf{A}\mathbf{B}_1)^{-1}\mathbf{B}_1'$$

is a positive semidefinite matrix of rank $= p-$ rank $\mathbf{B}$.

Theorem 12. If $\mathbf{A}$ and $\mathbf{B}$ are two $p \times p$ positive matrices, then

(i) $$\mathbf{A} - \mathbf{A}(\mathbf{A}+\mathbf{B})^{-1}\mathbf{A} = (\mathbf{A}^{-1} + \mathbf{B}^{-1})^{-1}, \tag{2.6.15}$$
meaning the l.h.s. equals the inverse of the sum of the inverse of $\mathbf{A}$ and $\mathbf{B}$.

(ii) $$\mathbf{I} - \mathbf{A}(\mathbf{A}+\mathbf{B})^{-1} = \mathbf{B}(\mathbf{A}+\mathbf{B})^{-1}, \tag{2.6.16}$$
meaning the l.h.s. equals the product of B and the inverse of the sum of A and B.

For more results on matrix theory see Anderson (1984), Rao (1973), and Srivastava and Khatri (1979) among others.

2.7 Large Sample Theory

Finite sample theory is generally based on the assumption of normality. In practice, this assumption may not hold and asymptotic theory is invoked. The results are parallel to normal theory. In the understanding of asymptotic theory, we look forward to four distinct types of convergence of sequence of random variables to a limit. We provide only a limited amount of results which achieve our goal in the book. For details, we refer the readers to Feller (1957), Renyi (1970), Rohatgi and Saleh(2001), Sen and Singer (1993), and Ferguson(1996).

In the next section we provide the definition of each type of convergence.

2.7.1 Four Types of Convergence

There are two basic forms of convergence (1) sequences of distribution functions $F_n(x)$ converging to a distribution function $F(x)$ for all continuity points, x as $n \to \infty$, and (2) sequences of random variables $\{X_n | n = 1, 2, \dots\}$ converging to a random variable X as $n \to \infty$.

First we consider the convergence of sequences of distribution functions.

Definition 2.1. A sequence of random variables (r.v.'s) $\{T_n\}$ converges in distribution to T, meaning $T_n \xrightarrow{\mathcal{D}} T$, if $F_{T_n}(x) \to F_T(x)$ for all continuity points x of $F_T(x)$.

Convergence in distribution does not imply convergence of moments, nor does it imply convergence of probability density functions.

Definition 2.2. A sequence of r.v. $\{T_n\}$ converges in probability to a r.v. T if for every $\epsilon > 0$,

$$P\{|T_n - T| > \epsilon\} \to 0 \qquad \text{as } n \to \infty.$$

This is written as

$$T_n \xrightarrow{\mathrm{P}} T \qquad \text{or} \qquad T_n = T + O_p(1).$$

The convergence says that the sequence of probabilities $P\{|T_n - T|\} > \epsilon\}$ converges to 0 as $n \to \infty$. It does not say that $|T_n - T| < \epsilon$ for $n \geq N$ for a suitable sample size, N.

As a consequence of the definition we can easily see that

$$g(T_n) \xrightarrow{\mathrm{P}} g(T) \qquad \text{as } n \to \infty,$$

when g is a continuous function. Hence, $g(T_n) \xrightarrow{\mathrm{P}} g(c)$ when $T_n \xrightarrow{\mathrm{P}} c$, where c is a constant.

The following theorem is important:

Theorem 1. $T_n \xrightarrow{\mathrm{P}} T$ implies $T_n \xrightarrow{\mathcal{D}} T$ as $n \to \infty$.

Definition 2.3. A sequence of r.v.'s $\{T_n\}$ such that $E|T_n|^r < \infty$ for some $r > 0$ converges in the rth mean to the r.v. T if $E|T|^r < \infty$ and $E|T_n - T|^r \to 0$ as $n \to \infty$.

This is written as

$$T_n \xrightarrow{r^{\text{th}}} T \qquad \text{as } n \to \infty.$$

Convergence in r^{th} means implies convergence in probability, meaning $T_n \xrightarrow{r^{\text{th}}} T$ for some $r > 0$ implies $T_n \xrightarrow{\mathrm{P}} T$. If $r = 2$, then $\mathrm{Var}(T_n) \to \mathrm{Var}(T)$. In statistical methodology, convergence in the quadratic mean is very useful and will be used frequently in this book.

Finally, we consider the concept of almost sure convergence, sometimes called convergence of a sequence of r.v.'s $\{T_n\}$ to T with probability one or strong convergence of $\{T_n\}$ to T.

Definition 2.4. A sequence of r.v.'s $\{T_n\}$ converges almost sure to the r.v. T, meaning $T_n \overset{a.s.}{\to} T$, if $P\{\lim_{n\to\infty} T_n = T\} = 1$.

The relationship of the four types of convergence concepts is given by the following diagram:

$$T_n \overset{a.s.}{\to} T$$

$$\searrow$$

$$T_n \overset{\mathrm{P}}{\to} T \Longrightarrow F_{T_n}(x) \to F_T(x) \text{ for all continuity points of } F_T(x)$$

$$\nearrow$$

$$T_n \overset{r^{\text{th}}}{\to} T$$

Thus, we see that convergence in distribution is the weakest of all. We observe that this convergence in distribution is implied by the rest of the three and is most important for statistical inference since the limiting distribution is used to determine the asymptotic distributional bias, risks and confidence sets or the significance of tests of parameters of interest.

2.7.2 Law of Large Numbers

When $\{T_n\}$ convergences to a constant, say, c, then it falls in the domain of the **law of large numbers**. Accordingly, when $T_n \overset{\mathrm{P}}{\to} c$ or $T_n \overset{\mathcal{D}}{\to} c$, then we have the weak law of large numbers. If $T_n \overset{a.s.}{\to} c$, then we have the strong law of large numbers. If a distribution has a finite second moment, the law of large numbers is simplest and useful for statistical inference and the convergence is in quadratic mean.

Three laws of large numbers are related to the idea of consistency of estimators/tests. Weak laws are proved based on characteristic functions and the continuity theorem. The following theorem states the weak/strong law of large numbers with regard to the sample mean:

Theorem 2. Let $X_1, \cdots, X_n$ be i.i.d., and let $\bar{X}_n = \frac{1}{n}\sum_{i=1}^{n} X_i$ be the sample mean. Then:

(i) (Weak law) If $E|X| < \infty$, then $\bar{X}_n \overset{\mathrm{P}}{\to} E(\overline{X}) = \mu$.

(ii) (Strong law) $\bar{X}_n \overset{a.s.}{\to} \mu$ if and only if $E(X) < \infty$ and $\mu = E(X)$.

(iii) If $E|X|^2 < \infty$, then $\bar{X}_n \xrightarrow{2^{nd} \text{ mean}} \mu = E(X)$.

The proof may be obtained by (1) the characteristic function, (2) Chebycheff's inequality, or (3) the Borel-Cantelli Lemma (a.s.), which are stated below.

Chebycheff's Inequality If T is a nonnegative r.v. such that $E|T| < \infty$, then for every $c > 0$,

$$P\{T > cE(T)\} < \frac{1}{c}.$$

Borel-Cantelli Lemma. Let $\{A_n\}$ be a sequence of events such that $\sum_{n=1}^{\infty} P(A_n) < \infty$. Then

(1) $P(\lim A_n) = 0$,

(2) If $\{A_n\}$ are independent events, then $P(\lim A_n) = 1$.

Markov Inequality Let T be a nonnegative r.v. with finite r^{th} mean, meaning $\mu_r' = E(T)^r < \infty$, for some $r > 0$, Then for every $\epsilon > 0$,

$$P(T > \epsilon) \leq \epsilon^{-r}\mu_r'.$$

2.7.3 Central Limit Theorems

In determining confidence intervals or the level of significance α in test of a hypothesis, we need the exact distribution of the related statistics. In the nonnormal case, under certain regularity conditions, we may determine the asymptotic distribution that may be used to set up confidence intervals, and determine type I and power of tests based on the statistic in question.

It is well known that if $X_1, \cdots, X_n$ are i.i.d. r.v. from a cdf $F(x)$ with finite mean μ and variance σ^2, then for the sample mean $\bar{X}_n$ we have

$$\frac{\sqrt{n}}{\sigma}(\bar{X}_n - \mu) \sim \mathcal{N}(0,1) \qquad \text{as } n \to \infty.$$

This is the basic central limit theorem (CLT). However, there are situations where $X_1, \cdots, X_n$ are not i.i.d. r.v. but satisfy

$$E(X_i) = 0, V(X_i) = \sigma_i^2 < \infty, \quad (i = 1, \cdots, n).$$

Now, let $T_n = X_1 + \cdots + X_n$ and $\sigma_n^2 = \sigma_1^2 + \cdots + \sigma_n^2$. Then we may ask the question:

When is $T_n/\sigma_n \sim N(0,1)$ as $n \to \infty$?

The answer lies in the Lindeberg-Feller theorem, which gives the uniform integrability condition: If for every $\epsilon > 0$,

$$\frac{1}{\sigma_n^2}\sum_{i=1}^{n} E\Big\{X_i^2 I\Big(|X_i| \geq \epsilon\sigma_n\Big)\Big\} \to 0 \qquad \text{as } n \to \infty,$$

then

$$\frac{T_n}{\sigma_n} \sim \mathcal{N}(0,1).$$

The next theorem called the **Hajek-Sidak CLT** deals with the *asymptotic normality* of a linear combination of i.i.d. r.v.'s $X_1, \ldots, X_n$ meaning of $T_n = a_1X_1 + \cdots + a_nX_n$, where $E(X_i) = \mu$ and $E(X_i - \mu)^2$ are both finite.

Hajek-Sidak CLT. Let $X_1, \ldots, X_n$ be i.i.d. r.v. such that $E(X_i) = \mu$ and $\operatorname{Var} X_i = \sigma^2$ are both finite. Let

$$T_n = a_1X_1 + \cdots + a_nX_n.$$

Then

$$\frac{T_n - \mu \sum_{i=1}^{n} a_i}{\sigma\sqrt{(\sum_{i=1}^{n} a_i^2)}} \overset{\mathcal{D}}{\simeq} N(0,1)$$

whenever the Noether condition

$$\frac{\max_{1\le i\le n} a_i^2}{\sum_{i=1}^{n} a_i^2} \longrightarrow 0 \quad \text{as } n \to \infty$$

is satisfied.

Sometimes we have to deal with functions of two or more statistics, say, $g(T_n, S_n)$ where $F_{T_n}(x) \to F(x)$ and $S_n \overset{\mathrm{P}}{\to} c$ (constant). Then we need the limiting distribution of $g(T_n, S_n)$.

In this case, we can use the **Slusky theorem**. Stated in a simplest form the Slutsky theorem is as follows:
Let $X_1, \cdots, X_n$ be a sequence of r.v. with distributions $F_1, \cdots, F_n$, and suppose $F_n(x) \to F(x)$ as $n \to \infty$. Further, let $Y_1, \cdots, Y_n$ be another set of r.v.'s such that $Y_n \overset{\mathrm{P}}{\to} c$. Set $V_n = X_n + Y_n$, $W_n = X_nY_n$ and $Z_n = X_n/Y_n$. Then, as $n \to \infty$, the distribution of

(1) V_n tends to $F(x + c)$

(2) W_n tends to $F(cx)$

(3) Z_n tends to $F(x/c)$ for $c > 0$ as $n \to \infty$.

An important concept that dominates the asymptotic theory of rank-statistics is that of "contiguity of probability measures." We define the concept in the following:

Definition 2.5. Let $\{p_n\}$ and $\{q_n\}$ be the sequence of simple hypothesis densities defined in a measure space $(\mathcal{Y}, \mathcal{B}_n, \mu_n)$. If

$$P_n(A_n) \to 0 \text{ implies } Q_n(A_n) \to 0 \text{ as } n \to \infty,$$

then $\{Q_n\}$ are contiguous to $\{P_n\}$. Here $\partial P(A_n)/\partial\mu_n = p_n$ and $\partial Q(A_n)/\partial\mu_n = q_n$.

Generally, we are interested in the asymptotic distribution of statistics $\{T_n(Y)\}$. Then convergence of $\{T_n(Y)\} \to 0$ under $\{P_n\}$ implies $\{T_n(Y)\} \to 0$ under $\{Q_n\}$ if $\{Q_n\}$ is contiguous to $\{P_n\}$.

In this context the likelihood ratio statistics plays an important part. Consider the likelihood ratio statistic

$$\begin{aligned} L_n(Y) &= \tfrac{q_n}{p_n} \quad \text{if } p_n > 0 \\ &= 1 \quad \text{if } p_n > 0 \\ &= \infty \quad \text{if } 0 = p_n < q_n. \end{aligned}$$

If $\log L_n(Y)$ is asymptotically $N(-\sigma^2/2, \sigma^2)$ under $\{P_n\}$, then $\{Q_n\}$ is contiguous to $\{P_n\}$ (LeCam's lemma 1).

Let $T_n(Y)$ be a statistic. Assume that under $\{P_n\}$, $(T_n(Y), \log L_n(Y))$ has asymptotically bivariate normal distribution with mean (θ_1, θ_2) and covariance matrix $\begin{pmatrix} \sigma_{11} & \sigma_{12} \\ \sigma_{21} & \sigma_{22} \end{pmatrix}$. Then, under $\{Q_n\}$, the asymptotic distribution of $T_n(Y)$ is $N(\mu_1 + \sigma_{12}, \sigma_{11})$. (LeCam's lemma 3). LeCam's lemma 2 gives conditions when $\log L_n(Y)$ is asymptotically $N(-\frac{1}{2}\sigma^2, \sigma^2)$.

For further information on contiguity see Hajek and Sidak (1967), Hajek, Sidak, and Sen (1999), Puri and Sen (1986), and Jurečkova and Sen (1996) among others.

2.8 Nonparametric Theory: Preliminaries

In this section we present some of the basic preliminaries of nonparametric methods which are relevant in the discussion of the R-estimators of the parameters of the models in various chapters. The proofs and details of technicalities are available in Hajek (1969), Hajek and Sidak (1967), Hajek, Sidak, and Sen (1999), Puri and Sen (1986), Randles and Wolfe (1979), and Sen (1981) among others.

2.8.1 Order-Statistics, Ranks, and Sign Statistics

Let $Y_1, Y_2, \ldots, Y_n$ be n i.i.d. r.v. with absolutely continuous cdf $F(y)$ with absolutely continuous pdf $f(y)$. Let $\mathbf{Y}_{(\cdot)} = (Y_{(1)}, \cdots, Y_{(n)})'$ be the order-statistic vector corresponding to the i.i.d. r.v. $\mathbf{Y} = (Y_1, \ldots, Y_n)'$, which we write as

$$\mathbf{Y} = (Y_1, \ldots, Y_n)' = (Y_{(R_1)}, \cdots, Y_{(R_n)})' \tag{2.8.1}$$

where R_j is the rank of Y_j among $Y_1, \ldots, Y_n$. Thus, the joint distribution of $\mathbf{Y}_{(\cdot)}$ is

$$n! \prod_{i=1}^{n} f(y_{(i)}) \quad \text{on} \quad -\infty < y_{(1)} < \cdots < y_{(n)} < \infty, \tag{2.8.2}$$

and the joint distribution of $\mathbf{R} = (R_1, \ldots, R_n)'$ of the rank-vector is

$$\frac{1}{n!} \qquad \text{for } \mathbf{R} \in \mathcal{R}_n, \tag{2.8.3}$$

where $\mathcal{R}_n$ is the set of $n!$ permutations of the integer-vector $(1, 2, \ldots, n)$.

If the pdf of Y_j $(j = 1, \ldots, n)$ is **symmetric** about 0, then we may define four statistics relevant to the distribution:

(i) The absolute-value statistics

$$|\mathbf{Y}| = (|Y_1|, \ldots, |Y_n|)'. \tag{2.8.4a}$$

(ii) The absolute-value order-statistics

$$|\mathbf{Y}|_{(\cdot)} = (|Y_1|_{(R_1^+)}, \ldots, |Y_n|_{(R_n^+)})'. \tag{2.8.4b}$$

(iii) The rank-statistic, $\mathbf{R}^+ = (R_1^+, \ldots, R_n^+)'$, where

$$R_j^+ \text{ is the rank of } |Y_j| \text{ among } |Y_1|, \ldots, |Y_n|. \tag{2.8.4c}$$

(iv) The sign-statistic $\mathbf{S} = (\operatorname{sign} Y_1, \ldots, \operatorname{sign} Y_n)$. Clearly,

$$\mathbf{Y}_j = |Y_j| \operatorname{sign} Y_j, \quad j = 1, \ldots, n. \tag{2.8.4d}$$

The distributions of $|\mathbf{Y}|_{(\cdot)}$, $\mathbf{R}^+$, and $\mathbf{S}$ are independent with distributions given by

(i) $2^n n! \prod_{k=1}^n f(z_i)$, $0 < z_i < \cdots < z_n < \infty$,

(ii) $\frac{1}{n!}$ for $\mathbf{R}^+ \in \mathcal{R}_n$,

(iii) $\frac{1}{2^n}$ for $S \in \mathcal{S}_n$, the set of 2^n points. (2.8.5)

2.8.2 Linear rank-statistics (LRS)

Let $(Y_1, \ldots, Y_n)$ be a sample of size n from the cdf $f(y)$/pdf $f(y)$ that satisfies the following conditions:

(i) $F \in \mathcal{F}$, a class of absolutely continuous cdf's with absolutely continuous pdf $f(x)$ such that $f(x) = f(-x)$ for all $x \in (-\infty, \infty)$.

(ii) $$I(f) = \int_{-\infty}^{\infty} \left[\frac{f'(z)}{f(z)}\right]^2 f(z)dz < \infty, \tag{2.8.6}$$

meaning f has finite Fisher's information.

Let $\{\varphi(u);\ - < u < 1\}$ be the class of nonconstant, nondecreasing, and square integrable functions. Further, let

$$\varphi^+(u) = \varphi\left(\frac{1+u}{2}\right), \quad \varphi(u) + \varphi(1-u) = 0. \tag{2.8.7}$$

Now, consider a sample of size n from the uniform distribution $U(0,1)$, and define the scores for every n (≥ 1)

$$a_n(i) = E[\varphi(U_{in})] \ \text{ or } \ \varphi\Big(\frac{i}{n+1}\Big),$$

$$a_n^+(i) = E[\varphi^+(U_{in})] \ \text{ or } \ \varphi^+\Big(\frac{i}{n+1}\Big), \tag{2.8.8}$$

for $i = 1, 2, \ldots, n$, when $0 < U_{1n} < U_{2n} < \cdots < U_{nn} < 1$ are the order-statistics of the sample $(U_1, \ldots, U_n)'$.

In this book we are interested in the *linear rank-statistics* (related to location and regression parameter) of the type:

$$T_n(0) = n^{-1/2} \sum_{j=1}^{n} a_n^+(R_j^+) \ \text{sign}\ Y_j, \tag{2.8.9}$$

$$L_n(0) = n^{-1/2} \sum_{j=1}^{n} (x_j - \bar{x}_n) a_n(R_j), \quad \bar{x}_n = n^{-1}\Big(\sum_{i=1}^{n} x_i\Big). \tag{2.8.10}$$

Some results are given in the following theorems (see also Hajek and Sidak (1967), Chapter V). First, the mean and variance of the statistics are given by:

Theorem 1. Under the assumed regularity conditions,

(i) $E[T_n(0)] = 0$,

(ii) $\text{Var}[T_n(0)] = \frac{1}{4} \sum_{i=1}^{n} [a_n^+(i)]^2$,

and

(i) $E[L_n(0)] = 0$,

(ii) $\text{Var}[L_n(0)] = \frac{1}{n(n-1)} \Big[\sum_{i=1}^{n} (x_i - \bar{x}_n)^2\Big] \Big[\sum_{i=1}^{n} (a_n(i) - \bar{a}_n)^2\Big]$. $\qquad$ (2.8.11)

Standard Scores. For the symmetric location distributions $F(x - \theta)$ and for the test of $\theta = 0$, the scores are

$$a_n^+(i, f) = E\left[-\frac{f'(F^{-1}(\frac{1}{2} + \frac{1}{2}U_{in}))}{f(F^{-1}\frac{1}{2} + \frac{1}{2}U_{in}))}\right], \quad 0 < u < 1, \tag{2.8.12}$$

where $F^{-1}(U)$, $(0 < u < 1)$ is the quantile of order u of the cdf $F(y)$ and the score generating function is defined by

$$\psi(u, f) = -\frac{f'(F^{-1}(\frac{1}{2} + \frac{1}{2}u_{in}))}{f(F^{-1}(\frac{1}{2} + \frac{1}{2}u_{in}))}, \quad 0 < u < 1. \tag{2.8.13}$$

The Wilcoxon test is associated with the score function

$$\psi(u) = u, \quad 0 < u < 1, \tag{2.8.14a}$$

and the normal score test is associated with

$$\psi(u) = \Phi^{-1}\left(\frac{1}{2} + \frac{1}{2}u\right), \quad 0 < u < 1. \tag{2.8.14b}$$

For the regression model, the scores are defined by

$$a_n(i, f) = E\left[-\frac{f'(F^{-1}(U_{in}))}{f(F^{-1}(U_{in}))}\right], \quad 0 < u < 1, \tag{2.8.15a}$$

with the score generating function

$$\psi(u, f) = -\frac{f'(F^{-1}(u))}{f(F^{-1}(u))}, \quad 0 < u < 1. \tag{2.8.15b}$$

Theorem 2. Let

$$T_n(0) = n^{-1/2}\sum_{i=1}^{n} a_n^+(R_i^+) \text{ sign} Y_i, \tag{2.8.16}$$

where the scores $\{a_n^+(i)|i = 1, \dots, n\}$ converge to some square integrable function $\varphi^+(u)$ such that $\int_0^1 [\varphi^+(u)]^2 du > 0$. Then $T_n(0)$ follows approximately the normal distribution with mean 0 and variance $\int_0^1 [\varphi^+(u)]^2 du$.

If $\varphi^+(u) = \psi^+(u, f)$, then the variance is $I(f)$.

Theorem 3. Let

$$L_n(0) = n^{-1/2}\sum_{i=1}^{n}(x_i - \bar{x}_n)a_n(R_i), \tag{2.8.17}$$

where the scores $\{a_n(i)|i = 1, \dots, n\}$ converge to some square integrable function $\varphi(u)$ such that $A_\varphi^2 = \int_0^1 [\varphi(u) - \bar{\varphi}]^2 du > 0$, $\bar{\varphi} = \int_0^1 \varphi(u) du$. Then $L_n(0)$ follows approximately the normal distribution with mean 0 and variance

$$A_\varphi^2 Q = \left(\lim_{n\to\infty} n^{-1}Q_n\right)\int_0^1 [\varphi(u) - \bar{\varphi}]^2 du, \quad Q_n = \sum_{i=1}^{n}(x_i - \bar{x}_n)^2$$

and $Q = \lim n^{-1}Q_n$.

If $\varphi(u) = \psi(u, f)$, then $\bar{\varphi} = 0$ and

$$A_\varphi^2 = \int_0^1 \psi^2(u, f) du = I(f). \tag{2.8.18}$$

Theorem 4. (Saleh and Sen, 1978). Let

$$T_n(\mathbf{0}) = n^{-1/2}\sum_{i=1}^{n} a_n^+(R_1^+) \text{ sign } Y_i,$$

$$L_n(\mathbf{0}) = n^{-1/2}\sum_{i=1}^{n}(x_i - \bar{x}_n)a_n(R_i), \tag{2.8.19}$$

where $\{a_n^+(i)\}$ and $\{a_n(i)\}$ converge to some square integrable functions $\varphi^+(u)$ and $\varphi(u)$ such that $A_\varphi^2 = \int_0^1[\varphi(u) - \bar{\varphi}]^2du > 0$, $\bar{\varphi} = \int_0^1 \varphi(u)du$ and $\int_0^1[\varphi^+(u)]^2du < \infty$. Then

$$\left[\sqrt{n}T_n(0), \sqrt{n}L_n(0)\right]' \simeq N_2\left\{\begin{pmatrix} 0 \\ 0 \end{pmatrix}, A_\varphi^2 \text{ Diag}(1, Q)\right\}. \tag{2.8.20}$$

Now we consider the following linear rank-statistics related to multiple regression models:

$$T_n(\mathbf{0}) = n^{-1/2}\sum_{i=1}^{n} a_n^+(R_i^+) \text{ sign } Y_i,$$

$$\mathbf{L}_n(\mathbf{0}) = n^{-1/2}\sum_{i=1}^{n}(\mathbf{x}_i - \bar{\mathbf{x}}_n)a_n(R_i)$$

$$= (L_{1n}(\mathbf{0}), \ldots, L_{q,n}(\mathbf{0}))', \quad \bar{x}_{in} = n^{-1}\sum_{j=1}^{n} x_{ijn}, \tag{2.8.21}$$

where

$$\bar{\mathbf{x}}_i = n^{-1}\sum_{i=1}^{n}\mathbf{x}_i \quad \text{and} \quad \mathbf{Q}_n = \sum_{i=1}^{n}(\mathbf{x}_i - \bar{\mathbf{x}}_n)(\mathbf{x}_i - \bar{\mathbf{x}}_{in})'$$

is a $q \times q$ matrix satisfying the general Noether condition given by

$$\max_{1\le i\le n}(\mathbf{x}_i - \bar{\mathbf{x}}_{in})'\mathbf{Q}_n^{-1}(\mathbf{x}_i - \bar{\mathbf{x}}_{in}) \to 0 \text{ as } n \to \infty, \tag{2.8.22}$$

along with the generalized inverse $\mathbf{Q}_n^{-1}$ of the matrix $\mathbf{Q}_n$. Then

$$\left\{\begin{matrix} \sqrt{n} & T_n(\mathbf{0}) \\ 1\sqrt{n} & \mathbf{L}_n(\mathbf{0}) \end{matrix}\right\} \sim N_{p+1}\left\{\begin{pmatrix} \mathbf{0} \\ \mathbf{0} \end{pmatrix} A_\varphi^2 \begin{pmatrix} 1 & \mathbf{0} \\ \mathbf{0} & \mathbf{Q} \end{pmatrix}\right\}, \tag{2.8.23}$$

where $\mathbf{Q} = \lim_{n\to\infty} n^{-1}\mathbf{Q}_n$. Let $A_n^2 = (n-1)^{-1}\sum_{i=1}^{n}[a_n(i) - \bar{a}_n]^2$. Then $\mathcal{L}_n = A_n^{-2}\mathbf{L}_n'(\mathbf{0})\mathbf{Q}_n^{-1}\mathbf{L}_n(\mathbf{0})$ follows a central chi-square distribution with q degrees of freedom

2.8.3 Rank Estimators of the Parameters of Various Models

Location Model. Consider the location model

$$\mathbf{Y}_n = \theta \mathbf{1}_n + \mathbf{e}, \tag{2.8.24}$$

where $\mathbf{Y}_n = (Y_1, \ldots, Y_n)'$, $\mathbf{1}_n = (1, \ldots, 1)'$ and $\mathbf{e} = (e_1, \ldots, e_n)$ and the error-distribution is the absolutely continuous and symmetric cdf $F(e)$ with absolutely continuous symmetric pdf $f(e)$ such that $f(e) = f(-e)$ and $I(f) < \infty$ (see equation 2.8.6). Then consider the linear rank-statistics

$$T_n(a) = n^{-1/2} \sum_{i=1}^{n} a_n^+(R_i^+) \operatorname{sign}(Y_i - a), \tag{2.8.25}$$

where R_i^+ is the rank of $|Y_i - a|$ among $|Y_i - a|, \ldots, |Y_n - a|$ and

$$a_n^+(i) = E\varphi^+(U_{in}) \quad \text{or} \quad \varphi^+\left(\frac{1}{2} + \frac{1}{2}\frac{i}{n+1}\right), \tag{2.8.26}$$

as defined in Section 2.8.2. For every given $\mathbf{Y}$, $T_n(a)$ is a decreasing in a. Also, if $a = 0$, then $T_n(0)$ is symmetrically distributed about 0. Thus, following Adichie (1967) and Hodges and Lehmann (1963) among others, we define the R-estimator of the location parameter θ as

$$\tilde{\theta}_n = \frac{1}{2}\left\{\tilde{\theta}_n^{(1)} + \tilde{\theta}_n^{(2)}\right\}, \tag{2.8.27}$$

where $\tilde{\theta}_n^{(1)} = \sup\{a : T_n(a) > 0\}$ and $\tilde{\theta}_n^{(2)} = \inf\{a : T_n(a) < 0\}$. Then, based on Puri and Sen (1986) we state the following theorem:

Theorem 5. Under the assumed regularity condition as $n \to \infty$,

$$\sqrt{n}(\tilde{\theta}_n - \theta) \sim N(\mathbf{0}, \sigma^2), \quad \sigma^2 = A_\varphi^2/\gamma^2(\varphi, \psi), \tag{2.8.28}$$

where

$$\psi(u, f) = -\frac{f'(F^{-1}(\frac{1}{2} + \frac{1}{2}u))}{f(F^{-1}(\frac{1}{2} + \frac{1}{2}u))}, \quad 0 < u < 1,$$

$$\gamma(\varphi, \psi) = \int_0^1 \varphi^+(u)\psi(u, f)du. \tag{2.8.29}$$

Location and Regression Model. Consider the location and regression model

$$\mathbf{Y}_n = \theta \mathbf{1}_n + \beta \mathbf{x} + \mathbf{e}, \tag{2.8.30}$$

where $\mathbf{Y}_n = (Y_1, \ldots, Y_n)'$, $\mathbf{x} = (x_1, \ldots, x_n)'$, and $\mathbf{e} = (e_1, \ldots, e_n)$ and the error distribution satisfy the same condition as in Section 2.8.4. Then, consider the linear rank-statistics

$$T_n(a, b) = n^{-1/2} \sum_{i=1}^{n} a_n^+(R_i^+) \operatorname{sign}(Y_i - a - bx_i),$$

where R_i^+ is the rank of $|Y_i - a - bx_i|$ among $|Y_1 - a - bx_1|, \ldots, |Y_n - a - bx_n|$, and

$$L_n(\mathbf{b}) = n^{-1/2} \sum_{i=1}^{n} (x_i - \bar{x}_n) a_n(R_i), \tag{2.8.31}$$

where R_i is the rank of $Y_i - bx_i$ and $Y_1 - bx_1, \ldots, Y_n - b_n x_n$. Then the R-estimators of (θ, β) are given by

$$\tilde{\beta}_n = \frac{1}{2}(\tilde{\beta}_n^{(1)} + \tilde{\beta}_n^{(2)}), \quad \text{where} \quad \tilde{\beta}_n^{(1)} = \sup\{b : L_n(b) > 0\},$$
$$\tilde{\beta}_n^{(2)} = \inf\{b : L_n(b) < 0\},$$

and

$$\tilde{\theta}_n^{(1)} = \frac{1}{2}(\tilde{\theta}_n^{(1)} + \tilde{\theta}_n^{(2)}), \quad \text{where} \quad \tilde{\theta}_n^{(1)} = \sup\{a : T_n(a, \tilde{\beta}_n) > 0\}$$
$$\tilde{\theta}_n^{(2)} = \inf\{a : T_n(a, \tilde{\beta}_n) < 0\}. \tag{2.8.32}$$

By the fact that $T_n(0,0)$ and $L_n(0)$ are symmetrically distributed around the origin, 0, the asymptotic distribution of the estimators is as given in the following theorem:

Theorem 6. (Adichie, 1967). Under the assumed regularity condition and $\dfrac{\max_{1 \le i \le n}(x_i - \bar{x}_n)^2}{\Sigma(x_i - \bar{x}_n)^2} \to 0$ as $n \to \infty$, we have

$$\begin{pmatrix} \sqrt{n}(\tilde{\theta}_n - \theta) \\ \sqrt{n}(\tilde{\beta}_n - \beta) \end{pmatrix} \sim N_2 \left\{ \begin{pmatrix} 0 \\ 0 \end{pmatrix}, \sigma^2 \begin{pmatrix} 1 + \frac{\bar{x}}{Q} & -\frac{\bar{x}}{Q} \\ -\frac{\bar{x}}{Q} & \frac{1}{Q} \end{pmatrix} \right\}, \tag{2.8.33}$$

where $Q = \lim n^{-1} Q_n$, $Q_n = \sum_{i=1}^{n} (x_i - \bar{x}_n)^2$ and $\lim \bar{x}_n = \bar{x}$.

Multiple Regression Model. Consider the regression model

$$\mathbf{Y}_n = \theta \mathbf{1}_n + \mathbf{X}\boldsymbol{\beta} + \mathbf{e}, \tag{2.8.34}$$

where $\mathbf{Y}_n = (Y_1, \ldots, Y_n)'$, $\mathbf{X}$ is a $n \times p$ matrix of constants, and $\mathbf{e} = (e_1, \ldots, e_n)'$. Regression parameters are $(\theta, \boldsymbol{\beta}')'$ when $\boldsymbol{\beta} = (\beta_1, \ldots, \beta_p)'$. Consider the LRS defined by

$$T_n(a, \mathbf{b}) = n^{-1/2} \sum_{i=1}^{n} a_n^+(R_i^+) \operatorname{sign}(Y_i - a - \mathbf{b}'\mathbf{x}_i)$$

and

$$\mathbf{L}_n(\mathbf{b}) = n^{-1/2}\sum_{j=1}^{n}(\mathbf{x}_i - \bar{\mathbf{x}}_n)a_n(R_i), \quad i = 1, \dots, p, \tag{2.8.35}$$

where R_i^+ is the rank of $|Y_i - a - \mathbf{b}'\mathbf{x}_i|$ among $|Y_1 - a - \mathbf{b}'\mathbf{x}_1|, \dots, |Y_n - a - \mathbf{b}'\mathbf{x}_n|$ and R_i is the rank of $(Y_1 - a - \mathbf{b}'\mathbf{x}_i)$ among $(y_1 - a - \mathbf{b}'\mathbf{x}), \dots, (Y_n - a - \mathbf{b}'\mathbf{x}_n)$. To obtain the point estimates of $(\theta, \boldsymbol{\beta}')'$, we define the estimator of $\boldsymbol{\beta}$ as the center of gravity of the set $\mathbf{D}_n$ given by

$$\mathbf{D}_n = \{\mathbf{b}_n : ||\mathbf{L}_n(\mathbf{b})|| \text{ is minimum}\}, \tag{2.8.36a}$$

namely

$$\tilde{\boldsymbol{\beta}}_n = \text{ center of gravity of } \mathbf{D}_n. \tag{2.8.36b}$$

The R-estimator of θ is defined by

$$\tilde{\theta}_n = \frac{1}{2}\left(\tilde{\theta}_n^{(1)} + \tilde{\theta}_n^{(2)}\right),$$

where

$$\tilde{\theta}_n^{(1)} = \sup\left\{a : T_n(a, \tilde{\boldsymbol{\beta}}_n) > 0\right\}$$

and

$$\tilde{\theta}_n^{(2)} = \inf\left\{a : T_n(a, \tilde{\boldsymbol{\beta}}_n) < 0\right\}. \tag{2.8.37}$$

The asymptotic distribution of the R-estimators of $(\theta, \boldsymbol{\beta}')'$ is given by

$$\begin{pmatrix} \sqrt{n}(\tilde{\theta} - \theta) \\ \sqrt{n}(\tilde{\boldsymbol{\beta}}_n - \boldsymbol{\beta}) \end{pmatrix} \sim N_{p+1}\left\{\begin{pmatrix} 0 \\ \mathbf{0} \end{pmatrix}, \sigma^2 \begin{pmatrix} 1 + \bar{\mathbf{x}}'\mathbf{Q}^{-1}\bar{\mathbf{x}} & -\bar{\mathbf{x}}'\mathbf{Q}^{-1} \\ \mathbf{Q}^{-1}\bar{\mathbf{x}} & \mathbf{Q}^{-1} \end{pmatrix}\right\}, \tag{2.8.38}$$

where $\mathbf{Q} = \lim n^{-1}\mathbf{Q}_n$ and $\mathbf{Q}_n$ is defined after (2.8.21).

Asymptotic Linearity Results for Linear Rank-Statistics. We observed that R-estimators are not explicit functionals of ranks. The usual methodology fails to obtain an asymptotic distribution theory of these R-estimators. Jurečkova (1969, 1971) opened the door toward asymptotic linearity results in rank theory. Rank estimators have thus been popularized since 1970 as a means of proving the asymptotic theory of the R-estimators (see Saleh and Sen, 1978). Here, we present only the results related to the multiple regression model in Section 2.8.3 because they have broad application to the topics covered in this book.

Theorem 7. Consider the statistics $T_n(a, \mathbf{b})$ and $\mathbf{L}_n(\mathbf{b})$ defined by (2.8.35) and the rank estimators of θ and $\boldsymbol{\beta}$ given by (2.8.36a, b) and (2.8.37). Assume the *following conditions* together with (2.8.6) through (2.8.9):

(i) $\lim_{n\to\infty} n^{-1}\mathbf{Q}_n \to \mathbf{Q}$.

(ii) $\max_{1\leq i\leq n}(\mathbf{x}_i-\bar{\mathbf{x}}_n)'\mathbf{Q}_n^{-1}(\mathbf{x}_i-\bar{\mathbf{x}}_n)=o(n)$,

where $\mathbf{x}_i$ is the ith row of $\mathbf{X}$ and $\bar{\mathbf{x}}_n=n^{-1}\sum_{i=1}^n \mathbf{x}_i$.

$$\text{(iii)}\ I(f)=\int_{-\infty}^{\infty}\left[\frac{f'(t)}{f(t)}\right]^2 f(t)dt<\infty. \tag{2.8.39}$$

Let $\|\mathbf{a}\|=\sum_{k=1}^p |a^{(k)}|$, where $\mathbf{a}=(a^{(1)},\ldots,a^{(p)})'$. Then the R-estimator of $\boldsymbol{\beta}$ is defined by any central point of the set

$$\mathbf{S_b}=\{\mathbf{b}:\|\mathbf{L}_n(\mathbf{b})\|^2\ \text{ is minimum}\} \tag{2.8.40}$$

denoted by $\tilde{\boldsymbol{\beta}}_n$. Under $H_0:\theta=0$, $\boldsymbol{\beta}=\mathbf{0}$, the linearity results are as follows:

$$\text{(i)}\ \lim_{n\to\infty}P_0\left\{\sup_{\|\boldsymbol{\delta}\|<k}\|\mathbf{L}_n(n^{-\frac{1}{2}}\boldsymbol{\delta})-\mathbf{L}_n(\mathbf{0})+\gamma(\psi,\phi)\boldsymbol{\delta}'\mathbf{Q}\|>\varepsilon\right\}\xrightarrow{P}0, \tag{2.8.41}$$

where $k>0$ and $\boldsymbol{\delta}\in R^p$ and $\|\cdot\|$ denotes the p-dimensional Euclidean norm with

$$\gamma(\psi,\phi)=\int_0^1\psi(u)\phi(u)du. \tag{2.8.42}$$

$$\text{(ii)}\ \lim_{n\to\infty}P_0\left\{\sup_{|\delta_0|<k_0,\|\boldsymbol{\delta}\|<k}\left|T_n(n^{-1/2}\delta_0,n^{-1/2}\boldsymbol{\delta})-T_n(0,\mathbf{0})\right.\right.$$
$$\left.\left.+n^{-1/2}(\delta_0+\boldsymbol{\delta}'\bar{\mathbf{x}}_n)\gamma(\psi,\phi)\right|>\varepsilon\right\}\to 0. \tag{2.8.43}$$

These two results will help us to obtain asymptotic properties of the R-estimators.

2.9 Problems

1. If $X_1,\ldots,X_n$ are i.i.d. r.v. with $E(X_i)=\mu$ and $\text{Var}(X_i)=\sigma^2$
   (a) Show that the asymptotic distribution of $\sqrt{n}\frac{(\bar{X}-\mu)}{\sigma}$ is $N(0,1)$.
   (b) Show that the asymptotic distribution of $\sqrt{n}\frac{(\bar{X}-\mu)}{s_x}$ is $N(0,1)$, where $\bar{X}=n^{-1}(X_1+\ldots+X_n)$ and $s_x^2=n^{-1}\sum_{i=1}^n(X_i-\bar{X})^2$.
2. Prove the Stein identities (2.1.2) through (2.1.3b).
3. Verify the expressions for the cumulants of the noncentral chi-square distribution with ν degrees of freedom and noncentrality parameter $\Delta^2/2$ given by (2.2.2).
4. Prove Theorems 1 through 8.

5. Verify the identities given at (2.2.13a) through (2.2.13h).

6. Refer to Section 2.3. If $\mathbf{V} \sim W_p(\boldsymbol{\Sigma}, n)$, then (a) $E[\mathbf{V}] = n\boldsymbol{\Sigma}$ (b) $E[\mathbf{VAV}] = n(n+1)\boldsymbol{\Sigma}\mathbf{A}\boldsymbol{\Sigma} + n\,\mathrm{tr}(\mathbf{A}\boldsymbol{\Sigma})\boldsymbol{\Sigma}$, where $\mathbf{A}$ is a symmetric matrix, and (c) $E[\mathbf{V}^{-1}] = (n-p-1)^{-1}\boldsymbol{\Sigma}^{-1}$.

7. Prove Theorem 2 of Section 2.3.

8. Refer to Section 2.5. Prove that $\mathbf{Z} = \mathbf{V}_{\theta}^{-1/2}(\mathbf{x} - n\boldsymbol{\theta})$ converges to k-variate normal distribution with mean $\mathbf{0}$ and covariance matrix $\mathbf{B}$ as $n \to \infty$.

9. Refer to Section 2.7.1. Prove Theorem 1.

10. Refer to Section 2.6.2. Prove Theorem 2.

11. Prove the Lindeberg-Feller theorem.

12. Refer to Section 2.6.3. Prove the Hajek-Sidak central limit theorem.

13. Prove Slusky's theorem.

14. Prove LeCam's lemma 2.

15. Prove Theorem 2.8.1.

16. Prove Theorems 2.8.2 and 2.8.3.

17. Prove Theorem 1 of Section 2.8.4.

18. Prove that

$$\begin{pmatrix} \sqrt{n}(\tilde{\theta} - \theta) \\ \sqrt{n}(\tilde{\boldsymbol{\beta}}_n - \boldsymbol{\beta}) \end{pmatrix} \sim N_{p+1}\left\{ \begin{pmatrix} 0 \\ \mathbf{0} \end{pmatrix}, \sigma^2 \begin{pmatrix} 1 + \bar{\mathbf{x}}'\mathbf{Q}^{-1}\bar{\mathbf{x}} & -\bar{\mathbf{x}}'\mathbf{Q}^{-1} \\ \mathbf{Q}^{-1}\bar{\mathbf{x}} & \mathbf{Q}^{-1} \end{pmatrix} \right\},$$

where $\sigma^2 = A_{\varphi}^2/\gamma^2(\varphi, \psi)$.

Chapter 3

Preliminary Test Estimation

Outline

In the statistical literature preliminary test estimation of parameters was introduced by Bancroft (1944, 1964, 1965) to estimate the parameters of a model when it is suspected that some "uncertain prior information" on the parameter of interest is available. The method involves a statistical test of the "uncertain prior information" based on an appropriate statistic and a decision on whether the model based sample estimate or the prior information based estimate of the model parameters should be taken.

In this chapter we consider the simple linear model. We introduce the preliminary test estimation (PTE) of the *intercept parameter* when it is a priori suspected that the *slope parameter* has a pre-specified value. This problem was studied by Ahsanullah and Saleh (1972) and extended by Ahmed and Saleh (1988). We analytically evaluate the performance of the preliminary test estimator of the intercept parameter relative to the standard estimate (least squares or maximum likelihood) with respect to the mean-square error (mse) criterion.

Next, as a special case of the simple linear model, we consider the classical problem of *pooling means* in a two-sample situation due to Han and Bancroft (1968). Given two samples, the problem is to estimate the means when it is a priori suspected that the two population means may be equal. Accordingly we evaluate the preliminary test estimator against the traditional sample mean, using the mse criterion.

Finally, we consider the one-sample problem of estimating the mean based on a previous experimental result when it is suspected that the population mean has a pre-specified value. In this case we also assess the preliminary test estimator relative to the sample mean based on the mse criterion.

In addition, we consider the one-sample problem of *preliminary test estimation of variance* when it is a priori suspected that the mean has a pre-specified value. An extended problem is discussed by Stein (1964).

We also provide an alternative method of estimation that competes well, and discuss its merit compared to the preliminary test method of estimation.

All these studies are carried out assuming normal distribution of the errors with unknown variance. Also, we provide an asymptotic theory under nonnormal errors, and nonparametric methods of estimation due to Adichie (1967) and Saleh and Sen (1978). Lastly, problems are added to expand the results in various forms.

3.1 Simple Linear Model, Estimators, and Tests

3.1.1 Simple Linear Model

Consider the simple linear model with slope β, and intercept θ, given by

$$\mathbf{Y} = \theta \mathbf{1}_n + \beta \mathbf{x} + \mathbf{e}, \tag{3.1.1}$$

where $\mathbf{Y} = (Y_1, \ldots, Y_n)'$ is a vector of n observations, $\mathbf{x} = (x_1, \ldots, x_n)'$ is a vector of n known constants, $\mathbf{1}_n = (1, \ldots, 1)'$ is a vector of n 1's, and $\mathbf{e} = (e_1, \ldots, e_n)'$ is a vector of n independent errors such that $\mathbf{e} \sim \mathcal{N}_n(\mathbf{0}, \sigma^2 \mathbf{I}_n)$ with $\mathbf{I}_n$ the identity matrix of order n.

3.1.2 Estimation of the Intercept and Slope Parameter

Using the model (3.1.1) and the sample information from a normal distribution, we obtain the maximum likelihood estimator (MLE)/least squares estimator (LSE) of $(\theta, \beta)'$ as

$$\begin{pmatrix} \tilde{\theta}_n \\ \tilde{\beta}_n \end{pmatrix} = \begin{pmatrix} \bar{Y} - \tilde{\beta}_n \bar{x} \\ \frac{1}{Q}[\mathbf{x}'\mathbf{Y} - \frac{1}{n}(\mathbf{1}_n'\mathbf{x})(\mathbf{1}_n'\mathbf{Y})] \end{pmatrix}, \tag{3.1.2}$$

where

$$Q = \mathbf{x}'\mathbf{x} - \frac{1}{n}(\mathbf{1}_n'\mathbf{x})^2. \tag{3.1.3}$$

The exact distribution of $(\tilde{\theta}_n, \tilde{\beta}_n)'$ is a bivariate normal with mean $(\theta, \beta)'$ and covariance matrix

$$\frac{\sigma^2}{n}\begin{pmatrix} 1 + \frac{n\bar{x}^2}{Q} & -\frac{n\bar{x}}{Q} \\ -\frac{n\bar{x}}{Q} & \frac{n}{Q} \end{pmatrix}. \tag{3.1.4}$$

An unbiased estimator of the variance σ^2 is given by

$$s_e^2 = \frac{1}{n-2}\left(\mathbf{Y} - \tilde{\theta}_n\mathbf{1}_n - \tilde{\beta}_n\mathbf{x}\right)'\left(\mathbf{Y} - \tilde{\theta}_n\mathbf{1}_n - \tilde{\beta}_n\mathbf{x}\right), \tag{3.1.5}$$

which is independent of $(\tilde{\theta}_n, \tilde{\beta}_n)'$, and $(n-2)(s_e^2/\sigma^2)$ follows a central chi-square distribution with $n-2$ d.f.

3.1.3 Test for the Slope Parameter

Suppose that we want to test the null hypothesis $H_0 : \beta = \beta_0$ against $H_A : \beta \neq \beta_0$. Then we use the likelihood ratio (LR) test-statistic

$$\mathcal{L}_n = \frac{(\tilde{\beta}_n - \beta_0)^2 Q}{s_e^2}, \tag{3.1.6}$$

which follows a noncentral F-distribution with $(1, m = n-2)$ d.f. and non-centrality parameter $\Delta^2/2$ where

$$\Delta^2 = \frac{(\beta - \beta_0)^2 Q}{\sigma^2}. \tag{3.1.7}$$

Under H_0, $\mathcal{L}_n$ follows a central F-distribution. At the α-level of significance we obtain the critical value $F_{1,m}(\alpha)$ from this distribution and reject H_0 if $\mathcal{L}_n \geq F_{1,m}(\alpha)$; otherwise, we accept H_0.

3.2 PTE of the Intercept Parameter

This section deals with the problem of estimation of the **intercept parameter** θ when it is suspected that the slope parameter β may be β_0. As a special case it covers the two-sample problem of estimating one mean when it is suspected that the two means may be equal. Also, one-sample estimation of mean is obtained by letting $\mathbf{x} = \mathbf{0}$ and prior information $\theta = \theta_0$.

3.2.1 Unrestricted, Restricted, and Preliminary Test Estimators of the Intercept Parameter

From Section 3.1.2., we know that the unrestricted estimator (UE) of θ is

$$\tilde{\theta}_n = \bar{Y} - \tilde{\beta}_n \bar{x}. \tag{3.2.1}$$

If we suspect β to be β_0, then the restricted estimator (RE) of θ is given by

$$\hat{\theta}_n = \bar{Y} - \beta_0 \bar{x}. \tag{3.2.2}$$

In practice, the prior information that $\beta = \beta_0$ is **uncertain**. The doubt on this prior information can be removed by using "Fisher's recipe" of testing the hypothesis $H_0 : \beta = \beta_0$ against the alternative $H_A : \beta \neq \beta_0$. As a result of this test we choose $\tilde{\theta}_n$ or $\hat{\theta}_n$ based on the rejection or acceptance of H_0. Accordingly, we write the estimator as

$$\hat{\theta}_n^{\text{PT}} = \hat{\theta}_n I(\mathcal{L}_n < F_{1,m}(\alpha)) + \tilde{\theta}_n I(\mathcal{L}_n \geq F_{1,m}(\alpha)), \quad m = n-2, \tag{3.2.3}$$

called the preliminary test estimator (PTE), where $F_{1,m}(\alpha)$ is the α-level upper critical value of a central F-distribution with $(1, n-2)$ d.f. and $I(A)$ is the indicator function of the set A. For more details on PTE, see Judge and Bock (1978), Ahmed and Saleh (1988), and Ahsanullah and Saleh (1972). We can rewrite $\hat{\theta}_n^{\text{PT}}$ as

$$\hat{\theta}_n^{\text{PT}} = \tilde{\theta}_n + (\tilde{\beta}_n - \beta_0)\bar{x} I(\mathcal{L}_n < F_{1,m}(\alpha)). \tag{3.2.4}$$

If $\alpha = 1$, $\tilde{\theta}_n$ is always chosen, if $\alpha = 0$, $\hat{\theta}_n$ is chosen. Since $0 < \alpha < 1$, $\hat{\theta}_n^{\text{PT}}$ in repeated samples will result in a combination of $\tilde{\theta}_n$ and $\hat{\theta}_n$. Note that the PTE procedure leads to the choice of one of the two values, namely, either $\tilde{\theta}_n$ or $\hat{\theta}_n$. Also, the PTE procedure depends on the level of significance α.

3.2.2 Bias and MSE Expressions of the Estimators of the Intercept Parameter

Since our interest is to compare the UE, RE, and PTE of θ with respect to a bias and the mse, we obtain the expressions of these quantities in the following theorems and proofs. First we consider the theorem for the bias and the quadratic bias expressions.

Theorem 1.

(i) $b_1(\tilde{\theta}_n) = 0, \qquad B_1(\tilde{\theta}_n) = 0,$ (3.2.5a)

(ii) $b_2(\hat{\theta}_n) = (\beta - \beta_0)\bar{x}, \qquad B_2(\hat{\theta}_n) = \frac{\bar{x}^2}{Q}\Delta^2,$ (3.2.5b)

(iii) $b_3(\hat{\theta}_n^{\text{PT}}) = (\beta - \beta_0)\bar{x} G_{3,m}(\frac{1}{3}F_{1,m}(\alpha); \Delta^2)$, $m = n-2$

$$B_3(\hat{\theta}_n^{\text{PT}}) = \frac{\bar{x}^2}{Q}\Delta^2\Big\{G_{3,m}\Big(\tfrac{1}{3}F_{1,m}(\alpha); \Delta^2\Big)\Big\}^2, \tag{3.2.5c}$$

where $G_{m_1,m_2}(\cdot;\Delta^2)$ is the cdf of a noncentral F-distribution with (m_1, m_2) d.f. and noncentrality parameter $\Delta^2/2$. Here B_1, B_2, and B_3 are quadratic bias expressions for $\tilde{\theta}_n$, $\hat{\theta}_n$ or $\hat{\theta}_n^{PT}$.

Proof. The expressions for $b_1(\tilde{\theta}_n)$ and $B_1(\tilde{\theta}_n)$ are obvious. For $b_2(\hat{\theta}_n)$, we have

$$b_2(\hat{\theta}_n) = E(\hat{\theta}_n - \theta)$$
$$= E[\tilde{\theta}_n - \theta + (\tilde{\beta}_n - \beta_0)\bar{x}] = (\beta - \beta_0)\bar{x}$$

$$B_2(\hat{\theta}_n) = \frac{(\beta - \beta_0)^2\bar{x}^2}{\sigma^2} = \frac{\bar{x}^2}{Q}\Delta^2. \tag{3.2.6a}$$

Next,

$$b_3(\hat{\theta}_n^{\mathrm{PT}}) = E\Big[\tilde{\theta}_n + \Big(\tilde{\beta}_n - \beta_0)\bar{x}I(\mathcal{L}_n < F_{1,m}(\alpha)) - \theta\Big]$$
$$= \bar{x}E\big[(\tilde{\beta}_n - \beta_0)I(\mathcal{L}_n < F_{1,m}(\alpha))\big]$$
$$= \frac{\bar{x}\sigma}{\sqrt{Q}}E\left\{ZI\left(\frac{Z^2}{s_e^2/\sigma^2} < F_{1,m}(\alpha)\right)\right\}, \quad Z = \frac{(\tilde{\beta}_n - \beta_0)}{\sigma}\sqrt{Q},$$

where $Z \sim N(\Delta, 1)$ and Z is *independent* of s_e^2. We can evaluate the expectation conditionally on ms_e^2/σ^2.

Thus,

$$b_3\left(\hat{\theta}_n^{\mathrm{PT}}\right) = \frac{\sigma\bar{x}}{\sqrt{Q}}\Delta\, G_{3,m}\left(\frac{1}{3}F_{1,m}(\alpha);\Delta^2\right) \qquad \text{by Theorem 2.2.4}$$

and

$$B_3\left(\hat{\theta}_n^{\mathrm{PT}}\right) = \frac{\bar{x}^2}{Q}\Delta^2\left\{G_{3,m}\left(\frac{1}{3}F_{1,m}(\alpha);\Delta^2\right)\right\}^2. \tag{3.2.6b}$$

Next, we consider expressions for the mean square errors of $\tilde{\theta}_n$, $\hat{\theta}_n$, and $\hat{\theta}_n^{\mathrm{PT}}$.

Theorem 2. The mse expressions for $\tilde{\theta}_n$, $\hat{\theta}_n$, and $\hat{\theta}_n^{\mathrm{PT}}$ are given by

$$M_1(\tilde{\theta}_n) = \frac{\sigma^2}{n}\left(1 + \frac{n\bar{x}^2}{Q}\right), \quad M_2(\hat{\theta}_n) = \frac{\sigma^2}{n}\left(1 + \frac{n\bar{x}^2}{Q}\Delta^2\right) \tag{3.2.7a}$$

and

$$M_3\left(\hat{\theta}_n^{\mathrm{PT}}\right) = \frac{\sigma^2}{n}\left(1 + \frac{n\bar{x}^2}{Q}\right) - \sigma^2\frac{\bar{x}^2}{Q}G_{3,m}\left(\frac{1}{3}F_{1,m}(\alpha);\Delta^2\right)$$
$$+ \sigma^2\frac{\bar{x}^2}{Q}\Delta^2\left\{2G_{3,m}\left(\frac{1}{3}F_{1,m}(\alpha);\Delta^2\right) - G_{5,m}\left(\frac{1}{5}F_{1,m}(\alpha);\Delta^2\right)\right\}, \tag{3.2.7b}$$

respectively.

Proof. $M_1(\tilde{\theta}_n)$ follows from (3.1.4). As for $M_2(\hat{\theta}_n)$ we have

$$\begin{aligned}
M_2(\hat{\theta}_n) &= E(\hat{\theta}_n - \theta)^2 = E\{(\tilde{\theta}_n - \theta) + (\tilde{\beta}_n - \beta_0)\bar{x}\}^2 \\
&= \frac{\sigma^2}{n}\Big(1 + \frac{n\bar{x}^2}{Q}\Big) + \bar{x}^2 E(\tilde{\beta}_n - \beta_0)^2 + 2\bar{x}E\big[(\tilde{\theta}_n - \theta)(\tilde{\beta}_n - \beta_0)\big] \\
&= \frac{\sigma^2}{n}\Big(1 + \frac{n\bar{x}^2}{Q}\Big) + \frac{\bar{x}^2\sigma^2}{Q}(1 + \Delta^2) - \frac{2\bar{x}^2}{Q}\sigma^2 \qquad (3.2.8a) \\
&= \frac{\sigma^2}{n} + \frac{\bar{x}^2}{Q}\Delta^2 = \frac{\sigma^2}{n}\Big(1 + \frac{n\bar{x}^2}{Q}\Delta^2\Big). \qquad (3.2.8b)
\end{aligned}$$

Finally,

$$\begin{aligned}
M_3\big(\hat{\theta}_n^{\mathrm{PT}}\big) &= E\big(\hat{\theta}_n^{\mathrm{PT}} - \theta\big)^2 = E\Big\{(\tilde{\theta}_n - \theta) + (\tilde{\beta}_n - \beta_0)\bar{x}I(\mathcal{L}_n < F_{1,m}(\alpha))\Big\}^2 \\
&= E(\tilde{\theta}_n - \theta)^2 + \bar{x}^2 E\Big\{(\tilde{\beta}_n - \beta_0)^2 I(\mathcal{L}_n < F_{1,m}(\alpha))\Big\} \\
&\quad + 2\bar{x}E\Big\{(\tilde{\theta}_n - \theta)(\tilde{\beta}_n - \beta_0)I(\mathcal{L}_n < F_{1,m}(\alpha))\Big\}. \qquad (3.2.9a)
\end{aligned}$$

Let

$$Z = (\tilde{\beta}_n - \beta_0)\frac{\sqrt{Q}}{\sigma}.$$

Then, using the fact that

$$E\Big\{(\tilde{\theta}_n - \theta)|(\tilde{\beta}_n - \beta_0)\Big\} = -\bar{x}\Big\{(\tilde{\beta}_n - \beta_0) - (\beta - \beta_0)\Big\}, \qquad (3.2.9b)$$

We write the third term as equaling

$$\begin{aligned}
&- 2\frac{\bar{x}^2}{Q}\sigma^2 E\Big\{(\tilde{\beta}_n - \beta_n)^2\frac{Q}{\sigma^2}I(\mathcal{L}_n < F_{1,m}(\alpha))\Big\} \\
&+ 2\sigma^2\frac{\bar{x}^2}{Q}E\Big\{(\tilde{\beta}_n - \beta_0)(\beta - \beta_0)\frac{Q}{\sigma^2}I(\mathcal{L}_n < F_{1,m}(\alpha))\Big\} \\
&= -\frac{2\bar{x}^2}{Q}\sigma^2 E\Big\{Z^2 I\Big(\frac{Z^2}{s_e^2/\sigma^2} < F_{1,m}(\alpha)\Big)\Big\} \\
&+ 2\frac{\bar{x}^2}{Q}\sigma^2\Delta^2 E\Big\{ZI\Big(\frac{Z^2}{s_e^2/\sigma^2} < F_{1,m}(\alpha)\Big)\Big\} \qquad (3.2.9c)
\end{aligned}$$

Hence, using Theorems 2.2.4 and 2.2.5, we have

$$
\begin{aligned}
M_3\left(\hat{\theta}_n^{\mathrm{PT}}\right) &= \sigma^2\left(\frac{1}{n}+\frac{\bar{x}^2}{Q}\right) - \sigma^2\frac{\bar{x}^2}{Q}G_{3,m}\left(\frac{1}{3}F_{1,m}(\alpha);\Delta^2\right) \\
&\quad + \sigma^2\Delta^2\frac{\bar{x}^2}{Q}\left\{2G_{3,m}\left(\frac{1}{3}F_{1,m}(\alpha);\Delta^2\right) - G_{5,m}\left(\frac{1}{5}F_{1,m}(\alpha);\Delta^2\right)\right\} \\
&= \frac{\sigma^2}{n} + \frac{\bar{x}^2}{Q}\sigma^2\left[1 - G_{3,m}\left(\frac{1}{3}F_{1,m}(\alpha);\Delta^2\right)\right. \\
&\quad \left. + \Delta^2\left\{2G_{3,m}\left(\frac{1}{3}F_{1,m}(\alpha);\Delta^2\right) - G_{5,m}\left(\frac{1}{5}F_{1,m}(\alpha);\Delta^2\right)\right\}\right].
\end{aligned}
\tag{3.2.9d}
$$

3.2.3 Comparison of bias and mse functions

Since the bias and mse expressions are known to us, we may now compare them for the three estimators, namely, $\tilde{\theta}_n$, $\hat{\theta}_n$, and $\hat{\theta}_n^{\mathrm{PT}}$. Note that all the expressions are functions of Δ^2, which is the noncentrality parameter of the noncentral F-distribution. Also, Δ^2 is the standardized squared distance between β and β_0. First, we compare the bias based on the quadratic bias (QB-)functions as in Theorem 3.2.1. The graphs of the quadratic bias (QB) are shown in Figure 3.2.1. For $\bar{x}=0$ or under H_0,

$$
B_1(\tilde{\theta}_n) = B_2(\hat{\theta}_n) = B_3(\hat{\theta}_n^{\mathrm{PT}}) = 0; \tag{3.2.10}
$$

otherwise,

$$
B_1(\tilde{\theta}_n) = 0 \le B_3(\hat{\theta}_n^{\mathrm{PT}}) = \frac{\bar{x}^2}{Q}\Delta^2\left\{G_{3,m}\left(\frac{1}{3}F_{1,m}(\alpha);\Delta^2\right)\right\}^2 \le B_2(\hat{\theta}_n) \tag{3.2.11}
$$

for all $\Delta^2 > 0$ and $\bar{x} \neq 0$.

The quadratic bias of $\hat{\theta}_n$ is linear in Δ^2, while the quadratic bias of $\hat{\theta}_n^{\mathrm{PT}}$ increases to a maximum as Δ^2 moves away from the origin, and then decreases toward zero as $\Delta^2 \to \infty$.

Now we compare the MSE functions of the restricted and preliminary test estimators with respect to the traditional estimator, $\tilde{\theta}_n$. The mean-square relative efficiency (MRE) of $\hat{\theta}_n$ compared to $\tilde{\theta}_n$ may be written as

$$
\mathrm{MRE}(\hat{\theta}_n;\tilde{\theta}_n) = \left(1+\frac{n\bar{x}^2}{Q}\right)\left(1+\frac{n\bar{x}^2}{Q}\Delta^2\right)^{-1}. \tag{3.2.12}
$$

The efficiency is a decreasing function of Δ^2. Under H_0 it has the maximum value

$$
\mathrm{MRE}(\hat{\theta}_n;\tilde{\theta}_n) = \left(1+\frac{n\bar{x}^2}{Q}\right) \ge 1 \tag{3.2.13}
$$

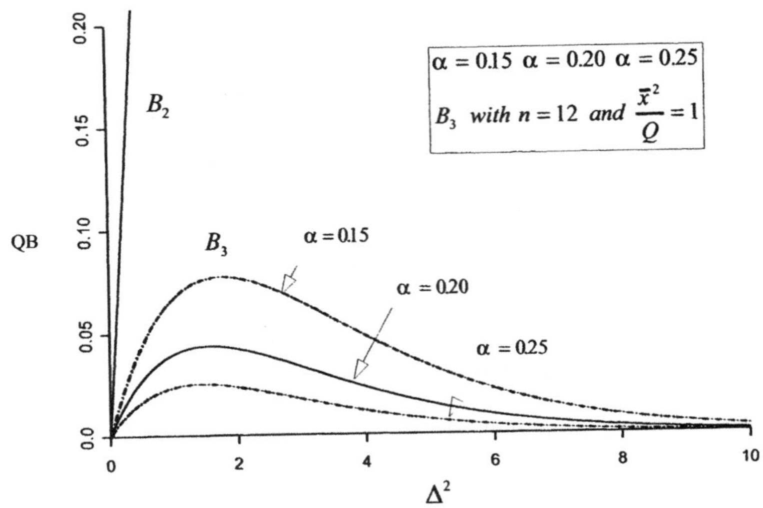

Figure 3.2.1 Graph of quadratic bias functions of the estimators

and

$$\text{MRE}(\hat{\theta}_n; \tilde{\theta}_n) \gtreqless 1 \text{ according as } \Delta^2 \lesseqgtr 1. \tag{3.2.14}$$

Thus, $\hat{\theta}_n$ performs better than $\tilde{\theta}_n$ whenever $\Delta^2 < 1$; otherwise, $\tilde{\theta}_n$ performs better.

The relative efficiency of $\hat{\theta}_n^{\text{PT}}$ compared to $\tilde{\theta}_n$ is given by

$$\text{MRE}(\hat{\theta}_n^{\text{PT}}; \tilde{\theta}_n) = [1 + g(\Delta^2)]^{-1}, \tag{3.2.15}$$

where

$$g(\Delta^2) = -\frac{\bar{x}^2}{Q}\left(\frac{1}{n} + \frac{\bar{x}^2}{Q}\right)^{-1} G_{3,m}\left(\frac{1}{3}F_{1,m}(\alpha); \Delta^2\right)$$

$$+ \Delta^2 \frac{\bar{x}^2}{Q}\left(\frac{1}{n} + \frac{\bar{x}^2}{Q}\right)^{-1} \left\{2G_{3,m}\left(\frac{1}{3}F_{1,m}(\alpha); \Delta^2\right) - G_{5,m}\left(\frac{1}{5}F_{1,m}(\alpha); \Delta^2\right)\right\}. \tag{3.2.16}$$

Under H_0, it has the maximum value

$$\text{MRE}\left(\hat{\theta}_n^{\text{PT}}; \tilde{\theta}_n\right) = \left\{1 - \frac{\bar{x}^2}{Q}\left(\frac{1}{n} + \frac{\bar{x}^2}{Q}\right)^{-1} G_{3,m}\left(\frac{1}{3}F_{1,m}(\alpha); 0\right)\right\}^{-1} (\geq 1) \tag{3.2.17}$$

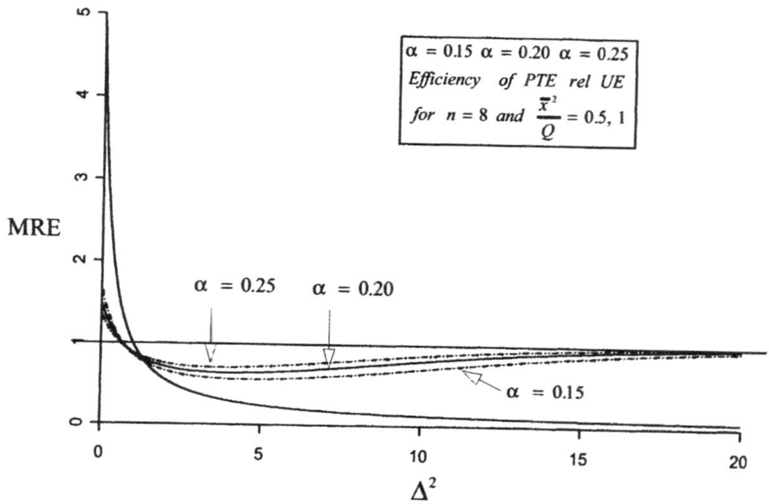

Figure 3.2.2 Graph of $\text{MRE}(\hat{\theta}_n; \tilde{\theta}_n)$ and $\text{MRE}(\hat{\theta}_n^{\text{PT}}; \tilde{\theta}_n)$

and $\text{MRE}(\hat{\theta}_n^{\text{PT}}; \tilde{\theta}_n) \geq 11$ according as

$$\Delta^2 \lesseqgtr K(F_{1,m}(\alpha); \Delta^2) = \frac{G_{3,m}(\frac{1}{3}F_{1,m}(\alpha); \Delta^2)}{2G_{3,m}(\frac{1}{3}F_{1,m}(\alpha); \Delta^2) - G_{5,m}(\frac{1}{5}F_{1,m}(\alpha); \Delta^2)}. \tag{3.2.18}$$

Hence, $\hat{\theta}_n^{\text{PT}}$ performs better than $\tilde{\theta}_n$ if $\Delta^2 \leq K(F_{1,m}(\alpha); \Delta^2)$; otherwise, $\tilde{\theta}_n$ is better than $\hat{\theta}_n^{\text{PT}}$. Since

$$2G_{3,m}\left(\frac{1}{3}F_{1,m}(\alpha); \Delta^2\right) - G_{5,m}\left(\frac{1}{5}F_{1,m}(\alpha); \Delta^2\right) > 0, \tag{3.2.19}$$

we obtain

$$K(F_{1,m}(\alpha); \Delta^2) \leq 1. \tag{3.2.20}$$

The graphs of the $\text{MRE}(\hat{\theta}_n : \tilde{\theta}_n)$ and $\text{MRE}(\hat{\theta}_n^{PT} : \tilde{\theta}_n)$ are shown in Figure 3.2.2.

3.2.4 Optimum Level of Significance of Preliminary Test

Consider the relative efficiency of $\hat{\theta}_n^{\text{PT}}$ compared to $\tilde{\theta}_n$. Denoting it by $\text{MRE}(\alpha, \Delta^2)$, we have

$$\text{MRE}(\alpha, \Delta^2) = [1 + g(\Delta^2)]^{-1}. \tag{3.2.21}$$

The graph of $\text{MRE}(\alpha, \Delta^2)$, as a function of Δ^2 for fixed α, *is decreasing, crossing the 1-line to a minimum at* $\Delta^2 = \Delta_0^2(\alpha)$ (say); *then it increases toward the 1-line as* $\Delta^2 \to \infty$. The maximum value of $\text{MRE}(\alpha, \Delta^2)$ occurs at $\Delta^2 = 0$ with the value

$$\max_{\Delta^2} \text{MRE}(\alpha, \Delta^2) =$$

$$\text{MRE}(\alpha, 0) = \left[1 - \frac{\bar{x}^2}{Q}\left(\frac{1}{n} + \frac{\bar{x}^2}{Q}\right)^{-1} G_{3,m}\left(\frac{1}{3}F_{1,m}(\alpha); 0\right)\right]^{-1} \geq 1 \quad (3.2.22)$$

for all $\alpha \in A$, the set of possible values of α. The value of $\text{MRE}(\alpha, 0)$ decreases as α increases. On the other hand, if $\alpha = 0$ and Δ^2 varies, the graphs of $\text{MRE}(0, \Delta^2) = 1$ and $\text{MRE}(1, \Delta^2)$ intersect at $\Delta^2 = 1$. In general, $\text{MRE}(\alpha_1, \Delta^2)$ and $\text{MRE}(\alpha_2, \Delta^2)$ intersect within the interval $0 \leq \Delta^2 \leq 1$; the value of Δ^2 at the intersection increases as α increases. Therefore, for two different α-values, $\text{MRE}(\alpha_1, \Delta^2)$ and $\text{MRE}(\alpha_2, \Delta^2)$ will always intersect below the 1-line.

In order to obtain an estimator (PTE) with a minimum guaranteed efficiency E_0, we adopt the following procedure: If $0 \leq \Delta^2 \leq 1$, we always choose $\tilde{\theta}_n$ because $\text{MRE}(\alpha, \Delta^2) \geq 1$ in this interval. However, in general, Δ^2 is unknown, so there is no way to choose an estimator that is uniformly best. For this reason, we select an estimator with minimum guaranteed efficiency, such as E_0, and look for a suitable α from the set $A_0 = \{\alpha | \text{MRE}(\alpha, \Delta^2) \geq E_0\}$. The estimator chosen maximizes $\text{MRE}(\alpha, \Delta^2)$ over all $\alpha \in A_0$ and Δ^2. Thus, we solve the following equation for the optimum α^*:

$$\min_{\Delta^2} \text{MRE}(\alpha, \Delta^2) = E(\alpha, \Delta_0^2(\alpha)) = E_0. \quad (3.2.23)$$

The solution α^* obtained this way gives the PTE with minimum guaranteed efficiency E_0, which may increase toward $\text{MRE}(\alpha^*, 0)$ given by (3.2.22).

Tables 3.2.1 and 3.2.2 give selected values of $\bar{x}^2/Q$ and $\alpha = 0.05(0.05)0.50$ for the procedure of choosing the level (α^*) of significance for $n = 8$ and $\bar{x}^2/Q = 1(.5)5$, $n = 12$, and $\bar{x}^2/Q = 0.1(0.1)0.9$.

Table 3.2.1 Maximum and Minimum Guaranteed Efficiencies for $n = 8$ and $\bar{x}^2/Q = 1.0(0.5)5.0$

α	$\bar{x}^2/Q$	1.0	1.5	2.0	2.5	3.0	3.5	4.0	4.5	5.0
0.05	$E_{\max}$	3.29753	3.16716	3.81283	3.94493	4.04011	4.11195	4.16810	4.21320	4.25021
	$E_{\min}$	0.32597	0.31773	0.31353	0.31099	0.30929	0.30806	0.30714	0.30643	0.30585
	$\Delta^2_{\min}$	7.05937	7.05937	5.53023	7.05937	7.05937	7.05937	7.05937	7.05937	7.05937
0.10	$E_{\max}$	2.27544	2.39283	2.46001	2.50352	2.53400	2.55654	2.57389	2.58764	2.59883
	$E_{\min}$	0.44789	0.43855	0.43378	0.43087	0.42892	0.42752	0.42646	0.42564	0.42497
	$\Delta^2_{\min}$	5.32801	5.32801	5.32801	5.32801	5.32801	5.32801	5.32801	5.32801	5.32801
0.15	$E_{\max}$	1.82656	1.88654	1.91991	1.94117	1.95589	1.96670	1.97496	1.98149	1.98678
	$E_{\min}$	0.54227	0.53289	0.52805	0.52510	0.52311	0.52168	0.52061	0.51976	0.51909
	$\Delta^2_{\min}$	4.50883	4.50883	4.50883	4.50883	4.50883	4.50883	4.50883	4.50883	4.50883
0.20	$E_{\max}$	1.40934	1.56154	1.64097	1.68976	1.72277	1.74659	1.76458	1.77866	1.78997
	$E_{\min}$	0.67415	0.62561	0.60605	0.59549	0.58887	0.58434	0.58104	0.57853	0.57656
	$\Delta^2_{\min}$	4.00897	4.00897	4.00897	4.00897	4.00897	4.00897	4.00897	4.00897	4.00897
0.25	$E_{\max}$	1.38335	1.40405	1.41526	1.42230	1.42712	1.43063	1.43330	1.43540	1.43710
	$E_{\min}$	0.70912	0.70127	0.69719	0.69469	0.69299	0.69177	0.69085	0.69013	0.68955
	$\Delta^2_{\min}$	0.60838	0.60838	0.60838	0.60838	0.60838	0.60838	0.60838	0.60838	0.60838
0.30	$E_{\max}$	1.30073	1.31595	1.32416	1.32929	1.33280	1.33535	1.33729	1.3882	1.34005
	$E_{\min}$	0.74827	0.74109	0.73735	0.73505	0.73349	0.73237	0.73152	0.73086	0.73033
	$\Delta^2_{\min}$	3.41621	3.41621	3.41621	3.41621	3.41621	3.41621	3.41621	3.41621	3.41621
0.35	$E_{\max}$	1.22121	1.23169	1.23731	1.24082	1.24321	1.24495	1.24627	1.24731	1.24815
	$E_{\min}$	0.79911	0.79298	0.78978	0.78781	0.78647	0.78551	0.78478	0.78421	0.78375
	$\Delta^2_{\min}$	3.22498	3.22498	3.22498	3.22498	3.22498	3.22498	3.22498	3.22498	3.22498
0.40	$E_{\max}$	1.09488	1.12019	1.13192	1.13868	1.14309	1.14618	1.14847	1.15014	1.15165
	$E_{\min}$	0.89949	0.87846	0.86936	0.86427	0.86102	0.85877	0.85712	0.85585	0.85485
	$\Delta^2_{\min}$	3.07468	3.07468	3.07468	3.07468	3.07468	3.07468	3.07468	3.07468	3.07468
0.45	$E_{\max}$	1.11847	1.12359	1.12632	1.12802	1.12917	1.13001	1.13065	1.13115	1.13155
	$E_{\min}$	0.87938	0.87532	0.87318	0.87187	0.87097	0.87033	0.86984	0.86946	0.86915
	$\Delta^2_{\min}$	2.95417	2.95417	2.95417	2.95417	2.95417	2.95417	2.95417	2.95417	2.95417
0.50	$E_{\max}$	1.08512	1.08868	1.09058	1.09176	1.09256	1.09314	1.09358	1.09393	1.09420
	$E_{\min}$	0.90992	0.90677	0.90512	0.90410	0.90340	0.90290	0.90252	0.90223	0.90199
	$\Delta^2_{\min}$	2.85621	2.85621	2.85621	2.85621	2.85621	2.85621	2.85621	2.85621	2.85621

Table 3.2.2 Maximum and Minimum Guaranteed Efficiencies for $n = 12$ and $\frac{\bar{x}^2}{Q} = 0.1(0.2)0.9$

α	$\bar{x}^2/Q$	0.1	0.2	0.3	0.4	0.5	0.6	0.7	0.8	0.9
0.05	E_{max}	1.85798	2.33498	2.63864	2.84893	3.00317	3.12114	3.21428	3.28968	3.35198
	E_{min}	0.47754	0.42471	0.40464	0.39408	0.38755	0.38313	0.37992	0.37750	0.37560
	Δ^2_{min}	5.53022	5.53022	5.53023	5.53022	5.53022	5.53022	5.53022	5.53022	5.53022
0.10	E_{max}	1.57358	1.82247	1.96155	2.05035	2.11197	2.15723	2.19189	2.21927	2.24145
	E_{min}	1.40934	1.56154	1.64097	1.68976	1.72277	1.74659	1.76458	1.77866	1.78997
	Δ^2_{min}	0.67415	0.62561	0.60605	0.59549	0.58887	0.58434	0.58104	0.57853	0.57656
0.15	E_{max}	1.40934	1.56154	1.64097	1.68976	1.72277	1.74659	1.76458	1.77866	1.78997
	E_{min}	0.67415	0.62561	0.60605	0.59549	0.58887	0.58434	0.58104	0.57853	0.57656
	Δ^2_{min}	3.94090	3.94090	3.94090	3.94090	3.94090	3.94090	3.94090	3.94090	3.94090
0.20	E_{max}	1.30133	1.40191	1.45223	1.48234	1.50256	1.51695	1.52774	1.53613	1.54284
	E_{min}	0.73816	0.69495	0.67714	0.66744	0.66133	0.65713	0.65406	0.65173	0.64989
	Δ^2_{min}	3.60055	3.60055	3.60055	3.60055	3.60055	3.60055	3.60055	3.60055	3.60055
0.25	E_{max}	1.22209	1.29033	1.32345	1.34302	1.35594	1.36510	1.37195	1.37725	1.38148
	E_{min}	0.79359	0.75642	0.74086	0.73231	0.72691	0.72318	0.72046	0.71838	0.71674
	Δ^2_{min}	3.35268	3.35268	3.35268	3.35268	3.35268	3.35268	3.35268	3.35268	3.35268
0.30	E_{max}	1.16931	1.21843	1.24181	1.25548	1.26446	1.27080	1.27552	1.27917	1.28207
	E_{min}	0.83380	0.80206	0.78860	0.78116	0.77644	0.77318	0.77079	0.76897	0.76753
	Δ^2_{min}	3.18236	3.18236	3.18236	3.18236	3.18236	3.18236	3.18236	3.18236	3.18236
0.35	E_{max}	1.12716	1.16235	1.17885	1.18842	1.19468	1.19908	1.20235	1.20487	1.20688
	E_{min}	0.86973	0.84357	0.83234	0.82610	0.82213	0.81938	0.81737	0.81582	0.81461
	Δ^2_{min}	3.04378	3.04378	3.04378	3.04378	3.04378	3.04378	3.04378	3.04378	3.04378
0.40	E_{max}	1.09488	1.12019	1.13192	1.13868	1.14309	1.14618	1.14847	1.15014	1.15165
	E_{min}	0.89949	0.87846	0.86936	0.86427	0.86102	0.85877	0.85712	0.85585	0.85485
	Δ^2_{min}	2.93352	2.93352	2.93352	2.93352	2.93352	2.93352	2.93352	2.93352	2.93352
0.45	E_{max}	1.06997	1.08809	1.09642	1.10120	1.10431	1.10648	1.10810	1.10934	1.11032
	E_{min}	0.92393	0.90750	0.90031	0.89629	0.89372	0.89193	0.88961	0.88961	0.88881
	Δ^2_{min}	2.84422	2.84422	2.84422	2.84422	2.84422	2.84422	2.84422	2.84422	2.84422
0.50	E_{max}	1.05070	1.06354	1.06940	1.07276	1.07493	1.07645	1.07758	1.07844	1.07913
	E_{min}	0.94378	0.93131	0.92584	0.92275	0.92077	0.91940	0.91838	0.91761	0.91700
	Δ^2_{min}	2.77103	2.77103	2.77103	2.77103	2.77103	2.77103	2.77103	2.77103	2.77103

To illustrate the use of the tables, we consider $n = 12$ and $\overline{x}^2/Q = 0.40$. We want a PTE with minimum guaranteed efficiency $E_0 = 0.7811$ (say). Then, using the table for $n = 12$ and $E_0 = 0.7811$ we find that $\alpha^* = 0.30$ with maximum efficiency $E^* = 1.2548$.

3.3 Two-Sample Problem and Pooling of Means

3.3.1 Model

Consider again the simple linear model (3.2.1)

$$\mathbf{Y} = \theta \mathbf{1}_n + \beta \mathbf{x} + \mathbf{e}, \tag{3.3.1}$$

where $\mathbf{x} = (0, \dots, 0, 1, \dots, 1)'$ is an $n \times 1$ vector of n_1 zeros and n_2 one's such that $n = n_1 + n_2$. Also, $\mathbf{e} \sim \mathcal{N}_n(\mathbf{0}, \sigma^2 \mathbf{I}_n)$, and in this case

$$\mu_1 = \theta, \quad \mu_2 = \theta + \beta \quad \text{and} \quad \beta = \mu_2 - \mu_1. \tag{3.3.2}$$

Further,

$$\overline{x} = \frac{n_2}{n_1 + n_2}, \quad Q = \frac{n_1 n_2}{n_1 + n_2} \quad \text{and} \quad \frac{\overline{x}^2}{Q} = \frac{n_2}{n_1(n_1 + n_2)}. \tag{3.3.3}$$

3.3.2 Estimation and Test of the Difference between Two Means

The maximum likelihood estimators of μ_1, μ_2, and σ^2 are

$$\tilde{\mu}_1 = \overline{y}_1, \quad \tilde{\mu}_2 = \overline{y}_2, \quad \text{and} \quad \tilde{\sigma}_n^2 = \frac{n_1 + n_2 - 2}{n_1 + n_2} s_p^2,$$

where

$$s_p^2 = \frac{1}{n_1 + n_2 - 2} \left\{ \sum_{j=1}^{n_1} (y_j - \overline{y}_1)^2 + \sum_{j=n_1+1}^{n_2} (y_j - \overline{y}_2)^2 \right\}. \tag{3.3.4}$$

Here $\overline{y}_1$ is the sample mean of the first n_1 y-observations while $\overline{y}_2$ is the sample mean of the last n_2 y-observations, and s_p^2 is the pooled sample variance of the y–observations. For testing the null hypothesis $H_0 : \mu_2 = \mu_1$ against $H_A : \mu_2 \neq \mu_1$, the likelihood ratio test-statistic is given by

$$\mathcal{L}_n = \frac{n_1 n_2}{n_1 + n_2} \frac{(\overline{y}_2 - \overline{y}_1)^2}{s_p^2}. \tag{3.3.5}$$

$\mathcal{L}_n$ follows a noncentral F-distribution with $(1, n_1 + n_2 - 2)$ d.f. and noncentrality parameter $\Delta^2/2$, where

$$\Delta^2 = \frac{n_1 n_2}{n_1 + n_2} \left(\frac{\mu_2 - \mu_1}{\sigma} \right)^2. \tag{3.3.6}$$

Under H_0, $\mathcal{L}_n$ follows a central F-distribution. Finally, the unrestricted, restricted, and preliminary test estimators of μ_1 are given by

$$\tilde{\mu}_1 = \overline{y}_1, \quad \hat{\mu}_1 = \frac{n_1\overline{y}_1 + n_2\overline{y}_2}{n_1 + n_2}$$

and

$$\hat{\mu}_1^{\mathrm{PT}} = \overline{y}_1 + \frac{n_2}{n_1 + n_2}(\overline{y}_2 - \overline{y}_1)I(\mathcal{L}_n < F_{1,m}(\alpha)), \tag{3.3.7}$$

where $I(A)$ is the indicator function of the set A and $F_{1,m}(\alpha)$ is the upper α-level critical value of the central F-distribution with $(1, m = n_1 + n_2 - 2)$ d.f.

3.3.3 Bias and mse Expression of the Three Estimators of a Mean

To obtain the bias, quadratic bias, and mse expressions of the estimators of μ_1, we simply substitute in the expressions of Theorems 3.2.1 and 3.2.2 corresponding values of $\overline{x}$ and Q given by (3.3.3). Thus, for the bias and quadratic bias we have the following theorem:

Theorem 1. (i) $b_1(\tilde{\mu}_1) = 0$ and $B_1(\tilde{\mu}_1) = 0$;

(ii) $b_2(\hat{\mu}_1) = (\mu_2 - \mu_1)\frac{n_2}{n_1+n_2}$ and $B_2(\hat{\mu}_1) = \frac{n_2}{n_1(n_1+n_2)}\Delta^2$;

(iii) $b_3(\hat{\mu}_1^{\mathrm{PT}}) = (\mu_2 - \mu_1)\frac{n_2}{n_1+n_2}G_{3,m}\left(\frac{1}{3}F_{1,m}(\alpha); \Delta^2\right)$, $m = n_1 + n_2 - 2$

$$\text{and } B_3(\hat{\mu}_1^{\mathrm{PT}}) = \frac{n_2}{n_1(n_1+n_2)}\Delta^2 \left\{ G_{3,m}\left(\tfrac{1}{3}F_{1,m}(\alpha); \Delta^2\right)\right\}^2. \tag{3.3.8}$$

For the mse expressions we have

Theorem 2.

$$M_1(\tilde{\mu}_1) = \frac{\sigma^2}{n_1}, \quad M_2(\hat{\mu}_1) = \frac{\sigma^2}{n_1 + n_2}\left(1 + \frac{n_2}{n_1}\Delta^2\right)$$

and

$$\begin{aligned} M_3(\hat{\mu}_1^{\mathrm{PT}}) = {} & \frac{\sigma^2}{n_1} - \frac{\sigma^2 n_2}{n_1(n_1 + n_2)}G_{3,m}\left(\frac{1}{3}F_{1,m}(\alpha); \Delta^2\right) \\ & + \frac{\sigma^2\Delta^2 n_2}{n_1(n_1 + n_2)}\left\{2G_{3,m}\left(\frac{1}{3}F_{1,m}(\alpha); \Delta^2\right)\right. \\ & \left. - G_{5,m}\left(\frac{1}{5}F_{1,m}(\alpha); \Delta^2\right)\right\}. \end{aligned} \tag{3.3.9}$$

It is clear from the expressions of the mse's of $\hat{\mu}_1$ and $\hat{\mu}_1^{\text{PT}}$ that the relative efficiencies of $\hat{\mu}_1$ and $\hat{\mu}_1^{\text{PT}}$ compared to $\tilde{\mu}_1$ are given by

$$\text{MRE}(\hat{\mu}_1; \tilde{\mu}_1) = \left\{ \frac{n_1}{n_1 + n_2} + \frac{n_2 \Delta^2}{n_1 + n_2} \right\}^{-1} \tag{3.3.10a}$$

and

$$\text{MRE}(\hat{\mu}_1^{\text{PT}}; \tilde{\mu}_1) = \{1 + g(\Delta^2)\}^{-1} \tag{3.3.10b}$$

respectively, where

$$\begin{aligned} g(\Delta^2) = & -\frac{n_2}{n_1 + n_2} G_{3,m}\left(\frac{1}{3}F_{1,m}(\alpha); \Delta^2\right) + \Delta^2 \frac{n_2}{n_1 + n_2} \\ & \times \left\{ 2G_{3,m}\left(\frac{1}{3}F_{1,m}(\alpha); \Delta^2\right) - G_{5,m}\left(\frac{1}{5}F_{1,m}(\alpha); \Delta^2\right) \right\}. \end{aligned} \tag{3.3.10c}$$

Note that $\text{MRE}(\hat{\mu}_1; \tilde{\mu}_1)$ is a decreasing function of Δ^2. The maximum value of $\text{MRE}(\hat{\mu}_1; \tilde{\mu}_1)$ occurs at $\Delta^2 = 0$ with the value $1 + n_2/n_1$; and as $\Delta^2 \to \infty$, $\text{MRE}(\hat{\mu}_1; \tilde{\mu}_1)$ tends to zero. Further, $\hat{\mu}_1$ performs better than $\tilde{\mu}_1$ for Δ^2 in the interval $[0, 1]$, while $\tilde{\mu}_1$ performs better than $\hat{\mu}_1$ outside this interval. Next, we compare $\hat{\mu}_1^{\text{PT}}$ relative to $\tilde{\mu}_1$. In this case, the efficiency has the maximum value

$$\text{MRE}(\hat{\mu}_1^{\text{PT}}; \tilde{\mu}_1) = \left[1 - \frac{n_2}{n_1 + n_2}(G_{3,m}\left(\frac{1}{3}F_{1,m}(\alpha); 0\right)\right]^{-1} \quad (\geq 1), \tag{3.3.11}$$

so it decreases monotonically, crossing the 1–line at $\Delta^2 = \Delta^2_{\text{min}}$, and then increases toward the 1–line as Δ^2 tends to infinity. The graphs of the $MRE(\hat{\mu}_1 : \tilde{\mu}_1)$ and the $MRE(\hat{\mu}_1^{PT} : \tilde{\mu}_1)$ are shown in Figure 3.3.1.

Based on the analysis of Section 3.2.4, the optimum level of significance α^* with minimum guaranteed efficiency E_0 is obtained by solving the following equation for α:

$$\min_{\Delta^2} \text{MRE}(\alpha, \Delta^2) = E(\alpha, \Delta_0^2(\alpha)) = E_0 \tag{3.3.12}$$

as in the case of (3.2.23). Tables 3.3.1 through 3.3.3 of maximum and minimum guaranteed values are provided for $\alpha = 0.05(0.05)0.50$ selected pair of integers (n_1, n_2). To illustrate the use of the table, let us choose $n_1 = 8$, $n_2 = 16$, and $E_0 = 0.66$. Then using the corresponding tabular values, we obtain $\alpha^* = 0.15$. The efficiency can go up to 1.442 if $\Delta^2 \to 0$.

Table 3.3.1 Maximum and Minimum Guaranteed Efficiencies

$\alpha\backslash(n_1, n_2)$		(4, 4)	(6, 4)	(8, 4)	(10, 4)	(4, 8)	(6, 8)	(8, 8)	(10, 8)
0.05	E^*	1.6445	1.4451	1.3400	1.2750	2.0301	1.7587	1.6005	1.4969
	E_0	0.4623	0.5477	0.6092	0.6558	0.4380	0.4878	0.5294	0.5647
	Δ_0	7.0594	6.3232	5.9338	5.6934	5.9338	5.6934	5.5302	5.4123
0.10	E^*	1.4605	1.3258	1.2522	1.2058	1.6746	1.5182	1.4208	1.3542
	E_0	0.5905	0.6647	0.7157	0.7530	0.5572	0.6039	0.6415	0.6726
	Δ_0	5.3280	4.9274	4.7087	4.5712	4.7087	4.5712	4.4767	4.4078
0.15	E^*	1.3415	1.2458	1.1921	1.1577	1.4755	1.3745	1.3089	1.2629
	E_0	0.6781	0.7410	0.7830	0.8131	0.6434	0.6851	0.7180	0.7446
	Δ_0	4.5088	4.2448	4.0983	4.0053	4.0983	4.0053	3.9409	3.8938
0.20	E^*	1.2578	1.1879	1.1479	1.1220	1.3471	1.2779	1.2317	1.1988
	E_0	0.7447	0.7973	0.8317	0.8560	0.7119	0.7482	0.7764	0.7988
	Δ_0	4.0090	3.8206	3.7150	3.5475	3.7150	3.5475	3.5005	3.5661
0.25	E^*	1.1962	1.1443	1.1142	1.0946	1.2579	1.2089	1.1756	1.1514
	E_0	0.7978	0.8411	0.8690	0.8884	0.7683	0.7993	0.8229	0.8415
	Δ_0	3.5667	3.5266	3.4474	3.3965	3.4474	3.3965	3.3611	3.3350
0.30	E^*	1.1495	1.1107	1.0880	1.0731	1.1931	1.1577	1.1333	1.1155
	E_0	0.8409	0.8760	0.8983	0.9137	0.8154	0.8412	0.8606	0.8758
	Δ_0	3.4162	3.3095	3.2488	3.2097	3.2488	3.2097	3.1824	3.1622
0.35	E^*	1.1135	1.0845	1.0674	1.0561	1.1445	1.1188	1.1009	1.0877
	E_0	0.8761	0.9042	0.9217	0.9338	0.8548	0.8758	0.8915	0.9037
	Δ_0	3.2250	3.1426	3.0955	3.0651	3.0955	3.0651	3.0438	3.0280
0.40	E^*	1.0854	1.0638	1.0510	1.0425	1.1076	1.0889	1.0757	1.0660
	E_0	0.9049	0.9269	0.9405	0.9498	0.8877	0.9044	0.9168	0.9263
	Δ_0	3.0747	3.0107	2.9739	2.9501	2.9739	2.9501	2.9335	2.9211
0.45	E^*	1.0634	1.0475	1.0381	1.0318	1.0792	1.0657	1.0561	1.0490
	E_0	0.9284	0.9451	0.9555	0.9625	0.9148	0.9278	0.9373	0.9446
	Δ_0	2.9542	2.9045	2.8758	2.8573	2.8758	2.8573	2.8442	2.8345
0.50	E^*	1.0462	1.0347	1.0279	1.0233	1.0573	1.0477	1.0408	1.0357
	E_0	0.9472	0.9598	0.9674	0.9726	0.9369	0.9467	0.9538	0.9593
	Δ_0	2.8562	2.8178	2.7956	2.7812	2.7956	2.7812	2.7710	2.7635

Table 3.3.2 Maximum and Minimum Guaranteed Efficiencies

$\alpha\backslash(n_1,n_2)$		(4, 12)	(6, 12)	(8, 12)	(10, 12)	(4, 16)	(6, 16)	(8, 16)	(10, 16)
0.05	E^*	2.2873	1.9918	1.8067	1.6798	2.4712	2.1720	1.9740	1.8332
	E_0	0.4286	0.4637	0.4948	0.5224	0.4234	0.4506	0.4753	0.4979
	Δ_0	5.5302	5.4123	5.3232	5.2534	5.3232	5.2534	5.1973	5.1512
0.10	E^*	1.7993	1.6456	1.5416	1.4664	1.8811	1.7363	1.6324	1.5542
	E_0	0.5440	0.5780	0.6072	0.6325	0.5369	0.5635	0.5873	0.6085
	Δ_0	4.4767	4.4078	4.3554	4.3142	4.3554	4.3142	4.2809	4.2534
0.15	E^*	1.5480	1.4540	1.3876	1.3381	1.5934	1.5081	1.4442	1.3947
	E_0	0.6293	0.6603	0.6866	0.7090	0.6216	0.6463	0.6679	0.6871
	Δ_0	3.9409	3.8938	3.8578	3.8294	3.8578	3.8294	3.8064	3.7874
0.20	E^*	1.3932	1.3310	1.2859	1.2516	1.4213	1.3661	1.3238	1.2902
	E_0	0.6983	0.7258	0.7487	0.7680	0.6908	0.7129	0.7320	0.7487
	Δ_0	3.5005	3.5661	3.5397	3.5188	3.5397	3.5188	3.5019	3.4880
0.25	E^*	1.28887	1.2458	1.2140	1.1895	1.3072	1.2696	1.2403	1.2167
	E_0	0.7559	0.7797	0.7992	0.8155	0.7490	0.7683	0.7847	0.7990
	Δ_0	3.3611	3.3350	3.3149	3.2991	3.3149	2.2991	3.2862	3.2756
0.30	E^*	1.2143	1.1839	1.1610	1.1432	1.2269	1.2006	1.1797	1.1628
	E_0	0.8045	0.8246	0.8409	0.8544	0.7985	0.8148	0.8287	0.8406
	Δ_0	3.1824	3.1622	3.1467	3.1344	3.1467	3.1344	3.1245	3.1163
0.35	E^*	1.1594	1.1376	1.1210	1.1080	1.1681	1.1494	1.1345	1.1222
	E_0	0.8456	0.8621	0.8754	0.8863	0.8405	0.8540	0.8654	0.8751
	Δ_0	3.0438	3.0280	3.0159	3.0063	3.0159	3.0063	2.9986	2.9921
0.40	E^*	1.1181	1.1024	1,0903	1.0809	1.1242	1.1108	1.1000	1.0911
	E_0	0.8802	0.8934	0.9040	0.9126	0.8759	0.8868	0.8959	0.9037
	Δ_0	2.9335	2.9211	2.9117	2.9041	2.9117	2.9041	2.8980	2.8929
0.45	E^*	1.0866	1.0753	1.0666	1.0598	1.0909	1.0813	1.0736	1.0672
	E_0	0.9088	0.9192	0.9274	0.9341	0.9055	0.9140	0.9211	0.9271
	Δ_0	2.8442	2.8345	2.8270	2.8212	2.8270	2.8212	2.8164	2.8124
0.50	E^*	1.0625	1.0545	1.0483	1.0434	1.0655	1.0587	1.9532	1.9487
	E_0	0.9323	0.9401	0.9463	0.9514	0.9297	0.9362	0.9416	0.9462
	Δ_0	2.7710	2.7635	2.7577	2.7530	2.7577	2.7530	2.7493	2.7462

Table 3.3.3 Maximum and Minimum Guaranteed Efficiencies

$\alpha\backslash(n_1, n_2)$		(3, 4)	(7, 8)	(11, 12)	(13, 14)	(15, 16)
0.05	E^*	1.8113	1.6673	1.6291	1.6189	1.6114
	E_0	0.4293	0.5133	0.5365	0.5428	0.5474
	Δ_0	7.0594	5.5302	5.1973	5.1127	5.0520
0.10	E^*	1.5633	1.4618	1.4351	1.4280	1.4228
	E_0	0.5579	0.6266	0.6451	0.6501	0.6538
	Δ_0	5.3280	4.4767	4.2809	4.2304	4.1941
0.15	E^*	1.4102	1.3364	1.3171	1.3119	1.3081
	E_0	0.6482	0.7048	0.7199	0.7240	0.7269
	Δ_0	4.5088	3.9409	3.8064	3.7715	3.7462
0.20	E^*	1.3059	1.2511	1.2367	1.2328	1.2301
	E_0	0.7158	0.7650	0.7773	0.7806	0.7830
	Δ_0	4.0090	3.5005	3.5019	3.4760	3.4574
0.25	E^*	1.2307	1.1895	1.1787	1.1758	1.1738
	E_0	0.7754	0.8133	0.8233	0.8259	0.8279
	Δ_0	3.5667	3.3611	3.2862	3.2667	3.2525
0.30	E^*	1.1746	1.1435	1.1354	1.1332	1.1316
	E_0	0.8222	0.8527	0.8607	0.8629	0.8644
	Δ_0	3.4162	3.1824	3.1245	3.1093	3.0983
0.35	E^*	1.1318	1.1084	1.1022	1.1006	1.0994
	E_0	0.8609	0.8851	0.8915	0.8932	0.8944
	Δ_0	3.2250	3.0438	2.9986	2.9867	2.9780
0.40	E^*	1.0987	1.0812	1.0766	1.0754	1.0745
	E_0	0.8928	0.9117	0.9167	0.9180	0.9189
	Δ_0	3.0747	2.9335	2.8980	2.8887	2.8819
0.45	E^*	1.0731	1.0601	1.0567	1.0558	1.0551
	E_0	0.9190	0.9334	0.9372	0.9382	0.9389
	Δ_0	2.9542	2.8442	2.8164	2.8090	2.8037
0.50	E^*	1.0531	1.0436	1.4120	1.0405	1.0400
	E_0	0.9402	0.9509	0.9537	0.9545	0.9550
	Δ_0	2.8562	2.7710	2.7493	2.7436	2.7394

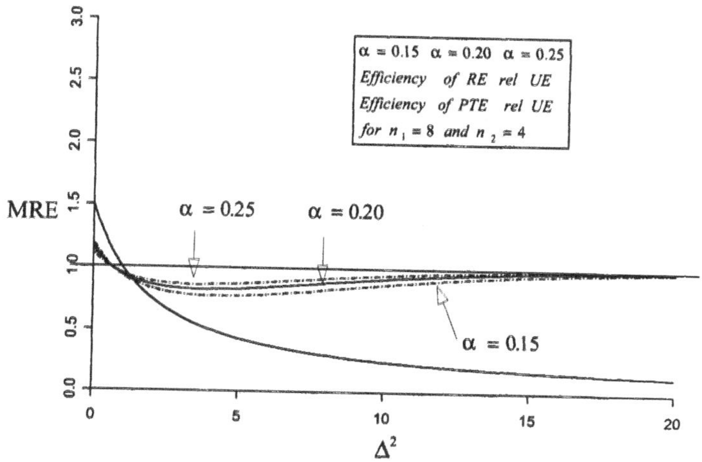

Figure 3.3.1 Graph of $\mathrm{MRE}(\hat{\mu}_1; \tilde{\mu}_1)$ and $\mathrm{MRE}(\hat{\mu}_1^{\mathrm{PT}}; \tilde{\mu}_1)$

3.4 One-Sample Problem: Estimation of Mean

3.4.1 Model

Consider the simple linear model

$$\mathbf{Y} = \theta \mathbf{1}_n + \beta \mathbf{x} + \mathbf{e}$$

where $\mathbf{e} \sim N_n(\mathbf{0}, \sigma^2 \mathbf{I}_n)$ and $\mathbf{x} = \mathbf{0}$. Then we have the reduced model for the mean given by

$$\mathbf{Y} = \theta \mathbf{1}_n + \mathbf{e}, \tag{3.4.1}$$

and the variance of the errors is σ^2. Our problem is to estimate θ when it is suspected that θ may be equal to θ_0. As before, we define the unrestricted, restricted, and preliminary test estimators of θ in the next section.

3.4.2 Unrestricted, Restricted, and Preliminary Test Estimators

Using the sample information and (3.4.1), we minimize the quadratic form

$$(\mathbf{Y} - \theta \mathbf{1}_n)'(\mathbf{Y} - \theta \mathbf{1}_n), \tag{3.4.2}$$

to obtain the unrestricted estimator of θ as

$$\tilde{\theta}_n = (\mathbf{1}_n'\mathbf{1}_n)^{-1}\mathbf{1}_n'\mathbf{Y} = \overline{Y} \quad \text{(sample mean)}. \tag{3.4.3}$$

Clearly, $E(\tilde{\theta}_n) = \theta$ and $\text{Var}(\tilde{\theta}_n) = \sigma^2/n$. Further, the unbiased estimator of σ^2 is given by

$$s_y^2 = \frac{1}{m}(\mathbf{Y} - \tilde{\theta}_n \mathbf{1}_n)'(\mathbf{Y} - \tilde{\theta}_n \mathbf{1}_n), \qquad m = n-1. \tag{3.4.4}$$

In order to test the null hypothesis $H_0 : \theta = \theta_0$ against $H_A : \theta \neq \theta_0$, we use the likelihood ratio test-statistic

$$\mathcal{L}_n = \frac{n(\tilde{\theta}_n - \theta_0)^2}{s_y^2}. \tag{3.4.5}$$

The exact distribution of $\mathcal{L}_n$ is the noncentral F-distribution with $(1, m)$ d.f. and noncentrality parameter $\Delta^2/2$, where $\Delta^2 = n(\theta - \theta_0)^2/\sigma^2$. Under H_0, it follows a central F-distribution. Thus, at the α-level of significance we reject H_0 if $\mathcal{L}_n \geq F_{1,m}(\alpha)$, where $F_{1,m}(\alpha)$ is the upper α-level critical value of the central F-distribution with $(1, m)$ d.f.

The restricted estimator of θ is simply θ_0, though we could combine $\tilde{\theta}_n$ and θ_0 as $d\tilde{\theta}_n + (1-d)\theta_0$, where $0 < d < 1$ is the coefficient of distrust on the value $\theta = \theta_0$ (type II error in testing H_0). However, in this case we will assume $d = 0$, meaning no distrust on $\theta = \theta_0$. So we define the preliminary test estimator of θ as

$$\begin{aligned} \hat{\theta}_n^{\text{PT}} &= \theta_0 I(\mathcal{L}_n < F_{1,m}(\alpha)) + \tilde{\theta}_n I(\mathcal{L}_n \geq F_{1,m}(\alpha)) \\ &= \tilde{\theta}_n - (\tilde{\theta}_n - \theta_0) I(\mathcal{L}_n < F_{1,m}(\alpha)), \end{aligned} \tag{3.4.6}$$

where $F_{1,m}(\alpha)$ is the α-level critical value.

3.4.3 Bias, mse, and Analysis of Efficiency

The bias expressions of $\tilde{\theta}_n$ and $\hat{\theta}_n^{\text{PT}}$ are

$$b_1(\tilde{\theta}_n) = 0, \quad B_1(\tilde{\theta}_n) = 0, \tag{3.4.7a}$$

$$b_2(\hat{\theta}_n^{\text{PT}}) = -(\theta - \theta_0) G_{3,m}\left(\frac{1}{3}F_{1,m}(\alpha); \Delta^2\right), \quad m = n-1,$$

and

$$B_2(\hat{\theta}_n^{\text{PT}}) = \Delta^2 \left\{ G_{3,m}\left(\frac{1}{3}F_{1,m}(\alpha); \Delta^2\right)\right\}^2. \tag{3.4.7b}$$

Similarly,

$$M_1(\tilde{\theta}_n) = \frac{\sigma^2}{n} \tag{3.4.8a}$$

and

$$M_2(\hat{\theta}_n^{\rm PT}) = \frac{\sigma^2}{n}\left[1 - G_{3,m}\left(\frac{1}{3}F_{1,m}(\alpha);\Delta^2\right)\right.$$
$$\left. +\Delta^2\left\{2G_{3,m}\left(\frac{1}{3}F_{1,m}(\alpha);\Delta^2\right) - G_{5,m}\left(\frac{1}{5}F_{1,m}(\alpha);\Delta^2\right)\right\}\right]. \tag{3.4.8b}$$

Thus, the efficiency expression for $\hat{\theta}_n^{\rm PT}$ to $\tilde{\theta}_n$ is given by,

$$\text{MRE}(\hat{\theta}_n^{\rm PT};\tilde{\theta}_n) = [1 + g(\Delta^2)]^{-1}, \tag{3.4.9}$$

where

$$g(\Delta^2) = -\ G_{3,m}\left(\frac{1}{3}F_{1,m}(\alpha);\Delta^2\right)$$
$$+\ \Delta^2\left\{2G_{3,m}\left(\frac{1}{3}F_{1,m}(\alpha);\Delta^2\right) - G_{5,m}\left(\frac{1}{5}F_{1,m}(\alpha);\Delta^2\right)\right\} \tag{3.4.10}$$

Hence, $\hat{\theta}_n^{\rm PT}$ performs better than $\tilde{\theta}_n$ whenever

$$0 \le \Delta^2 \le K(F_{1,m}(\alpha);\Delta^2), \tag{3.4.11}$$

where

$$K(F_{1,m}(\alpha);\Delta^2) = \frac{G_{3,m}(\frac{1}{3}F_{1,m}(\alpha);\Delta^2)}{2G_{3,m}(\frac{1}{3}F_{1,m}(\alpha);\Delta^2) - G_{5,m}(\frac{1}{5}F_{1,m}(\alpha);\Delta^2)}. \tag{3.4.12}$$

To obtain a PTE with minimum guaranteed efficiency E_0, we solve the equation

$$\min_{\Delta^2} MRE(\alpha,\Delta^2) = E(\alpha,\Delta_0^2(\alpha)) = E_0 \tag{3.4.13}$$

to find the optimum α-value α^* as before. Table 3.4.1 gives the maximum and minimum efficiency of the PTE of θ for various α and sample sizes.

Table 3.4.1 Minimum and Maximum Efficiency of PTE for $n = 8$ and $\overline{x}^2/Q = 0.1(0.1)0.9$

$\alpha\backslash n$		10	15	20	25	30	35	40
0.05	E^*	4.2577	4.0063	3.8912	3.8252	3.7825	3.7525	3.7304
	E_0	0.3350	0.3600	0.3720	0.3790	0.3836	0.3868	0.3893
	Δ_0	6.1009	5.5303	5.2857	5.1514	5.0656	5.0063	4.9629
0.10	E^*	2.5564	2.4529	2.4052	2.3778	2.3600	2.3475	2.3383
	E_0	0.4500	0.4722	0.4828	0.4889	0.4929	0.4958	0.4979
	Δ_0	4.8042	4.4762	4.3337	4.2535	4.2022	4.2657	4.1397
0.15	E^*	0.6827	0.6892	0.6924	0.6943	0.6955	0.6964	0.6971
	E_0	0.5405	0.5601	0.5693	0.5747	0.5782	0.5807	0.5825
	Δ_0	4.1626	2.9403	3.8429	3.7875	3.7520	3.7272	2.7090
0.20	E^*	1.6406	1.6033	1.5860	1.5760	1.5695	1.5649	1.5616
	E_0	0.6174	0.6345	0.6425	0.6471	0.6502	0.6523	0.6539
	Δ_0	3.7616	3.6006	3.5278	3.4873	3.4612	3.4430	3.4295
0.25	E^*	1.4514	1.4259	1.4141	1.4073	1.4029	1.3998	1.3975
	E_0	0.6844	0.6990	0.7059	0.7098	0.7124	0.7134	0.7156
	Δ_0	3.4811	3.3612	3.3066	3.2757	3.2557	3.2417	3.2314
0.30	E^*	1.3256	1.3077	1.2994	1.2946	1.2914	1.2892	1.2876
	E_0	0.7430	0.7553	0.7611	0.7645	0.7666	0.7682	0.7693
	Δ_0	3.2756	3.1824	3.1402	3.1150	3.0997	3.0890	3.0811
0.35	E^*	1.2374	1.2245	1.2185	1.2150	1.2128	1.2112	1.2100
	E_0	0.7941	0.8043	0.8090	0.8117	0.8135	0.8148	0.8157
	Δ_0	3.1163	3.0429	3.0110	2.9922	2.9801	2.9716	2.9653
0.40	E^*	1.1732	1.1639	1.1596	1.1571	1.1555	1.1543	1.1535
	E_0	0.8381	0.8463	0.8501	0.8523	0.8538	0.8548	0.8555
	Δ_0	2.9903	2.9335	2.9077	2.8930	2.8835	2.8768	2.8719
0.45	E^*	1.1256	1.1189	1.1158	1.1140	1.1128	1.1120	1.1114
	E_0	0.8756	0.8820	0.8850	0.8867	0.8878	0.8887	0.8892
	Δ_0	2.8885	2.8442	2.8240	2.8125	2.8050	2.7979	2.7941
0.50	E^*	1.0899	1.0851	1.0829	1.0816	1.0808	1.0802	1.0797
	E_0	0.9068	0.9118	0.9140	0.9153	0.9162	0.9168	0.9172
	Δ_0	2.8054	2.7695	2.7541	2.7461	2.7403	2.7362	2.7332

3.5 An Alternative Approach to Preliminary Test Estimation

3.5.1 Introduction

In the study of the properties of the preliminary test estimators, we note that the optimum preliminary test estimators have basically two characteristics:

(1) They depend heavily on the level of significance (generally $\alpha \geq 0.20$), which is a nonstandard choice of level of significance.

(2) They provide only two choices for the estimator, namely the restricted estimator or the unrestricted estimator on the result of the preliminary test.

To eliminate the dependency on the level of significance and the extreme choices for the estimator, we devise an **alternative** estimator that will **shrink** toward a targeted prior value of the parameter under consideration and provide interpolated values of the estimator between the extreme choices. The interpolated values depend on the value (not the test result) of the test-statistic. The targeted value of the parameter is called the **natural origin**. This approach was introduced by Rodrigues (1987) and elaborated on by Bolfarine and Zacks (1991) in connection with the prediction problem in finite populations. We elaborate the methodology in detail in our context. We first consider the one-sample problem of Section 3.4 to introduce the idea of such an alternative estimator.

3.5.2 One-Sample Problem

Consider the one-sample model of Section 3.4, which is

$$Y = \theta \mathbf{1}_n + \mathbf{e}, \tag{3.5.1}$$

where $\mathbf{e} \sim \mathcal{N}_n(\mathbf{0}, \sigma^2 \mathbf{I}_n)$.

It is suspected that θ may equal θ_0 (natural origin). So we wish to estimate θ based on a random sample $(y_1, \cdots, y_n)$ and the uncertain prior value θ_0. Then we have the preliminary test estimator of θ given by (3.4.6) based on θ_0, the unrestricted estimator, $\tilde{\theta}_n$ and the test-statistic $\mathcal{L}_n$, as

$$\hat{\theta}^{PT} = \tilde{\theta}_n - (\tilde{\theta} - \theta_0) I(\mathcal{L}_n < F_{1,m}(\alpha)); \ (m = n-1), \tag{3.5.2}$$

where $\mathcal{L}_n$ is the test-statistic given by (3.4.5)

$$\mathcal{L}_n = \frac{n(\tilde{\theta}_n - \theta_0)^2}{s_y^2}. \tag{3.5.3}$$

$F_{1,m}(\alpha)$ is the upper α-level critical value from a central F-distribution with $(1, m)$ d.f. The mean-square error of $\hat{\theta}_n^{PT}$ is

$$M_2(\hat{\theta}_n^{PT}) = \frac{\sigma^2}{n}\Big[1 - G_{3,m}\Big(\frac{1}{3}F_{1,m}(\alpha); \Delta^2\Big)$$
$$+\Delta^2\Big\{2G_{3,m}\Big(\frac{1}{3}F_{1,m}(\alpha); \Delta^2\Big) - G_{5,m}\Big(\frac{1}{5}F_{1,m}(\alpha); \Delta^2\Big)\Big\}\Big], \tag{3.5.4}$$

while the mean-square error of $\tilde{\theta}_n$ is $M_1(\tilde{\theta}_n) = \sigma^2/n$.

The expression (3.5.4) depends heavily on the level of significance. Thus applying the maxmin rule given at (3.4.13), we resolve the problem of choosing optimum α. To avoid the dependency of the estimator on α, we devise the following **shrinkage** estimator(SE) based on θ_0 and the sample $(y_1, \cdots, y_n)$

via $\tilde{\theta}_n$ and $\mathcal{L}_n$:

$$\hat{\theta}_n^S = \theta_0 + \left\{1 - c|\mathcal{L}_n^{1/2}|^{-1}\right\}\left(\tilde{\theta}_n - \theta_0\right) \tag{3.5.5a}$$

$$= \theta_0 + \left\{1 - \frac{cs_y}{|\sqrt{n}(\tilde{\theta}_n - \theta_0)|}\right\}\left(\tilde{\theta}_n - \theta_0\right) \tag{3.5.5b}$$

$$= \tilde{\theta}_n - \frac{cs_y(\tilde{\theta}_n - \theta_0)}{|\sqrt{n}(\tilde{\theta}_n - \theta_0)|}, \tag{3.5.5c}$$

where c is an appropriate nonnegative bounded constant. This estimator was introduced by Khan and Saleh (2001). Notice that $\hat{\theta}_n^S$ is similar to $\hat{\theta}_n^{PT}$ (see 3.5.2) where we have replaced the indicator function $I(\mathcal{L}_n < F_{1,m}(\alpha))$ by a continuous decreasing function $c|\mathcal{L}_n^{1/2}|^{-1}$ of $\mathcal{L}_n$. Thus, instead of two extreme choices, namely θ_0 or $\tilde{\theta}_n$, $\hat{\theta}_n^S$ provides the choice for all values between θ_0 and $\tilde{\theta}_n$ depending on the value of $\mathcal{L}_n$ for a given sample. $\hat{\theta}_n^S$ is a smooth version of $\hat{\theta}_n^{PT}$, where $\hat{\theta}_n^S \to \tilde{\theta}_n$ and $\hat{\theta}_n^{PT} \to \tilde{\theta}_n$ as $\mathcal{L}_n \to \infty$ and $\hat{\theta}_n^S \to \theta_0$ when $|\mathcal{L}_n^{1/2}|^{-1} \to c^{-1}$, while $\hat{\theta}_n^{PT} \to \theta_0$ for small values of $\mathcal{L}_n$.

The bias of $\hat{\theta}_n^S$ is seen to be

$$E[\hat{\theta}_n^S - \theta] = -cE[s_y]E\left[\frac{(\tilde{\theta}_n - \theta_0)}{|\sqrt{n}(\tilde{\theta}_n - \theta_0)|}\right]$$

$$= -\frac{c}{\sqrt{n}}E[s_y]E\left[\frac{Z}{|Z|}\right], \tag{3.5.6}$$

where $Z \sim \mathcal{N}(\Delta, 1)$ with $\Delta = \sqrt{n}(\theta - \theta_0)/\sigma$.

Theorem 1. If $Z \sim \mathcal{N}(\Delta, 1)$, then

$$E\left[\frac{Z}{|Z|}\right] = 1 - 2\Phi(-\Delta), \tag{3.5.7}$$

where Φ is the cdf of the standard normal distribution.

The proof is straightforward (see the problems at the end of the chapter; see also Rodrigues, 1987, and Bolfarine and Zacks, 1991). Hence the bias expression (3.5.6) becomes

$$-\frac{c\sigma}{n}K_n\left(2\Phi(\Delta) - 1\right), \quad \text{where} \quad K_n = \sqrt{\frac{2}{n-1}}\frac{\Gamma(n/2)}{\Gamma(n-1/2)}.$$

The quadratic bias function of $\hat{\theta}_n^S$ is further given by

$$B_4(\hat{\theta}_n^S) = c^2K_n^2\left\{2\Phi(\Delta) - 1\right\}^2. \tag{3.5.8}$$

The quadratic bias expression for the PTE is

$$B_3(\hat{\theta}_n^{PT}) = \Delta^2\left\{G_{3,m}\left(\frac{1}{3}F_{1,m}(\alpha); \Delta^2\right)\right\}^2, \tag{3.5.9}$$

which is dependent on α.

$B_4(\hat{\theta}_n^S)$ is a nondecreasing function of Δ^2 with initial value zero and increases monotonically to $c^2K_n^2$ as $\Delta^2 \to \infty$. On the other hand, $B_3(\hat{\theta}_n^{PT})$ is a function of (α, Δ) and, for fixed α, begins with the initial value zero and increases to a maximum at some point $\Delta = \Delta_0$; then it decreases gradually to zero again. Thus, except for small and large values of Δ, the bias of $\hat{\theta}_n^S$ is smaller than the bias of $\hat{\theta}_n^{PT}$.

Next, we consider the mean-square error of $\hat{\theta}_n^S$, which is

$$E(\hat{\theta}_n^S - \theta)^2 = E(\tilde{\theta}_n - \theta)^2 + c^2E(s_y^2)E\left\{\frac{(\tilde{\theta}_n - \theta_0)^2}{[\sqrt{n}(\tilde{\theta}_n - \theta_0)]^2}\right\}$$

$$- 2cE(s_y)E\left\{\frac{(\tilde{\theta}_n - \theta_0)(\tilde{\theta}_n - \theta)}{|\sqrt{n}(\tilde{\theta}_n - \theta_0)|}\right\} \tag{3.5.10}$$

$$= \frac{\sigma^2}{n} + c^2\frac{\sigma^2}{n} - 2c\frac{\sigma^2}{n}K_n\left\{E(|Z|) - \Delta E\left(\frac{Z}{|Z|}\right)\right\}. \tag{3.5.11}$$

To evaluate (3.5.11), we use the following theorem:

Theorem 2. If $Z \sim \mathcal{N}(\Delta, 1)$, then

$$E(|Z|) = \sqrt{\frac{2}{\pi}}\, e^{-\Delta^2/2} + \Delta\{2\Phi(\Delta) - 1\}. \tag{3.5.12}$$

Proof. The pdf of $|Z|$ is

$$\phi_{|Z|}(z) = \phi(z - \Delta) + \phi(z + \Delta), \tag{3.5.13}$$

and we have

$$E(|Z|) = \int_0^\infty z\phi(z - \Delta)dz + \int_0^\infty z\phi(z + \Delta)dz$$

$$= \int_\Delta^\infty z\phi(z)dz + \int_{-\Delta}^\infty z\phi(z)dz + \Delta\left\{\int_{-\Delta}^\infty \phi(z)dz - \int_\Delta^\infty \phi(z)dz\right\}$$

$$= \int_\Delta^\infty z\phi(z)dz + \int_{-\Delta}^\infty z\phi(z)dz + \Delta\{2\Phi(\Delta) - 1\}$$

$$= \sqrt{\frac{2}{\pi}}\, e^{-\Delta^2/2} + \Delta\{2\Phi(\Delta) - 1\}. \tag{3.5.14}$$

Hence, the mse of $\hat{\theta}_n^S$ is given by

$$M_4(\hat{\theta}_n^S) = \frac{\sigma^2}{n}\left\{1 + c^2 - 2cK_n\sqrt{\frac{2}{\pi}}\, e^{-\Delta^2/2}\right\}. \tag{3.5.15}$$

The value of c that minimizes $M_4(\hat{\theta}_n^S)$ is

$$c^* = K_n\sqrt{\frac{2}{\pi}}\, e^{-\Delta^2/2}, \tag{3.5.16}$$

which depends on Δ^2. To make c^* independent of Δ^2, we choose c as

$$c_0 = K_n\sqrt{\frac{2}{\pi}}. \tag{3.5.17}$$

Thus, the optimum value of $M_4(\hat{\theta}_n^S)$ reduces to

$$M_4(\hat{\theta}_n^S) = \frac{\sigma^2}{n}\left\{1 - \frac{2}{\pi}K_n^2\left[2\, e^{-\Delta^2/2} - 1\right]\right\}. \tag{3.5.18}$$

3.5.3 Comparison of PTE, $\hat{\theta}_n^{PT}$ and SE $\hat{\theta}_n^S$

The relative efficiency of $\hat{\theta}_n^S$ compared to $\tilde{\theta}_n$ is given by

$$RE(\hat{\theta}_n^S : \tilde{\theta}_n) = \left\{1 - \frac{2}{\pi}K_n^2\left[2\, e^{-\Delta^2/2} - 1\right]\right\}^{-1}, \tag{3.5.19}$$

while that of $\hat{\theta}_n^{PT}$ is given by

$$RE(\hat{\theta}_n^{PT} : \tilde{\theta}_n) = \Bigg[1 - G_{3,m}\Big(\frac{1}{3}F_{1,m}(\alpha); \Delta^2\Big)$$
$$+\Delta^2\left\{2G_{3,m}\Big(\frac{1}{3}F_{1,m}(\alpha); \Delta^2\Big) - G_{5,m}\Big(\frac{1}{5}F_{1,m}(\alpha); \Delta^2\Big)\right\}\Bigg]^{-1}. \tag{3.5.20}$$

Under the null hypothesis $H_0 : \theta = \theta_0$, $\Delta^2 = 0$; hence,

$$RE(\hat{\theta}_n^S : \tilde{\theta}_n) = \left[1 - \frac{2}{\pi}K_n^2\right]^{-1} \geq 1, \tag{3.5.21}$$

while

$$RE(\hat{\theta}_n^{PT} : \tilde{\theta}_n) = \left[1 - G_{3,m}\Big(\frac{1}{3}F_{1,m}(\alpha); 0\Big)\right]^{-1} \geq 1 \tag{3.5.22}$$

depending on the size of α. As $\Delta^2 \to \infty$,

$$RE(\hat{\theta}_n^S : \tilde{\theta}_n) \to \left[1 + \frac{2}{\pi}K_n^2\right]^{-1} \tag{3.5.23}$$

and

$$RE(\hat{\theta}_n^{PT} : \tilde{\theta}_n) \to 1. \tag{3.5.24}$$

In general, $RE(\hat{\theta}_n^S : \tilde{\theta}_n)$ decreases from $\left[1 - \frac{2}{\pi}K_n^2\right]^{-1}$ at $\Delta^2 = 0$ and crosses the 1–line at $\Delta^2 = ln\,4$ $(= 1.38)$ then drops to the minimum value $\left[1 + \frac{2}{\pi}K_n^2\right]^{-1}(\leq$

1) as $\Delta^2 \to \infty$. The loss of efficiency is $1 - \left[1 + \frac{2}{\pi}K_n^2\right]^{-1}$, while the gain in efficiency is $\left[1 - \frac{2}{\pi}K_n^2\right]^{-1}$. Thus, for $0 \leq \Delta^2 \leq ln\,4$, $\hat{\theta}_n^S$ performs better than $\tilde{\theta}_n$; otherwise, $\tilde{\theta}_n$ performs better outside this interval.

In the case of $\hat{\theta}_n^{PT}$, the graph of $RE(\hat{\theta}_n^{PT}; \tilde{\theta}_n)$ decreases from (3.5.22) to a minimum at $\Delta = \Delta_0$, and then increases to 1 as $\Delta \to \infty$. It performs better than $\tilde{\theta}_n$ in the range $0 \leq \Delta^2 \leq 1$. Thus, the range of better performance of $\hat{\theta}_n^S$ compared to $\hat{\theta}_n^{PT}$ is increased by an amount 0.38.

The following Table 3.5.1 gives the performance of $\hat{\theta}_n^S$ and $\hat{\theta}_n^{PT}$ (for each selected level of significance and sample sizes, n).

Table 3.5.1 Maximum and Minimum Efficiencies of SE and Efficiency of PTE at Δ_0 for Selected α

$\alpha \backslash n$		10	15	20	25	30	35	40
	E^{max}	2.6261	2.6692	2.6903	2.7029	2.7112	1.7171	1.7215
	E^{min}	0.6176	0.6152	0.6141	0.6135	0.6131	0.6128	0.6125
0.05	E_{Δ_0}	0.6408	0.6466	0.6498	0.6518	0.6532	0.6542	0.6550
	E_0	0.3350	0.3600	0.3720	0.3790	0.3836	0.3868	0.3893
	Δ_0	6.1009	5.5303	5.2857	5.1514	5.0656	5.0063	4.9629
0.15	E_{Δ_0}	0.6827	0.6892	0.6924	0.6943	0.6955	0.6964	0.6971
	E_0	0.5405	0.5601	0.5693	0.5747	0.5782	0.5807	0.5825
	Δ_0	4,1626	2.9403	3,8429	3,7875	3.7520	3.7272	2.7090
0.25	E_{Δ_0}	0.7133	0.7182	9.7206	0.7220	0.7229	0.7236	0.7241
	E_0	0.6844	0.6990	0.7059	0.7098	0.7124	0.7134	0.7156
	Δ_0	3.4811	3.3612	3.3066	3.2757	3.2557	3.2417	3.2314
0.35	E_{Δ_0}	0.7361	0.7395	0.7410	0.7420	0.7426	0.7420	9.7433
	E_0	0.7941	0.8043	0.8090	0.8117	0.8135	0.8148	0.8157
	Δ_0	3.1163	3.0429	3.0110	2.9922	2.9801	2.9716	2.9653
0.45	E_{Δ_0}	0.7536	0.7555	0.7564	0.7569	0.7572	0.7576	0.7578
	E_0	0.8756	0.8820	0.8850	0.8867	0.8878	0.8887	0.8892
	Δ_0	2.8885	2.8442	2.8240	2.8125	2.8050	2.7979	2.7941

The first two rows of the table give the maximum and minimum relative efficiency of SE for selected sample sizes. The maximum relative efficiency of SE increases as the sample size n increases, and as $n \to \infty$, it tends to $\pi/\pi - 2$ while the minimum efficiency decreases as n increases. Finally, as $n \to \infty$, it tends to $\pi/\pi + 2$. The maximum relative efficiency at $\Delta^2 = 0$ is $\left[1 - \frac{2}{\pi}K_n^2\right]^{-1}$, which varies depending on the sample sizes.

The remaining rows of the Table 3.5.1 contain the minimum relative efficiency, E_0 of the PTE at $\Delta = \Delta_0$, which is recorded for each $\alpha = 0.05(0.10)0.45$ with the corresponding efficiency (E_{Δ_0}) of SE for $\Delta = \Delta_0$. The graph of the relative efficiency of SE against Δ_0 is plotted in Figure 3.5.1 along with the graph of relative efficiency of PTE showing the relative positions of the efficiency over the Δ_0-values.

From Figure 3.5.1 it is clear that the relative efficiency of PTE compared to UE is smaller for $\Delta > 1$ and that of SE relative to UE is smaller for $\Delta > 1.38$. Thus, SE dominates UE over the wider interval $[0, 1.38]$ relative to the PTE

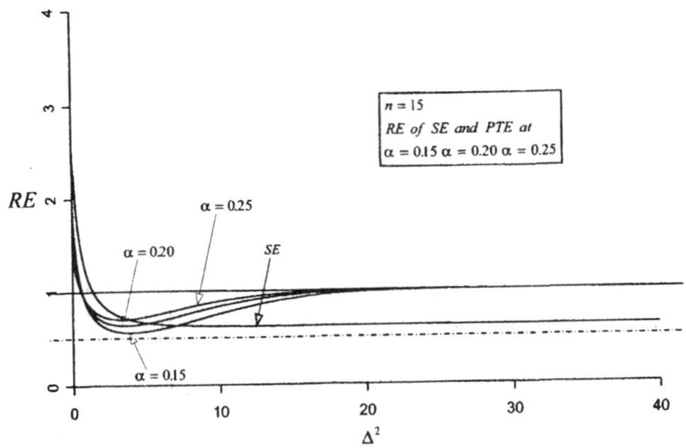

Figure 3.5.1 Graph of the relative efficiency of SE and PTE for different values of α

in the interval $[0, 1]$. Also from Figure 3.5.1 we see that SE has higher relative efficiency than PTE over all Δ at which efficiency of PTE is smallest.

In general, there is no uniform domination of SE over the PTE for all Δ and every α. When Δ is near the origin, SE over performs PTE for some α. Also, at $\Delta = 0$, the relative efficiency of SE is much higher than that of the PTE except for small α. Further, SE is independent of α. Thus, considering the overall performance of SE relative to PTE, SE is preferable to PTE because it produces interpolated estimators that are free of the level of significance, α.

3.5.4 Simple Linear Model and Shrinkage Estimation

Consider the simple model (3.1.1) again,

$$\mathbf{Y} = \theta \mathbf{1}_n + \beta \mathbf{x} + \mathbf{e}, \qquad \mathbf{e} \sim \mathcal{N}_n(\mathbf{0}, \sigma^2 \mathbf{I}_n), \tag{3.5.25}$$

and the estimator of (θ, β) as given by (3.1.2) and that for testing $\beta = \beta_0$ as given by (3.1.6). These estimators are repeated here as

$$\begin{pmatrix} \hat{\theta}_n \\ \hat{\beta}_n \end{pmatrix} = \begin{pmatrix} \bar{Y} - \tilde{\beta}_n \bar{\mathbf{x}} \\ \frac{1}{Q}[\mathbf{x}'\mathbf{Y} - \frac{1}{n}(\mathbf{1}_n'\mathbf{x})(\mathbf{1}_n'\mathbf{x})] \end{pmatrix} \tag{3.5.26}$$

and

$$\mathcal{L}_n = \frac{(\tilde{\beta}_n - \beta_0)^2 Q}{s_e^2}, \tag{3.5.27}$$

where $Q = \mathbf{x}'\mathbf{x} - \frac{1}{n}(\mathbf{1}_n'\mathbf{x})^2$ and

$$s_e^2 = (n-2)^{-1}(\mathbf{Y} - \tilde{\theta}_n \mathbf{1}_n - \tilde{\beta}_n \mathbf{x})'(\mathbf{Y} - \tilde{\theta}_n \mathbf{1}_n - \tilde{\beta}_n \mathbf{x}). \tag{3.5.28}$$

Consider the PTE and SE of β following Section 3.5.1 as

$$\hat{\beta}_n^{PT} = \tilde{\beta}_n - (\tilde{\beta}_n - \beta_0) I(\mathcal{L}_n < F_{1,m}(\alpha)), \qquad m = n-2, \tag{3.5.29}$$

and

$$\hat{\beta}_n^S = \tilde{\beta}_n - \frac{c s_e (\tilde{\beta}_n - \beta_0)}{\sqrt{Q}|\tilde{\beta}_n - \beta_0|}, \tag{3.5.30}$$

respectively, where c is the nonnegative shrinking constant. Similarly, the PTE and SE of θ are given by

$$\hat{\theta}_n^{PT} = \tilde{\theta}_n + (\tilde{\beta}_n - \beta_0)\bar{x} I(\mathcal{L}_n < F_{1,m}(\alpha)), \qquad m = n-2, \tag{3.5.31}$$

and

$$\hat{\theta}_n^S = \tilde{\theta}_n + \frac{c \bar{x} s_e (\tilde{\beta}_n - \beta_0)}{\sqrt{Q}|\tilde{\beta}_n - \beta_0|}, \tag{3.5.32}$$

respectively. The shrinkage estimators $(\hat{\theta}_n^S, \hat{\beta}_n^S)$ have been studied by Khan, Hoque, and Saleh (2002).

The bias expressions for the SE of β and θ may be obtained as

$$\begin{aligned} b_4(\hat{\beta}_n^S) &= -\frac{c}{\sqrt{Q}} E(s_e) E\left[\frac{\tilde{\beta}_n - \beta_0}{|\tilde{\beta}_n - \beta_0|}\right] \\ &= -\frac{c}{\sqrt{Q}} E(s_e) E\left[\frac{Z}{|Z|}\right], \end{aligned} \tag{3.5.33}$$

where $Z \sim \mathcal{N}(\Delta, 1)$ with $\Delta = (\beta - \beta_0) Q^{1/2}/\sigma$ and where

$$b_4(\hat{\theta}_n^S) = \frac{c\bar{x}}{\sqrt{Q}} E(s_e) E\left[\frac{Z}{|Z|}\right] = \frac{c\sigma\bar{x}}{\sqrt{Q}} K_n \Big(2\Phi(\Delta) - 1\Big) \tag{3.5.34}$$

with $K_n = \sqrt{\frac{2}{n-2}} \frac{\Gamma(\frac{n-1}{2})}{\Gamma(\frac{n-2}{2})}$. Similarly the mse expression for the SE of β and θ are, respectively, given by

$$\begin{aligned} M_4(\hat{\beta}_n^S) &= E\Big[(\tilde{\beta}_n - \beta)^2\Big] + \frac{c^2}{Q} E(s_e) - 2\frac{c}{\sqrt{Q}} E(s_e) E\left[\frac{(\tilde{\beta}_n - \beta_0)(\tilde{\beta}_n - \beta)}{|\tilde{\beta}_n - \beta_0|}\right] \\ &= \frac{\sigma^2}{Q} + \frac{c^2\sigma^2}{Q} - 2\frac{c\sigma^2}{Q} K_n \left\{ E(|Z|) - \Delta E\left[\frac{Z}{|Z|}\right] \right\} \\ &= \frac{\sigma^2}{Q} \left\{ 1 + c^2 - 2cK_n \sqrt{\frac{2}{\pi}}\, e^{-\Delta^2/2} \right\}, \\ &= \frac{\sigma^2}{Q} \left\{ 1 - \frac{2}{\pi} K_n^2 [2\, e^{-\Delta^2/2} - 1] \right\}, \quad \text{using } c = \sqrt{\frac{2}{\pi}} K_n, \end{aligned} \tag{3.5.35}$$

and

$$M_4(\hat{\theta}_n^S) = E\Big[(\tilde{\theta}_n - \theta)^2\Big] + \frac{c^2\bar{x}^2}{Q}E(s_e) + 2\frac{c\bar{x}^2}{\sqrt{Q}}E(s_e)E\left[\frac{(\tilde{\beta}_n - \beta_0)(\tilde{\theta}_n - \theta)}{|\tilde{\beta}_n - \beta_0|}\right]$$

$$= \sigma^2(\frac{1}{n} + \frac{\bar{x}^2}{Q}) + \frac{c^2\bar{x}^2\sigma^2}{Q} - 2K_n\frac{c\bar{x}^2\sigma^2}{Q}\left\{E(|Z|) - \Delta E\left[\frac{Z}{|Z|}\right]\right\}$$

$$= \frac{\sigma^2}{n} + \frac{\sigma^2\bar{x}^2}{Q}\left\{1 + c^2 - 2cK_n\sqrt{\frac{2}{\pi}}\, e^{-\Delta^2/2}\right\}$$

$$= \frac{\sigma^2}{n} + \frac{\sigma^2\bar{x}^2}{Q}\left\{1 - \frac{2}{\pi}K_n^2[2\, e^{-\Delta^2/2} - 1]\right\}, \qquad (3.5.36)$$

using the optimum value of c as before.

The relative efficiency of $\hat{\beta}_n^S$ with respect to $\tilde{\beta}$ and of $\hat{\theta}_n^S$ with respect to $\tilde{\theta}_n$ are

$$RE(\hat{\beta}_n^S : \tilde{\beta}_n) = \left[1 - \frac{2}{\pi}K_n^2\left(2\, e^{-\Delta^2/2} - 1\right)\right]^{-1} \qquad (3.5.37)$$

and

$$RE(\hat{\theta}_n^S : \tilde{\theta}_n) = \left[1 - \frac{2}{\pi}K_n^2\frac{\bar{x}^2}{Q}\left(\frac{1}{n} + \frac{\bar{x}^2}{Q}\right)^{-1}\left(2\, e^{-\Delta^2/2} - 1\right)\right]^{-1}, \qquad (3.5.38)$$

respectively. If $\Delta^2 = 0$,

$$RE(\hat{\beta}_n^S : \tilde{\beta}_n) = \left[1 - \frac{2}{\pi}K_n^2\right]^{-1}(\geq 1) \qquad (3.5.39)$$

and

$$RE(\hat{\theta}_n^S : \tilde{\theta}_n) = \left[1 - \frac{2}{\pi}K_n^2\frac{\bar{x}^2}{Q}\left(\frac{1}{n} + \frac{\bar{x}^2}{Q}\right)^{-1}\right]^{-1}(\geq 1) \qquad (3.5.40)$$

depending on the value of $\bar{x}^2/Q$. On the other hand, as $\Delta^2 \to \infty$,

$$RE(\hat{\beta}_n^S : \tilde{\theta}_n) = \left[1 + \frac{2}{\pi}K_n^2\right]^{-1}(\leq 1) \qquad (3.5.41)$$

and

$$RE(\hat{\theta}_n^S : \tilde{\theta}_n) = \left[1 + \frac{2}{\pi}K_n^2\frac{\bar{x}^2}{Q}\left(\frac{1}{n} + \frac{\bar{x}^2}{Q}\right)^{-1}\right]^{-1}(\leq 1). \qquad (3.5.42)$$

Note that both relative efficiencies are decreasing functions of Δ^2 with the highest efficiency at $\Delta^2 = 0$.

Some values of the efficiency $RE(\hat{\theta}_n^S : \tilde{\theta}_n)$ are given in Table 3.5.2 for chosen values of the coefficient of variation of the x-values. The efficiency $RE(\hat{\beta}_n^S : \tilde{\beta}_n)$ is available in Table 3.5.1, since it is similar to the one-sample problem.

Table 3.5.2 Minimum and Maximum Relative Efficiencies of SE and PTE for $n = 8$, $\alpha = 0.05(0.10)0.45$ and $\frac{\bar{x}^2}{Q} = 1(0.5)5$

$\alpha \backslash \bar{x}^2/Q$	1	1.5	2	2.5	3	3.5	4	4.5	5
E^{max}	2.08696	2.17801	2.22951	2.26263	2.28571	2.30273	2.31579	2.32613	2.33452
E^{min}	0.65756	0.64901	0.64457	0.64186	0.64002	0.63870	0.63770	0.63692	0.63630
0.05 E_{Δ_0}	0.67101	0.66262	0.65827	0.65560	0.65380	0.65250	0.65152	0.65076	0.65014
E_0	0.32597	0.31773	0.31353	0.31099	0.30929	0.30806	0.30714	0.30643	0.30585
Δ_0	7.05937	7.05937	7.05937	7.05937	7.05937	7.05937	7.05937	7.05937	7.05937
0.10 E_{Δ_0}	0.69048	0.68236	0.67814	0.67555	0.67380	0.67254	0.67159	0.67084	0.67024
E_0	0.44787	0.43855	0.43378	0.43087	0.42892	0.42752	0.42646	0.42564	0.42497
Δ_0	5.32801	5.32801	5.32801	5.32801	5.32801	5.32801	5.32801	5.32801	5.32801
0.15 E_{Δ_0}	0.70845	0.70060	0.69651	0.69400	0.69231	0.69108	0.69016	0.68944	0.68886
E_0	0.54227	0.53289	0.52805	0.52510	0.52311	0.52168	0.52061	0.51976	0.51909
Δ_0	4.50883	4.50883	4.50883	4.50883	4.50883	4.50883	4.50883	4.50883	4.50883
0.20 E_{Δ_0}	0.72438	0.71678	0.71282	0.71040	0.70875	0.70757	0.70667	0.70597	0.70541
E_0	0.62136	0.61244	0.60782	0.60499	0.60309	0.60172	0.60068	0.59987	0.59922
Δ_0	4.00897	4.00897	4.00897	4.00897	4.00897	4.00897	4.00897	4.00897	4.00897
0.25 E_{Δ_0}	0.74109	0.73378	0.72997	0.72763	0.72605	0.72491	0.72405	0.72337	0.72283
E_0	0.70912	0.70127	0.69719	0.69469	0.69299	0.69177	0.69085	0.69013	0.68955
Δ_0	3.60838	3.60838	3.60838	3.60838	3.60838	3.60838	3.60838	3.60838	3.60838
0.30 E_{Δ_0}	0.75070	0.74358	0.73986	0.73758	0.73603	0.73492	0.73407	0.73341	0.73288
E_0	0.74827	0.74109	0.73735	0.73505	0.73349	0.73237	0.73152	0.73086	0.73033
Δ_0	3.41621	3.41621	3.41621	3.41621	3.41621	3.41621	3.41621	3.41621	3.41621
0.35 E_{Δ_0}	0.76154	0.75461	0.75100	0.74878	0.74728	0.74620	0.74538	0.74474	0.74422
E_0	0.79911	0.79298	0.78978	0.78781	0.78647	0.78551	0.78478	0.78421	0.78375
Δ_0	3.22498	3.22498	3.22498	3.22498	3.22498	3.22498	3.22498	3.22498	3.22498
0.40 E_{Δ_0}	0.77106	0.76432	0.76081	0.75865	0.75719	0.75613	0.75533	0.75471	0.75421
E_0	0.84262	0.83755	0.83489	0.83326	0.83215	0.83134	0.83074	0.83026	0.82988
Δ_0	3.07468	3.07468	3.07468	3.07468	3.07468	3.07468	3.07468	3.07468	3.07468
0.45 E_{Δ_0}	0.77941	0.77286	0.76943	0.76732	0.76590	0.76487	0.76409	0.76348	0.76299
E_0	0.87938	0.87532	0.87318	0.87187	0.87097	0.87033	0.86984	0.86946	0.86915
Δ_0	2.95417	2.95417	2.95417	2.95417	2.95417	2.95417	2.95417	2.95417	2.95417
0.50 E_{Δ_0}	0.78673	0.78033	0.77699	0.77493	0.77354	0.77253	0.77177	0.77118	0.77070
E_0	0.90992	0.90677	0.90512	0.90410	0.90340	0.90290	0.90252	0.90223	0.90199
Δ_0	2.85621	2.85621	2.85621	2.85621	2.85621	2.85621	2.85621	2.85621	2.85621

Note: α is the level of significance, E^{max} is the maximum efficiency of SE, E_{min} is the minimum efficiency of SE, E_0 is the minimum efficiency of PTE, E_{Δ_0} is the minimum efficiency of SE at Δ_0, and Δ_0 is the value of Δ at which the minimum efficiency of PTE occurs.

3.5.5 The Two-Sample Problem and Shrinkage Estimation

Consider the linear method (3.5.25), where $\mathbf{x} = (0, \cdots, 0; 1, \cdots, 1)'$ is an $n \times 1$ vector with n_1 zero's and n_2 1's such that $n = n_1 + n_2$. Also, $\mathbf{e} \sim \mathcal{N}_n(\mathbf{0}, \sigma^2 \mathbf{I}_n)$, and in this case,

$$\bar{x} = \frac{n_2}{n_1 + n_2}, \; Q = \frac{n_1 n_2}{n_1 + n_2} \quad \text{and} \quad \frac{\bar{x}^2}{Q} = \frac{n_2}{n_1(n_1 + n_2)}. \tag{3.5.43}$$

For this model, $\mathbf{Y} = \theta \mathbf{1}_n + \beta \mathbf{x} + \mathbf{e}$,

$$\mu_1 = \theta \qquad \text{and} \qquad \mu_2 = \theta + \beta \tag{3.5.44}$$

with the estimators of μ_1 and μ_2 as

$$\tilde{\mu}_1 = \bar{y}_1 \qquad \text{and} \qquad \tilde{\mu}_2 = \bar{y}_2. \tag{3.5.45}$$

The unbiased estimator of σ^2 is s_p^2, where

$$s_p^2 = (n_1 + n_2 - 2)^{-1} \Big\{ \sum_{j=1}^{n_1} (y_j - \bar{y}_1)^2 + \sum_{j=n_1+1}^{n} (y_j - \bar{y}_2)^2 \Big\}. \tag{3.5.46}$$

For the test of the hypothesis H_0,

$$H_0 : \mu_1 = \mu_2 \;\; \text{against} \;\; \mu_1 \neq \mu_2 \tag{3.5.47}$$

we consider the statistic

$$\mathcal{L}_n = \frac{n_1 n_2}{n_1 + n_2} \frac{(\tilde{\mu}_2 - \tilde{\mu}_1)^2}{s_p^2}. \tag{3.5.48}$$

Now the shrinkage estimator of μ_1 may be written as

$$\hat{\mu}_1^S = \tilde{\mu}_1 - c|\mathcal{L}_n^{1/2}|^{-1}(\tilde{\mu}_1 - \hat{\mu}_n) = \tilde{\mu}_1 - c\sqrt{\frac{1}{n_1} + \frac{1}{n_2}} \frac{s_p(\tilde{\mu}_2 - \tilde{\mu}_1)}{|\tilde{\mu}_2 - \tilde{\mu}_1|}. \tag{3.5.49}$$

where $\hat{\mu}_n = n_1\tilde{\mu}_1 + n_2\tilde{\mu}_2 / n_1 + n_2$. The bias of $\hat{\mu}_1^S$ is given by

$$\begin{aligned} b(\hat{\mu}_1^S) &= -c\sqrt{\frac{1}{n_1} + \frac{1}{n_2}} E(s_p) E\left[\frac{Z}{|Z|}\right] \\ &= -c\sigma\sqrt{\frac{1}{n_1} + \frac{1}{n_2}} K_n \Big(2\Phi(\Delta) - 1\Big), \end{aligned} \tag{3.5.50}$$

where $Z \sim \mathcal{N}(\Delta, 1)$ with $\Delta^2 = n_1 n_2 (\mu_2 - \mu_1)^2 / (n_1 + n_2)\sigma^2$.

Similarly, the mse of $\hat{\mu}_1^S$ is then calculated as

$$\begin{aligned} M(\hat{\mu}_1^S) &= E\Big[(\hat{\mu}_1^S - \mu_1)^2\Big] \\ &= \frac{\sigma^2}{n_1} + \frac{n_2\sigma^2}{n_1(n_1 + n_2)} \Big\{ 1 + c^2 - 2cK_n\sqrt{\frac{2}{\pi}}\, e^{-\Delta^2/2} \Big\} \\ &= \frac{\sigma^2}{n_1} + \frac{n_2\sigma^2}{n_1(n_1 + n_2)} \Big\{ 1 - \frac{2}{\pi} K_n^2 \Big[2\, e^{-\Delta^2/2} - 1 \Big] \Big\}, \end{aligned} \tag{3.5.51}$$

with the optimum c-value as $\sqrt{(2/\pi)}K_n$ given earlier and K_n is defined by

$$K_n = \sqrt{\frac{2}{n_1+n_2-2}\,\frac{\Gamma(\frac{n_1+n_2-1}{2})}{\Gamma(\frac{n_1+n_2-2}{2})}}. \tag{3.5.52}$$

Hence, the $RE(\hat{\mu}_1^S; \tilde{\mu}_1)$ is

$$\left\{1 - \frac{2}{\pi}\frac{n_2}{(n_1+n_2)}K_n^2\left[2\,e^{-\Delta^2/2} - 1\right]\right\}^{-1}. \tag{3.5.53}$$

$RE(\hat{\mu}_1^S; \tilde{\mu}_1)$ is a decreasing function of Δ^2. At $\Delta^2 = 0$,

$$RE(\hat{\mu}_1^S; \tilde{\mu}_1) = \left\{1 - \frac{2}{\pi}\frac{n_2}{(n_1+n_2)}K_n^2\right\}^{-1} \quad (\geq 1), \tag{3.5.54}$$

while for $\Delta^2 \to \infty$,

$$RE(\hat{\mu}_1^S; \tilde{\mu}_1) = \left\{1 + \frac{2}{\pi}\frac{n_2}{(n_1+n_2)}K_n^2\right\}^{-1} \quad (\leq 1). \tag{3.5.55}$$

It is known from Section 3.3 that the relative efficiency of $\hat{\mu}_1^{PT}$ is

$$\begin{aligned}&\left[1 - \frac{n_2}{n_1+n_2}G_{3,m}\left(\frac{1}{3}F_{1,m}(\alpha); \Delta^2\right)\right.\\ &\left.+ \frac{n_2}{n_1+n_2}\Delta^2\left\{2G_{3,m}\left(\frac{1}{3}F_{1,m}(\alpha); \Delta^2\right) - G_{5,m}\left(\frac{1}{5}F_{1,m}(\alpha); \Delta^2\right)\right\}\right]^{-1}\\ &= \left[1 + g(\Delta^2)\right]^{-1};\end{aligned}$$

see (3.3.10b). The maximum value of $RE(\hat{\mu}_1^{PT}; \tilde{\mu}_1)$ is

$$\left[1 - \frac{n_2}{n_1+n_2}G_{3,m}\left(\frac{1}{3}F_{1,m}(\alpha); 0\right)\right]^{-1}, \qquad m = n_1+n_2-2, \tag{3.5.56}$$

and the minimum value for each α occurs at $\Delta^2 = \Delta_0^2$ with the value $[1 + g(\Delta_0^2)]^{-1}$ and Δ_0^2 depending on the size of α.

Table 3.5.3 provides the relative efficiencies of SE and PTE for each α at $\Delta^2 = \Delta_0^2$ to show that SE out forms the PTE for many α-values. Thus, SE is a preferred estimator of μ_1.

Table 3.5.3 Minimum and Maximum Relative Efficiency of SE and PTE for $\alpha = .05(.10).45$ and for selected samples

$\alpha \backslash (n_1, n_2)$		(4,4)	(4,8)	(4,12)	(4,16)	(15,16)
	E_{max}	1.4144	1.6771	1.8544	1.9816	1.4770
	E_{min}	0.7734	0.7124	0.6846	0.6688	0.7559
.05	E_{Δ_0}	0.7838	0.7341	0.7129	0.7012	0.7870
	E_0	0.4623	0.4380	0.4286	0.4234	0.5474
	Δ_0	7.0594	5.9338	5.5302	5.3232	5.0520
.10	E_{Δ_0}	0.7986	0.7535	0.7340	0.7230	0.8041
	E_0	0.5905	0.5572	0.5440	0.5309	0.6538
	Δ_0	5.3280	4.7087	4.4767	4.3554	4.1941
.15	E_{Δ_0}	0.8120	0.7694	0.7506	0.7400	0.8172
	E_0	0.6781	0.6434	0.6293	0.6216	0.7269
	Δ_0	4.5088	4.0983	3.9409	3.8578	3.7462
.20	E_{Δ_0}	0.8237	0.7826	0.7643	0.7538	0.8276
	E_0	0.7447	0.7119	0.6983	0.6908	0.7830
	Δ_0	4.0090	3.7150	3.6005	3.5397	3.4574
.25	E_{Δ_0}	0.8338	0.7939	0.7757	0.7654	0.8362
	E_0	0.7978	0.7883	0.7559	0.7490	0.8279
	Δ_0	3.5667	3.4474	3.3611	3.3149	3.2525
.30	E_{Δ_0}	0.8426	0.8034	0.7855	0.7752	0.8433
	E_0	0.8409	0.8154	0.8045	0.7985	0.8644
	Δ_0	3.4162	3.2488	3.1824	3.1467	3.0983
.35	E_{Δ_0}	0.8502	0.8117	0.7939	0.7837	0.8494
	E_0	0.8761	0.8548	0.8456	0.8405	0.8944
	Δ_0	3.2250	3.0955	3.0438	3.0159	2.9780
.40	E_{Δ_0}	0.8569	0.8189	0.8012	0.7909	0.8546
	E_0	0.9049	0.8877	0.8802	0.8759	0.9189
	Δ_0	3.0747	2.9739	2.9335	2.9117	2.8819
.45	E_{Δ_0}	0.8627	0.8251	0.8075	0.7972	0.8591
	E_0	0.9284	0.9148	0.9088	0.9055	0.9389
	Δ_0	2.9542	2.8758	2.8442	2.8270	2.8037
.50	E_{Δ_0}	0.8677	0.8304	0.8129	0.8027	0.8630
	E_0	0.9472	0.9369	0.9323	0.9297	0.9550
	Δ_0	2.8562	2.7956	2.7710	2.7577	2.7394

Note: α is the level of significance, n_1 and n_2 are the sample sizes, E_{max} is the maximum efficiency of SE, E_{min} is the minimum efficiency of SE, E_0 is the minimum efficiency of PTE, E_{Δ_0} is the minimum efficiency of SE at Δ_0, Δ_0 is the value of Δ at which the minimum efficiency of PTE occurs.

3.6 Estimation with Non-Normal Errors

In this section we consider the properties of the **unrestricted, restricted,** and **preliminary test and shrinkage estimators** in a simple linear model when the components of the error vector $\mathbf{e} = (e_1, \cdots, e_n)'$ in (3.2.1) are independent and

$$E(\mathbf{e}) = \mathbf{0} \quad \text{and} \quad E(\mathbf{e}\mathbf{e}') = \sigma^2\mathbf{I}_n \tag{3.6.1}$$

and the distribution of $\mathbf{e}$ is **nonnormal**, say, $F(\mathbf{e}) = \prod_{j=1}^n F_0(e_j)$.

3.6.1 Unrestricted, Restricted, Preliminary Test and Shrinkage Estimators, and the Test of Slope

Following the notations of Section 3.1 and 3.2, the unrestricted and the restricted estimators of θ and β are

$$\tilde{\theta}_n = \overline{y} - \tilde{\beta}_n\overline{x}, \quad \hat{\theta}_n = \overline{y} - \beta_0\overline{x} \quad (\text{when } \beta = \beta_0)$$

and

$$\tilde{\beta}_n = \frac{1}{Q}\left[\mathbf{x}'\mathbf{y} - \frac{1}{n}(\mathbf{1}_n'\mathbf{x})(\mathbf{1}_n'\mathbf{x})\right]. \tag{3.6.2}$$

As for the test of $H_0 : \beta = \beta_0$, the test-statistic $\mathcal{L}_n$ as in (3.1.6) may be used

$$\mathcal{L}_n = \frac{(\tilde{\beta}_n - \beta_0)^2 Q}{s_e^2}. \tag{3.6.3}$$

Therefore, the preliminary test estimator $\hat{\theta}_n^{\text{PT}}$ is simply,

$$\hat{\theta}_n^{\text{PT}} = \tilde{\theta}_n - (\tilde{\theta}_n - \hat{\theta}_n)I(\mathcal{L}_n < \mathcal{L}_{n,\alpha}) \tag{3.6.4}$$

where $\mathcal{L}_{n,\alpha}$ is the upper α-level value of the distribution of $\mathcal{L}_n$ under H_0. The shrinkage estimator $\hat{\theta}_n^S$ based on Section 3.5 is then defined by

$$\begin{aligned}\hat{\theta}_n^S &= \tilde{\theta}_n - c(\tilde{\theta}_n - \hat{\theta}_n)\frac{s_e}{\sqrt{Q}|\tilde{\beta}_n - \beta_0|} \\ &= \tilde{\theta}_n + \frac{cs_e\overline{x}}{\sqrt{Q}}\frac{\tilde{\beta}_n - \beta_0}{|\tilde{\beta}_n - \beta_0|}.\end{aligned} \tag{3.6.5}$$

3.6.2 Conditions for Asymptotic Normality of the Unrestricted Estimators of Intercept and Slope Parameters

We know that if the errors are independent normal with zero mean and variance σ^2, the exact distribution of $(\tilde{\theta}_n, \tilde{\beta}_n)'$ is a bivariate normal with mean $(\theta, \beta)'$ and covariance matrix

$$\sigma^2\begin{pmatrix} \frac{1}{n} + \frac{\overline{x}^2}{Q} & -\frac{\overline{x}}{Q} \\ -\frac{\overline{x}}{Q} & \frac{1}{Q} \end{pmatrix}. \tag{3.6.6}$$

Since $\mathbf{e} = (e_1, \ldots, e_n)'$ is **not** normally distributed, we need the following three regularity conditions (in this section and the next one) for the asymptotic normality of $(\tilde{\theta}_n, \tilde{\beta}_n)'$. See for example Sen and Singer (1993).

Theorem 1. Assume the following:

(i) $\lim_{n\to\infty} \overline{x} = \overline{x}_0$, $|\overline{x}_0| < \infty$.

(ii) Let $q_i = x_i - \overline{x}/\sqrt{Q}$ and $\max_{1\le i\le n} q_i^2 \to 0$ as $n \to \infty$. (3.6.7)

(iii) $\lim_{n\to\infty} n^{-1}Q = Q^0 < \infty$.

Then

$$\begin{pmatrix} \sqrt{n}(\tilde{\theta}_n - \theta) \\ \sqrt{n}(\tilde{\beta}_n - \beta) \end{pmatrix} \xrightarrow{\mathcal{D}} N_2\left\{ \begin{pmatrix} 0 \\ 0 \end{pmatrix}, \sigma^2 \begin{pmatrix} 1 + \frac{\overline{x}_0^2}{Q_0} & -\frac{\overline{x}}{Q^0} \\ -\frac{\overline{x}_0}{Q^0} & \frac{1}{Q^0} \end{pmatrix} \right\}, \tag{3.6.8}$$

where $\mathcal{D}$ stands for "in distribution."

As for the test of hypothesis, we see that by Sluskey's theorem as $n \to \infty$, $\mathcal{L}_n = (\tilde{\beta}_n - \beta_0)^2 Q/s_e^2$ converges to a central chi-square distribution with one d.f.

Under the fixed alternative hypothesis $A_\delta : \beta = \beta_0 + \delta$ we see that

$$\sqrt{n}(\tilde{\beta}_n - \beta_0) = \sqrt{n}(\tilde{\beta}_n - \beta) + \sqrt{n}\delta, \tag{3.6.9}$$

and under A_δ, that $\sqrt{n}(\tilde{\beta}_n - \beta_0) \sim \mathcal{N}(\sqrt{n}\delta, \sigma^2/Q_0)$. Also, $\sqrt{n}\delta \to \infty$, as $n \to \infty$ which implies that $\mathcal{L}_n = [(\tilde{\beta}_n - \beta_0)^2/Q]/s_e^2 \to \infty$ as $n \to \infty$. As a result, we have the following theorem:

Theorem 2. Under fixed alternatives

$$A_\delta : \beta = \beta_0 + \delta,$$

and the regularity conditions of Theorem 1, as $n \to \infty$,

(i) $\sqrt{n}(\hat{\beta}_n^{PT} - \beta) = \sqrt{n}(\tilde{\beta} - \beta) + o_p(1)$;

(ii) $\sqrt{n}(\hat{\beta}_n^{S} - \beta) = \sqrt{n}(\tilde{\beta} - \beta) + o_p(1)$. (3.6.10)

Proof. Under A_δ, we have

$$\begin{aligned} \lim_{n\to\infty} E\Big[n(\hat{\beta}_n^{PT} - \tilde{\beta}_n)^2\Big] &= \lim_{n\to\infty} E\Big[n(\tilde{\beta}_n - \beta_0)^2 I\Big(\mathcal{L}_n \le \mathcal{L}_{n,\alpha}\Big)\Big] \\ &= \lim_{n\to\infty} \Big\{\frac{\sigma^2}{Q} E\Big[\mathcal{L}_n I\Big(\mathcal{L}_n \le \mathcal{L}_{n,\alpha}\Big)\Big]\Big\} = \frac{\sigma^2}{Q^0} \lim_{n\to\infty} E\Big[\mathcal{L}_n I\Big(\mathcal{L}_n \le \mathcal{L}_{n,\alpha}\Big)\Big] \\ &\le \frac{\sigma^2}{Q^0}\chi_1^2(\alpha) \lim_{n\to\infty} E\Big[I\Big(\mathcal{L}_n \le \chi_1^2(\alpha)\Big)\Big] = 0, \end{aligned} \tag{3.6.11}$$

since $\mathcal{L}_n \to \infty$ as $n \to \infty$. Hence,

$$\sqrt{n}(\hat{\beta}_n^{PT} - \beta) = \sqrt{n}(\tilde{\beta}_n - \beta) + o_p(1).$$

Similarly,

$$\lim_{n\to\infty} E\Big[n(\hat{\beta}_n^S - \tilde{\beta}_n)^2\Big] = \lim_{n\to\infty} E\Big[\frac{c^2 s_e^2}{Q}\ \frac{\bar{x}(\tilde{\beta}_n - \beta)^2}{|\tilde{\beta}_n - \beta|^2}\Big] = 0, \tag{3.6.12}$$

since $n^{-1}Q \to Q^0$ as $n \to \infty$.

Thus, we have

$$\sqrt{n}(\hat{\beta}_n^S - \beta) = \sqrt{n}(\tilde{\beta}_n - \beta) + o_p(1).$$

Clearly, under A_δ, as $n \to \infty$, the asymptotic bias and mse expressions for $\hat{\beta}_n^{PT}$ and $\hat{\beta}_n^S$ are the same as for $\tilde{\beta}_n$, namely,

(i) $\lim_{n\to\infty} E\Big[\hat{\beta}_n^{PT} - \beta\Big] = 0$ and $\lim_{n\to\infty} E\Big[(\hat{\beta}_n^{PT} - \beta)^2\Big] = \frac{\sigma^2}{Q^0}$,

and

(ii) $\lim_{n\to\infty} E\Big[\hat{\beta}_n^S - \beta\Big] = 0$ and $\lim_{n\to\infty} E\Big[(\hat{\beta}_n^S - \beta)^2\Big] = \frac{\sigma^2}{Q^0}$.

Similarly, under A_δ,

(i) $\lim_{n\to\infty} E\Big[\hat{\theta}_n^{PT} - \theta\Big] = 0$ and $\lim_{n\to\infty} E\Big[(\hat{\theta}_n^{PT} - \theta)^2\Big] = \sigma^2\Big(1 + \frac{\bar{x}_0^2}{Q^0}\Big)$,

and

(ii) $\lim_{n\to\infty} E\Big[\hat{\theta}_n^S - \theta\Big] = 0$ and $\lim_{n\to\infty} E\Big[(\hat{\theta}_n^S - \theta)^2\Big] = \sigma^2\Big(1 + \frac{\bar{x}_0^2}{Q^0}\Big)$.

To overcome the difficulty of identical asymptotic distributions of different estimators in large samples under fixed alternatives, we consider the local alternatives

$$K_{(n)} : \beta_{(n)} = \beta_0 + n^{-1/2}\delta, \tag{3.6.13}$$

where δ is a fixed number. Then we see at once that the asymptotic distribution of $\mathcal{L}_n$ converges to a noncentral chi-square distribution, $H_1(\cdot;\Delta_0^2)$ with one d.f. and noncentrality parameter $\Delta_0^2/2$, where

$$\Delta_0^2 = \frac{\delta^2}{\sigma^2} Q^0. \tag{3.6.14}$$

Hence, the power function of an α-level test is given by

$$1 - H_1(\chi_1^2(\alpha); \Delta_0^2). \tag{3.6.15}$$

We will soon see that the asymptotic distributional bias and mse expressions of the various estimators of θ are different under $K_{(n)}$.

3.6.3 Asymptotic Distributional Bias and Mean Square Error Expressions, and Efficiency Analysis

The asymptotic distributional bias, quadratic bias, and mse expressions of the unrestricted, restricted, and preliminary test and shrinkage estimators of the slope and intercept parameters together with their asymptotic efficiency expressions are given in this section.

The following theorem gives the asymptotic distributional bias expressions of estimators of θ and β:

Theorem 3. Under $K_{(n)}$ and the regularity conditions of Theorem 3.6.1 the asymptotic distributional bias and quadratic bias expressions of the estimators of β and θ are given by

(i) $b_1(\tilde{\beta}_n) = 0$ and $B_1(\tilde{\beta}_n) = 0$;

(ii) $b_2(\hat{\beta}_n^{\rm PT}) = \delta H_3(\chi_1^2(\alpha);\Delta_0^2)$ and $B_2(\hat{\beta}_n^{\rm PT}) = \Delta_0^2\Big\{H_3(\chi_1^2(\alpha);\Delta_0^2)\Big\}^2$;

(iii) $b_3(\hat{\beta}_n^{\rm S}) = \dfrac{c\sigma}{\sqrt{Q^0}}\Big[2\Phi(\Delta_0)-1\Big]$ and $B_3(\hat{\beta}_n^{\rm S}) = c^2\Big[2\Phi(\Delta_0)-1\Big]^2$. (3.6.16)

For the bias expressions of the estimators of θ we respectively have

(i) $b_1(\tilde{\theta}_n) = 0$ and $B_1(\tilde{\theta}_n) = 0$;

(ii) $b_2(\hat{\theta}_n) = \delta\bar{x}_0$ and $B_2(\hat{\theta}_n) = \frac{\bar{x}_0^2}{Q^0}\Delta_0^2$;

(iii) $b_3(\hat{\theta}_n^{\rm PT}) = -\delta\bar{x}_0 H_3(\chi_1^2(\alpha);\Delta_0^2)$ and $B_3(\hat{\theta}_n^{\rm PT}) = \Delta_0^2\frac{\bar{x}_0^2}{Q^0}\Big\{H_3(\chi_1^2(\alpha);\Delta_0^2)\Big\}^2$;

(iv) $b_4(\hat{\theta}_n^{\rm S}) = \frac{c\sigma\delta\bar{x}_0}{\sqrt{Q^0}}\Big[2\Phi(\Delta_0)-1\Big]$ and $B_4(\hat{\theta}_n^{\rm S}) = c^2\frac{\bar{x}_0^2}{Q^0}\Delta_0^2\Big[2\Phi(\Delta_0)-1\Big]^2$. (3.6.17)

Proof. For $b_1(\tilde{\beta}_n) = B_1(\tilde{\beta}_n) = 0$ the proof is obvious. Consider $b_2(\hat{\beta}_n^{PT})$ and $b_3(\hat{\beta}_n^{S})$.

$$\begin{aligned} b_2(\hat{\beta}_n^{PT}) &= \lim_{n\to\infty} E\Big\{\sqrt{n}(\hat{\beta}_n^{PT} - \beta_{(n)})\Big\} \\ &= \lim_{n\to\infty} E\Big\{\sqrt{n}(\tilde{\beta}_n - \beta_0)I\Big(\mathcal{L}_n < \mathcal{L}_{n,\alpha}\Big)\Big\} \\ &= \delta H_3\Big(\chi_1^2(\alpha);\Delta_0^2\Big) \end{aligned}$$

and

$$B_2(\hat{\beta}_n^{PT}) = \frac{Q^0}{\sigma^2}b_2^2(\hat{\beta}_n^{PT}) = \Delta_0^2\Big\{H_3\Big(\chi_1^2(\alpha);\Delta_0^2\Big)\Big\}^2. \qquad (3.6.18)$$

Similarly,

$$b_3(\hat{\beta}_n^S) = \lim_{n\to\infty} E\Big\{\frac{c\sqrt{n}s_e}{\sqrt{Q}}\frac{\tilde{\beta}_n-\beta_0}{|\tilde{\beta}_n-\beta_0|}\Big\} = \frac{c\sigma}{\sqrt{Q^0}}E\Big[\frac{Z}{|Z|}\Big], \quad Z\sim\mathcal{N}(\Delta_0,1)$$
$$= \frac{c\sigma}{\sqrt{Q^0}}\Big[2\Phi(\Delta_0)-1\Big].$$

Hence,

$$B_2(\hat{\beta}_n^S) = c^2\Big[2\Phi(\Delta_0)-1\Big]^2. \tag{3.6.19}$$

Next, we consider the estimators of $\tilde{\theta}_n$, $\hat{\theta}_n$, $\hat{\theta}_n^{PT}$, and $\hat{\theta}_n^S$. Clearly, $b_1(\tilde{\theta}_n) = B_1(\tilde{\theta}_n) = 0$. Now, consider $\hat{\theta}_n$. We then have

$$b_2(\hat{\theta}_n) = \lim_{n\to\infty} E\{\sqrt{n}(\hat{\theta}_n-\theta)\} = \lim_{n\to\infty} E\{\sqrt{n}(\tilde{\beta}_n-\beta_0)\bar{x}\}$$
$$= \lim_{n\to\infty} E\left\{\sqrt{n}\frac{(\tilde{\beta}_n-\beta_0)}{\sigma}\frac{\bar{x}}{\sqrt{Q}}\sqrt{Q}\sigma\right\} = \sigma\frac{\bar{x}_0}{\sqrt{Q^0}}E\{Z\}, \quad Z\sim N(\Delta_0,1)$$
$$= \sigma\frac{\bar{x}_0}{\sqrt{Q^0}}\Delta_0$$

and normalizing, we have

$$B_2(\hat{\theta}_n) = \frac{\bar{x}_0^2}{Q^0}\Delta_0^2. \tag{3.6.20}$$

Now,

$$b_3(\hat{\theta}_n^{\rm PT}) = \lim_{n\to\infty} E\{\sqrt{n}(\hat{\theta}_n^{\rm PT}-\theta))\}$$
$$= -\lim_{n\to\infty} E\{\sqrt{n}(\tilde{\beta}_n-\beta_0)\bar{x}I(\mathcal{L}_n<\mathcal{L}_{n,\alpha})\}$$
$$= -\sigma\frac{\bar{x}_0}{\sqrt{Q_0}}E\left\{ZI(Z^2<\chi_1^2(\alpha))\right\}$$
$$= -\sigma\frac{\bar{x}_0}{Q_0}\Delta_0 H_3\left(\chi_1^2(\alpha);\Delta_0^2)\right\}.$$

Then, dividing by $\sigma\bar{x}_0/Q^0$ and squaring, we get

$$B_3(\hat{\theta}_n^{\rm PT}) = \frac{\bar{x}_0^2}{Q^0}\Delta_0^2\left\{H_3(\chi_1^2(\alpha);\Delta_0^2\right\}^2. \tag{3.6.21}$$

Finally,

$$b_4(\hat{\theta}_n^{\rm S}) = \lim_{n\to\infty} E\{\frac{\sqrt{n}c\bar{x}_0 s_e}{\sqrt{Q}}\frac{\tilde{\beta}_n-\beta_0}{|\tilde{\beta}_n-\beta_0|} = \frac{c\sigma\bar{x}_0}{\sqrt{Q_0}}[2\Phi(\Delta_0)-1].$$

Hence,

$$B_4(\hat{\theta}_n^{\rm S}) = c^2\frac{\bar{x}_0^2}{Q^0}[2\Phi(\Delta_0)-1]^2. \tag{3.6.22}$$

The next theorem gives the asymptotic distributional mse expressions.

Theorem 4. Under the regularity conditions of Theorem 1, the asymptotic mse expressions of the estimators of β are given by

$$M_1(\tilde{\beta}_n) = \frac{\sigma^2}{Q^0},$$

$$M_2(\hat{\beta}_n^{\mathrm{PT}}) = \frac{\sigma^2}{Q^0}\Big[1 - H_3(\chi_1^2(\alpha); \Delta_0^2) + \Delta_0^2\Big\{2H_3(\chi_1^2(\alpha); \Delta_0^2) - H_5(\chi_1^2(\alpha); \Delta_0^2)\Big]\Big\}$$

$$\text{and} \quad M_3(\hat{\beta}_n^{\mathrm{S}}) = \frac{\sigma^2}{Q^0}\Big\{1 - \frac{2}{\pi}\Big(2e^{-\Delta^2/2} - 1\Big)\Big\}. \tag{3.6.23}$$

That of the estimators of θ are given by

$$M_1(\tilde{\theta}_n) = \sigma^2\left(1 + \frac{\bar{x}_0^2}{Q^0}\right)$$

$$M_2(\hat{\theta}_n) = \sigma^2\left(1 + \frac{\bar{x}_0^2}{Q^0}\Delta_0^2\right)$$

$$M_3(\hat{\theta}_n^{\mathrm{PT}}) = \sigma^2\left(1 + \frac{\bar{x}_0^2}{Q^0}\right) - \sigma^2\frac{\bar{x}_0^2}{Q^0}H_3\left(\chi_1^2(\alpha); \Delta_0^2\right) + \sigma^2\frac{\bar{x}_0^2}{Q^0}\Delta_0^2\{2H_3(\chi_1^2(\alpha); \Delta_0^2) - H_5(\chi_1^2(\alpha); \Delta_0^2)\}.$$

$$\text{and} \quad M_4(\hat{\theta}_n^{\mathrm{S}}) = \sigma^2\Big[1 + \frac{\bar{x}_0^2}{Q^0}\Big\{1 - \frac{2}{\pi}\Big(2e^{-\Delta^2/2} - 1\Big)\Big\}\Big]. \tag{3.6.24}$$

Proof. Consider only the estimators of θ. In this case,

$$M_1(\tilde{\theta}_n) = \lim_{n\to\infty} E\Big[n(\tilde{\theta}_n - \theta)^2\Big] = \sigma^2\left(1 + \frac{\bar{x}_0^2}{Q^0}\right) \quad \text{by (3.5.7).} \tag{3.6.25}$$

Next, consider

$$\begin{aligned} M_2(\hat{\theta}_n) &= \lim_{n\to\infty} E\Big[n(\hat{\theta}_n - \theta)^2\Big] \\ &= \lim_{n\to\infty} En\Big[n(\tilde{\theta}_n - \theta) + (\tilde{\beta}_n - \beta_0)\bar{x}\Big]^2 \\ &= \lim E\Big[n(\tilde{\theta}_n - \theta)^2\Big] + \lim_{n\to\infty} E\Big[n(\tilde{\beta}_n - \beta_0)^2\bar{x}^2\Big] \\ &\quad + 2\lim_{n\to\infty}\Big\{\bar{x}E\Big[n(\tilde{\theta}_n - \theta)(\tilde{\beta}_n - \beta_0)\Big]\Big\} \end{aligned}$$

$$= \sigma^2\Big[\Big(1+\frac{\bar{x}_0^2}{Q^0}\Big)+\frac{\bar{x}_0^2\sigma^2}{Q^0}(1+\Delta_0^2)-2\frac{\bar{x}_0^2}{Q^0}\sigma^2\Big] \quad \text{using (3.2.9b)}$$

$$= \sigma^2\Big(1+\frac{\bar{x}_0^2}{Q^0}\Delta_0^2\Big). \tag{3.6.26}$$

Also,

$$M_3(\hat{\theta}_n^{\rm PT}) = \lim_{n\to\infty} E\Big[n(\hat{\theta}_n^{\rm PT}-\theta)^2\Big]$$

$$= \lim_{n\to\infty} En(\tilde{\theta}_n-\theta)^2 + \lim_{n\to\infty}\Big[\bar{x}^2E\{n(\tilde{\beta}_n-\beta_0)^2 I(\mathcal{L}_n < \mathcal{L}_{n,\alpha})\Big]$$

$$- 2\lim_{n\to\infty}\Big[\bar{x}E\Big\{n(\tilde{\theta}_n-\theta)(\tilde{\beta}_n-\beta_0)I(\mathcal{L}_n < \mathcal{L}_{n,\alpha}\Big\}\Big]$$

$$= \sigma^2\Big(1+\frac{\bar{x}_0^2}{Q^0}\Big) - \frac{\bar{x}_0^2\sigma^2}{Q^0}\{H_3(\chi_\alpha^2;\Delta_0^2) + \Delta_0^2H_5(\chi_1^2(\alpha);\Delta_0^2)\}$$

$$+ 2\frac{\bar{x}_0^2}{Q^0}\Delta_0^2\sigma^2H_3(\chi_1^2(\alpha);\Delta_0^2)$$

$$= \sigma^2\Big\{\Big(1+\frac{\bar{x}_0^2}{Q^0}\Big) - \frac{\bar{x}_0^2}{Q^0}H_3(\chi_1^2(\alpha);\Delta_0^2)$$

$$+ \frac{\bar{x}_0^2}{Q^0}\Delta_0^2\Big[2H_3(\chi_1^2(\alpha);\Delta_0^2) - H_5(\chi_1^2(\alpha);\Delta_0^2)\Big]\Big\}. \tag{3.6.27}$$

We used the limiting version of Theorems 2.2.4 and 2.2.5 and techniques in the proof of Theorem 3.3.2. Finally,

$$M_4(\hat{\theta}_n^{\rm S}) = \lim_{n\to\infty} E\Big[n(\hat{\theta}_n^{\rm S}-\theta)^2\Big]$$

$$= \lim_{n\to\infty} E\Big[n(\tilde{\theta}_n-\theta)^2\Big] + \frac{c^2}{Q^0}\lim_{n\to\infty} E[s_e^2]\bar{x}^2$$

$$+ 2c\lim_{n\to\infty} E\Big[\frac{(\tilde{\theta}_n-\theta)(\tilde{\beta}_n-\beta_0)\bar{x}_0}{\sqrt{Q}|\tilde{\beta}_n-\beta_0|}\Big]$$

$$= \sigma^2\Big(1+\frac{\bar{x}_0^2}{Q_0}\Big) + c^2\sigma^2\frac{\bar{x}_0^2}{Q_0} - 2c\sigma^2\sqrt{\frac{2}{\pi}\frac{\bar{x}_0^2}{Q_0}}\Big\{E[|Z|] - \Delta_0E\Big[\frac{Z}{|Z|}\Big]\Big\}$$

$$= \sigma^2\Big[1+\frac{\bar{x}_0^2}{Q^0}\Big\{1-\frac{2}{\pi}\Big[2e^{-\Delta^2/2}-1\Big]\Big\}\Big]. \tag{3.6.28}$$

The asymptotic efficiency expressions for $\hat{\theta}_n$, $\hat{\theta}_n^{\rm PT}$, and $\hat{\theta}_n^{\rm S}$ relative to $\tilde{\theta}_n$ are

$$\text{AMRE}(\hat{\theta}_n;\tilde{\theta}_n) = \Big(1+\frac{\bar{x}_0^2}{Q^0}\Big)\Big(1+\frac{\bar{x}_0^2}{Q^0}\Delta_0^2\Big)^{-1},$$

$$\text{AMRE}(\hat{\theta}_n^{\rm PT};\tilde{\theta}_n) = [1+g(\Delta_0^2)]^{-1},$$

and

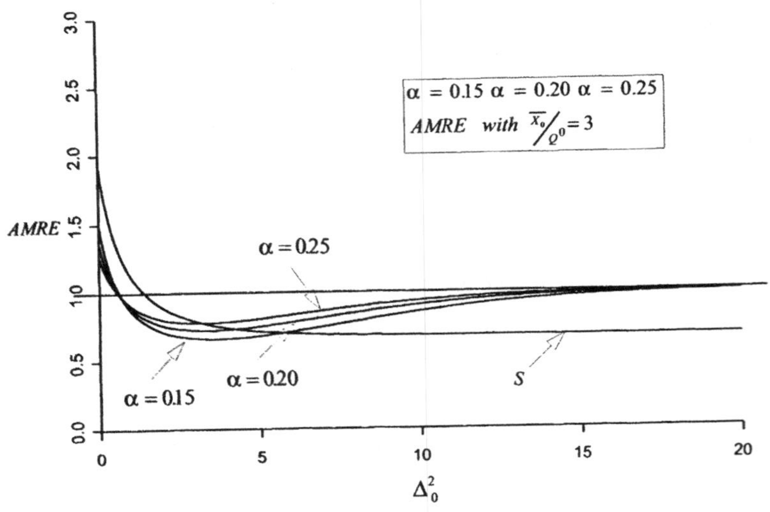

Figure 3.6.1 Graph of AMRE of $\hat{\theta}_n^{\text{PT}}$ and $\hat{\theta}_n^{\text{S}}$ relative to $\tilde{\theta}_n$

$$\text{AMRE}(\hat{\theta}_n^{\text{S}}; \tilde{\theta}_n) = \left\{1 - \frac{2}{\pi}\frac{\bar{x}_0^2}{Q^0}\left(1 + \frac{\bar{x}_0^2}{Q^0}\right)^{-1}\left[2e^{-\Delta_0^2/2} - 1\right]\right\}^{-1}, \quad (3.6.29)$$

where

$$g(\Delta_0^2) = -\frac{\bar{x}_0^2}{Q^0}\left(1 + \frac{\bar{x}_0^2}{Q^0}\right)^{-1} H_3(\chi_1^2(\alpha); \Delta_0^2)$$

$$+ \Delta_0^2\frac{\bar{x}_0^2}{Q^0}\left(1 + \frac{\bar{x}_0^2}{Q^0}\right)^{-1}\{2H_3(\chi_1^2(\alpha); \Delta_0^2) - H_5(\chi_1^2(\alpha); \Delta_0^2)\}. \quad (3.6.30)$$

Thus, $\hat{\theta}_n$ performs better than $\tilde{\theta}_n$ whenever

$$0 \le \Delta_0^2 \le 1; \quad (3.6.31)$$

otherwise, $\tilde{\theta}_n$ is better. Similarly, $\hat{\theta}_n^{\text{PT}}$ performs better than $\tilde{\theta}_n$ whenever

$$0 \le \Delta_0^2 \le \frac{H_3(\chi_1^2(\alpha); \Delta_0^2)}{2H_3(\chi_1^2(\alpha); \Delta_0^2) - H_5(\chi_1^2(\alpha); \Delta_0^2)} \le; \quad (3.6.32)$$

otherwise, $\tilde{\theta}_n$ is better. Figure 3.6.1 shows the graph of the AMRE of $\hat{\theta}_n^{PT}$ and $\hat{\theta}_n^{S}$ relative to $\tilde{\theta}_n$.

We present below Tables 3.6.1 and 3.6.2 for maximum and minimum MRE's for $\alpha = 0.05(0.05)0.50$ and various selected values of $\bar{x}_0^2/Q^0 = 0.1(0.1)0.9(0.5)5$.

Table 3.6.1 Maximum and Minimum Guaranteed Asymptotic Efficiencies of PTE

α	$\bar{x}^2/Q$	0.1	0.2	0.3	0.4	0.5	0.6	0.7	0.8	0.9
0.05	E_{max}	1.85798	2.33498	2.63864	2.84893	3.00317	3.12114	3.21428	3.28968	3.35198
	E_{min}	0.47754	0.42471	0.40464	0.39408	0.38755	0.38313	0.37992	0.37750	0.37560
	Δ^2_{min}	5.53022	5.53022	5.53023	5.53022	5.53022	5.53022	5.53022	5.53022	5.53022
0.1	E_{max}	1.57358	1.82247	1.96155	2.05035	2.11197	2.15723	2.19189	2.21927	2.24145
	E_{min}	0.59251	0.54011	0.51952	0.50851	0.50166	0.49699	0.49360	0.49102	0.48900
	Δ^2_{min}	4.47669	4.47669	4.47669	4.47669	4.47669	4.47669	4.47669	4.47669	4.47669
0.15	E_{max}	1.40934	1.56154	1.64097	1.68976	1.72277	1.74659	1.76458	1.77866	1.78997
	E_{min}	0.67415	0.62561	0.60605	0.59549	0.58887	0.58434	0.58104	0.57853	0.57656
	Δ^2_{min}	3.94090	3.94090	3.94090	3.94090	3.94090	3.94090	3.94090	3.94090	3.94090
0.2	E_{max}	1.30133	1.40191	1.45223	1.48243	1.50256	1.51695	1.52774	1.53613	1.54284
	E_{min}	0.73826	0.69495	0.67714	0.66744	0.66133	0.65713	0.65406	0.65173	0.64989
	Δ^2_{min}	3.60055	3.60055	3.60055	3.60055	3.60055	3.60055	3.60055	3.60055	3.60055
0.25	E_{max}	1.22209	1.29033	1.32345	1.34302	1.35594	1.36510	1.37195	1.37725	1.38148
	E_{min}	0.79359	0.75642	0.74086	0.73231	0.72691	0.72318	0.72046	0.71838	0.71674
	Δ^2_{min}	3.35268	3.35268	3.35268	3.35268	3.35268	3.35268	3.35268	3.35268	3.35268
0.3	E_{max}	1.16931	1.21843	1.24181	1.25548	1.26446	1.27080	1.27552	1.27917	1.28207
	E_{min}	0.83380	0.80206	0.78860	0.78116	0.77644	0.77318	0.77079	0.76897	0.76753
	Δ^2_{min}	3.18236	3.18236	3.18236	3.18236	3.18236	3.18236	3.18236	3.18236	3.18236
0.35	E_{max}	1.12716	1.16235	1.17885	1.18842	1.19468	1.19908	1.20235	1.20487	1.20688
	E_{min}	0.86973	0.84357	0.83234	0.82610	0.82213	0.81938	0.81737	0.81582	0.81461
	Δ^2_{min}	3.04378	3.04378	3.04378	3.04378	3.04378	3.04378	3.04378	3.04378	3.04378
0.4	E_{max}	1.09488	1.12019	1.13192	1.13868	1.14309	1.14618	1.14847	1.15024	1.15165
	E_{min}	0.89949	0.87846	0.86936	0.86427	0.86102	0.85877	0.85712	0.85585	0.85485
	Δ^2_{min}	2.93352	2.93352	2.93352	2.93352	2.93352	2.93352	2.93352	2.93352	2.93352
0.45	E_{max}	1.06997	1.08809	1.09642	1.10120	1.10431	1.10648	1.10810	1.10934	1.11032
	E_{min}	0.92393	0.90750	0.90032	0.89629	0.89372	0.89193	0.89061	0.88961	0.88881
	Δ^2_{min}	2.84422	2.84422	2.84422	2.84422	2.84422	2.84422	2.84422	2.84422	2.84422
0.5	E_{max}	1.05070	1.06354	1.06940	1.07276	1.07493	1.07645	1.07758	1.07844	1.07913
	E_{min}	0.94378	0.93131	0.92583	0.92275	0.92077	0.91940	0.91838	0.91761	0.91700
	Δ^2_{min}	2.77103	2.77103	2.77103	2.77103	2.77103	2.77103	2.77103	2.77103	2.77103

Table 3.6.2 Maximum and Minimum Guaranteed Asymptotic Efficiencies of PTE

α	$\bar{x}^2/Q$	1	1.5	2	2.5	3	3.5	4	4.5	5
0.05	E_{max}	1.56360	1.76224	1.92530	2.06155	2.17711	2.27635	2.36250	2.43800	2.50469
	E_{min}	0.57732	0.53232	0.50603	0.48878	0.47660	0.46753	0.46053	0.45495	0.45040
	Δ^2_{min}	4.68205	4.68205	4.68205	4.68205	4.68205	4.68205	4.68205	4.68205	4.68205
0.1	E_{max}	1.38958	1.50700	1.59696	1.66808	1.72573	1.77340	1.81347	1.84763	1.87709
	E_{min}	0.67745	0.63640	0.61169	0.59518	0.58337	0.57451	0.56761	0.56208	0.55756
	Δ^2_{min}	3.96923	3.96923	3.96923	3.96923	3.96923	3.96923	3.96923	3.96923	3.96923
0.15	E_{max}	1.28407	1.36142	1.41838	1.46207	1.49665	1.52469	1.54790	1.56742	1.58406
	E_{min}	0.74607	0.71001	0.68785	0.67285	0.66202	0.65384	0.64743	0.64229	0.63806
	Δ^2_{min}	3.58902	3.58902	3.58902	3.58902	3.58902	3.58902	3.58902	3.58902	3.58902
0.2	E_{max}	1.21225	1.26600	1.30455	1.33356	1.35618	1.37431	1.38917	1.40156	1.41206
	E_{min}	0.79855	0.76763	0.74830	0.73509	0.72548	0.71818	0.71244	0.70782	0.70401
	Δ^2_{min}	3.34095	3.34095	3.34095	3.34095	3.34095	3.34095	3.34095	3.34095	3.34095
0.25	E_{max}	1.16036	1.19880	1.22588	1.24599	1.26150	1.27384	1.28389	1.29222	1.29925
	E_{min}	0.84043	0.81444	0.79799	0.78664	0.77833	0.77199	0.76700	0.76296	0.75962
	Δ^2_{min}	3.16322	3.16322	3.16322	3.16322	3.16322	3.16322	3.16322	3.16322	3.16322
0.3	E_{max}	1.12151	1.14945	1.16886	1.18312	1.19406	1.20270	1.20971	1.21550	1.22037
	E_{min}	0.87447	0.85305	0.83935	0.82982	0.82282	0.81746	0.81322	0.80978	0.80694
	Δ^2_{min}	3.02878	3.02878	3.02878	3.02878	3.02878	3.02878	3.02878	3.02878	3.02878
0.35	E_{max}	1.09181	1.11223	1.12627	1.13652	1.14433	1.15048	1.15545	1.15955	1.16298
	E_{min}	0.90231	0.88502	0.87385	0.86605	0.86029	0.85586	0.85235	0.84950	0.84714
	Δ^2_{min}	2.92344	2.92344	2.92344	2.92344	2.92344	2.92344	2.92344	2.92344	2.92344
0.4	E_{max}	1.06881	1.08372	1.09390	1.10128	1.10689	1.11129	1.11483	1.11775	1.12019
	E_{min}	0.92507	0.91141	0.90252	0.89628	0.89166	0.88810	0.88527	0.88296	0.88105
	Δ^2_{min}	2.83892	2.83892	2.83892	2.83892	2.83892	2.83892	2.83892	2.83892	2.83892
0.45	E_{max}	1.05091	1.06172	1.06905	1.07435	1.07836	1.08149	1.08402	1.08609	1.08783
	E_{min}	0.94356	0.93303	0.92614	0.92128	0.91767	0.91488	0.91266	0.91085	0.90935
	Δ^2_{min}	2.76998	2.76998	2.76998	2.76998	2.76998	2.76998	2.76998	2.76998	2.76998
0.5	E_{max}	1.03698	1.04471	1.04992	1.05368	1.05652	1.05873	1.06051	1.06197	1.06319
	E_{min}	0.95845	0.95055	0.94536	0.94168	0.93894	0.93682	0.93514	0.93376	0.93262
	Δ^2_{min}	2.71315	2.71315	2.71315	2.71315	2.71315	2.71315	2.71315	2.71315	2.71315

To determine the optimum level of significance for the preliminary test estimator, we follow the same procedure as in Section 3.2.4, using Table 3.5.1. The AMRE$(\hat{\theta}_n^S;\tilde{\theta}_n)$ lies between $\left[1+\frac{2}{\pi}\frac{\bar{x}_0^2}{Q^0}\left(1+\frac{\bar{x}_0^2}{Q^0}\right)^{-1}\right]^{-1}$ and $\left[1-\frac{2}{\pi}\frac{\bar{x}_0^2}{Q^0}\left(1+\frac{\bar{x}_0^2}{Q^0}\right)^{-1}\right]^{-1}$. $\hat{\theta}_n^S$ performs better than $\tilde{\theta}_n$ in the interval $[0, ln\ 4]$. The graph of AMRE$(\hat{\theta}_n^S;\tilde{\theta}_n)$ shows the performance of $\hat{\theta}_n^S$ compared to $\tilde{\theta}_n$ and $\hat{\theta}_n^{PT}$ for all Δ_0^2 and selected α-values.

3.7 Two-Sample Problem and Estimation of the Mean

Consider the simple linear model as in Section 3.4 and Equation (3.4.1), where $E(\mathbf{e}) = \mathbf{0}$ and $E(\mathbf{ee}') = \sigma^2\mathbf{I}_n$, but the distribution of $\mathbf{e} \sim F(\mathbf{e})$ nonnormal. Further, let $\mathbf{x} = (0,\ldots,0;1,\ldots,1)'$ with n_1 zero's and n_2 1's be such that

$$\mu_1 = \theta \quad \text{and} \quad \mu_2 = \theta+\beta, \quad \beta = \mu_2-\mu_1$$

as in (3.4.2). Thus, referring back to the conditions of Theorem 1d (3.6.7), we have

$$\lim_{n\to\infty} \bar{x} = \lim \frac{n_2}{n_1+n_2} = \lambda, \tag{3.7.1a}$$

$$\lim_{n\to\infty} n^{-1}Q = \lambda(1-\lambda), \tag{3.7.1b}$$

$$\lim_{n\to\infty} \left[\frac{n\bar{x}^2}{Q}\right] = \lambda(1-\lambda)^{-1}. \tag{3.7.1c}$$

The unrestricted, restricted, and preliminary test estimators of μ_1 are

$$\tilde{\mu}_1 = \bar{y}_1, \quad \hat{\mu}_1 = \frac{n_1\bar{y}_1+n_2\bar{y}_2}{n_1+n_2}, \tag{3.7.2}$$

$$\hat{\mu}_1^{\text{PT}} = \bar{y}_1 - (\bar{y}_2-\bar{y}_1)I(\mathcal{L}_n < \mathcal{L}_{n\alpha}),$$

and

$$\hat{\mu}_1^{\text{S}} = \bar{y}_1 + \frac{\lambda(\bar{y}_2-\bar{y}_1)s_p}{|\bar{y}_2-\bar{y}_1|}. \tag{3.7.3}$$

The test-statistic for testing the null hypothesis $H_0 : \mu_2 = \mu_1$ against $H_A = \mu_2 \neq \mu_1$ is given by (3.4.4) as

$$\mathcal{L}_n = \frac{n_1n_2}{n_1+n_2}\left(\frac{\bar{y}_2-\bar{y}_1}{s_p}\right)^2. \tag{3.7.4}$$

Now,

$$\mathcal{L}_n \xrightarrow{\mathcal{D}} \chi_1^2(\Delta_0^2) \quad \text{as } n\to\infty. \tag{3.7.5}$$

Under the local alternatives,

$$K_{(n)} : \mu_2 = \mu_1 + n^{-1/2}\delta, \quad \delta \text{ fixed} \tag{3.7.6}$$

where $\Delta_0^2 = \lambda(1-\lambda)\frac{\delta^2}{\sigma^2}$, and approximation of the critical value

$$\mathcal{L}_{n,\alpha} \Longrightarrow \chi_1^2(\alpha) \quad (\alpha\text{-level critical value}). \tag{3.7.7}$$

Hence the asymptotic bias expressions of the three estimators are given by

$$b_1(\tilde{\mu}_1) = 0 \quad \text{and} \quad B_1(\tilde{\mu}_1) = 0,$$

$$b_2(\hat{\mu}_1) = \lim_{n\to\infty} \{\sqrt{n}(\hat{\mu}_1 - \mu_1)\} = \lambda\delta \quad \text{and} \quad B_2(\hat{\mu}_1) = \Delta_0^2$$

$$b_3(\hat{\mu}_1^{\text{PT}}) = \lambda\delta H_3(\chi_1^2(\alpha); \Delta_0^2) \text{ and } B_3(\hat{\mu}_1^{\text{PT}}) = \Delta_0^2\{H_3(\chi_1^2(\alpha); \sigma^2)\}^2. \tag{3.7.8}$$

Similarly, the mean square expressions for the estimators are

$$M_1(\tilde{\mu}_1) = \lim_{n\to\infty} E\{n(\tilde{\mu}_1 - \mu_1)^2\} = \sigma^2(1-\lambda)^{-1},$$

$$M_2(\hat{\mu}_1) = \lim_{n\to\infty} E\left\{n(\hat{\mu}_1 - \mu_1)^2\right\} = \sigma^2(1-\lambda)^{-1}, \{(1-\lambda) + \lambda\Delta_0^2\}$$

$$\begin{aligned} M_3(\hat{\mu}_1^{\text{PT}}) &= \lim_{n\to\infty} E\left\{n\left(\hat{\mu}^{\text{PT}} - \mu_1\right)^2\right\} \\ &= \sigma^2(1-\lambda)^{-1}\left[1 - \lambda H_3(\chi_1^2(\alpha); \Delta_0^2)\right. \\ &\quad \left. + \ \lambda\Delta_0^2\{2H_3(\chi_1^2(\alpha); \Delta_0^2) - H_5(\chi_1^2(\alpha); \Delta_0^2)\}\right], \end{aligned}$$

$$\begin{aligned} M_4(\hat{\mu}_1^S) &= \lim_{n\to\infty} E\left\{n\left(\hat{\mu}^{\text{S}} - \mu_1\right)^2\right\} \\ &= \sigma^2(1-\lambda)^{-1}\left[1 - \frac{2\lambda}{\pi}\left(2e^{-\Delta_0^2/2} - 1\right)\right]. \end{aligned} \tag{3.7.9}$$

The asymptotic efficiencies of $\hat{\mu}_1$ and $\hat{\mu}_1^{\text{PT}}$ relative to $\tilde{\mu}_1$ are

$$\text{AMRE}(\hat{\mu}_1 : \tilde{\mu}_1) = \left[1 - \lambda + \lambda\Delta_0^2\right]^{-1},$$

$$\text{AMRE}(\hat{\mu}_1^{\text{PT}} : \tilde{\mu}_1) = \left[1 + g(\Delta_0^2)\right]^{-1},$$

$$\text{AMRE}(\hat{\mu}_1^{\text{S}} : \tilde{\mu}_1) = \left[1 - \frac{2\lambda}{\pi}\left(2e^{-\Delta_0^2/2} - 1\right)\right]^{-1}, \tag{3.7.10}$$

where

$$g(\Delta_0^2) = -\lambda H_3(\chi_1^2(\alpha); \Delta_0^2) + \lambda\Delta_0^2\{2H_3(\chi_1^2(\alpha); \Delta_0^2) - H_5(\chi_1^2(\alpha); \Delta_0^2)\}. \tag{3.7.11}$$

Thus, $\hat{\mu}_1$ performs better than $\tilde{\mu}_1$ whenever $0 \le \Delta_0^2 \le 1$; otherwise, $\tilde{\mu}_1$ is better. Similarly, $\hat{\mu}_1^{\text{PT}}$ performs better than $\tilde{\mu}_1$ whenever $0 \le \Delta_0^2 \le \frac{H_3(\chi_1^2(\alpha);\Delta_0^2)}{2H_3(\chi_1^2(\alpha);\Delta_0^2) - H_5(\chi_1^2(\alpha);\Delta_0^2)} \le 1$; otherwise, $\tilde{\mu}_1$ is better.

Further, $\hat{\mu}_1^S$ is better than $\tilde{\mu}_1$ in the interval $[0, \ln 4]$. Like the estimation of the intercept when the slope is zero, $\hat{\mu}_1^S$ is comparable to the PTE of μ_1. $\hat{\mu}_1^S$ does not depend on the level of significance. Also, it interpolates between $\hat{\mu}_1$ and $\tilde{\mu}_1$. The minimum of guaranteed efficiency of $\hat{\mu}_1^S$ is $\left\{1+\frac{2\lambda}{\pi}\right\}^{-1}$ with a possible maximum relative efficiency $\left\{1-\frac{2\lambda}{\pi}\right\}^{-1}$.

3.8 One-Sample Problem and Estimation of the Mean

Consider, again, the model (3.7.1),

$$\mathbf{Y} = \theta \mathbf{1}_n + \boldsymbol{\varepsilon}, \tag{3.8.1}$$

where the components of $\boldsymbol{\varepsilon} = (\varepsilon_1, \dots, \varepsilon_n)'$ are independent and

$$E(\mathbf{e}) = \mathbf{0} \quad \text{and} \quad E(\mathbf{ee}') = \sigma^2 I_n.$$

Further $\mathbf{e} \sim F(\mathbf{e})$ is a nonnormal distribution. The LSE of $\boldsymbol{\theta}$ is $\tilde{\theta}_n = \bar{Y}$, the sample mean, and by the central limit theorem

$$\sqrt{n}(\tilde{\theta}_n - \theta) \overset{\mathcal{D}}{\simeq} N(0, \sigma^2 \mathbf{I}_n) \quad \text{as } n \to \infty. \tag{3.8.2}$$

Further, the estimate of σ^2 is

$$s_e^2 = \frac{1}{n-2}(Y - \overline{Y}\mathbf{1}_n)'(\mathbf{Y} - \overline{Y}1_n), \tag{3.8.3}$$

which converges in probability to σ^2. Hence, the test-statistic for the null hypothesis $H_0 : \theta = \theta_0$ against $H_A : \theta \neq \theta_0$ is

$$\mathcal{L}_n = \frac{n(\tilde{\theta}_n - \theta_0)^2}{s_e^2}, \tag{3.8.4}$$

which converges in distribution to $\chi_1^2(\Delta_0^2)$ as $n \to \infty$ under the local alternatives

$$K_{(n)} : \theta_{(n)} = \theta_0 + n^{-1/2}\delta, \quad \text{fixed } \delta, \tag{3.8.5}$$

where

$$\Delta_0^2 = \frac{\delta^2}{\sigma^2}. \tag{3.8.6}$$

Under H_0, $\mathcal{L}_{n,\alpha} \to \chi_1^2(\alpha)$.

The unrestricted, restricted, and the preliminary test and shrinkage estimators of θ are

$$\begin{aligned} \tilde{\theta}_n &= \overline{Y}, \quad \hat{\theta}_n = \theta_0, \\ \hat{\theta}_n^{\mathrm{PT}} &= \tilde{\theta}_n - (\tilde{\theta}_n - \theta_0) I(\mathcal{L}_n < \mathcal{L}_{n,\alpha}), \\ \hat{\theta}_n^S &= \tilde{\theta}_n - c s_e \frac{(\tilde{\theta}_n - \theta_0)}{|\tilde{\theta}_n - \theta_0|}, \end{aligned} \tag{3.8.7}$$

where $\mathcal{L}_{n,\alpha}$ is the α-level critical value of the $\mathcal{L}_n$-distribution. Since $\mathcal{L}_n$ converges in distribution to a central chi-square distribution with one d.f. as $n \to \infty$, we have $\mathcal{L}_{n,\alpha} \to \chi_1^2(\alpha)$.

The bias and mse expressions for $\tilde{\theta}_n$, $\hat{\theta}_n^{\mathrm{PT}}$ and $\hat{\theta}_n^{\mathrm{S}}$ are as follows: First, the bias expressions are given by

(i) $b_1(\tilde{\theta}_n) = \lim_{n\to\infty} E[\sqrt{n}(\tilde{\theta}_n - \theta_{(n)})] = 0$

(ii) $b_2(\hat{\theta}_n^{\mathrm{PT}}) = -\lim_{n\to\infty} E\left[\sqrt{n}(\tilde{\theta} - \theta_0) I\left(\frac{n(\tilde{\theta}_n - \theta_0)^2}{s_y^2} < \mathcal{L}_{n,\alpha}\right)\right].$

Under $K_{(n)}$, the r.h.s. of (ii) reduces to

$$
\begin{aligned}
&= -\sigma E\{Z I(Z^2 < \chi_1^2(\alpha))\}, \quad Z \sim N(\Delta_0, 1)\\
&= -\delta H_3(\chi_1^2(\alpha); \Delta_0^2).
\end{aligned}
$$

Similarly, we have

$$
\begin{aligned}
\text{(iii)}\quad b_3(\hat{\theta}_n^{\mathrm{S}}) &= \lim_{n\to\infty} E[\sqrt{n}(\hat{\theta}_n^S - \theta_{(n)})]\\
&= -c\sigma\left[2\Phi(\Delta_0^2) - 1\right].
\end{aligned}
$$

The mse expressions are given by

$$
\begin{aligned}
M_1(\tilde{\theta}_n) &= \lim_{n\to\infty} E[n(\tilde{\theta}_n - \theta)^2] = \sigma^2,\\
M_2(\hat{\theta}_n^{\mathrm{PT}}) &= \lim_{n\to\infty} E\{n(\hat{\theta}_n^{\mathrm{PT}} - \theta)^2\}\\
&= \sigma^2\left[1 - H_3(\chi_1^2(\alpha); \Delta_0^2) + \Delta_0^2\left\{2H_3(\chi_1^2(\alpha); \Delta_0^2) - H_5(\chi_1^2(\alpha); \Delta_0^2)\right\}\right]
\end{aligned}
$$

and

$$
\begin{aligned}
M_3(\hat{\theta}_n^{\mathrm{S}}) &= \lim_{n\to\infty} E\{n(\hat{\theta}_n^{\mathrm{S}} - \theta)^2\}\\
&= \sigma^2\left\{1 + c^2 - 2c\left(E[|Z|] - \Delta_0 E\left[\frac{Z}{|Z|}\right]\right)\right\}\\
&= \sigma^2\left\{1 - \frac{2}{\pi}\left[2e^{-\Delta_0^2} - 1\right]\right\}\\
&\text{for } c = \sqrt{\frac{2}{\pi}}, \quad Z \sim \mathcal{N}(\Delta_0, 1).
\end{aligned}
\tag{3.8.8}
$$

The asymptotic relative efficiency of $\hat{\theta}_n^{\mathrm{PT}}$ compared to $\tilde{\theta}_n$ is given by

$$
\mathrm{AMRE}(\hat{\theta}_n^{\mathrm{PT}}; \tilde{\theta}_n) = [1 + g(\Delta_0^2)]^{-1}, \tag{3.8.9}
$$

where

$$
g(\Delta_0^2) = -H_3(\chi_1^2(\alpha); \Delta_0^2) + \Delta_0^2\left\{2H_3(\chi_1^2(\alpha); \Delta_0^2) - H_5(\chi_1^2(\alpha); \Delta_0^2)\right\}, \tag{3.8.10}
$$

Hence, $\hat{\theta}_n^{\text{PT}}$ performs better than $\tilde{\theta}_n$ whenever

$$0 \leq \Delta_0^2 \leq \frac{H_3(\chi_1^2(\alpha); \Delta_0^2)}{2H_3(\chi_1^2(\alpha); \Delta_0^2) - H_5(\chi_1^2(\alpha); \Delta_0^2)} \leq 1; \tag{3.8.11}$$

otherwise, $\tilde{\theta}_n$ is better.

We can prepare a table for the maximum and minimum asymptotic relative efficiencies to determine the optimum level of significance for the PTE for application. Similarly,

$$\text{AMRE}(\hat{\theta}_n^S; \tilde{\theta}_n) = \left[1 - \frac{2}{\pi}\left(2e^{-\Delta_0^2/2} - 1\right)\right]^{-1}. \tag{3.8.12}$$

For $0 < \Delta_0^2 < ln\, 4$, $\hat{\theta}_n^S$ performs better than $\tilde{\theta}_n$; otherwise, $\tilde{\theta}_n$ better.

3.9 Stein Estimation of Variance: One-Sample Problem

Generally, we have seen that the preliminary test estimator performs better than the **unrestricted estimator** in a limited way. In this section we show that the **preliminary test estimator of variance** is uniformly better than the **unrestricted estimator.** An extended problem of variance is due to Stein (1964) with a review by Maatta and Casella (1990). We will be concerned with the one-sample model (3.4.1),

$$\mathbf{Y} = \theta \mathbf{1}_n + \mathbf{e}, \quad \mathbf{e} \sim N_n(\mathbf{0}, \sigma^2 \mathbf{I}_n), \tag{3.9.1}$$

and consider the problem of estimating σ^2 when it is suspected but not sure that θ may be equal to θ_0 (say). As before, we define various unrestricted, restricted, and preliminary test estimators of σ^2. First, we recall that the likelihood ratio test for the null hypothesis $H_0 : \theta = \theta_0$ against $H_A : \theta \neq \theta_0$ is given by

$$\mathcal{L}_n = \frac{n(\tilde{\theta}_n - \theta_0)^2}{s_y^2}, \tag{3.9.2}$$

where $\tilde{\theta}_n = \bar{Y}$ (sample mean) and

$$s_y^2 = \frac{1}{m}(\mathbf{Y} - \tilde{\theta}_n \mathbf{1}_n)'(\mathbf{Y} - \tilde{\theta}_n \mathbf{1}_n), \quad m = n - 1 \tag{3.9.3}$$

is the *unbiased unrestricted* estimator (UUE) of σ^2. The maximum likelihood estimator (MLE) of σ^2 is given by

$$\hat{\sigma}_n^{*2} = \frac{m s_y^2}{m+1}, \tag{3.9.4}$$

which is a *biased unrestricted* estimator (BUE) of σ^2.

The bias and mse of $\hat{\sigma}_n^{*2}$ are, respectively,

$$b_2(\hat{\sigma}_n^{*2}) = -\frac{\sigma^2}{m+1} \quad \text{and} \quad M_2(\hat{\sigma}_n^{*2}) = \frac{2(m+1)-1}{(m+1)^2}\sigma^4. \tag{3.9.5}$$

However, a better BUE of σ^2 may be obtained as

$$\hat{\sigma}_n^2 = \frac{ms_y^2}{m+2}. \tag{3.9.6}$$

The bias and mse of this estimator are

$$b_3(\hat{\sigma}_n^2) = -\frac{2\sigma^2}{m+1} \quad \text{and} \quad M_2(\hat{\sigma}_n^2) = \frac{2\sigma^4}{m+2}. \tag{3.9.7}$$

Thus, we may order the estimators as follows according to the mse criterion,

$$\hat{\sigma}_n^2 \succ \hat{\sigma}_n^{*2} \succ s_y^2 \quad (\succ \text{ stands for domination}). \tag{3.9.8}$$

Consider now that we have uncertain prior information on the mean of the normal distribution specified as $H_0 : \theta = \theta_0$. If this hypothesis is true, then we define the restricted unbiased estimator (RUE) of σ^2 as

$$\tilde{\sigma}_n^2 = \frac{1}{m+1}(\mathbf{Y} - \theta_0 \mathbf{1}_n)'(\mathbf{Y} - \theta_0 \mathbf{1}_n) = \frac{ms_y^2 + (m+1)(\tilde{\theta}_n - \theta_0)^2}{m+1}. \tag{3.9.9}$$

The bias and mse expressions of this estimator are given in the following theorem:

Theorem 1. Under $H_A : \theta \neq \theta_0$,

$$b_4(\tilde{\sigma}_n^2) = \frac{\sigma^2}{n}\Delta^2, \quad \Delta^2 = \frac{n}{\sigma^2}(\theta - \theta_0)^2 \tag{3.9.10}$$

and

$$M_4(\tilde{\sigma}_n^2) = \frac{\sigma^4}{n^2}\{2n + \Delta^2(\Delta^2 + 4)\}.$$

Proof.

$$\begin{aligned}\tilde{\sigma}_n^2 &= \frac{1}{n}(\mathbf{Y} - \theta_0 \mathbf{1}_n)'(\mathbf{Y} - \theta_0 \mathbf{1}_n) \\ &= \frac{1}{n}\{(n-1)S_n^2 + n(\tilde{\theta}_n - \theta_0)^2\}.\end{aligned} \tag{3.9.11}$$

Then we have

$$\tilde{\sigma}_n^2 \stackrel{\mathcal{D}}{=} \frac{\sigma^2}{n}\{\chi_{n-1}^2 + \chi_1^2(\Delta^2)\}, \tag{3.9.12}$$

where $\stackrel{\mathcal{D}}{=}$ stands for "equal in distribution to" and χ_{n-1}^2 is a central chi-square variable with $n-1$ d.f. and $\chi_1^2(\Delta^2)$ is a noncentral chi-square variable with degree of freedom and noncentrality parameter $\Delta^2/2$ and the two distributions are independent.

Clearly,

$$E(\tilde{\sigma}_n^2) = \frac{\sigma^2}{n}\{(n-1) + (1+\Delta^2)\} \tag{3.9.13}$$
$$= \frac{\sigma^2}{n}(n+\Delta^2).$$

Hence,

$$b_4(\tilde{\sigma}_n^2) = \frac{\sigma^2}{m+1}\Delta^2. \tag{3.9.14}$$

Similarly,

$$M_4(\tilde{\sigma}_n^2) = \text{ Var }(\tilde{\sigma}_n^2) + \left[b_4(\tilde{\sigma}_n^2)\right]^2, \tag{3.9.15}$$

where

$$\text{Var}(\tilde{\sigma}_n^2) = \frac{\sigma^4}{n^2}\{\text{Var}(\chi_{n-1}^2) + \text{Var}\big(\chi_1^2(\Delta^2)\big)\}. \tag{3.9.16}$$

Therefore,

$$\text{Var}(\tilde{\sigma}_n^2) = \frac{\sigma^4}{n^2}\{2(n-1) + 2(1+2\Delta^2)\}$$
$$= \frac{\sigma^4}{(m+1)^2}\left\{2(m+1) + 4\Delta^2\right\}. \tag{3.9.17}$$

This results in the mse expression to be

$$M_4(\tilde{\sigma}_n^2) = \frac{\sigma^4}{(m+1)^2}\left\{2(m+1) + \Delta^2(4+\Delta^2)\right\}. \tag{3.9.18}$$

Thus, under $H_0 : \theta = \theta_0$ we obtain

$$b_4(\tilde{\sigma}_n^2) = 0 \quad \text{and} \quad M_4(\tilde{\sigma}_n^2) = \frac{2\sigma^4}{m+1}. \tag{3.9.19}$$

It is clear from above that $\tilde{\sigma}_n^2$ is better than s_y^2 under the null-hypothesis. As the *departure constant* Δ^2 diverts from zero, the bias and mse values grow and become unbounded as $\Delta^2 \to \infty$. However, the two mse graphs intersect at

$$\Delta^2 = 2\left\{\sqrt{(1+\frac{1}{2}(m+1))} - 1\right\}. \tag{3.9.20}$$

This means that if $\Delta^2 \in \left[0, 2\{\sqrt{(1+\frac{1}{2}(m+1))} - 1\}\right]$, $\tilde{\sigma}_n^2$ performs better than s_y^2, whereas for $\Delta^2 \notin \left[0, 2\{\sqrt{(1+\frac{1}{2}(m+1))} - 1\}\right]$, s_y^2 is better. In this range the bias of $\tilde{\sigma}_n^2$ varies from 0 to $2\{\sqrt{(1+\frac{1}{2}(m+1))} - 1\}$.

In this context we may define another estimator $\tilde{\tilde{\sigma}}_n^2$, following Hodges and Lehmann (1951), namely

$$\tilde{\tilde{\sigma}}_n^2 = \frac{1}{m+3}(\mathbf{Y} - \theta_0\mathbf{1}_n)'(\mathbf{Y} - \theta_0\mathbf{1}_n) = \frac{ms_y^2 + (m+1)(\tilde{\theta}_n - \theta_0)^2}{m+3}. \quad (3.9.21)$$

Following the proof of Theorem 3.9.1, we find the bias to be

$$b_5(\tilde{\tilde{\sigma}}_n^2) = \frac{\sigma^2}{m+3}(\Delta^2 - 2), \quad (3.9.22)$$

and the mse of $\tilde{\sigma}_n^2$ is given by

$$M_5(\tilde{\tilde{\sigma}}_n^2) = \frac{\sigma^4}{(m+3)^2}\{2(m+3) + \Delta^4\}. \quad (3.9.23)$$

Now the mse difference is

$$M_4(\tilde{\sigma}_n^2) - M_5(\tilde{\tilde{\sigma}}_n^2) = \frac{4}{m+3} + \frac{4\Delta^2[(m+3)^2 + (m+2)\Delta^2]}{(m+1)^2(m+3)^2} > 0 \text{ for all } \Delta^2. \quad (3.9.24)$$

Similarly,

$$b_4(\tilde{\sigma}_n^2) - b_5(\tilde{\tilde{\sigma}}_n^2) = \frac{1}{m+3} + \frac{2\Delta^2}{(m+1)(m+3)} > 0 \text{ for all } \Delta^2. \quad (3.9.25)$$

Thus, with respect to the bias as well as the mse, $\tilde{\tilde{\sigma}}_n^2$ performs uniformly better than $\tilde{\sigma}_n^2$, meaning $M_5(\tilde{\tilde{\sigma}}_n^2) < M_4(\tilde{\sigma}_n^2)$ for all Δ^2.

Since the null hypothesis $H_0 : \theta = \theta_0$ is uncertain, we may consider several preliminary test estimators of σ^2. However, we consider only the estimator based on the *best of the unrestricted estimators* and *best of the restricted estimators*, namely based on $\hat{\sigma}_n^2$ and $\tilde{\tilde{\sigma}}_n^2$. Thus, we define the PTE of σ^2 as

$$\hat{\sigma}_n^{2PT} = \tilde{\tilde{\sigma}}_n^2 I(\mathcal{L}_n < F_{1,m}(\alpha)) + \hat{\sigma}_n^2 I(\mathcal{L}_n \geq F_{1,m}(\alpha)) \quad (3.9.26)$$

$$= \frac{ms_y^2}{m+1} I(\mathcal{L}_n \geq F_{1,m}(\alpha)) + \frac{(1 + \frac{1}{m}\mathcal{L}_n)ms_y^2}{m+3} I(\mathcal{L}_n < F_{1,m}(\alpha))$$

$$= \phi_\alpha(\mathcal{L}_n)ms_y^2, \quad (3.9.27)$$

where

$$\phi_\alpha(\mathcal{L}_n) = \frac{1}{m+2} I(\mathcal{L}_n \geq F_{1,m}(\alpha)) + \frac{(1 + \frac{1}{m}\mathcal{L}_n)}{m+3} I(\mathcal{L}_n < F_{1,m}(\alpha)). \quad (3.9.28)$$

Let us now set $F_{1,m}(\alpha) = m/(m+2)$ and define

$$\hat{\sigma}_n^{2[S]} = \phi_S(\mathcal{L}_n)ms_y^2, \quad (3.9.29)$$

where

$$\phi_S(\mathcal{L}_n) = \frac{1}{m+2} I\Big(\mathcal{L}_n \geq \frac{m}{m+2}\Big) + \frac{1+\frac{1}{m}\mathcal{L}_n}{m+3} I\Big(\mathcal{L}_n < \frac{m}{m+2}\Big). \quad (3.9.30)$$

Thus, $\hat{\sigma}_n^{2S}$ is the celebrated Stein estimator of σ^2.

The following theorem shows the optimality of $\hat{\sigma}_n^{2[S]}$ relative to $ms_y^2/(m+2)$ with respect to the mse criterion:

Theorem 2. $\hat{\sigma}^{2[S]}$ dominates $\hat{\sigma}_n^2$ uniformly with respect to mse criteria.

Proof. Consider the mse of $\hat{\sigma}_m^{2[S]}$:

$$\begin{aligned}
&E\Big[\hat{\sigma}_n^{2[S]} - \sigma^2\Big]^2 \\
&= E\Big[\hat{\sigma}^{2[S]}\phi_S(\mathcal{L}_n) m s_y^2 - \sigma^2\Big]^2 \\
&= \sigma^4 E\Big[\phi_S(\mathcal{L}_n)\frac{ms_y^2}{\sigma^2} - 1\Big]^2 \\
&= \sigma^4 E_{\mathcal{L}_n}\Big\{\big[\phi_S(\mathcal{L}_n)\big]^2 E(\chi_m^4|\mathcal{L}_n) - 2\phi_S(\mathcal{L}_n)E(\chi_m^2|\mathcal{L}_n) + 1\Big\}. \quad (3.9.31)
\end{aligned}$$

For fixed Δ^2 and each $\mathcal{L}_n$, this expression is a quadratic form in $\phi_S(\mathcal{L}_n)$ with a minimum at

$$\phi^*(\mathcal{L}_n) = \frac{E[\chi_m^2|\mathcal{L}_n]}{E[\chi_m^4|\mathcal{L}_n]}. \quad (3.9.32)$$

Using

$$\begin{aligned}
E[\chi_m^{2k}|\mathcal{L}_n] = \frac{1}{g_1(\mathcal{L}_n)} \sum_{n\geq 0} 2^k \Big(\frac{1}{m}\Big)^{(1/2)+r} \\
\times \frac{e^{-\Delta^2/2}\Big(\frac{\Delta^2}{2}\Big)^r \Gamma\big(\frac{m+1}{2}+r+k\big)\mathcal{L}_n^{(1/2)+r-1}}{\Gamma(r+1)\Gamma\frac{m}{2}\Gamma(\frac{1}{2}+r)(1+\frac{1}{m}\mathcal{L}_n)^{[(m+1)/2]+r+k}}, \quad (3.9.33)
\end{aligned}$$

where

$$g_1(\mathcal{L}_n) = \Sigma\Big(\frac{1}{m}\Big)^{\frac{1}{2}+r} \frac{e^{-\Delta^2/2}\Big(\frac{\Delta^2}{2}\Big)^r \Gamma\big(\frac{m+1}{2}+r+k\big)\mathcal{L}_n^{(1/2)+r-1}}{\Gamma(r+1)\Gamma\frac{m}{2}\Gamma(\frac{1}{2}+r)(1+\frac{1}{m}\mathcal{L}_n)^{[(m+1)/2]+r+k}} \quad (3.9.34)$$

and

$$E\left\{[\chi_m^{2k}|F_{1,m}(0)] = \mathcal{L}_n\right\} = 2^k \frac{\Gamma\left(\frac{m+1}{2}+k\right)}{\Gamma\left(\frac{m+1}{2}\right)\left(1+\frac{1}{m}\mathcal{L}_n\right)^k}. \quad (3.9.35)$$

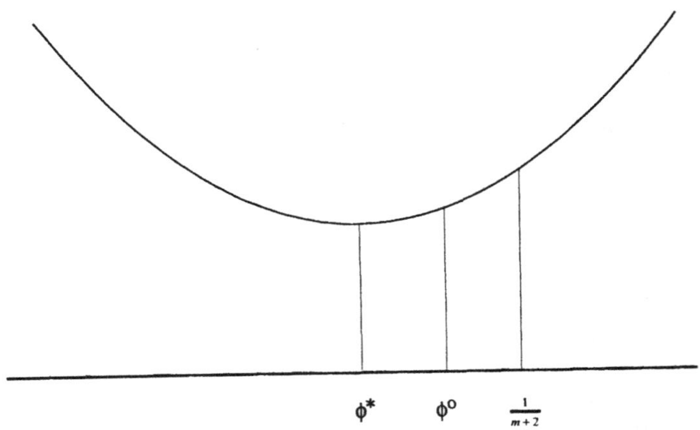

Figure 3.9.1 Graph of $[\phi_S(\mathcal{L}_n)]^2\{E(\chi_m^4|\mathcal{L}_n)\} - 2\phi_S(\mathcal{L}_n)E(\chi_m^2|\mathcal{L}_n) + 1$ as a function of $\phi_S(\mathcal{L}_n)$

We obtain

$$\begin{aligned}\phi_0(\mathcal{L}_n) &= \frac{E(\chi_m^2|\mathcal{L}_n)}{E(\chi_m^4|\mathcal{L}_n)} = \max_{\Delta^2} \phi^*(\mathcal{L}_n|\Delta^2 = 0) \\ &= \frac{1 + \frac{1}{m}\mathcal{L}_n}{m+3}. \end{aligned} \tag{3.9.36}$$

If $\mathcal{L}_n < m/(m+2)$, then $(1 + \frac{1}{m}\mathcal{L}_n)/(m+3) < \frac{1}{m+2}$, which implies

$$\phi^*(\mathcal{L}_n) \le \phi_0(\mathcal{L}_n) \le \frac{1}{m+2} \quad \text{for all } \Delta^2. \tag{3.9.37}$$

That is, $\phi_0(\mathcal{L}_n)$ is closer to the minimizing value than $1/(m+2)$. So, as shown in Figure 3.9.1, it is clear that for each Δ^2 and $\mathcal{L}_n$,

$$E\left\{\left[\phi_S(\mathcal{L}_n)\left(\frac{ms_y^2}{\sigma^2}\right) - 1\right]^2 \middle| \mathcal{L}_n\right\} \le E\left\{\left[\frac{1}{m+2}\left(\frac{ms_y^2}{\sigma^2}\right) - 1\right]^2 |\mathcal{L}_n\right\}. \tag{3.9.38}$$

It follows that $\phi_S(\mathcal{L}_n)ms_y^2$ dominates $ms_y^2/(m+2)$ uniformly in Δ^2. Further, it is clear that the PTE defined in (3.9.27) equals $\hat{\sigma}_n^{2[S]}$ whenever $F_{1,m}(\alpha) =$

$m(m+2)$. This may be obtained by minimizing the mse of $\hat{\sigma}_n^{2[PT]}$ for varying $F_{1,m}(\alpha)$. Thus, $\alpha^* = 1 - G_{1,m}(m/(m+2); 0)$ is the optimum level of significance for the PTE, $\hat{\sigma}_n^{2[PT]}$.

3.10 Nonparametric Methods: R-Estimation

In this section we consider nonparametric methods toward R-estimation of the intercept and the slope parameters of the simple linear model when it is suspected that the slope parameter may assume a pre-specified value. This procedure broadens the class of distributions and at the same time achieve robustness of the estimators against contamination in the data set. R-estimation of a location parameter after a preliminary test on regression is due to Saleh and Sen (1978), and estimation of a location parameter after a preliminary test is due to Tamura (1967) based on U-statistics.

We invoke asymptotic theory to develop the properties of the estimators we discuss. In the following sections we state the model, assumptions, and test of hypothesis, and define the estimators we consider.

3.10.1 Model and Assumptions

Let $y_1, \cdots, y_n$ be the response variables corresponding to preassigned values $x_1, \cdots, x_n$ satisfying the linear model

$$y_j = \theta + \beta x_j + e_j, \quad j = 1, \cdots, n. \tag{3.10.1}$$

The distribution function (d.f.) $F(e)$ of the errors belong to a class, $\mathcal{F}$, of symmetric absolutely continuous distribution function (cdf) with absolutely continuous density function (pdf), $f(e)$ such that the Fisher's information is finite,

$$I(f) = \int_{-\infty}^{\infty} \left(\frac{f'(z)}{f(z)}\right)^2 f(z)\, dz \ < \ \infty. \tag{3.10.2}$$

In addition, the covariates $x_1, \cdots, x_n$ satisfy the following conditions:

(i) $\lim_{n\to\infty} \bar{x}_n = \bar{x}_0, \quad |\bar{x}_0| \ < \ \infty.$ (3.10.3a)

(ii) $\lim_{n\to\infty} \frac{Q_n}{n} = Q^0 \ < \ \infty, \quad Q_n = \sum_{i=1}^n (x_i - \bar{x}_n)^2.$ (3.10.3b)

(iii) The x's are bounded and $\max_i \frac{(x_i - \bar{x}_n)^2}{Q_n} \ \to \ 0$ as $n \to \infty$. (3.10.3c)

The linear model (10.3.1) together with the assumptions (3.10.2) through (3.10.3a, b, c) is the basis of further investigations on test of the hypothesis $\theta = 0, \beta = 0$ or $\beta = 0$, and the estimate of (θ, β) using the sample $\{(x_1, y_1), \cdots, (x_n, y_n)\}$.

3.10.2 Test of Hypothesis

In nonparametric methods of statistics, one first develops test-statistic for testing the parameters of interest and in the second stage uses the test-statistic to define the estimators of the parameter by inversion technique. In the model (3.10.1) under consideration we consider the test of hypothesis, $H_0 : \theta = 0, \beta = 0$ against the alternatives that $\theta \neq 0, \beta \neq 0$ simultaneously. As such we consider the test-statistics well known in the literature (e.g., see Adichie, 1967; Hodges and Lehman, 1963; Puri and Sen, 1986) as follows:

Let $\phi = \{\phi(u), 0 < u < 1\}$ be a nondecreasing, skew-symmetric (i.e., $\phi(u) + \phi(1-u) = 0$, $0 < u < 1$) and square integrable score function $\phi^* = \{\phi^*(u) = \phi((1+u)/2), 0 < u < 1\}$, and for every $n(\geq 1)$, let

$$a_n(i) = E\phi(U_{ni}) \quad \text{or} \quad \phi\Big(\frac{i}{n+1}\Big),$$

$$a_n^*(i) = E\phi^*(U_{ni}) \quad \text{or} \quad \phi^*\Big(\frac{i}{n+1}\Big), \quad i = 1, \cdots, n, \tag{3.10.4}$$

where $U_{n1} < \cdots < U_{nn}$ are the ordered r.v's of a sample of size n from the rectangular cdf on (0.1). Finally, let $\mathbf{y}_n = (y_1, \cdots, y_n)'$ and for every real (a, b), and define $\mathbf{y}_n(a, b) = \mathbf{y}_n - a\mathbf{1}_n - b\mathbf{x}_n$, where $\mathbf{1}_n = (1, \cdots, 1)$ and $\mathbf{x}_n = (x_1, \cdots, x_n)'$. Consider then the statistics

$$T_n(a, b) = T(\mathbf{y}_n(a, b)) = \frac{1}{n}\sum_{i=1}^{n} a_n^*(R_{ni}^+(a, b))\text{sign}(Y_i - a - bx_i),$$

$$L_n(a, b) = L(\mathbf{y}_n(a, b)) = \frac{1}{n}\sum_{i=1}^{n}(x_i - \bar{x}_n)a_n(R_{ni}(a, b)), \tag{3.10.5}$$

where $R_{ni}(a, b)$ (or $R_{ni}^+(a, b)$) is the rank of $y_i - a - bx_i$ (or $|y_i - a - bx_i|$) among $y_i - a - bx_1, \cdots, y_n - a - bx_n$ (or $|y_1 - a - bx_1|, \cdots, |y_n - a - bx_n|$), for $i = 1, \cdots, n$. Note that $R_{ni}(a, b) = R_{ni}(0, b)$ for every real a, and hence, $R_{ni}(a, b)$ does not depend on a. We write it as $L_n(b)$; also, we write $R_{ni}(0, 0) = R_{ni}$ for $i = 1, \cdots, n$.

Note that for every given $\mathbf{y}_n$ and b, $T_n(a, b)$, is $\searrow$ in $a : -\infty < a < \infty$, and for every given $\mathbf{y}_n$, $L_n(b)$ is $\searrow$ in $b : -\infty < b < \infty$. Also, if in the model (3.10.1) we let $\theta = \beta = 0$, then $T_n(0, 0)$ and $L_n(0)$ both (marginally) have distributions symmetric about 0.

Thus for the one-sided test of $H_0 : \beta = 0$ against $H_A : \beta > 0$, our test consists in accepting or rejecting H_0 according as $L_n(0)$ is $<$ or $\geq L_{n,\alpha}$, where $P\{L_n(0) \geq L_{n,\alpha} | H_0\} \leq \alpha$, $0 < \alpha < 1$, and α is the level of significance of the test.

Similarly, for the test of $H_0 : \theta = 0$ against $H_A : \theta > 0$, the test consists of accepting and rejecting H_0 according as $T_n(0, 0) <$ or $\geq T_{n,\alpha}$, where $P_0\{T_n(0, 0) \geq T_{n,\alpha} | H_0\} \leq \alpha$.

Since we will be estimating θ and β when it is suspected that β is zero, we concentrate on the test of β that is, $H_0 : \beta = 0$ against $H_A : \beta > 0$, using the statistics $|L_n(0)|$ or $L_n^2(0)$. Let

$$A_\phi^2 = \int_0^1 \phi^2(u)du - (\int_0^1 \phi(u)du)^2, \tag{3.10.5a}$$

$$\psi(u) = -\frac{f'(F^{-1}(u))}{f(F^{-1}(u))}, \quad 0 < u < 1, \tag{3.10.5b}$$

$$A_\psi^2 = I(f) = \int_0^1 \psi^2(u)du, \tag{3.10.5c}$$

$$\gamma(\psi, \phi) = \int_0^1 \psi(u)\phi(u)du, \tag{3.10.5d}$$

and consider the estimate of A_ϕ^2 by

$$A_n^2 = \frac{1}{n-1}\sum_{i=1}^{n}[a_n(i) - \bar{a}_n]^2, \quad \bar{a}_n = \frac{1}{n}\sum_{i=1}^{n} a_n(i). \tag{3.10.6}$$

Let z_α be the upper α-level critical value from $\mathcal{N}(0,1)$. Then

$$\sqrt{n}Q_n^{-1/2}A_n^{-1}L_{n,\alpha} \;\to\; z_\alpha \text{ as } n \to \infty. \tag{3.10.7}$$

Alternatively, we can write

$$nQ_n^{-1}A_n^{-2}L_{n,\alpha}^2 \;\to\; \chi_1^2(\alpha) \text{ as } n \to \infty, \tag{3.10.8}$$

where $\chi_1^2(\alpha)$ is the α-level critical value from a central chi-square distribution with 1 d.f.

For small samples, $L_{n,\alpha}$, can be computed by direct enumeration of the null distribution of $L_n(0)$ generated by $n!$ equally likely permutation of the ranks $R_1, \cdots, R_n$.

These results follow from the fact that under the condition (3.10.3a, b, c) and under H_0 we have (see Hajek and Sidak, 1967; Hajek, Sidak, and Sen, 1999)

$$\begin{pmatrix} \sqrt{n}T_n(0,0) \\ \sqrt{n}L_n(0) \end{pmatrix} \sim \mathcal{N}_2\left\{\begin{pmatrix} 0 \\ 0 \end{pmatrix}, \; A_\phi^2 \text{ Diag}(1, Q^0)\right\}. \tag{3.10.9}$$

Consequently,

$$\lim_{n\to\infty} P\{nQ_n^{-1}A_n^{-2}L_n(0) \le x | H_0\} = H_1(x; 0). \tag{3.10.10}$$

3.10.3 Estimation of Intercept and Slope Parameters

We focus now on the estimation of θ and β based on the test-statistics $T_n(a, b)$ and $L_n(b)$ as follows (see Adichie, 1967; and Puri and Sen, 1986):

Let

$$\hat{\theta}_n^{(1)} = \sup\{a : T_n(a,0) > 0\} \text{ and } \hat{\theta}_n^{(2)} = \inf\{a : T_n(a,0) < 0\},$$

$$\hat{\theta}_n = \tfrac{1}{2}(\hat{\theta}_n^{(1)} + \hat{\theta}_n^{(2)}), \tag{3.10.11a}$$

$$\tilde{\beta}_n^{(1)} = \sup\{a : L_n(b) > 0\} \text{ and } \tilde{\beta}_n^{(2)} = \inf\{a : L_n(b) < 0\},$$

$$\tilde{\beta}_n = \tfrac{1}{2}(\tilde{\beta}_n^{(1)} + \tilde{\beta}_n^{(2)}), \tag{3.10.11b}$$

$$\tilde{\theta}_n^{(1)} = \sup\{a : T_n(a, \tilde{\beta}_n) > 0\} \text{ and } \tilde{\theta}_n^{(2)} = \inf\{a : T_n(a, \tilde{\beta}_n) < 0\},$$

$$\tilde{\theta}_n = \tfrac{1}{2}(\tilde{\theta}_n^{(1)} + \tilde{\theta}_n^{(2)}). \tag{3.10.11c}$$

Then $\hat{\theta}_n$ is a translation-invariant, robust, and consistent estimator of θ when $\beta = 0$, while $\tilde{\theta}_n$ is a similar estimator when β is unspecified.

For the preliminary test estimator of β and θ, we consider the statistic

$$\mathcal{L}_n = nQ_n^{-1}A_n^{-2}L_n^2(0) \overset{\mathcal{D}}{=} \chi_1^2 \quad \text{as } n \to \infty \tag{3.10.12}$$

from (3.10.10) under H_0. Let $\chi_1^2(\alpha)$ be the α-level upper critical value from a central chi-square distribution with 1 d.f., then the PTE of β is defined by

$$\hat{\beta}_n^{PT} = \tilde{\beta}_n - \tilde{\beta}_n I(\mathcal{L}_n < \chi_1^2(\alpha)). \tag{3.10.13}$$

The S-estimator of β is defined by

$$\hat{\beta}_n^S = \tilde{\beta}_n - \frac{cA_nQ_n^{1/2}\tilde{\beta}_n}{\sqrt{n}|L_n(0)|}. \tag{3.10.13}$$

Similarly, the PTE and S-estimator of θ are given by

$$\hat{\theta}_n^{PT} = \tilde{\theta}_n - (\tilde{\theta}_n - \hat{\theta}_n)I(\mathcal{L}_n < \chi_1^2(\alpha)) \tag{3.10.15}$$

and

$$\hat{\theta}_n^S = \tilde{\theta}_n - \frac{cA_nQ_n^{1/2}(\tilde{\theta}_n - \hat{\theta}_n)}{\sqrt{n}|L_n(0)|}, \tag{3.10.16}$$

respectively.

3.10.4 Asymptotic Distribution of Various Estimators and Their ADB and ADMSE Expressions

In order to obtain the asymptotic distributions of various statistics, we will primarily use (3.10.9) and the Jurečkova's (1969) linearity of rank-statistics under $H_0 : \beta = 0$ as $n \to \infty$. The following holds:

(i) $\sup\{\sqrt{n}|L_n(n^{-1/2}b) - L_n(0) + n^{-1/2}bQ^0\gamma(\psi,\phi)|; |b| \le k\} \xrightarrow{P} 0$

(ii) $\sup\{\sqrt{n}|T_n(n^{-\frac{1}{2}}a, n^{-1/2}b) - T_n(0,0)$

$$+n^{-1/2}(a + b\bar{x}_n)\gamma(\psi,\phi)|; |a| \le k, |b| \le k\} \xrightarrow{P} 0 \tag{3.10.17}$$

where k is a positive constant. Furthermore, under the stated conditions we have

$$\text{(i)} \begin{pmatrix} \sqrt{n}(\tilde{\theta}_n - \theta) \\ \sqrt{n}(\tilde{\beta} - \beta) \end{pmatrix} \sim \mathcal{N}_2\left\{\begin{pmatrix} 0 \\ 0 \end{pmatrix}; \left[\frac{A_\phi^2}{\gamma^2(\psi,\phi)}\right]\begin{pmatrix} 1 + \frac{\bar{x}_0^2}{Q^0} & -\frac{\bar{x}_0}{Q^0} \\ -\frac{\bar{x}_0}{Q^0} & \frac{1}{Q^0} \end{pmatrix}\right\}. \tag{3.10.18}$$

(ii) $\sqrt{n}(\tilde{\theta}_n - \theta) \sim \mathcal{N}(0, A_\phi^2/\gamma^2(\psi,\phi))$ when $\beta \neq 0$.

Note that $\mathcal{L}_n$ is a consistent test against $\beta \neq 0$. So, for **fixed alternatives**, the asymptotic distributions of $\sqrt{n}(\tilde{\beta}_n - \beta), \sqrt{n}(\hat{\beta}_n^{PT} - \beta)$ and $\sqrt{n}(\hat{\beta}_n^S - \beta)$ are all equivalent to $\mathcal{N}(0, A_\phi^2/\gamma(\psi,\phi)Q^0)$. Similar results hold for $\sqrt{n}(\tilde{\theta}_n - \theta), \sqrt{n}(\hat{\theta}_n^{PT} - \theta)$, and $\sqrt{n}(\hat{\theta}_n^S - \theta)$. Thus we consider the class of **local alternatives**

$$K_{(n)} : \beta_{(n)} = n^{-1/2}\delta \tag{3.10.19}$$

to obtain asymptotic distributions that provide fair comparisons of the properties of the estimators. Thus,

Theorem 1. (Saleh and Sen, 1978). Under $K_{(n)}$ and the assumed regularity conditions, as $n \to \infty$,

$$\text{(i)} \begin{pmatrix} \sqrt{n}(\tilde{\beta}_n - \beta) \\ \sqrt{n}L_n(0) \end{pmatrix} \sim \mathcal{N}_2\left\{\begin{pmatrix} 0 \\ \delta Q^0\gamma(\psi,\phi) \end{pmatrix}; A_\phi^2\begin{pmatrix} \frac{1}{\gamma^2(\psi,\phi)Q^0} & -\frac{1}{\gamma(\psi,\phi)} \\ -\frac{1}{\gamma(\psi,\phi)} & Q^0 \end{pmatrix}\right\}, \tag{3.10.20}$$

$$\text{(ii)} \begin{pmatrix} \sqrt{n}(\tilde{\theta}_n - \theta) \\ \sqrt{n}L_n(0) \end{pmatrix} \sim \mathcal{N}_2\left\{\begin{pmatrix} 0 \\ \delta Q^0\gamma(\psi,\phi) \end{pmatrix}; \boldsymbol{\Sigma}_1\right\},$$

$$\text{(iii)} \begin{pmatrix} \sqrt{n}(\hat{\theta}_n - \theta) \\ \sqrt{n}L_n(0) \end{pmatrix} \sim \mathcal{N}_2\left\{\begin{pmatrix} \delta\bar{x}_0 \\ \delta Q^0\gamma(\psi,\phi) \end{pmatrix}; \boldsymbol{\Sigma}_2\right\},$$

$$\text{where } \boldsymbol{\Sigma}_1 = A_\phi^2\begin{pmatrix} (1 + (\bar{x}_0/Q^0))/\gamma^2(\psi,\phi) & -\frac{\bar{x}_0}{\gamma(\psi,\phi)} \\ -\frac{\bar{x}_0}{\gamma(\psi,\phi)} & Q^0 \end{pmatrix},$$

$$\boldsymbol{\Sigma}_2 = A_\phi^2\begin{pmatrix} \frac{1}{\gamma^2(\psi,\phi)} & 0 \\ 0 & Q^0 \end{pmatrix},$$

$$\text{(iv)} \quad \lim P\{\mathcal{L}_n \leq x | K_{(n)}\} = H_1(x;\Delta^2), \quad \Delta^2 = \frac{\delta^2 Q^0\gamma^2(\psi,\phi)}{A_\phi^2}.$$

Proof. Note that by (3.10.11b), (3.10.9), and (3.10.17), we have $\sqrt{n}|\tilde{\beta} - \beta| = O_p(1)$, where under $\{K_{(n)}\}, n^{1/2}\beta = \delta = 0(1)$. Thus, under $\{K_{(n)}\}, \sqrt{n}|\tilde{\beta}| = O_p(1)$. Further under $\beta = 0$ by (3.10.17) we obtain $n^{1/2}L_n(0) = n^{1/2}\tilde{\beta}Q^0\gamma(\psi,\phi) + o_p(1)$. Hence, using contiguity of probability measures under $\{K_{(n)}\}$ and to that of $\beta = 0$, we obtain under $\{K_{(n)}\}$ as $n \to \infty$,

$$\begin{bmatrix} \sqrt{n}(\tilde{\beta} - \beta) \\ \sqrt{n}L_n(0) \end{bmatrix} \xrightarrow{\mathcal{D}} \mathcal{N}_2\left\{\begin{pmatrix} 0 \\ \delta\gamma(\psi,\phi)Q^0 \end{pmatrix}; A_\phi^2\begin{pmatrix} 1/\gamma^2(\psi,\phi)Q^0 & 1/\gamma(\psi,\phi) \\ 1/\gamma(\psi,\phi) & Q^0 \end{pmatrix}\right\}. \tag{3.10.21}$$

To prove (ii), we consider the hypothesis $\theta = \beta = 0$ together with the linearity results based on (3.10.17)

$$\sqrt{n}L_n(0) = \sqrt{n}\tilde{\beta}_n\sqrt{Q^0}\gamma(\psi,\phi) + o_p(1) \tag{3.10.22}$$

and

$$\sqrt{n}T_n(0,0) = \sqrt{n}(\tilde{\theta}_n + \tilde{\beta}\bar{x}_0)\gamma(\psi,\phi) + o_p(1).$$

Then we have

$$\sqrt{n}\theta_n\gamma(\psi,\phi) = \sqrt{n}T_n(0,0) - \frac{\bar{x}_0}{Q^*}\sqrt{n}L_n(0) + o_p(1). \tag{3.10.23}$$

Hence, from the contiguity of measures under $K^*_{(n)} : \theta = 0, \beta_{(n)} = n^{-1/2}\delta$, to those under $H^*_0 : \theta = \beta = 0$, we obtain from above that as $n \to \infty$,

$$\sqrt{n}\tilde{\theta}_n\gamma(\psi,\phi) = \sqrt{n}T_n(0,0) - \frac{\bar{x}_0}{Q^*}\sqrt{n}L_n(0) + o_p(1). \tag{3.10.24}$$

Finally, $(\sqrt{n}T_n(0,0), \sqrt{n}L_n(0))'$ under $\{K^*_{(n)}\}$ has the same distribution as $(\sqrt{n}T_n(0, n^{-1/2}\delta), \sqrt{n}L_n(n^{-1/2}\delta))$ under H^*_0. Hence, the asymptotic distribution of $(\sqrt{n}T_n(0,n^{-1/2}\delta), \sqrt{n}L_n(n^{-1/2}\delta))$ is bivariate normal with mean vector $\delta\gamma(\psi,\phi)(\bar{x}_0, Q^0)'$ and dispersion matrix $A^2_\phi\,\mathrm{Diag}(1, Q^0)$. Thus under $\{K_{(n)}\}$ as $n \to \infty$,

$$(\sqrt{n}T_n(0,0),\ \sqrt{n}L_n(0))' \sim \mathcal{N}_2\{\delta\gamma(\psi,\phi)(\bar{x}_0, Q^0)'; A^2_\phi\,\mathrm{Diag}(1,Q^0)\} \tag{3.10.25}$$

The proof of (3.10.20(ii)) follows from (3.10.23) and (3.10.25). Again, by noting (3.10.11a) and (3.10.17), under $\{K_{(n)}\}$, we have

$$\sqrt{n}\hat{\theta}_n\gamma(\psi,\phi) = \sqrt{n}T_n(0,0) + o_p(1), \tag{3.10.26}$$

and the proof of (3.10.20(iii)) follows. Clearly, $\mathcal{L}_n \xrightarrow{\mathcal{D}} \chi^2_1(\Delta^2)$ as $n \to \infty$.

Based on Theorem 3.10.1, we obtain the expressions for the asymptotic distributional bias (ADB) and the asymptotic distributional MSE (ADMSE) of various estimators of β and θ as given by the theorem below:

Theorem 2. Under $\{K_{(n)}\}$ and assumed regularity conditions, we have the ADB and ADMSE of the estimators of β as $n \to \infty$ given in (A), (B), and (C) below.

(A) UE, $\tilde{\beta}_n$

(i) $b_1(\tilde{\beta}_n) = 0$ and (ii) $M_1(\tilde{\beta}_n) = \sigma^2/Q^0$ with $\sigma^2 = A^2_\phi/\gamma^2(\psi,\phi)$,

(B) PTE, $\hat{\beta}^{PT}_n$

(i) $b_2(\hat{\beta}_n^{PT}) = -\delta H_3(\chi_1^2(\alpha); \Delta^2)$.

(ii) $M_2(\hat{\beta}_n^{PT}) = \frac{\sigma^2}{Q^0}[1 - H_3(\chi_1^2(\alpha); \Delta^2)$
$+\Delta^2\{2H_3(\chi_1^2(\alpha); \Delta^2) - H_5(\chi_1^2(\alpha); \Delta^2)\}]$.

(C) SE, $\hat{\beta}_n^S$

(i) $b_3(\hat{\beta}_n^S) = -\frac{c\sigma\delta}{Q^0}[2\Phi(\Delta^2) - 1]$.

(ii) $M_3(\hat{\beta}_n^S) = \frac{\sigma^2}{Q^0}\{1 - \sqrt{\frac{2}{\pi}}[2e^{-\Delta^2/2} - 1]\}$.

Proof. A(i) and A(ii) are obvious. We prove (B) and (C). The PTE of β may be written as $\hat{\beta}_n^{PT} = \tilde{\beta} - \tilde{\beta}I(\mathcal{L}_n < \chi_1^2(\alpha))$. Thus, the ADB may be written as

$$\begin{aligned}
&\lim_{n\to\infty} E[\sqrt{n}\tilde{\beta}_n I(\mathcal{L}_n < \chi_1^2(\alpha))] \\
&= -\lim_{n\to\infty} E[\sqrt{n}\beta_{(n)} I(n\tilde{\beta}_n^2 Q_n^{-1} A_n^{-2}\nu^2(\psi,\phi) < \chi_1^2(\alpha))] \\
&= -\delta H_3(\chi_1^2(\alpha); \Delta^2)
\end{aligned}$$

by Theorem 2.10.1 in Chapter 2.

Similarly, the ADMSE expression for PTE may be written as

$$\begin{aligned}
&\lim_{n\to\infty} E[n(\hat{\beta}_n^{PT} - \beta_{(n)})^2] \\
&= \lim_{n\to\infty}[n(\tilde{\beta}_n - \beta_{(n)})^2] + \lim_{n\to\infty}[n\tilde{\beta}_n^2 I(\mathcal{L}_n < \chi_1^2(\alpha))] \\
&\quad -2\lim_{n\to\infty} E[n(\tilde{\beta}_n - \beta_{(n)})\tilde{\beta}_n I(\mathcal{L}_n < \chi_1^2(\alpha))] \\
&= \sigma^2 - \lim_{n\to\infty} E[n\tilde{\beta}_n^2 I(\mathcal{L}_n < \chi_1^2(\alpha))] + 2\lim_{n\to\infty} E[n\beta_{(n)}\tilde{\beta}_n I(\mathcal{L}_n < \chi_1^2(\alpha))] \\
&= \frac{\sigma^2}{Q^0}[1 - H_3(\chi_1^2(\alpha); \Delta^2) + \Delta^2\{2H_3(\chi_1^2(\alpha); \Delta^2) - H_5(\chi_1^2(\alpha); \Delta^2)\}].
\end{aligned}$$

For C(i) and C(ii), we consider the bias

$$\begin{aligned}
\text{(i)} \lim_{n\to\infty} E[\sqrt{n}(\hat{\beta}_n^S - \beta_{(n)})] &= -\lim_{n\to\infty} E\Big[\frac{cA_n Q_n^{1/2}\sqrt{n}\tilde{\beta}_n}{Q^0\gamma(\psi,\phi)|\sqrt{n}\tilde{\beta}_n|}\Big] = -\frac{c\sigma}{\sqrt{Q^0}}E\Big[\frac{Z}{|Z|}\Big], \\
&= -\frac{c\sigma}{\sqrt{Q^0}}[2\Phi(\Delta) - 1], \quad \text{since } Z \sim \mathcal{N}(\Delta, 1)
\end{aligned}$$

The MSE of $\hat{\beta}_n^S$ is given by

$$\lim_{n\to\infty} E[\sqrt{n}(\hat{\beta}_n^S - \beta_{(n)})^2]$$

$$= - \lim_{n\to\infty} E[n(\tilde{\beta}_n - \beta)^2] + \lim_{n\to\infty} E\Big[\frac{c^2 A_n^2 Q_n}{[Q^0]^2\gamma^2(\psi,\phi)}\Big]$$

$$- \lim_{n\to\infty} E\Big[\frac{2cn(\tilde{\beta}_n - \beta_{(n)})\tilde{\beta}_n A_n Q_n^{1/2}}{Q^0\gamma(\psi,\phi)|\sqrt{n}\tilde{\beta}_n|}\Big]$$

$$= \frac{\sigma^2}{Q^0} + \ - 2c\frac{\sigma^2}{Q^0}E[|Z|] + 2\frac{c\sigma^2}{Q^0}\Delta E\Big[\frac{Z}{|Z|}\Big]$$

$$= \frac{\sigma^2}{Q^0}[1 + c^2 - 2c\sigma^2\{E[|Z|] - \Delta^2 E\Big[\frac{Z}{|Z|}\Big]\}]$$

$$= \frac{\sigma^2}{Q^0}\{1 - \frac{2}{\pi}(2e^{-\Delta^2/2} - 1)\}, \text{ by taking } c = \sqrt{\frac{2}{\pi}}.$$

The next theorem gives the ADB and ABMSE of the estimators of θ.

Theorem 3. Under $\{K_{(n)}\}$ and the assumed regularity conditions, we have as $n \to \infty$, the ADB and ADMSE of the estimators of θ as follows:

(A) UE, $\tilde{\theta}_n$

(i) $b_1(\tilde{\theta}_n) = 0$ and (ii) $M_1(\tilde{\theta}_n) = \sigma^2\Big(1 + \frac{\bar{x}_0^2}{Q^0}\Big)$.

(B) RE, $\hat{\theta}_n$

(i) $b_2(\hat{\theta}_n) = \delta\bar{x}_0$ and (ii) $M_2(\hat{\theta}_n) = \sigma^2\Big(1 + \frac{\bar{x}_0^2}{Q^0}\Delta^2\Big)$

(C) PTE, $\hat{\theta}_n^{PT}$

(i) $b_3(\hat{\theta}_n^{PT}) = \delta\bar{x}_0 H_3(\chi_1^2(\alpha);\Delta^2)$ and

(ii) $M_3(\hat{\theta}_n^{PT}) = \sigma^2\Big(1 + \frac{\bar{x}_0^2}{Q^0}\Big) - \frac{\sigma^2\bar{x}_0^2}{Q^0}H_3(\chi_1^2(\alpha);\Delta^2)$

$+\frac{\sigma^2\bar{x}_0^2}{Q^0}\Delta^2\{2H_3(\chi_1^2(\alpha);\Delta^2) - H_5(\chi_1^2(\alpha);\Delta^2)\}$.

(D) S-estimator, $\hat{\theta}_n^S$

(i) $b_4(\hat{\theta}_n^S) = \frac{c\sigma\bar{x}_0\delta}{\sqrt{Q^0}}[2\Phi(\Delta) - 1]$ and

(ii) $M_4(\hat{\theta}_n^S) = \sigma^2\Big[\Big(1 + \frac{\bar{x}_0^2}{Q^0}\Big)\Big\{1 - \frac{2}{\pi}\Big(2e^{-\Delta^2/2} - 1\Big)\Big\}\Big]$.

Proof. A(i) and (ii) are obvious. We consider B(i) and (ii). By the linearity results (3.10.26) and (3.10.22), we have $\sqrt{n}\hat{\theta}_n = \sqrt{n}(\tilde{\theta}_n + \tilde{\beta}\bar{x}_0) + o_p(1)$. Hence, the ADB and ADMSE are given as follows:.

(i) $\lim_{n\to\infty} E[\sqrt{n}(\hat{\theta}_n - \theta)] = \lim_{n\to\infty} E[\sqrt{n}\tilde{\beta}_n\bar{x}_0] = \delta\bar{x}_0$

(ii) $\lim_{n\to\infty} E[n(\hat{\theta}_n - \theta)^2]$

$$= \lim_{n\to\infty} E[n(\tilde{\theta}_n - \theta)^2] + \lim_{n\to\infty} E[n\tilde{\beta}_n^2\bar{x}_0^2] + 2\lim_{n\to\infty} E[n(\tilde{\theta}_n - \theta)\tilde{\beta}_n\bar{x}_0]$$

$$= \tfrac{\sigma^2}{Q^0}\left(1 + \tfrac{\bar{x}_0^2}{Q^0}\right) - \bar{x}_0^2 \lim_{n\to\infty} E[n\tilde{\beta}_n^2] + 2\delta\bar{x}_0^2$$

$$= \tfrac{\sigma^2}{Q^0}\left(1 + \tfrac{\bar{x}_0^2}{Q^0}\right) - \bar{x}_0^2\left(\tfrac{\sigma^2}{Q^0} + \delta^2\right) + 2\delta\bar{x}_0^2$$

$$= \tfrac{\sigma^2}{Q^0}\left(1 + \tfrac{\bar{x}_0^2}{Q^0}\right) - \tfrac{\bar{x}_0^2\sigma^2}{Q^0} + \delta^2\bar{x}_0^2$$

$$= \sigma^2\left(1 + \tfrac{\bar{x}_0^2}{Q^0}\Delta^2\right) \text{ from (3.2.9a).}$$

To prove C(i) and (ii), consider

(i) $\lim_{n\to\infty} E[\sqrt{n}(\hat{\theta}_n^{PT} - \theta)] = -\lim_{n\to\infty} E[\sqrt{n}\tilde{\beta}_n\bar{x}_0 I(\mathcal{L}_n < \chi_1^2(\alpha))]$

$$= \delta\bar{x}_0 H_3(\chi_1^2(\alpha); \Delta^2) \text{ by Theorem 3.10.1;}$$

(ii) $\lim_{n\to\infty} E[n(\hat{\theta}_n^{PT} - \theta)^2] = \lim_{n\to\infty} E[n(\tilde{\theta}_n - \theta)^2]$

$$+\bar{x}_0^2 \lim_{n\to\infty} E[n\tilde{\beta}_n^2 I(\mathcal{L}_n < \chi_1^2(\alpha))] + 2\bar{x}_0 \lim_{n\to\infty} E[n(\tilde{\theta}_n - \theta)\tilde{\beta} I(\mathcal{L}_n < \chi_1^2(\alpha))]$$

$$= \sigma^2\left(1 + \tfrac{\bar{x}_0^2}{Q^0}\right) - \bar{x}_0^2 \lim_{n\to\infty} E[n\tilde{\beta}_n^2 I(\mathcal{L}_n < \chi_1^2(\alpha))] + 2\delta^2\bar{x}_0^2 H_3(\chi_1^2(\alpha); \Delta^2)$$

$$= \sigma^2\left(1 + \tfrac{\bar{x}_0^2}{Q^0}\right) - \tfrac{\bar{x}_0^2}{Q^0}\sigma^2 H_3(\chi_1^2(\alpha); \Delta^2) - \bar{x}_0^2\delta^2 H_3(\chi_1^2(\alpha); \Delta^2)$$
$$+ 2\delta^2\bar{x}_0^2 H_5(\chi_1^2(\alpha); \Delta^2)$$

$$= \sigma^2\left(1 + \tfrac{\bar{x}_0^2}{Q^0}\right) - \tfrac{\bar{x}_0^2}{Q^0}\sigma^2 H_3(\chi_1^2(\alpha); \Delta^2)$$
$$+ \bar{x}_0^2\delta^2\{2H_3(\chi_1^2(\alpha); \Delta^2) - H_5(\chi_1^2(\alpha); \Delta^2)\}$$

$$= \sigma^2\Big\{\left(1 + \tfrac{\bar{x}_0^2}{Q^0}\right) - \tfrac{\bar{x}_0^2}{Q^0} H_3(\chi_1^2(\alpha); \Delta^2)$$
$$+ \tfrac{\bar{x}_0^2}{Q^0}\Delta^2\left[2H_3(\chi_1^2(\alpha); \Delta^2) - H_5(\chi_1^2(\alpha); \Delta^2)\right]\Big\}.$$

To prove D(i) and (ii), we consider

(i) $\lim_{n\to\infty} E[\sqrt{n}(\hat{\theta}_n^S - \theta)] = \lim_{n\to\infty} E\left[\sqrt{n}(\tilde{\theta}_n - \theta) - \frac{c\sqrt{n}(\tilde{\theta}_n - \hat{\theta}_n)A_nQ_n^{1/2}}{\sqrt{n}|L_n(0)|}\right]$

$$= -c\lim_{n\to\infty} E\left[\frac{\sqrt{n}(\tilde{\theta}_n - \hat{\theta}_n)A_nQ_n^{1/2}}{\sqrt{n}|L_n(0)|}\right]$$

$$= -c\bar{x}_0\sigma E\left[\tfrac{Z}{|Z|}\right],\ Z \sim \mathcal{N}(\Delta^2, 1)$$

$$= -c\bar{x}_0\sigma[2\Phi(\Delta) - 1];$$

$$\text{(ii)} \quad \lim_{n\to\infty} E[\sqrt{n}(\hat{\theta}_n^S - \theta)^2] = \lim_{n\to\infty} E\Big[\sqrt{n}(\tilde{\theta}_n - \theta) - \frac{c\sqrt{n}(\tilde{\theta}_n - \hat{\theta}_n)A_n Q_n^{1/2}}{\sqrt{n}|L_n(0)|}\Big]^2$$

$$= \lim_{n\to\infty} E[n(\tilde{\theta}_n - \theta)^2] + c^2 \lim_{n\to\infty} E\Big[\frac{n(\tilde{\theta}_n - \hat{\theta}_n)^2 A_n^2 Q_n}{n|L_n(0)|^2}\Big]$$

$$- 2c \lim_{n\to\infty} E\Big[\frac{n(\tilde{\theta}_n - \theta)(\tilde{\theta}_n - \hat{\theta}_n)A_n Q_n^{1/2}}{\sqrt{n}|L_n(0)|}\Big].$$

First note that $\sqrt{n}(\tilde{\theta}_n - \hat{\theta}_n) = \sqrt{n}\tilde{\beta}_n \bar{x}_0 + o_p(1)$, $\sqrt{n}L_n(0) = \sqrt{n}\tilde{\beta}_n Q^0 \gamma(\psi, \phi) + o_p(1)$ and $\lim_{n\to\infty} A_n^2 = A^2\phi$, $\lim_{n\to\infty} \frac{1}{n}Q_n = Q^0$. Hence, the above reduces to

$$\sigma^2\Big(1 + \frac{\bar{x}_0^2}{Q^0}\Big) + c^2\sigma^2\frac{\bar{x}_0^2}{Q^0} - 2c\bar{x}_0 \lim_{n\to\infty} E\Big[\frac{n(\tilde{\theta}_n - \theta)\tilde{\beta}_n A_\phi}{\sqrt{Q^0}\sqrt{n}|\tilde{\beta}_n|\gamma(\psi,\phi)}\Big]$$

$$= \sigma^2\Big(1 + \frac{\bar{x}_0^2}{Q^0}\Big) + c^2\sigma^2\frac{\bar{x}_0^2}{Q^0} - 2c\sigma^2\frac{\bar{x}_0^2}{Q^0}\sqrt{\frac{2}{\pi}}\Big\{E[Z] - \Delta^2 E\Big[\frac{Z}{|Z|}\Big]\Big\}, \quad Z \sim \mathcal{N}(\Delta, 1)$$

$$= \sigma^2\Big\{1 + \frac{\bar{x}_0^2}{Q^0}\Big[1 - \frac{2}{\pi}\Big(2e^{-\Delta^2/2} - 1\Big)\Big]\Big\} \text{ for } c = \sqrt{\frac{2}{\pi}}, \sigma^2 = A_\phi^2/\gamma^2(\psi, \phi).$$

From the result of Theorem 3.10.2, we may draw the same conclusion on the properties of the estimators of β and θ as in Section 3.6.

3.11 Conclusions

In this chapter we introduced the preliminary test estimation of the intercept parameter when the slope parameter has an a priori fixed value, and we compared its performance with the usual estimator under normal theory. In Section 3.5, we introduced an alternative estimator to PTE. Also, we included the Stein estimation of the variance as a PTE of σ^2.

In Section 3.6, we presented the asymptotic theory when the underlying error distribution is nonnormal. Similarly, two-sample and one-sample problems of estimating the mean were considered in Section 3.7 and 3.8. In this context the question of whether to pool or not to pool the sample is answered via preliminary test and shrinkage estimators. In Section 3.10, we included the theory of R-estimation of regression parameters and the PTE, together with the S-estimators.

The corresponding results when the errors follow a multivariate Student's t-distribution are presented in the problem set together with other results. In this way, a whole array of model assumptions are covered for the estimation of parameters for simple linear models.

3.12 Problems

1. Let $\mathbf{Y} = \theta\mathbf{1}_n + \mathbf{e}$, $\mathbf{e} \sim N_n(\mathbf{0}, \sigma^2\mathbf{I}_n)$. Show that

   (a) $\tilde{\theta}_n = (\mathbf{1}_n'\mathbf{1}_n)^{-1}\mathbf{1}_n\mathbf{Y}$,

(b) $E(\tilde{\theta}_n) = \theta$ and $E(\tilde{\theta}_n - \theta)^2 = \frac{\sigma^2}{n}$.

(c) Let $s_e^2 = \frac{1}{n-1}(\mathbf{Y} - \tilde{\theta}_n \mathbf{1}_n)'(\mathbf{Y} - \tilde{\theta}_n \mathbf{1}_n)$. Show that $E(s_e^2) = \sigma^2$.

(d) Show that $\mathcal{L}_n = \frac{n(\tilde{\theta}_n - \theta_0)^2}{s_e^2}$ follows the noncentral F-distribution with $(1, n-1)$ d.f. and noncentrality parameter $\Delta^2/2$ where $\Delta^2 = n(\theta - \theta)^2/\sigma^2$.

2. (a) Let $\hat{\theta}_n = d\tilde{\theta}_n + (1-d)\theta_0$. Show that (i) the bias of $\hat{\theta}_n$ is $-(1-d)(\theta - \theta_0)$ and (ii) the mean-square error of $\hat{\theta}_n$ is $\frac{\sigma^2}{n}[d^2 + (1-d)^2\Delta^2]$, where $\Delta^2 = \frac{n}{\sigma^2}(\theta - \theta_0)^2$. (iii) The relative efficiency of $\hat{\theta}_n$ compared to $\tilde{\theta}_n$ is

$$\text{MRE}(\hat{\theta}_n : \tilde{\theta}_n) = [d^2 + (1-d)^2\Delta^2]^{-1}.$$

(b) Let $\hat{\theta}_n^{\text{PT}} = \hat{\theta}_n I(\mathcal{L}_n < F_{1,m}(\alpha)) + \tilde{\theta}_n I(\mathcal{L}_n \geq F_{1,m}(\alpha))$, where $F_{1,m}(\alpha)$ is the α-level critical value from the central F-distribution with $(1, n-1)$ d.f. Show that the bias of $\hat{\theta}_n^{\text{PT}}$ is $b(\hat{\theta}_n^{\text{PT}}) = (1-d)(\theta - \theta_0)G_{3,m}(\frac{1}{3}F_{1,m}(\alpha); \Delta^2)$, $m = n-1$.

(c) The MSE of $\hat{\theta}_n^{\text{PT}}$ is

$$\begin{aligned} M(\hat{\theta}_n^{\text{PT}}) = \frac{\sigma^2}{n}\Big\{1 &- (1-d^2)G_{3,m}\Big(\frac{1}{3}F_{1,m}(\alpha); \Delta^2\Big) \\ &+ (1-d)\Delta^2\Big[2G_{3,m}\Big(\frac{1}{3}F_{1,m}(\alpha); \Delta^2\Big) \\ &- (1-d)G_{5,m}\Big(\frac{1}{5}F_{1,m}(\alpha); \Delta^2\Big)\Big]\Big\}. \end{aligned}$$

(d) Compare $\hat{\theta}_n^{\text{PT}}$ and $\tilde{\theta}_n$. Show the dominance region for $\hat{\theta}_n^{\text{PT}}$.

3. Let $\mathbf{Y} = \theta \mathbf{1}_n + \beta \mathbf{x} + \mathbf{e}$ as in Section 3.1. It is suspected that β may be equal to β_0. Define the shrinkage estimator $\hat{\beta}_n^{(d)} = d\tilde{\beta}_n + (1-d)\beta_0$ of β, $(0 < d < 1$ and d is known), where d is the *coefficient of distrust* on the prior $\beta = \beta_0$.

(a) Find the *bias, quadratic bias,* and *mse* of the estimator $\hat{\beta}_n^{(d)}$.

(b) Compare the estimator $\hat{\beta}_n^{(d)}$ with respect to $\tilde{\beta}_n$ using the bias and mse. (Find the efficiency of $\hat{\beta}_n^{(d)}$ with respect to $\tilde{\beta}_n$ and compare.)

(c) Let $\hat{\beta}_n^{\text{PT}} = \tilde{\beta}_n - \tilde{\beta}_n I(\mathcal{L}_n < F_{1,m}(\alpha))$ as in Section 3.2.1. Find the bias and mse of $\hat{\beta}_n^{\text{PT}}$.

4. Consider the problem of estimating the intercept parameter in the model

$$\mathbf{Y} = \theta \mathbf{1}_n + \beta \mathbf{x} + \mathbf{e}$$

when it is suspected that β may be β_0. Define the following estimators of θ:

(i) $\tilde{\theta}_n = \overline{Y} - \tilde{\beta}_n \overline{x}$ and $\hat{\theta}_n = \overline{Y} - \beta_0 \overline{x}$.

(ii) $\hat{\theta}_n^{(d)} = \overline{Y} - [d\tilde{\beta}_n + (1-d)\beta_0]\overline{x}$ $(0 < d < 1$, known).

(iii) $\hat{\theta}_n^{(\mathrm{d})\mathrm{pt}} = \hat{\theta}_n^{(d)} I(\mathcal{L}_n < F_{1,m}(\alpha)) + \tilde{\theta}_n I(\mathcal{L}_n \geq F_{1,m}(\alpha)), m = n-2,$

where $\mathcal{L}_n = (\tilde{\beta}_n - \beta_0)^2 \frac{Q}{s_e^2}$, where Q and s_e^2 are defined by (3.1.3) and (3.1.4), respectively, and $F_{1,m}(\alpha)$ is the α-level critical value of a central F-distribution with $(1, n-2)$ d.f.

(a) Find the bias, quadratic bias, and mse of the three estimators.

(b) Compare the estimators using the mse criterion.

5. Let $\mathbf{x} = (0, \dots, 0; 1, \dots, 1)'$ with n_1 zero's and n_2 1's. Then we have the two-sample problem, using the model $\mathbf{Y} = \theta \mathbf{1}_n + \beta\mathbf{x} + \mathbf{e}$. Then $\mu_1 = \theta$ and $\mu_2 = \theta + \beta$ and $\mu_2 - \mu_1 = \beta$. It is suspected that β may be equal to a known value β_0. Then three estimators of μ_1 may be

$$\text{(i)}\ \ \tilde{\mu}_1 = \overline{y}_1, \qquad \text{(ii)}\ \hat{\mu}_1^{(d)} = d\tilde{\mu}_1 + (1-d)(\overline{y}_2 - \beta_0),$$

where $\overline{y}_1$ and $\overline{y}_2$ are the means of the first n_1 y-values and n_2 y-values, respectively.

(a) Find the bias and mse of $\hat{\mu}_1^{(d)}$.

(b) Further, let $\hat{\mu}_1^{\mathrm{PT}} = \tilde{\mu}_1 I(\mathcal{L}_n \geq F_{1,m}(\alpha)) + \hat{\mu}_1^{s} I(\mathcal{L}_n < F_{1,m}(\alpha))$, where

$$\mathcal{L}_n = \frac{n_1 n_2}{n_1+n_2} \frac{(\overline{y}_2 - \overline{y}_1 - \beta_0)^2}{s_y^2}, \quad s_p^2 = \frac{1}{(n_1+n_2-2)} \sum_{i=1}^{2}\sum_{j=1}^{n_i} (y_{ij} - \overline{y}_i)^2,$$

and $m = n_1 + n_2 - 2$. Find the bias and mse of $\hat{\mu}_1^{(\mathrm{d})\mathrm{pt}}$ and compare with respect to $\tilde{\mu}_1$.

6. Consider the n-dimensional normal distribution $N_n(\mathbf{0}, \tau^2 \mathbf{I}_n)$ and the Inverted Gamma distribution $IG(\nu_0, \sigma^2)$ defined by

$$g(\nu_0, \sigma^2) = \frac{2}{\Gamma(\frac{\nu_0}{2})} \left(\frac{\nu_0\sigma^2}{2}\right)^{\frac{\nu_0}{2}} \tau^{-(\nu_0+1)} e^{-\frac{\nu_0\sigma^2}{2\tau^2}}; \quad \nu_0,\ \sigma,\ \tau > 0.$$

(a) Show that the mixture of these two distributions is the n-dimensional Student's t-distributions $M_t(\mathbf{0}, \sigma_e^2 I_n)$ defined by the density

$$f(\mathbf{e}, \nu_0, \sigma^2) = \frac{\Gamma(\frac{n+\nu_0}{2})}{(\pi\nu_0)^{n/2}\Gamma(\frac{\nu_0}{2})\sigma^n} \left(1 + \frac{\mathbf{e}'\mathbf{e}}{\nu_0\sigma^2}\right)^{-n+\nu_0/2},$$

$\nu_0,\ \sigma > 0, -\infty < e_i < \infty$.

(b) Show that

$$E(\mathbf{e}) = \mathbf{0} \quad \text{and} \quad E(\mathbf{e}\mathbf{e}') = \sigma_e^2 \mathbf{I}_n, \ \ \sigma_e^2 = \frac{\nu_0\sigma^2}{\nu_0 - 2} \quad (\nu_0 > 2).$$

7. Consider the simple linear model

$$\mathbf{Y} = \theta \mathbf{1}_n + \beta \mathbf{x} + \mathbf{e},$$

where $\mathbf{e} \sim M_t(\mathbf{0}, \sigma_e^2 \mathbf{I}_n)$.

(a) Show that the MLE of θ and β are the same as the least squares estimators, and the unbiased estimator of σ_e^2 is $s_e^2 = \frac{1}{n-2}(\mathbf{Y} - \tilde{\theta}_n \mathbf{1}_n - \tilde{\beta}_n \mathbf{x})'(\mathbf{Y} - \tilde{\theta}_n \mathbf{1}_n - \tilde{\beta}_n \mathbf{x})$.

(b) The restricted estimator of θ is $\hat{\theta}_n = \overline{Y} - \beta_0 \overline{x}$ when it is suspected that β may be equal to β_0. Find its bias and mse.

8. Suppose $\nu_0 > 2$ and $\mathbf{e} \sim M_t(\mathbf{0}, \sigma_e^2 \mathbf{I}_n)$ for the linear model in Problem 7.

(a) Show that the LR test of the null hypothesis $H_0 : \beta = \beta_0$ versus $H_A : \beta \neq \beta_0$ is given by test-statistic

$$\mathcal{L}_n = \frac{(\tilde{\beta}_n - \beta_0)^2 Q}{s_e^2}, \quad Q = \mathbf{x}'\mathbf{x} - \frac{1}{n}(\mathbf{1}_n'\mathbf{x}).$$

(b) Show that under H_0, $\mathcal{L}_n$ follows the central F-distribution with $(1, m)$ d.f. $(m = n - 2)$, while under H_A, it follows the distribution defined by the pdf

$$g(\mathcal{L}_n; \Delta^{*2}, \nu_0) = \sum_{r \geq 0} (n-2)^{r+(1/2)}$$
$$\times \frac{\Gamma(\frac{\nu_0}{2} + r)\Gamma(\frac{n-1}{2} + r)(\frac{\Delta^{*2}}{\nu_0 - 2})^r (\mathcal{L}_n)^{r-(1/2)}}{\Gamma(r+1)\Gamma(\frac{\nu_0}{2})\Gamma(\frac{1}{2} + r)\Gamma(\frac{n-2}{2})(1 + \frac{\Delta^{*2}}{\nu_0 - 2})^{\nu_0/2 + r}(1 + \frac{\mathcal{L}_n}{n-2})^{[(n-1)/2]+r}}$$

(which is not a noncentral F-distribution) where $\Delta^{*2} = \frac{\delta^2 Q}{\sigma_e^2}$.

9. Define the preliminary test estimator of the intercept parameter as in (3.2.4)

$$\hat{\theta}_n^{\rm PT} = \tilde{\theta}_n - (\tilde{\theta}_n - \hat{\theta}_n) I(\mathcal{L}_n < F_{1,m}(\alpha)).$$

(a) Show that the bias and MSE expressions are

(i) $b_1(\tilde{\theta}_n) = 0$, and $M_1(\tilde{\theta}_n) = \sigma_e^2 \left(\frac{1}{n} + \frac{\overline{x}^2}{Q}\right)$.

(ii) $b_2(\hat{\theta}_n) = -\delta \overline{x}$, and $M_2(\hat{\theta}_n) = \sigma_e^2 \left(\frac{1}{n} + \frac{\overline{x}^2}{Q}\Delta^{*2}\right)$, $\delta = \beta - \beta_0$.

(iii) $b_3(\hat{\theta}_n^{\rm PT}) = -\delta G_{3,m}^{(2)}\left(\frac{1}{3}F_{1,m}(\alpha); \Delta^{*2}\right)$, $m = n - 2$, and

$$M_3(\hat{\theta}_n) = \sigma_e^2 \left\{ \left(\frac{1}{n} + \frac{\overline{x}^2}{Q}\right) - \frac{\overline{x}^2}{Q} G_{3,m}^{(1)}\left(\frac{1}{3}F_{1,m}(\alpha); \Delta^{*2}\right) \right.$$
$$+ \Delta^{*2}\frac{\overline{x}^2}{Q}\left[2G_{3,m}^{(2)}\left(\frac{1}{3}F_{1,m}(\alpha); \Delta^{*2}\right)\right.$$
$$\left.\left. - G_{5,m}^{(2)}\left(\frac{1}{5}F_{1,m}(\alpha); \Delta^{*2}\right)\right]\right\},$$

where

$$G_{1+2i,m}^{(j)}\left(\frac{1}{1+2i}F_{1,m}(\alpha);\Delta^{*2}\right)$$

$$=\sum_{r\geq 0}\frac{\Gamma\left(\frac{\nu_0}{2}+r+j-2\right)}{\Gamma(r+1)\Gamma\left(\frac{\nu_0}{2}+j-2\right)}\frac{(\Delta^{*2}/\nu_0-2)^r}{\left(1+\frac{\Delta^{*2}}{\nu_0-2}\right)^{\frac{\nu_0}{2}+r+j-2}}I_{x_0}$$

$$\times\left[\frac{1}{2}(1+2i),\frac{m}{2}\right],$$

where $x_0=\frac{F_{1,m}(\alpha)}{m+F_{1,m}(\alpha)}$; and $j=1,2$.

(b) Compare the three estimators, namely, $\tilde{\theta}_n$, $\hat{\theta}_n$, and $\hat{\theta}_n^{\text{PT}}$ of θ and conclude that the performance of $\hat{\theta}_n$ and $\hat{\theta}_n^{\text{PT}}$ compared to $\tilde{\theta}_n$ is similar to that under the normal theory.

10. (Two-Sample Problem). Consider the simple linear model in Problem 2, and assume the vector $\mathbf{x}$ to be $(0,\ldots,0;1,\ldots,1)'$ with n_1 zero entries and n_2 entries of 1's. Assume $\mathbf{e}\sim M_t(\mathbf{0},\sigma_e^2 I_n)$, $n=n_1+n_2$.

(a) Show that the MLE of μ_1 and μ_2 are $\tilde{\mu}_1=\bar{y}_1$ and $\tilde{\mu}_2=\bar{y}_2$, respectively.

(b) Show that the LR test for $\beta=\mu_2-\mu_1=0$ is given by statistic

$$\mathcal{L}_n=\frac{n_1n_2(\bar{y}_2-\bar{y}_1)^2}{(n_1+n_2)s_p^2}.$$

(c) Show that $\mathcal{L}_n$ follows the central F-distribution with $(1,n_2+n_2-2)$ d.f. under H_0 and under $H_A:\mu_1\neq\mu_1$, it follows the distribution whose pdf is defined by $g(\mathcal{L}_n;\Delta^{*2},\nu_0)$ given in Problem 8(b) with

$$\mathcal{L}_n=\frac{n_1n_2}{n_1+n_2}\frac{(\bar{y}_2-\bar{y}_1)^2}{s_p^2},\quad\Delta^{*2}=\frac{n_1n_2}{n_1+n_2}\frac{(\mu_2-\mu_1)^2}{\sigma_e^2}.$$

(d) Consider the preliminary test estimator of μ_1 as

$$\hat{\mu}_1^{\text{PT}}=\tilde{\mu}_1-(\tilde{\mu}_1-\hat{\mu}_1)I(\mathcal{L}_n<F_{1,m}(\alpha))$$

where $F_{1,m}(\alpha)$ is the upper α-level critical value of the central F-distribution with $(1,n_1+n_2-2)$ d.f. Show that the bias and mse expressions of $\tilde{\mu}_1,\hat{\mu}_1$, and $\hat{\mu}_1^{\text{PT}}$ are given by

(i) $b_1(\tilde{\mu}_1)=0$, $B_1(\tilde{\mu}_1)=0$ and $M_1(\tilde{\mu}_1)=\frac{\sigma_e^2}{n_1}$, $\sigma_e^2=\frac{\nu_0\sigma^2}{\nu_0-2}$, $\nu_0>2$.

(ii) $b_2(\hat{\mu}_1)=-\frac{n_2}{n_1+n_2}(\mu_2-\mu_1)$, $B_2(\hat{\mu}_1)=\frac{n_2}{n_1(n_1+n_2)}\Delta^{*2}$, $\Delta^{*2}=\frac{n_1n_2}{n_1+n_2}\frac{(\mu_2-\mu_1)^2}{\sigma_e^2}$ and $M_2(\hat{\mu}_1)=\frac{\sigma_e^2}{n_1+n_2}\left(1+\frac{n_2}{n_1}\Delta^{*2}\right)$.

(iii) $b_3(\hat{\mu}_1^{\text{PT}} = -\frac{n_2}{n_1+n_2}(\mu_2 - \mu_1)G_{3,m}^{(2)}\left(\frac{1}{3}F_{1,m}(\alpha); \Delta^{*2}\right)$, $m = n_1 + n_2 - 2$,

$B_3(\hat{\mu}_1^{\text{PT}} = \frac{n_2}{n_1(n_1+n_2)}\Delta^{*2}\left\{G_{3,m}^{(2)}\left(\frac{1}{3}F_{1,m}(\alpha); \Delta^{*2}\right)\right\}^2$, and

$$\begin{aligned}M_3(\hat{\mu}_1^{\text{PT}}) = \frac{\sigma_e^2}{n_1} - \frac{\sigma_e^2 n_2}{n_1(n_1+n_2)}G_{3,m}^{(1)}\left(\frac{1}{3}F_{1,m}(\alpha); \Delta^{*2}\right) \\ + \sigma_e^2\Delta^{*2}\frac{n_2}{n_1+n_2}\left\{2G_{3,m}^{(2)}\left(\frac{1}{3}F_{1,m}(\alpha); \Delta^{*2}\right)\right. \\ \left. - G_{5,m}^{(2)}\left(\frac{1}{5}F_{1,m}(\alpha); \Delta^{*2}\right)\right\},\end{aligned}$$

where $G_{1+2i,m}^{(j)}\left(\frac{1}{1+2i}F_{1,m}(\alpha); \Delta^{*2}\right)$ has the same definition as in Problem 9a(iii).

11. (Application to Survey Sampling). Let $\mathcal{P} = \{1, \ldots, N\}$ denote a finite population of N units, where N is known. Associated with the kth unit of $\mathcal{P}$, there is a vector $(y_k, x_k)'$, $k = 1, \ldots, N$, assumed to be a random sample from the bivariate normal distribution. Assume that the population $\mathcal{P}$ was generated according to the model

$$\mathbf{Y} = \theta\mathbf{1}_N + \beta\mathbf{x} + \mathbf{e}, \quad E(\mathbf{e}_x) = \mathbf{0},$$

$$\text{Cov}(\mathbf{e}) = \sigma^2\mathbf{I}_N, \quad \text{and} \quad \mathbf{e} \sim \mathcal{N}_N(\mathbf{0}, \sigma_e^2\mathbf{I}_N).$$

To gain information about the population total $= \sum_{k=1}^N Y_k$, a sample s of n units is selected from $\mathcal{P}$ according to some specified plan. Thus, we are using a superpopulation model in survey sampling. Let $r = N - s$, the unobserved part of the population. The following predictor is considered:

$$\hat{T}_G = N\overline{Y}_s + (N - n)(\overline{x}_r - \overline{x}_s)\hat{\beta}_G,$$

where $\overline{Y}_s = \frac{1}{n}\sum_{i\in s} y_i$, $\overline{x}_r = \frac{1}{N-n}\sum_{j=s} x_j$, $\overline{x}_s = \frac{1}{n}\sum_{i\in s} x_i$ and $\hat{\beta}_G$ stands for the generic estimator. Let

(i) $$\tilde{\beta}_n = \frac{\mathbf{x}'\mathbf{y} - \frac{1}{n}(\mathbf{1}_n'\mathbf{x})(\mathbf{1}_n'\mathbf{y})}{\mathbf{x}'\mathbf{x} - \frac{1}{n}(\mathbf{1}_n\mathbf{x})^2};$$

(ii) $$\hat{\beta}^{\text{PT}} = \tilde{\beta}_n - (\tilde{\beta}_n - \beta_0)I(\mathcal{L}_n < F_{1,m}(\alpha))$$

$$\mathcal{L}_n = \frac{(\tilde{\beta}_n - \beta_0)^2 Q}{s_e^2}, \quad Q = \mathbf{x}'\mathbf{x} - \frac{1}{n}(\mathbf{1}_n\mathbf{x})^2,$$

$$s_e^2 = \frac{1}{n-1}(\mathbf{Y} - \tilde{\theta}_n\mathbf{1}_n - \tilde{\beta}_n\mathbf{x})'(\mathbf{Y} - \tilde{\theta}_n\mathbf{1}_n - \beta_n^2\mathbf{x}).$$

(a) Show that the bias and mse expressions of $\tilde{T}_n$ and $\hat{T}_n^{\rm PT}$ are

(i) $b(\tilde{T}_n) = 0$ and $b(\hat{T}_n^{\rm PT} = -(N-n)\beta_1 G_{3,m}\left(\frac{1}{3}F_{1,m}(\alpha); \Delta^2\right);$

(ii) $M_1(\tilde{T}_n) = N\frac{1}{f}(1-f)\sigma^2\left[1 + Nf(1-f)\frac{(\overline{x}_r - \overline{x}_s)^2}{Q}\right]$

and

$$M_2(\hat{T}_n^{\rm PT}) = N\frac{1}{f}(1-f)\sigma^2\left[1 + Nf(1-f)\frac{(\overline{x}_r - \overline{x}_s)^2}{Q}\right.$$
$$\times\left\{1 - G_{3,m}\left(\frac{1}{3}F_{1,m}(\alpha); \Delta^2\right)\right\}$$
$$\left. + \Delta^2\left\{2G_{3,m}(\frac{1}{3}F_{1,m}(\alpha); \Delta^2) - G_{5,m}(\frac{1}{5}F_{1,m}(\alpha); \Delta^2)\right\}\right].$$

(b) If, in the model

$$\mathbf{Y} = \theta\mathbf{1}_N + \beta\mathbf{x} + \mathbf{e},$$

$\mathbf{e}$ is distributed as $\mathcal{M}_t(\mathbf{0}, \sigma_e^2\mathbf{I}_N)$ as in Problem 6(a) show that

(i) $\mathbf{b}(\tilde{T}_n) = 0$ and $b(\hat{T}_n^{\rm PT}) = -(N-n)\beta_1 G_{3,m}^{(2)}(\frac{1}{3}F_{1,m}(\alpha); \Delta^{*2});$

(ii) $M_1(\tilde{T}_n) = N\frac{1}{f}(1-f)\sigma_e^2\left[1 + Nf(1-f)\frac{(\overline{x}_r - \overline{x}_s)^2}{Q}\right]$ and

$$M_2(\hat{T}_n^{\rm PT}) = N\frac{1}{f}(1-f)\sigma_e^2\left[1 + Nf(1-f)\frac{(\overline{x}_r - \overline{x}_s)^2}{Q}\right.$$
$$\times\left\{1 - G_{3,m}^{(1)}\left(\frac{1}{3}F_{1,m}(\alpha); \Delta^{*2}\right)\right\}$$
$$\left. + \Delta^{*2}\left\{2G_{3,m}^{(2)}\left(\frac{1}{3}F_{1,m}(\alpha); \Delta^{*2}\right) - G_{5,m}^{(2)}\left(\frac{1}{5}F_{1,m}(\alpha); \Delta^{*2}\right)\right\}\right],$$

where the G-functions are defined in Problem 9a(iii).

12. Prove Theorem 4 with respect to the estimate of β.

13. Refer to Section 3.10 (a) Show that the exact distribution of $T_n(0,0)$ and $L_n(0)$ is symmetric of 0. (b) Show that the R-estimators $\hat{\theta}_n$ and $\tilde{\beta}_n$ are invariant, robust, and consistent estimators of θ and β, respectively.

14. Refer to Section 3.10.3.

    (a) Find the asymptotic efficiency of $\tilde{\beta}_n$ and relative to the LSE of β.

    (b) Find the asymptotic efficiency of $\hat{\theta}_n$ and relative to the LSE of θ.

    (c) Find the joint asymptotic efficiency of $(\hat{\theta}_n, \tilde{\beta}_n)$ relative to the LSE of (θ, β).

Chapter 4

Stein-Type Estimation

Outline

Stein-type estimation (also called **shrinkage** estimation) was introduced by Stein(1956) and James and Stein (1961) in the statistical literature. The approach by Stein (1955, 1956) combines "uncertain prior information" on the parameters of interest and the sample observations from a Multiparameter statistical model. It is more sophisticated and improved than its pre-cursor, the **preliminary test estimation.** The application needs a parameter space with more than two dimensions. As pointed out in Section 3.5, the preliminary test estimation produces (1) only two values of the estimator, namely, the **unrestricted** estimator and the **restricted** estimator, and (2) depends heavily on the level of significance of the preliminary test. Stein-type estimators produce all possible values in between the unrestricted and restricted estimators depending on the sample values of the test-statistic used for the preliminary test, which shrinks toward the target vector parameter or its estimator. There is a substantial gain in the use of Stein-type estimators as the improvement carries over to the corresponding set-estimation problems.

In this chapter we introduce the Stein-type estimation of the mean vector of a standard p-variate normal distribution. The usual maximum likelihood estimation and test of hypothesis for the mean vector are presented. These results are then utilized to define (1) preliminary test estimator and (2) Stein-type estimators of the mean vector, which includes the James-Stein and the positive rule estimators. Risk difference and empirical Bayes approaches are presented to derive the Stein-type estimators, beginning with the preliminary test approach. Comparisons of the preliminary test and Stein-type estimators are made relative to the maximum likelihood estimator based on the MSE matrices and the weighted risks under quadratic loss functions. Some asymptotic theory is presented for nonnormal distributions together with the R-estimation in a nonparametric setup. For the preliminary test estimators, tables are presented to determine optimal size of the preliminary test. We also discuss recentered confidence sets with positive-rule Stein estimators for the normal model as well as for the nonnormal model and provide asymptotic domination of the recentered confidence set. Finite sample analysis of the efficiencies of the Stein-type estimators relative to the standard estimators are given along with the graphs of bias and efficiencies. Finally, problems are added to expand the results to multivariate t-distributions in addition to the related results.

4.1 Statistical Model, Estimation, and Tests

Consider the simple location model

$$\mathbf{Y}_i = \boldsymbol{\theta} + \mathbf{e}_i, \quad i = 1, \dots, n, \tag{4.1.1}$$

where $\mathbf{Y}_i = (Y_{1i}, \dots, Y_{pi})'$ is the ith observation vector, $\mathbf{e}_i = (e_{1i}, \dots, e_{pi})'$ is the ith error vector and $\boldsymbol{\theta} = (\theta_1, \dots, \theta_p)'$ is the vector parameter of the model. The primes mean transposition, as usual. It is assumed that

$$\mathbf{e}_i \sim \mathcal{N}_p(\mathbf{0}, \mathbf{I}_p), \quad i = 1, \dots, n. \tag{4.1.2}$$

Using the information (4.1.1) and (4.1.2), we obtain the maximum likelihood estimator (MLE) of $\boldsymbol{\theta}$ as

$$\tilde{\boldsymbol{\theta}}_n = \overline{\mathbf{Y}} = (\overline{Y_1}, \dots, \overline{Y_p})', \tag{4.1.3}$$

where $\overline{\mathbf{Y}}$ is the sample mean vector and

$$\tilde{\boldsymbol{\theta}}_n = \overline{\mathbf{Y}} \sim \mathcal{N}_p\left(\boldsymbol{\theta}, \frac{1}{n}\mathbf{I}_p\right). \tag{4.1.4}$$

Clearly, the bias vector, $\mathbf{b}_1(\tilde{\boldsymbol{\theta}}_n) = \mathbf{0}$ the quadratic bias, $B_1(\tilde{\boldsymbol{\theta}}_n) = 0$, and the normalized MSE matrix is $\mathbf{M}_1(\tilde{\boldsymbol{\theta}}_n) = E\{n(\tilde{\boldsymbol{\theta}}_n - \boldsymbol{\theta})(\tilde{\boldsymbol{\theta}}_n - \boldsymbol{\theta})'\} = \mathbf{I}_p$. Finally, under the quadratic loss function

$$L(\boldsymbol{\theta}^*, \boldsymbol{\theta}) = n\|\boldsymbol{\theta}^* - \boldsymbol{\theta}\|^2_{\mathbf{W}} = n(\boldsymbol{\theta}^* - \boldsymbol{\theta})'\mathbf{W}(\boldsymbol{\theta}^* - \boldsymbol{\theta}) \tag{4.1.5}$$

the weighted risk of $\tilde{\boldsymbol{\theta}}_n$ is

$$R_1(\tilde{\boldsymbol{\theta}}_n; \mathbf{W}) = \mathrm{tr}[\mathbf{M}_1(\tilde{\boldsymbol{\theta}}_n)\mathbf{W}], \quad \text{and} \quad \mathrm{tr}[\mathbf{M}_1(\tilde{\boldsymbol{\theta}}_n)\mathbf{W}] = p \text{ if } \mathbf{W} = \mathbf{I}_p. \tag{4.1.6}$$

Notice that $\tilde{\boldsymbol{\theta}}_n$ consists of the marginal sample means $\tilde{\theta}_{in} = \overline{Y}_i$, $i = 1, 2, \ldots, p$, as its components (which are independent).

For the test of the null hypothesis $H_0 : \boldsymbol{\theta} = \boldsymbol{\theta}_0$ against $H_A : \boldsymbol{\theta} \neq \boldsymbol{\theta}_0$, the likelihood ratio test yields the statistic

$$\mathcal{L}_n = n\|\tilde{\boldsymbol{\theta}}_n - \boldsymbol{\theta}_0\|^2. \tag{4.1.7}$$

Under H_0, $\mathcal{L}_n$ has a central chi-square distribution with p d.f. If $\boldsymbol{\theta} \neq \boldsymbol{\theta}_0$, $\mathcal{L}_n$ has the noncentral chi-square distribution with p d.f. and noncentrality parameter $\Delta^2/2$, where

$$\Delta^2 = n\|\boldsymbol{\theta} - \boldsymbol{\theta}_0\|^2. \tag{4.1.8}$$

The estimator $\tilde{\boldsymbol{\theta}}_n$ stands out because of the following optimality properties: it is a

(i) best unbiased estimator,
(ii) best equivariant estimator,
(iii) maximum likelihood estimator,
(iv) minimax estimator.

In this chapter we utilize the expressions of the MSE matrices and risks to evaluate the performance of various estimators. We also consider the MSE and risk based efficiencies. At this stage it is useful to remind the readers of the definition of inadmissibility of an estimator $\boldsymbol{\theta}_n^*$: An estimator $\boldsymbol{\theta}_n^*$ is inadmissible for $\boldsymbol{\theta}$ if there exists an estimator $\hat{\boldsymbol{\theta}}_n$ such that

$$R_1(\hat{\boldsymbol{\theta}}_n; \mathbf{W}) \leq R_2(\boldsymbol{\theta}_n^*; \mathbf{W})$$

for every value of $\boldsymbol{\theta}$ and strict inequality

$$R_1(\hat{\boldsymbol{\theta}}_n; \mathbf{W}) < R_2(\boldsymbol{\theta}_n^*; \mathbf{W}) \tag{4.1.9}$$

holds for some $\boldsymbol{\theta} \in \Omega$. In terms of MSE matrices we then have

$$\left|\mathbf{M}_1(\hat{\boldsymbol{\theta}}_n)\right| \leq |\mathbf{M}_2(\boldsymbol{\theta}_n^*)|$$

for every value of $\boldsymbol{\theta}$, and for some $\boldsymbol{\theta} \in \Omega$, strict inequality holds, meaning,

$$|\mathbf{M}_1(\boldsymbol{\theta}_n^*)| < |\mathbf{M}_2(\boldsymbol{\theta}_n^*)| . \tag{4.1.10}$$

In order to compute the bias vector, quadratic bias, MSE matrix, and the weighted risk of an estimator of the type

$$\boldsymbol{\theta}_n^* = \tilde{\boldsymbol{\theta}}_n \phi(n\|\tilde{\boldsymbol{\theta}}_n\|^2), \tag{4.1.11}$$

where ϕ is a real-valued function and $\tilde{\boldsymbol{\theta}}_n \sim \mathcal{N}_p(\boldsymbol{\theta}, n^{-1}\mathbf{I}_p)$, we use the Stein's (1956) formula given in the theorem below assuming $\boldsymbol{\theta}_0 = \mathbf{0}$ w.l.g.

Theorem 1. Under the assumed regularity conditions, we have the bias vector and the quadratic bias of $\boldsymbol{\theta}_n^*$ given by

(i) $\mathbf{b}(\boldsymbol{\theta}_n^*) = -\boldsymbol{\theta}\, E\big[1 - \phi(\chi^2_{p+2}(\Delta^2)\big]$

and

$$\text{(ii)}\quad \mathbf{B}(\boldsymbol{\theta}_n^*) = -\Delta^2 \left\{E\big[1 - \phi(\chi^2_{p+2}(\Delta^2)\big]\right\}^2, \quad \Delta^2 = n\|\boldsymbol{\theta}\|^2. \tag{4.1.12}$$

The MSE matrix and the weighted risk expressions under the loss function (4.1.5) are given by

$$\begin{aligned}\text{(iii)}\quad \mathbf{M}(\boldsymbol{\theta}_n^*) &= \mathbf{I}_p\, E\big[\phi^2(\chi^2_{p+2}(\Delta^2))\big] + n\boldsymbol{\theta}\boldsymbol{\theta}' E\big[\phi^2(\chi^2_{p+4}(\Delta^2))\big] \\ &\quad - 2n\boldsymbol{\theta}\boldsymbol{\theta}' E\big[\phi(\chi^2_{p+2}(\Delta^2))\big] + n\boldsymbol{\theta}\,\boldsymbol{\theta}'\end{aligned} \tag{4.1.13}$$

and

$$\begin{aligned}\text{(iv)}\quad R(\boldsymbol{\theta}_n^*; \mathbf{W}) &= \mathrm{tr}(\mathbf{W}) E\big[\phi^2(\chi^2_{p+2}(\Delta^2))\big] \\ &\quad + n\boldsymbol{\theta}'\mathbf{W}\boldsymbol{\theta}\Big\{1 - 2E\big[\phi^2(\chi^2_{p+2}(\Delta^2))\big] + E\big[\phi^2(\chi^2_{p+4}(\Delta^2))\big]\Big\} \\ &= pE\big[\phi^2(\chi^2_{p+2}(\Delta^2))\big] + \Delta^2\Big\{1 - 2E\big[\phi^2(\chi^2_{p+2}(\Delta^2))\big] \\ &\quad + E\big[\phi(\chi^2_{p+4}(\Delta^2))\big]\Big\} \quad \text{if } \mathbf{W} = \mathbf{I}.\end{aligned} \tag{4.1.14}$$

Proof.

$$\begin{aligned}\mathbf{b}(\boldsymbol{\theta}_n^*) &= E[\tilde{\boldsymbol{\theta}}_n \phi(n\|\tilde{\boldsymbol{\theta}}_n\|^2) - \boldsymbol{\theta}] \\ &= \boldsymbol{\theta}\, E\big[\phi(\chi^2_{p+2}(\Delta^2))\big] - \boldsymbol{\theta} \quad \text{by Theorem 2.1.6} \\ &= -\boldsymbol{\theta}\big\{1 - E\big[\phi(\chi^2_{p+2}(\Delta^2))\big]\big\}\end{aligned}$$

and

$$B(\boldsymbol{\theta}_n^*) = n\boldsymbol{\theta}'\boldsymbol{\theta}\Big\{1 - E\big[\phi(\chi^2_{p+2}(\delta^2))\big]\Big\}^2 = \Delta^2\{1 - E[\phi(\chi^2_{p+2}(\Delta^2))]\}^2. \tag{4.1.15}$$

Similarly,

$$\begin{aligned}\mathbf{M}(\boldsymbol{\theta}_n^*) &= nE\Big\{\big[\tilde{\boldsymbol{\theta}}_n\phi(n\|\tilde{\boldsymbol{\theta}}_n\|^2) - \boldsymbol{\theta}\big]\big[\tilde{\boldsymbol{\theta}}_n\phi(n\|\tilde{\boldsymbol{\theta}}_n\|^2) - \boldsymbol{\theta}\big]'\Big\} \\ &= E\big[n\tilde{\boldsymbol{\theta}}_n\tilde{\boldsymbol{\theta}}_n'\phi^2(n\|\tilde{\boldsymbol{\theta}}_n\|^2)\big] - 2n\boldsymbol{\theta}E\big[\tilde{\boldsymbol{\theta}}_n'\phi(n\|\tilde{\boldsymbol{\theta}}_n\|^2)\big] + n\boldsymbol{\theta}\boldsymbol{\theta}' \\ &= \mathbf{I}_p E\big[\phi^2(\chi^2_{p+2}(\Delta^2)\big] + n\boldsymbol{\theta}\boldsymbol{\theta}' E\big[\phi^2(\chi^2_{p+4}(\Delta^2))\big] \\ &\quad - 2n\boldsymbol{\theta}\boldsymbol{\theta}' E\big[\phi(\chi^2_{p+2}(\Delta^2))\big] + n\boldsymbol{\theta}\boldsymbol{\theta}' \\ &= \mathbf{I}_p E\big[\phi^2(\chi^2_{p+2}(\Delta^2))\big] \\ &\quad + n\boldsymbol{\theta}\boldsymbol{\theta}'\,\{1 - 2E[\phi(\chi^2_{p+2}(\Delta^2))] + E\big[\phi^2(\chi^2_{p+4}(\Delta^2))\big]\}\end{aligned} \tag{4.1.16}$$

by Theorems 2.1.6 and 2.1.7.

The weighted risk expression then follows as

$$
\begin{aligned}
R(\boldsymbol{\theta}_n^*; \mathbf{W}) &= \operatorname{tr}(\mathbf{W})E[\phi(\chi^2_{p+2}(\Delta^2))] \\
&\quad + n\boldsymbol{\theta}'\mathbf{W}\boldsymbol{\theta}\left\{1 - 2E[\phi(\chi^2_{p+2}(\Delta^2))] + E[\phi^2(\chi^2_{p+4}(\Delta^2))]\right\} \\
&= pE[\phi(\chi^2_{p+2}(\Delta^2))] + \Delta^2\{1 - 2E[\chi^2_{p+2}(\Delta^2)\} + E[\phi^2(\chi^2_{p+4}(\Delta^2)] \\
&\quad \text{if } \mathbf{W} = \mathbf{I}. \qquad (4.1.17)
\end{aligned}
$$

This theorem will be used in the sequel to obtain relevant results. Our choices of $\phi(n\|\tilde{\boldsymbol{\theta}}_n\|^2)$ will be the following:

$$
\begin{aligned}
\phi(n\|\tilde{\boldsymbol{\theta}}_n\|^2) &= I\left(n\|\tilde{\boldsymbol{\theta}}_n\|^2 > \chi^2_p(\alpha)\right) \\
&= \left(1 - \frac{(p-2)}{n\|\tilde{\boldsymbol{\theta}}_n\|^2}\right) \\
&= \left(1 - \frac{(p-2)}{n\|\tilde{\boldsymbol{\theta}}_n\|^2}\right) I(n\|\tilde{\boldsymbol{\theta}}_n\|^2 > \chi^2_p(\alpha))
\end{aligned}
$$

and

$$
= \left(1 - \frac{(p-2)}{n\|\tilde{\boldsymbol{\theta}}_n\|^2}\right) I(n\|\tilde{\boldsymbol{\theta}}_n\|^2 > p-2). \qquad (4.1.18)
$$

4.2 Preliminary Test Estimation

One aspect of $\tilde{\boldsymbol{\theta}}_n$ is that it ignores any relationship among the components of $\boldsymbol{\theta} = (\theta_1, \dots, \theta_p)'$. If this question is raised, then the entire estimation procedure may have to be modified. As a start, we consider that the relationship among the components of $\boldsymbol{\theta}$ is **defined** by the null hypothesis $H_0 : \boldsymbol{\theta} = \mathbf{0}$, meaning, $\boldsymbol{\theta}$ lies in the hyperplane defined by $\boldsymbol{\theta} = \mathbf{0}$. So the restricted estimator of $\boldsymbol{\theta}$ is $\hat{\boldsymbol{\theta}}_n = \mathbf{0}$. In this case, we consider the **preliminary test estimator** (PTE) of $\boldsymbol{\theta}$ as

$$
\begin{aligned}
\hat{\boldsymbol{\theta}}_n^{\text{PT}} &= \mathbf{0}\, I(n\|\tilde{\boldsymbol{\theta}}_n\|^2 < \chi^2_p(\alpha)) + \tilde{\boldsymbol{\theta}}_n I\left(n\|\tilde{\boldsymbol{\theta}}_n\|^2 > \chi^2_p(\alpha)\right) \\
&= \tilde{\boldsymbol{\theta}}_n - \tilde{\boldsymbol{\theta}}_n I\left(n\|\tilde{\boldsymbol{\theta}}_n\|^2 < \chi^2_p(\alpha)\right), \qquad (4.2.1)
\end{aligned}
$$

where $\mathbf{0}$ is the *restricted value* defined by H_0 and $I(A)$ is the indicator function of the set A and $\chi^2_p(\alpha)$ is the upper α-level critical value of a central chi-square distribution with p d.f.

As stated in Chapter 3, PTE was initiated by Bancroft (1944, 1964, 1965) and extended by Han and Bancroft (1968), among others, in a parametric setup. Saleh and Sen (1978) considered PTE within a nonparametric setup.

The PTE of $\boldsymbol{\theta}$ heavily depends on the level of significance, α, of the test and the value of $\hat{\boldsymbol{\theta}}_n^{\text{PT}}$ fluctuates between two values, namely $\tilde{\boldsymbol{\theta}}_n$ and $\mathbf{0}$. No intermediate values between $\mathbf{0}$ and $\tilde{\boldsymbol{\theta}}_n$ are available. It is clear that the bias vector of $\tilde{\boldsymbol{\theta}}_n$ is $\mathbf{b}_1(\tilde{\boldsymbol{\theta}}_n) = \mathbf{0}$ with the quadratic bias (QB), $B_1(\tilde{\boldsymbol{\theta}}_n) = 0$. The MSE matrix of $\tilde{\boldsymbol{\theta}}_n$ is $\mathbf{M}_1(\tilde{\boldsymbol{\theta}}_n) = \mathbf{I}_p$ with the risk $R_1(\tilde{\boldsymbol{\theta}}_n; \mathbf{W}) = \text{tr}(\mathbf{W}) = p$ if $\mathbf{W} = \mathbf{I}_p$.

Similarly, the bias vector $\mathbf{b}_2(\hat{\boldsymbol{\theta}}_n^{\text{PT}})$, the quadratic bias (QB) $B_2(\hat{\boldsymbol{\theta}}_n^{\text{PT}})$, the MSE matrix $\mathbf{M}_2(\hat{\boldsymbol{\theta}}_n^{\text{PT}})$, and the weighted risk $R_2(\hat{\boldsymbol{\theta}}_n^{\text{PT}}; \mathbf{W})$ are given by the following expressions, by the results of Theorem 1 of Section 4.1:

(i) $\mathbf{b}_2(\hat{\boldsymbol{\theta}}_n^{\text{PT}}) = -\boldsymbol{\theta} H_{p+2}(\chi_p^2(\alpha); \Delta^2)$ and

$$B_2(\hat{\boldsymbol{\theta}}_n^{\text{PT}}) = \Delta^2 \left\{H_{p+2}(\chi_p^2(\alpha); \Delta^2)\right\}^2, \tag{4.2.2}$$

where $H_\nu(\chi_p^2(\alpha); \Delta^2)$ is the cdf of a noncentral chi-square distribution with ν d.f. and noncentrality parameter $\Delta^2/2$ where $\Delta^2 = n\|\boldsymbol{\theta}\|^2$.

(ii)
$$\mathbf{M}_2(\hat{\boldsymbol{\theta}}_n^{\text{PT}}) = \mathbf{I}_p(1 - H_{p+2}(\chi_p^2(\alpha); \Delta^2)) + n\boldsymbol{\theta}\boldsymbol{\theta}'\{2H_{p+2}(\chi_p^2(\alpha); \Delta^2) - H_{p+4}(\chi_p^2(\alpha); \Delta^2)\}. \tag{4.2.3}$$

and

$$\begin{aligned} R_2(\hat{\boldsymbol{\theta}}_n^{\text{PT}}; \mathbf{W}) &= \text{tr}[\mathbf{M}_2(\hat{\boldsymbol{\theta}}_n^{\text{PT}})\mathbf{W}] \\ &= \text{tr}(\mathbf{W}) \left\{1 - H_{p+2}(\chi_p^2(\alpha); \Delta^2)\right\} \\ &\quad + (n\boldsymbol{\theta}'\mathbf{W}\boldsymbol{\theta}) \left\{2H_{p+2}(\chi_p^2(\alpha); \Delta^2) - H_{p+4}(\chi_p^2(\alpha); \Delta^2)\right\}, \quad (4.2.4) \\ &= p\{1 - H_{p+2}(\chi_p^2(\alpha); \Delta^2)\} \\ &\quad + \Delta^2 \left\{2H_{p+2}(\chi_p^2(\alpha); \Delta^2) - H_{p+4}(\chi_p^2(\alpha); \Delta^2)\right\}, \quad \text{if } \mathbf{W} = \mathbf{I}_p. \quad (4.2.5) \end{aligned}$$

Note that if $\Delta^2 = 0$, then $B_2(\hat{\boldsymbol{\theta}}_n^{\text{PT}}) = 0$; otherwise, it increases to a maximum and then drops down toward zero as $\Delta^2 \to \infty$. As regards the MSE matrix, we consider the pth root of the determinant of the matrix, that is,

$$\begin{aligned} &|\mathbf{M}_2(\hat{\boldsymbol{\theta}}_n^{\text{PT}})|^{1/p} \\ &= \left\{1 - H_{p+2}(\chi_p^2(\alpha); \Delta^2)\right\}^{(p-1)/p} \Big\{\left[1 - H_{p+2}(\chi_p^2(\alpha); \Delta^2)\right] \\ &\qquad + \Delta^2 \left[2H_{p+2}(\chi_p^2(\alpha); \Delta^2) - H_{p+4}(\chi_p^2(\alpha); \Delta^2)\right]\Big\}^{1/p}. \end{aligned} \tag{4.2.6}$$

Thus, $p|\mathbf{M}_2(\hat{\boldsymbol{\theta}}_n^{\text{PT}})|^{1/p} = R_2(\hat{\boldsymbol{\theta}}_n^{\text{PT}}; \mathbf{I}_p) = p\{1 - H_{p+2}(\chi_p^2(\alpha); 0)\}$ at $\Delta^2 = 0$, and as Δ^2 diverts from the origin, both $p|\mathbf{M}_2(\hat{\boldsymbol{\theta}}_n^{\text{PT}})|^{1/p}$ and $R_2(\hat{\boldsymbol{\theta}}_n^{\text{PT}}; \mathbf{I}_p)$ increase to a maximum crossing the p-line, and then decrease monotonically toward the value p as $\Delta^2 \to \infty$. See Figure 4.2.1. In general, the graph of $p|\mathbf{M}_2(\hat{\boldsymbol{\theta}}_n^{\text{PT}})|^{1/p}$ remains below the graph of $R_2(\hat{\boldsymbol{\theta}}_n^{\text{PT}}; \mathbf{I}_p)$ in the middle ($0 < \Delta^2 < \infty$), since these functions are drawn in the same scale.

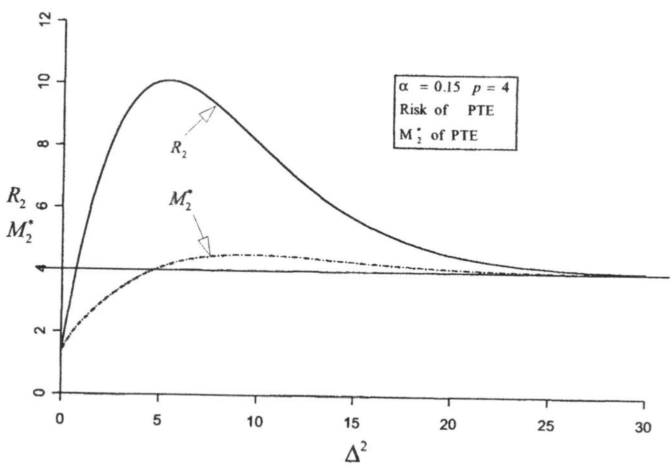

Figure 4.2.1 Graphs of $\mathbf{M}_2^* = p\left|\mathbf{M}_2(\hat{\boldsymbol{\theta}}_n^{\mathrm{PT}})\right|^{1/p}$ and $R_2 = R_2(\hat{\boldsymbol{\theta}}_n^{\mathrm{PT}};\mathbf{I}_p)$

The following theorem states that there are points in the parameter space where the PTE has risk larger than the MLE $\tilde{\boldsymbol{\theta}}_n$. It performs better in a small neighborhood of the origin. The theorem is due to Scolve, Morris, and Radhakrishnan (1972).

Theorem 1.

$$\begin{aligned} R_2(\hat{\boldsymbol{\theta}}_n^{\mathrm{PT}};\mathbf{W}) &> \mathrm{tr}(\mathbf{W}) \quad \text{if } (n\boldsymbol{\theta}'\mathbf{W}\boldsymbol{\theta}) > \mathrm{tr}(\mathbf{W}) \\ &< \mathrm{tr}(\mathbf{W}) \quad \text{if } (n\boldsymbol{\theta}'\mathbf{W}\boldsymbol{\theta}) < \mathrm{tr}(\mathbf{W}). \end{aligned} \tag{4.2.7}$$

Proof.

$$\begin{aligned} R_2(\hat{\boldsymbol{\theta}}_n^{\mathrm{PT}};\mathbf{W}) &= \mathrm{tr}(\mathbf{W})\{1 - H_{p+2}(\chi_p^2(\alpha);\Delta^2)\} \\ &\quad + (n\boldsymbol{\theta}'\mathbf{W}\boldsymbol{\theta})\{2H_{p+2}(\chi_p^2(\alpha);\Delta^2) - H_{p+4}(\chi_p^2(\alpha);\Delta^2)\} \\ &= [\mathrm{tr}(\mathbf{W}) - 2(n\boldsymbol{\theta}'\mathbf{W}\boldsymbol{\theta})]\{1 - H_{p+2}(\chi_p^2(\alpha);\Delta^2)\} \\ &\quad + (n\boldsymbol{\theta}'\mathbf{W}\boldsymbol{\theta})\{1 - H_{p+4}(\chi_p^2(\alpha);\Delta^2)\} + (n\boldsymbol{\theta}'\mathbf{W}\boldsymbol{\theta}) \\ &\geq [\mathrm{tr}(\mathbf{W}) - 2(n\boldsymbol{\theta}'\mathbf{W}\boldsymbol{\theta})]\{1 - H_{p+2}(\chi_p^2(\alpha);\Delta^2)\} \\ &\quad + (n\boldsymbol{\theta}'\mathbf{W}\boldsymbol{\theta})\{1 - H_{p+2}(\chi_p^2(\alpha);\Delta^2)\} + (n\boldsymbol{\theta}'\mathbf{W}\boldsymbol{\theta}), \end{aligned} \tag{4.2.8}$$

since χ^2_{p+4} is stochastically larger than χ^2_{p+2}. The r.h.s. of (4.2.8) equals

$$\begin{aligned}&\mathrm{tr}(\mathbf{W}) + [(n\boldsymbol{\theta}'\mathbf{W}\boldsymbol{\theta}) - \mathrm{tr}(\mathbf{W})]H_{p+2}(\chi^2_p(\alpha);\Delta^2) > \mathrm{tr}(\mathbf{W})\\ &\qquad \text{if } (n\boldsymbol{\theta}'\mathbf{W}\boldsymbol{\theta}) > \mathrm{tr}(\mathbf{W}).\end{aligned} \tag{4.2.9}$$

If $\mathbf{W} = \mathbf{I}_p$, then

$$R_2(\hat{\boldsymbol{\theta}}_n^{\mathrm{PT}};\mathbf{I}_p) > p \quad \text{if } \Delta^2 > p. \tag{4.2.10}$$

Now, choose $(n\boldsymbol{\theta}'\mathbf{W}\boldsymbol{\theta})$ such that $\mathrm{tr}(\mathbf{W}) - 2(n\boldsymbol{\theta}'\mathbf{W}\boldsymbol{\theta}) > 0$. Then

$$\begin{aligned}R_2(\hat{\boldsymbol{\theta}}_n^{\mathrm{PT}};\mathbf{W}) \le {}& [\mathrm{tr}(\mathbf{W}) - 2(n\boldsymbol{\theta}'\mathbf{W}\boldsymbol{\theta})]\{1 - H_{p+4}(\chi^2_p(\alpha);\Delta^2)]\\ &+(n\boldsymbol{\theta}'\mathbf{W}\boldsymbol{\theta})[1 - H_{p+4}(\chi^2_p(\alpha);\Delta^2)] + (n\boldsymbol{\theta}'\mathbf{W}\boldsymbol{\theta}),\end{aligned} \tag{4.2.11}$$

replacing $H_{p+2}(\chi^2_p(\alpha);\Delta^2)$ by $H_{p+4}(\chi^2_p(\alpha);\Delta^2)$. The r.h.s. of (4.2.11) after simplification equals

$$\begin{aligned}&\mathrm{tr}(\mathbf{W}) + [\mathrm{tr}(\mathbf{W}) - (n\boldsymbol{\theta}'\mathbf{W}\boldsymbol{\theta})]H_{p+4}(\chi^2_p(\alpha);\Delta^2) < \mathrm{tr}(\mathbf{W})\\ &\qquad \text{if } \ (n\boldsymbol{\theta}'\mathbf{W}\boldsymbol{\theta}) < \frac{1}{2}\mathrm{tr}(\mathbf{W}) < \mathrm{tr}(\mathbf{W}).\end{aligned} \tag{4.2.12}$$

If $\mathbf{W} = \mathbf{I}_p$, then $R_2(\hat{\boldsymbol{\theta}}_n^{\mathrm{PT}};\mathbf{I}_p) < p$ if $\Delta^2 < \frac{1}{2}p < p$.

Since the risk $R_1(\tilde{\boldsymbol{\theta}}_n;\mathbf{I}_p) = p$ and $\tilde{\boldsymbol{\theta}}_n$ is minimax, Theorem 4.2.1 states that no such PTE can be minimax. Further, as $\chi^2_p(\alpha) \to \infty$ (i.e., $\alpha \to 0$), the risk $R_2(\hat{\theta}_n;\mathbf{I}_p)$ at the origin (i.e., $\Delta^2 = 0$) corresponding to the PTE tends to be p. That is to say, we can find a PTE that does arbitrarily well at the origin at the cost of being poor elsewhere.

Consider the properties of $\hat{\boldsymbol{\theta}}_n^{PT}$ with respect to $\tilde{\boldsymbol{\theta}}_n$ based on the MSE criterion. For this consider the difference

$$\boldsymbol{\ell}'\mathbf{M}_2\left(\hat{\boldsymbol{\theta}}_n^{\mathrm{PT}}\right)\boldsymbol{\ell}' - \boldsymbol{\ell}'\boldsymbol{\ell} \tag{4.2.13}$$

for a given nonzero p-vector $\boldsymbol{\ell}$. Thus,

$$\begin{aligned}&\boldsymbol{\ell}'\boldsymbol{\ell}\Big\{1 - H_{p+2}(\chi^2_p(\alpha);\Delta^2)\Big\} + \boldsymbol{\ell}'n\boldsymbol{\theta}\boldsymbol{\theta}'\boldsymbol{\ell}\Big\{2H_{p+2}(\chi^2_p(\alpha);\Delta^2) - H_{p+4}(\chi^2_p(\alpha);\Delta^2)\Big\}\\ &\qquad - \boldsymbol{\ell}'\boldsymbol{\ell} \gtreqless 0\end{aligned} \tag{4.2.14}$$

according as

$$\frac{\boldsymbol{\ell}'n\boldsymbol{\theta}\boldsymbol{\theta}'\boldsymbol{\ell}}{\boldsymbol{\ell}'\boldsymbol{\ell}}\Big\{2H_{p+2}(\chi^2_p(\alpha);\Delta^2) - H_{p+4}(\chi^2_p(\alpha);\Delta^2)\Big\} \gtreqless H_{p+2}(\chi^2_p(\alpha);\Delta^2)$$

for all $\boldsymbol{\ell}$.

Since $\max_{\boldsymbol{\ell}} \frac{\boldsymbol{\ell}' n\boldsymbol{\theta}\boldsymbol{\theta}'\boldsymbol{\ell}}{\boldsymbol{\ell}'\boldsymbol{\ell}} = \Delta^2$, we have

$$\Delta^2 \gtrless \frac{H_{p+2}(\chi_p^2(\alpha);\Delta^2)}{\left\{2H_{p+2}(\chi_p^2(\alpha);\Delta^2) - H_{p+4}(\chi_p^2(\alpha);\Delta^2)\right\}}. \tag{4.2.15}$$

Thus, $\hat{\boldsymbol{\theta}}_n^{\text{PT}}$ is better than $\tilde{\boldsymbol{\theta}}_n$ whenever, $\Delta^2 < \frac{H_{p+2}(\chi_p^2(\alpha);\Delta^2)}{2H_{p+2}(\chi_p^2(\alpha);\Delta^2)-H_{p+4}(\chi_p^2(\alpha);\Delta^2)}$, whereas $\tilde{\boldsymbol{\theta}}_n$ is better than $\hat{\boldsymbol{\theta}}_n^{P_T}$ for $\Delta^2 > \frac{H_{p+2}(\chi_p^2(\alpha);\Delta^2)}{2H_{p+2}(\chi_p^2(\alpha);\Delta^2)-H_{p+4}(\chi_p^2(\alpha);\Delta^2)}$. (4.2.16)

We may define the MSE based relative efficiency (MRE) as:

$$\begin{aligned}
&\text{MRE}(\hat{\boldsymbol{\theta}}_n^{\text{PT}};\tilde{\boldsymbol{\theta}}_n)\\
&= \frac{\left|\mathbf{M}(\tilde{\boldsymbol{\theta}}_n)\right|^{1/p}}{\left|\mathbf{M}_2(\hat{\boldsymbol{\theta}}_n^{\text{PT}})\right|^{1/p}}\\
&= \Big[\{1 - H_{p+2}(\chi_p^2(\alpha);\Delta^2)\}^{p-1}\{(1 - H_{p+2}(\chi_p^2(\alpha);\Delta^2))\\
&\quad + \Delta^2(2H_{p+2}(\chi_p^2(\alpha);\Delta^2) - H_{p+4}(\chi_p^2(\alpha);\Delta^2))\}\Big]^{-1/p}\\
&= \{1 - H_{p+2}(\chi_p^2(\alpha);\Delta^2)\}^{-(p-1)/p}\Big[\{1 - H_{p+2}(\chi_p^2(\alpha);\Delta^2)\}\\
&\quad + \Delta^2\{2H_{p+2}(\chi_p^2(\alpha);\Delta^2) - H_{p+4}(\chi_p^2(\alpha);\Delta^2)\}\Big]^{-1/p}.
\end{aligned} \tag{4.2.17}$$

For each α, the efficiency of $\hat{\boldsymbol{\theta}}_n^{\text{PT}}$ relative to $\tilde{\boldsymbol{\theta}}_n$ attains the maximum value $E_{\max} = \{1 - H_{p+2}(\chi_p^2(\alpha);0)\}^{-1}$ at $\Delta^2 = 0$ and the minimum value $E_{\min}$ at $\Delta^2_{\min}(\alpha)$. The intersection of the graph of $\text{MRE}(\hat{\boldsymbol{\theta}}_n^{\text{PT}};\tilde{\boldsymbol{\theta}}_n)$ with the 1-line occurs at Δ_0^2. Thus, if $0 \le \Delta^2 \le \Delta^2$, then $\hat{\boldsymbol{\theta}}_n^{\text{PT}}$ is better than $\tilde{\boldsymbol{\theta}}_n$, and if $\Delta^2 > \Delta_0^2$, then $\tilde{\boldsymbol{\theta}}_n$ is better.

The cut-off point Δ_0^2 may be determined based on the inequality

$$0 \le \Delta^2 \le \frac{[\{1 - H_{p+2}(\chi_p^2(\alpha);\Delta^2)\}^{-p} - 1]\{1 - H_{p+2}(\chi_p^2(\alpha);\Delta^2)\}}{\{2H_{p+2}(\chi_p^2(\alpha);\Delta^2) - H_{p+4}(\chi_p^2(\alpha);\Delta^2)\}}. \tag{4.2.18}$$

The efficiency expression allows us to determine the optimum level of significance for minimum guaranteed efficiency, E_0, as the solution α^* of the equation

$$\min_{\Delta^2} \text{MRE}(\alpha,\Delta^2) = E(\alpha,\Delta_{\min}(\alpha)) = E_0\,, \tag{4.2.19}$$

where E_0 is a preassigned efficiency number for the PTE, and $\text{MRE}(\alpha,\Delta^2)$ stands for $\text{MRE}(\hat{\boldsymbol{\theta}}_n^{\text{PT}};\tilde{\boldsymbol{\theta}}_n)$ as a function of α and Δ^2. The efficiency based on the optimum level α^* may go up to $E_{\max} = \{1 - H_{p+2}(\chi_p^2(\alpha);0)\}^{-1}$.

Table 4.2.1. Maximum and Minimum Guaranteed MSE Based Efficiencies

α	p	4	6	8	10	12	14
0.05	E_{max}	6.75908	7.89292	8.72360	9.37375	9.90415	10.34988
	E_{min}	0.83930	0.90622	0.93840	0.95653	0.96781	0.97533
	Δ^2_{min}	12.55403	16.40307	19.66492	22.53881	25.14251	27.51153
0.1	E_{max}	3.92601	4.49118	4.89829	5.21311	5.46767	5.67991
	E_{min}	0.86685	0.92253	0.94938	0.96451	0.97391	0.98015
	Δ^2_{min}	10.62223	14.08485	17.06601	19.72510	22.15249	24.40138
0.15	E_{max}	2.89783	3.26722	3.53047	0.73248	3.89487	4.02965
	E_{min}	2.89783	3.26722	3.53047	3.73248	3.89487	4.01965
	Δ^2_{min}	9.47256	12.67623	15.46918	17.97908	20.28298	22.42997
0.2	E_{max}	2.35590	2.62511	2.81549	2.96075	3.07701	3.17317
	E_{min}	0.90103	0.94229	0.96248	0.97391	0.98100	0.98570
	Δ^2_{min}	8.64957	11.64973	14.29486	16.68815	18.89562	20.96061
0.25	E_{max}	2.01845	2.22631	2.37244	2.48345	2.57199	2.64502
	E_{min}	0.91385	0.94961	0.96728	0.97731	0.98354	0.98767
	Δ^2_{min}	8.00872	10.83690	13.35708	15.65218	17.77896	19.77572
0.3	E_{max}	1.78732	1.95346	2.06973	2.15774	2.22775	2.28535
	E_{min}	0.92509	0.95599	0.97144	0.98024	0.98572	0.93935
	Δ^2_{min}	7.48503	10.16181	12.57172	14.78051	16.83674	18.77429
0.35	E_{max}	1.61900	1.75471	1.84936	1.92081	1.97751	2.02408
	E_{min}	0.93513	0.96168	0.97513	0.98284	0.98764	0.99082
	Δ^2_{min}	7.04355	9.58252	11.89341	14.02413	16.01690	17.90156
0.4	E_{max}	1.49112	1.60350	1.68169	1.74059	1.78724	1.82549
	E_{min}	0.94421	0.96682	0.97845	0.98516	0.98935	0.99212
	Δ^2_{min}	6.66328	9.07737	11.29477	13.35346	15.28787	17.12444
0.45	E_{max}	1.39096	1.48475	1.54992	1.59894	1.63770	1.66944
	E_{min}	0.95247	0.97151	0.98148	0.98727	0.99089	0.99329
	Δ^2_{min}	6.33049	8.62734	10.75802	12.74924	14.62948	16.42136
0.5	E_{max}	1.31074	1.38937	1.44382	1.48480	1.51718	1.54367
	E_{min}	0.96000	0.97581	0.98425	0.98919	0.99229	0.99434
	Δ^2_{min}	6.03573	8.22239	10.27088	12.19820	14.02726	15.77741

Table 4.2.1 may be used to determine the optimum level of significance for the preliminary test (PT) when the MSE based criterion is used. For example, let $p = 6$, and suppose that we wish to obtain a PTE with minimum guaranteed efficiency $E_0 = 0.95$. Then Table 4.2.1 gives the value $\alpha^* = 0.25$ at the intersection of $\alpha = 0.25$ and $p = 6$. Hence, the optimum level for the PT in this case is 0.25 with a maximum possible efficiency 2.22631.

Now, consider the relative risk efficiency (RRE) based on the risk of $\hat{\boldsymbol{\theta}}_n^{PT}$

given by

$$\text{RRE}(\hat{\boldsymbol{\theta}}_n^{\text{PT}} : \tilde{\boldsymbol{\theta}}_n) = \Big[1 - H_{p+2}(\chi_p^2(\alpha); \Delta^2) + p^{-1}\Delta^2\{2H_{p+2}(\chi_p^2(\alpha); \Delta^2) - H_{p+4}(\chi_p^2(\alpha); \Delta^2)\}\Big]^{-1}. \tag{4.2.20}$$

The graph of $\text{RRE}(\hat{\boldsymbol{\theta}}_n^{\text{PT}}; \tilde{\boldsymbol{\theta}}_n)$ is similar to the graph of $\text{MRE}(\hat{\boldsymbol{\theta}}_n^{\text{PT}}; \hat{\boldsymbol{\theta}}_n)$. The maximum occurs at $\Delta^2 = 0$ with the same value $\{1 - H_{p+2}(\chi_p^2(\alpha); 0)\}^{-1}$. The efficiency decreases, crossing the 1-line at Δ^2_{min} to the value

$$\min_{\Delta^2} \text{RRE}(\alpha, \Delta^2_{\text{min}}(\alpha)) = \text{RRE}(\alpha, \Delta^2_{\text{min}}(\alpha)), \tag{4.2.21}$$

and then increases toward 1 as Δ^2 increases toward ∞. Here $\text{RRE}(\alpha, \Delta^2)$ stands for $\text{RRE}(\hat{\boldsymbol{\theta}}_n^{\text{PT}}; \tilde{\boldsymbol{\theta}}_n)$ as a function of α and Δ^2. Clearly, $\text{RRE}(\hat{\boldsymbol{\theta}}_n^{\text{PT}}; \tilde{\boldsymbol{\theta}}_n) \substack{\geq \\ <} 1$ according as

$$\Delta^2 \substack{\leq \\ >} \frac{pH_{p+2}(\chi_p^2(\alpha); \Delta^2)}{\{2H_{p+2}(\chi_p^2(\alpha); \Delta^2) - H_{p+1}(\chi_p^2(\alpha); \Delta^2)\}}. \tag{4.2.22}$$

The optimum value of α may be obtained by solving the equation

$$\min_{\Delta^2} \text{RRE}(\alpha, \Delta^2) = E(\alpha, \Delta_{\text{min}}(\alpha)) = E_0 \tag{4.2.23}$$

with a pre-assigned value E_0. Table 4.2.2 facilitates the determination of the optimum level of significance α^* based on MSE matrix and risk efficiencies.

As an application of Table 4.2.2, let $p = 6$. We wish to obtain a PTE with at least 95% efficiency. We find in the crossing of 0.40 and 0.95099 ($\equiv 0.95$) the desired α-value as $\alpha^* = 0.40$ with a maximum efficiency 1.60350 as in the case of MRE.

Table 4.2.2. Maximum and Minimum Guaranteed risk Based Efficiencies

α	p	4	6	8	10	12	14
0.05	E_{max}	6.75908	7.89292	8.72360	9.37375	9.90415	10.34988
	E_{min}	0.67741	0.76147	0.81707	0.85639	0.88538	0.90738
	Δ^2_{min}	8.70073	10.82463	12.79251	14.66773	16.48109	18.24932
0.1	E_{max}	3.92601	4.49118	4.89829	5.21311	5.46767	5.67991
	E_{min}	0.75506	0.82544	0.87014	0.90071	0.92260	0.93879
	Δ^2_{min}	7.73232	9.77526	11.68294	13.51002	15.28335	17.01821
0.15	E_{max}	2.89783	3.26722	3.53047	0.73248	3.89487	4.02965
	E_{min}	0.80429	0.86374	0.90061	0.92530	0.94267	4.95530
	Δ^2_{min}	7.16494	9.15150	11.01743	12.81104	14.55626	16.26681
0.2	E_{max}	2.35590	2.62511	2.81549	2.96075	3.07701	3.17317
	E_{min}	0.84032	0.89067	0.92141	0.94171	0.95579	0.96591
	Δ^2_{min}	6.76058	8.70106	10.53304	12.29940	14.02168	15.71231
0.25	E_{max}	2.01845	2.22631	2.37244	2.48345	2.57199	2.64502
	E_{min}	0.86844	0.91105	0.93681	0.95363	0.96518	0.97339
	Δ^2_{min}	6.44492	8.34493	10.14718	11.88969	13.59187	15.26500
0.3	E_{max}	1.78732	1.95346	2.06973	2.15774	2.22775	2.28535
	E_{min}	0.89120	0.92715	0.94874	0.96272	0.97225	0.97897
	Δ^2_{min}	6.18490	8.04786	9.82294	11.54366	13.22745	14.88455
0.35	E_{max}	1.61900	1.75471	1.84936	1.92081	1.97751	2.02408
	E_{min}	0.91005	0.94021	0.95827	0.96989	0.97776	0.98326
	Δ^2_{min}	5.96289	7.79095	9.54048	11.24068	12.90716	14.54916
0.4	E_{max}	1.49112	1.60350	1.68169	1.74059	1.78724	1.82549
	E_{min}	0.92589	0.95099	0.96604	0.97567	0.98215	0.98666
	Δ^2_{min}	5.76832	7.56281	9.28777	10.96826	12.61807	14.24553
0.45	E_{max}	1.39096	1.48475	1.54992	1.59894	1.63770	1.66944
	E_{min}	0.93933	0.96002	0.97245	0.98039	0.98571	0.98938
	Δ^2_{min}	5.59431	7.35598	9.05691	10.71809	12.35159	13.96478
0.5	E_{max}	1.31074	1.38937	1.44382	1.48480	1.51718	1.54367
	E_{min}	0.95079	0.96763	0.97781	0.98430	0.98862	0.99160
	Δ^2_{min}	5.43611	7.16522	8.84227	10.48427	12.10153	13.70053

4.3 Stein-Type Estimators

4.3.1 Introduction

Preliminary test estimation introduced by Bancroft (1944, 1964, 1965) was an early approach to the problem of estimation and test of hypothesis in cases where some uncertain prior information exists. We saw in Section 4.2 that the PTE of $\boldsymbol{\theta}$ fails to be minimax, although there is a small neighborhood near the origin where it performs better than the MLE, $\tilde{\boldsymbol{\theta}}_n$.

However, real breakthroughs came when Stein (1956) proved that there exist estimators that can improve on $\tilde{\boldsymbol{\theta}}_n$ under the loss function $n\|\tilde{\boldsymbol{\theta}}_n - \boldsymbol{\theta}\|^2$. He demonstrated that the estimator

$$\boldsymbol{\theta}_n^* = \left(1 - \frac{p}{n\|\tilde{\boldsymbol{\theta}}_n\|^2}\right)\tilde{\boldsymbol{\theta}}_n \tag{4.3.1}$$

improves on $\tilde{\boldsymbol{\theta}}_n$.

The genesis of this estimator stems out of a simple consideration: If $\tilde{\boldsymbol{\theta}}_n$ estimates $\boldsymbol{\theta}$ closely, does it mean $\|\tilde{\boldsymbol{\theta}}_n\|^2$ will estimate $\|\boldsymbol{\theta}\|^2$ closely? Clearly,

$$E[n\|\tilde{\boldsymbol{\theta}}_n\|^2] = n\|\boldsymbol{\theta}\|^2 + p. \tag{4.3.2}$$

Thus, $n\|\tilde{\boldsymbol{\theta}}_n\|^2$ has missed $n\|\boldsymbol{\theta}\|^2$ by an amount p. To obtain the estimator $\boldsymbol{\theta}_n^*$, we will pursue the following geometric approach together with the Pythagorean theorem: Since $\tilde{\boldsymbol{\theta}}_n$ is an estimator of $\boldsymbol{\theta}$ based on a sample of size n from $\mathcal{N}_p(\boldsymbol{\theta}, \mathbf{I}_p)$, we set

$$n\|\tilde{\boldsymbol{\theta}}_n - \boldsymbol{\theta}\|^2 = p \tag{4.3.3a}$$

and

$$n\|\tilde{\boldsymbol{\theta}}_n\|^2 = n\|\boldsymbol{\theta}\|^2 + p. \tag{4.3.3b}$$

Stein showed that instead of $\tilde{\boldsymbol{\theta}}_n$ the projection of $\boldsymbol{\theta}$ on $\tilde{\boldsymbol{\theta}}_n$ is a better estimate of $\boldsymbol{\theta}$. Denote this projection by $(1-\alpha)\tilde{\boldsymbol{\theta}}_n$. Then, from the geometrical representation given in Figure 4.3.1, we can conclude the following:

(i) $\|\mathbf{Z}\|^2 = n\|\tilde{\boldsymbol{\theta}}_n - \boldsymbol{\theta}\|^2 - n\alpha^2\|\tilde{\boldsymbol{\theta}}_n\|^2 = p - n\alpha^2\|\tilde{\boldsymbol{\theta}}_n\|^2$ (4.3.4)

and

(ii) $\|\mathbf{Z}\|^2 = n\|\boldsymbol{\theta}\|^2 - n(1-\alpha)^2\|\tilde{\boldsymbol{\theta}}_n\|^2 = n\|\tilde{\boldsymbol{\theta}}_n\|^2 - p - n(1-\alpha)^2\|\tilde{\boldsymbol{\theta}}_n\|^2$. (4.3.5)

Equating (4.3.4) with (4.3.5), we obtain

$$\hat{\alpha} = \frac{p}{n\|\tilde{\boldsymbol{\theta}}_n\|^2}. \tag{4.3.6}$$

Hence, the projection of $\boldsymbol{\theta}$ on $\tilde{\boldsymbol{\theta}}_n$ is given by (4.3.1) and repeated here:

$$\boldsymbol{\theta}_n^* = \left(1 - \frac{p}{n\|\tilde{\boldsymbol{\theta}}_n\|^2}\right)\tilde{\boldsymbol{\theta}}_n. \tag{4.3.7}$$

The bias vector and the quadratic bias (QB) of $\boldsymbol{\theta}_n^*$ are given by (using Theorem 1)

$$\mathbf{b}(\boldsymbol{\theta}_n^*) = -p\boldsymbol{\theta}E\left(\chi_{p+2}^{-2}(\Delta^2)\right) \quad \text{and} \quad B(\boldsymbol{\theta}_n^*) = p^2\Delta^2\{E\left(\chi_{p+2}^{-2}(\Delta^2)\right)\}^2, \tag{4.3.8}$$

where

$$E[\chi_p^{-2}(\Delta^2)] = e^{-\Delta^2/2}\sum_{r\geq 0}\frac{1}{r!}\left(\frac{1}{2}\Delta^2\right)^r(p-2+2r)^{-1}.$$

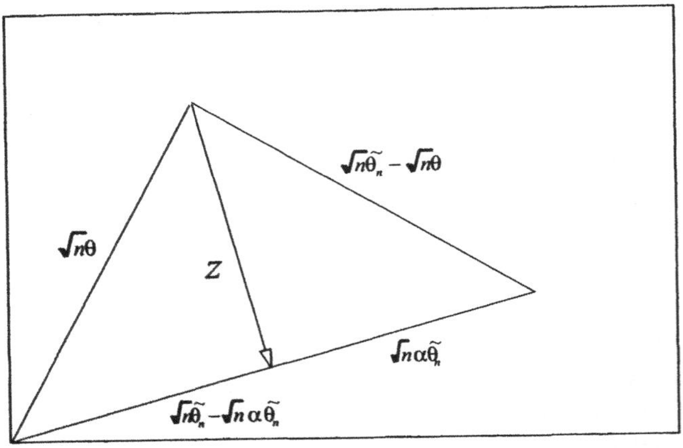

Figure 4.3.1 Geometrical representation of Stein's idea

In order to compare the performance of $\boldsymbol{\theta}_n^*$ relative to $\tilde{\boldsymbol{\theta}}_n$, we consider the MSE matrix and the weighted risk of $\boldsymbol{\theta}_n^*$ given by

$$\begin{aligned}\mathbf{M}(\boldsymbol{\theta}_n^*) = \mathbf{I}_p - p\mathbf{I}_p\Big[2E\Big(\chi_{p+2}^{-2}(\Delta^2)\Big) - pE\Big(\chi_{p+2}^{-4}(\Delta^2)\Big)\Big] \\ + p(n\boldsymbol{\theta}\boldsymbol{\theta}')\Big[2E\Big(\chi_{p+2}^{-2}(\Delta^2)\Big) - 2E\Big(\chi_{p+4}^{-2}(\Delta^2)\Big) \\ - pE\Big(\chi_{p+4}^{-4}(\Delta^2)\Big)\Big]\end{aligned} \tag{4.3.9}$$

and

$$\begin{aligned}&R(\boldsymbol{\theta}_n^*; \mathbf{W}) \\ &= \text{tr}[M(\boldsymbol{\theta}_n^*)\mathbf{W}] = \text{tr}(\mathbf{W}) - p\text{tr}(\mathbf{W})\Big[2E(\chi_{p+2}^{-2}(\Delta^2)) - pE(\chi_{p+2}^{-4}(\Delta^2))\Big] \\ &\quad + p(n\boldsymbol{\theta}'\mathbf{W}\boldsymbol{\theta})\Big[2E(\chi_{p+2}^{-2}(\Delta^2)) - 2E(\chi_{p+2}^{-4}(\Delta^2)) - pE(\chi_{p+4}^{-4}(\Delta^2))\Big] \\ &= p - p^2\Big[2E(\chi_{p+2}^{-2}(\Delta^2)) - pE(\chi_{p+2}^{-4}(\Delta^2))\Big] \\ &\quad + p\Delta^2\Big[2E(\chi_{p+2}^{-2}(\Delta^2)) - 2E(\chi_{p+4}^{-2}(\Delta^2)) + pE(\chi_{p+4}^{-4}(\Delta^2))\Big]\end{aligned} \tag{4.3.10}$$

if $\mathbf{W} = \mathbf{I}_p$.

By taking the risk difference $R(\tilde{\boldsymbol{\theta}}_n; I_p) - R(\boldsymbol{\theta}_n^*; I_p)$, we find that the risk of $\boldsymbol{\theta}_n^*$ is smaller than the MLE $\tilde{\boldsymbol{\theta}}_n$. Similarly, we consider the pth root of the determinant of $\mathbf{M}(\boldsymbol{\theta}_n^*)$:

$$
\begin{aligned}
&|\mathbf{M}(\boldsymbol{\theta}_n^*)|^{1/p} \\
&= \Big\{1 - p\Big[2E(\chi_{p+2}^{-2}(\Delta^2)) - pE(\chi_{p+2}^{-4}(\Delta^2))\Big]\Big\}^{(p-1)/p} \\
&\quad \Big\{1 - p\Big[2E(\chi_{p+2}^{-2}(\Delta^2)) - pE(\chi_{p+2}^{-4}(\Delta^2))\Big] \\
&\quad + p\Delta^2\Big[2E(\chi_{p+2}^{-2}(\Delta^2)) - 2E(\chi_{p+4}^{-2}(\Delta^2)) + pE(\chi_{p+4}^{-4}(\Delta^2))\Big]\Big\}^{1/p}. \quad (4.3.11)
\end{aligned}
$$

Thus, $p|M(\boldsymbol{\theta}_n^*)|^{1/p}$ is smaller than p for all Δ^2. Hence, the MLE $\tilde{\boldsymbol{\theta}}_n$ is inadmissible according to the MSE criterion. As $\Delta^2 \to \infty$, $p|M(\boldsymbol{\theta}_n^*)|^{1/p} \to p$, and for $\Delta^2 = 0$, we obtain $p|\mathbf{M}(\boldsymbol{\theta}_n^*)|^{1/p} = 2(1 - \frac{2}{p})^{-1}$. On the other hand, $R(\boldsymbol{\theta}_n^*; \mathbf{I}_p) \to p$ as $\Delta^2 \to \infty$ and at the origin $R(\boldsymbol{\theta}_n^*; \mathbf{I}_p) = 2(1 - \frac{2}{p})^{-1}$ is the same as the MSE case. Except at the endpoints the graph of $p|M(\boldsymbol{\theta}_n^*)|^{1/p}$ remains below the graph of $R(\boldsymbol{\theta}_n^*; \mathbf{I}_p)$ in the interval $0 < \Delta^2 < \infty$.

4.3.2 James-Stein Estimator (JSE)

Although $\boldsymbol{\theta}_n^*$ dominates the MLE $\tilde{\boldsymbol{\theta}}_n$, is it an optimum estimator of $\boldsymbol{\theta}$? To answer this question, consider the *shrinkage estimator*,

$$\hat{\boldsymbol{\theta}}_n^S(c) = \left(1 - \frac{c}{n\|\hat{\boldsymbol{\theta}}_n\|^2}\right)\tilde{\boldsymbol{\theta}}_n, \quad (4.3.12)$$

and determine the optimum constant c from the unweighted risk of $\hat{\boldsymbol{\theta}}_n^S(c)$. Now,

$$
\begin{aligned}
R(\hat{\boldsymbol{\theta}}_n^S(c); \mathbf{I}_p) &= p - 2c + 2cn\boldsymbol{\theta}'E\left[\frac{\tilde{\boldsymbol{\theta}}_n}{n\|\tilde{\boldsymbol{\theta}}_n\|^2}\right] + c^2E[n\|\tilde{\boldsymbol{\theta}}_n\|]^{-1} \\
&= p - 2c + 2c\Delta^2E(\chi_{p+2}^{-2}(\Delta^2)) + c^2E(\chi_p^{-2}(\Delta^2)). \quad (4.3.13)
\end{aligned}
$$

Minimizing $R(\hat{\boldsymbol{\theta}}_n^S(c); \mathbf{I}_p)$ with respect to c, we obtain

$$-2 + 2\Delta^2E\Big(\chi_{p+2}^{-2}(\Delta^2)\Big) + 2cE\Big(\chi_p^{-2}(\Delta^2)\Big) = 0 \quad (4.3.14a)$$

Now, writing $\Delta^2E[\chi_{p+2}^{-2}(\Delta^2)] = 1 - (p-2)E[\chi_p^{-2}(\Delta^2)]$, we obtain

$$-2 + 2\Big\{1 - (p-2)E\Big(\chi_p^{-2}(\Delta^2)\Big)\Big\} + 2cE\Big(\chi_p^{-2}(\Delta^2)\Big) = 0. \quad (4.3.14b)$$

This implies that the optimum value of c is

$$c^* = p - 2. \quad (4.3.14c)$$

Thus, the optimum Stein-type estimator of $\boldsymbol{\theta}$ is given by James and Stein (1961) as

$$\hat{\boldsymbol{\theta}}_n^S = \left(1 - \frac{(p-2)}{n\|\tilde{\boldsymbol{\theta}}_n\|^2}\right)\tilde{\boldsymbol{\theta}}_n. \tag{4.3.15}$$

Hence, the estimator $\hat{\boldsymbol{\theta}}_n^S$ is known as the *James-Stein estimator* (JSE) in the literature.

Note that if $n\|\tilde{\boldsymbol{\theta}}_n\|^2 \to \infty$, then $\hat{\boldsymbol{\theta}}_n^S = \tilde{\boldsymbol{\theta}}_n$, which matches with the PTE, $\hat{\boldsymbol{\theta}}_n^{PT}$. However, if $n\|\tilde{\boldsymbol{\theta}}_n\|^2 \to \mathbf{0}$, then $\hat{\boldsymbol{\theta}}_n^S$ becomes a negative estimator. In order to obtain the right value, one has to restrict $n\|\tilde{\boldsymbol{\theta}}_n\|^2 > p-2$. Stein (1966) defined the following estimator of $\boldsymbol{\theta}$:

$$\begin{aligned}\hat{\boldsymbol{\theta}}_n^{S+} &= \mathbf{0}\, I(n\|\tilde{\boldsymbol{\theta}}_n\|^2 \le p-2) + \hat{\boldsymbol{\theta}}_n^{S} I(n\|\tilde{\boldsymbol{\theta}}_n\|^2 > p-2) && (4.3.16)\\ &= \left(1 - \frac{p-2}{n\|\tilde{\boldsymbol{\theta}}_n\|^2}\right) I\left(n\|\tilde{\boldsymbol{\theta}}_n\|^2 > p-2\right)\tilde{\boldsymbol{\theta}}_n, && (4.3.17)\end{aligned}$$

which is known as the **positive-rule Stein estimator** (PRSE) of $\boldsymbol{\theta}$. The PRSE of $\boldsymbol{\theta}$ is a PTE. It combines $\mathbf{0}$ and $\hat{\boldsymbol{\theta}}_n^S$ via testing $H_0 : \boldsymbol{\theta} = \mathbf{0}$ based on $n\|\tilde{\boldsymbol{\theta}}_n\|^2$ with the critical value $p-2$.

Since the arrival of JS estimator, there was a tremendous serge of research in Stein-type estimation to fill the void in different directions. See, for example, Baranchik (1970), Berger (1980, 1985), Bock (1988), Brown (1966, 1988), Casella (1989), Cellier, Fourdrinier and Robert (1989), Cohen and Strawderman (1973), Efron and Morris (1973), Saleh and Sen (1978-1986), Hoffmann (1992), Brewster and Zidek (1974), Casella (1985), and Kubokawa (1991, 1994) among others. In the sections that follow we present various techniques to improve the estimator $\tilde{\boldsymbol{\theta}}_n$.

In the sequel we will discuss the MLE $\tilde{\boldsymbol{\theta}}_n$, the James-Stein estimator $\hat{\boldsymbol{\theta}}_n^S$, and the PRSE, $\hat{\boldsymbol{\theta}}_n^{S+}$ and variations thereof.

We have seen that the Stein estimator $\boldsymbol{\theta}_n^*$ given in (4.3.1) is a biased estimator. The bias vector and the QB of the James-Stein estimator are

$$b_3(\hat{\boldsymbol{\theta}}_n^S) = -(p-2)\boldsymbol{\theta} E\left(\chi_{p+2}^{-2}(\Delta^2)\right) \tag{4.3.18}$$

and $$B_3(\hat{\boldsymbol{\theta}}_n^S) = (p-2)^2\Delta^2\{E(\chi_{p+2}^{-2}(\Delta^2))\}^2, \tag{4.3.19}$$

respectively.

The MSE matrix and the weighted risk of $\hat{\boldsymbol{\theta}}_n^S$ may be obtained as

$$\begin{aligned}\mathbf{M}_3(\hat{\boldsymbol{\theta}}_n^S) = \mathbf{I}_p - (p-2)\mathbf{I}_p\{2E(\chi_{p+2}^{-2}(\Delta^2)) - (p-2)E(\chi_{p+2}^{-4}(\Delta^2))\}\\ + (p^2-4)n\boldsymbol{\theta}\boldsymbol{\theta}' E(\chi_{p+4}^{-4}(\Delta^2)) \qquad (4.3.20)\end{aligned}$$

and

$$\begin{aligned} R_3(\hat{\boldsymbol{\theta}}_n^S;\mathbf{W}) &= \mathrm{tr}[\mathbf{M}_3(\hat{\boldsymbol{\theta}}_n^S)\mathbf{W}] \\ &= \mathrm{tr}(\mathbf{W}) - (p-2)\mathrm{tr}(\mathbf{W})\left\{2E(\chi_{p+2}^{-2}(\Delta^2)) - (p-2)E(\chi_{p+2}^{-4}(\Delta^2))\right\} \\ &\quad + (p^2-4)(n\boldsymbol{\theta}'\mathbf{W}\boldsymbol{\theta})E(\chi_{p+4}^{-4}(\Delta^2)) \\ &= p - (p-2)^2E(\chi_p^{-2}(\Delta^2)) \text{ if } \mathbf{W} = \mathbf{I}_p. \end{aligned} \tag{4.3.21}$$

respectively.

The determinant of $\mathbf{M}_3(\hat{\boldsymbol{\theta}}_n^S)$ is given by

$$\begin{aligned} |\mathbf{M}_3(\hat{\boldsymbol{\theta}}_n^S)| = &\left\{1 - (p-2)\left[2E\big(\chi_{p+2}^{-2}(\Delta^2)\big) - (p-2)E\big(\chi_{p+2}^{-4}(\Delta^2)\big)\right]\right\}^{p-1} \times \\ &\left\{1 - (p-2)\left[2E\big(\chi_{p+2}^{-2}(\Delta^2)\big) - (p-2)E\big(\chi_{p+2}^{-4}(\Delta^2)\big)\right]\right. \\ &\left. + (p^2-4)\Delta^2E\big(\chi_{p+4}^{-4}(\Delta^2)\big)\right\}. \end{aligned} \tag{4.3.22}$$

At $\Delta^2 = 0$, $R_3(\hat{\boldsymbol{\theta}}_n^S;\mathbf{I}_p) = p|M_3(\hat{\boldsymbol{\theta}}_n^S)|^{1/p} = 2$ and as $\Delta^2 \to \infty$, $R_3(\hat{\boldsymbol{\theta}}_n^S;\mathbf{I}_p)$ and $p|\mathbf{M}_3(\hat{\boldsymbol{\theta}}_n^S)|^{1/p}$ converge to p. The graph of $p|\mathbf{M}_3(\hat{\boldsymbol{\theta}}_n^S)|^{1/p}$ remains below the graph of $R_3(\hat{\boldsymbol{\theta}}_n^S;\mathbf{I}_p)$ when $\Delta^2 \in (0,\infty)$. See Figure 4.3.2.

Consider the comparison of $\hat{\boldsymbol{\theta}}_n^S$ relative to $\tilde{\boldsymbol{\theta}}_n$. For a given nonzero vector $\boldsymbol{\ell}$ of p components we may evaluate the difference

$$\boldsymbol{\ell}'\mathbf{M}_3(\hat{\boldsymbol{\theta}}_n^S)\boldsymbol{\ell} - \boldsymbol{\ell}'\boldsymbol{\ell}. \tag{4.3.23}$$

Note that

$$\begin{aligned} \boldsymbol{\ell}'\mathbf{M}_3(\hat{\boldsymbol{\theta}}_n^S)\boldsymbol{\ell} = \boldsymbol{\ell}'\boldsymbol{\ell} &- (p-2)\boldsymbol{\ell}'\boldsymbol{\ell}\left\{2E\Big(\chi_{p+2}^{-2}(\Delta^2)\Big) - (p-2)E\Big(\chi_{p+2}^{-4}(\Delta^2)\Big)\right\} \\ &+ (p^2-4)\boldsymbol{\ell}'n\boldsymbol{\theta}\boldsymbol{\theta}'\boldsymbol{\ell}E\Big(\chi_{p+4}^{-4}(\Delta^2)\Big) \end{aligned} \tag{4.3.24}$$

and $\boldsymbol{\ell}'\mathbf{M}_3(\hat{\boldsymbol{\theta}}_n^S)\boldsymbol{\ell} - \boldsymbol{\ell}'\boldsymbol{\ell} < 0$ according as

$$\begin{aligned} &(p+2)\boldsymbol{\ell}'n\boldsymbol{\theta}\boldsymbol{\theta}'\boldsymbol{\ell}E\Big(\chi_{p+4}^{-4}(\Delta^2)\Big) \\ &< \ \boldsymbol{\ell}'\boldsymbol{\ell}\left\{2E\Big(\chi_{p+2}^{-2}(\Delta^2)\Big)\right\} - (p-2)E\Big(\chi_{p+2}^{-4}(\Delta^2)\Big)\Big\}. \end{aligned} \tag{4.3.25}$$

This implies

$$(p+2)\frac{\boldsymbol{\ell}'n\boldsymbol{\theta}\boldsymbol{\theta}'\boldsymbol{\ell}}{\boldsymbol{\ell}'\boldsymbol{\ell}}E\Big(\chi_{p+4}^{-4}(\Delta^2)\Big) < \left\{2E\Big(\chi_{p+2}^{-2}(\Delta^2)\Big) - (p-2)E\Big(\chi_{p+2}^{-4}(\Delta^2)\Big)\right\}. \tag{4.3.26}$$

Then

$$(p+2)\Delta^2E\Big(\chi_{p+4}^{-4}(\Delta^2)\Big) < \left\{2E\Big(\chi_{p+2}^{-2}(\Delta^2)\Big) - (p-2)E\Big(\chi_{p+2}^{-4}(\Delta^2)\Big)\right\} \tag{4.3.27}$$

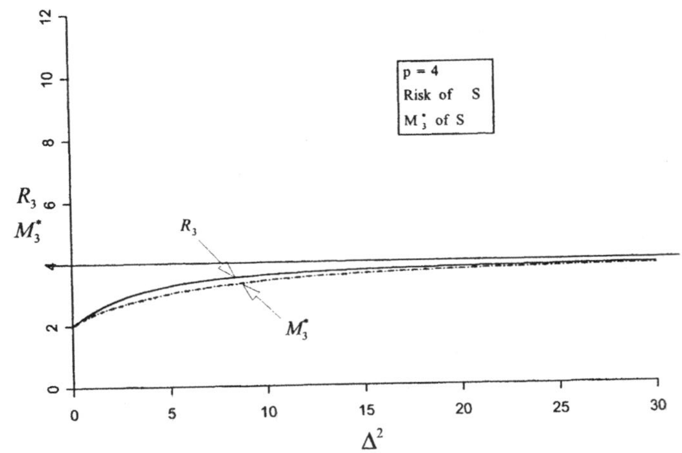

Figure 4.3.2 Graphs of $R_3 = R_3(\hat{\theta}_n^S; \mathbf{I}_p)$ and $M_3^* = p|\mathbf{M}_3(\hat{\boldsymbol{\theta}}_n^S)|^{1/p}$

holds for all Δ^2, which is equivalent to

$$pE\left[\chi_{p+2}^{-2}(\Delta^2)\right) < (p+1)(p-2)E\left[\chi_{p+2}^{-4}(\Delta^2)\right] \quad \text{for all } \Delta^2 \tag{4.3.28}$$

and may not always hold. Hence, $\mathbf{M}_3(\hat{\boldsymbol{\theta}}_n^S) - \mathbf{M}_1(\tilde{\boldsymbol{\theta}}_n)$ is not negative semi-definite (n.s.d.). Thus, $\hat{\boldsymbol{\theta}}_n^S$ does not dominate $\tilde{\boldsymbol{\theta}}_n$ uniformly with respect to the MSE criterion.

The efficiency of $\hat{\boldsymbol{\theta}}_n^S$ based on the MSE matrix is given by

$$\begin{aligned}
\text{MRE}(\hat{\boldsymbol{\theta}}_n^S; \tilde{\boldsymbol{\theta}}_n) &= \frac{|M_1(\tilde{\boldsymbol{\theta}}_n)|^{1/p}}{|M_3(\hat{\boldsymbol{\theta}}_n^S)|^{1/p}} \\
&= \left\{1 - (p-2)\left[2E(\chi_{p+2}^{-2}(\Delta^2)) - (p-2)E(\chi_{p+2}^{-4}(\Delta^2))\right]\right\}^{(p-1)/p} \times \\
&\quad \left\{1 - (p-2)\left[2E(\chi_{p+2}^{-2}(\Delta^2)) - (p-2)E(\chi_{p+2}^{-4}(\Delta^2))\right]\right. \\
&\quad \left. + (p^2-4)\Delta^2 E(\chi_{p+4}^{-4}(\Delta^2))\right\}^{-1/p}.
\end{aligned} \tag{4.3.29}$$

$\text{MRE}(\hat{\boldsymbol{\theta}}_n^S : \tilde{\boldsymbol{\theta}}_n)$ is a decreasing function of Δ^2. At $\Delta^2 = 0$, it assumes the maximum value $p/2$ and as $\Delta^2 \to \infty$, the efficiency tapers down to unity from above.

A consideration of the efficiency of $\hat{\boldsymbol{\theta}}_n^S$ relative to $\tilde{\boldsymbol{\theta}}_n$ based on the risk

($\mathbf{W} = \mathbf{I}_p$) expression leads to

$$\text{RRE}(\hat{\boldsymbol{\theta}}_n^S; \tilde{\boldsymbol{\theta}}_n) = \left[1 - p\left(1 - \frac{2}{p}\right)^2 E\left(\chi_p^{-2}(\Delta^2)\right)\right]^{-1}. \tag{4.3.30}$$

$\text{RRE}(\hat{\boldsymbol{\theta}}_n^S; \tilde{\boldsymbol{\theta}}_n)$ is also a decreasing function of Δ^2 with a maximum value $p/2$ and tends to 1 as $\Delta^2 \to \infty$ from below. See Figure 4.3.3 for both kinds of efficiencies.

Now we consider the risk analysis of $\hat{\boldsymbol{\theta}}_n^S$ relative to $\tilde{\boldsymbol{\theta}}_n$ for general $\mathbf{W}$. The weighted risk difference

$$\begin{aligned}
&R_1(\tilde{\boldsymbol{\theta}}_n; \mathbf{W}) - R_2(\hat{\boldsymbol{\theta}}_n^S; \mathbf{W}) \\
&= (p-2)\text{tr}(\mathbf{W})\left\{2E\left(\chi_{p+2}^{-2}(\Delta^2)\right) - (p-2)E\left(\chi_{p+2}^{-4}(\Delta^2)\right)\right\} \\
&\quad - (p^2-4)n\boldsymbol{\theta}'\mathbf{W}\boldsymbol{\theta}E\left(\chi_{p+4}^{-4}(\Delta^2)\right) \qquad (4.3.31) \\
&= (p-2)\text{tr}(\mathbf{W})\Bigg[(p-2)E\left(\chi_{p+2}^{-4}(\Delta^2)\right) \\
&\quad + \left\{1 - \frac{(p+2)(n\boldsymbol{\theta}'\mathbf{W}\boldsymbol{\theta})}{2\Delta^2\text{tr}(\mathbf{W})}\right\}2\Delta^2 E\left(\chi_{p+4}^{-4}(\Delta^2)\right)\Bigg]
\end{aligned}$$

≥ 0 if the matrix $\mathbf{W}$ satisfies the condition

$$\frac{\text{tr}(\mathbf{W})}{\text{Ch}_{\max}(\mathbf{W})} \geq \frac{p+2}{2}. \tag{4.3.32}$$

Thus, $\hat{\boldsymbol{\theta}}_n^S$ dominates $\tilde{\boldsymbol{\theta}}_n$ uniformly for all $\mathbf{W}$, satisfying (4.3.31). Further, as $\Delta^2 \to \infty$, the risk difference tends to zero from below.

4.3.3 Positive-Rule Stein Estimator (PRSE)

We have motivated the PRSE, $\hat{\boldsymbol{\theta}}_n^{S+}$ from the James-Stein estimator $\hat{\boldsymbol{\theta}}_n^S$ so that it may have negative coordinates. Clearly, $\hat{\boldsymbol{\theta}}_n^{S+}$ defined at (4.3.17) is a PTE. Also, this estimator can never change the sign of its coordinates, and hence the name. First, note that we may write

$$\hat{\boldsymbol{\theta}}_n^{S+} = \hat{\boldsymbol{\theta}}_n^S I\left(I(n\|\tilde{\boldsymbol{\theta}}_n\|^2 > p-2)\right) \tag{4.3.33}$$

in terms of the James-Stein estimator. Note also that $n\|\hat{\boldsymbol{\theta}}_n^{S+}\|^2 \leq \|\hat{\boldsymbol{\theta}}_n^S\|^2$.

The bias vector, QB, MSE matrix, and the risk expressions may be obtained as given below.

(i) $\quad \mathbf{b}_4(\hat{\boldsymbol{\theta}}_n^{S+}) = -\boldsymbol{\theta}\Big[H_{p+2}(p-2;\Delta^2) + (p-2)E\left(\chi_{p+2}^{-2}(\Delta^2)\right)$

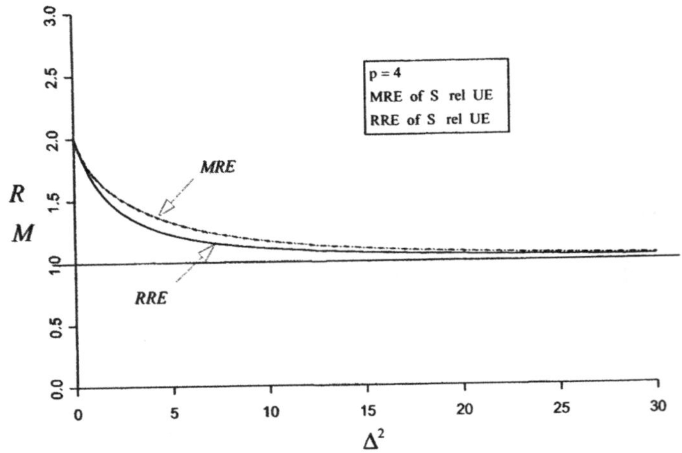

Figure 4.3.3 Graphs of MRE $= \text{MRE}(\hat{\boldsymbol{\theta}}_n^S; \tilde{\boldsymbol{\theta}}_n)$ and RRE $= \text{RRE}(\hat{\boldsymbol{\theta}}_n^S; \tilde{\boldsymbol{\theta}}_n)$

$$-(p-2)E\{\chi_{p+2}^{-2}(\Delta^2)I(\chi_{p+2}^2(\Delta^2) < p-2)\}\Big], \quad (4.3.34a)$$

(ii) $$B_4(\hat{\boldsymbol{\theta}}_n^{S+}) = \Delta^2\Big[H_{p+2}(p-2;\Delta^2) + (p-2)E(\chi_{p+2}^{-2}(\Delta^2)) \\ -(p-2)E\{\chi_{p+2}^{-2}(\Delta^2)I(\chi_{p+2}^2(\Delta^2) < p-2)\}\Big]^2, \quad (4.3.34b)$$

and
(iii) $\mathbf{M}_4(\hat{\boldsymbol{\theta}}_n^{S+})$

$$= \mathbf{M}_3(\hat{\boldsymbol{\theta}}_n^S) - \mathbf{I}_p E\Big[\Big\{1-(p-2)\chi_{p+2}^{-2}(\Delta^2)\Big\}^2 I(\chi_{p+2}^2(\Delta^2) < p-2)\Big] \\ +n\boldsymbol{\theta}\boldsymbol{\theta}'\Big\{2E\Big[(1-(p-2)\chi_{p+2}^{-2}(\Delta^2))I(\chi_{p+2}^2(\Delta^2) < p-2)\Big] \\ -E\Big[1-(p-2)\chi_{p+4}^{-2}(\Delta^2)\Big\}^2 I(\chi_{p+4}^2(\Delta^2) < p-2)\Big)\Big]\Big\}, \quad (4.3.35a)$$

where

$$\mathbf{M}_3(\hat{\boldsymbol{\theta}}_n^S) = \mathbf{I}_p - (p-2)\mathbf{I}_p\{2E(\chi_{p+2}^{-2}(\Delta^2)) - (p-2)E(\chi_{p+2}^{-4}(\Delta^2))\} \\ + (p-2)n\boldsymbol{\theta}\boldsymbol{\theta}'\{2E(\chi_{p+2}^{-2}(\Delta^2)) - 2E(\chi_{p+4}^{-2}(\Delta^2)) \\ + (p-2)E(\chi_{p+4}^{-4}(\Delta^2))\}. \quad (4.3.35b)$$

Therefore, the risk of $\hat{\boldsymbol{\theta}}_n^{S+}$ is

$$\begin{aligned}
\text{(iv)}\quad & R_4(\hat{\boldsymbol{\theta}}_n^{S+};\mathbf{W}) \\
&= \operatorname{tr}\left[\mathbf{M}_4(\hat{\boldsymbol{\theta}}_n^{S+})\mathbf{W}\right] \\
&= R_3(\hat{\boldsymbol{\theta}}_n^{S};\mathbf{W}) - \operatorname{tr}(\mathbf{W})E\Big[(1-(p-2)\chi_{p+2}^{-2}(\Delta^2))^2 I(\chi_{p+2}^{-2}(\Delta^2) < p-2)\Big] \\
&\quad +(\boldsymbol{\theta}'\mathbf{W}\boldsymbol{\theta})\Big\{2E\Big[(1-(p-2)\chi_{p+2}^{-2}(\Delta^2))I(\chi_{p+2}^{2}(\Delta^2) < p-2)\Big] \\
&\quad -E\Big[(1-(p-2)\chi_{p+4}^{-2}(\Delta^2))^2 I(\chi_{p+4}^{2}(\Delta^2) < p-2)\Big]\Big\} \qquad (4.3.36) \\
&= R(\hat{\boldsymbol{\theta}}_n^{S};\mathbf{I}_p) - pE\Big[(1-(p-2)\chi_{p+2}^{-2}(\Delta^2))^2 I(\chi_{p+2}^{2}(\Delta^2) < p-2)\Big] \\
&\quad +\Delta^2\Big\{2E\Big[(1-(p-2)\chi_{p+2}^{-2}(\Delta^2))I(\chi_{p+2}^{2}(\Delta^2) < p-2)\Big] \\
&\qquad -E\Big[(1-(p-2)\chi_{p+4}^{-2}(\Delta^2))^2 I(\chi_{p+4}^{-2}(\Delta^2) < p-2)\Big]\Big\} \qquad (4.3.37)
\end{aligned}$$

if $\mathbf{W} = \mathbf{I}_p$.

Next, we look at the risk comparison of $\hat{\boldsymbol{\theta}}_n^{S}$ and $\hat{\boldsymbol{\theta}}_n^{S+}$:

$$\begin{aligned}
& R_3(\hat{\boldsymbol{\theta}}_n^{S};\mathbf{W}) - R_4(\hat{\boldsymbol{\theta}}_n^{S+};\mathbf{W}) \\
&= \operatorname{tr}(\mathbf{W})E[(1-(p-2)\chi_{p+2}^{-2}(\Delta^2))^2 I(\chi_{p+2}^{2}(\Delta^2) < p-2)] \\
&\quad - (n\boldsymbol{\theta}'\mathbf{W}\boldsymbol{\theta})\Big\{2E[(1-(p-2)\chi_{p+2}^{-2}(\Delta^2))I(\chi_{p+2}^{2}(\Delta^2) < p-2)] \\
&\qquad - E[(1-(p-2)\chi_{p+4}^{-2}(\Delta^2))^2 I(\chi_{p+4}^{2}(\Delta^2) < p-2)]\Big\}. \qquad (4.3.38)
\end{aligned}$$

The term $2E[(1-(p-2)\chi_{p+2}^{-2}(\Delta^2))I(\chi_{p+2}^{2}(\Delta^2) < p-2)]$ is negative, whereas $E[(1-(p-2)\chi_{p+4}^{-2}(\Delta^2))^2 I(\chi_{p+4}^{2}(\Delta^2) < p-2)]$ is positive. Hence, the risk difference is positive. Hence, $\hat{\boldsymbol{\theta}}_n^{S+}$ dominates $\hat{\boldsymbol{\theta}}_n^{S}$ uniformly.

We know that $\tilde{\boldsymbol{\theta}}_n^{S}$ dominates $\tilde{\boldsymbol{\theta}}_n$. This comparison will show that $\hat{\boldsymbol{\theta}}_n^{S+}$ is better than $\tilde{\boldsymbol{\theta}}_n^{S}$, which may be written altogether as

$$\hat{\boldsymbol{\theta}}_n^{S+} \succ \hat{\boldsymbol{\theta}}_n^{S} \succ \tilde{\boldsymbol{\theta}}_n, \qquad (4.3.39)$$

where $\succ$ stands for domination under the risk criterion. In passing we mention the properties of the QB of the estimators shown in Figure 4.3.4.

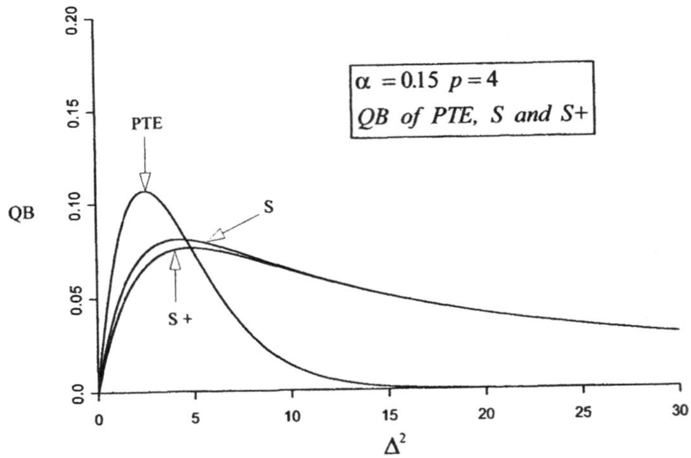

Figure 4.3.4 Graph of QB of estimators: PTE $= \hat{\boldsymbol{\theta}}_n^{PT}$, JSE $= \hat{\boldsymbol{\theta}}_n^{S}$, and PRSE $= \hat{\boldsymbol{\theta}}_n^{S+}$

Now, consider the MSE difference

$$\begin{aligned}
&\mathbf{M}_3(\hat{\boldsymbol{\theta}}_n^{S}) - \mathbf{M}_4(\hat{\boldsymbol{\theta}}_n^{S+}) \\
&= \mathbf{I}_p E\Big[(1-(p-2)\chi_{p+2}^{-2}(\Delta^2))^2 I(\chi_{p+2}^2(\Delta^2) < p-2)\Big] \\
&\quad -n\boldsymbol{\theta}\boldsymbol{\theta}'\Big\{2E\Big[(1-(p-2)\chi_{p+2}^{-2}(\Delta^2))I(\chi_{p+2}^2(\Delta^2) < p-2)\Big] \\
&\quad -E\big[(1-(p-2)\chi_{p+4}^{-2}(\Delta^2))^2 I(\chi_{p+4}^2(\Delta^2) < p-2)\big]\Big\}. \qquad (4.3.40)
\end{aligned}$$

For a nonzero vector, $\boldsymbol{\ell}$ we have

$$\begin{aligned}
&\boldsymbol{\ell}'\boldsymbol{\ell}E\Big[\big(1-(p-2)\chi_{p+4}^{-2}(\Delta^2)\big)^2 I\big(\chi_{p+2}^2(\Delta^2) < p-2\big)\Big] \\
&-(n\boldsymbol{\ell}'\boldsymbol{\theta}\boldsymbol{\theta}'\boldsymbol{\ell})\Big\{2E\Big[(1-(p-2)\chi_{p+2}^{-2}(\Delta^2))I\big(\chi_{p+2}^2(\Delta^2) < p-2\big)\Big] \\
&\qquad -E\Big[\big(1-(p-2)\chi_{p+4}^{-2}(\Delta^2)\big)^2 I\big(\chi_{p+4}^2(\Delta^2) < p-2\big)\Big]\Big\}. \qquad (4.3.41)
\end{aligned}$$

Since $\chi_{p+4}^2(\Delta^2) < p-2$, we have

$$1-(p-2)\chi_{p+2}^{-2}(\Delta^2) < 0, \qquad (4.3.42)$$

and the quantity in the second big bracket is negative. Thus, the whole expression is positive. Taking $\max(n\boldsymbol{\ell}'\boldsymbol{\theta}\boldsymbol{\theta}'\boldsymbol{\ell}/\boldsymbol{\ell}'\boldsymbol{\ell}) = \Delta^2$, we have

$$E\Big[\big(1-(p-2)\chi_{p+2}^{-2}(\Delta^2)\big)^2 I\big(\chi_{p+2}^2(\Delta^2) < p-2\big)\Big]$$
$$+\Delta^2\Big\{2E\Big[\big((p-2)\chi_{p+2}^{-2}(\Delta^2)-1\big)I\big(\chi_{p+2}^2(\Delta^2) < p-2\big)$$
$$+E\Big[\big(1-(p-2)\chi_{p+4}^{-2}(\Delta^2)\big)^2 I\big(\chi_{p+4}^2(\Delta^2) < p-2\big)\Big]\Big\} \geq 0. \quad (4.3.43)$$

This is positive for all Δ^2 and

$$\hat{\boldsymbol{\theta}}_n^{S+} \succ \hat{\boldsymbol{\theta}}_n^{S}. \quad (4.3.44)$$

The MSE based efficiency of $\hat{\boldsymbol{\theta}}_n^{S+}$ relative to $\tilde{\boldsymbol{\theta}}_n$ is given by

$$\text{MRE}(\hat{\boldsymbol{\theta}}_n^{S+};\tilde{\boldsymbol{\theta}}_n) = \Big(1-(p-2)\big[2E\big(\chi_{p+2}^{-2}(\Delta^2)\big)-(p-2)E\big(\chi_{p+2}^{-4}(\Delta^2)\big)\big]$$
$$E\big[\{1-(p-2)\chi_{p+2}^{-2}(\Delta^2)\}^2 I\big(\chi_{p+2}^2(\Delta^2) < p-2\big)\big]\Big)^{(p-1)/p}$$
$$\times\big[m_4(\Delta^2)\big]^{-1/p}. \quad (4.3.45)$$

where

$$m_4(\Delta^2) = \Big(1-(p-2)\big[2E\big(\chi_{p+2}^{-2}(\Delta^2)\big)-(p-2)E\big(\chi_{p+2}^{-4}(\Delta^2)\big)\big]$$
$$+(p^2-4)\Delta^2 E\big(\chi_{p+4}^{-4}(\Delta^2)\big)$$
$$+\Delta^2\Big\{2E\big[\big(1-(p-2)\chi_{p+2}^{-2}(\Delta^2)\big)I\big(\chi_{p+2}^2(\Delta^2) < p-2\big)\big]\Big)$$
$$-\ E\big[\big(1-(p-2)\chi_{p+4}^{-2}(\Delta^2)\big)^2 I\big(\chi_{p+4}^2(\Delta^2) < p-2\big)\big]\Big\}. \quad (4.3.46)$$

Similarly, the risk efficiency of $\hat{\boldsymbol{\theta}}_n^{S+}$ relative to $\tilde{\boldsymbol{\theta}}_n$ with $\mathbf{W} = \mathbf{I}_p$ may be obtained as

$$\text{RRE}(\hat{\boldsymbol{\theta}}_n^{S+};\tilde{\boldsymbol{\theta}}_n)$$
$$=\Big\{1-p^{-1}(p-2)^2E\big[\chi_p^{-2}(\Delta^2)\big]$$
$$-\ E\big[\{1-(p-2)\chi_{p+2}^{-2}(\Delta^2)\}^2 I\big(\chi_{p+2}^2(\Delta^2) < p-2\big)\big]$$
$$+\ p^{-1}\Delta^2\Big\{2E\big[\{1-(p-2)\chi_{p+2}^{-2}(\Delta^2)\}I\big(\chi_{p+2}^2(\Delta^2) < p-2\big)\big]$$
$$-\ E\big[\big(1-(p-2)\chi_{p+4}^{-2}(\Delta^2)\big)^2 I\big(\chi_{p+4}^2(\Delta^2) < p-2\big)\big]\Big\}^{-1}. \quad (4.3.47)$$

Table 4.3.1 presents some risk gains of $\hat{\boldsymbol{\theta}}_n^{S+}$ over $\hat{\boldsymbol{\theta}}_n^{S}$. Also, see Figure 4.3.5 for the risk of the four estimators, namely $\tilde{\boldsymbol{\theta}}_n, \hat{\boldsymbol{\theta}}_n^{\text{PT}}, \hat{\boldsymbol{\theta}}_n^{S}$, and $\hat{\boldsymbol{\theta}}_n^{S+}$.

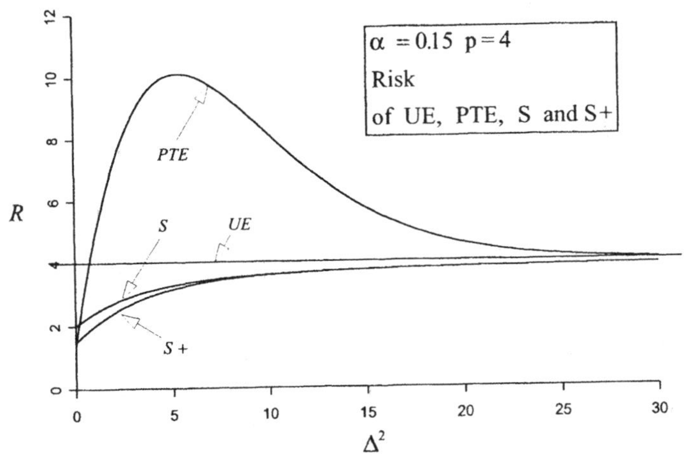

Figure 4.3.5 Graph of the risks of the four estimators ($\text{UE} = \tilde{\boldsymbol{\theta}}_n$, $\text{PTE} = \hat{\boldsymbol{\theta}}_n^{\text{PT}}$, $\text{S} = \hat{\boldsymbol{\theta}}_n^{S}$, and $S^+ = \hat{\boldsymbol{\theta}}_n^{S+}$)

Table 4.3.1 Risk Gain of PRSE over JSE

$p\backslash\Delta^2$	0	1	2	3	4	5	6	7	8	9	10
4	1.35914	1.20103	1.52558	1.08236	1.05536	1.03769	1.02583	1.01776	1.01222	1.00841	1.00578
6	1.47781	1.26169	1.16530	1.11089	1.07668	1.05389	1.03822	1.02724	1.01945	1.01390	1.00993
8	1.54504	1.29510	1.18779	1.12778	1.08993	1.06447	1.04672	1.03405	1.02490	1.01823	1.01335
10	1.59024	1.31722	1.20302	1.13959	1.09951	1.07239	1.05329	1.03949	1.02938	1.02190	1.01634
12	1.62348	1.33333	1.21430	1.14855	1.10698	1.07872	1.05866	1.04405	1.03322	1.02512	1.01901

4.3.4 Sclove-Morris-Radhakrishnan Modifications

It is interesting to note that the PRSE, $\hat{\boldsymbol{\theta}}_n^{S+}$ has two representations, namely

$$\hat{\boldsymbol{\theta}}_n^{S+} = \hat{\boldsymbol{\theta}}_n^{S} I(n\|\tilde{\boldsymbol{\theta}}_n\|^2 > p-2), \tag{4.3.48a}$$

which is a PTE based on $\boldsymbol{\theta}_n^S$ and $\mathbf{0}$ with the critical value $(p-2)$, and

$$\hat{\boldsymbol{\theta}}_n^{S+} = \boldsymbol{\theta}_n^{*\text{PT}}[1-(p-2)(n\|\tilde{\boldsymbol{\theta}}_n\|)^{-1}], \tag{4.3.48b}$$

which is a Stein-type estimator where $\tilde{\boldsymbol{\theta}}_n$ is replaced by

$$\boldsymbol{\theta}_n^{*\text{PT}} = \tilde{\boldsymbol{\theta}}_n I(n\|\tilde{\boldsymbol{\theta}}_n\|^2 > p-2), \tag{4.3.48c}$$

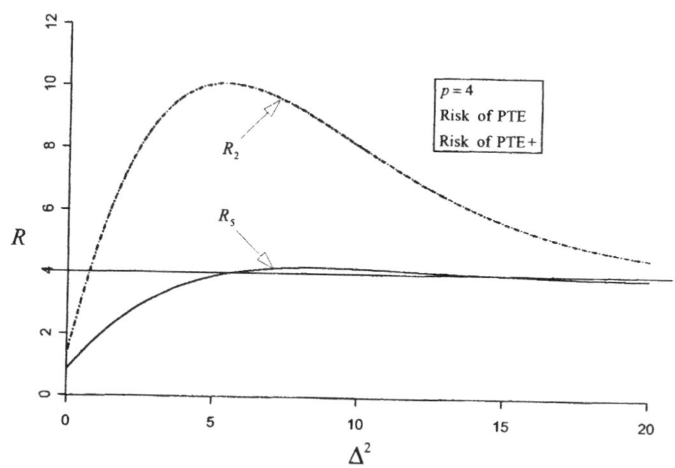

Figure 4.3.6 Graph of $R_5 = R_5(\hat{\boldsymbol{\theta}}_n^{\mathrm{PT}+}; \mathbf{I}_p)$ and $R_2 = R_2(\hat{\boldsymbol{\theta}}_n^{\mathrm{PT}}; \mathbf{I}_p)$

which is an ordinary PTE with critical value $(p-2)$. We can represent the risk expression of $\hat{\boldsymbol{\theta}}_n^{S+}$ in terms of the risk expression of $\boldsymbol{\theta}_n^{*\mathrm{PT}}$ as follows:

$$
\begin{aligned}
&R_4(\hat{\boldsymbol{\theta}}_n^{S+}; \mathbf{W}) = \\
&R_2(\boldsymbol{\theta}_n^{*\mathrm{PT}}; \mathbf{W}) - (p-2)\mathrm{tr}(\mathbf{W})\Big[2E(\chi_{p+2}^{-2}(\Delta^2)) - (p-2)E(\chi_{p+2}^{-4}(\Delta^2))\Big] \\
&+(p^2-4)(n\boldsymbol{\theta}'\mathbf{W}\boldsymbol{\theta})E(\chi_{p+4}^{-4}(\Delta^2)) \\
&+(p-2)\mathrm{tr}(\mathbf{W})\Big\{2E_P\Big[\frac{1}{p+2r}H_{p+2r}(p-2;0)\Big] \\
&\qquad -(p-2)E_P\Big[\frac{1}{(p+2r)(p-2+2r)}H_{p-2+2r}(p-2;0)\Big]\Big\} \\
&-(p-2)(n\boldsymbol{\theta}'\mathbf{W}\boldsymbol{\theta})\Big\{2E_P\Big[\frac{1}{p+2r}H_{p+2r}(p-2;0)\Big] \\
&\qquad -2E_P\Big[\frac{1}{(p+2r)(p-2+2r)}H_{p-2+2r}(p-2;0)\Big] \\
&\qquad +(p-2)E_P\Big[\frac{1}{(p+2+2r)(p+2r)}H_{p+2r}(p-2;0)\Big]\Big\}, \qquad (4.3.49)
\end{aligned}
$$

where E_P stands for the expectation with respect to the Poisson r.v. r with mean $\Delta^2/2$ and $H_\nu(\cdot;0)$ is the cdf of a central chi-square variable with ν d.f.

Also, we have the Sclove-Morris-Radhakrishnan (1972) modification of the PTE, $\hat{\boldsymbol{\theta}}_n^{\text{PT}}$ as given below. First we have the usual one

$$\hat{\boldsymbol{\theta}}_n^{\text{PT}} = \tilde{\boldsymbol{\theta}}_n I(n\|\tilde{\boldsymbol{\theta}}_n\|^2 > \chi_p^2(\alpha)), \tag{4.3.50}$$

and then we have the Stein-type modification of $\hat{\boldsymbol{\theta}}_n^{\text{PT}}$ given by

$$\hat{\boldsymbol{\theta}}_n^{\text{PT+}} = \hat{\boldsymbol{\theta}}_n^{\text{PT}} \left(1 - \frac{(p-2)}{n\|\tilde{\boldsymbol{\theta}}_n\|^2}\right). \tag{4.3.51}$$

The following theorem shows the uniform dominance of $\hat{\boldsymbol{\theta}}_n^{\text{PT+}}$ over $\hat{\boldsymbol{\theta}}_n^{\text{PT}}$. Let $R_5(\hat{\boldsymbol{\theta}}_n^{\text{PT+}}; \mathbf{I}_p)$ be the risk of $\hat{\boldsymbol{\theta}}_n^{\text{PT+}}$. See Figure 4.3.6.

Theorem 1. Let $\hat{\boldsymbol{\theta}}_n^{\text{PT}}$, $\hat{\boldsymbol{\theta}}_n^{\text{PT+}}$ and $\hat{\boldsymbol{\theta}}_n^{S+}$ be the estimators defined by (4.3.50), (4.3.51), and (4.3.17). Then, for $p \geq 3$,

$$R_2\left(\hat{\boldsymbol{\theta}}_n^{\text{PT}}; \mathbf{I}_p\right) \geq R_5\left(\hat{\boldsymbol{\theta}}_n^{\text{PT+}}; \mathbf{I}_p\right) \quad \text{for all } \Delta^2, \tag{4.3.52}$$

with strict inequality for at least one Δ^2.

Proof. Consider the risk difference $R_2(\hat{\boldsymbol{\theta}}_n^{\text{PT}}; \mathbf{I}_p) - R_5(\hat{\boldsymbol{\theta}}_n^{\text{PT+}}; \mathbf{I}_p)$, where $R_5(\hat{\boldsymbol{\theta}}_n^{\text{PT+}}; \mathbf{I}_p)$ is the risk expression (4.3.36) with the critical value $(p-2)$ replaced by $\chi_p^2(\alpha)$. Then the risk difference is equal to

$$\begin{aligned}
&p(p-2)\Big[2E\left(\chi_{p+2}^{-2}(\Delta^2)\right) - (p-2)E\left(\chi_{p+2}^{-4}(\Delta^2)\right)\Big] - (p^2-4)\Delta^2 E\left(\chi_{p+4}^{-4}(\Delta^2)\right)\\
&+p(p-2)\Big[(p-2)E_P\Big\{\frac{1}{(p+2r)(p-2+2r)} H_{p-2+2r}\left(\chi_p^2(\alpha); 0\right)\Big]\\
&\qquad -2E_P\Big[\frac{1}{p+2r} H_{p+2r}\left(\chi_p^2(\alpha); 0\right)\Big]\Big\}\\
&\qquad +\Delta^2(p-2)\Big\{2E_P\Big[\frac{1}{p+2r} H_{p+2r}\left(\chi_p^2(\alpha); 0\right)\Big]\\
&\qquad -2E_P\Big[\frac{1}{(p+2r)(p-2+2r)} H_{p-2+2r}\left(\chi_p^2(\alpha); 0\right)\Big]\\
&\qquad +(p-2)E_P\Big[\frac{1}{(p+2r)(p-2+2r)} H_{p+2r}\left(\chi_p^2(\alpha); 0\right)\Big]\Big\}.
\end{aligned} \tag{4.3.53}$$

All the terms in the brackets are positive, and the theorem follows.

Further, consider the risk difference

$$
\begin{aligned}
&R_5(\hat{\boldsymbol{\theta}}_n^{\mathrm{PT}+}; \mathbf{I}_p) - R_4(\hat{\boldsymbol{\theta}}_n^{S+}; \mathbf{I}_p) \\
&= pE\Big[\{1-(p-2)\chi_{p+2}^{-2}(\Delta^2)\}^2 \\
&\quad \times \{I(\chi_{p+2}^{-2}(\Delta^2) < p-2) - I(\chi_{p+2}^{-2}(\Delta^2) < \chi_p^2(\alpha))\}\Big] \\
&\quad -\Delta^2 E\Big[2\Big\{1-(p-2)\chi_{p+2}^{-2}(\Delta^2)\Big\} \\
&\quad \times \Big\{I(\chi_{p+2}^2(\Delta^2) < p-2) - I(\chi_{p+2}^2(\Delta^2) < \chi_p^2(\alpha))\Big\} \\
&\quad -\Big\{1-(p-2)\chi_{p+4}^{-2}(\Delta^2)\Big\}^2 \\
&\quad \times \Big\{I(\chi_{p+4}^2(\Delta^2) < p-2) - I(\chi_{p+4}^2(\Delta^2) < \chi_p^2(\alpha))\Big\}\Big]. \qquad (4.3.54)
\end{aligned}
$$

The r.h.s. of (4.3.54) is *positive* if $\chi_p^2(\alpha) < (p-2)$. Thus, $\hat{\boldsymbol{\theta}}_n^{S+}$ dominates $\hat{\boldsymbol{\theta}}_n^{\mathrm{PT}+}$ uniformly in Δ^2 if $\chi_p^2(\alpha) < (p-2)$. $\hat{\boldsymbol{\theta}}_n^{S+}$ is always minimax, and it is a minimax substitute of $\hat{\boldsymbol{\theta}}_n^{\mathrm{PT}}$ whenever $\chi_p^2(\alpha) \le 2(p-2)$. If $\chi_p^2(\alpha) > 2(p-2)$, then $\hat{\boldsymbol{\theta}}_n^{\mathrm{PT}+}$ dominates $\hat{\boldsymbol{\theta}}_n^{\mathrm{PT}}$ but $\hat{\boldsymbol{\theta}}_n^{S+}$ may not dominate $\hat{\boldsymbol{\theta}}_n^{\mathrm{PT}}$. In this case, $\hat{\boldsymbol{\theta}}_n^{\mathrm{PT}+}$ may not be minimax.

Thus, if $\chi_p^2(\alpha) < (p-2)$, we can order the estimators $\hat{\boldsymbol{\theta}}_n^{\mathrm{PT}}$, $\hat{\boldsymbol{\theta}}_n^{\mathrm{PT}+}$, and $\hat{\boldsymbol{\theta}}_n^{S+}$ as $\hat{\boldsymbol{\theta}}_n^{\mathrm{PT}+} > \hat{\boldsymbol{\theta}}_n^{\mathrm{PT}}$ for all $\chi_p^2(\alpha)$ and Δ^2 and $\hat{\boldsymbol{\theta}}_n^{S+} > \hat{\boldsymbol{\theta}}_n^{\mathrm{PT}+}$. See Figure 4.3.7.

4.4 Derivation of the Stein-Type Estimators

In this section, we present some general methods of obtaining Stein-type estimators. They are

(1) Risk difference representation approach due to Stein (1956) and James and Stein (1961).

(2) Empirical Bayes approach due to Efron and Morris (1973), and

(3) Preliminary test approach (or quasi-empirical Bayes approach) due to Saleh and Sen (1978-1986).

4.4.1 Risk Difference Representation Approach

Let $R_0(\tilde{\boldsymbol{\theta}}_n; \boldsymbol{\theta})$ be the risk of $\tilde{\boldsymbol{\theta}}_n$ under the loss $n\|\tilde{\boldsymbol{\theta}}_n - \boldsymbol{\theta}\|^2$, and let $R_1(\boldsymbol{\delta}_n, \theta)$ be the risk of another estimator $\boldsymbol{\delta}_n$. In order to determine an efficient estimator, consider the risk difference

$$\nabla R_0 = R_0(\tilde{\boldsymbol{\theta}}_n; \boldsymbol{\theta}) - R_1(\boldsymbol{\delta}_n; \boldsymbol{\theta}). \qquad (4.4.1)$$

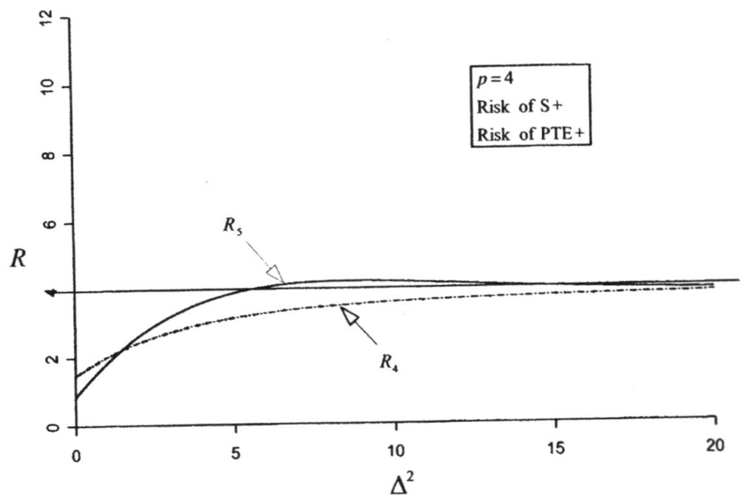

Figure 4.3.7 Graphs of $R_4 = R_4(\hat{\boldsymbol{\theta}}_n^{S+}; \mathbf{I}_p)$ and $R_5 = R_5(\hat{\boldsymbol{\theta}}_n^{\text{PT+}}; \mathbf{I}_p)$

The aim is to derive the estimator, $\boldsymbol{\delta}_n$ such that ∇R_0 as a function of $\boldsymbol{\theta}$ is **nonnegative and not identically zero.**

The Stein's identity (2.1.3) of Chapter 2 is an effective tool for **transforming risk differences** into an unbiased estimator of the "risk difference" considered as a parameter. This is given in the following theorem due to Hoffmann (1992):

Theorem 1. Let $\boldsymbol{\delta}_n = \tilde{\boldsymbol{\theta}}_n - \mathbf{g}(\tilde{\boldsymbol{\theta}}_n)$, where $\mathbf{g}(\tilde{\boldsymbol{\theta}}_n)$ is partially differentiable and $E\|\mathbf{g}(\tilde{\boldsymbol{\theta}}_n\|^2 < \infty$. Then

$$R_0(\tilde{\boldsymbol{\theta}}_n; \boldsymbol{\theta}) - R_1(\boldsymbol{\delta}_n; \boldsymbol{\theta}) = E_{\boldsymbol{\theta}}\left[2n \sum_{i=1}^{p} \frac{\partial g_i(\tilde{\boldsymbol{\theta}}_n)}{\partial \tilde{\theta}_{in}} - n\|\mathbf{g}(\tilde{\boldsymbol{\theta}}_n)\|^2\right]. \qquad (4.4.2)$$

Proof.

$$
\begin{aligned}
&R_0(\tilde{\boldsymbol{\theta}}_n;\boldsymbol{\theta}) - R_1(\boldsymbol{\delta}_n;\boldsymbol{\theta})\\
&= E_{\boldsymbol{\theta}}\big[n\|\tilde{\boldsymbol{\theta}}_n - \boldsymbol{\theta}\|^2\big] - E_{\boldsymbol{\theta}}\big[n\|\boldsymbol{\delta}_n - \boldsymbol{\theta}\|^2\big]\\
&= E_{\boldsymbol{\theta}}\big[n\|\tilde{\boldsymbol{\theta}}_n - \boldsymbol{\theta}\|^2\big] - E_{\boldsymbol{\theta}}\big[n\|\tilde{\boldsymbol{\theta}}_n - \boldsymbol{\theta}\|^2 - 2\{\mathbf{g}(\tilde{\boldsymbol{\theta}}_n)\}'n(\tilde{\boldsymbol{\theta}}_n - \boldsymbol{\theta}) + n\|\mathbf{g}(\tilde{\boldsymbol{\theta}}_n)\|^2\big]\\
&= 2E_{\boldsymbol{\theta}}\{[\mathbf{g}(\tilde{\boldsymbol{\theta}}_n)]'n(\tilde{\boldsymbol{\theta}}_n - \boldsymbol{\theta})\} - E_{\boldsymbol{\theta}}\{\|n\mathbf{g}(\tilde{\boldsymbol{\theta}}_n)\|^2\}\\
&= 2nE_{\boldsymbol{\theta}}\Big\{\sum_{i=1}^{p}\frac{\partial g_i(\tilde{\boldsymbol{\theta}}_n)}{\partial\tilde{\theta}_{in}}\Big\} - E_{\boldsymbol{\theta}}\{n\|\mathbf{g}(\tilde{\boldsymbol{\theta}}_n)\|^2\} \text{ by Stein's lemma 2 (2.1.3)}\\
&= E_{\boldsymbol{\theta}}\Big\{2n\sum_{i=1}^{p}\frac{\partial g_i(\tilde{\boldsymbol{\theta}}_n)}{\partial\tilde{\theta}_{in}} - n\|\mathbf{g}(\tilde{\boldsymbol{\theta}}_n)\|^2\Big\}\\
&= E_{\theta}\big[h(\tilde{\boldsymbol{\theta}}_n)\big],
\end{aligned}
\tag{4.4.3}
$$

where

$$h(\tilde{\boldsymbol{\theta}}_n) = 2n\sum_{i=1}^{p}\frac{\partial g_i(\tilde{\boldsymbol{\theta}}_n)}{\partial\tilde{\theta}_{in}} - n\|\mathbf{g}(\tilde{\boldsymbol{\theta}}_n)\|^2. \tag{4.4.4}$$

Then $h(\tilde{\boldsymbol{\theta}}_n)$ is an unbiased estimator of the risk difference. If we can show that $h(\tilde{\boldsymbol{\theta}}_n)$ is a positive-valued function for certain $\mathbf{g}(\tilde{\boldsymbol{\theta}}_n)$ satisfying the regularity conditions of Stein's lemma 2 (2.1.3), then Theorem 4.4.1 states that $\boldsymbol{\delta}_n^S = \tilde{\boldsymbol{\theta}}_n - \mathbf{g}(\tilde{\boldsymbol{\theta}}_n)$ has smaller quadratic risk than $\tilde{\boldsymbol{\theta}}_n$.

Now consider the class of estimators of $\boldsymbol{\theta}$ defined by

$$\boldsymbol{\delta}_n^{(c)} = \left(1 - \frac{c(p-2)}{n\|\tilde{\boldsymbol{\theta}}_n\|^2}\right)\tilde{\boldsymbol{\theta}}_n \tag{4.4.5}$$

based on a sample of size n from $\mathcal{N}_p(\boldsymbol{\theta}, \mathbf{I}_p)$, and let $\tilde{\boldsymbol{\theta}}_n$ be the MLE of $\boldsymbol{\theta}$. Then

$$
\begin{aligned}
&R_0(\tilde{\boldsymbol{\theta}}_n;\boldsymbol{\theta}) - R_1(\boldsymbol{\delta}_n;\boldsymbol{\theta})\\
&= p - \Big[p + c^2(p-2)^2E\big(\chi_p^{-2}(\Delta^2)\big) - 2c(p-2) + 2c(p-2)\Delta^2E\big(\chi_{p+2}^{-2}(\Delta^2)\big)\Big]\\
&= -2c(p-2)\{1-(p-2)E\big(\chi_p^{-2}(\Delta^2)\big)\} - c^2(p-2)^2E\big(\chi_p^{-2}(\Delta^2)\big) + 2c(p-2)\\
&= [2c(p-2)^2 - c^2(p-2)^2]E\big(\chi_p^{-2}(\Delta^2)\big)\\
&= (p-2)^2c(2-c)E\big(\chi_p^{-2}(\Delta^2)\big) > 0 \text{ for } 0 < c < 2,
\end{aligned}
\tag{4.4.6}
$$

where $E(\chi_p^{-2}(\Delta^2))$ is defined by (2.2.13) in Chapter 2. Here

$$\mathbf{g}(\tilde{\boldsymbol{\theta}}_n) = \frac{c(p-2)\tilde{\boldsymbol{\theta}}_n}{n\|\tilde{\boldsymbol{\theta}}_n\|^2}. \tag{4.4.7}$$

Then

(i) $E_\theta\{n\|\mathbf{g}(\tilde{\boldsymbol{\theta}}_n)\|^2\} = E_{\boldsymbol{\theta}}\left\{\frac{c^2(p-2)^2}{n\|\tilde{\boldsymbol{\theta}}_n\|^2}\right\} = c^2(p-2)^2 E\big(\chi_p^{-2}(\Delta^2)\big) < \infty$,

(ii) $h(\tilde{\boldsymbol{\theta}}_n) = \frac{c(2-c)(p-2)}{\|\tilde{\boldsymbol{\theta}}_n\|^2}$, $(0 < c < 2)$ is the unbiased estimator of the risk difference,

where

(iii) $\left|\frac{\partial g_i(\tilde{\boldsymbol{\theta}}_n)}{\partial \tilde{\theta}_{i,n}}\right| = E_\theta\left\{\frac{c(p-2)}{n\|\tilde{\boldsymbol{\theta}}_n\|^2} - \frac{2c(p-2)n\tilde{\theta}_{in}^2}{(n\|\tilde{\boldsymbol{\theta}}_n\|^2)^2}\right\} \le 3c(p-2)E\big(\chi_p^{-2}(\Delta^2)\big) < \infty.$

All conditions of Stein's lemma 2 (Section 2.1.3) are satisfied to justify the result (4.4.6). For $c = 1$, we obtain the JS estimator. Thus, the JS estimator has smaller risk than $\tilde{\boldsymbol{\theta}}_n$. As a matter of fact, the class of estimators defined by (4.4.5) has lesser risk than $\tilde{\boldsymbol{\theta}}_n$.

4.4.2 Empirical Bayes Estimation (EBE) Approach

Similar to Robbins (1955), Efron and Morris (1973) developed the empirical Bayes estimation approach to obtain the JS estimator of $\boldsymbol{\theta}$ as follows: Since $\tilde{\boldsymbol{\theta}}_n \sim \mathcal{N}_p(\boldsymbol{\theta}, \frac{1}{n}I_n)$, they incorporated the null hypothesis $H_0 : \boldsymbol{\theta} = \mathbf{0}$ into the description of the prior distribution π of $\boldsymbol{\theta}$. Thus, assuming $\boldsymbol{\theta} \sim \mathcal{N}_p(\mathbf{0}, \frac{\tau^2}{n}\mathbf{I}_p)$, they could obtain the joint distribution of $(\boldsymbol{\theta}', \tilde{\boldsymbol{\theta}}_n')$ as $\mathcal{N}_{2p}\left\{\begin{pmatrix}\mathbf{0}\\ \boldsymbol{\theta}\end{pmatrix}, n^{-1}\begin{pmatrix}\tau^2\mathbf{I} & \tau^2\mathbf{I}\\ \tau^2\mathbf{I} & (1+\tau^2)\mathbf{I}\end{pmatrix}\right\}$. Here the posterior distribution of $\boldsymbol{\theta}$ is given by

$$\boldsymbol{\theta}|\tilde{\boldsymbol{\theta}}_n \sim \mathcal{N}_p\left\{(1-B)\tilde{\boldsymbol{\theta}}_n, \frac{1}{n}(1-B)\mathbf{I}_p\right\}, \tag{4.4.8}$$

while the *marginal distribution* of $\sqrt{n}\,\tilde{\boldsymbol{\theta}}_n$ is given by

$$\mathcal{N}_p(\mathbf{0}, B^{-1}\mathbf{I}_p), \tag{4.4.9}$$

where

$$B = (1+\tau^2)^{-1}, \ 0 < B < 1. \tag{4.4.10}$$

The results (4.4.8) to (4.4.10) may be represented as an empirical Bayes tree given in Figure 4.4.1.

The p parameters are tied down together by the underlying common distribution of the θ's, but there are some differences in the θ's in general. Now, each $\theta_i\|\tilde{\theta}_{i,n} \sim N\left((1-B)\tilde{\theta}_{i,n}, (1-B)\right)$ so that

$$E\left(\theta_i\|\tilde{\theta}_{i,n}\right) = (1-B)\tilde{\theta}_{i,n}, \quad i = 1, \dots, p,$$

and the marginal distribution of $\sqrt{n}\,\tilde{\theta}_{i,n}$ is $N(\mathbf{0}, B^{-1})$. Then the Bayes estimator of $\boldsymbol{\theta}$ may be calculated as

$$\boldsymbol{\delta}_\pi = \boldsymbol{\delta}_\pi(\tilde{\boldsymbol{\theta}}_n) = (1-B)\tilde{\boldsymbol{\theta}}_n. \tag{4.4.11}$$

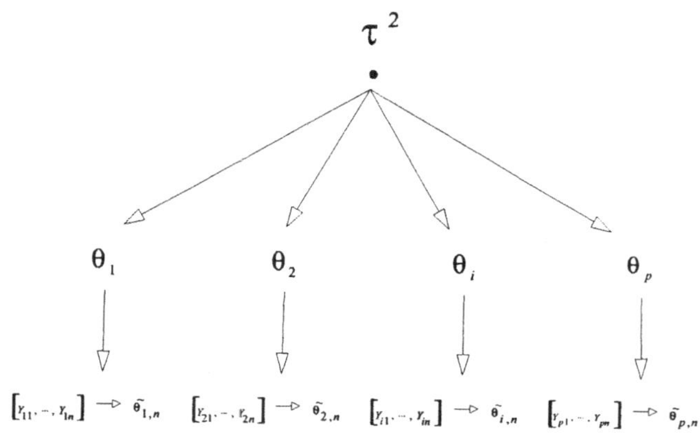

Figure 4.4.1 Empirical Bayes tree

The Bayes risk of $\boldsymbol{\delta}_\pi$ may be computed as

$$\begin{aligned}\rho(\boldsymbol{\delta}_\pi;\theta) &= E_{\boldsymbol{\theta}}\{n\|(\boldsymbol{\delta}_\pi-\boldsymbol{\theta})\|^2\}\\ &= E_{\boldsymbol{\theta}}\{n\|\tilde{\boldsymbol{\theta}}_n-\boldsymbol{\theta}\|^2\}+B^2E_{\boldsymbol{\theta}}\{n\|\tilde{\boldsymbol{\theta}}_n\|^2\}-2B\ E_{\boldsymbol{\theta}}\{n\tilde{\boldsymbol{\theta}}_n'(\tilde{\boldsymbol{\theta}}_n-\boldsymbol{\theta})\}\\ &= p-2Bp+Bp=p(1-B).\end{aligned} \tag{4.4.12}$$

However, the hyperparameter τ^2 or B in general is unknown. Therefore, we replace B by $\hat{B}_n$ given by

$$\hat{B}_n=\frac{(p-2)}{n\|\tilde{\boldsymbol{\theta}}_n\|^2} \tag{4.4.13}$$

to obtain the JS estimator

$$\hat{\boldsymbol{\theta}}_n^S=\left(1-\frac{(p-2)}{n\|\tilde{\boldsymbol{\theta}}_n\|^2}\right)\tilde{\boldsymbol{\theta}}_n.$$

To justify the use of $\hat{B}_n$, we note that

$$Bn\|\tilde{\boldsymbol{\theta}}_n\|^2\sim\chi_p^2. \tag{4.4.14}$$

Thus,

$$E_{\boldsymbol{\theta}}(n\|\tilde{\boldsymbol{\theta}}\|^2)^{-1}=\frac{B}{p-2}\quad\text{for } p>2, \tag{4.4.15}$$

and

$$E(\hat{B}_n) = (p-2)E_{\boldsymbol{\theta}}(n\|\tilde{\boldsymbol{\theta}}_n\|^2)^{-1} = B. \qquad (4.4.16)$$

This means that $\hat{B}_n$ is an unbiased estimator of B and the JS estimator arises from the Bayes estimator when we replace B by $\hat{B}_n$.

Notice that $\hat{B}_n$ can assume any value between 0 to ∞ but $B \in (0,1)$. Therefore, we may restrict the value of $n\|\tilde{\boldsymbol{\theta}}_n\|^2$ to more than $(p-2)$. This leads to the *positive-rule shrinkage estimator*, $\hat{\boldsymbol{\theta}}_n^{S+}$

$$\hat{\boldsymbol{\theta}}_n^{S+} = \left(1 - \frac{(p-2)}{n\|\tilde{\boldsymbol{\theta}}_n\|^2}\right) I\left(n\|\tilde{\boldsymbol{\theta}}_n\|^2 > p-2\right)\tilde{\boldsymbol{\theta}}_n. \qquad (4.4.17)$$

We may summarize the empirical Bayes estimation approach as follows:

(1) Obtain a Bayes estimator based on a class of prior distributions (here it is $N(\mathbf{0}, \frac{\tau^2}{n}I_p)$) that includes the "uncertain prior information" on the parameter of interest. (Here it is $\boldsymbol{\theta} = \mathbf{0}$.)

(2) Estimate the hyperparameter (here it is B) based on the marginal distribution of the MLE/BLUE of $\boldsymbol{\theta}$.

(3) Plug in the estimator of B into the Bayes estimator of $\boldsymbol{\theta}$, and obtain the EBE or Stein-type estimator. (Here it is $(1 - \hat{\mathrm{B}}_n)\tilde{\boldsymbol{\theta}}_n$.)

(4) If there is any restriction on B, adjust the estimator to set the estimator of B to fulfill this requirement. (See positive-rule shrinkage estimator in Section 4.3.3.)

For more information on EBE, see Deely and Lindley (1981).

For the EBE approach a major difficulty is faced when formulating a suitable **class of priors** for a given problem and the estimation of the hyperparameter. In Chapter 12, we will be presenting the estimation of binomial and multinomial probabilities to illustrate these difficulties with approximate solutions.

Finally, we consider the measure of improvement of an estimator $\boldsymbol{\delta}_n$ over $\boldsymbol{\delta}_0$ as the ratio

$$K(\boldsymbol{\delta}_n;\pi) = \frac{\rho(\boldsymbol{\delta}_0;\pi) - \rho(\boldsymbol{\delta}_n;\pi)}{\rho(\boldsymbol{\delta}_0;\pi) - \rho(\boldsymbol{\delta}_\pi;\pi)}, \qquad (4.4.18)$$

where the numerator is the Bayes risk difference between $\boldsymbol{\delta}_0$ and $\boldsymbol{\delta}_n$ and the denominator is the Bayes risk difference between $\boldsymbol{\delta}_0$ and $\boldsymbol{\delta}_\pi$. Clearly,

$$K(\boldsymbol{\delta}_n;\pi) \leq 1. \qquad (4.4.19)$$

If $\boldsymbol{\delta}_n$ is uniformly better than $\boldsymbol{\delta}_0$, then

$$0 \leq K(\boldsymbol{\delta}_n;\pi) \leq 1. \qquad (4.4.20)$$

An almost Bayes estimator is characterized by a value of $K(\boldsymbol{\delta}_n;\pi)$ near 1.

4.4.3 Quasi-empirical Bayes or Preliminary Test Estimation Approach

To obtain the Stein-type estimators using the PTE approach, the following steps may be followed for a given model with uncertain prior information:

(1) Obtain an optimum *unrestricted* estimator $\tilde{\boldsymbol{\theta}}_n$ of $\boldsymbol{\theta} \in \Omega$ and an optimum *restricted* estimator $\hat{\boldsymbol{\theta}}_n$ of $\boldsymbol{\theta} \in \omega$ (e.g., by likelihood/least squares method).

(2) Obtain an optimum (e.g., likelihood ratio type) test-statistic, $\mathcal{L}_n$ for testing the "uncertain prior information", say, $\boldsymbol{\theta} \in \omega$.

(3) Define the preliminary test estimator of $\boldsymbol{\theta}$ as

$$\hat{\boldsymbol{\theta}}_n^{\text{PT}} = \tilde{\boldsymbol{\theta}}_n - (\tilde{\boldsymbol{\theta}}_n - \hat{\boldsymbol{\theta}}_n) I(\mathcal{L}_n < \mathcal{L}_{n,\alpha}), \tag{4.4.21}$$

where $\mathcal{L}_{n,\alpha}$ is the α-level critical value of $\mathcal{L}_n$ from its distribution under $H_0 : \boldsymbol{\theta} \in \omega$.

(4) Replace the indicator function $I(\mathcal{L}_n < \mathcal{L}_{n,\alpha})$ by a smooth decreasing function $c\mathcal{L}_n^{-1}$, where c is a suitable constant. Then the **Stein-type estimator** is defined by

$$\hat{\boldsymbol{\theta}}_n^{S} = \tilde{\boldsymbol{\theta}}_n - c(\tilde{\boldsymbol{\theta}}_n - \hat{\boldsymbol{\theta}}_n)\mathcal{L}_n^{-1}, \tag{4.4.22a}$$

and the **positive-rule Stein-type estimator** by

$$\hat{\boldsymbol{\theta}}_n^{S+} = \hat{\boldsymbol{\theta}}_n + \{1 - c\mathcal{L}_n^{-1}\} I(\mathcal{L}_n > c)(\tilde{\boldsymbol{\theta}}_n - \hat{\boldsymbol{\theta}}_n) \text{ for the PRSE of } \boldsymbol{\theta}. \tag{4.4.22b}$$

For the James-Stein case, $\hat{\boldsymbol{\theta}}_n = \mathbf{0}$ and $\mathcal{L}_n = n\|\tilde{\boldsymbol{\theta}}_n\|^2$, and both the PTE approach and the EBE approach give identical results. Here, (4.4.22a) and (4.4.22b) are based on a simple decreasing function $c\mathcal{L}_n^{-1}$, but we will choose a more general function such as $c\phi(\mathcal{L}_n)$, where $\phi(\mathcal{L}_n)$ is a monotone decreasing function of the test-statistic $\mathcal{L}_n$. This allows us to cover various types of estimators. As a matter of fact we can write the four estimators compactly as (see Stein, 1966)

$$\boldsymbol{\theta}_n^* = \phi(\mathcal{L}_n)\tilde{\boldsymbol{\theta}}_n, \tag{4.4.23}$$

where

$$\begin{aligned} \phi(\mathcal{L}_n) &= 1, \\ &= I(\mathcal{L}_n > \chi_p^2(\alpha)), \\ &= 1 - c\mathcal{L}_n^{-1}, \\ \text{or } &= (1 - c\mathcal{L}_n^{-1}) I(\mathcal{L}_n > c), \quad c = p - 2 \end{aligned}$$

yields the estimators $\tilde{\boldsymbol{\theta}}_n, \hat{\boldsymbol{\theta}}_n^{\text{PT}}, \hat{\boldsymbol{\theta}}_n^{S}$, and $\hat{\boldsymbol{\theta}}_n^{S+}$, respectively. The quasi-empirical Bayes (or PTE) approach results in an identical, similar, or an *approximation* of the EBE approach, hence, the name "quasi-empirical Bayes estimation (QEBE)". We will consider this approach as a useful tool in many complicated situations in the chapters that follow.

(5) The constant c in the formulation of the Stein-type estimators is chosen to be

(i) $c = (\nu - 2)$ if $\mathcal{L}_n$ has a central chi-square distribution with ν d.f.

(ii) $c = \frac{(\nu_1 - 2)\nu_2}{\nu_1(\nu_2+2)}$ if $\mathcal{L}_n$ has a central F-distribution with (ν_1, ν_2) d.f.

In the James-Stein case here, $c = (p-2)$.

As an example consider the estimation of the mean vector $\boldsymbol{\theta}$ for the p-variate normal distribution $N_p(\boldsymbol{\theta}, \boldsymbol{\Sigma})$, where $\boldsymbol{\Sigma}$ is the covariance matrix based on the sample $\mathbf{Y}_1, \ldots, \mathbf{Y}_N$. Let $\overline{Y}$ and $\mathbf{S}$ be the sample mean vector, and let the sample covariance matrix have the Wishart distribution with $(N-1)$ d.f. and expectation $(N-1)\boldsymbol{\Sigma}$. Suppose that we need to estimate $\boldsymbol{\theta}$ with the loss function

$$L(\boldsymbol{\theta}_n^* \boldsymbol{\theta}) = N(\boldsymbol{\theta}_N^* - \boldsymbol{\theta})'\boldsymbol{\Sigma}^{-1}(\boldsymbol{\theta}_N^* - \boldsymbol{\theta}), \tag{4.4.24}$$

which is the Mahalonobis distance between $\boldsymbol{\theta}_N^*$ and $\boldsymbol{\theta}$.

Suppose that $\boldsymbol{\theta}_0$ is a prior uncertain value of $\boldsymbol{\theta}$. Then, using the PTE approach, we obtain the Stein-type estimator as

$$\hat{\boldsymbol{\theta}}_N^S = \boldsymbol{\theta}_0 + \left(1 - cT_N^{-2}\right)\left(\overline{Y} - \boldsymbol{\theta}_0\right), \tag{4.4.25}$$

where

$$T_N^2 = N(\overline{Y} - \boldsymbol{\theta}_0)'\mathbf{S}^{-1}(\overline{Y} - \boldsymbol{\theta}_0) \tag{4.4.26}$$

is the Hotellings T^2-statistic. Now, the distribution of $[T^2/(N-1)][(N-p)/p] \stackrel{\mathcal{D}}{=} F_{p,N-p}$, follows the F-distribution with $(p, N-p)$ d.f. Hence,

$$c = \frac{(p-2)m}{p(m+2)(N-1)}, \qquad m = N - p, \tag{4.4.27}$$

is the optimum shrinkage constant.

The EBE needs more analysis (see Efron and Morris, 1972) than this simple PTE approach as given below.

Assume that the prior distribution of $\boldsymbol{\theta}$ is $\mathcal{N}_p(\boldsymbol{\theta}_0, N^{-1}\mathbf{A})$ and we have $\tilde{\boldsymbol{\theta}}_N \sim \mathcal{N}_p(\boldsymbol{\theta}, N^{-1}\boldsymbol{\Sigma})$. It may be shown that the joint-distribution of $(\boldsymbol{\theta}', \tilde{\boldsymbol{\theta}}_N')'$ is given by $\mathcal{N}_{2p}\left\{\begin{pmatrix} \boldsymbol{\theta}_0 \\ \boldsymbol{\theta}_0 \end{pmatrix}, \begin{pmatrix} N^{-1}\mathbf{A} & N^{-1}\mathbf{A} \\ N^{-1}\mathbf{A} & N^{-1}(\mathbf{A}+\boldsymbol{\Sigma}) \end{pmatrix}\right\}$.

Consequently, the conditional distribution of $\boldsymbol{\theta}$ given $\tilde{\boldsymbol{\theta}}_N$ is

$$N_p\left\{\boldsymbol{\theta}_0 + \mathbf{A}(\mathbf{A}+\boldsymbol{\Sigma})^{-1}(\tilde{\boldsymbol{\theta}}_N - \boldsymbol{\theta}_0),\ N^{-1}\left(\mathbf{A}^{-1} + \boldsymbol{\Sigma}^{-1}\right)^{-1}\right\}.$$

Hence, the Bayes estimator of $\boldsymbol{\theta}$ is given by

$$\begin{aligned}\hat{\boldsymbol{\theta}}_N^B &= \boldsymbol{\theta}_0 + A(\mathbf{A}+\boldsymbol{\Sigma})^{-1}(\tilde{\boldsymbol{\theta}}_N - \boldsymbol{\theta}_0) \\ &= \boldsymbol{\theta}_0 + (I - \mathbf{B})(\tilde{\boldsymbol{\theta}}_N - \boldsymbol{\theta}_0),\ \mathbf{B} = \boldsymbol{\Sigma}(\mathbf{A}+\boldsymbol{\Sigma})^{-1},\end{aligned} \tag{4.4.28}$$

with

$$\operatorname{cov}(\hat{\boldsymbol{\theta}}_N^B) = N^{-1}\left(\mathbf{A}^{-1} + \boldsymbol{\Sigma}^{-1}\right)^{-1}.$$

Since $(\mathbf{A} + \boldsymbol{\Sigma})$ is unknown, so is $\mathbf{B}$, and we estimate $\mathbf{B}$ using the *marginal distribution* of $\mathbf{Y}_\alpha$ $(\alpha = 1, \ldots, N)$, which is $\mathcal{N}_p(\boldsymbol{\theta}_0, (\mathbf{A}+\boldsymbol{\Sigma}))$. Now, $\Sigma_{\alpha=1}^N(\mathbf{Y}_\alpha - \boldsymbol{\theta}_0)(\mathbf{Y}_\alpha - \boldsymbol{\theta}_0)'$ follows the Wishart distribution, $W_p(\mathbf{A} + \boldsymbol{\Sigma}), N)$, and we estimate $(\mathbf{A}+\boldsymbol{\Sigma})$ by $\left[\frac{1}{N}\sum_{\alpha=1}^N(\mathbf{Y}_\alpha - \boldsymbol{\theta}_0)(\mathbf{Y}_\alpha - \boldsymbol{\theta}_0)'\right] = \frac{1}{N}\{\mathbf{S}_N + N(\tilde{\boldsymbol{\theta}}_N - \boldsymbol{\theta}_0)(\tilde{\boldsymbol{\theta}}_N - \boldsymbol{\theta}_0)'\}$, where $\mathbf{S}_N = \sum_{\alpha=1}^N(\mathbf{Y}_\alpha - \tilde{\boldsymbol{\theta}}_N)(\mathbf{Y}_\alpha - \tilde{\boldsymbol{\theta}}_N)'$. Hence, an estimate of $\mathbf{B}$ is given by $\hat{\mathbf{B}}_N$,

$$\begin{aligned}\hat{\mathbf{B}}_N &= \mathbf{S}_N[\mathbf{S}_N + N(\tilde{\boldsymbol{\theta}}_N - \boldsymbol{\theta}_0)(\tilde{\boldsymbol{\theta}}_N - \boldsymbol{\theta}_0)']^{-1} \\ &= \left[\mathbf{I}_p + N\mathbf{S}_N^{-1}(\tilde{\boldsymbol{\theta}}_N - \boldsymbol{\theta}_0)(\tilde{\boldsymbol{\theta}}_N - \boldsymbol{\theta}_0)'\right]^{-1}.\end{aligned}$$

Then the empirical Bayes estimator of $\boldsymbol{\theta}$ is defined by

$$\hat{\boldsymbol{\theta}}_N^{EB} = \boldsymbol{\theta}_0 + (\mathbf{I} - \hat{\mathbf{B}}_N)(\tilde{\boldsymbol{\theta}}_N - \boldsymbol{\theta}_0). \tag{4.4.29}$$

Compare this expression with that of $\hat{\boldsymbol{\theta}}_N^S$ at (4.4.25).

4.4.4 How Close is the JS Estimator to the Bayes Estimator?

We consider the shrinkage estimator $\hat{\boldsymbol{\theta}}_n^S$. Based on the assumption that $\boldsymbol{\theta} \sim \mathcal{N}_p(\mathbf{0}, \frac{\tau^2}{n}\mathbf{I}_p) = \pi$ as in Section 4.3.2, we know that the a posteriori distribution of $\boldsymbol{\theta}$ is $\mathcal{N}_p\{(1 - B)\tilde{\boldsymbol{\theta}}_n; (1 - B)\mathbf{I}_p\}$, where $B = (1 + \tau^2)^{-1}$, $0 < B < 1$. Hence,

$$\rho(\boldsymbol{\delta}_\pi, \pi) = (1 - B)p, \tag{4.4.30}$$

$$\rho(\tilde{\boldsymbol{\theta}}_n, \pi) = p \quad \text{and} \quad \rho(\tilde{\boldsymbol{\theta}}_n; \pi) - \rho(\delta_\pi; \pi) = Bp. \tag{4.4.31}$$

On the other hand,

$$\rho(\tilde{\boldsymbol{\theta}}_n; \pi) - \rho(\hat{\boldsymbol{\theta}}_n^S; \pi) = E_\pi\left[R_1(\tilde{\boldsymbol{\theta}}_n) - R_3(\hat{\boldsymbol{\theta}}_n^S)\right] = (p-2)^2 E_\pi\left[E\left(\chi_p^{-2}(\Delta^2)\right)\right]. \tag{4.4.32}$$

Now, as in Hoffmann (1992), we have

$$E\left(\chi_p^{-2}(\Delta^2)\right) = E_{\boldsymbol{\theta}}\left[n\|\tilde{\boldsymbol{\theta}}_n\|^2\right]^{-1} = E_{\boldsymbol{\theta}}\left[\int_0^\infty \exp(-nu\|\tilde{\boldsymbol{\theta}}_n\|^2)du\right]. \tag{4.4.33}$$

The r.h.s. may be written following Hoffmann (1992) as

$$= \int_{-\infty}^\infty \cdots \int_{-\infty}^\infty \left[\int_0^\infty \exp(-nu\|\tilde{\boldsymbol{\theta}}_n\|^2)du\right] \prod_{j=1}^p \frac{\exp\left\{-\frac{n}{2}(\tilde{\theta}_{jn} - \theta_j)^2\right\}}{\sqrt{2\pi}} \sqrt{n}d\tilde{\theta}_{jn}. \tag{4.4.34}$$

Using the Fubini theorem, we obtain the r.h.s. as

$$= \int_0^\infty \prod_{j=1}^p \Big[\int_{-\infty}^{\infty} \exp\Big\{ -\frac{n}{2}(\tilde{\theta}_{jn} - \theta_j)^2 - nu\tilde{\theta}_{jn}^2 \Big\} (2\pi)^{-1/2} \sqrt{n} d\tilde{\theta}_{jn} \Big] du \quad (4.4.35)$$

$$= \int_0^\infty \prod_{j=1}^p \Big[\int_{-\infty}^{\infty} \exp\Big\{ -\frac{n}{2}(1+2u)\Big(\tilde{\theta}_{jn} - \frac{\theta_j}{1+2u}\Big)^2 \Big\} (2\pi)^{-1/2} \sqrt{n} d\tilde{\theta}_{jn} \Big]$$

$$\times \exp\Big(-\frac{nu}{1+2u}\theta_j^2 \Big) du \quad (4.4.36)$$

$$= \int_0^\infty (1+2u)^{-p/2} \exp\Big\{ -\frac{nu}{1+2u} \|\boldsymbol{\theta}\|^2 \Big\} du. \quad (4.4.37)$$

Setting $v = 2u/(1+2u)$ in (4.4.33), we get

$$= \frac{1}{2} \int_0^1 (1-v)^{\frac{p-2}{2}-1} \exp\Big\{ -\frac{v}{2}\Delta^2 \Big\} dv \quad \text{for } p \geq 3. \quad (4.4.38)$$

Thus, $E(\chi_p^{-2}(\Delta^2))$ has the integral representation (4.4.34) that restricts the shrinkage estimator $\hat{\boldsymbol{\theta}}_n^S$ for $p \geq 4$, and $E(\chi_p^{-2}(\Delta^2))$ is a decreasing function of Δ^2. The maximum value of $E(\chi_p^{-2}(\Delta^2))$ is $1/(p-2)$ which drops down to 0 as $\Delta^2 \to \infty$.

Thus, the risk difference $R_1(\tilde{\boldsymbol{\theta}}_n) - R_3(\hat{\boldsymbol{\theta}}_n^S)$ equals

$$\frac{(p-2)^2}{2} \int_0^1 (1-v)^{\frac{p-2}{2}-1} \exp\Big\{ -\frac{v}{2}\Delta^2 \Big\} dv, \quad (4.4.39)$$

which varies from $(p-2)$ (maximum) to 0 (minimum). Next, we compute the Bayes risk of $\hat{\boldsymbol{\theta}}_n^S$ given by

$$E_\pi\big[E(\chi_p^{-2}(\Delta^2))\big]$$

$$= \int_{R^p} \Big[\frac{(p-2)^2}{2} \int_0^1 (1-v)^{\frac{p-2}{2}-1} \exp\Big\{ -\frac{v}{2} n\|\boldsymbol{\theta}\|^2 \Big\} dv \Big]$$

$$\times \Big(\frac{1}{\sqrt{2\pi}} \Big)^p \exp\Big\{ -\frac{n}{2\pi^2} \|\boldsymbol{\theta}\|^2 \Big\} d\theta_1 \cdots d\theta_p$$

$$= \frac{(p-2)^2}{2} \int_0^1 \left(\frac{v + \frac{B}{1-B}}{1-v} \right)^{-p/2} (1-v)^{-2} dv \Big(\frac{B}{1-B} \Big)^{p/2}. \quad (4.4.40)$$

Substituting $z = \frac{v + [B/(1-B)]}{1-v}$ we obtain

$$= \frac{(p-2)^2}{2}(1-B) \int_{\frac{B}{1-B}}^\infty z^{-p/2} dz \Big(\frac{B}{1-B} \Big)^{p/2}$$

$$= B(p-2). \quad (4.4.41)$$

Thus, the value of $K(\hat{\boldsymbol{\theta}}_n^S;\pi)$ equals $(p-2)/p = 1-(2/p)$. For reasonably large values of p, the estimator $\hat{\boldsymbol{\theta}}_n^S$ closely resembles the Bayes estimator.

4.5 Stein-Type Estimation When the Variance is Unknown

4.5.1 Introduction: Model, Estimators, and Tests

Consider the model (4.1.1)

$$\mathbf{Y}_i = \boldsymbol{\theta} + \mathbf{e}_i, \quad i = 1, \dots, n,$$

where $\mathbf{Y}_i = (Y_{1i}, \dots, Y_{pi})'$ and $\mathbf{e}_i = (e_{1i}, \dots, e_{pi})'$ and $\boldsymbol{\theta} = (\theta_1, \dots, \theta_p)'$. Further,

$$\mathbf{e}_i \sim \mathcal{N}_p(\mathbf{0}, \sigma^2 \mathbf{I}_p). \tag{4.5.1}$$

The MLE/LSE of $\boldsymbol{\theta}$ is the same $\tilde{\boldsymbol{\theta}}_n$ as before, meaning $\tilde{\boldsymbol{\theta}}_n = (\overline{Y}_1, \dots, \overline{Y}_p)'$, but the unbiased estimator of σ^2 is given by

$$\begin{aligned} s_y^2 &= \frac{1}{p(n-1)} \operatorname{tr}\left\{\sum_{i=1}^{n} (\mathbf{Y}_i - \bar{\mathbf{Y}})(\mathbf{Y}_i - \bar{\mathbf{Y}})'\right\} \\ &= \frac{1}{p(n-1)} \sum_{i=1}^{p}\sum_{j=1}^{n} (Y_{ij} - \overline{Y}_i)^2. \end{aligned} \tag{4.5.2}$$

Further, $p(n-1)s_y^2/\sigma^2$ follows a central chi-square distribution with $m = p(n-1)$ d.f. In order to test the null hypothesis $H_0 : \boldsymbol{\theta} = \boldsymbol{\theta}_0$ against $H_A : \boldsymbol{\theta} \neq \boldsymbol{\theta}_0$, we use the LR-test

$$\mathcal{L}_n = \frac{n\|\tilde{\boldsymbol{\theta}}_n - \boldsymbol{\theta}_0\|^2}{p s_y^2} \tag{4.5.3}$$

that follows a central F-distribution with (p, m) d.f. under H_0 and a noncentral F-distribution with (p, m) d.f. and noncentrality parameter $\Delta^2/2$

$$\Delta^2 = \frac{n\|\boldsymbol{\theta} - \boldsymbol{\theta}_0\|^2}{\sigma^2} \tag{4.5.4}$$

under the alternative.

4.5.2 Preliminary Test and Stein-Type Estimators

If a priori it is suspected that $\boldsymbol{\theta}$ may be $\mathbf{0}$, then the preliminary test estimator of $\boldsymbol{\theta}$ is given by

$$\hat{\boldsymbol{\theta}}_n^{\mathrm{PT}} = \tilde{\boldsymbol{\theta}}_n - \tilde{\boldsymbol{\theta}}_n I(\mathcal{L}_n < F_{p,m}(\alpha)), \tag{4.5.5}$$

where $F_{p,m}(\alpha)$ is the α-level critical value of a central F-distribution with (p, m) d.f.

Similarly, the James-Stein estimator of $\boldsymbol{\theta}$ according to Section 4.4.3 is defined by

$$\hat{\boldsymbol{\theta}}_n^S = \left(1 - \frac{(p-2)m}{p(m+2)}\mathcal{L}_n^{-1}\right)\tilde{\boldsymbol{\theta}}_n, \tag{4.5.6}$$

and the positive-rule shrinkage estimator is defined by

$$\hat{\boldsymbol{\theta}}_n^{S+} = \left(1 - \frac{(p-2)m}{p(m+2)}\mathcal{L}_n^{-1}\right) I\left(\mathcal{L}_n > \frac{(p-2)m}{p(m+2)}\right)\tilde{\boldsymbol{\theta}}_n. \tag{4.5.7}$$

It may be noted that if we write $\hat{\boldsymbol{\theta}}_n^S$ as $(1 - c\mathcal{L}_n^{-1})\tilde{\boldsymbol{\theta}}_n$, then it is easy to show that the optimum c equals $(p-2)m/p(m+2)$.

4.5.3 Empirical Bayes Estimation When the Variance Is Unknown

To obtain the empirical estimator of $\boldsymbol{\theta}$, we consider the class of priors

$$\boldsymbol{\theta} \sim N_p\left(\mathbf{0}, \frac{\tau^2}{n}\mathbf{I}_p\right). \tag{4.5.8}$$

Hence,

$$\boldsymbol{\theta}|\tilde{\boldsymbol{\theta}}_n \sim N_p\{(1-B)\tilde{\boldsymbol{\theta}}_n, (1-B)\mathbf{I}_p\}, \tag{4.5.9}$$

where $B = \sigma^2(\sigma^2 + \tau^2)^{-1}$, $0 < B < 1$. Thus, the Bayes estimator of $\boldsymbol{\theta}$ is defined by

$$\boldsymbol{\delta}_\pi = (1-B)\tilde{\boldsymbol{\theta}}_n, \tag{4.5.10}$$

with Bayes risk $p(1-B)$.

In the expression for $\boldsymbol{\delta}_\pi$, B is *unknown* since σ^2 and τ^2 are both unknown. To estimate B, we follow the same procedure as (4.4.14) to (4.4.17) where we consider $\sigma^2(\sigma^2+\tau^2)^{-1}$ instead of $1+\tau^2$.

We note that the marginal distribution of $\sqrt{n}\tilde{\boldsymbol{\theta}}_n$ is $N_p(\mathbf{0}, B^{-1}\mathbf{I}_p)$. Thus, $Bn\|\tilde{\boldsymbol{\theta}}\|^2$ has the central chi-square distribution with p d.f. Accordingly, the uniformly minimum variance unbiased estimator of $(\sigma^2+\tau^2)^{-1}$ is $(p-2)/n\|\tilde{\boldsymbol{\theta}}_n\|^2$. Also, the best scale invariant estimator of σ^2 is $ms_y^2/(m+2)$. Thus, inserting the estimators of $(\sigma^2+\tau^2)^{-1}$ and σ^2, we obtain the empirical Bayes estimator of $\boldsymbol{\theta}$ as

$$\hat{\boldsymbol{\theta}}_n^{EB} = \left(1 - \frac{(p-2)ms_y^2}{(m+2)n\|\tilde{\boldsymbol{\theta}}_n\|^2}\right)\tilde{\boldsymbol{\theta}}_n \tag{4.5.11a}$$

$$= \left(1 - \frac{(p-2)m\mathcal{L}_n^{-1}}{p(m+2)}\right)\tilde{\boldsymbol{\theta}}_n. \tag{4.5.11b}$$

Thus, writing

$$\boldsymbol{\theta}_n^* = \phi(\mathcal{L}_n)\tilde{\boldsymbol{\theta}}_n, \tag{4.5.12}$$

where

$$\begin{aligned}\phi(\mathcal{L}_n) &= 1,\\ &= I(\mathcal{L}_n > F_{p,m}(\alpha)),\\ &= 1 - c\mathcal{L}_n^{-1},\end{aligned}$$

or

$$= (1 - c\mathcal{L}_n^{-1})I(\mathcal{L}_n > c), \quad c = \frac{(p-2)m}{p(m+2)},$$

yields the estimators $\tilde{\boldsymbol{\theta}}_n, \hat{\boldsymbol{\theta}}_n^{\mathrm{PT}}, \hat{\boldsymbol{\theta}}_n^{S}$, and $\hat{\boldsymbol{\theta}}_n^{S+}$, respectively. Here the PT approach coincides with the EB approach.

We may also consider the Stein estimator of σ^2 (see Section 3.8) in place of $ms_y^2/m+2$ in (4.5.11a, b) to obtain another estimator of $\boldsymbol{\theta}$ (see Berry, 1994), namely

$$\hat{\boldsymbol{\theta}}_n^{IS} = \{1 - c(m+2)\phi_S(\mathcal{L}_n)\mathcal{L}_n^{-1}\}\tilde{\boldsymbol{\theta}}_n, \tag{4.5.13}$$

where

$$\phi_S(\mathcal{L}_n) = \frac{1}{m+2}I\left(\mathcal{L}_n > \frac{m}{m+2}\right) + \frac{1+\frac{p}{m}\mathcal{L}_n}{m+p+2}I\left(\mathcal{L}_n < \frac{m}{m+2}\right),$$

and $c = (p-2)m/p(m+2)$. We study this estimator in Section 4.5.6. in terms of the test-statistic $\mathcal{L}_n$ as in (4.5.3).

We will see later that $\hat{\boldsymbol{\theta}}_n^{\mathrm{IS}}$ improves on $\hat{\boldsymbol{\theta}}_n^{S}$ (4.5.6) uniformly in Δ^2.

4.5.4 Bias, MSE Matrices, and Risk Expressions

The bias vector and quadratic bias expressions are

(i) $\mathbf{b}_1(\tilde{\boldsymbol{\theta}}_n) = \mathbf{0}$ and $B_1(\tilde{\boldsymbol{\theta}}_n) = 0.$ (4.5.14a)

(ii) $\mathbf{b}_2(\hat{\boldsymbol{\theta}}_n^{\mathrm{PT}}) = -\boldsymbol{\theta}G_{p+2,m}(\ell_\alpha;\Delta^2)$ and $B_2(\hat{\boldsymbol{\theta}}_n^{\mathrm{PT}}) = \Delta^2\left\{G_{p+2,m}(\ell_\alpha;\Delta^2)\right\}^2,$ (4.5.14b)

where $G_{\nu_1,\nu_2}(\cdot;\Delta^2)$ is the cdf of a noncentral F-distribution with (ν_1,ν_2) d.f. and noncentrality parameter $\Delta^2/2$ and

$$\ell_\alpha = \frac{p}{p+2}F_{p,m}(\alpha). \tag{4.5.14c}$$

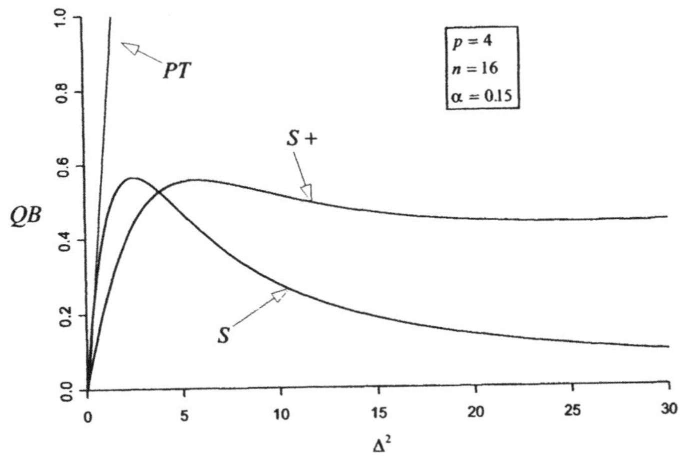

Figure 4.5.1 Graph of QB of estimators: PTE, JSE, and PRSE $\left(PT = \hat{\boldsymbol{\theta}}_n^{\text{PT}},\ S = \hat{\boldsymbol{\theta}}_n^S,\ \text{and}\ S^+ = \hat{\boldsymbol{\theta}}_n^{S+}\right)$

$$\text{(iii)}\quad \mathbf{b}_3(\hat{\boldsymbol{\theta}}_n^S) = -cp\boldsymbol{\theta}E\left(\chi_{p+2}^{-2}(\Delta^2)\right), \quad c = \frac{(p-2)m}{p(m+2)}$$

$$\text{and}\quad B_3\left(\hat{\boldsymbol{\theta}}_n^S\right) = c^2p^2\Delta^2\left\{E\left(\chi_{p+2}^{-2}(\Delta^2)\right)\right\}^2, \tag{4.5.14d}$$

$$\text{(iv)}\quad \mathbf{b}_4(\hat{\boldsymbol{\theta}}_n^{S+}) = -\boldsymbol{\theta}\Bigg[G_{p+2,m}\left(\frac{cp}{(p+2)};\Delta^2\right) + \frac{cp}{(p+2)}E\left(F_{p+2,m}^{-1}(\Delta^2)\right)$$
$$-\frac{cp}{(p+2)}E\left\{F_{p+2,m}^{-1}(\Delta^2)I\left(F_{p+2,m}(\Delta^2) < \frac{cp}{(p+2)}\right)\right\}\Bigg].$$

$$B_4\left(\hat{\boldsymbol{\theta}}_n^{S+}\right) = \Delta^2\Bigg[G_{p+2,m}\left(\frac{cp}{(p+2)};\Delta^2\right) + \frac{cp}{(p+2)}E\left(F_{p+2,m}^{-1}(\Delta^2)\right)$$
$$-\frac{cp}{(p+2)}E\left\{F_{p+2,m}^{-1}(\Delta^2)I\left(F_{p+2,m}(\Delta^2) < \frac{cp}{p+2}\right)\right\}\Bigg]^2. \tag{4.5.14e}$$

The expressions are obtained following the procedure of Section 4.3 and applying the theorems in Chapter 2. Some graphs of QB of estimators are given in Figure 4.5.1.

Similarly, the MSE matrices and the quadratic risk expressions may be obtained as

$$\text{(i)}\ \mathbf{M}_1(\tilde{\boldsymbol{\theta}}_n) = \sigma^2\mathbf{I}_p \text{ and } R_1(\tilde{\boldsymbol{\theta}}_n;\mathbf{W}) = \sigma^2\text{tr}(\mathbf{W}) = p \text{ if } \mathbf{W} = \sigma^{-2}\mathbf{I}_p. \tag{4.5.15a}$$

(ii) $\mathbf{M}_2(\hat{\boldsymbol{\theta}}_n^{\text{PT}}) = \sigma^2\mathbf{I}_p - \sigma^2\mathbf{I}_p\left\{G_{p+2,m}(\ell_\alpha;\Delta^2)\right\} + n\boldsymbol{\theta}\boldsymbol{\theta}'\{2G_{p+2,m}(\ell_\alpha;\Delta^2)$
$-G_{p+4,m}(\ell_\alpha^*;\Delta^2)\}$, (4.5.15b)
where $\ell_\alpha = \frac{p}{p+2}F_{p,m}(\alpha)$ and $\ell_\alpha^* = \frac{p}{p+4}F_{p,m}(\alpha)$.
Further, the risk is given by

$$\begin{aligned}R_2(\tilde{\boldsymbol{\theta}}_n^{\text{PT}};\mathbf{W}) &= \sigma^2\text{tr}(\mathbf{W})\left\{1 - G_{p+2,m}(\ell_\alpha;\Delta^2)\right\}\\ &\quad + (n\boldsymbol{\theta}'\mathbf{W}\boldsymbol{\theta})\left\{2G_{p+2,m}(\ell_\alpha;\Delta^2) - G_{p+4,m}(\ell_\alpha^*;\Delta^2)\right\}\\ &= p\left\{1 - G_{p+2,m}(\ell_\alpha;\Delta^2)\right\}\\ &\quad +\Delta^2\left\{2G_{p+2,m}(\ell_\alpha;\Delta^2) - G_{p+4,m}(\ell_\alpha^*;\Delta^2)\right\},\end{aligned}$$

if $\mathbf{W} = \sigma^{-2}\mathbf{I}_p$. (4.5.15c)

(iii) $$\mathbf{M}_3(\hat{\boldsymbol{\theta}}_n^S) = \sigma^2\mathbf{I}_p - cp\sigma^2\mathbf{I}_p\left\{2E\left(\chi_{p+2}^{-2}(\Delta^2)\right) - (p-2)E\left(\chi_{p+4}^{-4}(\Delta^2)\right)\right\}$$
$$+ cpm\boldsymbol{\theta}\boldsymbol{\theta}'\left\{2E\left(\chi_{p+2}^{-2}(\Delta^2)\right) - 2E\left(\chi_{p+2}^{-4}(\Delta^2)\right) + (p-2)E\left(\chi_{p+4}^{-4}(\Delta^2)\right)\right\} \quad (4.5.15\text{d})$$

and

$$\begin{aligned}&R_3(\hat{\boldsymbol{\theta}}_n^S;\mathbf{W})\\ &= \sigma^2\text{tr}(\mathbf{W}) - \sigma^2 cp\text{tr}(\mathbf{W})\left\{2E\left(\chi_{p+2}^{-2}(\Delta^2)\right) - (p-2)E\left(\chi_{p+2}^{-4}(\Delta^2)\right)\right\}\\ &\quad + cp(n\boldsymbol{\theta}'\mathbf{W}\boldsymbol{\theta})\left\{2E\left(\chi_{p+2}^{-2}(\Delta^2)\right) - 2E\left(\chi_{p+4}^{-2}(\Delta^2)\right) + (p-2)E\left(\chi_{p+4}^{-4}(\Delta^2)\right)\right\}\\ &= p - (p-2)^2\frac{m}{m+2}E\left(\chi_p^{-2}(\Delta^2)\right), \quad \text{if } \mathbf{W} = \sigma^{-2}\mathbf{I}_p. \qquad (4.5.15\text{e})\end{aligned}$$

(iv) $$\begin{aligned}&\mathbf{M}_4(\hat{\boldsymbol{\theta}}_n^{S+})\\ &= \mathbf{M}_3(\hat{\boldsymbol{\theta}}_n^S) - \sigma^2\mathbf{I}_pE\left[\left(1 - \frac{cp}{p+2}F_{p+2}^{-1}(\Delta^2)\right)^2 I\left(F_{p+2,m}(\Delta^2) < \frac{cp}{p+2}\right)\right]\\ &\quad + n\boldsymbol{\theta}\boldsymbol{\theta}'\left\{2E\left[\left(1 - \frac{cp}{p+2}F_{p+2,m}^{-1}(\Delta^2)\right)I\left(F_{p+2,m}(\Delta^2) < \frac{cp}{p+2}\right)\right]\right.\\ &\quad \left. - E\left[\left(1 - \frac{cp}{p+4}F_{p+4,m}^{-1}(\Delta^2)\right)^2 I\left(F_{p+4,m}(\Delta^2) < \frac{cp}{p+4}\right)\right]\right\} \qquad (4.5.15\text{f})\end{aligned}$$

and

$$\begin{aligned}&R_4\left(\hat{\boldsymbol{\theta}}_n^{S+};\mathbf{W}\right)\\ &= R_3(\hat{\boldsymbol{\theta}}_n^S;\mathbf{W}) - \sigma^2\text{tr}(\mathbf{W})E\left[\left(1-\frac{cp}{p+2}F_{p+2,m}^{-1}(\Delta^2)\right)^2 I\left(F_{p+2,m}(\Delta^2)<\frac{cp}{p+2}\right)\right]\\ &\quad +(n\boldsymbol{\theta}'\mathbf{W}\boldsymbol{\theta})\left\{2E\left[\left(1 - \frac{cp}{p+2}F_{p+2,m}^{-1}(\Delta^2)\right)I\left(F_{p+2,m}(\Delta^2) < \frac{cp}{p+2}\right)\right]\right.\end{aligned}$$

$$- E\left[\left(1 - \frac{cp}{p+4} F_{p+4,m}^{-1}(\Delta^2)\right)^2 I\left(F_{p+4,m}(\Delta^2) < \frac{cp}{p+4}\right)\right]$$

$$= R_3(\hat{\boldsymbol{\theta}}_n^S; \sigma^{-2}\mathbf{I}_p) - \sigma^{-2} pE\left[\left(1 - \frac{cp}{p+2} F_{p+2,m}^{-1}(\Delta^2)\right)^2 I\left(F_{p+2,m}(\Delta^2) < \frac{cp}{p+2}\right)\right]$$

$$+ \Delta^2 \Big\{ 2E\left[\left(1 - \frac{cp}{p+2} F_{p+2,m}^{-1}(\Delta^2)\right)\right] I\left(F_{p+2,m}(\Delta^2) < \frac{cp}{p+2}\right)$$

$$-E\left[\left(1 - \frac{cp}{p+4} F_{p+4,m}^{-1}(\Delta^2)\right)^2 I\left(F_{p+4,m}(\Delta^2) < \frac{cp}{p+4}\right)\right]\Big\}$$

if $\mathbf{W} = \sigma^{-2}\mathbf{I}_p$. (4.5.15g)

4.5.5 Risk Analysis of the Estimators

First compare the unrestricted estimator $\tilde{\boldsymbol{\theta}}_n$ and the PTE, $\hat{\boldsymbol{\theta}}_n^{\text{PT}}$. The risk difference in this case is given by

$$\begin{aligned} & R_1(\tilde{\boldsymbol{\theta}}_n; \mathbf{W}) - R_2(\hat{\boldsymbol{\theta}}_n^{\text{PT}}; \mathbf{W}) \\ &= \sigma^2 \text{tr}(\mathbf{W}) G_{p+2,m}(\ell_\alpha; \Delta^2) \\ &\quad -(n\boldsymbol{\theta}'\mathbf{W}\boldsymbol{\theta}) \left\{2G_{p+2,m}(\ell_\alpha; \Delta^2) - G_{p+4,m}(\ell_\alpha; \Delta^2)\right\}. \end{aligned} \quad (4.5.16)$$

Thus, the r.h.s. is nonnegative whenever

$$0 \le \Delta^2 \le \frac{\text{tr}(\mathbf{W}) G_{p+2,m}(\ell_\alpha; \Delta^2)}{Ch_{\max}(\mathbf{W})\{2G_{p+2,m}(\ell_\alpha; \Delta^2) - G_{p+4,m}(\ell_\alpha^*; \Delta^2)\}} \le \frac{\text{tr}(\mathbf{W})}{Ch_{\max}(\mathbf{W})}. \quad (4.5.17a)$$

In this interval, $\hat{\boldsymbol{\theta}}_n^{\text{PT}}$ performs better than $\tilde{\boldsymbol{\theta}}_n$, and $\tilde{\boldsymbol{\theta}}_n$ performs better than $\hat{\boldsymbol{\theta}}_n^{\text{PT}}$ whenever

$$\Delta^2 \ge \frac{\text{tr}(\mathbf{W}) G_{p+2,m}(\ell_\alpha; \Delta^2)}{Ch_{\min}(\mathbf{W})\{2G_{p+2,m}(\ell_\alpha; \Delta^2) - G_{p+4,m}(\ell_\alpha^*; \Delta^2)\}} \ge \frac{\text{tr}(\mathbf{W})}{Ch_{\min}(\mathbf{W})}. \quad (4.5.17b)$$

Next, we compare $\hat{\boldsymbol{\theta}}_n^S$ and $\tilde{\boldsymbol{\theta}}_n$. Here, the risk difference

$$\begin{aligned} & R_1(\tilde{\boldsymbol{\theta}}_n; \mathbf{W}) - R_3(\hat{\boldsymbol{\theta}}_n^S; \mathbf{W}) \\ &= \sigma^2 \text{tr}(\mathbf{W}) \Big\{ (p-2) E\left(\chi_{p+2}^{-4}(\Delta^2)\right) \\ &\qquad + \left[1 - \frac{(p+2) n\boldsymbol{\theta}'\mathbf{W}\boldsymbol{\theta}}{2\sigma^2\Delta^2 \text{tr}(\mathbf{W})}\right] 2\Delta^2 E\left(\chi_{p+4}^{-4}(\Delta^2)\right)\Big\}. \end{aligned} \quad (4.5.18)$$

This risk difference is positive for all $\mathbf{W}$ satisfying the condition

$$\frac{\text{tr}(\mathbf{W})}{Ch_{\max}(\mathbf{W})} \ge \frac{p+2}{2}, \quad (4.5.19)$$

and $\hat{\boldsymbol{\theta}}_n^S$ dominates $\tilde{\boldsymbol{\theta}}_n$ for all Δ^2 whenever $\mathbf{W}$ satisfies (4.5.19). The risk difference converges to zero as $\Delta^2 \to \infty$, and at $\Delta^2 = 0$, it is $2(p+m)/(m+2)$, which tends to 2 as $m \to \infty$.

If we compare $\hat{\boldsymbol{\theta}}_n^{\mathrm{PT}}$ and $\hat{\boldsymbol{\theta}}_n^S$, we notice that under $H_0 : \boldsymbol{\theta} = \mathbf{0}$,

$$R_3(\hat{\boldsymbol{\theta}}_n^S; \mathbf{W}) = R_2(\hat{\boldsymbol{\theta}}_n^{\mathrm{PT}}; \mathbf{W}) + \sigma^2 p\Big\{G_{p+2,m}(\ell_\alpha; 0) - c\Big\}$$

$$\geq R_2(\hat{\boldsymbol{\theta}}_n^{\mathrm{PT}}; \mathbf{W}), \quad \text{if } c > G_{p+2,m}(\ell_\alpha; 0). \tag{4.5.20}$$

Thus, the risk of $\hat{\boldsymbol{\theta}}_n^S$ is bigger than the risk of $\hat{\boldsymbol{\theta}}_n^{\mathrm{PT}}$ for all levels of significance α, satisfying

$$\left\{\alpha : \ell_\alpha > G_{p+2,m}^{-1}(c, 0)\right\}; \tag{4.5.21}$$

otherwise, the risk of $\hat{\boldsymbol{\theta}}_n^{\mathrm{PT}}$ is bigger.

The picture changes as Δ^2 moves away from the origin. As $\Delta^2 \to \infty$, the risk difference approaches zero. For intermediate value of Δ^2, the two risks intersect.

Comparison of $\hat{\boldsymbol{\theta}}_n^{S+}$ and $\hat{\boldsymbol{\theta}}_n^S$ shows that the risk difference is

$$\begin{aligned}
&R_4\left(\hat{\boldsymbol{\theta}}_n^{S+}; \mathbf{W}\right) - R_3\left(\hat{\boldsymbol{\theta}}_n^S; \mathbf{W}\right) \\
&= -\sigma^2 \mathrm{tr}(\mathbf{W}) E\left[\left\{1 - \frac{cp}{p+2} F_{p+2,m}^{-1}(\Delta^2)\right\}^2 I\left(F_{p+2,m}(\Delta^2) < \frac{cp}{p+2}\right)\right] \\
&\quad +(n\boldsymbol{\theta}'\mathbf{W}\boldsymbol{\theta})\Bigg\{E\left[2\left(1 - \frac{cp}{p+2} F_{p+2,m}^{-1}(\Delta^2)\right) I\left(F_{p+2,m}(\Delta^2) < \frac{cp}{p+2}\right)\right] \\
&\quad -E\left[\left(1 - \frac{cp}{p+4} F_{p+4,m}^{-1}(\Delta^2)\right)^2 I\left(F_{p+4,m}(\Delta^2) < \frac{cp}{p+4}\right)\right]\Bigg\}.
\end{aligned} \tag{4.5.22}$$

The r.h.s. is negative, since

$$\left(0 < F_{p+2,m}(\Delta^2) < \frac{cp}{p+2}\right) \Longleftrightarrow \left(\frac{cp}{p+2} F_{p+2,m}^{-1}(\Delta^2) - 1\right) \geq 0. \tag{4.5.23}$$

Thus, for all $\Delta^2 \in (0, \infty)$,

$$R_4\left(\hat{\boldsymbol{\theta}}_n^{S+}; \mathbf{W}\right) \leq R_3\left(\hat{\boldsymbol{\theta}}_n^S; \mathbf{W}\right), \tag{4.5.24}$$

$\hat{\boldsymbol{\theta}}_n^{S+}$ dominates $\hat{\boldsymbol{\theta}}_n^S$, and $\hat{\boldsymbol{\theta}}_n^{S+}$ is a simple modification of $\hat{\boldsymbol{\theta}}_n^S$.

Now, consider the modification of the PTE, $\hat{\boldsymbol{\theta}}_n^{\mathrm{PT}}$ as defined at (4.5.5):

$$\hat{\boldsymbol{\theta}}_n^{\mathrm{PT}+} = \hat{\boldsymbol{\theta}}_n^{\mathrm{PT}}(1 - c\mathcal{L}_n^{-1}). \tag{4.5.25}$$

The bias vector and QB of $\hat{\boldsymbol{\theta}}_n^{\text{PT+}}$ are then evaluated as

$$\begin{aligned}
&\mathbf{b}_5\left(\hat{\boldsymbol{\theta}}_n^{\text{PT+}}\right) \\
&= \mathbf{b}_2\left(\hat{\boldsymbol{\theta}}_n^{\text{PT}}\right) - \frac{cp}{p+2}\boldsymbol{\theta} E\left[F_{p+2,m}^{-1}(\Delta^2) I\left(F_{p+2,m}(\Delta^2) > \ell_\alpha\right)\right] \\
&= -\boldsymbol{\theta}\left[G_{p+2,m}(\ell_\alpha;\Delta^2) + \frac{cp}{p+2}E\left\{F_{p+2,m}^{-1}(\Delta^2) I\left(F_{p+2,m}(\Delta^2) < \ell_\alpha\right)\right\}\right]. \quad (4.5.26)
\end{aligned}$$

The QB of $\hat{\boldsymbol{\theta}}_n^{\text{PT+}}$ is given by

$$\begin{aligned}
B_5\left(\hat{\boldsymbol{\theta}}_n^{\text{PT+}}\right) = \Delta^2\Bigg[&G_{p+2,m}(\ell_\alpha;\Delta^2) \\
&+ \frac{cp}{p+2}E\left\{F_{p+2,m}^{-1}(\Delta^2) I\left(F_{p+2,m}(\Delta^2) < \ell_\alpha\right)\right\}\Bigg]^2. \quad (4.5.27)
\end{aligned}$$

The MSE matrix and weighted risk of $\hat{\boldsymbol{\theta}}_n^{\text{PT+}}$ is evaluated as given below for $p \geq 3$:

$$\begin{aligned}
&\mathbf{M}_5\left(\hat{\boldsymbol{\theta}}_n^{\text{PT+}}\right) = \mathbf{M}_2\left(\hat{\boldsymbol{\theta}}_n^{\text{PT}}\right) + \mathbf{M}_3\left(\hat{\boldsymbol{\theta}}_n^{S}\right) - \sigma^2\mathbf{I}_p \\
&+ \frac{cp}{p+2}\sigma^2\mathbf{I}_p E\left[F_{p+2,m}^{-1}(\Delta^2) I\left(F_{p+2,m}(\Delta^2) < \ell_\alpha\right)\left(2 - \frac{cp}{p+2}F_{p+2,m}^{-1}(\Delta^2)\right)\right] \\
&-(n\boldsymbol{\theta}\mathbf{W}\boldsymbol{\theta}')\Bigg\{\frac{2cp}{p+2}E\left[F_{p+2,m}^{-1}(\Delta^2) I\left(F_{p+2,m}(\Delta^2) < \ell_\alpha\right)\right] \\
&-\frac{cp}{p+4}E\left[F_{p+4,m}^{-1}(\Delta^2) I\left(F_{p+4,m}^{-1}(\Delta^2) < \ell_\alpha^*\right)\left(2 - \frac{cp}{p+4}F_{p+4,m}^{-1}(\Delta^2)\right)\right]\Bigg\}
\end{aligned} \quad (4.5.28)$$

and

$$\begin{aligned}
&R_5\left(\hat{\boldsymbol{\theta}}_n^{\text{PT+}} : \mathbf{W}\right) = \\
&R_2\left(\hat{\boldsymbol{\theta}}_n^{\text{PT}} : \mathbf{W}\right) + R_3\left(\hat{\boldsymbol{\theta}}_n^{S} : \mathbf{W}\right) - \sigma^2\text{tr}(\mathbf{W}) \\
&+ \frac{cp}{p+2}\sigma^2\text{tr}(\mathbf{W}) E\left[F_{p+2,m}^{-1}(\Delta^2) I\left(F_{p+2,m}(\Delta^2) < \ell_\alpha\right)\left(2 - \frac{cp}{p+2}F_{p+2,m}^{-1}(\Delta^2)\right)\right] \\
&-(n\boldsymbol{\theta}'\mathbf{W}\boldsymbol{\theta})\Bigg\{\frac{2cp}{p+2}E\left[F_{p+2,m}^{-1}(\Delta^2) I\left(F_{p+2,m}(\Delta^2) < \ell_\alpha\right)\right] \\
&-\frac{cp}{p+4}E\left[F_{p+4,m}^{-1}(\Delta^2) I\left(F_{p+4,m}^{-1}(\Delta^2) < \ell_\alpha^*\right)\left(2 - \frac{cp}{p+4}F_{p+4,m}^{-1}(\Delta^2)\right)\right]\Bigg\}. \quad (4.5.29)
\end{aligned}$$

Now, the risk difference $R_2\left(\hat{\boldsymbol{\theta}}_n^{\text{PT}};\mathbf{W}\right) - R_5\left(\hat{\boldsymbol{\theta}}_n^{\text{PT+}};\mathbf{W}\right)$ becomes

$$
\begin{aligned}
&+\sigma^2 cp\,\mathrm{tr}(\mathbf{W})\left\{2E\left(\chi_{p+2}^{-2}(\Delta^2)\right)-(p-2)E\left(\chi_{p+2}^{-4}(\Delta^2)\right)\right\}\\
&-cp(p+2)(n\boldsymbol{\theta}'\mathbf{W}\boldsymbol{\theta})E\left(\chi_{p+4}^{-4}(\Delta^2)\right)-\frac{cp}{p+2}\sigma^2\mathrm{tr}(\mathbf{W})\cdot\\
&E\Big[F_{p+2,m}^{-1}(\Delta^2)I\left(F_{p+2,m}(\Delta^2)<\ell_\alpha\right)\left(2-\frac{cp}{p+2}F_{p+2,m}^{-1}(\Delta^2)\right)\Big]\\
&+(n\boldsymbol{\theta}'\mathbf{W}\boldsymbol{\theta})\left\{\frac{2cp}{p+2}E\Big[F_{p+2,m}^{-1}(\Delta^2)I\left(F_{p+2,m}(\Delta^2)<\ell_\alpha\right)\Big]\right.\\
&\left.-\frac{cp}{p+4}E\Big[F_{p+4,m}^{-1}(\Delta^2)I\left(F_{p+4,m}(\Delta^2)<\ell_\alpha^*\right)\left(2-\frac{cp}{p+4}F_{p+4,m}^{-1}(\Delta^2)\right)\Big]\right\} \quad (4.5.30)
\end{aligned}
$$

≥ 0 for all (α,Δ^2) and $p\geq 3$. Hence,

$$
R_5\left(\hat{\boldsymbol{\theta}}_n^{\text{PT+}};\mathbf{W}\right)\leq R_2\left(\hat{\boldsymbol{\theta}}_n^{\text{PT}};\mathbf{W}\right)\quad\text{for all }(\alpha,\Delta^2), \qquad (4.5.31)
$$

where strict inequality holds for some (α,Δ^2). The risk of both $\hat{\boldsymbol{\theta}}_n^{\text{PT+}}$ and $\hat{\boldsymbol{\theta}}_n^{\text{PT}}$ approaches $\sigma^2\mathrm{tr}(\mathbf{W})$ as $\Delta^2\to\infty$. Both estimators share the property that as $\Delta^2\to\infty$, their risks converge to a common limit $\sigma^2\mathrm{tr}(\mathbf{W})$ but the risk of $\hat{\boldsymbol{\theta}}_n^{\text{PT+}}$ remains below that of the risk of $\hat{\boldsymbol{\theta}}_n^{\text{PT}}$. This shows clearly that $\hat{\boldsymbol{\theta}}_n^{\text{PT+}}$ is *superior* to $\hat{\boldsymbol{\theta}}_n^{\text{PT}}$. Furthermore, neither $\hat{\boldsymbol{\theta}}_n^{\text{PT+}}$ nor $\hat{\boldsymbol{\theta}}_n^{\text{PT}}$ is *superior* to $\tilde{\boldsymbol{\theta}}_n$, their risk functions intersect as Δ^2 increases leaving the origin. However, $\hat{\boldsymbol{\theta}}_n^{\text{PT+}}$ is a simple modification of $\hat{\boldsymbol{\theta}}_n^{\text{PT}}$ similar to the positive-rule shrinkage estimator and valid for $p\geq 3$.

Finally, we compare the PRSE, $\hat{\boldsymbol{\theta}}_n^{S+}$ and the improved PTE, $\hat{\boldsymbol{\theta}}_n^{\text{PT+}}$. Here the risk difference is

$$
\begin{aligned}
&R_5\left(\hat{\boldsymbol{\theta}}_n^{\text{PT+}}:\mathbf{W}\right)-R_4\left(\hat{\boldsymbol{\theta}}_n^{S+}:\mathbf{W}\right)\\
&=\sigma^2\mathrm{tr}(\mathbf{W})E\Big[\left(1-\frac{cp}{p+2}F_{p+2,m}^{-1}(\Delta^2)\right)^2\left\{I\left(F_{p+2,m}(\Delta^2)<\frac{cp}{p+2}\right)\right.\\
&\qquad\left.-I\left(F_{p+2,m}(\Delta^2)<\ell_\alpha\right)\right\}\Big]
\end{aligned}
$$

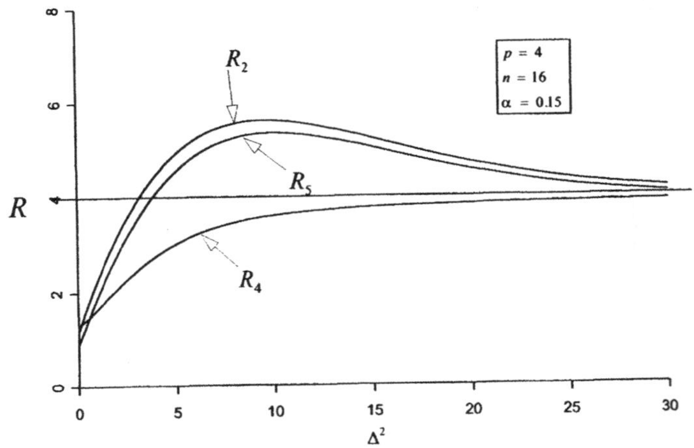

Figure 4.5.2 Graph of $R_2 = R_2(\hat{\boldsymbol{\theta}}_n^{\text{PT}}; \sigma^{-2}\mathbf{I}_p)$, $R_4 = R_4(\hat{\boldsymbol{\theta}}_n^{S+}; \sigma^{-2}\mathbf{I}_p)$, and $R_5 = R_5(\hat{\boldsymbol{\theta}}_n^{PT+}; \sigma^{-2}\mathbf{I}_p)$

$$+(n\boldsymbol{\theta}'\mathbf{W}\boldsymbol{\theta})\Bigg(E\bigg[2\bigg\{1 - \frac{cp}{p+2}F^{-1}_{p+2,m}(\Delta^2)\bigg\}$$

$$\bigg\{I\left(F_{p+2,m}(\Delta^2) < \ell_\alpha\right) - I\left(F_{p+2,m}(\Delta^2) < \frac{cp}{p+2}\right)\bigg\}\bigg]$$

$$- E\bigg[\bigg\{1 - \frac{cp}{p+4}F^{-1}_{p+4,m}(\Delta^2)\bigg\}^2\bigg[\bigg\{I\left(F_{p+4,m}(\Delta^2) < \ell^*_\alpha\right)$$

$$- I\left(F_{p+4,m}(\Delta^2) < \frac{cp}{p+4}\right)\bigg\}\bigg]\Bigg). \qquad (4.5.32)$$

Note that $\ell_\alpha < cp/(p+2)$ implies $F_{p,m}(\alpha) < c$ and similarly $\ell^*_\alpha < cp/(p+4)$ implies $F_{p,m}(\alpha) < c$.

The risk difference is *positive* for all $F_{p,m}(\alpha) < c$. Therefore, $\hat{\boldsymbol{\theta}}_n^{S+}$ is uniformly better than $\hat{\boldsymbol{\theta}}_n^{\text{PT}+}$. The estimator $\hat{\boldsymbol{\theta}}_n^{S+}$ is minimax, and therefore, $\hat{\boldsymbol{\theta}}_n^{S+}$ is a minimax substitute for the PTE whenever $F_{p,m}(\alpha) < 2c$. If $F_{p,m}(\alpha) > 2c$. Then $\hat{\boldsymbol{\theta}}_n^{\text{PT}+}$ dominates $\hat{\boldsymbol{\theta}}_n^{\text{PT}}$ as we saw before at (4.5.31), but $\hat{\boldsymbol{\theta}}_n^{S+}$ may not dominate $\hat{\boldsymbol{\theta}}_n^{\text{PT}}$. In this case, $\hat{\boldsymbol{\theta}}_n^{\text{PT}+}$ is not minimax. See Figure 4.5.2 for comparison of $\hat{\boldsymbol{\theta}}_n^{\text{PT}}$, $\hat{\boldsymbol{\theta}}_n^{S+}$, and $\hat{\boldsymbol{\theta}}_n^{PT+}$.

4.5.6 An Alternative Improved Estimator of $\boldsymbol{\theta}$

In Section 4.5.2 we considered various estimators of $\boldsymbol{\theta}$, and we defined another improved shrinkage estimator $\hat{\boldsymbol{\theta}}_n^{\mathrm{IS}}$ in (4.5.13) as

$$\hat{\boldsymbol{\theta}}_n^{IS} = \{1 - c(m+2)\phi_S(\mathcal{L}_n)\mathcal{L}_n^{-1}\}\tilde{\boldsymbol{\theta}}_n, \tag{4.5.33}$$

where

$$\phi_S(\mathcal{L}_n) = \frac{1}{m+2} I\left(\mathcal{L}_n > \frac{m}{m+2}\right) + \frac{1 + \frac{p}{m}\mathcal{L}_n}{m+p+2} I\left(\mathcal{L}_n < \frac{m}{m+2}\right), \tag{4.5.34}$$

(see Berry, 1994).

We now consider the bias vector and QB of $\hat{\boldsymbol{\theta}}_n^{IS}$ by expressing $\hat{\boldsymbol{\theta}}_n^{IS}$ in terms of the James-Stein estimator

$$\hat{\boldsymbol{\theta}}_n^{IS} = \hat{\boldsymbol{\theta}}_n^{S} - \frac{(p-2)}{(m+p+2)}\left(1 - \frac{p}{m+2}(F^*)^{-1}\right) I\left(F^* < \frac{p}{m+2}\right)\tilde{\boldsymbol{\theta}}_n, \tag{4.5.35}$$

where $F^* = n\|\tilde{\boldsymbol{\theta}}_n\|^2/ms_y^2$.

Then the bias vector and the QB are

$$\mathbf{b}_6\left(\hat{\boldsymbol{\theta}}_n^{IS}\right) = -\boldsymbol{\theta} h(F^*;\Delta^2) \quad \text{and} \quad B_6\left(\hat{\boldsymbol{\theta}}_n^{IS}\right) = \Delta^2 h^2(F^*,\Delta^2), \tag{4.5.36}$$

where

$$\begin{aligned} h(F^*,\Delta^2) &= \frac{(p-2)m}{m+2} E\left(\chi_{p+2}^{-2}(\Delta^2)\right) + \frac{(p-2)}{(m+p+2)} G_{p+2,m}(\ell_0;\Delta^2) \\ &- \frac{(p-2)m}{(m+2)(m+p+2)} \ell_0 E\left\{(F^*)_{p+2,m}^{-1}(\Delta^2) I\left(F_{p+2,m}^*(\Delta^2) < \ell_0\right)\right\}, \end{aligned} \tag{4.5.37}$$

with $\ell_0 = \frac{pm}{(p+2)(m+2)}$.

Next, we prove the following theorem showing $\hat{\boldsymbol{\theta}}_n^{IS}$ dominates $\hat{\boldsymbol{\theta}}_n^{S}$ uniformly over $\Delta^2 \in [0,\infty)$ via unweighted risk analysis:

Theorem 1. $R_6\left(\hat{\boldsymbol{\theta}}_n^{IS};\sigma^{-2}\mathbf{I}_p\right) \leq R_3\left(\hat{\boldsymbol{\theta}}_n^{S};\sigma^{-2}\mathbf{I}_p\right)$ for all $\Delta^2 \in [0,\infty)$ and strict inequality for at least one Δ^2.

Proof. First note that we can write $\hat{\boldsymbol{\theta}}_n^{IS}$ as

$$\hat{\boldsymbol{\theta}}_n^{IS} = \left(1 - \frac{(p-2)}{n\|\tilde{\boldsymbol{\theta}}_n\|^2} m s_Y^2 \phi_S\left(\frac{n\|\tilde{\boldsymbol{\theta}}_n\|^2}{p s_Y^2}\right)\right)\tilde{\boldsymbol{\theta}}_n. \tag{4.5.38}$$

We know from the proof of Theorem 3.9.2 in Section 3.9 that

$$E\left[\left\{\frac{ms_Y^2}{\sigma^2}\phi\left(\frac{n\|\tilde{\boldsymbol{\theta}}_n\|^2}{ps_Y^2}\right)-1\right\}^2\Big|\mathcal{L}_n\right] \le E\left[\left\{\frac{ms_Y^2}{(m+2)\sigma^2}-1\right\}^2\Big|\mathcal{L}_n\right] \text{ for all } \Delta^2.$$

The risk of $\hat{\boldsymbol{\theta}}_n^{IS}$ is given by

$$R_6\left(\hat{\boldsymbol{\theta}}_n^{IS};\sigma^{-2}\mathbf{I}_p\right) = p + E\left[\frac{(p-2)^2}{n\|\tilde{\boldsymbol{\theta}}_n^2\|^2}\frac{(ms_Y^2)^2}{\sigma^2}\phi_S^2\left(\frac{n\|\tilde{\boldsymbol{\theta}}_n^2\|^2}{ps_Y^2}\right)\right.$$
$$\left.-2\frac{(p-2)}{n\|\tilde{\boldsymbol{\theta}}_n\|^2}(\tilde{\boldsymbol{\theta}}_n-\boldsymbol{\theta})'\tilde{\boldsymbol{\theta}}_n\frac{ms_Y^2}{\sigma^2}\phi\left(\frac{n\|\tilde{\boldsymbol{\theta}}_n\|^2}{ps_y^2}\right)\right].$$

Now, by Stein's lemma, $E[(Z-U)h(Z)] = E[h'(Z)]$, where $h(Z)$ is an absolutely continuous function, we have

$$R_6\left(\hat{\boldsymbol{\theta}}_n^{IS};\sigma^{-2}\mathbf{I}_p\right)$$
$$= p + E_{\mathcal{L}_n}\left[\frac{(p-2)^2}{n\|\tilde{\boldsymbol{\theta}}_n\|^2}\frac{(ms_Y^2)^2}{\sigma^2}\phi_S^2\left(\frac{n\|\tilde{\boldsymbol{\theta}}_n\|^2}{ps_Y^2}\right) - 2\frac{(p-2)^2}{n\|\tilde{\boldsymbol{\theta}}_n\|^2}ms_Y^2\phi_S\left(\frac{n\|\tilde{\boldsymbol{\theta}}_n\|^2}{ps_Y^2}\right)\right]$$
$$= p - E\left[\frac{(p-2)^2\sigma^2}{n\|\tilde{\boldsymbol{\theta}}_n\|^2}\right] + E\left[\frac{(p-2)^2\sigma^2}{n\|\tilde{\boldsymbol{\theta}}_n\|^2}\right]E\left[\left\{\left(\frac{ms_Y^2}{\sigma^2}\right)\phi_S\left(\frac{n\|\tilde{\boldsymbol{\theta}}_n\|^2}{ps_Y^2}\right)-1\right\}^2\right],$$

while

$$R_3\left(\hat{\boldsymbol{\theta}}_n^S;\sigma^{-2}\mathbf{I}_p\right)$$
$$= p - E\left[\frac{(p-2)^2\sigma^2}{n\|\tilde{\boldsymbol{\theta}}_n\|^2}\right] + E\left[\frac{(p-2)^2\sigma^2}{n\|\tilde{\boldsymbol{\theta}}_n\|^2}\right]E\left[\left\{\frac{ms_Y^2}{(m+2)\sigma^2}-1\right\}^2\right].$$

Hence,

$$R_3\left(\hat{\boldsymbol{\theta}}_n^S;\sigma^{-2}\mathbf{I}_p\right) - R_6\left(\hat{\boldsymbol{\theta}}_n^{IS};\sigma^{-2}\mathbf{I}_p\right)$$
$$= E\left[\frac{(p-2)^2\sigma^2}{n\|\tilde{\boldsymbol{\theta}}_n\|^2}\right]\left(E\left[\left\{\frac{ms_Y^2}{(m+2)\sigma^2}-1\right\}^2\right]\right.$$
$$\left.-E\left[\left\{\left(\frac{ms_Y^2}{\sigma^2}\right)\phi_S\left(\frac{n\|\tilde{\boldsymbol{\theta}}_n\|^2}{ps_Y^2}\right)-1\right\}^2\right]\right) \ge 0 \text{ for all } \Delta^2.$$

Then,

$$R_6\left(\hat{\boldsymbol{\theta}}_n^{IS};\sigma^{-2}\mathbf{I}_p\right) \le R_3\left(\hat{\boldsymbol{\theta}}_n^S;\sigma^{-2}\mathbf{I}_p\right) \quad \text{for all } (\Delta^2, p\ge 3, m>1).$$

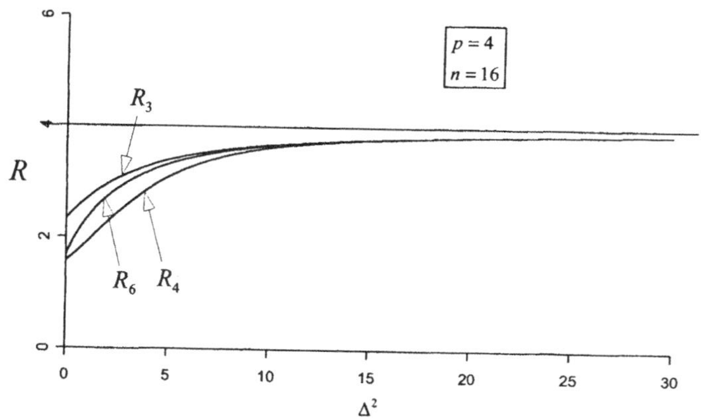

Figure 4.5.3 Graph of $R_6 = R_6(\hat{\boldsymbol{\theta}}_n^{IS}; \sigma^{-2}\mathbf{I}_p)$, $R_4 = R_4(\hat{\boldsymbol{\theta}}_n^{S+}; \sigma^{-2}\mathbf{I}_p)$, and $R_3 = R_3(\hat{\boldsymbol{\theta}}_n^{S}; \sigma^{-2}\mathbf{I}_p)$

Although $\hat{\boldsymbol{\theta}}_n^{IS}$ improves over $\hat{\boldsymbol{\theta}}_n^{S}$, the positive-rule version of $\hat{\boldsymbol{\theta}}_n^{IS}$ does not improve over $\hat{\boldsymbol{\theta}}_n^{S+}$.

We can write the expression of the risk of $\hat{\boldsymbol{\theta}}_n^{IS}$ as

$$\begin{aligned}
R_6\left(\hat{\boldsymbol{\theta}}_n^{IS}; \sigma^{-2}\mathbf{I}_p\right) &= R_3\left(\hat{\boldsymbol{\theta}}_n^{S}; \sigma^{-2}\mathbf{I}_p\right) \\
&- \frac{(p-2)^2}{(m+p+2)^2}\Bigg[\frac{m(p+2m+4)}{(m+2)} E\Big\{F_{p,m+2}^{-1}(\Delta^2) I\left(F_{p,m+2}(\Delta^2) < 1\right)\Big\} \\
&\qquad -2mG_{p,m+2}(1,\Delta^2) - \Big\{pG_{p,m}\Big(\frac{m}{m+2}; \Delta^2\Big) \\
&\qquad +\Delta^2 G_{p+2,m}\Big(\frac{mp}{(m+2)(p+2)}; \Delta^2\Big)\Big\}\Bigg] \\
&- \frac{2p(p-2)}{(m+p+2)} G_{p,m}\left(\frac{m}{m+2}; \Delta^2\right) + \frac{2m(p-2)^2}{(m+2)(m+p+2)} \\
&\times E\Big\{F_{p,m}^{-1}(\Delta^2) I\Big(F_{p,m}(\Delta^2) < \frac{m}{m+2}\Big)\Big\}. \qquad (4.5.39)
\end{aligned}$$

See Figure 4.5.3 for visual comparison of the estimators.

The MSE matrix of $\hat{\boldsymbol{\theta}}_n^{IS}$ can be obtained similarly.

4.6 Stein-Type Estimation: Nonnormal Distributions

4.6.1 Model, Estimation, and Test

Consider the model (4.1.1) again:

$$\mathbf{Y}_i = \boldsymbol{\theta} + \mathbf{e}_i, \quad i = 1, 2, \ldots, n,$$

where $\mathbf{Y}_i = (Y_{1i}, \ldots, y_{pi})'$, $\mathbf{e}_i = (e_{1i}, \ldots, e_{pi})' \sim \prod_{j=1}^{p} F_0(e_{ji})$ and $\boldsymbol{\theta} = (\theta_1, \ldots, \theta_p)'$ is the vector of means. Further, assume that

$$E(\mathbf{e}_i) = \mathbf{0} \quad \text{and} \quad E(\mathbf{e}_i\mathbf{e}_i') = \sigma^2\mathbf{I}_p. \tag{4.6.1}$$

It is well-known that the unbiased estimate of $\boldsymbol{\theta}$ is $\tilde{\boldsymbol{\theta}}_n = (\overline{Y}_1, \ldots, \overline{Y}_p)'$, and by the p-variate central limit theorem,

(i) $\sqrt{n}(\tilde{\boldsymbol{\theta}}_n - \boldsymbol{\theta}) \sim N_p(\mathbf{0}, \sigma^2\mathbf{I}_p)$ as $n \to \infty$; (4.6.2a)

(ii) The unbiased estimator of σ^2 is

$$s_e^2 = \{p(n-1)\}^{-1}\text{tr}\left\{\sum_{i=1}^{n}(\mathbf{Y}_i - \tilde{\boldsymbol{\theta}}_n)(\mathbf{Y}_i - \tilde{\boldsymbol{\theta}}_n)'\right\}, \tag{4.6.2b}$$

and $s_e^2 \xrightarrow{\text{P}} \sigma^2$ as $n \to \infty$.

In order to test the null hypothesis $H_0 : \boldsymbol{\theta} = \boldsymbol{\theta}_0$ against $H_A : \boldsymbol{\theta} \neq \boldsymbol{\theta}_0$, we consider the statistic

$$\mathcal{L}_n = \frac{n\|\tilde{\boldsymbol{\theta}}_n - \boldsymbol{\theta}_0\|^2}{s_e^2}. \tag{4.6.3}$$

Under H_0, as $n \to \infty$, $\mathcal{L}_n \xrightarrow{\mathcal{D}} \chi_p^2$, which is a central chi-square distribution with p d.f. Under the fixed alternative $H_\delta : \boldsymbol{\theta} = \boldsymbol{\theta}_0 + \boldsymbol{\delta}$,

$$\sqrt{n}(\tilde{\boldsymbol{\theta}}_n - \boldsymbol{\theta}_0) = \sqrt{n}(\tilde{\boldsymbol{\theta}}_n - \boldsymbol{\theta}) + \sqrt{n}\boldsymbol{\delta}, \tag{4.6.4}$$

and $n\|\boldsymbol{\delta}\|^2 \to \infty$ as $n \to \infty$. It follows that $\mathcal{L}_n \to \infty$ as $n \to \infty$. Consequently, for all fixed alternatives H_δ,

$$\lim_{n\to\infty} P\left\{\frac{n\|\tilde{\boldsymbol{\theta}}_n - \boldsymbol{\theta}_0\|^2}{s_e^2} \geq x \middle| H_\delta\right\} = 1 \tag{4.6.5}$$

for all $x \in (0, \infty)$.

Therefore, in order to obtain a meaningful distribution, we consider the class of **local alternatives** $K_{(n)}$ defined by

$$K_{(n)} : \boldsymbol{\theta}_{(n)} = \boldsymbol{\theta}_0 + n^{-1/2}\boldsymbol{\delta} \text{ for a fixed finite vector } \boldsymbol{\delta}. \tag{4.6.6}$$

Under $K_{(n)}$,

$$\lim_{n\to\infty} P\left(\frac{n\|\tilde{\boldsymbol{\theta}}_n - \boldsymbol{\theta}_0\|^2}{s_e^2} \le x \Big| K_{(n)}\right) = H_p(x;\Delta^2), \quad \Delta^2 = \frac{\|\boldsymbol{\delta}\|^2}{\sigma^2} \tag{4.6.7}$$

by Sluskey's theorem where $H_p(\cdot;\Delta^2)$ is the cdf of a chi-square distribution with p d.f. and noncentrality parameter $\Delta^2/2$.

4.6.2 Preliminary Test (or Quasi-empirical Bayes) Approach to Stein-Type Estimation of the Mean Vector

First, we consider $\tilde{\boldsymbol{\theta}}_n = (\overline{Y}_1, \ldots, \overline{Y}_p)'$ as the **unrestricted** estimator of $\boldsymbol{\theta}$. If we suspect that $\boldsymbol{\theta}$ is $\mathbf{0}$, then the **preliminary test estimator** of $\boldsymbol{\theta}$ is defined by

$$\hat{\boldsymbol{\theta}}_n^{\text{PT}} = \tilde{\boldsymbol{\theta}}_n - \tilde{\boldsymbol{\theta}}_n I(\mathcal{L}_n < \mathcal{L}_{n,\alpha}), \tag{4.6.8}$$

where $\mathcal{L}_{n,\alpha}$ is the upper α-level critical value from the exact distribution of $\mathcal{L}_n$, and under H_0 as $n \to \infty$, $\mathcal{L}_{n,\alpha}$ is approximated by $\chi_p^2(\alpha)$—an α-level chi-square value with p d.f. Notice that the difference between (4.2.1) and (4.6.8) where $\chi_p^2(\alpha)$ in (4.2.1) is replaced by $\mathcal{L}_{n,\alpha}$ though $\mathcal{L}_{n,\alpha} \to \chi_p^2(\alpha)$ as $n \to \infty$.

An empirical Bayes method to define the Stein-type estimator of $\boldsymbol{\theta}$ is not easy to formulate because we need to assign an appropriate prior distribution of $\boldsymbol{\theta}$ to obtain the a posterior distribution to use in developing the empirical Bayes estimator. In this case the preliminary test approach will overcome the difficulty by defining the Stein-type estimator as

$$\hat{\boldsymbol{\theta}}_n^{S} = \tilde{\boldsymbol{\theta}}_n - (p-2)\mathcal{L}_n^{-1}\tilde{\boldsymbol{\theta}}_n, \tag{4.6.9}$$

since $\mathcal{L}_n \neq 0$ with probability one. Here we have simply replaced $I(\mathcal{L}_n < \mathcal{L}_{n,\alpha})$ by $(p-2)\mathcal{L}_n^{-1}$. Further, following (4.2.3), we define the *positive-rule shrinkage estimator* as

$$\hat{\boldsymbol{\theta}}_m^{S+} = \{1 - (p-2)\mathcal{L}_n^{-1}\} I(\mathcal{L}_n > p-2)\tilde{\boldsymbol{\theta}}_n. \tag{4.6.10}$$

4.6.3 Asymptotic Distributional Bias Vector, Quadratic Bias, MSE Matrix, and Risk Expressions of the Estimators

Let $\boldsymbol{\theta}_n^*$ be any estimator of $\boldsymbol{\theta}$, let $\mathbf{W}$ be the p.s.d. matrix, and consider the quadratic loss function

$$\begin{aligned} L(\boldsymbol{\theta}_n^*;\boldsymbol{\theta}) &= n(\boldsymbol{\theta}_n^* - \boldsymbol{\theta})\mathbf{W}(\boldsymbol{\theta}_n^* - \boldsymbol{\theta}) \\ &= \operatorname{tr}[\mathbf{W}\{n(\boldsymbol{\theta}_n^* - \boldsymbol{\theta})(\boldsymbol{\theta}_n^* - \boldsymbol{\theta})'\}]. \end{aligned} \tag{4.6.11}$$

Let $\mathbf{M}(\boldsymbol{\theta}_n^*) = nE\{(\boldsymbol{\theta}_n^* - \boldsymbol{\theta})(\boldsymbol{\theta}_n^* - \boldsymbol{\theta})'\}$; then the risk is given by

$$R(\boldsymbol{\theta}_n^*; \mathbf{W}) = \text{tr}[\mathbf{W}\mathbf{M}(\boldsymbol{\theta}_n^*)]. \tag{4.6.12}$$

Consider the local alternatives

$$K_{(n)} : \boldsymbol{\theta}_{(n)} = \theta_0 \mathbf{1}_p + n^{-1/2}\boldsymbol{\delta} \quad (\boldsymbol{\delta} \text{ a fixed vector}), \tag{4.6.13}$$

and let the asymptotic distribution function (a.d.f.)

$$G_{\boldsymbol{\delta}}^*(\mathbf{x}) = \lim_{n\to\infty} P\left\{\sqrt{n}(\boldsymbol{\theta}_n^* - \boldsymbol{\theta}_{(n)}) \le \mathbf{x} \| K_{(n)}\right\} \tag{4.6.14}$$

if it exists. Further let the **asymptotic distributional MSE** (ADMSE) of $\boldsymbol{\theta}_n^*$ be denoted as

$$\mathbf{M}(\boldsymbol{\theta}_n^*) = \int \mathbf{x}\mathbf{x}' dG_{\boldsymbol{\delta}}^*(\mathbf{x}), \tag{4.6.15}$$

and let the asymptotic distributional bias vector (ADB) and the asymptotic distribution quadratic bias (ADQB) be

$$\mathbf{b}(\boldsymbol{\theta}_n^*) = \int \mathbf{x} dG_{\boldsymbol{\delta}}^*(\mathbf{x}) \quad \text{and} \quad B(\boldsymbol{\theta}_n^*) = [\mathbf{b}(\boldsymbol{\theta}_n^*)]'\boldsymbol{\Sigma}^{-1}[\mathbf{b}'(\boldsymbol{\theta}_n^*)], \tag{4.6.16}$$

where $\boldsymbol{\Sigma}$ is the covariance matrix of the unrestricted estimator. Finally, we have the asymptotic distributional risk (ADR) as

$$R(\boldsymbol{\theta}_n^*; \mathbf{W}) = \text{tr}[\mathbf{W}\mathbf{M}(\boldsymbol{\theta}_n^*)]. \tag{4.6.17}$$

Based on these definitions, we may obtain the ADB, ADQB, ADMSE, and ADR by the following two theorems:

Theorem 1. Under $\{K_{(n)}\}$ as $n \to \infty$, the ADB and ADQB are given by

(i) $\mathbf{b}_1(\tilde{\boldsymbol{\theta}}_n) = \mathbf{0}$ and $B_1(\tilde{\boldsymbol{\theta}}_n) = 0$,

(ii) $\mathbf{b}_2(\hat{\boldsymbol{\theta}}_n^{\text{PT}}) = -\boldsymbol{\delta} H_{p+2}(\chi_p^2(\alpha); \Delta^2)$ and $B_2(\hat{\boldsymbol{\theta}}_n^{\text{PT}}) = \Delta^2\{H_{p+2}(\chi_p^2(\alpha); \Delta^2)\}^2$,

(iii) $\mathbf{b}_3(\hat{\boldsymbol{\theta}}_n^S) = -(p-2)\boldsymbol{\delta} E(\chi_{p+2}^{-2}(\Delta^2))$ and

$$B_3(\hat{\boldsymbol{\theta}}_n^S) = (p-2)^2\Delta^2\{E(\chi_{p+2}^{-2}(\Delta^2))\}^2,$$

(iv) $$\mathbf{b}_4(\hat{\boldsymbol{\theta}}_n^{S+}) = -\boldsymbol{\delta}\Big[H_{p+2}(p-2; \Delta^2) + (p-2)E(\chi_{p+2}^{-2}(\Delta^2)) - (p-2)E\Big\{\chi_{p+2}^{-2}(\Delta^2))I\Big(\chi_{p+2}^{-2}(\Delta^2)\Big)\Big\}\Big]$$

and $$B_4(\hat{\boldsymbol{\theta}}_n^{S+}) = \Delta^2\Big[H_{p+2}(p-2, \Delta^2) + (p-2)E(\chi_{p+2}^{-2}(\Delta^2)) (p-2)E\Big\{\chi_{p+2}^{-2}(\Delta^2)I(\chi_{p+2}^{-2}(\Delta^2) < p-2)\Big\}\Big]^2.$$

Proof. Under $\{K_{(n)}\}$ as $n \to \infty$, we have (see Sen and Singer (1999))

(i) $\sqrt{n}\, s_e^{-1}(\tilde{\boldsymbol{\theta}}_n - \boldsymbol{\theta}_{(n)}) \xrightarrow{\mathcal{D}} \mathbf{Z} \sim N_p(\mathbf{0}, \mathbf{I}_p)$,

(ii) $\sqrt{n}\, s_e^{-1}(\hat{\boldsymbol{\theta}}_n^{\rm PT} - \boldsymbol{\theta}_{(n)}) \xrightarrow{\mathcal{D}} \mathbf{Z} - (\mathbf{Z} + \boldsymbol{\delta}^*) I(\|\mathbf{Z} + \boldsymbol{\delta}^*\|^2 < \chi_p^2(\alpha))$, $\boldsymbol{\delta}^* = \sigma^{-1}\boldsymbol{\delta}$,

(iii) $\sqrt{n}\, s_e^{-1}(\hat{\boldsymbol{\theta}}_n^{S} - \boldsymbol{\theta}_{(n)}) \xrightarrow{\mathcal{D}} \mathbf{Z} - c(\mathbf{Z} + \boldsymbol{\delta}^*)\|\mathbf{Z} + \boldsymbol{\delta}^*\|^{-2}$, $c = p-2$, and

(iv) $\sqrt{n} s_e^{-1}(\hat{\boldsymbol{\theta}}_n^{S+} - \boldsymbol{\theta}_{(n)}) \xrightarrow{\mathcal{D}} \mathbf{Z}(1 - c\|\mathbf{Z} + \boldsymbol{\delta}^*\|^{-2}) I(\|\mathbf{Z} + \boldsymbol{\delta}^*\|^2 > c) - \boldsymbol{\delta}^*$, where $c = (p-2)$,

Then

(i) $\mathbf{b}_1(\tilde{\boldsymbol{\theta}}_n) = \mathbf{0}$ and $B_1(\tilde{\boldsymbol{\theta}}_n) = 0$,

(ii) $\mathbf{b}_2(\hat{\boldsymbol{\theta}}_n^{\rm PT}) = \sigma E(\mathbf{U}_1)$, where

$$\mathbf{U}_1 = \mathbf{Z} - (\mathbf{Z} + \boldsymbol{\delta}^*) I(\|\mathbf{Z} + \boldsymbol{\delta}^*\|^2 < \chi_p^2(\alpha)) = \boldsymbol{\delta} H_{p+2}(\chi_p^2(\alpha); \Delta^2),$$

$$\Delta^2 = \|\boldsymbol{\delta}^*\|^2 \text{ and } B_2(\hat{\boldsymbol{\theta}}_n^{\rm PT}) = \Delta^2\{H_{p+2}(\chi_p^2(\alpha); \Delta^2)\}^2,$$

(iii) $B_3(\hat{\boldsymbol{\theta}}_n^{S}) = \sigma E(\mathbf{U}_2)$, where

$$\mathbf{U}_2 = \mathbf{Z} - c(\mathbf{Z} + \boldsymbol{\delta}^*)\|\mathbf{Z} + \boldsymbol{\delta}^*\|^{-2} = -c\boldsymbol{\delta} E[\chi_{p+2}^{-2}(\Delta^2)] \quad \text{and}$$

$$B_3(\hat{\boldsymbol{\theta}}_n^{S}) = c^2\Delta^2\Big\{E[\chi_{p+2}^{-2}(\Delta^2)]\Big\}^2.$$

(iv) $\mathbf{b}_4(\hat{\boldsymbol{\theta}}_n^{S+}) = \sigma E(\mathbf{U}_3)$, where

$$\begin{aligned}
\mathbf{U}_3 &= (\mathbf{Z} + \boldsymbol{\delta}^*)(1 - c\|\mathbf{Z} + \boldsymbol{\delta}^*\|^{-2}) I(\|\mathbf{Z} + \boldsymbol{\delta}^*\|^{-2} > c) - \boldsymbol{\delta}^* \\
&= \boldsymbol{\delta} E\Big[(1 - c\chi_{p+2}^{-2}(\Delta^2)) I(\chi_{p+2}^{2}(\Delta^2) > c)\Big] - \boldsymbol{\delta} \\
&= -c\boldsymbol{\delta} E[\chi_{p+2}^{-2}(\Delta^2)] - \boldsymbol{\delta} E[(1 - c\chi_{p+2}^{-2}(\Delta^2)) I(\chi_{p+2}^{-2}(\Delta^2) < c)], \; c = (p-2) \\
&= -\boldsymbol{\delta}\Big\{cE[\chi_{p+2}^{-2}(\Delta^2)] + H_{p+2}(c; \Delta^2) \\
&\qquad - cE[\chi_{p+2}^{-2}(\Delta^2) I(\chi_{p+2}^{-2}(\Delta^2) < c)]\Big\}
\end{aligned}$$

and

$$\begin{aligned}
B_4\Big(\hat{\boldsymbol{\theta}}_n^{S+}\Big) = \Delta^2\Big\{&cE[\chi_{p+2}^{-2}(\Delta^2)] + H_{p+2}(c; \Delta^2) \\
&- cE[\chi_{p+2}^{-2}(\Delta^2) I(\chi_{p+2}^{-2}(\Delta^2) < c)]\Big\}^2.
\end{aligned}$$

Note that these expressions are the same as in Theorem 4.4.1.

The next theorem gives the results on ADMSE and ADR expressions.

Theorem 2. Under $\{K_{(n)}\}$ as $n \to \infty$,

(i) $\mathbf{M}_1(\tilde{\boldsymbol{\theta}}_n) = \sigma^2\mathbf{I}_p$ and $R_1(\tilde{\boldsymbol{\theta}}_n; \mathbf{W}) = \sigma^2\mathrm{tr}(\mathbf{W}) = p$ if $\mathbf{W} = \sigma^{-2}\mathbf{I}_p$,

$$\text{(ii)}\ \mathbf{M}_2(\hat{\boldsymbol{\theta}}_n^{\text{PT}}) = \sigma^2\mathbf{I}_p\{1 - H_{p+2}(\chi_p^2(\alpha);\Delta^2)\} \\ +\boldsymbol{\delta}\boldsymbol{\delta}'\{2H_{p+2}(\chi_p^2(\alpha);\Delta^2) - H_{p+4}(\chi_p^2(\alpha);\Delta^2)\}$$

and

$$R_2(\hat{\boldsymbol{\theta}}_n^{\text{PT}};\mathbf{W}) = \sigma^2\text{tr}(\mathbf{W})\{1 - H_{p+2}(\chi_p^2(\alpha);\Delta^2)\} \\ +(\boldsymbol{\delta}'\mathbf{W}\boldsymbol{\delta})\{2H_{p+2}(\chi_p^2(\alpha);\Delta^2) - H_{p+4}(\chi_p^2(\alpha);\Delta^2)\} \\ = p\{1 - H_{p+2}(\chi_p^2(\alpha);\Delta^2)\} + \Delta^2\{2H_{p+2}(\chi_p^2(\alpha);\Delta^2) \\ -H_{p+4}(\chi_p^2(\alpha);\Delta^2)\} \quad \text{if } \mathbf{W} = \sigma^{-2}\mathbf{I}_p,$$

$$\text{(iii)}\ \mathbf{M}_3(\hat{\boldsymbol{\theta}}_n^S) = \sigma^2\mathbf{I}_p - \sigma^2(p-2)\mathbf{I}_p\{2E((\chi_{p+2}^{-2}(\Delta^2)) - (p-2)E((\chi_{p+2}^{-4}(\Delta^2))\} \\ +(p^2-4)\boldsymbol{\delta}\boldsymbol{\delta}'E((\chi_{p+4}^{-4}(\Delta^2))$$

and

$$R_3(\hat{\boldsymbol{\theta}}_n^S) \\ = \sigma^2\text{tr}(\mathbf{W}) - \sigma^2(p-2)\text{tr}(\mathbf{W})\Big\{2E\big[\chi_{p+2}^{-2}(\Delta^2)\big] - (p-2)E\big[\chi_{p+2}^{-4}(\Delta^2)\big]\Big\} \\ +(p^2-4)(\boldsymbol{\delta}'\mathbf{W}\boldsymbol{\delta})E\big[\chi_{p+4}^{-4}(\Delta^2)\big] \\ = p - (p-2)^2E((\chi_p^{-2}(\Delta^2)), \quad \text{if } \mathbf{W} = \sigma^{-2}\mathbf{I}_p,$$

$$\text{(iv)}\ \mathbf{M}_4(\hat{\boldsymbol{\theta}}_n^{S+}) = \mathbf{M}_3(\hat{\boldsymbol{\theta}}_n^S) \\ -\sigma^2\mathbf{I}_pE\Big[\big(1-(p-2)\chi_{p+2}^{-2}(\Delta^2)\big)^2 I\big(\chi_{p+2}^{-2}(\Delta^2) < p-2\big)\Big] \\ +\boldsymbol{\delta}\boldsymbol{\delta}'\Big\{2E\Big[\big(1-(p-2)\chi_{p+2}^{-2}(\Delta^2)\big)I\big(\chi_{p+2}^{-2}(\Delta^2) < p-2\big)\Big] \\ -E\Big[\big(1-(p-2)\chi_{p+4}^{-2}(\Delta^2)\big)^2 I\big(\chi_{p+4}^{-2}(\Delta^2) < p-2\big)\Big]\Big\}$$

and

$$R_4(\hat{\boldsymbol{\theta}}_n^{S+};\mathbf{W}) = R_3(\hat{\boldsymbol{\theta}}_n^S;\mathbf{W}) \\ -\sigma^2\text{tr}(\mathbf{W})E\big[\big(1-(p-2)\chi_{p+2}^{-2}(\Delta^2)\big)^2 I\big(\chi_{p+2}^{-2}(\Delta^2) < p-2\big)\big] \\ +(\boldsymbol{\delta}'\mathbf{W}\boldsymbol{\delta})\Big\{2E\big[\big(1-(p-2)\chi_{p+2}^{-2}(\Delta^2)\big)I\big(\chi_{p+2}^{-2}(\Delta^2) < p-2\big)\big] \\ -E\big[\big(1-(p-2)\chi_{p+4}^{-2}(\Delta^2)\big)^2 I\big(\chi_{p+4}^{2}(\Delta^2) < p-2\big)\big]\Big\}$$

$$
\begin{aligned}
&= p - (p-2)^2 E\big[\chi_p^{-2}(\Delta^2)\big] \\
&\quad -pE\big[\big(1-(p-2)\chi_{p+2}^{-2}(\Delta^2)I\big(\chi_{p+2}^{-2}(\Delta^2) < p-2\big)\big] \\
&\quad +\Delta^2\Big\{2E\big[1-(p-2)(\chi_{p+2}^{-2}(\Delta^2))I(\chi_{p+2}^{2}(\Delta^2)) < p-2)\big] \\
&\quad -E\Big[\big(1-(p-2)(\chi_{p+4}^{-2}(\Delta^2)\big)^2 I(\chi_{p+2}^{2}(\Delta^2)) < p-2)\big]\Big\} \\
&\qquad \text{if } \mathbf{W} = \sigma^{-2}\mathbf{I}_p.
\end{aligned}
$$

Proof. (i) is obvious.

(ii)
$$
\begin{aligned}
&\mathbf{M}_2(\hat{\boldsymbol{\theta}}_n^{\text{PT}}) \\
&= \sigma^2 E[\mathbf{U}_1\mathbf{U}_1'] \\
&= \sigma^2 E\big[\mathbf{Z} - (\mathbf{Z}+\boldsymbol{\delta}^*)I\big(\|\mathbf{Z}+\boldsymbol{\delta}^*\|^2 < \chi_p^2(\alpha)\big)\big] \\
&\quad \times\big[\mathbf{Z} - (\mathbf{Z}+\boldsymbol{\delta}^*)I(\|\mathbf{Z}+\boldsymbol{\delta}^*\|^2 < \chi_p^2(\alpha))\big]' \\
&= \sigma^2 E[\mathbf{Z}\mathbf{Z}'] + \sigma^2 E\big[(\mathbf{Z}+\boldsymbol{\delta}^*)(\mathbf{Z}+\boldsymbol{\delta}^*)'I(\|\mathbf{Z}+\boldsymbol{\delta}^*\|^2 < \chi_p^2(\alpha))\big] \\
&\quad -2\sigma^2 E\big[\mathbf{Z}(\mathbf{Z}+\boldsymbol{\delta}^*)'I(\|\mathbf{Z}+\boldsymbol{\delta}^*\|^2 < \chi_p^2(\alpha))\big] \\
&= \sigma^2\mathbf{I}_p - \sigma^2\{\mathbf{I}_p H_{p+2}(\chi_p^2(\alpha);\Delta^2) + \sigma^{-2}\boldsymbol{\delta}\boldsymbol{\delta}' H_{p+4}(\chi_p^2(\alpha);\Delta^2)\} \\
&\quad +2\boldsymbol{\delta}\boldsymbol{\delta}' H_{p+2}(\chi_p^2(\alpha);\Delta^2) \\
&= \sigma^2\mathbf{I}_p\{1 - H_{p+2}(\chi_p^2(\alpha);\Delta^2)\} \\
&\quad +\boldsymbol{\delta}\boldsymbol{\delta}'\{2H_{p+2}(\chi_p^2(\alpha);\Delta^2) - H_{p+4}(\chi_p^2(\alpha);\Delta^2)\}.
\end{aligned}
$$

Hence,

$$
\begin{aligned}
R_2(\hat{\boldsymbol{\theta}}_n^{\text{PT}};\mathbf{W}) &= \sigma^2 \text{tr}\{\mathbf{W}\}\Big\{1 - H_{p+2}(\chi_p^2(\alpha);\Delta^2)\Big\} \\
&\quad + (\boldsymbol{\delta}'\mathbf{W}\boldsymbol{\delta})\Big\{2H_{p+2}(\chi_p^2(\alpha);\Delta^2) - H_{p+4}(\chi_p^2(\alpha);\Delta^2)\Big\} \\
&= p\Big\{1 - H_{p+2}(\chi_p^2(\alpha);\Delta^2)\Big\} \\
&\quad + \Delta^2\Big\{2H_{p+2}(\chi_p^2(\alpha);\Delta^2) - H_{p+4}(\chi_p^2(\alpha);\Delta^2)\Big\} \text{ if } \mathbf{W} = \sigma^{-2}\mathbf{I}_p.
\end{aligned}
$$

(iii)
$$
\begin{aligned}
&\mathbf{M}_3(\hat{\boldsymbol{\theta}}_n^{S}) = \sigma^2 E[\mathbf{U}_2\mathbf{U}_2'] \\
&= \sigma^2 E\big[\mathbf{Z} - c(\mathbf{Z}+\boldsymbol{\delta}^*)\|\mathbf{Z}+\boldsymbol{\delta}^*\|^{-2}\big]\big[\mathbf{Z} - c(\mathbf{Z}+\boldsymbol{\delta}^*)\|\mathbf{Z}+\boldsymbol{\delta}^*\|^{-2}\big]', \; c = p-2 \\
&= \sigma^2 E[\mathbf{Z}\mathbf{Z}'] + \sigma^2 c^2 E\big[(\mathbf{Z}+\boldsymbol{\delta}^*)(\mathbf{Z}+\boldsymbol{\delta}^*)'(\|\mathbf{Z}+\boldsymbol{\delta}^*\|^{-4}\big] \\
&\quad - 2c\sigma^2 E\big[\mathbf{Z}'(\mathbf{Z}+\boldsymbol{\delta}^*)'\|\mathbf{Z}+\boldsymbol{\delta}^*\|^{-2}\big]
\end{aligned}
$$

$$= \sigma^2 E[\mathbf{ZZ}'] + \sigma^2 c^2 \Big\{ \mathbf{I}_p E[\chi_{p+2}^{-4}(\Delta^2)] + \boldsymbol{\delta}^*\boldsymbol{\delta}^{*\prime} E[\chi_{p+2}^{-4}(\Delta^2)] \Big\}$$

$$-2c\sigma^2 \Big\{ \mathbf{I}_p E\big[\chi_{p+2}^{-2}(\Delta^2)\big] + \boldsymbol{\delta}^*\boldsymbol{\delta}^{*\prime} E\big[\chi_{p+4}^{-2}(\Delta^2)\big] - \boldsymbol{\delta}^*\boldsymbol{\delta}^* E\big[\chi_{p+2}^{-2}(\Delta^2)\big] \Big\}$$

$$= \sigma^2 \mathbf{I}_p - \sigma^2 c \mathbf{I}_p \Big\{ 2E\big[\chi_{p+2}^{-2}(\Delta^2)\big] - cE(\chi_{p+2}^{-4}(\Delta^2)) \Big\}$$

$$+ c\boldsymbol{\delta\delta}' \Big\{ 2E\big[\chi_{p+2}^{-2}(\Delta^2)\big] - 2E\big[\chi_{p+4}^{-2}(\Delta^2)\big] + cE\big[\chi_{p+4}^{-4}(\Delta^2)\big] \Big\}$$

$$= \sigma^2 \mathbf{I}_p - \sigma^2 (p-2) \mathbf{I}_p \Big\{ 2E(\chi_{p+2}^{-2}(\Delta^2)) - (p-2)E(\chi_{p+2}^{-4}(\Delta^2)) \Big\}$$

$$+ (p^2-4)\boldsymbol{\delta\delta}' E(\chi_{p+4}^{-4}(\Delta^2)).$$

Hence,

$$R_3(\hat{\boldsymbol{\theta}}_n^S; \mathbf{W}) = \sigma^2 \mathrm{tr}(\mathbf{W}) - (p-2)\sigma^2 \mathrm{tr}(\mathbf{W}) \Big\{ 2E(\chi_{p+2}^{-2}(\Delta^2))$$

$$-(p-2)E(\chi_{p+2}^{-4}(\Delta^2)) \Big\} + (p^2-4)(\boldsymbol{\delta}'\mathbf{W}\boldsymbol{\delta})E(\chi_{p+4}^{-4}(\Delta^2))$$

$$= p - (p-2)^2 E\left(\chi_p^{-2}(\Delta^2)\right) \quad \text{if } \mathbf{W} = \sigma^{-2}\mathbf{I}_p.$$

Similarly, the asymptotic MSE and risk expressions may be verified for $\mathbf{M}_4(\hat{\boldsymbol{\theta}}_n^{S+})$ and $R_4(\hat{\boldsymbol{\theta}}_n^{S+}; \mathbf{W})$, respectively.

Clearly, all the analysis and conclusions of Section 4.4.3 continue to hold.

4.7 Improving James-Stein Estimator Toward Admissible Estimator

4.7.1 Introduction

In Section 4.3.2 we introduced the positive-rule Stein estimator, $\hat{\boldsymbol{\theta}}_n^{S+}$, defining it as a preliminary test estimator of the mean vector $\boldsymbol{\theta}$ with critical value $p-2$. A second improvement was done in Section 4.5.6 by substituting the Stein estimator of the variance to obtain the estimator, $\hat{\boldsymbol{\theta}}_n^{IS}$. We found that $\hat{\boldsymbol{\theta}}_n^{S+}$ improves $\hat{\boldsymbol{\theta}}_n^{S}$, and $\hat{\boldsymbol{\theta}}_n^{IS}$ improves only $\hat{\boldsymbol{\theta}}_n^{S}$ but not $\hat{\boldsymbol{\theta}}_n^{S+}$. In this section, we consider again an improvement of $\hat{\boldsymbol{\theta}}_n^{S}$ using the preliminary test estimator approach (Kubokawa, 1991) to obtain an admissible estimator that turns out to be the generalized Bayes estimator given by Strawderman (1971) and Berger (1976).

4.7.2 Improving $\hat{\boldsymbol{\theta}}_n^{S}$ via PTE

Let us again consider the model (4.1.1) where

$$\mathbf{Y}_i = \boldsymbol{\theta} + \mathbf{e}_i, \quad i = 1, \ldots, n, \tag{4.7.1}$$

with $\mathbf{e}_i \sim \mathcal{N}_p(\mathbf{0}, \mathbf{I}_p)$. Then $\tilde{\boldsymbol{\theta}}_n = (\overline{Y}_1, \ldots, \overline{Y}_p)'$ is the MLE/LSE of $\boldsymbol{\theta}$. Next, we consider two Stein-type estimators, namely

$$\hat{\boldsymbol{\theta}}_n^S = \left(1 - (p-2)(n\|\tilde{\boldsymbol{\theta}}_n\|)^{-1}\right)\tilde{\boldsymbol{\theta}}_n \tag{4.7.2}$$

and

$$\hat{\boldsymbol{\theta}}_n^S(c) = \left(1 - c(n\|\tilde{\boldsymbol{\theta}}_n\|)^{-1}\right)\tilde{\boldsymbol{\theta}}_n, \quad (c > 0). \tag{4.7.3}$$

Now, we combine the two estimators via a preliminary test procedure with the critical value k, as follows:

$$\begin{aligned}
\hat{\boldsymbol{\theta}}_n^S(c,k) &= \hat{\boldsymbol{\theta}}_n^S I(n\|\tilde{\boldsymbol{\theta}}_n\|^2 > k) + \hat{\boldsymbol{\theta}}_n^S(c) I(n\|\tilde{\boldsymbol{\theta}}_n\|^2 \le k) \\
&= \hat{\boldsymbol{\theta}}_n^S - \left(\hat{\boldsymbol{\theta}}_n^S - \hat{\boldsymbol{\theta}}_n^S(c)\right) I(n\|\tilde{\boldsymbol{\theta}}_n\|^2 \le k) \\
&= \tilde{\boldsymbol{\theta}}_n - (c-p+2)(n\|\tilde{\boldsymbol{\theta}}_n\|^2)^{-1} I(n\|\tilde{\boldsymbol{\theta}}_n\|^2 \le k)\tilde{\boldsymbol{\theta}}_n.
\end{aligned} \tag{4.7.4}$$

Minimizing the quadratic risk of $\hat{\boldsymbol{\theta}}_n^S(c,k)$ given by

$$\begin{aligned}
&p + (c-p+2)^2 E\left[\|\mathbf{Z}\|^{-2} I(\|\mathbf{Z}\|^2 < k)\right] \\
&-2(c-p+2) E\left[\|\mathbf{Z}\|^{-2}\mathbf{Z}'(\mathbf{Z}-\boldsymbol{\delta}) I(\|\mathbf{Z}\|^2 < k)\right]
\end{aligned} \tag{4.7.5}$$

with respect to c, we obtain the optimum value of c as

$$c_1^*(k, \Delta^2) = (p-2) + \frac{E\{\|\mathbf{Z}\|^{-2}\mathbf{Z}'(\mathbf{Z}-\boldsymbol{\delta}) I(\|\mathbf{Z}\|^2 < k)\}}{E\{\|\mathbf{Z}\|^{-2} I(\|\mathbf{Z}\|^2 < k)\}}, \tag{4.7.6}$$

where $\mathbf{Z} \sim \mathcal{N}_p(\boldsymbol{\delta}, \mathbf{I}_p)$, $\boldsymbol{\delta} = \sqrt{n}\boldsymbol{\theta}$.

Now, $c_1^*(k, \Delta^2)$ can be rewritten as

$$c_1^*(k; \Delta^2) = (p-2) + \frac{E\left\{\left(1 - \sqrt{2\Delta^2}\frac{U}{\|\mathbf{Z}\|^2}\right) I(\|\mathbf{Z}^2\| < k)\right\}}{E\left\{\frac{1}{\|\mathbf{Z}\|^2} I(\|\mathbf{Z}\|^2 < k)\right\}}, \quad \Delta^2 = \|\boldsymbol{\delta}\|^2, \tag{4.7.7}$$

where $U = \mathbf{Z}'\boldsymbol{\delta}/\|\boldsymbol{\delta}\|$. The explicit form of (4.7.7) is given by

$$c_1^*(k; \Delta^2) = (p-2) + \frac{H_p(k; \Delta^2) - E_P\left\{\frac{2r}{p-2+2r} H_{p-2+2r}(k; 0)\right\}}{E_P\left\{\frac{1}{p-2+2r} H_{p-2+2r}(k; 0)\right\}}, \tag{4.7.8}$$

and $H_\nu(x; \Delta^2)$ is the cdf. of a noncentral chi squared distribution with ν d.f. and noncentrality parameter $\Delta^2/2$, and E_P stands for the expectation operator with respect to Poisson distribution with mean $\Delta^2/2$.

Using (2.2.3), we can write

$$c_1^*(k; \Delta^2) = (p-2) - \frac{2E_P\{h_{p+2r}(k; 0)\}}{E_P\left\{\frac{1}{p-2+2r} H_{p-2+2r}(k; 0)\right\}}. \tag{4.7.9}$$

Now, rewriting (4.7.9) as

$$c_1^*(k;\Delta^2) = (p-2) - \frac{2h_p(k;\Delta^2)}{\left[\int_0^k t^{-1}h_p(t;\Delta^2)dt\right]}, \tag{4.7.10}$$

we see that $c_1^*(k;\Delta^2)$ is less than or equal to

$$c_1^*(k;0) = (p-2) - \frac{2h_p(k;0)}{\left[\int_0^k t^{-1}h_p(t;0)dt\right]}, \tag{4.7.11}$$

which is due to (2.2.7). Also, $c_1^*(k;0)$ is an increasing function of k and $0 < c_1^*(k;0) < p-2$. Further, we can write $c_1^*(k;0)$ as

$$c_1^*(k;0) = p-2-2\left[\int_0^1 t^{p/2-2}\exp\left\{\frac{1}{2}k(1-t)\right\}dt\right]^{-1}. \tag{4.7.12}$$

Then we have the following theorem:

Theorem 1. The risk function of $\hat{\boldsymbol{\theta}}_n^S(c,k)$ is quadratic in c and minimized at $c = c_1^*(k,\Delta^2)$ where $c_1^*(k;\Delta^2) \le c_1^*(k;0)$, and $c_1^*(k;0)$ is increasing in k, $(0 < c_1^*(k;0) < p-2)$.

The message of the theorem is that for all Δ^2, $c_1^*(k;0)$ is closer to the minimizing value of the risk of $\hat{\boldsymbol{\theta}}_n^S(c,k)$ than $p-2$. Hence, we obtain the main result.

Theorem 2. $\hat{\boldsymbol{\theta}}_n^S(c_1^*(k;0),k)$ dominates $\hat{\boldsymbol{\theta}}_n^S$.

4.7.3 Iterative PTE to Obtain an Admissible Estimator

Now we select another critical value k' such that $0 < k' < k$, and we define another (PTE)-type estimator

$$\begin{aligned}
&\hat{\boldsymbol{\theta}}_n^S(c_1^*(k;0),k',k) \\
&= \left\{1 - c_1^*(k';0)(n\|\tilde{\boldsymbol{\theta}}_n\|^2)^{-1}\right\}\tilde{\boldsymbol{\theta}}_n I(n\|\tilde{\boldsymbol{\theta}}_n\|^2 < k') \\
&\quad + \left\{1 - c_1^*(k;0)(n\|\tilde{\boldsymbol{\theta}}_n\|^2)^{-1}\right\}\tilde{\boldsymbol{\theta}}_n I(k' \le n\|\tilde{\boldsymbol{\theta}}_n\|^2 < k) \\
&\quad + \hat{\boldsymbol{\theta}}_n^S I(n\|\tilde{\boldsymbol{\theta}}_n\|^2 \ge k).
\end{aligned} \tag{4.7.13}$$

Then $\hat{\boldsymbol{\theta}}_n^S(c^*(k;0);k',k)$ dominates $\hat{\boldsymbol{\theta}}_n^S(c^*(k;0);k)$ by similar method as in Theorem 2.

Now, as in Brewster and Zidek (1974), we define a sequence of estimators of the form

$$\hat{\boldsymbol{\theta}}_n^S(c_1^*(k_{i,j};0),k_{i,j}) = \{1 - c_1^*(k_{i,j};0)(n\|\tilde{\boldsymbol{\theta}}_n\|^2)^{-1}\}\tilde{\boldsymbol{\theta}}_n, \tag{4.7.14}$$

if $k_{ij-1} < n\|\tilde{\boldsymbol{\theta}}\|^2 \leq k_{ij}$, for a sequence of a finite partition of $[0, \infty)$ represented by

$$0 = k_{i,0} < \cdots < k_{i,n_i-1} < k_{i,n_i} = \infty \tag{4.7.15}$$

such that $\max|k_{i,j} - k_{i,j-1}| \to 0$ and $k_{i,n_i-1} \to \infty$ as $i \to \infty$. The sequence of estimators $\hat{\boldsymbol{\theta}}_n^S(c^*(k_{ij};0), k_{ij})$ converges pointwise to $\boldsymbol{\theta}_n^{*S}$ defined by

$$\boldsymbol{\theta}_n^{*S} = \left\{1 - \frac{c_1^*(n\|\tilde{\boldsymbol{\theta}}_n\|^2;0)}{n\|\tilde{\boldsymbol{\theta}}_n\|^2}\right\}\tilde{\boldsymbol{\theta}}_n, \tag{4.7.16}$$

and this is the generalized Bayes estimator of $\boldsymbol{\theta}$ given by the theorem.

Theorem 3. The estimator $\boldsymbol{\theta}_n^{*S}$ is an admissible estimator of $\boldsymbol{\theta}$ dominating $\hat{\boldsymbol{\theta}}_n^S$ with respect to the prior

$$\pi^*(\boldsymbol{\theta}) = \int_0^1 (2\pi)^{-p/2}\lambda^{-2}\left(\frac{\lambda}{1-\lambda}\right)\exp\left\{-\frac{1}{2}\left(\frac{\lambda}{1-\lambda}\right)\Delta^2\right\}d\lambda \tag{4.7.17}$$

Proof. See Strawderman (1971), Berger (1976), and Brown and Hwang (1982).

4.7.4 Extension to the Case Where the Variance Is Unknown

In this section, we extend the results of Sections 4.7.1 to 4.7.3 to the case where the variance is unknown. For the unknown variance, the standard James-Stein estimator is given by

$$\hat{\boldsymbol{\theta}}_n^S = \left\{1 - \frac{(p-2)m}{(m+2)}\frac{s_y^2}{n\|\tilde{\boldsymbol{\theta}}_n\|^2}\right\}\tilde{\boldsymbol{\theta}}_n, \tag{4.7.18}$$

as shown by (4.5.6) where $\mathcal{L}_n = n\|\tilde{\boldsymbol{\theta}}_n\|^2/ps_y^2$.

Thus, we define the PTE by

$$\hat{\boldsymbol{\theta}}_n^S(c,k) = \hat{\boldsymbol{\theta}}_n^S I\left(\frac{n\|\tilde{\boldsymbol{\theta}}_n\|^2}{s_y^2} \geq k\right) + \hat{\boldsymbol{\theta}}_n^S(c) I\left(\frac{n\|\tilde{\boldsymbol{\theta}}_n\|^2}{s_y^2} < k\right), \tag{4.7.19}$$

where

$$\hat{\boldsymbol{\theta}}_n^S(c) = \left\{1 - c\frac{ms_y^2}{n\|\tilde{\boldsymbol{\theta}}_n\|^2}\right\}\tilde{\boldsymbol{\theta}}_n. \tag{4.7.20}$$

Define $c_2^*(k,0)$ by

$$c_2^*(k;0) = \frac{p-2}{m+2} - \frac{2}{m+2}\left[\int_0^1 \frac{(1+k)^{(p+m)/2}}{(1+kz)^{(p+m)/2+1}} z^{\frac{p}{2}-2}dz\right]^{-1}. \tag{4.7.21}$$

Then

Theorem 4. (Kubokawa, 1991) The estimator $\hat{\boldsymbol{\theta}}_n^S(c_2^*(k;0),k)$ dominates $\hat{\boldsymbol{\theta}}_n^S$.

Proof. The noncentrality parameter of the noncentral F-distribution in this case is $\Delta^2/2$, where $\Delta^2 = n\|\theta\|^2/\sigma^2$.

It may be shown that the minimum of the risk of $\hat{\boldsymbol{\theta}}_n^S(c,k)$ is given by

$$c_2^* = \frac{E\Big[\big(\frac{ms_y^2}{\sigma^2}(1-\mathbf{Z}'\boldsymbol{\delta}\|\mathbf{Z}\|^{-2})I\big(\frac{\|\mathbf{Z}\|^2}{ms_y^2}<k\big)\Big]}{E\Big[\Big(\frac{ms_y^2}{\sigma^2}\Big)^2 I\Big(\frac{n\|\tilde{\boldsymbol{\theta}}_n\|^2}{ps_y^2}<k\Big)\Big]}, \tag{4.7.22}$$

where $\mathbf{Z} = \sqrt{n}\tilde{\boldsymbol{\theta}}_n$, $\boldsymbol{\delta} = \sqrt{n}\boldsymbol{\theta}$, and $\Delta^2 = \|\boldsymbol{\delta}\|^2/\sigma^2$. Then (4.7.22) can be written as

$$\begin{aligned} c_2^*(k,\Delta^2) &= \frac{E_P\left[\int_0^\infty v^{m/2}e^{-v/2}dv\left\{H_{p+2r}(kv,0) - \frac{2r}{p-2+2r}H_{p-2+2r}(kv;0)\right\}\right]}{E_P\left[\int_0^\infty v^{(m+1)/2}e^{-v}\frac{1}{p-2+2r}H_{p-2+2r}(kv;0)\right]} \\ &= \frac{\int_0^\infty v^{m/2}e^{-v/2}\left\{(p-2)\int_0^{kv}t^{-1}h_p(t;\Delta^2)dt - 2h_p(kv;\Delta^2)\right\}dv}{\left[\int_0^\infty v^{m/2+1}e^{-v/2}\int_0^{kv}t^{-1}h_p(t;\Delta^2)dt\,dv\right]}. \end{aligned} \tag{4.7.23}$$

Integrating by parts we obtain

$$\begin{aligned} &\int_0^\infty e^{-v/2}\left\{v^{m/2+1}\int_0^{kv}t^{-1}h_p(t;\Delta^2)dt\right\}dv \\ &= (m+2)\int_0^\infty v^{m/2}e^{-v/2}\int_0^{kv}t^{-1}h_p(t;\Delta^2)dt\,dv \\ &\quad + 2\int_0^\infty v^{m/2}e^{-v/2}h_p(kv;\Delta^2)dv, \end{aligned} \tag{4.7.24}$$

so that

$$c_2^*(k;\Delta^2) = (p-2) - 2\frac{g(\Delta^2)}{[m+2+2g(\Delta^2)]}, \tag{4.7.25}$$

where

$$g(\Delta^2) = \frac{\int_0^\infty v^{m/2}e^{v/2}h_p(kv;\Delta^2)dv}{\left[\int_0^\infty v^{m/2}e^{v/2}\int_0^{kv}t^{-1}h_p(t;\Delta^2)dt\,dv\right]}. \tag{4.7.26}$$

Let $B(\nu) = 2^{-v/2}\{\Gamma(\nu/2)\}^{-1}$ and let

$$g^*(z,\Delta^2) = E_P\left[\frac{B(p+2r)}{B(m+p+2r)}z^{(p+2r)/2-1}(1+z)^{-(m+p+2r)/2}\right], \tag{4.7.27}$$

then $g(\Delta^2)$ may be rewritten as

$$g(\Delta^2) = \frac{g^*(k, \Delta^2)}{\int_0^k z^{-1} g^*(z; \Delta^2) dz}. \tag{4.7.28}$$

Similar to (4.7.11) we can show that $g(\Delta^2) \geq g(0)$. So from (4.7.16) we have

$$c_2^*(k; \Delta^2) \leq c_2^*(k; 0), \tag{4.7.29}$$

where $c_2^*(k; \Delta^2)$ is given by (4.7.25) and $c_2^*(k; 0)$ is increasing in k, and $0 < c_2^*(k; 0) < (p-2)/(m+2)$. Hence, the theorem follows.

As a limiting form corresponding to (4.7.14), we consider the estimator

$$\boldsymbol{\theta}_n^{*S} = \left\{1 - c_2^* \left(\frac{n\|\tilde{\boldsymbol{\theta}}_n\|^2}{ms_y^2}; 0 \right) \frac{ms_y^2}{n\|\tilde{\boldsymbol{\theta}}_n\|^2} \right\} \tilde{\boldsymbol{\theta}}_n, \tag{4.7.30}$$

which is identical to the generalized Bayes estimator derived by Lin and Tsai (1973).

4.8 Confidence Set Estimation Based on Stein-Type Estimators

4.8.1 Introduction

In the last few sections, we dealt with the problem of improving on the usual point estimator of a multivariate normal mean. In this section, we discuss the companion problem, that of the set estimation of the normal mean. The development of the set estimation was slow due to increased difficulty, and many of the techniques presented in the last few sections do not carry over to set estimation.

Classical confidence sets derived by Neyman-Pearson theory have been the subject of many criticisms, and alternative procedures have been proposed. The $(1-\gamma)$ confidence set for the mean vector $\boldsymbol{\theta}$ in the model (4.1.1) can be written as the Neyman-Pearson solution

$$C^0(\tilde{\boldsymbol{\theta}}_n) = \{\boldsymbol{\theta} : n\|\boldsymbol{\theta} - \tilde{\boldsymbol{\theta}}_n\|^2 < k_\gamma\}, \tag{4.8.1}$$

where $k_\gamma = \chi_p^2(\gamma)$ is the upper γ-level critical value from the central chi-square distribution with p d.f. The set $C^0(\tilde{\boldsymbol{\theta}}_n)$ is a sphere centered at $\tilde{\boldsymbol{\theta}}_n$, the sample mean vector, that has probability $1-\gamma$ of covering the point $\boldsymbol{\theta}$. $C^0(\tilde{\boldsymbol{\theta}}_n)$ enjoys many optimality properties such as, it is

(i) best unbiased,

(ii) best translation invariant,

(iii) minimax, meaning $C^0(\tilde{\boldsymbol{\theta}}_n)$ minimizes the maximum expected volume with coverage probability $1-\gamma$ among all other procedures.

Stein (1962) raised the following question: Is $C^0(\tilde{\boldsymbol{\theta}}_n)$ the unique minimax set, or do others exist? If so, since the coverage probability of $C^0(\tilde{\boldsymbol{\theta}}_n)$ is constant for all $\boldsymbol{\theta}$, there is room for improvement of the coverage probability without increasing the volume. He showed that there exists sets with more coverage probability than the set $C^0(\tilde{\boldsymbol{\theta}}_n)$. Later Brown (1966, 1988) and Joshi (1967) independently showed that there exists a dominating minimax set for $p \geq 3$. They showed that if $p \geq 3$, there exists another confidence set $C^{BJ}(\hat{\boldsymbol{\theta}}_n^{BJ})$, where $\hat{\boldsymbol{\theta}}_n^{BJ} = \left\{1 - \frac{a}{(b+n\|\tilde{\boldsymbol{\theta}}_n\|^2)}\right\}\tilde{\boldsymbol{\theta}}_n$ dominating $C^0(\tilde{\boldsymbol{\theta}}_n)$ in the sense that (1) $P_{\boldsymbol{\theta}}\{C^{BJ}(\hat{\boldsymbol{\theta}}_n^{BJ})\} \geq P_{\boldsymbol{\theta}}\{C^0(\tilde{\boldsymbol{\theta}}_n)\}$ and (2) Volume of $C^{BJ}(\hat{\boldsymbol{\theta}}_n^{BJ}) \leq$ Volume of $C^0(\tilde{\boldsymbol{\theta}}_n)$ with strict inequality holding in either (1) or (2) for a set of $\boldsymbol{\theta}$ or $\tilde{\boldsymbol{\theta}}_n$. A sequence of studies then followed by Chen and Hwang (1988), Cohen and Strawderman (1973), Faith (1976), Ki and Tsui (1985), Berger (1980b), Casella (1985), Shinazaki (1989), and Casella and Hwang (1986) among others. Finally, Hwang and Casella (1982, 1984) were able to show analytically the following theorem:

Theorem 1. For $p \geq 3$, and $0 < c < c^*_{k_\gamma}$, the confidence set

$$C^{S+}(\hat{\boldsymbol{\theta}}_n^{S+}(c)) = \{\boldsymbol{\theta} : n\|\boldsymbol{\theta} - \hat{\boldsymbol{\theta}}_n^{S+}(c)\|^2 < k_\gamma\} \tag{4.8.2}$$

that is recentered at the positive-rule Stein estimator of $\boldsymbol{\theta}$, given by

$$\hat{\boldsymbol{\theta}}_n^{S+}(c) = \left\{1 - \frac{c}{n\|\hat{\boldsymbol{\theta}}_n\|^2}\right\} I(n\|\tilde{\boldsymbol{\theta}}_n\|^2 > c)\tilde{\boldsymbol{\theta}}_n, \tag{4.8.3}$$

has a *higher* coverage probability than $C^0(\tilde{\boldsymbol{\theta}}_n)$ for all $\boldsymbol{\theta}$.

The proof needed to show that the derivative $\frac{\partial}{\partial\Delta^2}P_{\boldsymbol{\theta}}\{C^c(\hat{\boldsymbol{\theta}}_n^{S+}(c)\}$ is negative. To extend the bound on c, they needed the derivative $\frac{\partial}{\partial c}P_{\boldsymbol{\theta}}\{C^c(\hat{\boldsymbol{\theta}}_n^{S+}(c)\} > 0$ in a certain range $0 < c \leq c^*_{k_\gamma}$.

The number $c^*_{k_\gamma}$ is determined by solving the equation

$$\left\{\sqrt{k_\gamma} + \sqrt{(k_\gamma + c^*_{k_\gamma})}\right\}^{p-3} = c_{k_\gamma}^{(p-3)/2}\exp(\sqrt{c^*_{k_\gamma}k_\gamma}). \tag{4.8.4}$$

This result holds for $p \geq 4$. But to cover the case for $p = 3$, the number $c^*_{k_\gamma}$ is chosen as the *minimum* of the two solutions of the equations

$$\left\{\sqrt{k_\gamma/2} + \sqrt{k_\gamma 4 + c_1}\right\}^{p-2} = c_1^{(p-2)/2}\exp\left(\sqrt{k_\gamma c_{1/2}}\right) \tag{4.8.5a}$$

and

$$\left\{\sqrt{(k_\gamma + 4c_2)} - \sqrt{k_\gamma}\right\}\left\{\sqrt{k_\gamma} + \sqrt{k_\gamma + c_2}\right\}^{p-2} = 2c_1^{(p-2)/2}c_2^{1/2}\exp\left(\sqrt{k_\gamma c_2}\right). \tag{4.8.5b}$$

An approximate value of c is $0.8(p-2)$.

4.8.2 Properties of the Recentered Confidence Set Based on PRSE

As a set of $\boldsymbol{\theta}$'s, $C^{S+}(\hat{\boldsymbol{\theta}}_n^{S+}(c))$ is a spherical ball. Its shape is different when we look at the dual representation on the $\boldsymbol{\theta}$-section:

$$C^{S+}(\hat{\boldsymbol{\theta}}^{S+}(c)) = \left\{\tilde{\boldsymbol{\theta}}_n : n\|\boldsymbol{\theta} - \hat{\boldsymbol{\theta}}_n^{S+}(c)\|^2 < k_\gamma\right\}. \tag{4.8.6}$$

It is no longer a ball. Figures 4.8.1 and 4.8.2 give the representation of $C^{S+}(\boldsymbol{\theta})$ for $p = 5$, $k_\gamma = 9.24$ (corresponding to $\gamma = 0.10$) and $c = 2.536$ for two values of Δ^2, namely $\frac{1}{2}\sqrt{k_\gamma}$ and $\sqrt{k_\gamma}$.

We have followed the convention adopted by Hwang and Casella (1982): we represented the sets in the $(\boldsymbol{\theta}, \tilde{\boldsymbol{\theta}}_n)$-plane where they are invariant under rotation around the $(\mathbf{0}, \boldsymbol{\theta})$-axis.

First, we have the following theorem:

Theorem 2. If $\Delta^2 \leq k_\gamma$,

$$P_{\boldsymbol{\theta}}\left\{C^0(\tilde{\boldsymbol{\theta}}_n)\right\} \leq P_{\boldsymbol{\theta}}\left\{C^{S+}(\hat{\boldsymbol{\theta}}_n^{S+}(c))\right\}. \tag{4.8.7}$$

Proof. Let us write $\hat{\boldsymbol{\theta}}_n^{S+}(c) = \gamma(\tilde{\boldsymbol{\theta}}_n)\tilde{\boldsymbol{\theta}}_n$, where

$$\gamma(\tilde{\boldsymbol{\theta}}_n) = \left(1 - \frac{c}{n\|\tilde{\boldsymbol{\theta}}_n\|^2}\right) I\left(n\|\tilde{\boldsymbol{\theta}}_n\|^2 > c\right). \tag{4.8.8}$$

Then $0 \leq \gamma(\tilde{\boldsymbol{\theta}}_n) \leq 1$. Since $\tilde{\boldsymbol{\theta}}_n \in C^{S+}(\boldsymbol{\theta})$ if and only if $\boldsymbol{\theta} \in C^{S+}(\tilde{\boldsymbol{\theta}}_n)$, it follows that

$$P_{\boldsymbol{\theta}}\left\{C^{S+}(\boldsymbol{\theta})\right\} = P_{\boldsymbol{\theta}}\left\{C^{S+}(\tilde{\boldsymbol{\theta}}_n)\right\}. \tag{4.8.9}$$

Since $\Delta^2 < k_\gamma$, $\mathbf{0} \in C^0(\boldsymbol{\theta})$ and, hence, if $\tilde{\boldsymbol{\theta}}_n \in C^0(\boldsymbol{\theta})$, then $\gamma(\tilde{\boldsymbol{\theta}}_n)\tilde{\boldsymbol{\theta}}_n \in C^0(\boldsymbol{\theta})$ by convexity of $C^0(\boldsymbol{\theta})$. Hence, (4.8.7) holds for all $\Delta^2 \leq k_\gamma$, since $C^{S+}(\boldsymbol{\theta})$ contains $C^0(\boldsymbol{\theta})$ for convexity reasons.

We may also note that $C^{S+}(\boldsymbol{\theta})$ contains the ball with radius $\sqrt{c}$ (as the value of the PRSE in this ball is 0). When $\boldsymbol{\theta} = \mathbf{0}$, the recentered set $C^{S+}(\boldsymbol{\theta})$ is also a ball centered at $\mathbf{0}$ with radius $(\sqrt{k_\gamma} + \sqrt{k_\gamma + 4c})/2$. As seen in Figures 4.8.1 and 4.8.2, the size of the set $C^{S+}(\boldsymbol{\theta})$ is decreasing as Δ^2 goes toward k_γ and the boundaries of $C^{S+}(\boldsymbol{\theta})$ and $C^0(\boldsymbol{\theta})$ then get closer. When $n\|\tilde{\boldsymbol{\theta}}_n\|^2 = c$, there exists a discontinuity in the shape of $C^{S+}(\boldsymbol{\theta})$ provided that $\Delta^2 = k_\gamma$.

The *discontinuity* of the set $C^{S+}(\hat{\boldsymbol{\theta}}_{(c)}^{S+})$ is reflected by its coverage probability by the decomposition of $P_{\boldsymbol{\theta}}\{\boldsymbol{\theta} \in C^{S+}(\hat{\boldsymbol{\theta}}_n^{S+}(c))\}$ as

$$P_{\boldsymbol{\theta}}\left\{C^{S+}(\hat{\boldsymbol{\theta}}_n^{S+}(c))\right\} = P_{\boldsymbol{\theta}}\{n\|\tilde{\boldsymbol{\theta}}_n\|^2 \leq c\} I(\Delta^2 \leq k_\gamma)$$
$$+ P_{\boldsymbol{\theta}}\left\{n\|\boldsymbol{\theta} - \hat{\boldsymbol{\theta}}_n^{S}(c)\|^2 \leq k_\gamma;\ n\|\tilde{\boldsymbol{\theta}}_n\|^2 \geq c\right\}, \tag{4.8.10}$$

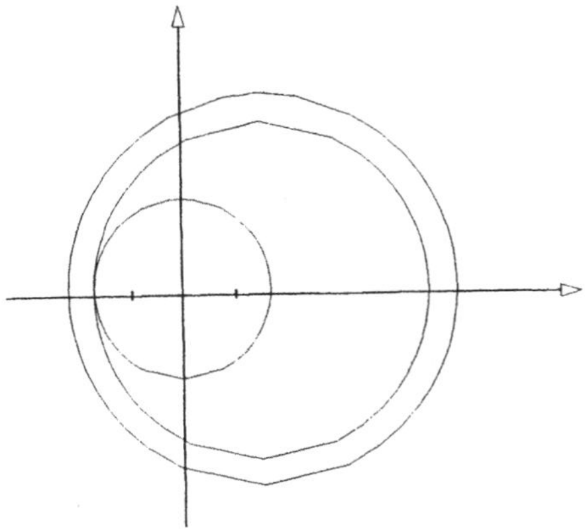

Figure 4.8.1 Projections of the sets $C^{S+}(\boldsymbol{\theta})$ (bigger) and $C^0(\boldsymbol{\theta})$ (smaller) on the $(\boldsymbol{\theta}, \tilde{\boldsymbol{\theta}}_n)$ plane for $\Delta = \sqrt{k_\gamma}/2$, $p = 5, k_\gamma = 9.24$, and $c^*_{k_\gamma} = 2.536$. (The ball of radius $\sqrt{c^*_{k_\gamma}}$ and center 0 are given for comparison.)

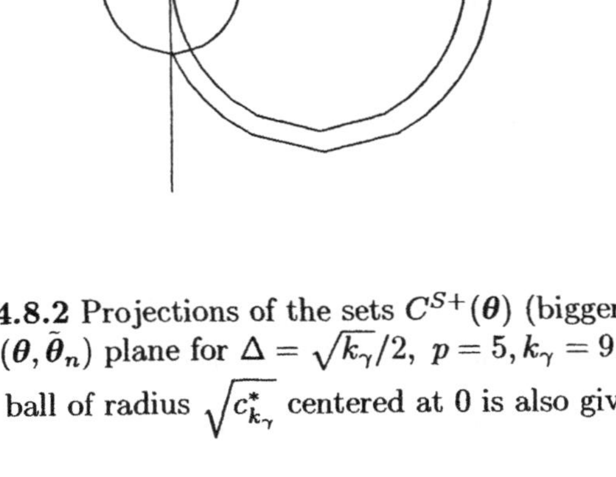

Figure 4.8.2 Projections of the sets $C^{S+}(\boldsymbol{\theta})$ (bigger) and $C^0(\boldsymbol{\theta})$ (smaller) on the $(\boldsymbol{\theta}, \tilde{\boldsymbol{\theta}}_n)$ plane for $\Delta = \sqrt{k_\gamma}/2$, $p = 5, k_\gamma = 9.24$, and $c^*_{k_\gamma} = 2.536$. (The ball of radius $\sqrt{c^*_{k_\gamma}}$ centered at 0 is also given.)

where

$$\hat{\boldsymbol{\theta}}_n^S(c) = \left\{1 - \frac{c}{n\|\tilde{\boldsymbol{\theta}}_n\|^2}\right\}\tilde{\boldsymbol{\theta}}_n. \tag{4.8.11}$$

Notice that when $\Delta^2 > k_\gamma$, the ball of radius $\sqrt{c}$ is excluded from $C^{S+}(\boldsymbol{\theta})$ and the sets $C^{S+}(\boldsymbol{\theta})$ and $C^0(\boldsymbol{\theta})$ intersect. Since $\hat{C}^{S+}(c)$ is the PRSE of $\boldsymbol{\theta}$, $C^{S+}(\boldsymbol{\theta})$ is farther away from the origin $\boldsymbol{\theta}$ than is $C^0(\boldsymbol{\theta})$.

The r.h.s. of (4.8.10) may be written as $\phi_1 + \phi_2$, where

$$\phi_1 = H_p(c; \Delta^2) I(\Delta^2 \leq k_\gamma)$$

and

$$\phi_2 = P_{\boldsymbol{\theta}}\left\{n\|\boldsymbol{\theta} - \hat{\boldsymbol{\theta}}_n^S(c)\|^2 \leq k_\gamma;\ n\|\tilde{\boldsymbol{\theta}}_n\|^2 \geq c\right\}. \tag{4.8.12}$$

Clearly, the probability is discontinuous at $\Delta^2 = k_\gamma$ since the first term vanishes for $\Delta^2 > k_\gamma$.

Before proving the next theorem, we consider the polar transformation to describe the set $C^{S+}(\boldsymbol{\theta})$. Let

$$r^2 = n\|\tilde{\boldsymbol{\theta}}_n\|^2 \quad \text{and} \quad \cos\beta = \frac{\boldsymbol{\theta}'\tilde{\boldsymbol{\theta}}_n}{\|\boldsymbol{\theta}\|\ \|\tilde{\boldsymbol{\theta}}_n\|}. \tag{4.8.13a}$$

Then, for $\Delta^2 = n\|\boldsymbol{\theta}\|^2 < k_\gamma$, we have

$$C^{S+}(\boldsymbol{\theta}) = \{(r, \beta) : r \leq r_+^0(\beta),\ \ \beta \in [-\pi, \pi]\}, \tag{4.8.13b}$$

where

$$r_+(\beta) = \frac{1}{2}\left\{r_+^0(\beta) + \sqrt{[r_+^0(\beta)]^2 + 4c}\right\} \tag{4.8.13c}$$

and

$$r_+(\beta) = \Delta\cos\beta + \sqrt{k_\gamma - \Delta^2\sin^2\beta}, \tag{4.8.13d}$$

whereas for $\Delta^2 > k_\gamma$, we have

$$C^{S+}(\boldsymbol{\theta}) = \{(r, \beta) : r_-(\beta) \leq r \leq r_+(\beta),\ \beta \in [-\beta_0, \beta_0]\}, \tag{4.8.14a}$$

where

$$\sin\beta_0 = \frac{k_\gamma}{\Delta},$$

$$r_-(\beta) = \frac{1}{2}\left\{r_-^0(\beta) + \sqrt{[r_-^0(\beta)]^2 + 4c}\right\}, \tag{4.8.14b}$$

and

$$r_-^0(\beta) = \Delta\cos\beta - \sqrt{k_\gamma - \Delta^2\sin\beta}. \tag{4.8.14c}$$

Further, note that $r_-(\beta) > c$. Also, $r_+(\beta)$, is a decreasing function of β, but the *distance* between $\boldsymbol{\theta}$ and the boundary of $C^{S+}(\boldsymbol{\theta})$ increases with β. The theorem that follows shows the discontinuity of the coverage probability.

Theorem 3. $\phi_2 = P_{\boldsymbol\theta}\{n\|\boldsymbol\theta - \hat{\boldsymbol\theta}_n^S(c)\|^2 \leq k_\gamma;\ n\|\tilde{\boldsymbol\theta}_n\|^2 > c\}$ is discontinuous at $\Delta^2 = k_\gamma$.

Proof. First, if $\Delta^2 < k_\gamma$ we have

$$\phi_2 = P_{\boldsymbol\theta}\{n\|\boldsymbol\theta - \hat{\boldsymbol\theta}_n^S(c)\|^2 \leq k_\gamma;\ n\|\tilde{\boldsymbol\theta}_n\|^2 > c\} = 2K\int_0^{\pi}\int_{\sqrt{c}}^{r_+(\beta)} h(r,\beta)dr d\beta, \tag{4.8.15}$$

where

$$h(r,\beta) = r^{p-1}\sin^{p-2}\beta \exp\left\{-\frac{1}{2}[r^2 - 2r\Delta\cos\beta + \Delta^2]\right\}$$

and $K = (2\pi)^{-(p-2)/2}\prod_{i=1}^{p-3}(\int_0^{\pi}\sin^i\beta\, d\beta)$, is the normalizing constant. As Δ increases to $\sqrt{k_\gamma}$, $r_+(\beta)$ goes to $\sqrt{k_\gamma}\left(\cos\beta + \sqrt{\cos\beta + \frac{c}{k_\gamma}}\right) = k_\gamma^*$. Therefore, for $\Delta^2 < k_\gamma$, the r.h.s. of (4.8.15) reduces to

$$2K\int_0^{\pi}\int_{\sqrt{c}}^{r_+(\beta)} h(r,\beta)dr\, d\beta. \tag{4.8.16}$$

Second, if $\Delta^2 > k_\gamma$, the r.h.s. of (4.8.15) becomes

$$2K\int_0^{\pi/2}\int_{\sqrt{c}}^{r_+(\beta)} h(r,\beta)dr\, d\beta, \tag{4.8.17}$$

since, as Δ decreases to $\sqrt{k_\gamma}$, $r_+(\beta) \to k_\gamma^*$ and $r_-(\beta) \to \sqrt{c}$ and $\beta_0 \to \pi/2$.

Combining (4.8.16) and (4.8.17), we can write

$$P\left(C^+(\hat{\boldsymbol\theta}_{(c)}^{S+})\right) = \begin{cases} H_p(c;\Delta^2)I(\Delta^2 < k_\gamma) \\ \quad +2K\int_0^{\pi}\int_{\sqrt{c}}^{r_+(\beta)} h(r,\beta)drd\beta & \text{if } \Delta^2 < k_\gamma, \\ H_p(c;\Delta^2)I(\Delta^2 < k_\gamma) \\ \quad +2K\int_0^{\pi/2}\int_{\sqrt{c}}^{r_+(\beta)} h(r,\beta)drd\beta & \text{if } \Delta^2 > k_\gamma. \end{cases}$$

Using the results given above, we present in Table 4.8.1 the components of the probability of coverage of $C^{S+}(\boldsymbol\theta_n)$ for $p = 3, 6$, and 9.

Table 4.8.1 Decomposition of the Coverage Probability.

(I) $p = 3$, $k_\gamma = 6.25$, and $c = 0.892$

Δ	0	0.25	0.50	0.75	1.0	1.25	1.5
ϕ_1	0.1727	0.1682	0.1556	0.1366	0.1138	0.0900	0.0675
ϕ_2	0.7799	0.7842	0.7964	0.8146	0.8364	0.8588	0.8795
$\phi_1 + \phi_2$	0.9526	0.9524	0.9520	0.9512	0.9502	0.9488	0.9470

(I) $p = 3$, $k_\gamma = 6.25$, and $c = 0.892$

Δ	1.75	2.0	2.25	2.5	4	6	8
$\phi_1 =$	0.0479	0.0323	0.0206	0.0125	–	–	–
ϕ_2	0.8969	0.9099	0.9182	,9212	0.9060	0.9022	0.9012
$\phi_1 + \phi_2$	0.9448	0.9422	0.9388	0.9337	0.9060	0.9022	0.9012

(II) $p = 6$, $k_\gamma = 10.64$, and $c = 3.666$

Δ	0	0.35	0.65	0.98	1.3	1.6	2
ϕ_1	0.2383	0.2306	0.2087	0.1766	0.1394	0.1025	0.0698
ϕ_2	0.7512	0.7588	0.7802	0.8116	0.8477	0.8833	0.9140
$\phi_1 + \phi_2$	0.9895	0.9894	0.9889	0.9882	0.9871	0.9858	0.9838

(II) $p = 6$, $k_\gamma = 10.64$, and $c = 3.666$

Δ	2.3	2.6	2.9	3.3	5	7	9
$\phi_1 =$	.0441	.0257	.0138	.0068	–	–	–
ϕ_2	0.9375	0.9530	0.9614	0.9637	0.9338	0.9169	0.9107
$\phi_1 + \phi_2$	0.9816	0.9787	0.9752	0.9705	0.9338	0.9169	0.9107

(III) $p = 9$, $k_\gamma = 14.68$, and $c = 5.872$

Δ	0	0.38	0.77	1.1	1.5	1.9	2.3
ϕ_1	0.2494	0.2380	0.2119	0.1741	0.1315	0.0909	0.0571
ϕ_2	0.7497	0.7590	0.7849	0.8223	0.8644	0.9042	0.9370
$\phi_1 + \phi_2$	.9991	0.9970	0.9968	0.9964	0.9959	0.9951	0.9941

(III) $p = 9$, $k_\gamma = 14.68$, and $c = 5.872$

Δ	2.7	3.1	3.4	3.8	5	7	9	11
$\phi_1 =$	0.0325	0.0166	0.0076	0.0031	-	-	-	-
ϕ_2	0.9602	0.9742	0.9808	0.9821	0.9639	0.9388	0.9255	0.9176
$\phi_1 + \phi_2$	.9927	0.9909	0.9884	0.9852	0.9639	0.9388	0.9255	0.9176

The table shows the decomposition of the coverage probability (4.8.10) where

$$\phi_1 = H_p(c; \Delta^2) I(\Delta^2 < k_\gamma) \tag{4.8.18a}$$

and

$$\phi_2 = P_{\boldsymbol{\theta}} \left\{ n\|\boldsymbol{\theta} - \hat{\boldsymbol{\theta}}_n^S(c)\|^2 < k_\gamma; \;\; n\|\tilde{\boldsymbol{\theta}}_n\|^2 > c \right\}. \tag{4.8.18b}$$

The different scales of Δ are intended to provide homogeneous coverage of $[0, \sqrt{k_\gamma}]$. Some combined values of c were obtained in Hwang and Casella (1984) and given in the Table 4.8.2.

Table 4.8.2 Some Upper Bounds of c for $\gamma = 0.10$ and 0.05

	3	4	5	6	7	8	9	10
$c^*_{.10}$	0.892	1.710	2.536	3.666	4.199	5.035	5.872	6.711
$c^*_{.05}$	0.810	1.602	2.408	3.223	4.042	4.866	5.692	6.521

Source: Adapted from Casella and Hwang, 1987

In general, we note that ϕ_1 is decreasing as Δ moves toward $\sqrt{k_\gamma}$ and ϕ_2 is increasing in Δ. From Table 4.8.1 (III) we find that the maximum of $\phi_1 + \phi_2$ is almost unity (0.9991) and the minimum value is 0.9176, which is more than 0.90. The confidence coefficient of the set $C^{S+}_{(c)}$ is thus always more than that of $C^0(\tilde{\boldsymbol{\theta}})$.

In concluding this section, we present without proof the asymptotic approximation of ϕ_2 as $\Delta^2 \to \infty$ motivated by the fact that $\Delta^2 > k_\gamma$.

Theorem 4. (Hwang and Casella, 1984). As $\Delta^2 \to \infty$, the asymptotic approximation of ϕ_2 is given by

$$\phi_2 \equiv (1-\gamma) - \frac{c}{2\Delta^2}[1-\gamma-h(\gamma)][c-2(p-2)] + O(\Delta^{-3/2}), \tag{4.8.19a}$$

where

$$h(\gamma) = 1 - \gamma - p^{-1}\left\{\frac{k_\gamma^{p/2} e^{-k_\gamma/2}}{2^{\frac{1}{2}(p-2)}\Gamma(\frac{p}{2})}\right\}. \tag{4.8.19b}$$

4.8.3 Confidence Set Based on Preliminary Test Estimator

Now, consider the PTE of $\boldsymbol{\theta}$:

$$\hat{\boldsymbol{\theta}}_n^{\text{PT}}(\alpha) = \tilde{\boldsymbol{\theta}}_n I(n\|\tilde{\boldsymbol{\theta}}_n\|^2 > d), d = \chi^2_p(\alpha),$$

where d is the critical value of the test-statistic $n|\tilde{\boldsymbol{\theta}}_n\|^2$ at some level of significance, say, α. The related confidence set is defined by

$$C^{\text{PT}}\left(\hat{\boldsymbol{\theta}}_n^{\text{PT}}(\alpha)\right) = \left\{\boldsymbol{\theta} : n\|\boldsymbol{\theta} - \hat{\boldsymbol{\theta}}_n^{\text{PT}}(\alpha)\|^2 \le k_\gamma\right\}. \tag{4.8.20}$$

From Section 4.2, we know that the quadratic risk of $\hat{\boldsymbol{\theta}}_n^{\text{PT}}(\alpha)$ is given by

$$R_2(\hat{\boldsymbol{\theta}}_n^{\text{PT}}(\alpha)) = p - pH_p(d;\Delta^2) + \Delta^2\{2H_{p+2}(d;\Delta^2) - H_{p+4}(d;\Delta^2)\}. \tag{4.8.21}$$

The graph of the efficiency of $\hat{\boldsymbol{\theta}}_n^{\text{PT}}(\alpha)$ relative to $\tilde{\boldsymbol{\theta}}_n$ as a function of Δ^2 may be described as follows: The maximum efficiency of $\hat{\boldsymbol{\theta}}_n^{\text{PT}}(\alpha)$ is $[1 - H_{p+2}(d;0)]^{-1} \ge$

1, which decreases, crossing the one-line at some point near $\Delta^2 = 1$, and goes to a minimum value at, say, $\Delta^2 = \Delta^2_{min}$. It then increases monotonically toward the 1-line as $\Delta^2 \to \infty$.

The properties of the confidence set $C^{\text{PT}}(\hat{\boldsymbol{\theta}}_n^{\text{PT}}(\alpha))$ are similar to the properties of the point estimator given by Theorem 3. First we note the decomposition of the coverage probability

$$P_{\boldsymbol{\theta}}\{\boldsymbol{\theta} \in C^{\text{PT}}(\hat{\boldsymbol{\theta}}_n^{\text{PT}}(\alpha))\}$$
$$= H_p(d; \Delta^2) I(\Delta^2 < k_\gamma) + P_{\boldsymbol{\theta}}\left\{n\|\tilde{\boldsymbol{\theta}}_n\|^2 > d;\ n\|\boldsymbol{\theta} - \tilde{\boldsymbol{\theta}}_n\|^2 \leq k_\gamma\right\}. \quad (4.8.22)$$

Then we conclude the inadmissibility of $C^{\text{PT}}(\hat{\boldsymbol{\theta}}_n^{\text{PT}}(\alpha))$ based on Theorem 5 given below.

Theorem 5. (Robert and Saleh, 2001; Chiou and Saleh, 2002).

(i) If $\Delta^2 < k_\gamma$, then $P_{\boldsymbol{\theta}}\{\boldsymbol{\theta} \in C^{\text{PT}}(\hat{\boldsymbol{\theta}}_n^{\text{PT}}(\alpha))\} \geq P_{\boldsymbol{\theta}}\{\boldsymbol{\theta} \in C^0(\tilde{\boldsymbol{\theta}}_n)\}$.

(ii) If $k_\gamma \leq \Delta^2 \leq (\sqrt{k_\gamma} + \sqrt{d})^2$, then $P_{\boldsymbol{\theta}}\{\boldsymbol{\theta} \in C^{\text{PT}}(\hat{\boldsymbol{\theta}}_n^{\text{PT}}(\alpha))\} < P_{\boldsymbol{\theta}}\{\boldsymbol{\theta} \in C^0_{k_\gamma}(\tilde{\boldsymbol{\theta}}_n)\}$.

(iii) If $\Delta^2 > (\sqrt{k_\gamma} + \sqrt{d})^2$, then $P_{\boldsymbol{\theta}}\{\boldsymbol{\theta} \in C^{\text{PT}}(\hat{\boldsymbol{\theta}}_n^{\text{PT}}(\alpha))\} = 1 - \gamma$. (4.8.23)

If $\Delta^2 < k_\gamma \leq d$, then $P_{\boldsymbol{\theta}}\{\boldsymbol{\theta} \in C^{\text{PT}}(\hat{\boldsymbol{\theta}}_n^{\text{PT}}(\alpha))\}$ is decreasing as a function of Δ^2. At $\Delta^2 = 0$, it equals $1 - \alpha$, corresponding to the critical value d. As Δ^2 increases to k_γ, the coverage probability of $C^{\text{PT}}(\hat{\boldsymbol{\theta}}_n^{\text{PT}}(\alpha))$ drops to a minimum below $1 - \gamma$ depending on the value of d, meaning the level of significance of the test for $\boldsymbol{\theta} = \mathbf{0}$, and then monotonically increases toward $1 - \gamma$ as $\Delta^2 \to (\sqrt{k_\gamma} + \sqrt{d})^2$. Finally, the coverage probability becomes equal to $1 - \gamma$ as $\Delta^2 > (\sqrt{k_\gamma} + \sqrt{d})^2$.

There is a discontinuity of the coverage probability at $\Delta^2 = k_\gamma$. Tabular values of the coverage probability are presented in Tables 4.8.3 and 4.8.4 for 90% confidence sets for $p = 5(2)15$ and the test levels $\alpha = 0.05$ and 0.10.

Proof. (i) $\Delta^2 < k_\gamma$. From (4.8.22) we have

$$P_{\Delta^2}\left(C^{\text{PT}}(\hat{\boldsymbol{\theta}}_n^{\text{PT}}(\alpha))\right)$$
$$= P_{\Delta^2}\{n\|\tilde{\boldsymbol{\theta}}_n\|^2 < d\} + P_{\Delta^2}\{n\|\tilde{\boldsymbol{\theta}}_n\|^2 > d; n\|\boldsymbol{\theta} - \tilde{\boldsymbol{\theta}}_n\|^2 \leq k_\gamma\}$$
$$\geq P_{\Delta^2}\{n\|\tilde{\boldsymbol{\theta}}_n\|^2 < d; n\|\boldsymbol{\theta} - \tilde{\boldsymbol{\theta}}_n\|^2 \leq k_\gamma\} + P_{\Delta^2}\{n\|\tilde{\boldsymbol{\theta}}_n\|^2 > d; n\|\boldsymbol{\theta} - \tilde{\boldsymbol{\theta}}_n\|^2 \leq k_\gamma\}$$
$$= P_0\{n\|\boldsymbol{\theta} - \tilde{\boldsymbol{\theta}}_n\|^2 < k_\gamma\} = 1 - \gamma.$$

(ii) $K_\gamma < \Delta^2 \le (\sqrt{k_\gamma} + \sqrt{d})^2$. From (4.8.22) we have

$$P_{\Delta^2}\Big(C^{\rm PT}(\hat{\theta}_n^{\rm PT}(\alpha))\Big) = P_{\Delta^2}\{n\|\tilde{\boldsymbol{\theta}}_n\|^2 > d; n\|\boldsymbol{\theta} - \tilde{\boldsymbol{\theta}}_n\|^2 \le k_\gamma\}$$
$$\le P_{\Delta^2}\{n\|\boldsymbol{\theta} - \tilde{\boldsymbol{\theta}}_n\|^2 < k_\gamma\} = 1 - \gamma.$$

(iii) $\Delta^2 > (\sqrt{k_\gamma} + \sqrt{d})^2$. Note that $(\sqrt{k_\gamma} + \sqrt{d})^2 < \Delta$. Now, $n\|\boldsymbol{\theta} - \tilde{\boldsymbol{\theta}}_n\|^2 \le k_\gamma \Longrightarrow \sqrt{n}\|\boldsymbol{\theta} - \tilde{\boldsymbol{\theta}}_n\| \le \sqrt{k_\gamma}$, which in turn implies

$$\Delta - \sqrt{n}\|\tilde{\boldsymbol{\theta}}_n\| \le \sqrt{n}\|\boldsymbol{\theta} - \tilde{\boldsymbol{\theta}}_n\| \le \sqrt{k_\gamma},$$

and we have $\Delta \le \sqrt{n}\|\tilde{\boldsymbol{\theta}}_n\| + \sqrt{k_\gamma}$. Thus, we conclude that

$$(\sqrt{k_\gamma} + \sqrt{d}) < \Delta < \sqrt{n}\|\tilde{\boldsymbol{\theta}}_n\| + \sqrt{k_\gamma}\,.$$

It follows that $\sqrt{d} < \sqrt{n}\|\tilde{\boldsymbol{\theta}}_n\|$ which is equivalent to $n\|\tilde{\boldsymbol{\theta}}_n\|^2 > d$. Thus, from (4.8.22),

$$P_{\Delta^2}\{n\|\tilde{\boldsymbol{\theta}}_n\|^2 > d;\; n\|\boldsymbol{\theta} - \tilde{\boldsymbol{\theta}}_n\|^2 < k_\gamma\}$$
$$= P_0\{n\|\boldsymbol{\theta} - \tilde{\boldsymbol{\theta}}_n\|^2 \le k_\gamma\} = 1 - \gamma.$$

This completes the proof.

Table 4.8.3 Coverage Probabilities for the Set $C^{\rm PT}(\hat{\boldsymbol{\theta}}_n^{\rm PT}(\alpha))$ with $\gamma = 0.10$ and $\alpha = 0.05$

Δ	$p = 5$	$p = 7$	$p = 9$	$p = 11$	$p = 13$	$p = 15$
0	0.9500	0.9500	0.9500	0.9500	0.9500	0.9500
1	0.9356	0.9350	0.9347	0.9347	0.9348	0.9350
2	0.9318	0.9304	0.9297	0.9293	0.9291	0.9289
3	0.9289	0.9271	0.9262	0.9256	0.9252	0.9249
4	0.9264	0.9245	0.9233	0.9226	0.9221	0.9218
5	0.9243	0.9222	0.9210	0.9202	0.9197	0.9193
6	0.9224	0.9202	0.9190	0.9181	0.9176	0.9171
7	0.9206	0.9184	0.9172	0.9163	0.9157	0.9153
8	0.9191	0.9169	0.9156	0.9147	0.9141	0.9137
9	0.9176	0.9154	0.9142	0.9133	0.9127	0.9123
10	0.5937	0.9141	0.9129	0.9121	0.9115	0.9110
15	0.7773	0.7296	0.6871	0.9074	0.9069	0.9066
20	0.8593	0.8336	0.8079	0.7826	0.7580	0.9039
25	0.8890	0.8777	0.8650	0.8514	0.8371	0.8223
50	0.9000	0.9000	0.9000	0.8999	0.8997	0.8995
100	0.9000	0.9000	0.9000	0.9000	0.9000	0.9000

Table 4.8.4 Coverage Probabilities for the Set $C^{\rm PT}(\hat{\boldsymbol{\theta}}_n^{\rm PT}(\alpha))$ with $\gamma = 0.10$ and $\alpha = 0.10$

Δ	$p = 5$	$p = 7$	$p = 9$	$p = 11$	$p = 13$	$p = 15$
0	0.9000	0.9000	0.9000	0.9000	0.9000	0.9000
1	0.9190	0.9162	0.9144	0.9131	0.9122	0.9114
2	0.9196	0.9167	0.9148	0.9135	0.9125	0.9117
3	0.9190	0.9160	0.9142	0.9129	0.9119	0.9111
4	0.9180	0.9151	0.9133	0.9120	0.9111	0.9103
5	0.9169	0.9141	0.9123	0.9111	0.9102	0.9095
6	0.9157	0.9130	0.9114	0.9102	0.9093	0.9087
7	0.9147	0.9120	0.9105	0.9094	0.9085	0.9079
8	0.9136	0.9111	0.9096	0.9085	0.9078	0.9072
9	0.9126	0.9102	0.9088	0.9078	0.9071	0.9065
10	0.6937	0.9094	0.9080	0.9071	0.9064	0.9059
15	0.8305	0.7975	0.7668	0.9044	0.9039	0.9036
20	0.8809	0.8657	0.8498	0.8336	0.8173	0.9021
25	0.8960	0.8901	0.8834	0.8758	0.8675	0.8588
50	0.9000	0.9000	0.9000	0.9000	0.8999	0.8999
100	0.9000	0.9000	0.9000	0.9000	0.9000	0.9000

4.8.4 Asymptotic Theory of Recentered Confidence Sets and Domination of Positive-Rule Coverage Probability

Consider the model (4.1.1) again, namely

$$\mathbf{Y}_i = \boldsymbol{\theta} + \mathbf{e}_i, \quad i = 1, \ldots, n, \tag{4.8.24}$$

where $\mathbf{e}_i = (e_{1i}, \ldots, e_{pi})' \sim \prod_{j=1}^{p} F_0(e_{ji})$ and $\boldsymbol{\theta} = (\theta_1, \ldots, \theta_p)'$ is the mean vector. Further, assume

$$E(\mathbf{e}_i) = \mathbf{0}, \quad \text{and} \quad E(\mathbf{e}_i\mathbf{e}_i') = \sigma^2\mathbf{I}_p. \tag{4.8.25}$$

The unbiased estimator of $\boldsymbol{\theta}$ is $\tilde{\boldsymbol{\theta}}_n = (\overline{Y}_1, \ldots, \overline{Y}_p)'$ and by the p-variate central limit theorem

$$\sqrt{n}(\tilde{\boldsymbol{\theta}}_n - \boldsymbol{\theta}) \sim \mathcal{N}_p(\mathbf{0}, \sigma^2\mathbf{I}_p) \ \text{ as } n \to \infty. \tag{4.8.26}$$

The unbiased estimator of σ^2 is

$$s_Y^2 = \{p(n-1)\}^{-1}\text{tr}\left\{\sum_{i=1}^{p}(\mathbf{Y}_i - \theta)(\mathbf{Y}_i - \theta)'\right\}, \tag{4.8.27}$$

and $s_Y^2 \to \sigma^2$ in probability as $n \to \infty$. For the test of $H_0 : \boldsymbol{\theta} = \boldsymbol{\theta}_0$ against $H_A : \boldsymbol{\theta} \neq \boldsymbol{\theta}_0$, we use the test-statistic

$$\mathcal{L}_n = \frac{n\|\tilde{\boldsymbol{\theta}}_n - \boldsymbol{\theta}_0\|^2}{s_Y^2}. \tag{4.8.28}$$

Under H_0, as $n \to \infty$, the distribution of $\mathcal{L}_n$ is a central chi-square distribution with p d.f.

For the asymptotic theory of confidence sets using PRSE, we assume that the prior uncertain information of $\boldsymbol{\theta}$ is equal to $\mathbf{0}$ (null vector). Then we define the Neyman-Pearson confidence set as follows:

$$C^0(\tilde{\boldsymbol{\theta}}_n) = \left\{\boldsymbol{\theta} : \frac{n\|\boldsymbol{\theta} - \tilde{\boldsymbol{\theta}}_n\|^2}{s_Y^2} \leq k_\gamma\right\}, \tag{4.8.29}$$

where $k_\gamma = \chi_p^2(\gamma)$ is the γ-level critical value of the chi-square distribution with p d.f. Correspondingly, the confidence set based on PRSE is defined by

$$C^{S+}(\hat{\boldsymbol{\theta}}_n^{S+}(c)) = \left\{\boldsymbol{\theta} : \frac{n\|\boldsymbol{\theta} - \hat{\boldsymbol{\theta}}_n^{S+}(c)\|^2}{s_Y^2} \leq k_\gamma\right\}, \tag{4.8.30}$$

where the positive-rule Stein estimator is given by

$$\hat{\boldsymbol{\theta}}_n^{S+}(c) = \left\{1 - \frac{cs_Y^2}{n\|\tilde{\boldsymbol{\theta}}_n\|^2}\right\} I\left(\frac{n\|\tilde{\boldsymbol{\theta}}_n\|^2}{s_Y^2} > c\right)\tilde{\boldsymbol{\theta}}_n \tag{4.8.31a}$$

and

$$\hat{\boldsymbol{\theta}}_n^{S}(c) = \left\{1 - \frac{cs_Y^2}{n\|\tilde{\boldsymbol{\theta}}_n\|^2}\right\}\tilde{\boldsymbol{\theta}}_n. \tag{4.8.31b}$$

Thus, we have the following theorem.

Theorem 6. Under the fixed alternative $\boldsymbol{\theta} = \boldsymbol{\delta}$,

$$\lim_{n\to\infty} P_{\boldsymbol{\theta}}\{\frac{n\|\boldsymbol{\theta} - \hat{\boldsymbol{\theta}}_n^{S+}(c)\|^2}{s_Y^2} \leq k_\gamma\} = \lim_{n\to\infty} P_{\boldsymbol{\theta}}\{\frac{n\|\boldsymbol{\theta} - \tilde{\boldsymbol{\theta}}_n\|^2}{s_Y^2} < k_\gamma\} = 1 - \gamma.$$

Proof. First consider the coverage probability

$$\lim_{n\to\infty} P_{\boldsymbol{\theta}}\{\frac{n\|\boldsymbol{\theta} - \tilde{\boldsymbol{\theta}}_n\|^2}{s_Y^2} \leq k_\gamma\} = P_{\boldsymbol{\delta}}\{\|\mathbf{Z}\|^2 \leq k_\gamma\} = 1 - \gamma,$$

since $\sqrt{n}(\tilde{\boldsymbol{\theta}}_n - \boldsymbol{\theta})/s_Y \xrightarrow{\mathcal{D}} \mathbf{Z} \sim \mathcal{N}_p(\mathbf{0}, \mathbf{I}_p)$ as $n \to \infty$, and $k_\gamma = \chi_p^2(\gamma)$, a γ-level critical value from the central chi-square distribution with p d.f.

Note that under fixed alternative $\mathcal{L}_n \to \infty$ as $n \to \infty$. Hence, $\mathcal{L}_n^{-1}$ converges to 0 in first mean and

$$\frac{\sqrt{n}(\hat{\boldsymbol{\theta}}_n^{S+}(c) - \boldsymbol{\theta})}{s_Y} \xrightarrow{\mathcal{D}} \frac{\sqrt{n}(\tilde{\boldsymbol{\theta}}_n - \boldsymbol{\theta})}{\sigma} \quad \text{as } n \to \infty,$$

which implies that

$$\lim P_{\boldsymbol{\theta}}\{\frac{n\|\boldsymbol{\theta} - \hat{\boldsymbol{\theta}}_n^{S+}(c)\|^2}{s_Y^2} \leq k_\gamma\} = \lim P_{\boldsymbol{\theta}}\{\frac{n\|\boldsymbol{\theta} - \tilde{\boldsymbol{\theta}}_n\|^2}{s_Y^2} < k_\gamma\} = 1 - \gamma.$$

Because of this equivalence of probability content of the two sets, we may consider a sequence of Pitman-type local alternatives and obtain the following theorem, which shows that the confidence set with PRSE as center of the sphere is locally better than the usual confidence set, $C^0(\tilde{\boldsymbol{\theta}}_n)$:

Theorem 7. Under the local alternatives $K_{(n)} : \boldsymbol{\theta}_{(n)} = n^{-1/2}\boldsymbol{\delta}$, $\boldsymbol{\delta}$ a fixed vector,

$$\lim_{n\to\infty} P_{K_{(n)}} \{C^{S+}(\boldsymbol{\theta}_{(n)})\} = H_p(c;\Delta^2)I(\Delta^2 < k_\gamma)$$
$$+P_{\boldsymbol{\delta}}\left\{\left\|\mathbf{Z} - \frac{c(\mathbf{Z}+\sigma^{-1}\boldsymbol{\delta})}{\|\mathbf{Z}+\sigma^{-1}\boldsymbol{\delta}\|^2}\right\|^2 \le k_\gamma; \|\mathbf{Z}+\sigma^{-1}\boldsymbol{\delta}\|^2 > c\right\}$$
$$\ge P_{\boldsymbol{\delta}}\{\|\mathbf{Z}\|^2 < k_\gamma\} = 1-\gamma. \tag{4.8.32}$$

Proof. First note that under $K_{(n)}$

$$\frac{\sqrt{n}(\tilde{\boldsymbol{\theta}}_n - \boldsymbol{\theta}_{(n)})}{s_Y} - \frac{c\sqrt{n}\tilde{\boldsymbol{\theta}}_n s_Y}{n\|\tilde{\boldsymbol{\theta}}_n\|^2} \xrightarrow{\mathcal{D}} \mathbf{Z} - \frac{c(\mathbf{Z}+\sigma^{-1}\boldsymbol{\delta})}{\|\mathbf{Z}+\sigma^{-1}\boldsymbol{\delta}\|^2},$$

where $\mathbf{Z} \sim \mathcal{N}_p(\mathbf{0}, \mathbf{I}_p)$. Therefore,

$$\lim_{n\to\infty} P_{K_{(n)}}\{\tilde{\boldsymbol{\theta}}_n \in C^{S+}(\boldsymbol{\theta}^{S+}_{(c)})\}$$
$$= \lim_{n\to\infty} P_{K_{(n)}}\left\{\frac{n\|\tilde{\boldsymbol{\theta}}_n\|^2}{s_Y^2} \le c\right\} I\left(\frac{n\|\tilde{\boldsymbol{\theta}}_n\|^2}{s_Y^2} \le k_\gamma\right)$$
$$+ \lim_{n\to\infty} P_{K_{(n)}}\left\{\frac{n\|\tilde{\boldsymbol{\theta}}^S_n(c)\|^2}{s_Y^2} \le k_\gamma; \frac{n\|\tilde{\boldsymbol{\theta}}_n\|^2}{s_Y^2} \le c\right\}$$
$$= H_p(c;\Delta^2)I(\Delta^2 \le k_\gamma) + P_{\boldsymbol{\delta}}\left\{\left\|\mathbf{Z} - \frac{c(\mathbf{Z}+\sigma^{-1}\boldsymbol{\delta})}{\|\mathbf{Z}+\sigma^{-1}\boldsymbol{\delta}\|^2}\right\|^2 < k_\gamma;\ \|\mathbf{Z}+\boldsymbol{\delta}/\sigma\|^2 > c\right\}$$
$$\ge P_{\boldsymbol{\delta}}\{\|\mathbf{Z}\|^2 < k_\gamma\} = 1-\gamma \text{ by Theorem 4.8.1.} \tag{4.8.33}$$

Thus, we can use the results of (4.8.10) on the decomposition of the coverage probability of positive-rule estimators in the normal case. Theorems 4.8.2 and 4.8.3 pertain to the nonnormal and asymptotic case. Thus, the domination of the probability of coverage in this case is given by Table 4.8.1, which holds for the asymptotic case.

4.9 Nonparametric Methods: *R*-Estimation

In this section, we consider the nonparametric methods of estimation of several location parameters $\boldsymbol{\theta} = (\theta_1, \ldots, \theta_p)'$ when it is suspected that $\boldsymbol{\theta}$ may be $\mathbf{0}$. This broadens the scope of the Stein-type estimators in a general class of symmetric distributions, and the robust Steins procedure ensues.

4.9.1 Model and Assumptions

Let $\mathbf{Y}_1, \mathbf{Y}_2, \ldots, \mathbf{Y}_n$ be n independent p-dimensional response vectors such that

$$\mathbf{Y}_j = \boldsymbol{\theta} + \mathbf{e}_j, \quad j = 1, \ldots, n, \tag{4.9.1}$$

holds. The error distribution $F(\mathbf{e}_j)$ is given by

$$F(\mathbf{e}_j) = \prod_{i=1}^{p} F_0(e_{ij}), \quad j = 1, \ldots, n, \tag{4.9.2}$$

where F_0 belongs to the class of absolutely continuous distributions, $\mathcal{F}$ with absolutely continuous pdf $f_0(e_{ij})$ such that

$$I(f_0) = \int_{-\infty}^{\infty} \left\{ \frac{f_0'(z)}{f_0(z)} \right\}^2 f_0(z)\, dz \; < \; \infty. \tag{4.9.3}$$

That is, the Fisher's information is bounded. Our objective is the estimation of $\boldsymbol{\theta}$ when it is suspected that $\boldsymbol{\theta}$ may be $\mathbf{0}$.

4.9.2 Test of Hypothesis

Consider the data set $\mathbf{Y}$ with independent rows and the corresponding ranks $R_{ij}^{+}(b_i)$ of the absolute deviations $|Y_{ij} - b_i|$ among $|Y_{i1} - b_i|, \ldots, |Y_{in} - b_i|$, $i = 1, \ldots, p$:

$$\mathbf{Y} = \begin{bmatrix} Y_{11} & \ldots & Y_{1n} \\ Y_{i1} & \ldots & Y_{in} \\ Y_{p1} & \ldots & Y_{pn} \end{bmatrix} \quad \text{and} \quad \mathbf{R}^{+} = \begin{bmatrix} R_{11}^{+}(b_1) & \ldots & R_{1n}^{+}(b_1) \\ R_{i1}^{+}(b_i) & \ldots & R_{in}^{+}(b_i) \\ R_{p1}^{+}(b_p) & \ldots & R_{pn}^{+}(b_p) \end{bmatrix}. \tag{4.9.4}$$

For every n (≥ 1) and $j = 1, \ldots, p$, we define a set of scores by setting

$$a_n^{+}(k) = E\big[\phi^{+}(U_{kn})\big] \quad \text{or} \quad \phi^{+}\Big(\frac{k}{n+1}\Big), \tag{4.9.5}$$

where $0 < U_{1n} \leq \cdots < U_{nn} < 1$ are the order statistics of a sample of size n from $U(0,1)$. For every $u \in (0,1)$,

$$\phi_j^{+}(u) = \phi_j\left(\frac{1+u}{2}\right), \quad \phi_j(u) + \phi_j(1-u) = 0, \tag{4.9.6}$$

the ϕ_j are all assumed to be nondecreasing, nonconstant and square integrable score-generating functions. Recall that $R_{ij}^{+}(b_i)$ is the rank of $|Y_{ij} - b_i|$ among $|Y_{i1} - b_i|$, $|Y_{i2} - b_i|, \ldots |Y_{in} - b_i|$ for $i = 1, \ldots, p$ and $j = 1, \ldots, n$.

Now, consider the vector of rank-statistics

$$\mathbf{T}_n(\mathbf{b}) = \big(T_{1n}(b_1), \ldots, T_{pn}(b_p)\big)', \tag{4.9.7a}$$

where

$$\mathbf{T}_{in}(b_i) = n^{-1}\sum_{j=1}^{n} a_N^+(R_{ij}^+(b_i))\text{sign}(Y_{ij} - b_i) \tag{4.9.7b}$$

for $i = 1, \dots, p$. Note that $T_i(b_i)$ is $\searrow$ in b_i ($i = 1, \dots, p$) (see Puri and Sen, 1971, Ch. 6). Let

$$A_\phi^2 = \int_0^1 \phi^2(u)du - \Big(\int_0^1 \phi(u)du\Big)^2, \quad A_\psi^2 = I(f) = \int_0^1 \psi^2(u)du,$$

$$\psi(u) = -\frac{f_0'(F_0^{-1}(u))}{f_0(F_0^{-1}(u))}, \quad 0 < u < 1, \quad \gamma(\psi, \phi) = \Big(\int_0^1 \psi(u)\phi(u)du\Big). \tag{4.9.8}$$

A consistent estimator of A_ϕ^2 may be defined as

$$A_n^2 = (n-1)^{-1}\sum_{i=1}^{n}[a_n^+(i) - \bar{a}_n^+]^2. \tag{4.9.9}$$

Then, as $n \to \infty$

$$\sqrt{n}\,\mathbf{T}_n(\mathbf{0}) \overset{\mathcal{D}}{\simeq} N_p(\mathbf{0}, A_\phi^2\mathbf{I}_p). \tag{4.9.10}$$

(See Hajek, Sidak, and Sen, 1999; Sen, 1986; Puri and Sen, 1971). Thus, for testing $H_0 : \boldsymbol{\theta}$ against $H_A : \boldsymbol{\theta} \neq \mathbf{0}$, we consider the rank-test $\mathcal{L}_N$, defined by

$$\mathcal{L}_n = nA_n^{-2}\left\|\mathbf{T}_n(\mathbf{0})\right\|^2, \tag{4.9.11}$$

which approximately follows a central chi-square distribution with p d.f. under H_0 as $n \to \infty$ (e.g., Puri and Sen, 1986).

4.9.3 Estimation of the Location Parameter

Now. we focus on the estimation of $\boldsymbol{\theta}$ based on the rank-statistic $T(\mathbf{b})$, following Adichie (1967) and Puri and Sen (1986). Let

$$\begin{aligned}\tilde{\boldsymbol{\theta}}_n^{(1)} &= \sup\{\mathbf{b} : \mathbf{T}_n(\mathbf{b}) > \mathbf{0}\},\\ \tilde{\boldsymbol{\theta}}_n^{(2)} &= \inf\{\mathbf{b} : \mathbf{T}_n(\mathbf{b}) < \mathbf{0}\}.\end{aligned} \tag{4.9.12}$$

Then, we define the UE of $\boldsymbol{\theta}$ as

$$\tilde{\boldsymbol{\theta}}_n = \frac{1}{2}\Big(\tilde{\boldsymbol{\theta}}_n^{(1)} + \tilde{\boldsymbol{\theta}}_n^{(2)}\Big). \tag{4.9.13}$$

The UE, $\tilde{\boldsymbol{\theta}}_n$ is a translation-invariant and consistent estimator of $\boldsymbol{\theta}$. The RE of $\boldsymbol{\theta}$ is of course, $\mathbf{0}$, which is prefixed by the null hypothesis H_0. It may be shown, following Puri and Sen (1971), Sen (1986), and Hajek, Sidak, and Sen (1999), that as $n \to \infty$,

$$\sqrt{n}(\tilde{\boldsymbol{\theta}}_n - \boldsymbol{\theta}) \sim N_p(\mathbf{0}, A_\phi^2/\gamma^2(\psi, \phi)\mathbf{I}_p). \tag{4.9.14}$$

(See also Section 3.10.)

Let $\chi_p^2(\alpha)$ be the α-level critical value for testing $H_0 : \boldsymbol{\theta} = \mathbf{0}$. Then, the PTE of $\boldsymbol{\theta}$ is defined by

$$\hat{\boldsymbol{\theta}}_n^{\mathrm{PT}} = \tilde{\boldsymbol{\theta}}_n - \tilde{\boldsymbol{\theta}}_n I(\mathcal{L}_n < \chi_p^2(\alpha)). \tag{4.9.15}$$

Similarly, the Stein-type estimator is defined by

$$\hat{\boldsymbol{\theta}}_n^{S} = \tilde{\boldsymbol{\theta}}_n - d\mathcal{L}_n^{-1}\tilde{\boldsymbol{\theta}}_n, \quad 0 < d < 2(p-2), \tag{4.9.16a}$$

but $\mathcal{L}_n$ takes the value 0 with positive probability. We define a modified Stein-type estimator as

$$\hat{\boldsymbol{\theta}}_n^{S}(\varepsilon_N) = \tilde{\boldsymbol{\theta}}_n - d\mathcal{L}_n^{-1} I(\mathcal{L}_n > \varepsilon_n)\tilde{\boldsymbol{\theta}}_n, \tag{4.9.16b}$$

where $\varepsilon_n \to 0$ as $n \to \infty$. A better estimator of $\boldsymbol{\theta}$ without the confusion above is the positive-rule Stein-type estimator $\hat{\boldsymbol{\theta}}_n^{S+}$, where

$$\hat{\boldsymbol{\theta}}_n^{S+} = (1 - d\mathcal{L}_n^{-1} I(\mathcal{L}_n > d)\tilde{\boldsymbol{\theta}}_n. \tag{4.9.17}$$

All these estimators parallel the estimators defined in Section 4.6.2.

4.9.4 ADB, ADQB, ADMSE, and ADQR of the Estimators of Location Parameters

It may be observed that the test-statistic $\mathcal{L}_n$ defined by (4.7.10) is a consistent test as such, when $n \to \infty$, all three estimators, namely $\hat{\boldsymbol{\theta}}_n^{\mathrm{PT}}$, $\hat{\boldsymbol{\theta}}_n^{S}$, and $\hat{\boldsymbol{\theta}}_n^{S+}$ that are equivalent in distribution to that of $\tilde{\boldsymbol{\theta}}_n$. Hence, the asymptotic distributional bias, MSE matrices, and the risk expressions are identical to that of $\sqrt{n}(\tilde{\boldsymbol{\theta}}_n - \boldsymbol{\theta})$. As we have seen before, to overcome this problem, we consider the class of local alternatives

$$K_{(n)} : \boldsymbol{\theta}_{(n)} = n^{-1/2}\boldsymbol{\delta}, \quad \boldsymbol{\delta} = (\delta_1, \ldots, \delta_p)'. \tag{4.9.18}$$

Then we have the following theorem under $K_{(n)}$, using the linearity result of Section 3.10:

$$\sup\left\{\sqrt{n}\left|T_{jn}(n^{-1/2}b_j) - T_{jn}(0) + n^{-1/2}b_j\gamma(\psi,\phi)\right|; |b_j| < k\right\} \xrightarrow{\mathrm{P}} 0 \tag{4.9.19}$$

for $j = 1, \ldots, p$.

Theorem 1. Under $\{K_{(n)}\}$ and assumed regularity conditions, as $n \to \infty$,
(i) $\sqrt{n}\gamma(\psi,\phi)A_n^{-1}(\hat{\boldsymbol{\theta}}_n^{\mathrm{PT}}(\alpha) - \boldsymbol{\theta}_{(n)}) \xrightarrow{\mathcal{D}} \mathbf{Z} - (\mathbf{Z} + \sigma^{-1}\boldsymbol{\delta})I(\|\mathbf{Z} + \sigma^{-1}\boldsymbol{\delta}\|^2 \le k_\alpha)$,
(ii) $\sqrt{n}\gamma(\psi,\phi)A_n^{-1}(\hat{\boldsymbol{\theta}}_n^{S}(c) - \boldsymbol{\theta}_{(n)}) \xrightarrow{\mathcal{D}} \mathbf{Z} - c(\mathbf{Z} + \boldsymbol{\delta}^{-1}\boldsymbol{\delta})\|\mathbf{Z} + \sigma^{-1}\boldsymbol{\delta}\|^{-2}$,
and
(iii) $\sqrt{n}\,\gamma(\psi,\phi)A_n^{-1}(\hat{\boldsymbol{\theta}}_n^{S+}(c) - \boldsymbol{\theta}_{(n)}) \xrightarrow{\mathcal{D}} \left(1 - \frac{c}{\|\mathbf{Z}+\sigma^{-1}\boldsymbol{\delta}\|^2}\right)(\mathbf{Z} + \sigma^{-1}\boldsymbol{\delta})I(\|\mathbf{Z} + \sigma^{-1}\boldsymbol{\delta}\|^2 > c)$.

Proof. Based on the linearity results given at (4.9.19), we have as $n \to \infty$,

$$\sqrt{n}\,\mathbf{T}_n(\mathbf{0}) = \sqrt{n}(\tilde{\boldsymbol{\theta}}_n - \boldsymbol{\theta}_{(n)})\gamma(\psi,\phi) + o_p(1). \tag{4.9.20}$$

Hence, using contiguity of probability measures under $\{K_{(n)}\}$ and to that of $\boldsymbol{\theta} = \mathbf{0}$, we obtain, as $n \to \infty$,

$$\begin{bmatrix} \sqrt{n}(\tilde{\boldsymbol{\theta}}_n - \boldsymbol{\theta}_{(n)} \\ \sqrt{n}\mathbf{T}_n(\mathbf{0}) \end{bmatrix} \overset{\mathcal{D}}{\simeq} N_{2p}\left\{\begin{pmatrix} \mathbf{0} \\ \boldsymbol{\delta}\gamma(\psi,\phi) \end{pmatrix}; \boldsymbol{\Sigma}\right\},$$

$$\boldsymbol{\Sigma} = \begin{pmatrix} A_\phi^2/\gamma^2(\psi,\phi)\mathbf{I}_p, & \gamma^{-1}(\psi,\phi)\mathbf{I}_p \\ \gamma^{-1}(\psi,\phi)\mathbf{I}_p & A_\phi^2\mathbf{I}_p \end{pmatrix}, \tag{4.9.21}$$

using (4.9.10). Now,

$$\begin{aligned}&\sqrt{n}\gamma(\psi,\phi)A_n^{-1}(\hat{\boldsymbol{\theta}}_n^{\mathrm{PT}}(\alpha) - \boldsymbol{\theta}_{(n)})\\ &= \sqrt{n}\gamma(\psi,\phi)A_n^{-1}(\tilde{\boldsymbol{\theta}}_n - \boldsymbol{\theta}_{(n)}) - \sqrt{n}\gamma(\psi,\phi)A_n^{-1}\tilde{\boldsymbol{\theta}}_n I(\mathcal{L}_n < \mathcal{L}_{n,\alpha}),\end{aligned} \tag{4.9.22a}$$

and as $n \to \infty$, the r.h.s. of (4.9.22a) converges in distribution to the r.v.

$$\mathbf{Z} - (\mathbf{Z} + \sigma^{-1}\boldsymbol{\delta})I(\|\mathbf{Z} + \sigma^{-1}\boldsymbol{\delta}\|^2 < k_\alpha), \tag{4.9.22b}$$

where $\mathcal{L}_{n,\alpha} \to \chi_p^2(\alpha) = k_\alpha$.

Similarly,

$$\begin{aligned}\sqrt{n}\gamma(\psi,\phi)A_n^{-1}(\hat{\boldsymbol{\theta}}_n^{S} - \theta_{(n)}) &= \sqrt{n}\gamma(\psi,\phi)A_n^{-1}(\tilde{\boldsymbol{\theta}}_n - \theta_{(n)})\\ &\quad - c\sqrt{n}\gamma(\psi,\phi)A_n^{-1}\tilde{\boldsymbol{\theta}}_n\mathcal{L}_n^{-1}I(\mathcal{L}_n > \varepsilon_n),\end{aligned} \tag{4.9.23}$$

and as $n \to \infty$, r.h.s. converges in distribution to the r.v.

$$\mathbf{Z} - c(\mathbf{Z} + \sigma^{-1}\boldsymbol{\delta})\|\mathbf{Z} + \sigma^{-1}\boldsymbol{\delta}\|^{-2}, \quad c = p - 2. \tag{4.9.24}$$

Finally,

$$\begin{aligned}&\sqrt{n}\gamma(\psi,\phi)A_n^{-1}(\hat{\boldsymbol{\theta}}_n^{S+} - \theta_{(n)})\\ &= \sqrt{n}\gamma(\psi,\phi)A_n^{-1}(\hat{\boldsymbol{\theta}}_n^{S+} - \boldsymbol{\theta}_{(n)}) - c(\sqrt{n}\gamma(\psi,\phi)A_n^{-1}\tilde{\boldsymbol{\theta}}_n\mathcal{L}_n^{-1}I(\mathcal{L}_n > \varepsilon_n),\end{aligned} \tag{4.9.25}$$

which converges in distribution to the r.v.

$$(\mathbf{Z} + \sigma^{-1}\boldsymbol{\delta})(1 - c\|\mathbf{Z} + \sigma^{-1}\boldsymbol{\delta}\|^{-2})I(\|\mathbf{Z} + \sigma^{-1}\|^2 > c). \tag{4.9.26}$$

As a result of Theorem 1, we obtain Theorem 2.

Theorem 2. Under $\{K_{(n)}\}$ and the assumed regularity conditions, as $n \to \infty$,

(i) $\sqrt{n}(\tilde{\boldsymbol{\theta}}_n - \boldsymbol{\theta}_{(n)}) \overset{\mathcal{D}}{\simeq} \mathcal{N}_p(\mathbf{0}, A_\phi^2/\gamma^2(\psi,\phi)\mathbf{I}_p)$, (4.9.27)

(ii) $\lim P_{K_{(n)}}\{nA_n^{-2}[\mathbf{T}_n(\mathbf{0})]'[\mathbf{T}_n(\mathbf{0})] \le x\} = H_p(x;\Delta^2)$, (4.9.28a)

where $H_p(\cdot;\Delta^2)$ is the cdf of a noncentral chi-square distribution with p d.f. and noncentrality parameter $\Delta^2/2$, where

$$\Delta^2 = \|\boldsymbol{\delta}\|^2/\sigma^2, \quad \sigma^2 = A_\phi^2/\gamma^2(\psi,\phi) \tag{4.9.28b}$$

(iii) The ADB, ADQB, ADMSE, and the ADQR of $\tilde{\boldsymbol{\theta}}_n$, $\tilde{\boldsymbol{\theta}}_n^{\mathrm{PT}}$, $\hat{\boldsymbol{\theta}}_n^{S}$, and $\hat{\boldsymbol{\theta}}_n^{S+}$ are given below: For UE, $(\tilde{\boldsymbol{\theta}}_n)$, we have

$$\begin{aligned}&\mathbf{b}_1(\tilde{\boldsymbol{\theta}}_n) = \mathbf{0} \text{ and } B_1(\tilde{\boldsymbol{\theta}}_n) = 0 \text{ and}\\ &\mathbf{M}_1(\tilde{\boldsymbol{\theta}}_n) = \sigma^2\mathbf{I}_p \text{ and } R_1(\tilde{\boldsymbol{\theta}}_n;\mathbf{W}) = \sigma^2\mathrm{tr}(\mathbf{W}).\end{aligned} \tag{4.9.29}$$

For PTE, $(\hat{\boldsymbol{\theta}}_n^{\mathrm{PT}})$, we have

$$\mathbf{b}_2(\hat{\boldsymbol{\theta}}_n^{\mathrm{PT}}) = -\boldsymbol{\delta}H_{p+2}(\chi_p^2(\alpha);\Delta^2) \text{ and } B_2(\hat{\boldsymbol{\theta}}_n^{\mathrm{PT}}) = \Delta^2\{H_{p+2}(\chi_p^2(\alpha);\Delta^2)\}^2; \tag{4.9.30}$$

$$\begin{aligned}\mathbf{M}_2(\hat{\boldsymbol{\theta}}_n) &= \sigma^2\mathbf{I}_p\{1 - H_{p+2}(\chi_p^2(\alpha);\Delta^2)\}\\ &+\boldsymbol{\delta}\boldsymbol{\delta}'\{2H_{p+2}(\chi_p^2(\alpha);\Delta^2) - H_{p+4}(\chi_p^2(\alpha);\Delta^2)\}\end{aligned}$$

and

$$\begin{aligned}R_2(\hat{\boldsymbol{\theta}}_n^{\mathrm{PT}};\mathbf{W}) &= \sigma^2\mathrm{tr}(\mathbf{W})\{1 - H_{p+2}(\chi_p^2(\alpha);\Delta^2)\}\\ &+(\boldsymbol{\delta}'\mathbf{W}\boldsymbol{\delta})\{2H_{p+2}(\chi_p^2(\alpha);\Delta^2) - H_{p+4}(\chi_p^2(\alpha);\Delta^2)\}.\end{aligned} \tag{4.9.31}$$

For SE, $(\hat{\boldsymbol{\theta}}_n^{S})$ we have

$$\mathbf{b}_3(\hat{\boldsymbol{\theta}}_n^{S}) = -(p-2)\boldsymbol{\delta}E[\chi_{p+2}^{-2}(\Delta^2)] \text{ and } B_3(\hat{\boldsymbol{\theta}}_n^{S}) = \Delta^2\{E[\chi_{p+2}^{-2}(\Delta^2)]\}^2;$$

$$\begin{aligned}\mathbf{M}_3(\hat{\boldsymbol{\theta}}_n^{S}) &= \sigma^2\mathbf{I}_p - \sigma^2(p-2)^2\mathbf{I}_p\{2E[\chi_{p+2}^{-2}(\Delta^2)] - (p-2)E[\chi_{p+2}^{-2}(\Delta^2)]\}\\ &+(p^2-4)\boldsymbol{\delta}\boldsymbol{\delta}'E[\chi_{p+4}^{-4}(\Delta^2)]\end{aligned}$$

and

$$\begin{aligned}\mathbf{R}_3(\hat{\boldsymbol{\theta}}_n^{S};\mathbf{W}) &= \sigma^2\mathrm{tr}(\mathbf{W}) - \sigma^2(p-2)^2\mathrm{tr}(\mathbf{W})\Big\{2E[\chi_{p+2}^{-2}(\Delta^2)\\ &-(p-2)E[\chi_{p+2}^{-4}(\Delta^2)]\Big\} + (p^2-4)(\boldsymbol{\delta}'\mathbf{W}\boldsymbol{\delta})E[\chi_{p+4}^{-4}(\Delta^2)].\end{aligned} \tag{4.9.32}$$

For the PRSE, $(\hat{\boldsymbol{\theta}}_n^{S+})$, we have

$$\begin{aligned}\mathbf{b}_4(\hat{\boldsymbol{\theta}}_n^{S+}) = -\ &\boldsymbol{\delta}\big\{(p-2)E[\chi_{p+2}^{-2}(\Delta^2)]\\ &+ E[(1-(p-2)\chi_{p+2}^{-2}(\Delta^2))I(\chi_{p+2}^2(\Delta^2)<p-2)]\big\}\end{aligned}$$

and

$$B_4(\hat{\boldsymbol{\theta}}_n^{S+}) = \Delta^2\{(p-2)E[\chi_{p+2}^{-2}(\Delta^2)] + E(1-(p-2)\chi_{p+2}^{-2}(\Delta^2)]I(\chi_{p+2}^2(\Delta^2)<p-2)\}^2;$$

$$\begin{aligned}\mathbf{M}_4(\hat{\boldsymbol{\theta}}_n^{S+}) = \mathbf{M}_3(\hat{\boldsymbol{\theta}}_n^{S}) &- \sigma^2\mathbf{I}_p E[(1-(p-2)\chi_{p+2}^{-2}(\Delta^2))^2 I(\chi_{p+2}^2(\Delta^2)<p-2)] \\ &+\boldsymbol{\delta}\boldsymbol{\delta}'\{2E[(1-(p-2)\chi_{p+2}^{-2}(\Delta^2))I(\chi_{p+2}^2(\Delta^2)<p-2)] \\ &-E[(1-(p-2)\chi_{p+4}^{-2}(\Delta^2))^2 I(\chi_{p+4}^2(\Delta^2)<p-2)]\}\end{aligned}$$

and

$$\begin{aligned}R_4(\hat{\boldsymbol{\theta}}_n^{S+};\mathbf{W}) = R_3(\hat{\boldsymbol{\theta}}_n^{S};\mathbf{W}) &- \sigma^2\mathrm{tr}(\mathbf{W})E[(1-(p-2)\chi_{p+2}^{-2}(\Delta^2))I(\chi_{p+2}^{-2}(\Delta^2))] \\ &+(\boldsymbol{\delta}'\mathbf{W}\boldsymbol{\delta})\{2E[(1-(p-2)\chi_{p+2}^{-2}(\Delta^2))I(\chi_{p+2}^2(\Delta^2)<p-2)] \\ &-E[(1-(p-2)\chi_{p+4}^{-2}(\Delta^2))^2 I(\chi_{p+4}^2(\Delta^2)<p-2)]\}. \quad (4.9.33)\end{aligned}$$

The expressions above are similar to the expressions in Theorems 1 and 2 of Section 4.6.3. Hence, the asymptotic properties of the four estimators are the same as given there.

Proof. (i) See Theorem 1. (ii) The proof follows from (4.9.21). For (iii), we let

$$\mathbf{U}_1 = \mathbf{Z} - (\mathbf{Z}+\sigma^{-1}\boldsymbol{\delta})I(\|\mathbf{Z}+\sigma^{-1}\boldsymbol{\delta})\|^2 < k_\alpha). \quad (4.9.34)$$

Then the bias vector is given by

$$\mathbf{b}_2(\hat{\boldsymbol{\theta}}_n^{\mathrm{PT}}) = E(\sigma\mathbf{U}_1) = -\boldsymbol{\delta}H_{p+2}(\chi_p^2(\alpha),\Delta^2).$$

Similarly,

$$\begin{aligned}\mathbf{M}_2(\hat{\boldsymbol{\theta}}_n^{\mathrm{PT}}) = \sigma^2 E[\mathbf{U}_1\mathbf{U}_1'] &= \sigma^2\mathbf{I}_p - \sigma^2\mathbf{I}_p H_{p+2}(\chi_p^2(\alpha);\Delta^2) \\ &+\boldsymbol{\delta}\boldsymbol{\delta}'\{2H_{p+2}(\chi_p^2(\alpha);\Delta^2) - H_{p+4}(\chi_p^2(\alpha);\Delta^2)\}. \quad (4.9.35)\end{aligned}$$

For the bias vector and MSE matrix of $\hat{\boldsymbol{\theta}}_n^{S}$, let

$$\mathbf{U}_2 = \mathbf{Z} - c(\mathbf{Z}+\sigma^{-1}\boldsymbol{\delta})\|\mathbf{Z}+\sigma^{-1}\boldsymbol{\delta}\|^{-2}.$$

Then compute

$$\mathbf{b}_3(\hat{\boldsymbol{\theta}}_n^{S}) = \sigma E(\mathbf{U}_2) \text{ and } \mathbf{M}_3(\hat{\boldsymbol{\theta}}_n^{S}) = \sigma^2 E[\mathbf{U}_2\mathbf{U}_2']. \quad (4.9.36)$$

Similarly, the bias vector and the MSE matrix of $\hat{\boldsymbol{\theta}}_n^{S+}$ can be computed by using the r.v.:

$$\mathbf{U}_3 = (\mathbf{Z} + \sigma^{-1}\boldsymbol{\delta})(1 - c\|\mathbf{Z} + \sigma^{-1}\boldsymbol{\delta}\|^{-2})I(\|\mathbf{Z} + \sigma^{-1}\boldsymbol{\delta}\|^2 > c).$$

Also, compute

$$\begin{aligned} \mathbf{b}_4(\hat{\boldsymbol{\theta}}_n^{S+}) &= \sigma E[\mathbf{U}_3], \\ \mathbf{M}_4(\hat{\boldsymbol{\theta}}_n^{S+} &= \sigma^2 E[\mathbf{U}_3\mathbf{U}_3']. \end{aligned} \tag{4.9.37}$$

4.9.5 Asymptotic Properties of Confidence Sets

In this section, we consider the asymptotic properties of the following $(1-\gamma)\%$ confidence sets:

$$C^0(\tilde{\boldsymbol{\theta}}_n) = \{n\|\boldsymbol{\theta} - \tilde{\boldsymbol{\theta}}_n\|^2 \le k_\gamma\}, \tag{4.9.38}$$

$$C^{\mathrm{PT}}(\hat{\boldsymbol{\theta}}_n^{\mathrm{PT}}(\alpha)) = \{n\|\boldsymbol{\theta} - \hat{\boldsymbol{\theta}}_n^{\mathrm{PT}}(\alpha)\|^2 \le k_\gamma\}, \tag{4.9.39}$$

and

$$C^{S+}(\hat{\boldsymbol{\theta}}_n^{S+}(c)) = \{n\|\boldsymbol{\theta} - \hat{\boldsymbol{\theta}}_n^{S+}(c)\|^2 \le k_\gamma\}, \tag{4.9.40}$$

where $k_\gamma = \chi_p^2(\gamma)$.

Similar to Theorem 4.8.6, we show that under fixed alternative, the asymptotic distributions of $\sqrt{n}(\hat{\boldsymbol{\theta}}_n^{\mathrm{PT}}(\alpha) - \boldsymbol{\theta})$ and $\sqrt{n}(\hat{\boldsymbol{\theta}}_n^{S+}(c) - \boldsymbol{\theta})$ are the same as that of $\sqrt{n}(\tilde{\boldsymbol{\theta}}_n - \boldsymbol{\theta})$. Hence, we consider the asymptotic properties of the confidence sets (4.9.26) through (4.9.28) under the local alternatives, $K_{(n)} : \boldsymbol{\theta}_{(n)} = n^{-1/2}\boldsymbol{\delta}$.

To begin, it is easy to see that

$$\lim_{n\to\infty} P_{K_{(n)}}\{n\|\boldsymbol{\theta}_{(n)} - \tilde{\boldsymbol{\theta}}_n\|^2 \le k_\gamma\} = P(\|\mathbf{Z}\|^2 < k_\gamma) = 1 - \gamma. \tag{4.9.41}$$

Next, we consider Theorem 4.9.1 and use the results (i) through (iii) to compute the probability of the different sets. Thus, the asymptotic coverage probabilities of $C^{\mathrm{PT}}(\hat{\boldsymbol{\theta}}_n^{\mathrm{PT}}(\alpha))$ and $C^{S+}(\hat{\boldsymbol{\theta}}_n^{S+}(c))$ reduce to

$$\begin{aligned} &\lim_{n\to\infty} P_{K_{(n)}}\{C^{\mathrm{PT}}(\hat{\theta}^{\mathrm{PT}}(\alpha))\} \\ &= H_p(k;\Delta^2)I(\Delta^2 < k_\gamma) + P_{\boldsymbol{\delta}}\{\|\mathbf{Z} + \sigma^{-1}\boldsymbol{\delta}\|^2 > k_\alpha; \|\mathbf{Z}\|^2 \le k_\gamma\} \end{aligned} \tag{4.9.42a}$$

and

$$\begin{aligned} &\lim_{n\to\infty} P_{K_{(n)}}\{C^{S+}(\hat{\theta}^{S+}(c))\}P_0\{\|\mathbf{Z}\|^2 < k_\gamma\} \\ &= H_p(c;\Delta^2)I(\Delta^2 < k_\gamma) + P_{\boldsymbol{\delta}}\left\{\left\|\mathbf{Z} - \frac{c(\mathbf{Z} + \sigma^{-1}\boldsymbol{\delta})}{\|\mathbf{Z} + \sigma^{-1}\boldsymbol{\delta}\|^2}\right\|^2 \le k_\gamma; \|\mathbf{Z} + \sigma^{-1}\boldsymbol{\delta}\|^2 > c\right\}. \end{aligned} \tag{4.9.42b}$$

The two expressions are similar to the expressions given in Theorem 4.8.7. Hence, the asymptotic coverage probabilities are similar to the finite sample coverage probabilities as given in Theorems 4.8.1 through 4.8.7.

4.10 Conclusions

In this chapter, we introduced the basic Stein-type estimators, in particular the James-Stein estimator together with the preliminary test estimator of the mean vector of a p-variate normal distribution with known and unknown covariance matrix. We have also presented three basic methods of obtaining Stein-type estimators. They are (1) the risk difference approach, (2) the empirical Bayes approach and the (3) the preliminary test (or quasi-empirical Bayes) approach. Of all these models, we particularly discussed the empirical Bayes and the preliminary test (or quasi-empirical Bayes) approach to obtain the Stein-type estimators. In general, we considered the (1) unrestricted (2) PTE, (3) Stein-type, and (4) positive-rule shrinkage estimators and evaluated their properties according to Table 4.10.1.

Table 4.10.1 Properties of Estimators

Type of Estimators	QB ↓	MSE ↓	Criterion Risk ↓	MRE ↓	RRE ↓
UE	$B_1(\tilde{\boldsymbol{\theta}}_n)$	$\mathbf{M}_1(\tilde{\boldsymbol{\theta}}_n)$	$R_1(\tilde{\boldsymbol{\theta}}_n;\mathbf{W})$	—	—
PTE	$B_1(\hat{\boldsymbol{\theta}}_n^{PT})$	$\mathbf{M}_2(\tilde{\boldsymbol{\theta}}_n^{PT})$	$R_1(\tilde{\boldsymbol{\theta}}_n^{PT};\mathbf{W})$	$\text{MRE}(\hat{\boldsymbol{\theta}}_n^{PT};\tilde{\boldsymbol{\theta}}_n)$	$\text{RRE}(\hat{\boldsymbol{\theta}}_n^{PT};\tilde{\boldsymbol{\theta}}_n)$
SE	$B_3(\hat{\boldsymbol{\theta}}_n^{S})$	$\mathbf{M}_3(\hat{\boldsymbol{\theta}}_n^{S})$	$R_3(\hat{\boldsymbol{\theta}}_n^{S};\mathbf{W})$	$\text{MRE}(\hat{\boldsymbol{\theta}}_n^{S};\tilde{\boldsymbol{\theta}}_n)$	$\text{RRE}(\hat{\boldsymbol{\theta}}_n^{S};\tilde{\boldsymbol{\theta}}_n)$
PRSE	$B_4(\hat{\boldsymbol{\theta}}_n^{S+})$	$\mathbf{M}_4(\hat{\boldsymbol{\theta}}_n^{S+})$	$R_4(\hat{\boldsymbol{\theta}}_n^{S+};\mathbf{W})$	$\text{MRE}(\hat{\boldsymbol{\theta}}_n^{S+};\tilde{\boldsymbol{\theta}}_n)$	$\text{RRE}(\hat{\boldsymbol{\theta}}_n^{S+};\tilde{\boldsymbol{\theta}}_n)$
IS	$B_5(\hat{\boldsymbol{\theta}}_n^{IS})$	$\mathbf{M}_5(\hat{\boldsymbol{\theta}}_n^{IS})$	$R_5(\hat{\boldsymbol{\theta}}_n^{IS};\mathbf{W})$	$\text{MRE}(\hat{\boldsymbol{\theta}}_n^{IS};\hat{\boldsymbol{\theta}}_n^{S})$	$\text{RRE}(\hat{\boldsymbol{\theta}}_n^{IS};\hat{\boldsymbol{\theta}}_n^{S})$

We showed the following among other results:

(i) $0 = B_1\left(\tilde{\boldsymbol{\theta}}_n\right) \le B_4\left(\hat{\boldsymbol{\theta}}_n^{S+}\right) \le B_3\left(\hat{\boldsymbol{\theta}}_n^{S}\right) \le B_2\left(\hat{\boldsymbol{\theta}}_n^{PT}\right) < \infty$, with some conditions on the level of significance, α

(ii) $R_1\left(\tilde{\boldsymbol{\theta}}_n;\mathbf{W}\right) \ge R_3\left(\hat{\boldsymbol{\theta}}_n^{S};\mathbf{W}\right) \ge R_4\left(\hat{\boldsymbol{\theta}}_n^{S+};\mathbf{W}\right)$ for all $\Delta^2 \in [0,\infty)$

(iii) $R_3\left(\hat{\boldsymbol{\theta}}_n^{S};\sigma^{-2}\mathbf{I}_p\right) \ge R_5\left(\hat{\boldsymbol{\theta}}_n^{IS};\sigma^{-2}\mathbf{I}_p\right)$ for all $\Delta^2 \in [0,\infty)$

(iv) $R_2\left(\hat{\boldsymbol{\theta}}_n^{PT};\mathbf{W}\right) \ge R_5\left(\hat{\boldsymbol{\theta}}_n^{PT+};\mathbf{W}\right)$ for all (α,Δ^2).

We concluded that for $p \ge 3$, $\hat{\boldsymbol{\theta}}_n^{S+}$ is preferable to $\hat{\boldsymbol{\theta}}_n^{S}$ and $\tilde{\boldsymbol{\theta}}_n$, and that $\hat{\boldsymbol{\theta}}_n^{PT+}$ is preferable to $\hat{\boldsymbol{\theta}}_n^{PT}$ for $p \le 2$. The estimator $\hat{\boldsymbol{\theta}}_n^{IS}$ is preferable to $\hat{\boldsymbol{\theta}}_n^{S}$ if $p \ge 3$ but not over $\hat{\boldsymbol{\theta}}_n^{S+}$. The positive-rule version of $\hat{\boldsymbol{\theta}}_n^{IS}$ is not preferable over $\hat{\boldsymbol{\theta}}_n^{S+}$ and should not be considered as a viable estimator. We discussed the recentered confidence set based on the PRSE and presented its properties in Section 4.8 with the asymptotic version under local alternatives around the uncertain prior information $\boldsymbol{\theta} = \mathbf{0}$. The confidence sets with recentered PSRE proved to dominate the usual set under local alternatives. The confidence set with unknown variance turned out to be a difficult problem that is open for solution and was hence avoided. We appealed to the asymptotic result for this problem. We also included the theory of Stein-type R-estimation of the location parameter $\boldsymbol{\theta}$ and the asymptotic properties of the point and the confidence set estimation of $\boldsymbol{\theta}$ under local alternatives.

4.11 Problems

1. Show that

   (i) $\left|\mathbf{M}_1(\tilde{\boldsymbol{\theta}}_n)\right| = 1.$

   (ii) $$\left|\mathbf{M}_2(\hat{\boldsymbol{\theta}}_n^{\text{PT}})\right| = \left\{1 - H_{p+2}(\chi_p^2(\alpha);\Delta^2)\right\}^{p-1} \times\left\{\left[1 - H_{p+2}(\chi_p^2(\alpha);\Delta^2)\right] +\Delta^2\left[2H_{p+2}(\chi_p^2(\alpha);\Delta^2) - H_{p+4}(\chi_p^2(\alpha);\Delta^2)\right]\right\}.$$

   (iii) $$\left|\mathbf{M}_3(\hat{\boldsymbol{\theta}}_n^{S})\right| = \left\{1-(p-2)\left[2E(\chi_{p+2}^{-2}(\Delta^2))-(p-2)E(\chi_{p+2}^{-4}(\Delta^2))\right]\right\}^{p-1} \times\left\{1-(p-2)\left[2E(\chi_{p+2}^{-2}(\Delta^2))-(p-2)E(\chi_{p+2}^{-4}(\Delta^2))\right] \times(p^2-4)\Delta^2E(\chi_{p+4}^{-4}(\Delta^2))\right\},$$

   and

   (iv) $$\left|\mathbf{M}_4(\hat{\boldsymbol{\theta}}_n^{S+})\right| = \Big(1-(p-2)[2E(\chi_{p+2}^{-2}(\Delta^2))-(p-2)E(\chi_{p+2}^{2}(\Delta^2))] -E[\{1-(p-2)\chi_{p+2}^{-2}(\Delta^2)\}^2 I(\chi_{p+2}^2(\Delta^2)<p-2)]\Big)^{p-1} m_4(\Delta^2),$$

   where $m_4(\Delta^2)$ is given by (4.4.59).

2. Refer to Section 4.4.2.

   (i) Compare $\hat{\boldsymbol{\theta}}_n^{\text{PT}}$ and $\hat{\boldsymbol{\theta}}_n^{S}$ with respect to MSE matrices and risk expressions, and show that neither dominates the other for some α, the level of significance.

   (ii) Compare $\hat{\boldsymbol{\theta}}_n^{\text{PT}}$ and $\hat{\boldsymbol{\theta}}_n^{S+}$ with respect to MSE matrices and risk expressions, and show that neither dominates the other for some α, the level of significance.

3. Let $\mathbf{Y}_i = \boldsymbol{\theta} + \mathbf{e}_i$; $i = 1,2,\ldots,n$, where $\mathbf{e}_i \sim N_p(\mathbf{0},\mathbf{I}_p)$. Let $\hat{\boldsymbol{\theta}}_n^{(d)} = d\tilde{\boldsymbol{\theta}}_n + (1-d)\theta_0\mathbf{1}_p$ where θ_0 is a fixed scalar parameter. Determine the bias, quadratic bias, MSE matrix, and quadratic risk based on the loss function $L(\boldsymbol{\theta}_n^*;\boldsymbol{\theta}) = (\boldsymbol{\theta}^*-\boldsymbol{\theta})'\mathbf{W}(\boldsymbol{\theta}^*-\boldsymbol{\theta})$ $(0<d<1)$ of the estimator, $\hat{\boldsymbol{\theta}}_n$.

4. Let $\mathbf{Y}_i = \boldsymbol{\theta} + \mathbf{e}_i$, $i = 1, 2, \ldots, n$, where $\mathbf{e}_i$ is distributed as

$$f(\mathbf{e}; \nu, \sigma^2) = \frac{\Gamma(\frac{np+v}{2})}{(\pi\nu)^{np/2}\Gamma(\frac{v}{2})}\left(1 + \frac{\sum_{i=1}^{n} \mathbf{e}_i'\mathbf{e}_i}{\nu\sigma^2}\right)^{(n+v)/2}, \quad \nu, \sigma^2 > 0.$$

Define

(i) $\tilde{\boldsymbol{\theta}}_n = \bar{\mathbf{Y}} = (\bar{Y}_1, \ldots, \bar{Y}_p)'$.

(ii) $\hat{\boldsymbol{\theta}}_n^{\text{PT}} = \tilde{\boldsymbol{\theta}}_n - \tilde{\boldsymbol{\theta}}_n I(n\|\tilde{\boldsymbol{\theta}}_n\|^2 < \chi_p^2(\alpha))$.

(iii) $\hat{\boldsymbol{\theta}}_n^{S} = \tilde{\boldsymbol{\theta}}_n - (p-2)(n\|\tilde{\boldsymbol{\theta}}_n\|^2)^{-1}\tilde{\boldsymbol{\theta}}_n$.

(iv) $\hat{\boldsymbol{\theta}}_n^{S+} = \tilde{\boldsymbol{\theta}}_n - \tilde{\boldsymbol{\theta}}_n I(n\|\tilde{\boldsymbol{\theta}}_n\|^2 < p-2) - (p-2)(n\|\tilde{\boldsymbol{\theta}}_n\|^2)\tilde{\boldsymbol{\theta}}_n + (p-2)(n\|\tilde{\boldsymbol{\theta}}_n\|^2)^{-1} I(n\|\tilde{\boldsymbol{\theta}}_n\|^2 < p-2)\tilde{\boldsymbol{\theta}}_n$ as in Sections 4.2 and 4.3. Here $\chi_p^2(\alpha)$ is the α-level upper critical value from a central chi-squared distribution with p d.f.

Show that the bias expressions are

(i) $\mathbf{b}_1(\tilde{\boldsymbol{\theta}}_n) = \mathbf{0}$ and $B_1(\tilde{\boldsymbol{\theta}}_n) = 0$.

(ii) $\mathbf{b}_2(\hat{\boldsymbol{\theta}}_n^{\text{PT}}) = -\boldsymbol{\theta} H_{p+2}^{(1)}(\chi_p^2(\alpha); \Delta^2)$ and $B_2(\hat{\boldsymbol{\theta}}_n^{\text{PT}}) = \Delta^2\{H_{p+2}^{(1)}(\chi_p^2(\alpha); \Delta^2)\}^2$, where

$$H_{p+2j}^{(1)}(\chi_p^2(\alpha); \Delta^2) = \sum_{r\geq 0} \frac{1}{r!} \frac{(\frac{\Delta^2}{\nu-2})^r}{(1+\frac{\Delta^2}{\nu-2})^{p+2j+2r/2}} H_{p+2j+2r}(\chi_p^2(\alpha); 0),$$

$j = 1, 2$.

(iii) $\mathbf{b}_3(\hat{\boldsymbol{\theta}}_n^{S}) = -(p-2)\boldsymbol{\theta} E^{(1)}(\chi_{p+2}^{-2}(\Delta^2))$ and $B_3(\hat{\boldsymbol{\theta}}_n^{S}) = (p-2)^2\Delta^2\{E^{(1)}(\chi_{p+2}^{-2}(\Delta^2)\}^2$, where

$$E^{(1)}\{\chi_{p+2j}^{-2}(\Delta^2)\} = \sum_{r\geq 0} \frac{1}{r!} \frac{(\frac{\Delta^2}{\nu-2})^r}{(1+\frac{\Delta^2}{\nu-2})^{p+2j+2r/2}} (p+2j+2r)^{-1},$$

$j = 1, 2$.

(iv)
$$\mathbf{b}_4(\hat{\boldsymbol{\theta}}_n^{S+}) = -\theta\Big[H_{p+2}^{(1)}(\chi_p^2(\alpha); \Delta^2) - (p-2)E^{(1)}(\chi_{p+2}^{-2}(\Delta^2)) + (p-2)E^{(2)}\Big\{\chi_{p+2}^{2}(\Delta^2) I(\chi_{p+2}^2(\Delta^2) < p-2)\Big\}\Big]$$

and

$$B_4(\hat{\boldsymbol{\theta}}_n^{S+}) = \Delta^2\Big[H_{p+2}^{(1)}(\chi_p^2(\alpha); \Delta^2) - (p-2)E^{(1)}(\chi_{p+2}^{-2}(\Delta^2)) + (p-2)E^{(2)}\Big\{\chi_{p+2}^{-2}(\Delta^2) I(\chi_{p+2}^2(\Delta^2) < p-2)\Big\}\Big]^2.$$

5. (Continuation of Problem 4.) Show that

(i) $\mathbf{M}_1(\tilde{\boldsymbol{\theta}}_n) = \mathbf{I}_p$ and $R_1(\tilde{\boldsymbol{\theta}}_n; \mathbf{W}) = \text{tr}(\mathbf{W}) \equiv p$ if $\mathbf{W} = \mathbf{I}_p$.

(ii) $$M_2(\hat{\boldsymbol{\theta}}_n^{\text{PT}}) = \mathbf{I}_p\{1 - H_{p+2}^{(1)}(\chi_p^2(\alpha); \Delta^2)\}$$
$$+n\boldsymbol{\theta}\boldsymbol{\theta}'\{2H_{p+2}^{(1)}(\chi_p^2(\alpha); \Delta^2) - H_{p+4}^{(1)}(\chi_p^2(\alpha); \Delta^2)\}$$

and

$$\begin{aligned} R_2(\hat{\boldsymbol{\theta}}_n^{\text{PT}}; \mathbf{W}) &= \text{tr}(\mathbf{W})\{1 - H_{p+2}^{(1)}(\chi_p^2(\alpha); \Delta^2)\} \\ &\quad + (n\boldsymbol{\theta}'\mathbf{W}\boldsymbol{\theta})\{2H_{p+2}^{(1)}(\chi_p^2(\alpha); \Delta^2) - H_{p+4}^{(1)}(\chi_p^2(\alpha); \Delta^2)\} \\ &= p\{1 - H_{p+2}^{(1)}(\chi_p^2(\alpha); \Delta^2)\} + \Delta^2\{2H_{p+2}^{(1)}(\chi_p^2(\alpha); \Delta^2) \\ &\quad - H_{p+2}^{(1)}(\chi_p^2(\alpha); \Delta^2)\} \text{ if } \mathbf{W} = \mathbf{I}_p. \end{aligned}$$

(iii) $$\mathbf{M}_3(\hat{\boldsymbol{\theta}}_n^S)$$
$$= \mathbf{I}_p - (p-2)\mathbf{I}_p\Big[2E^{(1)}(\chi_{p+2}^{-2}(\Delta^2)) - (p-2)E^{(1)}(\chi_{p+2}^{-4}(\Delta^2))\Big]$$
$$+ \ (p^2 - 4)n\boldsymbol{\theta}\boldsymbol{\theta}'E^{(1)}(\chi_{p+4}^{-4}(\Delta^2)),$$

where

$$E^{(1)}(\chi_{p+2j+2}^{-4}(\Delta^2))$$
$$= \sum \frac{1}{r!} \frac{\left(\frac{\Delta^2}{\nu-2}^r\right)}{(1 + \frac{\Delta^2}{\nu-2})^{p+2j+2r/2}} (p + 2j + 2r)^{-1}(p + 2j + 2r - 2)^{-1}$$

and

$$\begin{aligned} R_3(\hat{\boldsymbol{\theta}}_n^S; \mathbf{W}) &= \text{tr}(\mathbf{W}) - (p-2)\text{tr}\mathbf{W}\{2E^{(1)}(\chi_{p+2}^{2}(\Delta^2)) \\ &\quad - \ (p-2)E^{(1)}(\chi_{p+2}^{-4}(\Delta^2))\} \\ &\quad + \ (p^2 - r)n\boldsymbol{\theta}'\mathbf{W}\boldsymbol{\theta}E^{(1)}(\chi_{p+4}^{-4}(\Delta^2)) \\ &= p - (p-2)^2E^{(1)}(\chi_p^{-2}(\Delta^2)) \quad \text{if } \mathbf{W} = \mathbf{I}_p. \end{aligned}$$

(iv) $$\mathbf{M}_4(\hat{\boldsymbol{\theta}}_n^{S+})$$
$$= \mathbf{M}_3(\hat{\boldsymbol{\theta}}_n^S) - (p-2)\mathbf{I}_p\{2E^{(1)}(\chi_{p+2}^{-2}(\Delta^2)) - (p-2)E^{(1)}(\chi_{p+2}^{-4}(\Delta^2))\}$$
$$+ \ n\boldsymbol{\theta}\boldsymbol{\theta}'E^{(1)}\big[2\{1 - (p-2)(\chi_{p+2}^{-2}(\Delta^2)\}I(\chi_{p+2}^2(\Delta^2)) < p-2)$$
$$-\{1 - (p-2)(\chi_{p+4}^{-2}(\Delta^2)\}^2 I(\chi_{p+4}^2(\Delta^2)) < p-2)\big].$$

and

$$
\begin{aligned}
R_4(\hat{\boldsymbol{\theta}}_n^{S+};\mathbf{W})
&= R_3(\hat{\boldsymbol{\theta}}_n^{S};\mathbf{W})\\
&\quad -(p-2)\operatorname{tr}(\mathbf{W})\left[2E^{(1)}(\chi_{p+2}^{-2}(\Delta^2)) - (p-2)E^{(1)}(\chi_{p+2}^{-4}(\Delta^2))\right]\\
&\quad +(n\boldsymbol{\theta}'\mathbf{W}\boldsymbol{\theta})E\left[2\{1-(p-2)(\chi_{p+2}^{-2}(\Delta^2)\}I(\chi_{p+2}^{2}(\Delta^2)) < p-2)\right.\\
&\quad \left.-\{1-(p-2)(\chi_{p+4}^{-2}(\Delta^2)\}^2 I(\chi_{p+4}^{2}(\Delta^2)) < p-2)\right]\\
&= p-(p-2)^2E\left(\chi_p^{-2}(\Delta^2)\right)\\
&\quad -pE\left[\left\{1-(p-2)\chi_{p+2}^{-2}(\Delta^2)\right\}^2 I\left(\chi_{p+2}^{-2}(\Delta^2) < p-2\right)\right]\\
&\quad +\Delta^2E\left[2\{1-(p-2)(\chi_{p+2}^{-2}(\Delta^2)\}I(\chi_{p+2}^{2}(\Delta^2)) < p-2)\right.\\
&\quad \left.-\{1-(p-2)(\chi_{p+4}^{-2}(\Delta^2)\}^2 I(\chi_{p+4}^{2}(\Delta^2)) < p-2)\right] \quad \text{if } \mathbf{W} = \bar{\sigma}^2\mathbf{I}_p.
\end{aligned}
$$

6. Establish the following identities:

   (i) $E^{(1)}(\chi_{p+2j+2}^{-2}(\Delta^2))$

$$
= \sum_{r\geq 0} \frac{1}{r!} \frac{\left(\frac{\Delta^2}{\nu-2}\right)^r}{\left(1+\frac{\Delta^2}{\nu-2}\right)^{(p+2j+2r)/2}} (p+2j+2r)^{-1}.
$$

   (ii) $E^{(1)}(\chi_{p+2j+2}^{-4}(\Delta^2))$

$$
= \sum_{r\geq 0} \frac{1}{r!} \frac{\left(\frac{\Delta^2}{\nu-2}\right)^r}{(1+\frac{\Delta^2}{\nu-2})^{(p+2j+2r)/2}} (p+2j+2r)^{-1}(p+2j+2r-2)^{-1}.
$$

   (iii) $\Delta^2E^{(1)}(\chi_{p+2j+2}^{-4}(\Delta^2)) = 1-(p-2)E^{(1)}(\chi_p^{-2}(\Delta^2))$.

   (iv) $\Delta^2E^{(1)}(\chi_{p+4}^{-4}(\Delta^2)) = E^{(1)}(\chi_p^{-2}(\Delta^2)) - (p-2)E^{(1)}(\chi_{p+2}^{-4}(\Delta^2))$.

   (v) $2E^{(1)}(\chi_{p+2}^{-4}(\Delta^2)) = E^{(1)}(\chi_p^{-2}(\Delta^2)) = E^{(1)}(\chi_{p+2}^{-2}(\Delta^2))$.

7. Consider the model

$$
\mathbf{Y}_i = \boldsymbol{\theta} + \mathbf{e}_i, \quad i = 1, \ldots, n
$$

as in (4.1.1), and assume that the joint distribution $\mathbf{e}_1, \ldots, \mathbf{e}_n$ is given by multivariate t-distribution $M_t(0, \sigma^2\mathbf{I}_{np})$.

$$
f(\mathbf{e};\nu_0,\sigma^2) = \frac{\Gamma\left(\frac{np+\nu_0^2}{2}\right)}{(\pi\nu_0)^{np/2}\Gamma\left(\frac{\nu_0}{2}\right)\sigma^{np}} \left(1+\frac{\sum_{i=1}^n \mathbf{e}_i'\mathbf{e}_i}{\nu_0\sigma^2}\right)^{-(np+\nu_0)/2},
$$

$\nu_0, \sigma^2 > 0$. Let

(i) $\tilde{\boldsymbol{\theta}}_n = (\overline{Y}_1, \ldots, \overline{Y}_p)'$.

(ii) $\hat{\boldsymbol{\theta}}_n^{\text{PT}} = \tilde{\boldsymbol{\theta}}_n - \tilde{\boldsymbol{\theta}}_n I(\mathcal{L}_n < F_{p,m}(\alpha)), \quad \mathcal{L}_n = \frac{n\|\tilde{\boldsymbol{\theta}}_n\|^2}{ps_e^2}$.

(iii) $\hat{\boldsymbol{\theta}}_n^{S} = \left\{1 - \frac{(p-2)m}{(m+2)} \frac{s_e^2}{n\|\tilde{\boldsymbol{\theta}}_n\|^2}\right\} \tilde{\boldsymbol{\theta}}_n, \quad m = (n-1)p.$

(iv) $\hat{\boldsymbol{\theta}}_n^{S+} = \left\{1 - \frac{(p-2)m}{(m+2)} \frac{s_e^2}{n\|\tilde{\boldsymbol{\theta}}_n\|^2}\right\} I\left(\mathcal{L}_n > \frac{(p-2)m}{(m+2)}\right) \tilde{\boldsymbol{\theta}}_n.$

Show that the bias expressions of the estimators (i), (ii), (iii), and (iv) are

(i) $\mathbf{b}_1(\tilde{\boldsymbol{\theta}}_n) = \mathbf{0}$.

(ii) $\hat{\boldsymbol{\theta}}_n^{\text{PT}}) = -\boldsymbol{\theta} G_{p+2,m}^{(2)}(x'; \Delta^{*2})$.

(iii) $\mathbf{b}_3(\hat{\boldsymbol{\theta}}_n^{S}) = -cpE^{(2)}(\chi_{p+2}^{-2}(\Delta^{*2}))$, $c = \frac{(p-2)m}{p(m+2)}$.

(iv) $$\mathbf{b}_4(\hat{\boldsymbol{\theta}}_n^{S+}) = -\boldsymbol{\theta}\Big\{\frac{pc}{p+2} E^{(2)}(\chi_{p+2,m}^{-1}(\Delta^{*2})) + G_{p+2,m}^{(2)}(x'; \Delta^{*2}) - \frac{pc}{p+2} E^{(2)}\big[F_{p+2,m}^{-1}(\Delta^{*2}))I(F(\Delta^{*2}) \le pc)\big]\Big\},$$

where

$$G_{p+2i,m}^{(j)}(s_i'; \Delta^{*2}) = \sum_{r\ge 0} \frac{\Gamma\frac{(\nu_0+2r-2j-4)}{2}\left(\frac{\Delta^{*2}}{\nu_0-2}\right)^r I_{x_0}\left[\frac{1}{2}(p+2i)+r; \frac{1}{2}m\right]}{\Gamma(r+i)\Gamma\left(\frac{\nu_0+2j-4}{2}\right)\left(1+\frac{\Delta^{*2}}{\nu_0-2}\right)^{\frac{\nu_0+2r+2j-4}{2}}},$$

$$x_0 = \frac{pF_{p,m}(\alpha)}{m+pF_{p,m}(\alpha)}, \quad \Delta^{*2} = \frac{n\|\boldsymbol{\theta}\|^2}{\sigma_e^2}, \quad \sigma_e^2 = \frac{\nu_0\sigma^2}{\nu_0-2}$$

and

$$E^{(j)}(\chi_{p+s}^{-2}(\Delta^{*2})) = \sum_{r\ge 0} \frac{\Gamma(\frac{\nu_0}{2}+r+j-2)\left(\frac{\Delta^{*2}}{\nu_0-2}\right)^r (p+s+2r-2)^{-1}}{\Gamma(r+1)\Gamma(\frac{\nu_0}{2}+j-2)(1+\frac{\Delta^{*2}}{\nu_0-2})\frac{\nu_0}{2}+r+j-2},$$

$$E^{(j)}(\chi_{p+s}^{-2}(0)) = (p+s-2)^{-1}, \quad j = 1, 2,$$

and

$$E^{(j)}\big[F_{p+s,m}^{-1}(\Delta^{*2})I\big(F_{p+s,m}(\Delta^{*2}) < \frac{pc}{p+2}\big)\big]$$
$$= \sum_{r\ge 0} \frac{\Gamma(\frac{\nu_0}{2}+r+j-2)\left(\frac{\Delta^{*2}}{\nu_0-2}\right)^r (p+s)(p+s+2r-2)^{-1}}{\Gamma(r+1)\Gamma(\frac{\nu_0}{2}+j-2)(1+\frac{\Delta^{*2}}{\nu_0-2})^{(\nu_0/2)+r+j-2}} \times I_x\Big(\frac{1}{2}(p+s+2r-2); \frac{1}{2}(m+2)\Big),$$

where $x = \frac{pc}{m+pc}$, $\nu_0 > 2$.

(b) Show that the MSE matrices of the estimators are

(i) $\mathbf{M}_1(\tilde{\boldsymbol{\theta}}_n) = \sigma^2\mathbf{I}_p,$

(ii) $\mathbf{M}_2(\hat{\boldsymbol{\theta}}_n^{\mathrm{PT}}) = \sigma_e^2\mathbf{I}_p\{1 - G_{p+2,m}^{(1)}(x'; \Delta^{*2})\}$

$$+n\boldsymbol{\theta}\boldsymbol{\theta}'\{2G_{p+2,m}^{(1)}(x_1'; \Delta^{*2}) - G_{p+4,m}^{(1)}(x_2; \Delta^{*2})\}.$$

(iii) $\mathbf{M}_3(\hat{\boldsymbol{\theta}}_n^{S}) = \sigma_e^2\mathbf{I}_p - pc\sigma_e^2\mathbf{I}_p\Big\{(p-2)E^{(1)}(\chi_{p+2}^{-2}(\Delta^{*2}))$

$$+ 2\Delta^{*2}E^{(2)}(\chi_{p+4}^{-4}(\Delta^{*2}))\Big\}$$

$$+ npc(p+2)\boldsymbol{\theta}\boldsymbol{\theta}'E^{(2)}(\chi_{p+4}^{-4}(\Delta^{*2})).$$

8. (a) Find the expressions for the (i) bias vector, (ii) MSE matrix, and (iii) the risk of the estimator $\hat{\boldsymbol{\theta}}_n^{S}(c, k)$ given by (4.7.5).

   (b) Show that $c_1^*(k, \Delta^2)$ minimizes the risk of $\hat{\boldsymbol{\theta}}_n^{S}(c, k)$.

9. Prove Theorem 4.7.1.

10. Prove Theorem 4.8.3.

11. Prove Theorem 4.8.4.

12. Verify the expressions of ADB, ADQB, ADMSE, and ADQR of the R-estimators of $\boldsymbol{\theta}$ of Theorem 4.9.1.

Chapter 5

ANOVA Model

Outline

An important model belonging to the class of general linear hypotheses is the analysis of variance (ANOVA) model. In this model, p samples of differing sample sizes $n_1, \ldots, n_p$ from p normal distributions $N(\theta_1, \sigma^2), \ldots, N(\theta_p, \sigma^2)$, respectively, are collected, where $\boldsymbol{\theta} = (\theta_1, \ldots, \theta_p)'$ is the unknown treatment mean vector and σ^2 is the common unknown variance. It is usual to test the null hypothesis $H_0 : \theta_1 = \cdots = \theta_p = \theta_0$ (unknown), that is, the treatments are equal to an unknown mean θ_0 against the alternative hypothesis, H_A, that at least one pair of treatment means differ. Here $n = n_1 + \cdots + n_p$ is the total size of the samples collected from the p populations. The main object of this chapter is to study the properties of various improved estimators of $\boldsymbol{\theta} = (\theta_1, \ldots, \theta_p)'$. Accordingly, we consider the **unrestricted, restricted, preliminary test,** and **Stein-type estimators of $\boldsymbol{\theta}$.** We also present the Bayes and the empirical estimators of $\boldsymbol{\theta}$. We compare various estimators using the MSE matrix and the quadratic risk criterion. We also present the confidence set analysis based on the estimators as centers of the set with fixed volume. Finally, we discuss the asymptotic properties of the estimators when errors follows some nonnormal distributions, as well as the point and confidence set R-estimation of the treatment means to conclude the chapter.

5.1 Model, Estimation, and Tests

5.1.1 ANOVA model

Consider the ANOVA model

$$\mathbf{Y} = \mathbf{B}\boldsymbol{\theta} + \boldsymbol{\varepsilon}, \tag{5.1.1}$$

where

$$\begin{aligned}
\mathbf{Y} &= (y_{11}, \ldots, y_{1n_1}, \ldots, y_{p1}, \ldots, y_{pn_p})', \\
\boldsymbol{\theta} &= (\theta_1, \ldots, \theta_p)', \\
\boldsymbol{\varepsilon} &= (\varepsilon_{11}, \ldots, \varepsilon_{1n_1}, \ldots, \varepsilon_{p1}, \ldots, \varepsilon_{pn_p})' \\
\mathbf{B} &= \text{Block diagonal vectors} = \text{Diag}(\mathbf{1}_{n_1}, \ldots, \mathbf{1}_{n_p}), \\
\mathbf{1}_{n_i} &= (1, 1, \ldots, 1)', \text{ an } n_i\text{-tuple of 1's}, \\
n &= n_1 + n_2 + \cdots + n_p.
\end{aligned} \tag{5.1.2}$$

The distribution of the vector $\boldsymbol{\varepsilon}$ is $\mathcal{N}_n(\mathbf{0}, \sigma^2\mathbf{I}_n)$, where $\mathbf{I}_n$ is the identity matrix of order n and σ^2 is the variance of the errors.

Generally, the main object of ANOVA problems is the test of the null hypothesis $H_0 : \boldsymbol{\theta} = \theta_0\mathbf{1}_p$ against the alternative $H_A : \boldsymbol{\theta} \neq \theta_0\mathbf{1}_p$. But, in this chapter, our objective is to improve the usual estimators of the treatment means. In the next section, we consider the usual estimation and test of hypotheses in the ANOVA model, with an example and calculations of various related statistics together with the proof of likelihood statistics for the test of the hypothesis $H_0 : \boldsymbol{\theta} = \theta_0\mathbf{1}_p$ as the basis of developing the improved estimators.

5.1.2 Estimation of the Parameters of the One-Way ANOVA Model

Consider the ANOVA model (5.1.1) and (5.1.2). From the sample information and the model, the unrestricted estimator (UE) of $\boldsymbol{\theta}$ based on least squares (LS) or maximum likelihood (ML) method is given by

$$\begin{aligned}
\tilde{\boldsymbol{\theta}}_n &= (\mathbf{B}'\mathbf{B})^{-1}\mathbf{B}'\mathbf{Y} \\
&= \mathbf{N}^{-1}\mathbf{B}\mathbf{Y} = (\overline{Y}_1, \ldots, \overline{Y}_p)' = (\tilde{\theta}_1, \ldots, \tilde{\theta}_p)',
\end{aligned} \tag{5.1.3}$$

where $\mathbf{N} = \mathbf{B}'\mathbf{B} = \text{Diag}(n_1, \ldots, n_p)$. The corresponding *unrestricted unbiased estimator* (UUE) of σ^2 can be written as

$$s_e^2 = \frac{1}{m}(\mathbf{Y} - \mathbf{B}\tilde{\boldsymbol{\theta}}_n)'(\mathbf{Y} - \mathbf{B}\tilde{\boldsymbol{\theta}}_n), \quad m = n - p. \tag{5.1.4}$$

It is clear that $\tilde{\boldsymbol{\theta}}_n$ is distributed as $N_p(\boldsymbol{\theta}, \sigma^2\mathbf{N}^{-1})$ independent of the distribution of ms_e^2/σ^2, which has a central chi-square distribution with m degrees of freedom (d.f.).

Suppose that in addition to the sample information and the model, we have some additional information which consists of the null hypothesis $H_0 : \boldsymbol{\theta} = \theta_0\mathbf{1}_p$. If this hypothesis is true, then we estimate the common value of $\theta_0\mathbf{1}_p$ by

$$\hat{\boldsymbol{\theta}}_n = \frac{1}{n}\mathbf{1}_p\mathbf{1}_p'\mathbf{N}\tilde{\boldsymbol{\theta}}_n. \tag{5.1.5}$$

Thus, $\hat{\boldsymbol{\theta}}_n$ is the *restricted* estimator (RE) of $\boldsymbol{\theta}$ under the null hypothesis $H_0 : \boldsymbol{\theta} = \theta_0\mathbf{1}_p$.

Table 5.1.1 One-Way ANOVA Data

	1st Sample	Mean		pth Sample	Mean	
Data	y_{11}	θ_1	$\cdots$	y_{p1}	θ_p	
	y_{12}	θ_1	$\cdots$			
	y_{1n_1}	θ_1	$\cdots$	y_{pn_p}	θ_p	
Sample means	$\overline{y}_1 = \tilde{\theta}_1$	θ_1	$\cdots$	$\overline{y}_p = \tilde{\theta}_p$	θ_p	
Totals	$T_1 = n_1\overline{y}_1$	$n_1\theta_1$		$T_p = n_p\overline{y}_p$	$n_p\theta_p$	$\overline{\overline{y}} = \hat{\theta}_0$

The grand mean of the data set is $\overline{\overline{y}} = \hat{\theta}_{0n}$, given by

$$\hat{\theta}_{0n} = \frac{1}{n}(T_1 + \cdots + T_p). \tag{5.1.6}$$

5.1.3 Test of Equality of the Treatment Means

In this section, we consider the test of the null hypothesis in the ANOVA model, namely

$$H_0 : \boldsymbol{\theta} = \theta_0\mathbf{1}_p,$$

against the alternative

$$H_A : \boldsymbol{\theta} \neq \theta_0\mathbf{1}_p. \tag{5.1.7}$$

The test is given by the following theorem:

Theorem 1. The likelihood ratio (LR) statistic is given by

$$\mathcal{L}_n = \frac{\tilde{\boldsymbol{\theta}}_n'\mathbf{H}'\mathbf{N}\mathbf{H}\tilde{\boldsymbol{\theta}}_n}{(p-1)s_e^2}, \quad \mathbf{H} = \mathbf{I}_p - \frac{1}{n}\mathbf{1}_p\mathbf{1}_p'\mathbf{N}. \tag{5.1.8}$$

Under H_0, $\mathcal{L}_n$ has the central F-distribution with $(p-1, m)$ d.f., and under H_A it has the noncentral F-distribution with $(p-1, m)$ d.f. and noncentrality parameter $\Delta^2/2$, where

$$\Delta^2 = \frac{\boldsymbol{\theta}'\mathbf{H}'\mathbf{N}\mathbf{H}\boldsymbol{\theta}}{\sigma^2}. \tag{5.1.9}$$

Proof. The likelihood ratio (LR) statistic is obtained using the **maximum** of the likelihood functions under H_0 and H_A given by

$$L_0 = s_0^{-n}(\sqrt{2\pi})^{-n}\exp\left\{-\frac{n}{2}\right\}, \quad ns_0^2 = (\mathbf{Y} - \mathbf{B}\hat{\boldsymbol{\theta}}_n)'(\mathbf{Y} - \mathbf{B}\hat{\boldsymbol{\theta}}_n),$$

and

$$L_A = s_A^{-n}(\sqrt{2\pi})^{-n}\exp\left\{-\frac{n}{2}\right\}, \quad ns_A^2 = (\mathbf{Y} - \mathbf{B}\tilde{\boldsymbol{\theta}}_n)'(\mathbf{Y} - \mathbf{B}\tilde{\boldsymbol{\theta}}_n) = ms_e^2,$$

respectively. Hence, the LR statistic is the result of

$$\lambda^{2/n} = \left(\frac{L_0}{L_A}\right)^{2/n} = \frac{(\mathbf{Y} - \mathbf{B}\tilde{\boldsymbol{\theta}}_n)'(\mathbf{Y} - \mathbf{B}\tilde{\boldsymbol{\theta}}_n)}{(\mathbf{Y} - \mathbf{B}\hat{\boldsymbol{\theta}}_n)'(\mathbf{Y} - \mathbf{B}\hat{\boldsymbol{\theta}}_n)} = \frac{1}{1 + \frac{p-1}{m}\mathcal{L}_n}, \tag{5.1.10}$$

since $(\mathbf{Y} - \mathbf{B}\hat{\boldsymbol{\theta}}_n)'(\mathbf{Y} - \mathbf{B}\hat{\boldsymbol{\theta}}_n) = (\mathbf{Y} - \mathbf{B}\tilde{\boldsymbol{\theta}}_n)'(\mathbf{Y} - \mathbf{B}\tilde{\boldsymbol{\theta}}_n) + \tilde{\boldsymbol{\theta}}_n'\mathbf{H}'\mathbf{N}\mathbf{H}\tilde{\boldsymbol{\theta}}_n$. Thus we take $\mathcal{L}_n = \tilde{\boldsymbol{\theta}}_n'\mathbf{H}'\mathbf{N}\mathbf{H}\tilde{\boldsymbol{\theta}}_n/(p-1)s_e^2$ as the LR statistic for testing H_0 against H_A.

To obtain the distribution $\mathcal{L}_n$ under H_A (H_0), we consider an orthogonal matrix $\boldsymbol{\Gamma} = (\boldsymbol{\Gamma}_1, \boldsymbol{\Gamma}_2)$, where $\boldsymbol{\Gamma}_1$ is a $p \times (p-1)$ matrix and $\boldsymbol{\Gamma}_2$ is a p-vector such that $\boldsymbol{\Gamma}_2'\boldsymbol{\Gamma}_1 = \mathbf{0}$ and $\boldsymbol{\Gamma}_1\boldsymbol{\Gamma}_1' + \boldsymbol{\Gamma}_2\boldsymbol{\Gamma}_2' = \mathbf{I}_p$. Also, it diagonalizes the symmetric idempotent matrix $\mathbf{N}^{-1/2}\mathbf{H}'\mathbf{N}\mathbf{H}\mathbf{N}^{-1/2}$ of rank $(p-1)$, meaning $\boldsymbol{\Gamma}'\mathbf{N}^{-1/2}\mathbf{H}'\mathbf{N}\mathbf{H}\mathbf{N}^{-1/2}\boldsymbol{\Gamma}' = \begin{pmatrix} \mathbf{I}_{p-1} & 0 \\ 0 & 0 \end{pmatrix}$. Then, defining $\boldsymbol{\omega} = \sigma^{-1}\boldsymbol{\Gamma}N^{1/2}\tilde{\boldsymbol{\theta}}_n$ and $\boldsymbol{\eta} = \sigma^{-1}\boldsymbol{\Gamma}\mathbf{N}^{-1/2}\boldsymbol{\theta}$, we have $\boldsymbol{\omega} \sim N_p(\boldsymbol{\eta}, \mathbf{I}_p)$. Further, we partition $\boldsymbol{\omega} = (\omega_1', \omega_2')'$ and $\boldsymbol{\eta} = (\eta_1', \eta_2')'$, where $\boldsymbol{\omega}_1$ is a subvector of order $(p-1)$ and ω_2 is a scalar random variables such that

$$\begin{pmatrix} \omega_1 \\ \omega_2 \end{pmatrix} \sim \mathcal{N}_p\left\{\begin{pmatrix} \eta_1 \\ \eta_2 \end{pmatrix}; \begin{pmatrix} \mathbf{I}_{p-1} & 0 \\ 0 & 1 \end{pmatrix}\right\}.$$

Hence, $\|\omega_1\|^2$ follows the noncentral chi-square distribution with $(p-1)$ d.f. and noncentrality parameter $\Delta^2/2$, where

$$\Delta^2 = \frac{1}{\sigma^2}[\boldsymbol{\theta}'\mathbf{H}'\mathbf{N}\mathbf{H}\boldsymbol{\theta}] = \|\eta_1\|^2. \tag{5.1.11}$$

Then, under H_A, $\mathcal{L}_n$ has the noncentral F-distribution with $[(p-1), m]$ d.f. and noncentrality parameter $\Delta^2/2$, since ms_e^2/σ^2 is independently distributed as a central chi-square variable with m d.f.

The derivation of the LR test is given only to see the required transformations that come in the picture, which will be used later in subsequent discussions.

Some of the distributional results are given in the following theorem that are easy to prove.

Theorem 2. Under the assumed conditions,

(i) $\mathbf{X}_{(n)} = (\tilde{\boldsymbol{\theta}}_n - \boldsymbol{\theta}) \sim \mathcal{N}_p(\mathbf{0}; \sigma^2\mathbf{N}^{-1})$

(ii) $\mathbf{Y}_{(n)} = (\tilde{\boldsymbol{\theta}}_n - \hat{\boldsymbol{\theta}}_n) \sim \mathcal{N}_p(\mathbf{H}\boldsymbol{\theta}; \sigma^2\mathbf{H}\mathbf{N}^{-1})$;

(iii) $\mathbf{Z}_{(n)} = (\hat{\boldsymbol{\theta}}_n - \theta_0\mathbf{1}_p) \sim \mathcal{N}_p\left((\bar{\theta} - \theta_0)\mathbf{1}_p; \frac{\sigma^2}{n}\mathbf{B}\right)$, $\mathbf{B} = \mathbf{1}_p\mathbf{1}_p'$, and $\bar{\theta}\mathbf{1}_p = \frac{1}{n}\mathbf{1}_p\mathbf{1}_p'\mathbf{N}\boldsymbol{\theta}$;

(iva) $$\begin{bmatrix} \mathbf{X}_{(n)} \\ \mathbf{Y}_{(n)} \end{bmatrix} \sim \mathcal{N}_{2p}\left\{\begin{pmatrix} \mathbf{0} \\ \mathbf{H}\boldsymbol{\theta} \end{pmatrix} \sigma^2 \begin{pmatrix} \mathbf{N}^{-1} & \mathbf{H}\mathbf{N}^{-1} \\ \mathbf{N}^{-1}\mathbf{H}' & \mathbf{H}\mathbf{N}^{-1} \end{pmatrix}\right\};$$

(ivb) $$\begin{bmatrix} \mathbf{Z}_{(n)} \\ \mathbf{Y}_{(n)} \end{bmatrix} \sim \mathcal{N}_{2p}\left\{\begin{pmatrix} (\bar{\theta} - \theta_0)\mathbf{1}_p \\ \mathbf{H}\boldsymbol{\theta} \end{pmatrix}; \sigma^2 \begin{pmatrix} \frac{1}{n}\mathbf{B} & \mathbf{0} \\ \mathbf{0} & \mathbf{H}\mathbf{N}^{-1} \end{pmatrix}\right\}. \quad (5.1.12)$$

We now present an example to illustrate the test of the null hypothesis of equality of means in several samples. (Here it is four samples.)

Example 1. Suppose that 48 discs were allotted to four laboratories, each with 12 discs, to measure tin-coating thickness (in mm) in order to find uniformity in tin-coating in the overall production. Accordingly, the data in Table 5.1.2 are obtained.

Table 5.1.2 Amount of Tin-Coating (mm)

Labs	< - - - - Samples - - - - >	Total
1	0.25, 0.27, 0.22, 0.30, 0.27, 0.28, 0.32, 0.24, 0.31, 0.26, 0.21, 0.28	3.21
2	0.18, 0.28, 0.21, 0.23, 0.25, 0.20, 0.27, 0.19, 0.24, 0.22, 0.29, 0.16	2.72
3	0.19, 0.25, 0.27, 0.24, 0.18, 0.26, 0.28, 0.24, 0.25, 0.20, 0.21, 0.19	2.76
4	0.23, 0.30, 0.28, 0.28, 0.24, 0.34, 0.20, 0.18, 0.24, 0.28, 0.22, 0.21	3.00

Note that the estimate of common mean θ_0 is $\hat{\theta}_0 = 0.242$. The computed value of $\mathcal{L}_n$ is $2.87 > 2.82$, the 5% critical value with (3.44) d.f. Hence, the hypothesis of equality of the means is not accepted.

5.2 Preliminary Test Approach and Stein-Type Estimators

In this section, we discuss the preliminary test (or quasi-empirical Bayes) approach to shrinkage estimation in addition to the Bayes and empirical Bayes approach.

5.2.1 Preliminary Test Approach (or Quasi-empirical Bayes Approach)

Following Chapter 3, we write the preliminary test estimator (PTE) of $\boldsymbol{\theta}$ as

$$\hat{\boldsymbol{\theta}}_n^{PT} = \tilde{\boldsymbol{\theta}}_n - (\tilde{\boldsymbol{\theta}}_n - \hat{\boldsymbol{\theta}}_n) I(\mathcal{L}_n < F_{p-1,m}(\alpha)), \tag{5.2.1}$$

where $F_{p-1,m}(\alpha)$ is the upper α-level critical value of the F-distribution with $(p-1, m)$ d.f. PTE of $\boldsymbol{\theta}$ depends on the level of significance and the choice of estimators remains between the two values $\tilde{\boldsymbol{\theta}}_n$ and $\hat{\boldsymbol{\theta}}_n$. To overcome this difficulty we define the James-Stein-type estimator

$$\begin{aligned} \hat{\boldsymbol{\theta}}_n^{S} &= \hat{\boldsymbol{\theta}}_n + \{1 - c\mathcal{L}_n^{-1}\}(\tilde{\boldsymbol{\theta}}_n - \hat{\boldsymbol{\theta}}_n) \\ &= \tilde{\boldsymbol{\theta}}_n - c\mathcal{L}_n^{-1}(\tilde{\boldsymbol{\theta}}_n - \hat{\boldsymbol{\theta}}_n), \end{aligned} \tag{5.2.2}$$

where $c = \frac{(p-3)m}{(p-1)(m+2)}$. Notice that we have replaced $I(\mathcal{L}_n < F_{p-1,m}(\alpha))$ by $c\mathcal{L}_n^{-1}$ to obtain this estimator. The constant c is defined based on the degrees of freedom $[(p-1), m]$. The quantity $(p-3)$ is associated with the expected value of a reciprocal chi-square variable with $(p-1)$ d.f.

The estimator, $\hat{\boldsymbol{\theta}}_n^{S}$ may take values past $\hat{\boldsymbol{\theta}}_n$. Thus, we consider the positive-rule version (PRSE) of $\hat{\boldsymbol{\theta}}_n^{S}$ given by

$$\hat{\boldsymbol{\theta}}_n^{S+} = \hat{\boldsymbol{\theta}}_n + \left(1 - c\mathcal{L}_n^{-1}\right) I(\mathcal{L}_n > c)(\tilde{\boldsymbol{\theta}}_n - \hat{\boldsymbol{\theta}}_n), \tag{5.2.3}$$

which is a PTE based on $\hat{\boldsymbol{\theta}}_n^{S}$ and $\hat{\boldsymbol{\theta}}_n$ with critical value c.

The estimators above were discussed by Ali and Saleh (1991c) and Khan, Bashir (1997).

Now, using Theorem 2, we can write

(i) $(\hat{\boldsymbol{\theta}}_n^{PT} - \boldsymbol{\theta}) = \mathbf{X}_{(n)} - \mathbf{Y}_{(n)} I\left(\frac{||\mathbf{Y}_{(n)}||^2_{\mathbf{N}}}{(p-1)s_e^2} < F_{p-1,m}(\alpha)\right)$,

(ii) $(\hat{\boldsymbol{\theta}}_n^{S} - \boldsymbol{\theta}) = \mathbf{X}_{(n)} - c(p-1)s_e^2 \mathbf{Y}_{(n)} ||_{\mathbf{N}} \mathbf{Y}_{(n)}||^{-2}_{\mathbf{N}}$,

(iii) $(\hat{\boldsymbol{\theta}}_n^{S+} - \boldsymbol{\theta}) = \mathbf{X}_{(n)} - c(p-1)s_e^2 \mathbf{Y}_{(n)} ||\mathbf{Y}_{(n)}||^{-2}_{\mathbf{N}}$

$$-\mathbf{Y}_{(n)}(1 - c(p-1)s_e^2 ||\mathbf{Y}_{(n)}||^{-2}_{\mathbf{N}}) I\left(\frac{||\mathbf{Y}_{(n)}||^2_{\mathbf{N}}}{(p-1)s_e^2} \le c\right). \tag{5.2.4}$$

These expressions can be used to compute the expressions for bias, MSE matrices, and the quadratic risks of the different estimators.

5.2.2 Bayes and Empirical Bayes Estimators of Treatment Means

To obtain a Bayes estimator of $\boldsymbol{\theta}$, we consider the distribution of $\mathbf{Y} \sim N_n(\mathbf{B}\boldsymbol{\theta}, \sigma^2\mathbf{I}_n)$ together with the prior distribution of $\boldsymbol{\theta}$ as $N_p(\theta_0\mathbf{1}_p, \tau^2\mathbf{N}^{-1})$, which includes the null hypothesis $H_0 : \boldsymbol{\theta} = \theta_0\mathbf{1}_n$ (θ_0 unknown) into the prior of $\boldsymbol{\theta}$. Then, by Theorem 3 of Section 2.3, the posterior distribution of $\boldsymbol{\theta}$ given $\tilde{\boldsymbol{\theta}}_n$ is $N_p\{\theta_0\mathbf{1}_p + (1-d)(\tilde{\boldsymbol{\theta}}_n - \theta_0\mathbf{1}_p), \tau^2(1-d)\mathbf{N}^{-1}\}$, where $\mathbf{N} = \mathbf{B}'\mathbf{B}$ and $d = \frac{\sigma^2}{\sigma^2+\tau^2}$. Thus, under the quadratic loss $L(\boldsymbol{\theta}^*, \boldsymbol{\theta}) = \frac{n}{\sigma^2}||\boldsymbol{\theta}^* - \boldsymbol{\theta}||^2$, the Bayes estimator of $\boldsymbol{\theta}$ is given by

$$\hat{\boldsymbol{\theta}}_n^B = \theta_0\mathbf{1}_p + (1-d)(\tilde{\boldsymbol{\theta}}_n - \theta_0\mathbf{1}_p) \tag{5.2.5}$$

with covariance matrix

$$\tau^2(1-d)\mathbf{N}^{-1}. \tag{5.2.6}$$

The marginal distribution of $\tilde{\boldsymbol{\theta}}_n$ is $N_p\big(\theta_0\mathbf{1}_p, \mathbf{N}^{-1}(\sigma^2+\tau^2)^{-1}\big)$. Accordingly, minimizing $(\tilde{\boldsymbol{\theta}}_n - \theta_0\mathbf{1}_p)'\mathbf{N}(\tilde{\boldsymbol{\theta}}_n - \theta_0\mathbf{1}_p)$ w.r.t. θ_0, we obtain the estimate of θ_0 as

$$\hat{\theta}_{0n} = (\mathbf{1}_p'\mathbf{N}\mathbf{1}_p)^{-1}\mathbf{1}_p'\mathbf{N}'\tilde{\boldsymbol{\theta}}_n. \tag{5.2.7}$$

Let $\hat{\boldsymbol{\theta}}_n = \hat{\theta}_{0n}\mathbf{1}_p$. Clearly, $E^m(\hat{\boldsymbol{\theta}}_n) = \theta_0\mathbf{1}_p$, where E^m stands for the expectation based on *marginal* distribution of $\tilde{\boldsymbol{\theta}}_n$. Furthermore, $(\tilde{\boldsymbol{\theta}}_n - \hat{\boldsymbol{\theta}}_n) = \mathbf{H}\tilde{\boldsymbol{\theta}}_n$ is marginally $N_p(\mathbf{0}, (\sigma^2+\tau^2)^{-1}\mathbf{H}\mathbf{N}^{-1}\mathbf{H}')$ and

$$\tilde{\boldsymbol{\theta}}_n'\mathbf{H}'\mathbf{N}\mathbf{H}\tilde{\boldsymbol{\theta}}_n \overset{\mathcal{D}}{=} (\sigma^2+\tau^2)\chi^2_{p-1}. \tag{5.2.8}$$

Therefore,

$$E^m\left[\frac{1}{\tilde{\boldsymbol{\theta}}_n'\mathbf{H}'\mathbf{N}\mathbf{N}\hat{\boldsymbol{\theta}}_n}\right] = \frac{1}{(p-3)(\sigma^2+\tau^2)}, \tag{5.2.9}$$

and the unbiased estimator of $(\sigma^2+\tau^2)^{-1}$ is given by

$$\frac{(p-3)}{(\tilde{\boldsymbol{\theta}}_n'\mathbf{H}'\mathbf{N}\mathbf{H}\tilde{\boldsymbol{\theta}}_n)}. \tag{5.2.10}$$

Notice that no prior information is available on the variance σ^2. To obtain a Bayes estimate of σ^2, we have to use the so-called *noninformative* prior $\pi(\sigma^2) = 1/\sigma^2$ and estimate σ^2 from the marginal distribution of s_e^2, namely, $(n-p)s_e^2 \simeq \sigma^2\chi^2_{n-p}$. This leads to s_e^2 as the Bayes estimate of σ^2. But $(ms_e^2)/m+2$ is a scale-invariant better estimator of σ^2. Thus, the estimator of d is $\hat{d} = \frac{(p-3)m\mathcal{L}_n^{-1}}{(p-1)(m+2)}$. Accordingly, substituting the estimators of the extra parameters, we have the following *empirical Bayes estimator* (EBE) of $\boldsymbol{\theta}$:

$$\hat{\boldsymbol{\theta}}_n^{EB} = \hat{\boldsymbol{\theta}}_n + \left(1 - \frac{m(p-3)s_e^2}{(m+2)\tilde{\boldsymbol{\theta}}_n'\mathbf{H}'\mathbf{N}\mathbf{H}\tilde{\boldsymbol{\theta}}_n}\right)(\tilde{\boldsymbol{\theta}}_n - \hat{\boldsymbol{\theta}}_n). \tag{5.2.11}$$

This is the estimated version of the Bayes estimator (5.2.4(ii)), with weights depending logically on the test-statistics $\mathcal{L}_n$ defined by (5.1.8). We proposed the same estimator, $\hat{\boldsymbol{\theta}}_n^S$, using the preliminary test approach.

Now, if $\mathcal{L}_n \to 0$, $\mathcal{L}_n^{-1} \to \infty$, this creates a problem in the definition (5.2.11) of the empirical Bayes estimation. To keep the estimator of d bounded below 1, we restrict $\mathcal{L}_n$ to be greater than c as given in Chapter 4. Thus, we obtain the positive-rule shrinkage estimator (PRSE) of $\boldsymbol{\theta}$ defined by $\hat{\boldsymbol{\theta}}_n^{S+}$ in (5.2.3) and repeated again here:

$$\hat{\boldsymbol{\theta}}_n^{S+} = \hat{\boldsymbol{\theta}}_n + \left(1 - c\mathcal{L}_n^{-1}\right) I(\mathcal{L}_n > c)(\tilde{\boldsymbol{\theta}}_n - \hat{\boldsymbol{\theta}}_n). \tag{5.2.12}$$

We proceed to write the quasi-empirical Bayes estimators as

$$\boldsymbol{\theta}_n^* = \hat{\boldsymbol{\theta}}_n + (\tilde{\boldsymbol{\theta}}_n - \hat{\boldsymbol{\theta}}_n)\phi(\mathcal{L}_n) \tag{5.2.13}$$

where, by choosing

$$\begin{aligned}
\phi(\mathcal{L}_n) &= 1,\\
&= 0,\\
&= I(\mathcal{L}_n > F_{p,m}(\alpha)),\\
&= 1 - c\mathcal{L}_n^{-1},\\
\text{or} \quad &= (1 - c\mathcal{L}_n^{-1})I(\mathcal{L}_n > c), \quad c = \frac{(p-3)m}{(m+2)(p-1)},
\end{aligned}$$

we obtain the estimators $\tilde{\boldsymbol{\theta}}_n$, $\hat{\boldsymbol{\theta}}_n$ $\hat{\boldsymbol{\theta}}_n^{PT}$, $\hat{\boldsymbol{\theta}}_n^S$, and $\hat{\boldsymbol{\theta}}_n^{S+}$, respectively. The following example illustrates the computation of the various estimators:

Example 1. Computation of the preliminary test and shrinkage estimators. Consider the data in Table 5.1.2. For these we have $\mathcal{L}_n = 2.87$ and $c = \frac{1}{3}\left(\frac{44}{46}\right) = 0.319$. Let $\alpha = 0.15$. Then the preliminary test based on $\mathcal{L}_n$ **rejects** H_0, and $\hat{\boldsymbol{\theta}}_n^{PT} = (0.268, 0.227, 0.230, 0.250)' = \tilde{\boldsymbol{\theta}}_n$. To compute $\hat{\boldsymbol{\theta}}_n^S$ and $\hat{\boldsymbol{\theta}}_n^{S+}$, we have

$$\begin{aligned}
\hat{\theta}_1^S &= \tilde{\theta}_1 - c(\tilde{\theta}_1 - \hat{\theta}_0)\mathcal{L}_n^{-1} = 0.268 - (0.319)(0.268 - 0.242) = 0.210,\\
\hat{\theta}_2^S &= 0.227 - (0.319)(0.227 - 0.242) = 0.232,\\
\hat{\theta}_3^S &= 0.230 - (0.319)(0.230 - 0.242) = 0.234,\\
\hat{\theta}_4^S &= 0.250 - (0.319)(0.250 - 0.242) = 0.252.
\end{aligned}$$

Hence, $\hat{\boldsymbol{\theta}}_{48}^S = (0.210, 0.232, 0.234, 0.252)$.

Now, $\mathcal{L}_n = 2.87 > 0.319$. Therefore, $\hat{\boldsymbol{\theta}}_{48}^S = \hat{\boldsymbol{\theta}}_{48}^{S+}$. We note that the PTE of $\boldsymbol{\theta}$ is $\hat{\boldsymbol{\theta}}_{48}^{PT} = (0.268, 0.227, 0.230, 0.250)$, while $\hat{\boldsymbol{\theta}}_{48}^{S+} = (0.210, 0.232, 0.234, 0.252)$ in this case.

5.3 Bias, Quadratic Bias, MSE, and Risk Expressions

In this section, we obtain the bias, quadratic bias (QB), MSE matrix, and risk expressions for the unrestricted, restricted, preliminary test, James-Stein and the positive-rule Stein estimators.

5.3.1 Bias Expressions

First, we note that

$$\mathcal{L}_n = \frac{\tilde{\boldsymbol{\theta}}'_n \mathbf{H}'\mathbf{N}\mathbf{H}\tilde{\boldsymbol{\theta}}_n}{(p-1)s_e^2} = \frac{m\mathbf{Z}_1'\mathbf{Z}_1}{(p-1)\chi_m^2} \stackrel{\mathcal{D}}{=} F_{p-1,m}(\Delta^2), \quad \Delta^2 = \frac{\boldsymbol{\theta}'\mathbf{H}'\mathbf{N}\mathbf{H}\boldsymbol{\theta}}{\sigma^2}.$$

Therefore, we obtain the bias and the quadratic bias expressions as given in Theorem 1 below:

Theorem 1.

(i) $\mathbf{b}_1(\tilde{\boldsymbol{\theta}}_n) = \mathbf{0}$ and $B_1(\tilde{\boldsymbol{\theta}}_n) = 0,$ (5.3.1a)

(ii) $\mathbf{b}_1(\hat{\boldsymbol{\theta}}_n) = -\mathbf{H}\boldsymbol{\theta} = -\boldsymbol{\delta}$ and $B_2(\hat{\boldsymbol{\theta}}_n) = (\boldsymbol{\delta}'\mathbf{N}\boldsymbol{\delta})\sigma^{-2} = \Delta^2,$ (5.3.1b)

(iii) $\mathbf{b}_3(\hat{\boldsymbol{\theta}}_n^{PT}) = -\mathbf{H}\boldsymbol{\theta} G_{p+1,m}(\ell_\alpha;\Delta^2) = -\boldsymbol{\delta} G_{p+1,m}(\ell_\alpha;\Delta^2),$

$\ell_\alpha = \frac{p-1}{p+1}F_{p-1,m}(\alpha)$ and $B_3(\hat{\boldsymbol{\theta}}_n^{PT}) = \Delta^2\{G_{p+1,m}(\ell_\alpha;\Delta^2)\}^2.$ (5.3.1c)

(iv) $\mathbf{b}_4(\hat{\boldsymbol{\theta}}_n^{S}) = -c(p-1)\boldsymbol{\delta} E[\chi_{p+1}^{-2}(\Delta^2)]$

and $B_4(\hat{\boldsymbol{\theta}}_n^{S}) = c^2(p-1)^2\Delta^2\Big\{E\Big[\chi_{p+1}^{-2}(\Delta^2)\Big]\Big\}^2.$ (5.3.1d)

(v) $$\mathbf{b}_5(\hat{\boldsymbol{\theta}}_n^{S+}) = -\boldsymbol{\delta}\Big\{G_{p+1,m}\Big(c\tfrac{(p-1)}{(p+1)};\Delta^2\Big) + c\tfrac{(p-1)}{(p+1)}E\Big[F_{p+1,m}^{-1}(\Delta^2)\Big] - c\tfrac{(p-1)}{(p+1)}E\Big[F_{p+1,m}^{-1}(\Delta^2)I\Big(F_{p+1,m}(\Delta^2) < c\tfrac{(p-1)}{(p+1)}\Big)\Big]\Big\}. \quad (5.3.1e)$$

(vi) $$B_5(\hat{\boldsymbol{\theta}}_n^{S+}) = \Delta^2\Big\{G_{p+1,m}\Big(\tfrac{c(p-1)}{(p+1)};\Delta^2\Big) + \tfrac{c(p-1)}{(p+1)}E\Big[F_{p+1,m}^{-1}(\Delta^2)\Big] - \tfrac{c(p-1)}{(p+1)}E\Big[F_{p+1,m}^{-1}(\Delta^2)I\Big(F_{p+1,m}(\Delta^2) < \tfrac{c(p-1)}{(p+1)}\Big)\Big]\Big\}^2. \quad (5.3.1f)$$

Here $E[F_{p+1,m}^{-1}(\Delta^2]$ is the expectation of the reciprocal of the noncentral F-variable with $(p+1,m)$ d.f. and noncentrality parameter $\Delta^2/2$. Similarly,

$$E\left[F_{p+1,m}^{-1}(\Delta^2)I\left(F_{p+1,m}(\Delta^2) < c\frac{(p-1)}{(p+1)}\right)\right] \quad (5.3.2)$$

is the truncated expectation of the reciprocal of a noncentral F-variable with $(p+1, m)$ d.f. and noncentrality parameter $\Delta^2/2$. Further, $G_{\nu_1,\nu_2}(\cdot;\Delta^2)$ is the cdf of a noncentral F-distribution with (ν_1, ν_2) d.f. and noncentrality parameter $\Delta^2/2$.

Proof.

(i) The proof is obvious.

(ii) The bias vector of $\hat{\boldsymbol{\theta}}_n$ may be calculated as

$$\begin{aligned}\mathbf{b}_2(\hat{\boldsymbol{\theta}}_n) = E(\hat{\boldsymbol{\theta}}_n - \boldsymbol{\theta}) &= \frac{1}{n}\mathbf{1}_p\mathbf{1}_p'\mathbf{N}\boldsymbol{\theta} - \boldsymbol{\theta}\\ &= -(\mathbf{I}_p - \frac{1}{n}\mathbf{1}_p\mathbf{1}_p'\mathbf{N})\boldsymbol{\theta} = -\mathbf{H}\boldsymbol{\theta} = -\boldsymbol{\delta}.\end{aligned}$$

Hence,

$$B_2(\hat{\boldsymbol{\theta}}_n) = (\boldsymbol{\delta}'\mathbf{N}\boldsymbol{\delta})\sigma^{-2} = \Delta^2.$$

(iii) For $\mathbf{b}_3(\hat{\boldsymbol{\theta}}_n^{PT})$, we have

$$\begin{aligned}E(\hat{\boldsymbol{\theta}}_n^{PT} - \boldsymbol{\theta}) &= -E\{(\tilde{\boldsymbol{\theta}}_n - \hat{\boldsymbol{\theta}}_n)I(\mathcal{L}_n < F_{p-1,m}(\alpha))\}\\ &= -E\left\{\mathbf{H}\tilde{\boldsymbol{\theta}}_n I\left(\frac{\tilde{\boldsymbol{\theta}}'\mathbf{H}'\mathbf{N}\mathbf{H}\tilde{\boldsymbol{\theta}}_n}{(p-1)s_e^2} < F_{p-1,m}(\alpha)\right)\right\}\\ &= \sigma E\left\{\mathbf{N}^{-1/2}\boldsymbol{\Gamma}_1\mathbf{Z}_1 I\left(\frac{||\mathbf{Z}_1||^2}{s_e^2/\sigma^2} < (p-1)F_{p-1,m}(\alpha)\right)\right\}\end{aligned}$$

by using the transformations in the proof of Theorem 5.3.1

$$\begin{aligned}&= \sigma\mathbf{N}^{-1/2}\boldsymbol{\Gamma}_1\left(\frac{1}{\sigma}\boldsymbol{\Gamma}_1'\mathbf{N}^{1/2}\boldsymbol{\theta}\right)G_{p+1,m}(\ell_\alpha, \Delta^2) \text{ by Theorem 2.1.16}\\ &= \mathbf{H}\boldsymbol{\theta}G_{p+1,m}(\ell_\alpha,\Delta^2) = -\boldsymbol{\delta}G_{p+1,m}(\ell_\alpha,\Delta^2).\end{aligned}$$

Since $\mathbf{H}\tilde{\boldsymbol{\theta}}_n \sim \mathcal{N}(\mathbf{H}\boldsymbol{\theta}, \sigma^2\mathbf{H}\mathbf{N}^{-1})$ and $\tilde{\boldsymbol{\theta}}'\mathbf{H}'\mathbf{N}\mathbf{H}\tilde{\boldsymbol{\theta}} \sim \chi^2_{p-1}(\Delta^2)$ with $p-1$ d.f. and mse^2/σ^2 is a χ^2_m with m d.f. independent of $\chi^2_{p-1}(\Delta^2)$.

Consequently,

$$B_3(\hat{\boldsymbol{\theta}}_n^{PT}) = \Delta^2\{G_{p+1,m}(\ell_\alpha,\Delta^2)\}^2.$$

(iv) $\mathbf{b}_4(\hat{\boldsymbol{\theta}}_n^S) = E(\hat{\boldsymbol{\theta}}_n - \boldsymbol{\theta}) + E\big[(\tilde{\boldsymbol{\theta}}_n - \hat{\boldsymbol{\theta}}_n)(1 - c\mathcal{L}_n^{-1})\big]$

$$= -\mathbf{H}\boldsymbol{\theta} + \mathbf{H}\boldsymbol{\theta} - c(p-1)E\left\{\frac{\mathbf{H}\tilde{\boldsymbol{\theta}}_n s_e^2}{\tilde{\boldsymbol{\theta}}_n'\mathbf{HNH}\tilde{\boldsymbol{\theta}}_n}\right\}$$

$$= -c(p-1)\sigma E\left\{\frac{\mathbf{N}^{-1/2}\boldsymbol{\Gamma}_1\mathbf{Z}_1}{||\mathbf{Z}_1||^2}\right\}$$

$$= -c(p-1)\sigma\mathbf{N}^{-1/2}\boldsymbol{\Gamma}_1\left(\frac{1}{\sigma}\boldsymbol{\Gamma}_1'\mathbf{N}^{1/2}\boldsymbol{\theta}\right)E[\chi_{p+1}^{-2}(\Delta^2)] \quad \text{by Theorem 2.1.16}$$

$$= -c(p-1)(\mathbf{N}^{-1/2}\boldsymbol{\Gamma}_1\boldsymbol{\Gamma}_1'\mathbf{N}^{1/2}\boldsymbol{\theta})E[\chi_{p+1}^{-2}(\Delta^2)]$$

$$= -c(p-1)\mathbf{H}\boldsymbol{\theta}E[\chi_{p+1}^{-2}(\Delta^2)].$$

Consequently,

$$B(\hat{\boldsymbol{\theta}}_n^S) = c^2(p-1)^2\Delta^2\{E[\chi_{p+1}^{-2}(\Delta^2)]\}^2.$$

(v) $\mathbf{b}_5(\hat{\boldsymbol{\theta}}_n^{S+}) = E(\hat{\boldsymbol{\theta}}_n^{S+} - \boldsymbol{\theta}) = E(\hat{\boldsymbol{\theta}}_n^{S} - \boldsymbol{\theta}) - E\{(1 - c\mathcal{L}_n^{-1})I(\mathcal{L}_n < c)(\tilde{\boldsymbol{\theta}}_n - \hat{\boldsymbol{\theta}}_n)\}.$

$$= b_4(\hat{\boldsymbol{\theta}}_n^S) - E\{(1 - c\mathcal{L}_n^{-1})I(\mathcal{L}_n < c)(\tilde{\boldsymbol{\theta}}_n - \hat{\boldsymbol{\theta}}_n)\}.$$

We evaluate the second term above as follows:

$$E\{(\tilde{\boldsymbol{\theta}}_n - \hat{\boldsymbol{\theta}}_n)(1 - c\mathcal{L}_n^{-1})I(\mathcal{L}_n < c)\}$$

$$= \sigma E\left\{(p-1)s_e^2\mathbf{N}^{-1/2}\boldsymbol{\Gamma}_1\mathbf{Z}_1(1 - c||\mathbf{Z}_1||^{-2})I\left(\frac{||\mathbf{Z}_1||^2}{s_e^2/\sigma^2} \le (p-1)c\right)\right\}$$

$$= \left(\frac{p-1}{p+1}\right)\sigma\mathbf{N}^{-1/2}\boldsymbol{\Gamma}_1\left(\frac{1}{\sigma}\boldsymbol{\Gamma}_1'\mathbf{N}^{1/2}\boldsymbol{\theta}\right)$$

$$\times E\left\{\left(1 - c\,\frac{p-1}{p+1}F_{p+1,m}^{-1}(\Delta^2)\right)I\left(F_{p+1,m}(\Delta^2) < \frac{p-1}{p+1}c\right)\right\}$$

$$= \left(\frac{p-1}{p+1}\right)\mathbf{H}\boldsymbol{\theta}E\left\{\left(1 - c\,\frac{p-1}{p+1}F_{p+1,m}^{-1}(\Delta^2)\right)I\left(F_{p+1,m}(\Delta^2) < \frac{p-1}{p+1}c\right)\right\}$$

$$= \left(\frac{p-1}{p+1}\right)\boldsymbol{\delta}E\left\{\left(1 - c\,\frac{p-1}{p+1}F_{p+1,m}^{-1}(\Delta^2)\right)I\left(F_{p+1,m}^{-1}(\Delta^2) < \frac{p-1}{p+1}c\right)\right\}.$$

Substituting the above expression, we obtain the bias vector $\mathbf{b}_5(\hat{\boldsymbol{\theta}}_n^{S+})$ and $B_5(\hat{\boldsymbol{\theta}}_n^{S+})$ as in the theorem.

Since bias expressions are vectors, we use the quadratic bias expressions to study the properties of the biases of the five estimators. The quadratic bias of $\tilde{\boldsymbol{\theta}}_n$ is zero and that of $\hat{\boldsymbol{\theta}}_n$ is Δ^2, which is an unbounded line starting from the origin. The quadratic bias of PTE, SE, and PRSE are bounded in Δ^2, satisfying the relation

$$B_1(\tilde{\boldsymbol{\theta}}_n) = 0 \le B_5(\hat{\boldsymbol{\theta}}_n^{S+}) \le B_4(\hat{\boldsymbol{\theta}}_n^{S}) \le B_3(\hat{\boldsymbol{\theta}}_n^{PT}) \le B_2(\hat{\boldsymbol{\theta}}_n) \qquad (5.3.3)$$

whenever

$$G_{p+2,m}\left(\frac{p}{p+2}F_{p,m}(\alpha);\Delta^2\right) \ge \frac{(p-3)m}{(m+2)}E\left[\chi_{p+2}^{-2}(\Delta^2)\right].$$

5.3.2 MSE Matrix and Risk Expressions

The MSE matrix and the risk expression under the quadratic loss function

$$\begin{aligned} L(\boldsymbol{\theta}^*;\boldsymbol{\theta}) &= (\boldsymbol{\theta}^*-\boldsymbol{\theta})'\mathbf{W}(\boldsymbol{\theta}^*-\boldsymbol{\theta}) \\ &= \|\boldsymbol{\theta}^*-\boldsymbol{\theta}\|_{\mathbf{W}}^2 \end{aligned} \qquad (5.3.4)$$

are given by the following theorem:

Theorem 2. (i) $\mathbf{M}_1(\tilde{\boldsymbol{\theta}}_n) = \sigma^2\mathbf{N}^{-1}$ and $R_1(\tilde{\boldsymbol{\theta}}_n;\mathbf{W}) = \sigma^2\text{tr}(\mathbf{W}\mathbf{N}^{-1})$. (5.3.5a)

(ii) $\mathbf{M}_2(\hat{\boldsymbol{\theta}}_n) = \frac{\sigma^2}{n}\mathbf{1}_p\mathbf{1}_p' + \boldsymbol{\delta}\boldsymbol{\delta}'$ and $R_2(\hat{\boldsymbol{\theta}}_n;\mathbf{W}) = \frac{\sigma^2}{n}\mathbf{1}_p'\mathbf{W}\mathbf{1}_p + \boldsymbol{\delta}'\mathbf{W}\boldsymbol{\delta} = 1+\Delta^2$ if $\mathbf{W} = \sigma^{-2}\mathbf{N}$. (5.3.5b)

(iii) $$\begin{aligned}\mathbf{M}_3(\hat{\boldsymbol{\theta}}_n^{PT}) &= \sigma^2\mathbf{N}^{-1} - \sigma^2(\mathbf{H}\mathbf{N}^{-1}\mathbf{H}')G_{p+1,m}(\ell_\alpha;\Delta^2) \\ &\quad +(\boldsymbol{\delta}\boldsymbol{\delta}')\{2G_{p+1,m}(\ell_\alpha;\Delta^2) - G_{p+3,m}(\ell_\alpha^*;\Delta^2)\}\end{aligned}$$

where $\ell_\alpha = \frac{p-1}{p+1}F_{p-1,m}(\alpha)$, $\ell_\alpha^* = \frac{p-1}{p+3}F_{p-1,m}(\alpha)$, and $\mathbf{H}\mathbf{N}^{-1}\mathbf{H}' = \mathbf{H}\mathbf{N}^{-1}$.

Further,

$$\begin{aligned} R_3(\hat{\boldsymbol{\theta}}_n^{PT};\mathbf{W}) &= \sigma^2\text{tr}(\mathbf{W}\mathbf{N}^{-1}) - \sigma^2\text{tr}(\mathbf{W}\mathbf{H}\mathbf{N}^{-1})G_{p+1,m}(\ell_\alpha;\Delta^2) \\ &\quad +(\boldsymbol{\delta}'\mathbf{W}\boldsymbol{\delta})\{2G_{p+1,m}(\ell_\alpha;\Delta^2) - G_{p+3,m}(\ell_\alpha;\Delta^2)\}. \end{aligned} \qquad (5.3.5c)$$

(iv) $$\begin{aligned}\mathbf{M}_4(\hat{\boldsymbol{\theta}}_n^{S}) &= \sigma^2\mathbf{N}^{-1} - c(p-1)\sigma^2(\mathbf{H}\mathbf{N}^{-1})\{(p-3)E[\chi_{p+1}^{-2}(\Delta^2)] \\ &\quad +2\Delta^2E[\chi_{p+3}^{-4}(\Delta^2)]\} + c(p^2-1)\boldsymbol{\delta}\boldsymbol{\delta}'E[\chi_{p+3}^{-4}(\Delta^2)]\end{aligned}$$

and

$$\begin{aligned} R_4(\hat{\boldsymbol{\theta}}_n^{S};\mathbf{W}) &= \sigma^2\text{tr}(\mathbf{W}\mathbf{N}^{-1}) - c(p-1)\sigma^2\text{tr}(\mathbf{W}\mathbf{H}\mathbf{N}^{-1}) \\ &\times\left\{(p-3)E[\chi_{p+1}^{-2}(\Delta^2)] + 2\Delta^2E[\chi_{p+3}^{-4}(\Delta^2)]\left[1-\frac{(p+1)(\boldsymbol{\delta}'\mathbf{W}\boldsymbol{\delta})}{2(\boldsymbol{\delta}'\mathbf{N}\boldsymbol{\delta})\text{tr}(\mathbf{W}\mathbf{H}\mathbf{N}^{-1})}\right]\right\}. \end{aligned} \qquad (5.3.5d)$$

(v) $\mathbf{M}_5(\hat{\boldsymbol{\theta}}_n^{S+}) = \mathbf{M}_4(\hat{\boldsymbol{\theta}}_n^{S})$

$$-\sigma^2(\mathbf{HN}^{-1})E\Big[\Big(1-\tfrac{c(p-1)}{(p+1)}F^{-1}_{p+1,m}(\Delta^2)\Big)^2 I\Big(F_{p+1,m}(\Delta^2) < c\tfrac{(p-1)}{(p+1)}\Big)\Big]$$
$$+\ \boldsymbol{\delta\delta}'\Big\{2E\Big[\Big(1-\tfrac{c(p-1)}{(p+1)}F^{-1}_{p+1,m}(\Delta^2)\Big) I\Big(F_{p+1,m}(\Delta^2) < \tfrac{c(p-1)}{(p+1)}\Big)\Big]$$
$$-E\Big[\Big(1-\tfrac{c(p-1)}{(p+3)}F^{-1}_{p+3,m}(\Delta^2)-1\Big)^2 I\Big(F_{p+3,m}(\Delta^2) < \tfrac{c(p-1)}{(p+3)}\Big)\Big]\Big\}. \tag{5.3.5e}$$

and

$$R_5(\hat{\boldsymbol{\theta}}_n^{S+};\mathbf{W}) = R_4(\hat{\boldsymbol{\theta}}_n^{S};\mathbf{W})$$
$$-\sigma^2\mathrm{tr}(\mathbf{WHN}^{-1})E\Big[\Big(1-\tfrac{c(p-1)}{(p+1)}F^{-1}_{p+1,m}(\Delta^2)\Big)^2 I(F_{p+1,m}(\Delta^2) < \tfrac{c(p-1)}{(p+1)}\Big)\Big]$$
$$+(\boldsymbol{\delta}'\mathbf{W}\boldsymbol{\delta})E\Big\{2E\Big[\Big(1-\tfrac{c(p-1)}{(p+1)}F^{-1}_{p+1,m}(\Delta^2)\Big) I(F_{p+1,m}(\Delta^2) < \tfrac{c(p-1)}{(p+1)}\Big)\Big]$$
$$-E\Big[\Big(1-\tfrac{c(p-1)}{(p+3)}F^{-1}_{p+3,m}(\Delta^2)\Big)^2 I\Big(F_{p+3,m}(\Delta^2) < \tfrac{c(p-1)}{(p+3)}\Big)\Big]\Big\}. \tag{5.3.5f}$$

Proof.

(i) $\mathbf{M}_1(\tilde{\boldsymbol{\theta}}_n) = E(\tilde{\boldsymbol{\theta}}_n-\boldsymbol{\theta})(\tilde{\boldsymbol{\theta}}_n-\boldsymbol{\theta})' = \sigma^2\mathbf{N}^{-1}$

$$R_1(\tilde{\boldsymbol{\theta}}_n;\mathbf{W}) = E(\tilde{\boldsymbol{\theta}}_n-\boldsymbol{\theta})'\mathbf{W}(\tilde{\boldsymbol{\theta}}_n-\boldsymbol{\theta}) = \sigma^2\mathrm{tr}(\mathbf{WN}^{-1}). \tag{5.3.6a}$$

The risk $R_1(\tilde{\boldsymbol{\theta}}_n;\mathbf{W})$ equals p if $\mathbf{W} = \sigma^{-2}\mathbf{N}$.

(ii) $\mathbf{M}_2(\hat{\boldsymbol{\theta}}_n) = E(\hat{\boldsymbol{\theta}}_n-\boldsymbol{\theta})(\hat{\boldsymbol{\theta}}_n-\boldsymbol{\theta})' = \frac{\sigma^2}{n}\mathbf{1}_p\mathbf{1}_p' + \boldsymbol{\delta\delta}' = \sigma^2\mathbf{N}^{-1} - \sigma^2(\mathbf{HN}^{-1}) + \boldsymbol{\delta\delta}'$

and

$$R_2(\hat{\boldsymbol{\theta}}_n;\mathbf{W})$$
$$= E(\hat{\boldsymbol{\theta}}_n-\boldsymbol{\theta})'\mathbf{W}(\hat{\boldsymbol{\theta}}_n-\boldsymbol{\theta}) = \mathrm{tr}[\mathbf{WM}_2(\hat{\boldsymbol{\theta}}_n)]$$
$$= \tfrac{\sigma^2}{n}\mathbf{1}_p'\mathbf{W}\mathbf{1}_p + \boldsymbol{\delta}'\mathbf{W}\boldsymbol{\delta} = \sigma^2\mathrm{tr}(\mathbf{WN}^{-1}) - \sigma^2\mathrm{tr}(\mathbf{WHN}^{-1}) + \boldsymbol{\delta}'\mathbf{W}\boldsymbol{\delta}, \tag{5.3.6b}$$

which reduces to

$$R_2(\hat{\boldsymbol{\theta}}_n, \mathbf{N}\sigma^{-2}) = 1+\Delta^2 \ \text{ if } \ \mathbf{W} = \mathbf{N}\sigma^{-2}.$$

(iii) $\mathbf{M}_3(\hat{\boldsymbol{\theta}}_n^{PT}) = E(\hat{\boldsymbol{\theta}}_n^{PT}-\boldsymbol{\theta})(\hat{\boldsymbol{\theta}}_n^{PT}-\boldsymbol{\theta})'$

$$= E(\tilde{\boldsymbol{\theta}}_n-\boldsymbol{\theta})(\tilde{\boldsymbol{\theta}}_n-\boldsymbol{\theta})' - 2E(\tilde{\boldsymbol{\theta}}_n-\boldsymbol{\theta})\tilde{\boldsymbol{\theta}}_n'\mathbf{H}'I(\mathcal{L}_n < F_{p-1,m}(\alpha))$$
$$+\ (\mathbf{H}\tilde{\boldsymbol{\theta}}_n\boldsymbol{\theta}_n'\mathbf{H})I(\mathcal{L}_n < F_{p-1,m}(\alpha)).$$

Now, $\mathbf{H}\tilde{\boldsymbol{\theta}}_n = \mathbf{N}^{-1/2}\boldsymbol{\Gamma}_1\boldsymbol{\Gamma}_1'\mathbf{N}^{1/2}\tilde{\boldsymbol{\theta}}_n = \sigma\mathbf{N}^{-1/2}\boldsymbol{\Gamma}_1\boldsymbol{\Gamma}_1'\mathbf{N}^{1/2}\mathbf{N}^{-1/2}\boldsymbol{\Gamma}_1\mathbf{Z}_1 = \sigma\mathbf{N}^{-1/2}\boldsymbol{\Gamma}_1\mathbf{Z}_1$, and $\tilde{\boldsymbol{\theta}} - \boldsymbol{\theta} = \sigma\mathbf{N}^{-1/2}\boldsymbol{\Gamma}(\mathbf{Z} - \boldsymbol{\eta})$, $\boldsymbol{\Gamma} = (\boldsymbol{\Gamma}_1, \boldsymbol{\Gamma}_2)$ and $\chi_m^2 = ms_e^2/\sigma^2$.

Thus, we write the expression for $\mathbf{M}_3(\hat{\boldsymbol{\theta}}_n^{PT})$ as

$$\sigma^2\mathbf{N}^{-1} - 2\sigma^2 E\left\{\left(\mathbf{N}^{-1/2}\boldsymbol{\Gamma}_1(\mathbf{Z}_1 - \boldsymbol{\eta}_1)\mathbf{Z}_1'\boldsymbol{\Gamma}_1'\mathbf{N}^{-1/2}\right) I\left(\frac{\|\mathbf{Z}_1\|^2}{\chi_m^2} < \frac{(p-1)}{m}F_{p-1,m}(\alpha)\right)\right\}$$
$$+\sigma^2 E\left\{\left(\mathbf{N}^{-1/2}\boldsymbol{\Gamma}_1\mathbf{Z}_1\mathbf{Z}_1'\boldsymbol{\Gamma}_1\mathbf{N}^{-1/2}\right) I\left(\frac{\|\mathbf{Z}_1\|^2}{\chi_m^2} < \frac{(p-1)}{m}F_{p-1,m}(\alpha)\right)\right\},$$

$$\begin{aligned}
&= \sigma^2\mathbf{N}^{-1} - \left\{\sigma^2(\mathbf{N}^{-1/2}\boldsymbol{\Gamma}_1\boldsymbol{\Gamma}_1'\mathbf{N}^{-1/2})G_{p+1,m}(\ell_\alpha;\Delta^2)\right.\\
&\quad \left.+ (\mathbf{N}^{-1/2}\boldsymbol{\Gamma}_1\boldsymbol{\eta}_1\boldsymbol{\eta}_1'\boldsymbol{\Gamma}_1'N^{-1/2})G_{p+3,m}(\ell_\alpha^*;\Delta^2)\right\}\\
&\quad + 2\sigma^2(\mathbf{N}^{-1/2}\boldsymbol{\Gamma}_1\boldsymbol{\eta}_1\boldsymbol{\eta}_1'\boldsymbol{\Gamma}_1'N^{-1/2})G_{p+1,m}(\ell_\alpha;\Delta^2)\\
&= \sigma^2\mathbf{N}^{-1} - \sigma^2(\mathbf{H}\mathbf{N}^{-1})G_{p+1,m}(\ell_\alpha;\Delta^2)\\
&\quad + \boldsymbol{\delta}\boldsymbol{\delta}'\left\{2G_{p+1,m}(\ell_\alpha;\Delta^2) - G_{p+3,m}(\ell_\alpha^*;\Delta^2)\right\}.
\end{aligned}$$

Hence, the risk of $\hat{\boldsymbol{\theta}}_n^{PT}$ is given by

$$\begin{aligned}
&R_3(\hat{\boldsymbol{\theta}}_n^{PT};\mathbf{W})\\
&= \sigma^2\text{tr}(\mathbf{W}\mathbf{N}^{-1}) - \sigma^2\text{tr}(\mathbf{W}\mathbf{H}\mathbf{N}^{-1})G_{p+1,m}(\ell_\alpha;\Delta^2)\\
&\quad + (\boldsymbol{\delta}'\mathbf{W}\boldsymbol{\delta})\left\{2G_{p+1,m}(\ell_\alpha;\Delta^2) - G_{p+3,m}(\ell_\alpha^*;\Delta^2)\right\}\\
&= p - (p-1)G_{p+1,m}(\ell_\alpha;\Delta^2) + \Delta^2\left\{2G_{p+1,m}(\ell_\alpha;\Delta^2) - G_{p+3,m}(\ell_\alpha^*;\Delta^2)\right\}
\end{aligned}$$

if $\mathbf{W} = \sigma^{-2}\mathbf{N}$. (5.3.6c)

(iv) $\mathbf{M}_4(\hat{\boldsymbol{\theta}}_n^S) = E\{(\hat{\boldsymbol{\theta}}_n^S - \boldsymbol{\theta})(\hat{\boldsymbol{\theta}}_n^S - \boldsymbol{\theta})'\}$

$$\begin{aligned}
&= E\{(\tilde{\boldsymbol{\theta}}_n - \boldsymbol{\theta})(\tilde{\boldsymbol{\theta}}_n - \boldsymbol{\theta})' - 2cE\left\{(\tilde{\boldsymbol{\theta}}_n - \boldsymbol{\theta})\tilde{\boldsymbol{\theta}}_n'\mathbf{H}'\left[\frac{(p-1)s_e^2}{(\tilde{\boldsymbol{\theta}}_n'\mathbf{H}'\mathbf{N}\mathbf{H}\tilde{\boldsymbol{\theta}}_n)}\right]\right\}\\
&\quad + c^2E\left\{(\mathbf{H}\tilde{\boldsymbol{\theta}}_n\tilde{\boldsymbol{\theta}}_n'\mathbf{H})')\left[\frac{(p-1)s_e^2}{(\tilde{\boldsymbol{\theta}}_n'\mathbf{H}'\mathbf{N}\mathbf{H}\tilde{\boldsymbol{\theta}}_n)}\right]^2\right\}\\
&= \sigma^2\mathbf{N}^{-1} - 2c\sigma^2E\left\{\left(\mathbf{N}^{-1/2}\boldsymbol{\Gamma}_1(\mathbf{Z}_1 - \boldsymbol{\eta}_1)\mathbf{Z}_1'\boldsymbol{\Gamma}_1'\mathbf{N}^{-1/2}\right)\left[\frac{(p-1)s_e^2}{\sigma^2(||\mathbf{Z}_1||^2)}\right]\right\}\\
&\quad + c^2\sigma^2\left\{\left(N^{-1/2}\boldsymbol{\Gamma}_1\mathbf{Z}_1\mathbf{Z}_1'\boldsymbol{\Gamma}_1'\mathbf{N}^{-1/2}\right)\left[\frac{(p-1)s_e^2}{\sigma^2(||\mathbf{Z}_1||^2)}\right]^2\right\}
\end{aligned}$$

$$= \sigma^2\mathbf{N}^{-1} - 2\sigma^2 c(p-1)E\Big\{\big(\mathbf{N}^{-1/2}\boldsymbol{\Gamma}_1\mathbf{Z}_1\mathbf{Z}_1'\boldsymbol{\Gamma}_1'\mathbf{N}^{-1/2}\big)(\|\mathbf{Z}_1\|)^{-2}\Big\}$$

$$+ 2\sigma^2 c(p-1)\Big\{\big(N^{-1/2}\boldsymbol{\Gamma}_1\boldsymbol{\eta}_1\mathbf{Z}_1'\boldsymbol{\Gamma}_1'\mathbf{N}^{-1/2}\big)(\|\mathbf{Z}_1\|)^{-2}\Big\}$$

$$+ \sigma^2 c^2(p-1)^2\Big(\frac{m+2}{m}\Big)E\Big\{\big(N^{-1/2}\boldsymbol{\Gamma}_1\mathbf{Z}_1\mathbf{Z}_1'\boldsymbol{\Gamma}_1'\mathbf{N}^{-1/2}\big)(\|\mathbf{Z}_1\|)^{-4}\Big\}$$

$$= \sigma^2\mathbf{N}^{-1} - 2\sigma^2 c(p-1)\Big\{\big(\mathbf{N}^{-1/2}\boldsymbol{\Gamma}_1\boldsymbol{\Gamma}_1'\mathbf{N}^{-1/2}\big)E(\chi_{p+1}^{-2}(\Delta^2))$$

$$+ \big(N^{-1/2}\boldsymbol{\Gamma}_1\boldsymbol{\eta}_1\boldsymbol{\eta}_1'\boldsymbol{\Gamma}_1'\mathbf{N}^{-1/2}\big)E\big(\chi_{p+3}^{-2}(\Delta^2)\big)\Big\}$$

$$+ 2\sigma^2 c(p-1)\big(\mathbf{N}^{-1/2}\boldsymbol{\Gamma}_1\boldsymbol{\eta}_1\boldsymbol{\eta}_1'\boldsymbol{\Gamma}_1'\mathbf{N}^{-1/2}\big)E\big(\chi_{p+1}^{-2}(\Delta^2)\big)$$

$$+ \sigma^2 c^2(p-1)^2\Big(\frac{m+2}{m}\Big)\Big\{\big(N^{-1/2}\boldsymbol{\Gamma}_1\boldsymbol{\Gamma}_1'\mathbf{N}^{-1/2}\big)E\big(\chi_{p+1}^{-4}(\Delta^2)\big)$$

$$+ \big(\mathbf{N}^{1/2}\boldsymbol{\Gamma}_1\boldsymbol{\eta}_1\boldsymbol{\eta}_1'\boldsymbol{\Gamma}_1'\mathbf{N}^{-1/2}\big)E\big(\chi_{p+3}^{-4}(\Delta^2)\big)\Big\}$$

$$= \sigma^2\mathbf{N}^{-1} - \sigma^2 c(p-1)(\mathbf{H}\mathbf{N}^{-1})\{2E\big(\chi_{p+1}^{-2}(\Delta^2)\big) - (p-3)E\big(\chi_{p+1}^{-4}(\Delta^2)\big)\}$$

$$+ c(p-1)\boldsymbol{\delta}\boldsymbol{\delta}'\Big\{2E\big(\chi_{p+1}^{-2}(\Delta^2)\big) - 2E\big(\chi_{p+3}^{-4}(\Delta^2)\big) + (p-3)E\big(\chi_{p+3}^{-4}(\Delta^2)\big)\Big\}$$

$$= \sigma^2\mathbf{N}^{-1} - \sigma^2 c(p-1)(\mathbf{H}\mathbf{N}^{-1})\{2E\big(\chi_{p+1}^{-2}(\Delta^2)\big) - (p-3)E\big(\chi_{p+1}^{-4}(\Delta^2)\big)\}$$

$$+ c(p^2-1)\boldsymbol{\delta}\boldsymbol{\delta}'E\big(\chi_{p+3}^{-4}(\Delta^2)\big)$$

$$= \sigma^2\mathbf{N}^{-1} - \sigma^2\frac{(p-3)m}{(m+2)}(\mathbf{H}\mathbf{N}^{-1})\{2E\big(\chi_{p+1}^{-2}(\Delta^2)\big) - (p-3)E\big(\chi_{p+1}^{-4}(\Delta^2)\big)\}$$

$$+ \frac{(p-3)(p+1)m}{(m+2)}(\mathbf{H}\boldsymbol{\theta}\boldsymbol{\theta}'\mathbf{H}')E\big(\chi_{p+3}^{-4}(\Delta^2)\big) \quad \text{since} \quad c = \frac{(p-3)m}{(p-1)(m+2)}.$$

The risk of $\hat{\boldsymbol{\theta}}_n^S$ is given by

$$R_4(\hat{\boldsymbol{\theta}}_n^S;\mathbf{W}) = \sigma^2\mathrm{tr}(\mathbf{W}\mathbf{N}^{-1}) - \sigma^2\frac{(p-3)m}{(m+2)}\mathrm{tr}(\mathbf{W}\mathbf{N}^{-1}\mathbf{H})$$

$$\times\{2E\big(\chi_{p+1}^{-2}(\Delta^2)\big) - (p-3)E\big(\chi_{p+1}^{-4}(\Delta^2)\big)\}$$

$$+ \frac{(p-3)(p+1)m}{(m+2)}(\boldsymbol{\delta}\mathbf{W}\boldsymbol{\delta})E\big(\chi_{p+3}^{-4}(\Delta^2)\big)$$

$$= p - \frac{(p-3)(p-1)m}{(m+2)}\{2E\big(\chi_{p+1}^{-2}(\Delta^2)\big) - (p-3)E\big(\chi_{p+1}^{-4}(\Delta^2)\big)\}$$

$$+ \frac{(p-3)(p+1)m}{(m+2)}\Delta^2 E\big(\chi_{p+3}^{-4}(\Delta^2)\big)$$

if $\mathbf{W} = \sigma^{-2}\mathbf{N}$. (5.3.6d)

Rewriting this formula for $R_4(\hat{\boldsymbol{\theta}}_n^S : \mathbf{W})$, we obtain the expression of the theorem.

(ii) $\mathbf{M}_5(\hat{\boldsymbol{\theta}}_n^{S+})$

$$
\begin{aligned}
&= E(\hat{\boldsymbol{\theta}}_n^{S+} - \boldsymbol{\theta})(\hat{\boldsymbol{\theta}}_n^{S+} - \boldsymbol{\theta})' \\
&= E(\hat{\boldsymbol{\theta}}_n^{S} - \boldsymbol{\theta})(\hat{\boldsymbol{\theta}}^{S} - \boldsymbol{\theta})' + c^2 E[(\mathbf{H}\tilde{\boldsymbol{\theta}}_n\tilde{\boldsymbol{\theta}}_n'\mathbf{H}')I(\mathcal{L}_n < c)(1 - c\mathcal{L}_n^{-1})^2] \\
&\quad - 2E[\{(\hat{\boldsymbol{\theta}}^{S} - \boldsymbol{\theta})(\tilde{\boldsymbol{\theta}}_n'\mathbf{H}'\}(1 - c\mathcal{L}_n^{-1})I(\mathcal{L}_n < c)] \\
&= \mathbf{M}_4(\hat{\boldsymbol{\theta}}_n^{S}) + c^2 E[(\mathbf{H}\tilde{\boldsymbol{\theta}}_n\tilde{\boldsymbol{\theta}}_n'\mathbf{H}')\mathcal{L}_n^{-2}I(\mathcal{L}_n < c)(1 - c\mathcal{L}_n^{-1})^2] \\
&\quad -2cE[\{(\tilde{\boldsymbol{\theta}}_n - \boldsymbol{\theta})\tilde{\boldsymbol{\theta}}_n'\mathbf{H}'\}\mathcal{L}_n^{-1}(1 - c\mathcal{L}_n^{-1})I(\mathcal{L}_n < c)] \\
&\quad +2cE[(\mathbf{H}\tilde{\boldsymbol{\theta}}_n\tilde{\boldsymbol{\theta}}_n'\mathbf{H}')\mathcal{L}_n^{-1}(1 - c\mathcal{L}_n^{-1})I(\mathcal{L}_n < c)] \\
&= \mathbf{M}_4(\hat{\boldsymbol{\theta}}_n^{S}) + \sigma^2 c^2 (p-1)^2 E\Bigg[(\mathbf{N}^{-1/2}\boldsymbol{\Gamma}_1\mathbf{Z}_1\mathbf{Z}_1'\boldsymbol{\Gamma}_1'\mathbf{N}^{-1/2})\frac{\chi_m^4}{||\mathbf{Z}_1||^4} \\
&\qquad \times I\left(\frac{||\mathbf{Z}_1||^2}{\chi_m^2} < \frac{(p-1)}{m}c\right)\left(1 - c\frac{(p-1)}{m}\frac{\chi_m^2}{||\mathbf{Z}_1||^2}\right)^2\Bigg] \\
&\quad -2c(p-1)\sigma^2 E\Bigg[(\mathbf{N}^{-1/2}\boldsymbol{\Gamma}_1(\mathbf{Z}_1 - \boldsymbol{\eta}_1)\mathbf{Z}_1'\boldsymbol{\Gamma}_1'\mathbf{N}^{-1/2})\left(\frac{\chi_m^2}{||\mathbf{Z}_1||^2}\right) \\
&\qquad \times \left(1 - c\frac{(p-1)}{m}\frac{\chi_m^2}{||\mathbf{Z}_1||^2}\right) I\left(\frac{||\mathbf{Z}_1||^2}{\chi_m^2} < \frac{(p-1)}{m}c\right)\Bigg] \\
&\quad + 2\sigma^2 c(p-1)E\Bigg[\left(n^{-1/2}\boldsymbol{\Gamma}_1\mathbf{Z}_1\mathbf{Z}_1'\boldsymbol{\Gamma}_1'\mathbf{N}^{-1/2}\right)\left(\frac{\chi_m^2}{||\mathbf{Z}_1||^2}\right) \\
&\qquad \times \left(1 - c\frac{(p-1)}{m}\frac{\chi_m^2}{||\mathbf{Z}_1||^2}\right) I\left(\frac{||\mathbf{Z}_1||^2}{\chi_m^2} < \frac{(p-1)}{m}c\right)\Bigg].
\end{aligned}
$$

Evaluating the terms as before, we have

$$
\begin{aligned}
&\mathbf{M}_5(\hat{\boldsymbol{\theta}}_n^{S+}) = \mathbf{M}_4(\hat{\boldsymbol{\theta}}_n^{S}) - \sigma^2(\mathbf{H}\mathbf{N}^{-1}) \\
&\times E\Bigg[\left(1 - \frac{c(p-1)}{(p+1)}F_{p+1,m}^{-1}(\Delta^2)\right)^2 I\left(F_{p+1,m}(\Delta^2) < \frac{c(p-1)}{(p+1)}\right)\Bigg] \\
&+ \boldsymbol{\delta}\boldsymbol{\delta}'\Bigg\{2E\Bigg[\left(1 - \frac{c(p-1)}{(p+1)}F_{p+1,m}^{-1}(\Delta^2) - 1\right)I\left(F_{p+1,m}(\Delta^2) < \frac{p-1}{p+1}c\right)\Bigg] \\
&\quad -E\Bigg[\left(1 - \frac{c(p-1)}{(p+3)}F_{p+3,m}^{-1}(\Delta^2)\right)^2 I\left(F_{p+3,m}(\Delta^2) < \frac{c(p-1)}{p+3}\right)\Bigg]\Bigg\}.
\end{aligned}
$$

The risk of $\hat{\boldsymbol{\theta}}_n^{S+}$ is then easily computed as

$$
\begin{aligned}
&R_5(\hat{\boldsymbol{\theta}}_n^{S+};\mathbf{W})\\
&= R_4(\hat{\boldsymbol{\theta}}_n^{S};\mathbf{W}) - \sigma^2\mathrm{tr}(\mathbf{WHN}^{-1})E\Big[\Big(1-\frac{c(p-1)}{p+1}F_{p+1,m}^{-1}(\Delta^2)\Big)^2\\
&\qquad\qquad \times I\Big(F_{p+1,m}(\Delta^2) < \frac{c(p-1)}{p+1}\Big)\Big]\\
&\quad +(\boldsymbol{\delta}'\mathbf{W}\boldsymbol{\delta})\Big\{2E\Big[\Big(1-\frac{c(p-1)}{p+1}F_{p+1,m}^{-1}(\Delta^2)\Big)\\
&\qquad\qquad \times I\Big(F_{p+1,m}(\Delta^2) < \frac{c(p-1)}{p+1}c\Big)\Big]\\
&\quad - E\Big[\Big(\frac{c(p-1)}{p+3}F_{p+3,m}^{-1}(\Delta^2)-1\Big)^2 I\Big(F_{p+3,m}(\Delta^2) < \frac{c(p-1)}{p+3}\Big)\Big]\Big\}\\
&= R_3(\hat{\boldsymbol{\theta}}_n^{S};\sigma^{-2}\mathbf{I}_p) - \sigma^2(p-1)E\Big[\Big(1-\frac{c(p-1)}{p+1}F_{p+1,m}^{-1}(\Delta^2)\Big)^2\\
&\qquad\qquad \times I\Big(F(\Delta^2) < \frac{c(p-1)}{p+1}\Big)\Big]\\
&\quad + \Delta^2\Big\{2E\Big[\Big(1-\frac{c(p-1)}{p+1}F_{p+1,m}^{-1}(\Delta^2)-1\Big)I\Big(F_{p+1,m}(\Delta^2)<\frac{c(p-1)}{p+1}\Big)\Big]\\
&\quad -E\Big[\Big(1-\frac{c(p-1)}{p+3}F_{p+3,m}^{-1}(\Delta^2)\Big)^2 I\Big(F_{p+3,m}(\Delta^2) < \frac{c(p-1)}{p+3}\Big)\Big]\Big\},
\end{aligned}
$$

if $\mathbf{W} = \sigma^{-2}\mathbf{N}$. (5.3.6e)

5.4 Risk Analysis and Risk Efficiency

5.4.1 Comparison of $\hat{\boldsymbol{\theta}}_n$ and $\tilde{\boldsymbol{\theta}}_n$.

The risk-difference $R_1(\tilde{\boldsymbol{\theta}}_n;\mathbf{W}) - R_2(\hat{\boldsymbol{\theta}}_n;\mathbf{W})$ shows that $\hat{\boldsymbol{\theta}}_n$ performs better than $\tilde{\boldsymbol{\theta}}_n$ if

$$\sigma^{-2}(\boldsymbol{\delta}'\mathbf{W}\boldsymbol{\delta}) \le \mathrm{tr}(\mathbf{WHN}^{-1}), \tag{5.4.1a}$$

and $\tilde{\boldsymbol{\theta}}_n$ performs better than $\hat{\boldsymbol{\theta}}_n$ if

$$\sigma^{-2}(\boldsymbol{\delta}'\mathbf{W}\boldsymbol{\delta}) \ge \mathrm{tr}(\mathbf{WHN}^{-1}). \tag{5.4.1b}$$

If $\mathbf{W} = \sigma^{-2}\mathbf{N}$, then $\hat{\boldsymbol{\theta}}_n$ performs better than $\tilde{\boldsymbol{\theta}}_n$ when $\Delta^2 < p-1$; otherwise, $\tilde{\boldsymbol{\theta}}_n$ performs better than $\hat{\boldsymbol{\theta}}_n$.

The risk-based efficiency of $\hat{\boldsymbol{\theta}}_n$ compared to $\tilde{\boldsymbol{\theta}}_n$ is given by

$$\text{RRE}(\hat{\boldsymbol{\theta}}_n : \tilde{\boldsymbol{\theta}}_n) = \frac{R_1(\tilde{\boldsymbol{\theta}}_n : \sigma^{-2}\mathbf{N})}{R_2(\hat{\boldsymbol{\theta}}_n : \sigma^{-2}\mathbf{N})} = \frac{p}{1+\Delta^2}. \tag{5.4.2}$$

$\text{RRE}(\hat{\boldsymbol{\theta}}_n : \tilde{\boldsymbol{\theta}}_n)$ is a decreasing function of Δ^2. At $\Delta^2 = 0$, the maximum efficiency of p is attained. However, $\hat{\boldsymbol{\theta}}_n$ performs better when $\tilde{\boldsymbol{\theta}}_n$ is the interval $[0, p-1)$, and outside this interval, $\tilde{\boldsymbol{\theta}}_n$ performs better than $\hat{\boldsymbol{\theta}}_n$.

5.4.2 Comparison of $\hat{\boldsymbol{\theta}}_n^{PT}$ and $\tilde{\boldsymbol{\theta}}_n(\hat{\boldsymbol{\theta}}_n)$.

Here the risk-difference shows that $\hat{\boldsymbol{\theta}}_n^{PT}$ performs better than $\tilde{\boldsymbol{\theta}}_n$ if

$$\Delta^2 < \frac{\text{tr}[\mathbf{WHN}^{-1}]G_{p+1,m}(\ell_\alpha;\Delta^2)}{Ch_{\max}(\mathbf{WN}^{-1})\{2G_{p+1,m}(\ell_\alpha;\Delta^2) - G_{p+3,m}(\ell_\alpha^*;\Delta^2)\}}, \tag{5.4.3a}$$

while $\tilde{\boldsymbol{\theta}}_n$ performs better than $\hat{\boldsymbol{\theta}}_n^{PT}$ if

$$\Delta^2 > \frac{\text{tr}[\mathbf{WHN}^{-1}]G_{p+1,m}(\ell_\alpha;\Delta^2)}{Ch_{\min}(\mathbf{WN}^{-1})\{2G_{p+1,m}(\ell_\alpha;\Delta^2) - G_{p+3,m}(\ell_\alpha^*;\Delta^2)\}}. \tag{5.4.3b}$$

If $\mathbf{W} = \sigma^{-2}\mathbf{N}$, then $\hat{\boldsymbol{\theta}}_n^{\text{PT}}$ performs better than $\tilde{\boldsymbol{\theta}}_n$ when

$$\Delta^2 < \frac{(p-1)G_{p+1,3}(\ell_\alpha;\Delta^2)}{[2G_{p+1,m}(\ell_\alpha;\Delta^2) - G_{p+1,m}(\ell_\alpha;\Delta^2)]}; \tag{5.4.4}$$

otherwise, $\tilde{\boldsymbol{\theta}}_n$ performs better than $\hat{\boldsymbol{\theta}}_n$.

Now, consider the null hypothesis H_0 and the risk-difference of $\hat{\boldsymbol{\theta}}_n$ and $\hat{\boldsymbol{\theta}}_n^{\text{PT}}$, namely

$$\begin{aligned}
&R_2(\hat{\boldsymbol{\theta}}_n;\mathbf{W}) - R_3(\hat{\boldsymbol{\theta}}_n^{PT};\mathbf{W}) \\
&= \frac{\sigma^2}{n}\mathbf{1}_p'\mathbf{W}\mathbf{1}_n - \sigma^2\text{tr}(\mathbf{WN}^{-1}) + \sigma^2\text{tr}(\mathbf{WHN}^{-1})G_{p+1,m}(\ell_\alpha;0) \\
&= -\ \sigma^2\text{tr}(\mathbf{WHN}^{-1})(1 - G_{p+1,m}(\ell_\alpha;0)) < 0.
\end{aligned} \tag{5.4.5}$$

Thus, under H_0, $\hat{\boldsymbol{\theta}}_n$ performs better than $\hat{\boldsymbol{\theta}}_n^{\text{PT}}$. In general, $\hat{\boldsymbol{\theta}}_n$ performs better than $\hat{\boldsymbol{\theta}}_n^{\text{PT}}$ if

$$\Delta^2 < \frac{\text{tr}(\mathbf{WHN}^{-1})(1 - G_{p+1,m}(\ell_\alpha;\Delta^2))}{\{1 - 2G_{p+1,m}(\ell_\alpha;\Delta^2) + G_{p+3,m}(\ell_\alpha^*;\Delta^2)\}}, \tag{5.4.6}$$

and $\hat{\boldsymbol{\theta}}_n^{\text{PT}}$ performs better than $\hat{\boldsymbol{\theta}}_n$ when

$$\Delta^2 > \frac{\text{tr}(\mathbf{WHN}^{-1})(1 - G_{p+1,m}(\ell_\alpha;\Delta^2))}{\{1 - 2G_{p+1,m}(\ell_\alpha;\Delta^2) + G_{p+3,m}(\ell_\alpha^*;\Delta^2)\}}. \tag{5.4.7}$$

If $\mathbf{W} = \sigma^{-2}\mathbf{N}$, then the r.h.s. of (5.4.6) and (5.4.7) reduces to

$$\frac{(p-1)(1-G_{p+1,m}(\ell_\alpha;\Delta^2))}{\{1-2G_{p+1,m}(\ell_\alpha;\Delta^2)+G_{p+3,m}(\ell_\alpha^*;\Delta^2)\}}. \quad (5.4.8)$$

Now, consider the risk-efficiency of $\hat{\boldsymbol{\theta}}_n^{\mathrm{PT}}$ relative to $\tilde{\boldsymbol{\theta}}_n$ for $\mathbf{W} = \sigma^{-2}\mathbf{N}$, given by

$$\begin{aligned}\mathrm{RRE}(\hat{\boldsymbol{\theta}}_n^{\mathrm{PT}} : \tilde{\boldsymbol{\theta}}_n)) &= \left[1-\left(1-\frac{1}{p}\right)G_{p+1,m}(\ell_\alpha;\Delta^2)\right.\\ &\quad \left.+\Delta^2 p^{-1}\left\{2G_{p+1,m}(\ell_\alpha;\Delta^2)-G_{p+3,m}(\ell_\alpha^*;\Delta^2)\right\}\right]^{-1}.\end{aligned} \quad (5.4.9)$$

We note the following properties of $\mathrm{RRE}(\hat{\boldsymbol{\theta}}_n^{\mathrm{PT}};\tilde{\boldsymbol{\theta}}_n)$:

(i) $\mathrm{RRE}(\hat{\boldsymbol{\theta}}_n^{\mathrm{PT}} : \tilde{\boldsymbol{\theta}}_n) \gtreqless 1$ according as

$$\Delta^2 \lesseqgtr \frac{(p-1)G_{p+1,m}(\ell_\alpha;\Delta^2)}{\{2G_{p+1,m}(\ell_\alpha;\Delta^2)-G_{p+3,m}(\ell_\alpha^*;\Delta^2)\}} \quad (\le p-1) \quad (5.4.10)$$

from (5.4.3a) and (5.4.3b)

(ii) The maximum efficiency is attained at $\Delta^2 = 0$ with the value

$$\left\{1-\left(1-\frac{1}{p}\right)G_{p+1,m}(\ell_\alpha;0)\right\}^{-1}. \quad (5.4.11)$$

(iii) $\mathrm{RRE}(\hat{\boldsymbol{\theta}}_n^{\mathrm{PT}};\tilde{\boldsymbol{\theta}}_n)$ decreases as a function of Δ^2 for fixed $\alpha \in (0,1)$ crossing the 1-line to a minimum, say, at $\Delta^2 = \Delta^2_{\min}$; then monotonically increases toward unity as $\Delta^2 \to \infty$. Since PTE is not an estimator with uniform dominance over $\tilde{\boldsymbol{\theta}}_n$, we can obtain a PTE with minimum guaranteed efficiency, say, E_0 by solving the inequality

$$\max_{\alpha\in\mathcal{A}} \min_{\Delta^2} \mathrm{RRE}(\hat{\boldsymbol{\theta}}_n^{\mathrm{PT}};\tilde{\boldsymbol{\theta}}_n) = E(\alpha,\Delta^2_{\min}(\alpha)) = E_0. \quad (5.4.12)$$

Tables of maximum and minimum efficiency may be obtained as before.

(iv) As $\alpha \to 0$, $\hat{\boldsymbol{\theta}}_n^{\mathrm{PT}} \to \hat{\boldsymbol{\theta}}_n$, while $\hat{\boldsymbol{\theta}}_n^{\mathrm{PT}} \to \tilde{\boldsymbol{\theta}}_n$ as $\alpha \to 1$.

Next, we consider the risk efficiency of $\hat{\boldsymbol{\theta}}_n^{\mathrm{PT}}$ relative to $\hat{\boldsymbol{\theta}}_n$ for $\mathbf{W} = \mathbf{N}$ given by

$$\begin{aligned}\mathrm{RRE}(\hat{\boldsymbol{\theta}}_n^{\mathrm{PT}} : \hat{\boldsymbol{\theta}}_n)) &= p^{-1}(1+\Delta^2)\left[1-\left(1-\frac{1}{p}\right)G_{p+1,m}(\ell_\alpha;\Delta^2)\right.\\ &\quad \left.+\Delta^2 p^{-1}\left\{2G_{p+1,m}(\ell_\alpha;\Delta^2)-G_{p+3,m}(\ell_\alpha^*;\Delta^2)\right\}\right]^{-1}. \end{aligned} \quad (5.4.13)$$

We note the following properties of $\text{RRE}(\hat{\boldsymbol{\theta}}_n^{\text{PT}};\hat{\boldsymbol{\theta}}_n)$:

(i) $\text{RRE}(\hat{\boldsymbol{\theta}}_n^{\text{PT}};\hat{\boldsymbol{\theta}}_n)\gtreqless 1$ according as (5.4.14)

$$\Delta^2 \leq \frac{(p-1)\{1-G_{p+1,m}(\ell_\alpha;\Delta^2)\}}{\{2G_{p+1,m}(\ell_\alpha;\Delta^2)-G_{p+3,m}(\ell_\alpha^*;\Delta^2)\}}.$$

(ii) Under H_0, the efficiency is given by

$$p^{-1}\left[1-\left(1-\frac{1}{p}\right)G_{p+1,m}(\ell_\alpha;0)\right]^{-1} \quad (\geq p^{-1}),$$

while that of $\hat{\boldsymbol{\theta}}_n^{\text{PT}}$ rel $\tilde{\boldsymbol{\theta}}_n$ is given by

$$\left[1-\left(1-\frac{1}{p}\right)G_{p+1,m}(\ell_\alpha;0)\right]^{-1} \quad (\geq 1)$$

Thus, under H_0,

$$p^{-1} \leq \text{RRE}(\hat{\boldsymbol{\theta}}_n^{\text{PT}};\hat{\boldsymbol{\theta}}_n) \leq \text{RRE}(\hat{\boldsymbol{\theta}}_n^{\text{PT}};\tilde{\boldsymbol{\theta}}_n).$$

(iii) As $\Delta^2 \to \infty$, $\text{RRE}(\hat{\boldsymbol{\theta}}_n^{\text{PT}};\hat{\boldsymbol{\theta}}_n) \to \infty$.

(iv) As $\alpha \to 0$, $\hat{\boldsymbol{\theta}}_n^{\text{PT}} \to \hat{\boldsymbol{\theta}}_n$, and as $\alpha \to 1$, $\hat{\boldsymbol{\theta}}_n^{\text{PT}} \to \tilde{\boldsymbol{\theta}}_n$.

5.4.3 Comparison of $\hat{\boldsymbol{\theta}}_n^S$, $\hat{\boldsymbol{\theta}}_n^{S+}$, and $\tilde{\boldsymbol{\theta}}_n$.

First note that the risk-difference is

$$\begin{aligned}
&R_1(\tilde{\boldsymbol{\theta}}_n;\mathbf{W}) - R_4(\hat{\boldsymbol{\theta}}_n^S;\mathbf{W}) \\
&= c(p-1)\sigma^2\text{tr}(\mathbf{WHN}^{-1})\Big\{(p-3)E[\chi_{p+1}^{-1}(\Delta^2)] \\
&\quad + 2\Delta^2E(\chi_{p+3}^{-4}(\Delta^2))\left[1-\frac{(p+1)(\boldsymbol{\delta}'\mathbf{W}\boldsymbol{\delta})}{2(\boldsymbol{\delta}'\mathbf{N}\boldsymbol{\delta})\text{tr}(\mathbf{WHN}^{-1})}\right]\Big\} \geq 0,
\end{aligned} \tag{5.4.15}$$

for all Δ^2 and all matrices $\mathbf{W}$ such that

$$\frac{\text{tr}(\mathbf{WHN}^{-1})}{Ch_{\max}(\mathbf{WN}^{-1})} \geq \frac{p+1}{2}. \tag{5.4.16}$$

Hence, $\hat{\boldsymbol{\theta}}_n^S$ dominates $\tilde{\boldsymbol{\theta}}_n$ uniformly for all $\mathbf{W}$ satisfying (5.4.16).

Next, we consider the comparison of $\hat{\boldsymbol{\theta}}_n^{S+}$ and $\hat{\boldsymbol{\theta}}_n^{S}$. Again, the risk-difference

$$
\begin{aligned}
&R_4\left(\hat{\boldsymbol{\theta}}_n^{S};\mathbf{W}\right)-R_5\left(\hat{\boldsymbol{\theta}}_n^{S+};\mathbf{W}\right)\\
&=\sigma^2\operatorname{tr}\left(\mathbf{WHN}^{-1}\right)E\Big[\Big(1-\frac{c(p-1)}{p+1}F_{p+1,m}^{-1}(\Delta^2)\Big)^2 I\Big(F_{p+1,m}(\Delta^2)<\frac{p-1}{p+1}c\Big)\Big]\\
&\quad+(\boldsymbol{\delta}'\mathbf{W}\boldsymbol{\delta})\Big\{2E\Big[\Big(\frac{c(p-1)}{(p+1)}F_{p+1,m}^{-1}(\Delta^2)-1\Big)^2 I\Big(F_{p+1,m}(\Delta^2)<\frac{p-1}{p+1}c\Big)\Big]\\
&\quad-E\Big[\Big(\frac{c(p-1)}{(p+3)}F_{p+3,m}^{-1}(\Delta^2)-1\Big)^2 I\Big(F_{p+3,m}(\Delta^2)<\frac{p-1}{p+3}c\Big)\Big]\Big\}
\end{aligned}
$$

which is nonnegative for all $\boldsymbol{\delta}$.

If $\mathbf{W}=\sigma^{-2}\mathbf{N}$, the r.h.s. of risk-difference reduces to

$$
\begin{aligned}
&(p-1)E\Big[\Big(\frac{c(p-1)}{(p+1)}F_{p+1,m}^{-1}(\Delta^2)-1\Big)I\Big(F_{p+1,m}(\Delta^2)<\frac{p-1}{p+1}c\Big)\Big]\\
&\quad+\Delta^2\Big\{2E\Big[\Big(\frac{c(p-1)}{(p+1)}F_{p+1,m}^{-1}(\Delta^2)-1\Big)I\Big(F_{p+1,m}(\Delta^2)<\frac{p-1}{p+1}c\Big)\Big]\\
&\quad-E\Big[\Big(\frac{c(p-1)}{(p+3)}F_{p+3,m}^{-1}(\Delta^2)-1\Big)^2 I\Big(F_{p+3,m}(\Delta^2)<\frac{p-1}{p+3}c\Big)\Big]\Big\}.
\end{aligned}
$$

Thus, we obtain the order in the risk functions of $\hat{\boldsymbol{\theta}}_n$, $\hat{\boldsymbol{\theta}}_n^{S}$, and $\hat{\boldsymbol{\theta}}_n^{S+}$ as

$$
R_1(\tilde{\boldsymbol{\theta}}_n;\mathbf{W})\geq R_4(\hat{\boldsymbol{\theta}}_n^{S};\mathbf{W})\geq R_5(\hat{\boldsymbol{\theta}}_n^{S+};\mathbf{W})\quad\text{for all }\Delta^2,\tag{5.4.17}
$$

and dominance of $\hat{\boldsymbol{\theta}}_n^{S+}$ over $\hat{\boldsymbol{\theta}}_n^{S}$ as well as over $\tilde{\boldsymbol{\theta}}_n$ is established.

It may be verified that $\mathrm{RRE}(\hat{\boldsymbol{\theta}}_n^{S};\tilde{\boldsymbol{\theta}}_n)$ and $\mathrm{RRE}(\hat{\boldsymbol{\theta}}_n^{S+};\hat{\boldsymbol{\theta}}_n^{S})$ are decreasing functions of Δ^2 bounded below by the 1-line.

The risk-efficiency of $\hat{\boldsymbol{\theta}}_n^{S}$ compared to $\tilde{\boldsymbol{\theta}}_n$ is given by

$$
\begin{aligned}
&\mathrm{RRE}(\hat{\boldsymbol{\theta}}_n^{S};\tilde{\boldsymbol{\theta}}_n)\\
&=\Big(1-\frac{(p-3)(p+1)m}{p(m+2)}\Big\{2E\Big[X_{p+1}^{-2}(\Delta^2)\Big]-(p-3)E\Big[X_{p+1}^{-4}(\Delta^2)\Big]\Big\}\\
&\quad+\frac{(p-3)(p+1)^m}{p(m+2)}\Delta^2E\Big[X_{p+1}^{-4}(\Delta^2)\Big]\Big)^{-1}.
\end{aligned}\tag{5.4.18}
$$

Similarly, the risk-efficiency of $\hat{\boldsymbol{\theta}}_n^{S}$ compared to $\hat{\boldsymbol{\theta}}_n^{S}$ may be written as

$$\begin{aligned}
&\mathrm{RRE}(\hat{\boldsymbol{\theta}}_n^{S+};\hat{\boldsymbol{\theta}}_n^{S})\\
&=\Bigg[1-\{R_3(\hat{\boldsymbol{\theta}}_n^{S};\sigma^{-2}I_p)\}^{-1}\Big\{(p-1)\\
&\times E\Big[\Big(1-\frac{c(p-1)}{p+1}F_{p+1,m}^{-1}(\Delta^2(\Big)^2 I\Big(F_{p+1,m}(\Delta^2)<\frac{c(p-1)}{p+1}\Big)\Big]\\
&+\Delta^2 E\Big[\Big(\frac{c(p-1)}{(p+3)}F_{p+3,m}^{-1}(\Delta^2)-1\Big)^2 I\Big(F_{p+3,m}(\Delta^2)<\frac{c(p-1)}{p+3}\Big)\Big\}\\
&-2\{R_3(\hat{\boldsymbol{\theta}}_n^{S};\sigma^{-2}\mathbf{I}_p)\}\Big]^{-1}\Delta^2 E\Big[\Big(\frac{c(p-1)}{p+1}F_{p+1,m}^{-1}(\Delta^2)-1\Big)\\
&\times I\Big(F_{p+1,m}(\Delta^2)<\frac{c(p-1)}{p+1}\Big)\Big]\Bigg]^{-1}. \qquad (5.4.19)
\end{aligned}$$

5.5 MSE Matrix Analysis and Efficiency

In this section, we consider the MSE matrix analysis of the various estimators and look at the efficiency based on the MSE matrices.

5.5.1 Comparison of $\hat{\boldsymbol{\theta}}_n$ and $\tilde{\boldsymbol{\theta}}_n$

In this case, consider the matrix-difference

$$M_1(\tilde{\boldsymbol{\theta}}_n)-M_2(\hat{\boldsymbol{\theta}}_n)=\sigma^2\mathbf{N}^{-1}-\frac{\sigma^2}{n}\mathbf{1}_p\mathbf{1}_p'-\boldsymbol{\delta}\boldsymbol{\delta}'=\sigma^2(\mathbf{H}\mathbf{N}^{-1})-\boldsymbol{\delta}\boldsymbol{\delta}'. \qquad (5.5.1)$$

This MSE-difference is positive definite whenever for a given nonzero vector $\boldsymbol{\ell}=(\ell_1,\dots,\ell_p)'$ we have

$$\sigma^2\boldsymbol{\ell}'\mathbf{H}\mathbf{N}^{-1}\boldsymbol{\ell}-\boldsymbol{\ell}'\boldsymbol{\delta}\boldsymbol{\delta}'\boldsymbol{\ell}>0. \qquad (5.5.2)$$

That is,

$$\boldsymbol{\ell}'\mathbf{H}\mathbf{N}^{-1}\boldsymbol{\ell}\geq\sigma^{-2}\boldsymbol{\ell}'\boldsymbol{\delta}\boldsymbol{\delta}\boldsymbol{\ell}. \qquad (5.5.3)$$

Since $\boldsymbol{\ell}'\mathbf{N}^{-1}\boldsymbol{\ell}>0$, we consider

$$\max_{\boldsymbol{\ell}}\frac{\boldsymbol{\ell}'[\mathbf{H}\mathbf{N}^{-1}]\boldsymbol{\ell}}{\boldsymbol{\ell}'\mathbf{N}^{-1}\boldsymbol{\ell}}\geq\sigma^{-2}\max_{\boldsymbol{\ell}}\frac{\boldsymbol{\ell}'\boldsymbol{\delta}\boldsymbol{\delta}'\boldsymbol{\ell}}{\boldsymbol{\ell}'\mathbf{N}^{-1}\boldsymbol{\ell}}$$

or

$$Ch_{\max}[\mathbf{H}]\geq\Delta^2. \qquad (5.5.4)$$

Hence, $\hat{\theta}_n$ performs better than $\tilde{\theta}_n$ if $0 \le \Delta^2 \le p-1$; otherwise, $\tilde{\theta}_n$ performs better than $\hat{\theta}_n$.

The efficiency of $\hat{\theta}_n$ relative to $\tilde{\theta}_n$ is then given by

$$\begin{aligned}\text{MRE}(\hat{\theta}_n;\tilde{\theta}_n) &= \frac{|\mathbf{M}_1(\tilde{\theta}_n)|^{1/p}}{|\mathbf{M}_2(\hat{\theta}_n)|^{1/p}} \\ &= \frac{|\sigma^2\mathbf{N}^{-1}|^{1/p}}{|\frac{\sigma^2}{n}\mathbf{1}_p\mathbf{1}_p' + \delta\delta'|^{1/p}} \\ &= \left|\frac{1}{n}\mathbf{1}_p\mathbf{1}_p'\mathbf{N} + \sigma^{-2}\delta\delta'\mathbf{N}\right|^{-1/p}. \end{aligned} \tag{5.5.5}$$

But, $|\frac{1}{n}\mathbf{1}_p\mathbf{1}_p'\mathbf{N} + \sigma^{-2}\delta\delta'\mathbf{N}| = 0$, $\text{MRE}(\hat{\theta}_n;\tilde{\theta}_n)$ is meaningless. Thus, $\text{MRE}(\hat{\theta}_n;\tilde{\theta}_n)$ is not the suitable expression for it. We must depend on $\text{RRE}(\hat{\theta}_n;\tilde{\theta}_n)$.

5.5.2 Comparison of $\hat{\theta}_n^{\text{PT}}$ Relative to $\tilde{\theta}_n$ and $\hat{\theta}_n$

The MSE matrix difference is given by

$$\begin{aligned}\mathbf{M}_3(\hat{\theta}_n^{\text{PT}}) - M_1(\tilde{\theta}_n) &= -\sigma^2(\mathbf{H}\mathbf{N}^{-1})G_{p+1,m}(\ell_\alpha;\Delta^2) \\ &\quad +\delta\delta'\{2G_{p+1,m}(\ell_\alpha;\Delta^2) - G_{p+3,m}(\ell_\alpha^*;\Delta^2)\}\end{aligned} \tag{5.5.6}$$

It can be shown by the method in Section 5.5.1 that $\hat{\theta}_n^{\text{PT}}$ performs better than $\tilde{\theta}_n$ if

$$\Delta^2 < \frac{G_{p+1,m}(\ell_\alpha;\Delta^2)}{\{2G_{p+1,m}(\ell_\alpha;\Delta^2) - G_{p+3,m}(\ell_\alpha^*;\Delta^2)\}}; \tag{5.5.7}$$

otherwise, $\tilde{\theta}_n$ performs better than $\hat{\theta}_n^{\text{PT}}$. Similarly, we can show that $\hat{\theta}_n$ performs better than $\hat{\theta}_n^{\text{PT}}$ if

$$\Delta^2 < \frac{\{1 - G_{p+1,m}(\ell_\alpha;\Delta^2)\}}{\{1 - 2G_{p+1,m}(\ell_\alpha;\Delta^2) + G_{p+3,m}(\ell_\alpha^*;\Delta^2)\}}. \tag{5.5.8}$$

Otherwise, $\hat{\theta}_n^{\text{PT}}$ performs better than $\hat{\theta}_n$. Thus, none of the estimators dominate the other.

Now, the MSE-based efficiency of $\hat{\theta}_n^{\text{PT}}$ relative to $\tilde{\theta}_n$ and $\hat{\theta}_n$ is given respectively by

$$\begin{aligned}\text{MRE}(\hat{\theta}_n^{\text{PT}};\tilde{\theta}_n) &= \frac{|\mathbf{M}_1(\tilde{\theta}_n)|^{1/p}}{|\mathbf{M}_3(\hat{\theta}_n^{\text{PT}})|^{1/p}} \\ &= \Big|\mathbf{I}_p - \mathbf{H}G_{p+1,m}(\ell_\alpha;\Delta^2) + \sigma^{-2}(\delta\delta'\mathbf{N})\{2G_{p+1,m}(\ell_\alpha;\Delta^2) \\ &\qquad -G_{p+3,m}(\ell_\alpha^*;\Delta^2)\}\Big|^{-1/p}.\end{aligned} \tag{5.5.9}$$

Table 5.5.1 Maximum and Minimum Guaranteed Efficiencies

$\alpha\backslash(n,p)$		(15,4)	(15,6)	(15,8)	(15,10)	(15,12)
0.05	E^*	7.5774	10.0840	12.2909	14.4987	17.0014
	E_0	0.7986	0.8450	0.8661	0.8633	1.0528
	Δ_0	17.7443	31.0244	53.0571	98.2937	205.7908
0.10	E^*	4.2183	5.4015	6.4131	7.4080	8.5414
	E_0	0.7004	0.7535	0.7703	0.7584	0.7494
	Δ_0	15.0004	25.7483	43.9494	85.8566	212.9081
0.15	E^*	3.0422	3.7906	4.4183	5.0280	5.7225
	E_0	0.8653	0.8999	0.9117	0.9054	0.8624
	Δ_0	11.6013	19.6068	31.9353	58.4570	169.2741
0.20	E^*	2.4357	2.9685	3.4089	3.8323	4.3139
	E_0	0.8856	0.9146	0.9248	0.9197	0.8808
	Δ_0	10.2014	17.0557	27.2477	47.9643	133.8645
0.25	E^*	2.0642	2.4680	2.7979	3.1126	3.4696
	E_0	0.9022	0.9265	0.9353	0.9311	0.8974
	Δ_0	9.1734	15.1878	23.8846	40.7270	106.7507
0.30	E^*	1.8130	2.1308	2.3881	2.6317	2.9074
	E_0	0.9165	0.9367	0.9442	0.9406	0.9116
	Δ_0	8.3718	13.7302	21.2998	35.3589	86.3478
0.35	E^*	1.6321	1.8883	2.0941	2.2879	2.5065
	E_0	0.9289	0.9455	0.9520	0.9489	0.9238
0.40	E^*	1.4961	1.7057	1.8732	2.0300	2.2065
	E_0	0.9399	0.9534	0.9588	0.9565	0.9346
	Δ_0	7.1811	11.5522	17.4971	27.7770	59.8843
0.45	E^*	1.3906	1.56360	1.7013	1.8297	1.9738
	E_0	0.9497	0.9604	0.9649	0.9627	0.9442
	Δ_0	6.7219	10.7033	16.0321	24.9568	51.0341
0.50	E^*	1.3069	1.4502	1.5641	1.6698	1.7883
	E_0	0.9584	0.9668	0.9705	0.9685	0.9528
	Δ_0	6.3262	9.9655	14.7650	22.5630	43.9497

Since there exists an orthogonal matrix $\mathbf{\Gamma}$ such that $\mathbf{\Gamma H \Gamma}' = \text{Diag}(1,1,\ldots,1,0)$ and $\mathbf{\Gamma}\boldsymbol{\delta}\boldsymbol{\delta}'\mathbf{N}\mathbf{\Gamma}'\text{Diag}(\Delta^2, 0\ldots,0)$, we write (5.5.10) as

$$
\begin{aligned}
&= \left\{1 - G_{p+1,m}(\ell_\alpha;\Delta^2)\right\}^{-p-2/p} \times \\
&\quad \Big[1 - G_{p+1,m}(\ell_\alpha;\Delta^2) + \Delta^2\{2_{p+1,m}(\ell_\alpha;\Delta^2) \\
&\quad -G_{p+3,m}(\ell^*_\alpha;\Delta^2)\}\Big]^{-1/p},
\end{aligned}
\tag{5.5.10}
$$

Note that under H_0,

$$
\text{MRE}\Big(\hat{\boldsymbol{\theta}}_n^{\text{PT}};\tilde{\boldsymbol{\theta}}_n\Big) = \Big\{1 - G_{p+1,m}(\ell_\alpha;0)\Big\}^{-1+(1/p)} \quad (\geq 1),
$$

and under the alternatives, that is, $\Delta^2 > 0$, $\text{MRE}(\hat{\boldsymbol{\theta}}_n^{\text{PT}};\tilde{\boldsymbol{\theta}}_n)$ is decreasing function of Δ^2 for fixed α until $\Delta^2 = \Delta^2_{\min}$. Then it increases toward 1 as $\Delta^2 \to \infty$.

To determine optimum level of significance α to obtain a minimum guaranteed efficiency E_0, we solve the equation for α:

$$\min_{\Delta^2} \text{MRE}(\hat{\boldsymbol{\theta}}_n^{\text{PT}}; \tilde{\boldsymbol{\theta}}_n) = E(\alpha, \Delta^2_{\text{min}}(\alpha)) = E_0. \tag{5.5.11}$$

Some tabular values are given in Table 5.5.1 for various p and n values.

5.5.3 Comparison of $\tilde{\boldsymbol{\theta}}_n$ and $\hat{\boldsymbol{\theta}}_n^S$ ($\hat{\boldsymbol{\theta}}_n^S$ and $\hat{\boldsymbol{\theta}}_n^{S+}$)

First, consider the comparison of $\tilde{\boldsymbol{\theta}}_n$ and $\hat{\boldsymbol{\theta}}_n^S$. In this case, the MSE-matrix difference is given by

$$\begin{aligned}
&\mathbf{M}_1(\tilde{\boldsymbol{\theta}}_n) - \mathbf{M}_4(\hat{\boldsymbol{\theta}}_n^S) \\
&= \sigma^2 c\frac{(p-3)}{m+2}(\mathbf{HN}^{-1})\Big\{2E[\chi^{-2}_{p+1}(\Delta^2)] - (p-3)E[\chi^{-4}_{p+1}(\Delta^2)]\Big\} \\
&\quad - c\frac{(p-3)(p+1)}{m+2}(\boldsymbol{\delta\delta}')E[\chi^{-4}_{p+3}(\Delta^2)].
\end{aligned} \tag{5.5.12}$$

For $\hat{\boldsymbol{\theta}}_n^S$ to dominate $\tilde{\boldsymbol{\theta}}_n$ based on MSE matrices, we must show that (5.5.13) is nonnegative definite. Thus, consider the quadratic form

$$\begin{aligned}
&\sigma^2 \boldsymbol{\ell}'(\mathbf{HN}^{-1}\boldsymbol{\ell})\left\{2E[\chi^{-2}_{p+1}(\Delta^2)] - (p-3)E[\chi^{-4}_{p+1}(\Delta^2)]\right\} \\
&\quad -(p+1)\boldsymbol{\ell}'(\boldsymbol{\delta\delta}')\boldsymbol{\ell}\, E[\chi^{-4}_{p+3}(\Delta^2)]
\end{aligned} \tag{5.5.13}$$

for a given nonzero vector $\boldsymbol{\ell} = (\ell_1, \dots, \ell_p)'$. Dividing by $\boldsymbol{\ell}'\mathbf{N}^{-1}\boldsymbol{\ell} (> 0)$, and maximizing over all ℓ, we have

$$\begin{aligned}
&\max_{\boldsymbol{\ell}} \frac{\boldsymbol{\ell}'\mathbf{HN}^{-1}\boldsymbol{\ell}}{\boldsymbol{\ell}'\mathbf{N}^{-1}\boldsymbol{\ell}}\Big\{2E[\chi^{-2}_{p+1}(\Delta^2)] - (p-3)E[\chi^{-4}_{p+1}(\Delta^2)]\Big\} \\
&\qquad - (p+1)\max_{\boldsymbol{\ell}} \frac{\boldsymbol{\ell}'\boldsymbol{\delta\delta}'\boldsymbol{\ell}}{\boldsymbol{\ell}'\mathbf{N}^{-1}\boldsymbol{\ell}} E[\chi^{-4}_{p+1}(\Delta^2)] \qquad (5.5.14)\\
&= \Big\{E[\chi^{-2}_{p+1}(\Delta^2)] + \Delta^2 E[\chi^{-4}_{p+3}(\Delta^2)]\Big\} - (p+1)\Delta^2 E[\chi^{-4}_{p+3}(\Delta^2)] \\
&= E[\chi^{-2}_{p+1}(\Delta^2)] - p\Delta^2 E[\chi^{-4}_{p+3}(\Delta^2)] \geq 0 \quad \text{for all } \Delta^2, \qquad (5.5.15)
\end{aligned}$$

which does not always hold. Hence, $\hat{\boldsymbol{\theta}}_n^S$ does not dominate $\tilde{\boldsymbol{\theta}}_n$ uniformly.

The MSE-efficiency of $\hat{\boldsymbol{\theta}}_n^S$ compared to $\tilde{\boldsymbol{\theta}}_n$ is given by

$$\begin{aligned}
\text{MRE}(\hat{\boldsymbol{\theta}}_n^S;\tilde{\boldsymbol{\theta}}_n) &= \frac{|\mathbf{M}_1(\tilde{\boldsymbol{\theta}}_n)|^{1/p}}{|\mathbf{M}_4(\hat{\boldsymbol{\theta}}_n^S)|^{1/p}} \\
&= \Big|\mathbf{I}_p - c\frac{(p-1)}{m+2}\mathbf{H}\Big\{2E[\chi_{p+1}^{-2}(\Delta^2)] - (p-3)E[\chi_{p+1}^{-4}(\Delta^2)]\Big\} \\
&\quad + c\frac{(p-3)(p+1)}{m+2}(\boldsymbol{\delta}\boldsymbol{\delta}'\mathbf{N})E[\chi_{p+3}^{-4}(\Delta^2)]\Big|^{-1/p} \\
&= \Big[1 - c\frac{(p-1)}{m+2}\Big\{2E[\chi_{p+1}^{-2}(\Delta^2)] - (p-3)E[\chi_{p+1}^{-4}(\Delta^2)]\Big\}^{(p+2)/p} \\
&\quad \times\Big[1 - c\frac{(p-1)}{m+2}\Big\{2E[\chi_{p+1}^{-2}(\Delta^2)] - (p-3)E[\chi_{p+1}^{-4}(\Delta^2)]\Big\} \\
&\quad + c(p-3)(p+1)m + 2\Delta^2 E[\chi_{p+3}^{-4}(\Delta^2)]\Big]^{-1/p}. \qquad (5.5.16)
\end{aligned}$$

Under H_0, the MRE $(\hat{\boldsymbol{\theta}}_n^S;\tilde{\boldsymbol{\theta}}_n)$ becomes

$$\text{MRE}\left(\hat{\boldsymbol{\theta}}_n^S;\tilde{\boldsymbol{\theta}}_n\right) = \left\{1 - \frac{(p-3)m}{(p-1)(m+2)}\right\}^{-1+(1/p)} \ (\geq 1), \qquad (5.5.17)$$

and as $\Delta^2 \to \infty$, $\text{MRE}(\hat{\boldsymbol{\theta}}_n^S;\tilde{\boldsymbol{\theta}}_n) \to 1$.

In order to compare $\hat{\boldsymbol{\theta}}_n^{S+}$ and $\hat{\boldsymbol{\theta}}_n^S$, we note from (5.3.6e) that

$$\begin{aligned}
&\mathbf{M}_4(\hat{\boldsymbol{\theta}}_n^S) - M_5(\hat{\boldsymbol{\theta}}_n^{S+}) \\
&= \sigma^2(\mathbf{H}\mathbf{N}^{-1})E\left[\left(1 - \frac{c(p-1)}{p+1}F_{p+1,m}^{-1}(\Delta^2)\right)^2 I\left(F_{p+1,m}(\Delta^2) < \frac{c(p-1)}{p+1}\right)\right] \\
&-\boldsymbol{\delta}\boldsymbol{\delta}'\left\{2E\left[\left(1 - \frac{c(p-1)}{p+1}F_{p+1,m}^{-1}(\Delta^2\right) I\left(F_{p+1,m}(\Delta^2) < \frac{c(p-1)}{p+1}\right)\right]\right. \\
&\left.-E\Big[\Big(1 - \frac{c(p-1)}{p+1}F_{p+3,m}^{-1}(\Delta^2\Big)^2 I\Big(F_{p+3,m}(\Delta^2) < \frac{c(p-1)}{p+3}\Big)\Big]\right\}. \qquad (5.5.18)
\end{aligned}$$

The r.h.s. of (5.5.18) is positive-semidefinite, since

$$0 < F(\Delta^2) < \frac{c(p-1)}{(p+3)} < \frac{c(p-1)}{p+1}.$$

Thus, the MSE-based efficiency of $\hat{\boldsymbol{\theta}}_n^{S+}$ relative to $\hat{\boldsymbol{\theta}}_n^S$ is given by

$$MRE\left(\hat{\boldsymbol{\theta}}_n^{S+};\hat{\boldsymbol{\theta}}_n^S\right) = \frac{|\mathbf{M}_4(\hat{\boldsymbol{\theta}}_n^S)|^{1/p}}{|\mathbf{M}_5(\hat{\boldsymbol{\theta}}_n^{S+})|^{1/p}} = \frac{\gamma_{41}^{(p-2)/p}\cdot\gamma_{42}^{1/p}}{\gamma_{51}^{(p-2)/p}\cdot\gamma_{52}^{1/p}}, \qquad (5.5.19)$$

where

$$\gamma_{41} = \Big[1 - \tfrac{c(p-1)}{m+2}\{2E[\chi_{p+1}^{-2}(\Delta^2)] - (p-3)E[\chi_{p+1}^{4}(\Delta^2)]\}\Big],$$

$$\gamma_{42} = \Big[1 - \tfrac{c(p-1)}{m+2}\{2E[\chi_{p+1}^{-2}(\Delta^2)] - (p-3)E[\chi_{p+1}^{4}(\Delta^2)]\} + \tfrac{c(p^2-1)}{m+2}\Delta^2 E[\chi_{p+3}^{-4}(\Delta^2)]\Big], \tag{5.5.20a}$$

$$\gamma_{51} = \left\{\gamma_{41} - E\left[\left(1 - \frac{c(p-1)}{p+1}F_{p+2,m}^{-1}(\Delta^2)\right)^2 I\left(F_{p+1,m}(\Delta^2) < \frac{c(p-1)}{p+1}\right)\right]\right\}, \tag{5.5.20b}$$

$$\gamma_{52} = \Bigg\{\gamma_{42} - E\Bigg\{\left(1 - \frac{c(p-1)}{p+1}F_{p+1,m}^{-1}(\Delta^2)\right)^2 + \Delta^2 E\Bigg\{2E\left[\left(1 - \frac{c(p-1)}{p+1}F_{p+1,m}^{-1}(\Delta^2)\right) I\left(F_{p+1,m}(\Delta^2) < \frac{c(p-1)}{p+1}\right)\right.$$
$$\left. -E\left[\left(1 - \frac{c(p-1)}{p+3}F_{p+3,m}^{-1}(\Delta^2)\right)^2 I\left(F_{p+3,m}(\Delta^2) < \frac{c(p-1)}{p+3}\right)\right]\right]\Bigg\}\Bigg\}. \tag{5.5.20c}$$

Under H_0, $\mathrm{MRE}(\hat{\boldsymbol{\theta}}_n^{S+}; \hat{\boldsymbol{\theta}}_n^{S}) = \frac{\gamma_{41}^{(p-1)/p}}{\gamma_{51}^{(p-1)/p}}$ (≥ 1), and as $\Delta^2 \to \infty$, $\mathrm{MRE}(\hat{\boldsymbol{\theta}}_n^{S+}; \hat{\boldsymbol{\theta}}_n^{S})$ $\to 1$ from above.

The MSE-efficiency of $\hat{\boldsymbol{\theta}}_n^{S+}$ compared to $\hat{\boldsymbol{\theta}}_n^{S}$ may be written as

$$\mathrm{MRE}\left(\hat{\boldsymbol{\theta}}_n^{S+}; \tilde{\boldsymbol{\theta}}_n^{S}\right) = \frac{|\mathbf{M}_4(\hat{\boldsymbol{\theta}}_n^{S})|^{1/p}}{|\mathbf{M}_5(\hat{\boldsymbol{\theta}}_n^{S+})|^{1/p}}$$

$$= \Bigg|\mathbf{I}_p - \sigma^2\left(\mathbf{M}_4(\hat{\boldsymbol{\theta}}_n^{S})\right)^{-1}(\mathbf{HN}^{-1})$$
$$\times E\left[\left(1 - \frac{c(p-1)}{(p+1)}F_{p+1,m}^{-1}(\Delta^2)\right)^2 I\left(F_{p+1,m}(\Delta^2) < \frac{c(p-1)}{(p+1)}\right)\right]$$
$$-(\boldsymbol{\delta\delta}')E\left[\left(\frac{c(p-1)}{(p+3)}F_{p+3,m}^{-1}(\Delta^2) - 1\right)^2 I\left(F_{p+3,m}(\Delta^2) < \frac{c(p-1)}{(p+3)}\right)\right]$$
$$+(\mathbf{M}_4(\hat{\boldsymbol{\theta}}_1^{S}))^{-1}(\boldsymbol{\delta\delta}')2E\Bigg[\left(c\frac{c(p-1)}{(p+1)}F_{p+3,m}^{-1}(\Delta^2) - 1\right)$$
$$\times I\left(F_{p+1,m}(\Delta^2) < \frac{c(p-1)}{(p+1)}\right)\Bigg]\Bigg|^{-1}. \tag{5.5.21}$$

It may be shown that under H_0, the MRE $(\hat{\boldsymbol{\theta}}_n^{S+}; \hat{\boldsymbol{\theta}}_n^{S}) \geq 1$, and as $\Delta^2 \to \infty$, MRE $(\hat{\boldsymbol{\theta}}_n^{S+}; \hat{\boldsymbol{\theta}}_n^{S})$ tends to 1 from above.

5.6 Improving the PTE

As in Chapter 4, we consider the improvement of the PTE $\hat{\boldsymbol{\theta}}_n^{\mathrm{PT}}$ of $\boldsymbol{\theta}$ for $p \geq 4$. We recall that the PTE of $\boldsymbol{\theta}$ is given by

$$\hat{\boldsymbol{\theta}}_n^{\mathrm{PT}} = \tilde{\boldsymbol{\theta}}_n - (\tilde{\boldsymbol{\theta}}_n - \hat{\boldsymbol{\theta}}_n) I(\mathcal{L}_n \leq F_{p-1,m}(\alpha)) \tag{5.6.1}$$

based on $\tilde{\boldsymbol{\theta}}_n$ and $\hat{\boldsymbol{\theta}}_n$. Now, for $p \geq 4$, we have the Stein-type estimator of $\boldsymbol{\theta}$ given by

$$\hat{\boldsymbol{\theta}}_n^{S} = \tilde{\boldsymbol{\theta}}_n - c(\tilde{\boldsymbol{\theta}}_n - \hat{\boldsymbol{\theta}}_n)\mathcal{L}_n^{-1}, \quad c = \frac{(p-3)m}{(p-1)(m+2)}. \tag{5.6.2}$$

Thus, for $p \geq 4$, we may consider the estimator $\hat{\boldsymbol{\theta}}_n^{PT+}$ by combining $\hat{\boldsymbol{\theta}}_n^{S}$ and $\hat{\boldsymbol{\theta}}_n$ via the test on $\boldsymbol{\theta} = \theta_0 \mathbf{1}_p$. We may write

$$\begin{aligned}
\hat{\boldsymbol{\theta}}_n^{PT+} &= \hat{\boldsymbol{\theta}}_n I(\mathcal{L}_n < F_{p-1,m}(\alpha)) + \hat{\boldsymbol{\theta}}_n^{S} I(\mathcal{L}_n \geq F_{p-1,m}(\alpha)) \\
&= \hat{\boldsymbol{\theta}}_n I(\mathcal{L}_n < F_{p-1,m}(\alpha)) + \hat{\boldsymbol{\theta}}_n^{S} - \hat{\boldsymbol{\theta}}_n^{S} I(\mathcal{L}_n < F_{p-1,m}(\alpha)) \\
&= \hat{\boldsymbol{\theta}}_n^{S} - (\hat{\boldsymbol{\theta}}_n^{S} - \hat{\boldsymbol{\theta}}_n) I(\mathcal{L}_n < F_{p-1,m}(\alpha)) \\
&= \hat{\boldsymbol{\theta}}_n^{PT} - c(\tilde{\boldsymbol{\theta}}_n - \hat{\boldsymbol{\theta}}_n)\mathcal{L}_n^{-1} I(\mathcal{L} \geq F_{p-1,m}(\alpha)).
\end{aligned} \tag{5.6.3}$$

The bias vector, $\mathbf{b}_6(\hat{\boldsymbol{\theta}}_n^{PT+})$ and the quadratic bias of $\hat{\boldsymbol{\theta}}_n^{PT+}$ are then given by

$$\begin{aligned}
\mathbf{b}_6\left(\hat{\boldsymbol{\theta}}_n^{PT+}\right) &= \mathbf{b}_3\left(\hat{\boldsymbol{\theta}}_n^{PT}\right) + \frac{c(p-1)}{(p+1)}\boldsymbol{\delta} E\left[F_{p+1,m}^{-1}(\Delta^2) I(F_{p+1,m}(\Delta^2) < \ell_\alpha)\right] \\
&\quad - \frac{(p-3)m}{(m+2)}\boldsymbol{\delta} E\left[\chi_{p+1}^{-2}(\Delta^2)\right] \\
&= \mathbf{b}_3\left(\hat{\boldsymbol{\theta}}_n^{PT}\right) + \mathbf{b}_4\left(\hat{\boldsymbol{\theta}}_n^{S}\right) + \frac{c(p-1)}{(p+1)}\boldsymbol{\delta} \\
&\quad \times E\left[F_{p+1,m}(\Delta^2) I(F_{p+1,m}(\Delta^2) < \ell_\alpha)\right]
\end{aligned} \tag{5.6.4}$$

and

$$\begin{aligned}
b_6\left(\hat{\boldsymbol{\theta}}_n^{PT+}\right) = \Delta^2 \Big\{ &\frac{c(p-1)}{(p+1)} E\left[F_{p+1,m}^{-1}(\Delta^2) I(F_{p+1,m}(\Delta^2) < \ell_\alpha)\right] \\
&- \frac{(p-3)m}{(m+2)} E\left[\chi_{p+1}^{-2}(\Delta^2)\right] - G_{p+1,m}(\ell_\alpha; \Delta^2) \Big\}^2
\end{aligned} \tag{5.6.5}$$

respectively. Note that (5.6.5) and (5.3.1f) are similar except for ℓ_α.

The corresponding MSE-matrix and the risk expressions are given by

$$\mathbf{M}_6(\hat{\boldsymbol{\theta}}_n^{\text{PT+}}) = M_4(\hat{\boldsymbol{\theta}}_n^S)$$
$$- \sigma^2(\mathbf{HN}^{-1})E\Big[\Big(1 - \frac{c(p-1)}{(p+1)}F_{p+1,m}^{-1}(\Delta^2)\Big)^2 I\Big(F_{p+1,m}^{-1}(\Delta^2) < \ell_\alpha\Big)\Big]$$
$$+ \boldsymbol{\delta\delta}'\Big\{2E\Big[\Big(1 - \frac{c(p-1)}{(p+1)}F_{p+1,m}^{-1}(\Delta^2)\Big)I\Big(F_{p+1,m}^{-1}(\Delta^2) < \ell_\alpha\Big)\Big]$$
$$- E\Big[\Big(1 - \frac{c(p-1)}{p+3}F_{p+3,m}^{-1}(\Delta^2)\Big)^2 I(F_{p+3,m}(\Delta^2) < \ell_\alpha^*)\Big]\Big\} \quad (5.6.6)$$

and

$$R_6\Big(\hat{\boldsymbol{\theta}}_n^{PT+}; \mathbf{W}\Big) = R_4(\hat{\boldsymbol{\theta}}_n^S; \mathbf{W}) - \sigma^2\text{tr}(\mathbf{WHN}^{-1})$$
$$\times E\Big[\Big(1 - \frac{c(p-1)}{(p+1)}F_{p+1,m}^{-1}(\Delta^2)\Big)I\Big(F_{p+1,m}(\Delta^2) < \ell_\alpha\Big)\Big]$$
$$+ (\boldsymbol{\delta}'\mathbf{W}\boldsymbol{\delta})\Big\{2E\Big[\Big(1 - \frac{c(p-1)}{(p+3)}F_{p+3,m}^{-1}(\Delta^2)\Big)^2 I\Big(\Big(F_{p+3,m}(\Delta^2) < \ell_\alpha^*\Big)\Big]\Big\}, \quad (5.6.7)$$

respectively. The percentage improvements of $\hat{\boldsymbol{\theta}}_n^{PT+}$ over $\hat{\boldsymbol{\theta}}_n^{\text{PT}}$ under the null hypothesis H_0 for $\alpha = 0.05$ and 0.10 are given in Table 5.6.1.

Similarly, we can make an improvement on $\hat{\boldsymbol{\theta}}_n^S$ by using the Stein-type-estimator of the variance σ^2, as in Section 4.5.6 of Chapter 4. We write estimator of $\boldsymbol{\theta}$ as

$$\hat{\boldsymbol{\theta}}_n^{IS} = \hat{\boldsymbol{\theta}}_n + \Big\{1 - c(m+2)\mathcal{L}_n^{-1}\phi_s(\mathcal{L}_n)\Big\}(\tilde{\boldsymbol{\theta}}_n - \hat{\boldsymbol{\theta}}_n), \quad (5.6.8)$$

where

$$\phi_s(\mathcal{L}_n) = \frac{1}{m+2}I\Big(\mathcal{L}_n \geq \frac{m}{m+2}\Big) + \frac{1 + \frac{p-1}{m}\mathcal{L}_n}{m+p+1}I\Big(\mathcal{L}_n < \frac{m}{m+2}\Big). \quad (5.6.9)$$

This estimator improves over $\hat{\boldsymbol{\theta}}_n^S$ uniformly. The risk function of $\hat{\boldsymbol{\theta}}_n^{IS}$ is given by

$$R_7\Big(\hat{\boldsymbol{\theta}}_n^{IS} : \sigma^{-2}\mathbf{I}_p\Big) = R_4\Big(\hat{\boldsymbol{\theta}}_n^S : \sigma^{-2}\mathbf{I}_p\Big) - \frac{(p-3)^2}{(p+m+1)^2}\Big[\frac{m(p+2m+3)}{m+2}$$
$$\times E\Big\{F_{p-1,m+2}^{-1}(\Delta^2)I\Big(F_{p-1,m+2}(\Delta^2) < 1\Big)\Big\}$$
$$- 2mG_{p-1,m+2}(1; \Delta^2) - \Big\{(p-1)G_{p-1,m}\Big(\frac{m(p-1)}{(m+2)(p+1)}; \Delta^2\Big)$$
$$+ \Delta^2 G_{p+1,m}\Big(\frac{m(p-1)}{(m+2)(p+3)}; \Delta^2\Big)\Big\}\Big] - \frac{2(p-1)(p-3)}{m+p+1}G_{p-1,m}\Big(\frac{m}{m+2}; \Delta^2\Big)$$
$$+ \frac{2(p-1)(p-3)^2}{(m+2)(m+p+1)}E\Big\{F_{p-1,m}^{-1}(\Delta^2)I\Big(F_{p-1,m}(\Delta^2) < \frac{m}{m+2}\Big)\Big\}. \quad (5.6.10)$$

Table 5.6.1 Tables for Percentage Improvement of $\hat{\boldsymbol{\theta}}_n^{\mathrm{PT+}}$ over $\hat{\boldsymbol{\theta}}_n^{\mathrm{PT}}$

α	$\frac{n}{p}$	18	20	22	24	26	28	30	32
0.05	4	19.62	20.16	20.59	20.95	21.25	21.50	21.72	21.91
	5	36.20	37.29	38.15	38.85	39.43	39.91	40.32	40.68
	6	46.82	48.27	49.39	50.28	51.01	51.61	52.11	52.54
	7	52.87	54.54	55.78	56.74	57.51	58.13	58.65	59.09
	8	56.14	57.97	59.28	60.28	61.06	61.68	62.20	62.63
	9	57.82	59.86	61.29	62.34	63.15	63.80	64.32	64.76
	10	58.52	60.90	62.52	63.39	64.59	65.29	65.87	66.34
	11	58.49	61.39	63.30	64.66	65.70	66.51	67.16	67.70
	12	57.74	61.41	63.74	65.37	66.59	67.54	68.30	68.90
	13	56.08	60.92	63.83	65.82	67.82	68.40	69.29	70.01

α	$\frac{n}{p}$	18	20	22	24	26	28	30	32
0.10	4	22.59	23.12	23.54	23.88	24.17	24.42	24.63	24.81
	5	38.94	39.94	40.73	41.37	41.89	42.33	42.71	43.03
	6	48.39	49.67	50.66	51.45	52.09	52.62	53.07	53.45
	7	53.40	54.85	55.94	56.79	57.47	58.03	58.49	58.89
	8	56.02	57.64	58.83	59.73	60.45	61.03	61.51	61.92
	9	57.40	59.27	60.60	61.61	62.40	63.03	63.55	63.98
	10	58.01	60.26	61.83	63.00	63.90	64.62	65.21	65.69
	11	58.00	60.80	62.70	64.09	65.16	65.99	66.67	67.24
	12	57.32	60.89	63.25	64.93	66.19	67.18	67.97	68.62
	13	55.79	60.48	63.43	65.48	66.99	68.15	69.07	69.82

Source: Bashir Khan Ph.D. Thesis, 1997

5.7 ANOVA Model: Nonnormal Errors

In this section, once again, we consider the ANOVA model

$$\mathbf{Y} = \mathbf{B}\boldsymbol{\theta} + \boldsymbol{\varepsilon} \tag{5.7.1}$$

as in (5.1.1) and (5.1.2) except that $\boldsymbol{\varepsilon} \sim \prod_{i=1}^{p}\prod_{j=1}^{n_i} F(\varepsilon_{ij})$ where the cdf F is unknown but satisfies the conditions

$$\text{(i)}\quad E(\boldsymbol{\varepsilon}) = \mathbf{0} \text{ and (ii) } E(\boldsymbol{\varepsilon}\boldsymbol{\varepsilon}') = \sigma^2\mathbf{I}_n,\ (0 < \sigma^2 < \infty); \tag{5.7.2}$$

$$\text{(iii)}\quad \boldsymbol{\Lambda}_n = \mathrm{Diag}(\lambda_{1n}, \cdots, \lambda_{p,n}),\quad \lambda_{j,n} = \frac{n_j}{n}\quad (j = 1, \cdots, p),$$

$$\lim_{n\to\infty}(\lambda_{jn}) = \lim_{n\to\infty}\left(\frac{n_j}{n}\right) = \lambda_j\quad (0 < \lambda_j < 1), \tag{5.7.3}$$

$$\lim_{n\to\infty}(\boldsymbol{\Lambda}_n) = \boldsymbol{\Lambda} \quad\text{and}\quad \mathbf{J}_n = \mathbf{I}_p - \mathbf{1}_p\mathbf{1}_p'\boldsymbol{\Lambda}_n. \tag{5.7.4}$$

Further,

$$\lim_{n\to\infty}\mathbf{J}_n = \mathbf{J} = \mathbf{I}_p - \mathbf{1}_p\mathbf{1}_p'\boldsymbol{\Lambda}. \tag{5.7.5}$$

5.7.1 Estimation and Test of Hypothesis

We define the **unrestricted** and **restricted** estimator of $\boldsymbol{\theta}$ as before, as $\tilde{\boldsymbol{\theta}}_n = (\overline{Y}_1, \ldots, \overline{Y}_p)'$ and $\hat{\boldsymbol{\theta}}_n = \frac{1}{n}\mathbf{1}_p\mathbf{1}_p'\mathbf{N}\tilde{\boldsymbol{\theta}}_n$, respectively. Also, the unbiased estimator of σ^2 is defined as

$$s_e^2 = (n-p)^{-1}\sum_{i=1}^{p}\sum_{j=1}^{n_i}(Y_{ij} - \overline{Y}_i)^2, \tag{5.7.6}$$

and s_e^2 is a consistent estimator of σ^2.

In order to test the null hypothesis $H_0 : \boldsymbol{\theta} = \theta_0\mathbf{1}_p$, we consider the statistic

$$\mathcal{L}_n = \frac{n\tilde{\boldsymbol{\theta}}_n'\mathbf{J}_n'\boldsymbol{\Lambda}_n\mathbf{J}_n\tilde{\boldsymbol{\theta}}_n}{s_e^2}. \tag{5.7.7}$$

Under H_0, $\mathcal{L}_n$ is distributed approximately as a central chi-square variable with $(p-1)$ d.f. On the other hand, under fixed alternatives

$$H_{\boldsymbol{\delta}} : \boldsymbol{\theta} = \theta_0\mathbf{1}_p + \boldsymbol{\delta}, \tag{5.7.8}$$

we can write

$$\sqrt{n}s_e^{-1}(\tilde{\boldsymbol{\theta}}_n - \hat{\boldsymbol{\theta}}_n) = \sqrt{n}s_e^{-1}(\tilde{\boldsymbol{\theta}}_n - \boldsymbol{\theta}) - \sqrt{n}s_e^{-1}(\hat{\boldsymbol{\theta}}_n - \theta_0\mathbf{1}_p) + \sqrt{n}s_e^{-1}\boldsymbol{\delta}$$

and

$$\mathcal{L}_n = ns_e^{-2}||\tilde{\boldsymbol{\theta}}_n - \hat{\boldsymbol{\theta}}_n||^2 = ||\sqrt{n}s_e^{-1}(\tilde{\boldsymbol{\theta}}_n - \boldsymbol{\theta}) - \sqrt{n}s_e^{-1}(\hat{\boldsymbol{\theta}}_n - \theta_0\mathbf{1}_p) + \sqrt{n}s_e^{-1}\boldsymbol{\delta}||^2. \tag{5.7.9}$$

Now, by the standard central limit theorem together with the Slusky's theorem, we have

$$\begin{aligned}
&\text{(i)}\ \ \sqrt{n}s_e^{-1}(\tilde{\boldsymbol{\theta}}_n - \boldsymbol{\theta}) \xrightarrow{\mathcal{D}} \mathcal{N}_p(\mathbf{0}, \boldsymbol{\Lambda}^{-1}),\\
&\text{(ii)}\ \ n\sigma^{-2}||\boldsymbol{\delta}||^2 \to \infty \quad \text{as} \quad n \to \infty.
\end{aligned} \tag{5.7.10}$$

It follows that $\mathcal{L}_n = ns_e^{-1}\|\tilde{\boldsymbol{\theta}}_n - \hat{\boldsymbol{\theta}}_n\|^2 \to \infty$ as $n \to \infty$. Consequently, for all **fixed alternatives** $H_{\boldsymbol{\delta}}$,

$$P_{\boldsymbol{\delta}}\{\mathcal{L}_n > \kappa\} \to 1 \quad \text{for all} \quad \kappa. \tag{5.7.11}$$

Therefore, to obtain a meaningful asymptotic distribution of $\mathcal{L}_n$ as $n \to \infty$, we consider a class of local alternatives $\{K_{(n)}\}$ defined by

$$K_{(n)} : \boldsymbol{\theta}_{(n)} = \theta_0\mathbf{1}_p + n^{-1/2}\boldsymbol{\delta}. \tag{5.7.12}$$

Then we have the following theorem:

Theorem 1. Under $\{K_{(n)}\}$ and the assumed regularity conditions, we obtain the following as $n \to \infty$:

(i) $\mathbf{X}_{(n)} = \sqrt{n}s_e^{-1}(\tilde{\boldsymbol{\theta}}_n - \theta_0\mathbf{1}_p) \sim \mathcal{N}_p(\boldsymbol{\delta}^*, \boldsymbol{\Lambda}^{-1}), \quad \boldsymbol{\delta}^* = \sigma^{-1}\boldsymbol{\delta}.$

(ii) $\mathbf{Y}_{(n)} = \sqrt{n}s_e^{-1}(\tilde{\boldsymbol{\theta}}_n - \hat{\boldsymbol{\theta}}_n) \sim \mathcal{N}_p(\mathbf{J}\boldsymbol{\delta}^*, \mathbf{J}\boldsymbol{\Lambda}^{-1}).$

(iii) $\mathbf{Z}_{(n)} = \sqrt{n}s_e^{-1}(\hat{\boldsymbol{\theta}}_n - \theta_0\mathbf{1}_p) \sim \mathcal{N}_p(\mathbf{0}, \mathbf{B}); \quad \mathbf{B} = \mathbf{1}_p\mathbf{1}_p'.$

(iv) $\begin{pmatrix} \mathbf{X}_{(n)} \\ \mathbf{Y}_{(n)} \end{pmatrix} \sim \mathcal{N}_{2p}\left\{\begin{pmatrix} \boldsymbol{\delta}^* \\ \mathbf{J}\boldsymbol{\delta}^* \end{pmatrix}, \begin{pmatrix} \boldsymbol{\Lambda}^{-1} & \mathbf{J}\boldsymbol{\Lambda}^{-1} \\ \boldsymbol{\Lambda}^{-1}\mathbf{J}' & \mathbf{J}\boldsymbol{\Lambda}^{-1} \end{pmatrix}\right\}$;and

$$\begin{pmatrix} \mathbf{Z}_{(n)} \\ \mathbf{Y}_{(n)} \end{pmatrix} \sim \mathcal{N}_{2p}\left\{\begin{pmatrix} \mathbf{0} \\ \mathbf{J}\boldsymbol{\delta}^* \end{pmatrix}, \begin{pmatrix} \mathbf{B} & \mathbf{0} \\ \mathbf{0} & \mathbf{J}\boldsymbol{\Lambda}^{-1} \end{pmatrix}\right\}. \tag{5.7.13}$$

(v) $\mathcal{L}_n$ is asymptotically distributed as a noncentral chi-square with $p-1$ d.f. and noncentral parameter $\Delta^2/2$, $\Delta^2 = \sigma^{-2}(\boldsymbol{\delta}'\mathbf{J}'\boldsymbol{\Lambda}\mathbf{J}\boldsymbol{\delta}) = \boldsymbol{\delta}^*\mathbf{J}'\boldsymbol{\Lambda}\mathbf{J}\boldsymbol{\delta}^*$. The proof follows by defining $\boldsymbol{\Gamma}_2 = \boldsymbol{\Lambda}_n^{1/2}\mathbf{1}_p$, then $\boldsymbol{\Gamma}_1\boldsymbol{\Gamma}_1' = \mathbf{I}_p - \boldsymbol{\Lambda}_n^{1/2}\mathbf{1}_p\mathbf{1}_p'\boldsymbol{\Lambda}_n^{1/2}$.

5.7.2 Preliminary Test and Stein-Type Estimators

We combine the **unrestricted** and the **restricted** estimators of $\boldsymbol{\theta}$ to obtain the PTE and Stein-type estimators as follows:

(i) $\hat{\boldsymbol{\theta}}_n^{PT} = \tilde{\boldsymbol{\theta}}_n - (\tilde{\boldsymbol{\theta}}_n - \hat{\boldsymbol{\theta}}_n)I(\mathcal{L}_n < \mathcal{L}_{n,\alpha}),$ (5.7.15a)

where $\mathcal{L}_{n,\alpha}$ is the upper α-level critical value of $\mathcal{L}_n$ for finite samples,

(ii) $\hat{\boldsymbol{\theta}}_n^{S} = \tilde{\boldsymbol{\theta}}_n - c^*(\tilde{\boldsymbol{\theta}}_n - \hat{\boldsymbol{\theta}}_n)\mathcal{L}_n^{-1}, \quad c^* = \frac{(p-3)m}{m+2},$ (5.7.15b)

and

$$\hat{\boldsymbol{\theta}}_n^{S+} = \hat{\boldsymbol{\theta}}_n + (1 - c^*\mathcal{L}_n^{-1})I(\mathcal{L}_n > c^*)(\tilde{\boldsymbol{\theta}}_n - \hat{\boldsymbol{\theta}}_n) \tag{5.7.15c}$$

are the usual **Stein-rule** estimator and **positive-rule** Stein-estimator, respectively, since $\mathcal{L}_n \neq 0$ with probability one.

5.8 ADB, ADQB, ADMSE, and ADQR of the Estimators

In this section, we present the asymptotic distributional bias (ADB), quadratic bias (ADQB), MSE-matrices (ADMSE), and the quadratic risks (ADQR) of the estimators of $\boldsymbol{\theta}$. Accordingly, consider the estimator $\boldsymbol{\theta}_n^*$, and let $\mathbf{W}$ be a p.s.d. matrix and the quadratic loss function

$$\begin{aligned} L(\boldsymbol{\theta}_n^*, \boldsymbol{\theta}) &= n(\boldsymbol{\theta}_n^* - \boldsymbol{\theta})'\mathbf{W}(\boldsymbol{\theta}_n^* - \boldsymbol{\theta}) \\ &= \text{tr}[\mathbf{W}\{n(\boldsymbol{\theta}_n^* - \boldsymbol{\theta})(\boldsymbol{\theta}_n^* - \boldsymbol{\theta})'\}]. \end{aligned} \tag{5.8.1}$$

Let the ADM be $\mathbf{M}(\boldsymbol{\theta}_n^*) = \lim_{n\to\infty} E\{n(\boldsymbol{\theta}_n^* - \boldsymbol{\theta})(\boldsymbol{\theta}_n^* - \boldsymbol{\theta})'\}$. Then the ADQR of $\boldsymbol{\theta}_n^*$ is given by

$$R(\boldsymbol{\theta}_n^*; \mathbf{W}) = \text{tr}[\mathbf{W}\mathbf{M}(\boldsymbol{\theta}_n^*)]. \tag{5.8.2}$$

In many situations, the asymptotic distribution of $\sqrt{n}s_e^{-1}(\boldsymbol{\theta}_n^* - \boldsymbol{\theta})$ is equivalent to $\sqrt{n}\sigma^{-1}(\tilde{\boldsymbol{\theta}}_n - \boldsymbol{\theta})$ under fixed alternatives $\boldsymbol{\theta} = \theta_0\mathbf{1}_p + \boldsymbol{\delta}$, and thus to obtain appropriate asymptotic distribution of $\sqrt{n}s_e^{-1}(\boldsymbol{\theta}_n^* - \boldsymbol{\theta})$, we consider the class of local alternatives

$$K_{(n)} : \boldsymbol{\theta}_{(n)} = \theta_0\mathbf{1}_p + n^{-1/2}\boldsymbol{\delta} \tag{5.8.3}$$

and define the asymptotic cdf of $\sqrt{n}s_e^{-1}(\boldsymbol{\theta}_n^* - \boldsymbol{\theta}_{(n)})$ as

$$G_{\boldsymbol{\delta}}^*(\mathbf{x}) = \lim_{n\to\infty} P\left\{\sqrt{n}s_e^{-1}(\boldsymbol{\theta}_n^* - \boldsymbol{\theta}_{(n)}) \le \mathbf{x}|K_{(n)}\right\} \tag{5.8.4}$$

if it exists. Then the ABD and ADQB are defined by

$$\mathbf{b}(\boldsymbol{\theta}_n^*) = \int \mathbf{x}dG_{\boldsymbol{\delta}}^*(\mathbf{x}), \tag{5.8.5}$$

$$B(\boldsymbol{\theta}_n^*) = [\mathbf{b}(\boldsymbol{\theta}_n^*)]'\mathbf{M}(\tilde{\boldsymbol{\theta}}_n)[\mathbf{b}(\boldsymbol{\theta}_n^*)], \tag{5.8.6}$$

where

$$\mathbf{M}(\tilde{\boldsymbol{\theta}}_n) = \lim_{n\to\infty} E\{n(\tilde{\boldsymbol{\theta}}_n - \boldsymbol{\theta}_{(n)})(\tilde{\boldsymbol{\theta}}_n - \boldsymbol{\theta}_{(n)})'\}. \tag{5.8.7}$$

Finally, the ADQR of $\sqrt{n}s_e^{-1}(\boldsymbol{\theta}_n^* - \boldsymbol{\theta}_{(n)})$ is given by (5.8.2) as

$$\mathbf{M}(\boldsymbol{\theta}_n^*) = \int \mathbf{x}\mathbf{x}'dG_{\boldsymbol{\delta}}^*(\mathbf{x}). \tag{5.8.8}$$

5.8.1 Asymptotic Distribution of the Estimators under Fixed Alternatives

First, note that under $\boldsymbol{\theta} = \theta_0\mathbf{1}_p + \boldsymbol{\delta}$,

$$\sqrt{n}s_e^{-1}(\tilde{\boldsymbol{\theta}}_n - \boldsymbol{\theta}) \sim \mathcal{N}_p(\mathbf{0}, \boldsymbol{\Lambda}^{-1}). \tag{5.8.9}$$

Now, consider the quadratic difference

$$\begin{aligned} ns_e^{-2}||\hat{\boldsymbol{\theta}}_n^{PT} - \tilde{\boldsymbol{\theta}}_n||^2 &= ns_e^{-2}||(\tilde{\boldsymbol{\theta}}_n - \hat{\boldsymbol{\theta}}_n)||^2 I(\mathcal{L}_n < \mathcal{L}_{n,\alpha}) \\ &= \mathcal{L}_n I(\mathcal{L}_n < \mathcal{L}_{n,\alpha}) \le \mathcal{L}_{n,\alpha} I(\mathcal{L}_n < \mathcal{L}_{n,\alpha}). \end{aligned} \tag{5.8.10}$$

Thus, by the consistency of the test based on $\mathcal{L}_n$,

$$E\{\mathcal{L}_n I(\mathcal{L}_n < \mathcal{L}_{n,\alpha})\} \to 0 \text{ as } n \to \infty. \tag{5.8.11}$$

Hence, as $n \to \infty$,

$$\sqrt{n}s_e^{-1}(\hat{\boldsymbol{\theta}}_n^{PT} - \boldsymbol{\theta}) = \sqrt{n}\sigma^{-1}(\tilde{\boldsymbol{\theta}}_n - \boldsymbol{\theta}) + o_p(1). \tag{5.8.12}$$

Similarly, consider the quadratic difference

$$ns_e^{-2}||\hat{\boldsymbol{\theta}}_n^S - \tilde{\boldsymbol{\theta}}_n||^2 = ns_e^{-2}||(\tilde{\boldsymbol{\theta}}_n - \hat{\boldsymbol{\theta}}_n)||^2 c^{*2}\mathcal{L}_n^{-2} = c^{*2}\mathcal{L}_n^{-1}. \quad (5.8.13)$$

Since $\mathcal{L}_n \geq 0$ and $\mathcal{L}_n \xrightarrow{\text{P}} \infty$ as $n \to \infty$, we have $E[\mathcal{L}_n^{-1}] \to 0$ as $n \to \infty$ while $c^* \to (p-3)$. Hence,

$$\sqrt{n}s_e^{-1}(\hat{\boldsymbol{\theta}}_n^S - \boldsymbol{\theta}) = \sqrt{n}\sigma^{-1}(\tilde{\boldsymbol{\theta}}_n - \boldsymbol{\theta}) + o_p(1). \quad (5.8.14)$$

Finally, we show that

$$\sqrt{n}s_e^{-1}(\hat{\boldsymbol{\theta}}_n^{S+} - \boldsymbol{\theta}) = \sqrt{n}\sigma^{-1}(\tilde{\boldsymbol{\theta}}_n - \boldsymbol{\theta}) + o_p(1) \quad (5.8.15)$$

as $n \to \infty$. In this case,

$$\begin{aligned}
&ns_e^{-2}||\hat{\boldsymbol{\theta}}_n^{S+} - \tilde{\boldsymbol{\theta}}_n||^2 \\
&= ns_e^{-2}||\hat{\boldsymbol{\theta}}_n^S - \tilde{\boldsymbol{\theta}}_n||^2 + ns_e^{-2}||\tilde{\boldsymbol{\theta}}_n - \hat{\boldsymbol{\theta}}_n||^2(1 - c^*\mathcal{L}_n^{-1})^2 I(\mathcal{L}_n < c^*) \\
&\quad -2ns_e^{-2}(\hat{\boldsymbol{\theta}}_n^S - \tilde{\boldsymbol{\theta}}_n)(\tilde{\boldsymbol{\theta}}_n - \hat{\boldsymbol{\theta}}_n)(1 - c^*\mathcal{L}_n^{-1})I(\mathcal{L}_n < c^*) \\
&= c^{*2}\mathcal{L}_n^{-1} + \mathcal{L}_n(1 - c^*\mathcal{L}_n^{-1})^2 I(\mathcal{L}_n < c^*) + 2c * \mathcal{L}_n^{-1}(1 - c^*\mathcal{L}_n^{-1})I(\mathcal{L}_n < c^*) \\
&\leq c^{*2}\mathcal{L}_n^{-1} + c^*(1 - c^*\mathcal{L}_n^{-1})^2 I(\mathcal{L}_n < c^*) + 2c^*\mathcal{L}_n^{-1}(1 - c^*\mathcal{L}_n^{-1})I(\mathcal{L}_n < c^*) \\
&\to 0 \quad \text{as} \quad n \to \infty. \qquad (5.8.16)
\end{aligned}$$

Hence,

$$\lim_{n\to\infty} E\left\{ns_e^{-2}||\hat{\boldsymbol{\theta}}_n^{S+} - \tilde{\boldsymbol{\theta}}_n||^2\right\} = 0 \quad (5.8.17)$$

and

$$\sqrt{n}s_e^{-1}(\hat{\boldsymbol{\theta}}_n^{S+} - \boldsymbol{\theta}_n) = \sqrt{n}\sigma^{-1}(\tilde{\boldsymbol{\theta}}_n - \boldsymbol{\theta}) + o_p(1). \quad (5.8.18)$$

We conclude that under fixed alternatives $\boldsymbol{\theta} = \theta_0\mathbf{1}_p + \boldsymbol{\delta}$, asymptotic distribution of $\sqrt{n}s_e^{-1}(\hat{\boldsymbol{\theta}}_n^{PT} - \boldsymbol{\theta})$, $\sqrt{n}s_e^{-1}(\hat{\boldsymbol{\theta}}_n^S - \boldsymbol{\theta})$, and $\sqrt{n}s_e^{-1}(\hat{\boldsymbol{\theta}}_n^{S+} - \boldsymbol{\theta})$ is the same normal distribution $\mathcal{N}_p(\mathbf{0}, \boldsymbol{\Lambda}^{-1})$. Hence, the ADB, ADQB, AMSE, and ADQR of the estimators are the same as that of $\sqrt{n}s_e^{-1}(\tilde{\boldsymbol{\theta}}_n - \boldsymbol{\theta})$, while the asymptotic distribution of $\sqrt{n}s_e^{-1}(\hat{\boldsymbol{\theta}}_n - \boldsymbol{\theta})$ is degenerate.

5.8.2 Asymptotic Distribution of the Estimators under Local Alternatives

In this section, we consider the local alternatives $\{K_{(n)}\}$ given at (5.8.3) and Theorem 5.7.1 to obtain the ADB, ADQB, ADM, and ADQR of various estimators of $\boldsymbol{\theta}$.

Clearly,

(i) $\mathbf{b}_1(\tilde{\boldsymbol{\theta}}_n) = \mathbf{0}$ and $B_1(\tilde{\boldsymbol{\theta}}_n) = 0$

and

(ii) $\mathbf{b}_2(\hat{\boldsymbol{\theta}}_n) = \boldsymbol{\delta}^*$ and $B_2(\hat{\boldsymbol{\theta}}_n) = \Delta^2 \qquad (\boldsymbol{\delta}^* = \sigma^{-1}\boldsymbol{\delta}).$ (5.8.19)

(iii) From Theorem 5.7.1, we write

$$\sqrt{n}s_e^{-1}(\hat{\boldsymbol{\theta}}_n^{PT} - \theta_0\mathbf{1}_p) = \mathbf{X}_{(n)} - \mathbf{Y}_{(n)}I(\mathcal{L}_n < \mathcal{L}_{n,\alpha}) \tag{5.8.20a}$$

$$\xrightarrow{\mathcal{D}}(\mathbf{X} + \boldsymbol{\delta}^*) - (\mathbf{Y} + \mathbf{J}\boldsymbol{\delta}^*)I(\|\mathbf{Y} + \mathbf{J}\boldsymbol{\delta}^*\|^2 < \chi^2_{p-1}(\alpha)), \tag{5.8.20b}$$

where (since $\mathcal{L}_{n,\alpha} \to \chi^2_{p-1}(\alpha)$) as $n \to \infty$),

$$\begin{pmatrix} \mathbf{X} \\ \mathbf{Y} \end{pmatrix} \sim \mathcal{N}_{2p}\left\{\begin{pmatrix} -\boldsymbol{\delta}^* \\ \mathbf{0} \end{pmatrix}; \begin{pmatrix} \boldsymbol{\Lambda}^{-1}, & \mathbf{J}\boldsymbol{\Lambda}^{-1} \\ \boldsymbol{\Lambda}^{-1}\mathbf{J}', & \mathbf{J}\boldsymbol{\Lambda}^{-1} \end{pmatrix}\right\}. \tag{5.8.21}$$

so we arrive at the theorem

Theorem 1. Under $\{K_{(n)}\}$ and assumed regularity conditions, the asymptotic distribution of $\sqrt{n}s_e^{-1}(\hat{\boldsymbol{\theta}}_n^{PT} - \theta_0\mathbf{1}_p)$ is given by

$$G_p^{PT}(\mathbf{x}) = \Phi_p(\mathbf{x}; \mathbf{0}, \mathbf{B})H_{p-1}(\chi^2_{p-1}(\alpha); \Delta^2) + \int_{\Gamma(\boldsymbol{\delta}^*)}\!\!\cdots\!\int \Phi_p(\mathbf{x} - \mathbf{Y}; \mathbf{0}, \mathbf{B})d\Phi_p(\mathbf{Y}; \mathbf{0}, \mathbf{J}\boldsymbol{\Lambda}^{-1}), \tag{5.8.22}$$

where

$$\Gamma(\boldsymbol{\delta}^*) = \left\{\mathbf{z} : (\mathbf{Y} + \mathbf{J}\boldsymbol{\delta}^*)'(\mathbf{Y} + \mathbf{J}\boldsymbol{\delta}^*) \geq \chi^2_{p-1}(\alpha)\right\}, \tag{5.8.23}$$

$\boldsymbol{\Phi}_p(\mathbf{x}; \boldsymbol{\Sigma})$ cdf of a p-variate normal distribution with mean vector $\boldsymbol{\mu}$ and covariance matrix, $\boldsymbol{\Sigma}$.

Similarly, we write $\sqrt{n}s_e^{-1}(\hat{\boldsymbol{\theta}}_n^{S} - \theta_0\mathbf{1}_p)$ as

$$\sqrt{n}s_e^{-1}(\hat{\boldsymbol{\theta}}_n^{S} - \theta_0\mathbf{1}_p) = \mathbf{X}_{(n)} - \frac{(p-3)m}{m+2}\mathbf{Y}_{(n)}\{\mathbf{Y}_n'\boldsymbol{\Lambda}\mathbf{Y}_{(n)}\}^{-1}$$

$$\xrightarrow{\mathcal{D}}(\mathbf{X} + \boldsymbol{\delta}^*) - (p-3)(\mathbf{Y} + \mathbf{J}\boldsymbol{\delta}^*)\{\|\mathbf{Y} + \mathbf{J}\boldsymbol{\delta}^*\|^{-2}\}. \tag{5.8.24}$$

The r.h.s. of (5.8.24) is the asymptotic representation of $\sqrt{n}s_e^{-1}(\hat{\boldsymbol{\theta}}_n - \theta_0\mathbf{1}_p)$. A similar representation of $\sqrt{n}s_e^{-1}(\hat{\boldsymbol{\theta}}_n^{S+} - \theta_0\mathbf{1}_p)$ can be easily obtained by following such simple steps as

$$\sqrt{n}s_e^{-1}\left(\hat{\boldsymbol{\theta}}_n^{S+} - \theta_0\mathbf{1}_p\right)$$

$$\xrightarrow{\mathcal{D}}(\mathbf{X} + \boldsymbol{\delta}^*) - (p-3)(\mathbf{Y} + \mathbf{J}\boldsymbol{\delta}^*)\|\mathbf{Y} + \mathbf{J}\boldsymbol{\delta}^*\|^2\}I(\|\mathbf{Y} + \mathbf{J}\boldsymbol{\delta}^*\|^2 > p-3) - (\mathbf{Y} + \mathbf{J}\boldsymbol{\delta}^*)I(\|\mathbf{Y} + \mathbf{J}\boldsymbol{\delta}^*\|^2 < p-3). \tag{5.8.25}$$

5.8.3 ADB, ADQB, MSE-Matrices, and ADQR of $\hat{\boldsymbol{\theta}}_n^{PT}$ $\hat{\boldsymbol{\theta}}_n^{S}$ and $\hat{\boldsymbol{\theta}}_n^{S+}$

First, we consider $\hat{\boldsymbol{\theta}}_n^{PT}$. In this case, we note that

$$\sqrt{n}s_e^{-1}\left(\boldsymbol{\theta}_{(n)}^{PT} - \theta_0\mathbf{1}_p\right) \xrightarrow{\mathcal{D}} \mathbf{U}_1 = \mathbf{X} + \boldsymbol{\delta}^* - (\mathbf{Y} + \mathbf{J}\boldsymbol{\delta}^*)I(||\mathbf{Y} + \mathbf{J}\boldsymbol{\delta}^*||^2 < \chi_{p-1}^2(\alpha)). \tag{5.8.26}$$

Then

$$\begin{aligned}\mathbf{b}_3\left(\hat{\boldsymbol{\theta}}_n^{PT}\right) &= \lim_{n\to\infty} E\left\{\sqrt{n}s_e^{-1}\left(\hat{\boldsymbol{\theta}}_n^{PT} - \theta_0\mathbf{1}_p^*\right) - s_e^{-1}\boldsymbol{\delta}\right\}\\ &= E(\mathbf{X} + \boldsymbol{\delta}^*) - \mathbf{J}\boldsymbol{\delta}^* H_{p+1}(\chi_{p-1}^2(\alpha); \Delta^2) - \boldsymbol{\delta}^*\\ &= -\mathbf{J}\boldsymbol{\delta}^* H_{p+1}(\chi_{p-1}^2(\alpha); \Delta^2).\end{aligned} \tag{5.8.27}$$

Thus,

$$B_3\left(\hat{\boldsymbol{\theta}}_n^{PT}\right) = \Delta^2\left\{H_{p+1}(\chi_{p-1}^2(\alpha); \Delta^2\right\}^2. \tag{5.8.28}$$

The MSE-matrix for $\hat{\boldsymbol{\theta}}_n^{PT}$ may be obtained as

$$\begin{aligned}\mathbf{M}_3\left(\hat{\boldsymbol{\theta}}_n^{PT}\right) &= E(\mathbf{U}_1\mathbf{U}_1') = \boldsymbol{\Lambda}^{-1} - (\mathbf{J}\boldsymbol{\Lambda}^{-1})H_{p+1}(\chi_{p-1}^2(\alpha); \Delta^2) + (\mathbf{J}\boldsymbol{\delta}^*\boldsymbol{\delta}^{*'}\mathbf{J}')\\ &\quad\times\left\{2H_{p+1}(\chi_{p-1}^2(\alpha); \Delta^2) - H_{p+3}(\chi_{p-1}^2(\alpha); \Delta^2)\right\},\end{aligned} \tag{5.8.29}$$

and the quadratic risk as

$$\begin{aligned}R_3\left(\hat{\boldsymbol{\theta}}_n^{PT} : \mathbf{W}\right) &= \operatorname{tr}(\boldsymbol{\Lambda}^{-1}\mathbf{W}) - \operatorname{tr}(\mathbf{J}\boldsymbol{\Lambda}^{-1}\mathbf{W})H_{p+1}(\chi_{p-1}^2(\alpha); \Delta^2) + (\boldsymbol{\delta}^{*'}\mathbf{J}'\mathbf{W}\mathbf{J}\boldsymbol{\delta}^*)\\ &\quad\times\left\{2H_{p+1}(\chi_{p-1}^2(\alpha); \Delta^2) - H_{p+3}(\chi_{p-1}^2(\alpha); \Delta^2)\right\}.\end{aligned} \tag{5.8.30}$$

Next, we consider $\hat{\boldsymbol{\theta}}_n^{S}$. In this case,

$$\sqrt{n}s_e^{-1}\left(\hat{\boldsymbol{\theta}}_n^{S} - \boldsymbol{\theta}_{(n)}\mathbf{1}_p\right) \xrightarrow{\mathcal{D}} \mathbf{U}_2 = (\mathbf{X} + \boldsymbol{\delta}^*) - (p-3)(\mathbf{Y} + \mathbf{J}\boldsymbol{\delta}^*)||\mathbf{Y} + \mathbf{J}\boldsymbol{\delta}^*||^{-2}.$$

Thus,

$$E(\mathbf{U}_2) = \mathbf{b}_4\left(\hat{\boldsymbol{\theta}}_n^{S}\right) = -(p-3)\mathbf{J}\boldsymbol{\delta}^* E\left[\chi_{p+1}^{-2}(\Delta^2)\right] \tag{5.8.31a}$$

$$B_4\left(\hat{\boldsymbol{\theta}}_n^{S}\right) = -(p-3)^2\Delta^2\left\{E\left[\chi_{p+1}^{-2}(\Delta^2)\right]\right\}^2. \tag{5.8.31b}$$

The expressions for $\mathbf{M}_4\left(\hat{\boldsymbol{\theta}}_n^{S}\right)$ and $\mathbf{R}_4\left(\hat{\boldsymbol{\theta}}_n^{S}; \mathbf{W}\right)$ are computed as

$$\begin{aligned}\mathbf{M}_4\left(\hat{\boldsymbol{\theta}}_n^{S}\right) &= E(\mathbf{U}_2\mathbf{U}_2')\\ &= \boldsymbol{\Lambda}^{-1} - (p-3)(\mathbf{J}\boldsymbol{\Lambda}^{-1})\left\{2E\left[\chi_{p+1}^{-2}(\Delta^2)\right] - (p-3)E\left[\chi_{p+1}^{-4}(\Delta^2)\right]\right\}\\ &\quad + (p-3)(p+1)(\mathbf{J}\boldsymbol{\delta}^*\boldsymbol{\delta}^{*'}\mathbf{J}')E\left[\chi_{p+3}^{-4}(\Delta^2)\right],\end{aligned} \tag{5.8.32}$$

and

$$R_4\left(\hat{\boldsymbol{\theta}}_n^S;\mathbf{W}\right)$$
$$= \operatorname{tr}(\boldsymbol{\Lambda}^{-1}\mathbf{W}) - (p-3)\operatorname{tr}(\mathbf{J}\Lambda^{-1}\mathbf{W})\left\{2E\left[\chi_{p+1}^{-2}(\Delta^2)\right] - (p-3)E\left[\chi_{p+1}^{-4}(\Delta^2)\right]\right\}$$
$$+ (p-3)(p+1)(\boldsymbol{\delta}^*\mathbf{J}'\mathbf{W}\mathbf{J}\boldsymbol{\delta}^*)E\left[\chi_{p+3}^{-4}(\Delta^2)\right] \tag{5.8.33}$$

respectively.

The corresponding expressions for $\hat{\boldsymbol{\theta}}_n^{S+}$ are given by

$$\mathbf{b}_5\left(\hat{\boldsymbol{\theta}}_n^{S+}\right) = -\mathbf{J}\boldsymbol{\delta}^*\Big\{(p-3)E\left[\chi_{p+1}^{-2}(\Delta^2)I\left(\chi_{p+1}^{-2}(\Delta^2) > p-3\right)\right] + H_{p+1}(p-3;\Delta^2)\Big\}, \tag{5.8.34}$$

$$B_5\left(\hat{\boldsymbol{\theta}}_n^{S+}\right) = \Delta^2\Big\{(p-3)E\left[\chi_{p+1}^{-2}(\Delta^2)I\left(\chi_{p+1}^{-2}(\Delta^2) > p-3\right)\right] + H_{p+1}(p-3;\Delta^2)\Big\}^2, \tag{5.8.35}$$

$$\mathbf{M}_5\left(\hat{\boldsymbol{\theta}}_n^{S+}\right)$$
$$= \mathbf{M}_4\left(\hat{\boldsymbol{\theta}}_n^S\right) - (\mathbf{J}\boldsymbol{\Lambda}^{-1})E\Big[\left(1-(p-3)\chi_{p+1}^{-2}(\Delta^2)\right)^2 I\left(\chi_{p+1}^{2}(\Delta^2) < p-3\right)\Big]$$
$$+ (\mathbf{J}\boldsymbol{\delta}^*\boldsymbol{\delta}^{*\prime}\mathbf{J}')\Big\{2E\left[\left((p-3)\chi_{p+1}^{-2}(\Delta^2) - 1\right)I\left(\chi_{p+1}^{2}(\Delta^2) < p-3\right)\right]$$
$$- E\left[((p-3)\chi_{p+3}^{-2}(\Delta^2) - 1)^2 I(\chi_{p+3}^{2}(\Delta^2) < p-3)\right]\Big\} \tag{5.8.36}$$

and the ADQR of $\hat{\boldsymbol{\theta}}_n^{S+}$ is given by

$$R_5\left(\hat{\boldsymbol{\theta}}_n^{S+};\mathbf{W}\right)$$
$$= R_4\left(\hat{\boldsymbol{\theta}}_n^S;\mathbf{W}\right) - \operatorname{tr}\left(\mathbf{J}\boldsymbol{\Lambda}^{-1}\mathbf{W}\right)E\Big[\left(1-(p-3)\chi_{p+1}^{-2}(\Delta^2)\right)^2$$
$$\times I\left(\chi_{p+1}^{2}(\Delta^2) < p-3\right)\Big]$$
$$+ (\boldsymbol{\delta}^{*\prime}\mathbf{J}'\mathbf{W}\mathbf{J}\boldsymbol{\delta}^*)\Big\{2E\left[(p-3)\chi_{p+1}^{-2}(\Delta^2) - 1)I\left(\chi_{p+1}^{2}(\Delta^2) < p-3\right)\right].$$
$$- E\Big[\left((p-3)\chi_{p+3}^{-2}(\Delta^2) - 1\right)^2 I\left(\chi_{p+3}^{2}(\Delta^2) < p-3\right)\Big] \tag{5.8.37}$$

It can be shown by risk-difference analysis that

$$R_1\left(\tilde{\boldsymbol{\theta}}_n;\mathbf{W}\right) \geq R_4\left(\hat{\boldsymbol{\theta}}_n^S;\mathbf{W}\right) \geq R_5\left(\hat{\boldsymbol{\theta}}_n^{S+};\mathbf{W}\right) \tag{5.8.38}$$

for all Δ^2, and none of the estimators $\tilde{\boldsymbol{\theta}}_n$, $\hat{\boldsymbol{\theta}}_n^S$, $\hat{\boldsymbol{\theta}}_n^{S+}$, and $\hat{\boldsymbol{\theta}}_n$ dominate $\hat{\boldsymbol{\theta}}_n^{PT}$ uniformly. This is due to the size of the preliminary test of H_0 and unboundedness of the risk of $\hat{\boldsymbol{\theta}}_n$.

5.9 Confidence Set Estimation

In this section, we consider the confidence set estimation of the treatment means $\boldsymbol{\theta}$ in the model (5.1.1) and (5.1.2)

$$\mathbf{Y} = \mathbf{B}\boldsymbol{\theta} + \boldsymbol{\varepsilon}, \quad \boldsymbol{\varepsilon} \sim \mathcal{N}(\mathbf{0}, \sigma^2\mathbf{I}_n), \tag{5.9.1}$$

where $\mathbf{B}$ is the block-diagonal matrix of the vectors $\mathbf{1}_{n_1}, \ldots, \mathbf{1}_{n_p}$ and σ^2 is known. We have defined several estimators of $\boldsymbol{\theta}$ repeated here again:

(i) $\tilde{\boldsymbol{\theta}}_n = (\mathbf{B}'\mathbf{B})^{-1}\mathbf{B}'\mathbf{Y} = (\overline{Y}_1, \ldots, \overline{Y}_p)'$.

(ii) $\hat{\boldsymbol{\theta}}_n = \frac{1}{n}\mathbf{1}_p\mathbf{1}_p'\mathbf{N}\tilde{\boldsymbol{\theta}}_n$ when $\boldsymbol{\theta} = \theta_0\mathbf{1}_p$.

(iii) $\hat{\boldsymbol{\theta}}_n^{PT} = \tilde{\boldsymbol{\theta}}_n - \left(\tilde{\boldsymbol{\theta}}_n - \hat{\boldsymbol{\theta}}_n\right) I(\mathcal{L}_n < \chi_q^2(\alpha)), \quad q = p-1.$

(iv) $\hat{\boldsymbol{\theta}}_n^{S} = \tilde{\boldsymbol{\theta}}_n - c\left(\tilde{\boldsymbol{\theta}}_n - \hat{\boldsymbol{\theta}}_n\right) \mathcal{L}_n^{-1}, \quad 0 < c < 2(p-3).$

(v) $\hat{\boldsymbol{\theta}}_n^{S+} = \hat{\boldsymbol{\theta}}_n^{S} - (1 - c\mathcal{L}_n^{-1}) I(\mathcal{L}_n < c)\left(\tilde{\boldsymbol{\theta}}_n - \hat{\boldsymbol{\theta}}_n\right),$

where

$$\mathcal{L}_n = \sigma^{-2}\left(\tilde{\boldsymbol{\theta}}_n\mathbf{H}'\mathbf{N}\mathbf{H}\tilde{\boldsymbol{\theta}}_n\right). \tag{5.9.2}$$

Since $\tilde{\boldsymbol{\theta}}_n \sim \mathcal{N}_p(\boldsymbol{\theta}, \sigma^2\mathbf{N}^{-1})$, the $(1-\gamma)\%$ confidence set $C^0(\boldsymbol{\theta})$ is defined as

$$C^0\left(\tilde{\boldsymbol{\theta}}_n\right) = \left\{\boldsymbol{\theta}; \sigma^{-2}\left\|\left(\boldsymbol{\theta} - \tilde{\boldsymbol{\theta}}_n\right)\right\|_{\mathbf{N}}^2 \le \chi_p^2(\gamma)\right\}, \tag{5.9.3}$$

where $\chi_p^2(\gamma)$ is chosen such that $P_0\{\chi_p^2 \le \chi_p^2(\gamma)\} = 1-\gamma$. Thus, $C^0(\tilde{\boldsymbol{\theta}}_n)$ has coverage probability $1-\gamma$. The set $C^0(\tilde{\boldsymbol{\theta}}_n)$ is minimax in the sense that among all the confidence sets with coverage probability, at least $1-\gamma$. Also, $C^0(\tilde{\boldsymbol{\theta}}_n)$ minimizes the maximum volume. As we saw in Chapter 4, there is a scope to improve upon $C^0(\tilde{\boldsymbol{\theta}}_n)$ for $p \ge 4$ in the sense that (1) $P_{\boldsymbol{\theta}}\{\boldsymbol{\theta} \in C^*(\boldsymbol{\theta}_n^*)\} \ge P_{\boldsymbol{\theta}}(\boldsymbol{\theta} \in C^0(\tilde{\boldsymbol{\theta}}_n))$ and (2) The volume of $C^*(\boldsymbol{\theta}_n^*) \le$ volume of $C^0(\tilde{\boldsymbol{\theta}}_n)$ with strict inequality holding in either case for a set of positive Lebesgue measure of $\boldsymbol{\theta}$ or $\mathbf{Y}$, respectively.

In this section, we first consider the confidence sets that are centered at the estimators of the form

$$\boldsymbol{\theta}_n^* = \mathbf{A}\tilde{\boldsymbol{\theta}}_n + (\mathbf{I}-\mathbf{A})\tilde{\boldsymbol{\theta}}_n g(\mathcal{L}_n), \tag{5.9.4}$$

where $(\mathbf{I}-\mathbf{A})$ is an idempotent matrix of rank $p-1$ and $g(\mathcal{L}_n)$ is a non-decreasing function of the test-statistic $\mathcal{L}_n = \sigma^{-2}(\tilde{\boldsymbol{\theta}}_n'\mathbf{H}'\mathbf{N}\mathbf{H}\tilde{\boldsymbol{\theta}}_n)$. Particularly, we take $\mathbf{A} = \frac{1}{n}\mathbf{1}_p\mathbf{1}_p'\mathbf{N}$ and then $\mathbf{H} = \mathbf{I}_p - \mathbf{A}$.

(i) If $g(\mathcal{L}_n) = 1$, then $\boldsymbol{\theta}_n^* = \tilde{\boldsymbol{\theta}}_n$. (5.9.5)

(ii) If $g(\mathcal{L}_n) = 0$, then $\boldsymbol{\theta}_n^* = \hat{\boldsymbol{\theta}}_n$.

(iii) If $g(\mathcal{L}_n) = I(\mathcal{L}_n > \chi_q^2(\alpha))$, then $\boldsymbol{\theta}_n^* = \hat{\boldsymbol{\theta}}_n^{PT}$ $(q = p-1)$.

(iv) If $g(\mathcal{L}_n) = (1 - c\mathcal{L}_n^{-1})$; $0 < c < 2(p-3)$, then $\hat{\boldsymbol{\theta}}_n^* = \hat{\boldsymbol{\theta}}_n^S$.

(v) If $g(\mathcal{L}_n) = (1 - c\mathcal{L}_n^{-1})I(\mathcal{L}_n > c)$, then $\boldsymbol{\theta}_n^* = \hat{\boldsymbol{\theta}}_n^{S+}(c)$, which is the Lindley (1962) estimator of $\boldsymbol{\theta}$.

The estimator $\boldsymbol{\theta}_n^*$ shrinks toward the linear subspace defined by $\boldsymbol{\theta} = \theta_0 \mathbf{1}_p$.

5.9.1 Confidence Sets and Coverage Probabilities

We consider the class of confidence sets given by

$$C^*(\boldsymbol{\theta}_n^*) = \{\boldsymbol{\theta}; \sigma^{-2}||\boldsymbol{\theta} - \boldsymbol{\theta}_n^*||_{\mathbf{N}}^2 \leq \chi_p^2(\gamma)\}. \tag{5.9.6}$$

More specifically, we consider the following confidence sets:

(i) $C^0(\tilde{\boldsymbol{\theta}}_n) = \{\boldsymbol{\theta} : \sigma^{-2}||\boldsymbol{\theta} - \tilde{\boldsymbol{\theta}}_n||_{\mathbf{N}}^2 < \chi_p^2(\gamma)\}$.

(ii) $C^R(\hat{\boldsymbol{\theta}}_n) = \{\boldsymbol{\theta} : \sigma^{-2}||\boldsymbol{\theta} - \hat{\boldsymbol{\theta}}_n||_{\mathbf{N}}^2 < \chi_p^2(\gamma)\}$.

(iii) $C^{PT}(\hat{\boldsymbol{\theta}}_n^{PT}(\alpha)) = \{\boldsymbol{\theta} : \sigma^{-2}||\boldsymbol{\theta} - \hat{\boldsymbol{\theta}}_n^{PT}(\alpha)||_{\mathbf{N}}^2 < \chi_p^2(\gamma)\}$.

(iv) $C^S(\hat{\boldsymbol{\theta}}_n^S) = \{\boldsymbol{\theta} : \sigma^{-2}||\boldsymbol{\theta} - \hat{\boldsymbol{\theta}}_n^S||_{\mathbf{N}}^2 < \chi_p^2(\gamma)\}$.

(v) $C^{S+}(\hat{\boldsymbol{\theta}}_n^{S+}(c)) = \{\boldsymbol{\theta} : \sigma^{-2}||\boldsymbol{\theta} - \hat{\boldsymbol{\theta}}_n^{S+}(c)||_{\mathbf{N}}^2 < \chi_p^2(\gamma)\}$. (5.9.7)

Note that $N^{-1/2}(\mathbf{I} - \mathbf{A})'\mathbf{N}(\mathbf{I} - \mathbf{A})\mathbf{N}^{-1/2}$, $\mathbf{A} = \frac{1}{n}\mathbf{1}_p\mathbf{1}_p'\mathbf{N}$ is a symmetric idempotent matrix of rank $(p-1)$. An orthogonal matrix, $\boldsymbol{\Gamma} = (\Gamma_1, \Gamma_2)$ exists such that

$$\boldsymbol{\Gamma}\mathbf{N}^{-1/2}(\mathbf{I} - \mathbf{A})'\mathbf{N}(\mathbf{I} - \mathbf{A})\mathbf{N}^{-1/2}\boldsymbol{\Gamma}' = \begin{pmatrix} \mathbf{I}_{p-1} & \mathbf{0} \\ \mathbf{0} & 0 \end{pmatrix}$$

(see Section 5.1.3).

Defining $\boldsymbol{\omega}_1 = \frac{1}{\sigma}\boldsymbol{\Gamma}_1'\mathbf{N}^{1/2}\tilde{\boldsymbol{\theta}}_n$ and $\omega_2 = \frac{1}{\sigma}\boldsymbol{\Gamma}_2'\mathbf{N}^{1/2}\tilde{\boldsymbol{\theta}}_n$ we have

$$\begin{pmatrix} \boldsymbol{\omega}_1 \\ \omega_2 \end{pmatrix} \sim \mathcal{N}_p\left\{\begin{pmatrix} \frac{1}{\sigma}\boldsymbol{\Gamma}_1'\mathbf{N}^{1/2}\boldsymbol{\theta} \\ \frac{1}{\sigma}\boldsymbol{\Gamma}_2'\mathbf{N}^{1/2}\boldsymbol{\theta} \end{pmatrix}, \begin{pmatrix} \mathbf{I}_{p-1} & \mathbf{0} \\ \mathbf{0} & 1 \end{pmatrix}\right\}$$

$$= \mathcal{N}_p\left\{\begin{pmatrix} \boldsymbol{\eta}_1 \\ \eta_2 \end{pmatrix}; \begin{pmatrix} \mathbf{I}_{p-1} & \mathbf{0} \\ \mathbf{0} & 1 \end{pmatrix}\right\}. \tag{5.9.8}$$

As a result we write

$$\sigma^{-2}\|\boldsymbol{\theta}-\boldsymbol{\theta}_n^*\|_{\mathbf{N}}^2=(\eta_2-\omega_2)^2+\left\|\boldsymbol{\eta}_1-\boldsymbol{\omega}_1 g(\|\boldsymbol{\omega}_1\|^2)\right\|^2. \tag{5.9.9}$$

Thus we can write the coverage probability of (5.9.9) as

$$P_{\boldsymbol{\eta}}\left\{\left[(\eta_2-\omega_2)^2+\left\|\boldsymbol{\eta}_1-\boldsymbol{\omega}_1 g(\|\boldsymbol{\omega}_1\|^2)\right\|^2\right]\le\chi_p^2(\gamma)\right\}$$
$$=\int_0^{\chi_p^2(\gamma)}P_{\boldsymbol{\eta}}\left\{\left\|\boldsymbol{\eta}_1-\boldsymbol{\omega}_1 g(\|\boldsymbol{\omega}_1\|^2)\right\|^2\le(\chi_p^2(\gamma)-t)^+\right\}dH_1(t,0). \tag{5.9.10}$$

where $H_1(t;0)$ is the cdf of a central chi-squared distribution with one d.f. That is to say

(i) If $g(\|\boldsymbol{\omega}_1\|^2)=1$, we have

$$\begin{aligned}&P_0\left\{(\eta_2-\omega_2)^2+\|\boldsymbol{\eta}_1-\boldsymbol{\omega}_1\|^2<\chi_p^2(\gamma)\right\}\\&=P_0\left(\chi_p^2\le\chi_p^2(\gamma)\right)\\&=H_p(\chi_p^2(\gamma);0)=1-\gamma.\end{aligned} \tag{5.9.11}$$

(ii) If $g(\|\boldsymbol{\omega}_1\|^2)=0$, we have

$$P_{\boldsymbol{\eta}}\left\{(\eta_2-\omega_2)^2+\|\boldsymbol{\eta}_1\|^2\le\chi_p^2(\gamma)\right\}=H_1(\chi_p^2(\gamma)-\Delta^2;0). \tag{5.9.12}$$

(iii) If $g(\|\boldsymbol{\omega}_1\|^2)=I(\|\boldsymbol{\omega}_1\|^2>\chi_q^2(\alpha))$ with $q=p-1$, we have

$$\begin{aligned}&P_{\boldsymbol{\eta}}\left\{(\eta_2-\omega_2)^2+\|\boldsymbol{\eta}_1-\boldsymbol{\omega}_1 I(\|\boldsymbol{\omega}_1\|^2>\chi_q^2(\alpha))\|^2<\chi_p^2(\gamma)\right\}\\&=P_{\boldsymbol{\eta}}\left\{\chi_1^2+\chi_q^2\le\chi_p^2(\gamma);\|\boldsymbol{\omega}_1\|^2>\chi_q^2(\alpha)\right\}\\&\quad+P_{\boldsymbol{\eta}}\left\{\chi_1^2\le\chi_p^2(\gamma)-\Delta^2;\|\boldsymbol{\omega}_1\|^2<\chi_q^2(\alpha)\right\}\\&=H_1(\chi_p^2(\gamma)-\Delta^2;0)H_q(\chi_q^2(\alpha);\Delta^2)\\&\quad+\int_0^{\chi_p^2(\gamma)}P_{\boldsymbol{\eta}_1}\{\|\boldsymbol{\eta}_1-\boldsymbol{\omega}_1\|^2\le(\chi_p^2(\gamma)-t)^+;\|\boldsymbol{\omega}_1\|^2>\chi_q^2(\alpha)\}dH_1(t;0).\end{aligned} \tag{5.9.13}$$

(iv) If $g(\|\boldsymbol{\omega}_1\|^2)=(1-c\|\boldsymbol{\omega}_1\|^{-2})$, $0<c<2(p-3)$, we have

$$\begin{aligned}&P_{\boldsymbol{\eta}}\left\{(\eta_2-\omega_2)^2+\|\boldsymbol{\eta}_1-\boldsymbol{\omega}_1(1-c\|\boldsymbol{\omega}_1\|^{-2})\|^2<\chi_p^2(\gamma)\right\}\\&=\int_0^{\chi_p^2(\gamma)}P_{\boldsymbol{\eta}}\left\{\|\boldsymbol{\eta}_1-\boldsymbol{\omega}_1(1-c\|\boldsymbol{\omega}_1\|^{-2})\|^2<(\chi_p^2(\gamma)-t)^+\right\}dH_1\,(t;0).\end{aligned} \tag{5.9.14}$$

(v) If $g(||\omega_1||^2) = (1 - c||\omega_1||^{-2})I(||\omega_1||^2 > c)$, then

$$\begin{aligned}
&P_\eta\left\{(\eta_2 - \omega_2)^2 + ||\eta_1 - \omega_1(1 - c||\omega_1||^{-2})I(||\omega_1||^2 > c)||^2 \le \chi_p^2(\gamma)\right\} \\
&= P_\eta\{(\eta_2 - \omega_2)^2 + ||\eta_1||^2 \le \chi_p^2(\gamma); ||\omega_1||^2 < c\} \\
&\quad + P_\eta\left\{(\eta_2 - \omega_2)^2 + ||\eta_1 - \omega_1(1 - c||\omega_1||^{-2})||^2 < \chi_p^2(\gamma); ||\omega_1||^2 > c\right\} \\
&= H_1(\chi_p^2(\gamma) - \Delta^2; 0)H_q(c; \Delta^2) \\
&\quad + \int_0^{\chi_p^2(\gamma)} P_\eta\Big\{||\eta_1 - \omega_1(1 - c||\omega_1||^{-2})||^2 \\
&\qquad\qquad \le (\chi_p^2(\gamma) - t)^+; ||\omega_1||^2 > c\Big\} dH_1\ (t; 0). \qquad (5.9.15)
\end{aligned}$$

For further developments, see Casella and Hwang (1987).

5.9.2 Analysis of the Confidence Sets

We first note that $P(C^0(\tilde{\theta}_n)) = 1 - \gamma$, which is constant for all Δ^2. Next, we see that

$$P\left(C^R\left(\hat{\theta}_n\right)\right) = H_1\left(\chi_p^2(\gamma) - \Delta^2; 0\right), \qquad (5.9.16)$$

which is a decreasing function of Δ^2 with a maximum $H_1(\chi_q^2(\alpha) : 0)$ at $\Delta^2 = 0$ and zero when $\Delta^2 \to \chi_p^2(\gamma)$. The coverage probabilities of $C^0(\tilde{\theta}_n)$ and $C^R(\hat{\theta}_n)$ are equal when $\Delta^2 = \Delta_\gamma^2$, which is given by

$$\Delta_\gamma^2 = \chi_p^2(\gamma) - H_1^{-1}(1 - \gamma). \qquad (5.9.17)$$

Now, consider the confidence set $C^{PT}(\theta_n^{PT}(\alpha))$. In this case,

$$\begin{aligned}
P\left\{C^{PT}\left(\hat{\theta}_n^{PT}(\alpha)\right)\right\} &= H_1(\chi_p^2(\gamma) - \Delta^2; 0)H_q(\chi_q^2(\alpha); \Delta^2) \\
&+ \int_0^{\chi_p^2(\gamma)} P_{\eta_1}\left\{||\eta_1 - \omega_1||^2 \le (\chi_p^2(\gamma) - t)^+; ||\omega_1||^2 > \chi_q^2(\alpha)\right\} dH_1(t, 0),
\end{aligned} \qquad (5.9.18)$$

and we have the following theorem:

Theorem 1.

(i) If $\Delta^2 \le \chi_p^2(\gamma)$,

$$P\left(C^{PT}\left(\hat{\theta}_n^{PT}(\alpha)\right)\right) \ge 1 - \gamma, \qquad (5.9.19)$$

(ii) If $\chi_p^2(\gamma) \le \Delta^2 \le (\chi_p(\gamma) + \chi_q(\alpha))^2$, $q = p - 1$,

$$P\left(C^{PT}\left(\hat{\theta}_n^{PT}(\alpha)\right)\right) \le 1 - \gamma, \qquad (5.9.20)$$

(iii) If $\Delta^2 > (\chi_p(\gamma) + \chi_q(\alpha))^2$,

$$P\left(C^{PT}\left(\hat{\boldsymbol{\theta}}_n^{PT}(\alpha)\right)\right) = 1 - \gamma. \tag{5.9.21}$$

The coverage probability (5.9.18) drops to

$$\int_0^{\chi_p^2(\gamma)} P_{\eta_1}\{\chi_q^2 \leq (\chi_p^2(\gamma) - t)^+; \chi_q^2(\Delta^2) \geq \chi_q^2(\alpha)\}\, dH_1(t, 0) \tag{5.9.22}$$

as $\Delta^2 \to \chi_p^2(\gamma)$ and then increases to $1 - \gamma$ as $\Delta^2 > (\chi_p(\gamma) + \chi_q(\alpha))^2$.

Proof. (i) $\Delta^2 < \chi_p^2(\gamma)$. From (5.9.18), we have

$$\begin{aligned}
P(C^{\mathrm{PT}}(\hat{\boldsymbol{\theta}}_n^{\mathrm{PT}}(\alpha))) &= P\{(\eta_2 - \omega_2)^2 + \Delta^2 \leq \chi_p^2(\gamma);\ \|\boldsymbol{\omega}_1\|^2 < \chi_q^2(\alpha)\} \\
&\quad + \int_0^{\chi_p^2(\gamma)} P\{\|\eta_1 - \omega_1\|^2 \leq (\chi_p^2(\gamma) - t)^+;\ \|\boldsymbol{\omega}_1\|^2 > \chi_q^2(\alpha)\} dH_1(t, 0) \\
&\geq \int_0^{\chi_p^2(\gamma)} P_{\boldsymbol{\eta}_1}\{\|\boldsymbol{\eta}_1 - \boldsymbol{\omega}_1\|^2 \leq (\chi_p^2(\gamma) - t)^+;\ \|\boldsymbol{\omega}_1\|^2 < \chi_q^2(\alpha)\} d_p H_1(t, 0) \\
&\quad + \int_0^{\chi_p^2(\gamma)} P_{\boldsymbol{\eta}_1}\{\|\boldsymbol{\eta}_1 - \boldsymbol{\omega}_1\|^2 \leq (\chi_p^2(\gamma) - t)^+;\ \|\boldsymbol{\omega}_1\|^2 > \chi_q^2(\alpha)\} dH_1(t, 0) \\
&= \int_0^{\chi_p^2(\gamma)} P\{\|\boldsymbol{\eta}_1 - \boldsymbol{\omega}_1\|^2 \leq (\chi_p^2(\gamma) - t)^+\} dH_1(t, 0) \\
&= P_0(\|\eta_1 - \omega_1\|^2 + (\eta_2 - \omega_2)^2 \leq \chi_p^2(\gamma)) = 1 - \gamma.
\end{aligned}$$

(ii) $\chi_p^2(\gamma) \leq \Delta^2 \leq (\chi_p(\gamma) + \chi_q)^2$. In this case,

$$\begin{aligned}
&P(C^{\mathrm{PT}}(\hat{\boldsymbol{\theta}}_n^{\mathrm{PT}}(\alpha))) \\
&= \int_0^{\chi_p^2(\gamma)} P_{\boldsymbol{\eta}_1}\{\|\boldsymbol{\eta}_1 - \boldsymbol{\omega}_1\|^2 \leq (\chi_p^2(\gamma) - t)^+;\ \|\boldsymbol{\omega}_1\|^2 > \chi_q^2(\alpha)\} dH_1(t, 0) \\
&\leq \int_0^{\chi_p^2(\gamma)} P_{\boldsymbol{\eta}_1}\{\|\boldsymbol{\eta}_1 - \boldsymbol{\omega}_1\|^2 \leq (\chi_p^2(\gamma) - t)^+\} dH_1(t, 0) \\
&= P_0(\|\boldsymbol{\eta}_1 - \boldsymbol{\omega}_1\|^2 + (\eta_2 - \omega_2)^2 \leq \chi_p^2(\gamma)) \\
&= 1 - \gamma.
\end{aligned}$$

(iii) $\Delta^2 > (\chi_p(\gamma) + \chi_q(\alpha))^2$. In this case, $\chi_p(\gamma) + \chi_q(\alpha) < \Delta$ and $\|\boldsymbol{\eta}_1 - \boldsymbol{\omega}_1\|^2 < (\chi_p^2(\gamma) - t)^+$ imply $\|\boldsymbol{\eta}_1 - \boldsymbol{\omega}_1\| < \sqrt{\chi_p^2(\gamma) - t)^+}$, which in turn implies $\Delta - \|\boldsymbol{\omega}_1\|^2 \leq \|\boldsymbol{\eta}_1 - \boldsymbol{\omega}_1\| \leq \sqrt{\chi_p^2(\gamma) - t)^+}$. Thus, $\Delta < \|\boldsymbol{\omega}_1\|^2 + \sqrt{\chi_p(\gamma) - t)^+}$.

As a result, we have

$$\left\{\sqrt{\chi_p^2(\gamma) - t)^+} + \chi_q(\alpha)\right\} \leq \Delta \leq \|\boldsymbol{\omega}_1\|^2 + \sqrt{\chi_p(\gamma) - t)^+}$$

which implies that $\|\boldsymbol{\omega}_1\|^2 > (\chi_p^2(\gamma) - t)^+$. Thus,

$$\int_0^{\chi_p^2(\gamma)} P_{\boldsymbol{\eta}_1}\{\|\boldsymbol{\eta}_1 - \boldsymbol{\omega}_1\|^2 \leq (\chi_p^2(\gamma) - t)^+;\ \|\boldsymbol{\omega}_1\|^2 > \chi_q^2(\alpha)\} dH_1(t, 0)$$

$$= \int_0^{\chi_p^2(\gamma)} P_{\boldsymbol{\eta}_1}\{\|\eta_1 - \omega_1\|^2 \le (\chi_p^2(\gamma) - t)^+\} dH_1(t, 0)$$
$$= P_0\{\|\eta_1 - \omega_1)^2 + (\eta_2 - \omega_2)^2 \le \chi_p^2(\gamma)\} = 1 - \gamma.$$

Next, we consider the coverage probability of $C^{S+}(\hat{\boldsymbol{\theta}}_n^{S+}(c))$. In this case,

$$P\left\{C^{S+}(\hat{\boldsymbol{\theta}}_n^{S+}(c))\right\} = H_1(\chi_p^2(\gamma) - \Delta^2; 0) H_q(c; \Delta^2)$$
$$+ \int_0^{\chi_p^2(\gamma)} P_{\eta_1}\left\{\|\eta_1 - \omega_1(1 - c\|\omega_1\|^{-2})\|^2 \le \chi_p^2(\gamma) - t; \|\omega_1\|^2 > c\right\} dH_1(t, 0). \tag{5.9.23}$$

Again, we note that (5.9.23) is always greater than or equal to $1-\gamma$. To show this, we write

$$\int_0^{\chi_p^2(\gamma)} P_{\eta_1}\left\{\|\boldsymbol{\eta}_1 - \omega_1(1 - c\|\omega_1\|^{-2})\|^2 \le (\chi_p^2(\gamma) - t)^+; \|\omega_1\|^2 > c\right\} dH_1(t; 0)$$
$$\ge (1 - \gamma) - H_1(\chi_p^2(\gamma) - \Delta^2; 0) H_q(\chi_q^2(\alpha); \Delta^2). \tag{5.9.24}$$

From Chapter 4 and Hwang and Casella (1984), the condition for dominance of confidence set based on the **positive-rule Stein estimator** over the **unrestricted estimator** is that

$$\mathcal{G}_{(1)}(c; \chi_p^2(\gamma)) = \left[\frac{\frac{\chi_p(\gamma)}{2} + \sqrt{\left(\frac{\chi_p(\gamma)}{2}\right)^2 + c}}{\sqrt{c}}\right]^{p-3} \exp\left(-\frac{1}{2}\chi_p(\gamma)\sqrt{c}\right) \ge 1 \tag{5.9.25a}$$

and

$$\mathcal{G}^{(2)}(c; \chi_p^2(\gamma))$$
$$= \left[\frac{\chi_p(\gamma) + \sqrt{\chi_p^2(\gamma) + c}}{\sqrt{c}}\right]^{p-2} e^{-\chi_p(\gamma)\sqrt{c}} \left[\frac{\sqrt{(\chi_p^2(\gamma) + 4c)} - \chi_p(\gamma)}{2\sqrt{c}}\right] \ge 1 \tag{5.9.25b}$$

for $p \ge 3$.

Note that $\mathbf{Z}_1 \sim N_{p-1}(\eta_1, \mathbf{I}_{p-1})$, so it is sufficient to establish that $\mathcal{G}^{(1)}(c, b) \ge 1$ and $\mathcal{G}^{(2)}(c, b) \ge 1$ for $0 < c < c_0$ and $0 < b^2 \le \chi_p^2(\gamma)$.

Let us prove $\mathcal{G}^{(1)}(c, b) \ge 1$ for $0 < b^2 < \chi_p^2(\gamma)$. Note that for every value of b, $\mathcal{G}^{(1)}(c, b)$ is decreasing in c. Hence, it is sufficient to establish that there exists a c^* such that $\mathcal{G}^{(1)}(c^*, b) \ge 1$ and $\mathcal{G}^{(1)}(c^*, b) = 1$. Also, $\mathcal{G}^{(1)}(c, b)$ is strictly decreasing in b. Hence, $\mathcal{G}^{(1)}(c^*, b)$ either (1) strictly decreases to zero in b or (2) strictly increases to a unique maximum and then decreases to zero. Since $\mathcal{G}^{(1)}(c^*, 0) = \mathcal{G}^{(1)}(c^*; \chi_p(\gamma)) = 1$, then case 1 is not true, and in the case

2 we get $\mathcal{G}^{(1)}(c^*, 0) > 1$ for $0 < b^2 < \chi_p^2(\gamma)$. The proof with $\mathcal{G}^{(2)}(c, \chi_p(\gamma))$ is similar. In our case, $c^* = p - 3$, and computational results show that there is no significant difference in coverage probabilities. Now,

$$\begin{aligned}
&\int_0^{\chi_p^2(\gamma)} P_{\eta_1}\left\{\|\omega_1(1 - c\|\omega_1\|^{-2})\|^2 \leq (\chi_p^2(\gamma) - t)^+; \|\omega_1\|^2 > c\right\} dH_1(t, 0) \\
&\geq \int_0^{\chi_p^2(\gamma)} P_{\eta_1}\{\|\eta_1 - \omega_1\|^2 \leq \chi_p^2(\gamma) - t\} dH_1(t, 0) \\
&\qquad -H_1(\chi_p^2(\gamma) - \Delta^2; 0) H_q(\chi_q^2(\alpha); \Delta^2) \\
&\geq (1 - \gamma) - H_1(\chi_p^2(\gamma) - \Delta^2; 0) H_q(\chi_q^2(\alpha); \Delta^2). \qquad (5.9.26)
\end{aligned}$$

Consider further the solution of (5.9.25a) and (5.9.25b) for equality to 1 and let c_0 be the minimum of two solutions. Table 5.9.1 gives the values of c_0 is given from Casella and Hwang (1987).

Table 5.9.1 Values of c_0 from (5.9.25a) and (5.9.25b) for $p = 4(1)25$ and $\gamma = 0.05$ and 0.10

$\Delta^2 \backslash p$	$\gamma = 0.10$	$\gamma = 0.05$	p	$\gamma = 0.10$	$\gamma = 0.05$
4	0.669	0.633	15	9.888	9.675
5	1.517	1.458	16	10.731	10.510
6	2.392	2.281	17	11.574	11.345
7	3.211	3,983	18	12.418	12.182
8	4.036	3.893	19	13,263	13,020
9	4.865	4.710	20	14.109	13.859
10	5.697	5.531	21	14.955	14.699
11	6.532	6.355	22	15.802	15.539
12	7.369	7.182	23	16.650	16.381
13	8.207	8.011	24	17.498	17.223
14	9.047	8.842	25	18.346	18.065

The coverage probability of $C^{S+}\left(\hat{\boldsymbol{\theta}}_n^{S+}(c)\right)$ is given in Table 5.9.2 for $p = 4(4)24$ and $\gamma = 0.10$.

Table 5.9.2 Coverage Probabilities of the Set $C^{S+}\left(\hat{\boldsymbol{\theta}}_n^{S+}(c)\right)$ with $c = c_0$, $\gamma = 0.10$

$\Delta^2 \backslash p$	4	8	12	16	20	24
0	0.940	0.991	0.998	0.999	0.999	0.999
1	0.936	0.990	0.998	0.999	0.999	0.999
2	0.927	0.985	0.997	0.999	0.999	0.999
3	0.910	0.976	0.994	0.998	0.999	0.999
4	0.905	0.956	0.988	0.997	0.999	0.999
6	0.902	0.930	0.962	0.983	0.994	0.998
8	0.901	0.918	0.941	0.962	0.978	0.989
10	0.901	0.912	0.928	0.946	0.961	0.974
20	0.900	0.903	0.908	0.914	0.920	0.927
50	0.900	0.901	0.901	0.902	0.904	0.905
100	0.900	0.900	0.900	0.901	0.901	0.901

Table 5.9.3: Coverage Probabilities of the Set $C^{S+}\left(\hat{\boldsymbol{\theta}}_n^{S+}(c)\right)$ with $c_0 = p - 3$, $\gamma = 0.10$

$\Delta^2 \backslash p$	4	8	12	16	20	24
0	0.952	0.995	0.999	0.999	0.999	0.999
1	0.948	0.994	0.999	0.999	0.999	0.999
2	0.936	0.990	0.999	0.999	0.999	0.999
3	0.911	0.982	0.997	0.998	0.999	0.999
4	0.905	0.958	0.991	0.998	0.999	0.999
6	0.902	0.931	0.963	0.985	0.995	0.998
8	0.901	0.919	0.942	0.964	0.980	0.990
10	0.901	0.912	0.929	0.947	0.963	0.978
20	0.900	0.903	0.908	0.914	0.921	0.928
50	0.900	0.901	0.901	0.902	0.904	0.905
100	0.900	0.900	0.900	0.901	0.901	0.901

The value c_0 is almost near $p - 3$, which is our choice, in the Stein-rule estimation. Table 5.9.3 shows that by this choice, $c_0 = (p - 3)$. There is not much difference in coverage probability. (Tables 5.9.1 through 5.9.3 are due to Casella and Hwang (1987) reproduced with permission from Elsevier).

Thus, from the analysis of the coverage probabilities, we conclude that under H_0,

$$C^R\left(\hat{\boldsymbol{\theta}}_n\right) \geq C^{PT}\left(\hat{\boldsymbol{\theta}}_n^{PT}\right) \geq C^{S+}\left(\hat{\boldsymbol{\theta}}_n^{S+}\right) \geq C^0\left(\tilde{\boldsymbol{\theta}}_n\right), \qquad (5.9.27)$$

depending on the size of the test given by $\chi_q^2(\alpha)$. This may give rise to the ordering

$$C^R\left(\hat{\boldsymbol{\theta}}_n\right) \geq C^{S+}\left(\hat{\boldsymbol{\theta}}_n^{S+}\right) \geq C^{PT}\left(\hat{\boldsymbol{\theta}}_n^{PT}\right) \geq C^0\left(\tilde{\boldsymbol{\theta}}_n\right). \qquad (5.9.28)$$

As Δ^2 diverts from the origin, the dominance ordering changes drastically, leaving room for only $C^{S+}\left(\hat{\boldsymbol{\theta}}_n^{S+}\right)$, which dominates $C^0\left(\tilde{\boldsymbol{\theta}}_n\right)$ uniformly. Thus, for everyday use, we consider $C^{S+}\left(\hat{\boldsymbol{\theta}}_n^{S+}\right)$, which gives the guarantee that the coverage probability is at least $1-\gamma$.

5.10 Asymptotic Theory of Confidence Set Estimation

In this section, we develop the asymptotic theory of confidence sets for the treatment means. We recall the conditions on the ANOVA model (5.7.1) as given by (5.7.2) through (5.7.5). The confidence sets corresponding to the $(1-\gamma)\%$ confidence coefficient as given below.

$$C^0\left(\tilde{\boldsymbol{\theta}}_n\right)=\left\{\boldsymbol{\theta}: ns_e^{-2}\|\boldsymbol{\theta}-\tilde{\boldsymbol{\theta}}_n\|^2_{\boldsymbol{\Lambda}_n}\le\kappa_\gamma\right\}, \tag{5.10.1a}$$

where $P\left(C^0(\tilde{\boldsymbol{\theta}}_n)\right)=1-\gamma$ for $\kappa_\gamma=\chi_p^2(\gamma)$, since $P(\chi_p^2\le\chi_p^2(\gamma))=1-\gamma$.

$$C^R\left(\hat{\boldsymbol{\theta}}_n\right)=\left\{\boldsymbol{\theta}: ns_e^{-2}\|\boldsymbol{\theta}-\hat{\boldsymbol{\theta}}_n\|^2_{\boldsymbol{\Lambda}_n}\le\kappa_\gamma\right\} \tag{5.10.1b}$$

$$C^{PT}\left(\hat{\boldsymbol{\theta}}_n^{PT}(\alpha)\right)=\left\{\boldsymbol{\theta}: ns_e^{-2}\|\boldsymbol{\theta}-\hat{\boldsymbol{\theta}}_n^{PT}(\alpha)\|^2_{\boldsymbol{\Lambda}_n}\le\kappa_\gamma\right\} \tag{5.10.1c}$$

$$C^S\left(\hat{\boldsymbol{\theta}}_n^S\right)=\left\{\boldsymbol{\theta}: ns_e^{-2}\|\boldsymbol{\theta}-\hat{\boldsymbol{\theta}}_n^S\|^2_{\boldsymbol{\Lambda}_n}\le\kappa_\gamma\right\} \tag{5.10.1d}$$

$$C^{S+}\left(\hat{\boldsymbol{\theta}}_n^{S+}(c)\right)=\left\{\boldsymbol{\theta}: ns_e^{-2}\|\boldsymbol{\theta}-\hat{\boldsymbol{\theta}}_n^{S+}(c)\|^2_{\boldsymbol{\Lambda}_n}\le\kappa_\gamma\right\}. \tag{5.10.1e}$$

We will study the properties of these five $(1-\gamma)\%$ confidence sets as $n\to\infty$.

5.10.1 Asymptotic Representations of Normalized Estimators under Fixed Alternatives

From Section 5.8.1 we know that under fixed alternatives $\boldsymbol{\theta}=\theta_0\mathbf{1}_p+\boldsymbol{\delta}$:

$$\sqrt{n}s_e^{-1}\boldsymbol{\Lambda}_n^{1/2}\left(\tilde{\boldsymbol{\theta}}_n-\boldsymbol{\theta}\right)=\sigma^{-1}\boldsymbol{\Lambda}^{1/2}\left(\tilde{\boldsymbol{\theta}}_n-\boldsymbol{\theta}\right)+o_p(1), \tag{5.10.2a}$$

$$\sqrt{n}s_e^{-1}\boldsymbol{\Lambda}_n^{1/2}\left(\hat{\boldsymbol{\theta}}_n^{PT}-\boldsymbol{\theta}\right)=\sigma^{-1}\boldsymbol{\Lambda}^{1/2}\left(\tilde{\boldsymbol{\theta}}_n-\boldsymbol{\theta}\right)+o_p(1), \tag{5.10.2b}$$

$$\sqrt{n}s_e^{-1}\boldsymbol{\Lambda}_n^{1/2}\left(\hat{\boldsymbol{\theta}}_n^{S}-\boldsymbol{\theta}\right)=\sigma^{-1}\boldsymbol{\Lambda}^{1/2}\left(\tilde{\boldsymbol{\theta}}_n-\boldsymbol{\theta}\right)+o_p(1), \tag{5.10.2c}$$

and

$$\sqrt{n}s_e^{-1}\boldsymbol{\Lambda}_n^{1/2}\left(\hat{\boldsymbol{\theta}}_n^{S+}-\boldsymbol{\theta}\right)=\sigma^{-1}\boldsymbol{\Lambda}^{1/2}\left(\tilde{\boldsymbol{\theta}}_n-\boldsymbol{\theta}\right)+o_p(1). \tag{5.10.2d}$$

As in section 4.8.1 the asymptotic distribution of $\sqrt{n}s_e^{-1}\boldsymbol{\Lambda}_n^{1/2}(\hat{\boldsymbol{\theta}}_n-\boldsymbol{\theta})$ degenerates. Hence, $P(C^0(\tilde{\boldsymbol{\theta}}_n))$, $P(C^{PT}(\hat{\boldsymbol{\theta}}_n^{PT}(\alpha)))$, $P(C^S(\tilde{\boldsymbol{\theta}}_n))$, and $P(C^{S+}(\hat{\boldsymbol{\theta}}_n^{S+}(c)))$ tend to $1-\gamma$ as $n\to\infty$ under fixed alternatives.

5.10.2 Asymptotic Coverage Probability of the Confidence Sets under Local Alternatives

In order to obtain meaningful asymptotic coverage probabilities of the confidence sets in (5.10.1a-e), we consider the local alternatives $K_{(n)} : \boldsymbol{\theta}_{(n)} = \theta_0 \mathbf{1}_p + n^{-1/2}\boldsymbol{\delta}$ as given by (5.7.12). Accordingly, we start with the unrestricted estimator based confidence set $C^0(\tilde{\boldsymbol{\theta}}_n)$. In this case,

$$\sqrt{n}s_e^{-1}\boldsymbol{\Lambda}_n^{1/2}\left(\tilde{\boldsymbol{\theta}}_n - \boldsymbol{\theta}_{(n)}\right) \sim \mathbf{Z} \sim \mathcal{N}_p(\mathbf{0}, \mathbf{I}_p).$$

Hence,

$$\lim_{n\to\infty} P_{K_{(n)}}\left\{ns_e^{-1}\boldsymbol{\Lambda}_n^{1/2}\|\boldsymbol{\theta}_{(n)} - \tilde{\boldsymbol{\theta}}_n\|^2_{\boldsymbol{\Lambda}_n} \le \kappa_\gamma\right\} = P_0\left\{\|\mathbf{Z}\|^2 \le \chi_p^2(\gamma)\right\} = 1-\gamma, \tag{5.10.3}$$

which is constant for all Δ^2.

Next, we proceed to the set $C^R\left(\hat{\boldsymbol{\theta}}_n\right)$. In this case,

$$\begin{aligned}
&\lim_{n\to\infty} P_{K_{(n)}}\left\{ns_e^{-2}\|\boldsymbol{\theta}_{(n)} - \hat{\boldsymbol{\theta}}_n\|^2_{\boldsymbol{\Lambda}_n} \le \kappa_\gamma\right\}\\
&= P_{\eta_1}\left\{(\eta_2-\omega_2)^2 + \|\boldsymbol{\eta}_1\|^2 \le \chi_p^2(\gamma)\right\}\\
&= H_1(\chi_p^2(\gamma) - \Delta^2; 0)
\end{aligned} \tag{5.10.4}$$

by similar transformations as in Section 5.9.1 and equation (5.9.12). We note that the asymptotic coverage probability is the same as in the finite case and analysis similar to that of Section 5.9.2.

Accordingly,

$$\begin{aligned}
&\lim_{n\to\infty} P_{K_{(n)}}\left\{ns_e^{-2}\|\boldsymbol{\theta}_{(n)} - \hat{\boldsymbol{\theta}}_n^{PT}\|^2_{\boldsymbol{\Lambda}_n} \le \kappa_\gamma\right\}\\
&= H_1(\chi_p^2(\gamma) - \Delta^2; 0)H_q(\chi_q^2(\alpha); \Delta^2)\\
&\quad + \int_0^{\chi_p^2(\gamma)} P_{\eta_1}\left\{\|\boldsymbol{\eta}_1 - \boldsymbol{\omega}_1\|^2 \le (\chi_p^2(\gamma) - t)^+; \|\boldsymbol{\omega}_1\|^2 > \chi_q^2(\alpha)\right\} dH_1(t;0),
\end{aligned} \tag{5.10.5}$$

In the case positive-rule Stein-estimator, we obtain

$$\begin{aligned}
&\lim_{n\to\infty} P_{K_{(n)}}\left\{ns_e^{-2}\|\boldsymbol{\theta}_{(n)} - \hat{\boldsymbol{\theta}}_n^{S+}(c)\|^2_{\boldsymbol{\Lambda}_n} \le \kappa_\gamma\right\}\\
&= H_1(\chi_p^2(\gamma) - \Delta^2; 0)H_q(\chi_q^2(\alpha); \Delta^2)\\
&\quad + \int_0^{\chi_p^2(\gamma)} P_{\eta_1}\{\|\boldsymbol{\eta}_1 - \boldsymbol{\omega}_1\|^2(1 - c\|\boldsymbol{\omega}_1\|^{-2})\|^2 \le (\chi_p^2(\gamma) - t)^+; \|\boldsymbol{\omega}_1\|^2 >\}\\
&\qquad \times dH_1(t;0),
\end{aligned} \tag{5.10.6}$$

which is the same as (5.9.15). Hence, $C^{S+}\left(\hat{\boldsymbol{\theta}}_n^{S+}(c)\right)$ dominates $C^0\left(\tilde{\boldsymbol{\theta}}_n\right)$ for all Δ^2.

5.11 Nonparametric Methods: R-Estimation

In this section, we consider the nonparametric methods of estimation of several location parameters $\boldsymbol{\theta} = (\theta_1, \theta_2, \dots, \theta_p)'$ when it is suspected but not certain that $\boldsymbol{\theta} = \theta_0 \mathbf{1}_p$, $\mathbf{1}_p = (1 \dots 1)'$ is a p-tuple of ones and θ_0 is a scalar. This methodology broadens the scope of applications of the five estimators defined in earlier sections.

5.11.1 Model, Assumptions, and Linear Rank Statistics (LRS)

Consider the ANOVA model (5.1.1) in explicit form, as given below where S_i denotes the ith sample:

$$S_i : \mathbf{Y}_i = \theta_i \mathbf{1}_{n_i} + \boldsymbol{\varepsilon}_i, \quad i = 1, \dots, p, \tag{5.11.1}$$

where

$$\mathbf{Y}_i = (Y_{i1}, \dots, Y_{in_i})',$$
$$\boldsymbol{\varepsilon}_i = (\varepsilon_{i1}, \dots, \varepsilon_{in_i})',$$

with the cdf given by

$$F(\boldsymbol{\varepsilon}_i) = \prod_{j=1}^{n_i} F_0(Y_{ij} - \theta_i)$$

and

$$F(\boldsymbol{\varepsilon}_1, \dots, \boldsymbol{\varepsilon}_p) = \prod_{i=1}^{p} \prod_{j=1}^{n_i} F_0(Y_{ij} - \theta_i). \tag{5.11.2}$$

Further, let $\mathbf{D}_n = \text{Diag}\left(\frac{n_1}{n}, \dots, \frac{n_p}{n}\right)$ and $n = n_1 + \cdots + n_p$. (5.11.3)

It is assumed that $\{\varepsilon_{ij}\}$ are mutually independent and identically distributed with the cdf $F(\boldsymbol{\varepsilon}_1, \dots, \boldsymbol{\varepsilon}_p)$ defined by (5.11.2), where $F_0(\cdot)$ belongs to the class $\mathcal{F}$ of absolutely continuous c.d.f. with absolutely continuous p.d.f. $f_0(\cdot)$ having finite "Fisher's information", that is,

(i) $\displaystyle\int_{-\infty}^{\infty} \left\{\frac{f_0'(z)}{f_0(z)}\right\}^2 f_0(z)dz < \infty.$ (5.11.4)

(ii) $\displaystyle\lim_{n\to\infty} \frac{n_i}{n} = \lambda_i \quad (0 < \lambda_i < 1).$

(iii) Score function $a_n(\cdot)$ is generated by a function $\phi(u)$ that is nonconstant, nondecreasing, and square integrable and

$$a_n(k) = E[\phi(U_{k,n})], \quad \text{or} \quad \phi\left(\frac{k}{n+1}\right), \quad k = 1, \dots, n, \tag{5.11.5}$$

where U_{kn} is the order statistics with rank k in a random sample of size n from $U(0,1)$. Finally, let $R_{ij}(b_i)$ be the rank of $(Y_{ij} - b_i)$ among $\{(Y_{i1} - b_i), \dots, (Y_{in_i} - b_i) | i = 1, \dots, p\}$, and set

$$T_{n_i}(b_i) = \sum_{i=1}^{n_i} \left[a_{n_i}(R_{ij}(b_i)) - \bar{a}_{n_i}\right], \quad \bar{a}_{n_i} = n_i^{-1} \sum_{j=1}^{n_i} a_{n_i}(R_{ij}(b_i)). \tag{5.11.6}$$

Denote by $T_n^*(\mathbf{b}_i)$, the sum of $T_{n_i}(b_i)$,

$$T_n^*(\mathbf{b}) = \sum_{i=1}^{p} T_{n_i}(b_i) = \mathbf{1}_p' \mathbf{T}_n(\mathbf{b}), \tag{5.11.7a}$$

where $\quad \mathbf{T}_n(\mathbf{b}) = (T_{n_1}(b_1), \dots, T_{n_p}(b_p))'$. (5.11.7b)

As in Chapter 4, we define

$$A_\phi^2 = \left(\int_0^1 \phi^2(u)du\right) - \left(\int_0^1 \phi(u)du\right)^2$$

$$A_\psi^2 = \int_0^1 \psi^2(u)du, \quad \psi(u) = -\frac{f_0'(F_0^{-1}(u))}{f_0(F_0^{-1}(u))}, \quad 0 < u < 1,$$

and

$$\gamma(\psi, \phi) = \int_0^1 \psi(u)\phi(u)du. \tag{5.11.8}$$

Further, let

$$A_{n_i}^2 = (n_i - 1) \sum_{j=1}^{n_i} \left(a_{n_i}(R_{ij}) - \bar{a}_{n_i}\right)^2, \tag{5.11.9}$$

where R_{ij} stands for $R_{ij}(0)$ and $\bar{a}_{n_i} = n_i^{-1} \sum_{k=1}^{n_i} a_{n_i}(k)$. Then, by the basic theorems of Chapter 4 of Hajek and Sidak (1967) under $\boldsymbol{\theta} = \mathbf{0}$,

$$n^{-1/2}\mathbf{T}_n(\mathbf{0}) \sim N_p(\mathbf{0}, A_\phi^2 \boldsymbol{\Lambda}), \tag{5.11.10}$$

where $\boldsymbol{\Lambda} = \lim_{n\to\infty} \mathbf{D}_n$.

Note that $T_{n_i}(b_i) \searrow$ in b_i and $T_n^*(b_0) \searrow$ in b_0 under the model (5.11.1) and under $H_0 : \boldsymbol{\theta} = \theta_0 \mathbf{1}_p$ respectively. Also, under $\boldsymbol{\theta} = \mathbf{0}$, $T_{n_i}(0)$ and $T_n^*(0)$ are symmetric about 0. Thus, we define the **unrestricted** and the **restricted** estimators of $\boldsymbol{\theta}$ as follows:

Let $\tilde{\boldsymbol{\theta}}_n = (\tilde{\theta}_{n_1}, \dots, \tilde{\theta}_{n_p})'$ be the estimate of $\boldsymbol{\theta} = (\theta_1, \dots, \theta_p)'$. Then the components of $\tilde{\boldsymbol{\theta}}_n$ are given by

$$\tilde{\theta}_{n_i} = \frac{1}{2}\left[\sup\{b_i : T_{n_i}(b_i) > 0\} + \inf\{b_i : T_{n_i}(b_i) < 0\}\right], \quad i = 1, \dots, p, \tag{5.11.11a}$$

and under $\boldsymbol{\theta} = \theta_0 \mathbf{1}_p$ we define

$$\hat{\theta}_{0n} = \frac{1}{2}\Big[\sup\{b : T^*_{n_i}(b) > 0\} + \inf\{b : T^*_n(b) < 0\}\Big], \tag{5.11.11b}$$

as the estimate of θ_0. Then $\hat{\boldsymbol{\theta}}_n = \hat{\boldsymbol{\theta}}_{0n}\mathbf{1}_p$.

Note that both $\tilde{\boldsymbol{\theta}}_n$ and $\hat{\boldsymbol{\theta}}_n$ are translation-invariant robust and consistent estimators of $\boldsymbol{\theta}$ under the general model (5.11.1) and the restricted model with $\boldsymbol{\theta} = \theta_0\mathbf{1}_p$. For the preliminary test of the hypothesis, $H_0 : \boldsymbol{\theta} = \theta_0\mathbf{1}_p$, we use the nonparametric test-statistic

$$\mathcal{L}_n = A_n^{-2}\big[\mathbf{T}_n(\hat{\boldsymbol{\theta}}_n)\big]'\mathbf{D}_n^{-1}\big[\mathbf{T}_n(\hat{\boldsymbol{\theta}}_n)\big]', \quad \hat{\boldsymbol{\theta}}_n = \hat{\theta}_{0n}\mathbf{1}_p, \tag{5.11.12}$$

where

$$A_n^2 = (n-1)^{-1}\sum_{k=1}^{n}[a_n(k) - \bar{a}_n]^2 \tag{5.11.13}$$

and

$$\mathbf{T}_n(\hat{\boldsymbol{\theta}}_n) = \Big(T_{n_1}(\hat{\theta}_{0n}), \ldots, T_{n_p}(\hat{\theta}_{0n})\Big)'. \tag{5.11.14}$$

Also, Theorem 3.1 of Jurečkova (1969) implies that under $\boldsymbol{\theta} = \mathbf{0}$ and for any finite k (> 0),

$$\sup\Big\{n^{-1/2}\big|T_{n_i}(n^{-1/2}b_i) - T_{n_i}(0) + \lambda_i n^{1/2} b_i \lambda(\psi, \phi)\big| : |b_i| < k\Big\} \xrightarrow{P} 0 \tag{5.11.15a}$$

so that

$$\sup\left\{n^{-1/2}\big|T^*_n(n^{-1/2}\mathbf{b}) - T^*_n(0) + n^{1/2}\left(\sum_{i=1}^{p}\lambda_i b_i\right)\gamma(\psi,\phi)\big| : |\mathbf{b}| < k\right\} \xrightarrow{P} 0, \tag{5.11.15b}$$

where $|\mathbf{b}| = \max(|b_1|, \ldots, |b_p|)$. Thus, under $\boldsymbol{\theta} = \mathbf{0}$ and relative compactness of $n^{-1/2}\hat{\theta}_{0n}$, we obtain the relations

(i) $n^{-1/2}T^*_n(0)\gamma^{-1}(\psi,\phi) - n^{-1/2}\hat{\theta}_{0n} \xrightarrow{P} 0$

(ii) $n^{-1/2}T_{n_i}(\hat{\theta}_{0n}) - \Big(n^{-1/2}T_{n_i}(0) - \lambda_i n^{1/2}\hat{\theta}_{0n}\gamma(\psi,\phi)\Big) \xrightarrow{P} 0$

$$\text{or} \quad n^{-1/2}T_{n_i}(\hat{\theta}_{0n}) - \big(n^{-1/2}T_{ni}(0) - \lambda_i n^{-1/2}T^*_n(0)\big) \xrightarrow{P} 0. \tag{5.11.16}$$

In vector forms we may write

$$n^{-1/2}\mathbf{T}_n(\hat{\boldsymbol{\theta}}_n) - (\mathbf{I}_p - \mathbf{1}_p\mathbf{1}'_p\boldsymbol{\Lambda})n^{-1/2}\mathbf{T}_n(0) \xrightarrow{P} 0. \tag{5.11.17}$$

Then, we may write

$$\mathcal{L}_n = A_\phi^{-2}[\mathbf{T}_n(\mathbf{0})]'\mathbf{J}\boldsymbol{\Lambda}^{-1}\mathbf{J}'[\mathbf{T}_n(\mathbf{0})] + o_p(1), \tag{5.11.18}$$

where $\mathbf{J} = \mathbf{I}_p - \mathbf{1}_p\mathbf{1}'_p\boldsymbol{\Lambda}$. Hence, under $\boldsymbol{\theta} = \mathbf{0}$, as $n \to \infty$, $\mathcal{L}_n$ approximately follows the central chi-square distribution with $(p-1)$ d.f., since $\mathbf{J}\boldsymbol{\Lambda}^{-1}\mathbf{J}'$ is an idempotent matrix with rank $(p-1)$.

5.11.2 Preliminary Test and Stein-Type Estimators

We combine the **unrestricted** and the **restricted** estimators of $\boldsymbol{\theta}$ to obtain the preliminary test estimator and the Stein-type estimators as in Section 5.7.2, which are given by

$$\text{(i)}\quad \hat{\boldsymbol{\theta}}_n^{\text{PT}} = \tilde{\boldsymbol{\theta}}_n - (\tilde{\boldsymbol{\theta}}_n - \hat{\boldsymbol{\theta}}_n) I(\mathcal{L}_n < \mathcal{L}_{n,\alpha}), \tag{5.11.19}$$

where $\mathcal{L}_{n,\alpha}$ is the α-level critical value from the distribution of $\mathcal{L}_n$ under H_0 which converges to $\chi^2_{p-1}(\alpha)$ as $n \to \infty$. The Stein-type estimators are given by

$$\text{(ii)}\quad \hat{\boldsymbol{\theta}}_n^{S} = \tilde{\boldsymbol{\theta}}_n - c(\tilde{\boldsymbol{\theta}}_n - \hat{\boldsymbol{\theta}}_n)\mathcal{L}_n^{-1} I(\mathcal{L}_n > \varepsilon_n) \tag{5.11.20}$$

where $\varepsilon_n \to 0$ as $n \to \infty$ and $c = (p-3)$ and

$$\text{(ii)}\quad \hat{\boldsymbol{\theta}}_n^{S+} = \hat{\boldsymbol{\theta}}_n + (1 - c\mathcal{L}_n^{-1}) I(\mathcal{L}_n > c)(\tilde{\boldsymbol{\theta}}_n - \hat{\boldsymbol{\theta}}_n). \tag{5.11.21}$$

as in Section 5.7.2. Note that (5.11.20) involves $I(\mathcal{L}_n > \varepsilon_n)$, since $\mathcal{L}_n$ assumes the value 0 with positive probability.

5.11.3 Asymptotic Distributional Properties of R-Estimators

First of all we know from the basic theorems of Chapter 4 of Hajek and Sidak (1967), that as $n \to \infty$,

$$n^{-1/2}\mathbf{T}_n(\mathbf{0}) \sim N_p(\mathbf{0}, A_\phi^2 \boldsymbol{\Lambda}) \tag{5.11.22}$$

and under fixed alternatives $\mathcal{L}_n \to \infty$ as $n \to \infty$. Also, under fixed alternatives all the estimators $\sqrt{n}(\hat{\boldsymbol{\theta}}_n^{\text{PT}} - \boldsymbol{\theta})$, $\sqrt{n}(\hat{\boldsymbol{\theta}}_n^{S} - \boldsymbol{\theta})$ and $\sqrt{n}(\hat{\boldsymbol{\theta}}_n^{S+} - \boldsymbol{\theta})$ are equivalent in distribution to that of $\sqrt{n}(\tilde{\boldsymbol{\theta}}_n - \boldsymbol{\theta})$. Hence, we consider the class of local alternatives

$$K_{(n)} : \boldsymbol{\theta}_{(n)} = \theta_0 \mathbf{1}_p + n^{-1/2}\boldsymbol{\delta}. \tag{5.11.23}$$

It may be mentioned that this class of alternatives is contiguous to that of $\boldsymbol{\theta} = \theta_0 \mathbf{1}_p$. Thus, we have the following theorem:

Theorem 1. Under $\{K_{(n)}\}$ and the assumed regularity conditions w.r.t. the model (5.11.1), as $n \to \infty$, we have

(i) $\sqrt{n}(\tilde{\boldsymbol{\theta}}_n - \theta_0 \mathbf{1}_p) \sim N_p(\boldsymbol{\delta}, \sigma^2 \boldsymbol{\Lambda}^{-1})$; $\sigma^2 = A_\phi^2/\gamma^2(\psi, \phi)$,

(ii) $\sqrt{n}(\tilde{\boldsymbol{\theta}}_n - \hat{\boldsymbol{\theta}}_n) \sim N_p(\mathbf{J}\boldsymbol{\delta}, \sigma^2 \mathbf{J}\boldsymbol{\Lambda}^{-1})$, $\mathbf{J} = (\mathbf{I}_p - \mathbf{1}_p \mathbf{1}_p' \boldsymbol{\Lambda})$,

(iii) $\sqrt{n}(\hat{\boldsymbol{\theta}}_n - \boldsymbol{\theta}_0 \mathbf{1}_p) \sim N_p(\boldsymbol{\delta}, \sigma^2 \mathbf{B})$, $\mathbf{B} = \mathbf{1}_p \mathbf{1}_p'$,

(iv) $$\begin{pmatrix} \sqrt{n}(\tilde{\boldsymbol{\theta}}_n - \theta_0 \mathbf{1}_p) \\ \sqrt{n}(\tilde{\boldsymbol{\theta}}_n - \hat{\boldsymbol{\theta}}_n) \end{pmatrix} \sim N_{2p}\left\{\begin{pmatrix} \boldsymbol{\delta} \\ \mathbf{J}\boldsymbol{\delta} \end{pmatrix}; \sigma^2 \begin{pmatrix} \boldsymbol{\Lambda}^{-1} & \mathbf{J}\boldsymbol{\Lambda}^{-1} \\ \boldsymbol{\Lambda}^{-1}\mathbf{J}' & \mathbf{J}\boldsymbol{\Lambda}^{-1} \end{pmatrix}\right\},$$

(v) $$\begin{pmatrix} \sqrt{n}(\hat{\boldsymbol{\theta}}_n - \theta_0 \mathbf{1}_p) \\ \sqrt{n}(\tilde{\boldsymbol{\theta}}_n - \hat{\boldsymbol{\theta}}_n) \end{pmatrix} \sim N_{2p}\left\{\begin{pmatrix} \boldsymbol{\delta} \\ \mathbf{J}\boldsymbol{\delta} \end{pmatrix}; \sigma^2 \begin{pmatrix} \mathbf{B} & \mathbf{0} \\ \mathbf{0} & \mathbf{J}\boldsymbol{\Lambda}^{-1} \end{pmatrix}\right\}$$

and

(vi) $$\lim_{n\to\infty} P\{\mathcal{L}_n \le x | K_{(n)}\} = H_{p-1}(x;\Delta^2),\ \Delta^2 = \sigma^{-2}(\boldsymbol{\delta}'\mathbf{J}'\boldsymbol{\Lambda}\mathbf{J}\boldsymbol{\delta}). \tag{5.11.24}$$

Proof. Note that both $\tilde{\boldsymbol{\theta}}_n$ and $\hat{\boldsymbol{\theta}}_n$ (or $\bar{\boldsymbol{\theta}}_n$) are translation-invariant estimators. Thus, without loss of generality, we may assume $\boldsymbol{\theta} = \mathbf{0}$. Also, by the definition of the estimators (5.11.11a) and (5.11.10) and (5.11.15a, b) we have $n^{1/2}|\tilde{\boldsymbol{\theta}}_n - \theta| = O_p(1)$, while under $\{K_{(n)}\}$ $n^{1/2}(\theta - \theta_0 \mathbf{1}_p) = \boldsymbol{\delta} = O(1)$. Thus, $n^{1/2}|\tilde{\boldsymbol{\theta}}_n|$ (and $n^{1/2}|\hat{\boldsymbol{\theta}}_n|) = O_p(1)$. Observe also that under $\boldsymbol{\theta} = \mathbf{0}$,

$$n^{-1/2}\mathbf{T}_n(\mathbf{0}) = \boldsymbol{\Lambda} n^{1/2}\tilde{\boldsymbol{\theta}}_n \gamma(\psi,\phi) + o_p(1). \tag{5.11.25}$$

Hence, utilizing contiguity under $\{K_{(n)}\}$ to those under $\boldsymbol{\theta} = \mathbf{0}$, we obtain the distribution of $n^{1/2}(\tilde{\boldsymbol{\theta}}_n - \theta_0 \mathbf{1}_p)$ as $N_p(\boldsymbol{\delta}, \sigma^2\boldsymbol{\Lambda}^{-1})$, $\sigma^2 = A_\phi^2/\gamma^2(\psi,\phi)$. Similarly, the asymptotic distribution of $\sqrt{n}(\tilde{\boldsymbol{\theta}}_n - \hat{\boldsymbol{\theta}}_n)$ is $N_p(\mathbf{J}\boldsymbol{\delta}, \sigma^2\mathbf{J}\boldsymbol{\Lambda}^{-1})$ under $\{K_{(n)}\}$, while under $\boldsymbol{\theta} = \mathbf{0}$,

$$(\boldsymbol{\Lambda}^{-1} - \mathbf{1}_p\mathbf{1}_p')n^{-1/2}\mathbf{T}_n(\mathbf{0}) = \sqrt{n}(\tilde{\boldsymbol{\theta}}_n - \hat{\boldsymbol{\theta}}_n) + o_p(1). \tag{5.11.26}$$

Also,

$$(\mathbf{1}_p\mathbf{1}_p')n^{-1/2}\mathbf{T}_n(\mathbf{0}) = n^{-1/2}(\hat{\boldsymbol{\theta}}_n - \theta_0\mathbf{1}_p) + o_p(1), \tag{5.11.27}$$

and under $\{K_{(n)}\}$,

$$n^{1/2}(\hat{\boldsymbol{\theta}}_n - \theta_0\mathbf{1}_p) \sim N_p(\mathbf{0}, \sigma^2\mathbf{B}).$$

As for the asymptotic distribution of $\mathcal{L}_n$, as $n \to \infty$, we consider $\mathcal{L}_n$ under $\boldsymbol{\theta} = \mathbf{0}$ as

$$\mathcal{L}_n = A_\phi^{-2}[\mathbf{T}_n(\mathbf{0})]'\mathbf{J}'\boldsymbol{\Lambda}^{-1}\mathbf{J}[\mathbf{T}_n(\mathbf{0})] + o_p(1), \tag{5.11.28}$$

which follows a central chi-square with $(p-1)$ d.f. Under $\{K_{(n)}\}$, $\mathcal{L}_n$ is approximately distributed as a noncentral chi-square with $(p-1)$ d.f. and noncentrality parameter $\frac{1}{2}\Delta^2$, $\Delta^2 = A_\phi^{-2}\gamma^2(\psi,\phi)(\boldsymbol{\delta}'\mathbf{J}'\boldsymbol{\Lambda}\mathbf{J}\boldsymbol{\delta})$.

As a consequence of Theorem 1, we obtain ADB, ADQB, ADMSE, and ADQR of the five estimators, as given below following Section 5.8.3.

5.11.4 ADB, ADQB, ADMSE, and ADQR

For UE, $(\tilde{\boldsymbol{\theta}})$,

(i) $\mathbf{b}_1(\tilde{\boldsymbol{\theta}}_n) = \mathbf{0}$ and $B_1(\tilde{\boldsymbol{\theta}}_n) = 0$;

(ii) $\mathbf{M}_1(\tilde{\boldsymbol{\theta}}_n) = \sigma^2\boldsymbol{\Lambda}^{-1}$ and $R_1(\tilde{\boldsymbol{\theta}}_n; \mathbf{W}) = \sigma^2\text{tr}(\mathbf{W}\boldsymbol{\Lambda}^{-1})$. (5.11.28a)

For RE, $(\hat{\boldsymbol{\theta}}_n)$,

(i) $\mathbf{b}_2(\hat{\boldsymbol{\theta}}_n) = -\boldsymbol{\delta}$ and $B_2(\hat{\boldsymbol{\theta}}_n) = \Delta^2$, $\Delta^2 = \sigma^{-2}(\boldsymbol{\delta}'\mathbf{J}'\boldsymbol{\Lambda}\mathbf{J}\boldsymbol{\delta})$;

(ii) $M_2(\hat{\boldsymbol{\theta}}_n) = \sigma^2\mathbf{B} + \boldsymbol{\delta}\boldsymbol{\delta}'$ and $R_2(\hat{\boldsymbol{\theta}}_n; \mathbf{W}) = \sigma^2\text{tr}(\mathbf{W}\mathbf{B}) + \boldsymbol{\delta}'\mathbf{W}\boldsymbol{\delta}$. (5.11.28b)

For PTE, $(\hat{\boldsymbol{\theta}}_n^{\text{PT}})$,

(i) $\mathbf{b}_3(\hat{\boldsymbol{\theta}}_n^{\text{PT}}) = -\mathbf{J}\boldsymbol{\delta}H_{p+2}(\chi^2_{p-1}(\alpha); \Delta^2)$ and

$$B_3(\hat{\boldsymbol{\theta}}_n^{\text{PT}}) = \Delta^2\left\{H_{p+1}(\chi^2_{p-1}(\alpha); \Delta^2)\right\}^2;$$

(ii)
$$\mathbf{M}_3(\hat{\boldsymbol{\theta}}_n^{\text{PT}}) = \sigma^2\boldsymbol{\Lambda}^{-1} - \sigma^2\mathbf{J}\boldsymbol{\Lambda}^{-1}H_{p+1}(\chi^2_{p-1}(\alpha); \Delta^2) + (\mathbf{J}\boldsymbol{\delta}\boldsymbol{\delta}'\mathbf{J}')\left\{2H_{p+1}(\chi^2_{p-1}(\alpha); \Delta^2) - H_{p+3}(\chi^2_{p-1}(\alpha); \Delta^2)\right\}$$

and

$$R_3(\hat{\boldsymbol{\theta}}_n^{\text{PT}}; \mathbf{W}) = \sigma^2\text{tr}(\mathbf{W}\boldsymbol{\Lambda}^{-1}) - \sigma^2\text{tr}(\mathbf{W}\mathbf{J}\boldsymbol{\Lambda}^{-1})H_{p+1}(\chi^2_{p-1}(\alpha); \Delta^2) + (\boldsymbol{\delta}'\mathbf{J}\mathbf{W}\mathbf{J}'\boldsymbol{\delta})\{2H_{p+1}(\chi^2_{p-1}(\alpha); \Delta^2) - H_{p+3}(\chi^2_{p-1}(\alpha); \Delta^2)\}.$$

For SE, $(\hat{\boldsymbol{\theta}}_n^S)$,

(i) $\mathbf{b}_4(\hat{\boldsymbol{\theta}}_n^S) = -(p-3)\mathbf{J}\boldsymbol{\delta}E[\chi^2_{p-1}(\alpha); \Delta^2)]$ and

$$B_4(\hat{\boldsymbol{\theta}}_n^S) = (p-3)\Delta^2\{E[\chi^{-2}_{p+1}(\Delta^2)]\}^2;$$

(ii)
$$\mathbf{M}_4(\hat{\boldsymbol{\theta}}_n^S) = \sigma^2\text{tr}(\mathbf{W}\boldsymbol{\Lambda}^{-1}) - \sigma^2\text{tr}(\mathbf{W}\mathbf{J}\boldsymbol{\Lambda}^{-1}\{2E[\chi^{-2}_{p+1}(\Delta^2)], - (p-3)E[\chi^{-4}_{p+1}(\Delta^2)]\} + (p-3)(p+1)(\boldsymbol{\delta}'\mathbf{J}'\mathbf{W}\mathbf{J}\boldsymbol{\delta})E[\chi^{-4}_{p+3}(\Delta^2)]$$

$$R_4(\hat{\boldsymbol{\theta}}_n^S; \mathbf{W}) = \sigma^2\text{tr}(\mathbf{W}\boldsymbol{\Lambda}^{-1}) - \sigma^2\text{tr}(\mathbf{W}\mathbf{J}\boldsymbol{\Lambda}^{-1}) \times \{2E[\chi^{-2}_{p+1}(\Delta^2)] - (p-3)E[\chi^{-4}_{p+1}(\Delta^2)]\} + (p-3)(p+1)(\boldsymbol{\delta}'\mathbf{J}'\mathbf{W}\mathbf{J}\boldsymbol{\delta})E[\chi^{-4}_{p+3}(\Delta^2)]. \quad (5.11.28d)$$

For PRSE, $(\hat{\boldsymbol{\theta}}_n^{S+})$,

(i) $b_5(\hat{\boldsymbol{\theta}}_n^{S+}) = -\mathbf{J}\boldsymbol{\delta}\{(p-3)E[\chi_{p+1}^{-2}(\Delta^2)I(\chi_{p-1}^2(\Delta^2) > p-3)]$

$+ H_{p+1}(p-3;\Delta^2)\}$

and

$B_5(\hat{\boldsymbol{\theta}}_n^{S+}) = \Delta^2\{(p-3)E[\chi_{p+1}^{-2}(\Delta^2)I(\chi_{p-1}^2(\Delta^2) > p-3)]$

$+ H_{p+1}(p-3;\Delta^2)\}^2;$

(ii) $\mathbf{M}_5(\hat{\boldsymbol{\theta}}_n^{S+})$

$= \mathbf{M}_4(\hat{\theta}_n^S) - \sigma^2(\mathbf{J}\boldsymbol{\Lambda}^{-1})E[(1-(p-3)\chi_{p+1}^{-2}(\Delta^2)I(\chi_{p+1}^2(\Delta^2) < p-3)]$

$+ (\mathbf{J}\boldsymbol{\delta}\boldsymbol{\delta}'\mathbf{J}')\{2E[(1-(p-3)\chi_{p+1}^{-2}(\Delta^2))I(\chi_{p+1}^2(\Delta^2) < p-3)]$

$- E[(1-(p-3)\chi_{p+3}^{-2}(\Delta^2))^2 I(\chi_{p+3}^2(\Delta^2) < p-3))]\}$

and

$R_5(\hat{\boldsymbol{\theta}}_n^{S+};\mathbf{W})$

$= \mathrm{tr}(\mathbf{W}\mathbf{M}_4(\hat{\theta}_n^S)) - \sigma^2\mathrm{tr}(\mathbf{W}\mathbf{J}\boldsymbol{\Lambda}^{-1})$

$\times E[(1-(p-3)\chi_{p+1}^{-2}(\Delta^2))^2 I(\chi_{p+1}^2(\Delta^2) < p-3)]$

$+ (\boldsymbol{\delta}'\mathbf{J}'\mathbf{W}\mathbf{J}\boldsymbol{\delta})\{2E[(1-(p-3)\chi_{p+1}^{-2}(\Delta^2))I(\chi_{p+1}^2(\Delta^2) < p-3))]$

$$- E[(1-(p-3)\chi_{p+3}^{-2}(\Delta^2))^2 I(\chi_{p+3}^2(\Delta^2) < p-3))]\}. \qquad (5.11.28e)$$

A proof of the expressions above follows by considering

$$\mathbf{U}_1 = \mathbf{X} - (\mathbf{Y}+\mathbf{J}\boldsymbol{\delta})I(\|\mathbf{Y}+\mathbf{J}\boldsymbol{\delta}\|^2 < \chi_{p-1}^2(\alpha)), \qquad (5.11.29a)$$

$$\mathbf{U}_2 = \mathbf{X} - d(\mathbf{Y}+\mathbf{J}\boldsymbol{\delta})\|\mathbf{Y}+\mathbf{J}\boldsymbol{\delta}\|^{-2}, \quad d = p-3, \qquad (5.11.29b)$$

$$\mathbf{U}_3 = (\mathbf{X}+\boldsymbol{\delta}) - d(\mathbf{Y}+\mathbf{J}\boldsymbol{\delta})\|\mathbf{Y}+\mathbf{J}\boldsymbol{\delta}\|^{-2}I(\|\mathbf{Y}+\mathbf{J}\boldsymbol{\delta}\|^2 > d) - (\mathbf{Y}+\mathbf{J}\boldsymbol{\delta})I(\|\mathbf{Y}+\mathbf{J}\boldsymbol{\delta}\|^2 < d), \qquad (5.11.29c)$$

where

$$\begin{pmatrix} \mathbf{X} \\ \mathbf{Y} \end{pmatrix} \sim N_{2p}\left\{\begin{pmatrix} 0 \\ 0 \end{pmatrix}; \sigma^2\begin{pmatrix} \boldsymbol{\Lambda}^{-1} & \mathbf{J}\boldsymbol{\Lambda}^{-1} \\ \boldsymbol{\Lambda}^{-1}\mathbf{J}' & \mathbf{J}\boldsymbol{\Lambda}^{-1} \end{pmatrix}\right\}. \qquad (5.11.30)$$

By computing

(i) $E[\mathbf{U}_1]$, $\sigma^2 E[\mathbf{U}_1\mathbf{U}_1']$ and $\sigma^2 E[\mathbf{U}_1'\mathbf{W}\mathbf{U}_1]$,

(ii) $E[\mathbf{U}_2]$, $\sigma^2 E[\mathbf{U}_2\mathbf{U}_2']$ and $\sigma^2 E[\mathbf{U}_2'\mathbf{W}\mathbf{U}_2]$,

and

(iii) $E[\mathbf{U}_3]$, $\sigma^2 E[\mathbf{U}_3\mathbf{U}_3']$ and $\sigma^2 E[\mathbf{U}_3'\mathbf{W}\mathbf{U}_3]$, (5.11.31)

we obtain the bias, MSE matrix, and the quadratic risks of the PTE, SE, and PRSE. Next, we discuss the recentered confidence sets based on the five R-estimators.

First, we consider the usual $(1-\gamma)\%$ confidence set, namely $C^0(\tilde{\boldsymbol{\theta}}_n)$ defined by

$$C^0(\tilde{\boldsymbol{\theta}}_n) = \{\boldsymbol{\theta} : n\sigma_n^{-2}\|\boldsymbol{\theta} - \tilde{\boldsymbol{\theta}}_n\|^2_{\mathbf{D}_n} \le k_\gamma\}, \tag{5.11.32}$$

where $k_\gamma = \chi_p^2(\gamma)$, and $\sigma_n^2 = A_n^2/\hat{\gamma}_n^2$. For some positive a $(< \infty)$, we define a consistent estimator of $\gamma(\psi, \phi)$ by

$$\hat{\gamma}_n = \frac{1}{p}\sum_{j=1}^{p} \frac{T_{n_j}(\tilde{\boldsymbol{\theta}}_n + n^{-1/2}a\mathbf{1}_p) - T_{n_j}(\tilde{\boldsymbol{\theta}}_n - n^{-1/2}a\mathbf{1}_p)}{2n_j a_n}. \tag{5.11.33}$$

Then, the asymptotic coverage probability of the set, $C^0(\tilde{\boldsymbol{\theta}}_n)$ may be written as

$$\begin{aligned}\lim_{n\to\infty} P_{K_{(n)}}(C^0(\tilde{\boldsymbol{\theta}}_n)) &= \lim_{n\to\infty} P_{K_{(n)}}\{A_n^{-2}n^{-1}\|\mathbf{T}_n(0)\|^2_{\mathbf{D}_n^{-1}} \le k_\gamma\}\\ &= P_0(\chi_p^2 \le k_\gamma) = 1 - \gamma. \end{aligned}\tag{5.11.34}$$

We define the confidence sets by

$$C^*(\boldsymbol{\theta}_n^*) = \{\boldsymbol{\theta} : n\sigma_n^{-2}\|\boldsymbol{\theta} - \boldsymbol{\theta}_n^*\|^2_{\mathbf{D}_n} \le k_\gamma\}, \tag{5.11.35}$$

where

$$\boldsymbol{\theta}_n^* = \mathbf{A}\tilde{\boldsymbol{\theta}}_n + (I - \mathbf{A})\tilde{\boldsymbol{\theta}}_n g(\mathcal{L}_n), \quad \mathbf{A} = \mathbf{1}_p\mathbf{1}_p'\mathbf{D}_n, \tag{5.11.36}$$

as before (see Section 5.9.4).

Now, using the transformations as in Section 5.9.1, we have

$$\omega_1 = n^{1/2}\sigma_n^{-1}\boldsymbol{\Gamma}_1'\mathbf{D}_n^{1/2}\tilde{\boldsymbol{\theta}}_n \quad \text{and} \quad \omega_2 = n^{1/2}\sigma_n^{-1}\boldsymbol{\Gamma}_2'\mathbf{D}_n^{1/2}\tilde{\boldsymbol{\theta}}_n, \tag{5.11.37}$$

where $\boldsymbol{\Gamma}_2 = \mathbf{D}_n^{1/2}\mathbf{1}_p$. Then, as $n \to \infty$,

$$\begin{pmatrix}\omega_1\\ \omega_2\end{pmatrix} \sim N_p\left\{\begin{pmatrix}\eta_1\\ \eta_2\end{pmatrix}\begin{pmatrix}\mathbf{I}_{p-1} & 0\\ 0 & 1\end{pmatrix}\right\}. \tag{5.11.38}$$

Hence, as $n \to \infty$,

$$n\sigma_n^{-2}\|\boldsymbol{\theta} - \boldsymbol{\theta}_n^*\|^2_{\mathbf{D}_n} \stackrel{\mathcal{D}}{=} (\eta_2 - \omega_2)^2 + \|\omega_1 - \omega_1 g(\|\omega_1\|^2)\|^2. \tag{5.11.39}$$

The asymptotic coverage probability of the set $C^*(\boldsymbol{\theta}_n^*)$ is then given by

$$\begin{aligned}&\lim_{n\to\infty} P(C^*(\boldsymbol{\theta}_n^*))\\ &= P_{\boldsymbol{\eta}}\left\{[(\eta_2 - \omega_2)^2 + \|\boldsymbol{\eta}_1 - \omega_1 g(\|\omega_1\|^2)\|^2] \le k_\gamma\right\}\\ &= \int_0^{k_\gamma} P_{\boldsymbol{\eta}}\|\boldsymbol{\eta}_1 - \omega_1 g(\|\omega_1\|^2)\|^2] \le (k_\gamma - t)^+]dH_1(t, 0),\end{aligned}$$

which is the same as the expression at (5.9.10).

The rest of the analysis follows based on the material in Section 5.9.1 and 5.9.2 and is not repeated.

5.12 Conclusions

In this chapter, we considered the simple ANOVA problem with regard to the improved estimation of the parameters of the model (5.1.1). Accordingly, we discussed the unrestricted, restricted, PTE, SE, and the PRSE of the parameters based on normal theory as well as the nonnormal theory in an asymptotic setup. We added the nonparametric R-estimation methods to enhance the scope of the methodology. In addition, we discussed the recentered confidence set estimation and their finite sample as well as asymptotic properties of the coverage probabilities under local alternatives.

5.13 Problems

1. Prove Theorem 2 of Section 5.1.3.
2. Determine the bias-vector of $\hat{\boldsymbol{\theta}}_n^{PT}$, $\hat{\boldsymbol{\theta}}_n^{S}$ and $\hat{\boldsymbol{\theta}}_n^{S+}$ using the expressions at (5.2.4i-iii).
3. Determine the MSE matrices of $\hat{\boldsymbol{\theta}}_n^{PT}$, $\hat{\boldsymbol{\theta}}_n^{S}$, and $\hat{\boldsymbol{\theta}}_n^{S+}$ using the expressions at (5.2.4i-iii).
4. Show that $\hat{\boldsymbol{\theta}}_n^{S+}$ is the convex combination of $\hat{\boldsymbol{\theta}}_n$ and $\hat{\boldsymbol{\theta}}_n^{S}$, and hence $\hat{\boldsymbol{\theta}}_n^{S+}$ is a preliminary test estimator of $\boldsymbol{\theta}$ for $p \geq 4$.
5. Determine $|\mathbf{M}_1(\tilde{\boldsymbol{\theta}}_n)|$, $\mathbf{M}_2(\hat{\boldsymbol{\theta}}_n)|$, $\mathbf{M}_3(\hat{\boldsymbol{\theta}}_n^{PT})|$ $|\mathbf{M}_4(\hat{\boldsymbol{\theta}}_n^{S})|$, and $\mathbf{M}_5(\hat{\boldsymbol{\theta}}_n^{S+})|$, and verify the expressions for the relative efficiencies based on MSE matrices.
6. Prove Theorem 1 of Section 5.7.1.
7. Consider (5.8.16), and show that $E\{ns_e^{-2}\|\hat{\boldsymbol{\theta}}_n^{S+} - \boldsymbol{\theta}\|^2\} \to 0$ as $n \to \infty$ under fixed alternatives $\theta = \theta_0 \mathbf{1}_p + \boldsymbol{\delta}$.
8. Show that $\sqrt{n}s_e^{-1}\left(\hat{\boldsymbol{\theta}}_n^{PT} - \theta_0\mathbf{1}_p\right) \stackrel{\mathcal{D}}{=}$ $[(\mathbf{X} + \boldsymbol{\delta}^*) - (\mathbf{Y} + \mathbf{J}\boldsymbol{\delta}^*)I(\|\mathbf{Y} + \mathbf{J}\boldsymbol{\delta}^*\|^2 < \chi^2_{p-1}(\alpha))]$ under local alternatives $\boldsymbol{\theta}_{(n)} = \theta_0\mathbf{1}_p + n^{-1/2}\boldsymbol{\delta}$ where $\begin{pmatrix} \mathbf{X} \\ \mathbf{Y} \end{pmatrix} \sim \mathcal{N}_{2p}\left\{\begin{pmatrix} \mathbf{0} \\ \mathbf{0} \end{pmatrix}; \begin{pmatrix} \boldsymbol{\Lambda}^{-1} & \mathbf{J}\boldsymbol{\Lambda}^{-1} \\ \boldsymbol{\Lambda}^{-1}\mathbf{J}' & \mathbf{J}\boldsymbol{\Lambda}^{-1} \end{pmatrix}\right\}$ (Section 5.8.2).
9. Verify that $\sqrt{n}s_e^{-1}\left(\hat{\boldsymbol{\theta}}_n^{S} - \theta_0\mathbf{1}_p\right) \stackrel{\mathcal{D}}{=} X - (p-3)(\mathbf{Y} + \mathbf{J}\boldsymbol{\delta}^*)\|\mathbf{Y} + \mathbf{J}\boldsymbol{\delta}^*\|^{-2}$ given at (5.8.24).
10. Prove Theorem 5.8.1, equation (5.8.22).
11. Determine $\left|\mathbf{M}_3\left(\hat{\boldsymbol{\theta}}_n^{PT}\right)\right|$, $\left|\mathbf{M}_4\left(\hat{\boldsymbol{\theta}}_n^{S}\right)\right|$ and $\left|\mathbf{M}_5\left(\hat{\boldsymbol{\theta}}_n^{S+}\right)\right|$ using the expressions given at (5.8.29), (5.8.32), and (5.8.36).

12. Let $\mathbf{Y}_i = \boldsymbol{\theta} + \mathbf{e}_i$, $i = 1, \ldots, n$, where $\mathbf{e}_i$ is distributed as $f(\mathbf{e}_i; \nu, \sigma^2) = C\left(1 + \frac{\sum_{i=1}^{n} \mathbf{e}_i'\mathbf{e}_i}{\nu\sigma^2}\right)^{(n+\nu)/2}$; $\nu, \sigma^2 > 0$. Define $\tilde{\boldsymbol{\theta}}_n$, $\hat{\boldsymbol{\theta}}_n$, $\hat{\boldsymbol{\theta}}_n^{PT}$, $\hat{\boldsymbol{\theta}}_n^{S}$, and $\hat{\boldsymbol{\theta}}_n^{S+}$ as given in Section 5.2.1. Determine the bias vector, MSE matrices and weighted risk expression for the five estimators.
13. Prove Theorem 1 of Section 5.9.2.
14. Prove (5.20.2a–d) of Section 5.10.1.
15. Prove the asymptotic linearity results given by (5.11.15a, b).
16. Show that

$$n^{1/2}\hat{\theta}_n = n^{1/2}\bar{\theta}_n + o_p(1),$$

where $\bar{\theta}_n = \lambda_1\tilde{\theta}_{n_1} + \cdots + \lambda_p\tilde{\theta}_{n_p}$ using the asymptotic linearity results.
17. Verify the expressions for ADR and ADMSE matrices and for ADQR given by (5.11.28a–e) using (5.11.31) as the computational step.
18. Prove that $\mathcal{L}_n$ at (5.11.12) approximately follows the central chi-square distribution with $(p-1)$ d.f. under $H_0 : \boldsymbol{\theta} = \theta_0\mathbf{1}_p$.
19. Show that $\mathcal{L}_n = n(\tilde{\boldsymbol{\theta}}_n - \hat{\boldsymbol{\theta}}_n)'\boldsymbol{\Lambda}(\tilde{\boldsymbol{\theta}}_n - \hat{\boldsymbol{\theta}}_n) + o_p(1)$.
20. Verify the expressions for the asymptotic coverage probabilities of the confidence sets given by (5.11.35) explicitly with that of the Section 5.9.

Chapter 6

Parallelism Model

Outline

In this chapter, we consider another important model, namely the **parallelism model**, belonging to the class of linear hypotheses useful in the analysis of bioassay data, shelf-life determination of pharmaceutical products, and profitability analysis of factory products in terms of costs and outputs, among other applications. In this model, as in the ANOVA model, p independent bivariate samples $\{(x_{\alpha_1}, y_{\alpha_1}), (x_{\alpha_2}, y_{\alpha_2}), \ldots, (x_{\alpha n_\alpha}, y_{\alpha n_\alpha}) | \alpha = 1, \ldots, p\}$ are considered such that $y_{\alpha j} \sim \mathcal{N}(\theta_\alpha + \beta_\alpha x_{\alpha j}, \sigma^2)$ for each pair (α, j) with fixed $x_{\alpha j}$.

The parameters $\boldsymbol{\theta} = (\theta_1, \ldots, \theta_p)'$ and $\boldsymbol{\beta} = (\beta_1, \ldots, \beta_p)'$ are the intercept and slope vectors of the p lines, respectively, and σ^2 is the common unknown variance. In this model, it is common to test the parallelism hypothesis H_0 : $\boldsymbol{\beta} = \beta_0 \mathbf{1}_p$ (where β_0 is an unknown scalar) against the alternative H_A: at least one pair of the components of $\boldsymbol{\beta}$ differ. Hence, we designate this model as the "parallelism model". The main objective of this chapter is to study the properties of various estimators of $\boldsymbol{\theta}$ and $\boldsymbol{\beta}$ when it is suspected but not certain that the parallelism hypothesis $H_0 : \boldsymbol{\beta} = \beta_0 \mathbf{1}_p$ holds. Accordingly, we consider (1) the unrestricted estimator (UE), (2) the restricted estimator (RE), (3) the preliminary test estimator (PTE), (4) the James-Stein-type estimator (JSE) and (5) the positive-rule Stein estimator (PRSE) of $\boldsymbol{\theta}$ and $\boldsymbol{\beta}$. In addition, we consider the properties of the recentered confidence sets based on these estimators as the center of the sets with fixed volume. Both small and large sample cases are treated for the problems stated above. In addition, we include the R-estimation theory of point and confidence set estimation. Interested readers may refer to Lambert and Saleh (1985), Akritus, Saleh, and Sen (1985), Saleh and Sen (1985a), Saleh and Hassanein (1986), and Robert and Saleh (1991) for more information.

6.1 Model, Estimation, and Test of Hypothesis

6.1.1 Parallelism Model

Consider the p simple linear models

$$\mathbf{Y}_\alpha = \theta_\alpha \mathbf{1}_{n_\alpha} + \beta_\alpha \mathbf{x}_\alpha + \varepsilon_\alpha, \quad \alpha = 1, \ldots, p, \tag{6.1.1}$$

where

$$\mathbf{Y}_\alpha = (y_{\alpha 1}, \ldots y_{\alpha n_\alpha})', \ \mathbf{1}_{n_\alpha} = (1, \ldots, 1)' - \text{a } n_\alpha\text{-tuple of 1's},$$
$$\mathbf{x}_\alpha = (x_{\alpha 1}, \ldots, x_{\alpha n_\alpha})', \tag{6.1.2}$$

and $\varepsilon_\alpha \sim \mathcal{N}_{n_\alpha}(\mathbf{0}, \sigma^2 \mathbf{I}_{n_\alpha})$, $\mathbf{I}_{n_\alpha}$ being the identity matrix of order n_α.

Generally, the main objective of this model is the test of the parallelism hypothesis $H_0 : \boldsymbol{\beta} = \beta_0 \mathbf{1}_p$ (β_0 is an unknown scalar) against the alternative $H_A : \boldsymbol{\beta} \neq \beta_0 \mathbf{1}_p$. In this chapter, we pursue the problem of improving the usual estimators of $\boldsymbol{\theta}$ and $\boldsymbol{\beta}$ when it is suspected but not evident that H_0 holds.

6.1.2 Estimation of the Intercept and Slope Parameters

Consider (6.1.1) and (6.1.2) using the sample information and the model. The **unrestricted** estimator of $\boldsymbol{\theta}$ and $\boldsymbol{\beta}$ are given by the LSE/MLE as follows:

$$\text{(a)} \ \tilde{\boldsymbol{\theta}}_n = \bar{\mathbf{Y}} - \mathbf{T}_n \tilde{\boldsymbol{\beta}}_n \quad \text{and} \quad \text{(b)} \ \tilde{\boldsymbol{\beta}}_n = (\tilde{\beta}_{1n_1}, \ldots, \tilde{\beta}_{pn_p})', \tag{6.1.3}$$

where (i) $\bar{\mathbf{Y}} = (\bar{Y}_{1n_1}, \ldots, \bar{Y}_{pn_p})'$, (ii) $\mathbf{T}_n = \text{Diag}(\bar{x}_{1n_1}, \ldots, \bar{x}_{n_p})$, (iii) $\tilde{\beta}_{\alpha n_\alpha} = \dfrac{\mathbf{x}'_\alpha \mathbf{y}_\alpha - \frac{1}{n_\alpha}(\mathbf{1}'_{n_\alpha}\mathbf{x}_\alpha)(\mathbf{1}'_{n_\alpha}\mathbf{y}_\alpha)}{n_\alpha Q_\alpha}$, (iv) $n_\alpha Q_\alpha = \mathbf{x}'_\alpha \mathbf{x}_\alpha - \frac{1}{n_\alpha}(\mathbf{1}'_{n_\alpha}\mathbf{x}_\alpha)^2$, (v) $\bar{x}_{\alpha n_\alpha} = \frac{1}{n_\alpha}(\mathbf{1}'_{n_\alpha}\mathbf{x}_\alpha)$ and $\bar{Y}_{\alpha n_\alpha} = \frac{1}{n_\alpha}(\mathbf{1}_{n_\alpha}\mathbf{Y}_\alpha)$. (6.1.4)

Finally, the unbiased estimator of σ^2 is given by

$$s_e^2 = (n - 2p)^{-1} \sum_{\alpha=1}^{p} \|\mathbf{y}_\alpha - \tilde{\theta}_\alpha \mathbf{1}_{n_\alpha} - \tilde{\beta}_{\alpha, n_\alpha} \mathbf{x}_\alpha\|^2. \tag{6.1.5}$$

In the case where the slopes are equal, meaning $\boldsymbol{\beta} = \beta_0 \mathbf{1}_p$, the **restricted** estimators of $\boldsymbol{\theta}$ and $\boldsymbol{\beta}$ are given by

$$\hat{\boldsymbol{\theta}}_n = \tilde{\boldsymbol{\theta}}_n + \mathbf{T}_n \mathbf{H} \tilde{\boldsymbol{\beta}}_n \quad \text{and} \quad \hat{\boldsymbol{\beta}}_n = \frac{\mathbf{1}_p \mathbf{1}_p' \mathbf{D}_{22}^{-1} \tilde{\boldsymbol{\beta}}_n}{nQ} = \hat{\beta}_{0n} \mathbf{1}_p$$

where (i) $\hat{\boldsymbol{\beta}}_n = (\hat{\beta}_{0n}, \dots, \hat{\beta}_{0n})' = \hat{\beta}_{0n} \mathbf{1}_p$, (ii) $nQ = \sum_{\alpha=1}^{p} n_\alpha Q_\alpha$, (iii) $\mathbf{H} = \mathbf{I}_p - \frac{\mathbf{1}_p \mathbf{1}_p' \mathbf{D}_{22}^{-1}}{nQ}$, (iv) $\mathbf{D}_{22}^{-1} = \text{Diag}(n_1 Q_1, \dots, n_p Q_p)$. (6.1.6)

The following theorem gives the exact distribution of the estimators, whose proof is straightforward.

Theorem 1. Under the assumed conditions

$$\begin{pmatrix} \tilde{\boldsymbol{\theta}}_n - \boldsymbol{\theta} \\ \tilde{\boldsymbol{\beta}}_n - \boldsymbol{\beta} \end{pmatrix} \sim \mathcal{N}_{2p} \left\{ \begin{pmatrix} \mathbf{0} \\ \mathbf{0} \end{pmatrix}; \sigma^2 \begin{pmatrix} \mathbf{D}_{11} & -\mathbf{T}_n \mathbf{D}_{22} \\ -\mathbf{T}_n \mathbf{D}_{22} & \mathbf{D}_{22} \end{pmatrix} \right\}, \tag{6.1.7a}$$

$$\begin{pmatrix} \hat{\boldsymbol{\theta}}_n - \boldsymbol{\theta} \\ \hat{\boldsymbol{\beta}}_n - \boldsymbol{\beta} \end{pmatrix} \sim \mathcal{N}_{2p} \left\{ \begin{pmatrix} \mathbf{T}_n \mathbf{H} \boldsymbol{\beta} \\ \mathbf{0} \end{pmatrix}; \sigma^2 \begin{pmatrix} \mathbf{D}_{11}^* & \mathbf{D}_{12}^* \\ \mathbf{D}_{21}^* & \mathbf{D}_{22} \end{pmatrix} \right\}, \tag{6.1.7b}$$

$$\begin{pmatrix} \tilde{\boldsymbol{\beta}}_n - \boldsymbol{\beta} \\ \tilde{\boldsymbol{\beta}}_n - \hat{\boldsymbol{\beta}}_n \end{pmatrix} \sim \mathcal{N}_{2p} \left\{ \begin{pmatrix} \mathbf{0} \\ \mathbf{H} \boldsymbol{\beta} \end{pmatrix}; \sigma^2 \begin{pmatrix} \mathbf{D}_{22} & \mathbf{H} \mathbf{D}_{22} \\ \mathbf{D}_{22} \mathbf{H}' & \mathbf{H} \mathbf{D}_{22} \end{pmatrix} \right\}, \tag{6.1.7c}$$

$$\begin{pmatrix} \hat{\boldsymbol{\beta}}_n - \beta_0 \mathbf{1}_p \\ \tilde{\boldsymbol{\beta}}_n - \hat{\boldsymbol{\beta}}_n \end{pmatrix} \sim \mathcal{N}_{2p} \left\{ \begin{pmatrix} (\bar{\beta} - \beta_0) \mathbf{1}_p \\ \mathbf{H} \boldsymbol{\beta} \end{pmatrix}; \sigma^2 \begin{pmatrix} \frac{\mathbf{1}_p \mathbf{1}_p'}{nQ} & \mathbf{0} \\ \mathbf{0} & \mathbf{H} \mathbf{D}_{22} \end{pmatrix} \right\}, \tag{6.1.7d}$$

where (i) $\bar{\beta} \mathbf{1}_p = \frac{\mathbf{1}_p \mathbf{1}_p' \mathbf{D}_{22}^{-1} \boldsymbol{\beta}}{nQ}$, (ii) $\mathbf{D}_{11} = \mathbf{N}^{-1} + \mathbf{T}_n \mathbf{D}_{22} \mathbf{T}_n$; and $\mathbf{N} = \text{Diag}(n_1, \dots, n_p)$, (iii) $\mathbf{D}_{11}^* = \mathbf{N}^{-1} + \frac{\mathbf{T}_n \mathbf{1}_p \mathbf{1}_p' \mathbf{T}_n}{nQ}$ and $\mathbf{D}_{12}^* = -\frac{1}{nQ} \mathbf{1}_p \mathbf{1}_p' \mathbf{T}_n$, (iv) $\mathbf{H}$ is defined by (6.1.5 (iii)) and $\mathbf{D}_{22}$ is defined by (6.1.5iv). (6.1.8)

We will need these distributions later to compute the bias, the MSE matrices, and the weighted quadratic risks of the estimators of the form

$$\boldsymbol{\beta}_n^* = \mathbf{A} \tilde{\boldsymbol{\beta}}_n + (\mathbf{I} - \mathbf{A}) \tilde{\boldsymbol{\beta}}_n g(\mathcal{L}_n), \mathbf{A} = \frac{\mathbf{1}_p \mathbf{1}_p' \mathbf{D}_{22}^{-1}}{nQ}, \tag{6.1.9}$$

where $\mathbf{A}$ is an idempotent matrix of rank 1 and $g(\mathcal{L}_n)$ is a nondecreasing function of the test-statistic $\mathcal{L}_n$ for testing the null hypotheses $H_0 : \boldsymbol{\beta} = \beta_0 \mathbf{1}_p$ against the alternative $H_A : \boldsymbol{\beta} \neq \beta_0 \mathbf{1}_p$ given in Section 6.1.3 below.

Table 6.1.1 Parallelism Data

	First Sample	Second Sample		pth Sample
Data	(x_{11}, y_{11})	(x_{21}, y_{21})	$\cdots$	(x_{p1}, y_{p1})
	$\vdots$	$\vdots$	$\vdots$	$\vdots$
	(x_{1n_1}, y_{1n_1})	(x_{2n_2}, y_{2n_2})	$\cdots$	(x_{pn_p}, y_{pn_p})
Estimate	$\tilde{\theta}_{1n_1}, \tilde{\beta}_{1n_1}$	$\tilde{\theta}_{2n_2}, \tilde{\beta}_{2n_2}$	$\cdots$	$\tilde{\theta}_{n_\alpha} \tilde{\beta}_{n_\alpha}$
Variance Estimate	$\frac{1}{n_1-2}\|\mathbf{Y}_1 - \tilde{\theta}_{1n_1}\mathbf{1}_p - \tilde{\beta}_{1n_1}\mathbf{x}_1\|^2$	$\frac{1}{n_2-2}\|\mathbf{Y}_2 - \tilde{\theta}_{2n_2}\mathbf{1}_p - \tilde{\beta}_{2n_2}\mathbf{x}_2\|^2$	$\cdots$	$\frac{1}{n_p-2}\|\mathbf{Y}_p - \tilde{\theta}_{pn_p}\mathbf{1}_p - \tilde{\beta}_{pn_p}\mathbf{x}_p\|^2$

In the sequel, we will need the following statistics:

(a) Pooled estimate of slope: $\hat{\beta}_{0n} = \frac{n_1 Q_1 \tilde{\beta}_{1n_1} + \cdots + n_p Q_p \tilde{\beta}_{pn_p}}{nQ}$.

(b) Pooled estimate of variance: $s_e^2 = (n-2p)^{-1} \sum_{\alpha=1}^{p} \|\mathbf{y}_\alpha - \tilde{\theta}_\alpha \mathbf{1}_{n_\alpha} - \tilde{\beta}_{\alpha,n_\alpha}\mathbf{x}_\alpha\|^2$.

6.1.3 Test of Parallelism

In this section, we consider the testing of the parallelism hypothesis $H_0 : \boldsymbol{\beta} = \beta_0 \mathbf{1}_p$. The test-statistic is given by the following theorem:

Theorem 2. The likelihood ratio(LR) statistic for the null hypothesis H_0 is given by

$$\mathcal{L}_n = \frac{\tilde{\boldsymbol{\beta}}_n' \mathbf{H}' \mathbf{D}_{22}^{-1} \mathbf{H} \tilde{\boldsymbol{\beta}}_n}{(p-1)s_e^2}. \tag{6.1.10}$$

Under H_0, $\mathcal{L}_n$ follows the central F-distribution with $(p-1, m)$ d.f., where $m = n - 2p$ and under H_A, it follows the noncentral F-distribution with $(p-1, m)$ d.f. and noncentrality parameter $\Delta^2/2$, where

$$\Delta^2 = \frac{\boldsymbol{\beta}' \mathbf{H}' \mathbf{D}_{22}^{-1} \mathbf{H} \boldsymbol{\beta}}{\sigma^2}. \tag{6.1.11}$$

Proof. The likelihood ratio statistic is obtained by using the maximum of the likelihood function under H_0 and H_A, which are given by

$$L_0 = s_0^{-n} (\sqrt{2\pi})^{-n} \exp\left\{-\frac{n}{2}\right\}, \quad n s_0^2 = \sum_{\alpha=1}^{p} \|\mathbf{y}_\alpha - \hat{\theta}_\alpha \mathbf{1}_{n_\alpha} - \hat{\beta}_\alpha \mathbf{x}_\alpha\|^2 \tag{6.1.12}$$

and

$$L_A = s_A^{-n}(\sqrt{2\pi})^{-n}\exp\Big\{-\frac{n}{2}\Big\},\quad ns_A^2 = \sum_{\alpha=1}^{p}\|\mathbf{y}_\alpha - \tilde{\theta}_\alpha \mathbf{1}_{n_\alpha} - \tilde{\beta}_\alpha \mathbf{x}_\alpha\|^2 = ms_e^2, \tag{6.1.13}$$

respectively. Hence, the likelihood ratio statistic is the result of

$$\lambda^{2/n} = \left(\frac{L_0}{L_A}\right)^{2/n} = \frac{\sum_{\alpha=1}^{p}\|\mathbf{y}_\alpha - \tilde{\theta}_\alpha \mathbf{1}_{n_\alpha} - \tilde{\beta}_\alpha \mathbf{x}_\alpha\|^2}{\sum_{\alpha=1}^{p}\|\mathbf{y}_\alpha - \hat{\theta}_\alpha \mathbf{1}_{n_\alpha} - \hat{\beta}_\alpha \mathbf{x}_\alpha\|^2} = \frac{1}{1+\frac{p-1}{m}\mathcal{L}_n} \tag{6.1.14}$$

since

$$\sum_{\alpha=1}^{p}\|\mathbf{y}_\alpha - \hat{\theta}_\alpha \mathbf{1}_{n_\alpha} - \hat{\beta}_\alpha \mathbf{x}_\alpha\|^2 = \sum_{\alpha=1}^{p}\|\mathbf{y}_\alpha - \tilde{\theta}_\alpha \mathbf{1}_{n_\alpha} - \tilde{\beta}_\alpha \mathbf{x}_\alpha\|^2 + \tilde{\boldsymbol{\beta}}_n' \mathbf{H}'\mathbf{D}_{22}^{-1}\mathbf{H}\tilde{\boldsymbol{\beta}}_n. \tag{6.1.15}$$

Thus, we take $\mathcal{L}_n = \frac{\tilde{\boldsymbol{\beta}}_n' \mathbf{H}'\mathbf{D}_{22}^{-1}\mathbf{H}\tilde{\boldsymbol{\beta}}_n}{(p-1)s_e^2}$ as the LR statistic for testing H_0 against H_A.

To obtain the distribution of $\mathcal{L}_n$ under H_A (or H_0), we consider the orthogonal matrix $\boldsymbol{\Gamma} = (\boldsymbol{\Gamma}_1, \boldsymbol{\Gamma}_2)$. Here $\boldsymbol{\Gamma}_1$ is a $p \times p-1$ matrix, and $\boldsymbol{\Gamma}_2$ is a p-vector such that $\boldsymbol{\Gamma}_2'\boldsymbol{\Gamma}_1 = \mathbf{0}$ and $\boldsymbol{\Gamma}_1\boldsymbol{\Gamma}_1' + \boldsymbol{\Gamma}_2\boldsymbol{\Gamma}_2' = \mathbf{I}_p$ so that $\boldsymbol{\Gamma}_1\boldsymbol{\Gamma}_1' = \mathbf{I}_p - \boldsymbol{\Gamma}_2\boldsymbol{\Gamma}_2'$. Also, $\boldsymbol{\Gamma}$ diagonalizes the symmetric idempotent matrix $\mathbf{D}_{22}^{1/2}\mathbf{H}'\mathbf{D}_{22}^{-1}\mathbf{H}\mathbf{D}_{22}^{1/2}$ of rank $(p-1)$, that is,

$$\boldsymbol{\Gamma}\mathbf{D}_{22}^{1/2}\mathbf{H}'\mathbf{D}_{22}^{1/2}\boldsymbol{\Gamma}' = \begin{pmatrix} \mathbf{I}_{p-1} & \mathbf{0} \\ \mathbf{0} & 0 \end{pmatrix}.$$

Defining $\boldsymbol{\omega} = \sigma^{-1}\boldsymbol{\Gamma}\mathbf{D}_{22}^{1/2}\tilde{\boldsymbol{\beta}}_n$ and $\boldsymbol{\eta} = \sigma^{-1}\boldsymbol{\Gamma}\mathbf{D}_{22}^{1/2}\boldsymbol{\beta}$, we have $\boldsymbol{\omega} \approx N_p(\boldsymbol{\eta}, \mathbf{I}_p)$. Now partition $\boldsymbol{\omega} = (\boldsymbol{\omega}_1', \omega_2)'$ and $\boldsymbol{\eta} = (\boldsymbol{\eta}_1', \eta_2)'$, where $\boldsymbol{\omega}_1$ is a $(p-1)$-vector and ω_2 is a scalar random variable. Then $\begin{pmatrix} \boldsymbol{\omega}_1 \\ \omega_2 \end{pmatrix} \sim \mathcal{N}_p\left\{\begin{pmatrix} \boldsymbol{\eta}_1 \\ \eta_2 \end{pmatrix}, \begin{pmatrix} \mathbf{I}_{p-1} & \mathbf{0} \\ \mathbf{0} & 1 \end{pmatrix}\right\}$. Hence, $\|\boldsymbol{\omega}_1\|^2$ follows the noncentral chi-square distribution with $(p-1)$ d.f. and noncentrality parameter $\Delta^2/2$ with $\Delta^2 = \|\boldsymbol{\eta}_1\|^2 = \frac{\boldsymbol{\beta}'\mathbf{H}'\mathbf{D}_{22}^{-1}\mathbf{H}\boldsymbol{\beta}}{\sigma^2}$. Then, $\mathcal{L}_n$ under H_A follows the noncentral F-distribution with $(p-1, m)$ d.f. and noncentrality parameter $\Delta^2/2$ since ms_e^2/σ^2 follows the central chi-square distribution with m d.f. independent of $\tilde{\boldsymbol{\beta}}_n$. Clearly, under H_0, $\mathcal{L}_n$ follows the central F-distribution with $(p-1, m)$ d.f.

6.2 Preliminary Test and Stein-Type Estimators

Following Chapter 5, we now consider various estimators of the intercept $\boldsymbol{\theta}$ and slope vectors $\boldsymbol{\beta}$.

6.2.1 The Estimators of Intercepts and Slopes

The **unrestricted** and **restricted** estimators of $\boldsymbol{\theta}$ and $\boldsymbol{\beta}$ are given in Section 6.1.2. We consider the preliminary test estimators of $\boldsymbol{\theta}$ and $\boldsymbol{\beta}$ as follows:

$$\hat{\boldsymbol{\beta}}_n^{PT} = \tilde{\boldsymbol{\beta}}_n - \mathbf{H}\tilde{\boldsymbol{\beta}}_n I(\mathcal{L}_n < F_{p-1,m}(\alpha)) \tag{6.2.1a}$$

and

$$\hat{\boldsymbol{\theta}}_n^{PT} = \tilde{\boldsymbol{\theta}}_n + \mathbf{T}_n\mathbf{H}\tilde{\boldsymbol{\beta}}_n I(\mathcal{L}_n < F_{p-1,m}(\alpha)) \tag{6.2.1b}$$

respectively, where $F_{p-1,m}(\alpha)$ is the upper α-quantile of the central F-distribution with $(p-1,m)$ d.f.

To overcome the **dependency** on α, we consider the James-Stein-type estimators of $\boldsymbol{\theta}$ and $\boldsymbol{\beta}$, namely

$$\hat{\boldsymbol{\beta}}_n^{S} = \tilde{\boldsymbol{\beta}}_n - c\mathbf{H}\tilde{\boldsymbol{\beta}}_n\mathcal{L}_n^{-1} \tag{6.2.2a}$$

and

$$\hat{\boldsymbol{\theta}}_n^{S} = \tilde{\boldsymbol{\theta}}_n + c\mathbf{T}_n\mathbf{H}\tilde{\boldsymbol{\beta}}_n\mathcal{L}_n^{-1}, \tag{6.2.2b}$$

where $c = \frac{(p-3)m}{(p-1)(m+2)}$. Notice that we have replaced $I(\mathcal{L}_n < F_{p-1,m}(\alpha))$ by $c\mathcal{L}_n^{-1}$ in (6.2.1a,b) to obtain (6.2.2a,b).

The estimators $\hat{\boldsymbol{\theta}}_n^{S}$ and $\hat{\boldsymbol{\beta}}_n^{S}$ may exceed the values $\hat{\boldsymbol{\theta}}_n$ and $\hat{\boldsymbol{\beta}}_n$, respectively. Thus, we consider the convex combinations of $\hat{\boldsymbol{\theta}}_n$ and $\hat{\boldsymbol{\theta}}_n^{S}$, and $\hat{\boldsymbol{\beta}}_n$ and $\hat{\boldsymbol{\beta}}_n^{S}$, via the preliminary test procedure with critical value c to obtain the **positive-rule Stein-type estimators** of $\boldsymbol{\theta}$ and $\boldsymbol{\beta}$, respectively, as

$$\begin{aligned}
\hat{\boldsymbol{\theta}}_n^{S+} &= \hat{\boldsymbol{\theta}}_n I(\mathcal{L}_n < c) + \hat{\boldsymbol{\theta}}_n^{S} I(\mathcal{L}_n > c) \\
&= \hat{\boldsymbol{\theta}}_n - \{\hat{\boldsymbol{\theta}}_n - \hat{\boldsymbol{\theta}}_n^{S}\} I(\mathcal{L}_n > c) \\
&= \hat{\boldsymbol{\theta}}_n - (1 - c\mathcal{L}_n^{-1}) I(\mathcal{L}_n > c)\mathbf{T}_n\mathbf{H}\tilde{\boldsymbol{\beta}}_n \\
&= \tilde{\boldsymbol{\theta}}_n + \mathbf{T}_n\mathbf{H}\tilde{\boldsymbol{\beta}}_n\{1 - (1 - c\mathcal{L}_n^{-1}) I(\mathcal{L}_n > c)\}
\end{aligned} \tag{6.2.3a}$$

and

$$\begin{aligned}
\hat{\boldsymbol{\beta}}_n^{S+} &= \hat{\boldsymbol{\beta}}_n I(\mathcal{L}_n < c) + \hat{\boldsymbol{\beta}}_n^{S} I(\mathcal{L}_n > c) \\
&= \hat{\boldsymbol{\beta}}_n + (1 - c\mathcal{L}_n^{-1}) I(\mathcal{L}_n > c)\mathbf{H}\tilde{\boldsymbol{\beta}}_n
\end{aligned} \tag{6.2.3b}$$

with $p \geq 4$. We present the following data set related to shelf-life, to illustrate the summary statistics and the test-statistics. The tabular values contain results of six experiments conducted to determine the shelf-life of a drug by setting fixed concentrations of a component of the drug and observing the times(years) to destroy the concentration.

Table 6.2.1 Shelf-life Data

Batch	Time (years)	Concentration (%)	Batch	Time (years)	Concentration (%)
1	0.014	100.4	4	0.066	100.4
	0.230	100.3		0.343	100.0
	0.514	99.7		0.533	99.5
	0.769	99.2		0.802	99.3
	1.074	98.9		1.033	99.3
	1.533	98.2		1.538	98.2
	2.030	97.3			
	3.071	95.7	5	0.011	100.5
	4.649	94.5		0.310	99.8
				0.624	99.1
2	0.022	100.7		1.063	98.4
	0.118	100.6			
	0.272	100.3	6	0.011	100.1
	0.566	99.4		0.310	99.5
	1.165	98.6		0.624	98.5
	2.022	97.6		1.063	98.4
	3.077	96.4			
3	0.025	100.2			
	0.275	99.7			
	0.547	99.2			
	0.797	99.0			
	1.041	98.8			
	2.058	96.4			
	2.519	96.2			

Adapted from Roberg and Stegman (1991)

Table 6.2.2 contains the summary statistics related to Table 6.2.1.

Table 6.2.2 Individual Batch Regression Estimates

Batch	Intercept (%)	Slope (%/year)	MSE	Weight $n_i Q_i$
1	100.49	-1.515	0.019	14.63
2	100.66	-1.449	0.043	7.83
3	100.25	-1.682	0.062	5.14
4	100.45	-1.393	0.035	1.21
5	100.45	-1.999	0.011	0.61
6	99.98	-1.701	0.124	0.61
Pooled batch regression with equivalent slopes				
	99.90	-1.534	0.047	

The value of $\mathcal{L}_n = 1.64$ with p-value 0.186.

6.2.2 Bayes and Empirical Bayes Estimators of Intercepts and Slopes

Let $\boldsymbol{\delta} = (\boldsymbol{\theta}', \boldsymbol{\beta}')'$ be the intercept and slope parameter, and let $\tilde{\boldsymbol{\delta}}_n = (\tilde{\boldsymbol{\theta}}_n', \tilde{\boldsymbol{\beta}}_n')'$ be the **unrestricted** estimator and $\hat{\boldsymbol{\delta}}_n = (\hat{\boldsymbol{\theta}}_n', \hat{\boldsymbol{\beta}}_n')'$ the **restricted** estimator of $(\boldsymbol{\theta}', \boldsymbol{\beta}')'$. We then write

$$\hat{\boldsymbol{\delta}}_n = \begin{pmatrix} \mathbf{I}_p & \mathbf{T}_n\mathbf{H} \\ \mathbf{0} & \mathbf{I}_p - \mathbf{H} \end{pmatrix} \begin{pmatrix} \tilde{\boldsymbol{\theta}}_n \\ \tilde{\boldsymbol{\beta}}_n \end{pmatrix}. \tag{6.2.4}$$

By sufficiency considerations, we can reduce the model to the observation of the statistic, $\tilde{\boldsymbol{\delta}}_n \sim \mathcal{N}_{2p}\left\{\boldsymbol{\delta}, \sigma^2 \begin{pmatrix} \mathbf{D}_{11} & \mathbf{D}_{12} \\ \mathbf{D}_{21} & \mathbf{D}_{22} \end{pmatrix}\right\}$ and $m^* s_e^2 \sim \sigma^2 \chi^2_{n-2p}$ where $m^* = n - 2p + 2$ and

$$s_e^2 = (n - 2p + 2)^{-1} \sum_{\alpha=1}^{p} \|\mathbf{y}_\alpha - \tilde{\theta}_\alpha \mathbf{1}_{n_\alpha} - \tilde{\beta}_\alpha \mathbf{x}_\alpha\|^2. \tag{6.2.5}$$

The estimator s_e^2 of σ^2 is the best scale invariant under the loss

$$\mathcal{L}(\sigma^2, d) = \left(1 - \frac{d}{\sigma^2}\right)^2 \tag{6.2.6}$$

and dominates the mle of σ^2. Since our null hypothesis H_0 is $\boldsymbol{\beta} = \beta_0 \mathbf{1}_p$, we select the prior

$$\boldsymbol{\delta} \sim \mathcal{N}_{2p}\left\{\begin{pmatrix} \boldsymbol{\delta}_1^* \\ \boldsymbol{\delta}_2^* \end{pmatrix}; \tau^2 \begin{pmatrix} \mathbf{D}_{11} & -\mathbf{T}_n\mathbf{D}_{22} \\ -\mathbf{T}_n\mathbf{D}_{22} & \mathbf{D}_{22} \end{pmatrix}\right\}, \tag{6.2.7}$$

which gives the posterior distribution of $\boldsymbol{\delta}$ conditionally on σ^2 as

$$\boldsymbol{\delta}|\tilde{\boldsymbol{\delta}}_n \sim \mathcal{N}_{2p}\left\{\left(\frac{\sigma^2\boldsymbol{\delta}^* + \tau^2\tilde{\boldsymbol{\delta}}_n}{\sigma^2 + \tau^2}\right); \frac{\sigma^2\tau^2}{\sigma^2 + \tau^2} \begin{pmatrix} \mathbf{D}_{11} & \mathbf{D}_{12} \\ \mathbf{D}_{21} & \mathbf{D}_{22} \end{pmatrix}\right\}, \tag{6.2.8}$$

where $\boldsymbol{\delta}^* = \begin{pmatrix} \boldsymbol{\delta}_1^* \\ \boldsymbol{\delta}_2^* \end{pmatrix}$. Conditionally on σ^2, the associated posterior mean of $\boldsymbol{\delta}|\tilde{\boldsymbol{\delta}}_n$ may be written as

$$E(\boldsymbol{\delta}|\tilde{\boldsymbol{\delta}}_n) = \frac{\sigma^2\boldsymbol{\delta}^* + \tau^2\tilde{\boldsymbol{\delta}}_n}{\sigma^2 + \tau^2} = \boldsymbol{\delta}^* + \left(1 - \frac{\sigma^2}{\sigma^2 + \tau^2}\right)\left(\tilde{\boldsymbol{\delta}}_n - \boldsymbol{\delta}^*\right). \tag{6.2.9}$$

Now, using the noninformative prior on σ^2, i.e., $\pi(\sigma^2) = 1/\sigma^2$, we find that the Bayes estimate of σ^2 is s_e^2. Similarly, the empirical Bayes estimator of $\boldsymbol{\delta}^*$ and $(\sigma^2 + \tau^2)^{-1}$ based on the marginal distribution of $\tilde{\boldsymbol{\delta}}_n$, namely

$$\tilde{\boldsymbol{\delta}}_n \sim \mathcal{N}_{2p}\left\{\boldsymbol{\delta}^*, (\sigma^2 + \tau^2) \begin{pmatrix} \mathbf{D}_{11} & -\mathbf{T}_n\mathbf{D}_{22} \\ -\mathbf{D}_{22}\mathbf{T}_n & \mathbf{D}_{22} \end{pmatrix}\right\}, \tag{6.2.10}$$

is given by $\hat{\boldsymbol{\beta}}_n$ and $\hat{\boldsymbol{\theta}}_n$, since

$$E^m(\hat{\boldsymbol{\beta}}_n) = \beta_0 \mathbf{1}_p \qquad \text{and} \qquad E(\hat{\boldsymbol{\theta}}_n) = \boldsymbol{\theta}, \tag{6.2.11}$$

where $E^m[\cdot]$ stands for the marginal expectation of the r.v. in $[\cdot]$. Thus, $\hat{\boldsymbol{\beta}}_n$ and $\hat{\boldsymbol{\theta}}_n$ are empirical Bayes estimators of $\beta_0 \mathbf{1}_p$ and $\boldsymbol{\theta}$. Further,

$$\text{Cov}[(\tilde{\boldsymbol{\beta}}_n - \hat{\boldsymbol{\beta}}_n)] = (\sigma^2 + \tau^2)\mathbf{H}\mathbf{D}_{22}\mathbf{H}' \tag{6.2.12}$$

and

$$(\tilde{\boldsymbol{\beta}}_n - \hat{\boldsymbol{\beta}}_n)'\mathbf{D}_{22}^{-1}(\tilde{\boldsymbol{\beta}}_n - \hat{\boldsymbol{\beta}}_n) \overset{\mathcal{D}}{=} (\sigma^2 + \tau^2)\chi^2_{p-1}. \tag{6.2.13}$$

Therefore,

$$E^m\left\{\frac{1}{(\tilde{\boldsymbol{\beta}}_n - \hat{\boldsymbol{\beta}}_n)'\mathbf{D}_{22}^{-1}(\tilde{\boldsymbol{\beta}}_n - \hat{\boldsymbol{\beta}}_n)}\right\} = \frac{1}{(p-3)(\sigma^2 + \tau^2)}, \tag{6.2.14}$$

and the estimator of $1/(\sigma^2 + \tau^2)$ is given by

$$\frac{p-3}{(\tilde{\boldsymbol{\beta}}_n - \hat{\boldsymbol{\beta}}_n)'\mathbf{D}_{22}^{-1}(\tilde{\boldsymbol{\beta}}_n - \hat{\boldsymbol{\beta}}_n)}. \tag{6.2.15}$$

Hence, the empirical Bayes estimator of $\boldsymbol{\delta}$ may be written as

$$\begin{aligned} \hat{\boldsymbol{\delta}}_n^E &= \hat{\boldsymbol{\delta}}_n + \left\{1 - \frac{(p-3)s_e^2}{(\tilde{\boldsymbol{\beta}}_n - \hat{\boldsymbol{\beta}}_n)'\mathbf{D}_{22}^{-1}(\tilde{\boldsymbol{\beta}}_n - \hat{\boldsymbol{\beta}}_n)}\right\}(\tilde{\boldsymbol{\delta}}_n - \hat{\boldsymbol{\delta}}_n) \\ &= \hat{\boldsymbol{\delta}}_n + \left(1 - c\mathcal{L}_n^{-1}\right)(\tilde{\boldsymbol{\delta}}_n - \hat{\boldsymbol{\delta}}_n), \end{aligned} \tag{6.2.16}$$

where

$$c = \frac{(p-3)m}{(p-1)(m+2)} \quad \text{and} \quad \mathcal{L}_n = \frac{(\tilde{\boldsymbol{\beta}}_n - \hat{\boldsymbol{\beta}}_n)'\mathbf{D}_{22}^{-1}(\tilde{\boldsymbol{\beta}}_n - \hat{\boldsymbol{\beta}}_n)}{(p-1)s_e^2}, \tag{6.2.17}$$

with $m = n - 2p$ being the LR test for $H_0 : \boldsymbol{\beta} = \beta_0 \mathbf{1}_p$. The corresponding positive-rule empirical Bayes estimator of $\boldsymbol{\delta}$ is given by

$$\hat{\boldsymbol{\delta}}_n^{S+} = \hat{\boldsymbol{\delta}}_n + \left(1 - c\mathcal{L}_n^{-1}\right) I\left(\mathcal{L}_n > c\right)(\tilde{\boldsymbol{\delta}}_n - \hat{\boldsymbol{\delta}}_n), \tag{6.2.18}$$

which coincides with the PTE approach for the shrinkage estimators (see Section 6.1.4).

Example 1. Consider the data of Tables 6.1.1 through 6.1.3. We illustrate the computation of estimators of the intercept vector $\boldsymbol{\theta}$. The computation of the estimators of $\boldsymbol{\beta}$ may be obtained to verify these computations. Table 6.1.2 shows that $\mathcal{L}_n = 1.64$ with p-value 0.186. At the 0.25 level of significance, the hypothesis of equality of the slopes is **rejected**. The formulas for $\hat{\boldsymbol{\theta}}_n^{\text{PT}}$, $\hat{\boldsymbol{\theta}}_n^{S}$, and $\hat{\boldsymbol{\theta}}_n^{S+}$ are given by (6.2.1a) through

(6.2.3b). Accordingly, we have (i) The unrestricted estimator of $\boldsymbol{\theta}$ is $\tilde{\boldsymbol{\theta}}_{37} = (100.49, 100.60, 100.25, 100.45, 100.45, 99.98)'$. (ii) The restricted estimator of $\boldsymbol{\theta}$ is $\hat{\boldsymbol{\theta}}_{37} = (99.50, 99.90, 99.90, 99.90, 99.90, 99.90)'$. (iii) The PTE of $\boldsymbol{\theta}$ is $\hat{\boldsymbol{\theta}}_{37}^{PT} = \tilde{\boldsymbol{\theta}}_{37} = (100.49, 100.60, 100.25, 100.45, 100.45, 99.98)'$. (iv) The JSE of $\boldsymbol{\theta}$ is $\hat{\boldsymbol{\theta}}_{37}^{S} = (100.50, 100.62, 100.23, 100.47, 100.37, 99.48)'$, here $c = 0.556$ and $\mathcal{L}_n = 1.64$, and (v) The PRSE of $\boldsymbol{\theta}$ is then $\hat{\boldsymbol{\theta}}_{37}^{S+} = \hat{\boldsymbol{\theta}}_{37}^{S} = (100.50, 100.62, 100.23, 100.47, 100.37, 99.48)'$.

6.3 Bias, Quadratic Bias, MSE Matrices, and Risk Expressions

In this section, we present the expressions for **bias vector, quadratic bias,** MSE matrices, and **weighted** quadratic risks of various estimators. They are given in the sequel based on Theorem 6.1.1.

6.3.1 Unrestricted Estimators of $\boldsymbol{\beta}$ and $\boldsymbol{\theta}$

(i) $\mathbf{b}_1(\tilde{\boldsymbol{\beta}}_n) = \mathbf{0}$, and $B_1(\tilde{\boldsymbol{\beta}}_n) = 0$;

(ii) $\mathbf{M}_1(\tilde{\boldsymbol{\beta}}_n) = \sigma^2\mathbf{D}_{22}$, and $R_1(\tilde{\boldsymbol{\beta}}_n; \mathbf{W}) = \sigma^2 \operatorname{tr}(\mathbf{W}\mathbf{D}_{22})$. (6.3.1a)

Similarly,

(i) $\mathbf{b}_1(\tilde{\boldsymbol{\theta}}_n) = \mathbf{0}$, and $B_1(\tilde{\boldsymbol{\theta}}_n) = 0$

(ii) $\mathbf{M}_1(\tilde{\boldsymbol{\theta}}_n) = \sigma^2\mathbf{D}_{11}$, and $R_1(\tilde{\boldsymbol{\theta}}_n; \mathbf{W}) = \sigma^2 \operatorname{tr}(\mathbf{W}\mathbf{D}_{11})$. (6.3.1b)

6.3.2 Restricted Estimators of $\boldsymbol{\beta}$ and $\boldsymbol{\theta}$

(i) $\mathbf{b}_2(\hat{\boldsymbol{\beta}}_n) = -\mathbf{H}\boldsymbol{\beta}$, and $B_2(\hat{\boldsymbol{\beta}}_n) = \Delta^2 = \sigma^{-2}\boldsymbol{\beta}'\mathbf{H}'\mathbf{D}_{22}^{-1}\mathbf{H}\boldsymbol{\beta}$

(ii) $\mathbf{M}_2(\hat{\boldsymbol{\beta}}_n) = \frac{\sigma^2\mathbf{1}_p\mathbf{1}_p'}{nQ} + \mathbf{H}\boldsymbol{\beta}\boldsymbol{\beta}'\mathbf{H}'$ and $R_2(\hat{\boldsymbol{\beta}}_n; \mathbf{W}) = \frac{\sigma^2\mathbf{1}_p'\mathbf{W}\mathbf{1}_p}{nQ} + \boldsymbol{\beta}'\mathbf{H}'\mathbf{W}\mathbf{H}\boldsymbol{\beta}$. (6.3.2a)

Similarly,

(i) $\mathbf{b}_2(\hat{\boldsymbol{\theta}}_n) = \mathbf{T}_n\mathbf{H}\boldsymbol{\beta}$, and $B_2(\hat{\boldsymbol{\theta}}_n) = \sigma^{-2}\boldsymbol{\beta}'\mathbf{H}'\mathbf{T}_n'\mathbf{D}_{22}^{-1}\mathbf{T}_n\mathbf{H}\boldsymbol{\beta}$;

(ii) $\mathbf{M}_2(\hat{\boldsymbol{\theta}}_n) = \sigma^2\mathbf{D}_{11}^* + \mathbf{T}_n\mathbf{H}\boldsymbol{\beta}\boldsymbol{\beta}'\mathbf{H}'\mathbf{T}_n$ and $R_2(\hat{\boldsymbol{\theta}}_n; \mathbf{W}) = \sigma^2 \operatorname{tr}(\mathbf{W}\mathbf{D}_{11}^*) + \boldsymbol{\beta}'\mathbf{H}'\mathbf{T}_n'\mathbf{W}\mathbf{T}_n\mathbf{H}\boldsymbol{\beta}$, where $\mathbf{D}_{11}^* = \mathbf{N}^{-1} + \frac{\mathbf{T}_n'\mathbf{1}_p\mathbf{1}_p'\mathbf{T}_n}{nQ}$. (6.3.2b)

6.3.3 Preliminary Test Estimators of $\boldsymbol{\beta}$ and $\boldsymbol{\theta}$

(i) $\mathbf{b}_3(\hat{\boldsymbol{\beta}}_n^{\mathrm{PT}}) = -\mathbf{H}\boldsymbol{\beta}G_{p+1,m}(\ell_\alpha;\Delta^2)$ and $B_3(\hat{\boldsymbol{\beta}}_n^{\mathrm{PT}}) = \Delta^2\{G_{p+1,m}(\ell_\alpha;\Delta^2)\}^2$, where $\ell_\alpha = \frac{p-1}{p+1}F_{p-1,m}(\alpha)$

(ii) $\mathbf{M}_3(\hat{\boldsymbol{\beta}}_n^{\mathrm{PT}}) = \sigma^2\mathbf{D}_{22} - \sigma^2(\mathbf{H}\mathbf{D}_{22})G_{p+1,m}(\ell_\alpha;\Delta^2) + (\mathbf{H}\boldsymbol{\beta}\boldsymbol{\beta}'\mathbf{H}') \times \{2G_{p+1,m}(\ell_\alpha;\Delta^2) - G_{p+3,m}(\ell_\alpha^*;\Delta^2)\}$

and

$$
\begin{aligned}
R_3(\hat{\boldsymbol{\beta}}_n^{\mathrm{PT}};\mathbf{W}) &= \sigma^2\ \mathrm{tr}(\mathbf{W}\mathbf{D}_{22}) - \sigma^2\ tr(\mathbf{W}\mathbf{H}\mathbf{D}_{22})G_{p+1,m}(\ell_\alpha;\Delta^2)\\
&\quad +\boldsymbol{\beta}'\mathbf{H}'\mathbf{W}\mathbf{H}\boldsymbol{\beta}\{2G_{p+1,m}(\ell_\alpha;\Delta^2) - G_{p+3,m}(\ell_\alpha^*;\Delta^2)\},
\end{aligned}
$$

where $G_{\nu_1,\nu_2}(\cdot;\Delta^2)$ is the cdf of a noncentral F-distribution with (ν_1,ν_2) d.f. and noncentrality parameter $\frac{1}{2}\Delta^2$ and $\ell_\alpha^* = \frac{p-1}{p+3}F_{p-1,m}(\alpha)$. (6.3.3a)

Similarly,

(i) $\mathbf{b}_3(\hat{\boldsymbol{\theta}}_n^{\mathrm{PT}}) = \mathbf{T}_n\mathbf{H}\boldsymbol{\beta}G_{p+1,m}(\ell_\alpha;\Delta^2)$ and
$B_3(\hat{\boldsymbol{\theta}}_n^{\mathrm{PT}}) = \sigma^{-2}(\boldsymbol{\beta}'\mathbf{H}'\mathbf{T}_n\mathbf{D}_{22}^{-1}\mathbf{T}_n\mathbf{H}\boldsymbol{\beta})\{G_{p+1,m}(\ell_\alpha;\Delta^2)\}^2$;

(ii) $\mathbf{M}_3(\hat{\boldsymbol{\theta}}_n^{\mathrm{PT}}) = \sigma^2\mathbf{D}_{11} - \sigma^2(\mathbf{D}_{11} - \mathbf{D}_{11}^*)G_{p+1,m}(\ell_\alpha;\Delta^2) + (\mathbf{T}_n\mathbf{H}\boldsymbol{\beta}\boldsymbol{\beta}'\mathbf{H}'\mathbf{T}_n)\{2G_{p+1,m}(\ell_\alpha;\Delta^2) - G_{p+3,m}(\ell_\alpha^*;\Delta^2)\}$

and

$$
\begin{aligned}
R_3(\hat{\boldsymbol{\theta}}_n^{\mathrm{PT}};\mathbf{W}) &= \sigma^2\ \mathrm{tr}(\mathbf{W}\mathbf{D}_{11}) - \sigma^2\ tr[\mathbf{W}(\mathbf{D}_{11} - \mathbf{D}_{11}^*))]G_{p+1,m}(\ell_\alpha;\Delta^2)\\
&+ \boldsymbol{\beta}'\mathbf{H}'\mathbf{T}_n'\mathbf{W}\mathbf{T}_n\mathbf{H}\boldsymbol{\beta}\{2G_{p+1,m}(\ell_\alpha;\Delta^2) - G_{p+3,m}(\ell_\alpha^*;\Delta^2)\}. \qquad (6.3.3b)
\end{aligned}
$$

6.3.4 James-Stein-type Estimators of $\boldsymbol{\beta}$ and $\boldsymbol{\theta}$

(i) $\mathbf{b}_4(\hat{\boldsymbol{\beta}}_n^S) = -c(p-1)\mathbf{H}\boldsymbol{\beta}E[\chi_{p+1}^{-2}(\Delta^2)]$ and
$B_4(\hat{\boldsymbol{\beta}}_n^S) = c^2(p-1)^2\Delta^2\{E[\chi_{p+1}^{-2}(\Delta^2)]\}^2$;

(ii) $\mathbf{M}_4(\hat{\boldsymbol{\beta}}_n^S) = \sigma^2\mathbf{D}_{22} - c(p-1)\sigma^2(\mathbf{H}\mathbf{D}_{22})\{2E[\chi_{p+1}^{-2}(\Delta^2)] -(p-3)E[\chi_{p+1}^{-4}(\Delta^2)]\} + c(p^2-1)(\mathbf{H}\boldsymbol{\beta}\boldsymbol{\beta}'\mathbf{H}')E[\chi_{p+3}^{-4}(\Delta^2)]$

and

$$
\begin{aligned}
R_4(\hat{\boldsymbol{\beta}}_n^S;\mathbf{W}) &= \sigma^2\mathrm{tr}(\mathbf{W}\mathbf{D}_{22}) - c(p-1)\sigma^2\mathrm{tr}(\mathbf{W}\mathbf{H}\mathbf{D}_{22})\{2E[\chi_{p+1}^{-2}(\Delta^2)] -\\
&(p-3)E[\chi_{p+1}^{-4}(\Delta^2)]\} + c(p^2-1)(\boldsymbol{\beta}'\mathbf{H}'\mathbf{W}\mathbf{H}\boldsymbol{\beta})E[\chi_{p+3}^{-4}(\Delta^2)]. \qquad (6.3.4a)
\end{aligned}
$$

Similarly,

(i) $\mathbf{b}_4(\hat{\boldsymbol{\theta}}_n^S) = c(p-1)\mathbf{T}_n\mathbf{H}\boldsymbol{\beta}E[\chi_{p+1}^{-2}(\Delta^2)]$ and
$B_4(\hat{\boldsymbol{\theta}}_n^S) = c^2(p-1)^2\sigma^{-2}(\boldsymbol{\beta}'\mathbf{H}'\mathbf{T}_n\mathbf{D}_{22}^{-1}\mathbf{T}_n\mathbf{H}\boldsymbol{\beta})\{E[\chi_{p+1}^{-2}(\Delta^2)]\}^2$;

$$\text{(ii) } \mathbf{M}_4(\hat{\boldsymbol{\theta}}_n^S) = \sigma^2\mathbf{D}_{11} - \sigma^2 c(p-1)(\mathbf{T}_n\mathbf{H}\mathbf{D}_{22}\mathbf{H}'\mathbf{T}_n') \times \left\{(p-3)E\left[\chi_{p+1}^{-4}(\Delta^2)\right] + 2\Delta^2 E\left[\chi_{p+3}^{-4}(\Delta^2)\right]\right\} + c(p^2-1)(\mathbf{T}_n\mathbf{H}\boldsymbol{\beta}\boldsymbol{\beta}'\mathbf{H}'\mathbf{T}_n')E\left[\chi_{p+3}^{-4}(\Delta^2)\right]$$

and

$$R_4(\hat{\boldsymbol{\theta}}_n^S) = \sigma^2 \operatorname{tr}(\mathbf{W}\mathbf{D}_{11}) - \sigma^2 c(p-1)\operatorname{tr}(\mathbf{W}\mathbf{T}_n\mathbf{H}\mathbf{D}_{22}\mathbf{H}'\mathbf{T}_n') \times\left\{(p-3)E\left[\chi_{p+1}^{-4}(\Delta^2)\right] + 2\Delta^2 E\left[\chi_{p+3}^{-4}(\Delta^2)\right]\right\} + c(p^2-1)(\boldsymbol{\beta}'\mathbf{H}'\mathbf{T}_n'\mathbf{W}\mathbf{T}_n\mathbf{H}\boldsymbol{\beta})E\left[\chi_{p+3}^{-4}(\Delta^2)\right]. \tag{6.3.4b}$$

6.3.5 Positive-Rule Stein Estimators of β and θ

$$\text{(i) } \mathbf{b}_5(\hat{\boldsymbol{\beta}}_n^{S+}) = -\mathbf{H}\boldsymbol{\beta}\left\{G_{p+1,m}(c_1;\Delta^2) + c_1 E\left[F_{p+1,m}^{-1}(\Delta^2) I\left(F_{p+1,m}(\Delta^2) > c_1\right)\right]\right\}$$

and

$$B_5(\hat{\boldsymbol{\beta}}_n^{S+}) = \Delta^2\left\{G_{p+1,m}(c_1;\Delta^2) + c_1 E\left[F_{p+1,m}^{-1}(\Delta^2) I\left(F_{p+1,m}(\Delta^2) > c_1\right)\right]\right\}^2.$$

$$\text{(ii) } \mathbf{M}_5(\hat{\boldsymbol{\beta}}_n^{S+}) = \mathbf{M}_4(\hat{\boldsymbol{\beta}}_n^S) - \sigma^2(\mathbf{H}\mathbf{D}_{22})E\left[\left(1 - c_1 F_{p+1,m}^{-1}(\Delta^2)\right)^2 I\left(F_{p+1,m}(\Delta^2) < c_1\right)\right] + \mathbf{H}\boldsymbol{\beta}\boldsymbol{\beta}'\mathbf{H}' 2E\left[\left(1 - c_1 F_{p+1,m}^{-1}(\Delta^2)\right) I\left(F_{p+1,m}(\Delta^2) < c_1\right)\right] - E\left[\left(1 - c_2 F_{p+3,m}^{-1}(\Delta^2)\right)^2 I\left(F_{p+3,m}(\Delta^2) < c_2\right)\right]\}$$

and

$$R_5(\hat{\boldsymbol{\beta}}_n^{S+};\mathbf{W}) = R_4(\hat{\boldsymbol{\beta}}_n^S;\mathbf{W}) - \sigma^2\operatorname{tr}(\mathbf{W}\mathbf{H}\mathbf{D}_{22})E\left[\left(1 - c_1 F_{p+1,m}^{-1}(\Delta^2)\right)^2 I\left(F_{p+1,m}(\Delta^2) < c_1\right)\right]\} + \mathbf{H}\boldsymbol{\beta}\mathbf{W}\boldsymbol{\beta}'\mathbf{H}\left\{2E\left[\left(1 - c_1 F_{p+1,m}^{-1}(\Delta^2)\right) I\left(F_{p+1,m}(\Delta^2) < c_1\right)\right] - E\left[\left(1 - c_2 F_{p+3,m}^{-1}(\Delta^2)\right)^2 I\left(F_{p+3,m}(\Delta^2) < c_2\right)\right]\right\}, \tag{6.3.5a}$$

where $c_1 = \frac{c(p-1)}{p+1}$ and $c_2 = \frac{c(p-1)}{p+3}$ with $\boldsymbol{\delta} = \mathbf{H}\boldsymbol{\beta}$.
Similarly,

$$\text{(i) } \mathbf{b}_5(\hat{\boldsymbol{\theta}}_n^{S+}) = c_1(\mathbf{T}_n\mathbf{H}\boldsymbol{\beta})\left\{E\left[F_{p+1,m}^{-1}(\Delta^2)\right] - E\left[F_{p+1,m}^{-1}(\Delta^2) I\left(F_{p+1,m}(\Delta^2) < c_1\right)\right]\right\} + (\mathbf{T}_n\mathbf{H}\boldsymbol{\beta})G_{p+1,m}(c_1;\Delta^2) \text{ and}$$

$$B_5(\hat{\boldsymbol{\beta}}_n^{S+}) = \sigma^{-2}\left(\boldsymbol{\beta}'\mathbf{H}'\mathbf{T}_n'\mathbf{D}_{22}^{-1}\mathbf{T}_n\mathbf{H}\boldsymbol{\beta}\right)\left\{c_1 E\left[F_{p+1,m}^{-1}(\Delta^2)\right] - c_1 E\left[F_{p+1,m}^{-1}(\Delta^2) I\left(F_{p+1,m}(\Delta^2) < c_1\right)\right] + G_{p+1,m}(c_1;\Delta^2)\right\}^2;$$

(ii) $\mathbf{M}_5(\hat{\boldsymbol{\theta}}_n^{S+}) = \mathbf{M}_4(\hat{\boldsymbol{\theta}}_n^{S}) - \sigma^2(\mathbf{T}_n\mathbf{H}\mathbf{D}_{22}\mathbf{H}'\mathbf{T}_n')$

$$\times E\big[\big(1 - c_1 F_{p+1,m}^{-1}(\Delta^2)\big)^2 I\big(F_{p+1,m}(\Delta^2) < c_1\big)\big] + (\mathbf{T}_n\mathbf{H}\mathbf{D}_{22}\mathbf{H}'\mathbf{T}_n')$$
$$\times \Big\{2E\big[\big(1 - c_1 F_{p+1,m}^{-1}(\Delta^2)\big) I\big(F_{p+2,m}(\Delta^2) < c_1\big)\big]$$
$$- E\big[\big(1 - c_2 F_{p+3,m}^{-1}(\Delta^2)\big)^2 I\big(F_{p+3,m}(\Delta^2) < c_2\big)\big]\Big\}.$$

and

$$R_5(\hat{\boldsymbol{\theta}}_n^{S+}; \mathbf{W}) = R_4(\hat{\boldsymbol{\theta}}_n^{S}; \mathbf{W}) - \sigma^2 \operatorname{tr}(\mathbf{W}\mathbf{T}_n\mathbf{H}\mathbf{D}_{22}\mathbf{H}'\mathbf{T}_n')$$
$$\times E\big[\big(1 - c_1 F_{p+1,m}^{-1}(\Delta^2)\big)^2 I\big(F_{p+1,m}(\Delta^2) < c_1\big)\big] + (\boldsymbol{\beta}'\mathbf{H}'\mathbf{T}_n'\mathbf{W}\mathbf{T}_n\mathbf{H}\boldsymbol{\beta})$$
$$+ \Big\{2E\big[\big(1 - c_1 F_{p+1,m}^{-1}(\Delta^2)\big) I\big(F_{p+1,m}(\Delta^2) < c_1\big)\big]$$
$$- E\big[\big(1 - c_2 F_{p+3,m}^{-1}(\Delta^2)\big)^2 I\big(F_{p+3,m}(\Delta^2) < c_2\big)\big]\Big\} \qquad (6.3.5b)$$

where $E[F_{p+1,m}^{-1}(\Delta^2]$ is the expectation of the reciprocal of the noncentral F-variable with $(p+1, m)$ d.f. and noncentrality parameter $\Delta^2/2$. Similarly,

$$E\left[F_{p+1,m}^{-1}(\Delta^2) I\big(F_{p+1,m}(\Delta^2) < c_1\big)\right]$$

is the truncated expectation of the reciprocal of a noncentral F-variable with $(p+1, m)$ d.f. and noncentrality parameter $\Delta^2/2$. Further, $G_{\nu_1,\nu_2}(\cdot; \Delta^2)$ is the CDF of a noncentral F-distribution with (ν_1, ν_2) d.f. and noncentrality parameter $\Delta^2/2$.

6.4 Comparison of the Estimators of the Intercept Parameter

In this section, we discuss the comparison of the estimators of the intercept parameters, $\boldsymbol{\theta}$. The comparison of the estimators of the slope parameters $\boldsymbol{\beta}$ is similar to that of the treatments means in the ANOVA model given in Chapter 5.

6.4.1 Bias Comparison of the Estimators of the Intercept Parameter

The bias comparisons of the estimators of $\boldsymbol{\theta}$ is made via quadratic bias expressions as they are scalars. These expressions depend on $\mathbf{T}_n$ except for the unrestricted estimator $\tilde{\boldsymbol{\theta}}_n$. Hence, if $\mathbf{T}_n = \mathbf{0}$,

$$B_1(\tilde{\boldsymbol{\theta}}_n) = B_2(\hat{\boldsymbol{\theta}}_n) = B_3(\hat{\boldsymbol{\theta}}_n^{PT}) = B_4(\hat{\boldsymbol{\theta}}_n^{S}) = B_5(\hat{\boldsymbol{\theta}}_n^{S+}) = 0. \qquad (6.4.1)$$

If $\mathbf{T}_n \neq \mathbf{0}$, $B_1(\tilde{\boldsymbol{\theta}}_n)$ is independent of $\boldsymbol{\beta}$, and all the other expressions depend on $\boldsymbol{\beta}$ and thereby on Δ^2, the departure parameter from the null hypothesis.

First, note that $B_2(\hat{\boldsymbol{\theta}}_n)$ is **unbounded** as a function of Δ^2, since

$$\Delta^2 Ch_{\min}\big[\mathbf{T}_n^2\mathbf{D}_{22}\big] \le B_2(\hat{\boldsymbol{\theta}}_n) \ \le\ \Delta^2 Ch_{\max}\big[\mathbf{T}_n^2\mathbf{D}_{22}\big]. \qquad (6.4.2)$$

Further, under H_0,

$$B_1(\tilde{\boldsymbol{\theta}}_n) = B_2(\hat{\boldsymbol{\theta}}_n) = B_3(\hat{\boldsymbol{\theta}}_n^{\rm PT}) = B_4(\hat{\boldsymbol{\theta}}_n^{S}) = B_5(\hat{\boldsymbol{\theta}}_n^{S+}) = 0,$$

and in general,

$$B_5(\hat{\boldsymbol{\theta}}_n^{S+}) \le B_4(\hat{\boldsymbol{\theta}}_n^{S}) \le B_1(\tilde{\boldsymbol{\theta}}_n) \qquad (6.4.3)$$

and

$$0 \le B_3(\hat{\boldsymbol{\theta}}_n^{\rm PT}) \ \le\ B_2(\hat{\boldsymbol{\theta}}_n),$$

depending on the magnitude of the level of significance, α.

6.4.2 MSE-matrix Comparisons

Notice that all MSE-matrix expressions of the estimators of $\boldsymbol{\theta}$ involve the matrix $\mathbf{T}_n$. Thus, if $\mathbf{T}_n = \mathbf{0}$, then

$$\mathbf{M}_1(\tilde{\boldsymbol{\theta}}_n) = \mathbf{M}_2(\hat{\boldsymbol{\theta}}_n) = \mathbf{M}_3(\hat{\boldsymbol{\theta}}_n^{\rm PT}) = \mathbf{M}_4(\hat{\boldsymbol{\theta}}_n^{S}) = \mathbf{M}_5(\hat{\boldsymbol{\theta}}_n^{S+}) = \sigma^2\mathbf{N}^{-1}. \quad (6.4.4)$$

Clearly, for $\mathbf{T}_n = \mathbf{0}$, $\tilde{\boldsymbol{\theta}}_n$, $\hat{\boldsymbol{\theta}}_n$, $\hat{\boldsymbol{\theta}}_n^{\rm PT}$, $\mathbf{M}_4(\hat{\boldsymbol{\theta}}_n^{S})$ and $\mathbf{M}_5(\hat{\boldsymbol{\theta}}_n^{S+})$ are MSE equivalent. Further, we have the following theorem: on the properties of the MSE matrices under $H_0 : \boldsymbol{\beta} = \beta_0\mathbf{1}_p$.

Theorem 1. Under H_0 and assumed conditions, the quantities $\mathbf{M}_1(\tilde{\boldsymbol{\theta}}_n) - \mathbf{M}_2(\hat{\boldsymbol{\theta}}_n)$, $\mathbf{M}_3(\hat{\boldsymbol{\theta}}_n^{\rm PT}) - \mathbf{M}_2(\hat{\boldsymbol{\theta}}_n)$, $\mathbf{M}_1(\tilde{\boldsymbol{\theta}}_n) - \mathbf{M}_3(\hat{\boldsymbol{\theta}}_n^{\rm PT})$, $\mathbf{M}_1(\tilde{\boldsymbol{\theta}}_n) - \mathbf{M}_4(\hat{\boldsymbol{\theta}}_n^{S})$, and $\mathbf{M}_4(\hat{\boldsymbol{\theta}}_n^{S}) - \mathbf{M}_5(\hat{\boldsymbol{\theta}}_n^{S+})$ are positive semidefinite.

Proof. Under H_0, we may write

(i) $\mathbf{M}_1(\tilde{\boldsymbol{\theta}}_n) - \mathbf{M}_2(\hat{\boldsymbol{\theta}}_n) = \sigma^2(\mathbf{T}_n\mathbf{H}\mathbf{D}_{22}\mathbf{H}'\mathbf{T}_n)$,

(ii) $\mathbf{M}_1(\tilde{\boldsymbol{\theta}}_n) - \mathbf{M}_3(\hat{\boldsymbol{\theta}}_n^{\rm PT}) = \sigma^2(\mathbf{T}_n\mathbf{H}\mathbf{D}_{22}\mathbf{H}'\mathbf{T}_n)G_{p+1,m}(\ell_\alpha;0)$,

(iii) $\mathbf{M}_3(\hat{\boldsymbol{\theta}}_n^{\rm PT}) - \mathbf{M}_2(\hat{\boldsymbol{\theta}}_n) = \sigma^2(\mathbf{T}_n\mathbf{H}\mathbf{D}_{22}\mathbf{H}'\mathbf{T}_n)(1 - G_{p+1,m}(\ell_\alpha;0))$,

(iv) $\mathbf{M}_1(\tilde{\boldsymbol{\theta}}_n) - \mathbf{M}_4(\hat{\boldsymbol{\theta}}_n^{S}) = c\sigma^2(\mathbf{T}_n\mathbf{H}\mathbf{D}_{22}\mathbf{H}'\mathbf{T}_n)$,

(v) $\mathbf{M}_4(\hat{\boldsymbol{\theta}}_n^{S}) - \mathbf{M}_5(\hat{\boldsymbol{\theta}}_n^{S+}) = \sigma^2(\mathbf{T}_n\mathbf{H}\mathbf{D}_{22}\mathbf{H}'\mathbf{T}_n)$
$\times E\big[(1-c_1F_{p+1,m}^{-1}(0))^2 I(F_{p+1,m}(0) < c_1)\big]$, (6.4.5)

Expressions (i) through (v) are positive semidefinite matrices (p.s.d.) by the fact that

$$\frac{1}{nQ}(\mathbf{T}_n\mathbf{H}\mathbf{D}_{22}\mathbf{H}'\mathbf{T}_n) = \frac{1}{nQ}\mathbf{T}_n\mathbf{D}(\mathbf{I}_p - \mathbf{D}^{-1}\mathbf{1}_p\mathbf{1}_p'\mathbf{D}^{-1})\mathbf{D}\mathbf{T}_n, \tag{6.4.6}$$

where

$$\mathbf{D} = \sqrt{nQ}\,\mathrm{Diag}\big((n_1Q_1)^{-1/2}, \dots, (n_pQ_p)^{-1/2}\big) \tag{6.4.7}$$

and $\mathbf{D}^{-1}\mathbf{1}_p\mathbf{1}_p'\mathbf{D}^{-1}$ is of rank 1 and its only positive root is the trace of the matrix, the others being zero. Thus, we may write

$$\boldsymbol{\Gamma}(\mathbf{I}_p - \mathbf{D}^{-1}\mathbf{1}_p\mathbf{1}_p'\mathbf{D}^{-1})\boldsymbol{\Gamma}' = \begin{pmatrix} 0 & 0 & \cdots & \\ 0 & 1 & & \\ 0 & & \ddots & \\ 0 & & & 1 \end{pmatrix}$$

for some orthogonal matrix $\boldsymbol{\Gamma}$, which proves the claim. Further,

$$0 < G_{p+1,m}(\ell_\alpha; 0) < G_{p-1,m}(\ell_\alpha; 0) = 1 - \alpha \ \ (\leq 1).$$

Now, consider the MSE based comparisons in general

Comparison of $\hat{\theta}_n$ Relative to $\tilde{\theta}_n$. The MSE-matrix difference of the estimators is given by

$$\sigma^2(\mathbf{T}_n\mathbf{H}\mathbf{D}_{22}\mathbf{H}'\mathbf{T}_n) - (\mathbf{T}_n\mathbf{H}\boldsymbol{\beta}\boldsymbol{\beta}'\mathbf{H}'\mathbf{T}_n). \tag{6.4.8}$$

In order for (6.4.8) to be p.s.d, we must have

$$\boldsymbol{\ell}'(\mathbf{T}_n\mathbf{H}\mathbf{D}_{22}\mathbf{H}'\mathbf{T}_n)\boldsymbol{\ell} \geq \sigma^{-2}\boldsymbol{\ell}'(\mathbf{T}_n\mathbf{H}\boldsymbol{\beta}\boldsymbol{\beta}'\mathbf{H}'\mathbf{T}_n)\boldsymbol{\ell} \tag{6.4.9}$$

for a given nonzero vector $\boldsymbol{\ell}$. Hence, we must have

$$\frac{\sigma^{-2}\,\boldsymbol{\ell}'(\mathbf{T}_n\mathbf{H}\boldsymbol{\beta}\boldsymbol{\beta}'\mathbf{H}'\mathbf{T}_n)\boldsymbol{\ell}}{\boldsymbol{\ell}'(\mathbf{T}_n\mathbf{D}_{22}\mathbf{T}_n)\boldsymbol{\ell}} \leq \frac{\boldsymbol{\ell}'(\mathbf{T}_n\mathbf{H}\mathbf{D}_{22}\mathbf{H}'\mathbf{T}_n)\boldsymbol{\ell}}{\boldsymbol{\ell}'(\mathbf{T}_n\mathbf{D}_{22}\mathbf{T}_n)\boldsymbol{\ell}}. \tag{6.4.10}$$

But

$$\max_{\boldsymbol{\ell}} \frac{\sigma^{-2}\boldsymbol{\ell}'(\mathbf{T}_n\mathbf{H}\boldsymbol{\beta}\boldsymbol{\beta}'\mathbf{H}'\mathbf{T}_n)\boldsymbol{\ell}}{\boldsymbol{\ell}'(\mathbf{T}_n\mathbf{D}_{22}\mathbf{T}_n)\boldsymbol{\ell}} = \sigma^{-2}(\boldsymbol{\beta}'\mathbf{H}'\mathbf{D}_{22}^{-1}\mathbf{H}\boldsymbol{\beta}) = \Delta^2. \tag{6.4.11}$$

Hence,

$$\Delta^2 \leq \max_{\boldsymbol{\ell}} \frac{\boldsymbol{\ell}'\mathbf{T}_n(\mathbf{H}\mathbf{D}_{22})\mathbf{T}_n\boldsymbol{\ell}}{\boldsymbol{\ell}'\mathbf{T}_n(\mathbf{D}_{22})\mathbf{T}_n\boldsymbol{\ell}} \leq Ch_{\max}(\mathbf{H}) = 1. \tag{6.4.12}$$

Thus, for $\Delta^2 \leq 1$, $\hat{\boldsymbol{\theta}}_n$ performs better than $\tilde{\boldsymbol{\theta}}_n$; otherwise, $\tilde{\boldsymbol{\theta}}_n$ performs better. The efficiency of $\hat{\boldsymbol{\theta}}_n$ relative to $\tilde{\boldsymbol{\theta}}_n$ is given by

$$\begin{aligned}
\text{MRE}(\hat{\boldsymbol{\theta}}_n : \tilde{\boldsymbol{\theta}}_n) &= \frac{|\mathbf{M}_1(\tilde{\boldsymbol{\theta}}_n)|^{1/p}}{|\mathbf{M}_2(\hat{\boldsymbol{\theta}}_n)|^{1/p}} \\
&= \frac{|\mathbf{N}^{-1} + \mathbf{T}_n\mathbf{D}_{22}\mathbf{T}_n|^{1/p}}{|\mathbf{N}^{-1} + \frac{1}{nQ}\mathbf{T}_n\mathbf{1}_p\mathbf{1}_p'\mathbf{T}_n + \mathbf{T}_n\mathbf{H}\boldsymbol{\beta}\boldsymbol{\beta}'\mathbf{H}'\mathbf{T}_n|^{1/p}} \\
&= \left\{\prod_{\alpha=1}^{p}\left(\frac{1}{n_\alpha} + \frac{\bar{x}_\alpha^2}{n_\alpha Q_\alpha}\right)\right\}^{1/p} \\
&\quad \times \left|\mathbf{N}^{-1} + \frac{1}{nQ}\mathbf{T}_n\mathbf{1}_p\mathbf{1}_p'\mathbf{T}_n + \sigma^{-2}(\mathbf{T}_n\mathbf{H}\boldsymbol{\beta}\boldsymbol{\beta}'\mathbf{H}'\mathbf{T}_n)\right|^{-1/p}.
\end{aligned} \tag{6.4.13}$$

Now, we may write

$$\begin{aligned}
&\left|\mathbf{N}^{-1} + \frac{1}{nQ}\mathbf{T}_n\mathbf{1}_p\mathbf{1}_p'\mathbf{T}_n + \sigma^{-2}(\mathbf{T}_n\mathbf{H}\boldsymbol{\beta}\boldsymbol{\beta}'\mathbf{H}'\mathbf{T}_n)\right|^{-1/p} \\
&\quad = \left|\sigma^2\mathbf{D}_{11} - \mathbf{T}_n\mathbf{D}_{22}\mathbf{H}'\mathbf{T}_n + \mathbf{T}_n\mathbf{H}\boldsymbol{\beta}\boldsymbol{\beta}'\mathbf{H}'\mathbf{T}_n\right|^{-1/p}.
\end{aligned}$$

Then

$$\begin{aligned}
\text{MRE}(\hat{\boldsymbol{\theta}}_n : \tilde{\boldsymbol{\theta}}_n) &= \frac{|\sigma^2\mathbf{D}_{11}|^{1/p}}{|\sigma^2\mathbf{D}_{11} - \mathbf{T}_n\mathbf{D}_{22}\mathbf{H}'\mathbf{T}_n + \mathbf{T}_n\mathbf{H}\boldsymbol{\beta}\boldsymbol{\beta}'\mathbf{H}'\mathbf{T}_n|^{1/p}} \\
&= \left|\mathbf{I}_p - \mathbf{D}_{11}^{-1}\mathbf{T}_n\mathbf{D}_{22}(\mathbf{H}' - \sigma^{-2}\mathbf{D}_{22}^{-1}\mathbf{H}\boldsymbol{\beta}\boldsymbol{\beta}'\mathbf{H}')\mathbf{T}_n\right|^{-1/p},
\end{aligned} \tag{6.4.14}$$

where $\mathbf{H}$ is positive semidefinite with rank $p-1$ and $\mathbf{D}_{22}^{-1}\mathbf{H}\boldsymbol{\beta}\boldsymbol{\beta}'\mathbf{H}'$ is positive semidefinite with rank 1 and its only positive characteristic root is $\Delta^2 = \sigma^{-2}(\boldsymbol{\beta}'\mathbf{H}'\mathbf{D}_{22}^{-1}\mathbf{H}\boldsymbol{\beta})$. Since $\mathbf{D}_{22}$ is positive definite, $\mathbf{H}' - \sigma^{-2}\mathbf{D}_{22}^{-1}\mathbf{H}\boldsymbol{\beta}\boldsymbol{\beta}'\mathbf{H}'$ is positive definite. Thus, if Δ^2 is small, $\text{MRE}(\boldsymbol{\theta}_n : \tilde{\boldsymbol{\theta}}_n) \geq 1$, and for large values of Δ^2, $\text{MRE}(\hat{\boldsymbol{\theta}}_n : \tilde{\boldsymbol{\theta}}_n) < 1$. Further, as $\Delta^2 \to \infty$, the $\text{MRE}(\hat{\boldsymbol{\theta}}_n : \boldsymbol{\theta}_n)$ goes to zero. Thus, $\hat{\boldsymbol{\theta}}_n$ loses efficiency for large Δ^2.

Comparison of $\hat{\boldsymbol{\theta}}_n^{\text{PT}}$ Relative to $\tilde{\boldsymbol{\theta}}_n$. In this case, the MSE matrix difference of the estimators is given by

$$\begin{aligned}
&\mathbf{M}_1(\tilde{\boldsymbol{\theta}}_n) - \mathbf{M}_3(\hat{\boldsymbol{\theta}}_n^{\text{PT}}) \\
&= \sigma^2(\mathbf{T}_n\mathbf{H}\mathbf{D}_{22}\mathbf{H}'\mathbf{T}_n)G_{p+1,m}(\ell_\alpha; \Delta^2) \\
&\quad - \left(\mathbf{T}_n\mathbf{H}\boldsymbol{\beta}\boldsymbol{\beta}'\mathbf{H}\mathbf{T}_n\right)\left\{2G_{p+1,m}(\ell_\alpha; \Delta^2) - G_{p+3,m}(\ell_\alpha^*; \Delta^2)\right\}.
\end{aligned} \tag{6.4.15}$$

Thus, for a given nonzero vector $\boldsymbol{\ell}$ we obtain by setting (6.4.15) $\geq \mathbf{0}$ the expression below:

$$\sigma^{-2}\boldsymbol{\ell}'(\mathbf{T}_n\mathbf{H}\boldsymbol{\beta}\boldsymbol{\beta}'\mathbf{H}'\mathbf{T}_n)\boldsymbol{\ell} \leq \frac{\boldsymbol{\ell}'(\mathbf{T}_n\mathbf{H}\mathbf{D}_n\mathbf{H}'\mathbf{T}_n)\boldsymbol{\ell}G_{p+1,m}(\ell_\alpha; \Delta^2)}{\{2G_{p+1,m}(\ell_\alpha; \Delta^2) - G_{p+3,m}(\ell_\alpha^*; \Delta^2)\}}. \tag{6.4.16}$$

Dividing by $\boldsymbol{\ell}'(\mathbf{T}_n\mathbf{D}_{22}\mathbf{T}_n)\boldsymbol{\ell}$, and maximizing over all $\boldsymbol{\ell}$, we obtain

$$\Delta^2 \leq \frac{G_{p+1,m}(\ell_\alpha;\Delta^2)}{\{2G_{p+1,m}(\ell_\alpha;\Delta^2) - G_{p+3,m}(\ell_\alpha^*;\Delta^2)\}}. \tag{6.4.17}$$

Thus, $\hat{\boldsymbol{\theta}}_n^{\mathrm{PT}}$ performs better than $\tilde{\boldsymbol{\theta}}_n$ if Δ^2 satisfies (6.4.17); otherwise, $\tilde{\boldsymbol{\theta}}_n$ performs better than $\hat{\boldsymbol{\theta}}_n^{\mathrm{PT}}$. The MRE($\hat{\boldsymbol{\theta}}_n^{\mathrm{PT}} : \tilde{\boldsymbol{\theta}}_n$) is given by

$$\begin{aligned}
\mathrm{MRE}(\hat{\boldsymbol{\theta}}_n^{\mathrm{PT}} : \tilde{\boldsymbol{\theta}}_n) &= \frac{|\mathbf{M}_1(\tilde{\boldsymbol{\theta}}_n)|^{1/p}}{|\mathbf{M}_3(\hat{\boldsymbol{\theta}}_n^{\mathrm{PT}})|^{1/p}} \\
&= \Big|\mathbf{I}_p - \mathbf{D}_{11}^{-1}(\mathbf{T}_n\mathbf{H}\mathbf{D}_{22}\mathbf{H}'\mathbf{T}_n)G_{p+1,m}(\ell_\alpha;\Delta^2) \\
&\quad + \sigma^{-2}\mathbf{D}_{11}^{-1}(\mathbf{T}_n\mathbf{H}\boldsymbol{\beta}\boldsymbol{\beta}'\mathbf{H}'\mathbf{T}_n) \\
&\quad \times \Big\{2G_{p+1,m}(\ell_\alpha;\Delta^2) - G_{p+3,m}(\ell_\alpha^*;\Delta^2)\Big\}\Big|^{-1/p}.
\end{aligned} \tag{6.4.18}$$

Clearly, as a function of Δ^2, MRE($\hat{\boldsymbol{\theta}}_n^{\mathrm{PT}} : \tilde{\boldsymbol{\theta}}_n$) has a maximum at $\Delta^2 = 0$, decreases crossing the 1-line to a minimum, and then increases toward 1-line as $\Delta \to \infty$. The same conclusion will be reached if we consider the trace of the matrices instead of the determinant.

Comparison of $\hat{\boldsymbol{\theta}}_n^S$ Relative to $\tilde{\boldsymbol{\theta}}_n$. In this case, the MSE-matrix difference is given by

$$\begin{aligned}
\mathbf{M}_1(\tilde{\boldsymbol{\theta}}_n) - \mathbf{M}_4(\hat{\boldsymbol{\theta}}_n^S) &= c(p-1)(\mathbf{T}_n\mathbf{H}\mathbf{D}_{22}\mathbf{H}'\mathbf{T}_n) \\
&\quad \{(p-3)E[\chi_{p+1}^{-4}(\Delta^2)] + 2\Delta^2 E[\chi_{p+3}^{-4}(\Delta^2)]\} \\
&\quad -c(p^2-1)(\mathbf{T}_n\mathbf{H}\boldsymbol{\beta}\boldsymbol{\beta}'\mathbf{H}'\mathbf{T}_n E[\chi_{p+3}^{-4}(\Delta^2)].
\end{aligned} \tag{6.4.19}$$

In order that the MSE difference is p.s.d., we must have for every nonzero vector $\boldsymbol{\ell}$,

$$\begin{aligned}
&(p+1)(\boldsymbol{\ell}'\mathbf{T}_n\mathbf{H}\boldsymbol{\beta}\boldsymbol{\beta}'\mathbf{H}'\mathbf{T}_n\boldsymbol{\ell})E[\chi_{p+3}^{-4}(\Delta^2)] \\
&\leq (\boldsymbol{\ell}'\mathbf{T}_n\mathbf{H}\mathbf{D}_{22}\mathbf{H}'\mathbf{T}_n\boldsymbol{\ell})\{(p-3)E[\chi_{p+1}^{-4}(\Delta^2)] + 2\Delta^2E[\chi_{p+3}^{-4}(\Delta^2)]\}.
\end{aligned} \tag{6.4.20}$$

Dividing by $\boldsymbol{\ell}'\mathbf{T}_n\mathbf{D}_{22}\mathbf{T}_n\boldsymbol{\ell}$ and maximizing over $\boldsymbol{\ell}$, we have

$$(p+1)\Delta^2E[\chi_{p+3}^{-4}(\Delta^2)] \leq (p-3)E[\chi_{p+1}^{-4}(\Delta^2)] + 2\Delta^2E[\chi_{p+3}^{-4}(\Delta^2)],$$

or

$$(p-1)\Delta^2E[\chi_{p+3}^{-4}(\Delta^2)] \leq (p-3)E[\chi_{p+1}^{-4}(\Delta^2)], \tag{6.4.21}$$

since

$$\Delta^2E[\chi_{p+3}^{-4}(\Delta^2)] = E[\chi_{p+1}^{-2}(\Delta^2)] - (p-3)E[\chi_{p+1}^{-4}(\Delta^2)].$$

The inequality (6.4.21) does not hold for all Δ^2. Hence, $\hat{\boldsymbol{\theta}}_n^S$ does not dominate $\tilde{\boldsymbol{\theta}}_n$ uniformly. Here we have the efficiency expression as

$$
\begin{aligned}
\mathrm{MRE}(\hat{\boldsymbol{\theta}}_n^S : \tilde{\boldsymbol{\theta}}_n) = {} & |\mathbf{D}_{11}|^{1/p}\Big|\mathbf{D}_{11} - c(p-1)(\mathbf{T}_n\mathbf{H}\mathbf{D}_{22}\mathbf{H}'\mathbf{T}_n) \\
& \times\{(p-3)E[\chi_{p+1}^{-4}(\Delta^2)] + 2\Delta^2E[\chi_{p+3}^{-4}(\Delta^2)]\} \\
& +c(p^2-1)(\mathbf{T}_n\mathbf{H}\boldsymbol{\beta}\boldsymbol{\beta}'\mathbf{H}'\mathbf{T}_n)E[\chi_{p+3}^{-4}(\Delta^2)]\Big|^{-1/p} \\
= {} & \Big|\mathbf{I}_p - c(p-1)\mathbf{D}_{11}^{-1}(\mathbf{T}_n\mathbf{H}\mathbf{D}_{22}\mathbf{H}'\mathbf{T}_n) \\
& \times\{(p-3)E[\chi_{p+1}^{-4}(\Delta^2)] + 2\Delta^2E[\chi_{p+3}^{-4}(\Delta^2)]\} \\
& +\sigma^{-2}c(p^2-1)\mathbf{D}_{22}^{-1}(\mathbf{T}_n\mathbf{H}\boldsymbol{\beta}\boldsymbol{\beta}'\mathbf{H}'\mathbf{T}_n)E[\chi_{p+3}^{-4}(\Delta^2)]\Big|^{-1/p}.
\end{aligned}
\tag{6.4.22}
$$

Comparison of $\hat{\boldsymbol{\theta}}_n^{S+}$ Relative to $\tilde{\boldsymbol{\theta}}_n^S$. In this case, the MSE difference $\mathbf{M}_4(\tilde{\boldsymbol{\theta}}_n^S) - \mathbf{M}_5(\hat{\boldsymbol{\theta}}_n^{S+})$ equals

$$
\begin{aligned}
& \sigma^2(\mathbf{T}_n\mathbf{H}\mathbf{D}_{22}\mathbf{H}'\mathbf{T}_n')E[(1-c_1F_{p+1,m}^{-1}(\Delta^2))^2I(F_{p+1,m}(\Delta^2) < c_1)] \\
& +(\mathbf{T}_n\mathbf{H}\boldsymbol{\beta}\boldsymbol{\beta}'\mathbf{H}'\mathbf{T}_n')\Big\{E[(1-c_2F_{p+3,m}^{-1}(\Delta^2))^2I(F_{p+3,m}(\Delta^2) < c_2)] \\
& +2E[(1-c_1F_{p+1,m}^{-1}(\Delta^2))I(F_{p+1,m}(\Delta^2) < c_1)]\Big\}.
\end{aligned}
\tag{6.4.23}
$$

The R.H.S. is p.s.d. whenever for every nonzero vector $\boldsymbol{\ell}$,

$$
\begin{aligned}
& \sigma^2(\boldsymbol{\ell}'\mathbf{T}_n\mathbf{H}\mathbf{D}_{22}\mathbf{H}'\mathbf{T}_n'\boldsymbol{\ell})E[(1-c_1F_{p+1,m}^{-1}(\Delta^2))^2I(F_{p+1,m}(\Delta^2) < c_1)] \\
& +(\boldsymbol{\ell}'\mathbf{T}_n\mathbf{H}\boldsymbol{\beta}\boldsymbol{\beta}'\mathbf{H}'\mathbf{T}_n'\boldsymbol{\ell})\{E[(1-c_2F_{p+3,m}^{-1}(\Delta^2))^2I(F_{p+3,m}(\Delta^2) < c_2)] \\
& +2E[(1-c_1F_{p+1,m}^{-1}(\Delta^2))I(F_{p+1,m}(\Delta^2) < c_1)]\} \geq \mathbf{0}.
\end{aligned}
\tag{6.4.24}
$$

Dividing by $\boldsymbol{\ell}'\mathbf{T}_n\mathbf{D}_{22}\mathbf{T}_n\boldsymbol{\ell}$ and maximizing over $\boldsymbol{\ell}'$, we have

$$
\begin{aligned}
& E[(1-c_1F_{p+1,m}^{-1}(\Delta^2))^2I(F_{p+1,m}(\Delta^2) < c_1)] \\
& + \Delta^2\{E[(1-c_2F_{p+3,m}^{-1}(\Delta^2))^2I(F_{p+3,m}(\Delta^2) < c_2)] \\
& + 2E[(1-c_1F_{p+1,m}^{-1}(\Delta^2))I(F_{p+1,m}(\Delta^2) < c_1)]\},
\end{aligned}
\tag{6.4.25}
$$

which is positive. Hence, $\hat{\boldsymbol{\theta}}_n^{S+}$ dominates $\hat{\boldsymbol{\theta}}_n^S$ uniformly under the MSE criteria.

Here we have the efficiency expression as

$$\text{MRE}(\hat{\boldsymbol{\theta}}_n^{S+} : \tilde{\boldsymbol{\theta}}_n^{S}) = \frac{|\mathbf{M}_4(\tilde{\boldsymbol{\theta}}_n^{S})|^{1/p}}{|\mathbf{M}_5(\hat{\boldsymbol{\theta}}_n^{S+})|^{1/p}} = \Big|\mathbf{I}_p - \sigma^2\{\mathbf{M}_4(\tilde{\boldsymbol{\theta}}_n^{S})\}^{-1}\{(\mathbf{T}_n\mathbf{H}\mathbf{D}_{22}\mathbf{H}'\mathbf{T}_n)$$
$$\times E\big[(1 - c_1F_{p+1,m}^{-1}(\Delta^2))^2 I(F_{p+1,m}(\Delta^2) < c_1)\big] - (\mathbf{T}_n\mathbf{H}\boldsymbol{\beta}\boldsymbol{\beta}'\mathbf{H}'\mathbf{T}_n)$$
$$\times E\big[(1 - c_2F_{p+3,m}^{-1}(\Delta^2))^2 I(F_{p+3,m}(\Delta^2) < c_2)\big] - 2(\mathbf{T}_n\mathbf{H}\boldsymbol{\beta}\boldsymbol{\beta}'\mathbf{H}'\mathbf{T}_n)$$
$$\times E\big[(c_1F_{p+1,m}^{-1}(\Delta^2) - 1) I(F_{p+1,m}(\Delta^2) < c_1)\big]\}\Big|^{-1/p}, \tag{6.4.26}$$

where $c_1 = \frac{c(p-1)}{p+1}$ and $c_2 = \frac{c(p-1)}{p+3}$. $\text{MRE}(\hat{\boldsymbol{\theta}}_n^{S+} : \tilde{\boldsymbol{\theta}}_n^{S})$ is a decreasing function of Δ^2 with maximum at $\Delta^2 = 0$, and as $\Delta \to \infty$, it converges to 1 from above. Hence, $\hat{\boldsymbol{\theta}}_n^{S+}$ dominates $\tilde{\boldsymbol{\theta}}_n^{S}$ uniformly with respect to MSE matrices.

6.4.3 Weighted Risk Comparisons of the Estimators

In this section, we compare the estimators of $\boldsymbol{\theta}$ based on the weighted quadratic risk functions.

Comparison of $\hat{\boldsymbol{\theta}}_n$ Relative to $\tilde{\boldsymbol{\theta}}_n$. The risk-difference of the two estimators In this case, is given by

$$R_1(\tilde{\boldsymbol{\theta}}_n : \mathbf{W}) - R_2(\hat{\boldsymbol{\theta}}_n : \mathbf{W}) = \sigma^2\text{tr}\big[\mathbf{W}(\mathbf{D}_{11} - \mathbf{D}_{11}^*)\big] - (\boldsymbol{\beta}'\mathbf{H}'\mathbf{T}_n\mathbf{W}\mathbf{T}_n\mathbf{H}\boldsymbol{\beta}). \tag{6.4.27}$$

Thus, the risk-difference is $\lesseqgtr 0$ according as

$$\sigma^{-2}(\boldsymbol{\beta}'\mathbf{H}'\mathbf{T}_n\mathbf{W}\mathbf{T}_n\mathbf{H}\boldsymbol{\beta}) \gtreqless \text{tr}\big[\mathbf{W}(\mathbf{D}_{11} - \mathbf{D}_{11}^*)\big]. \tag{6.4.28}$$

Now,

$$\sigma^{-2}(\boldsymbol{\beta}'\mathbf{H}'\mathbf{T}_n\mathbf{W}\mathbf{T}_n\mathbf{H}\boldsymbol{\beta}) \le \Delta^2 Ch_{\max}(\mathbf{D}_{22}\mathbf{T}_n\mathbf{W}\mathbf{T}_n)$$
$$= \Delta^2 Ch_{\max}\big[(\mathbf{D}_{11} - \mathbf{N}^{-1})\mathbf{W}\big] \le \Delta^2 Ch_{\max}(\mathbf{D}_{11}\mathbf{W})$$
$$\le \Delta^2\text{tr}(\mathbf{D}_{11}\mathbf{W}) \quad \text{and} \quad \text{tr}\big[\mathbf{W}(\mathbf{D}_{11} - \mathbf{D}_{11}^*)\big] \le \text{tr}(\mathbf{D}_{11}\mathbf{W}). \tag{6.4.29}$$

Hence, for $\Delta^2 \le \frac{\text{tr}(\mathbf{D}_{11}\mathbf{W})}{Ch_{\max}(\mathbf{D}_{11}\mathbf{W})}$, $\hat{\boldsymbol{\theta}}_n$ performs better than $\tilde{\boldsymbol{\theta}}_n$, and if $\Delta^2 \ge \frac{\text{tr}(\mathbf{D}_{11}\mathbf{W})}{Ch_{\min}(\mathbf{D}_{11}\mathbf{W})}$, $\tilde{\boldsymbol{\theta}}_n$ performs better than $\hat{\boldsymbol{\theta}}_n$. (6.4.30)

Thus, the risk-efficiency of $\hat{\boldsymbol{\theta}}_n$ relative to $\tilde{\boldsymbol{\theta}}_n$ is

$$RRE(\hat{\boldsymbol{\theta}}_n : \tilde{\boldsymbol{\theta}}_n) = \frac{R_1(\tilde{\boldsymbol{\theta}}_n : \mathbf{W})}{R_2(\hat{\boldsymbol{\theta}}_n : \mathbf{W})}$$
$$= \left\{1 - \frac{\text{tr}(\mathbf{W}\mathbf{T}_n\mathbf{D}_{22}\mathbf{H}'\mathbf{T}_n')}{\text{tr}(\mathbf{D}_{11}\mathbf{W})} + \sigma^{-2}\frac{\boldsymbol{\beta}'\mathbf{H}'\mathbf{T}_n\mathbf{W}\mathbf{T}_n\mathbf{H}\boldsymbol{\beta}}{\text{tr}(\mathbf{D}_{11}\mathbf{W})}\right\}^{-1}. \tag{6.4.31}$$

Comparison of $\hat{\boldsymbol{\theta}}_n^{\text{PT}}$ Relative to $\tilde{\boldsymbol{\theta}}_n$ ($\hat{\boldsymbol{\theta}}_n$). In this case, by considering the risk-difference, we obtain

$$R_1(\tilde{\boldsymbol{\theta}}_n : \mathbf{W}) - R_3(\hat{\boldsymbol{\theta}}_n^{\text{PT}} : \mathbf{W}) \gtreqless 0 \text{ according as}$$

$$\sigma^{-2}(\boldsymbol{\beta}'\mathbf{H}'\mathbf{T}_n\mathbf{W}\mathbf{T}_n\mathbf{H}\boldsymbol{\beta}) \gtrless \frac{\operatorname{tr}\left[\mathbf{W}(\mathbf{D}_{11} - \mathbf{D}_{11}^*)\right] G_{p+1,m}(\ell;\Delta^2)}{\left\{2G_{p+1,m}(\ell;\Delta^2) - G_{p+3,m}(\ell^*;\Delta^2)\right\}}. \quad (6.4.32)$$

$$R_1(\hat{\boldsymbol{\theta}}_n : \mathbf{W}) - R_3(\hat{\boldsymbol{\theta}}_n^{\text{PT}} : \mathbf{W}) \gtreqless 0 \text{ according as}$$

$$\sigma^{-2}(\boldsymbol{\beta}'\mathbf{H}'\mathbf{T}_n\mathbf{W}\mathbf{T}_n\mathbf{H}\boldsymbol{\beta}) \lessgtr \frac{\operatorname{tr}\left[\mathbf{W}(\mathbf{D}_{11} - \mathbf{D}_{11}^*)\right]\left(1 - G_{p+1,m}(\ell;\Delta^2)\right)}{\left\{1 - 2G_{p+1,m}(\ell;\Delta^2) + G_{p+3,m}(\ell^*;\Delta^2)\right\}}. \quad (6.4.33)$$

We note that $\mathbf{D}_{11} - \mathbf{D}_{11}^*$ is a positive semidefinite matrix of rank of at most $p-1$. Therefore, for $\mathbf{T}_n \neq \mathbf{0}$, $\operatorname{tr}\left[\mathbf{W}(\mathbf{D}_{11} - \mathbf{D}_{11}^*)\right]$ is a nonnegative number. Hence, $\hat{\boldsymbol{\theta}}_n^{\text{PT}}$ performs better than $\tilde{\boldsymbol{\theta}}_n$ if

$$\Delta^2 \leq \frac{\operatorname{tr}(\mathbf{D}_{11}\mathbf{W})G_{p+1,m}(\ell_\alpha;\Delta^2)}{Ch_{\max}(\mathbf{D}_{11}\mathbf{W})\left\{2G_{p+1,m}(\ell_\alpha;\Delta^2) - G_{p+3,m}(\ell_\alpha^*;\Delta^2)\right\}}, \quad (6.4.34a)$$

and $\tilde{\boldsymbol{\theta}}_n$ performs better than $\hat{\boldsymbol{\theta}}_n^{\text{PT}}$ if

$$\Delta^2 \geq \frac{\operatorname{tr}(\mathbf{D}_{11}\mathbf{W})G_{p+1,m}(\ell_\alpha;\Delta^2)}{Ch_{\min}(\mathbf{D}_{11}\mathbf{W})\left\{2G_{p+1,m}(\ell_\alpha;\Delta^2) - G_{p+3,m}(\ell_\alpha^*;\Delta^2)\right\}}, \quad (6.4.34b)$$

Similarly, $\hat{\boldsymbol{\theta}}_n$ performs better than $\hat{\boldsymbol{\theta}}_n^{\text{PT}}$ if

$$\Delta^2 \leq \frac{\operatorname{tr}(\mathbf{D}_{11}\mathbf{W})\left[1 - G_{p+1,m}(\ell_\alpha;\Delta^2)\right]}{Ch_{\max}(\mathbf{D}_{11}\mathbf{W})\left\{1 - 2G_{p+1,m}(\ell_\alpha;\Delta^2) + G_{p+3,m}(\ell_\alpha^*;\Delta^2)\right\}}, \quad (6.4.35a)$$

and $\hat{\boldsymbol{\theta}}_n^{\text{PT}}$ performs better than $\hat{\boldsymbol{\theta}}_n$ if

$$\Delta^2 \geq \frac{\operatorname{tr}(\mathbf{D}_{11}\mathbf{W})\left[1 - G_{p+1,m}(\ell_\alpha;\Delta^2)\right]}{Ch_{\max}(\mathbf{D}_{11}\mathbf{W})\left\{1 - 2G_{p+1,m}(\ell_\alpha;\Delta^2) + G_{p+3,m}(\ell_\alpha^*;\Delta^2)\right\}}. \quad (6.4.35b)$$

The risk-efficency of $\hat{\boldsymbol{\theta}}_n^{\text{PT}}$ Relative to $\tilde{\boldsymbol{\theta}}_n$ is given by

$$\begin{aligned}
&RRE(\hat{\boldsymbol{\theta}}_n^{\text{PT}} : \tilde{\boldsymbol{\theta}}_n) \\
&= \frac{R_1(\tilde{\boldsymbol{\theta}}_n : \mathbf{W})}{R_2(\hat{\boldsymbol{\theta}}_n^{\text{PT}} : \mathbf{W})} \\
&= \left[1 - \frac{\operatorname{tr}[\mathbf{W}(\mathbf{D}_{11} - \mathbf{D}_{11}^*)]}{\operatorname{tr}(\mathbf{D}_{11}\mathbf{W})} G_{p+1,m}(\ell_\alpha;\Delta^2)\right. \\
&\quad \left. + \sigma^{-2}\frac{\boldsymbol{\beta}'\mathbf{H}'\mathbf{T}_n\mathbf{W}\mathbf{T}_n\mathbf{H}\boldsymbol{\beta}}{\operatorname{tr}(\mathbf{D}_{11}\mathbf{W})}\left\{2G_{p+1,m}(\ell_\alpha;\Delta^2) - G_{p+3,m}(\ell_\alpha^*;\Delta^2)\right\}\right]^{-1}.
\end{aligned}$$

(6.4.36)

Comparison of $\hat{\boldsymbol{\theta}}_n^S$ Relative to $\tilde{\boldsymbol{\theta}}_n$. In this case, the risk-difference is given by

$$\begin{aligned} R_1(\tilde{\boldsymbol{\theta}}_n : \mathbf{W}) - R_4(\hat{\boldsymbol{\theta}}_n^S : \mathbf{W}) &= \sigma^2 c(p-1)\mathrm{tr}\big[\mathbf{W}\mathbf{T}_n\mathbf{H}\mathbf{D}_{22}\mathbf{H}'\mathbf{T}_n\big] \\ &\times\{(p-3)E\big[\chi_{p+3}^{-4}(\Delta^2)\big] + 2\Delta^2 E\big[\chi_{p+3}^{-4}(\Delta^2)\big]\} \\ &-c(p^2-1)(\boldsymbol{\beta}'\mathbf{H}'\mathbf{T}_n\mathbf{W}\mathbf{T}_n\mathbf{H}\boldsymbol{\beta})E\big[\chi_{p+3}^{-4}(\Delta^2)\big] \text{ for } p \geq 4. \end{aligned} \tag{6.4.37}$$

Thus, the risk-difference is a decreasing function of $\sigma^{-2}(\boldsymbol{\beta}'\mathbf{H}'\mathbf{T}_n\mathbf{W}\mathbf{T}_n\mathbf{H}\boldsymbol{\beta})$.

Hence, replacing $\sigma^{-2}(\boldsymbol{\beta}'\mathbf{H}'\mathbf{T}_n\mathbf{W}\mathbf{T}_n\mathbf{H}\boldsymbol{\beta})$ by $\Delta^2\mathrm{tr}(\mathbf{D}_{11}\mathbf{W})$ in (6.4.37), we note that the risk-difference satisfies the equality

$$\begin{aligned} R_1(\tilde{\boldsymbol{\theta}}_n : \mathbf{W}) - R_4(\hat{\boldsymbol{\theta}}_n^S : \mathbf{W}) &\geq \sigma^2 c(p-1)\mathrm{tr}\big[\mathbf{W}'\mathbf{T}_n\mathbf{H}\mathbf{D}_{22}\mathbf{H}'\mathbf{T}_n\big] \\ &\times\Big\{(p-3)E\big[\chi_{p+3}^{-4}(\Delta^2)\big] + 2\Delta^2 E\big[\chi_{p+3}^{-4}(\Delta^2)\big]\Big\} \\ &-c(p^2-1)\Delta^2\mathrm{tr}(\mathbf{D}_{11}\mathbf{W})E\big[\chi_{p+3}^{-4}(\Delta^2)\big], \end{aligned} \tag{6.4.38}$$

and the R.H.S. is a decreasing function of Δ^2 with a maximum value

$$\sigma^2 c(p-1)^2(p-3)\mathrm{tr}\big[\mathbf{W}'\mathbf{T}_n\mathbf{H}\mathbf{D}_{22}\mathbf{H}'\mathbf{T}_n\big] \tag{6.4.39}$$

at $\Delta^2 = 0$. Hence, $\hat{\boldsymbol{\theta}}_n^S$ dominates $\tilde{\boldsymbol{\theta}}_n$ uniformly.

The risk-efficiency of $\hat{\boldsymbol{\theta}}_n^S$ relative to $\tilde{\boldsymbol{\theta}}_n$ is given by the expression

$$\begin{aligned} &RRE(\hat{\boldsymbol{\theta}}_n^S : \tilde{\boldsymbol{\theta}}_n) \\ &= \Bigg[1 - c(p-1)\frac{\mathrm{tr}[\mathbf{W}\mathbf{T}_n\mathbf{H}\mathbf{D}_{22}\mathbf{H}'\mathbf{T}_n]}{\mathrm{tr}[\mathbf{W}\mathbf{D}_{11}]} \\ &\quad \Big\{(p-3)E[\chi_{p+3}^{-4}(\Delta^2)] + 2\Delta^2 E[\chi_{p+3}^{-4}(\Delta^2)]\Big\} \\ &\quad -c(p^2-1)\frac{\sigma^{-2}(\boldsymbol{\beta}'\mathbf{H}'\mathbf{T}_n\mathbf{W}\mathbf{T}_n\mathbf{H}\boldsymbol{\beta})}{\mathrm{tr}[\mathbf{W}\mathbf{D}_{11}]}E[\chi_{p+3}^{-4}(\Delta^2)]\Bigg]^{-1}. \end{aligned} \tag{6.4.40}$$

Comparison of $\hat{\boldsymbol{\theta}}_n^{S+}$ Relative to $\tilde{\boldsymbol{\theta}}_n^S$. In this case, the risk-difference is

$$\begin{aligned} R_4(\tilde{\boldsymbol{\theta}}_n^S : \mathbf{W}) - R_5(\hat{\boldsymbol{\theta}}_n^{S+} : \mathbf{W}) &= \sigma^2\mathrm{tr}\big[\mathbf{W}'\mathbf{T}_n\mathbf{H}\mathbf{D}_{22}\mathbf{H}'\mathbf{T}_n\big] \\ \times E\Big[\big(1 - c_1F_{p+1,m}^{-1}(\Delta^2)\big)^2 I\big(F_{p+1,m}(\Delta^2) < c_1\big)\Big] &+ (\boldsymbol{\beta}'\mathbf{H}'\mathbf{T}_n\mathbf{W}\mathbf{T}_n\mathbf{H}\boldsymbol{\beta}) \\ \times E\Big[\big(1 - c_2F_{p+3,m}^{-1}(\Delta^2)\big)^2 I\big(F_{p+3,m}(\Delta^2) < c_2\big)\Big] &+ 2(\boldsymbol{\beta}'\mathbf{H}'\mathbf{T}_n\mathbf{W}\mathbf{T}_n\mathbf{H}\boldsymbol{\beta}) \\ \times E\Big[\big(1 - c_1F_{p+1,m}^{-1}(\Delta^2)\big)I\big(F_{p+1,m}(\Delta^2) < c_1\big)\Big] &\geq 0. \end{aligned} \tag{6.4.41}$$

Thus, $\hat{\boldsymbol{\theta}}_n^{S+}$ dominates $\hat{\boldsymbol{\theta}}_n^S$ uniformly. Hence, we have

$$R_1(\tilde{\boldsymbol{\theta}}_n : \mathbf{W}) \geq R_4(\hat{\boldsymbol{\theta}}_n^S : \mathbf{W}) \geq R_5(\hat{\boldsymbol{\theta}}_n^{S+} : \mathbf{W}) \text{ for all } \Delta^2. \tag{6.4.42}$$

The risk-efficiency of $\hat{\boldsymbol{\theta}}_n^{S+}$ relative to $\hat{\boldsymbol{\theta}}_n^{S}$ is the ratio of two risks.

Comparison of $\hat{\boldsymbol{\theta}}_n^{S}$ Relative to $\hat{\boldsymbol{\theta}}_n$. The risk of $\hat{\boldsymbol{\theta}}_n$ is unbounded while the risk of $\hat{\boldsymbol{\theta}}_n^{S}$ is bounded. Also under H_0,

$$\begin{aligned} R_2(\hat{\boldsymbol{\theta}}_n : \mathbf{W}) &= \sigma^2 \operatorname{tr}\big[\mathbf{W}\mathbf{D}_{11}^*\big] \\ R_4(\hat{\boldsymbol{\theta}}_n^{S} : \mathbf{W}) &= \sigma^2 \operatorname{tr}\big[\mathbf{W}\mathbf{D}_{11}\big] - c\sigma^2 \operatorname{tr}\big[\mathbf{W}(\mathbf{D}_{11} - \mathbf{D}_{11}^*)\big]. \end{aligned} \tag{6.4.43}$$

Hence, the risk-difference $R_4(\hat{\boldsymbol{\theta}}_n^{S} : \mathbf{W}) - R_4(\hat{\boldsymbol{\theta}}_n : \mathbf{W})$ In this case, is

$$\begin{aligned} &\sigma^2 \operatorname{tr}\big[\mathbf{W}\mathbf{D}_{11}\big] - c\sigma^2 \operatorname{tr}\big[\mathbf{W}(\mathbf{D}_{11} - \mathbf{D}_{11}^*)\big] - \sigma^2 \operatorname{tr}\big[\mathbf{W}\mathbf{D}_{11}^*\big] \\ &\quad = \sigma^2(1-c) \operatorname{tr}\big[\mathbf{W}\mathbf{D}_{11}\big] - \sigma^2(1-c) \operatorname{tr}\big[\mathbf{W}\mathbf{D}_{11}^*\big] \\ &\quad = \sigma^2(1-c) \operatorname{tr}\big[\mathbf{W}(\mathbf{D}_{11} - \mathbf{D}_{11}^*)\big] \geq 0. \end{aligned} \tag{6.4.44}$$

Hence, $\hat{\boldsymbol{\theta}}_n^{S}$ does not dominates $\hat{\boldsymbol{\theta}}_n$ at $\Delta^2 = 0$. On the other hand as Δ^2 moves away from the origin, $E\big[\chi_{p+1}^{-2}(\Delta^2)\big]$, $E\big[\chi_{p+3}^{-2}(\Delta^2)\big]$, $E\big[\chi_{p+1}^{-4}(\Delta^2)\big]$, and $E\big[\chi_{p+3}^{-4}(\Delta^2)\big]$ all decrease, so the risk-difference is less than zero. Thus, neither $\hat{\boldsymbol{\theta}}_n^{S}$ nor $\hat{\boldsymbol{\theta}}_n$ dominates the other.

Comparison of $\hat{\boldsymbol{\theta}}_n^{S}$ Relative to $\hat{\boldsymbol{\theta}}_n^{\mathrm{PT}}$. The risk of $\hat{\boldsymbol{\theta}}_n^{\mathrm{PT}}$ depends on the level of significance, while that of $\hat{\boldsymbol{\theta}}_n^{S}$ is free of this restriction. First, consider the risk-difference under H_0,

$$\begin{aligned} &R_4(\hat{\boldsymbol{\theta}}_n^{S} : \mathbf{W}) = \sigma^2 \operatorname{tr}\big[\mathbf{W}\mathbf{D}_{11}\big] - \sigma^2 c \operatorname{tr}\big[\mathbf{W}(\mathbf{D}_{11} - \mathbf{D}_{11}^*)\big] \\ &= R_3(\hat{\boldsymbol{\theta}}_n^{\mathrm{PT}} : \mathbf{W}) + \sigma^2 \operatorname{tr}\big[\mathbf{W}(\mathbf{D}_{11} - \mathbf{D}_{11}^*)\big]\big[G_{p+1,m}(\ell_\alpha; 0) - c\big] \geq R_3(\hat{\boldsymbol{\theta}}_n^{\mathrm{PT}} : \mathbf{W}) \\ &\text{whenever} \quad G_{p+1,m}(\ell_\alpha; 0) \geq c. \end{aligned} \tag{6.4.45}$$

Thus, $\hat{\boldsymbol{\theta}}_n^{S}$ does not dominate $\hat{\boldsymbol{\theta}}_n^{\mathrm{PT}}$ under H_0 whenever

$$G_{p+1,m}(\ell_\alpha; 0) \geq c, \tag{6.4.46}$$

which implies for certain values of α that neither $\hat{\boldsymbol{\theta}}_n^{S}$ nor $\hat{\boldsymbol{\theta}}_n^{\mathrm{PT}}$ dominate one another.

6.5 Estimation of the Regression Parameters: Nonnormal Errors

In this section, we consider the estimation of the intercept and slope parameters in the parallelism model when the error distribution is unknown. For each α $(\alpha = 1, \ldots, p)$, let $y_{\alpha 1}, \ldots, y_{\alpha n_\alpha}$ be independent random variables with distribution functions $F_{\alpha_1}, \ldots, F_{\alpha_{n_\alpha}}$, respectively, all defined on the real line $\mathcal{R} = (-\infty, \infty)$. It is assumed that

$$F_{\alpha_j}(y) = F\big((y_{\alpha_j} - \theta_\alpha - \beta_\alpha x_{\alpha_j})/\sigma\big), \quad y_{\alpha_j} \in \mathcal{R}, \tag{6.5.1}$$

for $\alpha = 1, \ldots, p$ and $j = 1, \ldots, n_\alpha$. Without loss of generality, assume

$$\int y dF_{\alpha_j}(y) = 0 \quad \text{and} \quad \int y^2 dF_{\alpha_j}(y) = 1. \tag{6.5.2}$$

6.5.1 Unrestricted, Restricted, Preliminary Test, James-Stein and Positive-Rule Stein Estimators and Test of Hypothesis

As in Section 6.1.2, the **unrestricted** and **restricted** estimators of $\boldsymbol{\theta}$ and $\boldsymbol{\beta}$ are as follows:

$$\tilde{\boldsymbol{\theta}}_n = \bar{\mathbf{Y}} - \mathbf{T}_n \tilde{\boldsymbol{\beta}}_n \quad \text{and} \quad \hat{\boldsymbol{\theta}}_n = \tilde{\boldsymbol{\theta}}_n + \mathbf{T}_n \mathbf{H}_n \tilde{\boldsymbol{\beta}}_n$$

where $\tilde{\boldsymbol{\beta}}_n = (\tilde{\beta}_{1n_1}, \ldots, \tilde{\beta}_{pn_p})'$, and

$$\tilde{\beta}_{\alpha n_\alpha} = \frac{\mathbf{x}'_\alpha \mathbf{y}_\alpha - \frac{1}{n_\alpha}(\mathbf{1}'_{n_\alpha}\mathbf{x}_\alpha)(\mathbf{1}'_{n_\alpha}\mathbf{y}_\alpha)}{n_\alpha Q_\alpha}$$

with

$$\mathbf{T}_n = \text{Diag}(\bar{x}_{1n_1}, \ldots, \bar{x}_{n_p}), \quad \mathbf{H}_n = \mathbf{I}_p - \frac{\mathbf{1}_p \mathbf{1}'_p \mathbf{D}_{22n}^{-1}}{n\bar{Q}_n},$$

and

$$n\bar{Q}_n = \sum_{\alpha=1}^{p} n_\alpha Q_\alpha \quad \text{and} \quad \mathbf{D}_{22n}^{-1} = \text{Diag}(n_1 Q_1, \ldots, n_p Q_p).$$

The restricted estimator is given by

$$\hat{\boldsymbol{\beta}}_n = \frac{\mathbf{1}_p \mathbf{1}'_p \mathbf{D}_{22n}^{-1}}{nQ}.$$

Also, the estimate of σ^2 is given by

$$s_e^2 = (n-2p)^{-1} \sum_{\alpha=1}^{p} \|\mathbf{y}_\alpha - \tilde{\theta}_\alpha \mathbf{1}_{n_\alpha} - \tilde{\beta}_\alpha \mathbf{x}_\alpha\|^2. \tag{6.5.3}$$

For the test of $H_0 : \boldsymbol{\beta} = \beta_0 \mathbf{1}_p$, we consider the test-statistic $\mathcal{L}_n$ defined by

$$\mathcal{L}_n = s_e^{-2}\big(\tilde{\boldsymbol{\beta}}'_n \mathbf{H}'_n \mathbf{D}_{22n}^{-1} \mathbf{H}_n \tilde{\boldsymbol{\beta}}_n\big). \tag{6.5.4}$$

Therefore, the PTE of $\boldsymbol{\theta}$ and $\boldsymbol{\beta}$ are defined by

$$\text{(i)}\ \hat{\boldsymbol{\theta}}_n^{\mathrm{PT}} = \tilde{\boldsymbol{\theta}}_n + \mathbf{T}_n\mathbf{H}_n\tilde{\boldsymbol{\beta}}_n I(\mathcal{L}_n < \ell_{n,\alpha}) \tag{6.5.6a}$$

$$\text{(ii)}\ \hat{\boldsymbol{\beta}}_n^{\mathrm{PT}} = \tilde{\boldsymbol{\beta}}_n - \mathbf{H}_n\tilde{\boldsymbol{\beta}}_n I(\mathcal{L}_n < \ell_{n,\alpha}), \tag{6.5.6b}$$

where $\ell_{n,\alpha}$ is the upper α-level value of the distribution of $\mathcal{L}_n$ under H_0. Similarly, the corresponding James-Stein and positive-rule Stein estimators of $\boldsymbol{\theta}$ are defined by

$$\text{(i)}\ \hat{\boldsymbol{\theta}}_n^{S} = \tilde{\boldsymbol{\theta}}_n + c_n\mathcal{L}_n^{-1}\mathbf{T}_n\mathbf{H}_n\tilde{\boldsymbol{\beta}}_n, \tag{6.5.7}$$

$$\text{(ii)}\ \hat{\boldsymbol{\theta}}_n^{S+} = \tilde{\boldsymbol{\theta}}_n + \{1 - (1 - c_n\mathcal{L}_n^{-1})I(\mathcal{L}_n > c_n)\}\mathbf{T}_n\mathbf{H}_n\tilde{\boldsymbol{\beta}}_n,$$

and that of $\boldsymbol{\beta}$ by

$$\text{(iii)}\ \hat{\boldsymbol{\beta}}_n^{S} = \tilde{\boldsymbol{\beta}}_n - c_n\mathbf{H}_n\tilde{\boldsymbol{\beta}}_n\mathcal{L}_n^{-1},$$

$$\text{(iv)}\ \hat{\boldsymbol{\beta}}_n^{S+} = \hat{\boldsymbol{\beta}}_n + (1 - c_n\mathcal{L}_n^{-1})I(\mathcal{L}_n > c_n)\mathbf{H}_n\tilde{\boldsymbol{\beta}}_n, \tag{6.5.8}$$

where $c_n = \frac{(p-3)m}{m+2}$ and $c_n \to c = (p-3)$ as $n \to \infty$.

6.5.2 Conditions for Asymptotic Properties of the Estimators and Their Distributions

We know that if errors are independent normal with zero mean and variance σ^2, the exact distribution of $(\tilde{\boldsymbol{\theta}}_n', \tilde{\boldsymbol{\beta}}_n')'$ is a $2p$-variate normal with mean $(\boldsymbol{\theta}', \boldsymbol{\beta}')'$ and covariance matrix

$$\sigma^2\begin{pmatrix} \mathbf{D}_{11n} & -\mathbf{T}_n\mathbf{D}_{22n} \\ -\mathbf{T}_n\mathbf{D}_{22n} & \mathbf{D}_{22n} \end{pmatrix}, \tag{6.5.9}$$

where $\mathbf{D}_{11n} = \mathrm{Diag}\Big(\frac{1}{n_1} + \frac{\bar{x}_{1n_1}}{n_1Q_1}, \dots, \frac{1}{n_p} + \frac{\bar{x}_{pn_p}}{n_pQ_p}\Big)$.
Since the errors are independent but not normally distributed, we need the following regularity conditions:

(i) For some r (≥ 2),

$$\mu_r = \int_{-\infty}^{\infty} |x|^r dF(x) < \infty. \tag{6.5.10}$$

This means $\sigma^2 = \int_{-\infty}^{\infty} x^2 dF(x) - \big(\int_{-\infty}^{\infty} x dF(x)\big)^2 < \infty$.

(ii) For each α $(= 1, 2, \dots, p)$, as $n \to \infty$,

$$\lambda_{\alpha n} = \frac{n_\alpha}{n} \to \lambda_\alpha, \quad 0 < \lambda_\alpha < 1,$$

$$\bar{x}_{\alpha n_\alpha} \to \bar{x}_\alpha, \quad |\bar{x}_\alpha| < \infty,$$

$$Q_{n_\alpha} \to Q_\alpha, \quad 0 < Q_\alpha < 1$$

$$\bar{Q}_n \to \bar{Q} = \sum_{\alpha=1}^{p} \lambda_\alpha Q_\alpha, \quad 0 < \bar{Q} < 1.$$

(iii) (Noether Condition). As $n \to \infty$,

$$\max_{1<\alpha<p}\left\{\max_{1\le j\le n_\alpha}\frac{\left(x_{\alpha j}-\bar{x}_{\alpha n_\alpha}\right)^2}{n_\alpha Q_\alpha}\right\}\to 0.$$

6.5.3 Asymptotic Distributions of the Estimators

Note that the estimators $\tilde{\theta}_n$ and $\tilde{\beta}_n$, $\hat{\theta}_n$ and $\hat{\beta}_n$ are all linear estimators. Thus, under $\sigma^2 < \infty$ and the Noether conditions above, the classical central limit theorem remains applicable. Further, $s_e^2 \xrightarrow{P} \sigma^2$ as $n \to \infty$.

Theorem 1. Under $H_0 : \boldsymbol{\beta} = \beta_0\mathbf{1}_p$ and assumed the regularity conditions, we have as $n \to \infty$,

$$\text{(i) } \sqrt{n}\begin{pmatrix}\tilde{\boldsymbol{\theta}}_n-\boldsymbol{\theta}\\ \tilde{\boldsymbol{\beta}}_n-\boldsymbol{\beta}\end{pmatrix}\sim\mathcal{N}_{2p}\left\{\begin{pmatrix}\mathbf{0}\\ \mathbf{0}\end{pmatrix},\ \sigma^2\begin{pmatrix}\boldsymbol{\Lambda}_{11} & -\mathbf{T}_0\boldsymbol{\Lambda}_{22}\\ -\mathbf{T}_0\boldsymbol{\Lambda}_{22} & \boldsymbol{\Lambda}_{22}\end{pmatrix}\right\}, \quad (6.5.11)$$

where $\boldsymbol{\Lambda}_{22}^{-1} = \text{Diag}(\lambda_1 Q_1,\ldots,\lambda_p Q_p)'$, $\mathbf{T}_0 = \lim_{n\to\infty}\mathbf{T}_n = \text{Diag}(\bar{x}_1,\ldots,\bar{x}_p)$, $\boldsymbol{\Lambda}_{11} = \boldsymbol{\Lambda}_0 + \mathbf{T}_0\boldsymbol{\Lambda}_{22}\mathbf{T}_0$, $\boldsymbol{\Lambda}_{12} = \boldsymbol{\Lambda}_{21} = \mathbf{T}_0\boldsymbol{\Lambda}_{22}$;

$$\text{(ii) }\sqrt{n}\begin{pmatrix}\hat{\boldsymbol{\theta}}_n-\boldsymbol{\theta}\\ \tilde{\boldsymbol{\beta}}_n-\boldsymbol{\beta}\end{pmatrix}\sim\mathcal{N}_{2p}\left\{\begin{pmatrix}\mathbf{0}\\ \mathbf{0}\end{pmatrix};\ \sigma^2\begin{pmatrix}\boldsymbol{\Lambda}_{11}^* & \boldsymbol{\Lambda}_{12}^*\\ \boldsymbol{\Lambda}_{21}^* & \boldsymbol{\Lambda}_{22}\end{pmatrix}\right\};$$

$$\text{(iii) } \sqrt{n}\begin{pmatrix}\tilde{\boldsymbol{\beta}}_n-\boldsymbol{\beta}\\ \tilde{\boldsymbol{\beta}}_n-\hat{\boldsymbol{\beta}}_n\end{pmatrix}\sim\mathcal{N}_{2p}\left\{\begin{pmatrix}\mathbf{0}\\ \mathbf{0}\end{pmatrix};\ \sigma^2\begin{pmatrix}\boldsymbol{\Lambda}_{22} & \mathbf{J}\boldsymbol{\Lambda}_{22}\\ \mathbf{J}\boldsymbol{\Lambda}_{22} & \mathbf{J}\boldsymbol{\Lambda}_{22}\end{pmatrix}\right\};$$

$$\text{(iv) } \sqrt{n}\begin{pmatrix}\hat{\boldsymbol{\beta}}_n-\beta_0\mathbf{1}_p\\ \tilde{\boldsymbol{\beta}}_n-\hat{\boldsymbol{\beta}}_n\end{pmatrix}\sim\mathcal{N}_{2p}\left\{\begin{pmatrix}\mathbf{0}\\ \mathbf{0}\end{pmatrix};\sigma^2\begin{pmatrix}\frac{\mathbf{1}_p\mathbf{1}_p'}{Q} & \mathbf{0}\\ \mathbf{0} & \mathbf{J}\boldsymbol{\Lambda}_{22}\end{pmatrix}\right\}$$

with $\lim_{n\to\infty}\mathbf{H}_n = \mathbf{J} = \mathbf{I}_p - \frac{\mathbf{1}_p\mathbf{1}_p'\boldsymbol{\Lambda}_{22}^{-1}}{Q}$, $\boldsymbol{\Lambda}_0^{-1} = \text{Diag}\,(\lambda_1,\ldots,\lambda_p)$, $\boldsymbol{\Lambda}_{11}^* = \boldsymbol{\Lambda}_0 + \frac{\mathbf{T}_0\mathbf{1}_p\mathbf{1}_p'\mathbf{T}_0}{Q}$, $\boldsymbol{\Lambda}_{12}^* = \boldsymbol{\Lambda}_{21}^* = -\frac{1}{Q}\mathbf{1}_p\mathbf{1}_p'\mathbf{T}_0$.

The results above coupled with the well-known Cochran theorem on quadratic forms in normal variables ensures that

$$\mathcal{L}_n^* = \sigma^{-2}n\left(\tilde{\boldsymbol{\beta}}_n-\hat{\boldsymbol{\beta}}_n\right)'\mathbf{D}_{22n}^{-1}\left(\tilde{\boldsymbol{\beta}}_n-\hat{\boldsymbol{\beta}}_n\right)\xrightarrow{\mathcal{D}}\chi_{p-1}^2 \quad\text{as}\quad n\to\infty. \quad (6.5.12)$$

On the other hand, $s_e^2 \to \sigma^2$ in probability as $n \to \infty$ and using Sluskey's theorem, we conclude that $\mathcal{L}_n \xrightarrow{\mathcal{D}} \chi_{p-1}^2$ as $n \to \infty$, and $\ell_{n,\alpha} \to \chi_{p-1}^2(\alpha)$ as $n \to \infty$.

Now, consider the fixed alternatives $H_{\boldsymbol{\delta}} : \boldsymbol{\beta} = \beta_0\mathbf{1}_p + \boldsymbol{\delta}$. We see that

$$\sqrt{n}(\tilde{\boldsymbol{\beta}}_n - \beta_0\mathbf{1}_p) = \sqrt{n}(\tilde{\boldsymbol{\beta}}_n - \boldsymbol{\beta}) + \sqrt{n}\boldsymbol{\delta}. \quad (6.5.13a)$$

Hence,

$$\sqrt{n}(\tilde{\boldsymbol{\beta}}_n - \hat{\boldsymbol{\beta}}_n) = \mathbf{H}_n\sqrt{n}(\tilde{\boldsymbol{\beta}}_n - \boldsymbol{\beta}) + \mathbf{H}_n(\sqrt{n}\boldsymbol{\delta}), \quad (6.5.13b)$$

where

$$\mathbf{H}_n\sqrt{n}(\tilde{\boldsymbol{\beta}}_n - \boldsymbol{\beta}) \sim \mathcal{N}(\mathbf{0}, \sigma^2 \mathbf{J}\boldsymbol{\Lambda}_{22}), \tag{6.5.13c}$$

since $\mathbf{H}_n = \mathbf{J} + O(\frac{1}{n})$ and $\mathbf{J} = \mathbf{I}_p - \dfrac{\mathbf{1}_p\mathbf{1}_p'\boldsymbol{\Lambda}_{22}^{-1}}{\bar{Q}}$. Further, $n\|\mathbf{J}\boldsymbol{\delta}\|^2 \to \infty$ as $n \to \infty$; hence, it follows that $\mathcal{L}_n \to \infty$ as $n \to \infty$. Consequently, $P(\mathcal{L}_n > k) \to 1$ as $n \to \infty$ for all $k \in \mathcal{R}_1$.

Furthermore, under a fixed alternative $H_{\boldsymbol{\delta}} : \boldsymbol{\beta} = \beta_0 \mathbf{1}_p + \boldsymbol{\delta}$,

$$\begin{aligned}
\text{(i)} \quad & \lim_{n\to\infty} E\{n(\hat{\boldsymbol{\beta}}_n^{\mathrm{PT}} - \tilde{\boldsymbol{\beta}}_n)'\mathbf{D}_{22n}^{-1}(\hat{\boldsymbol{\beta}}_n^{\mathrm{PT}} - \tilde{\boldsymbol{\beta}}_n)\} \\
&= \lim_{n\to\infty} E\{n(\tilde{\boldsymbol{\beta}}_n - \hat{\boldsymbol{\beta}}_n)'\mathbf{D}_{22n}^{-1}(\tilde{\boldsymbol{\beta}}_n - \hat{\boldsymbol{\beta}}_n) I(\mathcal{L}_n < \ell_{n,\alpha})\} \\
&= \lim_{n\to\infty} E(s_e^2) \lim_{n\to\infty} E\{\mathcal{L}_n I(\mathcal{L}_n < \ell_{n,\alpha})\} \\
&= \sigma^2 \lim_{n\to\infty} E\{\mathcal{L}_n I(\mathcal{L}_n < \ell_{n,\alpha})\} = 0
\end{aligned} \tag{6.5.14a}$$

by the consistency of the test-statistic $\mathcal{L}_n$;

$$\begin{aligned}
\text{(ii)} \quad & \lim_{n\to\infty} E\{n(\hat{\boldsymbol{\beta}}_n^{S} - \tilde{\boldsymbol{\beta}}_n)'\mathbf{D}_{22n}^{-1}(\hat{\boldsymbol{\beta}}_n^{S} - \tilde{\boldsymbol{\beta}}_n)\} \\
&= c^2\sigma^2 \lim_{n\to\infty} E\{s_e^2 n(\tilde{\boldsymbol{\beta}}_n - \hat{\boldsymbol{\beta}}_n)'\mathbf{D}_{22n}^{-1}(\tilde{\boldsymbol{\beta}}_n - \hat{\boldsymbol{\beta}}_n)\mathcal{L}_n^{-2}\} \\
&= c^2\sigma^2 \lim_{n\to\infty} E(\mathcal{L}_n^{-1}) = 0, \quad \text{where } c = (p-3),
\end{aligned} \tag{6.5.14b}$$

since $\mathcal{L}_n \geq 0$ and $\mathcal{L}_n \to \infty$ in probability;

$$\begin{aligned}
\text{(iii)} \quad & \lim_{n\to\infty} E\{n s_e^2 s_e^{-2}(\hat{\boldsymbol{\beta}}_n^{S+} - \hat{\boldsymbol{\beta}}^{S})'\mathbf{D}_{22n}^{-1}(\hat{\boldsymbol{\beta}}_n^{S+} - \hat{\boldsymbol{\beta}}^{S})\} \\
&= \sigma^2 \lim_{n\to\infty} E(\mathcal{L}_n^{-2}(1 - c\mathcal{L}_n^{-1})^2 I(\mathcal{L}_n < c)) = 0.
\end{aligned} \tag{6.5.14c}$$

Hence, $\tilde{\boldsymbol{\beta}}_n$, $\hat{\boldsymbol{\beta}}_n^{\mathrm{PT}}$, $\hat{\boldsymbol{\beta}}_n^{S}$, and $\hat{\boldsymbol{\beta}}_n^{S+}$ are risk equivalent under fixed alternatives.

Further, computing the asymptotic dispersion matrix of $\hat{\boldsymbol{\beta}}_n$, we obtain

$$\lim_{n\to\infty} E\{n(\hat{\boldsymbol{\beta}}_n - \boldsymbol{\beta})(\hat{\boldsymbol{\beta}}_n - \boldsymbol{\beta})'\} = \frac{\mathbf{1}_p\mathbf{1}_p'}{\bar{Q}} + n\boldsymbol{\delta}\boldsymbol{\delta}'. \tag{6.5.15}$$

Note that $n\|\boldsymbol{\delta}\|^2 \to \infty$ as $n \to \infty$. $\sqrt{n}(\hat{\boldsymbol{\beta}}_n - \boldsymbol{\beta})$ has unbounded dispersion matrix. Hence, the asymptotic distribution of $\tilde{\boldsymbol{\beta}}_n$ is degenerate. Thus, for any fixed alternatives $H_{\boldsymbol{\delta}} : \boldsymbol{\beta} = \beta_0\mathbf{1}_p + \boldsymbol{\delta}$ asymptotic distribution of $\hat{\boldsymbol{\beta}}_n^{\mathrm{PT}}$, $\hat{\boldsymbol{\beta}}_n^{S}$ and $\hat{\boldsymbol{\beta}}_n^{S+}$ is the same as that of $\sqrt{n}(\tilde{\boldsymbol{\beta}}_n - \boldsymbol{\beta})$, and that of $\hat{\boldsymbol{\beta}}_n$ is degenerate.

To obtain a meaningful asymptotic distribution of the estimators, we confine ourselves to the **local alternatives** defined by

$$K_{(n)} : \boldsymbol{\beta}_{(n)} = \beta_0\mathbf{1}_p + n^{-1/2}\boldsymbol{\delta}, \tag{6.5.16}$$

where $\boldsymbol{\delta} = (\delta_1, \ldots, \delta_p)'$ is fixed, $\beta_0 \in \mathcal{R}_1$, and $\delta_1, \cdots, \delta_p$ are not equal such that $\sum_{\alpha=1}^{p} \lambda_\alpha Q_\alpha \delta_\alpha / Q = 0$.

The following theorem: gives the asymptotic distribution of various estimators under $\{K_{(n)}\}$:

Theorem 2. Under $\{K_{(n)}\}$ and assumed regularity conditions of Section 6.5.3, as $n \to \infty$,

(i) $\sqrt{n}\begin{pmatrix} \tilde{\boldsymbol{\theta}}_n - \boldsymbol{\theta} \\ \tilde{\boldsymbol{\beta}}_n - \boldsymbol{\beta}_{(n)} \end{pmatrix} \sim \mathcal{N}_{2p}\left\{\begin{pmatrix} \mathbf{0} \\ \mathbf{0} \end{pmatrix}, \sigma^2 \begin{pmatrix} \boldsymbol{\Lambda}_{11} & -\mathbf{T}_0\boldsymbol{\Lambda}_{22} \\ -\mathbf{T}_0\boldsymbol{\Lambda}_{22} & \boldsymbol{\Lambda}_{22} \end{pmatrix}\right\}$;

(ii) $\sqrt{n}\begin{pmatrix} \hat{\boldsymbol{\theta}}_n - \boldsymbol{\theta} \\ \tilde{\boldsymbol{\beta}}_n - \boldsymbol{\beta}_{(n)} \end{pmatrix} \sim \mathcal{N}_{2p}\left\{\begin{pmatrix} \mathbf{T}_0\mathbf{J}\boldsymbol{\delta} \\ \mathbf{0} \end{pmatrix}, \sigma^2 \begin{pmatrix} \boldsymbol{\Lambda}_{11}^* & \boldsymbol{\Lambda}_{12}^* \\ \boldsymbol{\Lambda}_{21}^* & \boldsymbol{\Lambda}_{22} \end{pmatrix}\right\}$;

(iii) $\sqrt{n}\begin{pmatrix} \tilde{\boldsymbol{\beta}}_n - \boldsymbol{\beta}_{(n)} \\ \tilde{\boldsymbol{\beta}}_n - \hat{\boldsymbol{\beta}}_n \end{pmatrix} \sim \mathcal{N}_{2p}\left\{\begin{pmatrix} \mathbf{0} \\ \mathbf{J}\boldsymbol{\delta} \end{pmatrix}, \sigma^2 \begin{pmatrix} \boldsymbol{\Lambda}_{22} & -\mathbf{J}\boldsymbol{\Lambda}_{22} \\ -\mathbf{J}\boldsymbol{\Lambda}_{22} & \mathbf{J}\boldsymbol{\Lambda}_{22} \end{pmatrix}\right\}$;

(iv) $\mathcal{L}_n$ has asymptotically a noncentral chi-square distribution with $p-1$ d.f. and noncentrality parameter $\Delta^2/2$, where

$$\Delta^2 = \sigma^{-2}(\boldsymbol{\delta}'\mathbf{J}'\boldsymbol{\Lambda}_{22}^{-1}\mathbf{J}\boldsymbol{\delta}) = \sigma^{-2}(\boldsymbol{\delta}'\boldsymbol{\Lambda}_{22}^{-1}\boldsymbol{\delta}), \tag{6.5.17}$$

since $\mathbf{J}\boldsymbol{\delta} = \boldsymbol{\delta}$ from (6.5.16).

Now, by virtue of the definition of the PTE of $\boldsymbol{\beta}$ and $\boldsymbol{\theta}$, we have

$$\sqrt{n}(\hat{\boldsymbol{\beta}}_n^{\mathrm{PT}} - \boldsymbol{\beta}) = \sqrt{n}(\tilde{\boldsymbol{\beta}}_n - \boldsymbol{\beta}) - \sqrt{n}(\tilde{\boldsymbol{\beta}}_n - \hat{\boldsymbol{\beta}}_n)I(\mathcal{L}_n < \ell_{n,\alpha})$$

and

$$\sqrt{n}(\hat{\boldsymbol{\theta}}_n^{\mathrm{PT}} - \boldsymbol{\theta}) = \sqrt{n}(\tilde{\boldsymbol{\theta}}_n - \boldsymbol{\theta}) + \mathbf{T}_n\sqrt{n}(\tilde{\boldsymbol{\beta}}_n - \hat{\boldsymbol{\beta}}_n)I(\mathcal{L}_n < \ell_{n,\alpha}). \tag{6.5.18}$$

Therefore, using the fact that $\ell_{n,\alpha} \to \chi^2_{p-1}(\alpha)$, $\mathbf{T}_n \to \mathbf{T}_0$ and Theorem 6.6.1, we obtain the following theorem:

Theorem 3. Under $\{K_{(n)}\}$ and assumed regularity conditions, the cdf of the estimates of $\boldsymbol{\beta}$ and $\boldsymbol{\theta}$ are given by

$$\begin{aligned}
\text{(i)}\quad G_p^{PT(\boldsymbol{\beta})}(\mathbf{x}, \boldsymbol{\delta}) &= \lim_{n\to\infty} P\{\sqrt{n}(\hat{\boldsymbol{\beta}}_n^{\mathrm{PT}} - \boldsymbol{\beta}) \le \mathbf{x} | K_{(n)}\} \\
&= \lim_{n\to\infty} P\{\sqrt{n}(\hat{\boldsymbol{\beta}}_n - \boldsymbol{\beta}) \le \mathbf{x}\,; \mathcal{L}_n < \ell_{n,\alpha} | K_{(n)}\} \\
&\quad \lim_{n\to\infty} P\{\sqrt{n}(\tilde{\boldsymbol{\beta}}_n - \boldsymbol{\beta}) \le \mathbf{x}\,; \mathcal{L}_n \ge \ell_{n,\alpha} | K_{(n)}\} \\
&= \Phi_p(\mathbf{x} - \boldsymbol{\delta}; \mathbf{0}, \sigma^2\mathbf{B})H_{p-1}(\chi^2_{p-1}(\alpha); \Delta^2) \\
&\quad + \int_{E(\boldsymbol{\delta})} \Phi_p(\mathbf{x} - \mathbf{J}\boldsymbol{\delta}; \mathbf{0}, \sigma^2\mathbf{J}\boldsymbol{\Lambda}_{22})dG_p(\mathbf{Z}; \mathbf{0}, \sigma^2\boldsymbol{\Lambda}_{22}),
\end{aligned} \tag{6.5.19a}$$

where $\mathbf{B} = \frac{1}{Q}\mathbf{1}_p\mathbf{1}_p'$;

$$\text{(ii)}\quad G_p^{PT(\boldsymbol{\theta})}(\mathbf{x},\boldsymbol{\delta}) = \lim_{n\to\infty} P\Big\{\sqrt{n}(\hat{\boldsymbol{\theta}}_n^{\rm PT} - \boldsymbol{\theta}) \le \mathbf{x}\Big|K_{(n)}\Big\}$$

$$= \Phi_p(\mathbf{x} - \mathbf{T}_0\mathbf{J}\boldsymbol{\delta}; \mathbf{0}, \sigma^2\boldsymbol{\Lambda}_{11}^*)H_{p-1}(\chi_{p-1}^2(\alpha);\Delta^2)$$

$$+ \int_{E(\boldsymbol{\delta})} \Phi_p(\mathbf{x} + \mathbf{J}\mathbf{z}; \mathbf{0}, \sigma^2\boldsymbol{\Lambda}_0)dG_p(\mathbf{Z}; \mathbf{0}, \sigma^2\boldsymbol{\Lambda}_{22}), \qquad (6.5.19b)$$

where $$E(\boldsymbol{\delta}) = \Big\{\mathbf{Z}; \sigma^{-2}(\mathbf{Z} + \mathbf{J}\boldsymbol{\delta})'\boldsymbol{\Lambda}_{22}^{-1}(\mathbf{Z} + \mathbf{J}\boldsymbol{\delta}) \ge \chi_{p-1}^2(\alpha)\Big\}. \qquad (6.5.19c)$$

$\Phi_p(\mathbf{x}; \boldsymbol{\mu}, \boldsymbol{\Sigma})$ is the p-variate normal distribution with mean $\boldsymbol{\mu}$ and covariance matrix $\boldsymbol{\Sigma}$, and $H_m(\cdot;\Delta^2)$ is the cdf of a noncentral chi-square distribution function with m d.f. and noncentrality parameter $\Delta^2/2$.

The asymptotic distributions of $\sqrt{n}(\hat{\boldsymbol{\beta}}_n^S - \boldsymbol{\beta})$ and $\sqrt{n}(\hat{\boldsymbol{\beta}}_n^{S+} - \boldsymbol{\beta})$ are difficult to obtain but the asymptotic bias, MSE matrix, and weighted quadratic risk function can be computed writing

$$\sqrt{n}\mathbf{D}_{22n}^{-1/2}(\hat{\boldsymbol{\beta}}_n^S - \beta_0\mathbf{1}_p) = \sqrt{n}\mathbf{D}_{22n}^{-1/2}(\tilde{\boldsymbol{\beta}}_n - \beta_0\mathbf{1}_p) - c\sqrt{n}\mathbf{D}_{22n}^{-1/2}(\tilde{\boldsymbol{\beta}}_n - \hat{\boldsymbol{\beta}}_n)\mathcal{L}_n^{-1}$$

and

$$\sqrt{n}\,\mathbf{D}_{22n}^{-1/2}(\hat{\boldsymbol{\beta}}_n^{S+} - \beta_0\mathbf{1}_p) = \sqrt{n}\,\mathbf{D}_{22n}^{-1/2}(\tilde{\boldsymbol{\beta}}_n - \beta_0\mathbf{1}_p) - c\sqrt{n}\mathbf{D}_{22n}^{-1/2}(\tilde{\boldsymbol{\beta}}_n - \hat{\boldsymbol{\beta}}_n)\mathcal{L}_n^{-1}$$

$$-\{(1 - c\mathcal{L}_n^{-1})I(\mathcal{L}_n < c)\}\sqrt{n}\,\mathbf{D}_{22n}^{-1/2}(\tilde{\boldsymbol{\beta}}_n - \hat{\boldsymbol{\beta}}_n). \qquad (6.5.20a)$$

Similarly, we can write expressions for

$$\sqrt{n}\,\mathbf{D}_{11}^{-1/2}(\hat{\boldsymbol{\theta}}_n^{\rm PT} - \boldsymbol{\theta}),\ \sqrt{n}\,\mathbf{D}_{11}^{-1/2}(\hat{\boldsymbol{\theta}}_n^S - \boldsymbol{\theta}) \text{ and } \sqrt{n}\,\mathbf{D}_{11}^{-1/2}(\hat{\boldsymbol{\theta}}_n^{S+} - \boldsymbol{\theta}). \qquad (6.5.20b)$$

6.5.4 Expressions for ADB, ADQB, ADMSE, and ADQR of the Estimators

Note that under (6.5.1) and (6.5.2) and the regularity conditions (6.5.10i-iii), we can obtain the asymptotic distributional bias, MSE matrices, and the weighted risk expressions as follows:

Unrestricted Estimators.

$$\text{(i)}\ \mathbf{b}_1(\tilde{\boldsymbol{\beta}}_n) = \lim_{n\to\infty} E\Big\{\sqrt{n}\Big(\tilde{\boldsymbol{\beta}}_n - \boldsymbol{\beta}\Big)|K_{(n)}\Big\} = \mathbf{0} \text{ and } B_1(\tilde{\boldsymbol{\beta}}_n) = 0$$

$$\text{(ii)}\ \mathbf{M}_1(\tilde{\boldsymbol{\beta}}_n) = \lim_{n\to\infty} E\Big\{n\Big(\tilde{\boldsymbol{\beta}}_n - \boldsymbol{\beta}\Big)\Big(\tilde{\boldsymbol{\beta}}_n - \boldsymbol{\beta}\Big)'|K_{(n)}\Big\} = \sigma^2\boldsymbol{\Lambda}_{22}$$

$$\text{and } R_1(\tilde{\boldsymbol{\beta}}_n; \mathbf{W}) = \sigma^2\text{tr}\Big[\mathbf{W}\boldsymbol{\Lambda}_{22}\Big]. \qquad (6.5.21a)$$

Similarly,

(i) $\mathbf{b}_1(\tilde{\boldsymbol{\theta}}_n) = \lim_{n\to\infty} E\Big\{\sqrt{n}\Big(\tilde{\boldsymbol{\theta}}_n - \boldsymbol{\theta}\Big)|K_{(n)}\Big\} = \mathbf{0}$ and $B_1(\tilde{\boldsymbol{\theta}}_n) = 0$;

(ii)
$$\mathbf{M}_1(\tilde{\boldsymbol{\theta}}_n) = \lim_{n\to\infty} E\Big\{n\Big(\tilde{\boldsymbol{\theta}}_n - \boldsymbol{\theta}\Big)\Big(\tilde{\boldsymbol{\theta}}_n - \boldsymbol{\theta}\Big)'|K_{(n)}\Big\}$$
$$= \sigma^2\Big(\boldsymbol{\Lambda}_0 + \mathbf{T}_0\boldsymbol{\Lambda}_{22}\mathbf{T}_0\Big) = \sigma^2\boldsymbol{\Lambda}_{11}$$

and $R_1(\tilde{\boldsymbol{\theta}}_n;\mathbf{W}) = \sigma^2\mathrm{tr}\Big[\mathbf{W}\Big(\boldsymbol{\Lambda}_0 + \mathbf{T}_0\boldsymbol{\Lambda}_{22}\mathbf{T}_0\Big)\Big] = \sigma^2\mathrm{tr}\Big[\mathbf{W}\boldsymbol{\Lambda}_{11}\Big]$. (6.5.21b)

Restricted Estimators.

(i) $\mathbf{b}_2(\hat{\boldsymbol{\beta}}_n) = \lim_{n\to\infty} E\Big\{\sqrt{n}\Big(\hat{\boldsymbol{\beta}}_n - \boldsymbol{\beta}\Big)|K_{(n)}\Big\} = -\mathbf{J}\boldsymbol{\delta}$

and $B_2(\hat{\boldsymbol{\beta}}_n) = \sigma^{-2}\Big(\boldsymbol{\delta}'\mathbf{J}'\boldsymbol{\Lambda}_{22}^{-1}\mathbf{J}\boldsymbol{\delta}\Big) = \Delta^2$;

(ii) $\mathbf{M}_2(\hat{\boldsymbol{\beta}}_n) = \lim_{n\to\infty} E\Big\{n\Big(\hat{\boldsymbol{\beta}}_n - \boldsymbol{\beta}\Big)\Big(\hat{\boldsymbol{\beta}}_n - \boldsymbol{\beta}\Big)'|K_{(n)}\Big\} = \sigma^2\frac{\mathbf{1}_p\mathbf{1}_p'}{\bar{Q}} + \boldsymbol{\delta}\boldsymbol{\delta}'$

and $R_2(\hat{\boldsymbol{\beta}}_n;\mathbf{W}) = \sigma^2\frac{\mathbf{1}_p\mathbf{W}\mathbf{1}_p'}{\bar{Q}} + \boldsymbol{\delta}'\mathbf{W}\boldsymbol{\delta}$. (6.5.22a)

Similarly,

(i) $\mathbf{b}_2(\hat{\boldsymbol{\theta}}_n) = \lim_{n\to\infty} E\Big\{\sqrt{n}(\hat{\boldsymbol{\theta}}_n - \boldsymbol{\theta})|K_{(n)}\Big\} = \mathbf{T}_0\mathbf{J}\boldsymbol{\delta}$

and $B_2(\hat{\boldsymbol{\theta}}_n) = \sigma^{-2}(\boldsymbol{\delta}'\mathbf{J}'\mathbf{T}_0\boldsymbol{\Lambda}_{22}^{-1}\mathbf{T}_0\mathbf{J}\boldsymbol{\delta}) = \Delta^{*2}$;

(ii)
$$\mathbf{M}_2(\hat{\boldsymbol{\theta}}_n) = \lim_{n\to\infty} E\{n(\hat{\boldsymbol{\theta}}_n - \boldsymbol{\theta})(\hat{\boldsymbol{\theta}}_n - \boldsymbol{\theta})'|K_{(n)}\}$$
$$= \sigma^2(\boldsymbol{\Lambda}_{11}^* + \sigma^{-2}\mathbf{T}_0\mathbf{J}\boldsymbol{\delta}\boldsymbol{\delta}'\mathbf{J}'\mathbf{T}_0)$$

and $R_2(\hat{\boldsymbol{\theta}}_n;\mathbf{W}) = \sigma^2\mathrm{tr}[\mathbf{W}\boldsymbol{\Lambda}_{11}^*] + (\boldsymbol{\delta}'\mathbf{J}'\mathbf{T}_0\mathbf{W}\mathbf{T}_0\mathbf{J}\boldsymbol{\delta})$. (6.5.22b)

Preliminary Test Estimators.

(i) $\mathbf{b}_3(\hat{\boldsymbol{\beta}}_n^{\mathrm{PT}}) = -\mathbf{J}\boldsymbol{\delta}H_{p+1}(\chi^2_{p-1}(\alpha);\Delta^2)$ and

$B_3(\hat{\boldsymbol{\beta}}_n^{\mathrm{PT}}) = \Delta^2\{H_{p+1}(\chi^2_{p-1}(\alpha);\Delta^2)\}^2$;

(ii)
$$\mathbf{M}_3(\hat{\boldsymbol{\beta}}_n^{\mathrm{PT}}) = \sigma^2\boldsymbol{\Lambda}_{22} - \sigma^2\mathbf{J}\boldsymbol{\Lambda}_{22}H_{p+1}(\chi^2_{p-1}(\alpha);\Delta^2)$$
$$+(\mathbf{T}_0\mathbf{J}\boldsymbol{\delta}\boldsymbol{\delta}'\mathbf{J}'\mathbf{T}_0)\{2H_{p+1}(\chi^2_{p-1}(\alpha);\Delta^2) - H_{p+3}(\chi^2_{p-1}(\alpha);\Delta^2)\};$$

and

$$R_3(\hat{\boldsymbol{\beta}}_n^{\mathrm{PT}};\mathbf{W}) = \sigma^2\mathrm{tr}[\mathbf{W}\boldsymbol{\Lambda}_{22}] - \sigma^2\mathrm{tr}[\mathbf{W}\mathbf{J}\boldsymbol{\Lambda}_{22}]H_{p+1}(\chi^2_{p-1}(\alpha);\Delta^2)$$
$$+ (\boldsymbol{\delta}'\mathbf{J}'\mathbf{T}_0\mathbf{W}\mathbf{T}_0\mathbf{J}\boldsymbol{\delta})\{2H_{p+1}(\chi^2_{p-1}(\alpha);\Delta^2) - H_{p+3}(\chi^2_{p-1}(\alpha);\Delta^2)\}. \quad (6.5.23a)$$

Similarly,

(i) $\mathbf{b}_3(\hat{\boldsymbol{\theta}}_n^{\mathrm{PT}}) = \mathbf{T}_0\mathbf{J}\boldsymbol{\delta}H_{p+1}(\chi^2_{p-1}(\alpha);\Delta^2)$ and

$$B_3(\hat{\boldsymbol{\theta}}_n^{\mathrm{PT}}) = \sigma^{-2}(\boldsymbol{\delta}'\mathbf{J}'\mathbf{T}_0\boldsymbol{\Lambda}_{22}^{-1}\mathbf{T}_0\mathbf{J}\boldsymbol{\delta})\{H_{p+1}(\chi^2_{p-1}(\alpha);\Delta^2)\}^2;$$

(ii) $$\mathbf{M}_3(\hat{\boldsymbol{\theta}}_n^{\mathrm{PT}}) = \sigma^2\boldsymbol{\Lambda}_{11} - \sigma^2(\boldsymbol{\Lambda}_{11} - \boldsymbol{\Lambda}_{11}^*)H_{p+1}(\chi^2_{p-1}(\alpha);\Delta^2) + (\mathbf{T}_0\mathbf{J}\boldsymbol{\delta}\boldsymbol{\delta}'\mathbf{J}'\mathbf{T}_0)\{2H_{p+1}(\chi^2_{p-1}(\alpha);\Delta^2) - H_{p+3}(\chi^2_{p-1}(\alpha);\Delta^2)\}$$

and

$$R_3(\hat{\boldsymbol{\theta}}_n^{\mathrm{PT}};\mathbf{W}) = \sigma^2\mathrm{tr}[\mathbf{W}\boldsymbol{\Lambda}_{11}] - \sigma^2\mathrm{tr}[\mathbf{W}(\boldsymbol{\Lambda}_{11} - \boldsymbol{\Lambda}_{11}^*)]H_{p+1}(\chi^2_{p-1}(\alpha);\Delta^2) + (\boldsymbol{\delta}'\mathbf{J}'\mathbf{T}_0\mathbf{W}\mathbf{T}_0\mathbf{J}\boldsymbol{\delta})\{2H_{p+1}(\chi^2_{p-1}(\alpha);\Delta^2) - H_{p+3}(\chi^2_{p-1}(\alpha);\Delta^2)\}. \quad (6.5.23b)$$

James-Stein-Type Estimators.

(i) $\mathbf{b}_4(\hat{\boldsymbol{\beta}}_n^S) = -c\boldsymbol{\delta}E[\chi^{-2}_{p+1}(\Delta^2)]$ and $B_4(\hat{\boldsymbol{\beta}}_n^S) = c^2\Delta^2\{E[\chi^{-2}_{p+1}(\Delta^2)]\}^2;$

(ii) $$\mathbf{M}_4(\hat{\boldsymbol{\beta}}_n^S) = \sigma^2\boldsymbol{\Lambda}_{22} - \sigma^2(p-3)\mathbf{J}\boldsymbol{\Lambda}_{22}\{2E[\chi^{-2}_{p+1}(\Delta^2)] - (p-3)E[\chi^{-4}_{p+1}(\Delta^2)]\} + (p-3)(p-1)(\mathbf{T}_0\mathbf{J}\boldsymbol{\delta}\boldsymbol{\delta}'\mathbf{J}'\mathbf{T}_0)E[\chi^{-4}_{p+3}(\Delta^2)];$$

and

$$\begin{aligned} R_4(\hat{\boldsymbol{\beta}}_n^S;\mathbf{W}) = {} & \sigma^2\mathrm{tr}[\mathbf{W}\boldsymbol{\Lambda}_{22}] - \sigma^2(p-3)\mathrm{tr}[\mathbf{W}\mathbf{J}\boldsymbol{\Lambda}_{22}] \\ & \times\{2E[\chi^{-2}_{p+1}(\Delta^2)] - (p-3)E[\chi^{-4}_{p+1}(\Delta^2)]\} \\ & +(p-3)(p-1)(\boldsymbol{\delta}'\mathbf{J}'\mathbf{T}_0\mathbf{W}\mathbf{T}_0\mathbf{J}\boldsymbol{\delta})E[\chi^{-4}_{p+3}(\Delta^2)]. \quad (6.5.24a) \end{aligned}$$

Similarly,

(i) $\mathbf{b}_4(\hat{\boldsymbol{\theta}}_n^S) = -c\mathbf{T}_0\mathbf{J}\boldsymbol{\delta}E[\chi^{-2}_{p+1}(\Delta^2)]$

and $B_4(\hat{\boldsymbol{\theta}}_n^S) = c^2(\boldsymbol{\delta}'\mathbf{J}'\mathbf{T}_0\boldsymbol{\Lambda}_{22}^{-1}\mathbf{T}_0\mathbf{J}\boldsymbol{\delta})\{E[\chi^{-2}_{p+1}(\Delta^2)]\}^2;$

(ii) $$\mathbf{M}_4(\hat{\boldsymbol{\theta}}_n^S) = \sigma^2\boldsymbol{\Lambda}_{11} - \sigma^2(p-3)(\mathbf{T}_0\mathbf{J}\boldsymbol{\Lambda}_{22}\mathbf{J}'\mathbf{T}_0)\{2E[\chi^{-2}_{p+1}(\Delta^2)] - (p-3)E[\chi^{-4}_{p+1}(\Delta^2)]\} + (p-3)(p-1)(\mathbf{T}_0\mathbf{J}\boldsymbol{\delta}\boldsymbol{\delta}'\mathbf{J}'\mathbf{T}_0)E[\chi^{-4}_{p+3}(\Delta^2)];$$

and

$$\begin{aligned} R_4(\hat{\boldsymbol{\theta}}_n^S;\mathbf{W}) &= \sigma^2\mathrm{tr}[\mathbf{W}\boldsymbol{\Lambda}_{11}] - \sigma^2(p-3)\mathrm{tr}[\mathbf{W}(\mathbf{T}_0\mathbf{J}\boldsymbol{\Lambda}_{22}\mathbf{J}'\mathbf{T}_0)] \\ & \times\{2E[\chi^{-2}_{p+1}(\Delta^2)] - (p-3)E[\chi^{-4}_{p+1}(\Delta^2)]\} \\ & +(p-3)(p-1)(\boldsymbol{\delta}'\mathbf{J}'\mathbf{T}_0\mathbf{W}\mathbf{T}_0\mathbf{J}\boldsymbol{\delta})E[\chi^{-4}_{p+3}(\Delta^2)]. \quad (6.5.24b) \end{aligned}$$

Positive-Rule Stein Estimators.

(i) $\mathbf{b}_5(\hat{\boldsymbol{\beta}}_n^{S+}) = -\mathbf{J}\boldsymbol{\delta}\{H_{p+1}(p-3;\Delta^2) + (p-3)E[\chi^{-2}_{p+1}(\Delta^2)I(\chi^2_{p+1}(\Delta^2) > p-3)]\}$

and $B_5(\hat{\boldsymbol{\beta}}_n^{S+}) = \Delta^2\{H_{p+1}(p-3;\Delta^2) + (p-3)E[\chi_{p+1}^{-2}(\Delta^2)I(\chi_{p+1}^2(\Delta^2) > p-3)]\}^2$;

$$\begin{aligned}
\text{(ii)}\ \ \mathbf{M}_5(\hat{\boldsymbol{\beta}}_n^{S+}) &= \mathbf{M}_4(\hat{\boldsymbol{\beta}}_n^{S})\\
&-\sigma^2\mathbf{J}\boldsymbol{\Lambda}_{22}E\big[\big(1-(p-3)\chi_{p+1}^{-2}(\Delta^2)\big)^2 I\big(\chi_{p+1}^2(\Delta^2) < p-3\big)\big]\\
&+\mathbf{T}_0\mathbf{J}\boldsymbol{\delta}\boldsymbol{\delta}'\mathbf{J}\mathbf{T}_n - \big\}2E\big[\big((p-3)\chi_{p+3}^{-2}(\Delta^2)-1\big)I\big(\chi_{p+3}^2(\Delta^2) < p-3\big)\big]\\
&-E\big[\big(1-(p-3)\chi_{p+3}^{-2}(\Delta^2)\big)^2 I\big(\chi_{p+3}^2(\Delta^2) < p-3\big)\big]\big\}
\end{aligned}$$

and

$$\begin{aligned}
&R_5(\hat{\boldsymbol{\beta}}_n^{S+};\mathbf{W}) = \mathbf{R}_4(\hat{\boldsymbol{\beta}}_n^{S})\\
&-\sigma^2\mathrm{tr}[\mathbf{W}\mathbf{J}\boldsymbol{\Lambda}_{22}]E\big[\big(1-(p-3)\chi_{p+1}^{-2}(\Delta^2)\big)^2 I\big(\chi_{p+1}^2(\Delta^2) < p-3\big)\big]\\
&+(\boldsymbol{\delta}'\mathbf{J}'\mathbf{T}_0\mathbf{W}\mathbf{T}_0\mathbf{J}\boldsymbol{\delta})\big\{2E\big[\big((p-3)\chi_{p+3}^{-2}(\Delta^2)-1\big)I\big(\chi_{p+3}^2(\Delta^2) < p-3\big)\big]\\
&-E\big[\big(1-(p-3)\chi_{p+3}^{-2}(\Delta^2)\big)^2 I\big(\chi_{p+3}^2(\Delta^2) < p-3\big)\big]\big\}. \qquad (6.5.25a)
\end{aligned}$$

Similarly,

(i) $\mathbf{b}_5(\hat{\boldsymbol{\theta}}_n^{S+}) = \mathbf{T}_0\mathbf{J}\boldsymbol{\delta}E\big[\big(1-(p-3)\chi_{p+1}^{-2}(\Delta^2)\big)I\big(\chi_{p+1}^2(\Delta^2) > p-3\big)\big]$

and
$$\begin{aligned}
B_5(\hat{\boldsymbol{\theta}}_n^{S+}) &= \sigma^{-2}(\boldsymbol{\delta}'\mathbf{J}'\mathbf{T}_0\boldsymbol{\Lambda}_{22}^{-1}\mathbf{T}_0\mathbf{J}\boldsymbol{\delta})\{H_{p+1}(p-3;\Delta^2)\\
&\qquad +E\big[\big(1-(p-3)\chi_{p+1}^{-2}(\Delta^2)\big)I\big(\chi_{p+1}^2(\Delta^2) > p-3\big)\big]\}^2;
\end{aligned}$$

$$\begin{aligned}
\text{(ii)}\ \mathbf{M}_5(\hat{\boldsymbol{\theta}}_n^{S+}) &= \mathbf{M}_4(\hat{\boldsymbol{\theta}}_n^{S}) - \sigma^2(\mathbf{T}_0\mathbf{J}\boldsymbol{\Lambda}_{22}\mathbf{J}\mathbf{T}_0)\\
&\times E\big[\big(1-(p-3)\chi_{p+1}^{-2}(\Delta^2)\big)^2 I\big(\chi_{p+1}^2(\Delta^2) > p-3\big)\big]\\
&+(\mathbf{T}_0\mathbf{J}\boldsymbol{\delta}\boldsymbol{\delta}'\mathbf{J}'\mathbf{T}_0)\big\{2E\big[\big(1-(p-3)\chi_{p+3}^{-2}(\Delta^2)\big)I\big(\chi_{p+3}^2(\Delta^2) > p-3\big)\big]\\
&\times E\big[\big(1-(p-3)\chi_{p+3}^{-2}(\Delta^2)\big)^2 I\big(\chi_{p+3}^2(\Delta^2) > p-3\big)\big]\big\}
\end{aligned}$$

and

$$\begin{aligned}
&R_5(\hat{\boldsymbol{\theta}}_n^{S+};\mathbf{W}) = \mathbf{R}_4(\hat{\boldsymbol{\theta}}_n^{S}) - \sigma^2\,\mathrm{tr}[\mathbf{W}(\mathbf{T}_0\mathbf{J}\boldsymbol{\Lambda}_{22}\mathbf{J}'\mathbf{T}_0')]\\
&\times E\big[\big(1-(p-3)\chi_{p+1}^{-2}(\Delta^2)\big)^2 I\big(\chi_{p+1}^2(\Delta^2) > p-3\big)\big]\\
&+(\boldsymbol{\delta}'\mathbf{J}'\mathbf{T}_0\mathbf{W}\mathbf{T}_0\mathbf{J}\boldsymbol{\delta})\big\{2E\big[\big(1-(p-3)\chi_{p+3}^{-2}(\Delta^2)\big)I\big(\chi_{p+3}^2(\Delta^2) > p-3\big)\big]\\
&-E\big[\big(1-(p-3)\chi_{p+3}^{-2}(\Delta^2)\big)^2 I\big(\chi_{p+3}^2(\Delta^2) > p-3\big)\big]\big\} \qquad (6.5.25b)
\end{aligned}$$

6.6 Asymptotic Distributional Risk Properties

In this section, we provide the risk analysis of the estimators of $\boldsymbol{\theta}$ with the general quadratic loss function using the matrix $\mathbf{W}$ so that James-Stein and positive-rule Stein estimators of $\boldsymbol{\beta}$ dominate. The properties of the estimators of $\boldsymbol{\beta}$ may be obtained following Chapter 5 with the ANOVA model.

6.6.1 Comparison of $\tilde{\boldsymbol{\theta}}_n$ and $\hat{\boldsymbol{\theta}}_n$

The risk-difference In this case, is given by

$$R_1(\tilde{\boldsymbol{\theta}}_n : \mathbf{W}) - R_2(\hat{\boldsymbol{\theta}}_n : \mathbf{W}) = \sigma^2\{\text{tr}[\mathbf{W}(\boldsymbol{\Lambda}_{11} - \boldsymbol{\Lambda}_{11}^*)] - \sigma^{-2}(\boldsymbol{\delta}'\mathbf{J}'\mathbf{T}_0\mathbf{W}\mathbf{T}_0\mathbf{J}\boldsymbol{\delta})\}.$$

By Courant's theorem,

$$Ch_{\min}[(\mathbf{T}_0\mathbf{W}\mathbf{T}_0)\boldsymbol{\Lambda}_{22}] \leq \frac{\boldsymbol{\delta}'\mathbf{J}'\mathbf{T}_0\mathbf{W}\mathbf{T}_0\mathbf{J}\boldsymbol{\delta}}{\boldsymbol{\delta}'\mathbf{J}'\boldsymbol{\Lambda}_{22}^{-1}\mathbf{J}\boldsymbol{\delta}} \leq Ch_{\max}[(\mathbf{T}_0\mathbf{W}\mathbf{T}_0)\boldsymbol{\Lambda}_{22}].$$

Hence,

$$\begin{aligned} R_1(\tilde{\boldsymbol{\theta}}_n : \mathbf{W}) &- \sigma^2\text{tr}[\mathbf{W}(\boldsymbol{\Lambda}_{11} - \boldsymbol{\Lambda}_{11}^*)] + \sigma^2\Delta^2 Ch_{\min}[(\mathbf{T}_0\mathbf{W}\mathbf{T}_0)\boldsymbol{\Lambda}_{22}] \\ &\leq R_2(\hat{\boldsymbol{\theta}}_n : \mathbf{W}) \leq R_1(\tilde{\boldsymbol{\theta}}_n : \mathbf{W}) - \sigma^2\text{tr}[\mathbf{W}(\boldsymbol{\Lambda}_{11} - \boldsymbol{\Lambda}_{11}^*)] \\ &+\sigma^2\Delta^2 Ch_{\max}[(\mathbf{T}_0\mathbf{W}\mathbf{T}_0)\boldsymbol{\Lambda}_{22}]. \end{aligned} \tag{6.6.1}$$

When $\Delta^2 = 0$, the bounds are equal. Thus, (6.6.1) means $\hat{\boldsymbol{\theta}}_n$ performs better than $\tilde{\boldsymbol{\theta}}_n$ whenever

$$\Delta^2 \leq \frac{\text{tr}[\mathbf{W}(\boldsymbol{\Lambda}_{11} - \boldsymbol{\Lambda}_{11}^*)]}{Ch_{\max}[(\mathbf{T}_0\mathbf{W}\mathbf{T}_0)\boldsymbol{\Lambda}_{22}]}, \tag{6.6.2a}$$

whereas $\tilde{\boldsymbol{\theta}}_n$ performs better than $\hat{\boldsymbol{\theta}}_n$ whenever

$$\Delta^2 \geq \frac{\text{tr}[\mathbf{W}(\boldsymbol{\Lambda}_{11} - \boldsymbol{\Lambda}_{11}^*)]}{Ch_{\min}[(\mathbf{T}_0\mathbf{W}\mathbf{T}_0)\boldsymbol{\Lambda}_{22}]}. \tag{6.6.2b}$$

For $\mathbf{W} = \boldsymbol{\Lambda}_{22}^{-1}$, $\hat{\boldsymbol{\theta}}_n$ performs better than $\tilde{\boldsymbol{\theta}}_n$ in the interval $\left(0, \dfrac{\text{tr}[\boldsymbol{\Lambda}_{22}^{-1}(\boldsymbol{\Lambda}_{11} - \boldsymbol{\Lambda}_{11}^*)]}{Ch_{\max}[(\mathbf{T}_0\boldsymbol{\Lambda}_{22}^{-1}\mathbf{T}_0)\boldsymbol{\Lambda}_{22})]}\right)$ and worse outside the interval.

6.6.2 Comparison of $\hat{\boldsymbol{\theta}}_n^{\text{PT}}$ and $\tilde{\boldsymbol{\theta}}_n(\hat{\boldsymbol{\theta}}_n)$

Here we compare $\hat{\boldsymbol{\theta}}_n^{\text{PT}}$ and $\tilde{\boldsymbol{\theta}}_n$. The risk-difference In this case, is as follows:

$$\begin{aligned} R_1(\tilde{\boldsymbol{\theta}}_n : \mathbf{W}) - R_3(\hat{\boldsymbol{\theta}}_n^{\text{PT}} : \mathbf{W}) &= \sigma^2\text{tr}[\mathbf{W}(\boldsymbol{\Lambda}_{11} - \boldsymbol{\Lambda}_{11}^*)]H_{p+1}(\chi_{p-1}^2(\alpha); \Delta^2) \\ -(\boldsymbol{\delta}'\mathbf{J}'\mathbf{T}_0\mathbf{W}\mathbf{T}_0\mathbf{J}\boldsymbol{\delta})\{2H_{p+1}(\chi_{p-1}^2(\alpha); \Delta^2) &- H_{p+3}(\chi_{p-1}^2(\alpha); \Delta^2)\}. \end{aligned} \tag{6.6.3}$$

Thus, the R.H.S of (6.6.3) is nonnegative whenever

$$\Delta^2 \leq \frac{\text{tr}\big[\mathbf{W}(\boldsymbol{\Lambda}_{11} - \boldsymbol{\Lambda}_{11}^*)\big] H_{p+1}(\chi^2_{p-1}(\alpha);\Delta^2)}{Ch_{\max}\big[(\mathbf{T}_0\mathbf{W}\mathbf{T}_0)\boldsymbol{\Lambda}_{22}\big]\big\{2H_{p+1}(\chi^2_{p-1}(\alpha);\Delta^2) - H_{p+3}(\chi^2_{p-1}(\alpha);\Delta^2)\big\}}. \tag{6.6.4a}$$

In this range, $\hat{\boldsymbol{\theta}}_n^{\text{PT}}$ performs better than $\tilde{\boldsymbol{\theta}}_n$, whereas $\tilde{\boldsymbol{\theta}}_n$ performs better whenever

$$\Delta^2 \geq \frac{\text{tr}\big[\mathbf{W}(\boldsymbol{\Lambda}_{11} - \boldsymbol{\Lambda}_{11}^*)\big] H_{p+1}(\chi^2_{p-1}(\alpha);\Delta^2)}{Ch_{\min}\big[(\mathbf{T}_0\mathbf{W}\mathbf{T}_0)\boldsymbol{\Lambda}_{22}\big]\big\{2H_{p+1}(\chi^2_{p-1}(\alpha);\Delta^2) - H_{p+3}(\chi^2_{p-1}(\alpha);\Delta^2)\big\}}. \tag{6.6.4b}$$

Under H_0, $\hat{\boldsymbol{\theta}}_n^{\text{PT}}$ is superior to $\tilde{\boldsymbol{\theta}}_n$, since the risk-difference is positive for all α. We can describe the graph of $R_3(\hat{\boldsymbol{\theta}}_n^{\text{PT}} : \mathbf{W})$ as follows: It begins with a value

$$\sigma^2\,\text{tr}\big[\mathbf{W}\boldsymbol{\Lambda}_{11}\big] - \sigma^2\,\text{tr}\big[\mathbf{W}(\boldsymbol{\Lambda}_{11} - \boldsymbol{\Lambda}_{11}^*)\big] H_{p+1}(\chi^2_{p-1}(\alpha);0) \tag{6.6.5}$$

at $\Delta^2 = 0$, then increases, crossing the risk of $\tilde{\boldsymbol{\theta}}_n$ to a maximum, and then drops gradually toward $\sigma^2\text{tr}\big[\mathbf{W}\boldsymbol{\Lambda}_{11}\big]$ as $\Delta^2 \to \infty$.

By setting $\alpha = 0$, it can be shown that

$$R_2(\hat{\boldsymbol{\theta}}_n : \mathbf{W}) = R_3(\hat{\boldsymbol{\theta}}_n^{\text{PT}} : \mathbf{W}). \tag{6.6.6a}$$

For $\alpha = 1$, however, we obtain

$$R_1(\tilde{\boldsymbol{\theta}}_n : \mathbf{W}) = R_3(\hat{\boldsymbol{\theta}}_n^{\text{PT}} : \mathbf{W}). \tag{6.6.6b}$$

Now, we compare $\hat{\boldsymbol{\theta}}_n$ and $\hat{\boldsymbol{\theta}}_n^{\text{PT}}$. Both are superior to $\tilde{\boldsymbol{\theta}}_n$ under $H_0 : \boldsymbol{\beta} = \beta_0\mathbf{1}_p$. In general, the risk-difference is given by

$$\begin{aligned} &R_2(\hat{\boldsymbol{\theta}}_n : \mathbf{W}) - R_3(\hat{\boldsymbol{\theta}}_n^{\text{PT}} : \mathbf{W}) \\ &= -\sigma^2\text{tr}\big[\mathbf{W}(\boldsymbol{\Lambda}_{11} - \boldsymbol{\Lambda}_{11}^*)\big]\big[1 - H_{p+1}(\chi^2_{p-1}(\alpha);\Delta^2)\big] + (\boldsymbol{\delta}'\mathbf{J}'\mathbf{T}_0\mathbf{W}\mathbf{T}_0\mathbf{J}\boldsymbol{\delta}) \\ &\quad\times\big\{1 - 2H_{p+1}(\chi^2_{p-1}(\alpha);\Delta^2) + H_{p+3}(\chi^2_{p-1}(\alpha);\Delta^2)\big\}. \end{aligned} \tag{6.6.7}$$

Thus, $\hat{\boldsymbol{\theta}}_n^{\text{PT}}$ performs better than $\hat{\boldsymbol{\theta}}_n$ whenever

$$\Delta^2 \leq \frac{\text{tr}\big[\mathbf{W}(\boldsymbol{\Lambda}_{11} - \boldsymbol{\Lambda}_{11}^*)\big]\big[1 - H_{p+1}(\chi^2_{p-1}(\alpha);\Delta^2)\big]}{Ch_{\max}\big[(\mathbf{T}_0\mathbf{W}\mathbf{T}_0)\boldsymbol{\Lambda}_{22}\big]\big\{1 - 2H_{p+1}(\chi^2_{p-1}(\alpha);\Delta^2) + H_{p+3}(\chi^2_{p-1}(\alpha);\Delta^2)\big\}}, \tag{6.6.8a}$$

while for

$$\Delta^2 \geq \frac{\text{tr}\big[\mathbf{W}(\boldsymbol{\Lambda}_{11} - \boldsymbol{\Lambda}_{11}^*)\big]\big[1 - H_{p+1}(\chi^2_{p-1}(\alpha);\Delta^2)\big]}{Ch_{\min}\big[(\mathbf{T}_0\mathbf{W}\mathbf{T}_0)\boldsymbol{\Lambda}_{22}\big]\big\{1 - 2H_{p+1}(\chi^2_{p-1}(\alpha);\Delta^2) + H_{p+3}(\chi^2_{p-1}(\alpha);\Delta^2)\big\}}, \tag{6.6.8b}$$

$\hat{\boldsymbol{\theta}}_n$ performs better than $\hat{\boldsymbol{\theta}}_n^{\text{PT}}$. Under H_0, the risk ordering is given by

$$R_2(\tilde{\boldsymbol{\theta}}_n : \mathbf{W}) > R_3(\hat{\boldsymbol{\theta}}_n^{\text{PT}} : \mathbf{W}) > R_1(\hat{\boldsymbol{\theta}}_n : \mathbf{W}), \tag{6.6.9}$$

depending on the size of α.

6.6.3 Comparison of $\hat{\boldsymbol{\theta}}_n^S$ and $\tilde{\boldsymbol{\theta}}_n(\hat{\boldsymbol{\theta}}_n)$

In order to compare $\hat{\boldsymbol{\theta}}_n$ and $\hat{\boldsymbol{\theta}}_n^S$, we consider the risk-difference given by

$$\begin{aligned}&R_1(\tilde{\boldsymbol{\theta}}_n : \mathbf{W}) - R_4(\hat{\boldsymbol{\theta}}_n^S : \mathbf{W})\\&= \sigma^2(p-3)\mathrm{tr}\big[\mathbf{W}(\mathbf{T}_0\mathbf{J}\boldsymbol{\Lambda}_{22}\mathbf{J}'\mathbf{T}_0)\big]\Big\{(p-3)E\big[\chi_{p+1}^{-4}(\Delta^2)\big]\\&+\Big[1-\frac{(p+1)(\boldsymbol{\delta}'\mathbf{J}'\mathbf{T}_0\mathbf{W}\mathbf{T}_0\mathbf{J}\boldsymbol{\delta}}{2\sigma^2\Delta^2\mathrm{tr}[\mathbf{W}(\mathbf{T}_0\mathbf{J}\boldsymbol{\Lambda}_{22}\mathbf{J}'\mathbf{T}_0)]}(2\Delta^2)E\big[\chi_{p+1}^{-4}(\Delta^2)\big]\Big]\Big\}. \quad (6.6.10)\end{aligned}$$

The risk-difference is positive for all $\mathbf{W}$ such that

$$\frac{\mathrm{tr}\big[\mathbf{W}(\mathbf{T}_0\mathbf{J}\boldsymbol{\Lambda}_{22}\mathbf{J}'\mathbf{T}_0)\big]}{Ch_{\max}\big[(\mathbf{T}_0\mathbf{W}\mathbf{T}_0)\boldsymbol{\Lambda}_{22}\big]} \geq \frac{p+1}{2}. \quad (6.6.11)$$

Thus, $\hat{\boldsymbol{\theta}}_n^S$ uniformly dominates $\tilde{\boldsymbol{\theta}}_n$. Further, as $\Delta^2 \to \infty$,

$$R_4(\hat{\boldsymbol{\theta}}_n^S : \mathbf{W}) \;\to\; R_1(\tilde{\boldsymbol{\theta}}_n : \mathbf{W}).$$

To compare $\hat{\boldsymbol{\theta}}_n$ and $\hat{\boldsymbol{\theta}}_n^S$, we can write

$$\begin{aligned}&R_4(\hat{\boldsymbol{\theta}}_n^S : \mathbf{W}) = R_2(\hat{\boldsymbol{\theta}}_n : \mathbf{W}) + \sigma^2\mathrm{tr}\big[\mathbf{W}(\mathbf{T}_0\mathbf{J}\boldsymbol{\Lambda}_{22}\mathbf{J}'\mathbf{T}_0)\big]\\&-(\boldsymbol{\delta}'\mathbf{J}'\mathbf{T}_0\mathbf{W}\mathbf{T}_0\mathbf{J}\boldsymbol{\delta}) - \sigma^2(p-3)\mathrm{tr}\big[\mathbf{W}(\mathbf{T}_0\mathbf{J}\boldsymbol{\Lambda}_{22}\mathbf{J}'\mathbf{T}_0)\big]\Big\{(p-3)E\big[\chi_{p+1}^{-4}(\Delta^2)\big]\\&+\Big[1-\frac{(p+1)(\boldsymbol{\delta}'\mathbf{J}'\mathbf{T}_0\mathbf{W}\mathbf{T}_0\mathbf{J}\boldsymbol{\delta})}{2\sigma^2\Delta^2\mathrm{tr}\big[\mathbf{W}(\mathbf{T}_0\mathbf{J}\boldsymbol{\Lambda}_{22}\mathbf{J}'\mathbf{T}_0)\big]}(2\Delta^2)E\big[\chi_{p+1}^{-4}(\Delta^2)\big]\Big]\Big\}. \quad (6.6.12)\end{aligned}$$

Under H_0, this becomes

$$\begin{aligned}R_4(\hat{\boldsymbol{\theta}}_n^S : \mathbf{W}) &= R_2(\hat{\boldsymbol{\theta}}_n : \mathbf{W}) + \sigma^2\mathrm{tr}\big[\mathbf{W}(\mathbf{T}_0\mathbf{J}\boldsymbol{\Lambda}_{22}\mathbf{J}'\mathbf{T}_0)\big]\\&\quad\times\{1-(p-3)^2E\big[\chi_{p+1}^{-4}(\Delta^2)\big] \geq R_2(\hat{\boldsymbol{\theta}}_n : \mathbf{W})\end{aligned}, \quad (6.6.13a)$$

while

$$R_2(\hat{\boldsymbol{\theta}}_n : \mathbf{W}) = R_1(\tilde{\boldsymbol{\theta}}_n : \mathbf{W}) - \sigma^2\mathrm{tr}\big[\mathbf{W}(\mathbf{T}_0\mathbf{J}\boldsymbol{\Lambda}_{22}\mathbf{J}'\mathbf{T}_0)\big] \leq R_1(\tilde{\boldsymbol{\theta}}_n : \mathbf{W}). \quad (6.6.13b)$$

Thus, $\hat{\boldsymbol{\theta}}_n$ performs better than $\hat{\boldsymbol{\theta}}_n^S$ under H_0. However, as $\boldsymbol{\delta}$ moves away from the origin, $\sigma^{-2}(\boldsymbol{\delta}'\mathbf{J}'\mathbf{T}_0\mathbf{W}\mathbf{J}\boldsymbol{\delta})$ increases, and the risk of $\hat{\boldsymbol{\theta}}_n$ becomes unbounded while the risk of $\hat{\boldsymbol{\theta}}_n^S$ remains below the risk of $\tilde{\boldsymbol{\theta}}_n$ and merges with it as $\Delta \to \infty$. Thus, $\hat{\boldsymbol{\theta}}_n^S$ dominates $\hat{\boldsymbol{\theta}}_n^S$ outside an interval around the origin.

6.6.4 Comparison of $\hat{\boldsymbol{\theta}}_n^S$ and $\hat{\boldsymbol{\theta}}_n^{\mathrm{PT}}$

Now, we compare $\hat{\boldsymbol{\theta}}_n^S$ and $\hat{\boldsymbol{\theta}}_n^{\mathrm{PT}}$. First note that under H_0,

$$R_4(\hat{\boldsymbol{\theta}}_n^S : \mathbf{W}) = R_3(\tilde{\boldsymbol{\theta}}_n^{\mathrm{PT}} : \mathbf{W}) + \sigma^2 \mathrm{tr}\big[\mathbf{W}(\mathbf{T}_0\mathbf{J}\boldsymbol{\Lambda}_{22}\mathbf{J}'\mathbf{T}_0)\big]$$
$$\times\big[H_{p+1}(\chi^2_{p-1}(\alpha);0) - (p-3)\big] \geq R_3(\tilde{\boldsymbol{\theta}}_n^{\mathrm{PT}} : \mathbf{W}) \quad (6.6.14)$$

if

$$H_{p+1}(\chi^2_{p-1}(\alpha);0) > (p-3) \quad (6.6.15)$$

for some α. The risk of $\hat{\boldsymbol{\theta}}_n^S$ is smaller than that of the risk of $\hat{\boldsymbol{\theta}}_n^{\mathrm{PT}}$ when the critical value $\chi^2_{p-1}(\alpha)$ satisfies the opposite relation in (6.6.15). This means that $\hat{\boldsymbol{\theta}}_n^S$ does not always dominate $\tilde{\boldsymbol{\theta}}_n^{\mathrm{PT}}$ under H_0. We can order the risks as follows:

$$R_2(\hat{\boldsymbol{\theta}}_n : \mathbf{W}) < R_3(\hat{\boldsymbol{\theta}}_n^{\mathrm{PT}} : \mathbf{W}) < R_4(\hat{\boldsymbol{\theta}}_n^S : \mathbf{W}) < R_1(\tilde{\boldsymbol{\theta}}_n : \mathbf{W}) \quad (6.6.16)$$

when χ^2_α satisfies (6.6.16).

This picture changes as Δ^2 moves away from 0. As $\Delta^2 \to \infty$, the risk of $\hat{\boldsymbol{\theta}}_n^{\mathrm{PT}}$ and $\hat{\boldsymbol{\theta}}_n^S$ converge to the risk of $\tilde{\boldsymbol{\theta}}_n$. For a reasonable α-value, the risk of $\hat{\boldsymbol{\theta}}_n^{\mathrm{PT}}$ is smaller than the risk of $\hat{\boldsymbol{\theta}}_n^S$ for α, satisfying (6.6.15) and $p \geq 4$. Thus, none of the estimators dominates the other uniformly.

6.6.5 Comparison of $\hat{\boldsymbol{\theta}}_n^{S+}$ and $\hat{\boldsymbol{\theta}}_n^S$, $\hat{\boldsymbol{\theta}}_n$, $\hat{\boldsymbol{\theta}}_n^{\mathrm{PT}}$

Finally, we compare $\hat{\boldsymbol{\theta}}_n^S$ and $\hat{\boldsymbol{\theta}}_n^{S+}$. The risk-difference is given by

$$R_5(\hat{\boldsymbol{\theta}}_n^{S+} : \mathbf{W}) - R_4(\tilde{\boldsymbol{\theta}}_n^S : \mathbf{W}) = -\sigma^2 \,\mathrm{tr}\big[\mathbf{W}(\mathbf{T}_0\mathbf{J}\boldsymbol{\Lambda}_{22}\mathbf{J}'\mathbf{T}_0)\big]$$
$$\times E\big[\big(1-(p-3)\chi^{-2}_{p+1}(\Delta^2)\big)^2 I\big(\chi^2_{p+1}(\Delta^2) < p-3\big)\big]$$
$$-(\boldsymbol{\delta}'\mathbf{J}'\mathbf{T}_0\mathbf{W}\mathbf{T}_0\mathbf{J}\boldsymbol{\delta})\Big\{2E\big[\big(1-(p-3)\chi^{-2}_{p+1}(\Delta^2)\big)I\big(\chi^2_{p+1}(\Delta^2) < p-3\big)\big]$$
$$-E\big[\big(1-(p-3)\chi^{-2}_{p+3}(\Delta^2)\big)^2 I\big(\chi^2_{p+3}(\Delta^2) < p-3\big)\big]\Big\}. \quad (6.6.17)$$

The R.H.S. of (6.6.17) is negative since the expectation of a positive random variable is positive, and we get

$$\big(0 < \chi^2_{p+j}(\Delta^2) < p-3\big) \Rightarrow \big((p-3)\chi^{-2}_{p+j}(\Delta^2) - 1\big) \geq 0. \quad (6.6.18)$$

Thus,

$$E\big[\big((p-3)\chi^{-2}_{p+1}(\Delta^2) - 1\big)I\big(\chi^2_{p+1} < p-3\big)\big] \geq 0. \quad (6.6.19)$$

Therefore, for all $(\boldsymbol{\theta}', \boldsymbol{\beta}')'$, $R_5(\hat{\boldsymbol{\theta}}_n^{S+} : \mathbf{W}) \le R_4(\tilde{\boldsymbol{\theta}}_n^{S} : \mathbf{W})$. Further, we can write

$$R_1(\tilde{\boldsymbol{\theta}}_n : \mathbf{W}) \ge R_4(\hat{\boldsymbol{\theta}}_n^{S} : \mathbf{W}) \ge R_5(\tilde{\boldsymbol{\theta}}_n^{S+} : \mathbf{W}) \quad \text{for all } \Delta^2 \in (0, \infty) \quad (6.6.20)$$

and for $p \ge 4$.

To compare $\hat{\boldsymbol{\theta}}_n$ and $\tilde{\boldsymbol{\theta}}_n^{S+}$, we first consider the risk of $\tilde{\boldsymbol{\theta}}_n^{S+}$ under H_0:

$$\begin{aligned} R_5(\hat{\boldsymbol{\theta}}_n^{S+} : \mathbf{W}) &= R_2(\hat{\boldsymbol{\theta}}_n : \mathbf{W}) + \sigma^2 \mathrm{tr}\big[\mathbf{W}(\mathbf{T}_0 \mathbf{J} \boldsymbol{\Lambda}_{22} \mathbf{J}' \mathbf{T}_0)\big] \\ &\quad \times \left\{ \frac{2}{p-1} - E\big[(1-(p-3)\chi_{p+1}^{-2})^2 I(\chi_{p+1}^2 < p-3)\big] \right\} \\ &\ge R_2(\hat{\boldsymbol{\theta}}_n : \mathbf{W}), \end{aligned} \quad (6.6.21)$$

since

$$\begin{aligned} &E\big[(1-(p-3)\chi_{p+1}^{-2})^2 I(\chi_{p+1}^2 < p-3)\big] \\ &\le E(1-(p-3)\chi_{p+1}^{-2})^2 = \frac{2}{p-1}. \end{aligned} \quad (6.6.22)$$

Thus, $\hat{\boldsymbol{\theta}}_n$ performs better than $\hat{\boldsymbol{\theta}}_n^{S+}$ under H_0. However, as $\boldsymbol{\delta}$ moves away from zero, the risk of $\hat{\boldsymbol{\theta}}_n$ becomes unbounded while the risk of $\hat{\boldsymbol{\theta}}_n^{S+}$ remains below the risk of $\tilde{\boldsymbol{\theta}}_n$ and merges with it as $\Delta^2 \to \infty$. Thus, $\hat{\boldsymbol{\theta}}_n^{S+}$ dominates $\hat{\boldsymbol{\theta}}_n$ outside an interval around the origin.

Finally, we compare $\hat{\boldsymbol{\theta}}_n^{S+}$ and $\hat{\boldsymbol{\theta}}_n^{\mathrm{PT}}$. Under H_0, we obtain

$$\begin{aligned} R_5(\hat{\boldsymbol{\theta}}_n^{S+} : \mathbf{W}) &= R_3(\tilde{\boldsymbol{\theta}}_n^{\mathrm{PT}} : \mathbf{W}) + \sigma^2 \mathrm{tr}\big[\mathbf{W}(\mathbf{T}_0 \mathbf{J} \mathbf{W} \mathbf{J}' \mathbf{T}_0)\big] \\ \times &\left\{ \left(H_{p+1}(\chi_{p-1}^2(\alpha); 0) - \frac{p-3}{p-1} \right) - E\big[(1-(p-3)\chi_{p+1}^{-2})^2 I(\chi_{p+1}^2 < p-3)\big] \right\} \\ &\ge R_3(\tilde{\boldsymbol{\theta}}_n^{\mathrm{PT}} : \mathbf{W}) \end{aligned} \quad (6.6.23)$$

for all α, satisfying the relation

$$\left\{ \alpha : H_{p+1}(\chi_{p-1}^2(\alpha); 0) > \frac{p-3}{p-1} + E\Big[(1-(p-3)\chi_{p+1}^{-2})^2 I(\chi_{p+1}^2 < p-3)\Big] \right\}. \quad (6.6.24)$$

The risk of $\hat{\boldsymbol{\theta}}_n^{S+}$ is smaller than the risk of $\hat{\boldsymbol{\theta}}_n^{\mathrm{PT}}$ when the critical value $\chi_{p-1}^2(\alpha)$ satisfies the opposite inequality in (6.6.24). Thus, $\hat{\boldsymbol{\theta}}_n^{S+}$ does not always dominate $\tilde{\boldsymbol{\theta}}_n^{\mathrm{PT}}$ when H_0 is true. We can order the risks under H_0 as

$$R_2(\hat{\boldsymbol{\theta}}_n : \mathbf{W}) < R_3(\hat{\boldsymbol{\theta}}_n^{\mathrm{PT}} : \mathbf{W}) < R_5(\hat{\boldsymbol{\theta}}_n^{S+} : \mathbf{W}) < R_4(\hat{\boldsymbol{\theta}}_n^{S} : \mathbf{W}) < R_1(\tilde{\boldsymbol{\theta}}_n : \mathbf{W}). \quad (6.6.25)$$

The position of $R_3(\hat{\boldsymbol{\theta}}_n^{\mathrm{PT}} : \mathbf{W})$ may shift in between $R_2(\hat{\boldsymbol{\theta}}_n : \mathbf{W})$ and $R_5(\hat{\boldsymbol{\theta}}_n^{S+} : \mathbf{W})$ to in between $R_4(\hat{\boldsymbol{\theta}}_n^{S} : \mathbf{W})$ and $R_1(\tilde{\boldsymbol{\theta}}_n : \mathbf{W})$. Thus, the ordering under H_0 may change to

$$R_2(\hat{\boldsymbol{\theta}}_n : \mathbf{W}) < R_5(\hat{\boldsymbol{\theta}}_n^{S+} : \mathbf{W}) < R_4(\hat{\boldsymbol{\theta}}_n^{S} : \mathbf{W}) < R_3(\hat{\boldsymbol{\theta}}_n^{\mathrm{PT}} : \mathbf{W}) < R_1(\tilde{\boldsymbol{\theta}}_n : \mathbf{W}), \tag{6.6.26}$$

depending on the size of α. The picture changes as $\boldsymbol{\delta}$ moves from the origin. Note that the risk of $\tilde{\boldsymbol{\theta}}_n$ is constant and the risk of $\hat{\boldsymbol{\theta}}_n$ becomes unbounded as $\boldsymbol{\delta}$ moves away from $\mathbf{0}$ while risk of $\hat{\boldsymbol{\theta}}_n^{\mathrm{PT}}$, $\hat{\boldsymbol{\theta}}_n^{S}$, and $\hat{\boldsymbol{\theta}}_n^{S+}$ converges to the risk of $\tilde{\boldsymbol{\theta}}_n$ as $\boldsymbol{\delta}$ moves farthest away from $\mathbf{0}$. Thus, neither $\hat{\boldsymbol{\theta}}_n^{\mathrm{PT}}$ nor $\hat{\boldsymbol{\theta}}_n^{S+}$ dominates each other uniformly.

6.7 Asymptotic Distributional MSE-matrix Properties

In this section, we discuss the property of the asymptotic distributional MSE matrices of the estimators.

Comparison of $\hat{\boldsymbol{\theta}}_n$ and $\tilde{\boldsymbol{\theta}}_n$. Consider the difference of the MSE matrices

$$\mathbf{M}_1(\tilde{\boldsymbol{\theta}}_n) - \mathbf{M}_2(\hat{\boldsymbol{\theta}}_n) = \sigma^2(\boldsymbol{\Lambda}_{11} - \boldsymbol{\Lambda}_{11}^*) - (\mathbf{T}_0\mathbf{J}\boldsymbol{\delta}\boldsymbol{\delta}'\mathbf{J}'\mathbf{T}_0). \tag{6.7.1}$$

This expression is positive semidefinite if for any nonzero vector $\boldsymbol{\ell}$ we have

$$\frac{\boldsymbol{\ell}'(\boldsymbol{\Lambda}_{11} - \boldsymbol{\Lambda}_{11}^*)\boldsymbol{\ell}}{\boldsymbol{\ell}'(\mathbf{T}_0\boldsymbol{\Lambda}_{22}\mathbf{T}_0)\boldsymbol{\ell}} \geq \sigma^{-2}\frac{\boldsymbol{\ell}'(\mathbf{T}_0\mathbf{J}\boldsymbol{\delta}\boldsymbol{\delta}'\mathbf{J}'\mathbf{T}_0)\boldsymbol{\ell}}{\boldsymbol{\ell}'(\mathbf{T}_0\boldsymbol{\Lambda}_{22}\mathbf{T}_0)\boldsymbol{\ell}}. \tag{6.7.2}$$

Now,

$$\max_{\boldsymbol{\ell}} \sigma^{-2}\frac{\boldsymbol{\ell}'(\mathbf{T}_0\mathbf{J}\boldsymbol{\delta}\boldsymbol{\delta}'\mathbf{J}'\mathbf{T}_0)\boldsymbol{\ell}}{\boldsymbol{\ell}'(\mathbf{T}_0\boldsymbol{\Lambda}_{22}\mathbf{T}_0)\boldsymbol{\ell}} = \Delta^2 \tag{6.7.3a}$$

and

$$\max_{\boldsymbol{\ell}} \sigma^{-2}\frac{\boldsymbol{\ell}'(\boldsymbol{\Lambda}_{11} - \boldsymbol{\Lambda}_{11}^*)\boldsymbol{\ell}}{\boldsymbol{\ell}'(\mathbf{T}_0\boldsymbol{\Lambda}_{22}\mathbf{T}_0)\boldsymbol{\ell}} \leq Ch_{\max}(\mathbf{J}) = 1. \tag{6.7.3b}$$

If $\Delta^2 < 1$, then $\hat{\boldsymbol{\theta}}_n$ performs better than $\tilde{\boldsymbol{\theta}}_n$; otherwise, $\tilde{\boldsymbol{\theta}}_n$ performs better than $\hat{\boldsymbol{\theta}}_n$. The asymptotic efficiency of $\hat{\boldsymbol{\theta}}_n$ relative to $\tilde{\boldsymbol{\theta}}_n$ is given by

$$\begin{aligned}\mathrm{AMRE}(\hat{\boldsymbol{\theta}}_n : \tilde{\boldsymbol{\theta}}_n) &= \left|\mathbf{I}_p - \boldsymbol{\Lambda}_{11}^{-1}\{\mathbf{T}_0\mathbf{J}\boldsymbol{\Lambda}_{22}\mathbf{J}'\mathbf{T}_0 - \sigma^{-2}\mathbf{T}_0\mathbf{J}\boldsymbol{\delta}\boldsymbol{\delta}'\mathbf{J}'\mathbf{T}_0\}\right|^{-1/p} \\ &= \left|\mathbf{I}_p - \boldsymbol{\Lambda}_{11}^{-1}\{\mathbf{T}_0\boldsymbol{\Lambda}_{22}(\mathbf{J} - \sigma^{-2}\boldsymbol{\Lambda}_{22}^{-1}\mathbf{J}\boldsymbol{\delta}\boldsymbol{\delta}'\mathbf{J}')\mathbf{T}_0\}\right|^{-1/p}\end{aligned} \tag{6.7.4}$$

where $\mathbf{J}$ is positive semidefinite with rank $p-1$ while $\sigma^{-2}\boldsymbol{\Lambda}_{22}^{-2}(\mathbf{J}\boldsymbol{\delta}\boldsymbol{\delta}'\mathbf{J}')$ is positive semidefinite with rank 1 and the nonzero characteristic root of this matrix is $\sigma^{-2}\boldsymbol{\delta}'\mathbf{J}'\boldsymbol{\Lambda}_{22}^{-1}\mathbf{J}\boldsymbol{\delta} = \Delta^2$.

Thus, (6.7.4) exceeds 1 whenever $\Delta^2 < 1$ and falls below 1, whenever $\Delta^2 > 1$. Further, if $\Delta^2 \to \infty$, (6.7.4) tends to zero, showing that $\hat{\boldsymbol{\theta}}_n$ loses its efficiency as Δ^2 moves away from the unit interval.

Comparison of $\hat{\boldsymbol{\theta}}_n^{\mathrm{PT}}$ and $\tilde{\boldsymbol{\theta}}_n$. In this case, the difference of the MSE matrices is given by

$$\mathbf{M}_1(\tilde{\boldsymbol{\theta}}_n) - \mathbf{M}_3(\hat{\boldsymbol{\theta}}_n^{\mathrm{PT}}) = \sigma^2(\mathbf{T}_0\mathbf{J}\boldsymbol{\Lambda}_{22}\mathbf{J}'\mathbf{T}_0)H_{p+1}(\chi^2_{p-1}(\alpha);\Delta^2)$$
$$-(\mathbf{T}_0\mathbf{J}\boldsymbol{\delta}\boldsymbol{\delta}'\mathbf{J}'\mathbf{T}_0)\Big\{2H_{p+1}(\chi^2_{p-1}(\alpha);\Delta^2) - H_{p+3}(\chi^2_{p-1}(\alpha);\Delta^2)\Big\}. \quad (6.7.5)$$

For a given nonzero vector $\boldsymbol{\ell}$, we obtain setting R.H.S. of (6.7.5) ≥ 0 as

$$\boldsymbol{\ell}'\Big(\mathbf{T}_0\mathbf{J}\boldsymbol{\delta}\boldsymbol{\delta}'\mathbf{J}'\mathbf{T}_0\Big)\boldsymbol{\ell} \leq \frac{\boldsymbol{\ell}'\Big(\mathbf{T}_0\mathbf{J}\boldsymbol{\Lambda}_{22}\mathbf{J}'\mathbf{T}_0\Big)H_{p+1}(\chi^2_{p-1}(\alpha);\Delta^2)\boldsymbol{\ell}}{\{2H_{p+1}(\chi^2_{p-1}(\alpha);\Delta^2) - H_{p+3}(\chi^2_{p-1}(\alpha);\Delta^2)\}}. \quad (6.7.6)$$

Dividing by $\boldsymbol{\ell}'\mathbf{T}_0\boldsymbol{\Lambda}_{22}\mathbf{T}_0\boldsymbol{\ell}$ and maximizing over all $\boldsymbol{\ell}$, we obtain

$$\Delta^2 \leq \frac{H_{p+1}(\chi^2_{p-1}(\alpha);\Delta^2)}{\{2H_{p+1}(\chi^2_{p-1}(\alpha);\Delta^2) - H_{p+3}(\chi^2_{p-1}(\alpha);\Delta^2)\}}.$$

Thus, $\hat{\boldsymbol{\theta}}_n^{\mathrm{PT}}$ performs better than $\tilde{\boldsymbol{\theta}}_n$ whenever

$$\Delta^2 \leq \frac{H_{p+1}(\chi^2_{p-1}(\alpha);\Delta^2)}{\{2H_{p+1}(\chi^2_{p-1}(\alpha);\Delta^2) - H_{p+3}(\chi^2_{p-1}(\alpha);\Delta^2)\}}; \quad (6.7.7)$$

otherwise, $\tilde{\boldsymbol{\theta}}_n$ performs better than $\hat{\boldsymbol{\theta}}_n^{\mathrm{PT}}$. The AMRE$(\hat{\boldsymbol{\theta}}_n^{\mathrm{PT}};\tilde{\boldsymbol{\theta}}_n)$ is given by

$$\mathrm{AMRE}(\hat{\boldsymbol{\theta}}_n^{\mathrm{PT}}:\tilde{\boldsymbol{\theta}}_n) = \Big|\mathbf{I}_p - \sigma^2\boldsymbol{\Lambda}_{11}^{-1}(\mathbf{T}_0\mathbf{J}\boldsymbol{\Lambda}_{22}\mathbf{J}'\mathbf{T}_0)H_{p+1}(\chi^2_{p-1}(\alpha);\Delta^2)$$
$$-\boldsymbol{\Lambda}_{11}^{-1}(\mathbf{T}_0\mathbf{J}\boldsymbol{\delta}\boldsymbol{\delta}'\mathbf{J}'\mathbf{T}_0)\{2H_{p+1}(\chi^2_{p-1}(\alpha);\Delta^2) - H_{p+3}(\chi^2_{p-1}(\alpha);\Delta^2)\}\Big|^{-1/p}. \quad (6.7.8)$$

This is a function of Δ^2. It has its maximum at $\Delta^2 = 0$, then decreases crossing the 1-line to a minimum, and then increases as Δ^2 increases toward 1-line.

Comparison of $\hat{\boldsymbol{\theta}}_n^{S}$ and $\tilde{\boldsymbol{\theta}}_n$. In this case, the MSE-matrix difference is given by

$$\mathbf{M}_1(\tilde{\boldsymbol{\theta}}_n) - \mathbf{M}_4(\hat{\boldsymbol{\theta}}_n^{S}) = (p-3)\sigma^2(\mathbf{T}_0\mathbf{J}\boldsymbol{\Lambda}_{22}\mathbf{J}'\mathbf{T}_0)\{2E[\chi^{-2}_{p+1}(\Delta^2)]$$
$$-(p-3)E[\chi^{-4}_{p+1}(\Delta^2)]\} - (p-1)(p-3)(\mathbf{T}_0\mathbf{J}\boldsymbol{\delta}\boldsymbol{\delta}'\mathbf{J}'\mathbf{T}_0)E[\chi^{-4}_{p+3}(\Delta^2)]. \quad (6.7.9)$$

For the MSE difference to be p.s.d., we must have for every nonzero vector $\boldsymbol{\ell}$,

$$(p-1)(\boldsymbol{\ell}'\mathbf{T}_0\mathbf{J}\boldsymbol{\delta}\boldsymbol{\delta}'\mathbf{J}'\mathbf{T}_0\boldsymbol{\ell})E[\chi^{-4}_{p+3}(\Delta^2)]$$
$$\leq \boldsymbol{\ell}'(\mathbf{T}_0\mathbf{J}\boldsymbol{\Lambda}_{22}\mathbf{J}'\mathbf{T}_0)\boldsymbol{\ell}\{2E[\chi^{-2}_{p+1}(\Delta^2)] - (p-3)E[\chi^{-4}_{p+1}(\Delta^2)]\}. \quad (6.7.10)$$

Dividing by $\boldsymbol{\ell}'(\mathbf{T}_0\boldsymbol{\Lambda}_{22}\mathbf{T}_0)\boldsymbol{\ell}$ and maximizing over $\boldsymbol{\ell}$, we have

$$(p-1)\Delta^2 E\big[\chi_{p+3}^{-4}(\Delta^2)\big] \leq 2E\big[\chi_{p+1}^{-2}(\Delta^2)\big] - (p-3)E\big[\chi_{p+1}^{-4}(\Delta^2)\big] \quad (6.7.11)$$

or

$$E\big[\chi_{p+1}^{-2}(\Delta^2)\big] \leq pE\big[\chi_{p+1}^{-4}(\Delta^2)\big],$$

which does not hold for all Δ^2.

Hence, $\hat{\boldsymbol{\theta}}_n^S$ does not dominate $\tilde{\boldsymbol{\theta}}_n$ uniformly, as we have seen with respect to risk analysis. The asymptotic mean-square relative efficiency AMRE of $\hat{\boldsymbol{\theta}}_n^S$ compared to $\tilde{\boldsymbol{\theta}}_n$ is given by

$$\begin{aligned}\text{AMRE}(\hat{\boldsymbol{\theta}}_n^S : \tilde{\boldsymbol{\theta}}_n) &= \Big|\mathbf{I}_p - (p-3)\boldsymbol{\Lambda}_{11}^{-1}\Big\{(\mathbf{T}_0\mathbf{J}\boldsymbol{\Lambda}_{22}\mathbf{J}'\mathbf{T}_0)\big[2E\big[\chi_{p+1}^{-2}(\Delta^2)\big]\\ &-(p-3)E\big[\chi_{p+1}^{-4}(\Delta^2)\big]\big] + (p-1)(\mathbf{T}_0\mathbf{J}\boldsymbol{\delta}\boldsymbol{\delta}'\mathbf{J}'\mathbf{T}_0)E\big[\chi_{p+3}^{-4}(\Delta^2)\big]\Big\}\Big|^{-1/p}. \quad (6.7.12)\end{aligned}$$

Comparison of $\hat{\boldsymbol{\theta}}_n^{S+}$ and $\hat{\boldsymbol{\theta}}_n^S$. In this case, the MSE-matrix difference equals

$$\begin{aligned}&\mathbf{M}_4(\hat{\boldsymbol{\theta}}_n^S) - \mathbf{M}_5(\hat{\boldsymbol{\theta}}_n^{S+})\\ &= \sigma^2(\mathbf{T}_0\mathbf{J}\boldsymbol{\Lambda}_{22}\mathbf{J}'\mathbf{T}_0)E\big[(1-(p-3)\chi_{p+1}^{-2}(\Delta^2))^2 I(\chi_{p+1}^2(\Delta^2) > p-3)\big]\\ &+(\mathbf{T}_0\mathbf{J}\boldsymbol{\delta}\boldsymbol{\delta}'\mathbf{J}'\mathbf{T}_0)\Big\{E\big[(1-(p-3)\chi_{p+3}^{-2}(\Delta^2))^2 I(\chi_{p+3}^2(\Delta^2) > p-3)\big]\\ &+2E\big[(1-(p-3)\chi_{p+1}^{-2}(\Delta^2))I(\chi_{p+1}^2(\Delta^2) > p-3)\big]\Big\} \geq \mathbf{0}. \quad (6.7.13)\end{aligned}$$

Hence, $\hat{\boldsymbol{\theta}}_n^{S+}$ dominates $\hat{\boldsymbol{\theta}}_n^S$ uniformly. Furthermore,

$$\mathbf{M}_1(\tilde{\boldsymbol{\theta}}_n) - \mathbf{M}_4(\hat{\boldsymbol{\theta}}_n^{S+}) \geq \mathbf{0}$$

and

$$\mathbf{M}_4(\hat{\boldsymbol{\theta}}_n^S) - \mathbf{M}_5(\hat{\boldsymbol{\theta}}_n^{S+}) \geq \mathbf{0}. \quad (6.7.14)$$

So we can order the MSE matrices as

$$\mathbf{M}_5(\hat{\boldsymbol{\theta}}_n^{S+}) \prec \mathbf{M}_4(\hat{\boldsymbol{\theta}}_n^S) \prec \mathbf{M}_1(\tilde{\boldsymbol{\theta}}_n), \quad (6.7.15)$$

where $\prec$ means "domination".

The AMRE$(\hat{\boldsymbol{\theta}}_n^{S+} : \hat{\boldsymbol{\theta}}_n^S)$ is then given by

$$\begin{aligned}\text{AMRE}(\hat{\boldsymbol{\theta}}_n^{S+} : \hat{\boldsymbol{\theta}}_n^S) &= \Big|\mathbf{I}_p - \sigma^2\{\mathbf{M}_4(\hat{\boldsymbol{\theta}}_n^S)\}^{-1}(\mathbf{T}_0\mathbf{J}\boldsymbol{\Lambda}_{22}\mathbf{J}'\mathbf{T}_0)\\ &\times\Big\{E\big[(1-(p-3)\chi_{p+1}^{-2}(\Delta^2))^2 I(\chi_{p+1}^2(\Delta^2) > p-3)\big] + \sigma^{-2}(\mathbf{T}_0\mathbf{J}\boldsymbol{\delta}\boldsymbol{\delta}'\mathbf{J}'\mathbf{T}_0)\\ &\times E\big[(1-(p-3)\chi_{p+3}^{-2}(\Delta^2))^2 I(\chi_{p+3}^2(\Delta^2) > p-3)\big]\Big\}\\ &-2\{\mathbf{M}_4(\hat{\boldsymbol{\theta}}_n^S)\}^{-1}(\mathbf{T}_0\mathbf{J}\boldsymbol{\delta}\boldsymbol{\delta}'\mathbf{J}'\mathbf{T}_0)E\big[(1-(p-3)\chi_{p+1}^{-2}(\Delta^2))I(\chi_{p+1}^2(\Delta^2) > p-3)\big]\Big|^{-1/p}.\end{aligned}$$

(6.7.16)

Comparison of $\hat{\boldsymbol{\theta}}_n^S$ and $\hat{\boldsymbol{\theta}}_n$. In this case, the MSE-matrix difference is given by

$$\begin{aligned}
&\mathbf{M}_4(\hat{\boldsymbol{\theta}}_n^S) - \mathbf{M}_2(\hat{\boldsymbol{\theta}}_n)\\
&= \sigma^2(\boldsymbol{\Lambda}_{11} - \boldsymbol{\Lambda}_{11}^*) - (p-3)\sigma^2(\mathbf{T}_0\mathbf{J}\boldsymbol{\Lambda}_{22}\mathbf{J}'\mathbf{T}_0)\{2E[\chi_{p+1}^{-2}(\Delta^2)]\\
&-(p-3)E[\chi_{p+1}^{-4}(\Delta^2)]\} + (\mathbf{T}_0\mathbf{J}\boldsymbol{\delta}\boldsymbol{\delta}'\mathbf{J}'\mathbf{T}_0)\{(p-1)(p-3)E[\chi_{p+3}^{-4}(\Delta^2)] - 1\}.
\end{aligned} \quad (6.7.17)$$

Under H_0, (6.7.17) becomes

$$\sigma^2(\boldsymbol{\Lambda}_{11} - \boldsymbol{\Lambda}_{11}^*) - \frac{p-3}{p-1}\sigma^2(\mathbf{T}_0\mathbf{J}\boldsymbol{\Lambda}_{22}\mathbf{J}'\mathbf{T}_0) = \frac{2}{p-1}\sigma^2(\boldsymbol{\Lambda}_{11} - \boldsymbol{\Lambda}_{11}^*). \quad (6.7.18)$$

The expression (6.7.18) is p.s.d. Hence, $\hat{\boldsymbol{\theta}}_n$ performed better than $\hat{\boldsymbol{\theta}}_n^S$ under H_0. In general, the R.H.S. of (6.7.17) is negative-semidefinite(n.s.d) if for every nonzero vector $\boldsymbol{\ell}$, we have

$$\begin{aligned}
&\boldsymbol{\ell}'\Big(\boldsymbol{\Lambda}_{11} - \boldsymbol{\Lambda}_{11}^*\Big)\boldsymbol{\ell}\\
&-(p-3)\boldsymbol{\ell}'\Big(\mathbf{T}_0\mathbf{J}\boldsymbol{\Lambda}_{22}\mathbf{J}'\mathbf{T}_0\Big)\boldsymbol{\ell}\Big\{2E\Big[\chi_{p+1}^{-2}(\Delta^2)\Big] - (p-3)E\Big[\chi_{p+1}^{-4}(\Delta^2)\Big]\Big\}\\
&-\boldsymbol{\ell}'\Big(\mathbf{T}_0\mathbf{J}\boldsymbol{\Lambda}_{22}\mathbf{J}'\mathbf{T}_0\Big)\boldsymbol{\ell}\Big\{(p-1)(p-3)E\Big[\chi_{p+3}^{-4}(\Delta^2)\Big] - 1\Big\} \le \mathbf{0}.
\end{aligned} \quad (6.7.19)$$

Dividing by $\boldsymbol{\ell}'\mathbf{T}_0\boldsymbol{\Lambda}_{22}\mathbf{T}_0\boldsymbol{\ell}$ and maximizing over all $\boldsymbol{\ell}$, we obtain

$$\begin{aligned}
&1 - (p-3)\Big\{2E[\chi_{p+1}^{-2}(\Delta^2)] - (p-3)E[\chi_{p+1}^{-4}(\Delta^2)]\Big\}\\
&-\Delta^2\{(p-1)(p-3)E[\chi_{p+1}^{-4}(\Delta^2)] - 1\} \le \mathbf{0}
\end{aligned}$$

or

$$\Delta^2 \ge \frac{1 - (p-3)\{2E[\chi_{p+1}^{-2}(\Delta^2)] - (p-3)E[\chi_{p+1}^{-4}(\Delta^2)]\}}{(p-1)(p-3)E[\chi_{p+1}^{-4}(\Delta^2)] - 1}. \quad (6.7.20)$$

Thus, $\hat{\boldsymbol{\theta}}_n^S$ dominates $\hat{\boldsymbol{\theta}}_n$ whenever (6.7.20) is satisfied; otherwise, $\hat{\boldsymbol{\theta}}_n$ dominates $\hat{\boldsymbol{\theta}}_n^S$. Thus, neither $\hat{\boldsymbol{\theta}}_n^S$ nor $\hat{\boldsymbol{\theta}}_n$ dominates the other. The asymptotic efficiency of $\hat{\boldsymbol{\theta}}_n^S$ relative to $\hat{\boldsymbol{\theta}}_n$ is then given by

$$\text{AMRE}(\hat{\boldsymbol{\theta}}_n^S : \hat{\boldsymbol{\theta}}_n) = \frac{|\mathbf{M}_2(\hat{\boldsymbol{\theta}}_n)|^{1/p}}{|\mathbf{M}_4(\hat{\boldsymbol{\theta}}_n^S)|^{1/p}} = \frac{\text{AMRE}(\hat{\boldsymbol{\theta}}_n^S : \tilde{\boldsymbol{\theta}}_n)}{\text{AMRE}(\hat{\boldsymbol{\theta}}_n : \tilde{\boldsymbol{\theta}}_n)}, \quad (6.7.21)$$

where $\text{AMRE}(\hat{\boldsymbol{\theta}}_n^S : \tilde{\boldsymbol{\theta}}_n)$ and $\text{AMRE}(\hat{\boldsymbol{\theta}}_n : \tilde{\boldsymbol{\theta}}_n)$ are given by (6.7.16) and (6.7.4), respectively.

Comparison of $\hat{\boldsymbol{\theta}}_n^S$ and $\hat{\boldsymbol{\theta}}_n^{\mathrm{PT}}$. We may write the MSE-matrix difference as

$$\begin{aligned}\mathbf{M}_4(\hat{\boldsymbol{\theta}}_n^S) - \mathbf{M}_3(\hat{\boldsymbol{\theta}}_n^{\mathrm{PT}}) &= \sigma^2(\boldsymbol{\Lambda}_{11} - \boldsymbol{\Lambda}_{11}^*)\big[H_{p+1}(\chi_{p-1}^2(\alpha);\Delta^2) \\ &-(p-3)\{2E[\chi_{p+1}^{-2}(\Delta^2)] - (p-3)E[\chi_{p+1}^{-4}(\Delta^2)]\}\big] \\ &+(\mathbf{T}_0\mathbf{J}\boldsymbol{\delta}\boldsymbol{\delta}'\mathbf{J}'\mathbf{T}_0)\big[(p-1)(p-3)E[\chi_{p+3}^{-4}(\Delta^2)] \\ &-\{2E[\chi_{p+1}^{-2}(\Delta^2)] - E[\chi_{p+3}^{-2}(\Delta^2)]\}\big]. \end{aligned} \tag{6.7.22}$$

The expression (6.7.22) is p.s.d. whenever

$$\Delta^2 \le \frac{H_{p+1}(\chi_{p-1}^2(\alpha);\Delta^2) - (p-3)\{2E[\chi_{p+1}^{-2}(\Delta^2)] - (p-3)E[\chi_{p+1}^{-4}(\Delta^2)]\}}{\{2E[\chi_{p+1}^{-2}(\Delta^2)] - E[\chi_{p+3}^{-2}(\Delta^2)]\} - (p-1)(p-3)E[\chi_{p+3}^{-4}(\Delta^2)]}. \tag{6.7.23}$$

Hence, $\hat{\boldsymbol{\theta}}_n^{\mathrm{PT}}$ is superior to $\hat{\boldsymbol{\theta}}_n^S$ whenever (6.7.23) is satisfied; otherwise, $\hat{\boldsymbol{\theta}}_n^S$ is superior to $\hat{\boldsymbol{\theta}}_n^{\mathrm{PT}}$.

The asymptotic efficiency of $\hat{\boldsymbol{\theta}}_n^S$ relative to $\hat{\boldsymbol{\theta}}_n^{\mathrm{PT}}$ is given by

$$\mathrm{AMRE}(\hat{\boldsymbol{\theta}}_n^S : \hat{\boldsymbol{\theta}}_n^{\mathrm{PT}}) = \frac{\mathrm{AMRE}(\hat{\boldsymbol{\theta}}_n^S : \tilde{\boldsymbol{\theta}}_n)}{\mathrm{AMRE}(\hat{\boldsymbol{\theta}}_n^{\mathrm{PT}} : \tilde{\boldsymbol{\theta}}_n)}, \tag{6.7.24}$$

where $\mathrm{AMRE}(\hat{\boldsymbol{\theta}}_n^S : \tilde{\boldsymbol{\theta}}_n)$ and $\mathrm{AMRE}(\hat{\boldsymbol{\theta}}_n^{\mathrm{PT}} : \tilde{\boldsymbol{\theta}}_n)$ are given by (6.7.12) and (6.7.8), respectively.

Similarly, we can compare $\hat{\boldsymbol{\theta}}_n^{\mathrm{PT}}$ and $\hat{\boldsymbol{\theta}}_n$, $\hat{\boldsymbol{\theta}}_n^S$ and $\hat{\boldsymbol{\theta}}_n$, $\hat{\boldsymbol{\theta}}_n^{S+}$ and $\hat{\boldsymbol{\theta}}_n^{\mathrm{PT}}$, which are given in the problems section 6.12.

6.8 Confidence Set Estimation: Normal Case

So far, we have considered point estimators of the intercept and slope parameters in the parallelism model from preliminary test and Stein's perspectives. However, along with the point estimators, it is important to provide some information on confidence sets. This section is set out to study this problem.

6.8.1 Confidence Sets for the Slope Parameters

In this section, we consider the confidence set estimation of the slope parameters, $\boldsymbol{\beta} = (\beta_1, \cdots, \beta_p)'$ when it is suspected that $\boldsymbol{\beta} = \beta_0\mathbf{1}_p$ holds in the parallelism model. Consider, again, the model

$$\mathbf{Y}_\alpha = \theta_\alpha\mathbf{1}_{n_\alpha} + \beta_\alpha\mathbf{x}_\alpha + \varepsilon_\alpha, \quad \alpha = 1, \ldots, p, \tag{6.8.1}$$

where $\boldsymbol{\varepsilon}_\alpha \sim \mathcal{N}_{n_\alpha}(\mathbf{0}, \sigma^2\mathbf{I}_{n_\alpha})$ as stated in Section 6.1.1 and σ^2 **is known.** Thus, we consider several estimators of $\boldsymbol{\beta}$ of the type

$$\boldsymbol{\beta}_n^* = \mathbf{A}\tilde{\boldsymbol{\beta}}_n + (\mathbf{I}_p - \mathbf{A})\tilde{\boldsymbol{\beta}}_n g(\mathcal{L}_n), \tag{6.8.2}$$

where

$$\mathcal{L}_n = \sigma^{-2}\tilde{\boldsymbol{\beta}}_n'(\mathbf{I}_p - \mathbf{A})'\mathbf{D}_{22}^{-1}(\mathbf{I}_p - \mathbf{A})\tilde{\boldsymbol{\beta}}_n \tag{6.8.3}$$

and $g(\mathcal{L}_n)$ is a nondecreasing function of the test-statistic for testing the null hypothesis $H_0 : \boldsymbol{\beta} = \beta_0 \mathbf{1}_p$ against the alternative $H_A : \boldsymbol{\beta} \neq \beta_0 \mathbf{1}_p$. Here we consider $\mathbf{A}$ to be

$$\mathbf{A} = \frac{\mathbf{1}_p \mathbf{1}_p' \mathbf{D}_{22}^{-1}}{nQ}; \tag{6.8.4}$$

$\mathbf{I}_p - \mathbf{A}$ is an idempotent matrix of rank $p-1$. We limit $g(\mathcal{L}_n)$ to the following selected functions:

(i) $g(\mathcal{L}_n) = 1$, then $\boldsymbol{\beta}_n^* = \tilde{\boldsymbol{\beta}}_n$;

(ii) $g(\mathcal{L}_n) = 0$, then $\boldsymbol{\beta}_n^* = \hat{\boldsymbol{\beta}}_n$;

(iii) $g(\mathcal{L}_n) = I(\mathcal{L}_n > \chi_{p-1}^2(\alpha))$, then $\boldsymbol{\beta}_n^* = \hat{\boldsymbol{\beta}}_n^{\mathrm{PT}}$;

(iv) $g(\mathcal{L}_n) = (1 - c\mathcal{L}_n^{-1})$, $0 < c < 2(p-3)$ then $\boldsymbol{\beta}_n^* = \hat{\boldsymbol{\beta}}_n^{S}$;

and

(v) $g(\mathcal{L}_n) = (1 - c\mathcal{L}_n^{-1})I(\mathcal{L}_n > c)$, then $\boldsymbol{\beta}_n^* = \hat{\boldsymbol{\beta}}_n^{S+}$. (6.8.5)

Following Chapter 5 (Section 5.9.1), we define the five confidence sets as

(i) $C^0(\tilde{\boldsymbol{\beta}}_n) = \left\{\boldsymbol{\beta} : \sigma^{-2}\|\boldsymbol{\beta} - \tilde{\boldsymbol{\beta}}_n\|^2_{\mathbf{D}_{22}^{-1}} \leq \chi_p^2(\gamma)\right\}$;

(ii) $C^R(\hat{\boldsymbol{\beta}}_n) = \left\{\boldsymbol{\beta} : \sigma^{-2}\|\boldsymbol{\beta} - \hat{\boldsymbol{\beta}}_n\|^2_{\mathbf{D}_{22}^{-1}} \leq \chi_p^2(\gamma)\right\}$;

(iii) $C^{\mathrm{PT}}(\hat{\boldsymbol{\beta}}_n^{\mathrm{PT}}(\alpha)) = \left\{\boldsymbol{\beta} : \sigma^{-2}\|\boldsymbol{\beta} - \hat{\boldsymbol{\beta}}_n^{\mathrm{PT}}(\alpha)\|^2_{\mathbf{D}_{22}^{-1}} \leq \chi_p^2(\gamma)\right\}$;

(iv) $C^S(\hat{\boldsymbol{\beta}}_n^{S}) = \left\{\boldsymbol{\beta} : \sigma^{-2}\|\boldsymbol{\beta} - \hat{\boldsymbol{\beta}}_n^{S}\|^2_{\mathbf{D}_{22}^{-1}} \leq \chi_p^2(\gamma)\right\}$;

(v) $C^{S+}(\hat{\boldsymbol{\beta}}_n^{S+}(c)) = \left\{\boldsymbol{\beta} : \sigma^{-2}\|\boldsymbol{\beta} - \hat{\boldsymbol{\beta}}_n^{S+}(c)\|^2_{\mathbf{D}_{22}^{-1}} \leq \chi_p^2(\gamma)\right\}$. (6.8.6)

We like to provide the properties of the coverage probabilities of these five sets.

As before, let $\boldsymbol{\Gamma} = (\boldsymbol{\Gamma}_1, \boldsymbol{\Gamma}_2)$ be an orthogonal matrix, where $\boldsymbol{\Gamma}_1$ is $p \times p-1$ and $\boldsymbol{\Gamma}_2$ is a p-vector such that $\boldsymbol{\Gamma}_1\boldsymbol{\Gamma}_1' = \mathbf{I}_p - \boldsymbol{\Gamma}_2\boldsymbol{\Gamma}_2'$. Also, $\boldsymbol{\Gamma}$ diagonizes the symmetric idempotent matrix $\mathbf{D}_{22}^{1/2}\mathbf{H}'\mathbf{D}_{22}^{1/2}\mathbf{H}\mathbf{D}_{22}^{1/2}$ of rank $(p-1)$, meaning $\boldsymbol{\Gamma}\mathbf{D}_{22}^{1/2}\mathbf{H}'\mathbf{D}_{22}^{1/2}\mathbf{H}\mathbf{D}_{22}^{1/2}\boldsymbol{\Gamma}' = \begin{pmatrix} \mathbf{I}_{p-1} & \mathbf{0} \\ \mathbf{0} & 0 \end{pmatrix}$. Further, choose $\boldsymbol{\Gamma}_2 =$

$\mathbf{D}_{22}^{-1/2}\mathbf{1}_p/\sqrt{nQ}$, then $\mathbf{H} = \mathbf{D}_{22}^{1/2}\mathbf{\Gamma}_1\mathbf{\Gamma}_1'\mathbf{D}_{22}^{-1/2}$. Define $\omega_1 = \frac{1}{\sigma}\mathbf{\Gamma}_1'\mathbf{D}_{22}^{-1/2}\tilde{\beta}_n$ and $\omega_2 = \frac{1}{\sigma}\mathbf{\Gamma}_2'\mathbf{D}_{22}^{-1/2}\tilde{\beta}_n$. Then

$$\begin{pmatrix} \omega_1 \\ \omega_2 \end{pmatrix} \sim \mathcal{N}_p\left\{\begin{pmatrix} \eta_1 \\ \eta_2 \end{pmatrix}; \begin{pmatrix} \mathbf{I}_{p-1} & \mathbf{0} \\ \mathbf{0} & 1 \end{pmatrix}\right\}, \tag{6.8.7}$$

where $\eta_1 = \frac{1}{\sigma}\mathbf{\Gamma}_1'\mathbf{D}_{22}^{-1/2}\beta$ and $\eta_2 = \frac{1}{\sigma}\mathbf{\Gamma}_2'\mathbf{D}_{22}^{-1/2}\beta$.

Hence,

$$\sigma^{-2}\|\beta - \beta_n^*\|^2_{\mathbf{D}_{22}^{-1}} = \left\|\eta_1 - \omega_1 g(\|\omega_1\|^2)\right\|^2 + (\eta_2 - \omega_2)^2. \tag{6.8.8}$$

We can then write the **coverage probability** of the set

$$C^*(\beta_n^*) = \{\beta : \sigma^{-2}\|\beta - \beta_n^*\|^2_{\mathbf{D}_{22}^{-1}} \le \chi_p^2(\gamma)\} \tag{6.8.9}$$

as

$$P_\eta\left\{\left[(\eta_2 - \omega_2)^2 + \left\|\eta_1 - \omega_1 g(\|\omega_1\|^2)\right\|^2\right] \le \chi_p^2(\gamma)\right\}$$
$$= \int_0^{\chi_p^2(\gamma)} P_\eta\left\{\left\|\eta_1 - \omega_1 g(\|\omega_1\|^2)\right\|^2 \le (\chi_p^2(\gamma) - t)^+\right\} dH_1(t, 0), \tag{6.8.10}$$

where $H_\nu(\cdot : \Delta^2)$ is the cdf of a noncentral chi-square distribution with ν d.f. and noncentrality parameter $\Delta^2/2$. Specifically, we may write the coverage probability as follows:

(i) If $g(\|\omega_1\|^2) = 1$, then by definition

$$\begin{aligned} 1 - \gamma &= P_0\{(\eta_2 - \omega_2)^2 + \|\eta_1 - \omega_1\|^2 \le \chi_p^2(\gamma)\} \\ &= \int_0^{\chi_p^2(\gamma)} P_{\eta_1}\{\|\eta_1 - \omega_1\|^2 \le (\chi_p^2(\gamma) - t)^+\} dH_1(t, 0) \\ &= H_p(\chi_p^2(\gamma); 0). \end{aligned} \tag{6.8.11}$$

(ii) If $g(\|\omega_1\|^2) = 0$, then we have

$$\begin{aligned} &= P_\eta\left\{(\eta_2 - \omega_2)^2 + \|\eta_1\|^2 \le \chi_p^2(\gamma)\right\} \\ &= H_1(\chi_p^2(\gamma) - \Delta^2; 0). \end{aligned} \tag{6.8.12}$$

(iii) If $g(\|\omega_1\|^2) = I(\|\omega_1\|^2 > \chi_{p-1}^2(\alpha))$, then we have

$$\begin{aligned} &= P_\eta\{(\eta_2 - \omega_2)^2 + \left\|\eta_1 - \omega_1 I(\|\omega_1\|^2 > \chi_{p-1}^2(\alpha))\right\|^2 \le \chi_p^2(\gamma)\} \\ &= P_\eta\{(\eta_2 - \omega_2)^2 \le \chi_p^2(\gamma) - \Delta^2; \|\omega_1\|^2 \le \chi_{p-1}^2(\alpha)\} \\ &\quad + P_\eta\{\|\eta_1 - \omega_1\|^2 \le (\chi_p^2(\gamma) - t)^+; \|\omega_1\|^2 > \chi_{p-1}^2(\alpha)\} \\ &= H_1(\chi_p^2(\gamma); 0) H_{p-1}(\chi_{p-1}^2(\alpha); \Delta^2) \\ &\quad + \int_0^{\chi_p^2(\gamma)} P_{\eta_1}\{\|\eta_1 - \omega_1\|^2 \le (\chi_p^2(\gamma) - t)^+; \|\omega_1\|^2 > \chi_{p-1}^2(\alpha)\} dH_1(t, 0). \end{aligned} \tag{6.8.13}$$

(iv) If $g(\|\omega_1\|^2) = (1 - c\|\omega_1\|^{-2}),\ 0 < c < 2(p-3)$ then we have

$$= P_{\boldsymbol{\eta}}\{(\eta_2 - \omega_2)^2 + \left\|\boldsymbol{\eta}_1 - \omega_1(1 - c\|\omega_1\|^{-2})\right\|^2 \le \chi_p^2(\gamma)\}$$

$$= \int_0^{\chi_p^2(\gamma)} P_{\boldsymbol{\eta}_1}\{\|\boldsymbol{\eta}_1 - \omega_1(1 - c\|\omega_1\|^{-2})\|^2 \le (\chi_p^2(\gamma) - t)^+\} dH_1(t, 0). \tag{6.8.14}$$

(v) Finally, if $g(\|\omega_1\|^2) = (1 - c\|\omega_1\|^{-2}) I(\|\omega_1\|^2 > c)$, then we have

$$\begin{aligned}
&= P_{\boldsymbol{\eta}}\{(\eta_2 - \omega_2)^2 + \left\|\boldsymbol{\eta}_1 - \omega_1(1 - c\|\omega_1\|^{-2}) I(\|\omega_1\|^2 > c)\right\|^2 \le \chi_p^2(\gamma)\} \\
&= P_{\boldsymbol{\eta}}\{(\eta_2 - \omega_2)^2 + \|\boldsymbol{\eta}_1\|^2 \le \chi_p^2(\gamma); \|\omega_1\|^2 \le c\} \\
&\quad + P_{\boldsymbol{\eta}}\{\|\eta_2 - \omega_2\|^2 + \left\|\boldsymbol{\eta}_1 - \omega_1(1 - c\|\omega_1\|^{-2})\right\|^2 \le \chi_p^2(\gamma); \|\omega_1\|^2 > c\} \\
&= H_1(\chi_p^2(\gamma) - \Delta^2; 0) H_{p-1}(c; \Delta^2) \\
&\quad + \int_0^{\chi_p^2(\gamma)} P_{\boldsymbol{\eta}_1}\{\left\|\boldsymbol{\eta}_1 - \omega_1(1 - c\|\omega_1\|^{-2})\right\|^2 \\
&\qquad \le (\chi_p^2(\gamma) - t)^+; \|\omega_1\|^2 > c\} dH_1(t, 0).
\end{aligned} \tag{6.8.15}$$

6.8.2 Analysis of Coverage Probabilities

Clearly, $P(C^0(\tilde{\boldsymbol{\beta}}_n)) = 1 - \gamma$, which is a constant for all Δ^2. Next, we note that

$$P(C^R(\hat{\boldsymbol{\beta}}_n)) = H_1(\chi_p^2(\gamma) - \Delta^2; 0), \tag{6.8.16}$$

which is a decreasing function of Δ^2, has a maximum ($\ge 1-\gamma$) at $\Delta^2 = 0$, and decreases to zero when $\Delta^2 \to \chi_p^2(\gamma)$. The coverage probabilities of $P(C^0(\tilde{\boldsymbol{\beta}}_n))$ and $P(C^R(\hat{\boldsymbol{\beta}}_n))$ are equal when $\Delta^2 = \Delta_\gamma^2$ where

$$\Delta_\gamma^2 = \chi_p^2(\gamma) - H_1^{-1}(1 - \gamma). \tag{6.8.17}$$

Now, consider the confidence set $C^{\mathrm{PT}}(\hat{\boldsymbol{\beta}}_n^{\mathrm{PT}})$. In this case, the coverage probability is given by (6.8.13),

$$\begin{aligned}
&H_1(\chi_p^2(\gamma) - \Delta^2; 0) H_{p-1}(\chi_{p-1}^2(\alpha); \Delta^2) \\
&+ \int_0^{\chi_p^2(\gamma)} P_{\boldsymbol{\eta}_1}\{\|\boldsymbol{\eta}_1 - \omega_1\|^2 \le (\chi_p^2(\gamma) - t)^+; \|\omega_1\|^2 > \chi_{p-1}^2(\alpha)\} dH_1(t, 0),
\end{aligned} \tag{6.8.18}$$

and we have the following theorem: similar to Theorem 5.9.1:

Theorem 1.

(i) If $\Delta^2 \leq \chi_p^2(\gamma)$, $P(C^{\rm PT}(\hat{\beta}_n^{\rm PT}(\alpha))) \geq 1-\gamma$.

(ii) If $\chi_p^2(\gamma) \leq \Delta^2 \leq (\chi_p(\gamma)+\chi_{p-1}(\alpha))^2$, $P(C^{\rm PT}(\hat{\beta}_n^{\rm PT}(\alpha))) \leq 1-\gamma$.

(iii) If $\Delta^2 \geq (\chi_p(\gamma)+\chi_{p-1}(\alpha))^2$, $P(C^{\rm PT}(\hat{\beta}_n^{\rm PT}(\alpha))) = 1-\gamma$.

Note that there is a discontinuity at $\Delta^2 = \chi_p^2(\gamma)$.

Proof. Case (i): If $\Delta^2 \leq \chi_p^2(\gamma)$, then

$$P\Big\{C^{\rm PT}(\hat{\beta}_n^{\rm PT}(\alpha))\Big\} = P_{\boldsymbol{\eta}_1}\Big\{(\eta_2-\omega_2)^2+\Delta^2 \leq \chi_p^2(\gamma); \|\boldsymbol{\omega}_1\|^2 < \chi_{p-1}^2(\alpha)\Big\}$$
$$+\int_0^{\chi_p^2(\gamma)} P_{\boldsymbol{\eta}_1}\Big\{\|\boldsymbol{\eta}_1-\boldsymbol{\omega}_1\|^2 \leq (\chi_p^2(\gamma)-t)^+; \|\boldsymbol{\omega}_1\|^2 > \chi_{p-1}^2(\alpha)\Big\}dH_1(t,0) \tag{6.8.19}$$
$$\geq \int_0^{\chi_p^2(\gamma)} P_{\boldsymbol{\eta}_1}\Big\{\|\boldsymbol{\eta}_1-\boldsymbol{\omega}_1\|^2 \leq (\chi_p^2(\gamma)-t)^+; \|\boldsymbol{\omega}_1\|^2 < \chi_{p-1}^2(\alpha)\Big\}dH_1(t,0)$$
$$+\int_0^{\chi_p^2(\gamma)} P_{\boldsymbol{\eta}_1}\Big\{\|\boldsymbol{\eta}_1-\boldsymbol{\omega}_1\|^2 \leq (\chi_p^2(\gamma)-t)^+; \|\boldsymbol{\omega}_1\|^2 > \chi_{p-1}^2(\alpha)\Big\}dH_1(t,0)$$
$$=\int_0^{\chi_p^2(\gamma)} P_{\eta}\{\|\boldsymbol{\eta}_1-\boldsymbol{\omega}_1\|^2 \leq (\chi_p^2(\gamma)-t)^+\}dH_1(t,0)$$
$$= P_0\{\|\boldsymbol{\eta}_1-\boldsymbol{\omega}_1\|^2+(\eta_2-\omega_2)^2 \leq \chi_p^2(\gamma)\} = 1-\gamma. \tag{6.8.20}$$

Case (ii): If $\chi_p^2(\gamma) \leq \Delta^2 \leq (\chi_p(\gamma)+\chi_{p-1}(\alpha))^2$, then

$$P\{C^{\rm PT}(\hat{\beta}_n^{\rm PT}(\alpha))\}$$
$$=\int_0^{\chi_p^2(\gamma)} P_{\boldsymbol{\eta}_1}\Big\{\|\boldsymbol{\eta}_1-\boldsymbol{\omega}_1\|^2 \leq (\chi_p^2(\gamma)-t)^+; \|\boldsymbol{\omega}_1\|^2 > \chi_{p-1}^2(\alpha)\Big\}dH_1(t,0) \tag{6.8.21}$$
$$\leq P_0\Big\{(\eta_2-\omega_2)^2+\|\boldsymbol{\eta}_1-\boldsymbol{\omega}_1\|^2 \leq \chi_p^2(\gamma)\Big\} = 1-\gamma. \tag{6.8.22}$$

Case (iii): If $\Delta^2 \geq (\chi_p(\gamma)+\chi_{p-1}(\alpha))^2$,

$$P\Big\{C^{\rm PT}(\hat{\beta}_n^{\rm PT}(\alpha))\Big\}$$
$$= P_{\boldsymbol{\eta}_1}\Big\{(\eta_1-\omega_1)^2 \leq (\chi_p^2(\gamma)-t)^+; \|\boldsymbol{\omega}_1\|^2 \geq \chi_{p-1}^2(\alpha)\Big\}. \tag{6.8.23}$$

Now, $\|\boldsymbol{\eta}_1-\boldsymbol{\omega}_1\| \leq |(\chi_p^2(\gamma)-t)^+|^{1/2}$ implies

$$\Delta - \|\boldsymbol{\omega}_1\| \leq \|\boldsymbol{\eta}_1-\boldsymbol{\omega}_1\| \leq |(\chi_p^2(\gamma)-t)^+|^{1/2}$$

or

$$\Delta \leq \|\omega_1\| + |(\chi_p^2(\gamma) - t)^+|^{1/2}. \tag{6.8.24}$$

We then have

$$\left|(\chi_p^2(\gamma) - t)^+\right|^{1/2} + \chi_{p-1}(\alpha) \leq \Delta < \|\omega_1\| + |(\chi_p^2(\gamma) - t)^+|^{1/2}. \tag{6.8.25}$$

Hence, $\|\omega_1\|^2 > \chi_{p-1}^2(\alpha)$. Thus, from (6.8.23) we have the R.H.S. as

$$\begin{aligned}
&P_{\eta_1}\left\{\|\eta_1 - \omega_1\|^2 < (\chi_p^2(\gamma) - t)^+; \|\omega_1\|^2 > \chi_{p-1}^2(\alpha)\right\} \\
&= P_{\eta_1}\left\{(\eta_2 - \omega_2)^2 + \|\eta_1 - \omega_1\|^2 \leq \chi_p^2(\gamma); \|\omega_1\|^2 > \chi_{p-1}^2(\alpha)\right\} \\
&= P_0\left\{(\eta_2 - \omega_2)^2 + \|\eta_1 - \omega_1\|^2 \leq \chi_p^2(\gamma)\right\} = 1 - \gamma.
\end{aligned} \tag{6.8.26}$$

This completes the proof.

The graph of the coverage probability of $C^{\text{PT}}(\hat{\beta}_n^{\text{PT}}(\alpha))$ as a function of Δ^2 may be described as follows: As function of Δ^2, the coverage probability is decreasing in the range $0 \leq \Delta^2 \leq \chi_p^2(\gamma)$ with a maximum at $\Delta^2 = 0$. At $\Delta^2 = \chi_p^2(\gamma)$ there is discontinuity, and it drops to

$$\int_0^{\chi_p^2(\gamma)} P_{\eta_1}\left\{\chi_{p-1}^2 \leq (\chi_p^2(\gamma) - t)^+; \chi_{p-1}^2(\Delta^2) \geq \chi_{p-1}^2(\alpha)\right\} dH_1(t, 0), \tag{6.8.27}$$

then increases as Δ^2 increases in the range

$$\chi_p^2(\gamma) < \Delta^2 < (\chi_p(\gamma) + \chi_{p-1}(\alpha))^2, \tag{6.8.28}$$

and eventually goes to $1 - \gamma$ as Δ exceeds the limit

$$(\chi_p(\gamma) + \chi_{p-1}(\alpha))^2.$$

Next, we consider the coverage probability of the set $C^{S+}(\hat{\beta}_n^{S+}(c))$. In this case, we have

$$\begin{aligned}
P\{C^{S+}(\hat{\beta}_n^{S+}(c))\} &= H_1(\chi_p^2(\gamma) - \Delta^2; 0) H_{p-1}(c; \Delta^2) \\
&\quad + \int_0^{\chi_p^2(\gamma)} P_{\eta_1}\{\|\eta_1 - \omega_1(1 - c\|\omega_1\|^{-2})\|^2 \\
&\quad \leq (\chi_p^2(\gamma) - t)^+; \|\omega_1\|^2 > c\} dH_1(t, 0) \ \geq \ (1 - \gamma).
\end{aligned} \tag{6.8.29}$$

The first term on the R.H.S. of (6.8.29) is decreasing as a function of Δ^2, and the second term is increasing in Δ^2 and for all values of Δ^2 the coverage probability is at least $(1 - \gamma)$. Hence, $C^{S+}(\hat{\beta}_n^{S+}(c))$ uniformly dominates $C^0(\tilde{\beta}_n)$ in coverage probability.

Before we prove the assertion (6.8.29), we define the functions

$$\mathcal{M}_1\left(c; \left(\chi_p^2(\gamma) - t\right)^+\right) = \left[\frac{\frac{1}{2}\left(\chi_p^2(\gamma) - t\right)^+ + \sqrt{\frac{1}{2}\left(\chi_p^2(\gamma) - t\right)^+ + c}}{\sqrt{c}}\right]^{p-2} \exp\left\{-\left(\chi_p^2(\gamma) - t\right)^+ \sqrt{\frac{c}{2}}\right\} \tag{6.8.30a}$$

and

$$\mathcal{M}_2\left(c; \left(\chi_p^2(\gamma) - t\right)^+\right) = \left[\frac{\left(\chi_p^2(\gamma) - t\right)^+ + \sqrt{\left(\chi_p^2(\gamma) - t\right)^+ + c}}{\sqrt{c}}\right]^{p-2} \times \exp\left\{-\left(\chi_p^2(\gamma) - t\right)^+ \sqrt{c}\left[\frac{\sqrt{\left(\chi_p^2(\gamma) - t\right)^+ + 4c} - \sqrt{\left(\chi_p^2(\gamma) - t\right)^+}}{2\sqrt{c}}\right]\right\}. \tag{6.8.30b}$$

Theorem 2. $C^{S+}(\hat{\beta}_n^{S+}(c))$ has higher coverage probability than $C^0(\tilde{\beta}_n)$ for all Δ^2 such that $0 < c < c_0$, where c_0 is the minimum of the two unique solution

$$\mathcal{M}_1\left(c; \left(\chi_p^2(\gamma) - t\right)^+\right) = 1$$

and

$$\mathcal{M}_2\left(c; \left(\chi_p^2(\gamma) - t\right)^+\right) = 1. \tag{6.8.31}$$

Proof. We have to show that for every $0 < b^2 < \chi_p^2(\gamma)$,

$$H_1(\chi_p^2(\gamma) - \Delta^2; 0) H_{p-1}(c; \Delta^2) + \int_0^{\chi_p^2(\gamma)} P_{\eta_1}\left\{\left\|\eta_1 - \omega_1(1 - c\|\omega_1\|^{-2})\right\|^2 \le b^2; \|\omega_1\|^2 > c\right\} dH_1(t, 0)$$
$$\ge \int_0^{\chi_p^2(\gamma)} P_{\eta_1}\left\{\|\eta_1 - \omega_1\|^2 \le (\chi_p^2(\gamma) - t)^+\right\} dH_1(t; 0) = 1 - \gamma. \tag{6.8.32}$$

It is sufficient to establish that $\mathcal{M}_1(c; b) \ge 1$ and $\mathcal{M}_2(c; b) \ge 1$ for all $c \in (0, c_0)$ and $b^2 \in (0, \chi_p^2(\gamma))$. First, we prove that $\mathcal{M}_1(c; b) \ge 1$ for $b \in (0, \chi_p^2(\gamma))$. Note that $\mathcal{M}_1(c; b)$ for each fixed value of b is decreasing in c. Hence, it is sufficient to establish $\mathcal{M}_1(c^*; b) \ge 1$ where c^* satisfies $\mathcal{M}_1(c^*; \chi_p(\gamma)) = 1$. Note that

$$\frac{\partial^2}{\partial b^2} \log \mathcal{M}_1(c^*; b) < 0.$$

This is to say, $\frac{\partial}{\partial b}\log \mathcal{M}_1\left(c;b\right)$ is strictly decreasing in b. Hence, $\mathcal{M}_1(c^*;b)$ is (1) either strictly decreases to zero in b or (2) strictly increases to a unique maximum and then strictly decreases to zero. The first case does not hold, since

$$\mathcal{M}_1(c^*;0) = \mathcal{M}_1(c^*;\chi_p^2(\gamma)) = 1.$$

Hence, by (ii),

$$\mathcal{M}_1(c^*;b) > 1 \quad \text{for } 0 < b^2 < \chi_p^2(\gamma).$$

Similarly, $\mathcal{M}_2(c;b) > 1$ for $0 < b^2 < \chi_p^2(\gamma)$. Therefore,

$$\begin{aligned}
&H_1(\chi_p^2(\gamma) - \Delta^2; 0)H_{p-1}(c;\Delta^2) \\
&+ \int_0^{\chi_p^2(\gamma)} P_{\boldsymbol{\eta}_1}\{\|\boldsymbol{\eta}_1 - \boldsymbol{\omega}_1(1 - c\|\boldsymbol{\omega}_1\|^{-2})\|^2 \le (\chi_p^2(\gamma) - t)^+; \|\boldsymbol{\omega}_1\|^2 > c\}dH_1(t,0) \\
&\ge \int_0^{\chi_p^2(\gamma)} P_{\boldsymbol{\eta}_1}\{\|\boldsymbol{\eta}_1 - \boldsymbol{\omega}_1\|^2 \le (\chi_p^2(\gamma) - t)^+\}dH_1(t,0) \\
&= 1 - \gamma. \qquad (6.8.33)
\end{aligned}$$

6.8.3 Confidence Sets for the Intercept Parameters when σ^2 is Known

In this section, we propose some confidence sets for the intercept parameters for the model (6.8.1) when it is suspected but not certain that $H_0 : \boldsymbol{\beta} = \beta_0\mathbf{1}_p$ holds. Following Section 6.8.1, we define the five confidence sets given by the expressions below, assuming σ^2 is known:

$$\text{(i)}\ C^0(\tilde{\boldsymbol{\theta}}_n) = \left\{\boldsymbol{\theta} : \sigma^{-2}\|\boldsymbol{\theta} - \tilde{\boldsymbol{\theta}}_n\|^2_{\mathbf{D}_{11}^{-1}} \le \chi_p^2(\gamma)\right\}. \qquad (6.8.34a)$$

$$\text{(ii)}\ C^R(\hat{\boldsymbol{\theta}}_n) = \left\{\boldsymbol{\theta} : \sigma^{-2}\|\boldsymbol{\theta} - \hat{\boldsymbol{\theta}}_n\|^2_{\mathbf{D}_{11}^{-1}} \le \chi_p^2(\gamma)\right\}. \qquad (6.8.34b)$$

$$\text{(iii)}\ C^{\text{PT}}(\hat{\boldsymbol{\theta}}_n^{\text{PT}}) = \left\{\boldsymbol{\theta} : \sigma^{-2}\|\boldsymbol{\theta} - \hat{\boldsymbol{\theta}}_n^{\text{PT}}\|^2_{\mathbf{D}_{11}^{-1}} \le \chi_p^2(\gamma)\right\}. \qquad (6.8.34c)$$

$$\text{(iv)}\ C^S(\hat{\boldsymbol{\theta}}_n^S) = \left\{\boldsymbol{\theta} : \sigma^{-2}\|\boldsymbol{\theta} - \hat{\boldsymbol{\theta}}_n^S\|^2_{\mathbf{D}_{11}^{-1}} \le \chi_p^2(\gamma)\right\}. \qquad (6.8.34d)$$

$$\text{(v)}\ C^{S+}(\hat{\boldsymbol{\theta}}_n^{S+}(c)) = \left\{\boldsymbol{\theta} : \sigma^{-2}\|\boldsymbol{\theta} - \hat{\boldsymbol{\theta}}_n^{S+}(c)\|^2_{\mathbf{D}_{11}^{-1}} \le \chi_p^2(\gamma)\right\}. \qquad (6.8.34e)$$

We can write the confidence sets compactly as

$$C^*(\boldsymbol{\theta}_n^*) = \{\boldsymbol{\theta} : \sigma^{-2}\|\boldsymbol{\theta} - \boldsymbol{\theta}_n^*\|^2_{\mathbf{D}_{11}^{-1}} \le \chi_p^2(\gamma)\}, \qquad (6.8.35)$$

where $\theta_n^* = \hat{\theta}_n + (\tilde{\theta}_n - \hat{\theta}_n)g(\mathcal{L}_n)$, $g(\mathcal{L}_n)$ is given by (6.8.2) and $\mathcal{L}_n = \sigma^{-2}\tilde{\beta}_n'\mathbf{H}'\mathbf{D}_{22}^{-1}\mathbf{H}\tilde{\beta}_n$. By a suitable transformation (see Section 6.1.3), we then find that

$$\sigma^{-2}\|\theta - \theta_n^*\|^2 = (\eta_2 - \omega_2) + \|\eta_1 - \omega_1 g(\|\omega_1\|^2)\|^2, \tag{6.8.36}$$

where $\begin{pmatrix} \omega_1 \\ \omega_2 \end{pmatrix} \sim \mathcal{N}_p\left\{\begin{pmatrix} \eta_1 \\ \eta_2 \end{pmatrix}; \begin{pmatrix} \mathbf{I}_{p-1} & \mathbf{0} \\ \mathbf{0} & 1 \end{pmatrix}\right\}$. The coverage probability of (6.8.35) is then given by

$$P(C^*(\theta_n^*)) = P\{\|\eta_1 - \omega_1 g(\|\omega_1\|^2)\|^2 + (\eta_2 - \omega_2)^2 \le \chi_p^2(\gamma)\}, \tag{6.8.37}$$

which is the same as (6.8.10). Hence the details of Sections 6.8.1 and 6.8.2 are applicable to evaluate *properties* of the confidence sets with center $\tilde{\theta}_n$, $\hat{\theta}_n$, $\hat{\theta}_n^{PT}$, $\hat{\theta}_n^{S}$, and $\hat{\theta}_n^{S+}$, respectively, which are described in Section 6.8.3 replacing β_n^* by θ_n^*.

6.9 Confidence Set Estimation: Nonnormal Case

As in Section 6.8.1, we consider the confidence sets for β as follows, based on the five estimators of β:

$$\text{(i)}\ C^0(\tilde{\beta}_n) = \left\{\beta : ns_e^{-2}\|\beta - \tilde{\beta}_n\|^2_{\mathbf{D}_{22}^{-1}} \le \chi_p^2(\gamma)\right\}. \tag{6.9.1}$$

$$\text{(ii)}\ C^R(\hat{\beta}_n) = \left\{\beta : ns_e^{-2}\|\beta - \hat{\beta}_n\|^2_{\mathbf{D}_{22}^{-1}} \le \chi_p^2(\gamma)\right\}.$$

$$\text{(iii)}\ C^{PT}(\hat{\beta}_n^{PT}) = \left\{\beta : ns_e^{-2}\|\beta - \hat{\beta}_n^{PT}\|^2_{\mathbf{D}_{22}^{-1}} \le \chi_p^2(\gamma)\right\}.$$

$$\text{(iv)}\ C^S(\hat{\beta}_n^S) = \left\{\theta : ns_e^{-2}\|\beta - \hat{\beta}_n^S\|^2_{\mathbf{D}_{22}^{-1}} \le \chi_p^2(\gamma)\right\}.$$

$$\text{(v)}\ C^{S+}(\hat{\beta}_n^{S+}(c)) = \left\{\beta : ns_e^{-2}\|\beta - \hat{\beta}_n^{S+}(c)\|^2_{\mathbf{D}_{22}^{-1}} \le \chi_p^2(\gamma)\right\}.$$

In the nonnormal situation, we clearly have

$$\begin{aligned} &\lim_{n\to\infty} P_{K_{(n)}}\left\{ns_e^{-2}\|\beta - \tilde{\beta}_n\|^2_{\mathbf{D}_{22}^{-1}} \le \chi_p^2(\gamma)\right\} \\ &= P_0\left\{n\sigma^{-2}\|\beta - \tilde{\beta}_n^{S+}(c)\|^2_{\mathbf{\Lambda}_{22}^{-1}} \le \chi_p^2(\gamma)\right\} = 1 - \gamma. \end{aligned} \tag{6.9.2}$$

Now, consider the five estimators in a compact form as in Section 6.8.1:

$$\beta_n^* = \mathbf{A}_n\tilde{\beta}_n + (\mathbf{I} - \mathbf{A}_n)\tilde{\beta}_n g(\mathcal{L}_n) \tag{6.9.3}$$

where $\mathbf{A}_n = \mathbf{1}_p\mathbf{1}_p'\mathbf{D}_{22}^{-1}/nQ \to \mathbf{1}_p\mathbf{1}_p'\mathbf{\Lambda}_{22}^{-1}/\bar{Q}$ as $n \to \infty$. Then, under $\{K_{(n)}\}$ as $n \to \infty$, by Theorem 1 of Section 6.5.3, we can write

$$\lim_{n\to\infty} P_{K_{(n)}}\left\{ns_e^{-2}\|\boldsymbol{\beta} - \boldsymbol{\beta}_n^*\|^2_{\mathbf{D}_{22}^{-1}} \le \chi_p^2(\gamma)\right\}$$
$$= P_0\left\{\left\|\boldsymbol{\eta}_1 - \boldsymbol{\omega}_1 g(\|\boldsymbol{\omega}_1\|^2)\right\|^2 + (\boldsymbol{\eta}_2 - \boldsymbol{\omega}_2)^2 \le \chi_p^2(\gamma)\right\}, \tag{6.9.4}$$

where

$$\begin{pmatrix} \boldsymbol{\omega}_1 \\ \omega_2 \end{pmatrix} \sim \mathcal{N}_p\left\{\begin{pmatrix} \boldsymbol{\eta}_1 \\ \eta_2 \end{pmatrix}, \begin{pmatrix} \mathbf{I}_{p-1} & \mathbf{0} \\ \mathbf{0} & 1 \end{pmatrix}\right\}. \tag{6.9.5}$$

Thus, under $\{K_{(n)}\}$ as $n \to \infty$, we obtain the following asymptotic coverage probabilities according to the choice of $g(\|\boldsymbol{\omega}_1\|^2)$:

1. If $g(\|\boldsymbol{\omega}_1\|^2) = 1$, then (6.9.4) becomes

$$P_0\{\|\eta_1 - \boldsymbol{\omega}_1\|^2 + (\eta_2 - \omega_2)^2 \le \chi_p^2(\gamma)\}. \tag{6.9.6}$$

2. If $g(\|\boldsymbol{\omega}_1\|^2) = 0$, then (6.9.4) becomes

$$P_{\boldsymbol{\eta}_1}\{\|\boldsymbol{\eta}_1\|^2 + (\eta_2 - \omega_2)^2 \le \chi_p^2(\gamma)\} = H_1(\chi_p^2(\gamma) - \Delta^2; 0). \tag{6.9.7}$$

3. If $g(\|\boldsymbol{\omega}_1\|^2) = I(\|\boldsymbol{\omega}_1\|^2 \ge \chi_{p-1}^2(\alpha))$, then (6.9.4) becomes

$$\begin{aligned} &P_{\boldsymbol{\eta}_1}\{\|\boldsymbol{\eta}_1 - \omega_2\|^2 + (\eta_2 - \omega_2)^2 \le \chi_p^2(\gamma); \|\boldsymbol{\omega}_1\|^2 \le \chi_{p-1}^2(\alpha)\} \\ &\quad + P_{\boldsymbol{\eta}_1}\{\|\boldsymbol{\eta}_1 - \boldsymbol{\omega}_1\|^2 + (\eta_2 - \omega_2)^2 \le \chi_p^2(\gamma); \|\boldsymbol{\omega}_1\| > \chi_{p-1}^2(\alpha)\} \\ &= H_{p-1}(\chi_{p-1}^2(\alpha); \Delta^2) H_1(\chi_p^2(\gamma) - \Delta^2; 0) \\ &\quad + \int_0^{\chi_p^2(\gamma)} P_{\eta_1}\{\|\boldsymbol{\eta}_1 - \boldsymbol{\omega}_1\|^2 \le (\chi_p^2(\gamma) - t)^+, \|\boldsymbol{\omega}_1\|^2 > \chi_{p-1}^2(\alpha)\} dH_1(t, 0). \end{aligned} \tag{6.9.8}$$

4. If $g(\|\boldsymbol{\omega}_1\|^2) = 1 - c\|\boldsymbol{\omega}_1\|^2$, then (6.9.4) becomes

$$\begin{aligned} &P_{\boldsymbol{\eta}_1}\{\|\boldsymbol{\eta}_1 - \omega_1(1 - c\|\boldsymbol{\omega}1\|^2 + (\eta_2 - \omega_2)^2 \le \chi_p^2(\gamma); \|\boldsymbol{\omega}_1\|^2 \le \chi_{p-1}^2(\alpha)\} \\ &\quad + P_{\boldsymbol{\eta}_1}\{\|\boldsymbol{\eta}_1 - \boldsymbol{\omega}_1\|^2 + (\eta_2 - \omega_2)^2 \le \chi_p^2(\gamma); \|\boldsymbol{\omega}_1\| > \chi_{p-1}^2(\alpha)\} \\ &= H_{p-1}(\chi_{p-1}^2(\alpha); \Delta^2) H_1(\chi_p^2(\gamma) - \Delta^2; 0) \\ &\quad + \int_0^{\chi_p^2(\gamma)} P_{\boldsymbol{\eta}_1}\{\|\boldsymbol{\eta}_1 - \boldsymbol{\omega}_1\|^2 \le (\chi_p^2(\gamma) - t)^+, \|\boldsymbol{\omega}_1\|^2 > \chi_{p-1}^2(\alpha)\} dH_1(t, 0). \end{aligned} \tag{6.9.9}$$

5. If $g(\|\omega_1\|^2) = (1 - c\|\omega_1\|^2)I(\|\omega_1\|^2 > c$, then (6.9.4) becomes

$$\begin{aligned}
&P_{\boldsymbol{\eta}_1}\{\|\boldsymbol{\eta}_1 - \omega_1(1-(p-3)\|\boldsymbol{\omega}1\|^2 + (\eta_2 - \omega_2)^2 \leq \chi_p^2(\gamma)\} \\
&= P_{\boldsymbol{\eta}_1}\{(\eta_2 - \omega_2)^2 + \Delta^2 \leq \chi_p^2(\gamma); \|\boldsymbol{\omega}_1\|^2 < (p-3)\} \\
&\quad + P_{\boldsymbol{\eta}_1}\{(\eta_2-\omega_2)^2 + \|\boldsymbol{\eta}_1 - \boldsymbol{\omega}_1(1-c\|\boldsymbol{\omega}_1\|^{-2})\|^2 \leq \chi_p^2(\gamma); \|\boldsymbol{\omega}_1\|^2 < p-3\} \\
&= H_1(\chi_p^2(\gamma) - \Delta^2; 0)H_{p-1}((p-3);\Delta^2) \\
&\quad + \int_0^{\chi_p^2(\gamma)} P_{\boldsymbol{\eta}_1}\{\|\boldsymbol{\eta}_1 - \boldsymbol{\omega}_1(1-c\|\boldsymbol{\omega}_1\|^{-2})\| \leq \chi_p^2(\gamma) - t)^+; \\
&\qquad\qquad \times \|\boldsymbol{\omega}_1\|^2 > c)\}dH_1(t,0).
\end{aligned} \tag{6.9.10}$$

These expressions are similar to that of the expressions in Section 6.8.1. Hence, the analysis of the coverage probabilities is similar to that given in Section 6.8.2.

6.10 Nonparametric Methods: *R*-Estimation

In this section, we consider the nonparametric methods of estimation of the **intercept** and **slope** parameters of several simple linear models. These applications enlarge the scope of the theory of shrinkage estimation bringing it to a broader class of estimators with robustness properties.

6.10.1 Model, Assumptions, and Linear Rank Statistics

Consider, again, the set of several simple linear models given by

$$\mathbf{Y}_\alpha = \theta_\alpha \mathbf{1}_{n_\alpha} + \beta_\alpha \mathbf{x}_\alpha + \boldsymbol{\varepsilon}_\alpha, \quad \alpha = 1, \ldots, p, \tag{6.10.1}$$

where

$$\begin{aligned}
&\mathbf{Y}_\alpha = (Y_{\alpha 1}, \ldots, Y_{\alpha n_\alpha})', \quad \mathbf{x}_\alpha = (x_{\alpha 1}, \ldots, x_{\alpha n_\alpha})', \\
&\boldsymbol{\varepsilon}_\alpha = (\varepsilon_{\alpha 1}, \ldots, \varepsilon_{\alpha n_\alpha})',
\end{aligned} \tag{6.10.2}$$

with the cdf defined by

$$F(\boldsymbol{\varepsilon}_\alpha) = \prod_{j=1}^{n_\alpha} F_0(Y_{\alpha_j} - \theta_\alpha - \beta_\alpha x_{\alpha_j})$$

and

$$F(\boldsymbol{\varepsilon}_1, \ldots, \boldsymbol{\varepsilon}_p) = \prod_{\alpha=1}^{p}\prod_{j=1}^{n_\alpha} F_0(Y_{\alpha_j} - \theta_\alpha - \beta_\alpha x_{\alpha_j}). \tag{6.10.3}$$

Let $n = n_1 + \cdots + n_p$ and $\mathbf{\Lambda}_n^{-1} = \text{Diag}\left(\frac{n_1}{n}, \ldots, \frac{n_p}{n}\right)$. It is assumed that $\{\varepsilon_{\alpha_j}\}$ are mutually independent and identically distributed with cdf $F(\varepsilon_1, \ldots, \varepsilon_p)$ defined by (6.10.3), where $F_0(\cdot)$ belongs to the class $\mathcal{F}$ of absolutely continuous cdf with absolutely continuous pdf, $f_0(\cdot)$ having finite "Fisher information" as follows:

$$\text{(i)} \int_{-\infty}^{\infty} \left\{\frac{f_0'(z)}{f_0(z)}\right\}^2 f_0(z)dz < \infty. \tag{6.10.4}$$

(ii) $\lim_{n\to\infty} n_\alpha^{-1} \sum_{j=1}^{n_\alpha} x_{\alpha_j} \to \bar{x}_\alpha \; (|\bar{x}_\alpha| < \infty)$, $\lim_{n\to\infty} n^{-1} \sum_{j=1}^{n_\alpha} (x_{\alpha_j} - \bar{x}_{n_\alpha})^2 \to \lambda_\alpha Q_\alpha$.

(iii) $\lim_{n\to\infty} \frac{n_i}{n} = \lambda_i \quad (0 < \lambda_i < 1)$, meaning, $\lim_{n\to\infty} \Lambda_n^{-1} = \Lambda_0^{-1} = \text{Diag}(\lambda_1, \ldots, \lambda_p)$.

(iv) Score functions $a_n(\cdot)$ (and $a_n^+(\cdot)$) are generated by a function $\phi(u)$, $u \in (0,1)$, which is a nondecreasing, skew symmetric (i.e., $\phi(u)+\phi(1-u) = 0$ for all $u \in (0,1)$), and square integrable.

Let $\phi^+(u) = \phi\left(\frac{1+u}{2}\right)$, $u \in (0,1)$, and set

$$a_n(k) = E[\phi(U_{kn})] \quad \text{or} \quad \phi\left(\frac{k}{n+1}\right)$$

and

$$a_n^+(k) = E[\phi^+(U_{kn})] \quad \text{or} \quad \phi^+\left(\frac{k}{n+1}\right), \quad k = 1, \ldots, n, \tag{6.10.5}$$

where $0 < U_{1n} \le \cdots \le U_{nn} < 1$ are the order-statistics of a sample of size n from $U(0,1)$. Let $R_{\alpha j}(a_\alpha, b_\alpha)$ be the rank of $(Y_{\alpha j} - a_\alpha - b_\alpha x_{\alpha j})$ among $(Y_{\alpha 1} - a_\alpha - b_\alpha x_{\alpha 1}), \ldots, (Y_{\alpha n_\alpha} - a_\alpha - b_\alpha x_{\alpha n_\alpha})$ and similarly, let $R_{\alpha j}^+(a_\alpha, b_\alpha)$ be the rank of $\{|Y_{\alpha j} - a_\alpha - b_\alpha x_{\alpha j}|$ among $|(Y_{\alpha 1} - a_\alpha - b_\alpha x_{\alpha 1})|, \ldots, |(Y_{\alpha n_\alpha} - a_\alpha - b_\alpha x_{\alpha n_\alpha})|$ for $\alpha = 1, \ldots, p$. Let the vector of linear rank-statistics (LRS) be

$$\mathbf{T}_n(\mathbf{a}, \mathbf{b}) = (T_{n_1}(a_1, b_1), \ldots, T_{n_p}(a_p, b_p))',$$

where

$$T_{n_\alpha}(a_\alpha, b_\alpha) = \sum_{j=1}^{n_\alpha} a_{n_\alpha}^+(R_{\alpha j}^+(a_\alpha, b_\alpha)) \, \text{sign}(Y_{\alpha j} - a_\alpha - b_\alpha x_{\alpha j}), \tag{6.10.6a}$$

and

$$\mathbf{L}_n(\mathbf{b}) = (L_{n_1}(b_1), \ldots, L_{n_p}(b_p))',$$

where

$$L_{n_\alpha}(b_\alpha) = \sum_{j=1}^{n_\alpha} (x_{\alpha j} - \bar{x}_\alpha) a_{n_\alpha}(R_{\alpha j}(a_\alpha, b_\alpha)). \tag{6.10.6b}$$

Further, let

$$L_n^*(\mathbf{b}) = \sum_{j=1}^{p} L_{n_\alpha}(b_\alpha) = \mathbf{1}_p' \mathbf{L}_n(\mathbf{b}). \tag{6.10.7}$$

We will use the LRS above for the estimation and test of hypothesis-related problems regarding the intercepts and the slopes.

6.10.2 *R*-Estimation and Test of Hypothesis

As in Chapters 4 and 5, we define

$$A_\phi^2 = \int_0^1 \phi^2(u)du - \left(\int_0^1 \phi(u)du\right)^2,$$

$$A_\psi^2 = \int_0^1 \psi^2(u)du, \quad \psi(u) = -\frac{f_0'(F_0^{-1}(u))}{f_0(F_0^{-1}(u))}, \ 0 < u < 1,$$

and

$$\gamma(\psi, \phi) = \int_0^1 \psi(u)\phi(u)du. \tag{6.10.8}$$

We let

$$A_{n_\alpha}^2 = (n_\alpha - 1)^{-1} \sum_{k=1}^{n_\alpha} (a_{n_\alpha}(k) - \bar{a}_{n_\alpha})^2 \tag{6.10.9a}$$

with $\bar{a}_{n_\alpha} = n_\alpha^{-1} \sum_{k=1}^{n_\alpha} (a_{n_\alpha}(k)$ and $A_n^2 = (n-p)^{-1} \sum_{\alpha=1}^{p} (n_\alpha - 1) A_{n_\alpha}^2$. Further,

$$Q_{n_\alpha} = \sum_{j=1}^{n_\alpha} (x_{\alpha_j} - \bar{x}_{n_\alpha})^2, \quad \alpha = 1, \dots, p. \tag{6.10.9b}$$

Note that given $\mathbf{x}_\alpha = (x_{\alpha 1}, \dots, x_{\alpha n_\alpha})'$, $L_{n_\alpha}(b_\alpha) \searrow b_\alpha$ for $b_\alpha \in (-\infty, \infty)$ and $T_{n_\alpha}(a_\alpha, b_\alpha) \searrow a_\alpha$ for fixed b_α. Under the model (6.10.1) with $\theta_\alpha = \beta_\alpha = 0$, $T_{n_\alpha}(0,0)$ and $L_{n_\alpha}(0)$, both have distributions symmetric about 0.

First, we note that from the basic theorems of Hajek and Sidak (1967) in Chapter 4, we have that under $\theta_\alpha = \beta_\alpha = 0$

$$\begin{pmatrix} n^{-1/2}\mathbf{T}_n(\mathbf{0},\mathbf{0}) \\ n^{-1/2}\mathbf{L}_n(\mathbf{0}) \end{pmatrix} \sim \mathcal{N}_{2p}\left\{\begin{pmatrix} \mathbf{0} \\ \mathbf{0} \end{pmatrix}; \ A_\phi^2 \begin{pmatrix} \mathbf{\Lambda}_0^{-1} & \mathbf{0} \\ \mathbf{0} & \mathbf{\Lambda}_{22}^{-1} \end{pmatrix}\right\}, \tag{6.10.10}$$

where $\mathbf{\Lambda}_0^{-1} = \text{Diag}(\lambda_1, \dots, \lambda_p)$ and $\mathbf{\Lambda}_{22}^{-1} = \text{Diag}(\lambda_1 Q_1, \dots, \lambda_p Q_p)$. (6.10.11)

6.10.3 Estimation of the Intercepts θ_α and the Slope β_α

We consider two **unrestricted** estimators of θ_α and β_α based on (6.10.6a) and (6.10.6b), respectively, as follows:

$$\tilde{\beta}_{n_\alpha} = \frac{1}{2}\left[\sup\{b_\alpha : L_{n_\alpha}(b_\alpha) > 0\} + \inf\{b_\alpha : L_{n_\alpha}(b_\alpha) < 0\}\right], \tag{6.10.12a}$$

$$\tilde{\theta}_{n_\alpha} = \frac{1}{2}\big[\sup\{a_\alpha : T_{n_\alpha}(a_\alpha, \tilde{\beta}_{n_\alpha}) > 0\} + \inf\{a_\alpha : T_{n_\alpha}(a_\alpha, \tilde{\beta}_{n_\alpha}) < 0\}\big]. \tag{6.10.12b}$$

The **restricted estimator** of $\boldsymbol{\beta}$ under $\boldsymbol{\beta} = \beta_0 \mathbf{1}_p$ is defined by

$$\hat{\beta}_{0n} = \frac{1}{2}\big[\sup\{b : L_n^*(b) > 0\} + \inf\{b : L_n^*(b) < 0\}\big]. \tag{6.10.12c}$$

Hence, we denote the **unrestricted estimators** of $\boldsymbol{\theta}$ and $\boldsymbol{\beta}$ by

$$\tilde{\boldsymbol{\theta}}_n = (\tilde{\theta}_{n_1}, \dots, \tilde{\theta}_{n_p})', \quad \text{and} \quad \tilde{\boldsymbol{\beta}}_n = (\tilde{\beta}_{n_1}, \dots, \tilde{\beta}_{n_p})', \quad \text{respectively.} \tag{6.10.13}$$

Let us denote the **restricted estimator** of $\boldsymbol{\theta}$ by

$$\hat{\boldsymbol{\theta}}_{n_\alpha} = (\hat{\theta}_{n_1}, \dots, \hat{\theta}_{n_p})', \tag{6.10.14a}$$

where

$$\hat{\theta}_{n_\alpha} = \frac{1}{2}\big[\sup\{a_\alpha : T_{n_\alpha}(a_\alpha; \hat{\beta}_{0n}) > 0\} + \inf\{a_\alpha : T_{n_\alpha}(a_\alpha; \hat{\beta}_{0n}) < 0\}\big] \tag{6.10.14b}$$

and that of $\boldsymbol{\beta} = \beta_0 \mathbf{1}_p$ by $\hat{\boldsymbol{\beta}}_n = \hat{\beta}_{0n} \mathbf{1}_p$.

Test of Hypothesis: $\boldsymbol{\beta} = \beta_0 \mathbf{1}_p$. For the test of the null hypothesis $H_0 : \boldsymbol{\beta} = \beta_0 \mathbf{1}_p$, we use the nonparametric test due to Sen (1969), which is based on the statistic

$$\mathcal{L}_n = \sum_{\alpha=1}^{p-1} n_\alpha^{-1} A_{n_\alpha}^{-2} Q_{n_\alpha}^{-1} \big[L_{n_\alpha}(\hat{\beta}_{0n})\big]^2, \tag{6.10.15}$$

where $A_{n_\alpha}^2$ and Q_{n_α} are given by (6.10.9a) and (6.10.9b), respectively. Under the null hypothesis $H_0 : \boldsymbol{\beta} = \beta_0 \mathbf{1}_p$, $\mathcal{L}_n$ is approximately distributed as a chi-square with $(p-1)$ d.f. This allows us to define the PTE, SE, and PRSE of $\boldsymbol{\beta}$ as follows:

(i) $$\hat{\boldsymbol{\beta}}_n^{\mathrm{PT}} = \tilde{\boldsymbol{\beta}}_n - (\tilde{\boldsymbol{\beta}}_n - \hat{\boldsymbol{\beta}}_n) I(\mathcal{L}_n < \chi_{p-1}^2(\alpha)), \tag{6.10.16a}$$

(ii) $$\hat{\boldsymbol{\beta}}_n^{S} = \tilde{\boldsymbol{\beta}}_n - d(\tilde{\boldsymbol{\beta}}_n - \hat{\boldsymbol{\beta}}_n)\mathcal{L}_n^{-1} I(\mathcal{L} > \varepsilon_n), \quad d = p-3 \tag{6.10.16b}$$

where $\varepsilon_n \to 0$ as $n \to \infty$, and

(iii) $$\hat{\boldsymbol{\beta}}_n^{S+} = \hat{\boldsymbol{\beta}}_n - (1 - d\mathcal{L}_n^{-1}) I(\mathcal{L}_n > d)(\tilde{\boldsymbol{\beta}}_n - \hat{\boldsymbol{\beta}}_n), \tag{6.10.16c}$$

respectively. Thus, we have defined five possible estimators of $\boldsymbol{\beta}$ to generate five possible estimators of the intercept parameter $\boldsymbol{\theta}$ of the model

(i) $$\text{UE} : \tilde{\boldsymbol{\theta}}_n = (\tilde{\theta}_{n_1}, \dots, \tilde{\theta}_{n_p})'; \tag{6.10.17a}$$

(ii) $$\text{RE} : \hat{\boldsymbol{\theta}}_n = (\hat{\theta}_{n_1}, \dots, \hat{\theta}_{n_p})',; \tag{6.10.17b}$$

(iii) $$\text{PTE} : \hat{\boldsymbol{\theta}}_n^{\mathrm{PT}} = (\hat{\theta}_{n_1}^{\mathrm{PT}}, \dots, \tilde{\theta}_{n_p}^{\mathrm{PT}})'; \tag{6.10.18a}$$

where

$$\hat{\theta}_{n_\alpha}^{\mathrm{PT}} = \frac{1}{2}\Big[\sup\{a_\alpha : T_{n_\alpha}(a_\alpha; \hat{\beta}_n^{\mathrm{PT}}) > 0\} + \inf\{a_\alpha : T_{n_\alpha}(a_\alpha; \hat{\beta}_{n_\alpha}^{\mathrm{PT}}) < 0\}\Big] \tag{6.10.18b}$$

for $\alpha = 1, 2, \ldots, p$;

(iv) SE : $\hat{\boldsymbol{\theta}}_n^S = (\hat{\theta}_{n_1}^S, \ldots, \hat{\theta}_{n_p}^S)'$, (6.10.19a)

where

$$\hat{\theta}_{n_\alpha}^{S} = \frac{1}{2}\Big[\sup\{a_\alpha : T_{n_\alpha}(a_\alpha; \hat{\beta}_{n_\alpha}^{S}) > 0\} + \inf\{a_\alpha : T_{n_\alpha}(a_\alpha; \hat{\beta}_{n_\alpha}^{S}) < 0\}\Big] \tag{6.10.19b}$$

for $\alpha = 1, 2, \ldots, p$.

(v) PRSE: $\hat{\boldsymbol{\theta}}_n^{S+} = (\hat{\theta}_{n_1}^{S+}, \ldots, \hat{\theta}_{n_p}^{S+})'$ (6.10.20a)
where

$$\hat{\theta}_{n_\alpha}^{S+} = \frac{1}{2}\Big[\sup\{a_\alpha : T_{n_\alpha}(a_\alpha; \hat{\beta}_{n_\alpha}^{S+}) > 0\} + \inf\{a_\alpha : T_{n_\alpha}(a_\alpha; \hat{\beta}_{n_\alpha}^{S+}) < 0\}\Big] \tag{6.10.20b}$$

for $\alpha = 1, 2, \ldots, p$.

6.10.4 Asymptotic Distribution of the R-Estimators of the Slope Vector

First, note that by Jurečkova's (1969) asymptotic linearity result (see Section 2.8 of Chapter 2) we have under $\boldsymbol{\beta} = \mathbf{0}$, as $n \to \infty$ and $\kappa > 0$,

$$\sup\Big\{n^{-1/2}\big|L_{n_\alpha}(n^{-1/2}b_\alpha) - L_{n_\alpha}(0) + \lambda_\alpha n^{1/2} b_\alpha Q_\alpha \gamma(\psi, \phi)\big|\,|b_\alpha| < \kappa\Big\} \xrightarrow{\mathrm{P}} 0 \tag{6.10.21}$$

so that

$$\sup\Big\{n^{-1/2}\big|L_n^*(n^{-1/2}\mathbf{b}) - L_n^*(0) + n^{1/2}\Big(\sum_{\alpha=1}^{p} \lambda_\alpha b_\alpha Q_\alpha\Big)\gamma(\psi, \phi)\big|\,|b| < \kappa\Big\} \xrightarrow{\mathrm{P}} 0. \tag{6.10.22}$$

Also,

$$\sup\Big\{n^{-1/2}\big|L_n^*(n^{-1/2}b_0\mathbf{1}_p) - L_n^*(0) + \bar{Q}n^{1/2}b_0\gamma(\psi, \phi)\big|\,|b_0| < \kappa\Big\} \xrightarrow{\mathrm{P}} 0, \tag{6.10.23}$$

where $\bar{Q} = \sum_{\alpha=1}^{p} \lambda_\alpha Q_\alpha$.

Further, under $\boldsymbol{\theta} = \boldsymbol{\beta} = \mathbf{0}$ as $n \to \infty$ for $k\ (> 0)$, we have

$$\sup\Big\{n^{-1/2}\big|T_{n_\alpha}(n^{-1/2}(a_\alpha, b_\alpha)) - T_{n_\alpha}(0, 0)$$
$$+\lambda_\alpha n^{1/2}(a_\alpha + b_\alpha \bar{x}_\alpha)\gamma(\psi, \phi)\big|\,|a_\alpha| \le \kappa, |b_\alpha| \le \kappa\Big\} \xrightarrow{\mathrm{P}} 0. \tag{6.10.24}$$

(See Chapter 3, Section 3.10).

From (6.10.21) and (6.10.23), we can write under $\boldsymbol{\beta} = \mathbf{0}$,

$$n^{-1/2}L_n^*(\mathbf{0}) = n^{1/2}\Big(\sum_{\alpha=1}^{p} \lambda_\alpha Q_\alpha \tilde{\beta}_{n_\alpha}\Big)\gamma(\psi,\phi) + o_p(1)$$
$$= n^{1/2}\bar{Q}\bar{\beta}_{0n}\gamma(\psi,\phi) + o_p(1),$$

where

$$\bar{\beta}_n = \sum_{\alpha=1}^{p} \lambda_\alpha Q_\alpha \tilde{\beta}_{n_\alpha}(\bar{Q})^{-1}, \tag{6.10.25}$$

$$n^{1/2}\bar{Q}\hat{\beta}_{0n})\gamma(\psi,\phi) + o_p(1). \tag{6.10.26}$$

As a consequence, we have

$$\hat{\beta}_{0n} = \bar{\beta}_n + o_p(1). \tag{6.10.27}$$

Further, from (6.10.21) we have under $\boldsymbol{\beta} = \mathbf{0}$,

$$n^{-1/2}L_n(\mathbf{0}) = \boldsymbol{\Lambda}_{22} n^{1/2}\tilde{\boldsymbol{\beta}}_n \gamma(\psi,\phi) + o_p(1). \tag{6.10.28}$$

Hence,

$$n^{-1/2}(\tilde{\boldsymbol{\beta}}_n - \boldsymbol{\beta}) \sim \mathcal{N}_p(\mathbf{0}, \sigma^2\boldsymbol{\Lambda}_{22}), \quad \sigma^2 = \frac{A_\phi^2}{\gamma^2(\psi,\phi)}. \tag{6.10.29}$$

Now, again under $\boldsymbol{\beta} = \mathbf{0}$ and relative compactness of $n^{1/2}\hat{\beta}_{0n}$ (see Sen 1969), we obtain

(i) $n^{-1/2}L_n^*(0)\gamma^{-1}(\psi,\phi) - n^{1/2}\hat{\beta}_{0n} \xrightarrow{P} 0$

(ii) $n^{-1/2}L_{n_\alpha}(\hat{\beta}_{0n}) - \Big(n^{-1/2}L_{n_\alpha}(0) - \frac{\lambda_\alpha Q_\alpha}{\bar{Q}} n^{1/2}\hat{\beta}_{0n}\gamma(\psi,\phi)\Big) \xrightarrow{P} 0$

or

$$n^{-1/2}L_{n_\alpha}(\hat{\beta}_{0n}) - \Big(n^{-1/2}L_{n_\alpha}(0) - \frac{\lambda_\alpha Q_\alpha}{\bar{Q}} n^{1/2}L_n^*(0)\Big) \xrightarrow{P} 0. \tag{6.10.30}$$

Thus, in vector form, we write

$$n^{-1/2}\mathbf{L}_n(\hat{\boldsymbol{\beta}}_n) - \Big(\mathbf{I}_p - \frac{\boldsymbol{\Lambda}_{22}^{-1}\mathbf{1}_p\mathbf{1}_p'}{\bar{Q}}\Big)n^{-1/2}\mathbf{L}_n(0) \xrightarrow{P} 0. \tag{6.10.31}$$

Hence, under $\boldsymbol{\beta} = \mathbf{0}$ we write

$$\mathcal{L}_n = n^{-1}A_n^{-2}[\mathbf{L}_n(\mathbf{0})]'\mathbf{J}\boldsymbol{\Lambda}_{22}\mathbf{J}'[\mathbf{L}_n(\mathbf{0})] + o_p(1), \tag{6.10.32}$$

where $\mathbf{J} = \mathbf{I}_p - \frac{\mathbf{1}_p\mathbf{1}_p'}{Q}\mathbf{\Lambda}_{22}^{-1}$ and $\mathbf{J\Lambda}_{22}\mathbf{J}'$ is an idempotent matrix of rank $(p-1)$. Thus, $\mathcal{L}_n$ asymptotically approximates the central chi-square distribution with $(p-1)$ d.f.

We recall that under $\boldsymbol{\beta} = \mathbf{0}$ as $n \to \infty$,

$$n^{-1/2}\mathbf{L}_n(\mathbf{0}) \overset{\mathcal{D}}{\simeq} N_p(\mathbf{0}, A_\phi^2\mathbf{\Lambda}_{22}^{-1}),$$

and under fixed alternatives, $\mathcal{L}_n \to \infty$ as $n \to \infty$ and $\sqrt{n}(\boldsymbol{\beta}_n^* - \boldsymbol{\beta}) = \sqrt{n}(\tilde{\boldsymbol{\beta}}_n - \boldsymbol{\beta}) + o_P(1)$ as $n \to \infty$, where $\boldsymbol{\beta}_n^* = \hat{\boldsymbol{\beta}}_n^{\mathrm{PT}}$, $\hat{\boldsymbol{\beta}}_n^{S}$ and $\hat{\boldsymbol{\beta}}_n^{S+}$. Hence, we consider a class of local alternatives

$$K_{(n)} : \boldsymbol{\beta}_{(n)} = \beta_0\mathbf{1}_p + n^{-1/2}\boldsymbol{\delta}, \tag{6.10.33}$$

such that $\boldsymbol{\delta}'\Lambda_{22}^{-1}\mathbf{1}_p = 0$. We further note that this class of alternatives is contiguous to that of $H_0 : \boldsymbol{\beta} = \beta_0\mathbf{1}_p$. Thus, we have the following theorem:

Theorem 1. Under $\{K_{(n)}\}$ and the assumed regularity conditions with respect to the model (6.10.1) as $n \to \infty$, we have

(i) $\sqrt{n}(\tilde{\boldsymbol{\beta}}_n - \beta_0\mathbf{1}_\mathbf{p}) \sim \mathcal{N}_\mathbf{p}(\boldsymbol{\delta}, \sigma^2\mathbf{\Lambda}_{22})$, $\sigma^2 = \mathbf{A}_\phi^2/\gamma^2(\psi,\phi)$;

(ii) $\sqrt{n}(\tilde{\boldsymbol{\beta}}_n - \hat{\boldsymbol{\beta}}_n) \sim \mathcal{N}_p(\mathbf{J}\boldsymbol{\delta}, \sigma^2\mathbf{J\Lambda}_{22})$;

(iii) $\sqrt{n}(\hat{\beta}_{0n} - \beta_0)\mathbf{1}_p \sim \mathcal{N}_p(\boldsymbol{\delta}, \sigma^2\mathbf{B})$, $\mathbf{B} = \mathbf{1}_p\mathbf{1}_p'$;

(iv) $\begin{pmatrix} \sqrt{n}(\tilde{\boldsymbol{\beta}}_n - \beta_0\mathbf{1}_\mathbf{p}) \\ \sqrt{n}(\tilde{\boldsymbol{\beta}}_n - \hat{\boldsymbol{\beta}}_n) \end{pmatrix} \sim \mathcal{N}_{2p}\left\{\begin{pmatrix} \boldsymbol{\delta} \\ \mathbf{J}\boldsymbol{\delta} \end{pmatrix}; \sigma^2\begin{pmatrix} \mathbf{\Lambda}_{22} & \mathbf{J\Lambda}_{22} \\ \mathbf{\Lambda}_{22}\mathbf{J}' & \mathbf{J\Lambda}_{22} \end{pmatrix}\right\}$;

(v) $\begin{pmatrix} \sqrt{n}(\hat{\beta}_{0n} - \beta_0)\mathbf{1}_\mathbf{p} \\ \sqrt{n}(\tilde{\boldsymbol{\beta}}_n - \hat{\boldsymbol{\beta}}_n) \end{pmatrix} \sim \mathcal{N}_{2p}\left\{\begin{pmatrix} \boldsymbol{\delta} \\ \mathbf{J}\boldsymbol{\delta} \end{pmatrix}; \sigma^2\begin{pmatrix} \mathbf{B} & \mathbf{0} \\ \mathbf{0} & \mathbf{J\Lambda}_{22} \end{pmatrix}\right\}$;

(vi) $\lim_{n\to\infty} \{\mathcal{L}_n \le x | K_{(n)}\} = H_{p-1}(x; \Delta^2)$;

$$\text{where}\quad \mathbf{J} = \mathbf{I}_p - \frac{\mathbf{1}_p\mathbf{1}_p'}{Q}\mathbf{\Lambda}_{22}^{-1} \quad\text{and}\quad \Delta^2 = \sigma^{-2}(\boldsymbol{\delta}'\mathbf{\Lambda}_{22}^{-1}\boldsymbol{\delta}). \tag{6.10.34}$$

Proof. First note that $\tilde{\boldsymbol{\beta}}_n$ and $\hat{\boldsymbol{\beta}}_n$ are both translation-invariant estimators, and hence, we can assume without loss of generality, $\boldsymbol{\beta} = \beta_0\mathbf{1}_p$. Also, by (6.10.12a) and (6.10.12c) together with (6.10.10) and the linearity results (6.10.21) through (6.10.23), we have $n^{1/2}|\tilde{\boldsymbol{\beta}}_n - \boldsymbol{\beta}| = O_p(1)$ and $n^{1/2}|\hat{\boldsymbol{\beta}}_n - \boldsymbol{\beta}| = O_p(1)$, while under $\{K_{(n)}\}$, $n^{1/2}\tilde{\boldsymbol{\beta}}_n = \beta_0\mathbf{1}_p + \delta = O(1)$. Thus under $\{K_{(n)}\}$, $n^{1/2}\tilde{\boldsymbol{\beta}}_n = \beta_0\mathbf{1}_p + O_p(1)$ and $n^{1/2}\hat{\boldsymbol{\beta}}_n = \beta_0\mathbf{1}_p + O_p(1)$. Observe that under $H^* : \boldsymbol{\theta} = \boldsymbol{\beta} = \mathbf{0}$ by (6.10.12a-c), $n^{-1/2}\mathbf{L}_n(\mathbf{0}) = \mathbf{\Lambda}_{22}^{-1}n^{1/2}\tilde{\boldsymbol{\beta}}_n\gamma(\psi,\phi) + o_p(1)$. Hence, utilizing the contiguity of probability measures under $K_{(n)}^* : \theta = \mathbf{0}$, $\boldsymbol{\beta} - \beta_0\mathbf{1}_p = n^{-1/2}\boldsymbol{\delta}$ to those under H^*, we obtain under $K_{(n)}^*$ as $n \to \infty$, that $n^{-1/2}\mathbf{L}_n(\mathbf{0})$ has the same distribution as $n^{-1/2}\mathbf{L}_n(\boldsymbol{\delta})$ under

H^*. Thus, by (6.10.10) $n^{-1/2}\mathbf{L}_n(n^{-1/2}(\boldsymbol{\delta})$ is asymptotically p-variate normal with mean-vector $\boldsymbol{\Lambda}_{22}^{-1}\boldsymbol{\delta}\gamma(\psi,\phi)$ and dispersion matrix $A_\phi^2\boldsymbol{\Lambda}_{22}^{-1}$. Thus, under $\{K_{(n)}\}$, $n^{-1/2}\mathbf{L}_n(\mathbf{0}) \sim \mathcal{N}_p\{\boldsymbol{\Lambda}_{22}^{-1}\boldsymbol{\delta}\gamma(\psi,\phi), A_\phi^2\boldsymbol{\Lambda}_{22})$. Hence, $n^{1/2}(\tilde{\boldsymbol{\beta}}_n - \beta_0\mathbf{1}_p) \sim \mathcal{N}_p(\boldsymbol{\delta}, \sigma^2\boldsymbol{\Lambda}_{22})$, $\sigma^2 = A_\psi^2/\gamma^2(\psi,\phi)$.

Similarly, by the contiguity of probability measures under $K_{(n)}^*$ to those under H^*, we obtain under $K_{(n)}^*$ as $n \to \infty$, $n^{1/2}\hat{\boldsymbol{\beta}}_n = \frac{\mathbf{1}_p}{\bar{Q}}\mathbf{1}_p' n^{-1/2}\mathbf{L}_n(0)\gamma^{-1}(\psi,\phi) + o_p(1)$. By a similar argument as before, we have $\sqrt{n}(\hat{\boldsymbol{\beta}}_n - \beta_0\mathbf{1}_p) \sim \mathcal{N}_p(\boldsymbol{\delta}; \sigma^2\mathbf{B})$, $\mathbf{B} = \mathbf{1}_p\mathbf{1}_p'/\bar{Q}$. By similar steps, we obtain under $\{K_{(n)}\}$,

$$\text{(iii)} \quad n^{1/2}(\tilde{\boldsymbol{\beta}}_n - \hat{\boldsymbol{\beta}}_n) = (\boldsymbol{\Lambda}_{22} - \tfrac{1}{\bar{Q}}\mathbf{1}_p\mathbf{1}_p')n^{1/2}\mathbf{L}_n(\mathbf{0})\gamma^{-1}(\psi,\phi) + o_p(1)$$

$$\sim \mathcal{N}_p(\mathbf{J}\boldsymbol{\delta}, \sigma^2\mathbf{J}\boldsymbol{\Lambda}_{22}). \tag{6.10.35}$$

Further, under $\{K_{(n)}\}$ as $n \to \infty$,

$$\text{(iv)} \begin{pmatrix} n^{1/2}(\tilde{\boldsymbol{\beta}}_n - \beta_0\mathbf{1}_p \\ n^{1/2}(\tilde{\boldsymbol{\beta}}_n - \hat{\boldsymbol{\beta}}_n) \end{pmatrix} = \begin{pmatrix} \boldsymbol{\Lambda}_{22}n^{1/2}\mathbf{L}_n(\mathbf{0})\gamma^{-1}(\psi,\phi) \\ (\boldsymbol{\Lambda}_{22} - \frac{1}{\bar{Q}}\mathbf{1}_p\mathbf{1}_p')n^{1/2}\mathbf{L}_n(\mathbf{0})\gamma^{-1}(\psi,\phi) \end{pmatrix} + o_p(1)$$

$$\simeq \mathcal{N}_{2p}\left\{\begin{pmatrix} \boldsymbol{\delta} \\ \mathbf{J}\boldsymbol{\delta} \end{pmatrix}, \sigma^2\begin{pmatrix} \boldsymbol{\Lambda}_{22} & \mathbf{J}\boldsymbol{\Lambda}_{22} \\ \boldsymbol{\Lambda}_{22}\mathbf{J}' & \mathbf{J}\boldsymbol{\Lambda}_{22} \end{pmatrix}\right\} \tag{6.10.36}$$

by the same argument as before. Similarly, under $\{K_{(n)}\}$ as $n \to \infty$,

$$\text{(v)} \begin{pmatrix} \sqrt{n}(\hat{\boldsymbol{\beta}}_n - \beta_0\mathbf{1}_p) \\ \sqrt{n}(\tilde{\boldsymbol{\beta}}_n - \hat{\boldsymbol{\beta}}_0) \end{pmatrix} \sim \mathcal{N}_{2p}\left\{\begin{pmatrix} \boldsymbol{\delta} \\ \mathbf{J}\boldsymbol{\delta} \end{pmatrix}, \sigma^2\begin{pmatrix} \mathbf{B} & \mathbf{0} \\ \mathbf{0} & \mathbf{J}\boldsymbol{\Lambda}_{22} \end{pmatrix}\right\}. \tag{6.10.37}$$

(vi) To prove $\lim_{n\to\infty} P\{\mathcal{L}_n \le x | K_{(n)}\} = H_{p-1}(x;\Delta^2)$, we note that under $(\boldsymbol{\beta} - \beta_0\mathbf{1}_p) = \mathbf{0}$, $\mathcal{L}_n = nA_\phi^2[n\mathbf{L}_n(\mathbf{0})]'\mathbf{J}\boldsymbol{\Lambda}_{22}\mathbf{J}'[\mathbf{L}_n(\mathbf{0})] + o_p(1)$, which approximately follows a central chi-square distribution with $(p-1)$ d.f. Also, under $\{K_{(n)}\}$, as $n \to \infty$, we can write

$$\mathcal{L}_n = n(\tilde{\boldsymbol{\beta}}_n - \hat{\boldsymbol{\beta}}_n)\mathbf{J}'\boldsymbol{\Lambda}_{22}^{-1}\mathbf{J}(\tilde{\boldsymbol{\beta}}_n - \hat{\boldsymbol{\beta}}_n) + o_p(1). \tag{6.10.38}$$

By (iii) the expression above follows approximately a noncentral chi-square distribution with $(p-1)$ d.f. and a noncentrality parameter $\Delta^2/2$ with $\Delta^2 = \sigma^{-2}(\boldsymbol{\delta}'\boldsymbol{\Lambda}_{22}^{-1}\boldsymbol{\delta})$.

As a consequence of Theorem 6.10.1, we obtain the ADB, ADQB, ADMSE, and ADQR of the five estimators as given below.

For UE, $\tilde{\boldsymbol{\beta}}_n$,

(i) $\mathbf{b}_1(\tilde{\boldsymbol{\beta}}_n) = \mathbf{0}$ and $B_1(\tilde{\boldsymbol{\beta}}_n) = 0$;

$$\text{(ii) } \mathbf{M}_1(\tilde{\boldsymbol{\beta}}_n) = \sigma^2\boldsymbol{\Lambda}_{22} \quad \text{and} \quad R_1(\tilde{\boldsymbol{\beta}}_n;\mathbf{W}) = \sigma^2\text{tr}(\mathbf{W}\boldsymbol{\Lambda}_{22}). \tag{6.10.39a}$$

For RE, $\hat{\boldsymbol{\beta}}_n$,

(i) $\mathbf{b}_2(\hat{\boldsymbol{\beta}}_n) = -\boldsymbol{\delta}$ and $B_2(\hat{\boldsymbol{\beta}}_n) = \Delta^2$;

(ii) $\mathbf{M}_2(\hat{\boldsymbol{\beta}}_n) = \sigma^2 \frac{\mathbf{1}_p \mathbf{1}_p'}{Q} + \boldsymbol{\delta}\boldsymbol{\delta}'$ and $R_2(\hat{\boldsymbol{\beta}}_n; \mathbf{W}) = \sigma^2 \mathbf{1}_p \mathbf{W} \mathbf{1}_p' + \boldsymbol{\delta}' \mathbf{W} \boldsymbol{\delta}$.

$$(6.10.39b)$$

For PTE, $\hat{\boldsymbol{\beta}}_n^{\text{PT}}$,

(i) $\mathbf{b}_3(\hat{\boldsymbol{\beta}}_n^{\text{PT}}) = -\mathbf{J}\boldsymbol{\delta} H_{p+1}\big(\chi^2_{p-1}(\alpha); \Delta^2\big)$ and
$B_3(\hat{\boldsymbol{\beta}}_n^{\text{PT}}) = \Delta^2 \big\{H_{p+1}\big(\chi^2_{p-1}(\alpha); \Delta^2\big)\big\}^2$,

(ii)
$$\mathbf{M}_3(\hat{\boldsymbol{\beta}}_n^{\text{PT}}) = \sigma^2 \boldsymbol{\Lambda}_{22} - \sigma^2 J \boldsymbol{\Lambda}_{22} H_{p+1}\big(\chi^2_{p-1}(\alpha); \Delta^2\big) + \mathbf{J}\boldsymbol{\delta}\boldsymbol{\delta}'\mathbf{J}'\big\{2H_{p+2}\big(\chi^2_{p-1}(\alpha); \Delta^2\big) - H_{p+3}\big(\chi^2_{p-1}(\alpha); \Delta^2\big)\big\}$$

and

$$R_3(\hat{\boldsymbol{\beta}}_n^{\text{PT}}; \mathbf{W}) = \sigma^2 \text{tr}(\mathbf{W}\boldsymbol{\Lambda}_{22}) - \sigma^2 \text{tr}(\mathbf{W}\mathbf{J}\boldsymbol{\Lambda}_{22}) H_{p+1}\big(\chi^2_{p-1}(\alpha); \Delta^2\big) + (\boldsymbol{\delta}'\mathbf{J}'\mathbf{W}\mathbf{J}\boldsymbol{\delta})\Big\{2H_{p+1}\big(\chi^2_{p-1}(\alpha); \Delta^2\big) - H_{p+3}\big(\chi^2_{p-1}(\alpha); \Delta^2\big)\Big\}. \quad (6.10.39c)$$

For SE, $\hat{\boldsymbol{\beta}}_n^{S}$,

(i) $\mathbf{b}_4(\hat{\boldsymbol{\beta}}_n^{S}) = -(p-3)\mathbf{J}\boldsymbol{\delta}\, E\big[\chi^{-2}_{p+1}(\Delta^2)\big]$ and
$B_4(\hat{\boldsymbol{\beta}}_n^{S}) = (p-3)^2 \Delta^2 \big\{E\big[\chi^{-2}_{p+1}(\Delta^2)\big]\big\}^2$;

(ii)
$$\mathbf{M}_4(\hat{\boldsymbol{\beta}}_n^{S}) = \sigma^2 \boldsymbol{\Lambda}_{22} - (p-3)\sigma^2 \mathbf{J}\boldsymbol{\Lambda}_{22}\big\{2E\big[\chi^{-2}_{p+1}(\Delta^2)\big] - (p-3)E\big[\chi^{-4}_{p+1}(\Delta^2)\big]\big\} + (p-3)(p+1)(\mathbf{J}\boldsymbol{\delta}\boldsymbol{\delta}'\mathbf{J}')E\big[\chi^{-4}_{p+3}(\Delta^2)\big]$$

and

$$\begin{aligned} R_4(\hat{\boldsymbol{\beta}}_n^{S}; \mathbf{W}) = {} & \sigma^2 \text{tr}(\mathbf{W}\boldsymbol{\Lambda}_{22}) - (p-3)\sigma^2 \text{tr}(\mathbf{W}\mathbf{J}\boldsymbol{\Lambda}_{22}) \\ & \times \big\{2E\big[\chi^{-2}_{p+1}(\Delta^2)\big] - (p-3)E\big[\chi^{-4}_{p+1}(\Delta^2)\big]\big\} \\ & + (p-3)(p+1)(\boldsymbol{\delta}'\mathbf{J}\mathbf{W}\mathbf{J}\boldsymbol{\delta})E\big[\chi^{-4}_{p+3}(\Delta^2)\big]. \end{aligned} \quad (6.10.39d)$$

Finally, for **PRSE,** $\hat{\boldsymbol{\beta}}_n^{S+}$,

(i)
$$\mathbf{b}_5(\hat{\boldsymbol{\beta}}_n^{S+}) = -\mathbf{J}\boldsymbol{\delta}\big\{H_{p+1}(p-3; \Delta^2) + (p-3)E\big[\chi^{-2}_{p+1}(\Delta^2)\big]I\big(\chi^{-2}_{p+1}(\Delta^2) > p-3\big)\big]\big\}$$

and

$$B_5(\hat{\boldsymbol{\beta}}_n^{S+}) = \Delta^2\big\{H_{p+1}(p-3; \Delta^2) + (p-3)E\big[\chi^{-2}_{p+1}(\Delta^2)\big]I\big(\chi^{-2}_{p+1}(\Delta^2) > p-3\big)\big]\big\}^2;$$

(ii) $\mathbf{M}_5(\hat{\boldsymbol{\beta}}_n^{S+}) = \mathbf{M}_4(\hat{\boldsymbol{\beta}}_n^{S}) - \sigma^2(\mathbf{J}\boldsymbol{\Lambda}_{22})E\big[\big(1 - (p-3)\chi^{-2}_{p+1}(\Delta^2)\big)^2$

$$\times I\big(\chi^2_{p+1}(\Delta^2) < p-3\big)\big]$$
$$+(\mathbf{J}\boldsymbol{\delta}\boldsymbol{\delta}'\mathbf{J}')\big\{2E\big[\big(1-(p-3)\chi^{-2}_{p+1}(\Delta^2)\big)I\big(\chi^2_{p+1}(\Delta^2) < p-3\big)\big]$$
$$-E\big[\big(1-(p-3)\big(\chi^{-2}_{p+3}(\Delta^2)\big)^2 I\big(\chi^2_{p+3}(\Delta^2) < p-3\big)\big]\big\}$$

and

$$R_5(\hat{\boldsymbol{\beta}}_n^{S+};\mathbf{W}) =$$
$$R_4(\hat{\boldsymbol{\beta}}_n;\mathbf{W})-\sigma^2\text{tr}(\mathbf{W}\mathbf{J}\boldsymbol{\Lambda}_{22})E\big[\big(1-(p-3)\chi^{-2}_{p+1}(\Delta^2)\big)^2 I\big(\chi^2_{p+1}(\Delta^2)<p-3\big)\big]$$
$$+(\boldsymbol{\delta}'\mathbf{J}'\mathbf{W}\mathbf{J}\boldsymbol{\delta})\big\{2E\big[\big(1-(p-3)\chi^{-2}_{p+1}(\Delta^2)\big)I\big(\chi^2_{p+1}(\Delta^2) < p-3\big)\big]$$
$$-E\big[\big(1-(p-3)\chi^{-2}_{p+3}(\Delta^2)\big)^2 I\big(\chi^2_{p+4}(\Delta^2) < p-3\big)\big]\big\}. \quad (6.10.39e)$$

It should be observed that these expressions are similar to those in Section 6.5.4, and the properties of the five R-Estimators of $\boldsymbol{\beta}$ are the same as those given in Sections 6.5 and 6.6.

6.10.5 Asymptotic Distributional Properties of the R-Estimators of Intercepts

Recall from Section 6.10.2 that the unrestricted and the restricted estimators of $\boldsymbol{\theta}$ are those given by

$$\tilde{\boldsymbol{\theta}}_n = (\tilde{\theta}_{n_1},\ldots,\tilde{\theta}_{n_p})' \text{ and } \hat{\boldsymbol{\theta}}_n = (\hat{\theta}_{n_1},\ldots,\hat{\theta}_{n_p})', \quad (6.10.40)$$

respectively, as defined by (6.10.13a) and (6.10.14a). We will consider the distributional properties of these estimators under the local alternatives $\{K_{(n)}\}$.

Theorem 2. Under $\{K_{(n)}\}$ and the regularity conditions for the model (6.10.1), as $n \to \infty$,

$$\begin{pmatrix} \sqrt{n}(\tilde{\boldsymbol{\theta}}_n - \boldsymbol{\theta}) \\ n^{1/2}\mathbf{L}_n(\mathbf{0}) \end{pmatrix} \sim \mathcal{N}_{2p}\left\{\begin{pmatrix} \mathbf{0} \\ \boldsymbol{\Lambda}_{22}^{-1}\boldsymbol{\delta}\gamma(\psi,\phi) \end{pmatrix};\begin{pmatrix} \boldsymbol{\Sigma}_{11} & \boldsymbol{\Sigma}_{12} \\ \boldsymbol{\Sigma}_{21} & \boldsymbol{\Sigma}_{22} \end{pmatrix}\right\}, \quad (6.10.41)$$

where

$$\mathbf{T}_0 = \text{Diag}(\bar{x}_1,\ldots,\bar{x}_p),$$
$$\boldsymbol{\Sigma}_{11} = \frac{A_\phi^2}{\gamma^2(\psi,\phi)}(\boldsymbol{\Lambda}_0 + \mathbf{T}_0\boldsymbol{\Lambda}_{22}\mathbf{T}_0),$$
$$\boldsymbol{\Sigma}_{22} = A_\phi^2\boldsymbol{\Lambda}_{22}^{-1},$$

and

$$\boldsymbol{\Sigma}_{12} = -\frac{A_\phi^2}{\gamma(\psi,\phi)}\mathbf{T}_0. \quad (6.10.42)$$

Proof. Assume, without loss of generality, that $\boldsymbol{\theta} = \mathbf{0}$. Relation (6.10.24) and the definition of $\tilde{\boldsymbol{\theta}}_n$ imply that under $\{K_{(n)}\}$, $n^{1/2}\tilde{\boldsymbol{\theta}}_n$ is asymptotically equivalent to

$$\Lambda_0 n^{-1/2}\mathbf{T}_n(\mathbf{0},\mathbf{0})\gamma^{-1}(\psi,\phi) - \mathbf{T}_0 n^{1/2}\tilde{\boldsymbol{\beta}}_n. \qquad (6.10.43)$$

Using relation (6.10.21) and the definition of $\tilde{\boldsymbol{\beta}}_n$, we have under $\{K_{(n)}\}$,

$$\boldsymbol{\Lambda}_0 n^{-1/2}\mathbf{T}_n(\mathbf{0},\mathbf{0})\gamma^{-1}(\psi,\phi) - \frac{1}{\bar{Q}}\mathbf{T}_0\,\boldsymbol{\Lambda}_{22}n^{1/2}\mathbf{L}_n(\mathbf{0})\gamma^{-1}(\psi,\phi), \qquad (6.10.44)$$

which under $K^*_{(n)}: \boldsymbol{\theta} = \mathbf{0}, (\boldsymbol{\beta} - \beta_0\mathbf{1}_p) = n^{-1/2}\boldsymbol{\delta}$ is the same as

$$\left[\boldsymbol{\Lambda}_0 n^{-1/2}\mathbf{T}_n(\mathbf{0}, n^{-1/2}\boldsymbol{\delta}) - \mathbf{T}_0\,\boldsymbol{\Lambda}_{22}n^{1/2}\mathbf{L}_n(n^{-1/2}\boldsymbol{\delta})\right]\gamma^{-1}(\psi,\phi). \qquad (6.10.45)$$

As in (6.10.21) through (6.10.24), under H_0 the above is asymptotically equivalent to the random vector

$$\boldsymbol{\Lambda}_0 n^{-1/2}\mathbf{T}_n(\mathbf{0},\mathbf{0})\gamma^{-1}(\psi,\phi) - \frac{1}{\bar{Q}}\mathbf{T}_0\,\boldsymbol{\Lambda}_{22}n^{-1/2}\mathbf{L}_n(\mathbf{0})\gamma^{-1}(\psi,\phi). \qquad (6.10.46)$$

Thus, the last expression has the same asymptotic distribution under both $\{K_{(n)}\}$ and H_0. Since the asymptotic mean is $\boldsymbol{\Lambda}_{22}^{-1}\boldsymbol{\delta}\gamma(\psi,\phi)$, Theorem 6.10.2 follows from (6.10.10).

The next theorem relates to the asymptotic distribution of $\sqrt{n}(\hat{\boldsymbol{\theta}}_n - \boldsymbol{\theta})$.

Theorem 3. Under $\{K_{(n)}\}$ and the assumed regularity conditions for the model (6.10.1) as $n \to \infty$, we have

$$\begin{pmatrix} \sqrt{n}(\hat{\boldsymbol{\theta}}_n - \boldsymbol{\theta}) \\ n^{1/2}\mathbf{L}_n(\hat{\boldsymbol{\beta}}_n) \end{pmatrix} \sim \mathcal{N}_{2p}\left\{\begin{pmatrix} \mathbf{T}_0\boldsymbol{\delta} \\ \boldsymbol{\Lambda}_{22}^{-1}\boldsymbol{\delta}\gamma(\psi,\phi) \end{pmatrix}; \begin{pmatrix} \boldsymbol{\Sigma}^*_{11} & 0 \\ 0 & \boldsymbol{\Sigma}^*_{22} \end{pmatrix}\right\}, \qquad (6.10.47)$$

where

$$\boldsymbol{\Sigma}^*_{11} = \frac{A^2_\phi}{\gamma^2(\psi,\phi)}\left(\boldsymbol{\Lambda}_0 + \frac{\mathbf{T}_0\mathbf{1}_p\mathbf{1}'_p\mathbf{T}_o}{\bar{Q}}\right)$$

and

$$\Sigma^*_{22} = A^2_\phi\Lambda_0^{-1}\mathbf{B}\boldsymbol{\Lambda}_0^{-1},$$

with

$$\mathbf{B} = \mathrm{Diag}\left(\frac{Q_1}{\lambda_1}, \dots, \frac{Q_p}{\lambda_p}\right) - \left(\frac{Q_iQ_j}{\bar{Q}}\right), \quad i,j = 1,\dots,p. \qquad (6.10.48)$$

Proof. Assume, without loss of generality, $\boldsymbol{\theta} = \mathbf{0}$. From the relative compactness of $n^{1/2}\hat{\boldsymbol{\beta}}_n$, the definition of $\hat{\boldsymbol{\theta}}_n$ and the relations (6.10.21) through (6.10.24), it follows that

(i) $n^{1/2}\hat{\boldsymbol{\theta}}_n - (\boldsymbol{\Lambda}_0 n^{-1/2}\mathbf{T}_n(\mathbf{0},\mathbf{0})\gamma^{-1}(\psi,\phi) - \mathbf{T}_0 n^{1/2}\hat{\boldsymbol{\beta}}_n) \xrightarrow{P} 0,$
(ii) $n^{1/2}\hat{\boldsymbol{\beta}}_n - \frac{1}{Q}\mathbf{1}_p n^{1/2}\boldsymbol{\ell}_n^*(\mathbf{0})\gamma^{-1}(\psi,\phi) \xrightarrow{P} 0,$
and
(iii) $\gamma^{-1}(\psi,\phi)n^{-1/2}\mathbf{L}_n(\hat{\boldsymbol{\beta}}_n) - (n^{-1/2}\mathbf{L}_n(\mathbf{0}) - \boldsymbol{\Lambda}_{22}^{-1}n^{1/2}\hat{\boldsymbol{\beta}}_n\gamma^{-1}(\psi,\phi)) \xrightarrow{P} 0.$ (6.10.49)

The relations (6.10.48i, ii) imply

$$n^{1/2}\hat{\boldsymbol{\theta}}_n - (\boldsymbol{\Lambda}_0 n^{-1/2}\mathbf{T}_n(\mathbf{0},\mathbf{0})\gamma(\psi,\phi) - \frac{1}{\bar{Q}}\mathbf{T}_0\mathbf{1}_p n^{-1/2}L_n^*(0)\gamma^{-1}(\psi,\phi)) \xrightarrow{P} 0, \tag{6.10.50}$$

while the relations (6.10.48ii, iii) imply

$$n^{-1/2}\mathbf{L}_n(\hat{\boldsymbol{\beta}}_n) - (n^{-1/2}\mathbf{L}_n(0) - \frac{1}{\bar{Q}}\mathbf{T}_0\boldsymbol{\Lambda}_{22}^{-1}\mathbf{1}_p n^{-1/2}L_n^*(0)) \xrightarrow{P} 0. \tag{6.10.51}$$

Next, the relations (6.10.49) and (6.10.50) imply that under $\{K_{(n)}\}$, $\begin{pmatrix} \sqrt{n}(\hat{\boldsymbol{\theta}}_n - \boldsymbol{\theta}) \\ n^{-1/2}\mathbf{L}_n(\hat{\boldsymbol{\beta}}_n) \end{pmatrix}$ is asymptotically equivalent to the random vector

$$\begin{pmatrix} \boldsymbol{\Lambda}_0\mathbf{1}_p n^{-1/2}\mathbf{T}_n(\mathbf{0},\mathbf{0})\gamma^{-1}(\psi,\phi) - \frac{1}{\bar{Q}}\mathbf{T}_0\mathbf{1}_p n^{-1/2}\mathbf{L}_n^*\gamma^{-1}(\psi,\phi) \\ n^{-1/2}\mathbf{L}_n(\mathbf{0}) - \frac{1}{\bar{Q}}n^{-1/2}\boldsymbol{\Lambda}_{22}^{-1}\mathbf{1}_p L_n^*(\mathbf{0})\gamma^{-1}(\psi,\phi) \end{pmatrix}. \tag{6.10.52}$$

But the asymptotic distribution of (6.10.52) under $\{K_{(n)}\}$ is the same as

$$\begin{pmatrix} \boldsymbol{\Lambda}_0^{-1}n^{-1/2}\mathbf{T}_n(0), n^{1/2}\boldsymbol{\delta})\gamma^{-1}(\psi,\phi) - \frac{1}{\bar{Q}}\mathbf{T}_0\mathbf{1}_p n^{-1/2}\mathbf{L}_n^* - n^{-1/2}\boldsymbol{\delta})\gamma^{-1}(\psi,\phi) \\ n^{-1/2}\mathbf{L}_n(-n^{-1/2}\boldsymbol{\delta}) - \frac{1}{\bar{Q}}n^{-1/2}\boldsymbol{\Lambda}_{22}^{-1}\mathbf{1}_p L_n^*(-n^{-1/2}\boldsymbol{\delta})\gamma^{-1}(\psi,\phi) \end{pmatrix} \tag{6.10.53}$$

under $H^* : \boldsymbol{\theta} = \boldsymbol{\beta} = \mathbf{0}$ by the fact that the distribution of $\mathbf{T}_n(\mathbf{a},\mathbf{b})$ under $\boldsymbol{\theta} = \mathbf{a}$, $\boldsymbol{\beta} = \mathbf{b}$ is the same as that of $\mathbf{T}_n(\boldsymbol{\theta} - \mathbf{a}, \boldsymbol{\beta} - \mathbf{b})$ under $\boldsymbol{\theta} = \mathbf{0}$, $\boldsymbol{\beta} = \mathbf{0}$, and similarly for $\mathbf{L}_n(\mathbf{0})$. Using again (6.10.21) through (6.10.25), we find that the row vectors under H^* have the same distribution as

$$\begin{pmatrix} \boldsymbol{\Lambda}_0^{-1}n^{-1/2}\mathbf{T}_n(\mathbf{0},\mathbf{0})\gamma^{-1}(\psi,\phi) - \frac{1}{\bar{Q}}\mathbf{T}_0\mathbf{1}_p n^{-1/2}L_n^*(a)\gamma^{-1}(\psi,\phi) \\ \qquad + \mathbf{T}_0\boldsymbol{\delta} - \frac{1}{\bar{Q}}(\mathbf{T}_0\mathbf{1}_p)(\boldsymbol{\delta}'\boldsymbol{\Lambda}_{22}^{-1}\mathbf{1}_p') \\ n^{-1/2}\mathbf{L}_n(\mathbf{0}) - \frac{1}{\bar{Q}}\boldsymbol{\Lambda}_{22}^{-1}\mathbf{1}_p n^{-1/2}L_n^*(\mathbf{0})\gamma^{-1}(\psi,\phi) + \boldsymbol{\Lambda}_{22}^{-1}\boldsymbol{\delta}\gamma(\psi,\phi) \\ \qquad - \frac{1}{\bar{Q}}(\boldsymbol{\Lambda}_{22}^{-1}\mathbf{1}_p')(\boldsymbol{\delta}'\boldsymbol{\Lambda}_{22}^{-1}\mathbf{1}_p)\gamma(\psi,\phi) \end{pmatrix}. \tag{6.10.54}$$

However, by the definition of $\{K_{(n)}\}$, we have $\boldsymbol{\delta}'\boldsymbol{\Lambda}_{22}^{-1}\mathbf{1}_p = 0$. Thus, it follows immediately that the constant term converges to $(\mathbf{T}_0\boldsymbol{\delta}, \gamma(\psi,\phi)\boldsymbol{\Lambda}_{22}^{-1}\boldsymbol{\delta})$ as in the theorem. Now, by (6.10.10), it follows after straightforward computations that the asymptotic covariance of the vector above is given by the expressions in the theorem, that is, $\boldsymbol{\Sigma}^* = \begin{pmatrix} \boldsymbol{\Sigma}_{11}^* & \mathbf{0} \\ \mathbf{0} & \boldsymbol{\Sigma}_{22}^* \end{pmatrix}$.

Corollary 1. The random vector $\sqrt{n}(\hat{\boldsymbol{\theta}}_n - \boldsymbol{\theta})$ and the test-statistic $\mathcal{L}_n$ are independent.

Corollary 2. The statistic under $\{K_{(n)}\}$ is asymptotically equivalent to

$$A_\phi^{-2} n^{-1} \mathbf{L}_n'(\mathbf{0}) \Big(\boldsymbol{\Lambda}_{22} - \frac{1}{\bar{Q}} \mathbf{1}_p \mathbf{1}_p'\Big) \mathbf{L}_n'(0) = A_\phi^{-2} n^{-1} \mathbf{L}_n'(\mathbf{0}) (\mathbf{J} \boldsymbol{\Lambda}_{22} \mathbf{J}') \mathbf{L}_n'(0) \tag{6.10.55}$$

as in the proof (vi) of Theorem 6.10.1.

Based on the procedure above, we have the following theorem:

Theorem 4. Under $\{K_{(n)}\}$ and the assumed regularity conditions, as $n \to \infty$ we have

(i) $$\begin{pmatrix} \sqrt{n}(\tilde{\boldsymbol{\theta}}_n - \boldsymbol{\theta}) \\ \sqrt{n}(\tilde{\boldsymbol{\beta}}_n - \boldsymbol{\beta}) \end{pmatrix} \sim \mathcal{N}_{2p} \left\{ \begin{pmatrix} \mathbf{0} \\ \mathbf{0} \end{pmatrix}; \sigma^2 \begin{pmatrix} \boldsymbol{\Lambda}_{11} & -\mathbf{T}_0 \boldsymbol{\Lambda}_{22} \\ -\mathbf{T}_0' \boldsymbol{\Lambda}_{22} & \boldsymbol{\Lambda}_{22} \end{pmatrix} \right\},$$

$$\sigma^2 = \frac{A_\phi^2}{\gamma^2(\psi, \phi)}, \quad \text{where } \boldsymbol{\Lambda}_{11} = \boldsymbol{\Lambda}_0 + \mathbf{T}_0 \boldsymbol{\Lambda}_{22} \mathbf{T}_0. \tag{6.10.56}$$

(ii) $$\begin{pmatrix} \sqrt{n}(\hat{\boldsymbol{\theta}}_n - \boldsymbol{\theta}) \\ \sqrt{n}(\tilde{\boldsymbol{\beta}}_n - \boldsymbol{\beta}) \end{pmatrix} \sim \mathcal{N}_{2p} \left\{ \begin{pmatrix} \mathbf{T}_0 \boldsymbol{\delta} \\ \mathbf{0} \end{pmatrix}; \sigma^2 \begin{pmatrix} \boldsymbol{\Lambda}_{11}^* & \boldsymbol{\Lambda}_{12}^* \\ \boldsymbol{\Lambda}_{21}^* & \boldsymbol{\Lambda}_{22} \end{pmatrix} \right\},$$

$$\text{where } \boldsymbol{\Lambda}_{11}^* = \boldsymbol{\Lambda}_0 + \tfrac{1}{\bar{Q}} (\mathbf{T}_0 \mathbf{1}_p \mathbf{1}_p' \mathbf{T}_0) \text{ and } \boldsymbol{\Lambda}_{12}^* = -\tfrac{1}{\bar{Q}} \mathbf{1}_p \mathbf{1}_p' \mathbf{T}_0. \tag{6.10.57}$$

Proof. We know from (6.10.42) that

$$\begin{aligned} n^{1/2} \tilde{\boldsymbol{\theta}}_n &= \boldsymbol{\Lambda}_0 n^{-1/2} \mathbf{T}_n(0,0) \gamma^{-1}(\psi, \phi) - \mathbf{T}_0 n^{1/2} \tilde{\boldsymbol{\beta}}_n + o_p(1) \\ &= \boldsymbol{\Lambda}_0 n^{-1/2} \mathbf{T}_n(0,0) \gamma^{-1}(\psi, \phi) - \mathbf{T}_0 \boldsymbol{\Lambda}_{22} n^{-1/2} \mathbf{L}_n(\mathbf{0}) \gamma^{-1}(\psi, \phi) \\ &\quad + o_p(1). \end{aligned} \tag{6.10.58}$$

Also $n^{1/2} \tilde{\boldsymbol{\beta}}_n = \boldsymbol{\Lambda}_{22} n^{-1/2} \mathbf{L}_n(0) \gamma^{-1}(\psi, \phi) + o_p(1)$.

Hence, under $\{K_{(n)}\}$,

$$\begin{pmatrix} n^{1/2}(\tilde{\boldsymbol{\theta}}_n - \boldsymbol{\theta}) \\ n^{1/2}(\tilde{\boldsymbol{\beta}}_n - \boldsymbol{\beta}) \end{pmatrix} \sim \mathcal{N}_{2p} \left\{ \begin{pmatrix} \mathbf{0} \\ \mathbf{0} \end{pmatrix}; \sigma^2 \begin{pmatrix} \boldsymbol{\Lambda}_{11} & -\mathbf{T}_0 \boldsymbol{\Lambda}_{22} \\ -\mathbf{T}_0' \boldsymbol{\Lambda}_{22} & \boldsymbol{\Lambda}_{22} \end{pmatrix} \right\}.$$

Similarly, we can write

$$\begin{aligned} n^{1/2} \hat{\boldsymbol{\theta}}_n &= \boldsymbol{\Lambda}_{22} n^{-1/2} \mathbf{T}_n(\mathbf{0}, \mathbf{0}) \gamma^{-1}(\psi, \phi) - \frac{\mathbf{T}_0}{\bar{Q}} \mathbf{1}_p \mathbf{1}_p' n^{-1/2} \mathbf{L}_n(0) \gamma^{-1}(\psi, \phi) + o_p(1) \\ &= n^{1/2} \tilde{\boldsymbol{\theta}}_n + \mathbf{T}_0 (\boldsymbol{\Lambda}_{22} - \frac{1}{\bar{Q}} \mathbf{1}_p \mathbf{1}_p') n^{-1/2} \mathbf{L}_n(\mathbf{0}) \gamma^{-1}(\psi, \phi) + o_p(1) \\ &= n^{1/2} \tilde{\boldsymbol{\theta}}_n + \mathbf{T}_0 \mathbf{J} (n^{1/2} \tilde{\boldsymbol{\beta}}_n) + o_p(1), \end{aligned} \tag{6.10.59}$$

where $\mathbf{J} = \mathbf{I}_p - \frac{\mathbf{1}_p \mathbf{1}_p'}{\bar{Q}} \boldsymbol{\Lambda}_{22}^{-1}$. Hence, straightforward computation under $\{K_{(n)}\}$ leads to the theorem.

The results above show us that the test-statistic, $\mathcal{L}_n$ can be represented as

$$\mathcal{L}_n = n\sigma^{-2}\tilde{\boldsymbol{\beta}}_n'\mathbf{J}'\boldsymbol{\Lambda}_{22}^{-1}\mathbf{J}\tilde{\boldsymbol{\beta}}_n + o_p(1), \tag{6.10.60}$$

where $\sigma^2 = A_\phi^2/\gamma^2(\psi,\phi)$. Consequently, we can write the PTE, SE, and PRSE of $\boldsymbol{\theta}$ as follows:

(i) $n^{1/2}\hat{\boldsymbol{\theta}}_n^{\mathrm{PT}} = n^{1/2}\tilde{\boldsymbol{\theta}}_n + \mathbf{T}_0\mathbf{J}(n^{1/2}\tilde{\boldsymbol{\beta}}_n)I(\mathcal{L}_n < \chi_{p-1}^2(\alpha)) + o_p(1)$,

(ii) $n^{1/2}\hat{\boldsymbol{\theta}}_n^{S} = n^{1/2}\tilde{\boldsymbol{\theta}}_n + (p-3)\mathbf{T}_0\mathbf{J}(n^{1/2}\tilde{\boldsymbol{\beta}}_n)\mathcal{L}_n^{-1} + o_p(1)$,

and

(iii) $n^{1/2}\hat{\boldsymbol{\theta}}_n^{S+} = n^{1/2}\hat{\boldsymbol{\theta}}_n^{S} + \mathbf{T}_0\mathbf{J}(n^{1/2}\tilde{\boldsymbol{\beta}}_n)[1-(p-3)\mathcal{L}_n^{-1}]I(\mathcal{L}_n < p-3) + o_p(1)$.
(6.10.61)

Using Theorems 6.10.3 and 6.10.4, we can determine the ADB, ADQB, ADMSE, and ADQR of the R-Estimators of $\boldsymbol{\theta}$ under $\{K_{(n)}\}$ as follows:

Unrestricted Estimator ($\tilde{\boldsymbol{\theta}}_n$)

(i) $\mathbf{b}_1(\tilde{\boldsymbol{\theta}}_n) = \mathbf{0}$ and $B_1(\tilde{\boldsymbol{\theta}}_n) = 0$,

(ii) $\mathbf{M}_1(\tilde{\boldsymbol{\theta}}_n) = \sigma^2\boldsymbol{\Lambda}_{11}$ and $R_1(\tilde{\boldsymbol{\theta}}_n;\mathbf{W}) = \sigma^2\mathrm{tr}[\mathbf{W}\boldsymbol{\Lambda}_{11}]$.

Restricted Estimator ($\hat{\boldsymbol{\theta}}_n$)

(i) $\mathbf{b}_2(\hat{\boldsymbol{\theta}}_n) = \mathbf{T}_0\mathbf{J}\boldsymbol{\delta}$ and $B_2(\hat{\boldsymbol{\theta}}_n) = \sigma^{-2}(\boldsymbol{\delta}'\mathbf{J}'\mathbf{T}_0\boldsymbol{\Lambda}_{22}^{-1}\mathbf{T}_0\mathbf{J}\boldsymbol{\delta}) = \Delta^{*2}$;

(ii) $\mathbf{M}_2(\hat{\boldsymbol{\theta}}_n) = \sigma^2(\boldsymbol{\Lambda}_{11}^* + \sigma^{-2}\mathbf{T}_0\mathbf{J}\boldsymbol{\delta}\boldsymbol{\delta}'\mathbf{J}'\mathbf{T}_0)$ and $R_2(\hat{\boldsymbol{\theta}}_n;\mathbf{W}) = \sigma^2[\mathbf{W}\boldsymbol{\Lambda}_{11}^*] + (\boldsymbol{\delta}'\mathbf{J}'\mathbf{T}_0\mathbf{W}\mathbf{T}_0\mathbf{J}\boldsymbol{\delta})$.

Preliminary Test Estimator ($\hat{\boldsymbol{\theta}}_n^{\mathrm{PT}}$)

(i) $\mathbf{b}_3(\hat{\boldsymbol{\theta}}_n^{\mathrm{PT}}) = \mathbf{T}_0\mathbf{J}\boldsymbol{\delta}H_{p+1}(\chi_{p-1}^2(\alpha);\Delta^2)$ and

$$B_3(\hat{\boldsymbol{\theta}}_n^{\mathrm{PT}}) = \sigma^{-2}(\boldsymbol{\delta}'\mathbf{J}'\mathbf{T}_0\boldsymbol{\Lambda}_{22}^{-1}\mathbf{T}_0\mathbf{J}\boldsymbol{\delta})\{H_{p+1}(\chi_{p-1}^2(\alpha);\Delta^2)\}^2;$$

(ii) $\mathbf{M}_3(\hat{\boldsymbol{\beta}}_n^{\mathrm{PT}}) = \sigma^2\boldsymbol{\Lambda}_{11} - \sigma^2(\boldsymbol{\Lambda}_{11} - \boldsymbol{\Lambda}_{11}^*)H_{p+1}(\chi_{p-1}^2(\alpha);\Delta^2)$

$$+(\mathbf{T}_0\mathbf{J}\boldsymbol{\delta}\boldsymbol{\delta}'\mathbf{J}'\mathbf{T}_0)\{2H_{p+1}(\chi_{p-1}^2(\alpha);\Delta^2) - H_{p+3}(\chi_{p-1}^2(\alpha);\Delta^2)\};$$

and

$$R_3(\hat{\boldsymbol{\beta}}_n^{\mathrm{PT}};\mathbf{W}) = \sigma^2\mathrm{tr}[\mathbf{W}\boldsymbol{\Lambda}_{11}] - \sigma^2\mathrm{tr}[\mathbf{W}(\boldsymbol{\Lambda}_{11} - \boldsymbol{\Lambda}_{11}^*)]H_{p+1}(\chi_{p-1}^2(\alpha);\Delta^2)$$

$$+(\boldsymbol{\delta}'\mathbf{J}'\mathbf{T}_0\mathbf{W}\mathbf{T}_0\mathbf{J}\boldsymbol{\delta})\{2H_{p+1}(\chi_{p-1}^2(\alpha);\Delta^2) - H_{p+3}(\chi_{p-1}^2(\alpha);\Delta^2)\}. \tag{6.5.23a}$$

James-Stein-Type Estimator ($\hat{\boldsymbol{\theta}}_n^S$)

(i) $\mathbf{b}_4(\hat{\boldsymbol{\theta}}_n^S) = -(p-3)\mathbf{T}_0\mathbf{J}\boldsymbol{\delta}E[\chi_{p+1}^{-2}(\Delta^2)]$ and

$$B_4(\hat{\boldsymbol{\theta}}_n^S) = (p-3)^2(\boldsymbol{\delta}'\mathbf{J}'\mathbf{T}_0\boldsymbol{\Lambda}_{22}^{-1}\mathbf{T}_0\boldsymbol{\delta})\{E[\chi_{p+1}^{-2}(\Delta^2)]\}^2;$$

(ii) $\mathbf{M}_4(\hat{\boldsymbol{\theta}}_n^S) = \sigma^2\boldsymbol{\Lambda}_{11} - \sigma^2(p-3)(\mathbf{T}_0\mathbf{J}\boldsymbol{\Lambda}_{22}\mathbf{J}'\mathbf{T}_0)\{2E[\chi_{p+1}^{-2}(\Delta^2)]$

$-(p-3)E[\chi_{p+1}^{-4}(\Delta^2)]\} + (p-3)(p+1)(\mathbf{T}_0\mathbf{J}\boldsymbol{\delta}\boldsymbol{\delta}'\mathbf{J}'\mathbf{T}_0)E[\chi_{p+3}^{-4}(\Delta^2)]$

and $R_4(\hat{\boldsymbol{\theta}}_n^S;\mathbf{W}) = \sigma^2\text{tr}[\mathbf{W}\boldsymbol{\Lambda}_{22}] - \sigma^2(p-3)\text{tr}[\mathbf{W}(\mathbf{T}_0\mathbf{J}\boldsymbol{\Lambda}_{22}\mathbf{J}'\mathbf{T}_0)]\{2E[\chi_{p+1}^{-2}(\Delta^2)] - (p-3)E[\chi_{p+1}^{-4}(\Delta^2)]\} + (p-3)(p+1)(\boldsymbol{\delta}'\mathbf{J}'\mathbf{T}_0\mathbf{W}\mathbf{T}_0\mathbf{J}\boldsymbol{\delta})E[\chi_{p+3}^{-4}(\Delta^2)]$.

Positive-Rule Stein Estimator ($\hat{\boldsymbol{\theta}}_n^{S+}$)

(i) $\mathbf{b}_5(\hat{\boldsymbol{\theta}}_n^{S+}) = -\mathbf{T}_0\mathbf{J}\boldsymbol{\delta}\{E[(1-(p-3)\chi_{p+1}^{-2}(\Delta^2))I(\chi_{p+1}^2(\Delta^2) < p-3)]$

$+(p-3)E[\chi_{p+1}^{-2}(\Delta^2)]\}$ and

$$B_5(\hat{\boldsymbol{\theta}}_n^{S+}) = \sigma^{-2}(\boldsymbol{\delta}'\mathbf{J}'\mathbf{T}_0\boldsymbol{\Lambda}_{22}^{-1}\mathbf{T}_0\mathbf{J}\boldsymbol{\delta})\{(p-3)E[\chi_{p+1}^{-2}(\Delta^2)]$$

$$+E[1-(p-3)\chi_{p+1}^{-2}(\Delta^2)I(\chi_{p+1}^2(\Delta^2) < p-3)]\}^2,$$

(ii) $\mathbf{M}_5(\hat{\boldsymbol{\theta}}_n^{S+}) = \mathbf{M}_4(\hat{\boldsymbol{\theta}}_n^S) - \sigma^2\{(\mathbf{T}_0\mathbf{J}\boldsymbol{\Lambda}_{22}\mathbf{J}\mathbf{T}_0)$

$\times E[(1-(p-3)\chi_{p+1}^{-2}(\Delta^2))^2 I(\chi_{p+1}^2(\Delta^2) < p-3)] + \sigma^{-2}(\mathbf{T}_0\mathbf{J}\boldsymbol{\delta}\boldsymbol{\delta}'\mathbf{J}'\mathbf{T}_0)$

$\times E[(1-(p-3)\chi_{p+3}^{-2}(\Delta^2))^2 I(\chi_{p+3}^2(\Delta^2) < p-3)]\} - 2(\mathbf{T}_0\mathbf{J}\boldsymbol{\delta}\boldsymbol{\delta}'\mathbf{J}'\mathbf{T}_0)$

$\times E[((p-3)\chi_{p+3}^{-2}(\Delta^2) - 1)I(\chi_{p+3}^2(\Delta^2) < p-3)]$ and

$R_5(\hat{\boldsymbol{\theta}}_n^{S+};\mathbf{W}) = \mathbf{R}_4(\hat{\boldsymbol{\theta}}_n^S) - \sigma^2\{\text{tr}[\mathbf{W}(\mathbf{T}_0\mathbf{J}\boldsymbol{\Lambda}_{22}\mathbf{J}'\mathbf{T}_0)]$

$\times E[(1-(p-3)\chi_{p+1}^{-2}(\Delta^2))^2 I(\chi_{p+1}^2(\Delta^2) < p-3)] + \sigma^{-2}(\boldsymbol{\delta}'\mathbf{J}'\mathbf{T}_0\mathbf{W}\mathbf{T}_0\mathbf{J}\boldsymbol{\delta})$

$\times E[(1-(p-3)\chi_{p+3}^{-2}(\Delta^2))^2 I(\chi_{p+3}^2(\Delta^2) < p-3)]\} - 2(\boldsymbol{\delta}'\mathbf{J}'\mathbf{T}_0\mathbf{W}\mathbf{T}_0\mathbf{J}\boldsymbol{\delta})$

$\times E[(1-(p-3)\chi_{p+3}^{-2}(\Delta^2))I(\chi_{p+3}^2(\Delta^2) < p-3)]$.

These expressions are similar to those of Section 6.5. Hence, the conclusions on the properties of the estimators in Sections 6.6 and 6.7 hold.

6.10.6 Confidence Sets for Intercept and Slope Parameters

As in Section 6.8, we consider the confidence sets for $\boldsymbol{\beta}$ as follows, based on the estimators of $\boldsymbol{\beta}$ and $\boldsymbol{\theta}$:

(i) $C^0(\tilde{\boldsymbol{\beta}}_n) = \{\boldsymbol{\beta} : n\sigma_n^{-2}\|\boldsymbol{\beta} - \tilde{\boldsymbol{\beta}}_n\|^2_{\boldsymbol{\Lambda}_{22}^{-1}} \leq \chi_p^2(\gamma)\}$ and $C^0(\tilde{\boldsymbol{\theta}}_n) = \{\boldsymbol{\theta}; n\bar{\sigma}_n^2\|\boldsymbol{\theta} - \tilde{\boldsymbol{\theta}}_n\|^2_{\boldsymbol{\Lambda}_{11}} \leq \chi_p^2(\gamma)\}$ where $\sigma_n^2 = A_n^2/\hat{\gamma}^2(\psi, \phi)$ and $\hat{\gamma}(\psi, \phi)$ is a consistent estimator of $\gamma(\psi, \phi)$ defined by $\hat{\gamma}(\psi, \phi) = \frac{1}{p}\Sigma\hat{\gamma}_j(\psi, \phi)$, where

$$\begin{aligned}
\boldsymbol{\Lambda}_{22}^{(j)} &= j\text{th row-vector of } \boldsymbol{\Lambda}_{22},\\
\mathbf{e}_j &= \text{unit-vector with 1 in the } j\text{th place and 0 elsewhere},\\
\hat{\gamma}_j(\psi, \phi) &= \left\{L_{nj}\left(\tilde{\boldsymbol{\beta}}_n + \tfrac{a\mathbf{e}_j}{\sqrt{n}}\right) - L_{nj}(\tilde{\boldsymbol{\beta}}_n - \tfrac{a\mathbf{e}_j}{\sqrt{n}})\right\} \Big/ \left(2a\boldsymbol{\Lambda}_{22}^{(j)}\mathbf{e}_j\right) \text{ for }\\
j &= 1, \ldots, p.
\end{aligned}$$

(ii) $C^R(\hat{\boldsymbol{\beta}}_n) = \{\boldsymbol{\beta} : n\sigma_n^{-2}\|\boldsymbol{\beta} - \hat{\boldsymbol{\beta}}_n\|^2_{\boldsymbol{\Lambda}_{22}^{-1}} \leq \chi_p^2(\gamma)\}$ and

$$C^R(\hat{\boldsymbol{\theta}}_n) = \{\boldsymbol{\theta}; n\bar{\sigma}_n^2\|\boldsymbol{\theta} - \hat{\boldsymbol{\theta}}_n\|^2_{\boldsymbol{\Lambda}_{11}} \leq \chi_p^2(\gamma)\},$$

(iii) $C^{\text{PT}}(\hat{\boldsymbol{\beta}}_n^{\text{PT}}(\alpha)) = \{\boldsymbol{\beta} : n\sigma_n^{-2}\|\boldsymbol{\beta} - \hat{\boldsymbol{\beta}}_n^{\text{PT}}(\alpha)\|^2_{\boldsymbol{\Lambda}_{22}^{-1}} < \chi_p^2(\gamma)\}$ and

$$C^{\text{PT}}(\hat{\boldsymbol{\theta}}_n^{\text{PT}}) = \{\boldsymbol{\theta}; n\bar{\sigma}_n^2\|\boldsymbol{\theta} - \hat{\boldsymbol{\theta}}_n^{\text{PT}}(\alpha)\|^2_{\boldsymbol{\Lambda}_{11}} \leq \chi_p^2(\gamma)\}.$$

(iv) $C^S(\hat{\boldsymbol{\beta}}_n^S) = \{\boldsymbol{\beta} : n\sigma_n^{-2}\|\boldsymbol{\beta} - \hat{\boldsymbol{\beta}}_n\|^2_{\boldsymbol{\Lambda}_{22}^{-1}} \leq \chi_p^2(\gamma)\}$ and

$$C^S(\hat{\boldsymbol{\theta}}_n^S) = \{\boldsymbol{\theta}; n\bar{\sigma}_n^2\|\boldsymbol{\theta} - \hat{\boldsymbol{\theta}}_n^S\|^2_{\boldsymbol{\Lambda}_{11}} \leq \chi_p^2(\gamma)\}.$$

(v) $C^{S+}(\hat{\boldsymbol{\beta}}_n^{S+}) = \{\boldsymbol{\beta} : n\sigma_n^{-2}\|\boldsymbol{\beta} - \hat{\boldsymbol{\beta}}_n^{S+}\|^2_{\boldsymbol{\Lambda}_{22}^{-1}} \leq \chi_p^2(\gamma)\}$ and

$$C^{S+}(\hat{\boldsymbol{\theta}}_n^{S+}) = \{\boldsymbol{\theta}; n\bar{\sigma}_n^2\|\boldsymbol{\theta} - \hat{\boldsymbol{\theta}}_n^{S+}(\alpha)\|^2_{\boldsymbol{\Lambda}_{11}} \leq \chi_p^2(\gamma)\}.$$

Clearly,

$$\lim_{n\to\infty} P\left\{n\sigma_n^{-2}\|\boldsymbol{\beta} - \tilde{\boldsymbol{\beta}}_n\|^2 \leq \chi_p^2(\gamma)\right\} = 1 - \gamma$$

and

$$\lim P\{n\sigma_n^{-2}\|\boldsymbol{\theta} - \tilde{\boldsymbol{\theta}}_n\|^2_{\boldsymbol{\Lambda}_{11}} < \chi_p^2(\gamma)\} = 1 - \gamma.$$

Now, consider the five estimators in a compact form:

$$\boldsymbol{\beta}_n^* = \mathbf{A}\tilde{\boldsymbol{\beta}}_n + (\mathbf{I} - \mathbf{A})\tilde{\boldsymbol{\beta}}_n g(\mathcal{L}_n), \quad \boldsymbol{\theta}_n^* = \hat{\boldsymbol{\theta}}_n + (\tilde{\boldsymbol{\theta}}_n - \hat{\boldsymbol{\theta}}_n)g(\mathcal{L}_n),$$

where $\mathbf{A} = \mathbf{I}_p - \frac{\mathbf{1}_p\mathbf{1}_p'\mathbf{\Lambda}_{22}^{-1}}{Q}$. Then we can see that under $\{K_{(n)}\}$,

$$\begin{aligned}
&\lim_{n\to\infty} P_{K_{(n)}}\{n\sigma_n^{-2}\|\boldsymbol{\beta} - \boldsymbol{\beta}_n^*\|^2_{\mathbf{\Lambda}_{22}^{-1}} \le \chi_p^2(\gamma)\} \\
&= \lim P_{K_{(n)}}\{n\sigma_n^{-2}\|\boldsymbol{\theta} - \boldsymbol{\theta}_n^*\|^2_{\mathbf{\Lambda}_{11}^{-1}} \le \chi_p^2(\gamma)\} \\
&= P_0\{\|\eta_1 - \omega_1 g(\|\omega_1\|^2)\|^2 + (\eta_2 - \omega_2)^2 < \chi_p^2(\gamma)\},
\end{aligned}$$

which is similar to the expressions for the coverage probabilities given in Section 6.8. As such, the asymptotic coverage probabilities of each of the confidence sets are available in Section 6.8.1 with the analysis in Section 6.8.2 and are left to the readers to peruse.

6.11 Conclusions

In this chapter, we considered estimations of the *intercept* and *slope vectors* of *several simple linear models* to test the hypothesis of the equality of the slopes (i.e., the lines are *parallel*). We presented the normal theory of point estimation and confidence set estimation in Sections 6.1 through 6.4 and the nonnormal theory in Sections 6.5 through 6.7. In Section 6.8, we considered the confidence set estimation. Finally, we studied the nonparametric methods, namely the R-estimation by which we expanded the scope of the application of these estimators. In every case we considered the unrestricted, restricted, preliminary test, and Stein-type estimators of parameters. Asymptotic theory dominates the analysis of nonnormal and nonparametric methods under local alternative to parallelism hypothesis. Also, we discussed the confidence sets recentered at the five estimators.

6.12 Problems

1. Find the joint distribution of $((\hat{\boldsymbol{\theta}}_n - \boldsymbol{\theta})', (\tilde{\boldsymbol{\theta}}_n - \hat{\boldsymbol{\theta}}_n)')'$ based on the Theorem 6.1.1.
2. How would you test the hypothesis $H_{01} : \boldsymbol{\theta} = \theta_0\mathbf{1}_p$ against the alternative $H_{A1} : \boldsymbol{\theta} \neq \theta_0\mathbf{1}_p$. Define the test of H_0 after the test of $H_{02} : \boldsymbol{\beta} = \beta_0\mathbf{1}_p$ against $H_{A2} : \boldsymbol{\beta} \neq \beta_0\mathbf{1}_p$.
3. Verify the bias, quadratic bias, MSE matrices, and risk expressions of Section 6.3.
4. Define the improved estimation of the PTE, $\hat{\boldsymbol{\beta}}_n^{\mathrm{PT}}$, and $\hat{\boldsymbol{\theta}}_n^{\mathrm{PT}}$ and obtain their bias, quadratic bias, MSE, and risk expressions.
5. Verify the expressions of MRE$(\hat{\boldsymbol{\theta}}_n : \tilde{\boldsymbol{\theta}}_n)$, MRE$(\hat{\boldsymbol{\theta}}_n^{\mathrm{PT}} : \tilde{\boldsymbol{\theta}}_n)$, MRE$(\hat{\boldsymbol{\theta}}_n^{S} : \tilde{\boldsymbol{\theta}}_n)$, and MRE$(\hat{\boldsymbol{\theta}}_n^{S} : \tilde{\boldsymbol{\theta}}_n)$.
6. Consider the usual simple linear models

$$\mathbf{Y}_\alpha = \theta_\alpha\mathbf{1}_{n_\alpha} + \beta_\alpha\mathbf{x}_\alpha + \boldsymbol{\varepsilon}_\alpha, \quad \alpha = 1, \cdots, p,$$

where $\mathbf{Y}_\alpha$ is a $(n_\alpha \times 1)$-vector, $\mathbf{1}_{n_\alpha}$ is a 1-vector of n_α- tuples of ones, $\boldsymbol{\varepsilon}_\alpha = (\epsilon_{\alpha_1}, \cdots, \epsilon_{\alpha_{n_\alpha}})'$, $\mathbf{x}_\alpha = (x_{\alpha_1}, \cdots, x_{\alpha_{n_\alpha}})'$ and $\boldsymbol{\varepsilon}_\alpha$ are i.i.d. random variables $\mathcal{N}(0, \sigma^2 \mathbf{I}_{n_\alpha})$. Estimate $\boldsymbol{\theta}$ when it is suspected that $\boldsymbol{\theta}$ belongs to the subspace $\boldsymbol{\theta} = \theta_0 \mathbf{1}_p$. Define the unrestricted, restricted, PTE, SE, and PRSE. Determine the bias, quadratic bias, MSE, and risk expressions of these estimators.

7. Refer to the model in Problem 6. Now, assume that the distribution of $\boldsymbol{\varepsilon}_\alpha$ is

$$f(\boldsymbol{\varepsilon}_\alpha; \nu_\alpha, \sigma^2) = \frac{\Gamma(\frac{n_\alpha p + \nu_\alpha}{2})}{(\pi\nu_\alpha)^{\frac{n_\alpha p}{2}} \Gamma(\frac{p}{2})} \left(1 + \frac{\sum_{\alpha=1}^{n_\alpha} \boldsymbol{\varepsilon}'_\alpha \boldsymbol{\varepsilon}_\alpha}{\nu_\alpha \sigma^2}\right)^{n_\alpha + \nu_\alpha/2}, \quad \nu_\alpha, \sigma^2 > 0.$$

Define the unrestricted, restricted, PTE, SE, and PRSE when $\boldsymbol{\beta}$ belongs to the subspace $\boldsymbol{\beta} = \beta_0 \mathbf{1}_p$ $(\boldsymbol{\beta} = (\beta_1, \cdots, \beta_p)')$, and find their bias, quadratic bias, MSE matrices, and risk expressions.

8. Refer again to Problem 7. Define the confidence sets for $\boldsymbol{\beta}$ and $\boldsymbol{\theta}$. Can you find the expressions for the coverage probabilities of these sets?

9. Transform $\sigma^{-2}\|\boldsymbol{\theta}_n^* - \boldsymbol{\theta}\|^2_{\mathbf{D}_{11}^{-1}}$ to $\{(\eta_2 - \omega_2)^2 + \|\boldsymbol{\eta}_1 - \boldsymbol{\omega}_1 g(\|\boldsymbol{\omega}_1\|^2)\|^2\}$, and show that

$$\begin{aligned} &P\{\sigma^{-2}\|\boldsymbol{\theta}_n^* - \boldsymbol{\theta}\|^2_{\mathbf{D}_{11}^{-1}} \le \chi_p^2(\gamma)\} \\ &= P\{(\eta_2 - \omega_2)^2 + \|\boldsymbol{\eta}_1 - \boldsymbol{\omega}_1 g(\|\boldsymbol{\omega}_1\|^2)\|^2 \le \chi_p^2(\gamma)\}. \end{aligned}$$

10. Prove Theorem 6.8.1 for the confidence sets for $\boldsymbol{\theta}$.

11. Verify the expressions of the asymptotic biases, ADMSE, and ADQR of the R-Estimators of $\boldsymbol{\beta}$ and $\boldsymbol{\theta}$ under $\{K_{(n)}\}$.

12. Refer to (6.10.27). Show that the rank estimator of common slope β_0 and the average of the slope parameter $\tilde{\beta}_i$ $(i = 1, \ldots, p)$ are asymptotically equivalent in probability, meaning $\bar{\boldsymbol{\beta}}_n = \hat{\beta}_{0n} \mathbf{1}_p + o_p(1)$.

Chapter 7

Multiple Regression Model

Outline

The most important model belonging to the class of general linear hypotheses is the *multiple regression model.* It is useful in the analysis of scientific, engineering, agricultural, biostatistics, and econometric data among other applications. In this model, n independent $(p+1)$-variate samples $\{(x_{\alpha 1}, \ldots, x_{\alpha p})', y_\alpha | \alpha = 1, \ldots, n\}$ are considered such that $y_\alpha \sim \mathcal{N}(\beta_1 x_{\alpha 1} + \cdots + \beta_p x_{\alpha p}; \sigma^2)$ for each fixed p-vector $(x_{\alpha 1}, \ldots, x_{\alpha p})'$. The vector $\boldsymbol{\beta} = (\beta_1, \ldots \beta_p)'$ is the vector of *regression parameters* of the *model,* and $(x_{\alpha 1}, \ldots, x_{\alpha p})'$ are the *design/predictor* variables. The main use of the model is to estimate efficiently the regression parameter, $\boldsymbol{\beta} = (\beta_1, \ldots, \beta_p)'$ as well as to test the hypothesis $H_0 : \mathbf{H}\boldsymbol{\beta} = \mathbf{h}$ for a given matrix $\mathbf{H}_{q\times p}$ and $\mathbf{h}_{q\times 1}$ against the alternative $H_A : \mathbf{H}\boldsymbol{\beta} \neq \mathbf{h}$. Accordingly, we consider the *unrestricted, restricted, preliminary* test, *James-Stein-type,* and *positive-rule Stein-type* estimators. In addition, we consider the confidence set estimator of $\boldsymbol{\beta}$ based on the five estimators as the center of the confidence sets with fixed volume, and we study their properties. We also provide some asymptotic theory for nonnormal errors and R-estimators in a nonparametric setup.

7.1 Model, Estimation, and Tests

Consider the multiple regression model (MRM)

$$\mathbf{Y} = \mathbf{X}\boldsymbol{\beta} + \boldsymbol{\varepsilon}, \tag{7.1.1}$$

where $\mathbf{Y}$ is a $n \times 1$ vector of response variables, $\mathbf{X}$ is an $(n \times p)$ design matrix of full rank (i.e., $\mathbf{X}'\mathbf{X}$ is invertible), $\boldsymbol{\beta} = (\beta_1, \ldots, \beta_p)'$ is the vector of *regression* parameters, and $\boldsymbol{\varepsilon} = (\varepsilon_1, \ldots, \varepsilon_n)'$ is the $(n \times 1)$ vector of errors distributed according to normal distribution $N_n(\mathbf{0}, \sigma^2\mathbf{I}_n)$, $\mathbf{I}_n$ is the identity matrix of order n and σ^2 is the common variance of the error variables. For more details of MRM, see Graybill (1976) and Arnold (1981) among others.

Remark 1. This model includes the p-sample model to test the equality of the *population means.* This is done by choosing

$$\mathbf{X} = \begin{pmatrix} \mathbf{1}_{n_1} & \cdots & 0 & \cdots & & 0 \\ 0 & & \mathbf{1}_{n_2} & 0 & \cdots & 0 \\ \vdots & & \vdots & \vdots & & \vdots \\ 0 & & 0 & & \cdots & \mathbf{1}_{n_p} \end{pmatrix},$$

where $\mathbf{1}_{n_j} = (1, \cdots, 1)'$, a vector of n_j one's, $j = 1, 2, \ldots, p$.

Generally, the main objective of the multiple regression model is the estimation of the regression parameters and the prediction of response for a given design matrix. This is in addition to testing relevant hypotheses such as $H_0 : \mathbf{H}\boldsymbol{\beta} = \mathbf{h}$, where $\mathbf{H}$ is a $(q \times p)$ matrix of rank q and $\mathbf{h}$ is a $(q \times 1)$-vector of known constants. The hypothesis is relevant for many situations and includes the equality of the components of the vector $\boldsymbol{\beta}$ and the subhypothesis $H_0 : (\boldsymbol{\beta}_1', \boldsymbol{\beta}_2')' = (\boldsymbol{\beta}_1', \mathbf{0})'$. In this chapter,, we will pursue the objective of improving the traditional point estimator of $\boldsymbol{\beta}$ by combining the results on the test of $H_0 : \mathbf{H}\boldsymbol{\beta} = \mathbf{h}$ based on normal, nonnormal, and nonparametric theory. In addition, we will discuss properties of *recentered confidence sets* based on the improved estimators.

7.1.1 Estimation of Regression Parameters of the Model

Consider the MRM given by (7.1.1). On the basis of the sample information and the model, the *unrestricted estimator* (UE) of $\boldsymbol{\beta}$ by the least squares (LS) or maximum likelihood (ML) method is written as

$$\tilde{\boldsymbol{\beta}}_n = (\mathbf{X}'\mathbf{X})^{-1}\mathbf{X}'\mathbf{Y} = \mathbf{C}^{-1}\mathbf{X}'\mathbf{Y}, \quad \mathbf{C} = \mathbf{X}'\mathbf{X}. \tag{7.1.2}$$

The corresponding *unrestricted unbiased estimator* (UUE) of σ^2 is

$$s_e^2 = \frac{1}{m}(\mathbf{Y} - \mathbf{X}\tilde{\boldsymbol{\beta}}_n)'(\mathbf{Y} - \mathbf{X}\tilde{\boldsymbol{\beta}}_n), \quad m = n - p, \tag{7.1.3}$$

where $n = n_1 + n_2 + \cdots + n_p$. Clearly, $\tilde{\boldsymbol{\beta}}_n \sim \mathcal{N}_p(\boldsymbol{\beta}, \sigma^2\mathbf{C}^{-1})$ independent of the distribution of $\mathrm{m}s_e^2/\sigma^2$, which has a central chi-square distribution with m d.f.

Suppose now that in addition to the sample information and the model, we have some additional information described by the null hypothesis $H_0 : \mathbf{H}\boldsymbol{\beta} = \mathbf{h}$. If this hypothesis is true, then our revised least squares or ML estimator of $\boldsymbol{\beta}$ is given by

$$\hat{\boldsymbol{\beta}}_n = \tilde{\boldsymbol{\beta}}_n - \mathbf{C}^{-1}\mathbf{H}'(\mathbf{H}\mathbf{C}^{-1}\mathbf{H}')^{-1}(\mathbf{H}\tilde{\boldsymbol{\beta}}_n - \mathbf{h}). \tag{7.1.4}$$

Under H_0, $E(\hat{\boldsymbol{\beta}}_n) = \boldsymbol{\beta}$. Further, $\hat{\boldsymbol{\beta}}_n \sim \mathcal{N}_p(\boldsymbol{\beta}, \sigma^2\mathbf{A})$, where

$$\mathbf{A} = \mathbf{C}^{-1} - \mathbf{C}^{-1}\mathbf{H}'(\mathbf{H}\mathbf{C}^{-1}\mathbf{H}')^{-1}\mathbf{H}\mathbf{C}^{-1}. \tag{7.1.5}$$

Remark 2. For

$$\mathbf{X} = \begin{pmatrix} \mathbf{1}_{n_1} & 0 & & \cdots & 0 \\ 0 & \mathbf{1}_{n_2} & 0 & \cdots & 0 \\ \vdots & & \ddots & & \vdots \\ 0 & 0 & & \cdots & 0 \\ 0 & 0 & \cdots & \cdots & \mathbf{1}_{n_p} \end{pmatrix}, \quad \mathbf{X}'\mathbf{X} = \begin{pmatrix} n_1 & 0 & \cdots & 0 \\ 0 & n_2 & \cdots & \cdots \\ \vdots & & \ddots & \\ 0 & 0 & 0 & n_p \end{pmatrix} = \mathbf{N}.$$

$\tilde{\boldsymbol{\beta}}_n = \tilde{\boldsymbol{\mu}}_n = (\bar{Y}_1, \ldots, \bar{Y}_p)' \sim N_p(\boldsymbol{\mu}, \sigma^2\mathbf{N}^{-1})$, and $s_e^2 = m^{-1}\sum_{i=1}^{p}\sum_{j=1}^{n_i}(y_{ij} - \bar{y}_i)^2$. Further, for

$$\mathbf{H} = \begin{bmatrix} 1 & -1 & 0 & \cdots & & 0 \\ 1 & 0 & -1 & 0 & \cdots & 0 \\ & \vdots & \vdots & & \ddots & \vdots \\ 1 & 0 & \cdots & 0 & 0 & -1 \end{bmatrix} \quad \text{and} \quad \mathbf{h} = \mathbf{0},$$

$\hat{\boldsymbol{\beta}}_n = \hat{\mu}_n = n^{-1}(n_1\bar{y}_1 + \cdots + n_p\bar{y}_p) = \bar{\bar{y}}$. Under $H_0 : \mu_1 = \cdots = \mu_p = \mu$ we have $\hat{\mu}_n \sim N(\mu, \sigma^2/n)$.

7.1.2 Test of the Null Hypothesis, $\mathbf{H}\boldsymbol{\beta} = \mathbf{h}$

In this section, we consider the test of the null hypothesis $H_0 : \mathbf{H}\boldsymbol{\beta} = \mathbf{h}$ against the alternative $H_A : \mathbf{H}\boldsymbol{\beta} \neq \mathbf{h}$. The test is given by the following theorem:

Theorem 1. The likelihood ratio statistic for testing H_0 against H_A is given by

$$\mathcal{L}_n = \frac{(\mathbf{H}\tilde{\boldsymbol{\beta}}_n - \mathbf{h})'(\mathbf{H}\mathbf{C}^{-1}\mathbf{H}')^{-1}(\mathbf{H}\tilde{\boldsymbol{\beta}}_n - \mathbf{h})}{qs_e^2}. \tag{7.1.6}$$

Under H_A, $\mathcal{L}_n$ has a noncentral F-distribution with (q, m) d.f. and noncentrality parameter $\Delta^2/2$, where

$$\Delta^2 = \frac{(\mathbf{H}\beta - \mathbf{h})'(\mathbf{H}\mathbf{C}^{-1}\mathbf{H}')^{-1}(\mathbf{H}\beta - \mathbf{h})}{\sigma^2}. \tag{7.1.7}$$

Proof. The likelihood ratio test is given by L_0/L_A, where

$$L_0 = \hat{\sigma}_n^{-n}(\sqrt{2\pi})^{-n}\exp\left(-\frac{n}{2}\right) \quad \text{and} \quad L_A = \tilde{\sigma}_n^{-n}(\sqrt{2\pi})^{-n}\exp\left(-\frac{n}{2}\right) \tag{7.1.8}$$

with

$$n\hat{\sigma}_n^2 = (\mathbf{Y} - \mathbf{X}\hat{\beta}_n)'(\mathbf{Y} - \mathbf{X}\hat{\beta}_n) \quad \text{and} \quad n\tilde{\sigma}_n^2 = ms_e^2 = (\mathbf{Y} - \mathbf{X}\tilde{\beta}_n)'(\mathbf{Y} - \mathbf{X}\tilde{\beta}_n). \tag{7.1.9}$$

Now,

$$\begin{aligned}&(\mathbf{Y} - \mathbf{X}\hat{\beta}_n)'(\mathbf{Y} - \mathbf{X}\hat{\beta}_n)\\ &= (\mathbf{Y} - \mathbf{X}\tilde{\beta}_n)'(\mathbf{Y} - \mathbf{X}\tilde{\beta}_n) + (\mathbf{H}\tilde{\beta}_n - \mathbf{h})'(\mathbf{H}\mathbf{C}^{-1}\mathbf{H}')(\mathbf{H}\tilde{\beta}_n - \mathbf{h})\end{aligned}$$

and

$$\begin{aligned}\lambda^{2/n} &= \left(\frac{L_0}{L_A}\right)^{2/n}\\ &= \left\{\frac{(\mathbf{Y} - \mathbf{X}\tilde{\beta}_n)'(\mathbf{Y} - \mathbf{X}\tilde{\beta}_n)}{(\mathbf{Y} - \mathbf{X}\tilde{\beta}_n)'(\mathbf{Y} - \mathbf{X}\tilde{\beta}_n) + (\mathbf{H}\tilde{\beta}_n - \mathbf{h})'(\mathbf{H}\mathbf{C}^{-1}\mathbf{H}')^{-1}(\mathbf{H}\tilde{\beta}_n - \mathbf{h})}\right\}\end{aligned} \tag{7.1.10}$$

$$= a\left(\frac{1}{1 + \frac{q}{m}\mathcal{L}_n}\right), \quad a \text{ being a constant free of } \mathcal{L}_n.$$

Thus, $\mathcal{L}_n$ defined by (7.1.6) is the LR statistic for testing H_0 against H_A.

Since $\tilde{\beta}_n \sim \mathcal{N}_p(\beta, \sigma^2\mathbf{C}^{-1})$, $\mathbf{H}\tilde{\beta}_n \sim \mathcal{N}_q(\mathbf{H}\beta, \sigma^2(\mathbf{H}\mathbf{C}^{-1}\mathbf{H}'))$ so that $(\mathbf{H}\tilde{\beta}_n - \mathbf{h}) \sim \mathcal{N}_q(\mathbf{H}\beta - \mathbf{h}, \sigma^2(\mathbf{H}\mathbf{C}^{-1}\mathbf{H}'))$. Clearly, $(\mathbf{H}\tilde{\beta}_n - \mathbf{h})'(\mathbf{H}\mathbf{C}^{-1}\mathbf{H}')^{-1}(\mathbf{H}\tilde{\beta}_n - \mathbf{h})/\sigma^2$ follows a noncentral chi-squared distribution with q d.f. and noncentrality parameter $\Delta^2/2$. Hence, under H_A, $\mathcal{L}_n$ follows the noncentral F-distribution with (q, m) d.f. and noncentrality parameter $\Delta^2/2$. The theorem follows.

Remark 3. It is easy to see that the p-sample *test-statistic for the equality of means* is given by

$$\mathcal{L}_n = \frac{\sum_{j=1}^{p} n_j(\bar{y}_j - \bar{\bar{y}})^2/(p-1)}{\sum_{i=1}^{p}\sum_{j=1}^{n_j}(y_{ij} - \bar{y}_i)^2/m}.$$

Some distributional results involving the estimators, $\tilde{\beta}_n$ and $\hat{\beta}_n$ are given in the following theorem, which is easy to prove:

Theorem 2. Under the assumed conditions,

(i) $\mathbf{V}_{(n)}^{(1)} = (\tilde{\beta}_n - \beta) \sim \mathcal{N}_p(\mathbf{0}, \sigma^2\mathbf{C}^{-1})$;

(ii) $\mathbf{V}_{(n)}^{(2)} = (\hat{\beta}_n - \beta) \sim \mathcal{N}_p\{-\mathbf{C}^{-1}\mathbf{H}'(\mathbf{H}\mathbf{C}^{-1}\mathbf{H}')(\mathbf{H}\beta - \mathbf{h}), \sigma^2\mathbf{A}\}$ where $\mathbf{A} = \mathbf{C}^{-1} - \mathbf{C}^{-1}\mathbf{H}'(\mathbf{H}\mathbf{C}^{-1}\mathbf{H}')^{-1}\mathbf{H}\mathbf{C}^{-1}$;

(iii) $\mathbf{V}_{(n)}^{(3)} = (\tilde{\beta}_n - \hat{\beta}_n) \sim \mathcal{N}_p\{\mathbf{C}^{-1}\mathbf{H}'(\mathbf{H}\mathbf{C}^{-1}\mathbf{H}')^{-1}(\mathbf{H}\beta - \mathbf{h}), \sigma^2(\mathbf{C}^{-1} - \mathbf{A})\}$;

(iv) $$\begin{pmatrix} \mathbf{V}_n^{(1)} \\ \mathbf{H}\tilde{\beta}_n - \mathbf{h} \end{pmatrix} \sim \mathcal{N}_{p+q}\left\{\begin{pmatrix} \mathbf{0} \\ \mathbf{H}\beta - \mathbf{h} \end{pmatrix}, \sigma^2\begin{pmatrix} \mathbf{C}^{-1} & \mathbf{C}^{-1}\mathbf{H}' \\ \mathbf{H}\mathbf{C}^{-1} & \mathbf{H}\mathbf{C}^{-1}\mathbf{H}' \end{pmatrix}\right\};$$

(v) $$\begin{pmatrix} \mathbf{V}_n^{(1)} \\ \mathbf{V}_n^{(3)} \end{pmatrix} \sim \mathcal{N}_{2p}\left\{\begin{pmatrix} \mathbf{0} \\ \mathbf{C}^{-1}\mathbf{H}'(\mathbf{H}\mathbf{C}^{-1}\mathbf{H}')^{-1}(\mathbf{H}\beta - \mathbf{h}) \end{pmatrix}; \sigma^2\begin{pmatrix} \mathbf{C}^{-1} & \mathbf{C}^{-1} - \mathbf{A} \\ \mathbf{C}^{-1} - \mathbf{A} & \mathbf{C}^{-1} - \mathbf{A} \end{pmatrix}\right\};$$

(vi) $$\begin{pmatrix} \mathbf{V}_n^{(2)} \\ \mathbf{V}_n^{(3)} \end{pmatrix} \sim \mathcal{N}_{2p}\left\{\begin{pmatrix} -\mathbf{C}^{-1}\mathbf{H}'(\mathbf{H}\mathbf{C}^{-1}\mathbf{H}')^{-1}(\mathbf{H}\beta - \mathbf{h}) \\ \mathbf{C}^{-1}\mathbf{H}'(\mathbf{H}\mathbf{C}^{-1}\mathbf{H}')^{-1}(\mathbf{H}\beta - \mathbf{h}) \end{pmatrix}; \sigma^2\begin{pmatrix} \mathbf{A} & \mathbf{0} \\ \mathbf{0} & \mathbf{C}^{-1} - \mathbf{A} \end{pmatrix}\right\};$$

(vii) $$E\left\{\mathbf{V}_n^{(1)} | \mathbf{H}\tilde{\beta}_n - \mathbf{h}\right\} = \mathbf{C}^{-1}\mathbf{H}'(\mathbf{H}\mathbf{C}^{-1}\mathbf{H}')^{-1}(\mathbf{H}\tilde{\beta}_n - \mathbf{h}) - \mathbf{C}^{-1}\mathbf{H}'(\mathbf{H}\mathbf{C}^{-1}\mathbf{H}')^{-1}(\mathbf{H}\beta - \mathbf{h})$$
and
$$E\left\{\mathbf{V}_n^{(1)} | \mathbf{V}_n^{(3)}\right\} = \left[\mathbf{V}_n^{(3)} - \mathbf{C}^{-1}\mathbf{H}'(\mathbf{H}\mathbf{C}^{-1}\mathbf{H}')^{-1}(\mathbf{H}\beta - \mathbf{h})\right];$$

(viii) $P(\mathcal{L}_n \le x) = G_{q,m}(x; \Delta^2)$ where $G_{q,m}(x; \Delta^2)$ is the cdf of a noncentral F-distribution with (q, m) d.f. and noncentrality parameter $\Delta^2/2$.

7.2 Preliminary Test Approach and Stein-Type Estimation of Regression Parameters

In this section, we discuss the preliminary test approach to shrinkage estimation in addition to the empirical Bayes approach estimation of the regression parameters.

7.2.1 Preliminary Test (or Quasi-empirical Bayes) Approach

Following Chapter 5, we write the preliminary test estimate (PTE) of β as

$$\hat{\beta}_n^{\mathrm{PT}} = \tilde{\beta}_n - (\tilde{\beta}_n - \hat{\beta}_n) I(\mathcal{L}_n < F_{q,m}(\alpha)), \tag{7.2.1}$$

where $F_{q,m}(\alpha)$ is the upper α level critical value of the F-distribution with (q, m) d.f. PTE depends on the level of significance, α and to make it independent of α, we defined a James-Stein-type estimator, similar to the ANOVA and parallelism model as

$$\hat{\beta}_n^S = \tilde{\beta}_n - d(\tilde{\beta}_n - \hat{\beta}_n)\mathcal{L}_n^{-1}, \quad d = \frac{(q-2)m}{q(m+2)}. \tag{7.2.2}$$

Notice that the forms of $\hat{\beta}_n^{PT}$ and $\hat{\beta}_n^S$ are similar where (7.2.2) obtained from (7.2.1) by replacing $I(\mathcal{L}_n < F_{q,m}(\alpha))$ by $d\mathcal{L}_n^{-1}$ to make it independent of α, the level of significance.

The estimator $\hat{\beta}_n^S$ may go past the estimator $\hat{\beta}_n$. Thus, we consider the positive-rule Stein-type estimator given by

$$\hat{\beta}_n^{S+} = \hat{\beta}_n + \{1 - d\mathcal{L}_n^{-1}\} I(\mathcal{L}_n > d)(\tilde{\beta}_n - \hat{\beta}_n), \tag{7.2.3}$$

which is a PTE based on $\hat{\beta}_n$ and $\hat{\beta}_n^S$ with critical value d.

Now, consider the expressions for $\hat{\beta}_n^{PT}$, $\hat{\beta}_n^S$ and $\hat{\beta}_n^{S+}$ given below:

$$\text{(i)}\ \hat{\beta}_n^{PT} - \beta = \mathbf{V}_n^{(1)} - \mathbf{V}_n^{(3)} I\left(\frac{\|\mathbf{V}_n^{(3)}\|_{\mathbf{C}}^2}{q\, s_e^2} < F_{q,m}(\alpha)\right).$$

$$\text{(ii)}\ \hat{\beta}_n^S - \beta = \mathbf{V}_n^{(1)} - d\mathbf{V}_n^{(3)} \|\mathbf{V}_n^{(3)}\|_{\mathbf{C}}^{-2}. \tag{7.2.4}$$

$$\text{(iii)}\ \hat{\beta}_n^{S+} - \beta$$
$$= \mathbf{V}_n^{(1)} - d\mathbf{V}_n^{(3)}\|\mathbf{V}_n^{(3)}\|_{\mathbf{C}}^{-2} - \mathbf{V}_n^{(3)}\left(1 - d\|\mathbf{V}_n^{(3)}\|_{\mathbf{C}}^{-2}\right) I\left(\frac{\|\mathbf{V}_n^{(3)}\|_{\mathbf{C}}^{-2}}{q\, s_e^2} \le d\right).$$

These expressions will be used to compute the bias, MSE matrices, and quadratic risks of the estimators. For an application of PTE in the design of linear models, see Brenda (1996).

7.2.2 Bayes and Empirical Bayes Estimators of the Regression Parameters

Assume that $\mathbf{Y}|\beta \sim \mathcal{N}_n(\mathbf{X}\beta, \sigma^2\mathbf{I}_n)$ as given for the model (7.1.1) with the assumption that $\mathbf{C} = \mathbf{X}'\mathbf{X}$ is of full rank. Next, we assume $\beta \sim N_p(\nu, \tau^2\mathbf{V})$, where ν is a p-component vector and $\mathbf{V}$ is a $p \times p$ nonsingular matrix and τ^2 is a scalar. Then, we have the posterior distribution of β given $\mathbf{Y}$ in Theorem 1 that follows.

Theorem 1. The posterior distribution of β given $\mathbf{Y} = \mathbf{y}$ is

$$\mathcal{N}_p\left\{\nu + \left(\mathbf{C} + \frac{\sigma^2}{\tau^2}\mathbf{V}^{-1}\right)^{-1} \mathbf{X}'(\mathbf{Y} - \mathbf{X}\nu),\ \sigma^2\left(\mathbf{C} + \frac{\sigma^2}{\tau^2}V^{-1}\right)^{-1}\right\}. \tag{7.2.5}$$

Proof. First, note that

$$\begin{pmatrix} \mathbf{Y} \\ \beta \end{pmatrix} \sim N_{n+p}\left\{ \begin{pmatrix} X\nu \\ \nu \end{pmatrix}, \begin{bmatrix} \sigma^2\mathbf{I}_n + \tau^2\mathbf{XVX}' & \tau^2\mathbf{XV} \\ \tau^2\mathbf{VX}' & \tau^2\mathbf{V} \end{bmatrix} \right\}. \tag{7.2.6}$$

Hence, using the conditional formula, the distribution of β for $\mathbf{Y} = \mathbf{y}$ is given by

$$\begin{aligned} \mathcal{N}_p\{\nu + \tau^2\mathbf{VX}'(\sigma^2\mathbf{I}_n + \tau^2\mathbf{XVX}')(\mathbf{Y} - \mathbf{X}\nu), \\ \tau^2\mathbf{V} - \tau^4\mathbf{VX}'(\sigma^2\mathbf{I}_n + \tau^2\mathbf{XVX}')^{-1}\mathbf{XY}\}. \end{aligned} \tag{7.2.7}$$

Now, by Theorem 10 of Chapter 2, we have

$$\begin{aligned} (\sigma^2\mathbf{I}_n + \tau^2\mathbf{XVX}')^{-1} &= (\tau^{-2})\left[\frac{\sigma^2}{\tau^2}\mathbf{I}_n + \mathbf{XVX}'\right]^{-1} \\ &= (\tau^{-2})\left[\frac{\tau^2}{\sigma^2}\mathbf{I}_n - \frac{\tau^4}{\sigma^4}\mathbf{X}\left(\frac{\tau^2}{\sigma^2}\mathbf{X}'\mathbf{X} + \mathbf{V}^{-1}\right)^{-1}\mathbf{X}'\right] \\ &= \frac{1}{\sigma^2}\left[\mathbf{I}_n - X\left(\mathbf{X}'\mathbf{X} + \frac{\sigma^2}{\tau^2}\mathbf{V}^{-1}\right)^{-1}\mathbf{X}'\right]. \end{aligned} \tag{7.2.8}$$

We have from (7.2.7), the following expression:

$$\begin{aligned} &\tau^2\mathbf{VX}'(\sigma^2\mathbf{I}_n + \tau^2\mathbf{XVX}')^{-1} \\ &= \frac{\tau^2}{\sigma^2}\left[\mathbf{VX}' - \mathbf{VX}'\mathbf{X}(\mathbf{X}'\mathbf{X} + \frac{\sigma^2}{\tau^2}\mathbf{V}^{-1})^{-1}\mathbf{X}'\right] \\ &= \frac{\tau^2}{\sigma^2}\left\{\mathbf{VX}' - \left[\mathbf{V} - \frac{\sigma^2}{\tau^2}\left(\mathbf{X}'\mathbf{X} + \frac{\tau^2}{\sigma^2}\mathbf{V}^{-1}\right)^{-1}\mathbf{X}'\right]\right\} \\ &= \left(\mathbf{C} + \frac{\sigma^2}{\tau^2}\mathbf{V}^{-1}\right)^{-1}\mathbf{X}'. \end{aligned} \tag{7.2.9}$$

Hence,

$$\begin{aligned} &\tau^2\mathbf{VX}'(\sigma^2\mathbf{I}_n + \tau^2\mathbf{XVX}')^{-1}\mathbf{XV} \\ &= \left(\mathbf{C} + \frac{\sigma^2}{\tau^2}\mathbf{V}^{-1}\right)^{-1}\mathbf{CV} = \mathbf{V} - \frac{\sigma^2}{\tau^2}\left(\mathbf{C} + \frac{\sigma^2}{\tau^2}\mathbf{V}^{-1}\right)^{-1}. \end{aligned} \tag{7.2.10}$$

The conclusion of the theorem follows from (7.2.7) through (7.2.9).

Now, since $\mathbf{V} = \mathbf{C}^{-1}$, the Bayes estimator of β is given by

$$\hat{\beta}_n^B = E(\beta|\mathbf{Y} = \mathbf{y}) = \nu + (1 - B)(\tilde{\beta}_n - \nu), \tag{7.2.11}$$

where $B = \sigma^2/\sigma^2 + \tau^2$ and the a posterior distribution of β given $\mathbf{Y} = \mathbf{y}$ is

$$\mathcal{N}_p\left\{\nu + (1 - B)(\tilde{\beta}_n - \nu),\ \sigma^2(1 - B)\mathbf{C}^{-1}\right\}. \tag{7.2.12}$$

To obtain the empirical Bayes estimator of $\boldsymbol{\beta}$, the parameters $\boldsymbol{\nu}$, τ^2 and σ^2 have to be estimated because they are unknown. By taking $\mathbf{V} = \mathbf{C}^{-1}$, we obtain the marginal distribution of $\mathbf{Y}$ as

$$\mathbf{Y} \sim \mathcal{N}_n(\mathbf{X}\boldsymbol{\nu}, \boldsymbol{\Sigma}), \tag{7.2.13}$$

where

$$\boldsymbol{\Sigma} = \sigma^2\mathbf{I}_n + \tau^2 P_{\mathbf{X}}, \quad P_{\mathbf{X}} = \mathbf{X}\mathbf{C}^{-1}\mathbf{X}', \tag{7.2.14}$$

and $P_{\mathbf{X}}$ is the orthogonal projection on the subspace spanned by the columns of the design matrix, $\mathbf{X}$.

The next theorem provides the complete sufficient statistics for $(\boldsymbol{\nu}, \sigma^2 + \tau^2, \sigma^2)$ based on the marginal distribution of $\mathbf{Y}$ by direct computation.

Theorem 2. Based on the marginal distribution of $\mathbf{Y}$ and the assumption $\mathbf{H}\boldsymbol{\nu} = \mathbf{h}$,

$$\left\{\tilde{\boldsymbol{\beta}}_n, (\mathbf{H}\tilde{\boldsymbol{\beta}}_n - \mathbf{h})'(\mathbf{H}\mathbf{C}^{-1}\mathbf{H}')^{-1}(\mathbf{H}\tilde{\boldsymbol{\beta}}_n - \mathbf{h}), s_e^2\right\} \tag{7.2.15}$$

is a complete sufficient statistic for $(\boldsymbol{\nu}, \sigma^2 + \tau^2, \sigma^2)$.

Proof. First, note that

$$\begin{aligned}(\sigma^2 I_n + \tau^2 P_{\mathbf{X}})^{-1} &= \sigma^{-2}\left[\mathbf{I}_n - (1 - B)P_{\mathbf{X}}\right] \\ &= \sigma^{-2}\mathbf{C}^*, \quad B = \frac{\sigma^2}{\sigma^2 + \tau^2}.\end{aligned} \tag{7.2.16}$$

Hence, from (7.2.12), the probability density function of the marginal distribution of $\mathbf{Y}$ is proportional to

$$\exp\left\{-\frac{1}{2\sigma^2}\|\mathbf{Y} - \mathbf{X}\boldsymbol{\nu}\|^2_{\mathbf{C}^*}\right\}. \tag{7.2.17}$$

By the Pythagorean theorem,

$$\|\mathbf{Y} - \mathbf{X}\boldsymbol{\nu}\|^2_{\mathbf{C}^*} = \|\mathbf{Y} - \mathbf{X}\tilde{\boldsymbol{\beta}}_n\|^2_{\mathbf{C}^*} + \|\mathbf{X}\tilde{\boldsymbol{\beta}}_n - \mathbf{X}\boldsymbol{\nu}\|^2_{\mathbf{C}^*}. \tag{7.2.18}$$

Since $\mathbf{X}\tilde{\boldsymbol{\beta}}_n = \mathbf{X}(\mathbf{X}'\mathbf{X})^{-1}\mathbf{X}'\mathbf{Y} = P_{\mathbf{X}}\mathbf{Y}$, where $P_{\mathbf{X}} = \mathbf{X}(\mathbf{X}'\mathbf{X})^{-1}\mathbf{X}'$, and using the fact that

$$P_{\mathbf{X}}\mathbf{X} = \mathbf{X} \ \text{ and } \ \mathbf{Y} - \mathbf{X}\tilde{\boldsymbol{\beta}}_n = (\mathbf{I}_n - P_{\mathbf{X}})\mathbf{Y}, \tag{7.2.19}$$

we have

$$\begin{aligned}\|\mathbf{X}\tilde{\boldsymbol{\beta}}_n - \mathbf{X}\boldsymbol{\nu}\|^2_{\mathbf{C}^*} &= (\mathbf{X}\tilde{\boldsymbol{\beta}}_n - \mathbf{X}\boldsymbol{\nu})'[\mathbf{I}_n - (1 - B)P_{\mathbf{X}}](\mathbf{X}\tilde{\boldsymbol{\beta}}_n - \mathbf{X}\boldsymbol{\nu}) \\ &= (\tilde{\boldsymbol{\beta}}_n - \boldsymbol{\nu})'\mathbf{X}'[\mathbf{I}_n - (1 - B)P_{\mathbf{X}}](\mathbf{X}\tilde{\boldsymbol{\beta}}_n - \mathbf{X}\boldsymbol{\nu}) \\ &= B(\tilde{\boldsymbol{\beta}}_n - \boldsymbol{\nu})'\mathbf{C}(\tilde{\boldsymbol{\beta}}_n - \boldsymbol{\nu}) = B\|\tilde{\boldsymbol{\beta}}_n - \boldsymbol{\nu}\|^2_{\mathbf{C}}\end{aligned} \tag{7.2.20}$$

and

$$\begin{aligned}\|\mathbf{Y}-\mathbf{X}\tilde{\boldsymbol{\beta}}_n\|^2_{\mathbf{C}^*} &= (\mathbf{Y}-\mathbf{X}\tilde{\boldsymbol{\beta}}_n)'[\mathbf{I}_n-(1-B)P_{\mathbf{X}}](\mathbf{Y}-\mathbf{X}\tilde{\boldsymbol{\beta}}_n)\\ &= \mathbf{Y}'(\mathbf{I}_n-P_{\mathbf{X}})'[\mathbf{I}_n-(1-B)P_{\mathbf{X}}](\mathbf{I}_n-P_{\mathbf{X}})\mathbf{Y}\\ &= \mathbf{Y}'(\mathbf{I}_n-P_{\mathbf{X}})'(\mathbf{I}_n-P_{\mathbf{X}}\mathbf{Y}) = \|\mathbf{Y}-\mathbf{X}\tilde{\boldsymbol{\beta}}_n\|^2. \quad (7.2.21)\end{aligned}$$

Now, $\hat{\boldsymbol{\beta}}_n = [\mathbf{I}_p - \mathbf{C}^{-1}\mathbf{H}'(\mathbf{H}\mathbf{C}^{-1}\mathbf{H}')^{-1}\mathbf{H}]\tilde{\boldsymbol{\beta}}_n$, where $\mathbf{I}_p - \mathbf{C}^{-1}\mathbf{H}'(\mathbf{H}\mathbf{C}^{-1}\mathbf{H}')^{-1}\mathbf{H}$ is an idempotent matrix and $\mathbf{I}_n-(1-B)P_{\mathbf{X}}$ is the orthogonal projection on the subspace, R^q. Assume $\mathbf{h}=\mathbf{0}$. Then, by the Pythagorean theorem, we have

$$\|\tilde{\boldsymbol{\beta}}_n - \boldsymbol{\nu}\|^2_{\mathbf{C}} = \|\tilde{\boldsymbol{\beta}}_n - \hat{\boldsymbol{\beta}}_n\|^2_{\mathbf{C}} + \|\hat{\boldsymbol{\beta}}_n - \boldsymbol{\nu}\|^2, \quad (7.2.22)$$

where

$$\|\tilde{\boldsymbol{\beta}}_n - \hat{\boldsymbol{\beta}}_n\|^2_{\mathbf{C}} = (\mathbf{H}\tilde{\boldsymbol{\beta}}_n - \mathbf{h})'(\mathbf{H}\mathbf{C}^{-1}\mathbf{H}')^{-1}(\mathbf{H}\tilde{\boldsymbol{\beta}}_n - \mathbf{h}).$$

Therefore, from (7.2.16) through (7.2.21), it follows that

$$\sigma^{-2}\|\mathbf{Y}-\mathbf{X}\boldsymbol{\nu}\|^2_{\mathbf{C}^*} = \sigma^{-2}(n-p)s_e^2 + (\sigma^2+\tau^2)^{-1}\left\{\|\tilde{\boldsymbol{\beta}}_n - \hat{\boldsymbol{\beta}}_n\|^2_{\mathbf{C}} + \|\hat{\boldsymbol{\beta}}_n - \boldsymbol{\nu}\|^2_{\mathbf{C}}\right\}. \quad (7.2.23)$$

Now, by the condition of sufficiency and completeness for the regular exponential family, the theorem follows.

Note that

$$E(\tilde{\boldsymbol{\beta}}_n) = E(\mathbf{C}^{-1}\mathbf{X}'\mathbf{Y}) = \boldsymbol{\nu} \quad (7.2.24a)$$

$$\text{and} \qquad \text{Cov}(\tilde{\boldsymbol{\beta}}_n) = \text{Cov}(\mathbf{C}^{-1}\mathbf{X}'\mathbf{Y}) = (\sigma^2+\tau^2)\mathbf{C}^{-1}. \quad (7.2.24b)$$

Thus, imposing the condition $\mathbf{H}\boldsymbol{\nu} = \mathbf{h}$, we obtain

$$\begin{aligned}&E(\hat{\boldsymbol{\beta}}_n) = \boldsymbol{\nu}\\ \text{and} \qquad &(\mathbf{H}\tilde{\boldsymbol{\beta}}_n - \mathbf{h}) \sim N_q(\mathbf{0}, (\sigma^2+\tau^2)\mathbf{H}\mathbf{C}^{-1}\mathbf{H}'). \quad (7.2.25)\end{aligned}$$

Therefore, it follows from (7.2.25) that

$$\mathcal{L}_n^* = \frac{1}{\sigma^2+\tau^2}(\mathbf{H}\tilde{\boldsymbol{\beta}}_n - \mathbf{h})](\mathbf{H}\mathbf{C}^{-1}\mathbf{H}')^{-1}(\mathbf{H}\tilde{\boldsymbol{\beta}}_n - \mathbf{h}) \overset{\mathcal{D}}{\simeq} \chi^2_q. \quad (7.2.26)$$

Using Theorem 7.2.2 and (3.7.25), we have the UMVUE of $(\sigma^2+\tau^2)$ as $(q-2)\mathcal{L}_n^{-1}$, where

$$\mathcal{L}_n = \frac{(\mathbf{H}\tilde{\boldsymbol{\beta}}_n - \mathbf{h})'(\mathbf{H}\mathbf{C}^{-1}\mathbf{H})^{-1}(\mathbf{H}\tilde{\boldsymbol{\beta}}_n - \mathbf{h})}{qs_e^2} \quad (7.2.27)$$

and $\hat{\boldsymbol{\beta}}_n$ is UMVUE of $\boldsymbol{\nu}$. Further, the best invariant estimator of σ^2 is $ms_e^2/m+2$. Inserting the estimators above of $\boldsymbol{\nu}$, $(\sigma^2+\tau^2)^{-1}$, and σ^2 in the

Bayes estimator (7.2.10), we obtain the empirical Bayes estimator (EBE) of $\boldsymbol{\beta}$ as

$$\hat{\boldsymbol{\beta}}_n^{EB} = \hat{\boldsymbol{\beta}}_n + \left[1 - \frac{(q-2)m}{q(m+2)}\mathcal{L}_n^{-1}\right](\tilde{\boldsymbol{\beta}}_n - \hat{\boldsymbol{\beta}}_n). \tag{7.2.28}$$

Note that by the PTE approach, we have the same estimator of $\boldsymbol{\beta}$. Now, as $\mathcal{L}_n \to 0$, $\mathcal{L}_n^{-1} \to \infty$, the estimate will have values past $\hat{\boldsymbol{\beta}}_n$. Thus, we restrict $\mathcal{L}_n > \frac{(q-2)m}{q(m+2)}$ to obtain the positive-rule Stein estimator (PRSE),

$$\hat{\boldsymbol{\beta}}_n^{EB+} = \hat{\boldsymbol{\beta}}_n + \left[1 - \frac{(q-2)m}{q(m+2)}\mathcal{L}_n^{-1}\right] I\left(\mathcal{L}_n > \frac{(q-2)m}{q(m+2)}\mathcal{L}_n^{-1}\right)(\tilde{\boldsymbol{\beta}}_n - \hat{\boldsymbol{\beta}}_n). \tag{7.2.29}$$

Notice that (7.2.27) and (7.2.29) are the same as (7.2.3) and (7.2.4), respectively. The reader may refer to Ali and Saleh(1991b) and Ghosh, Saleh, and Sen (1989) for more information on EBE

7.3 Bias, Quadratic Bias, MSE and Quadratic Risk Expressions

In this section, we obtain the bias, quadratic bias, MSE matrices, and the weighted risk expressions of the five estimators of $\boldsymbol{\beta}$.

7.3.1 Bias Expressions

The bias and quadratic bias expressions of the five estimators of the regression coefficients are given in the following theorem:

Theorem 1.

(i) $\mathbf{b}_1(\tilde{\boldsymbol{\beta}}_n) = \mathbf{0}$ and $B_1(\tilde{\boldsymbol{\beta}}_n) = 0.$ (7.3.1)

(ii) $\mathbf{b}_2(\hat{\boldsymbol{\beta}}_n) = -\mathbf{C}^{-1}\mathbf{H}'(\mathbf{H}\mathbf{C}^{-1}\mathbf{H}')^{-1}(\mathbf{H}\boldsymbol{\beta} - \mathbf{h}) = \boldsymbol{\delta}$ (say) and

$B_2(\hat{\boldsymbol{\beta}}_n) = \frac{1}{\sigma^2}(\mathbf{H}\boldsymbol{\beta} - \mathbf{h})'(\mathbf{H}\mathbf{C}^{-1}\mathbf{H}')^{-1}(\mathbf{H}\boldsymbol{\beta} - \mathbf{h}) = \boldsymbol{\delta}'\mathbf{C}\boldsymbol{\delta}/\sigma^2 = \Delta^2.$

(iii) $\mathbf{b}_3(\hat{\boldsymbol{\beta}}_n^{PT}) = -\mathbf{C}^{-1}\mathbf{H}'(\mathbf{H}\mathbf{C}^{-1}\mathbf{H}')^{-1}(\mathbf{H}\boldsymbol{\beta} - \mathbf{h})G_{q+2,m}(\ell_\alpha; \Delta^2)$
$= -\boldsymbol{\delta}G_{q+2,m}(\ell_\alpha; \Delta^2),$

where $\ell_\alpha = \frac{q}{q+2}F_{q,m}(\alpha)$, and $B_3(\hat{\boldsymbol{\beta}}_n^{PT}) = \Delta^2\{G_{q+2,m}(\ell_\alpha; \Delta^2)\}^2.$

(iv) $\mathbf{b}_4(\hat{\boldsymbol{\beta}}_n^{S}) = -dq\mathbf{C}^{-1}\mathbf{H}'(\mathbf{H}\mathbf{C}^{-1}\mathbf{H}')^{-1}(\mathbf{H}\boldsymbol{\beta} - \mathbf{h})E[\chi_{q+2}^{-2}(\Delta^2)]$
$= -dq\boldsymbol{\delta}E[\chi_{q+2}^{-2}(\Delta^2)]$

and $B_4(\hat{\beta}_n^S) = d^2q^2\Delta^2\Big\{E\big[\chi_{q+2}^{-2}(\Delta^2)\big]\Big\}^2.$

(v) $$\mathbf{b}_5(\hat{\beta}_n^{S+}) = -\mathbf{C}^{-1}\mathbf{H}'(\mathbf{H}\mathbf{C}^{-1}\mathbf{H}')^{-1}(\mathbf{H}\beta - \mathbf{h})\Big\{d_1E\Big[F_{q+2,m}^{-1}(\Delta^2)I\big(F_{q+2,m}(\Delta^2) \le d_1\big)\Big] - d_1E[F_{q+2,m}^{-1}(\Delta^2)] - G_{q+2,m}(d_1;\Delta^2)\Big\}$$

$$= -\delta\Big\{d_1E\Big[F_{q+2,m}^{-1}(\Delta^2)I\Big(F_{q+2,m}(\Delta^2) \le d_1\Big)\Big] - d_1E[F_{q+2,m}^{-1}(\Delta^2)] - G_{q+2,m}(d_1;\Delta^2)\Big\}$$

and $$B_5(\hat{\beta}_n^{S+}) = \Delta^2\Big\{d_1E[F_{q+2,m}^{-1}(\Delta^2)I\Big(F_{q+2,m}(\Delta^2) \le d_1)\Big] - d_1E\big[F_{q+2,m}^{-1}(\Delta^2)\big] - G_{q+2,m}(\ell_\alpha;\Delta^2)\Big\}^2,$$

where $d_1 = \frac{q}{q+2}d$.

Proof. (i) is obvious. For others using Theorem 7.2.2, Theorem 2.2.6, and equation (7.2.4), we obtain the following:

(ii) $E(\hat{\beta}_n - \beta) = -\mathbf{C}^{-1}\mathbf{H}'(\mathbf{H}\mathbf{C}^{-1}\mathbf{H}')^{-1}(\mathbf{H}\beta - \mathbf{h})$ and $B_2(\hat{\beta}_n) = \Delta^2$. (7.3.2)

(iii) $$E(\hat{\beta}_n^{PT} - \beta) = -\mathbf{C}^{-1}\mathbf{H}'(\mathbf{H}\mathbf{C}^{-1}\mathbf{H}')^{-1}E\{(\mathbf{H}\tilde{\beta} - \mathbf{h})I(\mathcal{L}_n \le F_{q,m}(\alpha))\}$$
$$= -\mathbf{C}^{-1}\mathbf{H}'(\mathbf{H}\mathbf{C}^{-1}\mathbf{H}')^{-1}(\mathbf{H}\beta - \mathbf{h})G_{q+2,m}(\ell_\alpha;\Delta^2)$$

by Theorem 2.2.6. Thus, $B_3(\hat{\beta}_n^{PT}) = \Delta^2\{G_{q+2,m}(\ell_\alpha;\Delta^2)\}^2$.

(iv) $E(\hat{\beta}_n^S - \beta) = -dq\mathbf{C}^{-1}\mathbf{H}\mathbf{C}^{-1}\mathbf{H}')^{-1}(\mathbf{H}\beta - \mathbf{h})E\Big[\chi_{q+2}^{-2}(\Delta^2)\Big]$,

and $B_4(\hat{\beta}_n^S) = d^2q^2\Delta^2\left\{E\left[\chi_{q+2}^{-2}(\Delta^2)\right]\right\}^2$ by Theorem 2.1.16.

(v) $$E(\hat{\beta}_n^{S+} - \beta) = E\left[\hat{\beta}_n^S - \beta\right] - E\left[\tilde{\beta}_n - \hat{\beta}_n)I(\mathcal{L}_n \le d)\right] + dE\left[(\tilde{\beta}_n - \hat{\beta}_n)\mathcal{L}_n^{-1}I(\mathcal{L}_n \le d)\right].$$

The first term is given by (iv). The second term equals

$$-\delta(G_{q+2,m}(d_1;\Delta^2), \quad d_1 = \frac{q}{q+2}d.$$

Now, the third term can be written as

$$E\left[(\tilde{\beta}_n - \hat{\beta}_n)\mathcal{L}_n^{-1}I(\mathcal{L}_n \le d)\right] = d_1\delta E\left[F_{q+2,m}^{-1}I(F_{q+2,m}(\Delta^2) \le d_1)\right].$$

Collecting the expressions in (7.3.2) and $\mathbf{b}_4(\hat{\beta}_n^s)$, we obtain the expression $\mathbf{b}_5(\hat{\beta}_n^{S+})$. The expression for $B_5(\hat{\beta}_n^{S+})$ is easily verified.

In order to study the bias of the estimators, we consider the quadratic bias expressions $B_1(\tilde{\beta}_n)$, $B_2(\hat{\beta}_n)$, $B_3(\hat{\beta}_n^{PT})$, $B_4(\hat{\beta}_n^S)$, and $B_5(\hat{\beta}_n^{S+})$. Clearly, $\tilde{\beta}_n$ is unbiased and the bias of $\hat{\boldsymbol{\beta}}_n$ is unbounded. The bias of $\hat{\boldsymbol{\beta}}_n^{PT}$ depends on the size of α and Δ^2. The bias of $\hat{\boldsymbol{\beta}}_n^{S}$ and $\hat{\boldsymbol{\beta}}_n^{S+}$ depend on Δ^2 alone. Thus, we can establish the relation

$$0 = B_1(\tilde{\boldsymbol{\beta}}_n) \leq B_5(\hat{\boldsymbol{\beta}}_n^{S+}) \leq B_4(\hat{\boldsymbol{\beta}}_n^{S}) \leq B_3(\hat{\boldsymbol{\beta}}_n^{PT}) \leq B_2(\hat{\boldsymbol{\beta}}_n) \tag{7.3.3}$$

under certain conditions on α. Note that for $\alpha = 0$, $B_3(\hat{\boldsymbol{\beta}}_n^{PT}) = B_2(\hat{\boldsymbol{\beta}}_n)$, while for $\alpha = 1$, $B_3(\hat{\boldsymbol{\beta}}_n^{PT}) = B_1(\tilde{\boldsymbol{\beta}}_n)$. The position of $B_3(\hat{\boldsymbol{\beta}}_n^{PT})$ and $B_4(\hat{\boldsymbol{\beta}}_n^{S})$ switches, depending on the value of α.

7.3.2 MSE Matrices and Weighted Risks of the Estimators

The MSE matrices and the weighted risk based on the loss function

$$L(\boldsymbol{\beta}^*, \boldsymbol{\beta}) = (\boldsymbol{\beta}^* - \boldsymbol{\beta})'\mathbf{W}(\boldsymbol{\beta}^* - \boldsymbol{\beta}) = \|\boldsymbol{\beta}^* - \boldsymbol{\beta}\|_{\mathbf{W}}^2 \tag{7.3.4}$$

are given in the following theorem:

Theorem 2. The MSE matrices and the weighted risks of the estimators of $\boldsymbol{\beta}$ are given by

(i) $\mathbf{M}_1(\tilde{\boldsymbol{\beta}}_n) = \sigma^2\mathbf{C}^{-1}$ and $R_1(\tilde{\boldsymbol{\beta}}_n; \mathbf{W}) = \sigma^2 \operatorname{tr}(\mathbf{W}\mathbf{C}^{-1})$ (7.3.5)

(ii) $\mathbf{M}_2(\hat{\boldsymbol{\beta}}_n) = \sigma^2[\mathbf{C}^{-1} - \mathbf{C}^{-1}\mathbf{H}'(\mathbf{H}\mathbf{C}^{-1}\mathbf{H}')^{-1}\mathbf{H}\mathbf{C}^{-1}] + \boldsymbol{\delta}\boldsymbol{\delta}'$ and

$R_2(\hat{\boldsymbol{\beta}}_n; \mathbf{W}) = \sigma^2 \operatorname{tr}[\mathbf{W}\{\mathbf{C}^{-1} - \mathbf{C}^{-1}\mathbf{H}'(\mathbf{H}\mathbf{C}^{-1}\mathbf{H}')^{-1}\mathbf{H}\mathbf{C}^{-1}\}] + \boldsymbol{\delta}'\mathbf{W}\boldsymbol{\delta}$

where $\boldsymbol{\delta} = \mathbf{C}^{-1}\mathbf{H}'(\mathbf{H}\mathbf{C}^{-1}\mathbf{H}')^{-1}(\mathbf{H}\boldsymbol{\beta} - \mathbf{h})$.

(iii) $\mathbf{M}_3(\hat{\boldsymbol{\beta}}_n^{PT}) = \sigma^2\left[\mathbf{C}^{-1} - \{\mathbf{C}^{-1}\mathbf{H}'(\mathbf{H}\mathbf{C}^{-1}\mathbf{H}')^{-1}\mathbf{H}\mathbf{C}^{-1}\}G_{q+2,m}(\ell_\alpha; \Delta^2)\right]$

$+\boldsymbol{\delta}\boldsymbol{\delta}'\left\{2G_{q+2,m}(\ell_\alpha; \Delta^2) - G_{q+4,m}(\ell_\alpha^*; \Delta^2)\right\}$ and

$R_3(\hat{\boldsymbol{\beta}}_n^{PT}; \mathbf{W}) = \sigma^2 \operatorname{tr}(\mathbf{W}\mathbf{C}^{-1}) - \sigma^2 \operatorname{tr}[\mathbf{W}\mathbf{C}^{-1}\mathbf{H}'(\mathbf{H}\mathbf{C}^{-1}\mathbf{H}')^{-1}\mathbf{H}\mathbf{C}^{-1}]\times$

$G_{q+2,m}(\ell_\alpha; \Delta^2) + \boldsymbol{\delta}'\mathbf{W}\boldsymbol{\delta}\left\{2G_{q+2,m}(\ell_\alpha; \Delta^2) - G_{q+4,m}(\ell_\alpha^*; \Delta^2)\right\}$

where $\ell_\alpha = \frac{q}{q+2}F_{q,m}(\alpha)$ and $\ell_\alpha^* = \frac{q}{q+4}F_{q,m}(\alpha)$;

(iv) $\mathbf{M}_4(\hat{\boldsymbol{\beta}}_n^{S}) = \sigma^2\mathbf{C}^{-1} - dq\sigma^2\mathbf{C}^{-1}\mathbf{H}'(\mathbf{H}\mathbf{C}^{-1}\mathbf{H}')^{-1}\mathbf{H}\mathbf{C}^{-1}\Big\{(q-2)E[\chi_{q+2}^{-4}(\Delta^2)] +$

$2\Delta^2 E[\chi_{q+4}^{-4}(\Delta^2)]\Big\} + dq(q+2)\boldsymbol{\delta}\boldsymbol{\delta}'E[\chi_{q+4}^{-4}(\Delta^2)]$ and

$$\begin{aligned} R_4(\hat{\boldsymbol{\beta}}_n^{S}; \mathbf{W}) &= \sigma^2 \operatorname{tr}(\mathbf{W}\mathbf{C}^{-1}) - \sigma^2 \operatorname{tr}(\mathbf{W}\mathbf{C}^{-1}\mathbf{H}'(\mathbf{H}\mathbf{C}^{-1}\mathbf{H}')^{-1}\mathbf{H}\mathbf{C}^{-1}) \\ &\quad \times \{(q-2)E[\chi_{q+2}^{-4}(\Delta^2)] + 2\Delta^2 E[\chi_{q+4}^{-4}(\Delta^2)]\} \\ &\quad + dq(q+2)\boldsymbol{\delta}\mathbf{W}\boldsymbol{\delta}E[\chi_{q+4}^{-4}(\Delta^2)]; \end{aligned}$$

$$\text{(v)}\ \mathbf{M}_5\left(\hat{\beta}_n^{S+}\right) = \mathbf{M}_4\left(\hat{\beta}_n^{S}\right) - \sigma^2\left(\mathbf{C}^{-1}\mathbf{H}'(\mathbf{H}\mathbf{C}^{-1}\mathbf{H}')\mathbf{H}\mathbf{C}^{-1}\right)$$
$$E\Big[\left(1 - d_1 F_{q+2,m}^{-1}(\Delta^2)\right)^2 I\left(F_{q+2,m}(\Delta^2) < d_1\right)\Big]$$
$$- \boldsymbol{\delta}\boldsymbol{\delta}'\Big\{2E\Big[\left(d_1 F_{q+2,m}^{-1}(\Delta^2) - 1\right) I\left(F_{q+2,m}(\Delta^2) < d_1\right)\Big]$$
$$+E\Big[\left(1 - d_2 F_{q+4,m}^{-1}(\Delta^2)\right)^2 I\left(F_{q+4,m}(\Delta^2) < d_2\right)\Big]\Big\},$$

where $d_2 = \frac{q}{q+4}d$ and

$$R_5\left(\hat{\beta}_n^{S+};\mathbf{W}\right) = R_4\left(\hat{\beta}_n^{S};\mathbf{W}\right) - \sigma^2 \operatorname{tr}\Big[\mathbf{W}\mathbf{C}^{-1}\mathbf{H}'(\mathbf{H}\mathbf{C}^{-1}\mathbf{H}')^{-1}\mathbf{H}\mathbf{C}^{-1}\Big]$$
$$\times E\Big[\left(1 - d_1 F_{q+2,m}^{-1}(\Delta^2)\right)^2 I\left(F_{q+2,m}(\Delta^2) \le d_1\right)\Big]$$
$$+\boldsymbol{\delta}'\mathbf{W}\boldsymbol{\delta}\Big\{2E\Big[\left(1 - d_1 F_{q+2,m}^{-1}(\Delta^2)\right)^2 I\left(F_{q+4,m}(\Delta^2) < d_1\right)\Big]\Big\}$$
$$-E\Big[\left(1 - d_2 F_{q+4,m}^{-1}(\Delta^2)\right)^2 I\left(F_{q+4,m}(\Delta^2) < d_2\right)\Big]\Big\}$$

Proof. (i) is obvious. The remainder of the results follow by Theorems 7.2.2 and 7.2.6 and formula (7.2.4). First, consider

$$\begin{aligned}
\text{(ii)}\ \mathbf{M}_2(\hat{\beta}_n) &= E(\hat{\beta}_n - \beta)(\hat{\beta}_n - \beta)' \\
&= E\{(\tilde{\beta}_n - \beta) - \mathbf{C}^{-1}\mathbf{H}'(\mathbf{H}\mathbf{C}^{-1}\mathbf{H}')^{-1}(\mathbf{H}\tilde{\beta}_n - \mathbf{h})\} \\
&\quad \{(\tilde{\beta}_n - \beta) - \mathbf{C}^{-1}\mathbf{H}'(\mathbf{H}\mathbf{C}^{-1}\mathbf{H}')^{-1}(\mathbf{H}\tilde{\beta}_n - \mathbf{h})\}' \\
&= \sigma^2\mathbf{C}^{-1} + \mathbf{C}^{-1}\mathbf{H}'(\mathbf{H}\mathbf{C}^{-1}\mathbf{H}')^{-1}E\{(\mathbf{H}\tilde{\beta}_n - \mathbf{h})(\mathbf{H}\tilde{\beta}_n - \mathbf{h})'\} \times \\
&\quad (\mathbf{H}\mathbf{C}^{-1}\mathbf{H}')^{-1}\mathbf{H}\mathbf{C}^{-1} - 2\mathbf{C}^{-1}\mathbf{H}'(\mathbf{H}\mathbf{C}^{-1}\mathbf{H}')^{-1} \times \\
&\quad E\{(\mathbf{H}\tilde{\beta}_n - \mathbf{h})(\tilde{\beta}_n - \beta)'\} \\
&= \sigma^2\mathbf{C}^{-1} + \mathbf{C}^{-1}\mathbf{H}'(\mathbf{H}\mathbf{C}^{-1}\mathbf{H}')^{-1}\{\sigma^2(\mathbf{H}\mathbf{C}^{-1}\mathbf{H}') \\
&\quad + (\mathbf{H}\beta - \mathbf{h})(\mathbf{H}\beta - \mathbf{h})'\}(\mathbf{H}\mathbf{C}^{-1}\mathbf{H}')^{-1}\mathbf{H}\mathbf{C}^{-1} \\
&\quad - 2\sigma^2\mathbf{C}^{-1}\mathbf{H}'(\mathbf{H}\mathbf{C}^{-1}\mathbf{H}')^{-1}\mathbf{H}\mathbf{C}^{-1} \\
&= \sigma^2\mathbf{C}^{-1} - \sigma^2\mathbf{C}^{-1}\mathbf{H}'(\mathbf{H}\mathbf{C}^{-1}\mathbf{H}')^{-1}\mathbf{H}\mathbf{C}^{-1} + \boldsymbol{\delta}\boldsymbol{\delta}'. \qquad \text{(7.3.6a)}
\end{aligned}$$

The risk expression is given by

$$R_2(\hat{\beta}_n;\mathbf{W}) = \sigma^2 \operatorname{tr}[\mathbf{W}\mathbf{A}] + \boldsymbol{\delta}'\mathbf{W}\boldsymbol{\delta},\ \mathbf{A} = \mathbf{C}^{-1} - \mathbf{C}^{-1}\mathbf{H}'(\mathbf{H}\mathbf{C}^{-1}\mathbf{H}')^{-1}\mathbf{H}\mathbf{C}^{-1}. \qquad \text{(7.3.6b)}$$

$$
\begin{aligned}
\text{(iii)}\quad \mathbf{M}_3(\hat{\boldsymbol{\beta}}_n^{\mathrm{PT}}) &= E(\hat{\boldsymbol{\beta}}_n^{\mathrm{PT}} - \boldsymbol{\beta})(\hat{\boldsymbol{\beta}}_n^{\mathrm{PT}} - \boldsymbol{\beta})' \\
&= \{(\tilde{\boldsymbol{\beta}}_n - \boldsymbol{\beta}) - \mathbf{C}^{-1}\mathbf{H}'(\mathbf{H}\mathbf{C}^{-1}\mathbf{H}')^{-1}(\mathbf{H}\tilde{\boldsymbol{\beta}}_n - \mathbf{h})I(\mathcal{L}_n < F_{q,m}(\alpha)\} \\
&\quad \times \{(\tilde{\boldsymbol{\beta}} - \boldsymbol{\beta}) - \mathbf{C}^{-1}\mathbf{H}'(\mathbf{H}\mathbf{C}^{-1}\mathbf{H}')^{-1}(\mathbf{H}\tilde{\boldsymbol{\beta}}_n - \mathbf{h})I(\mathcal{L}_n < F_{q,m}(\alpha))\}' \\
&= E[(\tilde{\boldsymbol{\beta}}_n - \boldsymbol{\beta})(\tilde{\boldsymbol{\beta}}_n - \boldsymbol{\beta})'] - 2E[\mathbf{C}^{-1}\mathbf{H}'(\mathbf{H}\mathbf{C}^{-1}\mathbf{H}')^{-1}(\mathbf{H}\tilde{\boldsymbol{\beta}}_n - \mathbf{h}) \\
&\quad \times (\tilde{\boldsymbol{\beta}}_n - \boldsymbol{\beta})'I(\mathcal{L}_n < F_{q,m}(\alpha)] \\
&\quad + E[\mathbf{C}^{-1}\mathbf{H}'(\mathbf{H}\mathbf{C}^{-1}\mathbf{H}')^{-1}(\mathbf{H}\tilde{\boldsymbol{\beta}}_n - \mathbf{h})(\mathbf{H}\tilde{\boldsymbol{\beta}}_n - \mathbf{h})' \\
&\quad \times (\mathbf{H}\mathbf{C}^{-1}\mathbf{H}')^{-1}\mathbf{H}\mathbf{C}^{-1}I(\mathcal{L}_n < F_{q,m}(\alpha))]. \qquad (7.3.7)
\end{aligned}
$$

Now, we use the following general theorem to evaluate (7.3.7).

Theorem 3. Under the assumed conditions and continuous function $g(\mathcal{L}_n)$, we have

$$
\begin{aligned}
\text{(i)}\quad & E\{\mathbf{C}^{-1}\mathbf{H}'(\mathbf{H}\mathbf{C}^{-1}\mathbf{H}')^{-1}(\mathbf{H}\tilde{\boldsymbol{\beta}}_n - \mathbf{h})(\tilde{\boldsymbol{\beta}}_n - \boldsymbol{\beta})'g(\mathcal{L}_n)\} \\
&= \sigma^2[\mathbf{C}^{-1}\mathbf{H}'(\mathbf{H}\mathbf{C}^{-1}\mathbf{H}')^{-1}\mathbf{H}\mathbf{C}^{-1}]E\left[g\left(\frac{m}{q}\frac{\chi^2_{q+2}(\Delta^2)}{\chi^2_m}\right)\right] \\
&\quad + \boldsymbol{\delta}\boldsymbol{\delta}'\left\{E\left[g\left(\frac{m}{q}\frac{\chi^2_{q+4}(\Delta^2)}{\chi^2_m}\right)\right] - E\left[g\left(\frac{m}{q}\frac{\chi^2_{q+2}(\Delta^2)}{\chi^2_m}\right)\right]\right\}; \qquad (7.3.8)
\end{aligned}
$$

$$
\begin{aligned}
\text{(ii)}\quad & E\Big\{\Big[\mathbf{C}^{-1}\mathbf{H}'(\mathbf{H}\mathbf{C}^{-1}\mathbf{H}')^{-1}(\mathbf{H}\tilde{\boldsymbol{\beta}}_n - \mathbf{h})(\mathbf{H}\tilde{\boldsymbol{\beta}}_n - \mathbf{h})'(\mathbf{H}\mathbf{C}^{-1}\mathbf{H}')^{-1}\mathbf{H}\mathbf{C}^{-1}\Big] \\
&\quad \times g^2(\mathcal{L}_n)\Big\} \\
&= \sigma^2\left[\mathbf{C}^{-1}\mathbf{H}'(\mathbf{H}\mathbf{C}^{-1}\mathbf{H}')^{-1}\mathbf{H}\mathbf{C}^{-1}\right]E\left[g^2\left(\frac{m}{q}\frac{\chi^2_{q+2}(\Delta^2)}{\chi^2_m}\right)\right] \\
&\quad + \boldsymbol{\delta}\boldsymbol{\delta}'\left[g^2\left(\frac{m}{q}\frac{\chi^2_{q+4}(\Delta^2)}{\chi^2_m}\right)\right] \qquad (7.3.9)
\end{aligned}
$$

by Theorems 7.1.1 and 2.2.7.

Now, let $g(\mathcal{L}_n) = I(\mathcal{L}_n < F_{q,m}(\alpha))$. Then

$$Eg^s\left(\frac{m}{q}\frac{\chi^2_{q+2}(\Delta^2)}{\chi^2_m}\right) = E\Big\{I\left(F_{q+2,m}(\Delta^2) < \frac{q}{q+2}F_{q,m}(\alpha)\right)\Big\}$$
$$= G_{q+2,m}(\ell_\alpha : \Delta^2) \quad \text{and}$$
$$Eg^s\left(\frac{m}{q}\frac{\chi^2_{q+4}(\Delta^2)}{\chi^2_m}\right) = E\Big\{I\left(F_{q+4,m}(\Delta^2) < \frac{q}{q+4}F_{q,m}(\alpha)\right)\Big\}$$
$$= G_{q+4,m}(\ell^*_\alpha : \Delta^2)$$

for $s = 1, 2$. Applying Theorem 7.3.3 in (7.3.7) and choosing $g(\mathcal{L}_n) = I(\mathcal{L}_n < F_{q,m}(\alpha))$, we obtain

$$\mathbf{M}_3(\hat{\boldsymbol{\beta}}_n^{\text{PT}}) = \sigma^2\mathbf{C}^{-1} - \sigma^2(\mathbf{C}^{-1}\mathbf{H}'(\mathbf{H}\mathbf{C}^{-1}\mathbf{H}')^{-1}\mathbf{H}\mathbf{C}^{-1})G_{q+2,m}(\ell_\alpha;\Delta^2)$$
$$+ \boldsymbol{\delta}\boldsymbol{\delta}'\left\{2G_{q+2,m}(\ell_\alpha;\Delta^2) - G_{q+4,m}(\ell^*_\alpha;\Delta^2)\right\}$$

and

$$R_3\left(\hat{\beta}_n^{\text{PT}};\mathbf{W}\right) = \sigma^2\,\text{tr}(\mathbf{W}\mathbf{C}^{-1}) - \sigma^2\,\text{tr}[\mathbf{W}\mathbf{C}^{-1}\mathbf{H}'(\mathbf{H}\mathbf{C}^{-1}\mathbf{H}')\mathbf{H}\mathbf{C}^{-1}]G_{q+2}(\ell_\alpha;\Delta^2)$$
$$+\boldsymbol{\delta}'\mathbf{W}\boldsymbol{\delta}\left\{2G_{q+2,m}(\ell_\alpha;\Delta^2) - 2G_{q+4,m}(\ell^*_\alpha;\Delta^2)\right\}.$$

$$\mathbf{M}_4\left(\hat{\boldsymbol{\beta}}_n^S\right) = E\left(\hat{\boldsymbol{\beta}}_n^S - \boldsymbol{\beta}\right)\left(\hat{\boldsymbol{\beta}}_n^S - \boldsymbol{\beta}\right)'$$
$$= E(\tilde{\boldsymbol{\beta}} - \boldsymbol{\beta})(\tilde{\boldsymbol{\beta}}_n - \boldsymbol{\beta})' - 2dE[(\tilde{\boldsymbol{\beta}} - \boldsymbol{\beta})(\tilde{\boldsymbol{\beta}}_n - \hat{\boldsymbol{\beta}}_n)'\mathcal{L}_n^{-1}]$$
$$+ d^2E[(\tilde{\boldsymbol{\beta}}_n - \hat{\boldsymbol{\beta}}_n)(\tilde{\boldsymbol{\beta}}_n - \hat{\boldsymbol{\beta}}_n)'\mathcal{L}_n^{-2}].$$

Here $q(\mathcal{L}_n) = d\mathcal{L}_n^{-1}$ and $g^2(\mathcal{L}_n) = d^2\mathcal{L}_n^{-2}$. Also

$$\tilde{\boldsymbol{\beta}}_n - \hat{\boldsymbol{\beta}}_n = \mathbf{C}^{-1}\mathbf{H}'(\mathbf{H}\mathbf{C}^{-1}\mathbf{H}')^{-1}(\mathbf{H}\tilde{\boldsymbol{\beta}}_n - \mathbf{h}). \tag{7.3.10}$$

Then

(i) $$E\left[\frac{dq}{m}\frac{\chi^2_m}{\chi^2_{q+2}(\Delta^2)}\right] = \frac{(q-2)m}{m+2}E\left[\chi^{-2}_{q+2}(\Delta^2)\right],$$

(ii) $$E\left[\frac{dq}{m}\frac{\chi^2_m}{\chi^2_{q+4}(\Delta^2)}\right] = \frac{(q-2)m}{m+2}E\left[\chi^{-2}_{q+4}(\Delta^2)\right],$$

(iii) $$E\left[\frac{d^2q^2}{m^2}\frac{\chi^4_m}{\chi^4_{q+2}(\Delta^2)}\right] = \frac{(q-2)m}{m+2}E\left[\chi^{-4}_{q+2}(\Delta^2)\right],$$

(iv) $$E\left[\frac{d^2q^2}{m^2}\frac{\chi_m^4}{\chi_{q+4}^4(\Delta^2)}\right]=\frac{(q-2)^2m}{m+2}E\left[\chi_{q+4}^{-4}(\Delta^2)\right].$$

Based on (i) through (iv), we obtain, after simplification,

$$\begin{aligned}\mathbf{M}_4\left(\hat{\boldsymbol{\beta}}_n^S\right)&=\sigma^2\mathbf{C}^{-1}-dq\sigma^2\left[\mathbf{C}^{-1}\mathbf{H}'(\mathbf{H}\mathbf{C}^{-1}\mathbf{H}')^{-1}\mathbf{H}\mathbf{C}^{-1}\right]\\&\quad\times\left\{2E[\chi_{q+2}^{-2}(\Delta^2)]-(q-2)E[\chi_{q+2}^{-4}(\Delta^2)]\right\}\\&\quad+dq\boldsymbol{\delta}\boldsymbol{\delta}'\left\{2E[\chi_{q+2}^{-2}(\Delta^2)]-2E[\chi_{q+4}^{-2}(\Delta^2)]+(q-2)E[\chi_{q+4}^{-4}(\Delta^2)]\right\}.\end{aligned}$$

Hence,

$$\begin{aligned}R_4\left(\hat{\boldsymbol{\beta}}_n^S;\mathbf{W}\right)&=\sigma^2\operatorname{tr}(\mathbf{W}\mathbf{C}^{-1})-dq\sigma^2\operatorname{tr}[\mathbf{W}\mathbf{C}^{-1}\mathbf{H}'(\mathbf{H}\mathbf{C}^{-1}\mathbf{H}')\mathbf{H}\mathbf{C}^{-1}]\\&\quad\times\left\{2E[\chi_{q+2}^{-2}(\Delta^2)]-(q-2)E[\chi_{q+2}^{-4}(\Delta^2)]\right\}\\&\quad+dq\boldsymbol{\delta}\mathbf{W}\boldsymbol{\delta}'\left\{2E[\chi_{q+2}^{-2}(\Delta^2)]-2E[\chi_{q+4}^{-2}(\Delta^2)]\right.\\&\quad\left.+(q-2)E[\chi_{q+4}^{-4}(\Delta^2)]\right\}.\end{aligned}$$

Finally, if

(v) Using $g(\mathcal{L}_n)=1-(1-d\mathcal{L}_n^{-1})I(\mathcal{L}_n>d)$, we obtain by evaluating the terms in (7.3.8) the expression for $\mathbf{M}_5\left(\hat{\boldsymbol{\beta}}_n^{S+}\right)$. First, note that

$$\begin{aligned}\mathbf{M}_5\left(\hat{\boldsymbol{\beta}}_n^{S+}\right)&=E(\hat{\boldsymbol{\beta}}_n^{S+}-\boldsymbol{\beta})(\hat{\boldsymbol{\beta}}_n^{S+}-\boldsymbol{\beta})'\\&=E\left\{\left[(\hat{\boldsymbol{\beta}}_n^S-\boldsymbol{\beta})-(\tilde{\boldsymbol{\beta}}_n-\hat{\boldsymbol{\beta}}_n)(1-d\mathcal{L}_n^{-1})I(\mathcal{L}_n<d)\right]\right.\\&\quad\left.\times\left[(\hat{\boldsymbol{\beta}}_n^S-\boldsymbol{\beta})-(\tilde{\boldsymbol{\beta}}_n-\hat{\boldsymbol{\beta}}_n)(1-d\mathcal{L}_n^{-1})I(\mathcal{L}_n<d)\right]'\right\}\\&=E\left[(\hat{\boldsymbol{\beta}}^S-\boldsymbol{\beta})(\hat{\boldsymbol{\beta}}_n^S-\boldsymbol{\beta})'\right]\\&\quad+E\left[(\tilde{\boldsymbol{\beta}}_n-\hat{\boldsymbol{\beta}}_n)(\tilde{\boldsymbol{\beta}}_n-\hat{\boldsymbol{\beta}}_n)'(1-d\mathcal{L}_n^{-1})I(\mathcal{L}_n<d)\right]\\&\quad-2E\left[(\hat{\boldsymbol{\beta}}_n^S-\boldsymbol{\beta})(\tilde{\boldsymbol{\beta}}_n-\hat{\boldsymbol{\beta}}_n)'(1-d\mathcal{L}_n^{-1})I(\mathcal{L}_n<d)\right].\end{aligned}$$

By Theorem 7.3.3, we have

$$\begin{aligned}\mathbf{M}_5\left(\hat{\boldsymbol{\beta}}_n^{S+}\right)&=\mathbf{M}_4\left(\hat{\boldsymbol{\beta}}_n^S\right)-\sigma^2\left[\mathbf{C}^{-1}\mathbf{H}'(\mathbf{H}\mathbf{C}^{-1}\mathbf{H}')^{-1}\mathbf{H}\mathbf{C}^{-1}\right]\\&\quad\times E\left[\left(1-d_1F_{q+2,m}^{-1}(\Delta^2)\right)^2I\left(F_{q+2,m}(\Delta^2)\le d_1\right)\right]\\&\quad+\boldsymbol{\delta}\boldsymbol{\delta}'\left\{2E\left[\left(1-d_1F_{q+2,m}^{-1}(\Delta^2)\right)I\left(F_{q+2,m}(\Delta^2)<d_1\right)\right]\right.\\&\quad\left.-E\left[\left(1-d_2F_{q+4,m}^{-1}(\Delta^2)\right)I\left(F_{q+2,m}(\Delta^2)<d_1\right)\right]\right\},\end{aligned}$$

and the expression of $R_5\left(\hat{\beta}_n^{S+};\mathbf{W}\right)$ follows as we compute $\text{tr}\left(\mathbf{W}\mathbf{M}_5(\hat{\beta}_n^{S+})\right)$.

7.4 Risk Analysis of the Estimators

In this section, we provide the weighted risk analysis of the five estimators with the general loss function

$$L(\beta^*;\beta) = (\beta^* - \beta)'\mathbf{W}(\beta^* - \beta), \tag{7.4.1}$$

where $\mathbf{W}$ is a p.s.d. matrix. We impose conditions on $\mathbf{W}$ so that Stein-type estimators dominate the unrestricted estimator for all Δ^2.

Comparison of $\hat{\beta}_n$ and $\tilde{\beta}_n$

It is clear that the risk of $\tilde{\beta}_n$ is constant, meaning $\sigma^2\text{tr}(\mathbf{W}\mathbf{C}^{-1})$. However, the risk of $\hat{\beta}_n$ depends on $\delta'\mathbf{W}\delta$, since

$$R_2(\hat{\beta}_n;\mathbf{W}) = \sigma^2\,\text{tr}(\mathbf{W}\mathbf{C}^{-1}) - \sigma^2\,\text{tr}[\mathbf{W}\mathbf{C}^{-1}\mathbf{H}'(\mathbf{H}\mathbf{C}^{-1}\mathbf{H}')^{-1}\mathbf{H}\mathbf{C}^{-1}] + \delta'\mathbf{W}\delta, \tag{7.4.2}$$

where

$$\delta = \mathbf{C}^{-1}\mathbf{H}'(\mathbf{H}\mathbf{C}^{-1}\mathbf{H}')^{-1}(\mathbf{H}\beta - \mathbf{h}).$$

Note that $\mathbf{C}^{-1/2}\mathbf{H}'(\mathbf{H}\mathbf{C}^{-1}\mathbf{H}')^{-1}\mathbf{H}\mathbf{C}^{-1/2}$ is a symmetric idempotent matrix with rank q ($\leq p$). Therefore, there exists an orthogonal matrix $\mathbf{\Gamma}$ such that

$$\mathbf{\Gamma}\mathbf{C}^{-1/2}\mathbf{H}'(\mathbf{H}\mathbf{C}^{-1}\mathbf{H}')^{-1}\mathbf{H}\mathbf{C}^{-1/2}\mathbf{\Gamma}' = \begin{pmatrix} \mathbf{I}_q & \mathbf{0} \\ \mathbf{0} & \mathbf{0} \end{pmatrix}$$

and

$$\mathbf{\Gamma}\mathbf{C}^{-1/2}\mathbf{W}\mathbf{C}^{-1/2}\mathbf{\Gamma}' = \begin{pmatrix} \mathbf{A}_{11} & \mathbf{A}_{12} \\ \mathbf{A}_{12}' & \mathbf{A}_{22} \end{pmatrix}.$$

Then

$$\begin{aligned}
&\text{tr}\left[\mathbf{W}\{\mathbf{C}^{-1}\mathbf{H}'(\mathbf{H}\mathbf{C}^{-1}\mathbf{H}')^{-1}\mathbf{H}\mathbf{C}^{-1}\}\right] \\
&= \text{tr}\left[\{\mathbf{\Gamma}\mathbf{C}^{-1/2}\mathbf{W}\mathbf{C}^{-1/2}\mathbf{\Gamma}'\}\{\mathbf{\Gamma}\mathbf{C}^{-1/2}\mathbf{H}'(\mathbf{H}\mathbf{C}^{-1}\mathbf{H}')^{-1}\mathbf{H}\mathbf{C}^{-1/2}\mathbf{\Gamma}'\}\right] \\
&= \text{tr}\left\{\begin{bmatrix} \mathbf{A}_{11} & \mathbf{A}_{12} \\ \mathbf{A}_{21} & \mathbf{A}_{22} \end{bmatrix}\begin{bmatrix} \mathbf{I}_q & \mathbf{0} \\ \mathbf{0} & \mathbf{0} \end{bmatrix}\right\} = \text{tr}(\mathbf{A}_{11}),
\end{aligned} \tag{7.4.3}$$

and we may write

$$\begin{aligned}\boldsymbol{\delta}'\mathbf{W}\boldsymbol{\delta} &= (\mathbf{H}\boldsymbol{\beta}-\mathbf{h})'(\mathbf{H}\mathbf{C}^{-1}\mathbf{H}')^{-1}\mathbf{H}\mathbf{C}^{-1}\mathbf{W}\mathbf{C}^{-1}\mathbf{H}'(\mathbf{H}\mathbf{C}^{-1}\mathbf{H}')^{-1}(\mathbf{H}\boldsymbol{\beta}-\mathbf{h})\\ &= [\boldsymbol{\Gamma}\mathbf{C}^{1/2}\boldsymbol{\beta} - \boldsymbol{\Gamma}\mathbf{C}^{-1/2}\mathbf{H}'(\mathbf{H}\mathbf{C}^{-1}\mathbf{H}')^{-1}\mathbf{h}]'\\ &\quad \times[\boldsymbol{\Gamma}\mathbf{C}^{-1/2}\mathbf{H}'(\mathbf{H}\mathbf{C}^{-1}\mathbf{H}')^{-1}\mathbf{H}\mathbf{C}^{-1/2}\boldsymbol{\Gamma}'][\boldsymbol{\Gamma}\mathbf{C}^{-1/2}\mathbf{W}\mathbf{C}^{-1/2}\boldsymbol{\Gamma}]\\ &\quad \times[\boldsymbol{\Gamma}\mathbf{C}^{-1/2}\mathbf{H}'(\mathbf{H}\mathbf{C}^{-1}\mathbf{H}')^{-1}\mathbf{H}\mathbf{C}^{-1/2}\boldsymbol{\Gamma}'] \qquad (7.4.4)\\ &\quad \times[\boldsymbol{\Gamma}\mathbf{C}^{-1/2}\boldsymbol{\beta} - \boldsymbol{\Gamma}\mathbf{C}^{1/2}\mathbf{H}'(\mathbf{H}\mathbf{C}^{-1}\mathbf{H}')^{-1}\mathbf{h}]'\\ &= \boldsymbol{\eta}'\begin{pmatrix}\mathbf{I}_q & \mathbf{0}\\ \mathbf{0} & \mathbf{0}\end{pmatrix}\begin{pmatrix}\mathbf{A}_{11} & \mathbf{A}_{12}\\ \mathbf{A}_{12}' & \mathbf{A}_{22}\end{pmatrix}\begin{pmatrix}\mathbf{I}_q & \mathbf{0}\\ \mathbf{0} & \mathbf{0}\end{pmatrix}\boldsymbol{\eta}\\ &= \boldsymbol{\eta}_1'\mathbf{A}_{11}\boldsymbol{\eta}_1, \qquad (7.4.5)\end{aligned}$$

where

$$\boldsymbol{\eta} = \boldsymbol{\Gamma}\mathbf{C}^{1/2}\boldsymbol{\beta} - \boldsymbol{\Gamma}\mathbf{C}^{-1/2}\mathbf{H}'(\mathbf{H}\mathbf{C}^{-1}\mathbf{H}')^{-1}\mathbf{h} = \begin{pmatrix}\boldsymbol{\eta}_1\\ \boldsymbol{\eta}_2\end{pmatrix}.$$

Thus, using (7.4.3) and (7.4.4), we obtain

$$R_2(\hat{\boldsymbol{\beta}}_n;\mathbf{W}) = \sigma^2\operatorname{tr}(\mathbf{W}\mathbf{C}^{-1}) - \sigma^2\operatorname{tr}(\mathbf{A}_{11}) + \boldsymbol{\eta}_1'\mathbf{A}_{11}\boldsymbol{\eta}_1 \qquad (7.4.6)$$

By Courant's theorem (see Chapter 2), we may write

$$Ch_{\min}(\mathbf{A}_{11}) \le \frac{\boldsymbol{\eta}_1'\mathbf{A}_{11}\boldsymbol{\eta}_1}{\boldsymbol{\eta}_1'\boldsymbol{\eta}_1} \le Ch_{\max}(\mathbf{A}_{11}) \qquad (7.4.7)$$

or

$$\sigma^2\Delta^2 Ch_{\min}(\mathbf{A}_{11}) \le \boldsymbol{\eta}_1'\mathbf{A}_{11}\boldsymbol{\eta}_1 \le \sigma^2\Delta^2 Ch_{\max}(\mathbf{A}_{11}), \qquad (7.4.8)$$

where $Ch_{\min}(\mathbf{A}_{11})$ and $Ch_{\max}(\mathbf{A}_{11})$ are minimum and maximum characteristic roots of $\mathbf{A}_{11}$ and $\Delta^2 = \boldsymbol{\eta}_1'\boldsymbol{\eta}_1/\sigma^2$. Then, we write

$$\begin{aligned}&R_1(\tilde{\boldsymbol{\beta}}_n;\mathbf{W}) - \sigma^2\operatorname{tr}(\mathbf{A}_{11}) + Ch_{\min}(\mathbf{A}_{11})\\ &\le R_2(\hat{\boldsymbol{\beta}}_n;\mathbf{W})\\ &\le R_1(\tilde{\boldsymbol{\beta}}_n;\mathbf{W}) - \sigma^2\operatorname{tr}(\mathbf{A}_{11}) + Ch_{\max}(\mathbf{A}_{11}), \qquad (7.4.9)\end{aligned}$$

When $\Delta^2 = 0$, the bounds are equal. Thus, (7.4.9) means that $\hat{\boldsymbol{\beta}}_n$ performs better than $\tilde{\boldsymbol{\beta}}_n$ whenever

$$\Delta^2 \le \frac{\operatorname{tr}(\mathbf{A}_{11})}{Ch_{\max}(\mathbf{A}_{11})},$$

whereas $\tilde{\boldsymbol{\beta}}_n$ performs better than $\hat{\boldsymbol{\beta}}_n$ whenever

$$\Delta^2 \ge \frac{\operatorname{tr}(\mathbf{A}_{11})}{Ch_{\min}(\mathbf{A}_{11})}.$$

For $\mathbf{W} = \mathbf{C}$, we see $\hat{\boldsymbol{\beta}}_n$ performs better than $\tilde{\boldsymbol{\beta}}_n$ in the interval $[0, q]$ and outside this interval $\tilde{\boldsymbol{\beta}}_n$ performs better than $\hat{\boldsymbol{\beta}}_n$.

Comparison of $\hat{\boldsymbol{\beta}}_n^{\mathrm{PT}}$ and $\tilde{\boldsymbol{\beta}}_n$

Consider the risk-difference

$$
\begin{aligned}
&R_1(\tilde{\boldsymbol{\beta}}_n;\mathbf{W}) - R_3(\hat{\boldsymbol{\beta}}_n^{\mathrm{PT}};\mathbf{W}) \\
&= \sigma^2 \operatorname{tr}(\mathbf{A}_{11}) G_{q+2,m}(\ell_\alpha;\Delta^2) \\
&\quad -(\boldsymbol{\eta}_1'\mathbf{A}_{11}\boldsymbol{\eta}_1)\left\{2G_{q+2,m}(\ell_\alpha;\Delta^2) - G_{q+4,m}(\ell_\alpha^*;\Delta^2)\right\}. \qquad (7.4.10)
\end{aligned}
$$

The r.h.s. of (7.4.10) is nonnegative (≥ 0) whenever

$$
\Delta^2 \leq \frac{\operatorname{tr}(\mathbf{A}_{11})}{Ch_{\max}(\mathbf{A}_{11})}\, \frac{G_{q+2,m}(\ell_\alpha;\Delta^2)}{\{2G_{q+2,m}(\ell_\alpha;\Delta^2) - G_{q+4,m}(\ell_\alpha^*;\Delta^2)\}}. \qquad (7.4.11\text{a})
$$

In this range $\hat{\boldsymbol{\beta}}_n^{\mathrm{PT}}$ performs better than $\tilde{\boldsymbol{\beta}}_n$, whereas $\tilde{\boldsymbol{\beta}}_n$ performs better than $\tilde{\boldsymbol{\beta}}_n^{\mathrm{PT}}$ whenever

$$
\Delta^2 \geq \frac{\operatorname{tr}(\mathbf{A}_{11})}{Ch_{\min}(\mathbf{A}_{11})}\, \frac{G_{q+2,m}(\ell_\alpha;\Delta^2)}{\{2G_{q+2,m}(\ell_\alpha;\Delta^2) - G_{q+4,m}(\ell_\alpha^*;\Delta^2)\}}. \qquad (7.4.11\text{b})
$$

For $\mathbf{W} = \mathbf{C}$, $\operatorname{tr}(\mathbf{A}_{11}) = q$, and the required intervals follow from (7.4.11a) and (7.4.11b).

Under $H_0 : \mathbf{H}\boldsymbol{\beta} = \mathbf{h}$, $\hat{\boldsymbol{\beta}}_n^{\mathrm{PT}}$ is superior to $\tilde{\boldsymbol{\beta}}_n$, since (7.4.10) is positive for all α-values. We can describe the graph of $R_3(\hat{\boldsymbol{\beta}}_n^{\mathrm{PT}};\mathbf{W})$ as follows: At $\Delta^2 = 0$, the graph has a value equal to

$$
\sigma^2 \operatorname{tr}(\mathbf{W}\mathbf{C}^{-1}) - \sigma^2 \operatorname{tr}(\mathbf{A}_{11}) G_{q+2,m}(\ell_\alpha;0); \qquad (7.4.12)
$$

then it increases, crossing the line $\sigma^2 \operatorname{tr}(\mathbf{W}\mathbf{C}^{-1})$ to a maximum as Δ^2 increases from 0 to the value $\Delta^2_{\max}$. The graph proceeds to decrease towards $\sigma^2 \operatorname{tr}(\mathbf{W}\mathbf{C}^{-1})$ as $\Delta^2 \to \infty$. For $\alpha = 0$, $R_3(\hat{\boldsymbol{\beta}}_n^{\mathrm{PT}};\mathbf{W}) = R_2(\hat{\boldsymbol{\beta}}_n;\mathbf{W})$, and for $\alpha = 1$, $R_3(\hat{\boldsymbol{\beta}}_n^{\mathrm{PT}};\mathbf{W}) = R_1(\tilde{\boldsymbol{\beta}}_n;\mathbf{W})$. The efficiency of $\hat{\boldsymbol{\beta}}_n^{\mathrm{PT}}$ relative to $\tilde{\boldsymbol{\beta}}_n$ is $R_1(\tilde{\boldsymbol{\beta}}_n)[R_3(\hat{\boldsymbol{\beta}}_n^{\mathrm{PT}})]^{-1}$. Since $\hat{\boldsymbol{\beta}}_n^{\mathrm{PT}}$ is not uniformly better than $\tilde{\boldsymbol{\beta}}_n$, we must obtain a PTE, selecting an optimum level of significance α^* with a minimum guaranteed efficiency, say E_0, by solving the equation

$$
\max_{0<\alpha<1} \min_{\Delta^2 \geq 0} E(\alpha,\Delta^2) \geq E_0, \qquad (7.4.13\text{a})
$$

where

$$
E(\alpha,\Delta^2) = R_1(\tilde{\boldsymbol{\beta}}_1 : \mathbf{W})\left[R_3(\hat{\boldsymbol{\beta}}_n^{\mathrm{PT}} : \mathbf{W}\right]^{-1}. \qquad (7.4.13\text{b})
$$

Comparison of $\hat{\boldsymbol{\beta}}_n^{\rm PT}$ and $\hat{\boldsymbol{\beta}}_n$

Note that both $\boldsymbol{\beta}_n^{\rm PT}$ and $\boldsymbol{\beta}_n$ are superior to $\tilde{\boldsymbol{\beta}}_n$ under $H_0 : \mathbf{H}\boldsymbol{\beta} = \mathbf{h}$. In general, the risk-difference is given by

$$\begin{aligned} &R_2(\hat{\boldsymbol{\beta}}_n; \mathbf{W}) - R_3(\hat{\boldsymbol{\beta}}_n^{\rm PT}; \mathbf{W}) \\ &= -\sigma^2 \operatorname{tr}(\mathbf{A}_{11}) \left[1 - G_{q+2,m}(\ell_\alpha; \Delta^2)\right] + (\boldsymbol{\eta}_1' \mathbf{A}_{11} \boldsymbol{\eta}_1) \\ &\quad \times \left[1 - 2G_{q+2,m}(\ell_\alpha; \Delta^2) + G_{q+4,m}(\ell_\alpha^*; \Delta^2)\right]. \end{aligned} \tag{7.4.14}$$

Thus, $\hat{\boldsymbol{\beta}}_n$ is superior to $\hat{\boldsymbol{\beta}}_n^{\rm PT}$ whenever

$$\Delta^2 \leq \frac{\operatorname{tr}(\mathbf{A}_{11})[1 - G_{q+2,m}(\ell_\alpha; \Delta^2)]}{Ch_{\max}(\mathbf{A}_{11}[1 - 2G_{q+2,m}(\ell_\alpha; \Delta^2) + G_{q+4,m}(\ell_\alpha^*; \Delta^2)]}, \tag{7.4.15a}$$

and $\hat{\boldsymbol{\beta}}_n^{\rm PT}$ is superior to $\hat{\boldsymbol{\beta}}_n$ whenever

$$\Delta^2 \geq \frac{\operatorname{tr}(\mathbf{A}_{11})}{Ch_{\max}(\mathbf{A}_{11}} \; \frac{G_{q+2,m}(\ell_\alpha; \Delta^2)}{\{1 - 2G_{q+2,m}(\ell_\alpha; \Delta^2) + G_{q+4,m}(\ell_\alpha^*; \Delta^2)\}}. \tag{7.4.15b}$$

Under H_0, the risk of the three estimators may be ordered as

$$R_2(\hat{\boldsymbol{\beta}}_n; \mathbf{W}) \leq R_3(\hat{\boldsymbol{\beta}}_n^{\rm PT}; \mathbf{W}) \leq R_1(\tilde{\boldsymbol{\beta}}_n; \mathbf{W}). \tag{7.4.16}$$

Comparison of $\hat{\boldsymbol{\beta}}_n^{S}$ and $\tilde{\boldsymbol{\beta}}_n$

The risk-difference is given by

$$\begin{aligned} &R_1(\tilde{\boldsymbol{\beta}}_n; \mathbf{W}) - R_4(\hat{\boldsymbol{\beta}}_n^{S}; \mathbf{W}) \\ &= \sigma^2 dq \operatorname{tr}(\mathbf{A}_{11}) \Big\{ (q-2) E\left[\chi_{q+2}^{-4}(\Delta^2)\right] \Big\} \\ &\quad + \left[1 - \frac{(q+2)(\boldsymbol{\eta}_1' \mathbf{A}_{11} \boldsymbol{\eta}_1)}{2\sigma^2 \Delta^2 \operatorname{tr}(\mathbf{A}_{11})}\right] (2\Delta^2) E\left[\chi_{q+4}^{-4}(\Delta^2)\right]. \end{aligned} \tag{7.4.17}$$

The risk-difference is **positive** whenever

$$\frac{\operatorname{tr}(\mathbf{A}_{11})}{Ch_{\max}(\mathbf{A}_{11})} \geq \frac{q+2}{2}. \tag{7.4.18}$$

Note that $\mathbf{A}_{11}$ involves the matrix $\mathbf{W}$.

Thus, $\hat{\boldsymbol{\beta}}_n^{S}$ uniformly dominates $\tilde{\boldsymbol{\beta}}_n$. Further, as $\Delta^2 \to \infty$, the risk-difference tends to 0 from below.

Comparison of $\hat{\boldsymbol{\beta}}_n^S$ and $\hat{\boldsymbol{\beta}}_n$

In this case, the risk of $\hat{\boldsymbol{\beta}}_n^S$ may be written as

$$\begin{aligned}
&R_4(\hat{\boldsymbol{\beta}}_n^S; \mathbf{W}) \\
&= R_2(\hat{\boldsymbol{\beta}}_n; \mathbf{W}) + \sigma^2 \operatorname{tr}(\mathbf{A}_{11})\boldsymbol{\eta}_1'\mathbf{A}_{11}\boldsymbol{\eta}_1 + \sigma^2 dq \operatorname{tr}(\mathbf{A}_{11})\Big\{(q-2)E\left[\chi_{q+2}^{-4}(\Delta^2)\right]\Big\} \\
&\quad + \left[1 - \frac{(q+2)(\boldsymbol{\eta}_1'\mathbf{A}_{11}\boldsymbol{\eta}_1)}{2\sigma^2\Delta^2 \operatorname{tr}(\mathbf{A}_{11})}\right](2\Delta^2)E\left[\chi_{q+4}^{-4}(\Delta^2)\right]. \qquad (7.4.19)
\end{aligned}$$

Under H_0, this becomes

$$\begin{aligned}
R_4(\hat{\boldsymbol{\beta}}_n^S; \mathbf{W}) &= R_2(\hat{\boldsymbol{\beta}}_n; \mathbf{W}) + \sigma^2(1-d)\operatorname{tr}(\mathbf{A}_{11}) \\
&\geq R_2(\hat{\boldsymbol{\beta}}_n; \mathbf{W}), \qquad (7.4.20)
\end{aligned}$$

while

$$\begin{aligned}
R_2(\hat{\boldsymbol{\beta}}_n; \mathbf{W}) &= R_1(\tilde{\boldsymbol{\beta}}_n; \mathbf{W}) - \sigma^2 \operatorname{tr}(\mathbf{A}_{11}) \\
&\leq R_2(\tilde{\boldsymbol{\beta}}_n; \mathbf{W}). \qquad (7.4.21)
\end{aligned}$$

Thus, $\hat{\boldsymbol{\beta}}_n$ performs better than $\hat{\boldsymbol{\beta}}_n^S$ under H_0. However, as $\boldsymbol{\eta}_1$ moves away from the origin $\mathbf{0}$, meaning Δ^2 diverts from 0, and the risk of $\hat{\boldsymbol{\beta}}_n$ becomes unbounded. The risk of $\hat{\boldsymbol{\beta}}_n^S$, however, remains below the risk of $\tilde{\boldsymbol{\beta}}_n$ and merges with it as $\Delta^2 \to \infty$. Thus, $\hat{\boldsymbol{\beta}}_n^S$ dominates $\hat{\boldsymbol{\beta}}_n$ outside an interval around the origin.

Comparison of $\hat{\boldsymbol{\beta}}_n^S$ and $\hat{\boldsymbol{\beta}}_n^{\mathrm{PT}}$

First, under H_0 the risk of $\hat{\boldsymbol{\beta}}_n^S$ is given by

$$\begin{aligned}
R_4(\hat{\boldsymbol{\beta}}_n^S; \mathbf{W}) &= R_3(\hat{\boldsymbol{\beta}}_n^{\mathrm{PT}}; \mathbf{W}) + \sigma^2 \operatorname{tr}(\mathbf{A}_{11})[G_{q+2,m}(\ell_\alpha; 0) - d] \\
&\geq R_3(\hat{\boldsymbol{\beta}}_n^{\mathrm{PT}}; \mathbf{W}) \qquad (7.4.22)
\end{aligned}$$

whenever

$$G_{q+2,m}(\ell_\alpha; 0) > d. \qquad (7.4.23)$$

That is to say, (7.4.22) holds whenever there exists an α such that

$$\left\{\alpha : \frac{q}{q+2}F_{q,m}(\alpha) > G_{q+2,m}^{-1}(d, 0)\right\}. \qquad (7.4.24)$$

The risk of $\hat{\boldsymbol{\beta}}_n^S$ is smaller than that of the risk of $\hat{\boldsymbol{\beta}}_n^{\mathrm{PT}}$ when the critical value satisfies the opposite inequality, that is,

$$\left\{\alpha : G_{q+2,m}^{-1}(d, 0) > \frac{q}{q+2}F_{q,m}(\alpha)\right\}. \qquad (7.4.25)$$

This means that the James-Stein-type estimator $\hat{\boldsymbol{\beta}}_n^S$ does not always dominate the PTE, $\hat{\boldsymbol{\beta}}_n^{PT}$ under H_0, and we can order the risks as

$$R_2(\hat{\boldsymbol{\beta}}_n; \mathbf{W}) \leq R_3(\tilde{\boldsymbol{\beta}}_n^{PT}; \mathbf{W}) \leq R_1(\tilde{\boldsymbol{\beta}}_n; \mathbf{W}) \tag{7.4.26}$$

when α satisfies (7.4.24).

Comparison of $\hat{\boldsymbol{\beta}}_n^{S+}$ and $\hat{\boldsymbol{\beta}}_n^S$

The risk-difference of the estimators in this case is given by

$$\begin{aligned}
&R_4(\hat{\boldsymbol{\beta}}_n^S; \mathbf{W}) - R_5(\hat{\boldsymbol{\beta}}_n^{S+}; \mathbf{W}) \\
&= \sigma^2\Big\{\text{tr}(\mathbf{A}_{11}E\Big[\left(1 - d_1F_{q+2,m}^{-1}(\Delta^2)\right)^2 I(F_{q+2,m}(\Delta^2) < d_1)\Big] \\
&\quad + \sigma^{-2}(\boldsymbol{\eta}_1'\mathbf{A}_{11}\boldsymbol{\eta}_1)E\big[(1 - d_2F_{q+4,m}^{-1}(\Delta^2))^2 < d_2\big]\Big\} \\
&\quad - 2(\boldsymbol{\eta}_1'\mathbf{A}_{11}\boldsymbol{\eta}_1)E\left[(d_1F_{q+2,m}^{-1}(\Delta^2) - 1)I(F_{q+2,m}(\Delta^2) < d_1)\right].
\end{aligned} \tag{7.4.27}$$

The r.h.s. is positive semidefinite, since the expectation of a positive random variable is positive by the definition of an indicator function,

$$\left(0 < F_{q+2,m}(\Delta^2) < d_1\right) \longleftrightarrow \left(d_1F_{q+2,m}^{-1}(\Delta^2) - 1\right) \geq 0. \tag{7.4.28}$$

We get

$$E\left[\left(d_1F_{q+2,m}^{-1}(\Delta^2) - 1\right) I\left(F_{q+2,m}(\Delta^2) < d_1\right)\right] \geq 0. \tag{7.4.29}$$

Thus, for all Δ^2,

$$R_5(\hat{\boldsymbol{\beta}}_n^{S+}; \mathbf{W}) \leq R_4(\hat{\boldsymbol{\beta}}_n^S; \mathbf{W}), \tag{7.4.30}$$

and $\hat{\boldsymbol{\beta}}_n^{S+}$ not only confirms inadmissibility of $\hat{\boldsymbol{\beta}}_n^S$ but also provides a simple superior estimator. In passing, we remark that the risks of $\tilde{\boldsymbol{\beta}}_n$, $\hat{\boldsymbol{\beta}}_n^S$, and $\hat{\boldsymbol{\beta}}_n^{S+}$ can be ordered as

$$R_5(\hat{\boldsymbol{\beta}}_n^{S+}; \mathbf{W}) \leq R_4(\hat{\boldsymbol{\beta}}_n^S; \mathbf{W}) \leq R_1(\tilde{\boldsymbol{\beta}}_n; \mathbf{W}) \tag{7.4.31}$$

for all Δ^2, provided that

$$\left\{\mathcal{A} : \frac{\text{tr}(\mathbf{A}_{11})}{Ch_{\max}(\mathbf{A}_{11})} \geq \frac{q+2}{2}\right\} \quad \text{and } q > 3.$$

Comparison of $\hat{\beta}_n$ and $\hat{\beta}_n^{S+}$

First, consider the risk of $\hat{\beta}_n^{S+}$ under H_0, in terms of the risk of $\hat{\beta}_n$,

$$\begin{aligned} R_5(\hat{\beta}_n^{S+};\mathbf{W}) &= R_2(\hat{\beta}_n;\mathbf{W}) + \sigma^2 \operatorname{tr}(\mathbf{A}_{11})\{(1-d) \\ &\quad - E\left[\left(1 - d_1 F_{q+2,m}^{-1}(\Delta^2)\right)^2 I(F_{q+2,m}(0) < d_1)\right]\} \\ &\geq R_2(\hat{\beta}_n;\mathbf{W}), \end{aligned} \tag{7.4.32}$$

since

$$\begin{aligned} &E\left[\left(1 - d_1 F_{+2,m}^{-1}(0)\right)^2 I\left(F_{q+2,m}(0) < d_1\right)\right] \\ &\leq E\left[\left(1 - d_1 F_{q+2,m}^{-1}(0)\right)^2\right] = 1 - d. \end{aligned} \tag{7.4.33}$$

Clearly, $\hat{\beta}_n$ performs better than $\hat{\beta}_n^{S+}$ under H$_0$. However, as η_1 moves away from $\mathbf{0}$, Δ^2 increases. Then the risk of $\hat{\beta}_n$ becomes unbounded while the risk of $\hat{\beta}_n^{S+}$ remains below the risk of $\tilde{\beta}_n$ and merges with it as $\Delta^2 \to \infty$. Thus, $\hat{\beta}_n^{S+}$ dominates $\hat{\beta}_n$ outside an interval near the origin.

Comparison of $\hat{\beta}_n^{S+}$ and $\hat{\beta}_n^{\text{PT}}$

Under H_0, the risk of $\hat{\beta}_n^{S+}$ is written as

$$\begin{aligned} R_5(\hat{\beta}_n^{S+};\mathbf{W}) &= R_3(\hat{\beta}_n^{\text{PT}};\mathbf{W}) + \sigma^2 \operatorname{tr}(\mathbf{A}_{11})\{G_{q+2,m}(\ell_\alpha;0) - d \\ &\quad - E\left[\left(1 - d_1 F_{q+2,m}^{-1}(0)\right)^2 I(F_{q+2,m}(0) < d_1)\right]\} \\ &\geq R_3(\hat{\beta}_n^{\text{PT}};\mathbf{W}) \end{aligned} \tag{7.4.34}$$

for all α satisfying the condition

$$\left\{\alpha : G_{q+2,m}(\ell_\alpha;0) \geq d + E\left[\left(1 - d_1 F_{q+2,m}^{-1}(0)\right)^2 I(F_{q+2,m}(0) < d_1)\right]\right\}. \tag{7.4.35}$$

The risk of $\hat{\beta}_n^{S+}$ is smaller than that of $\hat{\beta}_n^{\text{PT}}$ when the critical value $\ell_\alpha = \frac{q}{q+2} F_{q,m}(\alpha)$ satisfies the opposite relation to (7.4.35). Thus, $\hat{\beta}_n^{S+}$ does not always dominate $\hat{\beta}_n^{\text{PT}}$ under H_0. We can order the risk of the estimators under H_0 as

$$R_2(\hat{\beta}_n;\mathbf{W}) \leq R_3(\hat{\beta}_n^{\text{PT}};\mathbf{W}) \leq R_4(\hat{\beta}_n^{S+};\mathbf{W}) \leq R_4(\hat{\beta}_n^{S};\mathbf{W}) \leq R_2(\tilde{\beta}_n;\mathbf{W}); \tag{7.4.36}$$

the position of the PTE $\hat{\boldsymbol{\beta}}_n^{\rm PT}$ shifts from "in between" $R_2(\hat{\boldsymbol{\beta}}_n;\mathbf{W})$ and $R_5(\hat{\boldsymbol{\beta}}_n^{S+};\mathbf{W})$ to "in between" $R_4(\hat{\boldsymbol{\beta}}_n^{S};\mathbf{W})$ and $R_1(\tilde{\boldsymbol{\beta}}_n;\mathbf{W})$. That is to say, the risk order is

$$R_2(\hat{\boldsymbol{\beta}}_n;\mathbf{W}) \leq R_5(\hat{\boldsymbol{\beta}}_n^{S+};\mathbf{W}) \leq R_4(\hat{\boldsymbol{\beta}}_n^{S};\mathbf{W}) \leq R_3(\hat{\boldsymbol{\beta}}_n^{\rm PT};\mathbf{W}) \leq R_1(\tilde{\boldsymbol{\beta}}_n;\mathbf{W}). \tag{7.4.37}$$

The picture changes when $\boldsymbol{\eta}_1$ moves away from the origin, $\mathbf{0}$. The risk of $\tilde{\boldsymbol{\beta}}_n$ remains constant, and the risk of $\hat{\boldsymbol{\beta}}_n$ unbounded, as Δ^2 moves away to infinity. Also, the risks of $\hat{\boldsymbol{\beta}}_n^{\rm PT}$, $\hat{\boldsymbol{\beta}}_n^{S}$, and $\hat{\boldsymbol{\beta}}_n^{S+}$ converge to the risk of $\tilde{\boldsymbol{\beta}}_n$ as $\Delta^2 \to \infty$. But for a reasonable value of Δ^2 near 0, the risk of $\hat{\boldsymbol{\beta}}_n^{\rm PT}$ is smaller than $\hat{\boldsymbol{\beta}}_n^{S+}$ as well as $\hat{\boldsymbol{\beta}}_n^{S}$ for α satisfying (7.4.35) for $q \geq 3$. Thus, neither $\hat{\boldsymbol{\beta}}_n^{\rm PT}$ nor $\hat{\boldsymbol{\beta}}_n^{S+}$ nor $\hat{\boldsymbol{\beta}}_n^{S}$ dominate each other. However, the risk of $\hat{\boldsymbol{\beta}}_n^{S+}$ is below the risk of $\hat{\boldsymbol{\beta}}_n^{S}$ as well as that of $\tilde{\boldsymbol{\beta}}_n$, while the risk of $\hat{\boldsymbol{\beta}}_n^{\rm PT}$ exceeds that risk of $\tilde{\boldsymbol{\beta}}$ at some intermediate values of Δ^2 depending on α.

7.5 MSE-Matrix Analysis of the Estimators

In this section, we study the MSE-matrix analysis of the estimators and determine their dominance properties.

Comparison of $\hat{\boldsymbol{\beta}}_n$ and $\tilde{\boldsymbol{\beta}}_n$

The MSE (matrix) difference of $\tilde{\boldsymbol{\beta}}_n$ and $\hat{\boldsymbol{\beta}}_n$ is

$$\mathbf{M}_1(\tilde{\boldsymbol{\beta}}_n) - \mathbf{M}_2(\hat{\boldsymbol{\beta}}_n) = \sigma^2\mathbf{C}^{-1}\mathbf{H}'(\mathbf{H}\mathbf{C}^{-1}\mathbf{H}')^{-1}\mathbf{H}\mathbf{C}^{-1} - \boldsymbol{\delta}\boldsymbol{\delta}'. \tag{7.5.1}$$

The MSE difference is p.s.d. whenever for a nonzero vector $\boldsymbol{\ell} = (\ell_1, \dots, \ell_p)'$ we have

$$\boldsymbol{\ell}'[\mathbf{M}_1(\tilde{\boldsymbol{\beta}}_n) - \mathbf{M}_2(\hat{\boldsymbol{\beta}}_n)]\boldsymbol{\ell} \geq 0. \tag{7.5.2}$$

That is to say,

$$\sigma^2\boldsymbol{\ell}'\{\mathbf{C}^{-1}\mathbf{H}'(\mathbf{H}\mathbf{C}^{-1}\mathbf{H}')^{-1}\mathbf{H}\mathbf{C}^{-1}\}\boldsymbol{\ell} \geq \boldsymbol{\ell}'\boldsymbol{\delta}\boldsymbol{\delta}'\boldsymbol{\ell}. \tag{7.5.3}$$

Hence, we have

$$\frac{\boldsymbol{\ell}'\boldsymbol{\delta}\boldsymbol{\delta}'\boldsymbol{\ell}}{\sigma^2\boldsymbol{\ell}'\mathbf{C}^{-1}\boldsymbol{\ell}} \leq \frac{\boldsymbol{\ell}'\{\mathbf{C}^{-1}\mathbf{H}'(\mathbf{H}\mathbf{C}^{-1}\mathbf{H}')^{-1}\mathbf{H}\mathbf{C}^{-1}\}\boldsymbol{\ell}}{\boldsymbol{\ell}'\mathbf{C}^{-1}\boldsymbol{\ell}} \quad \forall\, \boldsymbol{\ell}.$$

Therefore,

$$\Delta^2 = \max_{\boldsymbol{\ell}} \frac{\boldsymbol{\ell}'\boldsymbol{\delta}\boldsymbol{\delta}'\boldsymbol{\ell}}{\sigma^2\boldsymbol{\ell}'\mathbf{C}^{-1}\boldsymbol{\ell}} \leq \max_{\boldsymbol{\ell}} \frac{\boldsymbol{\ell}'\{\mathbf{C}^{-1}\mathbf{H}'(\mathbf{H}\mathbf{C}^{-1}\mathbf{H}')^{-1}\mathbf{H}\mathbf{C}^{-1}\}\boldsymbol{\ell}}{\boldsymbol{\ell}'\mathbf{C}^{-1}\boldsymbol{\ell}},$$

or

$$\Delta^2 \leq Ch_{\max}[\mathbf{H}'(\mathbf{H}\mathbf{C}^{-1}\mathbf{H}')^{-1}\mathbf{H}\mathbf{C}^{-1}] = 1, \tag{7.5.4}$$

since $\mathbf{C}^{-1/2}\mathbf{H}'(\mathbf{H}\mathbf{C}^{-1}\mathbf{H}')^{-1}\mathbf{H}\mathbf{C}^{-1/2}$ is an idempotent matrix with maximum characteristic value equal to 1. Thus, (7.5.1) is p.s.d. if and only if $\Delta^2 \le 1$. Therefore, $\hat{\boldsymbol{\beta}}_n$ performs better than $\tilde{\boldsymbol{\beta}}_n$ when $\Delta^2 \le 1$; otherwise, $\tilde{\boldsymbol{\beta}}_n$ performs better than $\hat{\boldsymbol{\beta}}_n$. The range of the domination of $\hat{\boldsymbol{\beta}}_n$ over $\tilde{\boldsymbol{\beta}}_n$ is **bigger** in the case of **risk analysis.** For $\mathbf{W} = \mathbf{C}$, the **range** of risk domination is $[0, q]$ and $q \ge 1$. The **MSE-based efficiency** of $\hat{\boldsymbol{\beta}}_n$ relative to $\tilde{\boldsymbol{\beta}}_n$ is meaningless, since

$$\mathrm{MRE}(\hat{\boldsymbol{\beta}}_n; \tilde{\boldsymbol{\beta}}_n) = \left|\mathbf{I}_p - \mathbf{H}'(\mathbf{H}\mathbf{C}^{-1}\mathbf{H}')^{-1}\mathbf{H}\mathbf{C}^{-1} - \sigma^{-2}\mathbf{C}\boldsymbol{\delta}\boldsymbol{\delta}'\right|^{-1/p} = 0, \quad (7.5.5)$$

while the *risk efficiency, which is meaningful,* is given by

$$\mathrm{RRE}(\hat{\boldsymbol{\beta}}_n; \tilde{\boldsymbol{\beta}}_n) = \left[1 - \frac{\mathrm{tr}(\mathbf{A}_{11})}{\mathrm{tr}(\mathbf{W}\mathbf{C}^{-1})} + \frac{\boldsymbol{\eta}_1'\mathbf{A}_{11}\boldsymbol{\eta}_1}{\mathrm{tr}(\mathbf{W}\mathbf{C}^{-1})}\right]^{-1}. \quad (7.5.6)$$

For $\mathbf{W} = \mathbf{C}$, $\mathrm{tr}(\mathbf{A}_{11}) = q$ and $\mathrm{tr}(\mathbf{W}\mathbf{C}^{-1}) = p$ so that (7.5.6) reduces to

$$p[(p-q) + \Delta^2]^{-1}. \quad (7.5.7)$$

Thus, the risk efficiency of $\hat{\boldsymbol{\beta}}_n$ is greatest when $\Delta^2 \le q$, while the range of Δ^2 is smaller for the MSE-matrix comparison.

Comparison of $\hat{\boldsymbol{\beta}}_n^{\mathrm{PT}}$, $\tilde{\boldsymbol{\beta}}_n$, and $\hat{\boldsymbol{\beta}}_n$

The MSE difference of $\tilde{\boldsymbol{\beta}}_n$ and $\hat{\boldsymbol{\beta}}_n^{\mathrm{PT}}$ is given by

$$\begin{aligned}\mathbf{M}_1(\tilde{\boldsymbol{\beta}}_n) - \mathbf{M}_3(\hat{\boldsymbol{\beta}}_n^{\mathrm{PT}}) &= \sigma^2\left(\mathbf{C}^{-1}\mathbf{H}'(\mathbf{H}\mathbf{C}^{-1}\mathbf{H}')^{-1}\mathbf{H}\mathbf{C}^{-1}\right)G_{q+2,m}(\ell_\alpha; \Delta^2)\\ &\quad - \boldsymbol{\delta}\boldsymbol{\delta}'\left\{2G_{q+2,m}(\ell_\alpha; \Delta^2) - G_{q+4,m}(\ell_\alpha^*; \Delta^2)\right\}. \quad (7.5.8)\end{aligned}$$

The MSE difference is p.s.d. whenever for any nonzero vector $\boldsymbol{\ell}$ we have

$$\begin{aligned}&\sigma^2\boldsymbol{\ell}'\left\{\mathbf{C}^{-1}\mathbf{H}'(\mathbf{H}\mathbf{C}^{-1}\mathbf{H})^{-1}\mathbf{H}\mathbf{C}^{-1}\right\}\boldsymbol{\ell}G_{q+2,m}(\ell_\alpha; \Delta^2)\\ &- (\boldsymbol{\ell}'\boldsymbol{\delta}\boldsymbol{\delta}'\boldsymbol{\ell})\left\{2G_{q+2,m}(\ell_\alpha; \Delta^2) - G_{q+4,m}(\ell_\alpha^*; \Delta^2)\right\} \ge 0, \quad (7.5.9)\end{aligned}$$

and consequently, we may write

$$\begin{aligned}\frac{\boldsymbol{\ell}'\boldsymbol{\delta}\boldsymbol{\delta}'\boldsymbol{\ell}}{\sigma^2\boldsymbol{\ell}'\mathbf{C}^{-1}\boldsymbol{\ell}} &\le \frac{G_{q+2,m}(\ell_\alpha; \Delta^2)}{\{2G_{q+2,m}(\ell_\alpha; \Delta^2) - G_{q+4,m}(\ell_\alpha^*; \Delta^2)}\\ &\quad - \frac{\boldsymbol{\ell}'\{\mathbf{C}^{-1}\mathbf{H}'(\mathbf{H}\mathbf{C}^{-1}\mathbf{H}')\mathbf{H}\mathbf{C}^{-1}\}\boldsymbol{\ell}}{\sigma^2\boldsymbol{\ell}'\mathbf{C}^{-1}\boldsymbol{\ell}}. \quad (7.5.10)\end{aligned}$$

Hence, taking the maximum over all nonzero vectors $\boldsymbol{\ell}$ we obtain

$$\Delta^2 \le \frac{G_{q+2,m}(\ell_\alpha; \Delta^2)}{\{2G_{q+2,m}(\ell_\alpha; \Delta^2) - G_{q+4,m}(\ell_\alpha^*; \Delta^2)\}}(\le 1). \quad (7.5.11)$$

This means $\hat{\boldsymbol{\beta}}_n^{\text{PT}}$ is superior to $\tilde{\boldsymbol{\beta}}_n$ in the range of Δ^2 is given by (7.5.11), otherwise $\tilde{\boldsymbol{\beta}}_n$ is superior to $\hat{\boldsymbol{\beta}}_n^{\text{PT}}$. The range of domination of $\hat{\boldsymbol{\beta}}_n^{\text{PT}}$ over $\tilde{\boldsymbol{\beta}}_n$ is bigger in the case of risk analysis. For $\mathbf{W} = \mathbf{C}$, the range of Δ^2 is $\leq q$. The risk-based efficiency of $\hat{\boldsymbol{\beta}}_n^{\text{PT}}$ relative to $\tilde{\boldsymbol{\beta}}_n$ is given by

$$\text{RRE}\left(\hat{\boldsymbol{\beta}}_n^{\text{PT}};\tilde{\boldsymbol{\beta}}_n\right) = \left[1 - \frac{\text{tr}(\mathbf{A}_{11})}{\text{tr}(\mathbf{WC}^{-1})}G_{q+2,m}(\ell_\alpha;\Delta^2) + \frac{\boldsymbol{\eta}_1'\mathbf{A}_{11}\boldsymbol{\eta}_1}{\text{tr}(\mathbf{WC}^{-1})}\right.$$
$$\left.\times \{2G_{q+2,m}(\ell_\alpha;\Delta^2) - G_{q+4,m}(\ell_\alpha^*;\Delta^2)\}\right]^{-1}. \quad (7.5.12)$$

For $\mathbf{W} = \mathbf{C}$, (7.5.12) equals

$$p\left[p - qG_{q+2,m}(\ell_\alpha;\Delta^2) + \Delta^2\left\{2G_{q+2,m}(\ell_\alpha;\Delta^2) - G_{q+4,m}(\ell_\alpha^*;\Delta^2)\right\}\right]^{-1}. \quad (7.5.13)$$

Thus, the risk efficiency of $\hat{\boldsymbol{\beta}}_n^{\text{PT}}$ is greatest when

$$\Delta^2 \leq \frac{qG_{q+2,m}(\ell_\alpha;\Delta^2)}{\{2G_{q+2,m}(\ell_\alpha;\Delta^2) - G_{q+4,m}(\ell_\alpha^*;\Delta^2)\}}. \quad (7.5.14)$$

The MSE-based efficiency of $\hat{\boldsymbol{\beta}}_n^{\text{PT}}$ relative to $\tilde{\boldsymbol{\beta}}_n$ is given by

$$\begin{aligned}
&\text{MRE}\left(\hat{\boldsymbol{\beta}}_n^{\text{PT}}:\tilde{\boldsymbol{\beta}}_n\right)\\
&= \Big|\mathbf{I}_p - \{\mathbf{H}'(\mathbf{HC}^{-1}\mathbf{H}')^{-1}\mathbf{HC}^{-1}\}G_{q+2,m}(\ell_\alpha;\Delta^2)\\
&\quad + \sigma^{-2}\mathbf{C}\boldsymbol{\delta}\boldsymbol{\delta}'\{2G_{q+2,m}(\ell_\alpha;\Delta^2) - G_{q+4,m}(\ell_\alpha^*;\Delta^2)\}\Big|^{-1/p}\\
&= \left[1 - G_{q+2,m}(\ell_\alpha;\Delta^2) + \Delta^2\{2G_{q+2,m}(\ell_\alpha;\Delta^2) - G_{q+4,m}(\ell_\alpha^*;\Delta^2)\}\right]^{-1}\\
&\quad \times \{1 - G_{q+2,m}(\ell_\alpha;\Delta^2)\}^{p-1/p}.
\end{aligned} \quad (7.5.15)$$

Thus, MRE of $\hat{\boldsymbol{\beta}}_n^{\text{PT}}$ is greatest when

$$\Delta^2 \leq \frac{\{1 - G_{q+2,m}(\ell_\alpha;\Delta^2)\}^{\frac{p-1}{p}} - \{1 - G_{q+2,m}(\ell_\alpha;\Delta^2)\}}{\{2G_{q+2,m}(\ell_\alpha;\Delta^2) - G_{q+4,m}(\ell_\alpha^*;\Delta^2)\}}, \quad (7.5.16)$$

which is smaller than (7.5.14).

To compare $\hat{\boldsymbol{\beta}}_n^{\text{PT}}$ and $\hat{\boldsymbol{\beta}}_n$ we consider the MSE difference

$$\begin{aligned}
&\mathbf{M}_2\left(\hat{\boldsymbol{\beta}}_n\right) - \mathbf{M}_3\left(\hat{\boldsymbol{\beta}}_n^{\text{PT}}\right)\\
&= -\sigma^2\left\{\mathbf{C}^{-1}\mathbf{H}'(\mathbf{HC}^{-1}\mathbf{H}')^{-1}\mathbf{HC}^{-1}\right\}\left\{1 - G_{q+2,m}(\ell_\alpha;\Delta^2)\right\}\\
&\quad + \boldsymbol{\delta}\boldsymbol{\delta}'\left\{1 - 2G_{q+2,m}(\ell_\alpha;\Delta^2) + G_{q+4,m}(\ell_\alpha^*;\Delta^2)\right\}.
\end{aligned} \quad (7.5.17)$$

The MSE difference is negative-semidefinite (n.s.d.) whenever for any nonzero vector $\boldsymbol{\ell}$ we have

$$\begin{aligned}\frac{(\boldsymbol{\ell}'\boldsymbol{\delta}\boldsymbol{\delta}'\boldsymbol{\ell})}{\sigma^2(\boldsymbol{\ell}'\mathbf{C}^{-1}\boldsymbol{\ell})}&\left[1-2G_{q+2,m}(\ell_\alpha;\Delta^2)+G_{q+4,m}(\ell_\alpha^*;\Delta^2)\right]\\ &\leq \frac{\boldsymbol{\ell}\{C^{-1}\mathbf{H}'(\mathbf{H}\mathbf{C}^{-1}\mathbf{H}')^{-1}\mathbf{H}\mathbf{C}^{-1}\}\boldsymbol{\ell}}{\boldsymbol{\ell}'\mathbf{C}^{-1}\boldsymbol{\ell}}\\ &\quad\times\frac{\{1-G_{q+2,m}(\ell_\alpha;\Delta^2)\}}{\{1-2G_{q+2,m}(\ell_\alpha;\Delta^2)+G_{q+4,m}(\ell_\alpha^*;\Delta^2)\}},\end{aligned}\tag{7.5.18}$$

which implies that the MSE difference is n.s.d. whenever

$$\Delta^2 \leq \frac{1-G_{q+2,m}(\ell_\alpha;\Delta^2)}{\{1-2G_{q+2,m}(\ell_\alpha;\Delta^2)+G_{q+4,m}(\ell_\alpha^*;\Delta^2)\}},\tag{7.5.19}$$

Thus, in this range $\hat{\boldsymbol{\beta}}_n$ is superior to $\hat{\boldsymbol{\beta}}_n^{\text{PT}}$; otherwise, $\hat{\boldsymbol{\beta}}_n^{\text{PT}}$ is superior. Here also the range of superiority of $\hat{\boldsymbol{\beta}}_n$ over $\hat{\boldsymbol{\beta}}_n^{\text{PT}}$ is smaller than the range for the risk comparison.

The MRE$(\hat{\boldsymbol{\beta}}_n^{\text{PT}};\hat{\boldsymbol{\beta}}_n)$ is given by

$$\text{MRE}(\hat{\boldsymbol{\beta}}_n^{\text{PT}};\hat{\boldsymbol{\beta}}_n)=$$

$$\frac{\left|\sigma^2\mathbf{C}^{-1}-\sigma^2\mathbf{C}^{-1}\mathbf{H}'(\mathbf{H}\mathbf{C}^{-1}\mathbf{H}')^{-1}\mathbf{H}\mathbf{C}^{-1}+\boldsymbol{\theta}\boldsymbol{\theta}'\right|^{1/p}}{\left|\sigma^2\mathbf{C}^{-1}-\sigma^2[\mathbf{C}^{-1}\mathbf{H}'(\mathbf{H}\mathbf{C}^{-1}\mathbf{H}')^{-1}\mathbf{H}\mathbf{C}^{-1}]G_{q+2,m}(\ell_\alpha;\Delta^2)+\boldsymbol{\theta}\boldsymbol{\theta}'\{2G_{q+2,m}(\ell_2;\Delta^2)-G_{q+4,m}(\ell_\alpha^*;\Delta^2)\}\right|^{1/p}}.$$

The MRE$(\hat{\boldsymbol{\beta}}_n^{\text{PT}};\hat{\boldsymbol{\beta}}_n)\geq 1$ whenever (7.5.19) is satisfied.

Comparison of $\hat{\beta}_n^S$, $\tilde{\boldsymbol{\beta}}_n$ and $\hat{\beta}_n$

We compare $\hat{\boldsymbol{\beta}}_n^S$ and $\tilde{\boldsymbol{\beta}}_n$. The MSE difference in this case is

$$\begin{aligned}\mathbf{M}_1\left(\tilde{\boldsymbol{\beta}}_n\right)-\mathbf{M}_4\left(\hat{\boldsymbol{\beta}}_n^S\right)&=dq\sigma^2\left\{\mathbf{C}^{-1}\mathbf{H}'(\mathbf{H}\mathbf{C}^{-1}\mathbf{H}')^{-1}\mathbf{H}\mathbf{C}^{-1}\right\}\\ &\quad\times\left\{(q-2)E\left[\chi_{q+2}^{-4}(\Delta^2)\right]+2\Delta^2E\left[\chi_{q+4}^{-4}(\Delta^2)\right]\right\}\\ &\quad-dq(q+2)\boldsymbol{\delta}\boldsymbol{\delta}'E\left[\chi_{q+4}^{-4}(\Delta^2)\right].\end{aligned}\tag{7.5.20}$$

The difference is p.s.d. whenever for a nonzero vector $\boldsymbol{\ell}$ we have

$$\begin{aligned}&dq\sigma^2\{\boldsymbol{\ell}'\{\mathbf{C}^{-1}\mathbf{H}'(\mathbf{H}\mathbf{C}^{-1}\mathbf{H}')^{-1}\mathbf{H}\mathbf{C}^{-1}\boldsymbol{\ell}\}\{(q-2)E[\chi_{q+4}^{-4}(\Delta^2)]\\ &+2\Delta^2E[\chi_{q+4}^{-4}(\Delta^2)]\}-dq(q+2)(\boldsymbol{\ell}'\boldsymbol{\delta}\boldsymbol{\delta}'\boldsymbol{\ell})E[\chi_{q+4}^{-4}(\Delta^2)]\geq 0.\end{aligned}\tag{7.5.21}$$

This implies that (after simplification)

$$q\Delta^2E\left[\chi_{q+4}^{-4}(\Delta^2)\right]\leq(q-2)E\left[\chi_{q+2}^{-4}(\Delta^2)\right],\tag{7.5.22}$$

which does not hold for all Δ^2. Thus, $\hat{\boldsymbol{\beta}}_n^S$ is not uniformly better than $\tilde{\boldsymbol{\beta}}_n$ under the MSE criterion.

The MRE$(\hat{\boldsymbol{\beta}}_n^S; \tilde{\boldsymbol{\beta}}_n)$ in this case is given by

$$\begin{aligned}
&\text{MRE}(\hat{\boldsymbol{\beta}}_n^S; \tilde{\boldsymbol{\beta}}_n) \\
&= \Big|\mathbf{I}_p - dq\mathbf{H}'(\mathbf{HC}^{-1}\mathbf{H}')^{-1}\mathbf{HC}^{-1}\Big\{(q-2)E\left[\chi_{q+2}^{-4}(\Delta^2)\right] \\
&\quad + 2\Delta^2 E\left[\chi_{q+4}^{-4}(\Delta^2)\right]\Big\} + dq(q+2)\sigma^{-2}\mathbf{C}\boldsymbol{\delta}\boldsymbol{\delta}'E\left[\chi_{q+4}^{-4}(\Delta^2)\right]\Big|^{-1/p} \\
&= \Big(1 - dq\Big\{(q-2)E\Big[\chi_{q+2}^{-4}(\Delta^2)\Big] + 2\Delta^2 E\big[\chi_{q+2}^{-4}(\Delta^2)\big]\Big\} \\
&\quad + dq(q+2)\Delta^2 E\left[\chi_{q+4}^{-4}(\Delta^2)\right]\Big)^{-1} \\
&\quad \times \left(1 - dq\left\{(q-2)E\left[\chi_{q+2}^{-4}(\Delta^2)\right] + 2\Delta^2 E\left[\chi_{q+4}^{-4}(\Delta^2)\right]\right\}\right)^{p-1/p}. \quad (7.5.23)
\end{aligned}$$

Similarly, the comparison of $\hat{\boldsymbol{\beta}}_n$ and $\hat{\boldsymbol{\beta}}_n^S$ is obtained through the MSE difference

$$\begin{aligned}
&\mathbf{M}_2\left(\hat{\boldsymbol{\beta}}_n\right) - \mathbf{M}_4\left(\hat{\boldsymbol{\beta}}_n^S\right) \\
&= -\sigma^2\mathbf{C}^{-1}\mathbf{H}'(\mathbf{HC}^{-1}\mathbf{H}')^{-1}\mathbf{HC}^{-1} \\
&\quad \times\Big\{1 - dq(q-2)E\left[\chi_{q+2}^{-4}(\Delta^2)\right] - 2dq\Delta^2 E\left[\chi_{q+4}^{-4}(\Delta^2)\right]\Big\} \\
&\quad + \boldsymbol{\delta}\boldsymbol{\delta}'\Big\{1 - dq(q+2)E\left[\chi_{q+4}^{-4}(\Delta^2)\right]\Big\}. \quad (7.5.24)
\end{aligned}$$

Under H_0, the MSE difference is n.s.d. Therefore, $\hat{\boldsymbol{\beta}}_n$ performs better than $\hat{\boldsymbol{\beta}}_n^S$ at $\Delta^2 = 0$. Except for small intervals around 0, $\hat{\boldsymbol{\beta}}_n^S$ performs better than $\hat{\boldsymbol{\beta}}_n$.

Comparison of $\hat{\boldsymbol{\beta}}_n^{\text{PT}}$, $\hat{\boldsymbol{\beta}}_n^S$, and $\hat{\boldsymbol{\beta}}_n^{S+}$

Consider the MSE differences

$$\begin{aligned}
&\mathbf{M}_4(\hat{\boldsymbol{\beta}}_n^S) - \mathbf{M}_3(\hat{\boldsymbol{\beta}}_n^{\text{PT}}) \\
&= \sigma^2\mathbf{C}^{-1}\mathbf{H}'(\mathbf{HC}^{-1}\mathbf{H}')^{-1}\mathbf{HC}^{-1} \\
&\quad \times\Big\{G_{q+2,m}(\ell_\alpha;\Delta^2) - dq(q-2)E\left[\chi_{q+2}^{-4}(\Delta^2)\right] - 2dq\Delta^2 E\left[\chi_{q+4}^{-4}(\Delta^2)\right]\Big\} \\
&\quad -\boldsymbol{\delta}\boldsymbol{\delta}'\Big\{2G_{q+2,m}(\ell_\alpha;\Delta^2) - G_{q+4,m}(\ell_\alpha^*;\Delta^2) - dq(q+2)E\left[\chi_{q+4}^{-4}(\Delta^2)\right]\Big\}
\end{aligned}$$

(7.5.25)

and

$$\begin{aligned}
&\mathbf{M}_5(\hat{\boldsymbol{\beta}}_n^{S+}) - \mathbf{M}_3(\hat{\boldsymbol{\beta}}_n^{\mathrm{PT}}) \\
&= \sigma^2\mathbf{C}^{-1}\mathbf{H}'(\mathbf{H}\mathbf{C}^{-1}\mathbf{H}')^{-1}\mathbf{H}\mathbf{C}^{-1} \\
&\quad \Big\{G_{q+2,m}(\ell_\alpha;\Delta^2) - dq(q-2)E\left[\chi_{q+2}^{-4}(\Delta^2)\right] - 2dq\Delta^2 E\left[\chi_{q+4}^{-4}(\Delta^2)\right] \\
&\qquad - E\big[(1-d_1F_{q+2,m}^{-1}(\Delta^2))^2 I(F_{q+2,m}(\Delta^2) < d_1)\big]\Big\} \\
&\quad + \boldsymbol{\delta}\boldsymbol{\delta}'\Big\{E[(1-d_2F_{q+4,m}^{-1}(\Delta^2))^2 I\left(F_{q+4,m}(\Delta^2) < d_2\right) \\
&\quad - \Big[2G_{q+2,m}(\ell_\alpha;\Delta^2) - G_{q+4,m}(\ell_\alpha^*;\Delta^2) - dq(q+2)E\big[\chi_{q+4}^{-4}(\Delta^2)\big]\Big] \\
&\quad - 2E\left[\left(d_1F_{q+2,m}^{-1}(\Delta^2) - 1\right) I\left(F_{q+2,m}(\Delta^2) < d_1\right)\right]\Big\}. \qquad (7.5.26)
\end{aligned}$$

The expression at (7.5.25) is p.s.d. whenever

$$\Delta^2 \le \frac{\{G_{q+2,m}(\ell_\alpha;\Delta^2) - dq(q-2)E\left[\chi_{q+2}^{-4}(\Delta^2)\right] - 2dq\Delta^2 E\left[\chi_{q+4}^{-4}(\Delta^2)\right]\}}{\{2G_{q+2,m}(\ell_\alpha;\Delta^2) - G_{q+4,m}(\ell_\alpha^*;\Delta^2) - dq(q-2)E\left[\chi_{q+2}^{-4}(\Delta^2)\right]\}}. \qquad (7.5.27)$$

Thus, $\hat{\boldsymbol{\beta}}_n^{\mathrm{PT}}$ performs better than $\hat{\boldsymbol{\beta}}_n^{S}$ whenever Δ^2 satisfies (7.5.27); otherwise, $\hat{\boldsymbol{\beta}}_n^{S}$ performs better than $\hat{\boldsymbol{\beta}}_n^{\mathrm{PT}}$. Neither estimator dominates the other.

Similarly, we can find the range of Δ^2 for which $\hat{\boldsymbol{\beta}}_n^{\mathrm{PT}}$ performs better than $\hat{\boldsymbol{\beta}}_n^{S+}$ using the expression (7.5.26) and observe that neither $\hat{\boldsymbol{\beta}}_n^{\mathrm{PT}}$ nor $\hat{\boldsymbol{\beta}}_n^{S+}$ dominates the other. MRE expressions can be obtained by straightforward computation. Next, we consider the MSE differences under H_0 for the comparison of $\hat{\boldsymbol{\beta}}_n^{S+}$ and $\hat{\boldsymbol{\beta}}_n$:

$$\mathbf{M}_5\left(\hat{\boldsymbol{\beta}}_n^{S+}\right) - \mathbf{M}_2\left(\hat{\boldsymbol{\beta}}_n\right) = \sigma^2\left[\mathbf{C}^{-1}\mathbf{H}'(\mathbf{H}\mathbf{C}^{-1}\mathbf{H}')^{-1}\mathbf{H}\mathbf{C}^{-1}\right]\Big\{1 - d - E\left[\left(1 - d_1F_{q+2,m}^{-1}(0)\right)^2 I\left(F_{q+2,m}(0) < d_1\right)\right]\Big\}. \qquad (7.5.28)$$

This is p.s.d., since

$$(1-d) \ge E\left[\left(1 - d_1F_{q+2,m}^{-1}(0)\right)^2 I\left(F_{q+2,m}(0) < d_1\right)\right]. \qquad (7.5.29)$$

This implies that $\hat{\boldsymbol{\beta}}_n$ is superior to $\hat{\boldsymbol{\beta}}_n^{S+}$ under H_0.

A similar comparison of $\hat{\boldsymbol{\beta}}_n^{S+}$ with $\hat{\boldsymbol{\beta}}_n^{\mathrm{PT}}$ under H_0 shows that

$$\begin{aligned}
&\mathbf{M}_5\left(\hat{\boldsymbol{\beta}}_n^{S+}\right) - \mathbf{M}_3\left(\hat{\boldsymbol{\beta}}_n^{\mathrm{PT}}\right) \\
&= \sigma^2\mathbf{C}^{-1}\mathbf{H}'(\mathbf{H}\mathbf{C}^{-1}\mathbf{H}')^{-1}\mathbf{H}\mathbf{C}^{-1}\Big\{G_{q+2,m}(\ell_\alpha;\Delta^2) - d \\
&\quad - E\left[\left(1 - d_1F_{q+2,m}^{-1}(0)\right)^2 I\left(F_{q+2,m}(0) < d_1\right)\right]\Big\} \qquad (7.5.30)
\end{aligned}$$

whenever there exists a set $\mathcal{A}$ of α such that

$$\mathcal{A} = \Big\{\alpha : G_{q+2,m}(\ell_\alpha; \Delta^2) \geq d + E\Big[\left(1 - d_1 F^{-1}_{q+2,m}(0)\right)^2 I\left(F_{q+2,m}(0) < d_1\right)\Big]\Big\}. \quad (7.5.31)$$

Thus, $\hat{\boldsymbol{\beta}}_n^{\mathrm{PT}}$ is superior to $\hat{\boldsymbol{\beta}}_n^{S+}$ whenever (7.5.31) is satisfied; otherwise, $\hat{\boldsymbol{\beta}}_n^{S+}$ is superior. Therefore, $\hat{\boldsymbol{\beta}}_n^{S+}$ does not always dominate $\hat{\boldsymbol{\beta}}_n^{\mathrm{PT}}$ when the hypothesis $H_0 : \mathbf{H}\boldsymbol{\beta} = \mathbf{h}$ holds. Hence, we can order the MSE matrices as

$$\mathbf{M}_1\left(\tilde{\boldsymbol{\beta}}_n\right) \geq \mathbf{M}_4\left(\hat{\boldsymbol{\beta}}_n^{S}\right) \geq \mathbf{M}_3\left(\hat{\boldsymbol{\beta}}_n^{\mathrm{PT}}\right) \geq \mathbf{M}_2\left(\hat{\boldsymbol{\beta}}_n\right), \quad (7.5.32)$$

where $\geq$ stands for $\mathbf{M}_2(\tilde{\boldsymbol{\beta}}_n) - \mathbf{M}_4(\hat{\boldsymbol{\beta}}_n^{S}) \geq \mathbf{0}$, $\mathbf{M}_4(\hat{\boldsymbol{\beta}}_n^{S}) - M_5(\hat{\boldsymbol{\beta}}_n^{S+}) \geq \mathbf{0}$, $\mathbf{M}_5(\hat{\boldsymbol{\beta}}_n^{S+}) - \mathbf{M}_3(\hat{\boldsymbol{\beta}}_n^{\mathrm{PT}}) \geq \mathbf{0}$, and $\mathbf{M}_3(\hat{\boldsymbol{\beta}}_n^{\mathrm{PT}}) - \mathbf{M}_2(\hat{\boldsymbol{\beta}}_n) \geq \mathbf{0}$.

The position of PTE may be changed to the order

$$\mathbf{M}_1\left(\tilde{\boldsymbol{\beta}}_n\right) \geq \mathbf{M}_3\left(\hat{\boldsymbol{\beta}}_n^{\mathrm{PT}}\right) \geq \mathbf{M}_4\left(\hat{\boldsymbol{\beta}}_n^{S}\right) \geq \mathbf{M}_5\left(\hat{\boldsymbol{\beta}}_n^{S+}\right) \geq \mathbf{M}_2\left(\hat{\boldsymbol{\beta}}_n\right), \quad (7.5.33)$$

depending on the size of α.

Comparison of $\hat{\boldsymbol{\beta}}_n^{S+}$, $\hat{\boldsymbol{\beta}}_n^{S}$ and $\tilde{\boldsymbol{\beta}}_n$

Finally, we compare $\hat{\boldsymbol{\beta}}_n^{S+}$ and $\hat{\boldsymbol{\beta}}_n^{S}$. In this case, MSE difference yields

$$\begin{aligned}
&\mathbf{M}_4\left(\hat{\boldsymbol{\beta}}_n^{S}\right) - \mathbf{M}_5\left(\hat{\boldsymbol{\beta}}_n^{S+}\right) \\
&= \sigma^2\Big\{ [\mathbf{C}^{-1}\mathbf{H}'(\mathbf{H}\mathbf{C}^{-1}\mathbf{H}')^{-1}\mathbf{H}\mathbf{C}^{-1}] \\
&\qquad \times E\left[\left(1 - d_1 F_{q+2,m}(\Delta^2)\right)^2 I\left(F_{q+2,m}(\Delta^2) < d_1\right)\right] \\
&\qquad + \sigma^{-2}\boldsymbol{\delta}\boldsymbol{\delta}' E\left[\left(1 - d_2 F^{-1}_{q+4,m}(\Delta^2)\right)^2 I\left(F_{q+4,m}(\Delta^2) < d_2\right)\right]\Big\} \\
&\quad + 2\boldsymbol{\delta}\boldsymbol{\delta}' E\left[\left(d_1^* F^{-1}_{q+2,m}(\Delta^2) - 1\right) I\left(F_{q+2,m}(\Delta^2) < d_1^*\right)\right] \geq \mathbf{0}. \quad (7.5.34)
\end{aligned}$$

The MSE difference is p.s.d. for all Δ^2, and hence, $\hat{\boldsymbol{\beta}}_n^{S+}$ dominates $\hat{\boldsymbol{\beta}}_n^{S}$ uniformly. But, $\hat{\boldsymbol{\beta}}_n^{S}$ does not dominate $\tilde{\boldsymbol{\beta}}_n$ uniformly. We can order the MSE of $\hat{\boldsymbol{\beta}}_n^{S}$ and $\hat{\boldsymbol{\beta}}_n^{S+}$ as

$$\mathbf{M}_4\left(\hat{\boldsymbol{\beta}}_n^{S}\right) \geq \mathbf{M}_5\left(\hat{\boldsymbol{\beta}}_n^{S+}\right). \quad (7.5.35)$$

7.6 Improving the PTE

As in Chapters 4, 5, and 6, we consider the improvement of the PTE, $\hat{\boldsymbol{\beta}}_n^{\text{PT}}$ of $\boldsymbol{\beta}$ when $q \geq 3$. We may write the improved estimator of $\hat{\boldsymbol{\beta}}_n^{\text{PT}}$ as

$$\hat{\boldsymbol{\beta}}_n^{\text{PT+}} = \hat{\boldsymbol{\beta}}_n^{\text{PT}} - d\left(\tilde{\boldsymbol{\beta}}_n - \hat{\boldsymbol{\beta}}_n\right)\mathcal{L}_n^{-1} I\left(\mathcal{L}_n \geq F_{q,m}(\alpha)\right) \qquad (7.6.1)$$

for $q \geq 3$.

The bias vector of $\hat{\boldsymbol{\beta}}_n^{\text{PT+}}$ and the quadratic bias of this estimator are given by

$$\mathbf{b}_5\left(\hat{\boldsymbol{\beta}}_n^{\text{PT+}}\right)$$
$$= \mathbf{b}_3\left(\hat{\boldsymbol{\beta}}_n^{\text{PT}}\right) + \mathbf{b}_4\left(\hat{\boldsymbol{\beta}}_n^{S}\right) + d_1\boldsymbol{\delta}E\left[F_{q+2,m}^{-1}(\Delta^2) I\left(F_{q+2,m}(\Delta^2) < \ell_\alpha\right)\right] \quad (7.6.2)$$

and

$$B_6\left(\hat{\boldsymbol{\beta}}_n^{\text{PT+}}\right) = \Delta^2\Big\{d_1 E\left[F_{q+2,m}^{-1}(\Delta^2) I\left(F_{q+2,m}(\Delta^2) < \ell_\alpha\right)\right]$$
$$- dqE\left[\chi_{q+2}^{-2}(\Delta^2)\right] - G_{q+2,m}(\ell_\alpha;\Delta^2)\Big\}^2. \quad (7.6.3)$$

The corresponding MSE-matrix and the weighted risk expressions are given by

$$\mathbf{M}_6\left(\hat{\boldsymbol{\beta}}_n^{\text{PT+}}\right)$$
$$= \mathbf{M}_3\left(\hat{\boldsymbol{\beta}}_n^{\text{PT}}\right) - \sigma^2 d_1[C^{-1}\mathbf{H}'(\mathbf{H}C^{-1}\mathbf{H}')^{-1}\mathbf{H}C^{-1}]$$
$$\times E\Big\{\left(2F_{q+2,m}^{-1}(\Delta^2) - d_1 F_{q+2,m}^{-2}(\Delta^2)\right) I\left(F_{q+2,m}(\Delta^2) > \ell_\alpha\right)\Big\}$$
$$- d_2\boldsymbol{\delta}\boldsymbol{\delta}'\Big[E\Big\{\left(2F_{q+4,m}^{-1}(\Delta^2) - d_2 F_{q+4,m}^{-2}(\Delta^2)\right) I\left(F_{q+4,m}(\Delta^2) > \ell_\alpha^*\right)\Big\}$$
$$-2E\left(F_{q+2,m}^{-1}(\Delta^2) I\left(F_{q+2,m}(\Delta^2) > \ell_\alpha\right)\right)\Big] \quad (7.6.4)$$

and

$$R_6\left(\hat{\boldsymbol{\beta}}_n^{\text{PT+}};\mathbf{W}\right)$$
$$= R_3\left(\hat{\boldsymbol{\beta}}_n^{\text{PT}};\mathbf{W}\right) - \sigma^2 d_1 \operatorname{tr}[\mathbf{W}\mathbf{C}^{-1}\mathbf{H}'(\mathbf{H}\mathbf{C}^{-1}\mathbf{H}')^{-1}\mathbf{H}\mathbf{C}^{-1}]$$
$$\times E\Big\{\left(2F_{q+2,m}^{-1}(\Delta^2) - d_1 F_{q+2,m}^{-2}(\Delta^2)\right) I\left(F_{q+2,m}(\Delta^2) > \ell_\alpha\right)\Big\}$$
$$- d_1(\boldsymbol{\delta}'\mathbf{W}\boldsymbol{\delta})\Big[E\Big\{\left(2F_{q+4,m}^{-1}(\Delta^2) - d_2 F_{q+4,m}^{-2}(\Delta^2)\right) I\left(F_{q+4,m}(\Delta^2) > \ell_\alpha^*\right)\Big\}$$
$$-2E\left(F_{q+2,m}^{-1}(\Delta^2) I\left(F_{q+2,m}(\Delta^2) > \ell_\alpha\right)\right)\Big]. \quad (7.6.5)$$

For $\mathbf{W} = \mathbf{C}$, the percentage improvements of $\hat{\boldsymbol{\beta}}_n^{\text{PT}+}$ over $\hat{\boldsymbol{\beta}}_n^{\text{PT}}$ are similar to the tabular values in Table 5.6.1 of Chapter 5.

7.7 Multiple Regression Model: Nonnormal Errors

7.7.1 Introduction

Consider the multiple regression model

$$\mathbf{Y}_n = \mathbf{X}_n\boldsymbol{\beta} + \mathbf{e}_n, \tag{7.7.1}$$

where $\mathbf{Y}_n = (y_1, \dots, y_n)'$ is the vector of observable response, $\mathbf{X}_n$ is an $n \times p$ matrix of known constants, $\boldsymbol{\beta} = (\beta_1, \dots, \beta_p)'$ is the vector of regression parameters, generally *unknown*, and $\mathbf{e}_n = (e_1, \dots, e_n)'$ is the vector of i.i.d. random errors assumed to follow a distribution, $F(e)$ such that $E(\mathbf{e}_n) = \mathbf{0}$ and $E(\mathbf{e}_n\mathbf{e}_n') = \sigma^2\mathbf{I}_n$ $(\sigma^2 < \infty)$. Further, assume that

$$\text{(i)} \qquad \max_{1\le i\le n} \mathbf{x}_i'(\mathbf{X}_n'\mathbf{X}_n)^{-1}\mathbf{x}_i \longrightarrow 0 \quad \text{as } n \to \infty, \tag{7.7.2}$$

where $\mathbf{x}_i'$ is the ith row of $\mathbf{X}_n$;

$$\text{(ii)} \qquad \lim_{n\to\infty} (n^{-1}(\mathbf{X}_n'\mathbf{X}_n)) = \mathbf{C}, \quad \text{finite and positive-definite matrix.}$$

The basic problem of this section is the estimation of the regression parameter $\boldsymbol{\beta}$ when it is suspected that $\boldsymbol{\beta}$ belongs to the subspace defined by $\mathbf{H}\boldsymbol{\beta} = \mathbf{h}$, where $\mathbf{H}$ is a *known* $q \times p$ matrix and $\mathbf{h}$ is a $q \times 1$ vector of known constants. Accordingly, we consider the *unrestricted, restricted,* preliminary test and Stein-type estimators of $\boldsymbol{\beta}$.

7.7.2 Estimation of Regression Parameters and Test of the Hypothesis

Based on the least squares principle, the unrestricted and the restricted estimators of $\boldsymbol{\beta}$ are given by

$$\tilde{\boldsymbol{\beta}}_n = (\mathbf{X}_n'\mathbf{X}_n)^{-1}\mathbf{X}_n'\mathbf{Y}_n \quad \text{and} \quad \hat{\boldsymbol{\beta}}_n = \tilde{\boldsymbol{\beta}}_n - \mathbf{C}_n^{-1}\mathbf{H}'(\mathbf{H}\mathbf{C}_n^{-1}\mathbf{H}')^{-1}(\mathbf{H}\tilde{\boldsymbol{\beta}}_n - \mathbf{h}), \tag{7.7.3}$$

where $\mathbf{C}_n = (\mathbf{X}_n'\mathbf{X}_n)$. We define the estimator of σ^2 by

$$s_e^2 = (n-p)^{-1}(\mathbf{Y}_n - \mathbf{X}_n\tilde{\boldsymbol{\beta}}_n)'(\mathbf{Y}_n - \mathbf{X}_n\tilde{\boldsymbol{\beta}}_n) \tag{7.7.4}$$

One may show that $s_e^2 \to \sigma^2$ almost surely (see for example Sen and Singer (1993, p. 281)).

For testing the null hypothesis $H_0 : \mathbf{H}\boldsymbol{\beta} = \mathbf{h}$, we consider test-statistic $\mathcal{L}_n$ defined by

$$\mathcal{L}_n = s_e^{-2}(\mathbf{H}\tilde{\boldsymbol{\beta}}_n - \mathbf{h})'(\mathbf{H}\mathbf{C}_n^{-1}\mathbf{H}')^{-1}(\mathbf{H}\tilde{\boldsymbol{\beta}}_n - \mathbf{h}). \tag{7.7.5}$$

Under H_0, $\mathcal{L}_n$ follows a central chi-square distribution with q d.f. as $n \to \infty$ (see Sections 7.8.1 and 7.8.2).

7.7.3 Preliminary Test and Stein-Type Estimation

As usual (see Chapters 5 and 6), we combine the *unrestricted* and the *restricted* estimators to obtain the PTE and Stein-type estimators of $\boldsymbol{\beta}$ as follows:

$$\hat{\boldsymbol{\beta}}_n^{\mathrm{PT}} = \tilde{\boldsymbol{\beta}}_n - (\tilde{\boldsymbol{\beta}}_n - \hat{\boldsymbol{\beta}}_n) I(\mathcal{L}_n < \mathcal{L}_{n,\alpha}), \tag{7.7.6}$$

where $\mathcal{L}_{n,\alpha}$ is the upper α-level critical value of the exact distribution of $\mathcal{L}_n$ under H_0. The James-Stein-type estimator (JSE) is defined by

$$\hat{\boldsymbol{\beta}}_n^{S} = \tilde{\boldsymbol{\beta}}_n - k_n(\tilde{\boldsymbol{\beta}}_n - \hat{\boldsymbol{\beta}}_n)\mathcal{L}_n^{-1}, \quad k_n = \frac{(q-2)(n-p)}{(n-p+2)}. \tag{7.7.7}$$

The positive-rule Stein-type estimator (PRSE) is defined by

$$\begin{aligned} \hat{\boldsymbol{\beta}}_n^{S+} &= \hat{\boldsymbol{\beta}}_n + (1 - k_n\mathcal{L}_n^{-1}) I(\mathcal{L}_n > k_n)(\tilde{\boldsymbol{\beta}}_n - \hat{\boldsymbol{\beta}}_n) \\ &= \hat{\boldsymbol{\beta}}_n^{S} - (1 - k_n\mathcal{L}_n^{-1}) I(\mathcal{L}_n < k_n)(\tilde{\boldsymbol{\beta}}_n - \hat{\boldsymbol{\beta}}_n), \end{aligned} \tag{7.7.8}$$

since $\mathcal{L}_n \neq 0$ with probability one.

7.8 Asymptotic Distribution of the Estimators

In order to find the asymptotic distributional bias (ADB), quadratic bias (ADQB), MSE-matrices (ADMSE), and quadratic risks (ADQR) of the estimators of $\boldsymbol{\beta}$, we need the asymptotic distributions of the various estimators and of the test-statistic, $\mathcal{L}_n$. As in Section 5.8, we consider the generic estimator $\boldsymbol{\beta}_n^*$ and a positive-semidefinite matrix $\mathbf{W}$ defining the loss function

$$\begin{aligned} L(\boldsymbol{\beta}_n^*; \boldsymbol{\beta}) &= n(\boldsymbol{\beta}_n^* - \boldsymbol{\beta})'\mathbf{W}(\boldsymbol{\beta}_n^* - \boldsymbol{\beta}) \\ &= \mathrm{tr}\big[\mathbf{W}\{n(\boldsymbol{\beta}_n^* - \boldsymbol{\beta})(\boldsymbol{\beta}_n^* - \boldsymbol{\beta})'\}\big]. \end{aligned} \tag{7.8.1}$$

Let $\mathbf{M}(\boldsymbol{\beta}_n^*)$ denote the $\lim_{n\to\infty} E\{n(\boldsymbol{\beta}_n^* - \boldsymbol{\beta})(\boldsymbol{\beta}_n^* - \boldsymbol{\beta})'\}$. Then the ADQR of $\boldsymbol{\beta}_n^*$ is given by

$$R(\boldsymbol{\beta}_n^*) = \mathrm{tr}[\mathbf{W}\mathbf{M}(\boldsymbol{\beta}_n^*)]. \tag{7.8.2}$$

In many situations, the asymptotic distribution of $\sqrt{n}\, s_e^{-1}(\boldsymbol{\beta}_n^* - \boldsymbol{\beta})$ is equivalent to the $\sqrt{n}\sigma^{-1}(\tilde{\boldsymbol{\beta}}_n - \boldsymbol{\beta})$ distribution as $n \to \infty$ under fixed alternatives

$$K_{\boldsymbol{\xi}} : \mathbf{H}\boldsymbol{\beta} = \mathbf{h} + \boldsymbol{\xi}. \tag{7.8.3}$$

Then, to obtain a meaningful asymptotic distribution of $\sqrt{n}s_e^{-1}(\boldsymbol{\beta}_n^* - \boldsymbol{\beta})$, we consider the class of local alternatives, $\{K_{(n)}\}$ defined by

$$K_{(n)} : \mathbf{H}\boldsymbol{\beta} = \mathbf{h} + n^{-1/2}\boldsymbol{\xi}. \tag{7.8.4}$$

Let the asymptotic cdf of $\sqrt{n}s_e^{-1}(\beta_n^* - \beta)$ under $\{K_{(n)}\}$ be

$$G_p(\mathbf{x}) = \lim_{n\to\infty} P_{K_{(n)}}\{\sqrt{n}s_e^{-1}(\beta_n^* - \beta) \leq \mathbf{x}\}. \tag{7.8.5}$$

If the asymptotic cdf exists, then the ADB and the ADQB are given by

$$\mathbf{b}(\beta_n^*) = \lim_{n\to\infty} E\left[\sqrt{n}(\beta_n^* - \beta)\right] = \int \mathbf{x}dG_p(\mathbf{x}), \tag{7.8.6a}$$

and

$$B(\beta_n^*) = \sigma^{-2}[\mathbf{b}(\beta_n^*)]'\mathbf{C}[\mathbf{b}(\beta_n^*)], \tag{7.8.6b}$$

respectively, where $\sigma^2\mathbf{C}^{-1}$ is the MSE-matrix of $\tilde{\beta}_n$ as $n \to \infty$. Defining

$$\mathbf{M}(\beta_n^*) = \int \mathbf{x}\mathbf{x}'dG_p(\mathbf{x}), \tag{7.8.7}$$

we have the weighted risk of β_n^* given by

$$R(\beta_n^*) = \text{tr}[\mathbf{W}\mathbf{M}(\beta_n^*)]. \tag{7.8.8}$$

7.8.1 Asymptotic Distribution of the Estimators under Fixed Alternatives

First, we consider the asymptotic distribution of $\sqrt{n}\, s_e^{-1}(\tilde{\beta}_n - \beta)$ under the regularity conditions given by 7.7.1(i) and (ii).

Theorem 1. Under the regularity conditions 7.7.1(i) and (ii) together with $\sigma^2 < \infty$, as $n \to \infty$,

$$\sqrt{n}\, s_e^{-1}(\tilde{\beta}_n - \beta) \overset{\mathcal{D}}{\sim} N_p(\mathbf{0}, \mathbf{C}^{-1}). \tag{7.8.9}$$

For proof readers are referred to Sen and Singer (1993, p. 280).

Next, we consider the asymptotic distribution of the test-statistic $\mathcal{L}_n$ for testing $H_0 : \mathbf{H}\beta = \mathbf{h}$ under some fixed alternative hypothesis of the form

$$K_{\xi} : \mathbf{H}\beta = \mathbf{h} + \xi, \tag{7.8.10}$$

where ξ is a fixed vector in $\mathcal{R}^q$. Then

$$\begin{aligned} s_e^{-1}(\mathbf{H}\tilde{\beta}_n - \mathbf{h}) &= s_e^{-1}\{\mathbf{H}(\tilde{\beta}_n - \beta) + (\mathbf{H}\beta - \mathbf{h})\} \\ &= s_e^{-1}\mathbf{H}(\tilde{\beta}_n - \beta) + s_e^{-1}\xi, \end{aligned}$$

and the statistic $\mathcal{L}_n$ can be written as

$$\begin{aligned} \mathcal{L}_n = s_e^{-2}(\tilde{\beta}_n - \beta)'\mathbf{H}'(\mathbf{H}\mathbf{C}_n^{-1}\mathbf{H}')^{-1}\mathbf{H}(\tilde{\beta}_n - \beta) + s_e^{-2}\xi'(\mathbf{H}\mathbf{C}_n^{-1}\mathbf{H}')^{-1}\xi \\ + 2s_e^{-2}(\tilde{\beta}_n - \beta)'\mathbf{H}'(\mathbf{H}\mathbf{C}_n^{-1}\mathbf{H}')^{-1}\xi. \end{aligned} \tag{7.8.11}$$

Under the assumed regularity conditions and the CLT, we obtain as $n \to \infty$,

$$\sqrt{n}\, s_e^{-1}(\mathbf{H}\tilde{\boldsymbol{\beta}}_n - \mathbf{h}) \overset{\mathcal{D}}{\sim} \mathcal{N}_q\left(\sqrt{n}\,\boldsymbol{\xi}\sigma^{-1}, (\mathbf{H}\mathbf{C}^{-1}\mathbf{H}')\right) \tag{7.8.12}$$

and $n\sigma_e^{-2}\boldsymbol{\xi}'(\mathbf{H}\mathbf{C}^{-1}\mathbf{H}')^{-1}\boldsymbol{\xi} \to \infty$. Similarly, the third term goes to infinity as $n \to \infty$. Hence, it follows that $\mathcal{L}_n \to \infty$ as $n \to \infty$. Consequently, for all hypotheses, $K_{\boldsymbol{\xi}} : \mathbf{H}\boldsymbol{\beta} = \mathbf{h} + \boldsymbol{\xi}$, $P\{\mathcal{L}_n > x\} \to 1$ as $n \to \infty$ for all $x \in R^+$. Then we have the following theorem:

Theorem 2. Under the regularity conditions and fixed alternatives, $K_{\boldsymbol{\xi}}$ as $n \to \infty$,

(i) $$\sqrt{n}s_e^{-1}\left(\hat{\boldsymbol{\beta}}_n^{\mathrm{PT}} - \tilde{\boldsymbol{\beta}}_n\right) = \sqrt{n}\sigma^{-1}(\tilde{\boldsymbol{\beta}}_n - \boldsymbol{\beta}) + o_P(1), \tag{7.8.13}$$

(ii) $$\sqrt{n}s_e^{-1}\left(\hat{\boldsymbol{\beta}}_n^{S} - \tilde{\boldsymbol{\beta}}_n\right) = \sqrt{n}\sigma^{-1}(\tilde{\boldsymbol{\beta}}_n - \boldsymbol{\beta}) + o_P(1),$$

(iii) $$\sqrt{n}s_e^{-1}\left(\hat{\boldsymbol{\beta}}_n^{S+} - \tilde{\boldsymbol{\beta}}_n\right) = \sqrt{n}\sigma^{-1}(\tilde{\boldsymbol{\beta}}_n - \boldsymbol{\beta}) + o_P(1).$$

Proof. Consider the quadratic-difference

$$s_e^{-2}\left\|\hat{\boldsymbol{\beta}}_n^{\mathrm{PT}} - \tilde{\boldsymbol{\beta}}_n\right\|_{\mathbf{C}_n}^2 = \mathcal{L}_n I(\mathcal{L}_n < \mathcal{L}_{n,\alpha}) \le \mathcal{L}_{n,\alpha} I(\mathcal{L}_n < \mathcal{L}_{n,\alpha}).$$

Then $\lim_{n\to\infty} E\{\mathcal{L}_{n,\alpha} I(\mathcal{L}_n < \mathcal{L}_{n,\alpha})\} = 0$. Hence, (i) is proved. Similarly,

$$s_e^{-2}\left\|\hat{\boldsymbol{\beta}}_n^{S} - \tilde{\boldsymbol{\beta}}_n\right\|_{\mathbf{C}_n}^2 = k_n^2 s_e^{-2}\left\|\tilde{\boldsymbol{\beta}}_n - \hat{\boldsymbol{\beta}}_n\right\|_{\mathbf{C}_n}^2 \mathcal{L}_n^{-2} = k_n^2\mathcal{L}_n^{-1}.$$

Since $\mathcal{L}_n \ge 0$ and $\mathcal{L}_n \to \infty$ as $n \to \infty$, we have $\lim_{n\to\infty} k_n \to k$ and

$$\lim_{n\to\infty} E(\mathcal{L}_n^{-1}) = 0 \Rightarrow \sqrt{n}\, s_e^{-1}(\hat{\boldsymbol{\beta}}_n^{S} - \boldsymbol{\beta}) = \sqrt{n}\sigma^{-1}(\tilde{\boldsymbol{\beta}}_n - \boldsymbol{\beta}) + o_P(1).$$

Finally,

$$\begin{aligned} s_e^{-2}\left\|\hat{\boldsymbol{\beta}}_n^{S+} - \tilde{\boldsymbol{\beta}}_n\right\|_{\mathbf{C}_n}^2 &= k_n^2\mathcal{L}_n^{-1} + \mathcal{L}_n(1 - k_n\mathcal{L}_n^{-1})^2 I(\mathcal{L}_n < k_n) \\ &\quad - 2k_n^2\mathcal{L}_n(1 - k_n\mathcal{L}_n^{-1})I(\mathcal{L}_n < k_n). \end{aligned}$$

Then $\lim_{n\to\infty} E\left\{s_e^{-2}\left\|\hat{\boldsymbol{\beta}}_n^{S+} - \tilde{\boldsymbol{\beta}}_n\right\|_{\mathbf{C}_n}^2\right\} = 0$. Hence,

$$\sqrt{n}\, s_e^{-1}(\hat{\boldsymbol{\beta}}_n^{S+} - \boldsymbol{\beta}) = \sqrt{n}\sigma^{-1}(\tilde{\boldsymbol{\beta}}_n - \boldsymbol{\beta}) + o_P(1). \qquad \text{Q.E.D.}$$

Theorem 7.8.2 shows that the asymptotic distribution of $\sqrt{n}(\hat{\boldsymbol{\beta}}_n^{\mathrm{PT}} - \boldsymbol{\beta})$, $\sqrt{n}(\hat{\boldsymbol{\beta}}_n^{S} - \boldsymbol{\beta})$, and $\sqrt{n}(\hat{\boldsymbol{\beta}}_n^{S+} - \boldsymbol{\beta})$ is given by $\mathcal{N}_p(\mathbf{0}, \sigma^2\mathbf{C}^{-1})$, and the ADB,

ADQB, ADMSE, and ADQR of the estimators $\hat{\boldsymbol{\beta}}_n^{PT}$, $\hat{\boldsymbol{\beta}}_n^{S}$, and $\hat{\boldsymbol{\beta}}_n^{S+}$ are all equal and are given by

$$\mathbf{b}_1(\tilde{\boldsymbol{\beta}}_n) = \mathbf{b}_3(\hat{\boldsymbol{\beta}}_n^{PT}) = \mathbf{b}_4(\hat{\boldsymbol{\beta}}_n^{S}) = \mathbf{b}_5(\hat{\boldsymbol{\beta}}_m^{S+}) = \mathbf{0},$$
$$\mathbf{M}_3(\hat{\boldsymbol{\beta}}_n^{PT}) = \mathbf{M}_4(\hat{\boldsymbol{\beta}}_n^{S}) = \mathbf{M}_5(\hat{\boldsymbol{\beta}}_n^{S+}) = \sigma^2\mathbf{C}^{-1},$$

and

$$R_1(\tilde{\boldsymbol{\beta}}_n : \mathbf{W}) = R_3(\hat{\boldsymbol{\beta}}_n^{PT} : \mathbf{W}) = R_4(\hat{\boldsymbol{\beta}}_n^{S} : \mathbf{W}) = R_5(\hat{\boldsymbol{\beta}}_n^{S+} : \mathbf{W}) = \sigma^2 \operatorname{tr}(\mathbf{W}\mathbf{C}^{-1}) \tag{7.8.14}$$

while

$$s_e^{-2}\left\|\hat{\boldsymbol{\beta}}_n - \tilde{\boldsymbol{\beta}}_n\right\|^2_{\mathbf{C}_n} = \mathcal{L}_n \to \infty \quad \text{as} \quad n \to \infty, \tag{7.8.15}$$

and $E(\mathcal{L}_n) \to \infty$ as $n \to \infty$.

Thus, the asymptotic distributions of $\sqrt{n}s_e^{-1}(\hat{\boldsymbol{\beta}}_n - \boldsymbol{\beta})$ and $\sqrt{n}s_e^{-1}(\tilde{\boldsymbol{\beta}}_n - \boldsymbol{\beta})$ are *not equivalent* as $n \to \infty$. Now,

$$\begin{aligned}
&\sqrt{n}s_e^{-1}(\hat{\boldsymbol{\beta}}_n - \boldsymbol{\beta})\\
&= \sqrt{n}s_e^{-1}(\tilde{\boldsymbol{\beta}}_n - \boldsymbol{\beta}) - \sqrt{n}s_e^{-1}\mathbf{C}^{-1}\mathbf{H}'(\mathbf{H}\mathbf{C}_n^{-1}\mathbf{H}')^{-1}(\mathbf{H}\tilde{\boldsymbol{\beta}}_n - \mathbf{h})\\
&= \sqrt{n}s_e^{-1}(\tilde{\boldsymbol{\beta}}_n - \boldsymbol{\beta}) - \sqrt{n}s_e^{-1}\mathbf{C}^{-1}\mathbf{H}'(\mathbf{H}\mathbf{C}_n^{-1}\mathbf{H}')^{-1}\{\mathbf{H}(\tilde{\boldsymbol{\beta}}_n - \boldsymbol{\beta}) + \boldsymbol{\xi}\}
\end{aligned} \tag{7.8.16}$$

implies that the asymptotic distribution of $\sqrt{n}s_e^{-1}(\hat{\boldsymbol{\beta}}_n - \boldsymbol{\beta})$ is degenerate under the fixed alternative $K_{\boldsymbol{\xi}}$.

7.8.2 Asymptotic Distribution of the Estimators under Local Alternatives, and ADB, ADQB, ADMSE, and ADQR

To obtain meaningful asymptotic distributions of the various estimators and the test-statistics, $\mathcal{L}_n$ we consider the following theorem:

Theorem 3. Under $K_{(n)}$: $\mathbf{H}\boldsymbol{\beta} = \mathbf{h} + n^{-1/2}\boldsymbol{\xi}$ and the assumed regularity conditions 7.7.2(i) and (ii), we have the following as $n \to \infty$:

(i) $\mathbf{V}_n^{(1)} = \sqrt{n}(\tilde{\boldsymbol{\beta}}_n - \boldsymbol{\beta}) \sim \mathcal{N}_p(\mathbf{0}, \sigma^2(\mathbf{C}^{-1}))$. (7.8.17)

(ii) $\mathbf{V}_n^{(2)} = \sqrt{n}(\hat{\boldsymbol{\beta}}_n - \boldsymbol{\beta}) \sim \mathcal{N}_p(-\boldsymbol{\delta}, \sigma^2\mathbf{A})$, where $\boldsymbol{\delta} = \mathbf{C}^{-1}\mathbf{H}'(\mathbf{H}\mathbf{C}^{-1}\mathbf{H}')^{-1}\boldsymbol{\xi}$, and $\mathbf{A} = \mathbf{C}^{-1} - \mathbf{C}^{-1}\mathbf{H}'(\mathbf{H}\mathbf{C}^{-1}\mathbf{H}')^{-1}\mathbf{H}\mathbf{C}^{-1}$.

(iii) $\mathbf{V}_n^{(3)} = \sqrt{n}(\tilde{\boldsymbol{\beta}}_n - \hat{\boldsymbol{\beta}}_n) \sim \mathcal{N}_p(\boldsymbol{\delta}, \sigma^2[\mathbf{C}^{-1} - \mathbf{A}])$.

(iv) $\begin{bmatrix} \mathbf{V}_n^{(1)} \\ \mathbf{V}_n^{(3)} \end{bmatrix} \sim \mathcal{N}_{2p}\left\{\begin{pmatrix} \mathbf{0} \\ \boldsymbol{\delta} \end{pmatrix}, \sigma^2\begin{pmatrix} \mathbf{C}^{-1} & \mathbf{C}^{-1} - \mathbf{A} \\ \mathbf{C}^{-1} - \mathbf{A} & \mathbf{C}^{-1} - \mathbf{A} \end{pmatrix}\right\}$.

(v) $\begin{bmatrix} \mathbf{V}_n^{(2)} \\ \mathbf{V}_n^{(3)} \end{bmatrix} \sim \mathcal{N}_{2p}\left\{\begin{pmatrix} -\boldsymbol{\delta} \\ \boldsymbol{\delta} \end{pmatrix}, \sigma^2\begin{pmatrix} \mathbf{A} & \mathbf{0} \\ \mathbf{0} & \mathbf{C}^{-1}-\mathbf{A} \end{pmatrix}\right\}.$

(vi) $\lim_{n\to\infty} P(\mathcal{L}_n \le x) = H_q(x; \Delta^2)$.

(vii) $\lim_{n\to\infty} P\{\sqrt{n}(\hat{\boldsymbol{\beta}}_n^{\mathrm{PT}} - \boldsymbol{\beta}) \le \mathbf{x}\} = H_q(\chi_q^2(\alpha); \Delta^2)\Phi_p(\mathbf{x} + \boldsymbol{\delta}, \mathbf{0}, \sigma^2\mathbf{A})$
$+ \int_{E(\boldsymbol{\delta})} \Phi_p(\mathbf{x} + \mathbf{Z}; \mathbf{0}, \sigma^2(\mathbf{C}^{-1} - \mathbf{A}))d\Phi_p(\mathbf{Z}, \mathbf{0}, \sigma^2(\mathbf{HC}^{-1}\mathbf{H}')),$

where $E(\boldsymbol{\delta}) = \{\mathbf{Z} : (\mathbf{Z} + \boldsymbol{\delta})'(\mathbf{HC}^{-1}\mathbf{H}')^{-1}(\mathbf{Z} + \boldsymbol{\delta}) \ge \chi_q^2(\alpha)\}$, and $\Phi_p(\cdot; \boldsymbol{\mu}, \boldsymbol{\Sigma})$ is the cdf of a p-variate normal distribution with mean $\boldsymbol{\mu}$ and covariance matrix, $\boldsymbol{\Sigma}$, and $H_\nu(\cdot; \Delta^2)$ is the cdf of a noncentral chi-square distribution with ν d.f. and noncentrality parameter $\Delta^2/2$.

It is difficult to obtain the asymptotic distributions of $\sqrt{n}(\hat{\boldsymbol{\beta}}_n^{S} - \boldsymbol{\beta})$ and $\sqrt{n}(\hat{\boldsymbol{\beta}}_n^{S+} - \boldsymbol{\beta})$, but we can obtain an asymptotic representation of these estimators under $\{K_{(n)}\}$ to facilitate the computation of ADB, ADQB, ADMSE and ADQR of these estimators. Hence, we have under $\{K_{(n)}\}$,

(viii) $\sqrt{n}\left(\hat{\boldsymbol{\beta}}_n^{S} - \boldsymbol{\beta}\right) \stackrel{\mathcal{D}}{=} \mathbf{Z} - k\left\{\dfrac{\mathbf{C}^{-1}\mathbf{H}'(\mathbf{HC}^{-1}\mathbf{H}')(\mathbf{HZ} + \boldsymbol{\xi})}{\sigma^{-2}(\mathbf{HZ} + \boldsymbol{\xi})'(\mathbf{HC}^{-1}\mathbf{H}')^{-1}(\mathbf{HZ} + \boldsymbol{\xi})}\right\};$

(ix) $\sqrt{n}\left(\hat{\boldsymbol{\beta}}_n^{S+} - \boldsymbol{\beta}\right) \stackrel{\mathcal{D}}{=} \mathbf{Z} - k\left\{\dfrac{\mathbf{C}^{-1}\mathbf{H}'(\mathbf{HC}^{-1}\mathbf{H}')(\mathbf{HZ} + \boldsymbol{\xi})}{\sigma^{-2}(\mathbf{HZ} + \boldsymbol{\xi})'(\mathbf{HC}^{-1}\mathbf{H}')^{-1}(\mathbf{HZ} + \boldsymbol{\xi})}\right\}$
$+[\mathbf{C}^{-1}\mathbf{H}'(\mathbf{HC}^{-1}\mathbf{H}')^{-1}(\mathbf{HZ} + \boldsymbol{\xi})]$
$\times\left\{1 - \dfrac{k}{\sigma^{-2}(\mathbf{HZ} + \boldsymbol{\xi})'(\mathbf{HC}^{-1}\mathbf{H}')^{-1}(\mathbf{HZ} + \boldsymbol{\xi})}\right\}$
$\times\, I\{\sigma^{-2}(\mathbf{HZ} + \boldsymbol{\xi})'(\mathbf{HC}^{-1}\mathbf{H}')(\mathbf{HZ} + \boldsymbol{\xi}) < k\},$

where $\mathbf{Z} \sim \mathcal{N}_p(\mathbf{0}, \sigma^2\mathbf{C}^{-1})$ and $k = q - 2$.

Proof. (i) See Theorem 7.8.1.

(ii) $\mathbf{V}_n^{(2)} = \sqrt{n}(\hat{\boldsymbol{\beta}}_n - \boldsymbol{\beta}) = \sqrt{n}\left\{(\tilde{\boldsymbol{\beta}}_n - \boldsymbol{\beta}) - \mathbf{C}_n^{-1}\mathbf{H}'(\mathbf{HC}_n^{-1}\mathbf{H}')^{-1}(\mathbf{H}\tilde{\boldsymbol{\beta}}_n - \mathbf{h})\right\}$
$= \sqrt{n}(\tilde{\boldsymbol{\beta}}_n - \boldsymbol{\beta}) - \mathbf{C}_n^{-1}\mathbf{H}'(\mathbf{HC}_n^{-1}\mathbf{H}')^{-1}\mathbf{H}[\sqrt{n}(\tilde{\boldsymbol{\beta}}_n - \boldsymbol{\beta}) + \boldsymbol{\xi}]$. By (i) $\mathbf{V}_n^{(2)} \sim \mathcal{N}_p(-\boldsymbol{\delta}, \sigma^2\mathbf{A})$ as $n \to \infty$.

(iii) $\mathbf{V}_n^{(3)} = \sqrt{n}(\tilde{\boldsymbol{\beta}}_n - \hat{\boldsymbol{\beta}}_n) = (\mathbf{C}_n^{-1}\mathbf{H}'(\mathbf{HC}_n^{-1}\mathbf{H}')^{-1}\mathbf{H}[\sqrt{n}(\tilde{\boldsymbol{\beta}}_n - \boldsymbol{\beta}) + \boldsymbol{\xi}]$
$\sim \mathcal{N}_p(\boldsymbol{\delta}, \sigma^2[\mathbf{C}^{-1} - \mathbf{A}])$ as $n \to \infty$.

(iv) and (v) follow similarly. To prove (vi), we note that $\sqrt{n}\, s_e^{-1}(\mathbf{H}\tilde{\boldsymbol{\beta}}_n - \mathbf{h}) \stackrel{\mathcal{D}}{\simeq} \mathcal{N}_q(\sigma^{-1}\boldsymbol{\xi}, (\mathbf{HC}^{-1}\mathbf{H}'))$ as $n \to \infty$. Hence,

$$n\, s_e^{-2}(\mathbf{H}\tilde{\boldsymbol{\beta}}_n - \mathbf{h})'(\mathbf{HC}_n^{-1}\mathbf{H}')^{-1}(\mathbf{H}\tilde{\boldsymbol{\beta}}_n - \mathbf{h}) \stackrel{\mathcal{D}}{=} \chi_q^2(\Delta^2)$$

as $n \to \infty$ where $\Delta^2 = \sigma^{-2}\xi'(\mathbf{HC}^{-1}\mathbf{H}')^{-1}\boldsymbol{\xi} = \sigma^{-1}\boldsymbol{\delta}'\mathbf{C}\boldsymbol{\delta}$.

(vi) $\lim_{n\to\infty} P\left\{\sqrt{n}\left(\hat{\boldsymbol{\beta}}_n^{\text{PT}} - \boldsymbol{\beta}\right) \leq \mathbf{x}\right\} = \lim P\left\{\sqrt{n}\left(\hat{\boldsymbol{\beta}}_n - \boldsymbol{\beta}\right) \leq \mathbf{x},\ \mathcal{L}_n \leq \mathcal{L}_{n,\alpha}\right\}$
$+ \lim_{n\to\infty} P\left\{\sqrt{n}\left(\tilde{\boldsymbol{\beta}}_n - \boldsymbol{\beta}\right) \leq \mathbf{x};\ \mathcal{L}_n > \mathcal{L}_{n,\alpha}\right\}.$

Since $\sqrt{n}(\hat{\boldsymbol{\beta}}_n - \boldsymbol{\beta})$ and $\sqrt{n}(\tilde{\boldsymbol{\beta}}_n - \hat{\boldsymbol{\beta}}_n)$ are independent, the first term reduces to

$$H_q(\chi_q^2(\alpha); \Delta^2)\Phi_p(\mathbf{x} + \boldsymbol{\delta}; \mathbf{0}, \sigma^2\mathbf{A})$$

as $n \to \infty$. The second term is obtained by conditional arguments as $n \to \infty$ and is given by

$$\int_{E(\boldsymbol{\delta})} \Phi_p\big(\mathbf{x} - \mathbf{C}^{-1}\mathbf{H}'(\mathbf{H}\mathbf{C}^{-1}\mathbf{H}')^{-1}\mathbf{Z}; \mathbf{0}, \sigma^2[\mathbf{C}^{-1} - \mathbf{A}]\big)$$
$$\times d\Phi_p\big(\mathbf{Z}; \mathbf{0}, \sigma^2(\mathbf{H}\mathbf{C}^{-1}\mathbf{H}')\big),$$

where $E(\boldsymbol{\delta}) = \left\{\mathbf{Z}; (\mathbf{Z} + \boldsymbol{\delta})'(\mathbf{H}\mathbf{C}^{-1}\mathbf{H})^{-1}(\mathbf{Z} + \boldsymbol{\delta}) \geq \chi_q^2(\alpha)\right\}.$

Proofs of (viii) and (ix) are obtained by writing the expressions in terms of $(\mathbf{H}\boldsymbol{\beta} - \mathbf{h}) = n^{-1/2}\boldsymbol{\xi}$ and then applying the distribution of the related statistics.

Based on the theorem above we can easily obtain the ADB, ADQB, ADMSE and ADQR of the estimators. First, we consider the ADB and ADQB of the estimators. Clearly,

(i) $\mathbf{b}_1(\tilde{\boldsymbol{\beta}}_n) = \lim_{n\to\infty} E(\sqrt{n}(\tilde{\boldsymbol{\beta}}_n - \boldsymbol{\beta})\} = \mathbf{0}$ and $B_1(\tilde{\boldsymbol{\beta}}_n) = 0,$ (7.8.18)

(ii) $\mathbf{b}_2(\hat{\boldsymbol{\beta}}_n) = \lim_{n\to\infty} E(\sqrt{n}(\hat{\boldsymbol{\beta}}_n - \boldsymbol{\beta})\} = -\boldsymbol{\delta}$ and $B_2(\hat{\boldsymbol{\beta}}_n) = \sigma^{-2}(\boldsymbol{\delta}'\mathbf{C}\boldsymbol{\delta}) = \Delta^2,$

(iii) $\mathbf{b}_3(\hat{\boldsymbol{\beta}}_n) = \lim_{n\to\infty} E(\sqrt{n}(\hat{\boldsymbol{\beta}}_n - \boldsymbol{\beta})\} = \lim_{n\to\infty} E(-\sqrt{n}(\tilde{\boldsymbol{\beta}}_n - \hat{\boldsymbol{\beta}}_n)\mathbf{I}(\mathcal{L}_n < \mathcal{L}_{n,\alpha})\}$
$= -\boldsymbol{\delta} H_{q+2}(\chi_q^2(\alpha); \Delta^2),$ since $\mathcal{L}_{n,\alpha} \to \chi_q^2(\alpha)$ as $n \to \infty$,

and

$$\mathbf{B}_3(\hat{\boldsymbol{\beta}}_n^{\text{PT}}) = \Delta^2 \left\{H_{q+2}(\chi_q^2(\alpha); \Delta^2)\right\}^2.$$

(iv) $\mathbf{b}_4(\hat{\boldsymbol{\beta}}_n^S) = \lim_{n\to\infty} E\{\sqrt{n}(\hat{\boldsymbol{\beta}}_n^S - \boldsymbol{\beta})\} = -k \lim_{n\to\infty} E\{\sqrt{n}(\tilde{\boldsymbol{\beta}}_n - \hat{\boldsymbol{\beta}}_n)\mathcal{L}_n^{-1}\}$

$= -k\boldsymbol{\delta} E[\chi_{q+2}^{-2}(\Delta^2)]\,, \qquad k = \lim_{n\to\infty} k_n = (q-2)$

and

$$B_4(\hat{\boldsymbol{\beta}}_n^S) = k^2\Delta^2\left\{E[\chi_{q+2}^{-2}(\Delta^2)]\right\}^2.$$

Similarly,

(v) $\mathbf{b}_5(\hat{\boldsymbol{\beta}}_n^{S+}) = \lim_{n\to\infty} E\{\sqrt{n}(\hat{\boldsymbol{\beta}}_n^{S+} - \boldsymbol{\beta})\} = \lim_{n\to\infty} E\{\sqrt{n}(\hat{\boldsymbol{\beta}}_n^{S} - \boldsymbol{\beta}) - \sqrt{n}(\tilde{\boldsymbol{\beta}}_n - \hat{\boldsymbol{\beta}}_n)$
$(1 - k_n\mathcal{L}_n^{-1})I(\mathcal{L}_n < k_n)\}$
$= \mathbf{b}_4(\hat{\boldsymbol{\beta}}_n^{S}) - \boldsymbol{\delta}E\{(1 - k\chi_{q+2}^{-2}(\Delta^2))I(\chi_{q+2}^2(\Delta^2) < k)\}$

and

$$B_5(\hat{\boldsymbol{\beta}}_n^{S+}) = \Delta^2\{kE[\chi_{q+2}^{-2}(\Delta^2)] - E[(1-k\chi_{q+2}^{-2}(\Delta^2))I(\chi_{q+2}^2(\Delta^2) < k)]\}^2.$$

We present the expressions for the ADQR and ADMSE of the estimators together with the ADMSE using Theorem 7.8.3. Notationally, they are similar to the expressions given in Section 7.3.2.

(i) $R_1(\tilde{\boldsymbol{\beta}}_n; \mathbf{W}) = \sigma^2 \operatorname{tr}(\mathbf{W}\mathbf{C}^{-1})$ and $\mathbf{M}_1(\tilde{\boldsymbol{\beta}}_n) = \sigma^2\mathbf{C}^{-1}$. (7.8.19)

(ii) $R_2(\hat{\boldsymbol{\beta}}_n; \mathbf{W}) = \sigma^2 \operatorname{tr}(\mathbf{W}\mathbf{C}^{-1}) - \sigma^2 \operatorname{tr}(\mathbf{A}_{11}) + \boldsymbol{\eta}_1'\mathbf{A}_{11}\boldsymbol{\eta}_1$ and

$\mathbf{M}_2(\hat{\boldsymbol{\beta}}_n) = \sigma^2(\mathbf{C}^{-1} - \mathbf{A}) + \boldsymbol{\delta}\boldsymbol{\delta}'$, $\mathbf{A}_{11} = \mathbf{W}\mathbf{C}^{-1}\mathbf{H}'(\mathbf{H}\mathbf{C}^{-1}\mathbf{H}')^{-1}\mathbf{H}\mathbf{C}^{-1}$.

(iii) $$R_3(\hat{\boldsymbol{\beta}}_n^{PT}; \mathbf{W}) = \sigma^2 \operatorname{tr}(\mathbf{W}\mathbf{C}^{-1}) - \sigma^2 \operatorname{tr}(\mathbf{A}_{11})H_{q+2}(\chi_q^2(\alpha); \Delta^2) + (\boldsymbol{\eta}_1'\mathbf{A}_{11}\boldsymbol{\eta}_1)\{2H_{q+2}(\chi_q^2(\alpha); \Delta^2) - H_{q+4}(\chi_q^2(\alpha); \Delta^2)\}$$

and

$$\mathbf{M}_3(\hat{\boldsymbol{\beta}}_n^{PT}; \mathbf{W}) = \sigma^2\mathbf{C}^{-1} - \sigma^2[\mathbf{C}^{-1} - \mathbf{A}]H_{q+2}(\chi_q^2(\alpha); \Delta^2) + \boldsymbol{\delta}\boldsymbol{\delta}'\{2H_{p+2}(\chi_q^2(\alpha); \Delta^2) - H_{p+1}(\chi_q^2(\alpha); \Delta^2)\}.$$

(iv) $$R_4(\hat{\boldsymbol{\beta}}_n^{S}; \mathbf{W}) = \sigma^2 \operatorname{tr}(\mathbf{W}\mathbf{C}^{-1}) - k\sigma^2 \operatorname{tr}(\mathbf{A}_{11})\Big\{2E[\chi_{q+2}^{-2}(\Delta^2] - kE[\chi_{q+2}^{-4}(\Delta^2)]\Big\} + k(k+4)(\boldsymbol{\eta}_1'\mathbf{A}_{11}\boldsymbol{\eta}_1)E[\chi_{q+4}^{-4}(\Delta^2)]$$

and

$$\mathbf{M}_4(\hat{\boldsymbol{\beta}}_n^{S}) = \sigma^2\mathbf{C}^{-1} - k\sigma^2[\mathbf{C}^{-1} - \mathbf{A}]\Big\{2E[\chi_{q+2}^{-2}(\Delta^2)] - kE[\chi_{q+4}^{-2}(\Delta^2)]\Big\} + k(k+4)\sigma^{-2}\boldsymbol{\delta}\boldsymbol{\delta}'E[\chi_{q+4}^{-4}(\Delta^2)],$$

where $k = q - 2$.

(v) $$R_5(\hat{\boldsymbol{\beta}}_n^{S+}; \mathbf{W}) = R_4(\hat{\beta}_n^S) - \sigma^2\Big\{\operatorname{tr}(\mathbf{A}_{11})E[(1-k\chi_{q+2}^{-2}(\Delta^2))^2 I(\chi_{q+2}^{-2}(\Delta^2) < k)] + \sigma^{-2}(\boldsymbol{\eta}_1'\mathbf{A}_{11}\boldsymbol{\eta}_1)E[(1-k)\,\chi_{q+4}^{-2}(\Delta^2))^2 I(\chi_{q+2}^2(\Delta^2) < k)]\Big\} - 2(\boldsymbol{\eta}_1'\mathbf{A}_{11}\boldsymbol{\eta}_1)E[(1 - k\chi_{q+2}^{-2}(\Delta^2))I(\chi_{q+2}^2(\Delta^2) < k]$$

and

$$\begin{aligned}\mathbf{M}_5(\hat{\beta}_n^{S+}) &= \mathbf{M}_4(\hat{\beta}_n^{S})\\ &\quad - \sigma^2\Big\{[\mathbf{C}^{-1} - \mathbf{A}]E\big[(1-k\chi_{q+2}^{-2}(\Delta^2))^2 I(\chi_{q+2}^2(\Delta^2)<k)\big]\\ &\quad + \sigma^{-2}\delta\delta' E\big[(1-k\chi_{q+4}^{-2}(\Delta^2))^2 I(\chi_{q+4}^2(\Delta^2)<k)\big]\Big\}\\ &\quad - 2[\delta\delta']E\big[(1-k\chi_{q+2}^{-2}(\Delta^2))I(\chi_{q+2}^2(\Delta^2)<k)\big].\end{aligned}$$

Proof. We provide the proofs for $R_3(\hat{\beta}_n^{\text{PT}};\mathbf{W})$ and $R_5(\hat{\beta}_n^{S+};\mathbf{W})$ only.

$$\begin{aligned}&R_3(\hat{\beta}_n^{\text{PT}};\mathbf{W})\\ &= \lim_{n\to\infty} E\big[n(\hat{\beta}_n^{\text{PT}} - \beta)'\mathbf{W}(\hat{\beta}_n^{\text{PT}} - \beta)\big]\\ &= \lim_{n\to\infty} E\big[n(\tilde{\beta}_n - \beta_n)'\mathbf{W}(\tilde{\beta}_n - \beta)\big]\\ &\quad + \lim_{n\to\infty} E\big[n(\tilde{\beta}_n - \hat{\beta}_n)'\mathbf{W}(\tilde{\beta}_n - \hat{\beta}_n)I(\mathcal{L}_n < \mathcal{L}_{n,\alpha})\\ &\quad - 2\lim_{n\to\infty} E\big[n(\tilde{\beta}_n - \beta)'\mathbf{W}(\tilde{\beta}_n - \hat{\beta}_n)I(\mathcal{L}_n < \mathcal{L}_{n,\alpha})\\ &= R_1(\tilde{\beta}_n;\mathbf{W}) - \lim_{n\to\infty} E\big[n(\tilde{\beta}_n - \hat{\beta})'\mathbf{W}(\tilde{\beta}_n - \hat{\beta})I(\mathcal{L}_n < \mathcal{L}_{n,\alpha})\\ &\quad + 2\eta'\mathbf{W}\Big\{\lim_{n\to\infty} E\big[\sqrt{n}(\tilde{\beta}_n - \hat{\beta}_n)I(\mathcal{L}_n > \mathcal{L}_{n,\alpha})\big]\Big\}\\ &= R_1(\tilde{\beta}_n;\mathbf{W}) - \sigma^2\,\text{tr}[\mathbf{C}^{-1}\mathbf{H}'(\mathbf{H}\mathbf{C}^{-1}\mathbf{H}')^{-1}\mathbf{H}\mathbf{C}^{-1}\mathbf{W}]H_{q+2}(\chi_q^2(\alpha);\Delta^2)\\ &\quad + (\delta'\mathbf{W}\delta)\Big\{2H_{q+2}(\chi_q^2(\alpha);\Delta^2) - H_{q+4}(\chi_q^2(\alpha);\Delta^2)\Big\}. \qquad (7.8.20)\end{aligned}$$

Since $\mathbf{C}^{-1/2}\mathbf{H}'(\mathbf{H}\mathbf{C}^{-1}\mathbf{H}')^{-1}\mathbf{H}\mathbf{C}^{-1/2}$ is a symmetric idempotent matrix with rank q ($\leq p$), there exists an orthogonal matrix $\mathbf{\Gamma}$ such that

$$\mathbf{\Gamma}\mathbf{C}^{-1/2}\mathbf{H}'(\mathbf{H}\mathbf{C}^{-1}\mathbf{H}')^{-1}\mathbf{H}\mathbf{C}^{-1/2}\mathbf{\Gamma}' = \begin{pmatrix} \mathbf{I}_q & \mathbf{0} \\ \mathbf{0} & \mathbf{0} \end{pmatrix},$$

$$\mathbf{\Gamma}\mathbf{C}^{-1/2}\mathbf{W}\mathbf{C}^{-1/2}\mathbf{\Gamma}' = \begin{pmatrix} \mathbf{A}_{11} & \mathbf{A}_{12} \\ \mathbf{A}_{12}' & \mathbf{A}_{22} \end{pmatrix}.$$

The matrices $\mathbf{A}_{11}$ and $\mathbf{A}_{12}$ are of order q and $p-q$, respectively. Hence,

$$\begin{aligned}&\text{tr}[\mathbf{C}^{-1}\mathbf{H}'(\mathbf{H}\mathbf{C}^{-1}\mathbf{H}')^{-1}\mathbf{H}\mathbf{C}^{-1}\mathbf{W}]\\ &= \text{tr}[\mathbf{\Gamma}\mathbf{C}^{-1/2}\mathbf{W}\mathbf{C}^{-1/2}\mathbf{\Gamma}')(\mathbf{\Gamma}\mathbf{C}^{-1/2}\mathbf{H}'(\mathbf{H}\mathbf{C}^{-1}\mathbf{H}')^{-1}\mathbf{H}\mathbf{C}^{-1/2}\mathbf{\Gamma}')]\\ &= \text{tr}\left[\begin{pmatrix} \mathbf{A}_{11} & \mathbf{A}_{12} \\ \mathbf{A}_{12}' & \mathbf{A}_{22} \end{pmatrix}\begin{pmatrix} \mathbf{I}_q & \mathbf{0} \\ \mathbf{0} & \mathbf{0} \end{pmatrix}\right]\\ &= \text{tr}(\mathbf{A}_{11}).\end{aligned}$$

Further, by Theorem 7.8.3(ii),

$$
\begin{aligned}
&\boldsymbol{\delta}'\mathbf{W}\boldsymbol{\delta}\\
&= \boldsymbol{\xi}'(\mathbf{HC}^{-1}\mathbf{H}')\mathbf{HC}^{-1}\mathbf{WC}^{-1}\mathbf{H}'(\mathbf{HC}^{-1}\mathbf{H}')^{-1}\boldsymbol{\xi}\\
&= \boldsymbol{\xi}'(\mathbf{HC}^{-1}\mathbf{H}')^{-1}\mathbf{HC}^{-1/2}\boldsymbol{\Gamma}'\boldsymbol{\Gamma}\mathbf{C}^{-1/2}\mathbf{WC}^{-1/2}\boldsymbol{\Gamma}'\boldsymbol{\Gamma}\mathbf{C}^{-1/2}\mathbf{H}'(\mathbf{HC}^{-1}\mathbf{H}')^{-1}\boldsymbol{\xi}\\
&= \boldsymbol{\xi}'(\mathbf{HC}^{-1}\mathbf{H}')^{-1}\mathbf{HC}^{-1/2}\boldsymbol{\Gamma}'\begin{pmatrix}\mathbf{I}_q & \mathbf{0}\\ \mathbf{0} & \mathbf{0}\end{pmatrix}\begin{pmatrix}\mathbf{A}_{11} & \mathbf{A}_{12}\\ \mathbf{A}_{12}' & \mathbf{A}_{22}\end{pmatrix}\begin{pmatrix}\mathbf{I}_q & \mathbf{0}\\ \mathbf{0} & \mathbf{0}\end{pmatrix}\\
&\qquad \times \boldsymbol{\Gamma}\mathbf{C}^{-1/2}\mathbf{H}'(\mathbf{HC}^{-1}\mathbf{H}')^{-1}\boldsymbol{\xi}. \qquad (7.8.21)
\end{aligned}
$$

Let $\boldsymbol{\eta} = \boldsymbol{\Gamma}\mathbf{C}^{-1/2}\mathbf{H}'(\mathbf{HC}^{-1}\mathbf{H}')^{-1}\boldsymbol{\xi}$; then, the r.h.s. of (7.8.21) becomes

$$
\boldsymbol{\eta}' = \begin{pmatrix}\mathbf{A}_{11} & \mathbf{0}\\ \mathbf{0} & \mathbf{0}\end{pmatrix}\boldsymbol{\eta} = \boldsymbol{\eta}_1'\mathbf{A}_{11}\boldsymbol{\eta}_1 \quad \text{where } \boldsymbol{\eta} = \begin{pmatrix}\boldsymbol{\eta}_1\\ \boldsymbol{\eta}_2\end{pmatrix}. \qquad (7.8.22)
$$

Thus, $R_3(\hat{\boldsymbol{\beta}}_n^{\text{PT}};\mathbf{W})$ reduces to

$$
\begin{aligned}
R_3(\hat{\boldsymbol{\beta}}_n^{\text{PT}};\mathbf{W}) &= R_1(\tilde{\boldsymbol{\beta}}_n;\mathbf{W}) - \sigma^2\operatorname{tr}(\mathbf{A}_{11})H_{q+2}(\chi_q^2(\alpha);\Delta^2)\\
&\quad + (\boldsymbol{\eta}_1'\mathbf{A}_{11}\boldsymbol{\eta}_1)\Big\{2H_{q+2}(\chi_q^2(\alpha);\Delta^2) - H_{q+4}(\chi_q^2(\alpha);\Delta^2)\Big\}.
\end{aligned}
$$

Next, we consider the risk expression for $\hat{\boldsymbol{\beta}}_n^S$:

$$
\begin{aligned}
&R_4\left(\hat{\boldsymbol{\beta}}_n^S;\mathbf{W}\right)\\
&= \lim_{n\to\infty} E\big[n(\hat{\boldsymbol{\beta}}_n^S - \boldsymbol{\beta})'\mathbf{W}\left(\hat{\boldsymbol{\beta}}_n^S - \boldsymbol{\beta}\right)\big]\\
&= \lim_{n\to\infty} En(\tilde{\boldsymbol{\beta}}_n - \boldsymbol{\beta})'\mathbf{W}(\tilde{\boldsymbol{\beta}}_n - \boldsymbol{\beta}) + k^2\lim E\big[n(\tilde{\boldsymbol{\beta}}_n - \hat{\boldsymbol{\beta}}_n)'\mathbf{W}(\tilde{\boldsymbol{\beta}}_n - \hat{\boldsymbol{\beta}}_n)\mathcal{L}_n^{-2}\big]\\
&\quad -2k\lim_{n\to\infty} E\big[n(\tilde{\boldsymbol{\beta}}_n - \boldsymbol{\beta})'\mathbf{W}(\tilde{\boldsymbol{\beta}}_n - \hat{\boldsymbol{\beta}}_n)\mathcal{L}_n^{-1}\big]\\
&= R_1(\tilde{\boldsymbol{\beta}}_n;\mathbf{W}) + k^2\lim_{n\to\infty} E\big[n(\tilde{\boldsymbol{\beta}}_n - \hat{\boldsymbol{\beta}}_n)'\mathbf{W}(\tilde{\boldsymbol{\beta}}_n - \hat{\boldsymbol{\beta}}_n)\mathcal{L}_n^{-2}\big]\\
&\quad -2k\lim_{n\to\infty} E\Big\{\big[n(\tilde{\boldsymbol{\beta}}_n - \hat{\boldsymbol{\beta}}_n)'\mathbf{W}(\tilde{\boldsymbol{\beta}}_n - \hat{\boldsymbol{\beta}}_n)\\
&\quad - (\mathbf{H}\boldsymbol{\beta} - \mathbf{h})'(\mathbf{HC}^{-1}\mathbf{H}')\mathbf{HC}^{-1}\mathbf{W}(\tilde{\boldsymbol{\beta}}_n - \hat{\boldsymbol{\beta}}_n)\big]\mathcal{L}_n^{-1}\Big\}\\
&= R_1(\tilde{\boldsymbol{\beta}}_n;\mathbf{W}) - 2k\sigma^2\operatorname{tr}\big[\mathbf{C}^{-1}\mathbf{H}'(\mathbf{HC}^{-1}\mathbf{H}')^{-1}\mathbf{HC}^{-1}\mathbf{W}\big]\\
&\quad - 2k\big[\boldsymbol{\xi}'(\mathbf{HC}^{-1}\mathbf{H}')^{-1}\mathbf{HC}^{-1}\mathbf{WC}^{-1}\mathbf{H}'(\mathbf{HC}^{-1}\mathbf{H}')^{-1}\boldsymbol{\xi}\big]E\big[\chi_{q+4}^{-2}(\Delta^2)\big]\\
&\quad + k^2\big[\boldsymbol{\xi}'(\mathbf{HC}^{-1}\mathbf{H}')^{-1}\mathbf{HC}^{-1}\mathbf{WC}^{-1}\mathbf{H}'(\mathbf{HC}^{-1}\mathbf{H}')^{-1}\boldsymbol{\xi}\big]E\big[\chi_{q+4}^{-2}(\Delta^2)\big]\\
&\quad + 2k\big[\boldsymbol{\xi}'(\mathbf{HC}^{-1}\mathbf{H}')^{-1}\mathbf{HC}^{-1}\mathbf{WC}^{-1}\mathbf{H}'(\mathbf{HC}^{-1}\mathbf{H}')^{-1}\boldsymbol{\xi}\big]E\big[\chi_{q+2}^{-2}(\Delta^2)\big]\\
&= R_1(\tilde{\boldsymbol{\beta}}_n;\mathbf{W}) - \sigma^2 k\operatorname{tr}(\mathbf{A}_{11})\{2E\big[\chi_{q+2}^{-2}(\Delta^2)\big] - kE\big[\chi_{q+2}^{-4}(\Delta^2)\big]\\
&\quad + k(k+4)(\boldsymbol{\eta}_1'\mathbf{A}_{11}\boldsymbol{\eta}_1)E\big[\chi_{q+4}^{-4}(\Delta^2)\big], \text{ by (7.8.22).} \qquad (7.8.23)
\end{aligned}
$$

Finally, we consider the risk expression for $\hat{\boldsymbol{\beta}}_n^{S+}$:

$$
\begin{aligned}
&R_5\left(\hat{\boldsymbol{\beta}}_n^{S+};\mathbf{W}\right)\\
&= \lim E\{n(\hat{\boldsymbol{\beta}}_n^{S+}-\boldsymbol{\beta})'\mathbf{W}(\hat{\boldsymbol{\beta}}_n^{S+}-\boldsymbol{\beta})\}\\
&= \lim E\Big\{n\big[(\hat{\boldsymbol{\beta}}^{S}-\boldsymbol{\beta})-(1-k_n\mathcal{L}_n^{-1})I(\mathcal{L}_n<k_n)(\tilde{\boldsymbol{\beta}}_n-\hat{\boldsymbol{\beta}}_n)\big]'\mathbf{W}\\
&\quad\times\big[(\hat{\boldsymbol{\beta}}_n^{S}-\boldsymbol{\beta})-(1-k_n\mathcal{L}_n^{-1})I(\mathcal{L}_n<k_n)(\tilde{\boldsymbol{\beta}}_n-\hat{\boldsymbol{\beta}}_n)\big]\Big\}\\
&= \lim_{n\to\infty} E\big[n(\hat{\boldsymbol{\beta}}_n^{S}-\boldsymbol{\beta})'\mathbf{W}(\hat{\boldsymbol{\beta}}^{S}-\boldsymbol{\beta})\big]\\
&\quad+\lim_{n\to\infty} E\big[(1-k_n\mathcal{L}_n^{-1})^2 I(\mathcal{L}_n<k_n)n(\tilde{\boldsymbol{\beta}}_n-\hat{\boldsymbol{\beta}}_n)'\mathbf{W}(\tilde{\boldsymbol{\beta}}_n-\hat{\boldsymbol{\beta}}_n)\big]\\
&\quad-2\lim_{n\to\infty} E\big[(1-k_n\mathcal{L}_n^{-1})I(\mathcal{L}_n<k_n)n(\hat{\boldsymbol{\beta}}_n^{S}-\boldsymbol{\beta})'\mathbf{W}(\tilde{\boldsymbol{\beta}}_n-\hat{\boldsymbol{\beta}}_n)\big]\\
&= R_4(\hat{\boldsymbol{\beta}}_n^{S};\mathbf{W})+\lim E\big[(1-k_n\mathcal{L}_n^{-1})^2 I(\mathcal{L}_n<k_n)n(\tilde{\boldsymbol{\beta}}_n-\hat{\boldsymbol{\beta}}_n)'\mathbf{W}(\tilde{\boldsymbol{\beta}}_n-\hat{\boldsymbol{\beta}}_n)\\
&\quad-2\lim_{n\to\infty} E\big[(1-k_n\mathcal{L}_n^{-1})I(\mathcal{L}_n<k_n)\{n(\tilde{\boldsymbol{\beta}}_n-\boldsymbol{\beta})\\
&\quad+(1-k_n\mathcal{L}_n^{-1})n(\tilde{\boldsymbol{\beta}}_n-\hat{\boldsymbol{\beta}}_n)-n(\tilde{\boldsymbol{\beta}}_n-\hat{\boldsymbol{\beta}}_n)\}\mathbf{W}(\tilde{\boldsymbol{\beta}}_n-\hat{\boldsymbol{\beta}}_n)\big]\\
&= R_4(\hat{\boldsymbol{\beta}}_n^{S};\mathbf{W})-\lim_{n\to\infty} E\Big\{(1-k_n\mathcal{L}_n^{-1})^2 I(\mathcal{L}_n<k)\big[n(\tilde{\boldsymbol{\beta}}_n-\hat{\boldsymbol{\beta}}_n)'\mathbf{W}(\tilde{\boldsymbol{\beta}}_n-\hat{\boldsymbol{\beta}}_n)\big]\Big\}\\
&\quad-2\lim_{n\to\infty} E\Big\{\big[n(\tilde{\boldsymbol{\beta}}_n-\hat{\boldsymbol{\beta}}_n)'\mathbf{W}(\tilde{\boldsymbol{\beta}}_n-\hat{\boldsymbol{\beta}}_n)\big](1-k_n\mathcal{L}_n^{-1})I(\mathcal{L}_n<k_n)\Big\}\\
&\quad+2\lim_{n\to\infty} E\Big\{\big[n(\tilde{\boldsymbol{\beta}}_n-\hat{\boldsymbol{\beta}}_n)'\mathbf{W}(\tilde{\boldsymbol{\beta}}_n-\hat{\boldsymbol{\beta}}_n)\big](1-k_n\mathcal{L}_n^{-1})I(\mathcal{L}_n<k_n)\Big\}\\
&= R_4(\hat{\boldsymbol{\beta}}_n^{S};\mathbf{W})-\sigma^2\operatorname{tr}(\mathbf{A}_{11})E\big[(1-k\chi_{q+2}^{-2}(\Delta^2))^2 I(\chi_{q+2}^2(\Delta^2)<k)\big]\\
&\quad+(\boldsymbol{\eta}_1'\mathbf{A}_{11}\boldsymbol{\eta}_1)\Big\{2E\big[(1-k\chi_{q+2}^{-2}(\Delta^2))I(\chi_{q+2}^2(\Delta^2)<k)\big]\\
&\quad-(\boldsymbol{\eta}_1'\mathbf{A}_{11}\boldsymbol{\eta}_1)E\big[(1-k\chi 1_{q+4}^{-2}(\Delta^2))^2 I(\chi_{q+4}^2(\Delta^2)<k)\big]\Big\}.
\end{aligned}
\tag{7.8.24}
$$

7.8.3 ADQR Analysis

First, compare $\hat{\boldsymbol{\beta}}_n$ and $\tilde{\boldsymbol{\beta}}_n$. The asymptotic risk efficiency of $\hat{\boldsymbol{\beta}}_n$ relative to $\tilde{\boldsymbol{\beta}}_n$ is given by

$$
\operatorname{ARRE}(\hat{\boldsymbol{\beta}}_n;\tilde{\boldsymbol{\beta}}_n)=\left[1-\frac{\operatorname{tr}(\mathbf{A}_{11})}{\operatorname{tr}(\mathbf{WC}^{-1})}+\frac{\boldsymbol{\eta}_1\mathbf{A}_{11}\boldsymbol{\eta}_1}{\operatorname{tr}(\mathbf{WC}^{-1})}\right]^{-1}. \tag{7.8.25}
$$

Now,

$$
\operatorname{Ch}_{min}(\mathbf{A}_{11})\Delta^2\le\frac{(\boldsymbol{\eta}_1'\mathbf{A}_{11}\boldsymbol{\eta}_1)}{\operatorname{tr}(\mathbf{WC}^{-1})}\le\operatorname{Ch}_{max}(\mathbf{A}_{11})\Delta^2;
$$

then

$$\left[1 - \frac{\mathrm{tr}(\mathbf{A}_{11})}{\mathrm{tr}(\mathbf{WC}^{-1})} + \mathrm{ch}_{max}(\mathbf{A}_{11})\Delta^2\right]^{-1}$$

$$\leq \mathrm{ARRE}(\hat{\boldsymbol{\beta}}_n; \tilde{\boldsymbol{\beta}}_n) \leq \left[1 - \frac{\mathrm{tr}(\mathbf{A}_{11})}{\mathrm{tr}(\mathbf{WC}^{-1})} + \mathrm{ch}_{max}(\mathbf{A}_{11})\Delta^2\right]^{-1}.$$

Under H_0, $\mathrm{ARRE}(\hat{\boldsymbol{\beta}}_n; \tilde{\boldsymbol{\beta}}_n)$ equals

$$\left[1 - \frac{\mathrm{tr}(\mathbf{A}_{11})}{\mathrm{tr}(\mathbf{WC}^{-1})}\right]^{-1} \geq 1.$$

However,

$$\mathrm{ARRE}(\hat{\boldsymbol{\beta}}_n; \tilde{\boldsymbol{\beta}}_n) \gtreqless 1$$

whenever

$$\boldsymbol{\eta}_1' \mathbf{A}_{11} \boldsymbol{\eta}_1 \gtreqless \mathrm{tr}(\mathbf{A}_{11}). \tag{7.8.26}$$

Next, the $\mathrm{ARRE}(\hat{\boldsymbol{\beta}}_n^{\mathrm{PT}}; \tilde{\boldsymbol{\beta}}_n)$ is given by

$$\mathrm{ARRE}(\hat{\boldsymbol{\beta}}_n^{\mathrm{PT}}; \tilde{\boldsymbol{\beta}}_n) = R_1(\tilde{\boldsymbol{\beta}}_n; \mathbf{W})\left[R_3(\hat{\boldsymbol{\beta}}_n^{\mathrm{PT}}; \mathbf{W})\right]^{-1} = [1 + g(\Delta^2)]^{-1}, \tag{7.8.27}$$

where

$$g(\Delta^2) = \sigma^2 \frac{(\boldsymbol{\eta}_1' \mathbf{A}_{11} \boldsymbol{\eta}_1)}{\mathrm{tr}(\mathbf{WC}^{-1})} \{2H_{q+2}(\chi_q^2(\alpha); \Delta^2) - H_{q+4}(\chi_q^2(\alpha); \Delta^2)\}$$
$$- \frac{\mathrm{tr}(\mathbf{A}_{11})}{\mathrm{tr}(\mathbf{WC}^{-1})} H_{q+2}(\chi_q^2(\alpha); \Delta^2). \tag{7.8.28}$$

Define

$$g_{\max}(\Delta^2) = \sigma^2 \mathrm{Ch}_{\max}(\mathbf{A}_{11})\Delta^2\{2H_{q+2}(\chi_q^2(\alpha); \Delta^2) - H_{q+4}(\chi_q^2(\alpha); \Delta^2)\}$$
$$- \frac{\mathrm{tr}(\mathbf{A}_{11})}{\mathrm{tr}(\mathbf{WC}^{-1})} H_{q+2}(\chi_q^2(\alpha); \Delta^2), \tag{7.8.29a}$$

$$g_{\min}(\Delta^2) = \sigma^2 \mathrm{Ch}_{\min}(\mathbf{A}_{11})\Delta^2\{2H_{q+2}(\chi_q^2(\alpha); \Delta^2) - H_{q+4}(\chi_q^2(\alpha); \Delta^2)\}$$
$$- \frac{\mathrm{tr}(\mathbf{A}_{11})}{\mathrm{tr}(\mathbf{WC}^{-1})} H_{q+2}(\chi_q^2(\alpha); \Delta^2). \tag{7.8.29b}$$

Then

$$[1 + g_{\max}(\Delta^2)]^{-1} \leq \mathrm{ARRE}(\hat{\boldsymbol{\beta}}_n^{\mathrm{PT}}; \tilde{\boldsymbol{\beta}}_n) \leq [1 + g_{\min}(\Delta^2)]^{-1}. \tag{7.8.30}$$

Under H_0, we have

$$\left[1 - \frac{\text{tr}(\mathbf{A}_{11})}{\text{tr}(\mathbf{WC}^{-1})}\right]^{-1} \geq 1,$$

and under non null situations, we have

$$(\boldsymbol{\eta}_1'\mathbf{A}_{11}\boldsymbol{\eta}_1)\sigma^{-2} \underset{<}{\geq} \frac{\text{tr}(\mathbf{A}_{11})H_{q+2}(\chi_q^2(\alpha);\Delta^2)}{\text{tr}(\mathbf{WC}^{-1})\{2H_{q+2}(\chi_q^2(\alpha);\Delta^2) - H_{q+4}(\chi_q^2(\alpha);\Delta^2)\}}. \tag{7.8.31}$$

Since PTE, $\hat{\boldsymbol{\beta}}_n^{\text{PT}}$ is not dominating $\tilde{\boldsymbol{\beta}}_n$ uniformly, we can select a PTE such that it has at least a pre-specified asymptotic efficiency, say, E_0, by selecting an optimum level of significance, α^{opt}. This is achieved by solving the equation

$$\min_{\Delta^2} E(\alpha,\Delta^2) = E(\alpha,\Delta^2_{\text{min}}) = E_0, \tag{7.8.32}$$

where $E(\alpha,\Delta^2) = [1 + g(\Delta^2)]^{-1}$. The resulting $(\alpha^{\text{opt}},\Delta^2)$ provide the minimum guaranteed ARRE for the application of PTE.

Now, compare $\hat{\boldsymbol{\beta}}_n^{S}$ and $\tilde{\boldsymbol{\beta}}_n$. We can rewrite $R_4(\hat{\boldsymbol{\beta}}_n^{S};\mathbf{W})$ as

$$\begin{aligned} R_4(\hat{\boldsymbol{\beta}}_n^{S};\mathbf{W}) &= R_1(\tilde{\boldsymbol{\beta}}_n;\mathbf{W}) - \sigma^2\text{tr}(\mathbf{A}_{11})\Big\{(q-2)E\big[\chi_{q+2}^{-2}(\Delta^2)\big] \\ &\quad + \Big[1 - \frac{(\boldsymbol{\eta}_1'\mathbf{A}_{11}\boldsymbol{\eta}_1)(q+2)}{2\Delta^2\,\text{tr}(\mathbf{A}_{11})}\Big]2\Delta^2 E\big[\chi_{q+4}^{-4}\Delta^2)\big]\Big\} \\ &\leq R_1(\tilde{\boldsymbol{\beta}}_n;\mathbf{W}) \text{ for all } (\Delta^2;\mathbf{W}), \text{ and } q \geq 3, \end{aligned}$$

such that

$$\frac{\text{Ch}_{\max}(\mathbf{A}_{11})}{\text{tr}(\mathbf{A}_{11})} \geq \frac{q+2}{2}. \tag{7.8.33}$$

The risk-difference is

$$\begin{aligned} &R_4(\hat{\boldsymbol{\beta}}_n^{S};\mathbf{W}) - R_5(\hat{\boldsymbol{\beta}}_n^{S+};\mathbf{W}) \\ &= \sigma^2\Big\{\text{tr}(\mathbf{A}_{11})E\big[(1 - k\chi_{q+2}^{-2}(\Delta^2))^2 I(\chi_{q+2}^2(\Delta^2) < k)\big] \\ &\quad + \sigma^{-2}(\boldsymbol{\eta}_1'\mathbf{A}_{11}\boldsymbol{\eta}_1)E\big[(1 - k\chi_{q+4}^{-2}(\Delta^2))^2\mathbf{I}(\chi_{q+4}^2(\Delta^2) < k)\big]\Big\} \\ &\quad - 2(\boldsymbol{\eta}_1'\mathbf{A}_{11}\boldsymbol{\eta}_1)E\big[(1 - k\chi_{q+2}^{-2}(\Delta^2))\mathbf{I}(\chi_{q+2}^2(\Delta^2) < k)\big] \geq 0, \end{aligned} \tag{7.8.34}$$

since $1 \leq k\chi_{q+2}^{-2}(\Delta^2)$ and $1 - k\chi_{q+2}^{-2}(\Delta^2) \leq 0$. Hence, for all $(\Delta^2,\mathbf{W})$ and $q \geq 3$,

$$R_5(\hat{\boldsymbol{\beta}}_n^{S+};\mathbf{W}) \leq R_4(\hat{\boldsymbol{\beta}}_n^{S};\mathbf{W}) \leq R_1(\tilde{\boldsymbol{\beta}}_n;\mathbf{W}) \tag{7.8.35}$$

holds.

Now, consider $R_4(\hat{\beta}_n^S; \mathbf{W})$ and $R_3(\hat{\beta}_n^{\text{PT}}; \mathbf{W})$. The ARRE $(\hat{\beta}_n; \tilde{\beta}_n)$ and ARRE $(\hat{\beta}_n^{\text{PT}}; \tilde{\beta}_n)$ both tend to 1 as Δ^2, for fixed α, while for a given α, they intersect at say, Δ_α^2, where Δ_α^2 is the solution of the equation

$$2E\left[\chi_{q+2}^{-2}(\Delta^2)\right] - kE\left[\chi_{q+2}^{-4}(\Delta^2)\right] - k(k+4)\frac{(\eta_1'\mathbf{A}_{11}\eta_1)}{\sigma^2 \operatorname{tr}(\mathbf{A}_{11})}E\left[\chi_{q+4}^{-4}(\Delta^2)\right]$$

$$= H_{q+2}\left(\chi_q^2(\alpha); \Delta^2\right) - \frac{\eta_1'\mathbf{A}_{11}\eta_1}{\sigma^2 \operatorname{tr}(\mathbf{A}_{11})}\left\{2H_{q+2}(\chi_q^2(\alpha); \Delta^2) - H_{q+4}(\chi_q^2(\alpha); \Delta^2)\right\}.$$

(7.8.36)

Note that for some α, there may not be any intersection. In this case, $\hat{\beta}_n^S$ is superior to $\hat{\beta}_n^{\text{PT}}$. When there is an intersection,

(i) $\hat{\beta}_n^{\text{PT}}$ is superior to $\hat{\beta}_n^S$ on $[0, \Delta_\alpha^2]$,

(ii) $\hat{\beta}_n^S$ is superior to $\hat{\beta}_n^{\text{PT}}$ on $(\Delta_\alpha^2, \infty)$,

(iii) $\hat{\beta}_n^{\text{PT}}$ and $\hat{\beta}_n^S$ are equally risk-efficient if $\Delta^2 = \Delta_\alpha^2$.

Clearly, neither $\hat{\beta}_n^{\text{PT}}$ nor $\hat{\beta}_n^S$ dominates each other. Finally, under H_0,

$$R_2(\hat{\beta}_n; \mathbf{W}) \le R_3(\hat{\beta}_n^{\text{PT}}; \mathbf{W}) \le R_4(\hat{\beta}_n^S; \mathbf{W}) \le R_1(\tilde{\beta}_n; \mathbf{W}). \qquad (7.8.37)$$

We conclude that if $q \ge 3$, then on the basis of ARRE ordering, namely

$$R_5(\hat{\beta}_n^{S+}; \mathbf{W}) \le R_4(\hat{\beta}_n^S; \mathbf{W}) \le R_1(\tilde{\beta}_n; \mathbf{W}).$$

We should use $\hat{\beta}_n^{S+}$, while for $q < 3$, we lean toward $\hat{\beta}_n^{\text{PT}}$ unless there is clear indication that H_0 is true to use $\hat{\beta}_n$, the restricted estimator.

Similarly, the ADMSE analysis may be carried out following Section 7.5.

7.9 Confidence Set Estimation

7.9.1 Preliminaries

In this section, we consider the confidence set estimation of the regression parameters β in the model

$$\mathbf{Y} = \mathbf{X}\beta + \varepsilon \qquad (7.9.1)$$

when it is suspected that β may belong to the subspace defined by $\mathbf{H}\beta = \mathbf{h}$, as before, where $\varepsilon \sim N_n(\mathbf{0}, \sigma^2\mathbf{I}_n)$ and σ^2 is known. As in Sections 7.1.1 and 7.2.1, we consider the five estimators:

(i) $\tilde{\beta}_n = (\mathbf{X}'\mathbf{X})^{-1}\mathbf{X}'\mathbf{Y}$.

(ii) $\hat{\boldsymbol{\beta}}_n = \tilde{\boldsymbol{\beta}}_n - \mathbf{C}^{-1}\mathbf{H}'(\mathbf{H}\mathbf{C}^{-1}\mathbf{H}')^{-1}(\mathbf{H}\tilde{\boldsymbol{\beta}}_n - \mathbf{h})$.

(iii) $\hat{\boldsymbol{\beta}}_n^{\mathrm{PT}} = \tilde{\boldsymbol{\beta}}_n - (\tilde{\boldsymbol{\beta}}_n - \hat{\boldsymbol{\beta}}_n)\mathbf{I}(\mathcal{L}_n < \chi_q^2(\alpha))$, (7.9.2)

(iv) $\hat{\boldsymbol{\beta}}_n^{S} = \tilde{\boldsymbol{\beta}}_n - d(\tilde{\boldsymbol{\beta}}_n - \hat{\boldsymbol{\beta}}_n)\mathcal{L}_n^{-1}$, where $d = q-2$.

(v) $\hat{\boldsymbol{\beta}}_n^{S+} = \hat{\boldsymbol{\beta}}_n^{S} - (1 - d\mathcal{L}_n^{-1})\mathbf{I}(\mathcal{L}_n < d)(\tilde{\boldsymbol{\beta}}_n - \hat{\boldsymbol{\beta}}_n)$.

where $\mathcal{L}_n$ is the test-statistic for testing $H_0 : \mathbf{H}\boldsymbol{\beta} = \mathbf{h}$, defined by

$$\mathcal{L}_n = \sigma^{-2}(\tilde{\boldsymbol{\beta}}_n - \hat{\boldsymbol{\beta}}_n)'(\mathbf{H}\mathbf{C}^{-1}\mathbf{H}')^{-1}(\tilde{\boldsymbol{\beta}}_n - \hat{\boldsymbol{\beta}}_n). \tag{7.9.3}$$

Since $\tilde{\boldsymbol{\beta}}_n \sim N_p(\boldsymbol{\beta}, \sigma^2\mathbf{C}^{-1})$, $\mathbf{C} = (\mathbf{X}'\mathbf{X})$, and the $(1-\gamma)\%$ confidence set $C_{\boldsymbol{\beta}}^0(\tilde{\boldsymbol{\beta}})$ is defined by

$$C_{\boldsymbol{\beta}}^0(\tilde{\boldsymbol{\beta}}_n) = \left\{\boldsymbol{\beta} : \sigma^{-2}\|\boldsymbol{\beta} - \tilde{\boldsymbol{\beta}}_n\|_{\mathbf{C}}^2 \le \chi_p^2(\gamma)\right\}, \tag{7.9.4}$$

where $\chi_p^2(\gamma)$ is chosen such that $P(\chi_p^2 \le \chi_p^2(\gamma)) = (1-\gamma)$. Thus, $C_{\boldsymbol{\beta}}^0(\tilde{\boldsymbol{\beta}}_n)$ has the coverage probability $1-\gamma$. The set $C_{\boldsymbol{\beta}}^0(\tilde{\boldsymbol{\beta}}_n)$ is minimax in the sense that among all the confidence sets, $C_{\boldsymbol{\beta}}^0(\tilde{\boldsymbol{\beta}}_n)$ has coverage probability with least $1-\gamma$. Also, $C_{\boldsymbol{\beta}}^0(\tilde{\boldsymbol{\beta}}_n)$ minimizes the maximum volume. As in Section 5.9, we notice that there is scope to improve on $C_{\boldsymbol{\beta}}^0(\tilde{\boldsymbol{\beta}}_n)$ for $q \ge 3$, in the sense that (i) $P_{\boldsymbol{\beta}}\{\boldsymbol{\beta} \in C_{\boldsymbol{\beta}}^*(\boldsymbol{\beta}_n^*)\} \ge P_{\boldsymbol{\beta}}\{\boldsymbol{\beta} \in C_{\boldsymbol{\beta}}^0(\tilde{\boldsymbol{\beta}}_n)\}$ and (ii) the volume of $C_{\boldsymbol{\beta}}^*(\boldsymbol{\beta}_n^*) \le$ volume of $C_{\boldsymbol{\beta}}^0(\tilde{\boldsymbol{\beta}}_n)$, where $\boldsymbol{\beta}_n^*$ stands for the generic notation for any of the five estimators.

In this section, we consider the confidence sets that are centered at the estimators of the form

$$\boldsymbol{\beta}_n^* = \hat{\boldsymbol{\beta}}_n + (\tilde{\boldsymbol{\beta}}_n - \hat{\boldsymbol{\beta}}_n)g(\mathcal{L}_n), \tag{7.9.5}$$

where $g(\mathcal{L}_n)$ is a nondecreasing function of the test-statistic $\mathcal{L}_n$. In particular, we consider the various functions as in Chapter 5, Section 5.9. Accordingly, we have

(i) $\tilde{\boldsymbol{\beta}}_n^* = \tilde{\boldsymbol{\beta}}_n$ if $g(\mathcal{L}_n) = 1$,

(ii) $\boldsymbol{\beta}_n^* = \hat{\boldsymbol{\beta}}_n$ if $g(\mathcal{L}_n) = 0$,

(iii) $\boldsymbol{\beta}_n^* = \hat{\boldsymbol{\beta}}_n^{\mathrm{PT}}$ if $g(\mathcal{L}_n) = I(\mathcal{L}_n \ge \chi_q^2(\alpha))$, where $\chi_q^2(\alpha)$ is the α-level critical value under H_0, (7.9.6)

(iv) $\boldsymbol{\beta}_n^* = \hat{\boldsymbol{\beta}}_n^{S}$ if $g(\mathcal{L}_n) = 1 - d\mathcal{L}_n^{-1}$,

(v) $\boldsymbol{\beta}_n^* = \hat{\boldsymbol{\beta}}_n^{S+}$ if $g(\mathcal{L}_n) = (1 - d\mathcal{L}_n^{-1})I(\mathcal{L}_n > d)$.

In the next section we consider the five confidence sets with centers $\tilde{\boldsymbol{\beta}}_n$, $\hat{\boldsymbol{\beta}}_n$, $\hat{\boldsymbol{\beta}}_n^{\mathrm{PT}}$, $\hat{\boldsymbol{\beta}}_n^{S}$, and $\hat{\boldsymbol{\beta}}_n^{S+}$, respectively.

7.9.2 Confidence Sets and the Coverage Probabilities

We consider the confidence sets on the β-section of the form

$$C^*_\beta(\beta^*_n) = \{\beta : \sigma^{-2}\|\beta - \beta^*_n\|^2 \leq \chi^2_p(\gamma)\}. \tag{7.9.7}$$

Thus, we arrive at the following confidence sets:

$$\begin{aligned}
&\text{(i) } C^0(\tilde{\beta}_n) = \{\beta : \sigma^{-2}\|\beta - \tilde{\beta}_n\|^2_{\mathbf{C}} \leq \chi^2_p(\gamma)\}.\\
&\text{(ii) } C^R(\hat{\beta}_n) = \{\beta : \sigma^{-2}\|\beta - \hat{\beta}_n\|^2_{\mathbf{C}} \leq \chi^2_p(\gamma)\}.\\
&\text{(iii) } C^{\text{PT}}(\hat{\beta}^{\text{PT}}_n) = \{\beta : \sigma^{-2}\|\beta - \hat{\beta}^{\text{PT}}_n\|^2_{\mathbf{C}} \leq \chi^2_p(\gamma)\}.\\
&\text{(iv) } C^S(\hat{\beta}^S_n) = \{\beta : \sigma^{-2}\|\beta - \hat{\beta}^S_n\|^2_{\mathbf{C}} \leq \chi^2_p(\gamma)\},\\
&\text{(v) } C^{S+}(\hat{\beta}^{S+}_n) = \{\beta : \sigma^{-2}\|\beta - \hat{\beta}^{S+}_n\|^2_{\mathbf{C}} \leq \chi^2_p(\gamma)\}.
\end{aligned} \tag{7.9.8}$$

Now, consider the orthogonal matrix $\mathbf{\Gamma}$ that diagonalizes the idempotent matrix $\mathbf{C}^{-1/2}\mathbf{H}(\mathbf{H}\mathbf{C}^{-1}\mathbf{H}')^{-1}\mathbf{C}^{-1/2}$ with rank q, meaning $\mathbf{\Gamma}\mathbf{C}^{-1/2}$ $\times\mathbf{H}(\mathbf{H}\mathbf{C}^{-1}\mathbf{H}')^{-1}\mathbf{C}^{-1/2}\mathbf{\Gamma}' = \begin{pmatrix} \mathbf{I}_q & 0 \\ 0 & 0 \end{pmatrix}$, and define the variable $\boldsymbol{\omega}$ and $\boldsymbol{\eta}$ as

$$\boldsymbol{\omega} = \sigma^{-1}\mathbf{\Gamma}_j\mathbf{C}^{1/2}\tilde{\beta}_n - \sigma^{-1}\mathbf{\Gamma}_j\mathbf{C}^{-1/2}\mathbf{H}'(\mathbf{H}\mathbf{C}^{-1}\mathbf{H}')^{-1}\mathbf{h}, \tag{7.9.9}$$

and

$$\boldsymbol{\eta} = \sigma^{-1}\mathbf{\Gamma}_j\mathbf{C}^{1/2}\beta - \sigma^{-1}\mathbf{\Gamma}_j\mathbf{C}^{-1/2}\mathbf{H}'(\mathbf{H}\mathbf{C}^{-1}\mathbf{H}')^{-1}\mathbf{h}. \tag{7.9.10}$$

Then, partitioning $\boldsymbol{\omega}$ and $\boldsymbol{\eta}$ as $(\boldsymbol{\omega}_1', \boldsymbol{\omega}_2')'$ and $(\boldsymbol{\eta}_1', \boldsymbol{\eta}_2')'$, we have

$$\begin{pmatrix} \boldsymbol{\omega}_1 \\ \boldsymbol{\omega}_2 \end{pmatrix} \sim N_p\left\{\begin{pmatrix} \boldsymbol{\eta}_1 \\ \boldsymbol{\eta}_2 \end{pmatrix}; \begin{pmatrix} \mathbf{I}_q & \mathbf{0} \\ \mathbf{0} & \mathbf{I}_{p-q} \end{pmatrix}\right\}. \tag{7.9.11}$$

where $\boldsymbol{\omega}_1$ is a q-vector and $\boldsymbol{\omega}_2$ is a $(p-q)$-vector respectively. Hence, we can show that

$$\sigma^{-2}\|\beta - \beta^*_n\|^2_{\mathbf{C}} = \|\boldsymbol{\eta}_1 - \boldsymbol{\omega}_1 g(\|\boldsymbol{\omega}_1\|^2)\|^2 + \|\boldsymbol{\eta}_2 - \boldsymbol{\omega}_2\|^2. \tag{7.9.12}$$

(See Problem 12 of Section 7.12). We can write the coverage probability of the set $C^*(\beta^*_n)$ as

$$\begin{aligned}
&P_\beta\{\sigma^{-2}\|\beta - \beta^*_n\|^2_{\mathbf{C}} \leq \chi^2_p(\gamma)\}\\
&= P_{\boldsymbol{\eta}}\{\|\boldsymbol{\eta}_1 - \boldsymbol{\omega}_1 g(\|\boldsymbol{\omega}_1\|^2)\|^2 + \|\boldsymbol{\eta}_2 - \boldsymbol{\omega}_2\|^2 \leq \chi^2_p(\gamma)\}\\
&= \int_0^{\chi^2_p(\gamma)} P_{\boldsymbol{\eta}_1}\{\|\boldsymbol{\eta}_1 - \boldsymbol{\omega}_1 g(\|\boldsymbol{\omega}_1\|^2)\|^2 \leq (\chi^2_p(\gamma) - t)^+\}dH_{p-q}(t, 0),
\end{aligned} \tag{7.9.13}$$

where $H_{p-q}(t;0)$ is the cdf of a central chi-square distribution with $(p-q)$ d.f. This allows us to write the coverage probabilities of the five confidence sets as given below:

If $g(\mathcal{L}_n) = 1$, then (7.9.14)

$$P_\beta\{\sigma^{-2}\|\beta - \tilde{\beta}_n\|_C^2 \le \chi_p^2(\gamma)\} = P_\eta\{\|\eta_1 - \omega_1\|^2 + \|\eta_2 - \omega_2\|^2 \le \chi_p^2(\gamma)\}$$
$$= P_0\{\chi_p^2 \le \chi_p^2(\gamma)\} = 1 - \gamma.$$

If $q(\mathcal{L}_n) = 0$, then

$$P_\beta\{\sigma^{-2}\|\beta - \hat{\beta}_n\|^2 \le \chi_p^2(\gamma)\} = P_\eta\{\|\eta_1\|^2 + \|\eta_2 - \omega_2\|^2 \le \chi_p^2(\gamma)\}$$
$$= H_{p-q}(\chi_p^2(\gamma) - \Delta^2; 0). \tag{7.9.15}$$

If $g(\mathcal{L}_n) = I(\mathcal{L}_n \ge \chi_q^2(\alpha))$, then

$$P_\beta\{\sigma^{-2}\|\beta - \hat{\beta}_n^{\text{PT}}\|_C^2 \le \chi_p^2(\gamma)\}$$
$$= P_\eta\{\|\eta_1 - \omega_1 I(\|\omega_1\|^2 \ge \chi_q^2(\alpha))\|^2 + \|\eta_2 - \omega_2\|^2 \le \chi_p^2(\gamma)\}$$
$$= P_\eta\{\|\eta_2 - \omega_2\|^2 + \|\eta_1\|^2 \le \chi_p^2(\gamma); \|\omega_1\|^2 \le \chi_q^2(\alpha))\}$$
$$+ P_\eta\{\|\eta_1 - \omega_1\|^2 + \|\eta_2 - \omega_2\|^2 < \chi_p^2(\gamma); \|\omega_1\|^2 > \chi_q^2(\alpha)\}$$
$$= H_q(\chi_q^2(\alpha); \Delta^2) H_{p-q}(\chi_p^2(\gamma) - \Delta^2; 0)$$
$$+ \int_0^{\chi_p^2(\gamma)} P_{\eta_1}\{\|\eta_1 - \omega_1)\|^2 \le (\chi_p^2(\gamma) - t)^+; \|\omega_1\|^2 > \chi_q^2(\alpha)\} dH_{p-q}(t;0). \tag{7.9.16}$$

If $g(\mathcal{L}_n) = 1 - d\mathcal{L}_n^{-1}$, $0 < d < 2(q-2)m/q(m+2)$, then

$$P_\beta\{\sigma^{-2}\|\beta - \hat{\beta}_n^S\|_C^2 \le \chi_p^2(\gamma)\}$$
$$= P_\eta\{\|\eta_1 - \omega_1(1 - d\|\omega_1\|^{-2})\|^2 + \|\eta_2 - \omega_2\|^2 \le \chi_p^2(\gamma)\}$$
$$= \int_0^{\chi_p^2(\gamma)} P_{\eta_1}\{\|\eta_1 - \omega_1(1 - d\|\omega_1\|^{-2})\|^2 \le (\chi_p^2(\gamma) - t)^+\} dH_{p-q}(t,0). \tag{7.9.17}$$

If $g(\mathcal{L}_n) = (1 - d\mathcal{L}_n^{-1})I(\mathcal{L}_n > d)$, then

$$\begin{aligned}
&P_\beta\{\sigma^{-2}\|\beta - \hat{\beta}_n^{S+}\|_C^2 \le \chi_p^2(\gamma)\} \\
&= P_\beta\{\|\eta_1 - \omega_1(1 - d\mathcal{L}_n^{-1})I(\|\omega_1\|^2 \ge d)\|^2 + \|\eta_2 - \omega_2\|^2 \le \chi_p^2(\gamma)\} \\
&= P_\eta\{\|\eta_1\|^2 + \|\eta_2 - \omega_2\|^2 \le \chi_p^2(\gamma); \|\omega_1\|^2 < d\} \\
&\quad + P_{\eta_1}\{\|\eta_1 - \omega_1(1 - d\|\omega_1\|^{-2})\|^2 + \|\eta_2 - \omega_2\|^2 \le \chi_p^2(\gamma); \|\omega_1\|^2 > d\} \\
&= H_{p-q}(\chi_p^2(\gamma) - \Delta^2; 0)H_q(d; \Delta^2) \\
&\quad + \int_0^{\chi_p^2(\gamma)} P_{\eta_1}\{\|\eta_1 - \omega_1(1 - d\|\omega_1\|^{-2}\|^2 \le (\chi_p^2(\gamma) - t)^+; \|\omega_1\|^2 > d\} \\
&\qquad \times dH_{p-q}(t; 0). \qquad (7.9.18)
\end{aligned}$$

7.9.3 Analysis of the Coverage Probabilities

$C^R(\hat{\beta}_n)$: Note that $P_0\{\sigma^{-2}\|\beta - \tilde{\beta}_n\|_C^2 \le \chi_p^2(\gamma)\} = 1 - \gamma$, which is independent of Δ^2. Thus, the coverage probability of $C_\beta^0(\tilde{\beta}_n)$ is constant as a function of Δ^2. Next, consider the coverage probability of $C^R(\hat{\beta}_n)$, which is given by

$$\begin{aligned}
&P_{\eta_1}\{\sigma^{-2}\|\beta - \hat{\beta}\|_C^2 \le \chi_p^2(\gamma)\} \\
&= P_{\eta_1}\{\|\eta_1\|^2 + \|\eta_2 - \omega_2\|^2 \le \chi_p^2(\gamma)\} \\
&= H_{p-q}(\chi_p^2(\gamma) - \Delta^2; 0). \qquad (7.9.19)
\end{aligned}$$

Now, $H_{p-q}(\chi_p^2(\gamma) - \Delta^2; 0)$ is a decreasing function of Δ^2. At $\Delta^2 = 0$, it attains the maximum value $H_{p-q}(\chi_p^2(\gamma); 0)$, and it tends to zero as $\Delta^2 \to \chi_p^2(\gamma)$. The coverage probability $P_0(C^0(\tilde{\beta}_n))$ is equal to $H_{p-q}(\chi_p^2(\gamma) - \Delta^2; 0)$ whenever

$$\Delta^2 = \chi_p^2(\gamma) - H_{p-q}^{-1}(1 - \gamma). \qquad (7.9.20)$$

$C^{PT}(\hat{\beta}_n^{PT})$: The coverage probability of this set is

$$\begin{aligned}
&P_\beta\{C^{PT}(\hat{\beta}_n^{PT})\} \\
&= H_{p-q}(\chi_p^2(\gamma) - \Delta^2; 0)H_q(\chi_q^2(\alpha); \Delta^2) \\
&\quad + \int_0^{\chi_p^2(\gamma)} P_{\eta_1}\{\|\eta_1 - \omega_1\|^2 \le (\chi_p^2(\gamma) - t)^+; \|\omega_1\|^2 > \chi_q^2(\alpha)\} \\
&\quad \times dH_{p-q}(t, 0). \qquad (7.9.21)
\end{aligned}$$

Then we have the following theorem:

Theorem.

(i) If $0 \le \Delta^2 < \chi_p^2(\gamma)$, then $P_{\boldsymbol{\beta}}\big(C^{\text{PT}}(\hat{\boldsymbol{\beta}}_n^{\text{PT}})\big) \ge 1-\gamma$. (7.9.22)

(ii) If $\chi_p^2(\gamma) < \Delta^2 < (\chi_p(\gamma)+\chi_q(\alpha))^2$, then $P_{\boldsymbol{\beta}}\big(C^{\text{PT}}(\hat{\boldsymbol{\beta}}_n^{\text{PT}})\big) < 1-\gamma$.

(iii) If $\Delta^2 \ge (\chi_p(\gamma)+\chi_q(\alpha))^2$, then $P_{\boldsymbol{\beta}}\big(C^{\text{PT}}(\hat{\boldsymbol{\beta}}_n^{\text{PT}})\big) = 1-\gamma$.

Proof. (i) If $\Delta^2 < \chi_p^2(\gamma)$, then

$$
\begin{aligned}
&P_{\boldsymbol{\beta}}\big(C_{\boldsymbol{\beta}}^{\text{PT}}(\hat{\boldsymbol{\beta}}_n^{\text{PT}})\big)\\
&= P_{\boldsymbol{\beta}}\{\|\boldsymbol{\omega}_1\|^2 < \chi_q^2(\alpha); \|\boldsymbol{\eta}_2-\boldsymbol{\omega}_2\|^2+\Delta^2 \le \chi_p^2(\gamma)\}\\
&\quad + \int_0^{\chi_p^2(\gamma)} P_{\boldsymbol{\eta}_1}\{\|\boldsymbol{\omega}_1\|^2 > \chi_q^2(\alpha); \|\boldsymbol{\eta}_1-\boldsymbol{\omega}_1\|^2 \le (\chi_p^2(\gamma)-t)^+\}dH_{p-q}(t,0)\\
&\ge \int_0^{\chi_p^2(\gamma)} P_{\boldsymbol{\eta}_1}\{\|\boldsymbol{\omega}_1\|^2 < \chi_q^2(\alpha); \|\boldsymbol{\eta}_1-\boldsymbol{\omega}_1\|^2 \le (\chi_p^2(\gamma)-t)^+\}dH_{p-q}(t,0)\\
&\quad + \int_0^{\chi_p^2(\gamma)} P_{\boldsymbol{\eta}_1}\{\|\boldsymbol{\omega}_1\|^2 > \chi_q^2(\alpha); \|\boldsymbol{\eta}_1-\boldsymbol{\omega}_1\|^2 \le (\chi_p^2(\gamma)-t)^+\}dH_{p-q}(t,0)\\
&= P_{\boldsymbol{\eta}_1}\{\|\boldsymbol{\eta}_1-\boldsymbol{\omega}_1\|^2+\|\boldsymbol{\eta}_2-\boldsymbol{\omega}_2\|^2 < \chi_p^2(\gamma)\} = 1-\gamma \qquad (7.9.23)
\end{aligned}
$$

(ii) If $\chi_p^2(\gamma) < \Delta^2 < (\chi_p(\gamma)+\chi_q(\alpha))^2$, then

$$
\begin{aligned}
&P_{\boldsymbol{\beta}}\big(C^{\text{PT}}(\hat{\boldsymbol{\beta}}_n^{\text{PT}})\big)\\
&= \int_0^{\chi_p^2(\gamma)} P_{\boldsymbol{\eta}_1}\{\|\boldsymbol{\omega}_1\|^2 > \chi_q^2(\alpha); \|\boldsymbol{\eta}_1-\boldsymbol{\omega}_1\|^2 \le (\chi_p^2(\gamma)-t)^+\}dH_{p-q}(t;0)\\
&< \int_0^{\chi_p^2(\gamma)} P_{\boldsymbol{\eta}_1}\{\|\boldsymbol{\eta}_1-\boldsymbol{\omega}_1\|^2 \le (\chi_p^2(\gamma)-t)^+\}dH_{p-q}(t,0)\\
&= P_0\big(\chi_p^2 \le \chi_p^2(\gamma)\big) = 1-\gamma. \qquad (7.9.24)
\end{aligned}
$$

(iii) If $\Delta^2 > (\chi_p(\gamma)+\chi_q(\alpha))^2$, then

$$
\begin{aligned}
&P_{\boldsymbol{\beta}}\big(C^{\text{PT}}(\hat{\boldsymbol{\beta}}_n^{\text{PT}})\big)\\
&= \int_0^{\chi_p^2(\gamma)} P_{\boldsymbol{\eta}_1}\{\|\boldsymbol{\omega}_1\|^2 > \chi_q^2(\alpha); \|\boldsymbol{\eta}_1-\boldsymbol{\omega}_1\|^2 \le (\chi_p^2(\gamma)-t)^+\}dH_{p-q}(t,0).
\end{aligned}
$$

Note that

$$\chi_p(\gamma)+\chi_q(\alpha) < \Delta = \|\boldsymbol{\eta}_1\|.$$

Since

$$\|\boldsymbol{\eta}_1 - \boldsymbol{\omega}_1\|^2 < (\chi_p^2(\gamma) - t)^+ \Rightarrow \|\boldsymbol{\eta}_1 - \boldsymbol{\omega}_1\| < \sqrt{(\chi_p^2(\gamma) - t)^+},$$

it implies

$$\begin{aligned}\Delta - \|\boldsymbol{\omega}_1\| \le \|\boldsymbol{\eta}_1 - \boldsymbol{\omega}_1\| &< \sqrt{(\chi_p^2(\gamma) - t)^+} \\ \Rightarrow \Delta \le \|\boldsymbol{\omega}_1\| &+ \sqrt{(\chi_p^2(\gamma) - t)^+}.\end{aligned}$$

Thus, we have

$$\sqrt{(\chi_p^2(\gamma) - t)^+} + \chi_q(\alpha) < \Delta < \|\boldsymbol{\omega}_1\| + \sqrt{(\chi_p^2(\gamma) - t)^+}.$$

It follows that

$$\|\boldsymbol{\omega}_1\|^2 > \chi_q^2(\alpha).$$

Thus,

$$\begin{aligned}&\int_0^{\chi_p^2(\gamma)} P_{\eta_1}\{\|\boldsymbol{\omega}_1\|^2 > \chi_q^2(\alpha); \|\boldsymbol{\eta}_1 - \boldsymbol{\omega}_1\|^2 \\ &\le (\chi_p^2(\gamma) - t)^+\} dH_{p-q}(t, 0) \\ &= \int_0^{\chi_p^2(\gamma)} P_{\eta_1}\{\|\boldsymbol{\eta}_1 - \boldsymbol{\omega}_1\|^2 \le (\chi_p^2(\gamma) - t)^+\} dH_{p-q}(t, 0) \\ &= P_0(\chi_p^2 < \chi_p^2(\gamma)) = 1 - \gamma. \qquad (7.9.25)\end{aligned}$$

The reader is referred to Tables 4.8.2a and 4.8.2b for the tabular values of the coverage probabilities against Δ^2-values for $p = 5(2)15$. Clearly, the coverage probability attains its maximum at $\Delta^2 = 0$. Then it decreases to $\Delta^2 = \chi_p^2(\gamma)$ and drops to a minimum value (there is a discontinuity at $\Delta^2 = \chi_p^2(\gamma)$). As $\Delta^2 > \chi_p^2(\gamma)$, the coverage probability increases monotone toward the value $1 - \gamma$ as $\Delta^2 \to \infty$. The picture of the coverage probability is similar to the risk of the PTE as a function of Δ^2.

<u>$C^S(\hat{\boldsymbol{\beta}}_n^S)$</u>: In this case, if $q \ge 3$ and $\Delta^2 > 0$, then

$$\begin{aligned}&P_{\boldsymbol{\beta}}(C^S(\hat{\boldsymbol{\beta}}_n^S)) \\ &= \int_0^{\chi_p^2(\gamma)} P_{\eta_1}\{\|\boldsymbol{\eta}_1 - \boldsymbol{\omega}_1(1 - d\|\boldsymbol{\omega}_1\|^{-2})\|^2 \le (\chi_p^2(\gamma) - t)^+\} dH_{p-q}(t; 0) \\ &\ge \int_0^{\chi_p^2(\gamma)} P_{\eta_1}\{\|\boldsymbol{\eta}_1 - \boldsymbol{\omega}_1\|^2 \le (\chi_p^2(\gamma) - t)^+\} dH_{p-q}(t; 0) = 1 - \gamma \qquad (7.9.26)\end{aligned}$$

for $0 \le d \le 2d$ by the fact that $\boldsymbol{\omega}_1(1 - d\|\boldsymbol{\omega}_1\|^{-2})$ is the Stein function, while $\boldsymbol{\omega}_1$ is the usual q-variate normal with mean $\boldsymbol{\eta}_1$ and covariance $\mathbf{I}_q$.

$C^{S+}(\hat{\beta}_n^{S+})$: In this case, we have the following theorem:

Theorem 1. If $q \geq 3$ and $\Delta^2 > 0$,

$$P_\beta\left\{C^+(\hat{\beta}_n^{S+})\right\} \geq P_0(C^0(\tilde{\beta}_n)) \quad \text{for all } \Delta^2 \in (0,\infty) \tag{7.9.27}$$

provided that $0 < d < d_0$ where d_0 is the minimum of the two unique

$$\mathcal{G}_q^{(1)}(d,\chi_p(\gamma)) = \left[\frac{\frac{1}{2}\chi_p(\gamma) + \sqrt{\chi_p^2(\gamma)/4 + d}}{\sqrt{d}}\right]^{p-q-2} \exp\left\{-\frac{1}{2}\chi_p(\gamma)\sqrt{d}\right\} = 1$$

and

$$\mathcal{G}_q^{(2)}(d,\chi_p(\gamma))$$
$$= \left[\frac{\chi_p(\gamma) + \sqrt{\chi_p^2(\gamma)+d}}{\sqrt{d}}\right]^{p-q-1} \exp\{-\chi_p(\gamma)\sqrt{d}\}\left[\frac{\chi_p^2(\gamma) + 4d - \chi_p(\gamma)}{2\sqrt{d}}\right] = 1. \tag{7.9.28}$$

Proof. The theorem is proved if we can show that for every $0 < b^2 \leq \chi_p^2(\gamma)$,

$$P_{\eta_1}\{\|\eta_1 - \omega_1(1 - d\|\omega_1\|^{-2})I(\|\omega_1\|^2 > d)\|^2 \leq b^2\} > P_{\eta_1}\{\|\eta_1 - \omega_1\|^2 \leq b^2\}. \tag{7.9.29}$$

Substituting the right-hand side of (7.9.28) for the integrand below, we get

$$\begin{aligned}
&P_\beta\left\{C^+(\hat{\beta}_n^{S+})\right\}\\
&= P_{\eta_1}\{\|\eta_2 - \omega_2\|^2 + \|\eta_1 - \omega_1(1 - d\|\omega_1\|^{-2})I(\|\omega_1\|^2 > d)\|^2 < \chi_p^2(\gamma)\}\\
&= \int_0^{\chi_p^2(\gamma)} P_{\eta_1}\{\|\eta_1 - \omega_1(1 - d\|\omega_1\|^{-2})I(\|\omega_1\|^2 > d)\|^2 \leq (\chi_p^2(\gamma) - t)^+\}\\
&\qquad \times dH_{p-q}(t;0)\\
&> \int_0^{\chi_p^2(\gamma)} P_{\eta_1}(\|\eta_1 - \omega_1\|^2 \leq (\chi_p^2(\gamma) - t)^+)dH_{p-q}(t;0)\\
&= P_0(\|\eta - \omega\|^2 \leq \chi_p^2(\gamma)) = 1 - \gamma.
\end{aligned} \tag{7.9.30}$$

Note that $\omega_1 \sim N_q(\eta_1, \mathbf{I}_q)$; hence, it is sufficient to establish that $\mathcal{G}_q^{(1)}(d,b) \geq 1$ and $\mathcal{G}_q^{(2)}(d,b) \geq 1$ for $0 < d < d_0$ and $0 < b^2 \leq \chi_p^2(\gamma)$.

Let us prove $\mathcal{G}_q^{(1)}(d,\chi_p(\gamma)) \geq 1$ for $0 < b^2 \leq \chi_p^2(\gamma)$. Note that for each b, $\mathcal{G}_q^{(1)}(d,b)$ is decreasing in d where d^* satisfies $G_q^{(1)}(d^*,\chi_p(\gamma)) = 1$. Hence, it is sufficient to prove that there exists a d^* such that $\mathcal{G}_q^{(1)}(d^*,b) \geq 1$. Note that

$\mathcal{G}_q^{(1)}(d,b)$ is strictly decreasing in b. As a result, $\mathcal{G}_q^{(1)}(d^*,b)$ either (1) strictly decreases to zero in b or (2) strictly increases to a unique maximum and then strictly decreases to zero. Since $\mathcal{G}_q^{(1)}(d^*,0) = \mathcal{G}_q^{(1)}(d;\chi_p(\gamma)) = 1$, then (1) is not true, and in the case of (2) we get $\mathcal{G}_q^{(1)}(d^*,b) > 1$ for $0 < b^2 < \chi_p^2(\gamma)$. The proof with $\mathcal{G}_q^{(2)}(d,\chi_p^2(\gamma))$ is similar. In our case where $d^* = q-2$, computational results show that there is no significant difference in the coverage probabilities.

Notice that

$$\|\omega_1(1-(q-2)\|\omega_1\|^{-2})\|^2 \geq \|\omega_1\{1-(q-2)\|\omega_1\|^{-2}\}I(\|\omega_1\|^2 > q-2)\|^2.$$

Hence,

$$\begin{aligned} &P_{\eta_1}\left[\|\eta_1 - \omega_1\{1-(q-2)\|\omega_1\|^{-2}\}\ I(\|\omega_1\|^2 > q-2)\|^2 < \chi_p^2(\gamma)\right] \\ &\geq\ P_{\eta_1}\left[\|\eta_1 - \omega_1\{1-(q-2)\|\omega_1\|^{-2}\} < \chi_p^2(\gamma)\right]. \end{aligned} \quad (7.9.31)$$

Thus, the confidence sets $C^S(\hat{\beta}_n^S)$ and $C^{S+}(\hat{\beta}_n^{S+})$ satisfy the dominance coverage probability

$$C^{S+}(\hat{\beta}_n^{S+}) \geq C^S(\hat{\beta}_n^S). \quad (7.9.32)$$

Consequently,

$$C^{S+}(\hat{\beta}_n^{S+}) \geq C^S(\tilde{\beta}_n) \geq C^0(\tilde{\beta}_n). \quad (7.9.33)$$

The reader is referred to Tables 5.9.2 and 5.9.3 for some values of coverage probabilities.

7.10 Asymptotic Theory of Confidence Sets

In Section 7.9 we discussed the finite sample properties of the confidence sets centered at the unrestricted, restricted, preliminary test, James-Stein-type, and positive-rule Stein estimators when σ^2 is known. When σ^2 is unknown, the closed form answer cannot be obtained. Hence, we consider the asymptotic theory, noting the fact that the estimator s_e^2 of σ^2 converges almost surely to σ^2 as $n \to \infty$. This leads us to use the results of Section 7.9 for moderate to large sample sizes. However, when the error distribution of the regression model is nonnormal, we consider the asymptotic theory as discussed in Section 7.7.

7.10.1 Confidence Sets

Again, consider the model

$$\mathbf{Y}_n = \mathbf{X}_n\boldsymbol{\beta} + \mathbf{e}_n \quad (7.10.1)$$

and the estimators $\tilde{\beta}_n$, $\hat{\beta}_n$, $\hat{\beta}_n^{\mathrm{PT}}$, $\hat{\beta}_n^S$, and $\hat{\beta}_n^{S+}$ of β given in Sections 7.7.1 and 7.7.2 and Sections 7.8.1 and 7.8.2.

In this section, we consider the asymptotic theory of confidence sets defined in Section 7.6 repeated below:

$$\begin{aligned}
&\text{(i) } C^0(\tilde{\beta}_n) = \{\beta : ns_e^{-2}\|\beta - \tilde{\beta}_n\|^2_{\mathbf{C}_n} \leq \chi^2_p(\gamma)\}.\\
&\text{(ii) } C^R(\hat{\beta}_n) = \{\beta : ns_e^{-2}\|\beta - \hat{\beta}_n\|^2_{\mathbf{C}_n} \leq \chi^2_p(\gamma)\}.\\
&\text{(iii) } C^{\mathrm{PT}}(\hat{\beta}_n^{\mathrm{PT}}) = \{\beta : ns_e^{-2}\|\beta - \hat{\beta}_n^{\mathrm{PT}}\|^2_{\mathbf{C}_n} \leq \chi^2_p(\gamma)\}.\\
&\text{(iv) } C^S(\hat{\beta}_n^S) = \{\beta : ns_e^{-2}\|\beta - \hat{\beta}_n^S\|^2_{\mathbf{C}_n} \leq \chi^2_p(\gamma)\}.\\
&\text{(v) } C^{S+}(\hat{\beta}_n^{S+}) = \{\beta : ns_e^{-2}\|\beta - \hat{\beta}_n^{S+}\|^2_{\mathbf{C}_n} \leq \chi^2_p(\gamma)\},
\end{aligned} \tag{7.10.2}$$

where $\chi^2_p(\gamma)$ is γ-level upper critical value of a central chi-square distribution with p d.f.

We can write the confidence sets compactly as

$$C^*(\beta_n^*) = \{\beta : ns_e^{-2}\|\beta - \beta_n^*\|^2_{\mathbf{C}_n} < \chi^2_p(\gamma)\}, \tag{7.10.3}$$

where

$$\beta_n^* = \tilde{\beta}_n + (\tilde{\beta}_n - \hat{\beta}_n)g(\mathcal{L}_n) \tag{7.10.4}$$

as before.

We study the properties of these sets in the next section and show that asymptotically they are similar to that of Section 7.9.

7.10.2 Asymptotic Properties of Confidence Sets

From Section 7.8.1, we observe that the asymptotic coverage probability of all the confidence sets is $1-\gamma$ under fixed alternative except $C_\beta^R(\hat{\beta}_n)$.

By Theorem 3 of Section 7.8.2, we obtain the asymptotic distribution of the statistics $\sqrt{n}s_e^{-1}(\beta_n^* - \beta)$, where $\beta_n^* = \tilde{\beta}_n, \hat{\beta}_n, \hat{\beta}_n^{\mathrm{PT}}, \hat{\beta}_n^S$, and $\hat{\beta}_n^{S+}$ under local alternatives

$$K_{(n)} : \mathbf{H}\beta = \mathbf{h} + n^{-1/2}\boldsymbol{\xi}. \tag{7.10.5}$$

First, we note that

$$\lim_{n\to\infty} P_{K_{(n)}}\{ns_e^{-2}\|\beta - \tilde{\beta}_n\|^2_{\mathbf{C}_n} \leq \chi^2_p(\gamma)\} = 1-\gamma, \tag{7.10.6}$$

The rest of the expressions are obtained using the basic transformation and technique given in Section 7.9.2.

Thus, we have

$$\begin{aligned}
&\lim_{n\to\infty} P_{K_{(n)}}\{ns_e^{-2}\|\boldsymbol{\beta}-\hat{\boldsymbol{\beta}}_n\|^2_{\mathbf{C}_n} \le \chi_p^2(\gamma)\} \\
&\qquad = P_{\eta_1}\{\|\boldsymbol{\eta}_2-\boldsymbol{\omega}_2\|^2+\|\boldsymbol{\eta}_1\|^2 \le \chi_p^2(\gamma)\} \\
&\qquad = H_{p-q}(\chi_p^2(\gamma)-\Delta^2;0), \quad \Delta^2 = \|\boldsymbol{\eta}_1\|^2. \qquad (7.10.7)
\end{aligned}$$

For the coverage probability of $C^{\mathrm{PT}}(\hat{\boldsymbol{\beta}}_n)$, we have

$$\begin{aligned}
&P_{K_{(n)}}\{ns_e^{-2}\|\boldsymbol{\beta}-\hat{\boldsymbol{\beta}}_n^{\mathrm{PT}}\|^2_{C_n} \le \chi_p^2(\gamma)\} \\
&= P_{\eta}\{\|\boldsymbol{\eta}_1-\boldsymbol{\omega}_2 g(\|\boldsymbol{\omega}_1\|^2)\|^2+\|\boldsymbol{\eta}_2-\boldsymbol{\omega}_2\|^2 \le \chi_p^2(\gamma)\}, \\
&\quad \text{where } g(\|\boldsymbol{\omega}_1\|^2) = I(\|\boldsymbol{\omega}_1\|^2 > \chi_q^2(\alpha)), \\
&= P_{\eta}\{\|\boldsymbol{\eta}_1\|^2+\|\boldsymbol{\eta}_2-\boldsymbol{\omega}_2\|^2 \le \chi_p^2(\gamma);\|\boldsymbol{\omega}_1\|^2 \le \chi_q^2(\alpha))\} \\
&\quad + P_{\eta}\{\|\boldsymbol{\eta}_1-\boldsymbol{\omega}_1\|^2+\|\boldsymbol{\eta}_2-\boldsymbol{\omega}_2\|^2 < \chi_p^2(\gamma);\|\boldsymbol{\omega}_1\|^2 > \chi_q^2(\alpha)\} \\
&= H_q(\chi_q^2(\alpha);\Delta^2)H_{p-q}(\chi_p^2(\gamma)-\Delta^2;0) \\
&\quad + \int_0^{\chi_p^2(\gamma)} P_{\eta_1}\{\|\boldsymbol{\eta}_1-\boldsymbol{\omega}_1)\|^2 \le (\chi_p^2(\gamma)-t)^+;\|\boldsymbol{\omega}_1\|^2 > \chi_q^2(\alpha)\}dH_{p-q}(t;0).
\end{aligned}$$

(7.10.8)

We get the expressions for $P_{\boldsymbol{\beta}}(C^S(\hat{\boldsymbol{\beta}}_n^S))$ and $P_{\boldsymbol{\beta}}(C^{S+}(\hat{\boldsymbol{\beta}}_n^{S+}))$ as

$$\begin{aligned}
&\lim_{n\to\infty} P_{K_{(n)}}\{ns_e^{-2}\|\boldsymbol{\beta}-\hat{\boldsymbol{\beta}}_n^S\|^2_{\mathbf{C}_n} \le \chi_p^2(\gamma)\} \\
&= P_{\eta_1}\{\|\boldsymbol{\eta}_1-\boldsymbol{\omega}_1(1-(q-2)\|\boldsymbol{\omega}_1\|^{-2})\|^2 \\
&\le (\chi_p^2(\gamma)-t)^+\}dH_{p-q}(t,0) \qquad (7.10.9)
\end{aligned}$$

and

$$\begin{aligned}
&\lim_{n\to\infty} P_{K_{(n)}}\{ns_e^{-2}\|\boldsymbol{\beta}-\hat{\boldsymbol{\beta}}_n^{S+}\|^2_{\mathbf{C}_n} \le \chi_p^2(\gamma)\} \\
&= H_{p-q}(\chi_p^2(\gamma)-\Delta^2;0)H_q(q-2;\Delta^2) \\
&\quad + \int_0^{\chi_p^2(\gamma)} P_{\eta_1}\{\|\boldsymbol{\eta}_1-\boldsymbol{\omega}_1(1-(q-2)\|\boldsymbol{\omega}_1\|^{-2})\|^2 \\
&\le (\chi_{p+q}^2(\gamma)-t)^+;\|\boldsymbol{\omega}_1\|^2 > q-2\}dH_{p-q}(t;0), \qquad (7.10.10)
\end{aligned}$$

respectively.

Since the asymptotic coverage probabilities of the five sets are similar to those given in Section 7.9, the properties are same as before.

7.11 Nonparametric Methods: R-Estimation

In this section, we consider nonparametric methods of estimation, namely the R-estimation of the regression parameters, $\beta = (\beta_1, \dots, \beta_p)'$ in the usual model

$$\mathbf{Y} = \mathbf{X}\beta + \varepsilon, \tag{7.11.1}$$

where $\mathbf{X}$ is a $n \times p$ design matrix of the known regression constants $(n \geq p \geq 1)$ and the components of the random vector ε are i.i.d.r.v., with continuous distribution function $F(\varepsilon_i)$ defined on the real line R^1. Further, we consider the following possible restriction on $\boldsymbol{\beta}$:

$$\mathbf{H}\beta = \mathbf{h}, \tag{7.11.2}$$

where $\mathbf{H}$ is a $q \times p$ matrix of known constants (rank q) and $\mathbf{h}$ is a $q \times 1$ vector of known constants as well.

Our problem is the *robust estimation* of $\boldsymbol{\beta}$ when it is suspected but one is not sure that (7.11.2) holds. We consider R-estimation of $\boldsymbol{\beta}$. Let $\tilde{\boldsymbol{\beta}}_n$ be the *unrestricted* and $\hat{\boldsymbol{\beta}}_n$ be the *restricted* (under (7.11.2)) *R-estimators* of $\boldsymbol{\beta}$ respectively. Further, let $\mathcal{L}_n$ be the rank-test of the null hypothesis $H_0 : \mathbf{H}\boldsymbol{\beta} = \mathbf{h}$ against the alternative $H_A : \mathbf{H}\boldsymbol{\beta} \neq \mathbf{h}$. Then following previous sections, we can consider three more R-estimators of $\boldsymbol{\beta}$, namely, PRE, SRE, and PRSRE given by

$$\text{PTRE} : \hat{\boldsymbol{\beta}}_n^{\text{PT}} = \tilde{\boldsymbol{\beta}}_n I(\mathcal{L}_n \geq \mathcal{L}_{n,\alpha}), \tag{7.11.3}$$

$$\text{SRE} : \hat{\boldsymbol{\beta}}_n^{S} = \tilde{\boldsymbol{\beta}}_n (1 - d\mathcal{L}_n^{-1} I(\mathcal{L}_n > c_n)) \quad (d = p-2), \tag{7.11.4}$$

$$\text{PRSRE} : \hat{\boldsymbol{\beta}}_n^{S+} = \hat{\boldsymbol{\beta}}_n + (\tilde{\boldsymbol{\beta}}_n - \hat{\boldsymbol{\beta}}_n)(1 - d\mathcal{L}_n^{-1}) I(\mathcal{L}_n > d), \tag{7.11.5}$$

where $c_n \to 0$ as $n \to \infty$, and $\mathcal{L}_{n,\alpha}$ is the α-level critical value from the null distribution of $\mathcal{L}_n$.

In this section, we compare and study various properties of the R-estimators of $\boldsymbol{\beta}$ in an asymptotic setup.

7.11.1 Linear Rank Statistics, R-Estimators and Confidence Sets

Let $R_i(b)$ be the rank of $Y_i - (\mathbf{x}_i - \bar{\mathbf{x}}_n)\mathbf{b}$ among $y_1 - (\mathbf{x}_1 - \bar{\mathbf{x}}_n)\mathbf{b}, \dots, y_n - (\mathbf{x}_n - \bar{\mathbf{x}}_n)\mathbf{b}$, where $\mathbf{x}_i$ is the ith row of $\mathbf{X} = (\mathbf{x}_1', \dots, \mathbf{x}_r')'$ and $\bar{\mathbf{x}}_n = n^{-1}\sum_{i=1}^n \mathbf{x}_i$. Now, consider the LRS,

$$\mathbf{L}_n(\mathbf{b}) = (L_{1n}(b_1), \dots, L_{pn}(b_p))' = \sum_{i=1}^n (\mathbf{x}_i - \bar{\mathbf{x}}_n) a_n(R_i(\mathbf{b})), \tag{7.11.6}$$

where (i) $a_n(k) = E\phi(U_{k,n})$ or $\phi(k/n+1)$ for square integrable and nondecreasing score generating function $\phi(\cdot)$ defined on $(0,1)$ as before (see Chapters 4, 5, and 6).

Let $\mathbf{C}_n = \sum_{i=1}^{n} (\mathbf{x}_i - \bar{\mathbf{x}}_n)(\mathbf{x}_i - \bar{\mathbf{x}}_n)'$, and assume that

(i) $\lim_{n\to\infty} \frac{1}{n}\mathbf{C}_n = \mathbf{C}$,

(ii) $\max_{1\le i\le n} \{(\mathbf{x}_i - \bar{\mathbf{x}}_n)\mathbf{C}_n^{-1}(\mathbf{x}_i - \bar{x}_n)\} = o(n)$.

$$I(f) = \int_{-\infty}^{\infty} \{f'(x)/f(x)\}^2 f(x)dx < \infty \text{ and } A_\phi^2 = \int_0^1 \phi^2(u) - \Big(\int_0^1 \phi(u)du\Big)^2.$$

$$\gamma(\psi,\phi) = \int_0^1 \psi(u)\phi(u)du \quad \text{and} \quad \int_0^1 \psi^2(u)du = I(f),$$

where

$$\psi(u) = -\frac{f'(F^{-1}(u))}{f(F^{-1}(u))}, \quad 0 < u < 1. \tag{7.11.7}$$

Further, if we put $\|\mathbf{a}\| = \sum_{m=1}^{p} a^{(m)}$, $\mathbf{a} = (a^{(1)}, \dots, a^{(p)})$, then the *unrestricted* R-estimator $\tilde{\boldsymbol{\beta}}_n$ of $\boldsymbol{\beta}$ is defined by any central point of the set

$$\mathbf{S} = \{\mathbf{b} : \|\mathbf{L}_n(\mathbf{b})\| = \text{ minimum}\}. \tag{7.11.8}$$

Using Jurečkova's (1969) linearity result (see Section 2.8.4), we have

$$\lim P\left\{ \sup_{\|\boldsymbol{\omega}\|<k} \left\|\mathbf{L}_n(\boldsymbol{\beta} + \frac{\boldsymbol{\omega}}{\sqrt{n}}) - \mathbf{L}_n(\boldsymbol{\beta}) + \gamma\mathbf{C}\boldsymbol{\omega}\right\| > \varepsilon \right\} = 0 \tag{7.11.9}$$

for any $k > 0$ and $\varepsilon > 0$, where $\boldsymbol{\omega}$ is a p-dimensional column vector and $\|\cdot\|$ denotes the p-dimensional Euclidean norm. We obtain

$$\sqrt{n}(\tilde{\boldsymbol{\beta}}_n - \boldsymbol{\beta}) \sim \mathcal{N}_p(\mathbf{0}, \sigma^2\mathbf{C}^{-1}), \quad \sigma^2 = A_\phi^2/\gamma^2(\psi,\phi). \tag{7.11.10}$$

To define the restricted R-estimator of $\boldsymbol{\beta}$, we mimic the least square estimator of β under $\mathbf{H}\boldsymbol{\beta} = \mathbf{h}$ as

$$\hat{\boldsymbol{\beta}}_n = \tilde{\boldsymbol{\beta}}_n - \mathbf{C}_n^{-1}\mathbf{H}'(\mathbf{H}\mathbf{C}_n^{-1}\mathbf{H}')(\mathbf{H}\tilde{\boldsymbol{\beta}}_n - \mathbf{h}). \tag{7.11.11}$$

In order to test the null hypothesis $H_0 : \mathbf{H}\boldsymbol{\beta} = \mathbf{h}$ against $H_A : \mathbf{H}\boldsymbol{\beta} \ne \mathbf{h}$, we consider the rank-test

$$\mathcal{L}_n = A_n^{-2}[\mathbf{L}_n(\hat{\boldsymbol{\beta}}_n)]'\mathbf{C}_n^{-1}[\mathbf{L}_n(\hat{\boldsymbol{\beta}}_n)], \tag{7.11.12}$$

where

$$A_n^2 = (n-1)^{-1}\sum_{k=1}^{n}(a_n(k) - \bar{a}_n)^2, \quad \bar{a}_n = n^{-1}\sum_{k=1}^{n} a_n(k). \tag{7.11.13}$$

We obtain the following theorem on the null distribution of $\mathcal{L}_n$:

Theorem 1. Assume (7.11.10). Then, under the null hypothesis $H_0 : \mathbf{H}\boldsymbol{\beta} = \mathbf{h}$ as $n \to \infty$, $\mathcal{L}_n$ approximately follows the central chi-square distribution with q d.f.

Proof. By the asymptotic linearity result (7.11.9) we have two relations:

$$\text{(i)}\quad \mathbf{L}_n(\hat{\boldsymbol{\beta}}_n) - \mathbf{L}_n(\boldsymbol{\beta}) + \gamma(\psi,\phi)\mathbf{C}\sqrt{n}(\hat{\boldsymbol{\beta}}_n - \boldsymbol{\beta})' \xrightarrow{\text{P}} \mathbf{0}$$

and $$\text{(ii)}\quad \mathbf{L}_n(\boldsymbol{\beta}) - \gamma(\psi,\phi)\mathbf{C}\sqrt{n}(\tilde{\boldsymbol{\beta}}_n - \boldsymbol{\beta})' \xrightarrow{\text{P}} \mathbf{0}. \tag{7.11.14}$$

From (i) and (ii) of (7.9.13), it follows that

$$\begin{aligned}\mathcal{L}_n &= \gamma^2(\psi,\phi)A_n^{-2}n(\tilde{\boldsymbol{\beta}}_n - \hat{\boldsymbol{\beta}}_n)'\mathbf{C}(\tilde{\boldsymbol{\beta}}_n - \hat{\boldsymbol{\beta}}_n) + o_P(1)\\ &= \gamma^2(\psi,\phi)A_n^{-2}n(\mathbf{H}\tilde{\boldsymbol{\beta}}_n - \mathbf{h})'(\mathbf{H}\mathbf{C}^{-1}\mathbf{H}')^{-1}(\mathbf{H}\tilde{\boldsymbol{\beta}}_n - \mathbf{h}) + o_P(1). \end{aligned}\tag{7.11.15}$$

After combining (7.11.10) and (7.11.14), we prove of the theorem.

We note that the test based on $\mathcal{L}_n$ is consistent against fixed $\boldsymbol{\beta}$ such that $\mathbf{H}\boldsymbol{\beta} \neq \mathbf{h}$. Thus $\lim_{n\to\infty} P(\mathcal{L}_n > k) \to 1$ for all $k \in R^1$, and the PTRE, SRE, and PRSRE of $\boldsymbol{\beta}$ is asymptotically equivalent to $\tilde{\boldsymbol{\beta}}_n$ under a fixed alternative. Hence, we consider the properties of the five estimators under the local alternatives:

$$K_{(n)} : \mathbf{H}\boldsymbol{\beta} = \mathbf{h} + n^{-1/2}\boldsymbol{\xi}, \quad \boldsymbol{\xi} = (\xi_1, \dots, \xi_q)'. \tag{7.11.16}$$

7.11.2 Asymptotic Distributional Properties of the *R*-estimators

By the asymptotic linearity results of Jurečkova (1969) (see also Puri and Sen (1986)), we have the following theorem:

Theorem 2. Under $\{K_{(n)}\}$ and the conditions (7.11.7) as $n \to \infty$, we conclude the following:

(i) $\sqrt{n}(\tilde{\boldsymbol{\beta}}_n - \boldsymbol{\beta}) \sim \mathcal{N}_p(\mathbf{0}, \sigma^2\mathbf{C}^{-1}), \quad \sigma^2 = A_\phi^2/\gamma^2(\psi,\phi),$ (7.11.17)

(ii) $\sqrt{n}(\hat{\boldsymbol{\beta}}_n - \boldsymbol{\beta}) \sim \mathcal{N}_p(-\boldsymbol{\delta}, \sigma^2\mathbf{A})$, where $\boldsymbol{\delta} = \mathbf{C}^{-1}\mathbf{H}'(\mathbf{H}\mathbf{C}^{-1}\mathbf{H}')\boldsymbol{\xi}$ and $\mathbf{A} = \mathbf{C}^{-1} - \mathbf{C}^{-1}\mathbf{H}'(\mathbf{H}\mathbf{C}^{-1}\mathbf{H}')^{-1}\mathbf{H}\mathbf{C}^{-1}$.

(iii) $\sqrt{n}(\tilde{\boldsymbol{\beta}}_n - \hat{\boldsymbol{\beta}}_n) \sim \mathcal{N}_p(\boldsymbol{\delta}, \sigma^2(\mathbf{C}^{-1} - \mathbf{A}))$,

(iv) $\lim P(\mathcal{L}_n \leq \kappa | K_{(n)}) = H_q(x; \Delta^2)$, where $\Delta^2 = \sigma^{-2}\boldsymbol{\delta}'\mathbf{C}\boldsymbol{\delta}' = \sigma^{-2}\boldsymbol{\xi}'(\mathbf{H}\mathbf{C}^{-1}\mathbf{H}')^{-1}\boldsymbol{\xi}$.

(v) $\lim P\{\sqrt{n}(\hat{\boldsymbol{\beta}}_n^{\text{PT}} - \boldsymbol{\beta}) \leq \kappa | K_{(n)}) = H_q(\chi_1^2(\alpha); \Delta^2)\Phi_p(\mathbf{x} + \boldsymbol{\delta}, \mathbf{0}, \sigma^2\mathbf{A}) + \int_{E(\boldsymbol{\xi})} \Phi_p(\mathbf{x} - \mathbf{C}^{-1}\mathbf{H}'(\mathbf{H}\mathbf{C}^{-1}\mathbf{H}')\mathbf{z}; \mathbf{0}, \sigma^2\mathbf{A})d\Phi_p(\mathbf{z}; 0, \sigma^2(\mathbf{H}\mathbf{C}^{-1}\mathbf{H}')^{-1})$ where $E(\boldsymbol{\xi}) = \{\mathbf{z}; \sigma^{-2}(\mathbf{H}\mathbf{Z} + \boldsymbol{\xi})'(\mathbf{H}\mathbf{C}^{-1}\mathbf{H}')^{-1}(\mathbf{H}\mathbf{Z} + \boldsymbol{\xi}) \geq \chi_q^2(\alpha)\}$.

(vi) $\sqrt{n}(\hat{\boldsymbol{\beta}}_n^{\text{PT}} - \boldsymbol{\beta}) = \mathbf{Z} - (\mathbf{H}\mathbf{Z} + \boldsymbol{\xi})I(\|\mathbf{H}\mathbf{Z} + \boldsymbol{\xi}\|^2_{(\mathbf{H}\mathbf{C}^{-1}\mathbf{H}')^{-1}} \leq \chi_q^2(\alpha)) + o_P(1)$, where $\mathbf{Z} \sim \mathcal{N}_p(\mathbf{0}, \sigma^2\mathbf{C}^{-1})$.

(vii) $\sqrt{n}(\hat{\beta}_n^S - \beta) = \mathbf{Z} - d\sigma^2\mathbf{C}^{-1}\mathbf{H}'(\mathbf{HZ} + \boldsymbol{\xi})\|\mathbf{HZ} + \boldsymbol{\xi}\|^{-2}_{(\mathbf{HC}^{-1}\mathbf{H}')^{-1}} + o_P(1).$

(viii) $\sqrt{n}(\hat{\beta}_n^{S+} - \beta) = \mathbf{Z} - d\sigma^2\mathbf{C}^{-1}\mathbf{H}'(\mathbf{HC}+\mathbf{H}')(\mathbf{HZ}+\boldsymbol{\xi})\|\mathbf{HZ}+\boldsymbol{\xi}\|^{-2}_{(\mathbf{HC}^{-1}\mathbf{H}')^{-1}}$
$- \sigma^2\mathbf{C}^{-1}\mathbf{H}'(\mathbf{HC}^{-1}\mathbf{H}')(\mathbf{HZ} + \boldsymbol{\xi})\{1 - d\|\mathbf{HZ} + \boldsymbol{\xi}\|^{-2}_{(\mathbf{HC}^{-1}\mathbf{H}')^{-1}}\}$
$\cdot I(\|\mathbf{HZ} + \xi\|^2_{(\mathbf{HC}^{-1}\mathbf{H}')^{-1}} < d) + o_P(1).$

Using Theorem 7.9.2, we compute the ADB, ADQR, ADMSE, and the ADQR of the five estimators given in the next two theorems:

Theorem 3. Under $\{K_{(n)}\}$ and conditions (7.11.7), as $n \to \infty$, the ADB, ADQB, ADMSE, and ADQR of the five estimators are given by

(i) URE($\tilde{\beta}_n$): (7.11.18)

$$b_1(\tilde{\beta}_n) = \mathbf{0} \quad \text{and} \quad B_1(\tilde{\beta}_n) = 0;$$

$$\mathbf{M}_1(\tilde{\beta}_n) = \sigma^2\mathbf{C}^{-1} \quad \text{and} \quad R_1(\tilde{\beta}_n; \mathbf{W}) = \sigma^2 \operatorname{tr}(\mathbf{C}^{-1}\mathbf{W}).$$

(ii) RRE($\hat{\beta}_n$):

$$\mathbf{b}_2(\hat{\beta}_n) = -\boldsymbol{\delta} \quad \text{and} \quad B_2(\hat{\beta}_n) = \Delta^2;$$

$$\mathbf{M}_2(\hat{\beta}_n) = \sigma^2(\mathbf{C}^{-1} - \mathbf{A}) + \boldsymbol{\delta\delta}' \quad \text{and}$$

$$R_2(\hat{\beta}_n; \mathbf{W}) = \sigma^2 \operatorname{tr}[(\mathbf{C}^{-1} - \mathbf{A})\mathbf{W}] + \boldsymbol{\delta}'\mathbf{W}\boldsymbol{\delta}$$

(iii) PTRE($\hat{\beta}_n^{\mathrm{PT}}$):

$$\mathbf{b}_3(\hat{\beta}_n^{\mathrm{PT}}) = -\boldsymbol{\delta} H_{q+2}(\chi_q^2(\alpha); \Delta^2) \text{ and}$$

$$B_3(\hat{\beta}_n^{\mathrm{PT}}) = \Delta^2 \left\{H_{q+2}(\chi_q^2(\alpha); \Delta^2)\right\}^2$$

$$\mathbf{M}_3(\hat{\beta}_n^{\mathrm{PT}}) = \sigma^2(\mathbf{C}^{-1} - \sigma^2(\mathbf{C}^{-1} - \mathbf{A})H_{q+2}(\chi_q^2(\alpha); \Delta^2)$$
$$+\boldsymbol{\delta\delta}' \left\{2H_{q+2}(\chi_q^2(\alpha); \Delta^2) - H_{q+4}(\chi_q^2(\alpha); \Delta^2)\right\}$$

and

$$R_3(\hat{\beta}_n^{\mathrm{PT}}) = \sigma^2 \operatorname{tr}(\mathbf{C}^{-1}\mathbf{W}) - \sigma^2 \operatorname{tr}[(\mathbf{C}^{-1} - \mathbf{A})\mathbf{W}]H_{q+2}(\chi_q^2(\alpha); \Delta^2)$$
$$+\boldsymbol{\delta}'\mathbf{W}\boldsymbol{\delta} \left\{2H_{q+2}(\chi_q^2(\alpha); \Delta^2) - H_{q+4}(\chi_q^2(\alpha); \Delta^2)\right\}.$$

(iv) SRE($\hat{\beta}_n^S$):

$$\mathbf{b}_4(\hat{\beta}_n^S) = -d\boldsymbol{\delta} E\left[\chi_{q+2}^{-2}(\Delta^2)\right] \text{ and } B_4(\hat{\beta}_n^S) = d^2\Delta^2 \left\{E\left[\chi_{q+2}^{-2}(\Delta^2)\right]\right\}^2$$

$$\mathbf{M}_4(\hat{\beta}_n^S) = \sigma^2\mathbf{C}^{-1} - d\sigma^2[\mathbf{C}^{-1} - \mathbf{A}]$$
$$\times \left\{2E\left[\chi_{q+2}^{-2}(\Delta^2)\right] - (q-2)E\left[\chi_{q+2}^{-4}(\Delta^2)\right]\right\}$$
$$+(q^2-4)\boldsymbol{\delta}\boldsymbol{\delta}'E\left[\chi_{q+2}^{-4}(\Delta^2)\right], \quad d = q-2,$$

and

$$R_4(\hat{\beta}_n^S; \mathbf{W}) = \sigma^2 \operatorname{tr}(\mathbf{C}^{-1}\mathbf{W}) - \sigma^2 \operatorname{tr}(\mathbf{A}_{11})$$
$$\times \left\{2E\left[\chi_{q+2}^{-2}(\Delta^2)\right] - (q-2)E\left[\chi_{q+2}^{-4}(\Delta^2)\right]\right\}$$
$$+ (q^2-4)\boldsymbol{\delta}'\mathbf{W}\boldsymbol{\delta}E\left[\chi_{q+2}^{-4}(\Delta^2)\right]$$

with $\mathbf{A}_{11} = [\mathbf{W}\mathbf{C}^{-1}\mathbf{H}'(\mathbf{H}\mathbf{C}^{-1}\mathbf{H}')^{-1}\mathbf{H}\mathbf{C}^{-1}]$,

(v) RRSRE$(\hat{\beta}_n^{S+})$

$$\mathbf{b}_5(\hat{\beta}_n^{S+}) = -\boldsymbol{\delta}\left\{dE[\chi_{q+2}^{-2}(\Delta^2)] + E[(1-d\chi_{q+2}^{-2}(\Delta^2))I(\chi_{q+2}^{-2}(\Delta^2)<d)]\right.$$

and

$$B_5(\hat{\beta}_n^{S+}) = \Delta^2\left\{dE[\chi_{q+2}^{-2}(\Delta^2)] + E[(1-d\chi_{q+2}^{-2}(\Delta^2))I(\chi_{q+2}^{-2}(\Delta^2) < d)]\right\}^2;$$

$$\mathbf{M}_5(\hat{\beta}_n^{S+})$$
$$= \mathbf{M}_4(\hat{\beta}_n^S) - \sigma^2(\mathbf{C}^{-1} - \mathbf{A})E\left[\left(1 - d\chi_{q+2}^{-2}(\Delta^2)\right)^2 I\left(\chi_{q+2}^{-2}(\Delta^2) < d\right)\right]$$
$$+ \boldsymbol{\delta}\boldsymbol{\delta}'\left\{2E\left[\left(1 - d\chi_{q+2}^{-2}(\Delta^2)\right)I\left(\chi_{q+2}^{-2}(\Delta^2) < d\right)\right]\right.$$
$$- E\left[\left(1 - d\chi_{q+2}^{-2}(\Delta^2)\right)^2 I\left(\chi_{q+4}^{-2}(\Delta^2) < d\right)\right]$$

and

$$\mathbf{R}_5(\hat{\beta}_n^{S+}; \mathbf{W}) = \operatorname{tr}(\mathbf{M}_4(\hat{\beta}_n^S)\mathbf{W}) - \sigma^2 \operatorname{tr}[(\mathbf{C}^{-1} - \mathbf{A})\mathbf{W}]$$
$$\times E\left[\left(1 - d\chi_{q+2}^{-2}(\Delta^2)\right)I\left(\chi_{q+2}^{2}(\Delta^2) < d\right)\right]$$
$$+ \boldsymbol{\delta}'\mathbf{W}\boldsymbol{\delta}\left\{2E\left[\left(1 - d\chi_{q+2}^{-2}(\Delta^2)\right)I\left(\chi_{q+2}^{2}(\Delta^2) < d\right)\right]\right.$$
$$\left. - E\left[\left(1 - d\chi_{q+4}^{-2}(\Delta^2)\right)^2 I\left(\chi_{q+4}^{-2}(\Delta^2) < d\right)\right]\right\},$$

where $d = q - 2$.

Notice that these expressions are similar to the expressions of Section 7.8.2. Hence, the properties of the five estimators are asymptotically similar.

7.11.3 Asymptotic Properties of the Recentered Confidence Sets Based on R-Estimators

We consider the recentered confidence sets based on the R-estimators, namely

(i) $$C^0(\tilde{\beta}_n) = \left\{\beta : n\sigma_n^{-2}\|\beta - \tilde{\beta}_n\|^2_{\mathbf{C}_n} \le \chi_p^2(\gamma)\right\},$$

(ii) $$C^R(\hat{\beta}_n) = \left\{\beta : n\sigma_n^{-2}\|\beta - \hat{\beta}_n\|^2_{\mathbf{C}_n} \le \chi_p^2(\gamma)\right\},$$

(iii) $$C^{\mathrm{PT}}(\hat{\beta}_n^{\mathrm{PT}}) = \left\{\beta : n\sigma_n^{-2}\|\beta - \hat{\beta}_n^{\mathrm{PT}}(\alpha)\|^2_{\mathbf{C}_n} \le \chi_p^2(\gamma)\right\},$$

(iv) $$C^S(\hat{\beta}_n^S) = \left\{\beta : n\sigma_n^{-2}\|\beta - \hat{\beta}_n^S\|^2_{\mathbf{C}_n} \le \chi_p^2(\gamma)\right\},$$

and

(v) $$C^{S+}(\hat{\beta}_n^{S+}(c)) = \left\{\beta : n\sigma_n^{-2}\|\beta - \hat{\beta}_n^{S+}(c)\|^2_{\mathbf{C}_n} \le \chi_p^2(\gamma)\right\}, \quad (7.11.19)$$

where $\sigma_n^2 = A_n^2/\hat{\gamma}(\psi,\phi)$ and $\hat{\gamma}(\psi,\phi)$ is a *consistent* estimator of $\gamma(\psi,\phi)$ given by

$$\hat{\gamma}(\psi,\phi) = p^{-1}\sum_{j=1}^{p}\hat{\gamma}_{jn}, \hat{\gamma}_{jn} = \frac{1}{2aC_n^{(j)}\mathbf{e}_j}\left\{L_{jn}\left(\tilde{\beta}_n - \frac{a\mathbf{e}_j}{\sqrt{n}}\right) - L_{jn}\left(\tilde{\beta}_n + \frac{a\mathbf{e}_j}{\sqrt{n}}\right)\right\}$$

$C_n^{(j)}$ is the jth row of $\mathbf{C}_n$, and $\mathbf{e}_j$ is the unit-vector with 1 at the jth place and zero elsewhere.

We can, again, write the confidence sets compactly as

$$C^*(\beta_n^*) = \left\{\beta : n\sigma_n^{-2}\|\beta - \beta_n^*\|^2_{\mathbf{C}_n} \le \chi_p^2(\gamma)\right\}, \quad (7.11.20)$$

where $\beta_n^* = \hat{\beta}_n + (\tilde{\beta}_n - \hat{\beta}_n)g(\mathcal{L}_n)$. We study the asymptotic ($n \to \infty$) properties of the confidence sets under the local alternative

$$K_{(n)} : \mathbf{H}\beta = \mathbf{h} + n^{-1/2}\boldsymbol{\xi}. \quad (7.11.21)$$

First, we note that

$$\lim_{n\to N} P_{K_{(n)}}\left\{n\sigma_n^{-2}\|\beta - \tilde{\beta}_n\|^2_{C_n} < \chi_p^2(\gamma)\right\} = 1 - \gamma. \quad (7.11.22)$$

By the linearity result (7.9.9),

$$n^{1/2}\mathbf{L}_n(0) = n^{1/2}\mathbf{C}(\tilde{\beta}_n - \beta)\gamma(\psi,\phi) + o_P(1)$$
$$n^{1/2}\mathbf{C}^{-1}\mathbf{L}_n(0) = n^{1/2}(\tilde{\beta}_n - \beta)\gamma(\psi,\phi) + o_P(1).$$

Hence,

$$\lim_{n\to\infty} n\sigma_n^{-2}\|\beta - \tilde{\beta}_n\|^2_{C_n} \xrightarrow{\mathrm{P}} \sigma_n^{-2}(\beta - \tilde{\beta}_n)\mathbf{C}(\beta - \tilde{\beta}_n) + o_P(1)$$
$$= A_\phi^{-2} n\left\|\mathbf{L}_n(\mathbf{0})\right\|^2,$$

$$\lim_{n\to\infty} P_0\{A_n^{-2}n\|\mathbf{L}_n(\mathbf{0})\|^2_{\mathbf{C}_n^{-1}} < \chi_p^2(\gamma)\} = P_0\{\chi_p^2 \le \chi_p^2(\gamma)\} = 1-\gamma. \quad (7.11.23)$$

Next, we consider the coverage probability

$$\begin{aligned}&\lim_{n\to\infty} P_{K_{(n)}}\Big\{n\sigma_n^{-2}\|\boldsymbol{\beta}-\tilde{\boldsymbol{\beta}}_n^*\|^2_{\mathbf{C}_n} < \chi_p^2(\gamma)\Big\}\\ &= P_{\boldsymbol{\eta}_1}\Big\{\|\boldsymbol{\eta}_1 - \boldsymbol{\omega}_1 g(\|\boldsymbol{\omega}_1\|^2)^2 + \|\boldsymbol{\eta}_2 - \boldsymbol{\omega}_2\|^2\|^2 \le \chi_p^2(\gamma)\Big\}\end{aligned}$$

by suitable transformation as in (7.9.9) and (7.9.10). Thus, we can write the asymptotic coverage probability as

$$\int_0^{\chi_p^2(\gamma)} P_{\eta_1}\Big\{\|\boldsymbol{\eta}_1 - \boldsymbol{\omega}_1 g(\|\boldsymbol{\omega}_1\|^2)\| \le (\chi_p^2(\gamma)-t)^+\Big\} dH_{p-q}(t;0). \quad (7.11.24)$$

This expression is the same as (7.9.13). Correspondingly, all conclusions derived in Section (7.9.3) hold. This expression is the same as (7.9.13). Hence, all analyses and conclusions are the same as given in Section 7.9.3 hold.

7.12 Conclusions

In this chapter, we studied the estimation of the regression parameters when it is suspected that the parameters belong to a subspace. Accordingly, the unrestricted, the restricted, the preliminary test, the Stein-type, and the positive-rule Stein-type estimators are defined, and their properties are studied in small as well as in large samples. In addition, we discuss the confidence set estimation of the parameters in both small and large samples, which include the nonparametric R-estimation methods.

7.13 Problems

1. Derive the expression for $\hat{\boldsymbol{\beta}}_n$ given in (7.1.4).
2. Verify remark 2 of Section 7.1.1.
3. Prove Theorem 2 of Section 7.1.2.
4. Verify the expression for $\text{MRE}(\hat{\boldsymbol{\beta}}_n^{\text{PT}}; \tilde{\boldsymbol{\beta}}_n)$.
5. Prove Theorem 3 of Section 7.3.
6. Find the expressions for the covariance matrices of $\hat{\boldsymbol{\beta}}_n$, $\hat{\boldsymbol{\beta}}_n^{\text{PT}}$, $\hat{\boldsymbol{\beta}}_n^{S}$, and $\hat{\boldsymbol{\beta}}_n^{S+}$.
7. Prove that the risk-difference of $\hat{\boldsymbol{\beta}}_n^{S}$ and $\tilde{\boldsymbol{\beta}}_n$ is given by (7.4.17).
8. Verify that the risk of $\hat{\boldsymbol{\beta}}_n^{S}$ can be expressed as the r.h.s. of (7.4.19).
9. Verify that the $\text{MRE}(\hat{\boldsymbol{\beta}}_n^{\text{PT}}; \tilde{\boldsymbol{\beta}}_n)$ is given by the R.H.S of (7.5.15).
10. Verify the expression for $\mathbf{M}_6(\hat{\boldsymbol{\beta}}_n^{\text{PT+}})$ and $R_6(\hat{\boldsymbol{\beta}}_n^{\text{PT+}} : \mathbf{W})$ given in Sections 7.6.4 and 7.6.5.

11. Prove that the risk $R_4(\hat{\beta}_n^S : \mathbf{W})$ may be expressed as

$$R_1(\tilde{\beta}_n : \mathbf{W}) - \sigma^2 \operatorname{tr}(\mathbf{A}_n)\Big\{(q-2)E[\chi_{q+2}^{-2}(\Delta^2)] + \Big[1 - \frac{(q+2)(\eta_1'\mathbf{A}_{11}\eta_1)}{2\Delta^2 \operatorname{tr}(\mathbf{A}_{11})}\Big]2\Delta^2 E[\chi_{q+4}^{-2}(\Delta^2)]\Big]$$

given by (7.8.33).

12. Prove that

$$\lim_{n\to\infty} P_{K_{(n)}}\Big\{ns_e^{-2}\|\beta - \beta_n^*\|_{\mathbf{C}_n}^2 < \chi_p^2(\gamma)\Big\}$$
$$= P_\eta\Big\{\|\eta_1 - \omega_1 g(\|\omega_1\|^2)\|^2 + \|\eta_2 - \omega_2\|^2 \le \chi_p^2(\gamma)\Big\}$$

by suitable transformations and analysis, and verify (7.10.2) through (7.10.5).

13. Prove that

$$\sigma^{-2}\|\beta - \beta_n^*\|_C^2 = \|\eta_1 - \omega_1 g(\|\omega_1\|^2)\|^2 + |\eta_2 - \omega_2\|^2.$$

14. Prove that $\sqrt{n}(\tilde{\beta}_n - \beta) \sim N_p(\mathbf{0}, \sigma^2(\mathbf{C}^{-1}))$ where $\sigma^2 = A_\phi^2/\gamma^2(\psi, \phi)$ under $K_{(n)}$.

15. Prove that $\sqrt{n}(\tilde{\beta}_n - \hat{\beta}_n) \sim N_p(-\delta,\ \sigma^2(\mathbf{C}^{-1} - \mathbf{A}))$ of Theorem 7.9.2 under $K_{(n)}$.

16. Prove that $\sqrt{n}(\hat{\beta}_n - \beta_n) \sim N_p(-\delta, \sigma^2\mathbf{A})$ under $K_{(n)}$.

17. Show that $\lim_{n\to\infty} P\{\mathcal{L}_n < x | K_{(n)}\} = H_q(x; \Delta^2)$.

18. Verify Theorem 7.9.3.

19. Numerically evaluate

$$\int_0^{\chi_p^2(\gamma)} P_{\eta_1}\{\|\eta_1 - \omega_1\|^2 \le (\chi_p^2(\gamma) - t)^+; \|\omega_1\|^2 > \chi_q^2(\alpha)\} dH_{p-q}(t, 0)$$

for $q = p - 1$.

Chapter 8

Multiple Regression Model: Stochastic Subspace Hypothesis

Outline

In Chapter 7, we considered the estimation of the regression parameters β in the multiple regression model $Y = \mathbf{X}\beta + \mathbf{e}$ when it is a priori suspected that β belongs to the subspace defined by $\mathbf{H}\beta = \mathbf{h}$. In this chapter, we consider the estimation of β when β belongs to the *stochastic subspace* defined by $\mathbf{h} = \mathbf{H}\beta + \mathbf{v}$, where $\mathbf{v}$ is normally distributed with mean δ and covariance matrix $\sigma^2\mathbf{\Omega}$, with $\mathbf{\Omega}$ being a known $(q \times q)$ positive-definite matrix. In addition, we consider the estimation of the regression parameters when $\mathbf{\Omega}$ is unknown based on a sample from previous study. Asymptotic theory is also provided for nonnormal errors in the models. Here we consider the idea of mixed estimation involving combined sample information from the model and an independent prior stochastic subspace information. The methodology is due to Theil and Goldberger (1961), Theil (1963), and Nagar and Kakwani (1964) among others. Further, we provide the confidence set analysis involving various estimators of $\boldsymbol{\beta}$ for finite as well as for large sample.

8.1 The Model, Estimation, and Test of Hypothesis

In this section, we formally present the model, estimation of the parameters, and related test of hypothesis results that will be used in the discussions of the multiple regression model.

8.1.1 The Model Formulation

Consider the estimation of the regression parameters in the model given by

$$\mathbf{Y} = \mathbf{X}\boldsymbol{\beta} + \mathbf{e} \tag{8.1.1}$$

when it is suspected that the parameter $\boldsymbol{\beta}$ belongs to the *stochastic subspace model*, namely

$$\mathbf{h} = \mathbf{H}\boldsymbol{\beta} + \mathbf{v}, \qquad \mathbf{v} \sim \mathcal{N}_q(\boldsymbol{\delta}, \sigma^2\boldsymbol{\Omega}), \tag{8.1.2}$$

where $\mathbf{h}$ is a known $q \times 1$ vector, $\mathbf{H}$ is a known $q \times p$ matrix of constants with rank q, and $\mathbf{v}$ is a $q \times 1$ vector of random errors. In the model above $\boldsymbol{\Omega}$ is a known matrix but σ^2 is unknown. This type of model appears in many econometric analyses. Reader may see Theil and Goldberger (1961) and Theil (1963).

For a suitable analysis, consider the sample information and model (8.1.1) together with the stochastic subspace model (8.1.2) to obtain the following mixed model:

$$\begin{bmatrix} \mathbf{Y} \\ \mathbf{h} \end{bmatrix} = \begin{bmatrix} \mathbf{X} \\ \mathbf{H} \end{bmatrix} \boldsymbol{\beta} + \begin{bmatrix} \mathbf{e} \\ \mathbf{v} \end{bmatrix}, \tag{8.1.3}$$

where

$$\begin{bmatrix} \mathbf{e} \\ \mathbf{v} \end{bmatrix} \sim \mathcal{N}_{n+q}\left\{ \begin{bmatrix} \mathbf{0} \\ \boldsymbol{\delta} \end{bmatrix}, \sigma^2 \begin{bmatrix} \mathbf{I}_n & \mathbf{0} \\ \mathbf{0} & \boldsymbol{\Omega} \end{bmatrix} \right\}. \tag{8.1.4}$$

Thus, our problem reduces to the estimation of $\boldsymbol{\beta}$ for the model (8.1.3) when it is suspected, but not certain, that

$$[-\mathbf{H}, \mathbf{I}_q] \begin{bmatrix} \boldsymbol{\beta} \\ \mathbf{H}\boldsymbol{\beta} + \boldsymbol{\delta} \end{bmatrix} = \boldsymbol{\delta} = \mathbf{0}. \tag{8.1.5}$$

Rewriting (8.1.3) and (8.1.5), we have the problem of estimating $\boldsymbol{\theta}$ in the model

$$\mathbf{Y}^* = \mathbf{Z}\boldsymbol{\theta} + \mathbf{E}, \tag{8.1.6}$$

when it is suspected that $\boldsymbol{\theta}$ belongs to the subspace $\mathbf{R}\boldsymbol{\theta} = \boldsymbol{\delta} = \mathbf{0}$, where

$$\mathbf{Y}^* = \begin{bmatrix} \mathbf{Y} \\ \boldsymbol{\Omega}^{-1/2}\mathbf{h} \end{bmatrix}, \quad \mathbf{Z} = \begin{bmatrix} \mathbf{X} & \mathbf{0} \\ \mathbf{0} & \boldsymbol{\Omega}^{-1/2} \end{bmatrix}, \quad \boldsymbol{\theta} = \begin{bmatrix} \boldsymbol{\beta} \\ \mathbf{H}\boldsymbol{\beta} + \boldsymbol{\delta} \end{bmatrix}, \quad \mathbf{R} = [-\mathbf{H}, \mathbf{I}_q], \tag{8.1.7}$$

and

$$\mathbf{E} = \begin{bmatrix} \mathbf{e} \\ \boldsymbol{\Omega}^{-1/2}(\mathbf{v} - \boldsymbol{\delta}) \end{bmatrix} \sim \mathcal{N}_{n+q}(\mathbf{0}, \sigma^2\mathbf{I}_{n+q}). \tag{8.1.8}$$

8.1.2 Mixed Model Estimation

Applying the least squares principle to the model (8.1.6), we have the *unrestricted mixed* estimator of $\boldsymbol{\theta}$ for

$$\tilde{\boldsymbol{\theta}}_n = (\mathbf{Z}'\mathbf{Z})^{-1}\mathbf{Z}'\mathbf{Y}^* = \begin{bmatrix} \mathbf{C}^{-1}\mathbf{X}'\mathbf{Y} \\ \mathbf{h} \end{bmatrix} = \begin{bmatrix} \tilde{\boldsymbol{\beta}}_n \\ \mathbf{h} \end{bmatrix}. \tag{8.1.9}$$

Similarly, the *mixed restricted estimator* of $\boldsymbol{\theta}$ subject to $\mathbf{R}\boldsymbol{\theta} = \mathbf{0}$ is given by

$$\begin{aligned} \hat{\boldsymbol{\theta}}_n &= \tilde{\boldsymbol{\theta}}_n - \mathbf{C}_z^{-1}\mathbf{R}'(\mathbf{R}\mathbf{C}_z^{-1}\mathbf{R}')^{-1}\mathbf{R}\tilde{\boldsymbol{\theta}}_n, \qquad \mathbf{C}_z = \mathbf{Z}'\mathbf{Z}, \\ &= \begin{bmatrix} \tilde{\boldsymbol{\beta}}_n \\ \mathbf{h} \end{bmatrix} - \begin{bmatrix} \mathbf{C}^{-1}\mathbf{H}'(\mathbf{H}\mathbf{C}^{-1}\mathbf{H}' + \boldsymbol{\Omega})^{-1}(\mathbf{H}\tilde{\boldsymbol{\beta}}_n - \mathbf{h}) \\ -\boldsymbol{\Omega}(\mathbf{H}\mathbf{C}^{-1}\mathbf{H}' + \boldsymbol{\Omega})^{-1}(\mathbf{H}\tilde{\boldsymbol{\beta}}_n - \mathbf{h}) \end{bmatrix}. \end{aligned} \tag{8.1.10}$$

Let $\hat{\boldsymbol{\beta}}_n$ denote the restricted estimator of $\boldsymbol{\beta}$; then

$$\hat{\boldsymbol{\beta}}_n = \tilde{\boldsymbol{\beta}}_n - \mathbf{C}^{-1}\mathbf{H}'(\mathbf{H}\mathbf{C}^{-1}\mathbf{H}' + \boldsymbol{\Omega})^{-1}(\mathbf{H}\tilde{\boldsymbol{\beta}}_n - \mathbf{h})$$

from (8.1.10). Let us now consider the estimation of $\boldsymbol{\beta}$ based on the formulations (8.1.3) and (8.1.4) using the generalized least squares principle (GLSE). The mixed restricted estimator of $\boldsymbol{\beta}$ of the mixed model (8.1.3) is given by

$$\begin{aligned} \hat{\boldsymbol{\beta}}_n^R &= \left([\mathbf{X}', \mathbf{H}'] \begin{pmatrix} \mathbf{I}_n & \mathbf{0} \\ \mathbf{0} & \boldsymbol{\Omega} \end{pmatrix}^{-1} \begin{bmatrix} \mathbf{X} \\ \mathbf{H} \end{bmatrix} \right)^{-1} \left([\mathbf{X}', \mathbf{H}'] \begin{pmatrix} \mathbf{I}_n & \mathbf{0} \\ \mathbf{0} & \boldsymbol{\Omega} \end{pmatrix}^{-1} \begin{bmatrix} \mathbf{Y} \\ \mathbf{h} \end{bmatrix} \right) \\ &= (\mathbf{C} + \mathbf{H}'\boldsymbol{\Omega}^{-1}\mathbf{H})^{-1}(\mathbf{C}\tilde{\boldsymbol{\beta}}_n + \mathbf{H}'\boldsymbol{\Omega}^{-1}\mathbf{h}). \end{aligned} \tag{8.1.11}$$

Since

$$(\mathbf{C} + \mathbf{H}'\boldsymbol{\Omega}^{-1}\mathbf{H})^{-1} = \mathbf{C}^{-1} - \mathbf{C}^{-1}\mathbf{H}'(\mathbf{H}\mathbf{C}^{-1}\mathbf{H}' + \boldsymbol{\Omega})^{-1}\mathbf{H}\mathbf{C}^{-1} \tag{8.1.12}$$

by Theorem 2.6.8 of Chapter 2, we have

$$\hat{\boldsymbol{\beta}}_n^R = (\tilde{\boldsymbol{\beta}}_n + \mathbf{C}^{-1}\mathbf{H}'\boldsymbol{\Omega}^{-1}\mathbf{h}) - \mathbf{C}^{-1}\mathbf{H}'(\mathbf{H}\mathbf{C}^{-1}\mathbf{H}' + \boldsymbol{\Omega})^{-1}\mathbf{H}(\tilde{\boldsymbol{\beta}}_n + \mathbf{C}^{-1}\mathbf{H}'\boldsymbol{\Omega}^{-1}\mathbf{h}). \tag{8.1.13}$$

Is there any difference between $\hat{\beta}$ and $\hat{\beta}_n^R$? To find the answer, consider the difference

$$\begin{aligned}
&\hat{\beta}_n - \hat{\beta}_n^R \\
&= [\tilde{\beta}_n - \mathbf{C}^{-1}\mathbf{H}'(\mathbf{H}\mathbf{C}^{-1}\mathbf{H}' + \boldsymbol{\Omega})^{-1}(\mathbf{H}\tilde{\beta}_n - \mathbf{h})] - [(\tilde{\beta}_n + \mathbf{C}^{-1}\mathbf{H}'\boldsymbol{\Omega}^{-1}\mathbf{h}) \\
&\quad - \mathbf{C}^{-1}\mathbf{H}'(\mathbf{H}\mathbf{C}^{-1}\mathbf{H}' + \boldsymbol{\Omega})^{-1}\mathbf{H}\mathbf{C}^{-1}(\mathbf{C}\tilde{\beta} + \mathbf{H}'\boldsymbol{\Omega}^{-1}\mathbf{h})] \\
&= \mathbf{C}^{-1}\mathbf{H}'[\boldsymbol{\Omega}^{-1} - (\mathbf{H}\mathbf{C}^{-1}\mathbf{H}' + \boldsymbol{\Omega})^{-1}(\mathbf{H}\mathbf{C}^{-1}\mathbf{H}' + \boldsymbol{\Omega})\boldsymbol{\Omega}^{-1}]\mathbf{h} = \mathbf{0}, \qquad (8.1.14)
\end{aligned}$$

which shows that $\hat{\beta} = \hat{\beta}_n^R$. Therefore, the *restricted mixed* estimator of $\boldsymbol{\theta}$ can be rewritten as

$$\hat{\boldsymbol{\theta}}_n = \begin{pmatrix} \hat{\beta}_n \\ \mathbf{H}\hat{\beta}_n \end{pmatrix}. \qquad (8.1.15)$$

If we premultiply $\tilde{\boldsymbol{\theta}}_n$ and $\hat{\boldsymbol{\theta}}_n$ by the $p \times (p+q)$ matrix $(\mathbf{I}_p, \mathbf{0}_{p\times q})$, we obtain $\tilde{\beta}_n$ and $\hat{\beta}_n$, respectively.

In the next section, we consider the test of the hypothesis $H_0 : \boldsymbol{\delta} = \mathbf{R}\boldsymbol{\theta} = \mathbf{0}$ against the alternative $H_A : \boldsymbol{\delta} \neq \mathbf{0}$.

8.1.3 Test of Hypothesis

The joint distribution of $(\tilde{\beta}_n', \hat{\beta}_n')'$ is given by

$$\begin{pmatrix} \tilde{\beta}_n \\ \hat{\beta}_n \end{pmatrix} \sim \mathcal{N}_{2p}\left\{\begin{bmatrix} \beta \\ \beta - \mathbf{C}^{-1}\mathbf{H}'(\mathbf{H}\mathbf{C}^{-1}\mathbf{H}' + \boldsymbol{\Omega})^{-1}\boldsymbol{\delta} \end{bmatrix}, \sigma^2\begin{bmatrix} \mathbf{C}^{-1} & \mathbf{A}_\Omega \\ \mathbf{A}_\Omega & \mathbf{A}_\Omega \end{bmatrix}\right\}, \qquad (8.1.16)$$

where $\mathbf{A}_\Omega = \mathbf{C}^{-1} - \mathbf{C}^{-1}\mathbf{H}'(\mathbf{H}\mathbf{C}^{-1}\mathbf{H}' + \boldsymbol{\Omega})^{-1}\mathbf{H}\mathbf{C}^{-1}$. For the estimation of σ^2, we consider

$$\begin{aligned}
s_e^2 &= m^{-1}(\mathbf{Y}^* - \mathbf{Z}\tilde{\boldsymbol{\theta}}_n)'(\mathbf{Y}^* - \mathbf{Z}\tilde{\boldsymbol{\theta}}_n) = \mathbf{Y}^*[\mathbf{I} - \mathbf{Z}(\mathbf{Z}'\mathbf{Z})^{-1}\mathbf{Z}']\mathbf{Y}^* \\
&= m^{-1}(\mathbf{Y} - \mathbf{X}\tilde{\beta}_n)'(\mathbf{Y} - \mathbf{X}\tilde{\beta}_n), \quad m = n - p. \qquad (8.1.17)
\end{aligned}$$

Further,

$$\begin{aligned}
(\tilde{\boldsymbol{\theta}}_n - \hat{\boldsymbol{\theta}}_n)'\mathbf{C}_z(\tilde{\boldsymbol{\theta}}_n - \hat{\boldsymbol{\theta}}_n) &= \tilde{\boldsymbol{\theta}}_n'\mathbf{R}'[\mathbf{R}\mathbf{C}_z^{-1}\mathbf{R}']^{-1}\mathbf{R}\tilde{\boldsymbol{\theta}}_n \\
&= (\mathbf{H}\tilde{\beta}_n - \mathbf{h})'(\mathbf{H}\mathbf{C}_z^{-1}\mathbf{H}' + \boldsymbol{\Omega})^{-1}(\mathbf{H}\tilde{\beta}_n - \mathbf{h}). \qquad (8.1.18)
\end{aligned}$$

Hence, we can define the statistics $\mathcal{L}_n^*$ for testing the hypothesis $H_0 : \mathbf{R}\boldsymbol{\theta} = \boldsymbol{\delta} = \mathbf{0}$ by

$$\begin{aligned}
\mathcal{L}_n^* &= \frac{\tilde{\boldsymbol{\theta}}_n'\mathbf{R}'(\mathbf{R}\mathbf{C}_z^{-1}\mathbf{R}')^{-1}\mathbf{R}\tilde{\boldsymbol{\theta}}_n}{qs_e^2} \\
&= \frac{(\mathbf{H}\tilde{\beta}_n - \mathbf{h})'(\mathbf{H}\mathbf{C}^{-1}\mathbf{H}' + \boldsymbol{\Omega})^{-1}(\mathbf{H}\tilde{\beta}_n - \mathbf{h})}{qs_e^2}, \qquad (8.1.19)
\end{aligned}$$

which is similar to the statistic $\mathcal{L}_n$ given in (7.1.6) except that it contains $\boldsymbol{\Omega}$.

The distribution of $\mathcal{L}_n^*$ under $H_0 : \boldsymbol{\delta} = \mathbf{R}\boldsymbol{\theta} = \mathbf{0}$ follows the central F-distribution with (q, m) d.f. Under $H_A : \boldsymbol{\delta} = \mathbf{R}\boldsymbol{\theta} \neq \mathbf{0}$, it follows the noncentral F-distribution with (q, m) d.f. and noncentrality parameter $\Delta^2/2$ where

$$\Delta^2 = \sigma^{-2}[\boldsymbol{\theta}'\mathbf{R}'(\mathbf{R}\mathbf{C}_z^{-1}\mathbf{R}')^{-1}\mathbf{R}\boldsymbol{\theta}] = \sigma^{-2}\boldsymbol{\delta}'\left(\mathbf{H}\mathbf{C}^{-1}\mathbf{H}' + \boldsymbol{\Omega}\right)^{-1}\boldsymbol{\delta}. \quad (8.1.20)$$

These statements are proved in the same way as in Theorem 7.1.1.

8.1.4 Preliminary Test and Stein-type Mixed Estimators

Following Section 7.2.1, we define the PTE of $\boldsymbol{\theta}$ as

$$\hat{\boldsymbol{\theta}}_n^{\text{PT}} = \tilde{\boldsymbol{\theta}}_n - (\tilde{\boldsymbol{\theta}}_n - \hat{\boldsymbol{\theta}}_n)I(\mathcal{L}_n^* < F_{q,m}(\alpha)), \quad (8.1.21)$$

where $F_{q,m}(\alpha)$ is the upper α-level critical value from the central F-distribution with (q, m) d.f.

Similarly, the James-Stein-type estimator of $\boldsymbol{\theta}$ is given by

$$\hat{\boldsymbol{\theta}}_n^{S} = \tilde{\boldsymbol{\theta}}_n - d(\tilde{\boldsymbol{\theta}}_n - \hat{\boldsymbol{\theta}}_n)\mathcal{L}_n^{*-1}, \quad (8.1.22)$$

and the PRSE of $\boldsymbol{\theta}$ is given by

$$\hat{\boldsymbol{\theta}}_n^{S+} = \hat{\boldsymbol{\theta}}_n + \{1 - d\mathcal{L}_n^{*-1}\}I(\mathcal{L}_n^* > d)(\tilde{\boldsymbol{\theta}}_n - \hat{\boldsymbol{\theta}}_n), \quad (8.1.23)$$

where

$$d = \frac{(q-2)m}{q(m+2)}. \quad (8.1.24)$$

Explicitly, the expressions for PTE, SE, and PRSE are given by

$$\hat{\boldsymbol{\theta}}_n^{\text{PT}} = \begin{bmatrix} \tilde{\boldsymbol{\beta}}_n \\ \mathbf{h} \end{bmatrix} - \begin{bmatrix} \tilde{\boldsymbol{\beta}}_n - \hat{\boldsymbol{\beta}}_n \\ \mathbf{h} - \mathbf{H}\hat{\boldsymbol{\beta}}_n \end{bmatrix} I(\mathcal{L}_n^* < F_{q,m}(\alpha)), \quad (8.1.25)$$

where $\hat{\boldsymbol{\beta}}_n = \tilde{\boldsymbol{\beta}}_n - \mathbf{C}^{-1}\mathbf{H}'(\mathbf{H}\mathbf{C}^{-1}\mathbf{H}' + \boldsymbol{\Omega})^{-1}\mathbf{H}\mathbf{C}^{-1}(\mathbf{H}\tilde{\boldsymbol{\beta}}_n - \mathbf{h})$.

$$\hat{\boldsymbol{\theta}}_n^{S} = \begin{bmatrix} \tilde{\boldsymbol{\beta}}_n \\ \mathbf{h} \end{bmatrix} - d \begin{bmatrix} \tilde{\boldsymbol{\beta}}_n - \hat{\boldsymbol{\beta}}_n \\ \mathbf{h} - \mathbf{H}\hat{\boldsymbol{\beta}}_n \end{bmatrix} \mathcal{L}_n^{*-1}, \quad (8.1.26)$$

and

$$\hat{\boldsymbol{\theta}}_n^{S+} = \begin{bmatrix} \hat{\boldsymbol{\beta}}_n \\ \mathbf{H}\hat{\boldsymbol{\beta}}_n \end{bmatrix} + \begin{bmatrix} \tilde{\boldsymbol{\beta}}_n - \hat{\boldsymbol{\beta}}_n \\ \mathbf{h} - \mathbf{H}\hat{\boldsymbol{\beta}}_n \end{bmatrix} (1 - d\mathcal{L}_n^{*-1})I(\mathcal{L}_n^* > d). \quad (8.1.27)$$

If we premultiply $\hat{\boldsymbol{\theta}}_n^{\text{PT}}$, $\hat{\boldsymbol{\theta}}_n^{S}$, and $\hat{\boldsymbol{\theta}}_n^{S+}$ by the $p \times (p+q)$ matrix $(\mathbf{I}_p, \mathbf{0}_{p\times q})$, we obtain the corresponding expressions for $\hat{\boldsymbol{\beta}}_n^{\text{PT}}$, $\hat{\boldsymbol{\beta}}_n^{S}$, and $\hat{\boldsymbol{\beta}}_n^{S+}$ (see also Khan, 1997).

8.2 Bias, Quadratic Bias, MSE Matrix, and the Quadratic Risk of the Estimators

In this section, we provide the bias, quadratic bias, MSE matrices, and risk expressions of the five estimators of $\boldsymbol{\theta}$.

8.2.1 Bias and Quadratic Bias Expressions

We consider the *bias* and *quadratic bias* expressions of the estimators of $\boldsymbol{\theta}$. They are given by

(i) $\mathbf{b}_1(\tilde{\boldsymbol{\theta}}_n) = \mathbf{0}$ and $B_1(\tilde{\boldsymbol{\theta}}_n) = 0$; (8.2.1)

(ii) $\mathbf{b}_2(\hat{\boldsymbol{\theta}}_n) = -\mathbf{C}_z^{-1}\mathbf{R}'(\mathbf{R}\mathbf{C}_z^{-1}\mathbf{R}')^{-1}\mathbf{R}\boldsymbol{\theta} = -\boldsymbol{\delta}_z$ (say), then

$$B_2(\hat{\boldsymbol{\theta}}_n) = \sigma^{-2}\boldsymbol{\delta}_z'\mathbf{C}_z\boldsymbol{\delta}_z = \Delta^2;$$

(iii) $\mathbf{b}_3(\hat{\boldsymbol{\theta}}_n^{\text{PT}}) = -\boldsymbol{\delta}_z G_{q+2,m}(\ell_\alpha;\Delta^2)$ and $B_3(\hat{\boldsymbol{\theta}}_n^{\text{PT}}) = \Delta^2\{G_{q+2,m}(\ell_\alpha;\Delta^2)\}^2$;

(iv) $\mathbf{b}_4(\hat{\boldsymbol{\theta}}_n^{S}) = -dq\boldsymbol{\delta}_z E[\chi_{q+2}^{-2}(\Delta^2)]$ and $B_4(\hat{\boldsymbol{\theta}}_n^{S}) = d^2q^2\Delta^2\{E[\chi_{q+2}^{-2}(\Delta^2)]\}^2$;

(v) $$\mathbf{b}_5(\hat{\boldsymbol{\theta}}_n^{S+}) = -\boldsymbol{\delta}_z\Big\{d_1 E[F_{q+2,m}^{-1}(\Delta^2)I(F_{q+2,m}(\Delta^2) < d_1)] - d_1 E[F_{q+2,m}^{-1}(\Delta^2)] - G_{q+2,m}(d_1;\Delta^2)\Big\}$$

and

$$B_5(\hat{\boldsymbol{\theta}}_n^{S+}) = \Delta^2\Big\{d_1 E[F_{q+2,m}^{-1}(\Delta^2)I(F_{q+2,m}(\Delta^2) < d_1)] - d_1 E[F_{q+2,m}^{-1}(\Delta^2)] - G_{q+2,m}(d_1;\Delta^2)\Big\}^2,$$

where $d_1 = qd/q+2$.

From the formulas above, we obtain the expressions for bias vector and quadratic bias of the estimators $\tilde{\boldsymbol{\beta}}_n$, $\hat{\boldsymbol{\beta}}_n$, $\hat{\boldsymbol{\beta}}_n^{\text{PT}}$, $\hat{\boldsymbol{\beta}}_n^{S}$, and $\hat{\boldsymbol{\beta}}_n^{S+}$ given below:

(i) $\mathbf{b}_1(\tilde{\boldsymbol{\beta}}_n) = \mathbf{0}$ and $B_1(\tilde{\boldsymbol{\beta}}_n) = 0$.

(ii) $\mathbf{b}_2(\hat{\boldsymbol{\beta}}_n) = -\mathbf{C}^{-1}\mathbf{H}'(\mathbf{H}\mathbf{C}^{-1}\mathbf{H}' + \boldsymbol{\Omega})^{-1}\boldsymbol{\delta}$ and $B_2(\hat{\boldsymbol{\beta}}_n) = \Delta^2$.

(iii) $\mathbf{b}_3(\hat{\boldsymbol{\beta}}_n^{\text{PT}}) = -\mathbf{C}^{-1}\mathbf{H}(\mathbf{H}\mathbf{C}^{-1}\mathbf{H}' + \boldsymbol{\Omega})^{-1}\boldsymbol{\delta}\, G_{q+2,m}(\ell_\alpha;\Delta^2)$ and
$B_3(\hat{\boldsymbol{\beta}}_n^{\text{PT}}) = \Delta^2\{(\ell_\alpha;\Delta^2)\}^2$.

(iv) $\mathbf{b}_4(\hat{\boldsymbol{\beta}}_n^{S}) = -dq\mathbf{C}^{-1}\mathbf{H}'(\mathbf{H}\mathbf{C}^{-1}\mathbf{H}' + \boldsymbol{\Omega})^{-1}\boldsymbol{\delta} E[\chi_{p+2}^{-2}(\Delta^2)]$ and
$B_4(\hat{\boldsymbol{\beta}}_n^{S}) = d^2q^2\Delta^2\{E[\chi_{p+2}^{-2}(\Delta^2)]\}^2$.

(v) $$\mathbf{b}_5(\hat{\boldsymbol{\beta}}_n^{S+}) = -\mathbf{C}^{-1}\mathbf{H}'(\mathbf{H}\mathbf{C}^{-1}\mathbf{H}' + \boldsymbol{\Omega})^{-1}\boldsymbol{\delta}\Big\{d_1 E\big[F_{q+2,m}^{-1}(\Delta^2) \times I(F_{q+2}(\Delta^2) < d_1)\big] - d_1 E\big[F_{q+2,m}^{-1}(\Delta^2)\big] - G_{q+2,m}(d_1;\Delta^2)\Big\}$$

and

$$B_5(\hat{\boldsymbol{\beta}}_n^{S+}) = d^2q^2\Delta^2\Big\{d_1 E\big[F_{q+2,m}^{-1}(\Delta^2) I(F_{q+2}(\Delta^2) < d_1)\big] - d_1 E\big[F_{q+2,m}^{-1}(\Delta^2)\big] - G_{q+2,m}(d_1;\Delta^2)\Big\}^2.$$

8.2.2 MSE Matrix and Risk Expressions

The MSE matrices and the quadratic risk expressions of $\tilde{\boldsymbol{\theta}}_n$, $\hat{\boldsymbol{\theta}}_n^{\text{PT}}$, $\hat{\boldsymbol{\theta}}_n^{S}$, and $\hat{\boldsymbol{\theta}}_n^{S+}$ are given by

(i) $$\mathbf{M}_1(\tilde{\boldsymbol{\theta}}_n) = \sigma^2\mathbf{C}_z^{-1} \text{ and } R_1(\tilde{\boldsymbol{\theta}}_{n+q};\mathbf{W}) = \sigma^2\operatorname{tr}(\mathbf{W}\mathbf{C}_z^{-1}); \qquad (8.2.2)$$

(ii) $$\mathbf{M}_2(\hat{\boldsymbol{\theta}}_n) = \sigma^2\mathbf{C}_z^{-1} - \sigma^2[\mathbf{C}_z^{-1}\mathbf{R}'(\mathbf{R}\mathbf{C}_z^{-1}\mathbf{R}')^{-1}\mathbf{R}\mathbf{C}_z^{-1} + \boldsymbol{\delta}_z\boldsymbol{\delta}_z' \text{ and}$$

$$R_2(\hat{\boldsymbol{\theta}}_n;\mathbf{W}) = \sigma^2\operatorname{tr}[\mathbf{W}\mathbf{C}_z^{-1}] - \sigma^2\operatorname{tr}[\mathbf{W}\mathbf{C}_z^{-1}\mathbf{R}'(\mathbf{R}\mathbf{C}_z^{-1}\mathbf{R}')^{-1}\mathbf{R}\mathbf{C}_z^{-1}] + \boldsymbol{\delta}_z'\mathbf{W}\boldsymbol{\delta}_z$$

(iii) $$\mathbf{M}_3(\hat{\boldsymbol{\theta}}_n^{\text{PT}}) = \sigma^2\mathbf{C}_z^{-1} - \sigma^2\mathbf{C}_z^{-1}\mathbf{R}'(\mathbf{R}\mathbf{C}_z^{-1}\mathbf{R}')^{-1}\mathbf{R}\mathbf{C}_z^{-1}G_{q+2,m}(\ell_\alpha;\Delta^2) \text{ and } + \boldsymbol{\delta}_z\boldsymbol{\delta}_z'\{2G_{q+2,m}(\ell_\alpha;\Delta^2) - G_{q+4,m}(\ell_\alpha^*;\Delta^2)\}$$

$$R_3(\hat{\boldsymbol{\theta}}_n^{\text{PT}}) = \sigma^2\operatorname{tr}[\mathbf{W}\mathbf{C}_z^{-1}] - \sigma^2\operatorname{tr}[\mathbf{W}\mathbf{C}_z^{-1}\mathbf{R}'(\mathbf{R}\mathbf{C}_z^{-1}\mathbf{R}')^{-1}\mathbf{R}\mathbf{C}_z^{-1}] \times G_{q+2,m}(\ell_\alpha;\Delta^2) + \boldsymbol{\delta}_z'\mathbf{W}\boldsymbol{\delta}_z\{2G_{q+2,m}(\ell_\alpha;\Delta^2) - G_{q+4,m}(\ell_\alpha^*;\Delta^2)\};$$

(iv) $$\mathbf{M}_4(\hat{\boldsymbol{\theta}}_n^{S}) = \sigma^2\mathbf{C}_z^{-1} - dq\sigma^2[\mathbf{C}_z^{-1}\mathbf{R}'(\mathbf{R}\mathbf{C}_z^{-1}\mathbf{R}')^{-1}\mathbf{R}\mathbf{C}_z^{-1}\Big\{2E[\chi_{q+2}^{-2}(\Delta^2)] - (q-2)E[\chi_{q+2}^{-4}(\Delta^2)]\Big\} + dq(q+2)\boldsymbol{\delta}_z\boldsymbol{\delta}_z'E[\chi_{q+4}^{-4}(\Delta^2)]$$

and

$$R_4(\hat{\boldsymbol{\theta}}_n^{S};\mathbf{W}) = \sigma^2\operatorname{tr}[\mathbf{W}\mathbf{C}_z^{-1}] - dq\sigma^2\operatorname{tr}[\mathbf{W}\mathbf{C}_z^{-1}\mathbf{R}'(\mathbf{R}\mathbf{C}_z^{-1}\mathbf{R}')^{-1}\mathbf{R}\mathbf{C}_z^{-1}] \times \Big\{2E[\chi_{q+2}^{-2}(\Delta^2)] - (q-2)E[\chi_{q+2}^{-4}(\Delta^2)]\Big\} + dq(q+2)\boldsymbol{\delta}_z'\mathbf{W}\boldsymbol{\delta}_z E[\chi_{q+4}^{-4}(\Delta^2)];$$

(v) $$\mathbf{M}_5(\hat{\boldsymbol{\theta}}_n^{S+}) = \mathbf{M}_4(\hat{\boldsymbol{\theta}}_n^{S}) - \sigma^2\mathbf{C}_z^{-1}\mathbf{R}'(\mathbf{R}\mathbf{C}_z^{-1}\mathbf{R}')^{-1}\mathbf{R}\mathbf{C}_z^{-1} \times E\left[(1 - d_1F_{q+2,m}^{-1}(\Delta^2))^2 I(F_{q+2,m}(\Delta^2) < d_1)\right] + \boldsymbol{\delta}_z\boldsymbol{\delta}_z'\left\{2E\left[(d_1F_{q+2,m}^{-1}(\Delta^2) - 1)I(F_{q+2,m}(\Delta^2) < d_1)\right] - E\left[(1 - d_2F_{q+4,m}^{-1}(\Delta^2))^2 I(F_{q+4,m}(\Delta^2) < d_2)\right]\right\},$$

and

$$
\begin{aligned}
R_5(\hat{\boldsymbol{\theta}}_n^{S+}) &= R_4(\hat{\boldsymbol{\theta}}_n^{S}) - \sigma^2 \operatorname{tr}\mathbf{W}\mathbf{C}_z^{-1}\mathbf{R}'(\mathbf{R}\mathbf{C}_z^{-1}\mathbf{R}')^{-1}\mathbf{R}\mathbf{C}_z^{-1} \\
&\times E\left[(1 - d_1 F_{q+2,m}^{-1}(\Delta^2))^2 I(F_{q+2,m}(\Delta^2) < d_1)\right] \\
&+ \boldsymbol{\delta}_z'\mathbf{W}\boldsymbol{\delta}_z\left\{2E\left[(d_1 F_{q+2,m}^{-1}(\Delta^2) - 1) I(F_{q+2,m}(\Delta^2) < d_1)\right]\right. \\
&\left.- E\left[(1 - d_2 F_{q+4,m}^{-1}(\Delta^2))^2 I(F_{q+4,m}(\Delta^2) < d_2)\right]\right\}.
\end{aligned}
$$

To obtain the MSE matrices of the estimators of $\tilde{\boldsymbol{\beta}}_n$, $\hat{\boldsymbol{\beta}}_n$, $\hat{\boldsymbol{\beta}}_n^{PT}$, $\hat{\boldsymbol{\beta}}_n^{S}$ and $\hat{\boldsymbol{\beta}}_n^{S+}$ we pre- and postmultiply the MSE matrices above by the $p \times (p+q)$ matrix $(\mathbf{I}_p, \mathbf{0}_{p+q})$ and its transpose, respectively. The resulting expressions are given below:

(i) $\mathbf{M}_1(\tilde{\boldsymbol{\beta}}_n) = \sigma^2\mathbf{C}^{-1}$ and $R_1(\tilde{\boldsymbol{\beta}}_n; \mathbf{W}) = \sigma^2 \operatorname{tr}(\mathbf{W}\mathbf{C}^{-1})$.

(ii)
$$
\begin{aligned}
\mathbf{M}_2(\hat{\boldsymbol{\beta}}_n) &= \sigma^2\mathbf{C}^{-1} - \sigma^2\mathbf{C}^{-1}\mathbf{H}'(\mathbf{H}\mathbf{C}^{-1}\mathbf{H}' + \boldsymbol{\Omega})^{-1}\mathbf{H}\mathbf{C}^{-1} \\
&+ \mathbf{C}^{-1}\mathbf{H}'(\mathbf{H}\mathbf{C}^{-1}\mathbf{H}' + \boldsymbol{\Omega})^{-1}\boldsymbol{\delta}\boldsymbol{\delta}'(\mathbf{H}\mathbf{C}^{-1}\mathbf{H}' + \boldsymbol{\Omega})^{-1}\mathbf{H}\mathbf{C}^{-1}
\end{aligned}
$$

and

$$
\begin{aligned}
R_2(\hat{\boldsymbol{\beta}}_n, \mathbf{W}) &= \sigma^2 \operatorname{tr}[\mathbf{W}\mathbf{C}^{-1}] - \sigma^2 \operatorname{tr}\left[\mathbf{W}\mathbf{C}^{-1}\mathbf{H}'(\mathbf{H}\mathbf{C}^{-1}\mathbf{H}' + \boldsymbol{\Omega})^{-1}\mathbf{H}\mathbf{C}^{-1}\right] \\
&+ \boldsymbol{\delta}'(\mathbf{H}\mathbf{C}^{-1}\mathbf{H}' + \boldsymbol{\Omega})^{-1}\mathbf{H}'\mathbf{C}^{-1}\mathbf{W}\mathbf{C}^{-1}\mathbf{H}'(\mathbf{H}\mathbf{C}^{-1}\mathbf{H}' + \boldsymbol{\Omega})^{-1}\mathbf{H}\mathbf{C}^{-1}\boldsymbol{\delta}.
\end{aligned}
$$

(iii) $\mathbf{M}_3(\hat{\boldsymbol{\beta}}_n^{PT}) = \sigma^2\mathbf{C}^{-1} - \sigma^2\left[\mathbf{C}^{-1}\mathbf{H}'(\mathbf{H}\mathbf{C}^{-1}\mathbf{H}' + \boldsymbol{\Omega})^{-1}\mathbf{H}\mathbf{C}^{-1}\right] G_{q+2,m}(\ell_\alpha; \Delta^2)$ $+\left[\mathbf{C}^{-1}\mathbf{H}'(\mathbf{H}\mathbf{C}^{-1}\mathbf{H}' + \boldsymbol{\Omega})^{-1}\boldsymbol{\delta}\boldsymbol{\delta}'(\mathbf{H}\mathbf{C}^{-1}\mathbf{H}' + \boldsymbol{\Omega})^{-1}\right]\left\{2G_{q+2,m}(\ell_\alpha; \Delta^2) - G_{q+4,m}(\ell_\alpha^*; \Delta^2)\right].$

(iv)
$$
\begin{aligned}
\mathbf{M}_4(\hat{\boldsymbol{\beta}}_n^{S}) &= \sigma^2\mathbf{C}^{-1} - dq\sigma^2\left[\mathbf{C}^{-1}\mathbf{H}'(\mathbf{H}\mathbf{C}^{-1}\mathbf{H}' + \boldsymbol{\Omega})^{-1}\mathbf{H}\mathbf{C}^{-1}\right] \\
&\times \left\{2E\left[\chi_{q+2}^{-2}(\Delta^2)\right] - (q-2)E\left[E_{q+2}^{-4}(\Delta^2)\right]\right\} \\
&+ dq(q+2)\left[\mathbf{C}^{-1}\mathbf{H}'(\mathbf{H}\mathbf{C}^{-1}\mathbf{H}' + \boldsymbol{\Omega})^{-1}\boldsymbol{\delta}\boldsymbol{\delta}'(\mathbf{H}\mathbf{C}^{-1}\mathbf{H}' + \boldsymbol{\Omega})\mathbf{H}\mathbf{C}^{-1}\right] E\left[\chi_{q+4}^{-4}(\Delta^2)\right]
\end{aligned}
$$

and

$$
\begin{aligned}
R_4(\hat{\beta}_n^{S}; \mathbf{W}) &= \sigma^2 \operatorname{tr}(\mathbf{W}\mathbf{C}^{-1}) - dq\sigma^2 \operatorname{tr}\left[\mathbf{W}\mathbf{C}^{-1}\mathbf{H}'(\mathbf{H}\mathbf{C}^{-1}\mathbf{H}' + \boldsymbol{\Omega})^{-1}\mathbf{H}\mathbf{C}^{-1}\right] \\
&\times \left\{2E\left[\chi_{q+2}^{-2}(\Delta^2)\right] - (q-2)E\left[\chi_{q+2}^{-4}(\Delta^2)\right]\right\} \\
&+ dq(q+2)[\delta'(\mathbf{H}\mathbf{C}^{-1}\mathbf{H}' + \boldsymbol{\Omega})^{-1} \\
&\times \mathbf{H}\mathbf{C}^{-1}\mathbf{W}\mathbf{C}^{-1}\mathbf{H}'(\mathbf{H}\mathbf{C}^{-1}\mathbf{H}' + \boldsymbol{\Omega})^{-1}\delta\big] E\left[\chi_{q+4}^{-4}(\Delta^2)\right].
\end{aligned}
$$

$$\text{(v)}\ \mathbf{M}_5(\hat{\beta}_n^{S+}) = \mathbf{M}_4(\hat{\beta}_n^{S}) - \sigma^2[\mathbf{C}^{-1}\mathbf{H}'(\mathbf{H}\mathbf{C}^{-1}\mathbf{H}'+\boldsymbol{\Omega})^{-1}\mathbf{H}\mathbf{C}^{-1}]$$
$$\times E\big[\big(1-d_1F_{q+2,m}^{-1}(\Delta^2)\big)^2 I\big(F_{q+2,m}(\Delta^2)<d_1\big)\big]$$
$$+\big[\mathbf{C}^{-1}\mathbf{H}'(\mathbf{H}\mathbf{C}^{-1}\mathbf{H}'+\boldsymbol{\Omega})^{-1}\boldsymbol{\delta}\boldsymbol{\delta}'(\mathbf{H}\mathbf{C}^{-1}\mathbf{H}'+\boldsymbol{\Omega})^{-1}\mathbf{H}\mathbf{C}^{-1}\big]$$
$$\times\big\{2E\big[\big(1-d_1F_{q+2,m}^{-1}(\Delta^2)\big)\, I\big(F_{q+2,m}(\Delta^2)<d_1\big)\big]$$
$$-E\big[\big(1-d_2F_{q+4,m}^{-1}(\Delta^2)\big)^2 I\big(F_{q+4,m}(\Delta^2)<d_2\big)\big]\big\}$$

and

$$R_5(\hat{\beta}_n^{S+};\mathbf{W})$$
$$= R_4(\hat{\beta}_n^{S};W) - \sigma^2\,\mathrm{tr}\big[\mathbf{W}\mathbf{C}^{-1}\mathbf{H}'(\mathbf{H}\mathbf{C}^{-1}\mathbf{H}'+\boldsymbol{\Omega})^{-1}\mathbf{H}\mathbf{C}^{-1}\big]$$
$$\times E\big[\big(1-d,F_{q+2,m}^{-1}(\Delta^2)\big)^2 I\big(F_{q+2,m}<d_1\big)\big]$$
$$-\ \boldsymbol{\delta}'(\mathbf{H}\mathbf{C}^{-1}\mathbf{H}'+\boldsymbol{\Omega})^{-1}\mathbf{H}\mathbf{C}^{-1}\mathbf{W}\mathbf{C}^{-1}\mathbf{H}'(\mathbf{H}\mathbf{C}^{-1}\mathbf{H}'+\boldsymbol{\Omega})^{-1}\boldsymbol{\delta}$$
$$\times\big\{2E\big[\big(1-d_1F_{q+2,m}^{-1}(\Delta^2)\big)\big)I\big(F_{q+2,m}(\Delta^2)<d_1\big)\big]$$
$$-E\big[\big(1-d_2F_{q+4,m}^{-1}(\Delta^2)\big)I\big(F_{q+4,m}(\Delta^2)<d_2\big)\big]\big\}.$$

8.2.3 MSE Matrix Comparisons of the Estimators

Comparison of $\tilde{\boldsymbol{\theta}}_n$ and $\hat{\boldsymbol{\theta}}_n$. First, we compare $\tilde{\boldsymbol{\theta}}_n$ and $\hat{\boldsymbol{\theta}}_n$. The difference between the MSE matrices is given by

$$M_1(\tilde{\boldsymbol{\theta}}_n) - M_2(\hat{\boldsymbol{\theta}}_n) = \sigma^2\mathbf{C}_z^{-1}\mathbf{R}'(\mathbf{R}\mathbf{C}_z^{-1}\mathbf{R}')^{-1}\mathbf{R}\mathbf{C}_z^{-1} - \boldsymbol{\delta}_z\boldsymbol{\delta}_z' \qquad (8.2.3)$$

The r.h.s. of (8.2.3) is positive-semidefinite whenever for a nonzero vector $\boldsymbol{\ell}=(l_1,\cdots,l_{n+q})'$ we have

$$\sigma^2\boldsymbol{\ell}'\mathbf{C}_z^{-1}\mathbf{R}'(\mathbf{R}\mathbf{C}_z^{-1}\mathbf{R}')^{-1}\mathbf{R}\mathbf{C}_z^{-1}\boldsymbol{\ell} \geq \boldsymbol{\ell}'\boldsymbol{\delta}_z\boldsymbol{\delta}_z'\boldsymbol{\ell}. \qquad (8.2.4)$$

Dividing by $\boldsymbol{\ell}'C_z\boldsymbol{\ell}$, we have

$$\frac{\boldsymbol{\ell}'\boldsymbol{\delta}_z\boldsymbol{\delta}_z'\boldsymbol{\ell}}{\sigma^2\boldsymbol{\ell}'\mathbf{C}_z^{-1}\boldsymbol{\ell}} \leq \frac{\boldsymbol{\ell}'\mathbf{C}_z^{-1}\mathbf{R}'(\mathbf{R}\mathbf{C}_z^{-1}\mathbf{R}')^{-1}\mathbf{R}\mathbf{C}_z^{-1}\boldsymbol{\ell}}{\boldsymbol{\ell}'\mathbf{C}_z^{-1}\boldsymbol{\ell}}.$$

Therefore,

$$\max_{\boldsymbol{\ell}}\frac{\boldsymbol{\ell}'\boldsymbol{\delta}_z\boldsymbol{\delta}_z'\boldsymbol{\ell}}{\sigma^2\boldsymbol{\ell}'\mathbf{C}_z^{-1}\boldsymbol{\ell}} = \Delta^2 \leq Ch_{max}\{\mathbf{R}'(\mathbf{R}\mathbf{C}_z^{-1}\mathbf{R}')^{-1}\mathbf{R}\mathbf{C}_z^{-1}\} = 1 \qquad (8.2.5)$$

Thus, the r.h.s. of (8.2.3) is a positive-semidefinite matrix whenever $\Delta^2 \leq 1$, and $\hat{\boldsymbol{\theta}}_n$ performs better than $\tilde{\boldsymbol{\theta}}_n$ in this range; otherwise, $\tilde{\boldsymbol{\theta}}_n$ is better than $\hat{\boldsymbol{\theta}}_n$. MSE-based efficiency of $\hat{\boldsymbol{\theta}}_n$ relative to $\tilde{\boldsymbol{\theta}}_n$ is meaningless, since

$$|\mathbf{I} - \mathbf{R}'(\mathbf{R}\mathbf{C}_z^{-1}\mathbf{R}')^{-1}\mathbf{R}\mathbf{C}_z^{-1} - \sigma^{-2}\mathbf{C}_z\boldsymbol{\delta}_z\boldsymbol{\delta}_z'| = 0.$$

Comparisons of $\hat{\boldsymbol{\theta}}_n^{\text{PT}}$, $\tilde{\boldsymbol{\theta}}_n$, and $\hat{\boldsymbol{\theta}}_n$. The difference between the MSE matrices $\tilde{\boldsymbol{\theta}}_n$ and $\hat{\boldsymbol{\theta}}_n^{\text{PT}}$ is given by

$$\mathbf{M}_1(\tilde{\boldsymbol{\theta}}_n) - \mathbf{M}_3(\hat{\boldsymbol{\theta}}_n^{\text{PT}}) = \sigma^2\mathbf{C}_z^{-1}\mathbf{R}'(\mathbf{R}\mathbf{C}_z^{-1}\mathbf{R}')^{-1}\mathbf{R}\mathbf{C}_z^{-1}G_{q+2,m}(\ell_\alpha;\Delta^2) - \boldsymbol{\delta}_z\boldsymbol{\delta}_z'\{2G_{q+2,m}(\ell_\alpha;\Delta^2) - G_{q+4,m}(\ell_\alpha^*;\Delta^2)\}. \quad (8.2.6)$$

Following Section 8.2.3a, we note that the r.h.s. of (8.2.6) is p.s.d. whenever

$$\Delta^2 \le \frac{G_{q+2,m}(\ell_\alpha;\Delta^2)}{\{2G_{q+2,m}(\ell_\alpha;\Delta^2) - G_{q+4,m}(\ell_\alpha^*;\Delta^2)\}}(\le 1). \quad (8.2.7)$$

Hence, $\hat{\boldsymbol{\theta}}_n^{\text{PT}}$ is superior to $\tilde{\boldsymbol{\theta}}_n$ in this range of Δ^2. For

$$\Delta^2 \ge \frac{G_{q+2,m}(\ell_\alpha;\Delta^2)}{\{2G_{q+2,m}(\ell_\alpha;\Delta^2) - G_{q+4,m}(\ell_\alpha^*;\Delta^2)\}}, \quad (8.2.8)$$

$\tilde{\boldsymbol{\theta}}_n$ is superior to $\hat{\boldsymbol{\theta}}_n^{\text{PT}}$.

Similarly, to compare $\hat{\boldsymbol{\theta}}_n^{\text{PT}}$ and $\hat{\boldsymbol{\theta}}_n$, we consider the difference between the MSE matrices:

$$\mathbf{M}_2(\hat{\boldsymbol{\theta}}_n) - \mathbf{M}_3(\hat{\boldsymbol{\theta}}_n^{\text{PT}}) = -\sigma^2\mathbf{C}_z^{-1}\mathbf{R}'(\mathbf{R}\mathbf{C}_z^{-1}\mathbf{R}')^{-1}\mathbf{R}\mathbf{C}_z^{-1}\{1 - G_{q+2,m}(\ell_\alpha;\Delta^2)\} + \boldsymbol{\delta}_z\boldsymbol{\delta}_z'\{1 - 2G_{q+2,m}(\ell_\alpha;\Delta^2) + G_{q+4,m}(\ell_\alpha^*;\Delta^2)\}, \quad (8.2.9)$$

which implies that $\hat{\boldsymbol{\theta}}_n$ is superior to $\hat{\boldsymbol{\theta}}_n^{\text{PT}}$ whenever

$$\Delta^2 \le \frac{1 - G_{q+2,m}(\ell_\alpha;\Delta^2)}{\{1 - 2G_{q+2,m}(\ell_\alpha;\Delta^2) + G_{q+4,m}(\ell_\alpha^*;\Delta^2)\}}; \quad (8.2.10)$$

otherwise, $\hat{\boldsymbol{\theta}}_n^{\text{PT}}$ is superior to $\hat{\boldsymbol{\theta}}_n$.

Comparisons of $\hat{\boldsymbol{\theta}}_n^{S}$, $\tilde{\boldsymbol{\theta}}_n$, and $\hat{\boldsymbol{\theta}}_n$. First, we compare $\tilde{\boldsymbol{\theta}}_n$ and $\hat{\boldsymbol{\theta}}_n^{S}$. In this case, the difference between the MSE matrices of $\tilde{\boldsymbol{\theta}}_n$ and $\hat{\boldsymbol{\theta}}_n^{S}$ is given by

$$\mathbf{M}_1(\tilde{\boldsymbol{\theta}}_n) - \mathbf{M}_4(\hat{\boldsymbol{\theta}}_n^{S}) = dq\sigma^2\mathbf{C}_z^{-1}\mathbf{R}'(\mathbf{R}\mathbf{C}_z^{-1}\mathbf{R}')^{-1}\mathbf{R}\mathbf{C}_z^{-1}\{2E[\chi_{q+2}^{-2}(\Delta^2)] - (q-2)E[\chi_{q+2}^{-4}(\Delta^2)]\} - dq(q+2)\boldsymbol{\delta}_z\boldsymbol{\delta}_z'E[\chi_{q+4}^{-4}(\Delta^2)]. \quad (8.2.11)$$

The r.h.s. of (8.2.11) is p.s.d. whenever for a given vector $\boldsymbol{\ell}$,

$$dq\sigma^2\{\boldsymbol{\ell}'\mathbf{C}_z^{-1}\mathbf{R}'(\mathbf{R}\mathbf{C}_z^{-1}\mathbf{R}')^{-1}\mathbf{R}\mathbf{C}_z^{-1}\boldsymbol{\ell}\}\{2E[\chi_{q+2}^{-2}(\Delta^2)] - (q-2)E[\chi_{q+2}^{-4}(\Delta^2)]\} - dq(q+2)\boldsymbol{\delta}_z\boldsymbol{\delta}_z'E[\chi_{q+4}^{-4}(\Delta^2)] \ge 0. \quad (8.2.12)$$

After simplification, as in Section 8.2.1, we obtain

$$q\Delta^2E[\chi_{q+4}^{-4}(\Delta^2)] \le (q-2)E[\chi_{q+2}^{-4}(\Delta^2)], \quad (8.2.13)$$

which does not hold for all Δ^2. Hence, $\hat{\boldsymbol{\theta}}_n^S$ is not uniformly better than $\tilde{\boldsymbol{\theta}}_n$. The $MRE(\hat{\boldsymbol{\theta}}_n^S;\tilde{\boldsymbol{\theta}}_n)$ is given by

$$\left(1-dq\{(q-2)E[\chi_{q+4}^{-4}(\Delta^2)]+2\Delta^2E[\chi_{q+2}^{-4}(\Delta^2)]\}+dq(q+2)\Delta^2E[\chi_{q+4}^{-4}(\Delta^2)]\right)^{-1} \times\left(1-dq\{(q-2)E[\chi_{q+2}^{-4}(\Delta^2)]+2\Delta^2E[\chi_{q+4}^{-4}(\Delta^2)]\}\right)^{\frac{p-1}{p}}. \quad (8.2.14)$$

Similarly, the comparison of $\hat{\boldsymbol{\theta}}_n$ and $\hat{\boldsymbol{\theta}}_n^S$ is obtained through the difference between the MSE matrices:

$$\begin{aligned}&\mathbf{M}_2(\hat{\boldsymbol{\theta}}_n)-\mathbf{M}_4(\hat{\boldsymbol{\theta}}_n^S)\\&=-\sigma^2\mathbf{C}_z^{-1}\mathbf{R}'(\mathbf{R}\mathbf{C}_z^{-1}\mathbf{R}')^{-1}\mathbf{R}\mathbf{C}_z^{-1}\{1-dq(q-2)E[\chi_{q+2}^{-4}(\Delta^2)]\\&\quad-2dq\Delta^2E[\chi_{q+4}^{-4}(\Delta^2)]\}+\boldsymbol{\delta}_z\boldsymbol{\delta}_z'\{1-dq(q+2)E[\chi_{q+4}^{-4}(\Delta^2)]\}. \quad (8.2.15)\end{aligned}$$

Under H_0, the MSE difference is n.s.d. Therefore, $\hat{\boldsymbol{\theta}}_n$ performs better than $\hat{\boldsymbol{\theta}}_n^S$ at $\Delta^2=0$. Further, $\hat{\boldsymbol{\theta}}_n$ performs better than $\hat{\boldsymbol{\theta}}_n^S$ in the interval given by

$$0\le\Delta^2\le\frac{1-dq(q-2)E[\chi_{q+2}^{-4}(\Delta^2)]}{1-dq^2E[\chi_{q+4}^{-4}(\Delta^2)]}. \quad (8.2.16)$$

Outside the interval, $\hat{\boldsymbol{\theta}}_n^S$ performs better than $\hat{\boldsymbol{\theta}}_n$. Thus, $\hat{\boldsymbol{\theta}}_n^S$ is not uniformly better than $\hat{\boldsymbol{\theta}}_n$.

Comparisons of $\hat{\boldsymbol{\theta}}_n^{S+}$, $\hat{\boldsymbol{\theta}}_n^S$, and $\hat{\boldsymbol{\theta}}_n^{\mathrm{PT}}$. Consider the difference between the MSE matrices of $\hat{\boldsymbol{\theta}}_n^S$ and $\hat{\boldsymbol{\theta}}_n^{S+}$:

$$\begin{aligned}&\mathbf{M}_4(\hat{\boldsymbol{\theta}}_n^S)-\mathbf{M}_5(\hat{\boldsymbol{\theta}}_n^{S+})\\&=\sigma^2\Big\{\mathbf{C}_z^{-1}\mathbf{R}'(\mathbf{R}\mathbf{C}_z^{-1}\mathbf{R}')^{-1}\mathbf{R}\mathbf{C}_z^{-1}E\big[(1-d_1F_{q+2,m}^{-1}(\Delta^2))^2I(F_{q+2,m}(\Delta^2)<d_1)\big]\\&\quad+\sigma^{-2}\boldsymbol{\delta}_z\boldsymbol{\delta}_z'E\big[(1-d_2F_{q+4,m}^{-1}(\Delta^2))^2I(F_{q+4,m}(\Delta^2)<d_2)\big]\Big\}\\&\quad+2\boldsymbol{\delta}_z\boldsymbol{\delta}_z'E\big[(d_1F_{q+2,m}^{-1}(\Delta^2)-1)I(F_{q+2,m}(\Delta^2)<d_1)\big]\ge\mathbf{0}. \quad (8.2.17)\end{aligned}$$

The difference between the MSE matrices is p.s.d. for all Δ^2. Hence, $\hat{\boldsymbol{\theta}}_n^{S+}$ dominates $\hat{\boldsymbol{\theta}}_n^S$ uniformly. But, $\hat{\boldsymbol{\theta}}_n^S$ does not dominate $\tilde{\boldsymbol{\theta}}_n$. Hence, we can write

$$\mathbf{M}_4(\hat{\boldsymbol{\theta}}_n^S)\ge\mathbf{M}_5(\hat{\boldsymbol{\theta}}_n^{S+}). \quad (8.2.18)$$

Next, we consider the difference between the MSE matrices, which is

$$\begin{aligned}\mathbf{M}_4(\hat{\boldsymbol{\theta}}_n^{S}) - \mathbf{M}_3(\hat{\boldsymbol{\theta}}_n^{\mathrm{PT}}) = \sigma^2\mathbf{C}_z^{-1}\mathbf{R}'(\mathbf{R}\mathbf{C}_z^{-1}\mathbf{R}')^{-1}\mathbf{R}\mathbf{C}_z^{-1}\Big\{&G_{q+2,m}(\ell_\alpha;\Delta^2)\\ &- dq(q-2)E[\chi_{q+2}^{-4}(\Delta^2)] - 2dq\Delta^2E[\chi_{q+4}^{-4}(\Delta^2)]\Big\}\\ &- \boldsymbol{\delta}_z\boldsymbol{\delta}_z'\Big\{2G_{q+2,m}(\ell_\alpha;\Delta^2) - G_{q+4,m}(\ell_\alpha^*;\Delta^2)\\ &- dq(q+2)E[\chi_{q+4}^{-4}(\Delta^2)]\Big\} \qquad (8.2.19)\end{aligned}$$

and

$$\begin{aligned}\mathbf{M}_5(\hat{\boldsymbol{\theta}}_n^{S+}) - \mathbf{M}_3(\hat{\boldsymbol{\theta}}_n^{\mathrm{PT}}) &= \sigma^2\mathbf{C}_z^{-1}\mathbf{R}'(\mathbf{R}\mathbf{C}_z^{-1}\mathbf{R}')^{-1}\mathbf{R}\mathbf{C}_z^{-1}\Big\{G_{q+2,m}(\ell_\alpha;\Delta^2)\\ &- dq(q-2)E[\chi_{q+2}^{-4}(\Delta^2)] - 2dq\Delta^2E[\chi_{q+4}^{-4}(\Delta^2)]\\ &- E\big[(1-d_1F_{q+2,m}^{-1}(\Delta^2))^2I(F_{q+2,m}(\Delta^2)<d_1)\big]\Big\}\\ &+ \boldsymbol{\delta}_z\boldsymbol{\delta}_z'\Big\{E\Big[(1-d_2F_{q+4,m}^{-1}(\Delta^2))^2I(F_{q+4,m}(\Delta^2)<d_2)\Big]\\ &-\big[2G_{q+2,m}(\ell_\alpha;\Delta^2) - G_{q+4,m}(\ell_\alpha^*;\Delta^2) - dq(q+2)E[\chi_{q+4}^{-4}(\Delta^2)]\big]\\ &- 2E\big[(1-d_1F_{q+2,m}^{-1}(\Delta^2))^2I(F_{q+2,m}(\Delta^2)<d_1)\big]\Big\}. \qquad (8.2.20)\end{aligned}$$

The r.h.s. of (8.2.19) is p.s.d. whenever

$$\Delta^2 \le \frac{G_{q+2,m}(\ell_\alpha;\Delta^2) - dq(q-2)E[\chi_{q+2}^{-4}(\Delta^2)] - 2dq\Delta^2E[\chi_{q+4}^{-4}(\Delta^2)]}{2G_{q+2,m}(\ell_\alpha;\Delta^2) - G_{q+4,m}(\ell_\alpha^*;\Delta^2) - dq(q+2)E[\chi_{q+2}^{-4}(\Delta^2)]}. \qquad (8.2.21)$$

Thus, $\hat{\boldsymbol{\theta}}_n^{\mathrm{PT}}$ performs better than $\hat{\boldsymbol{\theta}}_n^{S}$ whenever Δ^2 satisfies (8.2.21); otherwise, $\hat{\boldsymbol{\theta}}_n^{S}$ performs better than $\hat{\boldsymbol{\theta}}_n^{\mathrm{PT}}$. Neither $\hat{\boldsymbol{\theta}}_n^{S}$ nor $\hat{\boldsymbol{\theta}}_n^{\mathrm{PT}}$ dominates the other. Similarly, using (8.2.20), we note that neither $\hat{\boldsymbol{\theta}}_n^{S+}$ nor $\hat{\boldsymbol{\theta}}_n^{\mathrm{PT}}$ dominates each other.

Next, we consider the difference between the MSE matrices under H_0 to compare $\hat{\boldsymbol{\theta}}_n^{S+}$ and $\hat{\boldsymbol{\theta}}_n$:

$$\begin{aligned}\mathbf{M}_5(\hat{\boldsymbol{\theta}}_n^{S+}) - \mathbf{M}_2(\hat{\boldsymbol{\theta}}_n) = \sigma^2\mathbf{C}_z^{-1}\mathbf{R}'(\mathbf{R}\mathbf{C}_z^{-1}\mathbf{R}')^{-1}\mathbf{R}\mathbf{C}_z^{-1}\{&1-d\\ &-E\big[(1-d_1F_{q+2,m}^{-1}(\Delta^2))^2I(F_{q+2,m}(\Delta^2)<d_1)\big]\}. \qquad (8.2.22)\end{aligned}$$

This R.H.S of (8.2.22) is p.s.d. whenever

$$1-d \ge E\big[(1-d_1F_{q+2,m}^{-1}(\Delta^2))^2I(F_{q+2,m}(\Delta^2)<d_1)\big]. \qquad (8.2.23)$$

Hence, $\hat{\boldsymbol{\theta}}_n$ performs better than $\hat{\boldsymbol{\theta}}_n^{S+}$ under H_0, and $\hat{\boldsymbol{\theta}}_n^{S+}$ is not uniformly better than $\hat{\boldsymbol{\theta}}_n$. Similar conclusions hold on the properties of the five estimators of $\boldsymbol{\beta}$.

8.2.4 Risk Comparisons of the Estimations

In this section, we provide the weighted risk analysis of the five estimators with the loss function

$$L(\boldsymbol{\theta}^* : \boldsymbol{\theta}) = (\boldsymbol{\theta}^* - \boldsymbol{\theta})'\mathbf{W}(\boldsymbol{\theta}^* - \boldsymbol{\theta}), \tag{8.2.24}$$

where $\mathbf{W}$ is a p.s.d. matrix.

Comparisons of $\tilde{\boldsymbol{\theta}}_n$ and $\hat{\boldsymbol{\theta}}_n$. It is clear that the risk of $\tilde{\boldsymbol{\theta}}_n$ is constant, meaning $\sigma^2 \operatorname{tr}(\mathbf{W}\mathbf{C}_z^{-1})$, and the risk of $\hat{\boldsymbol{\theta}}_n$ depends on $\boldsymbol{\delta}_z'\mathbf{W}\boldsymbol{\delta}_z$, since

$$R_2(\hat{\boldsymbol{\theta}}_n : \mathbf{W}) = \sigma^2 \operatorname{tr}(\mathbf{W}\mathbf{C}_z^{-1}) - \sigma^2 \operatorname{tr}[\mathbf{W}\mathbf{C}_z^{-1}\mathbf{R}'(\mathbf{R}\mathbf{C}_z^{-1}\mathbf{R}')^{-1}\mathbf{R}\mathbf{C}_z^{-1}] + \boldsymbol{\delta}_z'\mathbf{W}\boldsymbol{\delta}_z. \tag{8.2.25}$$

Further, $\mathbf{C}_z^{-1/2}\mathbf{R}'(\mathbf{R}\mathbf{C}_z^{-1}\mathbf{R}')^{-1}\mathbf{R}\mathbf{C}_z^{-1/2}$ is a symmetric idempotent matrix with rank $q(\leq p+q)$. Therefore, there exists an orthogonal matrix $\boldsymbol{\Gamma}$ such that

$$\boldsymbol{\Gamma}\mathbf{C}_z^{-1/2}\mathbf{R}'(\mathbf{R}\mathbf{C}_z^{-1}\mathbf{R}')^{-1}\mathbf{R}\mathbf{C}_z^{-1/2}\boldsymbol{\Gamma}' = \begin{pmatrix} \mathbf{I}_q & \mathbf{0} \\ \mathbf{0} & \mathbf{0} \end{pmatrix} \tag{8.2.26}$$

and

$$\boldsymbol{\Gamma}\mathbf{C}_z^{-1/2}\mathbf{W}\mathbf{C}_z^{-1/2}\boldsymbol{\Gamma}' = \begin{pmatrix} \mathbf{A}_{11} & \mathbf{A}_{12} \\ \mathbf{A}_{12}' & \mathbf{A}_{22} \end{pmatrix}. \tag{8.2.27}$$

Thus,

$$\operatorname{tr}[\mathbf{R}\mathbf{C}_z^{-1}\mathbf{W}\mathbf{C}_z^{-1}\mathbf{R}'(\mathbf{R}\mathbf{C}_z^{-1}\mathbf{R}')^{-1}] = \operatorname{tr}(\mathbf{A}_{11}) \tag{8.2.28}$$

and

$$\boldsymbol{\delta}_z'\mathbf{W}\boldsymbol{\delta}_z = \boldsymbol{\eta}_1'\mathbf{A}_{11}\boldsymbol{\eta}_1, \quad \boldsymbol{\eta} = \boldsymbol{\Gamma}\mathbf{C}_z^{1/2}\boldsymbol{\theta} - \boldsymbol{\Gamma}\mathbf{C}_z^{-1/2}\mathbf{R}'(\mathbf{R}\mathbf{C}_z^{-1}\mathbf{R}')^{-1}\mathbf{R}\boldsymbol{\theta} = \begin{pmatrix} \boldsymbol{\eta}_1 \\ \boldsymbol{\eta}_2 \end{pmatrix}. \tag{8.2.29}$$

Hence,, we may write

$$R_2(\hat{\boldsymbol{\theta}}_n : \mathbf{W}) = \sigma^2 \operatorname{tr}(\mathbf{W}\mathbf{C}_z^{-1}) - \sigma^2 \operatorname{tr}(\mathbf{A}_{11}) + \boldsymbol{\eta}_1'\mathbf{A}_{11}\boldsymbol{\eta}_1. \tag{8.2.30}$$

We conclude that $\hat{\boldsymbol{\theta}}_n$ performs better than $\tilde{\boldsymbol{\theta}}_n$ whenever

$$\Delta^2 \leq \frac{\operatorname{tr}(\mathbf{A}_{11})}{Ch_{max}(\mathbf{A}_{11})}, \tag{8.2.31}$$

whereas $\tilde{\boldsymbol{\theta}}_n$ performs better than $\hat{\boldsymbol{\theta}}_n$ whenever

$$\Delta^2 \geq \frac{\operatorname{tr}(\mathbf{A}_{11})}{Ch_{min}(\mathbf{A}_{11})}. \tag{8.2.32}$$

If $\mathbf{W} = \mathbf{C}_z$, we see that $\hat{\boldsymbol{\theta}}_n$ performs better than $\tilde{\boldsymbol{\theta}}_n$ in the interval $[0, q]$ and $\tilde{\boldsymbol{\theta}}_n$ performs better than $\hat{\boldsymbol{\theta}}_n$ outside this interval.

Comparisons of $\hat{\boldsymbol{\theta}}_n^{\mathrm{PT}}$, $\tilde{\boldsymbol{\theta}}_n$ and $\hat{\boldsymbol{\theta}}_n$. Consider the risk-difference

$$R_1(\tilde{\boldsymbol{\theta}}_n : \mathbf{W}) - R_3(\hat{\boldsymbol{\theta}}_n^{\mathrm{PT}} : \mathbf{W}) = \sigma^2 \operatorname{tr}(\mathbf{A}_{11}) G_{q+2,m}(\ell_\alpha; \Delta^2)$$
$$- \boldsymbol{\eta}_1' \mathbf{A}_{11} \boldsymbol{\eta}_1 \Big\{ 2G_{q+2,m}(\ell_\alpha; \Delta^2) - G_{q+4,m}(\ell_\alpha^*; \Delta^2) \Big\}. \tag{8.2.33}$$

Thus, from risk-difference of (8.2.33), $\hat{\boldsymbol{\theta}}_n^{\mathrm{PT}}$ performs better than $\tilde{\boldsymbol{\theta}}_n$ whenever

$$\Delta^2 \leq \frac{\operatorname{tr}(\mathbf{A}_{11})}{Ch_{max}(\mathbf{A}_{11})} \frac{G_{q+2,m}(\ell_\alpha; \Delta^2)}{2G_{q+2,m}(\ell_\alpha; \Delta^2) - G_{q+4,m}(\ell_\alpha^*; \Delta^2)} \tag{8.2.34}$$

while $\tilde{\boldsymbol{\theta}}_n$ perform better than $\hat{\boldsymbol{\theta}}_n^{\mathrm{PT}}$ whenever

$$\Delta^2 \geq \frac{\operatorname{tr}(\mathbf{A}_{11})}{Ch_{min}(\mathbf{A}_{11})} \frac{G_{q+2,m}(\ell_\alpha; \Delta^2)}{2G_{q+2,m}(\ell_\alpha; \Delta^2) - G_{q+4,m}(\ell_\alpha^*; \Delta^2)}. \tag{8.2.35}$$

Since $\hat{\boldsymbol{\theta}}_n^{\mathrm{PT}}$ is not uniformly better than $\tilde{\boldsymbol{\theta}}_n$, we obtain a PTE by selecting an optimum level of significance α^* with a minimum guaranteed efficiency, say, E_0, by solving the equation

$$\min_{\Delta^2} E(\alpha, \Delta^2) = E(\alpha, \Delta^2_{\min}) = E_0, \tag{8.2.36}$$

where

$$E(\alpha, \Delta^2) = R_1(\tilde{\boldsymbol{\theta}}_n : \mathbf{W})[R_3(\hat{\boldsymbol{\theta}}_n^{\mathrm{PT}} : \mathbf{W})]^{-1}. \tag{8.2.37}$$

Next, we compare $\hat{\boldsymbol{\theta}}_n^{\mathrm{PT}}$ and $\hat{\boldsymbol{\theta}}_n$. Here

$$R_2(\hat{\boldsymbol{\theta}}_n : \mathbf{W}) - R_3(\hat{\boldsymbol{\theta}}_n^{\mathrm{PT}} : \mathbf{W}) = -\sigma^2 \operatorname{tr}(\mathbf{A}_{11}) \Big[1 - G_{q+2,m}(\ell_\alpha; \Delta^2) \Big]$$
$$+ \boldsymbol{\eta}_1' \mathbf{A}_{11} \boldsymbol{\eta}_1 \Big\{ 1 - 2G_{q+2,m}(\ell_\alpha; \Delta^2) + G_{q+4,m}(\ell_\alpha^*; \Delta^2) \Big\}. \tag{8.2.38}$$

Thus, $\hat{\boldsymbol{\theta}}_n$ performs better than $\hat{\boldsymbol{\theta}}_n^{\mathrm{PT}}$ whenever

$$\Delta^2 \leq \frac{\operatorname{tr}(\mathbf{A}_{11})}{Ch_{max}(\mathbf{A}_{11})} \frac{(1 - G_{q+2,m}(\ell_\alpha; \Delta^2))}{(1 - 2G_{q+2,m}(\ell_\alpha; \Delta^2) + G_{q+4,m}(\ell_\alpha^*; \Delta^2))} \tag{8.2.39}$$

and $\hat{\boldsymbol{\theta}}_n^{\mathrm{PT}}$ performs $\hat{\boldsymbol{\theta}}_n$ whenever

$$\Delta^2 \geq \frac{\operatorname{tr}(\mathbf{A}_{11})}{Ch_{min}(\mathbf{A}_{11})} \frac{(1 - G_{q+2,m}(\ell_\alpha; \Delta^2))}{(1 - 2G_{q+2,m}(\ell_\alpha; \Delta^2) + G_{q+4,m}(\ell_\alpha^*; \Delta^2))}. \tag{8.2.40}$$

Under H_0, the risks of the three estimators may be ordered as

$$R_2(\hat{\boldsymbol{\theta}}_n; \mathbf{W}) \leq R_3(\hat{\boldsymbol{\theta}}_n^{\mathrm{PT}}; \mathbf{W}) \leq R_1(\tilde{\boldsymbol{\theta}}_n; \mathbf{W}). \tag{8.2.41}$$

Comparisons of $\hat{\boldsymbol{\theta}}_n^{S+}$, $\hat{\boldsymbol{\theta}}_n^{S}$, and $\tilde{\boldsymbol{\theta}}_n$. The risk-difference of $\hat{\boldsymbol{\theta}}_n^{S}$ and $\tilde{\boldsymbol{\theta}}_n$ is given by

$$R_1(\tilde{\boldsymbol{\theta}}_n : \mathbf{W}) - R_4(\hat{\boldsymbol{\theta}}_n^{S} : \mathbf{W}) = \sigma^2 dq \operatorname{tr}(\mathbf{A}_{11})\Big\{(q-2)E[\chi_{q+2}^{-4}(\Delta^2)] + 2\Big[1 - \frac{(q+2)(\boldsymbol{\eta}_1'\mathbf{A}_{11}\boldsymbol{\eta}_1)}{2\sigma^2\Delta^2\operatorname{tr}(\mathbf{A}_{11})}\Big]\Delta^2 E[\chi_{q+4}^{-4}(\Delta^2)]\Big\}. \quad (8.2.42)$$

This risk-difference is p.s.d. for all $\mathcal{A}$ such that

$$\Big\{\mathcal{A} : \frac{\operatorname{tr}(\mathbf{A}_{11})}{Ch_{max}(\mathbf{A}_{11})} \geq \frac{q+2}{2}\Big\}. \quad (8.2.43)$$

Note that $\mathbf{A}_{11}$ involves $\mathbf{W}$. Thus, $\hat{\boldsymbol{\theta}}_n^{S}$ dominates $\tilde{\boldsymbol{\theta}}_n$ uniformly for all $\mathcal{A}$ given by (8.2.43).

Now, consider the risk-difference

$$\begin{aligned} R_4(\hat{\boldsymbol{\theta}}_n^{S} : \mathbf{W}) - R_5(\hat{\boldsymbol{\theta}}_n^{S+} : \mathbf{W}) &= \sigma^2\Big\{\operatorname{tr}(\mathbf{A}_{11}) \\ &\times E\Big[(1 - d_1F_{q+2,m}^{-1}(\Delta^2))^2 I(F_{q+2,m}(\Delta^2) < d_1)\Big] \\ &+ \sigma^{-2}\boldsymbol{\eta}_1'\mathbf{A}_{11}\boldsymbol{\eta}_1 E\Big[(1 - d_2F_{q+4,m}^{-1}(\Delta^2))^2 I(F_{q+4,m}(\Delta^2) < d_2)\Big]\Big\} \\ &- 2(\boldsymbol{\eta}_1'\mathbf{A}_{11}\boldsymbol{\eta}_1)E\Big[(d_1F_{q+2,m}^{-1}(\Delta^2) - 1)I(F_{q+2,m}(\Delta^2) < d_1)\Big]. \quad (8.2.44) \end{aligned}$$

The r.h.s. of (8.2.44) is positive, since the expectation of a positive random variable is positive, which follows from the fact that

$$\big(0 < F_{q+2,m}(\Delta^2) < d_1\big) \Longrightarrow (d_1F_{q+2,m}^{-1}(\Delta^2) - 1) \geq 0. \quad (8.2.45)$$

We thus obtain

$$E\Big[(d_1F_{q+2,m}^{-1}(\Delta^2) - 1)I(F_{q+2,m}(\Delta^2) < d_1)\Big] \geq 0. \quad (8.2.46)$$

For all $q \geq 3$,

$$R_5(\hat{\boldsymbol{\theta}}_n^{S+}; \mathbf{W}) \leq R_4(\hat{\boldsymbol{\theta}}_n^{S}; \mathbf{W}). \quad (8.2.47)$$

As a consequence, we obtain

$$R_5(\hat{\boldsymbol{\theta}}_n^{S+}; \mathbf{W}) \leq R_4(\hat{\boldsymbol{\theta}}_n^{S}; \mathbf{W}) \leq R_1(\tilde{\boldsymbol{\theta}}_n^{S}; \mathbf{W}) \quad (8.2.48)$$

for all Δ^2 provided

$$\frac{\operatorname{tr}(\mathbf{A}_{11})}{Ch_{max}(\mathbf{A}_{11})} \geq \frac{q+2}{2} \quad \text{and} \quad q \geq 3 \quad (8.2.49)$$

holds.

Comparisons of $\hat{\boldsymbol{\theta}}_n^S$ and $\hat{\boldsymbol{\theta}}_n^{\mathrm{PT}}$. Under H_0, the risk of $\hat{\boldsymbol{\theta}}_n^S$ may be written as

$$R_4(\hat{\boldsymbol{\theta}}_n^S;\mathbf{W}) = R_3(\hat{\boldsymbol{\theta}}_n^{\mathrm{PT}};\mathbf{W}) + \sigma^2 \operatorname{tr}(\mathbf{A}_{11})\Big[G_{q+2,m}(\ell_\alpha;0) - d\Big]$$
$$\geq R_3(\hat{\boldsymbol{\theta}}_n^{\mathrm{PT}};\mathbf{W}) \tag{8.2.50}$$

whenever

$$G_{q+2,m}(\ell_\alpha;0) \geq d. \tag{8.2.51}$$

That is to say, (8.2.35) holds whenever there exits a set of α such that

$$\mathcal{A}_\alpha = \left\{\alpha : \frac{q}{q+2}F_{q,m}(\alpha) \geq G_{q+2,m}^{-1}(d:0)\right\}. \tag{8.2.52}$$

The risk of $\hat{\boldsymbol{\theta}}_n^S$ is smaller than that of $\hat{\boldsymbol{\theta}}_n^{\mathrm{PT}}$ whenever the opposite inequality in (8.2.52) holds.

Further, under H_0, the risks may be ordered as

$$R_2(\hat{\boldsymbol{\theta}}_n;\mathbf{W}) \leq R_3(\hat{\boldsymbol{\theta}}_n^{\mathrm{PT}};\mathbf{W}) \leq R_1(\hat{\boldsymbol{\theta}}_n^S;\mathbf{W}) \tag{8.2.53}$$

when α satisfies (8.2.51).

As Δ^2 diverts from 0, the risk comparisons show that $\hat{\boldsymbol{\theta}}_n^{\mathrm{PT}}$ gets worse and so does $\hat{\boldsymbol{\theta}}_n$. Only $\tilde{\boldsymbol{\theta}}_n$ has stable risk. None of the estimators dominate each other.

Similar conclusions as in Sections 8.2.3 and 8.2.4 hold for the five estimators of $\boldsymbol{\beta}$.

8.3 Estimation of Regression Parameters with Prior Information from a Previous Sample

In the last two sections, we considered the estimation of the regression parameters $\boldsymbol{\beta}_1$ in the model

$$\mathbf{Y}_1 = \mathbf{X}_1\boldsymbol{\beta}_1 + \mathbf{e}_1 : \quad \mathbf{e}_1 \sim \mathcal{N}_{n_1}(\mathbf{0}, \sigma^2\mathbf{I}_{n_1}) \tag{8.3.1}$$

when it is suspected that $\boldsymbol{\beta}_1$ belongs to the stochastic subspace defined by

$$\mathbf{h} = \mathbf{H}_1\boldsymbol{\beta}_1 + \mathbf{v}, \tag{8.3.2}$$

where $\mathbf{v} \sim \mathcal{N}_q(\boldsymbol{\delta}, \sigma^2\boldsymbol{\Omega})$ and $\boldsymbol{\Omega}$ is *known.* In a practical situation, the assumption that $\boldsymbol{\Omega}$ is known may not be valid. However, one may gather information about $\boldsymbol{\Omega}$ based on a previous sample regression model

$$\mathbf{Y}_0 = \mathbf{X}_0\boldsymbol{\beta}_0 + \mathbf{e}_0 : \quad \mathbf{e}_0 \sim \mathcal{N}_{n_0}(\mathbf{0}, \sigma^2\mathbf{I}_{n_0}), \tag{8.3.3}$$

where $\mathbf{Y}_0$ is a $(n_0 \times 1)$ vector of responses, $\mathbf{X}_0$ is an $n_0 \times p_0$ known design matrix, $\boldsymbol{\beta}_0 = (\beta_{01}, \cdots, \beta_{0p_0})'$ is a $p_0 \times 1$ vector of unknown regression parameters, and $\mathbf{e}_0$ is a $n_0 \times 1$ vector distributed normally with mean $\mathbf{0}$ and covariance $\sigma^2 \mathbf{I}_{n_0}$. Further, in addition to the sample information, *it is suspected* that $\boldsymbol{\beta}_0$ and $\boldsymbol{\beta}_1$ are related through the constraint

$$\mathbf{H}_0 \boldsymbol{\beta}_0 = \mathbf{H}_1 \boldsymbol{\beta}_1, \tag{8.3.4}$$

where $\mathbf{H}_i$ is a $q \times p_i$ $(i = 0, 1)$ known matrices of constants.

Now, estimating $\boldsymbol{\beta}_0$ by $\tilde{\boldsymbol{\beta}}_{0n_0} = \mathbf{C}_0^{-1}\mathbf{X}_0'\mathbf{Y}_0$, $\mathbf{C}_0 = \mathbf{X}_0'\mathbf{X}_0$, we obtain the stochastic constraint defined by

$$\mathbf{h}_0 = \mathbf{H}_0 \tilde{\boldsymbol{\beta}}_{0n_0} = \mathbf{H}_1 \boldsymbol{\beta}_1 + \mathbf{v}_0, \tag{8.3.5}$$

where $\mathbf{v}_0 \sim \mathcal{N}_q(\boldsymbol{\delta}_0, \sigma^2 \boldsymbol{\Omega}_0)$, and $\boldsymbol{\Omega}_0 = \mathbf{H}_0 \mathbf{C}_0^{-1} \mathbf{H}_0'$, $\boldsymbol{\delta}_0 = \mathbf{H}_0 \boldsymbol{\beta}_0 - \mathbf{H}_1 \boldsymbol{\beta}_1$. Note that $\boldsymbol{\Omega}_0$ is known for this stochastic constraint.

Thus, our objective is to estimate $\boldsymbol{\beta}_1$ based on the model

$$\mathbf{Y}_1 = \mathbf{X}_1 \boldsymbol{\beta}_1 + \mathbf{e}_1$$

when it is suspected that $\boldsymbol{\beta}_1$ belongs to the stochastic subspace defined by

$$\mathbf{h}_0 = \mathbf{H}_1 \boldsymbol{\beta}_1 + \mathbf{v}_0.$$

The problem is discussed in Khan and Saleh (2005).

8.3.1 Estimation of $\boldsymbol{\beta}_1$ and Test of $\mathbf{H}_0\boldsymbol{\beta}_0 = \mathbf{H}_1\boldsymbol{\beta}_1$

First, we note that the LSE of $\boldsymbol{\beta}_1$ from (8.3.1) is $\tilde{\boldsymbol{\beta}}_{1n_1} = \mathbf{C}_1^{-1}\mathbf{X}_1'\mathbf{Y}_1$, $\mathbf{C}_1 = \mathbf{X}_1'\mathbf{X}_1$ and the restricted estimator of $\boldsymbol{\beta}_1$ subject to (8.3.5) is given by

$$\begin{aligned}\hat{\boldsymbol{\beta}}_{1n_1}^R &= (\mathbf{C}_1 + \mathbf{H}_1'\boldsymbol{\Omega}_0^{-1}\mathbf{H}_1)^{-1}(\mathbf{X}_1'\mathbf{Y}_1 + \mathbf{H}_1'\boldsymbol{\Omega}_0^{-1}\mathbf{h}_0) \\ &= \tilde{\boldsymbol{\beta}}_{1n_1} - \mathbf{C}_1^{-1}\mathbf{H}_1'(\mathbf{H}_1\mathbf{C}_1^{-1}\mathbf{H}_1' + \boldsymbol{\Omega}_0)^{-1}(\mathbf{H}_1\tilde{\boldsymbol{\beta}}_{1n_1} - \mathbf{h}_0) = \hat{\boldsymbol{\beta}}_{1n_1}. \end{aligned} \tag{8.3.6}$$

Under $\mathbf{H}_0\boldsymbol{\beta}_0 = \mathbf{H}_1\boldsymbol{\beta}_1$, we have $E(\hat{\boldsymbol{\beta}}_{1n_1}^R) = \boldsymbol{\beta}_1$ and

$$\mathrm{Cov}(\hat{\boldsymbol{\beta}}_{1n_1}) = \sigma^2(\mathbf{C}_1 + \mathbf{H}_1\boldsymbol{\Omega}_0^{-1}\mathbf{H}_1')^{-1}. \tag{8.3.7}$$

Further, the pooled estimator of σ^2 is given by

$$\tilde{\sigma}_n^2 = \frac{1}{m}\Big\{\mathbf{Y}_0'[\mathbf{I}_{n_0} - \mathbf{X}_0\mathbf{C}_0^{-1}\mathbf{X}_0']\mathbf{Y}_0 + \mathbf{Y}_1'[\mathbf{I}_{n_1} - \mathbf{X}_1\mathbf{C}_1^{-1}\mathbf{X}_1']\mathbf{Y}_1\Big\}, \tag{8.3.8}$$

where $m = n - (p_0 + p_1)$ and $n = n_0 + n_1$.

In order to test the null hypothesis $\mathbf{H}_0\boldsymbol{\beta}_0 = \mathbf{H}_1\boldsymbol{\beta}_1$ against the alternative $\mathbf{H}_0\boldsymbol{\beta}_0 \neq \mathbf{H}_1\boldsymbol{\beta}_1$, we consider the test-statistic, following Section 8.1.3, as

$$\mathcal{L}_n = \frac{(\mathbf{H}_1\tilde{\boldsymbol{\beta}}_{1n_1} - \mathbf{h}_0)'(\mathbf{H}_1\mathbf{C}_1^{-1}\mathbf{H}_1' + \boldsymbol{\Omega}_0)^{-1}(\mathbf{H}_1\tilde{\boldsymbol{\beta}}_{1n_1} - \mathbf{h}_0)}{q\tilde{\sigma}_n^2}, \tag{8.3.9}$$

which follows the noncentral F-distribution with (q, m) d.f. and noncentrality parameter $\Delta^2/2$ with

$$\Delta^2 = \frac{\boldsymbol{\delta}_0'(\mathbf{H}_1\mathbf{C}_1^{-1}\mathbf{H}_1' + \boldsymbol{\Omega}_0)^{-1}\boldsymbol{\delta}_0}{\sigma^2}. \tag{8.3.10}$$

8.3.2 The Mixed Estimators

Let

$$\boldsymbol{\theta} = \begin{bmatrix} \boldsymbol{\beta}_1 \\ \mathbf{H}_1\boldsymbol{\beta}_1 + \boldsymbol{\delta}_0 \end{bmatrix}. \tag{8.3.11}$$

Then the model (8.3.1) together with (8.3.5) can be written as

$$\begin{pmatrix} \mathbf{Y}_1 \\ \boldsymbol{\Omega}_0^{-1/2}\mathbf{h}_0 \end{pmatrix} = \begin{pmatrix} \mathbf{X}_1 & \mathbf{0} \\ \mathbf{0} & \boldsymbol{\Omega}_0^{-1/2} \end{pmatrix} \begin{pmatrix} \boldsymbol{\beta}_1 \\ \mathbf{H}_1\boldsymbol{\beta}_1 + \boldsymbol{\delta}_0 \end{pmatrix} + \begin{pmatrix} \mathbf{e}_1 \\ \boldsymbol{\Omega}_0^{-1/2}(\mathbf{v}_0 - \boldsymbol{\delta}_0) \end{pmatrix} \tag{8.3.12}$$

or

$$\mathbf{Y}^* = \mathbf{Z}\boldsymbol{\theta} + \mathbf{E}.$$

The *unrestricted* and the *restricted* estimators of $\boldsymbol{\theta}$ are given by

$$\tilde{\boldsymbol{\theta}}_n = \begin{pmatrix} \tilde{\boldsymbol{\beta}}_{1n_1} \\ \mathbf{h}_0 \end{pmatrix} \tag{8.3.13}$$

and

$$\begin{aligned} \hat{\boldsymbol{\theta}}_n &= \begin{pmatrix} \tilde{\boldsymbol{\beta}}_{1n_1} \\ \mathbf{h}_0 \end{pmatrix} - \begin{pmatrix} \mathbf{C}_1^{-1}\mathbf{H}_1'(\mathbf{H}_1\mathbf{C}_1^{-1}\mathbf{H}_1' + \boldsymbol{\Omega}_0)^{-1}(\mathbf{H}_1\tilde{\boldsymbol{\beta}}_{1n_1} - \mathbf{h}_0 \\ -\boldsymbol{\Omega}_0(\mathbf{H}_1\mathbf{C}_1^{-1}\mathbf{H}_1' + \boldsymbol{\Omega}_0)^{-1}(\mathbf{H}_1\tilde{\boldsymbol{\beta}}_{1n_1} - \mathbf{h}_0) \end{pmatrix} \\ &= \begin{pmatrix} \hat{\boldsymbol{\beta}}_{1n_1} \\ \mathbf{H}_1\hat{\boldsymbol{\beta}}_{1n_1} \end{pmatrix}, \end{aligned} \tag{8.3.14}$$

respectively. Similarly, the PTE, JSE, and PRSE of $\boldsymbol{\theta}$ are given by

$$\hat{\boldsymbol{\theta}}_n^{\mathrm{PT}} = \tilde{\boldsymbol{\theta}}_n - (\tilde{\boldsymbol{\theta}}_n - \hat{\boldsymbol{\theta}}_n)I(\mathcal{L}_n < F_{q,m}(\alpha)), \tag{8.3.15}$$

$$\hat{\boldsymbol{\theta}}_n^{S} = \tilde{\boldsymbol{\theta}}_n - d(\tilde{\boldsymbol{\theta}}_n - \hat{\boldsymbol{\theta}}_n)\mathcal{L}_n^{-1}, \; d = \frac{(q-2)m}{q(m+2)} \tag{8.3.16}$$

and

$$\hat{\boldsymbol{\theta}}_n^{S+} = \tilde{\boldsymbol{\theta}}_n + (\tilde{\boldsymbol{\theta}}_n - \hat{\boldsymbol{\theta}}_n)(1 - d\mathcal{L}_n^{-1})I(\mathcal{L}_n > d), \tag{8.3.17}$$

where $F_{q,m}(\alpha)$ is the αth-level upper critical value from the central F-distribution with (q, m) d.f.

Let

$$\mathbf{R} = [-\mathbf{H}_1, \mathbf{I}_{n_1}]. \tag{8.3.18}$$

Then we write the bias, MSE matrices, and risk of the estimators (8.3.13) through (8.3.17) as follows:

8.3.3 Bias Expressions

(i) $\mathbf{b}_1(\tilde{\boldsymbol{\theta}}_n) = \mathbf{0}$ and $B_1(\tilde{\boldsymbol{\theta}}_n) = 0.$ (8.3.19)

(ii) $\mathbf{b}_2(\hat{\boldsymbol{\theta}}_n) = -\mathbf{C}_z^{-1}\mathbf{R}(\mathbf{R}\mathbf{C}_z^{-1}\mathbf{R}')^{-1}\mathbf{R}\boldsymbol{\theta} = -\boldsymbol{\delta}_z$ and $B_2(\hat{\boldsymbol{\theta}}_n) = \Delta^2.$

(iii) $\mathbf{b}_3(\hat{\boldsymbol{\theta}}_n^{\mathrm{PT}}) = -\boldsymbol{\delta}_z G_{q+2,m}(\ell_\alpha;\Delta^2)$ and $B_3(\hat{\boldsymbol{\theta}}_n^{\mathrm{PT}}) = \Delta^2\{G_{q+2,m}(\ell_\alpha;\Delta^2)\}^2.$

(iv) $\mathbf{b}_4(\hat{\boldsymbol{\theta}}_n^{S}) = -dq\boldsymbol{\delta}_z E[\chi_{q+2}^{-2}(\Delta^2)]$ and $B_4(\hat{\boldsymbol{\theta}}_n^{S}) = d^2q^2\Delta^2\{E[\chi_{q+2}^{-2}(\Delta^2)]\}^2.$

(v) $$\mathbf{b}_5(\hat{\boldsymbol{\theta}}_n^{S+}) = -\boldsymbol{\delta}_z\Big\{d_1 E\big[F_{q+2,m}^{-1}(\Delta^2)I(F_{q+2,m}(\Delta^2) < d_1)\big] - d_1 E\big[F_{q+2,m}^{-1}(\Delta^2)\big] - G_{q+2,m}(d_1;\Delta^2)\Big\}$$

and

$$B_5(\hat{\boldsymbol{\theta}}_n^{S+}) = \Delta^2\Big\{d_1 E\big[F_{q+2,m}^{-1}(\Delta^2)I(F_{q+2,m}(\Delta^2) < d_1)\big] - d_1 E\big[F_{q+2,m}^{-1}(\Delta^2)\big] - G_{q+2,m}(d_1;\Delta^2)\Big\}^2.$$

8.3.4 MSE Matrix and Risk Expressions

(i) $\mathbf{M}_1(\tilde{\boldsymbol{\theta}}_n) = \sigma^2\mathbf{C}_z^{-1}$ and $R_1(\tilde{\boldsymbol{\theta}}_n:\mathbf{W}) = \sigma^2\,\mathrm{tr}(\mathbf{W}\mathbf{C}_z^{-1}).$ (8.3.20)

(ii) $\mathbf{M}_2(\hat{\boldsymbol{\theta}}_n) = \sigma^2\mathbf{C}_z^{-1} - \sigma^2\mathbf{C}_z^{-1}\mathbf{R}'(\mathbf{R}\mathbf{C}_z^{-1}\mathbf{R}')^{-1}\mathbf{R}\mathbf{C}_z^{-1} + \boldsymbol{\delta}_z\boldsymbol{\delta}_z'$

and

$$R_2(\hat{\boldsymbol{\theta}}_n:\mathbf{W}) = \sigma^2\mathrm{tr}(\mathbf{W}\mathbf{C}_z^{-1}) - \sigma^2\mathrm{tr}[\mathbf{W}\mathbf{C}_z^{-1}\mathbf{R}'(\mathbf{R}\mathbf{C}_z^{-1}\mathbf{R}')^{-1}\mathbf{R}\mathbf{C}_z^{-1}] + \boldsymbol{\delta}_z'\mathbf{W}\boldsymbol{\delta}_z,$$

(iii) $$\mathbf{M}_3(\hat{\boldsymbol{\theta}}_n^{\mathrm{PT}}) = \sigma^2\mathbf{C}_z^{-1} - \sigma^2\mathbf{C}_z^{-1}\mathbf{R}'(\mathbf{R}\mathbf{C}_z^{-1}\mathbf{R}')^{-1}\mathbf{R}\mathbf{C}_z^{-1}G_{q+2,m}(\ell_\alpha;\Delta^2) + \boldsymbol{\delta}_z\boldsymbol{\delta}_z'\Big\{2G_{q+2,m}(\ell_\alpha;\Delta^2) - G_{q+4,m}(\ell_\alpha^*;\Delta^2)\Big\}$$

and

$$R_3(\hat{\boldsymbol{\theta}}_n^{\mathrm{PT}}:\mathbf{W}) = \sigma^2\,\mathrm{tr}(\mathbf{W}\mathbf{C}_z^{-1}) - \sigma^2\mathrm{tr}[\mathbf{W}\mathbf{C}_z^{-1}\mathbf{R}'(\mathbf{R}\mathbf{C}_z^{-1}\mathbf{R}')^{-1}\mathbf{R}\mathbf{C}_z^{-1}] \times G_{q+2,m}(\ell_\alpha;\Delta^2) + \boldsymbol{\delta}_z'\mathbf{W}\boldsymbol{\delta}_z\Big\{2G_{q+2,m}(\ell_\alpha;\Delta^2) - G_{q+4,m}(\ell_\alpha^*;\Delta^2)\Big\}.$$

(iv) $$\mathbf{M}_4(\hat{\boldsymbol{\theta}}_n^{S}) = \sigma^2\mathbf{C}_z^{-1} - dq\,\sigma^2\mathbf{C}_z^{-1}\mathbf{R}'(\mathbf{R}\mathbf{C}_z^{-1}\mathbf{R}')^{-1}\mathbf{R}\mathbf{C}_z^{-1}\{2E[\chi_{q+2}^{-2}(\Delta^2)] - (q-2)E[\chi_{q+2}^{-4}(\Delta^2)]\} + dq(q+2)\boldsymbol{\delta}_z\boldsymbol{\delta}_z' E[\chi_{q+4}^{-4}(\Delta^2)]$$

and

$$R_4(\hat{\boldsymbol{\theta}}_n^{S}:\mathbf{W}) = \sigma^2\,\mathrm{tr}(\mathbf{W}\mathbf{C}_z^{-1}) - dq\,\sigma^2\mathrm{tr}[\mathbf{W}\mathbf{C}_z^{-1}\mathbf{R}'(\mathbf{R}\mathbf{C}_z^{-1}\mathbf{R}')^{-1}\mathbf{R}\mathbf{C}_z^{-1}] \times\{2E[\chi_{q+2}^{-2}(\Delta^2)] - (q-2)E[\chi_{q+2}^{-4}(\Delta^2)]\}$$

$$+dq(q+2)\boldsymbol{\delta}_z'\mathbf{W}\boldsymbol{\delta}_z E[\chi_{q+4}^{-4}(\Delta^2)].$$

(v) $$\mathbf{M}_5(\hat{\boldsymbol{\theta}}_n^{S+}) = \mathbf{M}_4(\hat{\boldsymbol{\theta}}_n^{S}) - \sigma^2\mathbf{C}_z^{-1}\mathbf{R}'(\mathbf{R}\mathbf{C}_z^{-1}\mathbf{R}')^{-1}\mathbf{R}\mathbf{C}_z^{-1}$$
$$\times E\left[\left(1 - d_1F_{q+2,m}^{-1}(\Delta^2)\right)^2 I(F_{q+2,m}(\Delta^2) < d_1)\right]$$
$$-\boldsymbol{\delta}_z\boldsymbol{\delta}_z'\{2E\left[(d_1F_{q+2,m}^{-1}(\Delta^2) - 1)I(F_{q+2,m}(\Delta^2) < d_1)\right]$$
$$-E\left[\left(1 - d_2F_{q+4,m}^{-1}(\Delta^2)\right)^2 I(F_{q+4,m}(\Delta^2) < d_2)\right]\}$$

and

$$R_5(\hat{\boldsymbol{\theta}}_n^{S+}:\mathbf{W}) = R_4(\hat{\boldsymbol{\theta}}_n^{S}:\mathbf{W}) - \sigma^2\,\mathrm{tr}\left[\mathbf{W}\mathbf{C}_z^{-1}\mathbf{R}'(\mathbf{R}\mathbf{C}_z^{-1}\mathbf{R}')^{-1}\mathbf{R}\mathbf{C}_z^{-1}\right]$$
$$\times E\left[\left(1 - d_1F_{q+2,m}^{-1}(\Delta^2)\right)^2 I(F_{q+2,m}(\Delta^2) < d_1)\right]$$
$$-\boldsymbol{\delta}_z'\mathbf{W}\boldsymbol{\delta}_z\Big\{2E\left[(d_1F_{q+2,m}^{-1}(\Delta^2) - 1)I\big(F_{q+2,m}(\Delta^2) < d_1\big)\right]$$
$$-E\left[\left(1 - d_2F_{q+4,m}^{-1}(\Delta^2)\right)^2 I(F_{q+4,m}(\Delta^2) < d_2)\right]\Big\}.$$

We note that the expressions above are similar to those in Sections 8.2.1 and 8.2.2. Hence, the properties of these estimators are similar to that of the estimators of Sections 8.1.2 and 8.1.4 given in Sections 8.2.3 and 8.2.4, respectively.

8.4 Multiple Regression Model and Stochastic Subspace Hypothesis: Nonnormal Errors

8.4.1 Introduction

Consider the multiple regression model

$$\mathbf{Y} = \mathbf{X}_n\boldsymbol{\beta} + \mathbf{e} \tag{8.4.1}$$

with the suspected prior information on $\boldsymbol{\beta}$ provided by the stochastic subspace constraint

$$\mathbf{h} = \mathbf{H}\boldsymbol{\beta} + \mathbf{v};\ E(\mathbf{v}) = \boldsymbol{\delta} \text{ and } E(\mathbf{v}\mathbf{v}') = \frac{\sigma^2}{n}\boldsymbol{\Omega} + \boldsymbol{\delta}\boldsymbol{\delta}', \tag{8.4.2}$$

where $\mathbf{e} = (e_1, \cdots, e_n)$ are i.i.d. r.v assumed to be distributed with the cdf $F(\mathbf{e})$ nonnormal such that $E(\mathbf{e}) = \mathbf{0}$ and $E(\mathbf{e}\mathbf{e}') = \sigma^2\mathbf{I}_n$. Further, assume that

(i) max $\mathbf{x}_i'\mathbf{C}_n^{-1}\mathbf{x}_i \to 0$ as $n \to \infty$, where $\mathbf{x}_i$ is ith row of $\mathbf{X}_n$, (8.4.3)

and

(ii) $\lim_{n\to\infty} [n^{-1}\mathbf{C}_n] = \mathbf{C}$, $\mathbf{C}_n = \mathbf{X}_n'\mathbf{X}_n$, meaning the generalized Noether condition is satisfied for asymptotic normality of the estimators to be defined.

With this setup our basic problem is the estimation of $\boldsymbol{\beta}$ when it is suspected that $\boldsymbol{\beta}$ belongs to the stochastic subspace defined by

$$\mathbf{h} = \mathbf{H}\boldsymbol{\beta} + \mathbf{v}. \tag{8.4.4}$$

Accordingly, we consider the *unrestricted, restricted, preliminary test, James-Stein-type*, and *positive rule Stein-type* estimators of $\boldsymbol{\beta}$.

8.4.2 Estimation of the Parameters and Test of Hypothesis

As in Section 8.1.1, we consider the *mixed model*,

$$\begin{bmatrix} \mathbf{Y} \\ n^{1/2}\boldsymbol{\Omega}^{-1/2}\mathbf{h} \end{bmatrix} = \begin{bmatrix} \mathbf{X}_n & \mathbf{0} \\ \mathbf{0} & n^{1/2}\boldsymbol{\Omega}^{-1/2} \end{bmatrix} \begin{bmatrix} \boldsymbol{\beta} \\ \mathbf{H}\boldsymbol{\beta}+\boldsymbol{\delta} \end{bmatrix} + \begin{bmatrix} \mathbf{e} \\ n^{1/2}\boldsymbol{\Omega}^{-1/2}(\mathbf{v}-\boldsymbol{\delta}) \end{bmatrix}, \tag{8.4.5}$$

where we assume that $\mathbf{e}$ and $\boldsymbol{\Omega}_n^{-1/2}(\mathbf{v}-\boldsymbol{\delta})$ are independent so that

$$\mathrm{Cov}\begin{bmatrix} \mathbf{e} \\ n^{1/2}\boldsymbol{\Omega}^{-1/2}(\mathbf{v}-\boldsymbol{\delta}) \end{bmatrix} = \begin{pmatrix} \mathbf{I}_n & \mathbf{0} \\ \mathbf{0} & \mathbf{I}_q \end{pmatrix}. \tag{8.4.6}$$

Defining

$$\boldsymbol{\theta} = \begin{bmatrix} \boldsymbol{\beta} \\ \mathbf{H}\boldsymbol{\beta}+\boldsymbol{\delta} \end{bmatrix}, \quad \mathbf{Z}_n = \begin{bmatrix} \mathbf{X}_n & \mathbf{0} \\ \mathbf{0} & n^{1/2}\boldsymbol{\Omega}^{-1/2} \end{bmatrix}, \quad \mathbf{Y}^* = \begin{bmatrix} \mathbf{Y} \\ n^{1/2}\boldsymbol{\Omega}^{-1/2}\mathbf{h} \end{bmatrix},$$

$$\mathbf{R} = [-\mathbf{H}, \mathbf{I}_q], \quad \text{and} \quad \mathbf{E} = \begin{bmatrix} \mathbf{e} \\ n^{1/2}\boldsymbol{\Omega}^{-1/2}(\mathbf{v}-\boldsymbol{\delta}) \end{bmatrix}, \tag{8.4.7}$$

we obtain the model

$$\mathbf{Y}^* = \mathbf{Z}_n\boldsymbol{\theta} + \mathbf{E}, \tag{8.4.8}$$

where we wish to estimate $\boldsymbol{\theta}$ when it is suspected that $\boldsymbol{\theta}$ belongs to the subspace $\mathbf{R}\boldsymbol{\theta} = \boldsymbol{\delta} = \mathbf{0}$. Thus, the *unrestricted estimator* of $\boldsymbol{\theta}$ is given by

$$\tilde{\boldsymbol{\theta}}_n = \begin{bmatrix} \tilde{\boldsymbol{\beta}}_n \\ \mathbf{h} \end{bmatrix}, \quad \tilde{\boldsymbol{\beta}}_n = \mathbf{C}_n^{-1}\mathbf{X}_n'\mathbf{Y}, \tag{8.4.9}$$

and the *restricted estimator* of $\boldsymbol{\theta}$ subject to $\mathbf{R}\boldsymbol{\theta} = \mathbf{0}$ is given by

$$\hat{\boldsymbol{\theta}}_n = \tilde{\boldsymbol{\theta}}_n - \mathbf{C}_{nz}^{-1}\mathbf{R}'(\mathbf{R}\mathbf{C}_{nz}^{-1}\mathbf{R}')^{-1}\mathbf{R}\tilde{\boldsymbol{\theta}}_n, \tag{8.4.10}$$

where

$$\mathbf{C}_{nz} = \mathbf{Z}_n'\mathbf{Z}_n = \begin{pmatrix} \mathbf{X}_n'\mathbf{X}_n & \mathbf{0} \\ \mathbf{0} & n\boldsymbol{\Omega}^{-1} \end{pmatrix} = \begin{pmatrix} \mathbf{C}_n & \mathbf{0} \\ \mathbf{0} & n\boldsymbol{\Omega}^{-1} \end{pmatrix} \tag{8.4.11}$$

and $n^{-1}\mathbf{C}_{nz} \to \mathbf{C}_z$. Then we assert that as $n \to \infty$ under the conditions (i) through (iii) of (8.4.3),

$$\sqrt{n}(\tilde{\boldsymbol{\theta}}_n - \boldsymbol{\theta}) \sim \mathcal{N}_{n+q}(\mathbf{0}, \sigma^2\mathbf{C}_z^{-1}), \tag{8.4.12}$$

where $\mathbf{C}_z = \begin{pmatrix} \mathbf{C} & \mathbf{0} \\ \mathbf{0} & \boldsymbol{\Omega}^{-1} \end{pmatrix}$. Similarly, as $n \to \infty$,

$$\sqrt{n}(\hat{\boldsymbol{\theta}}_n - \boldsymbol{\theta}) \sim \mathcal{N}_{n+q}(-\boldsymbol{\delta}_z, \sigma^2[\mathbf{C}_z^{-1} - \mathbf{C}_z^{-1}\mathbf{R}'(\mathbf{R}\mathbf{C}_z^{-1}\mathbf{R}')^{-1}\mathbf{R}\mathbf{C}_z^{-1}]), \tag{8.4.13}$$

where

$$\boldsymbol{\delta}_z = \mathbf{C}_z^{-1}\mathbf{R}'(\mathbf{R}\mathbf{C}_z^{-1}\mathbf{R}')^{-1}\mathbf{R}\boldsymbol{\theta}. \tag{8.4.14}$$

For the test of the null hypothesis, $\mathbf{R}\boldsymbol{\theta} = \mathbf{0}$, we consider the test-statistic

$$\mathcal{L}_n^* = \frac{\tilde{\boldsymbol{\theta}}_n'\mathbf{R}'(\mathbf{R}\mathbf{C}_{nz}^{-1}\mathbf{R}')^{-1}\mathbf{R}\tilde{\boldsymbol{\theta}}_n}{q\, s_e^2} = \frac{n(\mathbf{H}\tilde{\boldsymbol{\beta}}_n - \mathbf{h})'(n\mathbf{H}\mathbf{C}_{nz}^{-1}\mathbf{H}' + \boldsymbol{\Omega})^{-1}(\mathbf{H}\tilde{\boldsymbol{\beta}}_n - \mathbf{h})}{q\, s_e^2}. \tag{8.4.15}$$

where

$$s_e^2 = m^{-1}\mathbf{Y}'[\mathbf{I} - \mathbf{X}_n\mathbf{C}_n^{-1}\mathbf{X}_n]\mathbf{Y}, \quad m = n - p. \tag{8.4.16}$$

and $s_e^2 \xrightarrow{P} \sigma^2$ as $n \to \infty$. Hence, as $n \to \infty$, $\mathcal{L}_n^*$ converges to the central chi-square distribution with m d.f. under the null hypothesis $\boldsymbol{\delta} = \mathbf{0}$.

8.4.3 Preliminary Test and Stein-type Estimators

Following Section 8.1.4, we define PTE, JSE, and PRSE, respectively:

$$\hat{\boldsymbol{\theta}}_n^{\mathrm{PT}} = \tilde{\boldsymbol{\theta}}_n - (\tilde{\boldsymbol{\theta}}_n - \hat{\boldsymbol{\theta}}_n)I(\mathcal{L}_n^* < \chi_q^2(\alpha)), \tag{8.4.17}$$

$$\hat{\boldsymbol{\theta}}_n^{S} = \tilde{\boldsymbol{\theta}}_n - (q-2)(\tilde{\boldsymbol{\theta}}_n - \hat{\boldsymbol{\theta}}_n)\mathcal{L}_n^*, \tag{8.4.18}$$

$$\hat{\boldsymbol{\theta}}_n^{S+} = \hat{\boldsymbol{\theta}}_n + (\tilde{\boldsymbol{\theta}}_n - \hat{\boldsymbol{\theta}}_n)(1 - (q-2)\mathcal{L}_n^{*-1})I(\mathcal{L}_n^* > q-2). \tag{8.4.19}$$

8.5 Asymptotic Distribution of the Estimators

In order to obtain the asymptotic distributional quadratic bias (ADQB), MSE matrices (ADMSE), and the quadratic risks (ADQR) of $\boldsymbol{\theta}$, we need the asymptotic distributions of UE, RE, PTE, and PRSE of $\boldsymbol{\theta}$. As in Section 5.8, we consider the generic estimator $\boldsymbol{\theta}_n^*$ and a positive-semidefinite matrix $\mathbf{W}$ defining the loss function

$$\begin{aligned} L(\boldsymbol{\theta}_n^*; \boldsymbol{\theta}) &= n(\boldsymbol{\theta}_n^* - \boldsymbol{\theta})'\mathbf{W}(\boldsymbol{\theta}_n^* - \boldsymbol{\theta}) \\ &= \mathrm{tr}[\mathbf{W}\{n(\boldsymbol{\theta}_n^* - \boldsymbol{\theta})(\boldsymbol{\theta}_n^* - \boldsymbol{\theta})'\}]. \end{aligned} \tag{8.5.1}$$

Let $\mathbf{M}(\boldsymbol{\theta}_n^*)$ denote the asymptotic distributional MSE matrix of $\boldsymbol{\theta}_n^*$. Then the ADQR of $\boldsymbol{\theta}_n^*$ is given by

$$R(\boldsymbol{\theta}_n^* : \mathbf{W}) = \mathrm{tr}[\mathbf{W}\mathbf{M}(\boldsymbol{\theta}_n^*)]. \tag{8.5.2}$$

In many situations, the *asymptotic* distribution of $\sqrt{n}s_e^{-1}(\boldsymbol{\theta}_n^* - \boldsymbol{\theta})$ is equivalent to $\sqrt{n}s_e^{-1}(\tilde{\boldsymbol{\theta}}_n - \boldsymbol{\theta})$, as $n \to \infty$, under fixed alternatives of real values as given by

$$K_{\boldsymbol{\xi}} : \boldsymbol{\delta} = \boldsymbol{\xi}, \quad \text{where } \boldsymbol{\xi} \text{ is a nonzero vector.} \tag{8.5.3}$$

To obtain a meaningful asymptotic distribution of $\sqrt{n}s_e^{-1}(\boldsymbol{\theta}_n^* - \boldsymbol{\theta})$ we consider the class of local alternatives

$$K_{(n)} : \boldsymbol{\delta}_{(n)} = n^{-1/2}\boldsymbol{\xi}. \tag{8.5.4}$$

Let the asymptotic distribution of $\sqrt{n}s_e^{-1}(\boldsymbol{\theta}_n^* - \boldsymbol{\theta})$ under $K_{(n)}$ be

$$G_p(\mathbf{x}) = \lim_{n\to\infty} P_{K_{(n)}}\{\sqrt{n}s_e^{-1}(\boldsymbol{\theta}_n^* - \boldsymbol{\theta}) \le \mathbf{x}\} \tag{8.5.5}$$

if it exits, then the ADB and the ADQB are given by

$$\mathbf{b}(\boldsymbol{\theta}_n^*) = \lim_{n\to\infty} E[\sqrt{n}s_e^{-1}(\boldsymbol{\theta}_n^* - \boldsymbol{\theta})] = \int \mathbf{x}dG_{n+q}(\mathbf{x}) \tag{8.5.6}$$

and

$$B(\boldsymbol{\theta}_n^*) = [\mathbf{b}(\boldsymbol{\theta}_n^*)]'[\mathbf{M}_1(\tilde{\boldsymbol{\theta}}_n)]^{-1}[\mathbf{b}(\boldsymbol{\theta}_n^*)], \tag{8.5.7}$$

respectively, where

$$\mathbf{M}_1(\tilde{\boldsymbol{\theta}}_n) = \sigma^2\mathbf{C}_z^{-1}. \tag{8.5.8}$$

8.5.1 Asymptotic Distribution of the Estimators under Fixed Alternatives

First, we consider the asymptotic distribution of $\sqrt{n}s_e^{-1}(\tilde{\boldsymbol{\theta}}_n - \boldsymbol{\theta})$ under the regularity conditions (i) through (iii) of (8.4.3). From Theorem 7.8.1, we have

$$\sqrt{n}s_e^{-1}(\tilde{\boldsymbol{\theta}}_n - \boldsymbol{\theta}) \overset{\mathcal{D}}{=} \mathcal{N}_{n+q}(\mathbf{0}, \mathbf{C}_z^{-1}), \tag{8.5.9}$$

where $s_e^2 \to \sigma^2$ converges surely as $n \to \infty$. Next, we consider the asymptotic distribution of the test-statistic $\mathcal{L}_n^*$ under the fixed alternative $K_{\boldsymbol{\xi}} : \boldsymbol{\delta}_0 = \boldsymbol{\xi}$. Then

$$\sqrt{n}s_e^{-1}\mathbf{R}\tilde{\boldsymbol{\theta}}_n = \sqrt{n}s_e^{-1}\mathbf{R}(\tilde{\boldsymbol{\theta}}_n - \boldsymbol{\theta}) + \sqrt{n}s_e^{-1}\boldsymbol{\xi}. \tag{8.5.10}$$

So $\mathcal{L}_n^*$ can be written as

$$\begin{aligned}\mathcal{L}_n^* &= ns_e^{-2}(\tilde{\boldsymbol{\theta}}_n - \boldsymbol{\theta})'\mathbf{R}'(\mathbf{R}\mathbf{C}_{nz}^{-1}\mathbf{R}')^{-1}\mathbf{R}(\tilde{\boldsymbol{\theta}}_n - \boldsymbol{\theta}) \\ &\quad + ns_e^{-2}\boldsymbol{\xi}'(\mathbf{R}\mathbf{C}_{nz}^{-1}\mathbf{R}')^{-1}\boldsymbol{\xi} + 2ns_e^{-2}(\tilde{\boldsymbol{\theta}}_n - \boldsymbol{\theta})'\mathbf{R}'(\mathbf{R}\mathbf{C}_{nz}^{-1}\mathbf{R}')^{-1}\boldsymbol{\xi}.\end{aligned} \tag{8.5.11}$$

Under the conditions (i) through (iii) of (8.4.3) and CLT as $n \to \infty$, we obtain

$$\sqrt{n}\, s_e^{-1}\mathbf{R}\tilde{\boldsymbol{\theta}}_n \xrightarrow{\mathcal{D}} \mathcal{N}_q(\sqrt{n}\,\sigma^{-1}\boldsymbol{\xi}, \mathbf{R}\mathbf{C}_z^{-1}\mathbf{R}') \tag{8.5.12}$$

and $n\sigma^{-2}\boldsymbol{\xi}'(\mathbf{R}\mathbf{C}_z^{-1}\mathbf{R}')^{-1}\boldsymbol{\xi} \to \infty$, as $n \to \infty$, while the third term goes to zero as $n \to \infty$ in probability. Hence, under $K_{\boldsymbol{\xi}}$,

$$\lim_{n\to\infty} P(\mathcal{L}_n^* > x) = 1 \quad \text{for all } x \in R^+. \tag{8.5.13}$$

Thus, we obtain the following results:

(i) $\sqrt{n}s_e^{-1}(\hat{\boldsymbol{\theta}}_n^{\text{PT}} - \boldsymbol{\theta}) = \sqrt{n}\sigma^{-1}(\tilde{\boldsymbol{\theta}}_n - \boldsymbol{\theta}) + o_p(1).$ (8.5.14)

(ii) $\sqrt{n}s_e^{-1}(\hat{\boldsymbol{\theta}}_n^{S} - \boldsymbol{\theta}) = \sqrt{n}\sigma^{-1}(\tilde{\boldsymbol{\theta}}_n - \boldsymbol{\theta}) + o_p(1).$

(iii) $\sqrt{n}s_e^{-1}(\hat{\boldsymbol{\theta}}_n^{S+} - \boldsymbol{\theta}) = \sqrt{n}\sigma^{-1}(\tilde{\boldsymbol{\theta}}_n - \boldsymbol{\theta}) + o_p(1)$

by Theorem 7.8.2. The ADB and ADQB are given by

$$\mathbf{b}_1(\tilde{\boldsymbol{\theta}}_n) = \mathbf{b}_3(\hat{\boldsymbol{\theta}}_n^{\text{PT}}) = \mathbf{b}_4(\hat{\boldsymbol{\theta}}_n^{S}) = \mathbf{b}_5(\hat{\boldsymbol{\theta}}_n^{S+}) = \mathbf{0}$$

and

$$B_1(\tilde{\boldsymbol{\theta}}_n) = B_3(\hat{\boldsymbol{\theta}}_n^{\text{PT}}) = B_4(\hat{\boldsymbol{\theta}}_n^{S}) = B_5(\hat{\boldsymbol{\theta}}_n^{S+}) = 0. \tag{8.5.15}$$

Similarly,

$$\mathbf{M}_1(\tilde{\boldsymbol{\theta}}_n) = \mathbf{M}_3(\hat{\boldsymbol{\theta}}_n^{\text{PT}}) = \mathbf{M}_4(\hat{\boldsymbol{\theta}}_n^{S}) = \mathbf{M}_5(\hat{\boldsymbol{\theta}}_n^{S+}) = \sigma^2\mathbf{C}_z^{-1},$$

and the ADQR is given by

$$R_1(\tilde{\boldsymbol{\theta}}_n : \mathbf{W}) = R_3(\hat{\boldsymbol{\theta}}_n^{\text{PT}} : \mathbf{W}) = R_4(\hat{\boldsymbol{\theta}}_n^{S} : \mathbf{W}) = R_5(\hat{\boldsymbol{\theta}}_n^{S+} : \mathbf{W}) = \sigma^2 \operatorname{tr}[\mathbf{W}\mathbf{C}_z^{-1}]. \tag{8.5.16}$$

Now, for the restricted estimator, we have

$$ns_e^{-2}||\hat{\boldsymbol{\theta}}_n - \tilde{\boldsymbol{\theta}}_n||^2 = \mathcal{L}_n^* \to \infty, \tag{8.5.17}$$

and $E(\mathcal{L}_n^*) \to \infty$ as $n \to \infty$. The asymptotic distributions of $\sqrt{n}s_e^{-1}(\hat{\boldsymbol{\theta}}_n - \boldsymbol{\theta})$ and $\sqrt{n}s_e^{-1}(\tilde{\boldsymbol{\theta}}_n - \boldsymbol{\theta})$ are not equivalent. Further,

$$\begin{aligned}
&\sqrt{n}s_e^{-1}(\hat{\boldsymbol{\theta}}_n - \boldsymbol{\theta}) \\
&= \sqrt{n}s_e^{-1}(\tilde{\boldsymbol{\theta}}_n - \boldsymbol{\theta}) - \sqrt{n}s_e^{-1}\mathbf{C}_{nz}^{-1}\mathbf{R}'(\mathbf{R}\mathbf{C}_{nz}^{-1}\mathbf{R}')^{-1}\mathbf{R}\tilde{\boldsymbol{\theta}}_n \\
&= \sqrt{n}s_e^{-1}(\tilde{\boldsymbol{\theta}}_n - \boldsymbol{\theta}) - \sqrt{n}s_e^{-1}\mathbf{C}_{nz}^{-1}\mathbf{R}'(\mathbf{R}\mathbf{C}_{nz}^{-1}\mathbf{R}')^{-1}[\mathbf{R}(\tilde{\boldsymbol{\theta}}_n - \boldsymbol{\theta}) + \boldsymbol{\xi}],
\end{aligned} \tag{8.5.18}$$

and this implies that the asymptotic distribution of $\sqrt{n}s_e^{-1}(\hat{\boldsymbol{\theta}}_n - \boldsymbol{\theta})$ is degenerate under fixed alternatives $K_{\boldsymbol{\xi}}$, since $\sqrt{n}s_e^{2}\mathbf{C}_{nz}^{-1}\mathbf{R}'(\mathbf{R}\mathbf{C}_{nz}^{-1}\mathbf{R}')^{-1}\boldsymbol{\xi} \to \infty$ as $n \to \infty$.

8.5.2 Asymptotic Distribution of the Estimators under Local Alternatives

To obtain a meaningful asymptotic distribution of the estimators and the test-statistic $\mathcal{L}_n^*$, we consider the following theorem under $\{K_{(n)}\}$:

Theorem 1. Under $K_{(n)} : \boldsymbol{\delta}_{(n)} = n^{-1/2}\boldsymbol{\xi}$ and the assumed regularity conditions (i) through (iii) of (8.4.3), we have as $n \to \infty$ the following:

(i) $V_{n+q}^{(1)} = \sqrt{n}(\tilde{\boldsymbol{\theta}}_n - \boldsymbol{\theta}) \sim \mathcal{N}_{n+q}(\mathbf{0}, \sigma^2\mathbf{C}_z^{-1}).$ (8.5.19)

(ii) $V_{n+q}^{(2)} = \sqrt{n}(\hat{\boldsymbol{\theta}}_n - \boldsymbol{\theta}) \sim \mathcal{N}_{n+q}(-\boldsymbol{\delta}_z, \sigma^2\mathbf{A}_z),$

where $\mathbf{A}_z = \mathbf{C}_z^{-1} - \mathbf{C}_z^{-1}\mathbf{R}'(\mathbf{R}\mathbf{C}_z^{-1}\mathbf{R}')^{-1}\mathbf{R}\mathbf{C}_z^{-1}$ and

$\boldsymbol{\delta}_z = \mathbf{C}_z^{-1}\mathbf{R}'(\mathbf{R}\mathbf{C}_z^{-1}\mathbf{R}')\mathbf{R}\boldsymbol{\xi}.$

(iii) $V_{n+q}^{(3)} = \sqrt{n}(\tilde{\boldsymbol{\theta}}_n - \hat{\boldsymbol{\theta}}_n) \sim \mathcal{N}_{n+q}(\boldsymbol{\delta}_z, \sigma^2[\mathbf{C}_z^{-1} - \mathbf{A}_z]).$

(iv) $$\begin{bmatrix} V_{n+q}^{(1)} \\ V_{n+q}^{(3)} \end{bmatrix} \sim \mathcal{N}_{2(n+q)}\left\{ \begin{matrix} \mathbf{0} \\ \boldsymbol{\delta}_z \end{matrix} ; \sigma^2 \begin{pmatrix} \mathbf{C}_z^{-1} & \mathbf{C}_z^{-1} - \mathbf{A}_z \\ \mathbf{C}_z^{-1} - \mathbf{A}_z & \mathbf{C}_z^{-1} - \mathbf{A}_z \end{pmatrix}\right\}.$$

(v) $$\begin{bmatrix} V_{n+q}^{(2)} \\ V_{n+q}^{(3)} \end{bmatrix} \sim \mathcal{N}_{2(n+q)}\left\{ \begin{pmatrix} -\boldsymbol{\delta}_z \\ \boldsymbol{\delta}_z \end{pmatrix} ; \sigma^2 \begin{pmatrix} \mathbf{A}_z & \mathbf{0} \\ \mathbf{0} & \mathbf{C}_z^{-1} - \mathbf{A}_z \end{pmatrix}\right\}.$$

(vi) $\lim_{n\to\infty} P_{K_{(n)}}\{\mathcal{L}_n^* \le x\} = H_q(x; \Delta^2)$

(vii) $$\lim_{n\to\infty} P_{K_{(n)}}\{\sqrt{n}(\hat{\boldsymbol{\theta}}_n^{\text{PT}} - \boldsymbol{\theta}) \le x\} = H_q(\chi_q^2(\alpha) : \Delta^2)\Phi_{n+q}(\mathbf{x} + \boldsymbol{\delta}_z; \mathbf{0}; \sigma^2\mathbf{A}_z)$$
$$+ \int_{E(\boldsymbol{\delta}_z)} \Phi_{n+q}(\mathbf{x} + \mathbf{u}; \mathbf{0}; \sigma^2[\mathbf{C}_z^{-1} - \mathbf{A}_z])d\,\Phi_{n+q}(\mathbf{u}, \mathbf{0}; \sigma^2\mathbf{R}\mathbf{C}_z^{-1}\mathbf{R}').$$

where $E(\boldsymbol{\delta}_z) = \{\mathbf{U}; (\mathbf{U} + \boldsymbol{\delta}_z)'(\mathbf{R}\mathbf{C}_z^{-1}\mathbf{R}')^{-1}(\mathbf{U} + \boldsymbol{\delta}_z) \ge \chi_q^2(\alpha)\}$ and $\Phi_\nu(\mathbf{x}; \mathbf{U}; \boldsymbol{\Sigma})$ is the cdf of a ν-variate normal distribution with mean μ and covariance matrix $\boldsymbol{\Sigma}$ and $H_q(\cdot; \Delta^2)$ is the cdf of a noncentral chi-square distribution with q d.f. and the noncentrality parameter $\Delta^2/2$,

(viii) $$\sqrt{n}(\hat{\boldsymbol{\theta}}_n^S - \boldsymbol{\theta}) \overset{\mathcal{D}}{=} \mathbf{U} - k\left\{\frac{\mathbf{C}_z^{-1}\mathbf{R}'(\mathbf{R}\mathbf{C}_z^{-1}\mathbf{R}')^{-1}(\mathbf{R}\mathbf{U}+\boldsymbol{\xi})}{\sigma^{-2}(\mathbf{R}\mathbf{U}+\boldsymbol{\xi})'(\mathbf{R}\mathbf{C}_z^{-1}\mathbf{R}')^{-1}(\mathbf{R}\mathbf{U}+\boldsymbol{\xi})}\right\} \text{ as } n \to \infty.$$

(ix) $$\sqrt{n}(\hat{\boldsymbol{\theta}}_n^{S+} - \boldsymbol{\theta}) \overset{\mathcal{D}}{=} \mathbf{U} - k\left\{\frac{\mathbf{C}_z^{-1}\mathbf{R}'(\mathbf{R}\mathbf{C}_z^{-1}\mathbf{R}')^{-1}(\mathbf{R}\mathbf{U}+\boldsymbol{\xi})}{\sigma^{-2}(\mathbf{R}\mathbf{U}+\boldsymbol{\xi})'(\mathbf{R}\mathbf{C}_z^{-1}\mathbf{R}')^{-1}(\mathbf{R}\mathbf{U}+\boldsymbol{\xi})}\right\}$$
$$+\mathbf{C}_z^{-1}\mathbf{R}'(\mathbf{R}\mathbf{C}_z^{-1}\mathbf{R}')^{-1}(\mathbf{R}\mathbf{U}+\boldsymbol{\xi})\left\{1 - \frac{k}{\sigma^{-2}(\mathbf{R}\mathbf{U}+\boldsymbol{\xi})'(\mathbf{R}\mathbf{C}_z^{-1}\mathbf{R}')^{-1}(\mathbf{R}\mathbf{U}+\boldsymbol{\xi})}\right\}$$
$$\times I(\sigma^{-2}(\mathbf{R}\mathbf{U} + \boldsymbol{\xi})'(\mathbf{R}\mathbf{C}_z^{-1}\mathbf{R}')^{-1}(\mathbf{R}\mathbf{U} + \boldsymbol{\xi}) < k),$$

where $k = q - 2$ and $\mathbf{U} \sim \mathcal{N}_{n+q}(\mathbf{0}, \sigma^2\mathbf{C}_z^{-1}).$

Based on the distribution, we find the ADB, ADQB, and ADMSE matrices and the ADQR of the estimators. First, the ADB and the ADQB expressions are

$$\text{(i) } \mathbf{b}_1(\tilde{\boldsymbol{\theta}}_n) = \mathbf{0} \text{ and } B_1(\tilde{\boldsymbol{\theta}}_n) = 0 \quad \text{(ii) } \mathbf{b}_2(\hat{\boldsymbol{\theta}}_n) = -\boldsymbol{\delta}_z \text{ and } B_2(\hat{\boldsymbol{\theta}}_n) = \Delta^2, \tag{8.5.20}$$

$$\text{(iii) } \mathbf{b}_3(\hat{\boldsymbol{\theta}}_n^{PT}) = -\boldsymbol{\delta}_z G_{q+2}(\chi_q^2(\alpha);\Delta^2) \text{ and } B_3(\hat{\boldsymbol{\theta}}_n^{PT}) = \Delta^2\{G_{q+2}(\chi_q^2(\alpha);\Delta^2)\}^2,$$

$$\text{(iv) } \mathbf{b}_4(\hat{\boldsymbol{\theta}}_n^S) = -k\boldsymbol{\delta}_z E[\chi_{q+2}^{-2}(\Delta^2)],\ k = q-2 \text{ and } B_4(\hat{\boldsymbol{\theta}}_n^S) = \Delta^2\{E[\chi_{q+2}^{-2}(\Delta^2)]\}^2,$$

$$\text{(v) } \mathbf{b}_5(\hat{\boldsymbol{\theta}}_n^{S+}) = \mathbf{b}_4(\hat{\boldsymbol{\theta}}_n^S) - \boldsymbol{\delta}_z E[(1-k\chi_{q+2}^{-2}(\Delta^2))I(\chi_{q+2}^2(\Delta^2) < k)]$$

and

$$B_5(\hat{\boldsymbol{\theta}}_n^{S+}) = \Delta^2\{kE[\chi_{q+2}^{-2}(\Delta^2)] - E[(1-k\chi_{q+2}^{-2}(\Delta^2))I(\chi_{q+2}^2(\Delta^2) < k)]\}^2.$$

Next, we present the ADMSE and ADQR expressions as

$$\text{(i) } \mathbf{M}_1(\tilde{\boldsymbol{\theta}}_n) = \sigma^2\mathbf{C}_z^{-1} \text{ and } R_1(\tilde{\boldsymbol{\theta}}_n : \mathbf{W}) = \sigma^2 \operatorname{tr}[\mathbf{W}\mathbf{C}_z^{-1}], \tag{8.5.21}$$

$$\text{(ii) } \mathbf{M}_2(\hat{\boldsymbol{\theta}}_n) = \sigma^2[\mathbf{C}_z^{-1} - \mathbf{A}_z] + \boldsymbol{\delta}_z\boldsymbol{\delta}_z' \text{ and}$$

$$R_2(\hat{\boldsymbol{\theta}}_n : \mathbf{W}) = \sigma^2 \operatorname{tr}[\mathbf{W}\mathbf{C}_z^{-1}] - \sigma^2 \operatorname{tr}[\mathbf{W}(\mathbf{C}_z^{-1} - \mathbf{A}_z)] + \boldsymbol{\delta}_z'\mathbf{W}\boldsymbol{\delta}_z,$$

$$\begin{aligned}\text{(iii) } \mathbf{M}_3(\hat{\boldsymbol{\theta}}_n^{PT}) &= \sigma^2\mathbf{C}_z^{-1} - \sigma^2[\mathbf{C}_z^{-1} - \mathbf{A}_z]H_{q+2}(\chi_q^2(\alpha);\Delta^2) \\ &\quad +\boldsymbol{\delta}_z\boldsymbol{\delta}_z'\{2H_{q+2}(\chi_q^2(\alpha);\Delta^2) - H_{q+4}(\chi_q^2(\alpha);\Delta^2)\} \quad \text{and}\end{aligned}$$

$$\begin{aligned}R_3(\hat{\boldsymbol{\theta}}_n^{PT} : \mathbf{W}) &= \sigma^2 \operatorname{tr}(\mathbf{W}\mathbf{C}_z^{-1}) - \sigma^2 \operatorname{tr}[\mathbf{W}(\mathbf{C}_z^{-1} - \mathbf{A}_z)]H_{q+2}(\chi_q^2(\alpha);\Delta^2), \\ &\quad +\boldsymbol{\delta}_z'\mathbf{W}\boldsymbol{\delta}_z\{2H_{q+2}(\chi_q^2(\alpha);\Delta^2) - H_{q+4}(\chi_q^2(\alpha);\Delta^2)\}\end{aligned}$$

$$\begin{aligned}\text{(iv) } \mathbf{M}_4(\hat{\boldsymbol{\theta}}_n^S) &= \sigma^2\mathbf{C}_z^{-1} - k\sigma^2[\mathbf{C}_z^{-1} - \mathbf{A}_z]\{2E[\chi_{q+2}^{-2}(\Delta^2)] - (q-2)E[\chi_{q+2}^{-4}(\Delta^2)]\} \\ &\quad +k(k+2)\sigma^{-2}\boldsymbol{\delta}_z\boldsymbol{\delta}_z' E[\chi_{q+4}^{-4}(\Delta^2)] \text{ and}\end{aligned}$$

$$\begin{aligned}R_4(\hat{\boldsymbol{\theta}}_n^S : \mathbf{W}) &= \sigma^2 \operatorname{tr}(\mathbf{W}\mathbf{C}_z^{-1}) - k\sigma^2 \operatorname{tr}[\mathbf{W}(\mathbf{C}_z^{-1} - \mathbf{A}_z)] \\ &\quad \times\{2E[\chi_{q+2}^{-2}(\Delta^2)] - (q-2)E[\chi_{q+2}^{-4}(\Delta^2)]\} \\ &\quad +k(k+2)\sigma^{-2}\boldsymbol{\delta}_z'\mathbf{W}\boldsymbol{\delta}_z E[\chi_{q+4}^{-4}(\Delta^2)],\end{aligned}$$

$$\begin{aligned}\text{(v) } \mathbf{M}_5(\hat{\boldsymbol{\theta}}_n^{S+}) &= \mathbf{M}_4(\hat{\boldsymbol{\theta}}_n^S) - \sigma^2[\mathbf{C}_z^{-1} - \mathbf{A}_z]E[(1-k\chi_{q+2}^{-2}(\Delta^2))^2 I(\chi_{q+2}^2(\Delta^2) < k)] \\ &\quad +\boldsymbol{\delta}_z\boldsymbol{\delta}_z'\{2E[(1-k\chi_{q+2}^{-2}(\Delta^2))I(\chi_{q+2}^2(\Delta^2) < k)] \text{ and} \\ &\quad -E[(1-k\chi_{q+4}^{-2}(\Delta^2))^2 I(\chi_{q+2}^2(\Delta^2) < k)]\}\end{aligned}$$

$$R_5(\hat{\boldsymbol{\theta}}_n^{S+} : \mathbf{W}) = R_4(\hat{\boldsymbol{\theta}}_n^{S} : \mathbf{W}) - \sigma^2\{\text{tr}[\mathbf{W}(\mathbf{C}_z^{-1} - \mathbf{A}_z)]$$
$$\times E[(1 - k\chi_{q+2}^{-2}(\Delta^2))^2 I(\chi_{q+2}^2(\Delta^2) < k)]$$
$$-\boldsymbol{\delta}_z'\mathbf{W}\boldsymbol{\delta}_z\{2E[(1 - k\chi_{q+2}^{-2}(\Delta^2))I(\chi_{q+2}^2(\Delta^2) < k)]$$
$$-E[(1 - k\chi_{q+4}^{-2}(\Delta^2))^2 I(\chi_{q+2}^2(\Delta^2) < k)]\}.$$

The expressions above may be used to compare the properties of the five estimators as has been done in Section 8.2.3 and (8.2.4).

8.6 Confidence Set Estimation: Stochastic Hypothesis

In this section, we consider the confidence set estimation of the regression parameters, $\boldsymbol{\beta}$ in the model (8.1.) together with (8.1.2), namely

$$\mathbf{Y} = \mathbf{X}\boldsymbol{\beta} + \mathbf{e} \tag{8.6.1}$$

and

$$\mathbf{h} = \mathbf{H}\boldsymbol{\beta} + \boldsymbol{\delta} + \mathbf{v}.$$

Thus, we consider the five estimators

(i) $\tilde{\boldsymbol{\beta}}_n = (\mathbf{C}^{-1}\mathbf{X}'\mathbf{Y}.$

(ii) $\hat{\boldsymbol{\beta}}_n = \tilde{\boldsymbol{\beta}}_n - \mathbf{C}^{-1}\mathbf{H}'(\mathbf{H}\mathbf{C}^{-1}\mathbf{H}' + \boldsymbol{\Omega})^{-1}(\mathbf{H}\tilde{\boldsymbol{\beta}}_n - \mathbf{h}).$ (8.6.2)

(iii) $\hat{\boldsymbol{\beta}}_n^{\text{PT}} = \tilde{\boldsymbol{\beta}}_n - \mathbf{C}^{-1}\mathbf{H}'(\mathbf{H}\mathbf{C}^{-1}\mathbf{H}' + \boldsymbol{\Omega})^{-1}(\mathbf{H}\tilde{\boldsymbol{\beta}}_n - \mathbf{h})\mathbf{I}(\mathcal{L}_n < \chi_q^2(\alpha)),$

where $\mathcal{L}_n = \sigma^{-2}(\mathbf{H}\tilde{\boldsymbol{\beta}}_n - \mathbf{h})'(\mathbf{H}\mathbf{C}^{-1}\mathbf{H}' + \boldsymbol{\Omega})^{-1}(\mathbf{H}\tilde{\boldsymbol{\beta}}_n - \mathbf{h}).$

(iv) $\hat{\boldsymbol{\beta}}_n^{S} = \tilde{\boldsymbol{\beta}}_n - d\mathbf{C}^{-1}\mathbf{H}'(\mathbf{H}\mathbf{C}^{-1}\mathbf{H}' + \boldsymbol{\Omega})^{-1}(\mathbf{H}\tilde{\boldsymbol{\beta}}_n - \mathbf{h})\mathcal{L}_n^{-1},$

(v) $\hat{\boldsymbol{\beta}}_n^{S+} = \hat{\boldsymbol{\beta}}_n + \mathbf{C}^{-1}\mathbf{H}'(\mathbf{H}\mathbf{C}^{-1}\mathbf{H}' + \boldsymbol{\Omega})(\mathbf{H}\tilde{\boldsymbol{\beta}}_n - \mathbf{h})(1 - d\mathcal{L}_n^{-1})\mathbf{I}(\mathcal{L}_n < d).$

These estimators may be compactly written as

$$\boldsymbol{\beta}_n^* = \hat{\boldsymbol{\beta}}_n + (\tilde{\boldsymbol{\beta}}_n - \hat{\boldsymbol{\beta}}_n)g(\mathcal{L}_n) \tag{8.6.3}$$

as before.

Now, we write $\boldsymbol{\beta}_n^*$ as

$$\boldsymbol{\beta}_n^* = [\mathbf{I} - \mathbf{A}\mathbf{H}]\tilde{\boldsymbol{\beta}}_n - \mathbf{A}\mathbf{h} + \mathbf{A}(\mathbf{H}\tilde{\boldsymbol{\beta}}_n - \mathbf{h})g(\mathcal{L}_n), \tag{8.6.4}$$

where $\mathbf{A} = \mathbf{C}^{-1}\mathbf{H}'(\mathbf{H}\mathbf{C}^{-1}\mathbf{H}' + \boldsymbol{\Omega})^{-1}$. Then

$$\mathbf{A}'\mathbf{C}\mathbf{A} = (\mathbf{H}\mathbf{C}^{-1}\mathbf{H}' + \boldsymbol{\Omega})^{-1}(\mathbf{H}\mathbf{C}^{-1}\mathbf{H}')(\mathbf{H}\mathbf{C}^{-1}\mathbf{H}' + \boldsymbol{\Omega})^{-1} \tag{8.6.5}$$
$$(\mathbf{I} - \mathbf{A}\mathbf{H})'\mathbf{C}(\mathbf{I} - \mathbf{A}\mathbf{H}) = \mathbf{C} - \mathbf{H}'(\mathbf{H}\mathbf{C}^{-1}\mathbf{H}' + \boldsymbol{\Omega})^{-1}\mathbf{H} \tag{8.6.6}$$

Thus, we may consider the $(1-\gamma)\%$ confidence sets defined by

$$C^*(\beta_n^*) = \{\beta : \sigma^{-2}\|\beta - \beta_n^*\|_{\mathbf{C}}^2 \le \chi_p^2(\gamma)\}. \tag{8.6.7}$$

Accordingly, the five confidence sets of β are given by

(i) $C^0(\tilde{\beta}_n) = \{\beta : \sigma^{-2}\|\beta - \tilde{\beta}_n\|_{\mathbf{C}}^2 \le \chi_p^2(\gamma)\}$,

(ii) $C^R(\hat{\beta}_n) = \{\beta : \sigma^{-2}\|\hat{\beta} - \tilde{\beta}_n\|_{\mathbf{C}}^2 \le \chi_p^2(\gamma)\}$,

(iii) $C^{\mathrm{PT}}(\hat{\beta}_n^{\mathrm{PT}}(\alpha)) = \{\beta : \sigma^{-2}\|\beta - \hat{\beta}_n^{\mathrm{PT}}\|_{\mathbf{C}}^2 \le \chi_p^2(\gamma)\}$,

(iv) $C^S(\hat{\beta}_n^S) = \{\beta : \sigma^{-2}\|\beta - \hat{\beta}_n^S\|_{\mathbf{C}}^2 \le \chi_p^2(\gamma)\}$,

and

(v) $C^{S+}(\hat{\beta}_n^{S+}) = \{\beta : \sigma^{-2}\|\beta - \hat{\beta}_n^{S+}\|_{\mathbf{C}}^2 \le \chi_p^2(\gamma)\}$, respectively. (8.6.8)

Following Chapter 7, Section 7.9, we obtain the expressions after the coverage probabilities of these confidence sets by

$$\begin{aligned} &P\{\sigma^{-2}\|\beta - \beta_n^*|_{\mathbf{C}_z}^2 \le \chi_p^2(\gamma)\} \\ &= P_\eta\{|\eta_2 - \omega_2|^2 + |\eta_1 - \omega_1 g(\|\omega_1\|^2)|^2 \le \chi_p^2(\gamma)\}. \end{aligned} \tag{8.6.9}$$

As given in Section 7.9,

$$\begin{pmatrix} \omega_1 \\ \omega_2 \end{pmatrix} \sim N_p\left\{\begin{pmatrix} \eta_1 \\ \eta_2 \end{pmatrix}, \begin{pmatrix} \mathbf{I}_q & \mathbf{0} \\ \mathbf{0} & \mathbf{I}_{p-q} \end{pmatrix}\right\} \tag{8.6.10}$$

(see Problem 11).

Similar conclusions hold for the nonnormal case considering the asymptotic setup.

8.7 R-Estimation: Stochastic Hypothesis

In this section, we consider the problem of R-estimation of the regression parameters β in the model (as in Section 8.1)

$$\mathbf{Y}_n = \beta_0 \mathbf{1}_n + \mathbf{X}_n \beta + \mathbf{e}_n \tag{8.7.1}$$

when it is suspected but not certain that β belongs to the stochastic subspace defined by

$$\mathbf{h} = \mathbf{h}\beta + \boldsymbol{\delta} + \mathbf{v}_q, \tag{8.7.2}$$

where $\mathbf{h}$ is a q-vector of random variables, $\mathbf{H}$ is a $q \times p$ matrix of known constants, $\boldsymbol{\delta}$ is a q-vector of constants (unknown), and $\mathbf{v}_q$ is a q-vector of random variables such that $E(\mathbf{v}_q) = \mathbf{0}$ and $\text{Cov}(\mathbf{v}_q) = \frac{\sigma^2}{n}\boldsymbol{\Omega}$; here $\boldsymbol{\Omega}$ is a known $q \times q$ matrix of constants.

In order to obtain the R-estimator of $\boldsymbol{\beta}$, we first concentrate on the model (8.7.1) and consider the following toward estimating $\boldsymbol{\beta}$:

Let $R_i(\mathbf{b})$ be the rank of $Y_i - (\mathbf{x}_i - \bar{\mathbf{x}}_n)\mathbf{b}$ among $Y_1 - (\mathbf{x}_i - \bar{\mathbf{x}}_n)\mathbf{b}, \ldots, Y_n - (\mathbf{x}_n - \bar{\mathbf{x}}_n)\mathbf{b}$, where $\mathbf{x}_i$ is the ith row of $\mathbf{X} = (\mathbf{x}_1', \ldots, \mathbf{x}_n')'$ and $\bar{\mathbf{x}}_n = n^{-1}\sum_{i=1}^n \mathbf{x}_i$. Now, consider the LRS:

$$\mathbf{L}_n(\mathbf{b}) = (L_{1n}(b_1), \ldots, L_{pn}(b_p))' = \sum_{i=1}^n (\mathbf{x}_i - \bar{\mathbf{x}}_n)a_n(R_i(\mathbf{b})). \tag{8.7.3}$$

(i) For a square integrable and nondecreasing score generating function $\phi(\cdot)$ defined on $(0,1)$ as before (see Chapters 4, 5, and 6), $a_n(k) = E\phi(U_{k,n})$ or $\phi(\frac{k}{n+1})$.

(ii) Let $\mathbf{C}_n = \sum_{i=1}^n (\mathbf{x}_i - \bar{\mathbf{x}}_n)(\mathbf{x}_i - \bar{\mathbf{x}}_n)'$, and assume that $\lim_{n\to\infty} \mathbf{C}_n = \mathbf{C}$, and $\max_{1\le i\le n}\{(\mathbf{x}_i - \bar{\mathbf{x}}_n)C_n^{-1}(\mathbf{x}_i - \bar{\mathbf{x}}_n)\} = o(n)$.

(iii) $I(f) = \int_{-\infty}^{\infty}\left\{\frac{f'(x)}{f(x)}\right\}^2 f(x)dx < \infty$ and $A_\phi^2 = \int_0^1 \phi^2(u) - \left(\int_0^1 \phi(u)du\right)^2$.

(iv) $\gamma(\psi,\phi) = \int_0^1 \psi(u)\phi(u)du$.

(v) $\sigma^2 = A_\phi^2/\gamma^2(\psi,\phi)$, where $\psi(u) = -\dfrac{f'(F^{-1}(u))}{f(F^{-1}(u))}$, $0 < u < 1$, and

$$\int_0^1 \psi^2(u)du = I(f). \tag{8.7.4}$$

Further, if we put $\|\mathbf{a}\| = \sum_{m=1}^p a^{(m)}$, and $\mathbf{a} = (a^{(1)}, \ldots, a^{(p)})$, then the *unrestricted* R-estimator (URE), $\tilde{\boldsymbol{\beta}}_n$ of $\boldsymbol{\beta}$ is defined by any central point of the set

$$\mathbf{S} = \{\mathbf{b} : \|\mathbf{L}_n(\mathbf{b})\| = \text{ minimum}\}. \tag{8.7.5}$$

Then, by Jurečkova's (1969) linearity result (see Section 2.8.4), we have

$$\lim_{n\to\infty} P\left\{\sup_{\|\boldsymbol{\omega}\|<k}\left\|\mathbf{L}_n\left(\boldsymbol{\beta} + \frac{\boldsymbol{\omega}}{\sqrt{n}}\right) - \mathbf{L}_n(\boldsymbol{\beta}) + \gamma\mathbf{C}\boldsymbol{\omega}\right\| > \varepsilon\right\} = 0 \tag{8.7.6}$$

for any $k > 0$ and $\varepsilon > 0$, where $\boldsymbol{\omega}$ is a p-dimensional column vector and $\|\cdot\|$ denotes the p-dimensional Euclidean norm. Then we obtain

$$\sqrt{n}(\tilde{\boldsymbol{\beta}}_n - \boldsymbol{\beta}) \sim \mathcal{N}_p(\mathbf{0}, \sigma^2\mathbf{C}^{-1}), \quad \sigma^2 = A_\phi^2/\gamma^2(\psi,\phi). \tag{8.7.7}$$

To define the *restricted* R-estimator of $\boldsymbol{\beta}$, we mimic the least square estimator of $\boldsymbol{\beta}$ under (8.5.2) as the restricted rank estimator (RRE) of $\boldsymbol{\beta}$ given by

$$\hat{\boldsymbol{\beta}}_n = \tilde{\boldsymbol{\beta}}_n - \mathbf{C}_n^{-1}\mathbf{H}'\left(\mathbf{H}\mathbf{C}_n^{-1}\mathbf{H}' + \frac{1}{n}\boldsymbol{\Omega}\right)(\mathbf{H}\tilde{\boldsymbol{\beta}}_n - \mathbf{h}) \tag{8.7.8}$$

where $\text{Cov}(\mathbf{v}_q) = \frac{\sigma^2}{n}\mathbf{\Omega}$ and $\sigma^2 = A_\phi^2/\gamma^2(\psi,\phi)$ as defined in (8.7.4) and $\mathbf{\Omega}$ is a known $q \times q$ matrix of constants. In order to test the hypothesis $H_0 : \boldsymbol{\delta} = \mathbf{0}$ against the alternative $H_A : \boldsymbol{\delta} \neq \mathbf{0}$, we consider the rank-test given by

$$\mathcal{L}_n = \{\hat{\gamma}_n^2(\psi,\phi)A_n^{-2}\}n(\mathbf{H}\tilde{\boldsymbol{\beta}}_n - \mathbf{h})'\left(n\mathbf{H}\mathbf{C}_n^{-1}\mathbf{H}' + \mathbf{\Omega}\right)^{-1}(\mathbf{H}\tilde{\boldsymbol{\beta}}_n - \mathbf{h}), \quad (8.7.9)$$

where $A_n^2 = (n-1)^{-1}\sum_{k=1}^n (a_n(k) - \bar{a}_n)^2$, $\bar{a}_n = n^{-1}\sum_{k=1}^n a_n(k)$, and $\hat{\gamma}_n(\psi,\phi) = p^{-1}\sum_{j=}^p \hat{\gamma}_{jn}(\psi,\phi)$
and

$$\hat{\gamma}_{jn}(\psi,\phi) = \frac{L_{jn}\left(\tilde{\boldsymbol{\beta}}_n + \frac{a\mathbf{e}_j}{\sqrt{n}}\right) - L_{jn}\left(\tilde{\boldsymbol{\beta}}_n + \frac{a\mathbf{e}_j}{\sqrt{n}}\right)}{2aC^{(j)}\mathbf{e}_j}, \quad (8.7.10)$$

with the following two constraints:

(i) $C^{(j)}$ is the jth row of $\mathbf{C}$.

(ii) $\mathbf{e}_j$ is the unit-vector with 1 at the jth place and 0 elsewhere.

Here, $\hat{\gamma}_n(\psi,\phi)$ is a consistent estimator of $\gamma(\psi,\phi)$.

Thus, by (8.7.7) as $n \to \infty$, $\mathcal{L}_n$ converges in distribution to the central chi-square with q d.f. under $H_0 : \boldsymbol{\delta} = \mathbf{0}$. Further, under the *fixed* alternative, $H_\xi : \boldsymbol{\delta} = \boldsymbol{\xi}$, $\mathcal{L}_n \to \infty$ as $n \to \infty$, since

$$\mathcal{L}_n = \hat{\gamma}^2(\psi,\phi)A_n^{-2}n\tilde{\boldsymbol{\delta}}_n'(n\mathbf{H}\mathbf{C}_n^{-1}\mathbf{H}' + \mathbf{\Omega})^{-1}\tilde{\boldsymbol{\delta}}_n, \quad (8.7.11)$$

where $\tilde{\boldsymbol{\delta}}_n = \mathbf{H}\tilde{\boldsymbol{\beta}}_n - \mathbf{h}$.

Note that

(i) $\tilde{\sigma}_n^{-1}\sqrt{n}\tilde{\boldsymbol{\delta}}_n \sim \mathcal{N}\{n\sigma^{-1}\boldsymbol{\xi}, (\mathbf{H}\mathbf{C}^{-1}\mathbf{H}' + \mathbf{\Omega})\}$,

(ii) $n\xi'(\mathbf{H}\mathbf{C}^{-1}\mathbf{H}' + \mathbf{\Omega})^{-1}\boldsymbol{\xi} \to \infty$ as $n \to \infty$,

(iii) $n(\tilde{\boldsymbol{\delta}}_n - \boldsymbol{\xi})'(\mathbf{H}\mathbf{C}^{-1}\mathbf{H}' + \mathbf{\Omega})^{-1}\boldsymbol{\xi} \xrightarrow{\text{P}} 0$ as $n \to \infty$. (8.7.12)

Hence, under a fixed alternative, $\mathcal{L}_n \xrightarrow{\text{P}} \infty$ as $n \to \infty$ and $\lim_{n\to\infty} P(\mathcal{L}_n > k) \to 1$ for all $k \in R$. However, under the null hypothesis $H_0 : \boldsymbol{\delta} = \mathbf{0}$, $\mathcal{L}_n$ approximately follows a central chi-square distribution with q d.f.

Now, we consider the mixed rank estimators of $\boldsymbol{\theta} = \begin{pmatrix} \boldsymbol{\beta} \\ \mathbf{H}\boldsymbol{\beta} + \boldsymbol{\delta} \end{pmatrix}$. The URE of $\boldsymbol{\theta}$ is then given by $\tilde{\boldsymbol{\theta}}_n = \begin{pmatrix} \tilde{\boldsymbol{\beta}}_n \\ \mathbf{h} \end{pmatrix}$, and the RRE of $\boldsymbol{\theta}$ is given by $\hat{\boldsymbol{\theta}}_n = \begin{pmatrix} \tilde{\boldsymbol{\beta}}_n \\ \mathbf{h} - \frac{1}{n}\mathbf{\Omega}(\mathbf{H}\mathbf{C}_n^{-1}\mathbf{H}' + \frac{1}{n}\mathbf{\Omega})^{-1}(\mathbf{H}\tilde{\boldsymbol{\beta}}_n - \mathbf{h}) \end{pmatrix}$.

Let $\chi_q^2(\alpha)$ be the upper α-level critical value from a central chi-square distribution with q d.f. Then we define the PTRE of $\boldsymbol{\theta}$ as

$$\hat{\boldsymbol{\theta}}_n^{\text{PT}} = \tilde{\boldsymbol{\theta}}_n - (\tilde{\boldsymbol{\theta}}_n - \hat{\boldsymbol{\theta}}_n)I(\mathcal{L}_n < \chi_q^2(\alpha)). \quad (8.7.13\text{a})$$

Similarly, we define the SRE and PRSRE of $\boldsymbol{\theta}$ as the following:

$$\hat{\boldsymbol{\theta}}_n^S = \tilde{\boldsymbol{\theta}}_n - d(\tilde{\boldsymbol{\theta}}_n - \hat{\boldsymbol{\theta}}_n)\mathcal{L}_n^{-1}I(\mathcal{L}_n > \varepsilon_n), \tag{8.7.13b}$$

where $\varepsilon_n \to 0$ as $n \to \infty$, and $d = 1 - 3$ and

$$\hat{\boldsymbol{\theta}}_n^{S+} = \hat{\boldsymbol{\theta}}_n + (1 - d\mathcal{L}_n^{-1})I(\mathcal{L}_n > d)(\tilde{\boldsymbol{\theta}}_n - \hat{\boldsymbol{\theta}}_n) \tag{8.7.13c}$$

as in Section 8.4.3. Further, we can show, as in Section 8.7.1, that under the fixed alternative $H_3 : \boldsymbol{\delta} = \boldsymbol{\xi}$, we have

$$\begin{aligned}
&\text{(i)}\ \hat{\sigma}_n^{-1}\sqrt{n}(\tilde{\boldsymbol{\theta}}_n^{\text{PT}} = \sigma^{-1}\sqrt{n}(\tilde{\boldsymbol{\theta}}_n - \boldsymbol{\theta}) + o_P(1),\\
&\text{(ii)}\ \sigma_n^{-1}\sqrt{n}(\hat{\boldsymbol{\theta}}_n^S - \boldsymbol{\theta}) = \sigma^{-1}\sqrt{n}(\tilde{\boldsymbol{\theta}}_n - \boldsymbol{\theta}) + o_P(1),\\
&\text{(iii)}\ \hat{\sigma}_n^{-1}\sqrt{n}(\hat{\boldsymbol{\theta}}_n^{S+} - \boldsymbol{\theta}) = \sigma^{-1}\sqrt{n}(\tilde{\boldsymbol{\theta}}_n - \boldsymbol{\theta}) + o_P(1),
\end{aligned} \tag{8.7.14}$$

while the asymptotic distribution of $\sqrt{n}\hat{\sigma}_n(\hat{\boldsymbol{\theta}}_n - \boldsymbol{\theta})$ is degenerate as $n \to \infty$. Thus, as in Section 8.5.1, we consider the local alternatives

$$K_{(n)} : \boldsymbol{\delta}_{(n)} = n^{-1/2}\boldsymbol{\xi}, \quad \boldsymbol{\xi} \in R^n \tag{8.7.15}$$

to obtain the following theorem:

Theorem 1. Under $K_{(n)} : \boldsymbol{\delta}_{(n)} = n^{-1/2}\boldsymbol{\xi}$ and the assumed regularity conditions of (8.6.4), we have the following as $n \to \infty$:

(i) $V_{n+q}^{(1)} = \sqrt{n}(\tilde{\boldsymbol{\theta}}_n - \boldsymbol{\theta}) \sim \mathcal{N}_{n+q}(\mathbf{0}, \sigma^2\mathbf{C}_z^{-1}).$ (8.7.16)

(ii) $V_{n+q}^{(2)} = \sqrt{n}(\hat{\boldsymbol{\theta}}_n - \boldsymbol{\theta}) \sim \mathcal{N}_{n+q}(-\boldsymbol{\delta}_z, \sigma^2\mathbf{A}_z)$, where

$\mathbf{A}_z = \mathbf{C}_z^{-1} - \mathbf{C}_z^{-1}\mathbf{R}'(\mathbf{R}\mathbf{C}_z^{-1}\mathbf{R}')^{-1}\mathbf{R}\mathbf{C}_z^{-1}$ and $\boldsymbol{\delta}_z = \mathbf{C}_z^{-1}\mathbf{R}'(\mathbf{R}\mathbf{C}_z^{-1}\mathbf{R}')\mathbf{R}\boldsymbol{\xi}.$

(iii) $V_{n+q}^{(3)} = \sqrt{n}(\tilde{\boldsymbol{\theta}}_n - \hat{\boldsymbol{\theta}}_n) \sim \mathcal{N}_{n+q}(\boldsymbol{\delta}_z, \sigma^2[\mathbf{C}_z^{-1} - \mathbf{A}_z]).$

(iv) $$\begin{bmatrix} V_{n+q}^{(1)} \\ V_{n+q}^{(3)} \end{bmatrix} \sim \mathcal{N}_{2(n+q)}\left\{ \begin{matrix} \mathbf{0} \\ \boldsymbol{\delta}_z \end{matrix} ; \sigma^2 \begin{pmatrix} \mathbf{C}_z^{-1} & \mathbf{C}_z^{-1} - \mathbf{A}_z \\ \mathbf{C}_z^{-1} - \mathbf{A}_z & \mathbf{C}_z^{-1} - \mathbf{A}_z \end{pmatrix} \right\}.$$

(v) $$\begin{bmatrix} V_{n+q}^{(2)} \\ V_{n+q}^{(3)} \end{bmatrix} \sim \mathcal{N}_{2(n+q)}\left\{ \begin{pmatrix} -\boldsymbol{\delta}_z \\ \boldsymbol{\delta}_z \end{pmatrix} ; \sigma^2 \begin{pmatrix} \mathbf{A}_z & \mathbf{0} \\ \mathbf{0} & \mathbf{C}_z^{-1} - \mathbf{A}_z \end{pmatrix} \right\}.$$

(vi) $\lim_{n\to\infty} P_{K_{(n)}}\{\mathcal{L}_n^* \le x\} = H_q(x; \Delta^2).$

(vii) $$\lim_{n\to\infty} P_{K_{(n)}}\left\{\sqrt{n}(\hat{\boldsymbol{\theta}}_n^{\text{PT}} - \boldsymbol{\theta}) \le x\right\} = H_q(\chi_q^2(\alpha) : \Delta^2)\Phi_{n+q}(\mathbf{x} + \boldsymbol{\delta}_z; \mathbf{0}; \sigma^2\mathbf{A}_z)$$
$$+ \int_{E(\boldsymbol{\delta}_z)} \Phi_{n+q}(\mathbf{x} + \mathbf{u}; \mathbf{0}; \sigma^2[\mathbf{C}_z^{-1} - \mathbf{A}_z])d\,\Phi_{n+q}(\mathbf{u}, \mathbf{0}; \sigma^2\mathbf{R}\mathbf{C}_z^{-1}\mathbf{R}'),$$

where $E(\boldsymbol{\delta}_z) = \{\mathbf{U}; (\mathbf{U}+\boldsymbol{\delta}_z)'(\mathbf{R}\mathbf{C}_z^{-1}\mathbf{R}')^{-1}(\mathbf{U}+\boldsymbol{\delta}_z) \geq \chi_q^2(\alpha)\}$ and $\Phi_\nu(\mathbf{x}; \mathbf{U}; \boldsymbol{\Sigma})$ is the cdf of a ν-variate normal distribution with mean μ and covariance matrix $\boldsymbol{\Sigma}$ and $H_q(\cdot; \Delta^2)$ is the cdf of a noncentral chi-square distribution with q d.f. and noncentrality parameter $\Delta^2/2$.

(viii) $$\sqrt{n}(\hat{\boldsymbol{\theta}}_n^S - \boldsymbol{\theta}) \stackrel{\mathcal{D}}{=} \mathbf{U} - k\left\{\frac{\mathbf{C}_z^{-1}\mathbf{R}'(\mathbf{R}\mathbf{C}_z^{-1}\mathbf{R}')^{-1}(\mathbf{R}\mathbf{U}+\boldsymbol{\xi})}{\sigma^{-2}(\mathbf{R}\mathbf{U}+\boldsymbol{\xi})'(\mathbf{R}\mathbf{C}_Z^{-1}\mathbf{R}')^{-1}(\mathbf{R}\mathbf{U}+\boldsymbol{\xi})}\right\} \text{ as } n \to \infty.$$

(ix) $$\begin{aligned}\sqrt{n}(\hat{\boldsymbol{\theta}}_n^{S+} - \boldsymbol{\theta}) \stackrel{\mathcal{D}}{=} \mathbf{U} - k&\left\{\frac{\mathbf{C}_z^{-1}\mathbf{R}'(\mathbf{R}\mathbf{C}_z^{-1}\mathbf{R}')^{-1}(\mathbf{R}\mathbf{U}+\boldsymbol{\xi})}{\sigma^{-2}(\mathbf{R}\mathbf{U}+\boldsymbol{\xi})'(\mathbf{R}\mathbf{C}_z^{-1}\mathbf{R}')^{-1}(\mathbf{R}\mathbf{U}+\boldsymbol{\xi})}\right\}\\ &+ \mathbf{C}_z^{-1}\mathbf{R}'(\mathbf{R}\mathbf{C}_z^{-1}\mathbf{R}')^{-1}(\mathbf{R}\mathbf{U}+\boldsymbol{\xi})\\ &\times\left\{1 - \frac{k}{\sigma^{-2}(\mathbf{R}\mathbf{U}+\boldsymbol{\xi})'(\mathbf{R}\mathbf{C}_z^{-1}\mathbf{R}')^{-1}(\mathbf{R}\mathbf{U}+\boldsymbol{\xi})}\right\}\\ &\times I(\sigma^{-2}(\mathbf{R}\mathbf{U}+\boldsymbol{\xi})'(\mathbf{R}\mathbf{C}_z^{-1}\mathbf{R}')^{-1}(\mathbf{R}\mathbf{U}+\boldsymbol{\xi}) < k),\end{aligned}$$

where $k = q-2$ and $\mathbf{U} \sim \mathcal{N}_{n+q}(\mathbf{0}, \sigma^2\mathbf{C}_z^{-1})$.

Separating out the expressions for the properties of the estimators of the regression coefficients $\boldsymbol{\beta}$, we have as $n \to \infty$,

Theorem 2. Under $K_{(n)}$, we have

(i) $\sqrt{n}(\tilde{\boldsymbol{\beta}}_n - \boldsymbol{\beta}) \sim N_p(\mathbf{0}, \sigma^2\mathbf{C}^{-1}), \quad \sigma^2 = A_\phi^2/\gamma^2(\psi, \phi)$;

(ii) $\sqrt{n}(\hat{\boldsymbol{\beta}}_n - \boldsymbol{\beta}) \sim N_p(-\boldsymbol{\delta}, \sigma^2(\mathbf{C}^{-1} - \mathbf{A}))$ where $\mathbf{A} = \mathbf{C}^{-1}\mathbf{H}'(\mathbf{H}\mathbf{C}^{-1}\mathbf{H}' + \boldsymbol{\Omega})^{-1}\mathbf{H}\mathbf{C}^{-1}$, $\boldsymbol{\delta} = \mathbf{C}^{-1}\mathbf{H}'(\mathbf{H}\mathbf{C}^{-1}\mathbf{H}' + \boldsymbol{\Omega})^{-1}\mathbf{H}\boldsymbol{\xi}$;

(iii) $\sqrt{n}(\tilde{\boldsymbol{\beta}}_n - \hat{\boldsymbol{\beta}}_n) \sim N_p(\boldsymbol{\delta}, \sigma^2(\mathbf{C}^{-1} - \mathbf{A}))$;

(iv) $\lim P(\mathcal{L}_n \leq x|K_{(n)}) = H_q(x; \Delta^2)$; where $\Delta^2 = \sigma^{-2}\boldsymbol{\xi}'(\mathbf{H}\mathbf{C}^{-1}\mathbf{H}' + \boldsymbol{\Omega})^{-1}\boldsymbol{\xi}$.

(v) $\lim P\{\sqrt{n}(\hat{\boldsymbol{\beta}}_n^{\text{PT}} - \boldsymbol{\beta}) \leq \mathbf{x}|K_{(n)}\} = H_q(\chi_q^2(\alpha); \Delta^2)\Phi_p(\mathbf{x} + \mathbf{A}_\xi, \mathbf{0}, \sigma^2\mathbf{A}) + \int_{E(\boldsymbol{\xi})} \Phi_p(\mathbf{x} - \mathbf{C}^{-1}\mathbf{H}'(\mathbf{H}\mathbf{C}^{-1}\mathbf{H}' + \Omega)^{-1}\mathbf{Z}; (1, \sigma^2\mathbf{A})d\Phi_p(\mathbf{Z}; 0, \sigma^2(\mathbf{H}\mathbf{C}^{-1}\mathbf{H}' + \boldsymbol{\Omega}))$, where

$$E(\boldsymbol{\xi}) = \left\{\mathbf{Z}; \sigma^2(\mathbf{H}\mathbf{Z} + \boldsymbol{\xi})'(\mathbf{H}\mathbf{C}^{-1}\mathbf{H}' + \Omega)^{-1}(\mathbf{H}\mathbf{Z} + \boldsymbol{\xi}) \geq \chi_q^2(\alpha)\right\}.$$

Further,

(vi) $\sqrt{n}(\hat{\boldsymbol{\beta}}_n^{\text{PT}} - \boldsymbol{\beta}) = \mathbf{Z} - (\mathbf{H}\mathbf{Z} + \boldsymbol{\xi})I(\|\mathbf{H}\mathbf{Z} + \boldsymbol{\xi}\|^2_{(\mathbf{H}\mathbf{C}^{-1}\mathbf{H}'+\boldsymbol{\Omega})^{-1}} \leq \chi_q^2(\alpha)) + o_P(1)$, where $\mathbf{Z} \sim N_p(\mathbf{0}, \sigma^2\mathbf{C}^{-1})$;

(vii) $\sqrt{n}(\hat{\boldsymbol{\beta}}_n^S - \boldsymbol{\beta}) = \mathbf{Z} - d\sigma^2\mathbf{C}^{-1}\mathbf{H}'(\mathbf{H}\mathbf{C}^{-1}\mathbf{H}' + \boldsymbol{\Omega})^{-1}(\mathbf{H}\mathbf{Z} + \boldsymbol{\xi})$
$\times\|\mathbf{H}\mathbf{Z} + \boldsymbol{\xi}\|^{-2}_{\{(\mathbf{H}\mathbf{C}^{-1}\mathbf{H}'+\boldsymbol{\Omega})^{-1}} \leq \chi_q^2(\alpha)) + o_P(1)$;

(viii) $\sqrt{n}(\hat{\boldsymbol{\beta}}_n^{S+} - \boldsymbol{\beta}) = \mathbf{Z} - d\sigma^2\mathbf{C}^{-1}\mathbf{H}'(\mathbf{H}\mathbf{C}^{-1}\mathbf{H}' + \boldsymbol{\Omega})^{-1}(\mathbf{HZ} + \boldsymbol{\xi})$
$\times\|\mathbf{HZ} + \boldsymbol{\xi}\|^{-2}_{(\mathbf{HC}^{-1}\mathbf{H}'+\boldsymbol{\Omega})^{-1}} - \sigma^2\mathbf{C}^{-1}\mathbf{H}'(\mathbf{HC}^{-1}\mathbf{H}' + \boldsymbol{\Omega})^{-1}(\mathbf{HZ} + \boldsymbol{\xi})$
$\{1 - d\|\mathbf{HZ}+\boldsymbol{\xi}\|^{-2}_{(\mathbf{HC}^{-1}\mathbf{H}'+\boldsymbol{\Omega})^{-1}}\} I(\|\mathbf{HZ}+\boldsymbol{\xi}\|^2 < d)_{(\mathbf{HC}^{-1}\mathbf{H}'+\boldsymbol{\Omega})^{-1}} + o_P(1).$
(8.7.18)

Based on Theorem 8.7.2, we may compute the ADB, ADQR, ADMSE, and the ADQR of the five estimators given in the following two theorems:

Theorem 3. Under $\{K_{(n)}\}$ and conditions (8.1.4) as $n \to \infty$, the ADB, ADQB, ADMSE and ADQR of the five estimators are given as follows:

(i) URE($\tilde{\boldsymbol{\beta}}_n$):

$$\mathbf{b}_1(\tilde{\boldsymbol{\beta}}_n) = \mathbf{0} \quad \text{and} \quad B_1(\tilde{\boldsymbol{\beta}}_n) = 0 \tag{8.7.19}$$

$$\mathbf{M}_1(\tilde{\boldsymbol{\beta}}_n) = \sigma^2\mathbf{C}^{-1} \quad \text{and} \quad R_1(\tilde{\boldsymbol{\beta}}_n; \mathbf{W}) = \sigma^2 \operatorname{tr}(\mathbf{C}^{-1}\mathbf{W}).$$

(ii) RRE($\hat{\boldsymbol{\beta}}_n$):

$$\mathbf{b}_2(\hat{\boldsymbol{\beta}}_n) = -\boldsymbol{\delta} \quad \text{and} \quad B_2(\hat{\boldsymbol{\beta}}_n) = \Delta^2;$$

$$\mathbf{M}_2(\hat{\boldsymbol{\beta}}_n) = \sigma^2(\mathbf{C}^{-1} - \mathbf{A}) + \boldsymbol{\delta}\boldsymbol{\delta}' \quad \text{and}$$

$$R_2(\hat{\beta}_n; \mathbf{W}) = \sigma^2 \operatorname{tr}\big[(\mathbf{C}^{-1} - \mathbf{A})\mathbf{W}\big] + \boldsymbol{\delta}'\mathbf{W}\boldsymbol{\delta}.$$

(iii) PTRE($\hat{\boldsymbol{\beta}}_n^{PT}$):

$$\mathbf{b}_3(\hat{\boldsymbol{\beta}}_n^{PT}) = \boldsymbol{\delta}H_{q+2}(\chi_q^2(\alpha); \Delta^2) \text{ and } B_3(\hat{\boldsymbol{\beta}}_n^{PT}) = \Delta^2\{H_{q+2}(\chi_q^2(\alpha); \Delta^2)\}^2;$$

$$\mathbf{M}_3(\hat{\boldsymbol{\beta}}_n^{PT}) = \sigma^2\mathbf{C}^{-1} - \sigma^2(\mathbf{C}^{-1} - \mathbf{A})H_{q+2}(\chi_q^2(\alpha); \Delta^2) + (\boldsymbol{\delta}\boldsymbol{\delta}')$$
$$\{2H_{q+2}(\chi_q^2(\alpha); \Delta^2) - H_{q+4}(\chi_q^2(\alpha); \Delta^2)\}$$
and

$$R_3(\hat{\boldsymbol{\beta}}_n^{PT}) = \sigma^2 \operatorname{tr}(\mathbf{C}^{-1}\mathbf{W}) - \sigma^2 \operatorname{tr}[(\mathbf{C}^{-1} - \mathbf{A})\mathbf{W}]H_{q+2}(\chi_q^2(\alpha); \Delta^2)$$
$$+ (\boldsymbol{\delta}'\mathbf{W}\boldsymbol{\delta})\{2H_{q+2}(\chi_q^2(\alpha); \Delta^2) - H_{q+4}(\chi_q^2(\alpha); \Delta^2)\}.$$

(iv) SRE($\hat{\boldsymbol{\beta}}_n^S$):

$$\mathbf{b}_4(\hat{\boldsymbol{\beta}}_n^S) = -d\boldsymbol{\delta}E\big[\chi_{q+2}^{-2}(\Delta^2)\big] \text{ and } B_4(\hat{\boldsymbol{\beta}}_n^S) = d^2\Delta^2\{E\big[\chi_{q+2}^{-2}(\Delta^2)\big]\}^2;$$

$$\mathbf{M}_4(\hat{\boldsymbol{\beta}}_n^S) = \sigma^2\mathbf{C}^{-1} - d\sigma^2[\mathbf{C}^{-1} - \mathbf{A}]\{2E\big[\chi_{q+2}^{-2}(\Delta^2)\big]$$
$$-(q-2)E\big[\chi_{q+2}^{-4}(\Delta^2)\big]\} + (q^2 - 4)(\boldsymbol{\delta}\boldsymbol{\delta}')E\big[\chi_{q+4}^{-4}(\Delta^2)\big], \quad d = q - 2$$
and

$$R_1(\hat{\beta}_n^S; \mathbf{W}) = \sigma^2 \operatorname{tr}(\mathbf{C}^{-1}\mathbf{W}) - \sigma^2 \operatorname{tr}(\mathbf{A}_{11})\{2E\big[\chi_{q+2}^{-2}(\Delta^2)\big]$$
$$-(q-2)E\big[\chi_{q+2}^{-4}(\Delta^2)\big]\} + (q^2 - 4)\boldsymbol{\delta}'\mathbf{W}\boldsymbol{\delta}E\big[\chi_{q+4}^{-4}(\Delta^2)\big]$$

with $\mathbf{A}_{11} = [\mathbf{W}\mathbf{C}^{-1}\mathbf{H}'(\mathbf{H}\mathbf{C}^{-1}\mathbf{H}' + \boldsymbol{\Omega})^{-1}\mathbf{H}\mathbf{C}^{-1}]$.

(v) RRSRE($\hat{\beta}_n^{S+}$):

$$\mathbf{b}_5(\hat{\boldsymbol{\beta}}_n^{S+}) = \boldsymbol{\delta}\Big\{dE\big[\chi_{q+2}^{-2}(\Delta^2)\big] + E\big[(1-d\chi_{q+2}^{-2}(\Delta^2))I(\chi_{q+2}^{2}(\Delta^2)\leq d)\big]\Big\}$$

and

$$B_5(\hat{\boldsymbol{\beta}}_n^{S+}) = \Delta^2\Big\{dE\big[\chi_{q+2}^{-2}(\Delta^2)\big] + E\big[(1-d\chi_{q+2}^{-2}(\Delta^2))I(\chi_{q+2}^{-2} < d)\big]\Big\}^2;$$

$$\mathbf{M}_5(\hat{\boldsymbol{\beta}}_n^{S+}) = \mathbf{M}_4(\hat{\boldsymbol{\beta}}_n^{S}) = \sigma^2(\mathbf{C}^{-1}-\mathbf{A})E\big[(1-d\chi_{q+2}^{-2}(\Delta^2))I(\chi_{q+2}^{2}<d)\big]+(\boldsymbol{\delta}\boldsymbol{\delta}')$$
$$\times\Big\{2E\big[(1-d\chi_{q+2}^{-2}(\Delta^2))I(\chi_{q+2}^{2} < d)\big]-E\big[(1-d\chi_{q+4}^{-2}(\Delta^2))^2 I(\chi_{q+4}^{2} < d)\Big\}$$

and

$$R_5(\hat{\boldsymbol{\beta}}_n^{S+};\mathbf{W}) = \operatorname{tr}(\mathbf{M}_4(\hat{\boldsymbol{\beta}}_n^{S})\mathbf{W}) - \sigma^2\operatorname{tr}\big[(\mathbf{C}^{-1}-\mathbf{A})\mathbf{W}\big]$$
$$\times E\big[(1-d\chi_{q+2}^{-2}(\Delta^2))I(\chi_{q+2}^{2}(\Delta^2) < d\big]$$
$$+(\boldsymbol{\xi}'\mathbf{A}'\mathbf{W}\mathbf{A}\boldsymbol{\xi})\Big\{2E\big[(1-d\chi_{q+2}^{-2}(\Delta^2))I(\chi_{q+2}^{2} < d)\big]$$
$$-E\big[(1-d\chi_{q+4}^{-2}(\Delta^2))^2 I(\chi_{q+4}^{2} < d)\Big\}, \quad \text{where } d = q-2.$$

Notice that these expressions are similar to the expressions of Section 7.8.2. Hence, the properties of the five estimators are asymptotically similar.

8.8 Conclusions

In this chapter, we studied the estimation of regression parameters when it is suspected that the regression parameters belongs to a stochastic subspace under the assumption that the covariance of the errors in the description of the stochastic subspace is known. The problem of estimation of the regression parameters is also discussed when there exists prior sample information from a previous study. The usual five estimators are defined and their finite sample properties are discussed in detail. We find that the properties of these estimators are similar to the properties of estimators given in Chapter 7. We also obtain confidence sets based on these estimators and apply the developments of Chapter 7 to study them. The small sample as well as large sample properties of confidence sets are similar to that of Chapter 7.9. Some nonparametric methods leading to R-estimation are also discussed.

8.9 Problems

1. Show that (8.1.10) holds. First, show that $\mathbf{R}\mathbf{C}_z^{-1}\mathbf{R}' = \mathbf{H}\mathbf{C}^{-1}\mathbf{H}' + \boldsymbol{\Omega}$.
2. Verify (i) through (v) of (8.2.1). For (ii), write down the explicit form of $\boldsymbol{\delta}_z$. Show that $\boldsymbol{\delta}_z\mathbf{C}_z\boldsymbol{\delta}_z = \Delta^2$.

3. Verify (i) through (v) of (8.2.2).
4. Compare $\hat{\boldsymbol{\theta}}_n$ and $\hat{\boldsymbol{\theta}}_n^{S+}$ using MSE matrices and weighted risk expressions.
5. Find the expressions for

$$\text{MRE}(\hat{\boldsymbol{\theta}}_n^{\text{PT}}; \tilde{\boldsymbol{\theta}}_n), \quad \text{and} \quad \text{RRE}(\hat{\boldsymbol{\theta}}_n^{\text{PT}}; \tilde{\boldsymbol{\theta}}_n)$$

6. Find the expressions of the covariance matrices of $\hat{\boldsymbol{\theta}}_n, \hat{\boldsymbol{\theta}}_n^{\text{PT}}, \hat{\boldsymbol{\theta}}_n^{S}$, and $\hat{\boldsymbol{\theta}}_n^{S+}$.
7. Show that

$$\hat{\boldsymbol{\theta}}_{1n_1}^{R} = \tilde{\boldsymbol{\theta}}_{1n_1} - \mathbf{C}_1^{-1}\mathbf{H}_1'(\mathbf{H}_1\mathbf{C}_1^{-1}\mathbf{H}_1' + \boldsymbol{\Omega}_0)^{-1}(\mathbf{H}_1\tilde{\boldsymbol{\theta}}_{1n_1} - \mathbf{h})$$

as given in (8.3.6).
8. Verify (i) through (v) of (8.3.19).
9. Verify (i) through (v) of (8.3.20).
10. Consider the system of two equations and a set of stochastic constraints

$$\begin{bmatrix} \mathbf{Y}_1 \\ \mathbf{Y}_2 \\ \mathbf{r}_1 \end{bmatrix} = \begin{bmatrix} \mathbf{X}_1 & \mathbf{0} \\ \mathbf{0} & \mathbf{X}_2 \\ \mathbf{R}_1 & \mathbf{0} \end{bmatrix} \begin{bmatrix} \boldsymbol{\beta}_1 \\ \boldsymbol{\beta}_2 \end{bmatrix} + \begin{bmatrix} \mathbf{e}_1 \\ \mathbf{e}_2 \\ \mathbf{v}_1 \end{bmatrix}$$

where $(\mathbf{e}_1', \mathbf{e}_2', \mathbf{v}_1')$ is normally distributed with mean vector $(\mathbf{0}', \mathbf{0}', \boldsymbol{\delta}')$ and covariance matrix

$$\boldsymbol{\Sigma} = \begin{pmatrix} \sigma^2\mathbf{I} & \mathbf{0} & \mathbf{0} \\ \mathbf{0} & \sigma^2\mathbf{I} & \mathbf{0} \\ \mathbf{0} & \mathbf{0} & \sigma^2\boldsymbol{\Omega} \end{pmatrix}.$$

(a) Find the estimator $(\boldsymbol{\beta}_1', \boldsymbol{\beta}_2')'$.

(b) Find the mean vector and covariance matrix of the estimators.

(c) Determine the stochastically restricted estimator of $(\boldsymbol{\beta}_1', \boldsymbol{\beta}_2')'$ assuming $\boldsymbol{\Sigma}$ to be known.

(d) Determine the test-statistic for testing $\mathbf{R}_1\boldsymbol{\beta}_1 = \mathbf{r}_1$ and its distribution.

11. Verify the asymptotic expression of the coverage probabilities of $C^0(\tilde{\boldsymbol{\beta}}_n)$, $C^R(\hat{\boldsymbol{\beta}}_n)$, $C^{\text{PT}}(\hat{\boldsymbol{\beta}}_n^{\text{PT}})$, $C^S(\hat{\boldsymbol{\beta}}_n^{S})$, and $C^{S+}(\hat{\boldsymbol{\beta}}_n^{S+})$ from the developments of Section 8.7.3.
12. Prove Theorem 8.6.2.
13. Prove Theorem 8.6.3.
14. Consider the two sets of regression models with *nonnormal* errors

$$\mathbf{Y}_{n_1} = \mathbf{X}_1\boldsymbol{\beta}_1 + \mathbf{e}_{n_1},$$

and

$$\mathbf{Y}_{n_2} = \mathbf{X}_2\boldsymbol{\beta}_2 + \mathbf{e}_{n_2}$$

where $\boldsymbol{\beta}_1$ and $\boldsymbol{\beta}_2$ are p_1- and p_2-vectors of regression coefficients and $\mathbf{X}_1$ and $\mathbf{X}_2$ are $n_1 \times p_1$ and $n_2 \times p_2$ matrices of known constants and $\mathbf{Y}_{n_1}$ and Y_{n_2} are response vectors such that $n_i/n_1 + n_2 \to \lambda_i$ $(0 < \lambda_i < 1)$ as $n(= n_1 + n_2) \to \infty$. Let H_1 and H_2 be two $n_1 \times p_1$ and $n_2 \times p_2$ known matrices.
Formulate the (a) unrestricted, (b) restricted, (c) preliminary test estimators, (d) Stein-type estimators, and the (e) positive-rule Stein estimators when it is suspected but not certain that $\mathbf{H}_1\boldsymbol{\beta}_1 = \mathbf{H}_2\boldsymbol{\beta}_2$ may hold using (a) LSE procedure and (b) R-estimation procedure.

Chapter 9

Ridge Regression

Outline

In the last two chapters, we have studied the properties of five estimators of the regression parameters, namely the *unrestricted, the restricted, the preliminary test, the James-Stein-type*, and *the positive-rule Stein-type* estimators of the regression parameters $\boldsymbol{\beta}$ in the model

$$\mathbf{Y} = \mathbf{X}\boldsymbol{\beta} + \mathbf{e}; \quad \mathbf{e} \sim \mathbf{N}_n(\mathbf{0}, \sigma^2\mathbf{I}_n)$$

when it is suspected but not certain that $\boldsymbol{\beta}$ belongs to a nonstochastic subspace.

It is observed from above that the usual least squares estimator (LSE) of $\boldsymbol{\beta}$ depends heavily on the characteristics of the matrix $\mathbf{C} = \mathbf{X}'\mathbf{X}$. If the $\mathbf{C}$ matrix is ill-conditioned (near dependency among various columns of $\mathbf{C}$, called "multicollinearity"), then the least squares estimator (LSE) produces unduly large sampling variances. Moreover, some of the regression coefficients may be statistically insignificant with the wrong sign, and meaningful statistical inference becomes difficult for the researcher. Hoerl and Kennard (1970) found that "multicollinearity" is a common problem in the field of engineering and econometrics, among many other fields. To resolve this problem, they suggested the use of $\mathbf{C}(k) = \mathbf{X}'\mathbf{X} + k\mathbf{I}_p$ ($k \geq 0$), instead of $\mathbf{C}$ in the estimation of

β. Accordingly, we consider the estimating equation $(\mathbf{C} + k\mathbf{I}_p)\tilde{\beta}_n(k) = \mathbf{X}'\mathbf{Y}$ instead of $\mathbf{C}\tilde{\beta}_n = \mathbf{X}'\mathbf{Y}$. The solution $\tilde{\beta}_n(k) = (\mathbf{I}_p + k\mathbf{C}^{-1})^{-1}\tilde{\beta}_n$ is known as the *ridge regression estimator* (*RRE*) of β. Recent application of ridge regressions are given in Malthouse (1999) and the references there in. Ridge regression methods have been considered by many researchers, beginning with Hoerl and Kennard (1970) and followed by Gibbons (1981), Vinod and Ullah (1981), Sarker (1992), Saleh and Kibria (1993), Kibria (1996a, b), Gruber (1998), Singh and Tracy (1999), Tabatabaey (1995), Wencheko (2000), Inoue (2001), Montgomery et al. (2001), and very recently Kibria and Saleh (2003, 2004) among others.

The purpose of this chapter is twofold: (1) to present the basic ideas and properties of ridge regression estimators by considering the estimation of the regression parameter β when $\mathbf{C}$ is ill-conditioned and it is suspected that β may be restricted to the subspace $\mathbf{H}_0 : \beta = \mathbf{0}$, and (2) to discuss the estimation techniques of the ridge parameter k. First we consider the regression estimators and variations thereof, to show that these can be further improved via ridge regression (Section 9.1).

Consider the model given earlier. The usual *unrestricted* and the *restricted* estimators of β under the restriction $\beta = \mathbf{0}$ are given by

$$\tilde{\beta}_n = \mathbf{C}^{-1}\mathbf{X}'\mathbf{Y} \quad \text{and} \quad \hat{\beta}_n = \mathbf{0},$$

respectively. The usual preliminary test estimator (PTE) of β is defined by

$$\hat{\beta}_n^{\text{PT}} = \tilde{\beta}_n I(\mathcal{L}_n \geq F_{p,m}(\alpha)), \quad m = n - p$$

where $F_{p,m}(\alpha)$ is the α-level critical value from the null distribution of $\mathcal{L}_n$, $I(A)$ is the indicator function of the set A, $\mathcal{L}_n$ is the test-statistic for testing the null hypothesis $\beta = \mathbf{0}$ (against the alternative hypothesis $\beta \neq \mathbf{0}$) given by

$$\mathcal{L}_n = \frac{\tilde{\beta}_n' \mathbf{C} \tilde{\beta}_n}{p\tilde{\sigma}^2}$$

with

$$\tilde{\sigma}^2 = m^{-1}(\mathbf{Y} - \mathbf{X}\tilde{\beta}_n)'(\mathbf{Y} - \mathbf{X}\tilde{\beta}_n).$$

Similarly, the James-Stein-type estimator of β is given by

$$\hat{\beta}_n^S = \tilde{\beta}_n - d\tilde{\beta}_n\mathcal{L}_n^{-1}, \quad d = \frac{m(p-2)}{p(m+2)},$$

and the positive-rule Stein-type estimator of β is given by

$$\hat{\beta}_n^{S+} = (1 - d\mathcal{L}_n^{-1})I(\mathcal{L}_n > d)\tilde{\beta}_n.$$

In the next section, we define the ridge regression estimators of β corresponding to each of the estimators, beginning with the unrestricted ridge regression estimator (URRE).

9.1 Ridge Regression Estimators (RRE)

9.1.1 Ridge Regression with Normal Errors

In this section, we consider the ridge regression version of the four estimators of $\boldsymbol{\beta}$. Accordingly, the *unrestricted ridge regression estimator* (URRE) of $\boldsymbol{\beta}$ is defined by the *ridge matrix*, $\mathbf{R}(k)$ as follows:

$$\tilde{\boldsymbol{\beta}}_n(k) = (\mathbf{C} + k\mathbf{I}_p)^{-1}\mathbf{X}'\mathbf{Y} = \mathbf{C}^{-1}(k)\mathbf{X}'\mathbf{Y} = \mathbf{R}(k)\tilde{\boldsymbol{\beta}}_n, \qquad (9.1.1)$$

where $\mathbf{R}(k) = (\mathbf{I}_p + k\mathbf{C}^{-1})^{-1}$, $k \geq 0$, $\tilde{\boldsymbol{\beta}}_n$ is the unrestricted estimator of $\boldsymbol{\beta}$ and k is the ridge parameter. This is the well-known Hoerl and Kennard (1970) estimator of $\boldsymbol{\beta}$.

Further, due to Rao (1975), a generalized ridge regression estimator of $\boldsymbol{\beta}$ can be defined by choosing a positive-semidefinite matrix $\mathbf{G}$, instead of $k\mathbf{I}_p$, and writing the estimator as

$$\tilde{\boldsymbol{\beta}}_n(G) = (\mathbf{C} + \mathbf{G})^{-1}\mathbf{X}'\mathbf{Y} = (\mathbf{I}_p + \mathbf{G}\mathbf{C}^{-1})^{-1}\tilde{\boldsymbol{\beta}}_n. \qquad (9.1.2)$$

Under the null hypothesis $H_0 : \boldsymbol{\beta} = \mathbf{0}$, the *restricted ridge regression estimator* (RRRE) of $\boldsymbol{\beta}$ is defined by $\hat{\boldsymbol{\beta}}_n(k) = \mathbf{0}$, where $\mathbf{0}$ is the restricted estimator of $\boldsymbol{\beta}$.

Next, we consider the *preliminary test ridge regression estimator* (PTRRE) of $\boldsymbol{\beta}$ defined by

$$\hat{\boldsymbol{\beta}}_n^{\mathrm{PT}}(k) = \mathbf{R}(k)\hat{\boldsymbol{\beta}}_n^{\mathrm{PT}} = \mathbf{R}(k)\tilde{\boldsymbol{\beta}}_n - \mathbf{R}(k)\tilde{\boldsymbol{\beta}}_n I(\mathcal{L}_n \leq \mathcal{F}_{p,m}(\alpha)), \qquad (9.1.3)$$

where $\mathcal{L}_n$ is the test-statistic for testing $\boldsymbol{\beta} = \mathbf{0}$. These kinds of estimators were studied by Saleh and Kibria (1993) for the normal regression model and by Kibria (1996a) for the Student's t regression model. Parallel to PTRRE, we define the *Stein-type ridge regression estimator* (SRRE) of $\boldsymbol{\beta}$ as

$$\hat{\boldsymbol{\beta}}_n^{S}(k) = \mathbf{R}(k)\hat{\boldsymbol{\beta}}_n^{S} = \mathbf{R}(k)\tilde{\boldsymbol{\beta}}_n - d\mathbf{R}(k)\tilde{\boldsymbol{\beta}}_n\mathcal{L}_n^{-1}, \quad d = \frac{m(p-2)}{p(m+2)}. \qquad (9.1.4)$$

Finally, we consider the positive-rule Stein-type ridge regression estimator (PRSRRE) of $\boldsymbol{\beta}$ defined by

$$\hat{\boldsymbol{\beta}}_n^{S+}(k) = \mathbf{R}(k)\hat{\boldsymbol{\beta}}_n^{S+} = \mathbf{R}(k)\hat{\boldsymbol{\beta}}_n^{S} - (1 - d\mathcal{L}_n^{-1})I(\mathcal{L}_n < d)\mathbf{R}(k)\tilde{\boldsymbol{\beta}}_n. \quad (9.1.5)$$

These kinds of estimators are considered by Kibria and Saleh (2004a, b) for the normal regression model and by Tabatabaey, Kibria, and Saleh (2005) for the Student's t regression model.

The four estimators $\tilde{\boldsymbol{\beta}}_n$, $\hat{\boldsymbol{\beta}}_n^{\mathrm{PT}}$ $\hat{\boldsymbol{\beta}}_n^{S}$, and $\hat{\boldsymbol{\beta}}_n^{S+}$ therefore give rise to four corresponding ridge regression estimators that are of interest in practical work in applied statistics.

9.1.2 Nonparametric Ridge Regression Estimators

In this section, we consider ridge regression estimators based on rank statistics. As in Chapter 7, we consider the model

$$\mathbf{Y} = \mathbf{X}\boldsymbol{\beta} + \boldsymbol{\varepsilon}, \tag{9.1.6}$$

where $\mathbf{X}$ is a $n \times p$ matrix of known constants such that $\mathbf{C}_n = \sum_{i=1}^n (\mathbf{x}_i - \bar{\mathbf{x}}_n)(\mathbf{x}_i - \bar{\mathbf{x}}_n)'$ (where $\mathbf{x}_i$ is the ith row of $\mathbf{X}$ and $\bar{\mathbf{x}}_n = \frac{1}{n}\sum_{i=1}^n \mathbf{x}_i$) is ill-conditioned. This means that some of the characteristic roots of $\mathbf{C}_n$ are small, which makes some of the asymptotic variances very large, and as a result makes the inference problem difficult. Thus, as in Section 7.11.1, consider the linear rank statistics required for testing $H_0 : \boldsymbol{\beta} = \mathbf{0}$:

$$\mathbf{L}_n(\mathbf{b}) = \sum_{i=1}^n (\mathbf{x}_i - \bar{\mathbf{x}}_n) a_n(R_i(\mathbf{b})) \tag{9.1.7}$$

as given by (7.11.6). We assume (7.11.7i–iv). In particular, we mention $\text{limit}_{n\to\infty}(n^{-1}\mathbf{C}_n) = \mathbf{C}$ and $\max\{(\mathbf{x}_i - \bar{\mathbf{x}}_n)'\mathbf{C}_n^{-1}(\mathbf{x}_i - \bar{\mathbf{x}}_n)\} = o(n)$.

Let $\tilde{\boldsymbol{\beta}}_n^{(r)}$ be the unrestricted rank-estimator of $\boldsymbol{\beta}$ by choosing the central point of the set

$$\mathbf{S} = \{\mathbf{b} : ||L_n(\mathbf{b})|| = \text{minimum}\}. \tag{9.1.8}$$

Then, as in Section 7.11.7,

$$\sqrt{n}(\tilde{\boldsymbol{\beta}}_n^{(r)} - \boldsymbol{\beta}) \sim \mathcal{N}_p(\mathbf{0}, A_\phi^2/\gamma^2(\psi,\phi)\mathbf{C}^{-1}). \tag{9.1.9}$$

We define the ridge matrix $\mathbf{R}(k)$ as

$$\mathbf{R}(k) = \lim_{n\to\infty}(\mathbf{I}_p + k(n^{-1}\mathbf{C}_n)^{-1} = (\mathbf{I}_p + k\mathbf{C}^{-1})^{-1}. \tag{9.1.10}$$

Then the nonparametric ridge regression estimator of $\boldsymbol{\beta}$ is defined by $\tilde{\boldsymbol{\beta}}_n^{(r)}(k) = \mathbf{R}(k)\tilde{\boldsymbol{\beta}}_n^{(r)}$, and under the local alternatives, $\boldsymbol{\beta}_{(n)} = n^{-1/2}\boldsymbol{\delta}$, $\boldsymbol{\delta} = (\tilde{\delta}_1, \ldots, \boldsymbol{\delta}_p)'$,

$$\sqrt{n}(\tilde{\boldsymbol{\beta}}_{(k)}^{(r)} - \boldsymbol{\beta}) \sim \mathcal{N}_p\{-k\mathbf{C}^{-1}(k)\boldsymbol{\delta}, A_\phi^2/\gamma^2[\mathbf{R}(k)\mathbf{C}^{-1}\mathbf{R}(k)]\}. \tag{9.1.11}$$

Thus, the asymptotic distributional bias, covariance, and MSE matrix can be computed. We mention that the asymptotic distribution of the least square theory based ridge regression estimator $\mathbf{R}(k)\tilde{\boldsymbol{\beta}}_n$, where $\tilde{\boldsymbol{\beta}}_n$ is the LSE of $\boldsymbol{\beta}$, under the local alternative $K_{(n)} : \boldsymbol{\beta}_{(n)} = n^{-1/2}\boldsymbol{\delta}$ is again

$$\mathcal{N}_p\{-k\mathbf{C}^{-1}(k)\boldsymbol{\delta}; \sigma^2(\mathbf{R}(k)\mathbf{C}^{-1}\mathbf{R}(k))\}. \tag{9.1.12}$$

Hence, the asymptotic efficiency of the rank-based ridge regression estimator is

$$\sigma^2\gamma^2(\psi,\phi)A_\phi^{-2}, \tag{9.1.13}$$

which is the same as the efficiency of the R-estimator relative to the LSE of $\boldsymbol{\beta}$.

Based on the rank ridge regression estimator of $\boldsymbol{\beta}$, we can define the three more estimators, as before:

(i) $\hat{\boldsymbol{\beta}}_n^{\text{PT}}(k) = \tilde{\boldsymbol{\beta}}_n^{(r)}(k) I(\mathcal{L}_n \geq \chi_p^2(\alpha))$, (9.1.14)

(ii) $\hat{\boldsymbol{\beta}}_n^{S}(k) = \tilde{\boldsymbol{\beta}}_n^{(r)}(k)\{1 - (p-2)\mathcal{L}_n^{-1}\} I(\mathcal{L}_n > \varepsilon_n)$

where $\varepsilon_n \to 0$ as $n \to \infty$, and

(iii) $\hat{\boldsymbol{\beta}}_n^{S+}(k) = \tilde{\boldsymbol{\beta}}_n^{(r)}(k)\{1 - (p-2)\mathcal{L}_n^{-1}\} I(\mathcal{L}_n > (p-2))$,

respectively, where

$$\mathcal{L}_n = A_n^{-2}[\mathbf{L}_n'(0)\mathbf{C}_n^{-1}\mathbf{L}_n(\mathbf{0})] \tag{9.1.15}$$

with $A_n^2 = (n-1)^{-1}\sum_{k=1}^n (a_n(k) - \bar{a}_n)^2$; $\bar{a}_n = n^{-1}\sum_{k=1}^n a_n(k)$ is used to test the null hypothesis $H_0 : \boldsymbol{\beta} = \mathbf{0}$.

These estimators parallel those given in (9.1.3 through 9.1.5). The bias, covariance matrix, MSE matrix and risks of the estimators are given in Problems 10 and 11 and are left to the reader to verify.

9.2 Ridge Regression as Bayesian Regression Estimators

We justify the ridge regression estimator (RRE) as a Bayesian estimator of the regression parameter $\boldsymbol{\beta}$.

Let $\mathbf{Y}$ given $\boldsymbol{\beta}$ be $N_n(\mathbf{X}\boldsymbol{\beta}, \sigma^2\mathbf{I}_n)$, and let the prior distribution of $\boldsymbol{\beta}$ be $N_p(\boldsymbol{\gamma}, \tau^2\mathbf{V})$. Then the posterior distribution of $\boldsymbol{\beta}$ given $\mathbf{Y} = \mathbf{y}$ is

$$N_p\left\{\boldsymbol{\gamma} + (\mathbf{C} + \frac{\sigma^2}{\tau^2}\mathbf{V}^{-1})^{-1}\mathbf{X}'(\mathbf{y} - \mathbf{X}\boldsymbol{\gamma}),\ \sigma^2(\mathbf{C} + \frac{\sigma^2}{\tau^2}\mathbf{V}^{-1})^{-1}\right\}. \tag{9.2.1}$$

Thus, the Bayes estimator of $\boldsymbol{\beta}$ given $\mathbf{Y} = \mathbf{y}$ under a quadratic loss function is the posterior mean

$$E[\boldsymbol{\beta}|\mathbf{y}] = \boldsymbol{\gamma} + (\mathbf{C} + k\mathbf{V}^{-1})^{-1}\mathbf{X}'(\mathbf{y} - \mathbf{X}\boldsymbol{\gamma}), \tag{9.2.2}$$

where $k = \sigma^2/\tau^2$. This estimator is the *generalized ridge estimator* of $\boldsymbol{\beta}$. If $\mathbf{V} = \mathbf{I}_p$, and $\boldsymbol{\gamma} = \mathbf{0}$, then $E(\boldsymbol{\beta}|\boldsymbol{\gamma})$ is equal to the ridge estimator of $\boldsymbol{\beta}$ defined by Hoerl and Kennard (1970). Thus, the Bayes estimator of $\boldsymbol{\beta}$ in this case is given by

$$\tilde{\boldsymbol{\beta}}_n(k) = \mathbf{R}(k)\tilde{\boldsymbol{\beta}}_n, \qquad \mathbf{R}(k) = (\mathbf{I}_p + k\mathbf{C}^{-1})^{-1}, \quad k \geq 0. \tag{9.2.3}$$

The remaining three estimators are obtained by transforming the basic estimators given in the introduction. In addition, if k is estimated by some function,

$k(\mathbf{y})$ of the sample observations, then we obtain the so-called *adaptive ridge estimators* of $\boldsymbol{\beta}$. The parameter $k = \sigma^2/\tau^2$ is called the *ridge parameter* of the ridge estimator. There exists a number of methods in the literature for the estimation of k. The interested reader is referred to Newhouse and Oman (1971), Farebrother (1975), Hoerl and Kennard (1975), Hoerl, Kennard, and Baldwin (1975), Marquardt and Snee (1975), McDonald and Galarneau (1975), Obenchain (1975), Dempster et al. (1977), Lawless and Wang (1976), Lawless (1978), Hemmerle and Brantle (1978), and Vinod and Ullah (1981) among others. More discussion on the ridge regression is available in Gruber (1998) and Montgomery and Peck and Vining (2001). Hoerl and Kennard (1970) suggested use of the *ridge trace* to find an appropriate value of k. The *ridge trace* is a plot of $\tilde{\boldsymbol{\beta}}_n(k)$ versus k between 0 to 1. If the multicollinearity in the data is severe, the *ridge trace* will show the obvious instability in the regression coefficients. However, more work needs to be done on the *ridge trace*. An excellent review and study of this problem is given by Kibria (2003), details of which are provided in Section 9.6.

9.3 Bias Expression of Various Ridge Estimators

The bias vectors of various ridge estimators are given in this section.

Bias Vector of $\tilde{\boldsymbol{\beta}}_n(k)$. The bias vector of the unrestricted RRE is given by

$$\begin{aligned}\mathbf{b}_1(\tilde{\boldsymbol{\beta}}_n(k)) &= E(\tilde{\boldsymbol{\beta}}_n(k) - \boldsymbol{\beta}) = E[\mathbf{R}(k)\tilde{\boldsymbol{\beta}}_n - \boldsymbol{\beta}] \\ &= -[\mathbf{I}_p - \mathbf{R}(k)]\boldsymbol{\beta} = -k\mathbf{C}^{-1}(k)\boldsymbol{\beta}. \end{aligned} \tag{9.3.1}$$

9.3.1 Bias Vector of $\hat{\boldsymbol{\beta}}_n^{\text{PT}}(k)$

Similarly, the bias vector of PTRRE, $\hat{\boldsymbol{\beta}}_n^{\text{PT}}(k)$ can be written as

$$\begin{aligned}\mathbf{b}_2(\hat{\boldsymbol{\beta}}_n^{\text{PT}}(k)) &= E(\hat{\boldsymbol{\beta}}_n^{\text{PT}}(k) - \boldsymbol{\beta}) \\ &= E(\tilde{\boldsymbol{\beta}}_n(k) - \boldsymbol{\beta}) - E\{\mathbf{R}(k)\tilde{\boldsymbol{\beta}}_n I(\mathcal{L}_n \le F_{p,m}(\alpha))\} \\ &= -\,k\mathbf{C}^{-1}(k)\boldsymbol{\beta} - \mathbf{R}(k)\boldsymbol{\beta}G_{p+2,m}(\ell_\alpha;\Delta^2), \end{aligned} \tag{9.3.2}$$

where $\ell_\alpha = \frac{p}{p+2}F_{p,m}(\alpha)$ and $\Delta^2 = (\boldsymbol{\beta}'\mathbf{C}\boldsymbol{\beta})\sigma^{-2}$.

Bias Vector of $\hat{\boldsymbol{\beta}}_n^{S}(k)$. In a similar way, the bias vector of SRRE, $\hat{\boldsymbol{\beta}}_n^{S}(k)$ can be written as

$$\begin{aligned}\mathbf{b}_3\left(\hat{\boldsymbol{\beta}}_n^{S}(k)\right) &= E\left(\hat{\boldsymbol{\beta}}_n^{S}(k) - \boldsymbol{\beta}\right) = E(\tilde{\boldsymbol{\beta}}_n(k) - \boldsymbol{\beta}) - pd\mathbf{R}(k)E(\tilde{\boldsymbol{\beta}}_n\mathcal{L}_n^{-1}) \\ &= -k\mathbf{C}^{-1}(k)\boldsymbol{\beta} - pd\mathbf{R}(k)\boldsymbol{\beta}E\left[\chi_{p+2}^{-2}(\Delta^2)\right]. \end{aligned} \tag{9.3.3}$$

Bias Vector of $\hat{\beta}_n^{S+}(k)$. Finally, the bias vector of PRRRE, $\hat{\beta}_n^{S+}(k)$ can be written as

$$\begin{aligned}
\mathbf{b}_4(\hat{\beta}_n^{S+}(k)) &= E(\hat{\beta}_n^{S+}(k) - \beta) \\
&= E(\hat{\beta}_n^{S}(k) - \beta) - qd\mathbf{R}(k)E\{\tilde{\beta}_n(1 - d\mathcal{L}_n^{-1})I(\mathcal{L}_n < d)\} \\
&= -k\mathbf{C}^{-1}(k)\beta - d_1 E[F_{p+2,m}^{-1}(\Delta^2)]\beta - d_1\mathbf{R}(k)\beta G_{p+2,m}(d_1;\Delta^2) \\
&\quad + d_1\mathbf{R}(k)\beta E\left[F_{p+2,m}^{-1}(\Delta^2)I(F_{p+2,m}(\Delta^2) < d_1)\right] \\
&= -k\mathbf{C}^{-1}(k)\beta - \mathbf{R}(k)\beta\left\{E[(1 - d_1F_{p+2,m}^{-1}(\Delta^2))I(F_{p+2,m}(\Delta^2) < d_1)]\right. \\
&\quad \left. + d_1E[F_{p+2,m}^{-1}(\Delta^2)]\right\}, \qquad (9.3.4)
\end{aligned}$$

where $d_1 = \frac{p}{p+2}d$. The corresponding expressions of the rank estimators are given in Problem 12.

Bias Comparison. For $\alpha = 0$, the bias of $\hat{\beta}_n^{\text{PT}}(k)$ coincides with that of the restricted ridge regression estimator, $\mathbf{0}$, while for $\alpha = 1$, it coincides with that of $\tilde{\beta}_n(k)$, the unrestricted ridge regression estimator of β. Also, as the noncentrality parameter $\Delta^2 \to \infty$, $\mathbf{b}_2(\hat{\beta}_n^{\text{PT}}(k)) = \mathbf{b}_3(\hat{\beta}_n^{S}(k)) = \mathbf{b}_4(\hat{\beta}_n^{S+}(k)) = \mathbf{b}_1(\tilde{\beta}_n)$ are unbounded. Under $H_0: \beta = \mathbf{0}$ all the estimators are unbiased.

Now, we compare the biases of the proposed estimators under the alternative hypothesis. In order to present a clear-cut picture of the biases, we transform them in quadratic (scalar) form by defining $QB(\hat{\beta}^*) = \mathbf{b}(\hat{\beta}^*)'\mathbf{b}(\hat{\beta}^*)$ as the quadratic bias function of an estimator $\hat{\beta}^*$ of the parameter vector β. The quadratic bias functions of the estimators can be expressed by the following theorem:

Theorem 1. Quadratic bias of the URRE, PTRRE, SRRE, and PRSRRE estimators are given, respectively, by

$$QB_1(\tilde{\beta}_n(k)) = k^2\beta'\mathbf{C}^{-2}(k)\beta,$$

$$\begin{aligned}
QB_2(\hat{\beta}_n^{\text{PT}}(k)) &= \beta'\mathbf{R}'(k)\mathbf{R}(k)\beta\left[G_{p+2,m}(\ell_\alpha, \Delta^2)\right]^2 \\
&\quad + 2k\beta'\mathbf{C}^{-1}(k)\mathbf{R}(k)\beta G_{p+2,n-p}(\ell_\alpha;\Delta^2) + k^2\beta'\mathbf{C}^{-2}(k)\beta,
\end{aligned}$$

$$\begin{aligned}
QB_3(\hat{\beta}_n^{S}(k)) &= \beta'\mathbf{R}'(k)\mathbf{R}(k)\beta\left[d_1E[F_{p+2,m}^{-1}(\Delta^2)]\right]^2 \\
&\quad + 2kd_1\beta'\mathbf{C}^{-1}(k)\mathbf{R}(k)\beta E[F_{p+2,m}^{-1}(\Delta^2)] + k^2\beta'\mathbf{C}^{-2}(k)\beta,
\end{aligned}$$

$$\begin{aligned}
QB_4(\hat{\beta}_n^{S+}(k)) &= \beta'\mathbf{R}'(k)\mathbf{R}(k)\beta[\mathrm{F}(\Delta^2)]^2 + 2k\beta'\mathbf{C}^{-1}(k)\mathbf{R}(k)\beta\mathrm{F}(\Delta^2) \\
&\quad + k^2\beta'\mathbf{C}^{-2}(k)\beta, \qquad (9.3.5)
\end{aligned}$$

where

$$\mathbf{F}(\Delta^2) = -E\left[(1 - d_1 F_{q+2,m}^{-1}(\Delta^2)) I\left(F_{q+2,m}(\Delta^2) \le d_1\right)\right] - d_1 E[F_{p+2,m}^{-1}(\Delta^2)].$$

The quadratic bias of the PTRRE depends on α, the size of the test. The magnitude of the RRRE increases without a bound and tends to ∞ as $\Delta^2 \to \infty$. Since both $E[F_{p+2,m}^{-1}(\Delta^2)]$ and $G_{q+2,n-p}(\ell_\alpha, \Delta^2)$ are decreasing functions of Δ^2, the quadratic biases of the PTRRE, SRRE, and PRSRRE estimators start from 0 and increase to a point, and then decrease gradually to $k^2\boldsymbol{\beta}'\mathbf{C}^{-2}(k)\boldsymbol{\beta}$ as $\Delta^2 \to \infty$. Under the null hypothesis, the quadratic biases are the same for all estimators. However, under the alternative hypothesis, the bias of the PRSRRE estimator remains below the curve of SRRE and PTRRE for all $k > 0$. Based on the analysis of equation (9.3.5), we can establish the following inequality:

$$QB_1(\tilde{\boldsymbol{\beta}}_n(k)) \le QB_4(\hat{\boldsymbol{\beta}}_n^{S+}(k)) \le QB_3(\hat{\boldsymbol{\beta}}_n^{S}(k)) \le QB_2(\hat{\boldsymbol{\beta}}_n^{\mathrm{PT}}(k)) \le \boldsymbol{\beta}'\boldsymbol{\beta}$$

whenever

$$d_1 E[F_{p+2,m}^{-1}(\Delta^2)] \le G_{p+2,m}(\ell_\alpha; \Delta^2); \tag{9.3.6}$$

otherwise,

$$QB_1(\tilde{\boldsymbol{\beta}}_n(k)) \le QB_4(\hat{\boldsymbol{\beta}}_n^{S+}(k)) \le QB_2(\hat{\boldsymbol{\beta}}_n^{\mathrm{PT}}(k)) \le QB_3(\hat{\boldsymbol{\beta}}_n^{S}(k)) \le \boldsymbol{\beta}'\boldsymbol{\beta}$$

whenever

$$d_1 E[F_{p+2,m}^{-1}(\Delta^2)] \ge G_{p+2,n-p}(\ell_\alpha; \Delta^2). \tag{9.3.7}$$

9.4 Covariance, MSE Matrix, and Weighted Risk Functions

In this section, we obtain the expressions for covariance matrices, MSE matrices, and the weighted risk functions of the ridge regression estimators.

URRE, $\tilde{\boldsymbol{\beta}}_n(k)$. The covariance matrix of $\tilde{\boldsymbol{\beta}}_n(k)$ is given by

$$\begin{aligned}\mathbf{V}_1(k) &= E\big[\mathbf{R}(k)(\tilde{\boldsymbol{\beta}}_n - \boldsymbol{\beta})(\tilde{\boldsymbol{\beta}}_n - \boldsymbol{\beta})'\mathbf{R}'(k)\big] \\ &= \sigma^2\mathbf{R}(k)\mathbf{C}^{-1}\mathbf{R}'(k).\end{aligned} \tag{9.4.1}$$

As a result, the MSE matrix of $\tilde{\boldsymbol{\beta}}_n(k)$ is given by

$$\mathbf{M}_1(\tilde{\boldsymbol{\beta}}_n(k)) = \sigma^2\mathbf{R}(k)\mathbf{C}^{-1}\mathbf{R}'(k) + k^2\mathbf{C}^{-1}(k)\boldsymbol{\beta}\boldsymbol{\beta}'\mathbf{C}^{-1}(k). \tag{9.4.2}$$

The weighted risk function of $\tilde{\beta}_n(k)$ based on a quadratic loss function with the *p.s.d* matrix $\mathbf{W}$ is given by

$$\begin{aligned}R_1(\tilde{\beta}_n(k);\mathbf{W}) &= \sigma^2 \operatorname{tr}[\mathbf{W}\mathbf{R}(k)\mathbf{C}^{-1}\mathbf{R}'(k)] + k^2\beta'\mathbf{C}^{-1}(k)\mathbf{W}\mathbf{C}^{-1}(k)\beta \\ &= \sigma^2 \operatorname{tr}[\mathbf{R}(k)\mathbf{C}^{-1}\mathbf{R}'(k)] + k^2\beta'\mathbf{C}^{-2}(k)\beta, \quad \text{for} \quad \mathbf{W}=\mathbf{I}_p.\end{aligned} \tag{9.4.3}$$

PTRRE, $\hat{\beta}_n^{PT}(k)$. The covariance matrix of $\hat{\beta}_n^{PT}(k)$ is given by

$$\begin{aligned}&\mathbf{V}_2(\hat{\beta}_n^{PT}(k)) = \mathbf{R}(k)\mathbf{V}_2(\hat{\beta}_n^{PT})\mathbf{R}'(k)\\ &= E[\mathbf{R}(k)(\hat{\beta}_n^{PT}-\beta) + \mathbf{R}(k)\beta G_{p+2,m}(\ell_\alpha;\Delta^2)]\\ &\quad\times[\mathbf{R}(k)(\hat{\beta}_n^{PT}-\beta) + \mathbf{R}(k)\beta G_{p+2,m}(\ell_\alpha;\Delta^2)]'\\ &= \sigma^2[\mathbf{R}(k)\mathbf{C}^{-1}\mathbf{R}'(k)] - \sigma^2[\mathbf{R}(k)\mathbf{C}^{-1}\mathbf{R}'(k)]G_{p+2,m}(\ell_\alpha;\Delta^2)]\\ &\quad+\big[2G_{p+2,m}(\ell_\alpha;\Delta^2) - G_{p+4,m}(\ell_\alpha^*;\Delta^2) - \{G_{p+2,m}(\ell_\alpha;\Delta^2)\}^2\big]\mathbf{R}(k)\beta\beta'\mathbf{R}'(k).\end{aligned} \tag{9.4.4}$$

Thus, the MSE matrix is given by

$$\begin{aligned}\mathbf{M}_2(\hat{\beta}_n^{PT}(k)) &= \sigma^2\mathbf{R}(k)\mathbf{C}^{-1}\mathbf{R}'(k) - \sigma^2\left[\mathbf{R}(k)\mathbf{C}^{-1}\mathbf{R}'(k)\right]G_{p+2,m}(\ell_\alpha;\Delta^2)\\ &\quad+\mathbf{R}(k)\beta\beta'\mathbf{R}'(k)\{2G_{p+2,m}(\ell_\alpha;\Delta^2) - G_{p+4,m}(\ell_\alpha^*;\Delta^2)\}\\ &\quad+k\big[\mathbf{R}(k)\beta\beta'\mathbf{C}^{-1}(k) + \mathbf{C}^{-1}(k)\beta\beta'\mathbf{R}(k)\big]G_{p+2,m}(\ell_\alpha;\Delta^2)\\ &\quad+k^2\mathbf{C}^{-1}(k)\beta\beta'\mathbf{C}^{-1}(k),\end{aligned} \tag{9.4.5}$$

and the weighted risk expression is given by

$$\begin{aligned}R_2(\hat{\beta}_n^{PT}(k);\mathbf{W}) &= \sigma^2 \operatorname{tr}[\mathbf{W}\mathbf{R}(k)\mathbf{C}^{-1}\mathbf{R}'(k)]\\ &\quad-\sigma^2\operatorname{tr}[\mathbf{W}\mathbf{R}'(k)\mathbf{C}^{-1}\mathbf{R}(k)]G_{p+2,m}(\ell_\alpha;\Delta^2)\\ &\quad+\beta'\mathbf{R}'(k)\mathbf{W}\mathbf{R}(k)\beta\{2G_{p+2,m}(\ell_\alpha;\Delta^2) - G_{p+4,m}(\ell_\alpha^*;\Delta^2)\}\\ &\quad+k^2[\beta'\mathbf{C}^{-1}(k)\mathbf{W}\mathbf{C}^{-1}(k)\beta] + k[\beta'\mathbf{C}^{-1}(k)\mathbf{W}\mathbf{R}(k)\beta\\ &\quad+\beta'\mathbf{R}'(k)\mathbf{W}\mathbf{C}^{-1}(k)\beta]G_{p+2,m}(\ell_\alpha;\Delta^2),\end{aligned} \tag{9.4.6}$$

where $\ell_\alpha^* = \frac{p}{p+4}F_{p,m}(\alpha)$ and $G_{p+4,m}(\ell_\alpha;\Delta^2)$ is the *cdf* of a noncentral F-distribution with $(p+4,m)$ degrees of freedom.

For $\mathbf{W}=\mathbf{I}_p$,

$$\begin{aligned}&R_2(\hat{\beta}_n^{PT}(k);\mathbf{I}_p) =\\ &\sigma^2\operatorname{tr}[\mathbf{R}(k)\mathbf{C}^{-1}\mathbf{R}'(k)](1-G_{p+2,m}(\ell_\alpha;\Delta^2))\\ &+\beta'\mathbf{R}'(k)\mathbf{R}(k)\beta\{2G_{p+2,m}(\ell_\alpha;\Delta^2) - G_{p+4,m}(\ell_\alpha^*;\Delta^2)\}\\ &+k^2\beta'\mathbf{C}^{-2}(k)\beta + k\beta'[\mathbf{C}(k)\mathbf{R}(k) + \mathbf{R}(k)\mathbf{C}^{-1}(k)]\beta G_{p+2,m}(\ell_\alpha;\Delta^2).\end{aligned} \tag{9.4.7}$$

SRRE, $\hat{\beta}_n^S(k)$. In this case, the covariance matrix of $\hat{\beta}_n^S(k)$ is given by

$$\begin{aligned}\mathbf{V}_3(\hat{\beta}_n^S(k)) &= \sigma^2\mathbf{R}(k)\mathbf{C}^{-1}\mathbf{R}'(k) - \sigma^2 pd[\mathbf{R}(k)\mathbf{C}^{-1}\mathbf{R}'(k)] \\ &\quad \times \{2E(\chi_{p+2}^{-2}(\Delta^2)) - (p-2)E(\chi_{p+2}^{-4}(\Delta^2))\} \\ &\quad + dp\mathbf{R}(k)\beta\beta'\mathbf{R}(k)\left\{(p+2)E[\chi_{p+4}^{-4}(\Delta^2)] - dp\left(E[\chi_{p+2}^{-2}(\Delta^2)]\right)^2\right\}.\end{aligned} \tag{9.4.8}$$

The MSE matrix of $\hat{\beta}_n^S(k)$ is

$$\begin{aligned}\mathbf{M}_3(\hat{\beta}_n^S(k)) &= \sigma^2(\mathbf{R}(k)\mathbf{C}^{-1}\mathbf{R}'(k)) - \sigma^2 pd[\mathbf{R}(k)\mathbf{C}^{-1}\mathbf{R}'(k)] \\ &\quad \times \{2E(\chi_{p+2}^{-2}(\Delta^2)) - (p-2)E(\chi_{p+2}^{-4}(\Delta^2))\} \\ &\quad + pd(p+2)(\mathbf{R}(k)\beta\beta'\mathbf{R}'(k))E(\chi_{p+4}^{-4}(\Delta^2)) \\ &\quad + dpk[\mathbf{R}(k)\beta\beta'\mathbf{C}^{-1}(k) + \mathbf{C}^{-1}(k)\beta\beta'\mathbf{R}'(k)]E(\chi_{p+2}^{-2}(\Delta^2)) \\ &\quad + k^2\mathbf{C}^{-1}(k)\beta\beta'\mathbf{C}^{-1}(k).\end{aligned} \tag{9.4.9}$$

The weighted risk function of $\hat{\beta}_n^S(k)$ follows immediately by calculating the $\mathrm{tr}[\mathbf{W}\mathbf{M}_3(\hat{\beta}_n^S(k))]$ as given below:

$$\begin{aligned}&R_3(\hat{\beta}_n^S(k);\mathbf{W}) \\ &= \sigma^2\,\mathrm{tr}[\mathbf{W}\mathbf{R}(k)\mathbf{C}^{-1}\mathbf{R}'(k)] - \sigma^2 pd\,\mathrm{tr}[\mathbf{W}\mathbf{R}(k)\mathbf{C}^{-1}\mathbf{R}(k)]\{2E[\chi_{p+2}^{-2}(\Delta^2)] \\ &\quad - (p-2)E\chi_{p+2}^{-4}(\Delta^2)]\} + dp(p+2)(\beta'\mathbf{R}'(k)\mathbf{W}\mathbf{R}(k)\beta)E[\chi_{p+4}^{-4}(\Delta^2)] \\ &\quad + k^2\beta'\mathbf{C}^{-1}(k)\mathbf{W}\mathbf{C}^{-1}(k)\beta + pdk\{\beta'\mathbf{C}^{-1}(k)\mathbf{W}\mathbf{R}(k)\beta \\ &\quad + \beta'\mathbf{R}'(k)\mathbf{W}\mathbf{C}^{-1}(k)\beta\}E[\chi_{p+2}^{-2}(\Delta^2)]\end{aligned} \tag{9.4.10}$$

$$\begin{aligned}&= \sigma^2\,\mathrm{tr}[\mathbf{R}(k)\mathbf{C}^{-1}\mathbf{R}'(k)] - \sigma^2 pd\,\mathrm{tr}[\mathbf{R}(k)\mathbf{C}^{-1}\mathbf{R}'(k)] \\ &\quad \times\{2E[\chi_{p+2}^{-2}(\Delta^2)] - (p-2)E[\chi_{p+2}^{-4}(\Delta^2)]\} \\ &\quad + dp(p+2)(\beta'\mathbf{R}'(k)\mathbf{R}(k)\beta)E[\chi_{p+4}^{-4}(\Delta^2)] + k^2\beta'\mathbf{C}^{-2}(k)\beta \\ &\quad + pdk[\beta'\mathbf{C}^{-1}(k)\mathbf{R}(k)\beta + \beta'\mathbf{R}'(k)\mathbf{C}^{-1}(k)\beta]E[\chi_{p+2}^{-2}(\Delta^2)]\end{aligned} \tag{9.4.11}$$

for $\mathbf{W} = \mathbf{I}_p$.

Finally, we consider the covariance and MSE matrix to obtain the risk of the positive-rule Stein-type ridge regression estimator (PRSRRE). They are provided in the following subsection. For the expressions for the R-estimators see Problem 13.

PRSRRE, $\hat{\boldsymbol{\beta}}_n^{S+}(k)$. In this case, the covariance matrix of $\hat{\boldsymbol{\beta}}_n^{S+}(k)$ is given by

$$
\begin{aligned}
\mathbf{V}_4(\hat{\boldsymbol{\beta}}_n^{S+}(k)) &= \mathbf{R}(k)\mathbf{V}_3(\hat{\boldsymbol{\beta}}_n^{S})\mathbf{R}'(k) - \sigma^2[\mathbf{R}(k)\mathbf{C}^{-1}\mathbf{R}'(k)] \\
&\quad \times E\left[(1 - d_1 F_{p+2,m}^{-1}(\Delta^2))^2 I(F_{p+2,m}(\Delta^2) < d_1)\right] \\
&\quad + \mathbf{R}(k)\boldsymbol{\beta}\boldsymbol{\beta}'\mathbf{R}'(k)\big[\{2E[1 - d_1 F_{p+2,m}^{-1}(\Delta^2))I(F_{p+2,m}(\Delta^2) < d_1)] \\
&\quad - \{E[(1 - d_2 F_{p+4,m}^{-1}(\Delta^2))^2 I(F_{p+4,m}(\Delta^2) < d_2)]\}\} \\
&\quad - E[(1 - d_1 F_{p+2,m}^{-1}(\Delta^2))I(F_{p+2,m}(\Delta^2) < d_1)] \\
&\quad + d_1 E[F_{p+2,m}^{-1}(\Delta^2)]\}^2\big],
\end{aligned} \tag{9.4.12}
$$

where $d_1 = \frac{p}{p+2}d$. The MSE matrix of $\hat{\boldsymbol{\beta}}_n^{S+}(k)$ is

$$
\begin{aligned}
&\mathbf{M}_4(\hat{\boldsymbol{\beta}}_n^{S+}(k)) \\
&= \mathbf{M}_3(\hat{\boldsymbol{\beta}}_n^{S}(k)) - \sigma^2[\mathbf{R}(k)\mathbf{C}^{-1}\mathbf{R}'(k)] \\
&\quad \times E\big[(1 - d_1 F_{p+2,m}^{-1}(\Delta^2))^2 I(F_{p+2,m}(\Delta^2) < d_1)\big] \\
&\quad + (\mathbf{R}(k)\boldsymbol{\beta}\boldsymbol{\beta}'\mathbf{R}'(k))\{2E\big[(1 - d_1 F_{p+2,m}^{-1}(\Delta^2))(F_{p+2,m}(\Delta^2) < d_1)\big] \\
&\quad - E\big[(1 - d_2 F_{p+4,m}^{-1}(\Delta^2))^2 I(F_{p+4,m}(\Delta^2) < d_2)\big]\} \\
&\quad - k\big[\mathbf{R}(k)\boldsymbol{\beta}\boldsymbol{\beta}'\mathbf{C}^{-1}(k) + \mathbf{C}^{-1}(k)\boldsymbol{\beta}\boldsymbol{\beta}'\mathbf{R}'(k)\big] \\
&\qquad \times \{E\big[(1 - d_1 F_{p+2,m}^{-1}(\Delta^2))I(F_{p+2,m}(\Delta^2) < d_1)\big]\},
\end{aligned} \tag{9.4.13}
$$

where $d_2 = \frac{p}{p+4}d$. The risk function of $\hat{\boldsymbol{\beta}}_n^{S+}(k)$ is given by

$$
\begin{aligned}
&R_4(\hat{\boldsymbol{\beta}}_n^{S+}(k); \mathbf{W}) \\
&= R_3(\hat{\boldsymbol{\beta}}_n^{S}(k); \mathbf{W}) - \sigma^2 \operatorname{tr}[\mathbf{W}\mathbf{R}(k)\mathbf{C}^{-1}\mathbf{R}'(k)] \\
&\quad \times E\left[(1 - d_1 F_{p+2,m}^{-1}(\Delta^2))^2 I(F_{p+2,m}(\Delta^2) < d_1)\right] + (\boldsymbol{\beta}'\mathbf{R}'(k)\mathbf{W}\mathbf{R}(k)\boldsymbol{\beta}) \\
&\quad \times \left[\{2E[(1 - d_1 F_{p+2,m}^{-1}(\Delta^2))I(F_{p+2,m}(\Delta^2) < d_1)\right. \\
&\quad -E[(1 - d_2 F_{p+4,m}^{-1}(\Delta^2))^2 I(F_{p+4,m}(\Delta^2) < d_2)]\} \\
&\quad +k\big[\boldsymbol{\beta}'\mathbf{C}^{-1}(k)\mathbf{W}\mathbf{R}(k)\boldsymbol{\beta} + \boldsymbol{\beta}'\mathbf{R}'(k)\mathbf{W}\mathbf{C}^{-1}(k)\boldsymbol{\beta}\big] \\
&\quad \times\{E\big[(1 - d_1 F_{p+2,m}^{-1}(\Delta^2))I\big(F_{p+2,m}(\Delta^2) < d_1\big)\big]\}.
\end{aligned} \tag{9.4.14}
$$

These expressions are obtained by direct computation of the expectation of various related expressions. In the following sections, we will compare the performance of the estimators by using risk functions. Note that $E[F_{\nu_1,\nu_2}^{-k}(\Delta^2)I(F_{\nu_1,\nu_2}(\Delta^2) < d)]$ is the kth truncated moment of a noncentral F-distribution with (ν_1, ν_2) d.f.'s.

9.5 Performance of Estimators: Risk Analysis

Since $\mathbf{C}$ is a *p.s.d* matrix, there exists an orthogonal matrix $\mathbf{\Gamma}$ such that

$$\mathbf{\Gamma}'\mathbf{C}\mathbf{\Gamma} = \mathbf{\Lambda} = \text{Diag}(\lambda_1, \lambda_2, \dots, \lambda_p),$$

where $\lambda_1 \geq \lambda_2 \geq \dots \lambda_p > 0$ are the eigenvalues of $\mathbf{C}$. It is easy to see that the eigenvalues of $\mathbf{R}(k) = (\mathbf{I}_p + k\mathbf{C}^{-1})^{-1}$ and of $\mathbf{C}(k) = (\mathbf{C} + k\mathbf{I}_p)$ are $\left(\frac{\lambda_1}{\lambda_1+k}, \frac{\lambda_2}{\lambda_2+k}, \dots, \frac{\lambda_p}{\lambda_p+k}\right)$ and $(\lambda_1 + k, \lambda_2 + k, \dots, \lambda_p + k)$, respectively. Then we obtain the following identities:

$$\text{tr}(\mathbf{R}'(k)\mathbf{C}^{-1}\mathbf{R}(k)) = \sum_{i=1}^{p} \frac{\lambda_i}{(\lambda_i + k)^2}, \tag{9.5.1}$$

$$\boldsymbol{\beta}'\mathbf{C}^{-2}(k)\boldsymbol{\beta} = \sum_{i=1}^{p} \frac{\alpha_i^2}{(\lambda_i + k)^2}, \qquad \boldsymbol{\alpha} = \mathbf{\Gamma}'\boldsymbol{\beta}, \tag{9.5.2}$$

For the risk comparison, we let $\mathbf{W} = \mathbf{I}_p$.

Comparison of $\tilde{\boldsymbol{\beta}}_n$ and $\tilde{\boldsymbol{\beta}}_n(k)$. The risk function of $\tilde{\boldsymbol{\beta}}_n(k)$ is given by

$$R_1(\tilde{\boldsymbol{\beta}}_n(k); I_p) = \sigma^2 \sum_{i=1}^{p} \frac{\lambda_i}{(\lambda_i + k)^2} + k^2 \sum_{i=1}^{p} \frac{\alpha_i^2}{(\lambda_i + k)^2}. \tag{9.5.3}$$

If $k = 0$, then the first term equals $\sigma^2 \sum_{i=1}^{p} \lambda_i^{-1}$, and the second term equals 0. The first term is a continuous monotonically decreasing function of k, and its derivative w.r.t. k approaches ∞ as $k \to 0+$ and $\lambda_p \to 0+$ as $k \to 0$. The second term is also a continuous monotonically increasing function of k, and its derivative w.r.t. k approaches 0 as $k \to 0+$. Note that the second term approaches $||\boldsymbol{\beta}||^2$ as $k \to \infty$. Thus, differentiating (9.5.3) with respect to k, we get

$$\frac{\partial R_1(\tilde{\boldsymbol{\beta}}_n(k); I_p)}{\partial k} = 2\sum_{i=1}^{p} \frac{\lambda_i^2}{(\lambda_i + k)^3}(k\alpha_i^2 - \sigma^2). \tag{9.5.4}$$

Next we define

$$k_1 = \frac{\sigma^2}{\max_{1\leq i\leq p} \alpha_i^2} = \frac{\sigma^2}{\alpha_{\max}^2}, \tag{9.5.5}$$

where $\alpha_{\max}$ is the largest element of α. We see that a sufficient condition for (9.5.4) to be negative is that there exists a $k \in (0, k_1)$ such that URRE, $\tilde{\boldsymbol{\beta}}_n(k)$ has smaller risk than that of unrestricted estimator, $\tilde{\boldsymbol{\beta}}_n$. Similarly, we can show that a sufficient condition for $R_1(\tilde{\boldsymbol{\beta}}_n(k); \mathbf{I}_p)$ to be smaller than $R_1(\tilde{\boldsymbol{\beta}}_n; \mathbf{I}_p) = \sigma^2 \,\text{tr}\mathbf{C}^{-1}$ is that there exists a $\Delta^2 \in (0, \Delta^2(k)]$ where

$$\Delta^2(k) = \frac{\text{tr}[\mathbf{C}^{-1} - \mathbf{R}(k)\mathbf{C}^{-1}\mathbf{R}(k)]}{k^2 Ch_{\min}[\mathbf{C}^{-2}(k)\mathbf{C}^{-1}]}. \tag{9.5.6}$$

Combining (9.5.5) and (9.5.6), we have the following theorem:

Theorem 1. A sufficient condition for the superiority of $\tilde{\boldsymbol{\beta}}_n(k)$ over $\tilde{\boldsymbol{\beta}}_n$ is that there exists a pair (Δ^2, k) such that

$$(\Delta^2, k) \in (0, \Delta^2(k_1)] \times (0, k_1]. \tag{9.5.7}$$

Further, the MSE difference between $\tilde{\boldsymbol{\beta}}_n$ and $\tilde{\boldsymbol{\beta}}_n(k)$ is given by

$$\begin{aligned} &\sigma^2(\mathbf{C}^{-1} - \mathbf{R}(k)\mathbf{C}^{-1}\mathbf{R}'(k)) - k^2\mathbf{C}^{-1}(k)\boldsymbol{\beta}\boldsymbol{\beta}'\mathbf{C}^{-1}(k) \qquad (9.5.8)\\ &= \boldsymbol{\Gamma}(\boldsymbol{\Lambda} + k\mathbf{I}_p)^{-1}\{(\sigma^2(\boldsymbol{\Lambda} + k\mathbf{I}_p)\boldsymbol{\Lambda}^{-1}(\boldsymbol{\Lambda} + k\mathbf{I}) - \sigma^2\boldsymbol{\Lambda} - k^2\boldsymbol{\alpha}\boldsymbol{\alpha}'\}(\boldsymbol{\Lambda} + k\mathbf{I}_p)^{-1}\boldsymbol{\Gamma}'\\ &= \boldsymbol{\Gamma}(\boldsymbol{\Lambda} + k\mathbf{I}_p)^{-1}\left[k\sigma^2(2\mathbf{I}_p + k\boldsymbol{\Lambda}^{-1}) - k^2\boldsymbol{\alpha}\boldsymbol{\alpha}'\right](\boldsymbol{\Lambda} + k\mathbf{I}_p)^{-1}\boldsymbol{\Gamma}'. \qquad (9.5.9)\end{aligned}$$

The r.h.s. of (9.5.8) is an n.n.d. matrix if and only if (1) $\boldsymbol{\alpha}$ belongs to the range of $(2\mathbf{I}_p + k\boldsymbol{\Lambda}^{-1})$ and (2) $\boldsymbol{\alpha}'(2\mathbf{I}_p + k\boldsymbol{\Lambda}^{-1})^{-1}\boldsymbol{\alpha} \le \sigma^2/k$, holds true (see Baksalary and Kala, 1983). Using Courant's theorem, we have

$$\Delta^2 \le \frac{1}{k}\left[\max_{1\le i\le p}\left(\frac{1}{2\lambda_i + k}\right)\right]^{-1} = \Delta^2(k_1). \tag{9.5.10}$$

Comparison of PTRRE with PTE and URRE.
Case 1: Null hypothesis $\mathbf{H}_0 : \boldsymbol{\beta} = \mathbf{0}$. In this case, the risk-difference of the two estimators, namely $\hat{\boldsymbol{\beta}}_n^{\text{PT}}$ and $\hat{\boldsymbol{\beta}}_n^{\text{PT}}(k)$, is given by

$$\sigma^2 \operatorname{tr}[\mathbf{C}^{-1} - \mathbf{R}(k)\mathbf{C}^{-1}\mathbf{R}'(k)](1 - G_{p+2,m}(\ell_\alpha; 0)) \ge 0. \tag{9.5.11}$$

Hence, $\hat{\boldsymbol{\beta}}_n^{\text{PT}}(k)$ dominates $\hat{\boldsymbol{\beta}}_n^{\text{PT}}$ uniformly for all k (≥ 0) and ℓ_α.
Case 2: Alternative hypothesis $\mathbf{H}_A : \boldsymbol{\beta} \ne \mathbf{0}$. The risk-difference in this case is given by

$$\begin{aligned}&\sigma^2 \operatorname{tr}[\mathbf{C}^{-1} - \mathbf{R}(k)\mathbf{C}^{-1}\mathbf{R}'(k)](1 - G_{p+2,m}(\ell_\alpha; \Delta^2))\\ &-\boldsymbol{\beta}'\Big[\mathbf{I} - \mathbf{C}\mathbf{C}^{-2}(k)\mathbf{C}\{2G_{p+2,m}(\ell_\alpha; \Delta^2) - G_{p+4,m}(\ell_\alpha^*; \Delta^2)\} - k^2\mathbf{C}^{-2}(k)\\ &-k\{\mathbf{I} - \mathbf{C}\mathbf{C}^{-2}(k) + \mathbf{C}^{-2}(k)\mathbf{C}\}G_{p+2,m}(\ell_\alpha; \Delta^2)\Big]\boldsymbol{\beta}. \qquad (9.5.12)\end{aligned}$$

The expression (9.5.12) is nonnegative if and only if

$$\begin{aligned}&\boldsymbol{\beta}'\Big[k^2\mathbf{C}^{-2}(k) + \{I - \mathbf{C}\mathbf{C}^{-2}(k)\mathbf{C}\}\{2G_{p+2,m}(\ell_\alpha; \Delta^2) - G_{p+4,m}(\ell_\alpha^*; \Delta^2)\}\\ &+k\{\mathbf{C}\mathbf{C}^{-2}(k) + \mathbf{C}^{-2}(k)\mathbf{C}\}G_{p+2,m}(\ell_\alpha; \Delta^2)\Big]\boldsymbol{\beta}\\ &\le \sigma^2 \operatorname{tr}\big[\mathbf{C}^{-1} - \mathbf{R}(k)\mathbf{C}^{-1}\mathbf{R}'(k)\big](1 - G_{p+2,m}(\ell_\alpha; \Delta^2)). \qquad (9.5.13)\end{aligned}$$

By standard calculations using Courant's theorem, we obtain

$$\Delta^2 \le \frac{\sigma^2 \operatorname{tr}[\mathbf{C}^{-1} - \mathbf{R}(k)\mathbf{C}^{-1}\mathbf{R}'(k)](1 - G_{p+2,m}(\ell_\alpha; \Delta^2))}{Ch_{\min}[\mathbf{A}\mathbf{C}^{-1}]} = \Delta^2(k, \alpha), \tag{9.5.14}$$

where

$$\mathbf{A} = k^2\mathbf{C}^{-2}(k) + \{I - \mathbf{C}\mathbf{C}^{-2}(k)\}\{2G_{p+2,m}(\ell_\alpha;\Delta^2) - G_{p+4,m}(\ell^*_\alpha;\Delta^2)\}$$
$$+k\{\mathbf{C}\mathbf{C}^{-2}(k) + \mathbf{C}^{-2}(k)\mathbf{C}\}G_{p+2,m}(\ell_\alpha;\Delta^2). \quad (9.5.15)$$

Now, differentiating the risk of $\hat{\boldsymbol{\beta}}_n^{\text{PT}}(k)$ w.r.t. k, we obtain

$$\frac{\partial R_2(\hat{\boldsymbol{\beta}}_n^{\text{PT}}(k);\mathbf{I}_p)}{\partial k} = -2\sum_{i=1}^{p}\frac{\lambda_i}{(\lambda_i+k)^3}\Big\{k\alpha_i^2(1 - G_{p+2,m}(\ell_\alpha;\Delta^2))$$
$$-\sigma^2(1 - G_{p+2,m}(\ell_\alpha;\Delta^2)) - \lambda_i^2\alpha_i^2(2G_{p+2,m}(\ell_\alpha;\Delta^2)$$
$$-G_{p+4,m}(\ell^*_\alpha;\Delta^2))\Big\}. \quad (9.5.16)$$

A sufficient condition for (9.5.16) to be negative is that there exists a $k \in (0, k_2)$ where

$$k_2 = \frac{\min\limits_{1\le i\le p}\left[\sigma^2(1 - G_{p+2,m}(\ell_\alpha;\Delta^2)) + \lambda_i^2\alpha_i^2(2G_{p+2,m}(\ell_\alpha;\Delta^2) - G_{p+4,m}(\ell^*_\alpha;\Delta^2))\right]}{\max\limits_{1\le i\le p}\left[\alpha_i^2(1 - G_{p+2,m}(\ell_\alpha;\Delta^2))\right]}. \quad (9.5.17)$$

For $\ell_\alpha = 0$, we have $k_2 = k_1$ as in Section 9.5.1. Thus, $\hat{\boldsymbol{\beta}}_n^{\text{PT}}(k)$ is superior to $\hat{\boldsymbol{\beta}}_n^{\text{PT}}$ whenever $k \in [0, k_2]$ under H_A.

This leads us to the following theorem:

Theorem 2. A sufficient condition for $\hat{\boldsymbol{\beta}}_n^{\text{PT}}(k)$ to be superior to $\hat{\boldsymbol{\beta}}_n^{\text{PT}}$ is that there exists a pair (Δ^2, k) such that

$$(\Delta^2, k) \in (0, \Delta^2(k_2, \alpha)] \times (0, k_2]. \quad (9.5.18)$$

Consider the MSE comparison between $\hat{\boldsymbol{\beta}}_n^{\text{PT}}$ and $\hat{\boldsymbol{\beta}}_n^{\text{PT}}(k)$. Here the MSE difference is given by

$$\sigma^2[\mathbf{C}^{-1} - \mathbf{R}(k)\mathbf{C}^{-1}\mathbf{R}'(k)](1 - G_{p+2,m}(\ell_\alpha;\Delta^2))$$
$$+[\boldsymbol{\beta}\boldsymbol{\beta}' - \mathbf{R}(k)\boldsymbol{\beta}\boldsymbol{\beta}'\mathbf{R}'(k)]\{2G_{p+2,m}(\ell_\alpha;\Delta^2) - G_{p+4,m}(\ell^*_\alpha;\Delta^2)\}$$

$$-k^2\mathbf{C}^{-1}(k)\beta\beta'\mathbf{C}^{-1}(k) - k[\mathbf{R}(k)\beta\beta'\mathbf{C}^{-1}(k) + \mathbf{C}^{-1}(k)\beta\beta'\mathbf{R}'(k)] \times G_{p+2,m}(\ell_\alpha; \Delta^2) \qquad (9.5.19)$$

$$= \Gamma(\Lambda + k\mathbf{I}_p)^{-1}\Big\{\sigma^2\big[(\Lambda + k\mathbf{I}_p)\Lambda^{-1}(\Lambda + k\mathbf{I}_p) - \Lambda\big]\big(1 - G_{p+2,m}(\ell_\alpha; \Delta^2)\big)$$
$$+ \big[(\Lambda + k\mathbf{I}_p)\alpha\alpha'(\Lambda + k\mathbf{I}_p) - \Lambda\alpha\alpha'\Lambda\big]\big[(2G_{p+2,m}(\ell_\alpha; \Delta^2) - G_{p+4,m}(\ell_\alpha^*; \Delta^2)\big]$$
$$- k^2\alpha\alpha' - k[\alpha\alpha'\Lambda + \Lambda\alpha\alpha']G_{p+2,m}(\ell_\alpha; \Delta^2)\Big\}(\Lambda + k\mathbf{I}_p)^{-1}\Gamma'$$
$$= \Gamma(\Lambda + k\mathbf{I}_p)^{-1}\Big\{k\sigma^2(2\mathbf{I}_p + k\Lambda^{-1})(1 - G_{p+2,m}(\ell_\alpha; \Delta^2))$$
$$+\big[k(\alpha\alpha'\Lambda + \Lambda\alpha\alpha') + k^2\alpha\alpha'\big]\big(2G_{p+2,m}(\ell_\alpha; \Delta^2) - G_{p+4,m}(\ell_\alpha^*; \Delta^2)\big)$$
$$- k^2\alpha\alpha' - k\big[\alpha\alpha'\Lambda + \Lambda\alpha\alpha'\big]G_{p+2,m}(\ell_\alpha; \Delta^2)\Big\}(\Lambda + k\mathbf{I}_p)^{-1}\Gamma'$$
$$= \Gamma(\Lambda + k\mathbf{I}_p)^{-1}\Big[k\sigma^2(2\mathbf{I}_p + k\Lambda^{-1})\big(1 - G_{p+2,m}(\ell_\alpha; \Delta^2)\big)$$
$$- \big\{k^2\alpha\alpha'(1 - 2G_{p+2,m}(\ell_\alpha; \Delta^2) + G_{p+4,m}(\ell_\alpha^*; \Delta^2))$$
$$+k(\Lambda\alpha\alpha' + \alpha\alpha'\Lambda)\big(G_{p+2,m}(\ell_\alpha; \Delta^2)) - G_{p+4,m}(\ell_\alpha^*; \Delta^2)\big)\big\}\Big](\Lambda + k\mathbf{I}_p)^{-1}\Gamma'. \qquad (9.5.20)$$

The r.h.s. of (9.5.20) is an n.n.d. matrix if and only if

$$k\alpha'(2\mathbf{I}_p + k\Lambda^{-1})^{-1}\alpha\big[1 - 2G_{p+2,m}(\ell_\alpha; \Delta^2) + G_{p+4,m}(\ell_\alpha^*; \Delta^2)\big]$$
$$+ 2\alpha'\Lambda(2\mathbf{I}_p + k\Lambda^{-1})^{-1}\alpha\big[G_{p+2,m}(\ell_\alpha; \Delta^2) - G_{p+4,m}(\ell_\alpha^*; \Delta^2)\big]$$
$$\le \sigma^2\big(1 - G_{p+2,m}(\ell_\alpha; \Delta^2)\big) \quad \text{holds true} \qquad (9.5.21)$$

or

$$\Delta^2 \le \frac{\sigma^2}{k}(1 - G_{p+2,m}(\ell_\alpha; \Delta^2))$$
$$\times\Bigg[\max_{1\le i\le p}\left(\frac{1}{2\lambda_i + k}\right)\Big\{1 - 2G_{p+2,m}(\ell_\alpha; \Delta^2) + G_{p+4,m}(\ell_\alpha^*; \Delta^2)\Big\}$$
$$+ \max_{1\le i\le p}\left(\frac{\lambda_i}{2\lambda_i + k}\right)\Big\{G_{p+2,m}(\ell_\alpha; \Delta^2) + G_{p+4,m}(\ell_\alpha^*; \Delta^2)\Big\}\Bigg]^{-1} \qquad (9.5.22)$$

If $\ell_\alpha = 0$, (9.5.22) reduces to

$$\Delta^2 \le \frac{\sigma^2}{k}\left[\max_{1\le i\le p}\left(\frac{1}{2\lambda_i + k}\right)\right]^{-1}. \qquad (9.5.23)$$

Now, we consider the efficiency of PTRRE relative to URRE given by

$$E_k(\Delta^2; \alpha) = \big[1 - g_k(\Delta^2, \alpha)\big]^{-1}, \qquad (9.5.24)$$

where

$$g_k(\Delta^2,\alpha) = \frac{g_k^{(1)}(\Delta^2,\alpha)}{g_k^{(2)}(\Delta^2,\alpha)}$$

with

$$\begin{aligned} &g_k^{(1)}(\Delta^2,\alpha) \\ &= \big[\sigma^2 \operatorname{tr}(\mathbf{R}(k)\mathbf{C}^{-1}\mathbf{R}'(k))G_{p+2,m}(\ell_\alpha;\Delta^2) + \boldsymbol{\beta}'\mathbf{R}'(k)\mathbf{R}(k)\boldsymbol{\beta}\{2G_{p+2,m}(\ell_\alpha;\Delta^2) \\ &\quad - G_{p+4,m}(\ell_\alpha^*;\Delta^2)\} + 2\boldsymbol{\beta}'[\mathbf{C}^{-1}(k)\mathbf{R}(k) + \mathbf{R}'(k)\mathbf{C}^{-1}(k)]\boldsymbol{\beta}G_{p+2,m}(\ell_\alpha;\Delta^2)\big] \end{aligned} \tag{9.5.25a}$$

and

$$g_k^{(2)}(\Delta^2,\alpha) = \sigma^2 \operatorname{tr}(\mathbf{R}(k)\mathbf{C}^{-1}\mathbf{R}(k)) + k^2\boldsymbol{\beta}'\mathbf{C}^{-2}(k)\boldsymbol{\beta}. \tag{9.5.25b}$$

Note that $E_k(\Delta^2;\alpha)$ attains its maximum at $\Delta^2 = 0$ with the value

$$E_{\max}(k,0;\alpha) = \big[1 - G_{p+2,m}(\ell_\alpha;0)\big]^{-1}. \tag{9.5.26}$$

As Δ^2 increases, $E_k(\Delta^2;\alpha)$ decreases, crossing the 1–line to a minimum $E_k(\Delta^2_{\min}(\alpha);\alpha)$ at $\Delta^2 = \Delta^2_{\min}$, and then increases toward the 1–line as $\Delta^2 \to \infty$. Since PTRRE does not dominate URRE uniformly, **we select an optimum α** by prefixing an efficiency E_k^0 and solving for α satisfying the equation

$$E_k(\Delta^2_{\min}(\alpha);\alpha) \geq E_k^0. \tag{9.5.27}$$

The solution α^* obtained in this way is considered to be appropriate for the **preliminary test ridge regression estimator** (PTRRE) for $\boldsymbol{\beta}$.

Comparison of SRRE, $\hat{\boldsymbol{\beta}}_n^S(k)$ and SE, $\hat{\boldsymbol{\beta}}_n^S$.

Case 1: Null hypothesis $\mathbf{H}_0 : \boldsymbol{\beta} = \mathbf{0}$. The risk-difference function, in this case, turns out to be

$$R_3(\hat{\boldsymbol{\beta}}_n^S;\mathbf{I}_p) - R_3(\hat{\boldsymbol{\beta}}_n^S(k);\mathbf{I}_p) = \sigma^2 \operatorname{tr}[\mathbf{C}^{-1} - \mathbf{R}(k)\mathbf{C}^{-1}\mathbf{R}'(k)](1-d) > 0, \tag{9.5.28}$$

which is nonnegative for all k (> 0). Hence, SRRE dominates SE uniformly in k under H_0.

Case 2: Alternative hypothesis $\mathbf{H}_A : \boldsymbol{\beta} \neq \mathbf{0}$. In this case, the risk function is given by

$$\begin{aligned} &R_3(\hat{\boldsymbol{\beta}}_n^S(k);\mathbf{I}_p \\ &= \sigma^2 \operatorname{tr}[\mathbf{R}(k)\mathbf{C}^{-1}\mathbf{R}'(k)]\Big[1 - pd\Big\{(p-2)E[\chi_{p+2}^{-4}(\Delta^2)] \\ &\quad + \Big(1 - \frac{(p+2)\boldsymbol{\beta}'[\mathbf{I}_p - \mathbf{R}'(k)\mathbf{R}(k)]\boldsymbol{\beta}}{2\sigma^2\Delta^2 \operatorname{tr}[\mathbf{R}(k)\mathbf{C}^{-1}\mathbf{R}(k)]}\Big)\Big(2\Delta^2 E[\chi_{p+4}^{-4}(\Delta^2)]\Big)\Big\} \\ &\quad + k^2\boldsymbol{\beta}'\mathbf{C}^{-2}(k)\boldsymbol{\beta} + k\boldsymbol{\beta}'[\mathbf{C}^{-1}(k)\mathbf{R}(k) + \mathbf{R}(k)\mathbf{C}^{-1}(k)]\boldsymbol{\beta}E[\chi_{p+2}^{-2}(\Delta^2)]. \end{aligned} \tag{9.5.29}$$

Differentiating (9.5.29) with respect to k, we obtain

$$\frac{\partial R_3(\hat{\beta}_n^S(k); I_p)}{\partial k} = 2\sum_{i=1}^{p} \frac{1}{(\lambda_i + k)^3}\Big\{k\lambda_i\alpha_i^2(1 - pdE[\chi_{p+2}^{-2}(\Delta^2)])$$

$$- \sigma^2\Big(\lambda_i - pd\lambda_i\Big[(p-2)E[\chi_{p+2}^{-4}(\Delta^2)] + \Big(1 - \frac{(p+2)\lambda_i\alpha_i^2}{2\sigma^2\Delta^2}\Big)$$

$$(2\Delta^2)E[\chi_{p+4}^{-4}(\Delta^2)]\Big]\Big) - pd\lambda_i^2\alpha_i^2 E[\chi_{p+2}^{-2}(\Delta^2)]\Big\}. \qquad (9.5.30)$$

Hence, a sufficient condition for (9.5.30) to be negative is that there exists a k such that $0 < k < k_3$ where $k_3 = k_3^{(1)}(\Delta^2)/k_2^{(2)}(\Delta^2)$ with

$$k_3^{(1)}(\Delta^2) = \min_{1\le i\le p}\Big\{\sigma^2\Big\{\lambda_i - pd\lambda_i\Big[(p-2)E[\chi_{p+2}^{-4}(\Delta^2)] + \Big(1 - \frac{(p+2)\lambda_i\alpha_i^2}{2\sigma^2\Delta^2}\Big)$$

$$(2\Delta^2 E[\chi_{p+4}^{-4}(\Delta^2)]\Big]\Big\} - pd\lambda_i^2\alpha_i^2 E[\chi_{p+2}^{-2}(\Delta^2)]\Big\} \qquad (9.5.31a)$$

and

$$k_3^{(2)}(\Delta^2) = \max_{1\le i\le p} \lambda_i\alpha_i^2(1 - pdE[\chi_{p+2}^{-2}(\Delta^2)]). \qquad (9.5.31b)$$

Hence, $\hat{\beta}_n^S(k)$ has smaller risk than $\hat{\beta}_n^S$ for $k \in [0, k_3]$. Further, we can check the sufficient condition on Δ^2 by considering the risk-difference

$$R_3(\hat{\beta}_n^S; I_p) - R_3(\hat{\beta}_n^S(k); I_p)$$

$$= \sigma^2 \operatorname{tr}[\mathbf{C}^{-1} - \mathbf{R}(k)\mathbf{C}^{-1}\mathbf{R}(k)][1 - pd\{2E[\chi_{p+2}^{-2}(\Delta^2)] - (p-2)E[\chi_{p+2}^{-4}(\Delta^2)]\}]$$

$$+p(p+2)d\boldsymbol{\beta}'[\mathbf{I}_p - \mathbf{R}'(k)\mathbf{R}(k)]\boldsymbol{\beta}'E[\chi_{p+4}^{-4}(\Delta^2)] - k^2\boldsymbol{\beta}'\mathbf{C}^{-2}(k)\boldsymbol{\beta}$$

$$-k\boldsymbol{\beta}'[\mathbf{C}^{-1}(k)\mathbf{R}(k) + \mathbf{R}'(k)\mathbf{C}^{-1}(k)]\boldsymbol{\beta}E[\chi_{p+2}^{-2}(\Delta^2)]. \qquad (9.5.32)$$

Using Courant's theorem, we find that SRRE is superior to SE whenever $\Delta^2 < \Delta_3^2(k)$, defined by

$$\Delta_3^2(k) = \frac{\Delta_{31}^2(k)}{\Delta_{32}^2(k)}, \qquad (9.5.33)$$

where

$$\Delta_{31}^2(k) = \operatorname{tr}[\mathbf{C}^{-1} - \mathbf{R}(k)\mathbf{C}^{-1}\mathbf{R}'(k)][1 - pd\{2E[\chi_{p+2}^{-2}(\Delta^2)] - (p-2)E[\chi_{p+2}^{-4}(\Delta^2)] \qquad (9.5.34a)$$

and

$$\Delta_{31}^2(k) = Ch_{\max}(\mathbf{AC}^{-1}) \qquad (9.5.34b)$$

with

$$\mathbf{A} = k^2\mathbf{C}^{-2}(k) + k[\mathbf{C}^{-1}(k)\mathbf{R}'(k) + \mathbf{R}(k)\mathbf{C}^{-1}(k)]E[\chi_{p+2}^{-2}(\Delta^2)]$$
$$-p(p+2)d[\mathbf{I}_p - \mathbf{R}(k)\mathbf{R}'(k)]E[\chi_{p+4}^{-4}(\Delta^2)]. \qquad (9.5.35)$$

Based on the analysis above we state the following theorem:

Theorem 3. The SRRE is superior to SE in the sense of having smaller quadratic risk if there exists a pair (Δ^2, k) such that

$$(\Delta^2, k) \in (0, \Delta^2(k_3)] \times (0, k_3]. \qquad (9.5.36)$$

Comparison of SRRE with URRE. We consider the risk-difference

$$R_1(\tilde{\boldsymbol{\beta}}_n(k); \mathbf{I}_p) - R_3(\hat{\boldsymbol{\beta}}_n^S(k); \mathbf{I}_p)$$
$$= pd\sigma^2 \operatorname{tr}[\mathbf{R}(k)\mathbf{C}^{-1}\mathbf{R}(k)]\Big\{(p-2)E[\chi_{p+2}^{-4}(\Delta^2)]$$
$$+\left(1 - \frac{(p+2)\boldsymbol{\beta}'[\mathbf{I}_p - \mathbf{R}'(k)\mathbf{R}(k)]\boldsymbol{\beta}}{2\sigma^2\Delta^2 \operatorname{tr}[\mathbf{R}(k)\mathbf{C}^{-1}\mathbf{R}(k)]}\right)(2\Delta^2 E[\chi_{p+4}^{-4}(\Delta^2)])\Big\}$$
$$-k\boldsymbol{\beta}'[\mathbf{C}^{-1}(k)R(k) + \mathbf{R}(k)\mathbf{C}^{-1}(k)]\boldsymbol{\beta}E[\chi_{p+2}^{-2}(\Delta^2)]. \qquad (9.5.37)$$

The risk-difference is nonnegative for all pairs (Δ^2, k) whenever

$$\frac{\operatorname{tr}[\mathbf{R}(k)\mathbf{C}^{-1}\mathbf{R}'(k)]}{Ch_{\max}\{[\mathbf{I}_p - \mathbf{R}'(k)\mathbf{R}(k)]\mathbf{C}^{-1}\}} \geq \frac{p+2}{2}. \qquad (9.5.38)$$

On the other hand, the risk-difference is nonnegative whenever there exists a value of $k \in (0, k_4]$, where

$$k_4 = \frac{\sigma^2 \min\limits_{1\leq i\leq p}\left\{\lambda_i(p-2)E[\chi_{p+2}^{-4}] + \left(1 - \frac{(p+2)\lambda_i\alpha_i^2}{2\sigma^2\Delta^2}\right)(2\Delta^2 E[\chi_{p+4}^{-4}(\Delta^2)]\right\}}{\max\limits_{1\leq i\leq p}\{\lambda_i\alpha_i^2 E[\chi_{p+2}^{-2}(\Delta^2)]\}}. \qquad (9.5.39)$$

Hence, we obtain the following theorem:

Theorem 4. A sufficient condition for SRRE to have smaller quadratic risk than URRE is that there exists a pair $(\Delta^2, k) \in (0, \infty) \times (0, k_4]$ and the following holds:

$$\frac{\operatorname{tr}[\mathbf{R}(k)\mathbf{C}^{-1}\mathbf{R}(k)]}{Ch_{\max}\{[\mathbf{I}_p - \mathbf{R}(k)\mathbf{R}(k)]\mathbf{C}^{-1}\}} \geq \frac{p+2}{2}. \qquad (9.5.40)$$

Now, consider the MSE matrix-difference

$$\begin{aligned}
&\mathbf{M}_1(\tilde{\beta}_n(k)) - \mathbf{M}_3(\hat{\beta}_n^S(k)) \\
&= \sigma^2 pd[\mathbf{R}(k)\mathbf{C}^{-1}\mathbf{R}'(k)]\{2E[\chi_{p+2}^{-2}(\Delta^2)] - (p-2)E[\chi_{p+2}^{-4}(\Delta^2)]\} \\
&\quad -pd(p+2)[\mathbf{R}(k)\beta\beta'\mathbf{R}'(k)]E[\chi_{p+4}^{-4}(\Delta^2)] \\
&\quad -pdk[\mathbf{R}(k)\beta\beta'\mathbf{C}^{-1}(k) + \mathbf{C}^{-1}(k)\beta\beta'\mathbf{R}'(k)]E[\chi_{p+2}^{-2}(\Delta^2)] \qquad (9.5.41) \\
&= \mathbf{\Gamma}(\mathbf{\Lambda} + k\mathbf{I}_p)^{-1}\Big[\sigma^2 pd\mathbf{\Lambda}\{2E[\chi_{p+2}^{-2}(\Delta^2)] - (p-2)E[\chi_{p+2}^{-4}(\Delta^2)]\} \\
&\quad -pd(p+2)(\alpha\alpha')E[\chi_{p+4}^{-4}(\Delta^2)] - pdk[\alpha\alpha'\mathbf{\Lambda} + \mathbf{\Lambda}\alpha\alpha']E[\chi_{p+2}^{-2}(\Delta^2)]\Big] \\
&\quad (\mathbf{\Lambda} + k\mathbf{I}_p)^{-1}\mathbf{\Gamma}'. \qquad (9.5.42)
\end{aligned}$$

The expression (9.5.42) is an n.n.d. matrix if and only if for all k, we have

$$\begin{aligned}
&\Big\{(p+2)\alpha'\mathbf{\Lambda}^{-1}\alpha E[\chi_{p+4}^{-4}(\Delta^2)] + 2k\alpha'\alpha E[\chi_{p+2}^{-2}(\Delta^2)]\Big\} \\
&\quad \leq \sigma^2\{2E[\chi_{p+2}^{-2}(\Delta^2)] - (p-2)E[\chi_{p+2}^{-4}(\Delta^2)]\}. \qquad (9.5.43)
\end{aligned}$$

Comparison of SRRE with PTRRE. The risk difference in this case is

$$\begin{aligned}
&R_3(\hat{\beta}_n^S(k);\mathbf{I}_p) - R_2(\hat{\beta}_n^{\text{PT}}(k);\mathbf{I}_p) \\
&= \sigma^2 \operatorname{tr}(\mathbf{R}(k)\mathbf{C}^{-1}\mathbf{R}(k))\Big\{G_{p+2,m}(\ell_\alpha;\Delta^2) - pd\Big(2E[\chi_{p+2}^{-2}(\Delta^2)] \\
&\quad -(p-2)E[\chi_{p+2}^{-4}(\Delta^2)]\Big)\Big\} \\
&\quad -\beta'\mathbf{R}'(k)\mathbf{R}(k)\beta\Big\{(2G_{p+2,m}(\ell_\alpha;\Delta^2) - G_{p+4,m}(\ell_\alpha^*;\Delta^2)) \\
&\quad -pd(p+2)E[\chi_{p+2}^{-4}(\Delta^2)]\Big\} + k[\beta'\mathbf{C}^{-1}(k)\mathbf{R}(k)\beta + \beta'\mathbf{R}'(k)\mathbf{C}^{-1}\beta] \\
&\quad \{pdE[\chi_{p+2}^{-2}(\Delta^2)] - G_{p+2,m}(\ell_\alpha;\Delta^2)\}. \qquad (9.5.44)
\end{aligned}$$

The r.h.s. of (9.5.44) will be nonpositive whenever

$$\begin{aligned}
&\Delta^2 \leq \Delta^2(k,\alpha) = \\
&\frac{\sigma^2 \operatorname{tr}[\mathbf{R}'(k)\mathbf{C}^{-1}\mathbf{R}(k)]\{G_{p+2,m}(\ell_\alpha;\Delta^2) - (2E[\chi_{p+2}^{-2}(\Delta^2)] - (p-2)E[\chi_{p+2}^{-4}(\Delta^2)])\}}{Ch_{\max}\{\mathbf{A}\mathbf{C}^{-1}\}}
\end{aligned}$$

(9.5.45)

where

$$\begin{aligned}\mathbf{AC}^{-1} &= [\mathbf{R}'(k)\mathbf{R}(k)\mathbf{C}^{-1}] \\ &\quad\times\Big\{(2G_{p+2,m}(\ell_\alpha;\Delta^2) - G_{p+4,m}(\ell_\alpha^*;\Delta^2)) + pd(p+2)E\big[\chi_{p+4}^{-4}(\Delta^2)\big]\Big\} \\ &\quad -k\big[\mathbf{C}^{-1}(k)\mathbf{R}(k)\mathbf{C}^{-1} + \mathbf{R}'(k)\mathbf{C}^{-1}(k)\mathbf{C}^{-1}\big] \\ &\quad\times\Big\{pdE\big[\chi_{p+2}^{-2}(\Delta^2)\big] - G_{p+2,m}(\ell_\alpha;\Delta^2)\Big\}. \end{aligned} \tag{9.5.46}$$

On the other hand, the risk-difference is nonpositive whenever there exists a value $k \in [0, k_5]$, where

$$k_5 = \frac{k_5^{(1)}(\Delta^2,\alpha)}{k_5^{(2)}(\Delta^2,\alpha)} \tag{9.5.47}$$

with

$$\begin{aligned} k_5^{(1)}(\Delta^2,\alpha) &= \max_{1\le i\le p}\Big[\sigma^2\lambda_i - \alpha_i\Big\{G_{p+2,m}(\ell_\alpha;\Delta^2) - pd\Big[(p-2)E\big[\chi_{p+2}^{-4}(\Delta^2)\big] \\ &\quad +\Big(1 - \frac{(p+2)\lambda_i\alpha_i^2}{2\Delta^2}\Big)(2\Delta^2E\big[\chi_{p+4}^{-4}(\Delta^2)\big])\Big] \\ &\quad -\lambda_i^2\alpha_i^2\big[2G_{p+2,m}(\ell_\alpha;\Delta^2) - G_{p+4,m}(\ell_\alpha^*;\Delta^2)\big]\Big\}\Big] \end{aligned} \tag{9.5.48}$$

and

$$k_5^{(2)}(\Delta^2,\alpha) = \min_{1\le i\le p}\Big\{2\lambda_i^2\alpha_i^2\big(G_{p+2,m}(\ell_\alpha;\Delta^2) - pdE\big[\chi_{p+2}^{-2}(\Delta^2)\big)\Big\}. \tag{9.5.49}$$

Hence, we state the following theorem:

Theorem 5. A sufficient condition for SRRE to be superior to PTRRE is that there exists a pair $(\Delta^2, k) \in (0, \Delta^2(k,\alpha)] \times (0, k_5]$ for $\alpha \in (0,1)$.

Under H_0, the risk-difference becomes

$$\sigma^2\,\mathrm{tr}[\mathbf{R}(k)\mathbf{C}^{-1}\mathbf{R}'(k)]\{G_{p+2,m}(\ell_\alpha;0) - d\}, \tag{9.5.50}$$

which is nonnegative whenever

$$G_{p+2,m}(\ell_\alpha;0) > d. \tag{9.5.51}$$

$\hat{\boldsymbol{\beta}}_n^{\mathrm{PT}}(k)$ is superior to $\hat{\boldsymbol{\beta}}_n^{S}(k)$; otherwise, $\hat{\boldsymbol{\beta}}_n^{S}(k)$ is superior to $\hat{\boldsymbol{\beta}}_n^{\mathrm{PT}}(k)$.

Now, consider the MSE matrix difference of $\hat{\beta}_n^{\text{PT}}(k)$ and $\hat{\beta}_n^S(k)$, which is

$$
\begin{aligned}
&\mathbf{M}_2(\hat{\beta}_n^S(k)) - \mathbf{M}_3(\hat{\beta}_n^{\text{PT}}(k)) = \\
&\sigma^2 pd[\mathbf{R}(k)\mathbf{C}^{-1}\mathbf{R}'(k)]\{G_{p+2,m}(\ell_\alpha;\Delta^2) - 2E[\chi_{p+2}^{-2}(\Delta^2)] + (p-2)E[\chi_{p+2}^{-4}(\Delta^2)]\} \\
&-[\mathbf{R}(k)\mathbf{C}^{-1}\mathbf{R}'(k)]\{(2G_{p+2,m}(\ell_\alpha;\Delta^2) - G_{p+4,m}(\ell_\alpha^*;\Delta^2)) \\
&-pd(p+2)E[\chi_{p+4}^{-4}(\Delta^2)]\} - k[\mathbf{R}(k)\boldsymbol{\beta}\boldsymbol{\beta}'\mathbf{C}^{-1}(k) + \mathbf{C}^{-1}(k)\boldsymbol{\beta}\boldsymbol{\beta}'\mathbf{R}'(k)] \\
&\quad\times\{G_{p+2,m}(\ell_\alpha;\Delta^2) - E[\chi_{p+2}^{-4}(\Delta^2)]\} \qquad (9.5.52) \\
&= \boldsymbol{\Gamma}(\boldsymbol{\Lambda} + k\mathbf{I}_p)^{-1}\{\sigma^2 pd\boldsymbol{\Lambda}[G_{p+2,m}(\ell_\alpha;\Delta^2) - 2E[\chi_{p+2}^{-2}(\Delta^2)]E[\chi_{p+2}^{-4}(\Delta^2)]] \\
&-\boldsymbol{\alpha}\boldsymbol{\alpha}'[2G_{p+2,m}(\ell_\alpha;\Delta^2) - G_{p+4,m}(\ell_\alpha^*;\Delta^2) - pd(p+2)E[\chi_{p+4}^{-4}(\Delta^2)]] \\
&-k[\boldsymbol{\alpha}\boldsymbol{\alpha}'\boldsymbol{\Lambda} + \boldsymbol{\Lambda}\boldsymbol{\alpha}\boldsymbol{\alpha}']\{G_{p+2,m}(\ell_\alpha;\Delta^2) - E[\chi_{p+2}^{-2}(\Delta^2)]\}(\Lambda + k\mathbf{I}_p)^{-1}\boldsymbol{\Gamma}'. \qquad (9.5.53)
\end{aligned}
$$

The r.h.s. of (9.5.53) is an n.n.d. matrix if and only if

$$
\begin{aligned}
&\sigma^2 pd\{G_{p+2,m}(\ell_\alpha;\Delta^2) - 2E[\chi_{p+2}^{-2}(\Delta^2)] + (p-2)E[\chi_{p+2}^{-4}(\Delta^2)]\} \geq \\
&\boldsymbol{\alpha}'\boldsymbol{\Lambda}^{-1}\boldsymbol{\alpha}\{2G_{p+2,m}(\ell_\alpha;\Delta^2) - G_{p+4,m}(\ell_\alpha^*;\Delta^2) - pd(p+2)E[\chi_{p+4}^{-4}(\Delta^2)]\} \\
&+2k[\boldsymbol{\alpha}\boldsymbol{\alpha}'\{G_{p+2,m}(\ell_\alpha;\Delta^2) - E[\chi_{p+2}^{-2}(\Delta^2)]\}, \qquad (9.5.54)
\end{aligned}
$$

provided that

$$
\begin{aligned}
&G_{p+2,m}(\ell_\alpha;\Delta^2) > E[\chi_{p+2}^{-2}(\Delta^2)], \\
&2G_{p+2,m}(\ell_\alpha;\Delta^2) - G_{p+4,m}(\ell_\alpha^*;\Delta^2) \geq pd(p+2)E[\chi_{p+2}^{-2}(\Delta^2)],
\end{aligned}
$$

and

$$
G_{p+2,m}(\ell_\alpha;\Delta^2) \geq 2E[\chi_{p+2}^{-2}(\Delta^2)] - (p-2)E[\chi_{p+2}^{-4}(\Delta^2)]. \qquad (9.5.55)
$$

Under the null hypothesis $H_0 : \boldsymbol{\beta} = \mathbf{0}$, the expression (9.5.54) becomes

$$
\begin{aligned}
\sigma^2 pd\left(G_{p+2,m}(\ell_\alpha;0) - \frac{1}{p}\right) &\geq \boldsymbol{\alpha}'\boldsymbol{\Lambda}^{-1}\boldsymbol{\alpha}(2G_{p+2,m}(\ell_\alpha;0) - G_{p+4m}(\ell_\alpha^*;0) - d) \\
&+2k\boldsymbol{\alpha}'\boldsymbol{\alpha}\left(G_{p+2,m}(\ell_\alpha;0) - \frac{1}{p}\right) \qquad (9.5.56)
\end{aligned}
$$

whenever

$$
G_{p+2,m}(\ell_\alpha;0) > \frac{1}{p} \quad \text{and} \quad 2G_{p+2,m}(\ell_\alpha;0) - G_{p+4,m}(\ell_\alpha^*;0) > d. \qquad (9.5.57)
$$

Thus, $\hat{\beta}_n^{\text{PT}}(k)$ is superior to $\hat{\beta}_n^S(k)$ whenever (9.5.56) is satisfied under (9.5.57).

Comparison of PRSRRE and PRSE.

The risk-difference of the two estimators is given by

$$
\begin{aligned}
&R_4\left(\hat{\boldsymbol{\beta}}_n^{S+};\mathbf{I}_p\right) - R_4\left(\hat{\beta}_n^{S+}(k);\mathbf{I}_p\right) = R_3\left(\hat{\beta}_n^{S};\mathbf{I}_p\right) - R_3\left(\hat{\boldsymbol{\beta}}_n^{S}(k);\mathbf{I}_p\right) \\
&-\sigma^2 \operatorname{tr}\left[\mathbf{C}^{-1} - \mathbf{R}(k)\mathbf{C}^{-1}\mathbf{R}'(k)\right] E\left[\left((1 - d_1 F_{p+2,m}^{-1}(\Delta^2)\right)^2 I\left(F_{p+2,m}^{-1}(\Delta^2) < d_1\right)\right] \\
&-\{\boldsymbol{\beta}'\boldsymbol{\beta} - \boldsymbol{\beta}'\mathbf{R}'(k)\mathbf{R}(k)\boldsymbol{\beta}\}\left\{2E\left[((1 - d_1 F_{p+2,m}^{-1}(\Delta^2))\, I\left(F_{p+2,m}^{-1}(\Delta^2) < d_1\right)\right]\right. \\
&\left.-E\left[\left((1 - d_2 F_{p+4,m}^{-1}(\Delta^2)\right)^2 I\left(F_{p+4,m}(\Delta^2) < d_2\right)\right]\right\} \\
&-k\left[\boldsymbol{\beta}'\mathbf{R}'(k)\mathbf{C}^{-1}(k)\boldsymbol{\beta} + \boldsymbol{\beta}'\mathbf{C}^{-1}(k)\mathbf{R}(k)\boldsymbol{\beta}\right] \\
&\quad\times E\left[((1 - d_1 F_{p+2,m}^{-1}(\Delta^2))\, I\left(F_{p+2,m}^{-1}(\Delta^2) < d_1\right)\right]. \qquad (9.5.58)
\end{aligned}
$$

The risk-difference above is nonnegative whenever$\Delta^2 \leq \Delta^2(k)$, where

$$
\Delta^2(k) = \frac{\begin{array}{c} R_3(\hat{\boldsymbol{\beta}}_n^{S};\mathbf{I}_p) - \mathbf{R}_3(\hat{\beta}_n^S(k);I_p) - \sigma^2 \operatorname{tr}[\mathbf{C}^{-1} - \mathbf{R}(k)\mathbf{C}^{-1}\mathbf{R}'(k)] \\ \times E[(1 - d_1 F_{p+2,m}^{-1}(\Delta^2))^2 I(F_{p+2,m}^{-1}(\Delta^2) < d_1)] \end{array}}{Ch_{\max}\left[(\mathbf{A}^* + \mathbf{B}^*)\mathbf{C}^{-1}\right]} \qquad (9.5.59)
$$

with

$$
\begin{aligned}
\mathbf{A}^* = [\mathbf{I}_p - \mathbf{R}'(k(\mathbf{R}(k)]\{&2E[(1 - d_1 F_{p+2,m}^{-1}(\Delta^2))I(F_{p+2,m}(\Delta^2)) \\
&\times I(F_{p+2,m}(\Delta^2) < d_1)] - E[(1 - d_2 F_{p+4,m}^{-1}(\Delta^2))^2 I(F_{p+2,m}(\Delta^2) < d_2)]\}
\end{aligned}
$$

and

$$
\mathbf{B}^* = k[\mathbf{R}'(k)\mathbf{C}^{-1}(k) + \mathbf{C}^{-1}(k)\mathbf{R}(k)]E[(1 - d_1 F_{p+2,m}^{-1}(\Delta^2))I(F_{p+2,m}(\Delta^2) < d_1)]. \qquad (9.5.60)
$$

Further, differentiating $R_4(\hat{\boldsymbol{\beta}}_n^{S+}(k);\mathbf{I}_p)$ w.r.t. k yields

$$
\begin{aligned}
&\frac{\partial R_4(\hat{\beta}_n^{S+}(k);\mathbf{I}_p)}{\partial k} = \frac{\partial R_3(\hat{\beta}_n^S(k(;\mathbf{I}_p)}{\partial k} \\
&+2\sum_{i=1}^{p} \frac{1}{(\lambda_i + k)^3}\left\{k\alpha_i^2\lambda_i E\left[\left(d_1 F_{p+2,m}^{-1}(\Delta^2) - 1\right)^2 I\left(F_{p+2,m}(\Delta^2) < d_1\right)\right]\right. \\
&+\sigma^2\lambda_i E\left[\left(1 - d_1 F_{p+2,m}^{-1}(\Delta^2)\right)^2 I\left(F_{p+2,m}(\Delta^2) < d_1\right)\right] \\
&+\lambda_i^2\alpha_i^2 E\left[\left(1 - d_2 F_{p+4,m}^{-1}(\Delta^2)\right)^2 I\left(F_{p+4,m}(\Delta^2) < d_2\right)\right] \\
&\left.+\lambda_i^2\alpha_i^2 E\left[\left(d_1 F_{p+2,m}^{-1}(\Delta^2) - 1\right) I\left(F_{p+2,m}(\Delta^2) < d_1\right)\right]\right\}, \qquad (9.5.61)
\end{aligned}
$$

where $\partial R_3(\hat{\beta}_n^S(k);I_p)/\partial k$ is given by (9.5.30). Hence, sufficient condition for (9.5.60) to be negative is that there exists a $k \in (0, k_6]$, where

$$
k_6 = \frac{f_1(\Delta^2)}{f_2(\Delta^2)}, \qquad (9.5.62)
$$

with

$$f_1(\Delta^2) = \min_{1\le i\le p}\Big\{\sigma^2\Big(\lambda_i - pd\lambda_i\big[(p-2)E[\chi^{-4}_{p+2}(\Delta^2)]$$
$$+\left(1-\frac{(p+2)\lambda_i\alpha_i^2}{2\sigma^2\Delta^2}\right)\left(2\Delta^2E[\chi^{-4}_{p+4}(\Delta^2)]\right)\Big)$$
$$-\lambda_iE\big[(1-d_1F^{-1}_{p+2,m}(\Delta^2))^2I(F_{p+2,m}(\Delta^2)<d_1)\big]$$
$$-\lambda_i^2\alpha_i^2E\big[(1-d_1F^{-1}_{p+2,m}(\Delta^2)-1)I(F_{p+2,m}(\Delta^2)<d_1)\big]$$
$$+\,pd\alpha_i^2\lambda_i^2E[\chi^{-2}_{p+2}(\Delta^2)]\Big\} \tag{9.5.63}$$

and

$$f_2(\Delta^2) = \max_{1\le i\le p}\alpha_i^2\lambda_i\Big\{1+E[(d_1F^{-1}_{p+2,m}(\Delta^2)-1_I(F_{p+2,m}(\Delta^2)<d_1)]$$
$$-pdE[\chi^{-2}_{p+2}(\Delta^2)\Big\}. \tag{9.5.64}$$

Hence, $\hat{\boldsymbol{\beta}}_n^{S+}(k)$ dominates $\hat{\boldsymbol{\beta}}_n^{S}(k)$ for $k\in(0,k_6)$. Combining (9.5.59) and (9.5.62), we obtain the next theorem.

Theorem 6. A sufficient condition for $\hat{\boldsymbol{\beta}}_n^{S+}(k)$ to dominate $\hat{\boldsymbol{\beta}}_n^{S+}$ is that there exists a pair $(\Delta^2,k)\in(0,\Delta_7^2(k)]\times(0,k_6]$.

The MSE matrix comparison of $\hat{\boldsymbol{\beta}}_n^{S+}$ and $\hat{\boldsymbol{\beta}}_n^{S+}(k)$ can be carried out by considering the difference $\mathbf{M}_4(\hat{\boldsymbol{\beta}}_n^{S+})-\mathbf{M}_4(\hat{\boldsymbol{\beta}}_n^{S+}(k))$, and this is left to the reader as an exercise (Problem 7).

9.6 Estimation of the Ridge Parameter

In this section, we discuss variety of estimators of the ridge parameter k. We first define a simple generalization of the ridge regression estimator and consider the estimators based on this generalization. As before, we consider the diagonalization of the design matrix $\mathbf{C}$ as $\boldsymbol{\Lambda}=\text{Diag}(\lambda_1,\dots,\lambda_p)$ so that the canonical form of the linear model (9.0.1) is written as

$$\mathbf{Y}=\mathbf{X}^*\boldsymbol{\alpha}+\mathbf{e}, \tag{9.6.1}$$

where $\mathbf{X}^*=\mathbf{X}\boldsymbol{\Gamma}$ and $\boldsymbol{\alpha}=\boldsymbol{\Gamma}'\boldsymbol{\beta}$ and $\boldsymbol{\Gamma}$ is defined as in Section 9.5. This allows us to write one simple generalization of the ridge regression estimator as

$$\tilde{\boldsymbol{\alpha}}(\mathbf{k})=(\mathbf{X}^{*'}\mathbf{X}^*+\mathbf{K})^{-1}\mathbf{X}'^*\mathbf{Y}, \tag{9.6.2}$$

where $\mathbf{K}=\text{Diag}(k_1,k_2,\dots,k_p)$, $(k_i>0|i=1,\dots,p)$ while $\tilde{\boldsymbol{\alpha}}=(\mathbf{X}'^*\mathbf{X}^*)\mathbf{X}^{*'}\mathbf{Y}$ is the least squares estimator (LSE) of $\boldsymbol{\alpha}$. Clearly, the MSE expression for

$\tilde{\alpha}(\mathbf{k})$ is

$$\text{MSE}(\tilde{\alpha}(\mathbf{k})) = \text{E}(\tilde{\alpha}(\mathbf{k}) - \alpha)'(\tilde{\alpha}(\mathbf{k}) - \alpha) = \sigma^2 \sum_{i=1}^{p} \frac{\lambda_i}{(\lambda_i + k_i)^2} + \sum_{i=1}^{p} \frac{k_i^2 \alpha_i^2}{(\lambda_i + k_i)^2}. \tag{9.6.3}$$

Hoerl and Kennard (1970) showed that the optimal value of k_i is σ^2/α_i^2 and they subsequently came up with the estimator of k for the case $\mathbf{K} = k\mathbf{I}_p$ as

$$\hat{k}_{HK} = \frac{\sigma^2}{\tilde{\alpha}_{\max}^2}, \quad \tilde{\alpha}_{\max}(\tilde{\alpha}_1, \ldots, \tilde{\alpha}_p). \tag{9.6.4}$$

There are two variations of the estimator proposed by Hoerl and Kennard (1970). One is due to Hoerl, Kennard, and Baldwin (1975), given by

$$\hat{k}_{HKB} = \frac{p\tilde{\sigma}^2}{\sum_{i=1}^{p} \tilde{\alpha}_i^2} = \frac{p\tilde{\sigma}^2}{\tilde{\beta}_n' \tilde{\beta}_n}, \tag{9.6.5}$$

which is the harmonic mean of $\hat{k}_i = \tilde{\sigma}^2/\tilde{\alpha}_i^2$, $(i = 1, \ldots, p)$ and the other is due to Hocking, Speed, and Lynn (1976), given by

$$\hat{k}_{HSL} = \tilde{\sigma}^2 \frac{\sum_{i=1}^{p} (\lambda_i \tilde{\alpha}_i)^2}{\left(\sum_{i=1}^{p} (\lambda_i \tilde{\alpha}_i^2)\right)^2}. \tag{9.6.6}$$

In addition, a Bayesian approach by Lawless and Wang (1976) led to the estimator

$$\hat{k}_{LW} = \frac{p\tilde{\sigma}^2}{\tilde{\beta}_n' \, \mathbf{C} \, \tilde{\beta}_n}. \tag{9.6.7}$$

More recently, Kibria (2003) proposed the following three estimators of k, namely the arithmetic mean, geometric mean, and the median of $\tilde{\sigma}^2/\tilde{\alpha}_i^2$ $(i = 1, \ldots, p)$. Explicitly, they are

$$\text{(i)} \ \hat{k}_{\text{AM}} = \tfrac{1}{p} \sum_{i=1}^{p} \frac{\tilde{\sigma}^2}{\tilde{\alpha}_i^2}, \tag{9.6.8}$$

$$\text{(ii)} \ \hat{k}_{\text{GM}} = \frac{\tilde{\sigma}^2}{\left(\prod_{i=1}^{p} \tilde{\alpha}_i^2\right)^{1/p}}$$

(iii) $\hat{k}_{\text{med}} = \text{med}_{1 \le i \le p} \left\{ \frac{\tilde{\sigma}^2}{\tilde{\alpha}_i^2} \right\}$, respectively.

It is observed from equations (9.6.6) through (9.6.8) that for $\tilde{\alpha}_i = \tilde{\alpha}$ $(i = 1, \ldots, p)$, all proposed estimators except $\hat{k}_{\text{HSL}}$ become $\tilde{\sigma}^2/\tilde{\alpha}^2$. Also, all estimators except $\hat{k}_{\text{HLW}}$ and $\hat{k}_{\text{HSL}}$ are independent of $(\lambda_1, \ldots, \lambda_p)$. In order to study the properties of the estimators, we use the MSE criterian following Dempster et al. (1977), Lawless and Wang (1976), and Gibbons (1981) among others. All these authors used simulation studies to assess the finite sample properties of the estimators. It is reported that all the estimators have smaller MSE values than the LSE. It is also shown that $\hat{k}_{\mathrm{GM}}$ and $\hat{k}_{\mathrm{LW}}$ performed equivalently and little better than $\hat{k}_{\mathrm{HKB}}$. For the small correlation, $\hat{k}_{\mathrm{GM}}$ performed better than $\hat{k}_{\mathrm{LW}}$ and for high correlation $\hat{k}_{\mathrm{LW}}$ performed better than $\hat{k}_{\mathrm{GM}}$. Further, it is noted that $\hat{k}_{\mathrm{AM}}$ performed quite well compared to the others. Anyway, more studies are necessary to differentiate between the performances of the estimators, and as such the area remains open for research, since no mathematical analysis has yet been recorded.

An Example. In this section, we consider one real-life example to demonstrate the performance of the ridge estimators. The example was considered by Anderson et al. (1993, p. 579). This example contains data for 19 stocks on the following variables: profit margin, growth rate, type of industry and price to earning ratio. We consider the following linear regression model:

$$Y = \beta_0 + \beta_1 X_1 + \beta_2 X_2 + \beta_3 X_3 + e, \tag{9.6.9}$$

where $\mathbf{Y}$ is the price to earning ratio, X_1 is the profit margin, X_2 is the growth rate, and X_3 is the type of industry. The random error is assumed to follow the normal distribution with mean 0 and variance σ^2.

The correlation matrix of the variables in model (9.6.9) is presented in Table 9.1.

Table 9.1 Correlations among the Variables

	X_1	X_2	X_3	$\mathbf{Y}$
X_1	1.00000	0.98684	0.99701	0.99700
X_2	0.98684	1.00000	0.99472	0.99131
X_3	0.99701	0.99472	1.00000	0.99950
$\mathbf{Y}$	0.99700	0.99131	0.99950	1.00000

Source: Adapted from Kibria (2003)

As observed in Table 9.1, the explanatory variables are strongly correlated. Moreover, the ratio of the largest and the smallest root of the design matrix, $\lambda_1/\lambda_p = 8072.09207/15.83687 = 509.7025$, which implies the existence of multicollinearity in the data set. So it is adequate to compare the proposed ridge estimators with this real data set.

The estimated MSE of the proposed estimators along with the ridge regression coefficients are presented in Table 9.2.

Table 9.2 MSE and the Estimated Regression Coefficients

Estimators	LSE	$\hat{k}_{HK}$	$\hat{k}_{HKB}$	$\hat{k}_{LW}$	$\hat{k}_{HSL}$	$\hat{k}_{AM}$	$\hat{k}_{GM}$	$\hat{k}_{MED}$
MSE	10.308	0.319	0.312	0.666	0.319	0.384	0.332	0.311
$\hat{\beta}_1$	0.656	0.674	0.675	0.656	0.674	0.654	0.671	0.675
$\hat{\beta}_2$	0.336	0.374	0.386	0.336	0.374	0.421	0.399	0.381
$\hat{\beta}_3$	0.782	0.446	0.361	0.782	0.446	0.201	0.287	0.396

Source: Adapted from Kibria (2003)

From Table 9.2, we know that all the proposed estimators perform better than LSE in the sense of having smaller MSE. $\hat{k}_{MED}$ and $\hat{k}_{HKB}$ perform equivalently and little better than other estimators. However, $\hat{k}_{LW}$ performs worse compared to other estimators. Again, the performance of $\hat{k}_{GM}$ is between $\hat{k}_{LW}$ and $\hat{k}_{HKB}$.

9.7 Conclusions

In this chapter, we discussed the estimation procedures of the regression parameters for the ill-conditioned data. First, we analyzed some finite sample theory of four well-known ridge regression estimators of $\boldsymbol{\beta}$ that are a combination of the sample and nonsample information. The RRRE performs best compared to other estimators in the neighborhood of the null hypothesis; however, it performs worse when Δ^2 moves away from its origin. We may order the risk of the estimators under H_0 as

$$R_2(\hat{\boldsymbol{\beta}}_n(k); \mathbf{I}_p) \le R_3(\hat{\boldsymbol{\beta}}_n^{\text{PT}}(k); \mathbf{I}_p) \le R_5(\hat{\boldsymbol{\beta}}_n^{S+}(k); \mathbf{I}_p)$$
$$\le R_4(\hat{\boldsymbol{\beta}}_n^{S}(k); \mathbf{I}_p) \le R_1(\tilde{\boldsymbol{\beta}}_n(k); \mathbf{I}_p),$$

while the position of the PTRRE shifts from "in between" $R_5(\hat{\boldsymbol{\beta}}_n^{S+}(k); \mathbf{I}_p)$ and $R_2(\hat{\boldsymbol{\beta}}_n(k); \mathbf{I}_p)$ to "in between" $R_4(\hat{\boldsymbol{\beta}}_n^{S}(k); \mathbf{I}_p)$ and $R_1(\tilde{\boldsymbol{\beta}}_n(k); \mathbf{I}_p)$. That is to say, the risk order is

$$R_2(\hat{\boldsymbol{\beta}}_n(k); I_p) \le R_5(\hat{\boldsymbol{\beta}}_n^{S+}(k); I_p) \le \mathbf{R}_4(k)(\hat{\boldsymbol{\beta}}_n^{S}(k); I_p)$$
$$\le R_3(\hat{\boldsymbol{\beta}}_n^{\text{PT}}(k); I_p) \le R_1(\tilde{\boldsymbol{\beta}}_n(k); I_p).$$

The picture changes when $\boldsymbol{\beta}$ moves away from the origin $\mathbf{0}$. We found that there exists a value of k such that ridge regression estimators dominate each others and over the traditional estimators. Note that the application of PRRRE and SRRE is constrained by the requirement $q \ge 3$, while PTRRE does not need such constraint. However, the choice of the level of significance of the test has a dramatic impact on the nature of the risk function for the PT estimator. Thus, when $q \ge 3$, we use PRRRE; otherwise, PTRRE with some optimum size α.

Second, we considered various estimators of ridge parameter k and specified their relative properties based on documented simulation results.

9.8 Problems

1. Prove Theorem 1.

2. Show that $\mathbf{V}_1(0) - \mathbf{V}_1(k)$ is n.n.d. matrix for all $k > 0$, where $\mathbf{V}_1(k)$ is given by (9.4.1).

3. Show that $\mathbf{V}_2(0) - \mathbf{V}_2(k)$ is n.n.d. matrix for all $k > 0$, where $\mathbf{V}_2(k)$ is given by (9.4.4).

4. Show that $\mathbf{V}_3(0) - \mathbf{V}_3(k)$ is n.n.d. matrix for all $k > 0$, where $\mathbf{V}_3(k)$ is given by (9.4.8).

5. Show that $\mathbf{V}_4(0) - \mathbf{V}_4(k)$ is n.n.d. matrix for all $k > 0$, where $\mathbf{V}_2(k)$ is given by (9.4.12).

6. Determine the conditions when $\mathbf{M}_3(\hat{\boldsymbol{\beta}}_n^S) - \mathbf{M}_3(\hat{\boldsymbol{\beta}}_n^S(k))$ is n.n.d. matrix for all $k > 0$.

7. Determine the conditions when $\mathbf{M}_4(\hat{\boldsymbol{\beta}}_n^{S+}) - \mathbf{M}_4(\hat{\boldsymbol{\beta}}_n^{S+}(k))$ is n.n.d. matrix for all $k > 0$.

8. Show that $\mathbf{V}_1(0) - \mathbf{V}_2(k)$ is n.n.d. matrix for all $k > 0$.

9. Show that $\mathbf{V}_1(0) - \mathbf{V}_3(k)$ is n.n.d. matrix for all $k > 0$.

10. Consider the regression model (9.0.1)

$$\mathbf{Y} = \mathbf{X}\boldsymbol{\beta} + \varepsilon_n, \quad \boldsymbol{\varepsilon}_n \sim N_n(\mathbf{0}, \sigma^2\mathbf{I}_n)$$

and the estimation of $\boldsymbol{\beta}$ when the restriction $\mathbf{H}_{q\times p}\boldsymbol{\beta}_{p\times 1} = \mathbf{h}_{q\times 1}$ holds. Thus, we have the unrestricted and the restricted estimators $\tilde{\boldsymbol{\beta}}_n = \mathbf{C}^{-1}\mathbf{X}'\mathbf{Y}$ and $\hat{\boldsymbol{\beta}}_n = \tilde{\boldsymbol{\beta}}_n - \mathbf{C}^{-1}\mathbf{H}'(\mathbf{H}\mathbf{C}^{-1}\mathbf{H}')^{-1}(\mathbf{H}\tilde{\boldsymbol{\beta}}_n - \mathbf{h})$. Define the ridge estimators as follows:

(i) $\tilde{\boldsymbol{\beta}}_n(k) = \mathbf{R}(k)\tilde{\boldsymbol{\beta}}_n$,

(ii) $\hat{\boldsymbol{\beta}}_n(k) = \mathbf{R}(k)\hat{\boldsymbol{\beta}}_n$,

(iii) $\hat{\boldsymbol{\beta}}_n^{\text{PT}}(k) = \mathbf{R}(k)\hat{\boldsymbol{\beta}}_n^{\text{PT}}$, $\hat{\boldsymbol{\beta}}_n^{\text{PT}} = \tilde{\boldsymbol{\beta}}_n - (\tilde{\boldsymbol{\beta}}_n - \hat{\boldsymbol{\beta}}_n)I(\mathcal{L}_n \leq F_{q,m}(\alpha))$,

where

$$\mathcal{L}_n = \frac{(\mathbf{H}\tilde{\boldsymbol{\beta}}_n - \mathbf{h})'(\mathbf{H}\mathbf{C}^{-1}\mathbf{H}')^{-1}(\mathbf{H}\tilde{\boldsymbol{\beta}}_n - \mathbf{h})}{q\tilde{\sigma}_n^2},$$

$$\tilde{\sigma}_n^2 = m^{-1}(\mathbf{Y} - \mathbf{X}\tilde{\boldsymbol{\beta}}_n)'(\mathbf{Y} - \mathbf{X}\tilde{\boldsymbol{\beta}}_n), \quad m = n - p.$$

a) Find the bias vector of the estimators: $\tilde{\boldsymbol{\beta}}_n(k)$, $\hat{\boldsymbol{\beta}}_n(k)$, and $\hat{\boldsymbol{\beta}}_n^{\text{PT}}(k)$.

b) Find the covariance matrix and the MSE matrix of the estimators: $\tilde{\boldsymbol{\beta}}_n(k)$, $\hat{\boldsymbol{\beta}}_n(k)$, and $\hat{\boldsymbol{\beta}}_n^{\text{PT}}(k)$.

c) Compare the estimators: $\tilde{\boldsymbol{\beta}}_n$ vs. $\tilde{\boldsymbol{\beta}}_n(k)$, $\hat{\boldsymbol{\beta}}_n$ vs. $\hat{\boldsymbol{\beta}}_n(k)$, and $\hat{\boldsymbol{\beta}}_n^{\text{PT}}$ vs. $\hat{\boldsymbol{\beta}}_n^{\text{PT}}(k)$ using covariance matrices and using MSE matrices.

11. Define the Stein estimator in the ridge setup as

$$\hat{\boldsymbol{\beta}}_n^S(k) = \mathbf{R}(k)\hat{\boldsymbol{\beta}}_n^S \quad \text{and} \quad \hat{\boldsymbol{\beta}}_n^{S+}(k) = \mathbf{R}(k)\hat{\boldsymbol{\beta}}_n^{S+},$$

$$\hat{\boldsymbol{\beta}}_n^S = \tilde{\boldsymbol{\beta}}_n - d(\tilde{\boldsymbol{\beta}}_n - \hat{\boldsymbol{\beta}}_n)\mathcal{L}_n^{-1}, \quad d = \frac{(q-2)m}{q(m+2)},$$

$$\hat{\boldsymbol{\beta}}_n^{S+} = \hat{\boldsymbol{\beta}}_n + (\tilde{\boldsymbol{\beta}}_n - \hat{\boldsymbol{\beta}}_n)\{1 - d\mathcal{L}_n^{-1}\}I(\mathcal{L}_n > d),$$

a) Find the bias vectors of $\hat{\boldsymbol{\beta}}_n^S(k)$ and $\hat{\boldsymbol{\beta}}_n^{S+}(k)$.

b) Find the covariance and MSE matrices of $\hat{\boldsymbol{\beta}}_n^S(k)$ and $\hat{\boldsymbol{\beta}}_n^{S+}(k)$.

c) Compare the estimators:

$$\tilde{\boldsymbol{\beta}}_n(k) \text{ vs. } \hat{\boldsymbol{\beta}}_n^S(k),$$

$$\hat{\boldsymbol{\beta}}_n^S(k) \text{ vs. } \hat{\boldsymbol{\beta}}_n^{S+}(k),$$

$$\text{and} \qquad \tilde{\boldsymbol{\beta}}_n^{\text{PT}}(k) \text{ vs. } \hat{\boldsymbol{\beta}}_n^S(k),$$

using covariance and MSE matrices.

12. Consider the following rank estimators of $\boldsymbol{\beta}$ as given by (9.1.14a–c):

a) $\hat{\boldsymbol{\beta}}_n^{(r)}(k) = \mathbf{R}(k)\tilde{\boldsymbol{\beta}}_n^{(r)}$,

b) $\hat{\boldsymbol{\beta}}_n^{\text{PT}}(k) = \hat{\boldsymbol{\beta}}_n^{(r)}(k) = \hat{\boldsymbol{\beta}}_n^{(r)}(k)I(\mathcal{L}_n > \chi_p^2(\alpha))$

c) $\hat{\boldsymbol{\beta}}_n^S(k) = \hat{\boldsymbol{\beta}}_n^{(r)}(k)\{1 - (p-2)\mathcal{L}_n^{-1}\}I(\mathcal{L}_n > \varepsilon_n)$

d) $\hat{\boldsymbol{\beta}}_n^{S+}(k) = \hat{\boldsymbol{\beta}}_n^{(r)}(k)\{1 - (p-2)\mathcal{L}_n^{-1}\}I(\mathcal{L}_n > (p-2))$.

13. Show that asymptotic expressions of bias of the estimators are given as follows:

a) $\mathbf{b}_1(\tilde{\boldsymbol{\beta}}_n^{(r)}(k)) = -k\mathbf{C}^{-1}(k)\boldsymbol{\delta}$.

b) $\mathbf{b}_2(\hat{\boldsymbol{\beta}}_n^{\text{PT}}(k)) = -k\mathbf{C}^{-1}(k)\boldsymbol{\delta} - \mathbf{R}(k)\boldsymbol{\delta}H_{p+2}(\chi_p^2(\alpha); \Delta^2)$,
$\Delta^2 = \sigma^{-2}(\boldsymbol{\delta}'\mathbf{C}\boldsymbol{\delta})$ where $\sigma^2 = A_\phi^2/\gamma^2(\psi, \phi)$.

c) $\mathbf{b}_3(\hat{\boldsymbol{\beta}}_n^S(k)) = -k\mathbf{C}^{-1}(k)\boldsymbol{\delta} - (p-2)\mathbf{R}(k)\boldsymbol{\delta}E[\chi_{p+2}^{-2}(\Delta^2)]$.

d) $\mathbf{b}_4(\hat{\boldsymbol{\beta}}_n^{S+}(k)) = -k\mathbf{C}^{-1}(k)\boldsymbol{\delta} - \mathbf{R}(k)\boldsymbol{\delta}\{E(1 - (p-2)\chi_{p+2}^{-2}(\Delta^2))$.

$$I(\chi_{p+2}^2(\Delta^2) < p-2)\} + (p-2)E[\chi_{p+2}^{-2}(\Delta^2)].$$

14. Continuation of Problem 12. Asymptotic variance MSE matrix and risk expressions of the four rank based ridge regression estimators are given

by

(i) $\mathbf{M}_1(\tilde{\beta}_n^{(r)}(k)) = \sigma^2[\mathbf{R}(k)\mathbf{C}^{-1}\mathbf{R}(k)] + k^2\mathbf{C}^{-1}(k)\boldsymbol{\delta}\boldsymbol{\delta}'\mathbf{C}^{-1}(k),$

$\mathbf{V}_1(k) = \sigma^2[\mathbf{R}(k)\mathbf{C}^{-1}\mathbf{R}(k)],$

$R_1(\tilde{\beta}_n^{(r)}(k);\mathbf{Q}) = \sigma^2 \operatorname{tr}[\mathbf{Q}\mathbf{R}(k)\mathbf{C}^{-1}\mathbf{R}(k))] + k^2(\mathbf{C}^{-1}(k)\mathbf{Q}\mathbf{C}^{-1}(k)\boldsymbol{\delta});$

(ii) $\mathbf{M}_2(\hat{\beta}_n^{\text{PT}}(k))$

$$= \sigma^2[\mathbf{R}(k)\mathbf{C}^{-1}\mathbf{R}(k)] - \sigma^2[\mathbf{R}(k)\mathbf{C}^{-1}\mathbf{R}(k)]H_{p+2}(\chi_p^2(\alpha);\Delta^2)$$
$$-[\mathbf{R}(k)\boldsymbol{\delta}\boldsymbol{\delta}'\mathbf{R}(k)]\{2H_{p+2}(\chi_p^2(\alpha);\Delta^2) - H_{p+4}(\chi_p^2(\alpha);\Delta^2)\}$$
$$+k[\mathbf{R}(k) + \mathbf{C}^{-1}(k)\boldsymbol{\delta}\boldsymbol{\delta}'\mathbf{R}(k)]H_{p+2}(\chi_p^2(\alpha);\Delta^2)$$
$$+k^2[\mathbf{C}^{-1}(k)\boldsymbol{\delta}\boldsymbol{\delta}'\mathbf{C}^{-1}(k)],$$

$$\mathbf{V}_2(k) = \sigma^2[\mathbf{R}(k)\mathbf{C}^{-1}\mathbf{R}(k)] - \sigma^2[\mathbf{R}(k)\mathbf{C}^{-1}\mathbf{R}(k)]H_{p+2}(\chi_p^2(\alpha);\Delta^2)$$
$$+[\mathbf{R}(k)\boldsymbol{\delta}\boldsymbol{\delta}'\mathbf{R}(k)]\{2H_{p+4}(\chi_p^2(\alpha);\Delta^2) + (H_{p+2}(\chi_p^2(\alpha);\Delta^2))^2\},$$

and

$$R_2(\hat{\beta}_n^{\text{PT}}(k);\mathbf{Q}) = \sigma^2 \operatorname{tr}[\mathbf{Q}(\mathbf{R}(k)\mathbf{C}^{-1}\mathbf{R}(k))](1 - H_{p+2}(\chi_p^2(\alpha);\Delta^2)$$
$$+ [\mathbf{R}(k)\boldsymbol{\delta}\boldsymbol{\delta}'\mathbf{R}(k)]\{2H_{p+4}(\chi_p^2(\alpha);\Delta^2) + (H_{p+4}(\chi_p^2(\alpha);\Delta^2))^2\}$$
$$+ k^2\boldsymbol{\delta}'\mathbf{C}^{-1}(k)\mathbf{Q}\mathbf{C}^{-1}(k)\boldsymbol{\delta} + k[\boldsymbol{\delta}'\mathbf{C}^{-1}(k)\mathbf{Q}\mathbf{R}(k)\boldsymbol{\delta}$$
$$+ \boldsymbol{\delta}'\mathbf{R}(k)\mathbf{Q}\mathbf{C}^{-1}(k)\boldsymbol{\delta}]H_{p+2}(\chi_p^2(\alpha);\Delta^2);$$

(iii) $\mathbf{M}_3(\hat{\beta}_n^{S}(k)) = \sigma^2(\mathbf{R}(k)\mathbf{C}^{-1}\mathbf{R}(k))$

$$\times\Big[1 - pd\{2E[\chi_{p+2}^{-2}(\Delta^2)] - (p-2)E[\chi_{p+2}^{-4}(\Delta^2)]\}\Big]$$
$$+ pd(p+2)(\mathbf{R}(k)\boldsymbol{\delta}\boldsymbol{\delta}'\mathbf{R}'(k))E[\chi_{p+2}^{-4}(\Delta^2)] + k^2(\mathbf{C}^{-1}(k)\boldsymbol{\delta}\boldsymbol{\delta}'\mathbf{C}^{-1}(k))$$
$$+ dfk[\mathbf{R}(k)\boldsymbol{\delta}\boldsymbol{\delta}'\mathbf{C}^{-1}(k) + \mathbf{C}^{-1}(k)\boldsymbol{\delta}\boldsymbol{\delta}'\mathbf{R}'(k))]E[\chi_{p+2}^{-4}(\Delta^2)],$$

$\mathbf{V}_3(k) =$

$$\sigma^2(\mathbf{R}(k)\mathbf{C}^{-1}\mathbf{R}(k))\Big[1 - pd\{2E[\chi_{p+2}^{-2}(\Delta^2)] - (p-2)E[\chi_{p+2}^{-4}(\Delta^2)]\}\Big]$$
$$+ pd(\mathbf{R}(k)\boldsymbol{\delta}\boldsymbol{\delta}'\mathbf{R}'(k))\{(p+2)E[\chi_{p+4}^{-4}(\Delta^2)] - dp[(\chi_{p+2}^{-2}(\Delta^2)]^2\}$$

and

$$
\begin{aligned}
R_3(\hat{\beta}_n^S(k);\mathbf{Q}) = {} & \sigma^2 \operatorname{tr}[\mathbf{Q}(\mathbf{R}(k)\mathbf{C}^{-1}\mathbf{R}(k))](1 - pd\{2E[\chi_{p+2}^{-2}(\Delta^2)]\}] \\
& + dp(p+2)(\boldsymbol{\delta}\mathbf{R}(k)\mathbf{Q}\mathbf{R}(k)\boldsymbol{\delta})E[\chi_{p+4}^{-4}(\Delta^2)] \\
& + k^2(\boldsymbol{\delta}'\mathbf{C}^{-1}(k)\mathbf{Q}\mathbf{C}^{-1}\boldsymbol{\delta}) + pdk \\
& [\boldsymbol{\delta}'\mathbf{C}^{-1}(k)\mathbf{Q}\mathbf{R}(k)\boldsymbol{\delta} + \boldsymbol{\delta}'\mathbf{R}'(k)\mathbf{Q}\mathbf{C}^{-1}(k)\boldsymbol{\delta}]E[\chi_{p+2}^{-2}(\Delta^2)].
\end{aligned}
$$

Similarly, derive the expressions for (a) $\mathbf{M}_4(\hat{\beta}_n^{S+}(k))$, (b) $\mathbf{V}_4(k)$, and (c) $R_4(\hat{\beta}_n^{S+}(k);\mathbf{Q})$, respectively.

Chapter 10

Regression Models with Autocorrelated Errors

Outline

In Chapters 7 and 8, we considered the estimation of the regression parameters $\boldsymbol{\beta}$ when it is suspected that the regression parameters belongs to a (1) subspace (Chapter 7) or to a (2) stochastic subspace (Chapter 8) while the errors of the regression model are iid r.v. (normal) with mean *zero* and variance σ^2. In this chapter, estimation of the regression parameters concerns the case where the errors of the model are autocorrelated. These models are often satisfactory representation of time-series data in econometrics. We provide some limited nonparametric analyses for autoregressive models. With regard to a regression model with AR[1] errors, we attempt to provide only some initial results.

In Section 10.1, we extend the simple linear model of Chapter 3 to introduce autocorrelated errors, and in Sections 10.2 through 10.5, we introduce autocorrelated errors to the multiple regression model of Chapter 7. Section 10.6 contains the estimation of parameters in an autoregressive Gaussian process, and Section 10.7 deals with the R-estimation of regression parameters in an AR[p] model. We sketch the R-estimation for the regression models with AR[1] errors in Section 10.8.

10.1 Simple Linear Model with Autocorrelated Errors

Consider the simple linear regression model as in Chapter 3 with autocorrelated errors given by

$$\mathbf{Y}_n = \theta \mathbf{1}_n + \beta \mathbf{x} + \boldsymbol{\varepsilon}_n, \tag{10.1.1}$$

where $\boldsymbol{\varepsilon}_n = (\epsilon_1, \cdots, \epsilon_n)'$ and the components of $\boldsymbol{\varepsilon}_n$ satisfy the first-order stationary autocorrelated model defined by

$$\epsilon_i = \rho \epsilon_{i-1} + v_i, \quad |\rho| < 1; \quad i = 1, \cdots, n, \tag{10.1.2}$$

with $E(\mathbf{v}) = 0$, $E(\mathbf{v}\mathbf{v}') = \sigma_v^2 \mathbf{I}_n$, where $\mathbf{v} = (v_1, \cdots, v_n)'$. This specification for each ϵ_i implies that $\boldsymbol{\varepsilon}_n$ is a Gaussian vector with $E(\boldsymbol{\varepsilon}_n) = \mathbf{0}$ and

$$E(\boldsymbol{\varepsilon}_n \boldsymbol{\varepsilon}_n') = \sigma_v^2 \boldsymbol{\Sigma}_\rho, \tag{10.1.3}$$

where

$$\boldsymbol{\Sigma}_\rho = \begin{pmatrix} 1 & \rho & \cdots & \rho^{n-1} \\ \rho & 1 & \cdots & \rho^{n-2} \\ \vdots & \vdots & \vdots & \vdots \\ \rho^{n-1} & \rho^{n-2} & \cdots & 1 \end{pmatrix}, \tag{10.1.4}$$

with the inverse

$$\boldsymbol{\Sigma}_\rho^{-1} = (1-\rho^2)^{-1} \begin{pmatrix} 1 & -\rho & 0 & \cdots & 0 & 0 \\ -\rho & 1+\rho^2 & -\rho & 0 & \cdots & 0 \\ \vdots & \vdots & \vdots & \vdots & \vdots & \vdots \\ 0 & \cdots & 0 & 0 & -\rho & 1 \end{pmatrix},$$

Further, from (10.1.1) it is clear that

$$\mathbf{Y}_n \xrightarrow{\mathcal{D}} \mathcal{N}_n(\theta \mathbf{1}_n + \beta \mathbf{x},\ \sigma_v^2 \boldsymbol{\Sigma}_\rho). \tag{10.1.5}$$

10.1.1 Estimation of the Intercept and Slope Parameters when ρ is Known

The usual unbiased estimators of θ and β (when $\rho = 0$) are given by

$$\begin{pmatrix} \tilde{\theta}_n \\ \tilde{\beta}_n \end{pmatrix} = \begin{pmatrix} \bar{Y} - \tilde{\beta}_n \bar{x} \\ \frac{1}{Q}[\mathbf{x}'\mathbf{Y}_n - \frac{1}{n}(\mathbf{1}_n'\mathbf{x})(\mathbf{1}_n'\mathbf{Y}_n)] \end{pmatrix} \tag{10.1.6}$$

with the covariance matrix

$$\frac{\sigma_v^2}{n}\begin{pmatrix} 1+\frac{n\bar{x}_n}{Q} & -\frac{n\bar{x}}{Q} \\ -\frac{n\bar{x}_n}{Q} & \frac{1}{Q} \end{pmatrix}, \quad Q = \mathbf{x}'\mathbf{x} - \frac{1}{n}(\mathbf{1}_n'\mathbf{x})^2 \tag{10.1.7}$$

The unbiased estimator of σ_v^2 is then the usual s_e^2 given as

$$s_e^2 = \frac{1}{n-2}(\mathbf{Y}_n - \tilde{\theta}_n\mathbf{1}_n - \tilde{\beta}_n\mathbf{x})'(\mathbf{Y}_n - \tilde{\theta}_n\mathbf{1}_n - \tilde{\beta}_n\mathbf{x}). \tag{10.1.8}$$

Now for the model (10.1.1), the **unbiased estimators** of θ and β can be obtained based on the generalized least squares estimation (GLSE) procedure as

$$\begin{pmatrix} \tilde{\theta}_n(\rho) \\ \tilde{\beta}_n(\rho) \end{pmatrix} = \frac{1}{K_1K_2 - K_3^2}\begin{pmatrix} K_2\mathbf{1}_n'\mathbf{\Sigma}_\rho^{-1}\mathbf{Y}_n - K_3\mathbf{x}'\mathbf{\Sigma}_\rho^{-1}\mathbf{Y}_n \\ K_1\mathbf{x}'\mathbf{\Sigma}_\rho^{-1}\mathbf{Y}_n - K_3\mathbf{1}_n'\mathbf{\Sigma}_\rho^{-1}\mathbf{Y}_n \end{pmatrix} \tag{10.1.9}$$

with the covariance matrix

$$\text{cov}\begin{pmatrix} \tilde{\theta}_n(\rho) \\ \tilde{\beta}_n(\rho) \end{pmatrix} = \frac{\sigma_v^2}{K_1K_2 - K_3^2}\begin{pmatrix} K_2 & -K_3 \\ -K_3 & K_1 \end{pmatrix} = \sigma_v^2(\mathbf{X}'\mathbf{\Sigma}_\rho^{-1}\mathbf{X})^{-1} \tag{10.1.10}$$

where $\mathbf{X} = (\mathbf{1}_n', \mathbf{x}')'$,

$$K_1 = \mathbf{1}_n'\mathbf{\Sigma}_\rho^{-1}\mathbf{1}_n = \frac{n}{1+\rho}\left(1 - \frac{\rho}{n}\right), \tag{10.1.11a}$$

$$K_2 = \mathbf{x}'\mathbf{\Sigma}_\rho^{-1}\mathbf{x} = \frac{1}{1-\rho^2)}\left\{(1-\rho^2)x_1^2 + \sum_{i=2}^{n}(x_i - \rho x_{i-1})^2\right\} \tag{10.1.11b}$$

and

$$K_3 = \mathbf{1}_n'\mathbf{\Sigma}_\rho^{-1}\mathbf{x} = \frac{n}{1+\rho}\left[\bar{x}_n - \frac{1}{n}\rho\left(\sum_{i=1}^{n-1} x_i\right)\right] \tag{10.1.11c}$$

are **functions of the autocorrelation,** ρ **assumed known.** Here, we suppress ρ in the quantities K_1, K_2, and K_3.

To test the null hypothesis, $H_0 : \beta = \mathbf{0}$ against $H_A : \beta \neq 0$, we first consider the χ^2-statistic given by

$$\chi^2 = \frac{1}{\sigma_v^2 K_1}[\tilde{\beta}_n(\rho)]^2[K_1K_2 - K_3^2] \stackrel{\mathcal{D}}{=} \chi_1^2(\Delta^2), \tag{10.1.12}$$

since $\tilde{\beta}_n(\rho) \sim \mathcal{N}\left(\beta, \sigma_v^2\frac{K_1}{K_1K_2-K_3^2}\right)$ and $\Delta^2 = \frac{\beta^2[K_1K_2-K_3^2]}{\sigma_v^2 K_1}$.

On the other hand, the estimate of σ_v^2 is given by

$$\tilde{\sigma}_v^2 = \frac{1}{n-2}[\mathbf{Y}_n - \tilde{\theta}_n(\rho)\mathbf{1}_n - \tilde{\beta}_n(\rho)\mathbf{x}]'\mathbf{\Sigma}_\rho^{-1}[\mathbf{Y}_n - \tilde{\theta}_n(\rho)\mathbf{1}_n - \tilde{\beta}_n(\rho)\mathbf{x}]. \tag{10.1.13}$$

The exact distribution of $m\frac{\tilde{\sigma}_v^2}{\sigma_v^2} \stackrel{\mathcal{D}}{=} \chi_m^2$ (m=n-2) and independent of $\tilde{\beta}_n(\rho)$. Hence, we define the test-statistic, $\mathcal{L}_n$ as

$$\mathcal{L}_n = \frac{[\tilde{\beta}_n(\rho)]^2}{\tilde{\sigma}_v^2 K_1}[K_1 K_2 - K_3^2] \stackrel{\mathcal{D}}{=} F_{1,m}(\Delta^2) \tag{10.1.14}$$

for testing $H_0 : \beta = 0$, where $\mathcal{L}_n \stackrel{\mathcal{D}}{=} F_{1,m}$ under H_0. If $\beta = 0$ with certainty, the **restricted estimator** of θ can be written as

$$\hat{\theta}_n(\rho) = K_1^{-1}(\mathbf{1}_n'\boldsymbol{\Sigma}_\rho^{-1}\mathbf{Y}_n) = \tilde{\theta}_n(\rho) + \frac{K_3}{K_1}\tilde{\beta}_n(\rho) \tag{10.1.15}$$

with variance

$$\mathrm{Var}(\hat{\theta}_n(\rho)) = \frac{\sigma_v^2}{K_1}. \tag{10.1.16}$$

10.1.2 Preliminary Test and S-Estimation of β and θ

Let $F_{1,m}(\alpha)$ be the α-level critical value of $\mathcal{L}_n$ (under H_0), then the PTE of β and θ are given by

$$\hat{\beta}_n^{PT}(\rho) = \tilde{\beta}_n(\rho) - \tilde{\beta}_n(\rho) I(\mathcal{L}_n < F_{1,m}(\alpha)) \tag{10.1.17}$$

and

$$\begin{aligned}\hat{\theta}_n^{PT}(\rho) &= \tilde{\theta}_n(\rho) - [\tilde{\theta}_n(\rho) - \hat{\theta}_n(\rho)] I(\mathcal{L}_n < F_{1,m}(\alpha)) \\ &= \tilde{\theta}_n + \frac{K_3}{K_1}\tilde{\beta}_n I(\mathcal{L}_n < F_{1,m}(\alpha))\end{aligned} \tag{10.1.18}$$

respectively. Similarly, we may define the **S-estimators** of β and θ following Section 3.5 as

$$\hat{\beta}_n^S(\rho) = \left\{1 - \frac{c\tilde{\sigma}_v K_1^{1/2}}{(K_1K_2 - K_3^2)^{1/2}|\tilde{\beta}_n(\rho)|}\right\}\tilde{\beta}_n(\rho) \tag{10.1.19}$$

and

$$\hat{\theta}_n^S(\rho) = \tilde{\theta}_n(\rho) + \frac{cK_3\tilde{\sigma}_v\tilde{\beta}_n(\rho)}{K_1^{1/2}(K_1K_2 - K_3^2)^{1/2}|\tilde{\beta}_n(\rho)|} \tag{10.1.20}$$

respectively. Note that the **restricted estimator** of β is 0 and that of θ is $\hat{\theta}_n$ (since $\beta = 0$). It can be shown, following Section 3.5, that the bias and the mean-square error of the estimators of β are given by the following:

(i) Unrestricted estimator

$$b_1(\tilde{\beta}_n(\rho)) = 0 \ \text{ and } \ M_1(\tilde{\beta}_n(\rho)) = \frac{\sigma_v^2 K_1}{K_1K_2 - K_3^2}. \tag{10.1.21}$$

(ii) PTE

$$b_2(\hat{\beta}_n^{PT}(\rho)) = -\beta G_{3,m}\left(\frac{1}{3}F_{1,m}(\alpha);\Delta^2\right); \text{ and}$$

$$\begin{aligned} M_2(\hat{\beta}_n^{PT}(\rho)) = {} & \frac{\sigma_v^2 K_1}{K_1K_2 - K_3^2} - \frac{\sigma_v^2 K_1}{K_1K_2 - K_3^2} G_{3,m}\left(\frac{1}{3}F_{1,m}(\alpha);\Delta^2\right) \\ & + \frac{\sigma_v^2 K_1}{K_1K_2 - K_3^2}\Delta^2 \left\{2G_{3,m}\left(\frac{1}{3}F_{1,m}(\alpha);\Delta^2\right)\right. \\ & \left. - G_{5,m}\left(\frac{1}{5}F_{1,m}(\alpha);\Delta^2\right)\right\}. \end{aligned} \tag{10.1.22}$$

(iii) S-estimator

$$b_3(\hat{\beta}_n^S(\rho)) = -\frac{cE(\tilde{\sigma}_v)K_1}{K_1K_2 - K_3^2}\Big\{1 - 2\Phi(-\Delta)\Big\},\ E(\tilde{\sigma}_v) = \sigma_v K_n; \text{ and}$$

$$M_3(\hat{\beta}_n^S(\rho)) = \frac{\sigma_v^2 K_1}{K_1K_2 - K_3^2}\left\{1 - \frac{2}{\pi}K_n^2\left[2e^{-\Delta^2/2} - 1\right]\right\}, \tag{10.1.23}$$

using $c = \sqrt{\frac{2}{\pi}}K_n$, $K_n = \sqrt{\frac{2}{n}}\frac{\Gamma(\frac{n+1}{2})}{\Gamma(\frac{n}{2})}$.

Next we consider the bias, mean-square errors of $\tilde{\theta}_n(\rho), \hat{\theta}_n(\rho), \hat{\theta}_n^{PT}(\rho)$, and $\hat{\theta}_n^S(\rho)$. They are given as follows:

(i) Unrestricted estimator

$$b_1(\tilde{\theta}_n(\rho)) = 0 \text{ and } M_1(\tilde{\theta}_n(\rho)) = \tfrac{\sigma_v^2 K_2}{K_1K_2 - K_3^2}. \tag{10.1.24}$$

(ii) Restricted Estimator

$$b_2(\hat{\theta}_n(\rho)) = \beta\tfrac{K_3}{K_1} \text{ and } M_2(\hat{\theta}_n(\rho)) = \tfrac{\sigma_v^2}{K_1} + \beta^2\left(\tfrac{K_3}{K_1}\right)^2. \tag{10.1.25}$$

(iii) PTE

$$b_2(\hat{\theta}_n^{PT}(\rho)) = \beta\tfrac{K_3}{K_1}G_{3,m}\left(\tfrac{1}{3}F_{1,m}(\alpha);\Delta^2\right); \text{ and}$$

$$\begin{aligned} M_2(\hat{\theta}_n^{PT}(\rho)) = {} & \tfrac{\sigma_v^2 K_2}{K_1K_2 - K_3^2} - \tfrac{\sigma_v^2 K_3^2}{K_1(K_1K_2 - K_3^2)}G_{3,m}\left(\tfrac{1}{3}F_{1,m}(\alpha);\Delta^2\right) \\ + {} & \tfrac{\sigma_v^2 K_3^2}{K_1(K_1K_2 - K_3^2)}\Delta^2\Big\{2G_{3,m}\left(\tfrac{1}{3}F_{1,m}(\alpha);\Delta^2\right) - G_{5,m}\left(\tfrac{1}{5}F_{1,m}(\alpha);\Delta^2\right)\Big\}. \end{aligned} \tag{10.1.26}$$

(iv) S-estimator

$$b_4(\hat{\theta}_n^S(\rho)) = \frac{cK_3E(\tilde{\sigma}_v)}{K_1^{1/2}(K_1K_2 - K_3^2)^{1/2}}\{1 - 2\Phi(-\Delta)\}; \quad \text{and}$$

$$M_4(\hat{\theta}_n^S(\rho)) = \sigma_v^2\left\{\frac{K_2}{K_1K_2 - K_3^2} - \frac{2}{\pi}K_n\frac{K_3^2}{K_1(K_1K_2 - K_3^2)}\left(2e^{-\Delta^2/2} - 1\right)\right\}. \tag{10.1.27}$$

In general, for small samples, the variances of the estimators with known values of ρ are smaller than the traditional estimators given by (10.1.6). However, it can be shown that under the conditions of Theorem 3.6.1 of Chapter 3, as $n \to \infty$, the variance and covariance matrices of the two sets of estimators are equal. Hence, asymptotically they have the same efficiency.

10.1.3 Estimation of the Intercept and Slope Parameters When Autocorrelation Is Unknown

If ρ is unknown, the estimation of $(\theta, \beta)'$ becomes a bit cumbersome, since ρ is a nuisance parameter. Fisher's recipe tells us to estimate this parameter or test it out. Let us then consider the case where we want to estimate ρ consistently. We follow Durbin's (1960) method of estimating ρ conditionally as follows: Let us transform the responses of the model (10.1.1) to the model (see Fuller, 1976)

$$\begin{aligned} Y_1 &= \theta + \beta x_1 + v_1, \\ Y_t &= (1-\rho)\theta + \rho Y_{t-1} + \beta(x_t - \rho x_{t-1}) + v_t, \quad t = 2, 3, \cdots, n, \end{aligned} \tag{10.1.28}$$

where

$$\mathbf{v} = (v_1, \cdots, v_n)' \sim \mathcal{N}_n\left\{\mathbf{0}, \sigma_v^2 \text{Diag}[(1-\rho^2)^{-1}, 1, \cdots, 1]\right\}. \tag{10.1.29}$$

The **conditional** likelihood function of (θ, β) for fixed Y_1 is given by

$$-(n-1)\log(2\pi) - (n-1)\log\sigma_v^2 - \sigma_v^{-2}\sum_{t=2}^{n}[(Y_t - \rho Y_{t-1}) - (1-\rho)\theta - \beta(x_t - \rho x_{t-1})]^2. \tag{10.1.30}$$

We can use the likelihood function or the LSE method to obtain the estimator of ρ as

$$\text{(i)} \quad \tilde{\rho}_n = \frac{\sum_{t=2}^n (Y_t - \tilde{\theta}_n - \tilde{\beta}_n x_t)(Y_{t-1} - \tilde{\theta}_n - \tilde{\beta}_n x_{t-1})}{\sum_{t=2}^n (Y_{t-1} - \tilde{\theta}_n - \tilde{\beta}_n x_{t-1})^2}, \tag{10.1.31}$$

$$\text{(ii)} \quad (1 - \tilde{\rho}_n)\tilde{\theta}_n = (\bar{Y}_{(0)} - \tilde{\rho}_n\bar{Y}_{(-1)}) - \tilde{\beta}_n(\bar{x}_{(0)} - \tilde{\rho}_n\bar{x}_{(-1)}), \quad \text{and}$$

$$\text{(iii)} \quad \tilde{\beta}_n = \frac{\sum_{t=2}^n (x_t - \tilde{\rho}_n x_{t-1})\{(Y_t - \tilde{\rho}_n Y_{t-1}) - (1-\tilde{\rho}_n)\tilde{\theta}_n - \tilde{\beta}_n(x_t - \tilde{\rho}_n x_{t-1})\}}{\sum_{t=2}^n (x_t - \tilde{\rho}_n x_{t-1})^2},$$

where $\bar{Y}_{(-1)} = \frac{1}{n-1}(Y_1 + \cdots + Y_{n-1})$ and $\bar{Y}_{(0)} = \frac{1}{n-1}(Y_2 + \cdots + Y_n)$. Similarly, $\bar{x}_{(-1)} = \frac{1}{n-1}(x_1 + \cdots + x_{n-1})$ and $\bar{x}_{(0)} = \frac{1}{n-1}(x_2 + \cdots + x_{n-1})$.

In order to obtain the stable estimate $\tilde{\rho}_n$ of ρ, we follow the iterative procedure based on the initial estimates of θ and β given in (10.1.6) to obtain initial $\tilde{\rho}_n^{(1)}$, which helps to obtain the second stage estimates of θ and β, namely $\tilde{\theta}_n^{(1)}(\tilde{\rho}_n)$ and $\tilde{\beta}_n^{(1)}(\tilde{\rho}_n)$ based on (10.1.30). The method is repeated until stable estimates are obtained. The method produces consistent and efficient estimators of ρ, θ and β. Thus, the final estimates of θ and β can be written as

$$\begin{pmatrix} \tilde{\theta}_n(\tilde{\rho}_n) \\ \tilde{\beta}_n(\tilde{\rho}_n) \end{pmatrix} = \frac{1}{K_1(\tilde{\rho}_n)K_2(\tilde{\rho}_n) - K_3^2(\tilde{\rho}_n)} \begin{pmatrix} K_2(\tilde{\rho}_n)(\mathbf{x}'\boldsymbol{\Sigma}_{\tilde{\rho}_n}^{-1}\mathbf{Y}_n) - K_3(\tilde{\rho}_n)(\mathbf{1}_n'\boldsymbol{\Sigma}_{\tilde{\rho}_n}^{-1}\mathbf{Y}_n) \\ K_1(\tilde{\rho}_n)(\mathbf{x}'\boldsymbol{\Sigma}_{\tilde{\rho}_n}^{-1}\mathbf{Y}_n) - K_3(\tilde{\rho}_n)(\mathbf{1}_n'\boldsymbol{\Sigma}_{\tilde{\rho}_n}^{-1}\mathbf{Y}_n) \end{pmatrix}, \quad (10.1.32)$$

where $K_i(\tilde{\rho}_n), i = 1, 2, 3$, are functions of $\tilde{\rho}_n$ and the consistent estimate of σ_v^2 is given by

$$\tilde{\sigma}_v^2(\tilde{\rho}_n) = \frac{1}{m-1}(\mathbf{Y}_n - \tilde{\theta}_n(\tilde{\rho}_n)\mathbf{1}_n - \tilde{\beta}_n(\tilde{\rho}_n)\mathbf{x})'\boldsymbol{\Sigma}_{\tilde{\rho}_n}^{-1}(\mathbf{Y}_n - \tilde{\theta}_n(\tilde{\rho}_n)\mathbf{1}_n - \tilde{\beta}_n(\tilde{\rho}_n)\mathbf{x}). \quad (10.1.33)$$

Consequently, we can show that as $n \to \infty$,

$$\begin{pmatrix} \sqrt{n}(\tilde{\theta}_n(\tilde{\rho}_n) - \theta) \\ \sqrt{n}(\tilde{\beta}_n(\tilde{\rho}_n) - \beta) \\ \sqrt{n}(\tilde{\rho}_n - \rho) \end{pmatrix} \overset{D}{\simeq} \mathcal{N}_3\left\{\begin{pmatrix} 0 \\ 0 \\ 0 \end{pmatrix}; \frac{\sigma_v^2}{K_1K_2 - K_3^2}\begin{pmatrix} K_2 & -K_3 & 0 \\ -K_3 & K_1 & 0 \\ 0 & 0 & (1-\rho^2)(K_1K_2 - K_3^2) \end{pmatrix}\right\}. \quad (10.1.34)$$

There are two other methods of obtaining consistent estimators of ρ, namely (the Cochrane-Orcutt (1949) two stage procedure and the Prais-Winsten (1954) two-stage procedure. They are described below.

Cochrane-Orcutt Two-Stage Procedure. Here one begins with the first stage regression estimators of ρ given by

$$\hat{\rho}_n^{(1)} = \frac{\sum_{t=2}^n \hat{\varepsilon}_t\hat{\varepsilon}_{t-1}}{\sum_{t=1}^n \hat{\varepsilon}_t^2}, \quad (10.1.35)$$

where $\hat{\varepsilon}_t = Y_t - \tilde{\theta}_n - \tilde{\beta}_n x_t$ and $(\tilde{\theta}_n, \tilde{\beta}_n)'$ are the estimates based on (10.1.6). In the second stage, new variables are defined by

$$\mathbf{Y}^* = \boldsymbol{\Gamma}_{CO}\mathbf{Y} = \boldsymbol{\Gamma}_{CO}(\mathbf{1}_n, \mathbf{x})\begin{pmatrix} \theta \\ \beta \end{pmatrix} + \boldsymbol{\Gamma}_{CO}\,\boldsymbol{\varepsilon}, \quad (10.1.36)$$

where

$$\mathbf{\Sigma}_{CO} = \begin{pmatrix} -\rho & 1 & 0 & 0 & \cdots & 0 \\ 0 & -\rho & 1 & 0 & \cdots & 0 \\ \vdots & \vdots & \vdots & \vdots & & \\ 0 & 0 & \cdots & 0 & -\rho & 1 \end{pmatrix}_{(n-1)\times n} . \tag{10.1.37}$$

is a $(n-1) \times n$ matrix to be used as the basis of second-round residuals $Y_t - \hat{\beta}_n(\tilde{\rho}_n)x_t$ to obtain $\hat{\rho}_n^{(2)}$ using (10.1.31). The process is repeated until stable values $\tilde{\rho}_n$ is obtained and $\tilde{\rho}_n$ has the same asymptotic properties as the MLE as studied by Kmenta (1972).

Prais-Winsten (1954) Two-Stage Procedure. Durbin's method and Cochrane-Orcutt's method lose one observation that can have a significant effect in evaluating the estimates. For this reason, Prais and Winsten (1954) proposed $n \times n$ matrix $\mathbf{\Gamma}_{PW}$ defined by

$$\mathbf{\Gamma}_{PW} = \begin{pmatrix} \sqrt{1-\rho^2} & 0 & 0 & 0 & \cdots & 0 \\ -\rho & 1 & 0 & 0 & \cdots & 0 \\ \vdots & \vdots & \vdots & \vdots & & \\ 0 & 0 & \cdots & 0 & -\rho & 1 \end{pmatrix}_{n\times n} . \tag{10.1.38}$$

By way of a transformation, we obtain the model

$$\mathbf{Y}^* = \mathbf{\Gamma}_{PW}\mathbf{Y} = \mathbf{\Gamma}_{PW}(\mathbf{1}_n, \mathbf{x})\begin{pmatrix} \theta \\ \beta \end{pmatrix} + \mathbf{\Gamma}_{PW}\varepsilon, \tag{10.1.39}$$

and by the procedure of Cochrane-Orcutt, we obtain a consistent estimate of ρ. Note that $\tilde{\rho}_n \xrightarrow{P} \rho$ and $\mathbf{\Sigma}_{\tilde{\rho}_n} \xrightarrow{P} \mathbf{\Sigma}_\rho$ as $n \to \infty$. Hence, as $n \to \infty$,

$$\begin{pmatrix} \sqrt{n}(\tilde{\theta}_n(\tilde{\rho}_n) - \theta) \\ \sqrt{n}(\tilde{\beta}_n(\tilde{\rho}_n) - \beta) \end{pmatrix} \xrightarrow{\mathcal{D}} \mathcal{N}_2\left\{\begin{pmatrix} 0 \\ 0 \end{pmatrix}, \frac{\sigma_v^2}{K_1K_2 - K_3^2}\begin{pmatrix} K_2 & -K_3 \\ -K_3 & K_1 \end{pmatrix}\right\}. \tag{10.1.40}$$

Now, in order to test the hypothesis $H_0 : \beta = 0$, we can use the statistic

$$\mathcal{L}_n = \frac{[\tilde{\beta}_n(\tilde{\rho}_n)]^2\{K_1(\tilde{\rho}_n)K_2(\tilde{\rho}_n) - K_3^2(\tilde{\rho}_n)\}}{[\tilde{\sigma}_v(\tilde{\rho}_n)]^2K_1(\tilde{\rho}_n)} \xrightarrow{\mathcal{D}} \chi_1^2 \text{ as } n \to \infty \tag{10.1.41}$$

(under H_0). From the results, we can define the PTE and S-estimators of β as

$$\hat{\beta}_n^{PT}(\tilde{\rho}_n) = \tilde{\beta}_n(\tilde{\rho}_n) - \tilde{\beta}_n(\tilde{\rho}_n)I(\mathcal{L}_n < \chi_1^2(\alpha));$$

$$\hat{\beta}_n^{S}(\tilde{\rho}_n) = \left\{1 - \frac{c\tilde{\sigma}_n(\tilde{\rho}_n)K_1^{1/2}(\tilde{\rho}_n)}{[K_1(\tilde{\rho}_n)K_2(\tilde{\rho}_n) - K_3^2(\tilde{\rho}_n)]^{1/2}|\tilde{\beta}_n(\tilde{\rho}_n)|}\right\}\tilde{\beta}_n(\tilde{\rho}_n), \tag{10.1.42}$$

and that of θ as

$$\hat{\theta}_n^{PT}(\tilde{\rho}_n) = \tilde{\theta}_n(\tilde{\rho}_n) + \tilde{\beta}_n(\tilde{\rho}_n)\frac{K_3(\tilde{\rho}_n)}{K_1(\tilde{\rho}_n)} I(\mathcal{L}_n < \chi_1^2(\alpha))$$

$$\hat{\theta}_n^S(\tilde{\rho}_n) = \tilde{\theta}_n(\tilde{\rho}_n) + \frac{cK_3(\tilde{\rho}_n)\tilde{\sigma}_n(\tilde{\rho}_n)}{K_1^{1/2}(\tilde{\rho}_n)[K_1(\tilde{\rho}_n)K_2(\tilde{\rho}_n) - K_3^2(\tilde{\rho}_n)]^{1/2}} \frac{\tilde{\beta}_n(\tilde{\rho}_n)}{|\tilde{\beta}_n(\tilde{\rho}_n)|}. \tag{10.1.43}$$

The restricted estimator of β is 0 and that of θ is $\tilde{\theta}_n(\tilde{\rho}_n) + \frac{K_3(\tilde{\rho}_n)}{K_1(\tilde{\rho}_n)}\tilde{\beta}_n(\tilde{\rho}_n)$.

Now we may find the asymptotic expressions for the ADB and ADMSE of the estimators of β are given by the following:

(i) Unrestricted estimator of β

$$b_1(\tilde{\beta}_n(\tilde{\rho}_n)) \simeq 0 \quad \text{and} \quad M_1(\tilde{\beta}_n(\tilde{\rho}_n)) \simeq \frac{\sigma_v^2 K_1}{K_1K_2 - K_3^2}, \tag{10.1.44}$$

(ii) PTE of β

$$b_2(\hat{\beta}_n^{PT}(\tilde{\rho}_n)) \simeq \beta H_3(\chi_1^2(\alpha); \Delta^2), \ \Delta^2 = \frac{\delta^2[K_1K_2 - K_3^2]}{\sigma_v^2 K_1}$$

and

$$M_2(\hat{\beta}_n^{PT}(\tilde{\rho}_n)) \simeq \frac{\sigma_v^2 K_1}{K_1K_2 - K_3^2} - \frac{\sigma_v^2 K_1}{K_1K_2 - K_3^2} H_3(\chi_1^2(\alpha); \Delta^2)$$
$$+ \frac{\sigma_v^2 K_1}{K_1K_2 - K_3^2}\Delta^2 \left\{2H_3(\chi_1^2(\alpha); \Delta^2) - H_5(\chi_1^2(\alpha); \Delta^2)\right\}, \tag{10.1.45}$$

(iii) S-estimator of β

$$b_3(\hat{\beta}_n^S(\rho)) \simeq -\frac{cE(\tilde{\sigma}_v)K_1}{K_1K_2 - K_3^2}\{1 - 2\Phi(-\Delta)\}; \quad \text{and}$$

$$M_3(\hat{\beta}_n^S(\rho)) \simeq \frac{\sigma_v^2 K_1}{K_1K_2 - K_3^2}\left\{1 - \frac{2}{\pi}(e^{-\Delta^2/2} - 1)\right\}. \tag{10.1.46}$$

Similarly, the ADB and ADMSE expressions for the estimators of θ are given by

$$b_1(\tilde{\theta}_n(\tilde{\rho}_n)) \simeq 0 \quad \text{and} \quad M_1(\tilde{\theta}_n(\rho)) \simeq \frac{\sigma_v^2 K_2}{K_1 K_2 - K_3^2}, \tag{10.1.47}$$

$$b_2(\hat{\theta}_n(\tilde{\rho}_n)) \simeq \beta \frac{K_3}{K_1} \quad \text{and} \quad M_2(\hat{\theta}_n(\tilde{\rho}_n)) \simeq \frac{\sigma_v^2}{K_1} + \beta_1^2 \frac{K_3^2}{K_1} \tag{10.1.48}$$

$$b_3(\hat{\theta}_n^{PT}(\tilde{\rho}_n)) \simeq \beta \frac{K_3}{K_1} H_3(\chi_1^2(\alpha); \Delta^2) \quad \text{and}$$

$$M_3(\hat{\theta}_n^{PT}(\tilde{\rho}_n)) \simeq \frac{\sigma_v^2 K_2}{K_1 K_2 - K_3^2} - \frac{\sigma_v^2 K_3}{K_1(K_1 K_2 - K_3^2)} H_3(\chi_1^2(\alpha); \Delta^2)$$
$$+ \frac{\sigma_v^2 K_3^2}{K_1(K_1 K_2 - K_3^2)} \Delta^2 \Big\{ 2H_3(\chi_1^2(\alpha); \Delta^2) - H_5(\chi_1^2(\alpha); \Delta^2) \Big\},$$

$$b_4(\hat{\theta}_n^S(\rho)) \simeq \frac{c\sigma_v K_3}{K_1^{1/2}(K_1 K_2 - K_3^2)} \delta \Big\{ 1 - 2\Phi(-\Delta) \Big\} \quad \text{and}$$

$$M_4(\hat{\theta}_n^S(\rho)) \simeq \sigma_v^2 \left\{ \frac{K_2}{K_1 K_2 - K_3^2} - \frac{2}{\pi} \frac{K_3^2}{K_1(K_1 K_2 - K_3^2)} \left(2e^{-\Delta^2/2} - 1 \right) \right\}, \tag{10.1.49}$$

respectively.

Based on these results, the conclusion of Chapter 3 follows on the properties of the estimators of β and θ.

10.2 Multiple Regression Model with Autocorrelation

Consider the multiple regression model of Chapter 7 with autocorrelation ρ given by

$$\mathbf{Y} = \mathbf{X}\boldsymbol{\beta} + \boldsymbol{\varepsilon}, \tag{10.2.1}$$

where $\mathbf{Y} = (Y_1, \cdots, Y_n)'$, $\mathbf{X}$ is a $n \times p$ design matrix, and $\boldsymbol{\varepsilon} = (\epsilon_1, \cdots, \epsilon_n)'$ satisfies

$$\epsilon_i = \rho\epsilon_{i-1} + v_i, \quad i = 1, \cdots, n, \quad \text{and} \quad |\rho| < 1. \tag{10.2.2}$$

Thus, our problem is to estimate $\boldsymbol{\beta}$ when it is suspected that $\boldsymbol{\beta}$ belongs to the subspace defined by $\mathbf{H}\boldsymbol{\beta} = \mathbf{h}$ as in Chapter 7, and also to test the hypothesis $H_0 : \mathbf{H}\boldsymbol{\beta} = \mathbf{h}$ against $\mathbf{H}\boldsymbol{\beta} \neq \mathbf{h}$.

10.2.1 Estimation of β and Test of Hypothesis of $\mathbf{H}\beta = \mathbf{h}$

Assuming ρ to be known and using the **generalized least squares principle**, we minimize

$$(\mathbf{Y} - \mathbf{X}\beta)'\Sigma_\rho^{-1}(\mathbf{Y} - \mathbf{X}\beta) \tag{10.2.3}$$

w.r.t β to obtain the **unrestricted** estimator $\tilde{\beta}_n(\rho)$ as

$$\tilde{\beta}_n(\rho) = (\mathbf{X}'\Sigma_\rho^{-1}\mathbf{X})^{-1}(\mathbf{X}'\Sigma_\rho^{-1}\mathbf{Y}_n) \tag{10.2.4}$$

with $\mathrm{Cov}(\tilde{\beta}_n(\rho)) = \sigma_v^2(\mathbf{X}'\Sigma_\rho^{-1}\mathbf{X})^{-1}$.

The **restricted** estimator of β subject to $\mathbf{H}\beta = \mathbf{h}$ is then given by

$$\hat{\beta}_n(\rho) = \tilde{\beta}_n(\rho) - \mathbf{C}_\rho^{-1}\mathbf{H}'(\mathbf{H}\mathbf{C}_\rho^{-1}\mathbf{H}')^{-1}(\mathbf{H}\tilde{\beta}_n(\rho) - \mathbf{h}), \tag{10.2.5}$$

where $\mathbf{C}_\rho = (\mathbf{X}'\Sigma_\rho^{-1}\mathbf{X})$, and the covariance matrix of $\hat{\beta}_n(\rho)$ is given by

$$\mathrm{Cov}(\hat{\beta}_n(\rho)) = \sigma_v^2[\mathbf{C}_\rho^{-1} - \mathbf{C}_\rho^{-1}\mathbf{H}(\mathbf{H}\mathbf{C}_\rho^{-1}\mathbf{H}')^{-1}\mathbf{H}\mathbf{C}_\rho^{-1}]. \tag{10.2.6}$$

The estimator of σ_v^2 can be written as

$$\tilde{\sigma}_v^2 = \frac{1}{m}[\mathbf{Y}_n - \mathbf{X}\tilde{\beta}_n(\rho)]'\Sigma_\rho^{-1}[\mathbf{Y}_n - \mathbf{X}\tilde{\beta}_n(\rho)], \tag{10.2.7}$$

where $m = n - p$. Hence for the test of the hypothesis $H_0 : \mathbf{H}\beta = \mathbf{h}$ against $H_A : \mathbf{H}\beta \neq \mathbf{h}$, we can use the test-statistic

$$\mathcal{L}_n(\rho) = \frac{(\mathbf{H}\tilde{\beta}_n(\rho) - \mathbf{h})'(\mathbf{H}\mathbf{C}_\rho^{-1}\mathbf{H}')^{-1}(\mathbf{H}\tilde{\beta}_n(\rho) - \mathbf{h})}{q\tilde{\sigma}_v^2}. \tag{10.2.8}$$

Clearly, $\mathcal{L}_n(\rho) \stackrel{\mathcal{D}}{=} F_{q,m}(\Delta^2)$, where

$$\Delta^2 = \frac{1}{\sigma^2}(\mathbf{H}\beta - \mathbf{h})'(\mathbf{H}\mathbf{C}_\rho^{-1}\mathbf{H}')^{-1}(\mathbf{H}\beta - \mathbf{h}). \tag{10.2.9}$$

If ρ is unknown, we estimate ρ consistently by $\tilde{\rho}_n$ by the procedure of Section 10.1.3.

10.2.2 Preliminary Test, James-Stein and Positive-Rule Stein-Type Estimators of β

Following Chapter 7, Section 7.1.2, we obtain the PTE, JSE, and PRSE of β as follows when ρ is known:

(i) $\hat{\beta}_n^{PT}(\rho) = \tilde{\beta}_n(\rho) - (\tilde{\beta}_n(\rho) - \hat{\beta}_n(\rho))I(\mathcal{L}_n(\rho) < F_{q,m}(\alpha)),$ (10.2.10)

(ii) $\hat{\beta}_n^{S}(\rho) = \tilde{\beta}_n(\rho) - d(\tilde{\beta}_n(\rho) - \hat{\beta}_n(\rho))\mathcal{L}_n^{-1}(\rho), d = \frac{(q-2)m}{q(m+2)},$

(iii) $\hat{\boldsymbol{\beta}}_n^{S+}(\rho) = \hat{\boldsymbol{\beta}}_n(\rho) + (\tilde{\boldsymbol{\beta}}_n(\rho) - \hat{\boldsymbol{\beta}}_n(\rho))[1 - d\mathcal{L}_n^{-1}(\rho)]I(\mathcal{L}_n(\rho) > d)$,

respectively, when ρ is known. When ρ is not known, we define the estimators of $\boldsymbol{\beta}$ replacing ρ by $\tilde{\rho}_n$ in the expressions (10.2.10) and obtain

$$\text{(i) } \hat{\boldsymbol{\beta}}_n(\tilde{\rho}_n) = \tilde{\boldsymbol{\beta}}_n(\tilde{\rho}_n) - \mathbf{C}_{\tilde{\rho}_n}^{-1}\mathbf{H}'(\mathbf{H}\mathbf{C}_{\tilde{\rho}_n}^{-1}\mathbf{H}')^{-1}(\mathbf{H}\tilde{\boldsymbol{\beta}}_n(\tilde{\rho}_n) - \mathbf{h}), \tag{10.2.11}$$

(ii) $\hat{\boldsymbol{\beta}}_n^{PT}(\tilde{\rho}_n) = \tilde{\boldsymbol{\beta}}_n(\tilde{\rho}_n) - (\tilde{\boldsymbol{\beta}}_n(\tilde{\rho}_n) - \hat{\boldsymbol{\beta}}_n(\tilde{\rho}_n))I(\mathcal{L}_n^*(\tilde{\rho}_n) < \chi_q^2(\alpha))$,

(iii) $\hat{\boldsymbol{\beta}}_n^{S}(\tilde{\rho}_n) = \tilde{\boldsymbol{\beta}}_n(\tilde{\rho}_n) - (\tilde{\boldsymbol{\beta}}_n(\tilde{\rho}_n) - \hat{\boldsymbol{\beta}}_n(\tilde{\rho}_n))(q-2)\mathcal{L}_n^{*-1}(\tilde{\rho}_n)$,

(iv) $\hat{\boldsymbol{\beta}}_n^{S+}(\tilde{\rho}_n) = \hat{\boldsymbol{\beta}}_n(\tilde{\rho}_n) + (\tilde{\boldsymbol{\beta}}_n(\tilde{\rho}_n) - \hat{\boldsymbol{\beta}}_n(\tilde{\rho}_n))[1-(q-2)\mathcal{L}_n^{*-1}(\tilde{\rho}_n)]I(\mathcal{L}_n^*(\tilde{\rho}_n) > (q-2))$,

substituting $\tilde{\rho}_n$ in place of ρ, where

$$\mathcal{L}_n^*(\tilde{\rho}_n) = \frac{(\mathbf{H}\boldsymbol{\beta}_n(\tilde{\rho}_n) - \mathbf{h})'(\mathbf{H}\mathbf{C}_{\tilde{\rho}_n}^{-1}\mathbf{H}')^{-1}(\mathbf{H}\boldsymbol{\beta}_n(\tilde{\rho}_n) - \mathbf{h})}{\tilde{\sigma}_v^2(\tilde{\rho}_n)}. \tag{10.2.12}$$

Then $\tilde{\sigma}_v^2$ is defined by

$$\tilde{\sigma}_v^2(\tilde{\rho}_n) = \frac{1}{m-1}[\mathbf{Y} - \mathbf{X}\tilde{\boldsymbol{\beta}}_n(\tilde{\rho}_n)]'\hat{\boldsymbol{\Sigma}}_{\tilde{\rho}_n}^{-1}[\mathbf{Y} - \mathbf{X}\tilde{\boldsymbol{\beta}}_n(\tilde{\rho}_n)]. \tag{10.2.13}$$

Naturally, these estimators will have more errors than the previous ones given in (10.2.10) above.

10.3 Bias, MSE Matrices, and the Risk of Estimators When ρ Is Known

When ρ is known, the generalized least squares estimator of $\boldsymbol{\beta}$ is $\tilde{\boldsymbol{\beta}}_n(\rho)$ given by (10.1.5), and $(\tilde{\boldsymbol{\beta}}_n(\rho) - \boldsymbol{\beta})$ is distributed normally with mean $\mathbf{0}$ and covariance matrix $\sigma_v^2\mathbf{C}_\rho^{-1}$.
Hence, the bias and quadratic bias of $\tilde{\boldsymbol{\beta}}_n(\rho)$ are given by

$$b_1(\tilde{\boldsymbol{\beta}}_n(\rho)) = \mathbf{0} \quad \text{and} \quad B_1(\tilde{\boldsymbol{\beta}}_n(\rho)) = 0, \tag{10.3.1}$$

respectively. Similarly, we find the MSE matrix given by

$$M_1(\tilde{\boldsymbol{\beta}}_n(\rho)) = \sigma^2(\mathbf{X}'\boldsymbol{\Sigma}_\rho^{-1}\mathbf{X})^{-1} = \sigma^2\mathbf{C}_\rho^{-1}, \tag{10.3.2}$$

and the quadratic risk under the loss function

$$L(\boldsymbol{\beta}_n^*, \boldsymbol{\beta}) = (\boldsymbol{\beta}_n^* - \boldsymbol{\beta})'\mathbf{W}(\boldsymbol{\beta}_n^* - \boldsymbol{\beta}) \tag{10.3.3}$$

is given by

$$R_1(\tilde{\boldsymbol{\beta}}_n(\rho) : \mathbf{W}) = \sigma_v^2 \operatorname{tr}[\mathbf{W}\mathbf{C}_\rho^{-1}]. \tag{10.3.4}$$

We know that the usual LSE of $\boldsymbol{\beta}$ ignoring ρ (i.e., assuming $\rho = 0$) is $\tilde{\boldsymbol{\beta}}_n = (\mathbf{X}'\mathbf{X})^{-1}\mathbf{X}'\mathbf{Y}$, which is **unbiased**, having a positive-definite symmetric MSE matrix $(\sigma_v^2(\mathbf{C}^{-1}\mathbf{C}_\rho\mathbf{C}^{-1})$. Clearly, the MSE matrix difference of $(\mathbf{X}'\mathbf{X})^{-1}\mathbf{X}'\mathbf{Y}$ and of $\tilde{\boldsymbol{\beta}}_n(\rho)$ is given by

$$\sigma_v^2[\mathbf{C}^{-1}\mathbf{C}_\rho\mathbf{C}^{-1} - \mathbf{C}_\rho^{-1}] \tag{10.3.5}$$

and is a positive-semidefinite matrix showing that $\tilde{\boldsymbol{\beta}}_n(\rho)$ is better than $\tilde{\boldsymbol{\beta}}_n = (\mathbf{X}'\mathbf{X})^{-1}\mathbf{X}'\mathbf{Y}$ whenever $\rho \neq 0$. The bias of the restricted estimator, $\hat{\boldsymbol{\beta}}_n(\rho)$ can be shown to be

$$\mathbf{b}_2(\hat{\boldsymbol{\beta}}_n(\rho)) = -\mathbf{C}_\rho^{-1}\mathbf{H}'(\mathbf{H}\mathbf{C}_\rho^{-1}\mathbf{H}')(\mathbf{H}\boldsymbol{\beta} - \mathbf{h}) = -\boldsymbol{\delta}_\rho \tag{10.3.6}$$

and, say,

$$B_2(\hat{\boldsymbol{\beta}}_n(\rho)) = \sigma_v^{-2}\boldsymbol{\delta}_p'\mathbf{C}_\rho\boldsymbol{\delta}_p = \Delta_\rho^2. \tag{10.3.7}$$

The MSE matrix and the risk expressions of $\hat{\boldsymbol{\beta}}_n(\rho)$ under (10.3.3) are given by

$$\mathbf{M}_2(\hat{\boldsymbol{\beta}}_n(\rho)) = \sigma^2\mathbf{C}_\rho^{-1} + \boldsymbol{\delta}_\rho\boldsymbol{\delta}_\rho' \tag{10.3.8}$$

and

$$R_2(\hat{\boldsymbol{\beta}}_n(\rho) : \mathbf{W}) = \sigma^2 \operatorname{tr}[\mathbf{W}\mathbf{C}_\rho^{-1}] + \boldsymbol{\delta}_\rho'\mathbf{W}\boldsymbol{\delta}_\rho, \tag{10.3.9}$$

respectively.

Similarly, we find the bias, MSE matrices, and the risk of $\hat{\boldsymbol{\beta}}_n^{PT}(\rho), \hat{\boldsymbol{\beta}}_n^{S}(\rho)$ and $\hat{\boldsymbol{\beta}}_n^{S+}(\rho)$ as follows: First, we consider $\hat{\boldsymbol{\beta}}_n^{PT}(\rho)$, and we obtain

$$\mathbf{b}_3(\hat{\boldsymbol{\beta}}_n^{PT}(\rho)) = -\boldsymbol{\delta}_\rho G_{q+2,m}(\ell_\alpha : \Delta_\rho^2), \quad \ell_\alpha = \frac{q}{q+2}F_{q,m}(\alpha), \tag{10.3.10}$$

$$B_3(\hat{\boldsymbol{\beta}}_n^{PT}(\rho)) = \Delta_\rho^2\{G_{q+2,m}(\ell_\alpha : \Delta_\rho^2)\}^2,$$

$$\begin{aligned}\mathbf{M}_3(\hat{\boldsymbol{\beta}}_n^{PT}(\rho)) &= \sigma_v^2\mathbf{C}_\rho^{-1} - \sigma_v^2\mathbf{C}_\rho^{-1}\mathbf{H}'(\mathbf{H}\mathbf{C}_\rho^{-1}\mathbf{H}')\mathbf{H}\mathbf{C}_\rho^{-1}G_{q+2,m}(\ell_\alpha : \Delta_\rho^2),\\ &\quad +\boldsymbol{\delta}_\rho\boldsymbol{\delta}_\rho'\{2G_{q+2,m}(\ell_\alpha : \Delta_\rho^2) - G_{q+4,m}(\ell_\alpha^* : \Delta_\rho^2)\},\end{aligned}$$

$$\begin{aligned}R_3(\hat{\boldsymbol{\beta}}_n^{PT}(\rho); \mathbf{W}) &= \sigma_v^2 \operatorname{tr}[\mathbf{W}\mathbf{C}_\rho^{-1}] - \sigma_v^2 \operatorname{tr}[\mathbf{W}\mathbf{C}_\rho^{-1}\mathbf{H}'(\mathbf{H}\mathbf{C}_\rho^{-1}\mathbf{H}')\mathbf{H}\mathbf{C}_\rho^{-1}]\\ &\quad \times G_{q+2,m}(\ell_\alpha : \Delta_\rho^2) + \boldsymbol{\delta}_\rho'\mathbf{W}\boldsymbol{\delta}_\rho\{2G_{q+2,m}(\ell_\alpha : \Delta_\rho^2)\\ &\quad - G_{q+4,m}(\ell_\alpha^* : \Delta_\rho^2)\},\end{aligned}$$

with $\ell_\alpha^* = \frac{q}{q+4}F_{q,m}(\alpha)$.

Next, we consider the Stein-type estimator, $\hat{\boldsymbol{\beta}}_n^{S}(\rho)$ and obtain the bias, quadratic bias, MSE matrices, and risk expressions as given below:

$$\mathbf{b}_4(\hat{\boldsymbol{\beta}}_n^S(\rho)) = -dq\boldsymbol{\delta}_\rho E[\chi_{q+2}^{-2}(\Delta_\rho^2)], \tag{10.3.11}$$

$$B_4(\hat{\boldsymbol{\beta}}_n^S(\rho)) = d^2q^2\Delta_\rho^2\{E[\chi_{q+2}^{-2}(\Delta_\rho^2)]\}^2,$$

$$\begin{aligned}\mathbf{M}_4(\hat{\boldsymbol{\beta}}_n^S(\rho)) &= \sigma_v^2\mathbf{C}_\rho^{-1} - \sigma_v^2\mathbf{C}_\rho^{-1}\mathbf{H}'(\mathbf{H}\mathbf{C}_\rho^{-1}\mathbf{H}')\mathbf{H}\mathbf{C}_\rho^{-1}\{(q-2)E[\chi_{q+2}^{-2}(\Delta_\rho^2)]\\ &\quad +2\Delta_\rho^2E[\chi_{q+4}^{-4}(\Delta_\rho^2)]\} + dq(q+2)\boldsymbol{\delta}_\rho\boldsymbol{\delta}_\rho'E[\chi_{q+4}^{-4}(\Delta_\rho^2)],\end{aligned}$$

$$\begin{aligned}R_4(\hat{\boldsymbol{\beta}}_n^S(\rho);\mathbf{W}) &= \sigma_v^2\operatorname{tr}[\mathbf{W}\mathbf{C}_\rho^{-1}] - \sigma_v^2\operatorname{tr}[\mathbf{W}\mathbf{C}_\rho^{-1}\mathbf{H}'(\mathbf{H}\mathbf{C}_\rho^{-1}\mathbf{H}')\mathbf{H}\mathbf{C}_\rho^{-1}]\\ &\quad\times\{(q-2)E[\chi_{q+2}^{-2}(\Delta_\rho^2)] + 2\Delta_\rho^2E[\chi_{q+4}^{-4}(\Delta_\rho^2)]\}\\ &\quad +dq(q+2)\boldsymbol{\delta}_\rho'\mathbf{W}\boldsymbol{\delta}_\rho E[\chi_{q+4}^{-4}(\Delta_\rho^2)].\end{aligned}$$

Finally, for $\hat{\boldsymbol{\beta}}_n^{S+}(\rho)$, we obtain

$$\begin{aligned}\mathbf{b}_5(\hat{\boldsymbol{\beta}}_n^{S+}(\rho)) &= -\boldsymbol{\delta}_\rho\Big\{d_1E[F_{q+2,m}^{-1}(\Delta_\rho^2)I(F_{q+2,m}(\Delta_\rho^2) < d_1)] - d_1E[F_{q+2,m}^{-1}(\Delta_\rho^2)]\\ &\quad -G_{q+2,m}(\ell_\alpha:\Delta_\rho^2)\Big\},\quad \left(d_1 = \frac{qd}{q+2}\right);\end{aligned} \tag{10.3.12}$$

$$\begin{aligned}B_5(\hat{\boldsymbol{\beta}}_n^{S+}(\rho)) &= \Delta_\rho^2\Big\{d_1E[F_{q+2,m}^{-1}(\Delta_\rho^2)I(F_{q+2,m}(\Delta_\rho^2) < d_1)] - d_1E[F_{q+2,m}^{-1}(\Delta_\rho^2)]\\ &\quad -G_{q+2,m}(\ell_\alpha:\Delta_\rho^2)\Big\}^2,\end{aligned}$$

$$\begin{aligned}\mathbf{M}_5(\hat{\boldsymbol{\beta}}_n^{S+}(\rho)) &= \mathbf{M}_4(\hat{\boldsymbol{\beta}}_n^S(\rho)) - \sigma_v^2(\mathbf{C}_\rho^{-1}\mathbf{H}'(\mathbf{H}\mathbf{C}_\rho^{-1}\mathbf{H}')^{-1}\mathbf{H}\mathbf{C}_\rho^{-1})\\ &\quad\times E[(1-d_1F_{q+2,m}^{-1})^2I(F_{q+2,m}(\Delta_\rho^2) < d_1)]\\ &\quad +\boldsymbol{\delta}_\rho\boldsymbol{\delta}_\rho'\Big\{2E[(1-d_1F_{q+2,m}^{-1})I(F_{q+2,m}(\Delta_\rho^2) < d_1)]\}\\ &\quad -E[(1-d_2F_{q+4,m}^{-1})^2I(F_{q+4,m}(\Delta_\rho^2) < d_2)]\Big\}\quad\left(d_2 = \frac{qd}{q+4}\right),\end{aligned}$$

$$\begin{aligned}R_5(\hat{\boldsymbol{\beta}}_n^{S+}(\rho):\mathbf{W}) &= R_4(\hat{\boldsymbol{\beta}}_n^S(\rho);\mathbf{W}) - \sigma_v^2\operatorname{tr}\{\mathbf{W}[\mathbf{C}_\rho^{-1}\mathbf{H}'(\mathbf{H}\mathbf{C}_\rho^{-1}\mathbf{H}')^{-1}\mathbf{H}\mathbf{C}_\rho^{-1}]\}\\ &\quad\times E[(1-d_1F_{q+2,m}^{-1})^2I(F_{q+2,m}(\Delta_\rho^2) < d_1)]\\ &\quad +\boldsymbol{\delta}_\rho'\mathbf{W}\boldsymbol{\delta}_\rho\Big\{2E[(1-d_1F_{q+2,m}^{-1})I(F_{q+2,m}(\Delta_\rho^2) < d_1)]\}\\ &\quad -E[(1-d_2F_{q+4,m}^{-1})I(F_{q+4,m}(\Delta_\rho^2) < d_2)]\Big\}.\end{aligned} \tag{10.3.13}$$

10.4 ADB, ADMSE, and ADQR of the Estimators (ρ Unknown)

If ρ is **unknown**, we replace ρ by a consistent estimator, that is, the estimator $\tilde{\rho}_n$ that converges to ρ in probability ($\tilde{\rho}_n \xrightarrow{\text{P}} \rho$) as $n \to \infty$ to obtain the estimators given by (10.2.11). Then we can write (as $n \to \infty$) under fixed alternatives $H_\delta : \mathbf{H}\boldsymbol{\beta} = \mathbf{h} + \boldsymbol{\delta}$,

$$\sqrt{n}(\boldsymbol{\beta}_n^*(\tilde{\rho}_n) - \boldsymbol{\beta}) = \sqrt{n}(\tilde{\boldsymbol{\beta}}_n(\rho) - \boldsymbol{\beta}) + o_p(1), \tag{10.4.1}$$

where $\boldsymbol{\beta}_n^*(\cdot)$ stands for any one of $\tilde{\boldsymbol{\beta}}_n(\cdot), \hat{\boldsymbol{\beta}}_n(\cdot), \hat{\boldsymbol{\beta}}_n^{PT}(\cdot), \hat{\boldsymbol{\beta}}_n^{S}(\cdot)$ and $\hat{\boldsymbol{\beta}}_n^{S+}(\cdot)$. Under the **local alternatives** $H_A : \mathbf{H}\boldsymbol{\beta} = \mathbf{h} + n^{-1/2}\boldsymbol{\xi}$, $\mathcal{L}_n^*(\tilde{\rho}_n)$ follows the noncentral chi-square distribution with q D.F. and noncentrality parameter $\Delta_\rho^2/2$ as $n \to \infty$, where $\Delta_\rho^2 = \sigma_v^{-2}\boldsymbol{\delta}_\rho'\mathbf{C}_\rho\boldsymbol{\delta}_\rho$ and $\boldsymbol{\delta}_\rho = \mathbf{C}_\rho^{-1}\mathbf{H}'(\mathbf{H}\mathbf{C}_\rho^{-1}\mathbf{H}')^{-1}\boldsymbol{\xi}$.

These results allow us to calculate the ADB, ADMSE, and ADQR of the estimators as follows:

(i) For **ADB** and **ADQB** of $\tilde{\boldsymbol{\beta}}_n(\tilde{\rho}_n), \hat{\boldsymbol{\beta}}_n(\tilde{\rho}_n), \hat{\boldsymbol{\beta}}_n^{PT}(\tilde{\rho}_n), \hat{\boldsymbol{\beta}}_n^{S}(\tilde{\rho}_n)$, and $\hat{\boldsymbol{\beta}}_n^{S+}(\tilde{\rho}_n)$,

$$\mathbf{b}_1(\tilde{\boldsymbol{\beta}}_n(\tilde{\rho}_n)) = \mathbf{0} \quad \text{and} \quad B_1(\tilde{\boldsymbol{\beta}}_n(\tilde{\rho}_n)) = 0$$

$$\mathbf{b}_2(\hat{\boldsymbol{\beta}}_n(\tilde{\rho}_n)) = -\boldsymbol{\delta}_\rho \quad \text{and} \quad B_2(\hat{\boldsymbol{\beta}}_n(\tilde{\rho}_n)) = \Delta_\rho^2$$

$$\mathbf{b}_3(\hat{\boldsymbol{\beta}}_n^{PT}(\tilde{\rho}_n)) = -\boldsymbol{\delta}_\rho H_{q+2}(\chi_q^2(\alpha) : \Delta_\rho^2), \quad \text{and}$$

$$B_3(\hat{\boldsymbol{\beta}}_n^{PT}(\tilde{\rho}_n)) = \Delta_\rho^2\{H_{q+2}(\chi_q^2(\alpha) : \Delta_\rho^2)\}^2$$

$$\mathbf{b}_4(\hat{\boldsymbol{\beta}}_n^{S}(\tilde{\rho}_n)) = -(q-2)\boldsymbol{\delta}_\rho E[\chi_{q+2}^{-2}(\Delta_\rho^2)] \quad \text{and}$$

$$B_4(\hat{\boldsymbol{\beta}}_n^{S}(\tilde{\rho}_n)) = \Delta_\rho^2\{E[\chi_{q+2}^{-2}(\Delta_\rho^2)]\}^2$$

$$\mathbf{b}_5(\hat{\boldsymbol{\beta}}_n^{S+}(\tilde{\rho}_n)) = \mathbf{b}_4(\hat{\boldsymbol{\beta}}_n^{S}(\tilde{\rho}_n)) - \boldsymbol{\delta}_\rho E[(1-(q-2)\chi_{q+2}^{-2}(\Delta_\rho^2))I(\chi_{q+2}^2(\Delta_\rho^2) < q-2)]$$

$$B_5(1 - \hat{\boldsymbol{\beta}}_n^{S+}(\tilde{\rho}_n)) = \Delta_\rho^2\{(q-2)E[\chi_{q+2}^{-1}(\Delta_\rho^2)]E[(1-(q-2)\chi_{q+2}^{-2}(\Delta_\rho^2)) \times I(\chi_{q+2}^2(\Delta_\rho^2) < q-2)]\}^2$$

(ii) For **ADMSE** and **ADQR** of $\tilde{\boldsymbol{\beta}}_n(\tilde{\rho}_n), \hat{\boldsymbol{\beta}}_n(\tilde{\rho}_n), \hat{\boldsymbol{\beta}}_n^{PT}(\tilde{\rho}_n), \hat{\boldsymbol{\beta}}_n^{S}(\tilde{\rho}_n)$, and $\hat{\boldsymbol{\beta}}_n^{S+}(\tilde{\rho}_n)$,

$$\mathbf{M}_1(\tilde{\boldsymbol{\beta}}_n(\tilde{\rho}_n)) = \sigma_v^2\mathbf{C}_\rho^{-1} \quad \text{and} \quad R_1(\tilde{\boldsymbol{\beta}}_n(\tilde{\rho}_n)) = \sigma_v^2 \operatorname{tr}[\mathbf{W}\mathbf{C}_\rho^{-1}]; \tag{10.4.2}$$

$$\mathbf{M}_2(\hat{\boldsymbol{\beta}}_n(\tilde{\rho}_n)) = \sigma_v^2[\mathbf{C}_\rho^{-1} - \mathbf{C}_\rho^{-1}\mathbf{H}'(\mathbf{H}\mathbf{C}_\rho^{-1}\mathbf{H}')^{-1}\mathbf{H}\mathbf{C}_\rho^{-1}] + \boldsymbol{\delta}_\rho\boldsymbol{\delta}_\rho'$$

$$R_2(\hat{\boldsymbol{\beta}}_n(\tilde{\rho}_n) : \mathbf{W}) = \sigma_v^2 \operatorname{tr}[\mathbf{W}\mathbf{C}_\rho^{-1}] - \sigma_v^2 \operatorname{tr}[\mathbf{W}\mathbf{C}_\rho^{-1}\mathbf{H}'(\mathbf{H}\mathbf{C}_\rho^{-1}\mathbf{H}')^{-1}\mathbf{H}\mathbf{C}_\rho^{-1}] + \boldsymbol{\delta}_\rho'\mathbf{W}\boldsymbol{\delta}_\rho; \tag{10.4.3}$$

$$\mathbf{M}_3(\hat{\boldsymbol{\beta}}_n^{PT}(\tilde{\rho}_n)) = \sigma_v^2\mathbf{C}_\rho^{-1} - \sigma_v^2[\mathbf{C}_\rho^{-1}\mathbf{H}'(\mathbf{H}\mathbf{C}_\rho^{-1}\mathbf{H}')^{-1}]\mathbf{H}\mathbf{C}_\rho^{-1}H_{q+2}(\chi_q^2(\alpha);\Delta_\rho^2)]$$
$$+\boldsymbol{\delta}_p\boldsymbol{\delta}_p'[2H_{q+2}(\chi_q^2(\alpha);\Delta_\rho^2) - H_{q+4}(\chi_q^2(\alpha);\Delta_\rho^2)];$$

$$R_3(\hat{\boldsymbol{\beta}}_n^{PT}(\tilde{\rho}_n):\mathbf{W}) = \sigma_v^2\,\mathrm{tr}[\mathbf{W}\mathbf{C}_\rho^{-1}]$$
$$-\sigma_v^2\,\mathrm{tr}[\mathbf{W}\mathbf{C}_\rho^{-1}\mathbf{H}'(\mathbf{H}\mathbf{C}_\rho^{-1}\mathbf{H}')^{-1}\mathbf{H}\mathbf{C}_\rho^{-1}]H_{q+2}(\chi_q^2(\alpha);\Delta_\rho^2)$$
$$+\boldsymbol{\delta}_\rho'\mathbf{W}\boldsymbol{\delta}_\rho[2H_{q+2}(\chi_q^2(\alpha);\Delta_\rho^2) - H_{q+4}(\chi_q^2(\alpha);\Delta_\rho^2)], \quad (10.4.4)$$

$$\mathbf{M}_4(\hat{\boldsymbol{\beta}}_n^{S}(\tilde{\rho}_n)) = \sigma_v^2\mathbf{C}_\rho^{-1} - (q-2)\sigma_v^2[\mathbf{C}_\rho^{-1}\mathbf{H}'(\mathbf{H}\mathbf{C}_\rho^{-1}\mathbf{H}')^{-1}\mathbf{H}\mathbf{C}_\rho^{-1}]$$
$$\times\{2E[\chi_{q+2}^{-2}(\Delta_\rho^2)] - (q-2)E[\chi_{q+2}^{-4}(\Delta_\rho^2)]\}$$
$$+(q^2-4)\boldsymbol{\delta}_\rho\boldsymbol{\delta}_\rho'E[\chi_{q+4}^{-4}(\Delta_\rho^2)]$$

$$R_4(\hat{\boldsymbol{\beta}}_n^{S}(\tilde{\rho}_n):\mathbf{W}) = \sigma_v^2\,\mathrm{tr}[\mathbf{W}\mathbf{C}_\rho^{-1}] - (q-2)\sigma_v^2\,\mathrm{tr}[\mathbf{A}_{10}(\rho)]\{2E[\chi_{q+2}^{-2}(\Delta_\rho^2)]$$
$$-(q-2)E[\chi_{q+2}^{-4}(\Delta_\rho^2)]\}$$
$$+(q^2-4)\boldsymbol{\eta}_1\mathbf{A}_{10}(\rho)\boldsymbol{\eta}_1E[\chi_{q+4}^{-4}(\Delta_\rho^2)], \quad (10.4.5)$$

where $\mathbf{A}_{10}(\rho)$ is the $q \times q$ matrix in the partition of

$$\boldsymbol{\Gamma}\mathbf{C}_\rho^{-1/2}\mathbf{W}\mathbf{C}_\rho^{-1/2}\boldsymbol{\Gamma}' = \begin{pmatrix} \mathbf{A}_{10}(\rho) & \mathbf{A}_{12}(\rho) \\ \mathbf{A}_{12}'(\rho) & \mathbf{A}_{22}(\rho) \end{pmatrix} \text{ and } \boldsymbol{\delta}_\rho'\mathbf{W}\boldsymbol{\delta}_\rho = (\boldsymbol{\eta}_1'\mathbf{A}_{10}(\rho)\boldsymbol{\eta}_1) \quad (10.4.6)$$

similar to Section 7.4.1a except that $\mathbf{C}$ replaced by $\mathbf{C}_\rho$;

$$\mathbf{M}_5(\hat{\boldsymbol{\beta}}_n^{S+}(\tilde{\rho}_n)) = \mathbf{M}_4(\hat{\boldsymbol{\beta}}_n^{S}(\tilde{\rho}_n)) - \sigma_v^2[\mathbf{C}_\rho^{-1}\mathbf{H}'(\mathbf{H}\mathbf{C}_\rho^{-1}\mathbf{H}')^{-1}\mathbf{H}\mathbf{C}_\rho^{-1}]$$
$$\times E[(1-(q-2)\chi_{q+2}^{-2}(\Delta_\rho^2))I(\chi_{q+2}^2(\Delta_\rho^2) < q-2)]$$
$$+\boldsymbol{\delta}_\rho\boldsymbol{\delta}_\rho'\Big\{2E\big[(1-(q-2)\chi_{q+2}^{-2}(\Delta_\rho^2))I(\chi_{q+2}^2(\Delta_\rho^2) < q-2)\big]$$
$$-E[(1-(q-2)\chi_{q+4}^{-2}(\Delta_\rho^2))^2 I(\chi_{q+4}^2(\Delta_\rho^2) < q-2)]\Big\}$$

$$R_5(\hat{\boldsymbol{\beta}}_n^{S+}(\tilde{\rho}_n):\mathbf{W}) = R_4(\hat{\boldsymbol{\beta}}_n^{S}(\tilde{\rho}_n);\mathbf{W})$$
$$-\sigma_v^2\,\mathrm{tr}[\mathbf{A}_{10}(\rho)]E[(1-(q-2)\chi_{q+2}^{-2}(\Delta_\rho^2))I(\chi_{q+2}^2(\Delta_\rho^2)<q-2)]$$
$$+\boldsymbol{\eta}_1'\mathbf{A}_{10}(\rho)\boldsymbol{\eta}_1\Big\{2E[(1-(q-2)\chi_{q+2}^{-2}(\Delta_\rho^2))I(\chi_{q+2}^2(\Delta_\rho^2)<q-2)]$$
$$-E[(1-(q-2)\chi_{q+4}^{-2}(\Delta_\rho^2))^2 I(\chi_{q+4}^2(\Delta_\rho^2) < q-2)]\Big\}. \quad (10.4.7)$$

Based on the results of Sections 10.2 and 10.3, we can compare the five estimators as in Chapter 7, and the analysis reach the same conclusions as in Chapter 7.

10.5 Estimation of Regression Parameters When ρ Is Near Zero

Judge and Bock (1978) considered the properties of the preliminary test estimator of $\boldsymbol{\beta}$ for the model (10.2.1) with autocorrelation ρ, when it is suspected that ρ is zero. We attempt to improve on the procedure in this section. To define the preliminary test estimator of $\boldsymbol{\beta}$, Judge and Bock used the Durbin-Waston test (1950, 1951), $\hat{D}_W$, defined by

$$\hat{D}_W = \sum_{t=2}^{n} \left[\frac{(\hat{\epsilon}_t - \hat{\epsilon}_{t-1})^2}{\sum_{t=1}^{n} \hat{\epsilon}_t^2} \right], \tag{10.5.1}$$

where $\hat{\epsilon}_t$ is the residual error $Y_t - \mathbf{X}\tilde{\boldsymbol{\beta}}_n(\tilde{\rho}_n), t = 1, 2, \cdots, n$ for testing $H_0 : \rho = 0$ against $H_A : \rho > 0$. In this case, the critical region is of the form $\{\hat{D}_W | \hat{D}_W < D_L(\alpha)\}$, and the decision is reached as follows: If $\hat{D}_W < D_L(\alpha)$, H_0 is rejected, and if $\hat{D}_W > D_u(\alpha)$, the hypothesis H_0 is not rejected. If $D_L(\alpha) < \hat{D}_W < D_u(\alpha)$ then H_0 is inconclusive. Durbin and Watson provided the upper and lower critical values $D_u(\alpha)$ and $D_L(\alpha)$, respectively, with various sample sizes.

A second test due to Berenblutt and Webb (1973) is indicated by the $\mathcal{G}_1$-test, defined by

$$\mathcal{G}_1 = \frac{\mathbf{Y}'[\mathbf{B} - \mathbf{B}\mathbf{X}(\mathbf{X}'\mathbf{B}\mathbf{X})^{-1}\mathbf{X}'\mathbf{B}]\mathbf{Y}}{\mathbf{Y}'[\mathbf{I} - \mathbf{X}(\mathbf{X}'\mathbf{X})^{-1}\mathbf{X}']\mathbf{Y}}, \tag{10.5.2}$$

where $\mathbf{B}$ is the matrix

$$\mathbf{B} = \begin{pmatrix} 2 & -1 & 0 & 0 & \cdots & 0 \\ -1 & 2 & -1 & 0 & \cdots & 0 \\ 0 & 0 & 0 & 0 & \cdots & 0 \\ 0 & 0 & \cdots & -1 & 2 & -1 \\ 0 & 0 & \cdots & 0 & -1 & -1 \end{pmatrix}. \tag{10.5.3}$$

The power of the $\mathcal{G}_1$ test is generally higher than the power of the D–W-test for higher values of ρ; otherwise, $\mathcal{G}_1$ test is similar to the D–W-test. So the $\mathcal{G}_1$ test can be reduced to the same canonical form as the D–W-test.

Both tests have been obtained by approximating the likelihood function for ρ in two different ways. (Durbin and Watson, 1950, 1951, and Berenblutt and Webb, 1973, for details; for some related problems, see Kmenta and Gilbert 1968, 1970).

10.5.1 Preliminary Test and Stein-Type Estimators (Chen and Saleh, 1993)

If $\rho = 0$, then the LSE is $\tilde{\boldsymbol{\beta}}_n = (\mathbf{X}'\mathbf{X})^{-1}\mathbf{X}\mathbf{Y}$. The estimator of $\boldsymbol{\beta}$ is given by

$$\tilde{\boldsymbol{\beta}}_n(\tilde{\rho}_n) = (\mathbf{X}'\boldsymbol{\Sigma}_{\tilde{\rho}_n}^{-1}\mathbf{X})^{-1}\mathbf{X}'\boldsymbol{\Sigma}_{\tilde{\rho}_n}^{-1}\mathbf{Y}, \tag{10.5.4}$$

if ρ is unknown.

Let $\mathcal{L}_n$ be a suitable test-statistic for testing $H_0 : \rho = 0$ against $\rho > 0$. Also, let $\mathcal{L}_{n,\alpha}$ be the α-level critical value. Then the PTE of $\boldsymbol{\beta}$ is defined by

$$\hat{\boldsymbol{\beta}}_n^{PT}(\tilde{\rho}_n) = \tilde{\boldsymbol{\beta}}_n(\tilde{\rho}_n) - (\tilde{\boldsymbol{\beta}}_n(\tilde{\rho}_n) - \tilde{\boldsymbol{\beta}}_n)I(\mathcal{L}_n < \mathcal{L}_{n,\alpha}). \tag{10.5.5}$$

By the Durbin-Watson test, we can write the PTE of $\boldsymbol{\beta}$ as

$$\begin{aligned}\hat{\boldsymbol{\beta}}_n^{PT(D)}(\tilde{\rho}_n) = \tilde{\boldsymbol{\beta}}_n(\tilde{\rho}_n)I(-2 < \hat{D}_W - D_u(\alpha) < 0)\\ +\tilde{\boldsymbol{\beta}}_n I(0 < \hat{D}_W - D_L(\alpha) < 2)\end{aligned} \tag{10.5.6}$$

and by the $\mathcal{G}_1$ test,

$$\hat{\boldsymbol{\beta}}_n^{PT(\mathcal{G}_1)}(\tilde{\rho}_n) = \tilde{\boldsymbol{\beta}}_n(\tilde{\rho}_n)I(-2 < \mathcal{G}_1 - D_u(\alpha) < 0) + \tilde{\boldsymbol{\beta}}_n I(0 < \mathcal{G}_1 - D_L(\alpha) < 2). \tag{10.5.7}$$

Now, consider the Stein-type estimators of $\boldsymbol{\beta}$. For $\rho = 0$, the Stein-type estimator of $\boldsymbol{\beta}$ is defined as

$$\hat{\boldsymbol{\beta}}_n^S = \left(1 - \frac{(p-2)m\hat{\sigma}_v^2}{(m+2)\tilde{\boldsymbol{\beta}}_n'\mathbf{X}'\mathbf{X}\tilde{\boldsymbol{\beta}}_n}\right)\tilde{\boldsymbol{\beta}}_n, \quad m = n - p, \tag{10.5.8}$$

with $\hat{\sigma}_v^2 = \frac{1}{m}(\mathbf{Y} - \mathbf{X}\tilde{\boldsymbol{\beta}}_n)'(\mathbf{Y} - \mathbf{X}\tilde{\boldsymbol{\beta}}_n)$. However, for $\rho \neq 0$, we estimate ρ by $\tilde{\rho}_n$ and define the Stein-type estimator of $\boldsymbol{\beta}$ as

$$\hat{\boldsymbol{\beta}}_n^S(\tilde{\rho}_n) = \left(1 - \frac{(p-2)(m-1)\hat{\sigma}_v^2(\tilde{\rho}_n)\hat{\sigma}_v^2}{(m+1)\tilde{\boldsymbol{\beta}}_n'(\tilde{\rho}_n)\mathbf{X}'\boldsymbol{\Sigma}_{\tilde{\rho}_n}^{-1}\mathbf{X}\tilde{\boldsymbol{\beta}}_n(\tilde{\rho}_n)}\right)\tilde{\boldsymbol{\beta}}_n(\tilde{\rho}_n). \tag{10.5.9}$$

Thus, based on the Stein-type estimators, we define the preliminary test estimators of $\boldsymbol{\beta}$ as

$$\hat{\boldsymbol{\beta}}_n^{PT(1)}(\tilde{\rho}_n) = \tilde{\boldsymbol{\beta}}_n^S I(-2 < \hat{D}_W - D_u < 0) + \tilde{\boldsymbol{\beta}}_n^S(\tilde{\rho}_n)I(0 < \hat{D}_W - D_L < 2) \tag{10.5.10}$$

and based on D_W and $\mathcal{G}_1$ tests,

$$\hat{\boldsymbol{\beta}}_n^{PT(2)}(\tilde{\rho}_n) = \tilde{\boldsymbol{\beta}}_n^S I(-2 < \mathcal{G}_1 - D_u < 0) + \tilde{\boldsymbol{\beta}}_n^S(\tilde{\rho}_n)I(0 < \mathcal{G}_1 - D_L < 2). \tag{10.5.11}$$

For the multiple regression model $\mathbf{Y} = \mathbf{X}\boldsymbol{\beta} + \boldsymbol{\varepsilon}$, where $\epsilon_i = \rho\epsilon_{i-1} + v_i, |\rho| < 1, i = 1, 2, \cdots, n$, and $\mathbf{v} \sim \mathcal{N}_n(\mathbf{0}, \sigma_v^2\boldsymbol{\Sigma}_\rho)$, we consider the risk analysis of the four pairs of estimators: (1) two unrestricted estimators, (2) two PTEs, and (3) two Stein-type estimators, and (4) two PTEs based on the Stein-type estimators.

From our simulation study, we learn that the (1) shrinkage estimators performs better than the unrestricted estimators and the (2) the shrinkage PTE performs better than the usual PTEs. Further, the PTEs based on the $\mathcal{G}_1$ test perform better than the PTEs based on the D–W-test. In the next section, we describe the design of the Monte Carlo study.

10.5.2 Design of Monte Carlo Experiment

In the Monte Carlo study, two hundred samples of size 25 were generated using the orthonormal statistical model

$$\mathbf{Y} = \mathbf{X}\boldsymbol{\beta}_0 + \boldsymbol{\varepsilon}, \mathbf{X}'\mathbf{X} = \mathbf{I}, (n = 25), \tag{10.5.12}$$

where $\boldsymbol{\beta}_0 = (13.9, 10.79, 6.13, 3.01, 10.81)'$. The autocorrelated errors were generated by the first-order autocorrelated model $\epsilon_t = \rho\epsilon_{t-1} + v_t$, $|\rho| < 1$. The vector $\mathbf{v} = (v_1, \cdots, v_{25})$ was generated from $\mathcal{N}(0, 60.8)$ and was assumed to be iid. The generated v's are then used to create the ϵ_t's by varying the autocorrelation in steps of tenths from 0.0 to 0.9. We obtained the MLE of the parameters for the 200 samples in order to compute the test-statistics. We only choose ρ to be nonnegative. The tests are based on the 0.01 and 0.05 levels of significance for the D–W-test and $\mathcal{G}_1$ test.

10.5.3 Empirical Results and Conclusions

The simulation study reported by Judge and Bock (1978) gave performances of the PTE of $\boldsymbol{\beta}$ based on both the D–W-test and $\mathcal{G}_1$-test and they are the precursor of the study by Chen and Saleh (1993). In order to carry out the risk analysis of the estimators, we consider the unweighted and weighted loss function given by

$$L_I(\boldsymbol{\beta}^* - \boldsymbol{\beta}) = (\boldsymbol{\beta}^* - \boldsymbol{\beta})'(\boldsymbol{\beta}^* - \boldsymbol{\beta}) \quad \text{and} \quad R(\boldsymbol{\beta}^* : \mathbf{I}) = E[(\boldsymbol{\beta}^* - \boldsymbol{\beta})'(\boldsymbol{\beta}^* - \boldsymbol{\beta})], \tag{10.5.13}$$

$$L_W(\boldsymbol{\beta}^* - \boldsymbol{\beta}) = (\boldsymbol{\beta}^* - \boldsymbol{\beta})'\mathbf{W}(\boldsymbol{\beta}^* - \boldsymbol{\beta}), \quad \text{and}$$
$$R(\boldsymbol{\beta}^* : \mathbf{W}) = \text{tr}\{\mathbf{W}E[(\boldsymbol{\beta}^* - \boldsymbol{\beta})'(\boldsymbol{\beta}^* - \boldsymbol{\beta})]\}, \tag{10.5.14}$$

where we used the weight to be $\sigma_v^{-2}\mathbf{C}_\rho$. Based on these results we can calculate the performances of the estimators. The sampling experiment and the chosen model are as given in (10.3.10). Table 10.4.1 and Table 10.4.2 give the empirical risks based on (10.3.14). The major findings are as follows:

(i) The shrinkage estimators have uniformly lower risks than the unrestricted estimators for all ρ-values. Generalized least squares estimators have smaller risk than the ordinary least squares estimators. The **Durbin's two-stage** estimator has the lowest risk compared to the **Cochrane Orcutt** (CO) and **Prais-Winsten** (PW) estimators, in Table 10.4.1 and Table 10.4.2, respectively. The same results were observed in the studies by Griliches and Rao (1969) and Judge and Bock (1978).

Table 10.4.1 Empirical Risks for Different Estimators prior to Testing

ρ	OLS	GLS	Durbin	CO	PW
0.0	280.56	280.56	280.98	292.91	290.82
0.1	290.94	287.68	294.15	301.25	298.98
0.2	308.07	296.94	304.65	310.81	309.37
0.3	335.57	310.44	320.08	327.67	325.17
0.4	380.18	331.31	344.73	353.14	350.75
0.5	454.95	364.49	385.04	394.43	392.15
0.6	587.68	418.03	448.87	461.60	458.88
0.8	1436.42	643.91	716.10	777.24	768.53
0.9	3208.83	848.30	958.54	1244.43	1225.25

Table 10.4.2 Empirical Risks for Different Estimators prior to Testing Shrinkage Estimates

ρ	OLS	GLS	Durbin	CO	PW
0.0	225.91	225.80	230.67	236.30	234.57
0.1	234.84	233.23	238.86	244.57	242.52
0.2	248.73	242.82	249.18	254.66	252.29
0.3	270.24	254.83	262.27	267.65	265.24
0.4	303.86	269.61	278.98	284.93	282.83
0.5	357.15	290.40	303.12	310.22	308.28
0.6	449.66	322.93	341.52	350.36	348.02
0.7	639.65	375.14	403.59	418.13	414.54
0.8	1010.84	457.53	507.77	548.22	542.14
0.9	2609.99	571.12	652.46	865.44	852.51

In Tables 10.4.3 through 10.4.6, we present the empirical results the two PTEs. One is based on the D–W-test, and the other on the $\mathcal{G}_1$-test at 1% and 5% levels of significance. The results shows that the PTE with shrinkage estimators performs better than the PTE based on the unrestricted estimators.

Table 10.4.3 Empirical Risk Values for PTE Based on D–W and $\mathcal{G}_1$ statistic, $\alpha = 0.01$

E	Durbin		CO		PW	
ρ	D–W	$\mathcal{G}_1$	D–W	$\mathcal{G}_1$	D–W	$\mathcal{G}_1$
0.0	283.43	283.51	286.48	287.51	285.91	286.43
0.1	294.56	292.75	298.16	297.16	297.28	295.67
0.2	307.24	304.49	310.75	308.97	309.58	307.34
0.3	324.10	320.68	328.14	326.45	326.84	324.42
0.4	350.87	346.35	355.61	352.91	353.96	351.03
0.5	394.16	385.45	398.91	394.63	397.25	392.38
0.6	458.03	449.06	466.23	461.67	464.02	458.98
0.7	545.95	545.61	573.78	573.59	569.22	596.04
0.8	717.06	716.10	777.57	777.24	768.86	768.53
0.9	1009.81	958.54	1252.77	1244.43	1232.51	1225.25

Table 10.4.4 Empirical Risk Values for Shrinkage PTE Based on D–W and $\mathcal{G}_1$ Statistic, $\alpha = 0.01$

E	Durbin		CO		PW	
ρ	D–W	$\mathcal{G}_1$	D–W	$\mathcal{G}_1$	D–W	$\mathcal{G}_1$
0.0	271.15	260.54	273.59	263.43	273.17	262.68
0.1	274.03	262.93	276.62	266.14	276.10	265.07
0.2	280.15	268.05	282.67	271.06	281.78	269.84
0.3	289.07	277.00	291.59	280.29	290.63	278.41
0.4	298.38	286.41	301.16	290.28	299.77	288.74
0.5	326.04	305.72	329.05	312.55	327.66	310.69
0.6	357.54	343.72	362.50	352.42	360.52	350.10
0.7	407.99	407.07	422.07	421.34	418.51	417.78
0.8	508.92	507.77	548.93	548.22	542.85	542.14
0.9	707.72	652.46	881.06	865.44	866.75	852.51

Table 10.4.5 Empirical Risk Values for PTE
Based on D–W and $\mathcal{G}_1$ Statistic, $\alpha = 0.05$

E	Durbin		CO		PW	
ρ	D–W	$\mathcal{G}_1$	D–W	$\mathcal{G}_1$	D–W	$\mathcal{G}_1$
0.0	283.90	283.96	288.19	288.69	286.90	287.10
0.1	292.65	292.25	296.77	297.10	295.38	295.32
0.2	304.14	303.76	308.71	309.59	306.95	307.42
0.3	321.25	320.21	326.36	326.94	324.42	324.66
0.4	346.52	344.84	353.05	353.23	351.05	350.85
0.5	385.30	385.04	394.48	394.42	392.21	392.15
0.6	448.87	448.87	461.59	461.59	458.88	458.88
0.7	545.12	545.12	573.35	573.35	568.80	568.80
0.8	716.10	716.10	777.24	777.24	768.53	768.53
0.9	958.54	958.54	1244.43	1244.43	1225.25	1225.25

Table 10.4.6 Empirical Risk Values for Shrinkage PTE
Based on D–W and $\mathcal{G}_1$ Statistic, $\alpha = 0.05$

E	Durbin		CO		PW	
ρ	D–W	$\mathcal{G}_1$	D–W	$\mathcal{G}_1$	D–W	$\mathcal{G}_1$
0.0	258.32	249.92	261.46	253.44	260.66	252.33
0.1	260.79	252.10	263.78	255.61	262.82	254.07
0.2	265.52	255.71	268.70	259.65	267.01	257.65
0.3	275.13	266.14	278.29	270.51	276.41	268.34
0.4	286.18	279.71	290.04	285.61	288.32	283.52
0.5	305.37	303.93	312.30	310.01	310.37	309.08
0.6	341.81	341.81	350.64	350.64	348.29	348.29
0.7	403.59	403.59	418.13	418.13	414.54	414.54
0.8	507.77	507.77	548.22	548.22	542.14	542.14
0.9	652.46	652.46	865.44	865.44	852.51	852.51

The Tables 10.4.3 through 10.4.6 are prepared with the headers $\rho/\mathbf{E}$. These headers indicate that for various values of ρ, the estimators (E) are based on the Durbin, Cochrane–Orcutt and Prais–Winsten two-stage procedures by way of the tests (T), D–W and $\mathcal{G}_1$ for the PTEs, and risks are calculated by the formula (10.3.13). Table 10.4.7 through 10.4.12 give the risk tables for the same estimators based on the weighted loss function given by (10.3.15).

Table 10.4.7 Empirical Risks for Different Estimators prior to Testing

ρ	OLS	GLS	Durbin	CO	PW
0.0	5.15	5.15	5.76	6.02	5.98
0.1	5.32	5.19	5.77	6.03	5.99
0.2	5.51	5.23	5.79	6.03	5.99
0.3	5.75	5.28	5.82	6.02	5.99
0.4	6.10	5.33	5.84	6.02	5.99
0.5	6.69	5.39	5.86	6.03	6.00
0.6	7.68	5.43	5.87	6.07	6.04
0.7	9.48	5.47	5.92	6.17	6.15
0.8	13.10	5.50	6.02	6.47	6.46
0.9	21.31	5.56	6.04	7.38	7.39

Table 10.4.8 Empirical Risks for Different Estimators prior to Testing-Shrinkage Estimates

ρ	OLS	GLS	Durbin	CO	PW
0.0	3.84	3.81	4.32	4.53	4.50
0.1	3.98	3.82	4.30	4.51	4.48
0.2	4.12	3.85	4.32	4.50	4.48
0.3	4.28	3.90	4.34	4.49	4.47
0.4	4.51	3.95	4.36	4.48	4.46
0.5	4.88	4.02	4.38	4.48	4.46
0.6	5.52	4.09	4.41	4.50	4.48
0.7	6.80	4.18	4.48	4.57	4.56
0.8	9.61	4.26	4.59	4.80	4.79
0.9	16.60	4.33	4.61	5.54	5.55

Table 10.4.9 Empirical Risk Values for PTE Based on D–W and $\mathcal{G}_1$ Statistic, $\alpha = 0.01$

E	Durbin		CO		PW	
ρ	D–W	$\mathcal{G}_1$	D–W	$\mathcal{G}_1$	D–W	$\mathcal{G}_1$
0.0	5.40	5.39	5.54	5.56	5.54	5.54
0.1	5.46	5.44	5.62	5.62	5.61	5.60
0.2	5.52	5.49	5.68	5.66	5.67	5.64
0.3	5.56	5.54	5.71	5.69	5.69	5.67
0.4	5.61	5.57	5.74	5.71	5.72	5.69
0.5	5.64	5.57	5.78	5.73	5.76	5.70
0.6	5.66	5.58	5.82	5.76	5.80	5.74
0.7	5.64	5.63	5.87	5.87	5.85	5.85
0.8	5.74	5.72	6.15	6.14	6.14	6.13
0.9	5.84	5.73	7.06	7.01	7.07	7.02

Table 10.4.10 Empirical Risk Values for Shrinkage PTE Based on D–W and $\mathcal{G}_1$ Statistic, $\alpha = 0.01$

E	Durbin		CO		PW	
ρ	D–W	$\mathcal{G}_1$	D–W	$\mathcal{G}_1$	D–W	$\mathcal{G}_1$
0.0	4.23	4.21	4.34	4.33	4.34	4.32
0.1	4.27	4.24	4.40	4.38	4.40	4.37
0.2	4.32	4.29	4.44	4.42	4.44	4.40
0.3	4.37	4.36	4.47	4.47	4.47	4.45
0.4	4.42	4.38	4.50	4.47	4.50	4.46
0.5	4.47	4.38	4.54	4.48	4.53	4.46
0.6	4.48	4.41	4.56	4.50	4.54	4.48
0.7	4.50	4.49	4.58	4.57	4.57	4.56
0.8	4.60	4.59	4.80	4.80	4.80	4.79
0.9	4.68	4.61	5.61	5.54	5.63	5.55

Table 10.4.11 Empirical Risk Values for PTE Based on D–W and $\mathcal{G}_1$ Statistic, $\alpha = 0.05$

E	Durbin		CO		PW	
ρ	D–W	$\mathcal{G}_1$	D–W	$\mathcal{G}_1$	D–W	$\mathcal{G}_1$
0.0	5.40	5.40	5.57	5.58	5.55	5.55
0.1	5.43	5.43	5.61	5.63	5.59	5.60
0.2	5.48	5.48	5.66	5.68	5.63	5.65
0.3	5.53	5.53	5.69	5.70	5.67	5.68
0.4	5.58	5.55	5.71	5.72	5.69	5.69
0.5	5.57	5.56	5.73	5.73	5.70	5.70
0.6	5.58	5.58	5.76	5.76	5.74	5.74
0.7	5.63	5.63	5.86	5.86	5.84	5.84
0.8	5.72	5.72	6.14	6.14	6.13	6.13
0.9	5.73	5.73	7.01	7.01	7.02	7.02

Table 10.4.12 Empirical Risk Values for Shrinkage PTE Based on D–W and $\mathcal{G}_1$ Statistic, $\alpha = 0.05$

E	Durbin		CO		PW	
ρ	D–W	$\mathcal{G}_1$	D–W	$\mathcal{G}_1$	D–W	$\mathcal{G}_1$
0.0	4.22	4.21	4.34	4.35	4.33	4.33
0.1	4.24	4.23	4.38	4.38	4.37	4.36
0.2	4.29	4.29	4.42	4.44	4.40	4.42
0.3	4.35	4.34	4.46	4.47	4.45	4.46
0.4	4.40	4.36	4.48	4.48	4.47	4.46
0.5	4.38	4.38	4.48	4.48	4.46	4.46
0.6	4.41	4.41	4.50	4.50	4.48	4.48
0.7	4.48	4.48	4.57	4.57	4.56	4.56
0.8	4.59	4.59	4.80	4.80	4.79	4.79
0.9	4.61	4.61	5.54	5.54	5.55	5.55

(Tables 10.4.1 – 10.4.12 are due to Chen and Saleh (1993) reproduced here with permission of VSP publishers, an imprint of Brill Academic Publishers, Leiden, The Netherlands.)

10.6 Estimation of Parameters of an Autoregressive Gaussian Process

Consider the very often encountered autoregressive model in time-series defined by the Gaussian process $Y_t, t = \cdots, -1, 0, 1, 2, \cdots$, satisfying the difference equation

$$Y_t - \theta_1 Y_{t-1} - \cdots - \theta_p Y_{t-p} = v_t, \tag{10.6.1}$$

where $v_t, t = \cdots, -1, 0, 1, 2, \cdots$ is a sequence of iid Gaussian random errors such that $E(v_t) = 0$ and $E(v_t^2) = \sigma_v^2 > 0$. The coefficient vector $\boldsymbol{\theta} = (\theta_1, \cdots, \theta_p)'$ is such that all the roots of the characteristic equation

$$h_p(z) = 1 - \theta_1 z - \cdots - \theta_p z^p = 0 \tag{10.6.2}$$

exceed one in absolute value. This model is often used by economists for time-series analyses of data price index, unemployment, labor force, population, gross national product, and so forth. It can be verified that the parameters $(\boldsymbol{\theta}', \sigma^2)'$ of (10.6.2) are uniquely determined by the **spectral density function** (sp.d.f) of Y_t,

$$f_{\theta}(\lambda) = \frac{\sigma_v^2}{2\pi} |h_p(z)|^{-2}, \quad z = e^{i\lambda}. \tag{10.6.3}$$

Also, the $p \times p$ covariance matrix $\mathbf{B}$ of the autoregressive process Y_t of order p with sp.d.f (10.6.3) satisfies the so-called Yule-Walker equation (e.g., see Doob, 1953, and Hannan, 1970):

$$\beta(\tau) - \theta_1\beta(\tau-1) - \cdots - \theta_p\beta(\tau-p) = \delta_{0\tau}\sigma_v^2, \tau = 0, 1, \cdots, \tag{10.6.4}$$

where $\delta_{0\tau} = 1$ or 0 accordingly as $\tau = 0$ or $\neq 0$ (see also Brockwell and Davis, 1996).

Now, suppose that $\boldsymbol{\theta}$ belongs to the subspace $\mathbf{H}\boldsymbol{\theta} = \mathbf{h}$, where $\mathbf{H}$ is a $q \times p$-matrix and $\mathbf{h}$ is a p-vector of known constants. Our problem is to estimate $\boldsymbol{\theta}$ based on a sample $Y_1, Y_2, \cdots, Y_n$ when it is suspected that $\mathbf{H}\boldsymbol{\theta} = \mathbf{h}$ holds. Accordingly, we define (as in Chapter 7) five estimators of $\boldsymbol{\theta}$: (1) the unrestricted estimator $\tilde{\boldsymbol{\theta}}_n$, (2) the restricted estimator $\hat{\boldsymbol{\theta}}_n$, (3) the PTE, $\hat{\boldsymbol{\theta}}_n^{PT}$, (4) the Stein-type (SE) $\hat{\boldsymbol{\theta}}_n^{S}$ and (5) the positive-rule Stein estimator (PRSE) $\hat{\boldsymbol{\theta}}_n^{S+}$, respectively, and study their properties.

10.6.1 Estimation and Test of Hypothesis

Our proposed estimators of $\boldsymbol{\theta}$ are based on the "principal part" (PP) of the logarithm of the likelihood function, $L_n = \log p(Y_1, \cdots, Y_n)$ denoted by $\tilde{L}_n$ since $n^{-1/2}(L_n - \tilde{L}_n) \xrightarrow{P} 0$ as $n \to \infty$ (e.g., see Whittle, 1952, and Dzhaparidze, 1986). For the model (10.6.1) it has been shown that the approximation $\tilde{L}_n$ is given by

$$\begin{aligned}\tilde{L}_n &= -\frac{n}{2}\Big\{\log 2\pi\sigma_v^2 + \bar{\sigma}_v^2 \int_{-\pi}^{\pi} I_n(\lambda)|h_p(z)|^2 d\lambda\Big\} \\ &= -\frac{n}{2}\Big\{\log 2\pi\sigma_v^2 + \bar{\sigma}_v^2 \sum_{k,j=0}^{j} \theta_k\theta_j\beta_n^*(k-j)\Big\},\end{aligned} \tag{10.6.5}$$

where $\theta_0 = 1$ and $I_n(\lambda)$ is the periodogram of the process $\{Y_t\}$, while

$$\beta_n^*(\tau) = \int_{-\pi}^{\pi} I_n(\lambda)e^{i\lambda\tau}d\lambda = n^{-1}\sum_{t=1}^{n-\tau} Y_tY_{t+\tau}, \ \tau = 0, 1, \cdots, n-1, \tag{10.6.6}$$

are the elements of the estimated covariance matrix $\mathbf{B}_n^*$. Denoting by $\boldsymbol{\beta}_n^* = (\beta_n^*(1), \cdots, \beta_n^*(p))'$, we define the **unrestricted** principal part maximum likelihood estimator (UPPMLE) as $\tilde{\boldsymbol{\theta}}_n$ as

$$\tilde{\boldsymbol{\theta}}_n = \mathbf{B}_n^{*-1}\boldsymbol{\beta}_n^*. \tag{10.6.7}$$

If $\boldsymbol{\theta}$ belongs to the subspace $\mathbf{H}\boldsymbol{\theta} = \mathbf{h}$, then the **restricted** principal part MLE (RPPMLE) of $\boldsymbol{\theta}$ can be written as

$$\hat{\boldsymbol{\theta}}_n = \tilde{\boldsymbol{\theta}}_n - \mathbf{B}_n^{*-1}\mathbf{H}'(\mathbf{H}\mathbf{B}_n^{*-1}\mathbf{H}')^{-1}(\mathbf{H}\tilde{\boldsymbol{\theta}}_n - \mathbf{h}). \tag{10.6.8}$$

We consider here the test of the hypothesis $H_0 : \mathbf{H}\boldsymbol{\theta} = \mathbf{h}$ against $H_A : \mathbf{H}\boldsymbol{\theta} \neq \mathbf{h}$ based on the Wald-type statistic:

$$\mathcal{L}_n = n\tilde{\sigma}_n^{-2}(\mathbf{H}\tilde{\boldsymbol{\theta}}_n - \mathbf{h})'(\mathbf{H}\mathbf{B}_n^{*-1}\mathbf{H}')^{-1}(\mathbf{H}\tilde{\boldsymbol{\theta}}_n - \mathbf{h}), \tag{10.6.9}$$

where

$$\tilde{\sigma}_n^2 = \beta_n^*(0) - \beta_n^{*'}\tilde{\theta}_n. \tag{10.6.10}$$

In the next section we will show that as $n \to \infty$, $\mathcal{L}_n$ follows a central chi-squares distribution with q D.F. Then we can define the principal part preliminary test estimator (PPPTMLE) as

$$\hat{\theta}_n^{PT} = \tilde{\theta}_n - (\tilde{\theta}_n - \hat{\theta}_n)I(\mathcal{L}_n < \chi_q^2(\alpha)) \tag{10.6.11}$$

Similiarly, following Chapter 7, we define the Stein-type principal part estimator (SPPMLE) and the positive-rule principal part Stein estimator (PRPPMLE) as follows:

$$\hat{\theta}_n^S = \tilde{\theta}_n - d(\tilde{\theta}_n - \hat{\theta}_n)\mathcal{L}_n^{-1}, d = q - 2 \tag{10.6.12}$$

and

$$\hat{\theta}_n^{S+} = \hat{\theta}_n + (1 - d\mathcal{L}_n^{-1})I(\mathcal{L}_n > d)(\tilde{\theta}_n - \hat{\theta}_n), \tag{10.6.13}$$

respectively. See also Saleh (1992).

10.6.2 Asymptotic Theory of the Estimators and the Test-Statistics

Our main purpose is the comparison of the five estimators based on the asymptotic theory of the estimators and loss function

$$L(\theta_n^*, \theta) = n(\theta_n^* - \theta)'\mathbf{W}(\theta_n^* - \theta), \tag{10.6.14}$$

where $\mathbf{W}$ is a positive-semidefinite weight matrix.

It can be shown that $\mathcal{L}_n$ is a consistent test, and for any fixed alternative $H_\delta : \mathbf{H}\theta = \mathbf{h} + \delta$, all the estimators will be asymptotically risk equivalent to $\tilde{\theta}_n$ except that $\hat{\theta}_n$ tends to be degenerate as $n \to \infty$. Thus, we resolve this issue by considering the class of **local alternatives** (as in Chapter 7) to be

$$K_{(n)} : \mathbf{H}\theta = \mathbf{h} + n^{-1/2}\xi, \xi \neq \mathbf{0}. \tag{10.6.15}$$

First, we note that the following theorem holds as $n \to \infty$ under $\{K_{(n)}\}$:

Theorem 1. Under $\{K_{(n)}\}$ and the assumed regularity conditions for the model (10.6.1), we have as $n \to \infty$

(i) $\lim P_{K_{(n)}}\{\sqrt{n}(\tilde{\theta}_n - \theta) \le \mathbf{x}\} = \Phi_p(\mathbf{x}, 0; \sigma_v^2\mathbf{B})$,

$$\mathbf{B} = \begin{pmatrix} \beta(0) & \beta(1) & \cdots & \beta(p-1) \\ \beta(1) & \beta(2) & \cdots & \beta(p-2) \\ \vdots & \vdots & \vdots & \vdots \\ \beta(p-1) & \beta(p-2) & \cdots & \beta(0) \end{pmatrix}$$

is the Toeplitz matrix associated with sp.d.f.; (10.6.16)

(ii) $\lim P_{K_{(n)}}\{\sqrt{n}(\hat{\boldsymbol{\theta}}_n - \boldsymbol{\theta}) \le \mathbf{x}\} = \Phi_p(\mathbf{x},, \boldsymbol{\delta}, 0; \sigma_v^2 \mathbf{V}),$ (10.6.17)

where $\mathbf{V} = \mathbf{B}^{-1} - \mathbf{B}^{-1}\mathbf{H}'(\mathbf{H}\mathbf{B}^{-1}\mathbf{H}')^{-1}\mathbf{H}\mathbf{B}^{-1}$ and $\boldsymbol{\delta} = \mathbf{B}^{-1}\mathbf{H}'(\mathbf{H}\mathbf{B}^{-1}\mathbf{H}')^{-1}\boldsymbol{\xi}$;

(iii) $\lim P_{K_{(n)}}(\mathcal{L}_n \le \mathbf{x}) = H_q(\mathbf{x}; \Delta^2), \Delta^2 = \sigma_v^{-2}\boldsymbol{\delta}'(\mathbf{H}\mathbf{B}^{-1}\mathbf{H}')^{-1}\boldsymbol{\delta};$ (10.6.18)

(iv) $\lim P_{K_{(n)}}\{\sqrt{n}(\hat{\boldsymbol{\theta}}_n^{PT} - \boldsymbol{\theta}) \le \mathbf{x}\} = H_q(\chi_q^2(\alpha); \Delta^2)\Phi_p(\mathbf{x} - \boldsymbol{\delta}; \mathbf{0}; \sigma_v^2\mathbf{V})$

$$+ \int \Phi_p(\mathbf{x} - \mathbf{Z}; \mathbf{0}; \sigma_v^2\mathbf{V})d\Phi_p(\mathbf{Z}, \mathbf{0}; \sigma_v^2(\mathbf{H}\mathbf{B}^{-1}\mathbf{H}')^{-1}), \tag{10.6.19}$$

where $E(\boldsymbol{\xi}) = \{\mathbf{Z} : (\mathbf{H}\mathbf{Z} + \boldsymbol{\xi})'(\mathbf{H}\mathbf{B}^{-1}\mathbf{H}')^{-1}(\mathbf{H}\mathbf{Z} + \boldsymbol{\xi}) \ge \chi_q^2(\alpha)\}$ and where $\Phi_p(\mathbf{x}; \mathbf{0}; \boldsymbol{\Sigma})$ stands for the p-variate normal cdf with mean $\mathbf{0}$ and covariance matrix $\boldsymbol{\Sigma}$, and $H_q(\mathbf{x}; \Delta^2)$ is the cdf of a noncentral chi-square distribution with q-D.F. and noncentrality parameter $\Delta^2/2$.

(v) the asymptotic representation of $\sqrt{n}(\hat{\boldsymbol{\theta}}_n^S - \boldsymbol{\theta})$ given by

$$\sqrt{n}(\hat{\boldsymbol{\theta}}_n^S - \boldsymbol{\theta}) \xrightarrow{\mathcal{D}} \mathbf{Z} - d\left\{\frac{\mathbf{B}^{-1}\mathbf{H}'(\mathbf{H}\mathbf{B}^{-1}\mathbf{H}')^{-1}(\mathbf{H}\mathbf{Z}+\boldsymbol{\xi})}{\sigma_v^{-2}(\mathbf{H}\mathbf{Z}+\boldsymbol{\xi})'(\mathbf{H}\mathbf{B}^{-1}\mathbf{H}')^{-1}(\mathbf{H}\mathbf{Z}+\boldsymbol{\xi})}\right\}, \tag{10.6.20}$$

where $\mathbf{Z} \sim \mathcal{N}_p(\mathbf{0}, \sigma_v^2\mathbf{B}^{-1})$ and $d = q - 2$;

and

(vi) $$\sqrt{n}(\hat{\boldsymbol{\theta}}_n^{S+} - \boldsymbol{\theta}) \xrightarrow{\mathcal{D}} \mathbf{Z} - d\left\{\frac{\mathbf{B}^{-1}\mathbf{H}'(\mathbf{H}\mathbf{B}^{-1}\mathbf{H}')^{-1}(\mathbf{H}\mathbf{Z}+\boldsymbol{\xi})}{\sigma_v^{-2}(\mathbf{H}\mathbf{Z}+\boldsymbol{\xi})'(\mathbf{H}\mathbf{B}^{-1}\mathbf{H}')^{-1}(\mathbf{H}\mathbf{Z}+\boldsymbol{\xi})}\right\}$$
$$- \left[1 - \sigma_v^{-2}d\{(\mathbf{H}\mathbf{Z} + \boldsymbol{\xi})'(\mathbf{H}\mathbf{B}^{-1}\mathbf{H}')^{-1}(\mathbf{H}\mathbf{Z} + \boldsymbol{\xi})\}^{-1}\right.$$
$$\times I\{(\mathbf{H}\mathbf{Z} + \boldsymbol{\xi})'(\mathbf{H}\mathbf{B}^{-1}\mathbf{H}')^{-1}(\mathbf{H}\mathbf{Z} + \boldsymbol{\xi}) < d\}$$
$$\left.\times \{\mathbf{B}^{-1}\mathbf{H}'(\mathbf{H}\mathbf{B}^{-1}\mathbf{H}')^{-1}(\mathbf{H}\mathbf{Z} + \boldsymbol{\xi})\}\right]. \tag{10.6.21}$$

Proof. (i) Since $\mathbf{B}_n^* \xrightarrow{P} \mathbf{B}$ as $n \to \infty$, we can write

$$\tilde{\boldsymbol{\theta}}_n = \mathbf{B}_n^{*-1}\boldsymbol{\beta}_n^*(\tau) = \mathbf{B}^{-1}\boldsymbol{\beta}_n^*(\tau) + o_p(1). \tag{10.6.22}$$

Further as $n \to \infty$, $\sqrt{n}(\tilde{\boldsymbol{\theta}}_n - \boldsymbol{\theta}) \stackrel{\mathcal{D}}{=} \mathcal{N}_p(\mathbf{0}, \sigma_v^2\mathbf{B}^{-1})$, being a PPMLE (see Dzhaparidze, 1986). Hence, we set

$$\mathbf{Z} = \sqrt{n}(\tilde{\boldsymbol{\theta}}_n - \boldsymbol{\theta}) = \sqrt{n}\mathbf{B}^{-1}(\boldsymbol{\beta}_n^*(\tau) - \boldsymbol{\beta}(\tau)) \stackrel{\mathcal{D}}{=} \mathcal{N}_p(\mathbf{0}, \sigma_v^2\mathbf{B}^{-1}), \text{as } n \to \infty. \tag{10.6.23}$$

(ii) **Restricted estimator**

$$\sqrt{n}(\hat{\boldsymbol{\theta}}_n - \boldsymbol{\theta}) = \sqrt{n}\{(\tilde{\boldsymbol{\theta}}_n - \boldsymbol{\theta}) - \sqrt{n}\mathbf{B}_n^{*-1}\mathbf{H}'(\mathbf{H}\mathbf{B}_n^{*-1}\mathbf{H}')^{-1}(\mathbf{H}\tilde{\boldsymbol{\theta}}_n - \mathbf{h})\}$$
$$= \mathbf{Z} - \mathbf{B}^{-1}\mathbf{H}'(\mathbf{H}\mathbf{B}^{-1}\mathbf{H}')^{-1}(\mathbf{H}\mathbf{Z} + \boldsymbol{\xi}) + o_p(1). \tag{10.6.24}$$

As $n \to \infty$,

$$\sqrt{n}(\hat{\boldsymbol{\theta}}_n - \boldsymbol{\theta}) \stackrel{\mathcal{D}}{=} \mathcal{N}(-\mathbf{B}^{-1}\mathbf{H}'(\mathbf{H}\mathbf{B}^{-1}\mathbf{H}')^{-1}\boldsymbol{\xi}, \sigma_v^2\mathbf{V}) \tag{10.6.25}$$

where $\mathbf{V} = \mathbf{B}^{-1} - \mathbf{B}^{-1}\mathbf{H}'(\mathbf{H}\mathbf{B}^{-1}\mathbf{H}')^{-1}\mathbf{H}\mathbf{B}^{-1}$.

(iii) **Test-Statistic**

$$\begin{aligned}\mathcal{L}_n &= n\tilde{\sigma}_n^{-2}(\mathbf{H}\tilde{\boldsymbol{\theta}}_n - \mathbf{h})'(\mathbf{H}\mathbf{B}_n^{*-1}\mathbf{H}')^{-1}(\mathbf{H}\tilde{\boldsymbol{\theta}}_n - \mathbf{h})\\ &= \sigma_v^{-2}(\mathbf{H}\mathbf{Z} + \boldsymbol{\xi})'(\mathbf{H}\mathbf{B}^{-1}\mathbf{H}')^{-1}(\mathbf{H}\mathbf{Z} + \boldsymbol{\xi}) + o_p(1). \qquad (10.6.26)\end{aligned}$$

As $n \to \infty$,

$$\begin{aligned}\lim P_{K_{(n)}}\{\mathcal{L}_n \le \mathrm{x}\} &= P_{\Delta^2}\{\sigma_v^{-2}(\mathbf{H}\mathbf{X} + \boldsymbol{\xi})'(\mathbf{H}\mathbf{B}^{-1}\mathbf{H}')^{-1}(\mathbf{H}\mathbf{Z} + \boldsymbol{\xi}) \le \mathrm{x}\}\\ &= H_q(\mathrm{x};\Delta^2), \quad \Delta^2 = \sigma_v^{-2}\boldsymbol{\delta}'(\mathbf{H}\mathbf{B}^{-1}\mathbf{H}')^{-1}\boldsymbol{\delta} \qquad (10.6.27)\end{aligned}$$

Next (iv), (v) and (vi) follows from (i)-(iii) and noting that $\{\sqrt{n}(\hat{\boldsymbol{\theta}}_n - \boldsymbol{\theta})', \sqrt{n}(\tilde{\boldsymbol{\theta}}_n - \hat{\boldsymbol{\theta}}_n)'\}'$ are statistically independent as $n \to \infty$.

10.6.3 ADB, ADMSE Matrices, and ADQR of the Estimators

Using Theorem 10.6.1 we can find expressions of the asymptotic distribution bias (ADB), MSE (ADMSE) matrices, and the risks (ADQR), which are given below estimator by estimator.

(i) UPPMLE:

$$\begin{aligned}&\mathbf{b}_1(\tilde{\boldsymbol{\theta}}_n) = \mathbf{0} \quad \text{and} \quad B_1(\tilde{\boldsymbol{\theta}}_n) = 0\\ &\mathbf{M}_1(\tilde{\boldsymbol{\theta}}_n) = \sigma_v^2\mathbf{B}^{-1} \quad \text{and} \quad R_1(\tilde{\boldsymbol{\theta}}_n : \mathbf{W}) = \sigma_v^2 \operatorname{tr}(\mathbf{W}\mathbf{B}^{-1}). \qquad (10.6.28)\end{aligned}$$

(ii) RPPMLE:

$$\begin{aligned}&\mathbf{b}_2(\hat{\boldsymbol{\theta}}_n) = -\boldsymbol{\delta} \quad \text{and} \quad B_2(\hat{\boldsymbol{\theta}}_n) = \Delta^2,\\ &\mathbf{M}_2(\hat{\boldsymbol{\theta}}_n) = \sigma_v^2\mathbf{V} + \boldsymbol{\delta}\boldsymbol{\delta}', \text{ where } \mathbf{V} = \mathbf{B}^{-1} - \mathbf{B}^{-1}\mathbf{H}(\mathbf{H}\mathbf{B}^{-1}\mathbf{H}')^{-1}\mathbf{H}\mathbf{B}^{-1},\\ &R_2(\hat{\boldsymbol{\theta}}_n : \mathbf{W}) = \sigma_v^2 \operatorname{tr}(\mathbf{W}\mathbf{V}) + \boldsymbol{\delta}'\mathbf{W}\boldsymbol{\delta}. \qquad (10.6.29)\end{aligned}$$

(iii) PTPPMLE:

$$\begin{aligned}&\mathbf{b}_3(\hat{\boldsymbol{\theta}}_n^{PT} = -\boldsymbol{\delta}H_{q+2}(\chi_q^2(\alpha);\Delta^2),\\ &B_3(\hat{\boldsymbol{\theta}}_n^{PT}) = \Delta^2\{H_{q+2}(\chi_q^2(\alpha);\Delta^2)\}^2,\\ &\mathbf{M}_3(\hat{\boldsymbol{\theta}}_n^{PT}) = \sigma_v^2\mathbf{B}^{-1} - \sigma_v^2(\mathbf{B}^{-1}\mathbf{H}'(\mathbf{H}\mathbf{B}^{-1}\mathbf{H}')^{-1}\mathbf{H}\mathbf{B}^{-1})H_{q+2}(\chi_q^2(\alpha);\Delta^2)\\ &\qquad +\boldsymbol{\delta}\boldsymbol{\delta}'\{2H_{q+2}(\chi_q^2(\alpha);\Delta^2) - H_{q+4}(\chi_q^2(\alpha);\Delta^2)\},\\ &R_3(\hat{\boldsymbol{\theta}}_n^{PT} : \mathbf{W}) = \sigma_v^2 \operatorname{tr}(\mathbf{W}\mathbf{B}^{-1}) - \sigma_v^2 \operatorname{tr}\{\mathbf{W}[\mathbf{B}^{-1}\mathbf{H}'(\mathbf{H}\mathbf{B}^{-1}\mathbf{H}')^{-1}\mathbf{H}\mathbf{B}^{-1}]\}\\ &\times H_{q+2}(\chi_q^2(\alpha);\Delta^2) + \boldsymbol{\delta}'\mathbf{W}\boldsymbol{\delta}\{2H_{q+2}(\chi_q^2(\alpha);\Delta^2) - H_{q+4}(\chi_q^2(\alpha);\Delta^2)\}.\\ &\qquad (10.6.30)\end{aligned}$$

(iv) SPPMLE:

$$\mathbf{b}_4(\hat{\boldsymbol{\theta}}_n^S) = -(q-2)\boldsymbol{\delta}E[\chi_{q+2}^{-2}(\Delta^2)] \text{ and } B_4(\hat{\boldsymbol{\theta}}_n^S) = (q-2)^2\Delta^2\{E[\chi_{q+2}^{-2}(\Delta^2)]\}^2,$$

$$\mathbf{M}_4(\hat{\boldsymbol{\theta}}_n^S) = \sigma_v^2\mathbf{B}^{-1} - \sigma_v^2(q-2)\mathbf{B}^{-1}\mathbf{H}'(\mathbf{H}\mathbf{B}^{-1}\mathbf{H}')^{-1}\mathbf{H}\mathbf{B}^{-1}\{2E[\chi_{q+2}^{-2}(\Delta^2)] \\ -(q-2)E[\chi_{q+2}^{-4}(\Delta^2)]\} + (q^2-4)\boldsymbol{\delta}\boldsymbol{\delta}'E[\chi_{q+4}^{-4}(\Delta^2)],$$

$$R_4(\hat{\boldsymbol{\theta}}_n^S : \mathbf{W}) = \sigma_v^2 \operatorname{tr}[\mathbf{W}\mathbf{B}^{-1}] - \sigma_v^2(q-2)\operatorname{tr}[\mathbf{W}\mathbf{B}^{-1}\mathbf{H}'(\mathbf{H}\mathbf{B}^{-1}\mathbf{H}')^{-1}\mathbf{H}\mathbf{B}^{-1}] \\ \times\{2E[\chi_{q+2}^{-2}(\Delta^2)] - (q-2)E[\chi_{q+2}^{-4}(\Delta^2)]\} \\ +(q^2-4)\boldsymbol{\delta}'\mathbf{W}\boldsymbol{\delta}E[\chi_{q+4}^{-4}(\Delta^2)]. \tag{10.6.31}$$

(v) PRSPPMLE,

$$\mathbf{b}_5(\hat{\boldsymbol{\theta}}_n^{S+}) = \mathbf{b}_4(\hat{\boldsymbol{\theta}}_n^S) - \boldsymbol{\delta}E[(1-(q-2)\chi_{q+2}^{-2}(\Delta^2))I(\chi_{q+2}^2(\Delta^2) < q-2)]$$

$$B_5(\hat{\boldsymbol{\theta}}_n^{S+}) = \Delta^2\{(q-2)E[\chi_{q+2}^{-2}(\Delta^2)] \\ -E[(1-(q-2)\chi_{q+2}^{-2}(\Delta^2))I(\chi_{q+2}^2(\Delta^2) < q-2)]\}^2,$$

$$\mathbf{M}_5(\hat{\boldsymbol{\theta}}_n^{S+}) = \mathbf{M}_4(\hat{\boldsymbol{\theta}}_n^S) - \sigma_v^2\{\mathbf{B}^{-1}\mathbf{H}'(\mathbf{H}\mathbf{B}^{-1}\mathbf{H}')^{-1}\mathbf{H}\mathbf{B}^{-1} \\ \times E[(1-(q-2)\chi_{q+2}^{-2}(\Delta^2))^2 I(\chi_{q+2}^2(\Delta^2) < q-2)] \\ +\boldsymbol{\delta}\boldsymbol{\delta}'\{2E[(1-(q-2)\chi_{q+2}^{-2}(\Delta^2))I(\chi_{q+2}^2(\Delta^2) < q-2)] \\ -E[(1-(q-2)\chi_{q+4}^{-2}(\Delta^2))^2 I(\chi_{q+4}^2(\Delta^2) < q-2)]\} \tag{10.6.32}$$

$$R_5(\hat{\boldsymbol{\theta}}_n^{S+} : \mathbf{W}) = R_4(\hat{\boldsymbol{\theta}}_n^S) - \sigma_v^2\{\operatorname{tr}[\mathbf{W}\mathbf{B}^{-1}\mathbf{H}'(\mathbf{H}\mathbf{B}^{-1}\mathbf{H}')^{-1}\mathbf{H}\mathbf{B}^{-1}] \\ \times E[(1-(q-2)\chi_{q+2}^{-2}(\Delta^2))^2 I(\chi_{q+2}^2(\Delta^2) < q-2)] \\ +2\boldsymbol{\delta}'\mathbf{W}\boldsymbol{\delta}\{2E[(1-(q-2)\chi_{q+2}^{-2}(\Delta^2))I(\chi_{q+2}^2(\Delta^2) < q-2)] \\ -E[(1-(q-2)\chi_{q+4}^{-2}(\Delta^2))^2 I(\chi_{q+2}^2(\Delta^2) < q-2)]\}. \tag{10.6.33}$$

The expressions above are similar to the asymptotic expressions of the regression estimators of Chapter 7. Hence, the properties of the five estimators are similar to those of Chapter 7.

Similarly, we can develop the confidence sets similar to Section 7.9 of Chapter 7 and arrive at the same conclusion as given there.

10.7 R-Estimation of the Parameters of the AR[p]-Models

Let $\varepsilon_0, \varepsilon_{\pm 1}, \varepsilon_{i\pm 2}, \ldots,$ be i.i.d.r.v.'s with cdf F defined on $\mathbb{R}$. Let $\mathbf{Y}_0 = (Y_0, Y_1, \ldots, Y_{1-p})'$ be an observable random vector independent of $v_1, v_2, \ldots$

for some fixed integer p (≥ 1). Consider the pth order autoregressive, AR[p] model where the observations $Y_1, Y_2, \ldots, Y_n$ satisfy the relation

$$Y_i = \theta_1 Y_{i-1} + \cdots + \theta_p Y_{i-p} + v_i, \quad 1 \leq i \leq n, \quad (p \geq 1) \tag{10.7.1}$$

and assume that all roots of

$$x^p - \theta_1 x^{p-1} - \theta_2 x^{p-2} \cdots - \theta_p = 0 \tag{10.7.2}$$

are in $(-1, 1)$. Here $\boldsymbol{\theta} = (\theta_1, \ldots, \theta_p)' \in R^p$ is the vector of unknown autoregressive parameters. Our primary interest is to obtain R-estimators of $\boldsymbol{\theta}$ when it is suspected but not certain that $\boldsymbol{\theta}$ belongs to the subspace defined by $\mathbf{H}\boldsymbol{\theta} = \mathbf{h}$, where $\mathbf{H}$ is a $p \times q$ matrix and $\mathbf{h}$ is a q-vector of constants, respectively. For a simple subhypothesis problem, see also Koul and Saleh (1993).

10.7.1 R-Estimation of the Parameters of the AR[p] Model

In this section, we introduce a class of R-estimators of $\boldsymbol{\theta}$ for the AR[p] model (10.7.1) and discuss their asymptotic properties.

Let $\mathbf{Y}_i = (Y_i, \ldots, Y_{i-p+1})'$, $1 \leq i \leq n$, and define $R_i(\mathbf{b})$ as the rank of $Y_i - \mathbf{b}'\mathbf{Y}_{i-1}$ among $\{Y_j - \mathbf{b}'\mathbf{Y}_{j-1} | j = 1, \ldots, n\}$. Set $R_i(\mathbf{b}) = 0$ for $i \leq 0$. Let ϕ be a nondecreasing function from $[0, 1]$ to the real line, and define $\mathbf{L}_n(\mathbf{b}) = (L_1(\mathbf{b}), \ldots, L_p(\mathbf{b}))'$, where

$$L_j(\mathbf{b}) = n^{-1/2} \sum_{i=j+1}^{n} Y_{i-j} \phi\left(\frac{R_i(\mathbf{b})}{n+1}\right), \quad 1 \leq j \leq p, \tag{10.7.3}$$

with $\mathbf{b} = (b_1, \ldots, b_p)' \in R^p$. The class of rank-statistic $L_n(\mathbf{b})$, one for each ϕ, is similar to a class of similar rank-statistics for the linear regression model discussed in Chapter 7 where we replaced the weights $\{Y_{i-j}\}$ by appropriate design points as discussed in Hajek, Sidak, and Sen (1999) and Puri and Sen (1986) among others. For an alternative class of rank-statistics, see Hallin and Puri (1988).

It is natural to define $\tilde{\boldsymbol{\theta}}_n$ as the *unrestricted* (UE) R-estimator of $\boldsymbol{\theta}$ by the relationship

$$\inf_{b \in R^p} \|\mathbf{L}_n(\mathbf{b})\| = \|\mathbf{L}_n(\tilde{\boldsymbol{\theta}}_n)\|. \tag{10.7.4}$$

Then we can state the asymptotic linearity result as given below.

Theorem 1. Assume that (10.7.1) and (10.7.2) hold. In addition, assume that the following conditions hold.

A. (i) $E(v) = 0$, $0 < E(v^4) < \infty$ and $v_1, \ldots, v_n$ are i.i.d.r.v. with cdf, F,
 (ii) F has a uniformly continuous pdf f (> 0), almost everywhere.

B. ϕ is nondecreasing and differentiable with the derivative ϕ' being uniformly continuous on $[0, 1]$. Then, for every $0 < k < \infty$,

$$\sup_{\|\boldsymbol{\xi}\|<k} \left\|\mathbf{L}_n(\mathbf{b} + n^{-1/2}\boldsymbol{\xi}) - \hat{\mathbf{L}}_n + \boldsymbol{\xi}'\boldsymbol{\Sigma}\gamma\right\| = o_p(1), \tag{10.7.5}$$

where $\hat{\mathbf{L}}_n = (\hat{L}_1, \ldots, \hat{L}_p)'$ with

$$\hat{L}_j = n^{-1/2} \sum_{i=j+1}^{n} (Y_{i-j} - \bar{Y}_j)[\phi(F(v_i)) - \bar{\phi}], \quad 1 \le j \le p,$$

$$\bar{\phi} = \int_0^1 \phi(t)dt, \quad \bar{Y}_j = n^{-1} \sum_{i=j+1}^{n} Y_{i-j}$$

$\gamma = \int f d\phi(F)$ and $\boldsymbol{\Sigma} = ((\beta(i-j)))$ is the Toeplitz matrix with elements

$$\beta(k) = \text{Cov}(X_0, X_k), \quad 1 \le k \le p. \tag{10.7.6}$$

The theorem above covers Wilcoxon-type scores but not the normal score. For a proof, see Koul and Saleh (1993) and Problem 14.

As a consequence of Theorem 1, we can write

$$n^{1/2}(\tilde{\boldsymbol{\theta}}_n - \boldsymbol{\theta}) = \gamma\boldsymbol{\Sigma}^{-1}\hat{\mathbf{L}}_n(\mathbf{0}) + o_p(1). \tag{10.7.7}$$

Observe that $\hat{\mathbf{L}}_n$ is a vector of square integrable mean zero martingales with $E[\hat{\mathbf{L}}_n\hat{\mathbf{L}}_n'] = \sigma_\phi^2\boldsymbol{\Sigma}$, where $\sigma_\phi^2 = \text{Var}(\phi(U))$. Thus, by a routine Cramer-Wold device and Corollary 3.1 of Hall and Heyde (1980), we obtain

$$\hat{\mathbf{L}}_n \sim N_p(\mathbf{0}, \sigma_\phi^2\boldsymbol{\Sigma})$$

$$\text{and} \qquad n^{1/2}(\tilde{\boldsymbol{\theta}}_n - \boldsymbol{\theta}) \sim N_p(\mathbf{0}, \gamma^{-2}\sigma_\phi^2\boldsymbol{\Sigma}^{-1}). \tag{10.7.8}$$

The proof is left to the reader; see Problem 15.

10.7.2 Tests of Hypothesis and Improved R-Estimators of $\boldsymbol{\theta}$

Since our objective is to improve upon $\tilde{\boldsymbol{\theta}}_n$ when $\boldsymbol{\theta}$ belongs to subspace defined by $\mathbf{H}\boldsymbol{\theta} = \mathbf{h}$, we need to test the null hypothesis $H_0 : \mathbf{H}\boldsymbol{\theta} = \mathbf{h}$ against the alternative $H_A : \mathbf{H}\boldsymbol{\theta} \ne \mathbf{h}$. For this, we first consider the restricted R-estimators of $\boldsymbol{\theta}$ as

$$\hat{\boldsymbol{\theta}}_n = \tilde{\boldsymbol{\theta}}_n - \hat{\boldsymbol{\Sigma}}_n\mathbf{H}'(\mathbf{H}\hat{\boldsymbol{\Sigma}}_n^{-1}\mathbf{H})^{-1}(\mathbf{H}\tilde{\boldsymbol{\theta}}_n - \mathbf{h}), \tag{10.7.9}$$

where $\hat{\boldsymbol{\Sigma}}_n = ((\hat{\sigma}_{nij}))$ consistent estimator of $\boldsymbol{\Sigma}$ with

$$\hat{\sigma}_{nij} = \sum_{k=1}^{n-\max(i,j)} (Y_{k-i} - \bar{Y}_i)(Y_{k-j} - \bar{Y}_j), \quad 1 \le i, j \le n, \tag{10.7.10}$$

since $E(v_i^2) < \infty$ and the ergodic theorem implies $\frac{1}{n}\hat{\boldsymbol{\Sigma}}_n \xrightarrow{P} \boldsymbol{\Sigma}$ as $n \to \infty$.

Now, we consider the test-statistic $\mathcal{L}_n$, defined by

$$\mathcal{L}_n = n\hat{\sigma}_\phi^2 \mathbf{L}_n'(\hat{\boldsymbol{\theta}}_n)\hat{\boldsymbol{\Sigma}}_n^{-1}\mathbf{L}_n(\hat{\boldsymbol{\theta}}_n), \tag{10.7.11}$$

where $\hat{\sigma}_\phi^2 = n^{-1}\sum_{k=1}^n \left(\phi\left(\frac{k}{n+1}\right) - \bar{\phi}_n\right)^2$, and $\bar{\phi}_n = n^{-1}\sum_{k=1}^n \phi\left(\frac{k}{n+1}\right)$.

It can be shown (see Problem 16) that under H_0, $\mathcal{L}_n$ has asymptotically the central chi-square distribution with q D.F. Let $\chi_q^2(\alpha)$ be the upper $100\alpha\%$ critical value of the central chi-square distribution with q D.F. Then we define the preliminary test rank estimator (PTRE) as

$$\hat{\boldsymbol{\theta}}_n^{\mathrm{PT}} = \tilde{\boldsymbol{\theta}}_n - \left(\tilde{\boldsymbol{\theta}}_n - \hat{\boldsymbol{\theta}}_n\right) I\left(\mathcal{L}_n < \chi_q^2(\alpha)\right). \tag{10.7.12}$$

The Stein-type rank estimator (SRE) is defined by

$$\hat{\boldsymbol{\theta}}_n^{S} = \hat{\boldsymbol{\theta}}_n + (1 - c\mathcal{L}_n^{-1})(\tilde{\boldsymbol{\theta}}_n - \hat{\boldsymbol{\theta}}_n), \quad c = q-2, \tag{10.7.13}$$

and the positive-rule Stein-type rank estimator (PRSRE) is defined by

$$\hat{\boldsymbol{\theta}}_n^{S+} = \hat{\boldsymbol{\theta}}_n + (1 - c\mathcal{L}_n^{-1})I(\mathcal{L}_n > c)(\tilde{\boldsymbol{\theta}}_n - \hat{\boldsymbol{\theta}}_n), \quad q \geq 3. \tag{10.7.14}$$

10.7.3 Asymptotic Bias, MSE Matrix, and Risks of the *R*-Estimators

In order to obtain the expressions of the ADB, ADQB, ADMSE, and ADQR of the five *R*-estimators, we follow Saleh and Sen (1986), Sen and Saleh (1987), and Saleh (1992) and use the sequence of total alternatives. $K_{(n)} : \mathbf{H}\boldsymbol{\theta} = \mathbf{h} + n^{-1/2}\boldsymbol{\xi}$, $\boldsymbol{\xi} \neq \mathbf{0}$. We need the following additional assumptions (in addition to the assumptions in Theorem 1):

A* (ii) The error distribution F has an absolutely continuous pdf f such that $I(f) = \int_{-\infty}^{\infty}\left(\frac{f'(t)}{f(t)}\right)^2 f(t)dt < \infty$.

It can be shown (see Akritus and Johnson, 1982) that under A (i), (ii), B, and A* (ii), $K_{(n)}$ is contiguous to H_0. Using this argument and that of Theorem 3.2 of Sen and Saleh (1982), we obtain Theorem 2 as given below.

Theorem 2. Under (10.7.1) and (10.7.2), A(i), (ii), A*(ii), and B, the ADB, ADQB, ADMSE, and ADQR are given as follows:

A. **For URE $\tilde{\boldsymbol{\theta}}_n$,**

(i) $\mathbf{b}_1(\tilde{\boldsymbol{\theta}}_n) = \mathbf{0}$ and $B_1(\tilde{\boldsymbol{\theta}}_n) = 0$,

(ii) $\mathbf{M}_1(\tilde{\boldsymbol{\theta}}_n) = \sigma_\phi^2\boldsymbol{\Sigma}^{-1}$ and $R_1(\tilde{\boldsymbol{\theta}}_n; \mathbf{Q}) = \sigma_\phi^2\,\mathrm{tr}[\mathbf{Q}\boldsymbol{\Sigma}^{-1}]$.

B. **For RRE $\hat{\boldsymbol{\theta}}_n$,**

(i) $\mathbf{b}_2(\hat{\boldsymbol{\theta}}_n) = -\boldsymbol{\delta}$, $\boldsymbol{\delta} = \boldsymbol{\Sigma}^{-1}\mathbf{H}'(\mathbf{H}\boldsymbol{\Sigma}^{-1}\mathbf{H}')^{-1})\boldsymbol{\xi}$ and

$B_2(\hat{\boldsymbol{\theta}}_n) = \Delta^2 = \sigma_\phi^{-2}\boldsymbol{\delta}'(\mathbf{H}\boldsymbol{\Sigma}^{-1}\mathbf{H}')^{-1}\boldsymbol{\delta}$,

(ii) $\mathbf{M}_2(\hat{\boldsymbol{\theta}}_n) = \sigma_\phi^2[\boldsymbol{\Sigma}^{-1} - \boldsymbol{\Sigma}^{-1}\mathbf{H}(\mathbf{H}\boldsymbol{\Sigma}^{-1}\mathbf{H}')\mathbf{H}\boldsymbol{\Sigma}^{-1}] + \boldsymbol{\delta}\boldsymbol{\delta}'$ and

$R_2(\hat{\boldsymbol{\theta}}_n; \mathbf{Q}) = \sigma_\phi^2 \operatorname{tr}[\mathbf{Q}\{\boldsymbol{\Sigma}^{-1} - \boldsymbol{\Sigma}^{-1}\mathbf{H}'(\mathbf{H}\boldsymbol{\Sigma}^{-1}\mathbf{H}')^{-1}\mathbf{H}\boldsymbol{\Sigma}^{-1}\}] + \boldsymbol{\delta}\boldsymbol{\delta}'$.

C. **For PTRE,** $\hat{\boldsymbol{\theta}}_n^{\mathrm{PT}}$,

(i) $\mathbf{b}_3(\hat{\boldsymbol{\theta}}_n^{\mathrm{PT}}) = -\boldsymbol{\delta} H_{q+2}(\chi_q^2(\alpha); \Delta^2)$ and

$B_3(\hat{\rho}_n^{\mathrm{PT}}) = \Delta^2\{H_{q+2}(\chi_q^2(\alpha); \Delta^2)\}^2$

(ii) $$\mathbf{M}_3(\hat{\boldsymbol{\theta}}_n^{\mathrm{PT}}) = \sigma_\phi^2\boldsymbol{\Sigma}^{-1} - (\boldsymbol{\Sigma}^{-1}\mathbf{H}'(\mathbf{H}\boldsymbol{\Sigma}^{-1}\mathbf{H}')^{-1}\mathbf{H}\boldsymbol{\Sigma}^{-1})H_{q+2}(\chi_q^2(\alpha); \Delta^2) + \boldsymbol{\delta}\boldsymbol{\delta}'\{2H_{q+2}(\chi_q^2(\alpha); \Delta^2) - H_{q+4}(\chi_q^2(\alpha); \Delta^2)\}$$

and

$$R_3(\hat{\boldsymbol{\theta}}_n; \mathbf{Q}) = \sigma_\phi^2 \operatorname{tr}(\mathbf{Q}\boldsymbol{\Sigma}^{-1}) - \sigma_\phi^2 \operatorname{tr}[\mathbf{Q}\{\boldsymbol{\Sigma}^{-1}\mathbf{H}'(\mathbf{H}\boldsymbol{\Sigma}^{-1}\mathbf{H}')^{-1}\mathbf{H}\boldsymbol{\Sigma}^{-1}] \times H_{q+2}(\chi_q^2(\alpha); \Delta^2) + (\boldsymbol{\delta}'\mathbf{Q}\boldsymbol{\delta})\{2H_{q+2}(\chi_q^2(\alpha); \Delta^2) - H_{q+4}(\chi_q^2(\alpha); \Delta^2)\}$$

D. **For SRRE,** $\hat{\boldsymbol{\theta}}_n^{S}$,

(i) $\mathbf{b}_4(\hat{\boldsymbol{\theta}}_n^{S}) = -c\boldsymbol{\delta} E[\chi_{q+2}^{-2}(\Delta^2)]$ and $B_4(\hat{\boldsymbol{\theta}}_n^{S}) = c^2\Delta^2\{E[\chi_{q+2}^{-2}(\Delta^2)]\}^2$,

(ii) $$\mathbf{M}_4(\hat{\boldsymbol{\theta}}_n^{S}) = \sigma_\phi^2\boldsymbol{\Sigma}^{-1} - c\sigma_\phi^2\{\boldsymbol{\Sigma}^{-1}\mathbf{H}'(\mathbf{H}\boldsymbol{\Sigma}^{-1}\mathbf{H}')^{-1}\mathbf{H}\boldsymbol{\Sigma}^{-1}\} \times\{2E[\chi_{q+2}^{-2}(\Delta^2)] - cE[\chi_{q+2}^{-4}(\delta^2)]\} + c(c+4)\boldsymbol{\delta}\boldsymbol{\delta}'E[\chi_{q+4}^{-4}(\Delta^2)]$$

and

$$R_4(\hat{\boldsymbol{\theta}}_n^{S}; \mathbf{Q}) = \sigma_\phi^2 \operatorname{tr}(\mathbf{Q}\boldsymbol{\Sigma}^{-1}) - c\sigma_\phi^2 \operatorname{tr}[Q\{\boldsymbol{\Sigma}^{-1}\mathbf{H}'(\mathbf{H}\boldsymbol{\Sigma}^{-1}\mathbf{H}')^{-1}\mathbf{H}\boldsymbol{\Sigma}^{-1}\}] \times\{2E[\chi_{q+2}^{-2}(\Delta^2)] - cE[\chi_{q+2}^{-4}(\Delta^2)]\} + c(c+4)(\boldsymbol{\delta}'\mathbf{Q}b\delta)E[\chi_{q+4}^{-4}(\Delta^2)],$$

where $c = q - 2$.

E. **For PRSRE,** $\hat{\boldsymbol{\theta}}_n^{S+}$,

(i) $\mathbf{b}_5(\hat{\boldsymbol{\theta}}_n^{S+}) = \mathbf{b}_4(\hat{\boldsymbol{\theta}}_n^{S}) - \boldsymbol{\delta} E[(1 - c\chi_{q+2}^{-2}(\Delta^2))I(\chi_{q+2}^2(\Delta^2) < c)]$ and

$$B_5(\hat{\boldsymbol{\theta}}_n^{S+}) = \Delta^2\{cE[\chi_{q+2}^{-2}(\delta^2)] - E[(1 - c\chi_{q+2}^{-2}(\Delta^2))\, I(\chi_{q+2}^2(\Delta^2) < c)]\}^2,$$

(ii) $$\mathbf{M}_5(\hat{\boldsymbol{\theta}}_n^{S+}) = \mathbf{M}_4(\hat{\boldsymbol{\theta}}_n^{S}) - \sigma_\phi^2(\boldsymbol{\Sigma}^{-1}\mathbf{H}(\mathbf{H}\boldsymbol{\Sigma}^{-1}\mathbf{H}')^{-1}\mathbf{H}\boldsymbol{\Sigma}^{-1}) \times\{E[(1 - c\chi_{q+2}^{2}(\Delta^2))^2 I(\chi_{q+2}^2(\Delta^2) < c)]\} + \boldsymbol{\delta}\boldsymbol{\delta}'\{2E[(1 - c\chi_{q+2}^{-1}(\Delta^2))I(\chi_{q+2}^2(\Delta^2) < c)] - E[(1 - c\chi_{q+4}^{-2}(\Delta^2))^2 I(\chi_{q+2}^2(\Delta^2) < c)]\}$$

and

$$R_5(\hat{\boldsymbol{\theta}}_n^{S+};\mathbf{Q}) = R_4(\hat{\boldsymbol{\theta}}_n^{S};\mathbf{Q}) - \sigma_\phi^2 \operatorname{tr}\big[\mathbf{Q}\boldsymbol{\Sigma}^{-1}\mathbf{H}'(\mathbf{H}\boldsymbol{\Sigma}^{-1}\mathbf{H}')^{-1}\mathbf{H}\boldsymbol{\Sigma}^{-1}\big]$$
$$\times\big\{E\big[(1-c\chi_{q+2}^{-2}(\Delta^2))^2 I(\chi_{q+2}^2(\Delta^2)<c)\big]\big\}$$
$$+(\boldsymbol{\delta}'\mathbf{Q}\boldsymbol{\delta})\big\{2E\big[1-c\chi_{q+2}^{-2}(\Delta^2))I(\chi_{q+2}^2(\Delta^2)<c)$$
$$-E\big[(1-c\chi_{q+4}^{-2}(\Delta^2))^2 I(\chi_{q+2}^2(\Delta^2)<c)\big]\big\}.$$

The proof is left to the reader; see Problem 18.

The asymptotic dominance picture of the R-estimators are the same as that of Chapter 7 and Section 10.6. The asymptotic properties of the recentered confidence sets are similar to Section 7.9, and can be obtained by way of Problem 19.

10.8 R-Estimation of the Parameters with AR[1] Errors

Consider the simple linear regression model

$$Y_1 = \beta_0 + \beta_1 x_i + e_i, \quad e_i = \rho e_{i-1} + v_i, \quad |\rho| < 1, \quad i = 1, 2, \ldots, n, \quad (10.8.1)$$

where v_i are i.i.d. standardized r.v.'s with distribution function F, assumed to be symmetric around 0. The design variables x_i are real known constants and nonrandom. We focus on the asymptotic distribution of a number of signed rank estimators of β_0 and β_1.

Let ϕ be a nondecreasing function from $[0,1]$ to $\mathbb{R}$ with $\phi(u) = -\phi(1-u)$, for all $u \in [0,1]$. Let $\mathbf{b} = (b_0, b_1)' \in \mathbb{R}^2$, $e_{i,b} := Y_i - b_0 - b_1 x_i$, and $R_{i,b}^+$ denote the rank of $|e_{j,\mathbf{b}}|$ among $|e_{j,\mathbf{b}}|$. $1 \le j \le n$. Define $\phi^+(u) = \phi(u+1)/2)$ and

$$\mathbf{S}(\mathbf{b}) := n^{-1}\sum_{i=1}^{n}\begin{pmatrix}1\\ x_i\end{pmatrix}\phi^+\left(\frac{R_{i,\mathbf{b}}^+}{n+1}\right)\operatorname{sign}(e_{i,\mathbf{b}}),$$

$$\hat{\mathbf{S}} := n^{-1}\sum_{i=1}^{n}\begin{pmatrix}1\\ x_i\end{pmatrix}\phi(F|e_{i,\mathbf{b}}|)\operatorname{sign}(e_{i,\mathbf{b}}), \quad (10.8.2)$$

Let $\boldsymbol{\beta} := (\beta_0, \beta_1)'$. Then under the symmetry of f around zero, $E[\mathbf{S}(\boldsymbol{\beta})] = \mathbf{0}$. Define the signed rank estimator vector $\tilde{\boldsymbol{\beta}}_n$ to be a minimizer of $\|\mathbf{S}(\mathbf{b})\|$ with respect to $\mathbf{b} \in \mathbb{R}^2$ or a solution of the equations $\mathbf{S}(\mathbf{b}) = \mathbf{0}$. Let X denote the $n \times 2$ design matrix whose ith row is $(1, x_i)$, $1 \le i \le n$.

Then from Koul (1977) we obtain the following result:

Theorem 1. In addition to the assumptions above, suppose that the following holds: The d.f. F has a uniformly continuous Lebesgue density f, the score function ϕ is right-continuous and bounded, and the design matrix X

is of the full rank and satisfies $n^{1/2}\max_{1\le i\le n}|x_i|/\sqrt{\sum_{i=1}^n x_i^2} = O(1)$. Also, assume that the integral $\gamma := \int f d\phi(F) > 0$. Then

$$n^{1/2}(\beta_n^+ - \beta) = \gamma^{-1}(X'X)^{-1}n^{1/2}\hat{S} + o_p(1). \tag{10.8.3}$$

Now, let $\psi(e_i) := \phi(F(|e_i|)sign(e_i)$, and

$$C_n(\rho) := n^{-1}\sum_{i=1}^n\sum_{j=1}^n \begin{pmatrix} 1 & x_j \\ x_i & x_ix_j \end{pmatrix} E\psi(e_i)\psi(e_j). \tag{10.8.4}$$

Then the covariance matrix of the leading vector on the right-hand side can be written as

$$E\{(X'X)^{-1}n\hat{S}\hat{S}'(X'X)^{-1}\} = (X'X)^{-1}C_n(\rho)(X'X)^{-1}. \tag{10.8.5}$$

Assuming that this matrix has a positive definite limit, say, $\Sigma(\rho)$, we can further conclude that

$$n^{1/2}(\beta_n^+ - \beta) \Longrightarrow \mathcal{N}_2(0, \gamma^{-2}\Sigma(\rho)). \tag{10.8.6}$$

Further characterization of the covariance matrix $\Sigma(\rho)$ depends on the design, F and the score function. See, for example, Section 4 of Koul (1977) for the details involved in the case where e_i is a Gaussian process with mean zero, variance 1, and the lag k correlation $\rho^{|k|}$. It is really cumbersome.

The main difficulty is how to compute a reasonable approximation to the covariances $E[\psi(e_i)\psi(e_j)]$ as a function of $(i-j)$ in general when e_i's are a moving average of the i.i.d. r.v.'s.

If ρ if **not** known, following Durbin (1960), we can write the model (10.8.1) as

$$Y_i = \lambda + \rho Y_{i-1} + \beta_1(x_i - \rho x_{i-1}) + v_i, \quad i = 2, \ldots, n, \tag{10.8.7}$$

conditionally on Y_1, where $\lambda = (1-\rho)\beta_0$. Now, we consider the rank-statistics $\hat{S}_n(u)$ defined by

$$\hat{S}_n(u) = n^{-1}\sum_{i=2}^n \left(Y_{i-1} - \tilde{\beta}_{0n} - \tilde{\beta}_{1n}x_{i-1}\right)\phi\left(\frac{R_i(\tilde{\beta}_{0n}, \tilde{\beta}_{1n})}{n+1}\right), \tag{10.8.8}$$

where $R_i(\tilde{\beta}_{0n}, \tilde{\beta}_{1n})$ is the rank of residuals $Y_{i-1} - \tilde{\beta}_{0n} - \tilde{\beta}_{1n}x_{i-1}$ among $\{Y_j - \tilde{\beta}_{0n} - \tilde{\beta}_{1n}x_j | 2 \le j \le n\}$. The first residuals are computed based on the R-estimators of β_0 and β, assuming ρ to be zero as in Chapter 3. Then

$$\hat{\rho}_n^{(1)} = \frac{1}{2}\Big[\sup\{u : \hat{S}_n(u) > 0\} + \inf\{u : \hat{S}_n(u) < 0\}\Big]. \tag{10.8.9}$$

Next, we compute β_1 and β_0 as given below.

$$\hat{\beta}_{1n}^{(1)} = \frac{1}{2}\Big[\sup\{b : L_n(b|\hat{\rho}_n^{(1)}) > 0\} + \inf\{b : L_n(b|\hat{\rho}_n^{(1)}) < 0\}\Big]$$

and

$$\tilde{\beta}_{0n}^{(1)} = \frac{1}{2}\Big[\sup\{a : T_n(a, \hat{\beta}_{1n}^{(1)}|\hat{\rho}_n^{(1)}) > 0\} + \inf\{a : T_n(a, \hat{\beta}_{1n}^{(1)}|\hat{\rho}_n^{(1)}) < 0\}\Big], \tag{10.8.10}$$

where

$$T_n(a, b|\hat{\rho}_n^{(1)}) = n^{-1} \sum_{t=2}^{n} a_n^+ (R_t^+(a, b|\hat{\rho}_n^{(1)}) sign(Y_t^* - a - bx_t^*)$$

and

$$L_n(b|\hat{\rho}_n^{(1)}) = n^{-1} \sum_{t=2}^{n} (x_k^* - \bar{x}_n^*) a_n (R_t(b|\hat{\rho}_n^{(1)})), \tag{10.8.11}$$

respectively and $Y_t^* = Y_t - \rho Y_{t-1}$ and $x_t^* = x_t - \rho x_{t-1}$ for $t = 1, \ldots, n$ (see Chapter 3). The process is repeated until the R-estimators are stabilized. The asymptotic distribution of $\sqrt{n}\{(\tilde{\beta}_{0n} - \beta_0), (\tilde{\beta}_{1n} - \beta_1)\}$ can be proved to be an asymptotically bivariate normal covariance matrix as given by (10.8.6) (see Problem 20), so this is an open problem.

10.9 Conclusions

In this chapter, we considered the simple linear model and the multiple regression model with autocorrelated errors and discussed various estimators as in Chapter 7. If the autocorrelation in the model is unknown, we showed that by asymptotic theory consistent estimation of the autocorrelation parameter (ρ) and asymptotic properties of the various estimators can be obtained. Further, we discussed the asymptotic properties of the estimators of the parameters of an autoregressive model. In addition, we discussed R-estimation of the regression parameters, where autocorrelation may or may not be known, based on Durbin's (1960) method.

10.10 Problems

1. Verify that the inverse of the matrix, $\boldsymbol{\Sigma}_\rho$ defined by

$$\boldsymbol{\Sigma}_\rho = \begin{pmatrix} 1 & \rho & \rho^2 & \cdots & \rho^{n-1} \\ \rho & 1 & \rho & \cdots & \rho^{n-2} \\ \vdots & \vdots & \vdots & \vdots & \vdots \\ \rho^{n-1} & \rho^{n-2} & \cdots & \cdots & 1 \end{pmatrix}$$

to be

$$\frac{1}{1-\rho^2} \begin{pmatrix} 1 & -\rho & 0 & 0 & 0 \\ -\rho & 1+\rho^2 & -\rho & \cdots & 0 \\ \vdots & \vdots & \vdots & \vdots & \vdots \\ 0 & \cdots & -\rho & 1+\rho^2 & -\rho \\ 0 & \cdots & 0 & -\rho & 1 \end{pmatrix}$$

2. Show that if $\mathbf{Y}_n \sim \mathcal{N}(\theta\mathbf{1}_n + \beta\mathbf{x}, \sigma_v^2\mathbf{\Sigma}_\rho)$, where $\mathbf{Y}_n = (Y_1, \cdots, Y_n)'$ and $\mathbf{x} = (x_1, \cdots, x_n)'$, then the following hold:
   (a) $\bar{\mathbf{Y}} \sim \mathcal{N}(\theta + \beta\bar{x}, \frac{\sigma_v^2}{n}\mathbf{\Sigma}_\rho)$,
   (b) If ρ is known, the estimators of θ and β are given by (10.1.9) with covariance matrix given by (10.1.10).
   (c) The **unbiased estimator** of σ_v^2 $(m = n - 2)$

$$\tilde{\sigma}_v^2 = \frac{1}{m}[\mathbf{Y}_n - \tilde{\theta}_n(\rho)\mathbf{1}_n - \tilde{\beta}_n(\rho)\mathbf{x}]'\mathbf{\Sigma}_\rho^{-1}[\mathbf{Y}_n - \tilde{\theta}_n(\rho)\mathbf{1}_n - \tilde{\beta}_n'(\rho)\mathbf{x}],$$

   and $m\tilde{\sigma}_v^2/\sigma_v^2$ follows the central chi-square distribution with m D.F. independent of $\tilde{\theta}_n(\rho)$ and $\tilde{\beta}_n(\rho)$.
3. Show that the bias and MSE of the estimators $\tilde{\beta}_n(\rho), \hat{\beta}_n^{PT}(\rho)$ and $\hat{\beta}_n^S(\rho)$ are given by (10.1.20) through (10.1.22).
4. Show that the bias and MSE of the estimators $\tilde{\theta}_n(\rho), \hat{\theta}_n(\rho), \hat{\theta}_n^{PT}(\rho)$, and $\hat{\theta}_n^S(\rho)$ are given by (10.1.23) through (10.1.26).
5. Consider the model (10.1.1) and the prediction of Y value given $x = x_0$, where ρ is known,

$$\tilde{Y}_n(x_0) = \tilde{\theta}_n(\rho) + \tilde{\beta}_n(\rho)x_0.$$

   Find the bias and MSE of the following estimators:
   (i) $\tilde{Y}_n(x_0)$,
   (ii) $\hat{Y}_n(x_0) = \bar{Y}$,
   (iii) $\hat{Y}_n^{PT}(x_0) = \bar{Y}I(\mathcal{L}_n < F_{1,m}(\alpha)) + \tilde{Y}_n(x_0)I(\mathcal{L}_n \geq F_{1,m}(\alpha))$
   (iv) $\hat{Y}_n^S(x_0) = \bar{Y} + \left(1 - \frac{c\tilde{\sigma}_v K_1^{1/2}}{(K_1K_2 - K_3^2)^{1/2}|\tilde{\beta}_n(\rho)|}\right)\tilde{\beta}_n(\rho)x_0.$
6. Consider the model (10.1.1) when ρ is unknown. Show that
   (a) $\tilde{\rho}_n = \rho + O_p(n^{-1/2})$, holds as $n \to \infty$. (b) $\tilde{\theta}_n = \theta + O_p(n^{-1/2})$,
   (c) $\tilde{\beta}_n = \beta + O_p(n^{-1/2})$ and
   (d) $[\sqrt{n}(\tilde{\theta}_n - \theta), \sqrt{n}(\tilde{\beta}_n - \beta), \sqrt{n}(\tilde{\rho}_n - \rho)]'$

$$\simeq \mathcal{N}_3\left\{(0,0,0); \frac{\sigma_v^2}{K_1K_2 - K_3^2}\begin{pmatrix} K_2 & -K_3 & 0 \\ -K_3 & K_1 & 0 \\ 0 & 0 & (1-\rho^2)(K_1K_2 - K_3^2) \end{pmatrix}\right\}.$$

7. Verify that expressions for ADB and ADMSE of the estimators of ρ, β and θ as given by (10.1.20) through (10.1.26).
8. Consider the multiple regression model (10.2.1) when ρ is known. Show that

$$\mathbf{C}^{-1}\mathbf{C}_\rho\mathbf{C}^{-1} - \mathbf{C}^{-1} \geq \mathbf{0}.$$

9. Verify the expressions for the bias, quadratic bias, MSE matrices, and the risks for the estimators of $\boldsymbol{\beta}$ given by (10.3.6) through (10.3.13).
10. Following Chapter 7, Section 7.8.1, show that the estimators of $\boldsymbol{\beta}$ when ρ is unknown are asymptotically risk equivalent under fixed alternatives (refer to section 10.6).
11. Verify the expression for the ADB, ADQB, ADMSE, and ADQR of $\boldsymbol{\beta}$ when ρ is consistently estimated by $\tilde{\rho}_n$, given by (10.4.4) through (10.4.9).
12. Following Chapter 7, Section 7.8.1, show that the estimators of $\boldsymbol{\theta}$ are asymptotically risk equivalent under fixed alternatives.
13. Following Chapter 7, Section 7.9, develop the asymptotic properties of the five confidence sets, where the centers are $\tilde{\boldsymbol{\theta}}_n, \hat{\boldsymbol{\theta}}_n, \hat{\boldsymbol{\theta}}_n^{PT}, \hat{\boldsymbol{\theta}}_n^{S}$, and $\hat{\boldsymbol{\theta}}_n^{S+}$, respectively (Ref 10.6).
14. Prove Theorem 1 of Section 10.7.2.
15. Prove $n^{-1/2}\hat{\mathbf{L}}_n(\mathbf{0}) \sim \mathcal{N}_p(\mathbf{0}, \sigma_\phi^2 \boldsymbol{\Sigma})$ and $\sqrt{n}(\tilde{\boldsymbol{\theta}}_n - \boldsymbol{\theta}) \sim \mathcal{N}_p(\mathbf{0}, \gamma^{-2}\sigma_\phi^2 \boldsymbol{\Sigma})$.
16. Prove that as $n \to \infty$,

$$\mathcal{L}_n = \sigma_\phi^2 (\mathbf{H}\tilde{\boldsymbol{\theta}}_n - \mathbf{h})'(\mathbf{H}\boldsymbol{\Sigma}^{-1}\mathbf{H}')^{-1}(\mathbf{H}\tilde{\boldsymbol{\theta}}_n - \mathbf{h}) + o_p(1),$$

where $\mathcal{L}_n$ is defined by (10.7.11).
17. Prove that the local alternatives $\{K_{(n)}\}$ defined by $\mathbf{H}\boldsymbol{\theta} = h + n^{-1/2}\boldsymbol{\xi}$ is contiguous to $H_0 : \mathbf{H}\boldsymbol{\theta} = \mathbf{h}$.
18. Prove Theorem 2 of Section 10.7.4.
19. Formulate the recentered $(1-\alpha)\%$ confidence interval when $\boldsymbol{\Sigma}$ is known, and study the dominance properties as Section 7.9.
20. Develop the R-estimation of the intercept and multiple regression parameters in the model

$$\mathbf{Y} = \beta_0 \mathbf{1}_n + \mathbf{X}\boldsymbol{\beta} + \mathbf{e}$$

when (i) ρ is known and (ii) when ρ is unknown.

Chapter 11

Multivariate Models

Outline

Real-life data sets involve measurements of two or more characteristics of individual samples chosen for study. For example, height, weight, chest-girth, and the like, can be treated as vectors of observations, $\mathbf{Y} = (Y_1, Y_2, \cdots, Y_p)'$, of p-component characteristics. The component characteristics are not necessarily independent but correlated. The usual distribution that is used for analysis of a vector data set of fixed sample size is the multivariate normal distribution (MND). Thus, for statistical analysis we consider a random sample $\mathbf{Y}_1, \cdots, \mathbf{Y}_N$ of size N from the MND, $\mathcal{N}_p(\boldsymbol{\theta}, \boldsymbol{\Sigma})$ with mean vector $\boldsymbol{\theta}$ and the $p \times p$ covariance matrix, $\boldsymbol{\Sigma}$. In general, $(\boldsymbol{\theta}, \boldsymbol{\Sigma})$ is **unknown**, and the simplest statistical problem is the estimation of $(\boldsymbol{\theta}, \boldsymbol{\Sigma})$ based on the sample, $\mathbf{Y}_1, \mathbf{Y}_2, \cdots, \mathbf{Y}_N$. It is well-known (e.g., see Anderson, 1984; Srivastava and Khatri, 1986) that the maximum likelihood estimator of $(\boldsymbol{\theta}, \boldsymbol{\Sigma})$ is $(\overline{\mathbf{Y}}_N, \mathbf{S})$, where $\overline{\mathbf{Y}}_N = \frac{1}{N}\sum_{i=1}^{N} \mathbf{Y}_i$ and $(N-1)\mathbf{S} = \sum_{i=1}^{N}(\mathbf{Y}_i - \overline{\mathbf{Y}}_N)(\mathbf{Y}_i - \overline{\mathbf{Y}}_N)'$. Further, $(\overline{\mathbf{Y}}_N, \mathbf{S})$ satisfies the estimation criteria of Fisher, sufficiency, unbiasedness, uniform minimum variance unbiased estimation (UMVUE), as well as the decision theoretic criterion of "minimaxity".

As we mentioned in Chapter 4, Stein (1956) proposed that the usual estimator of $\boldsymbol{\theta}$, namely $\overline{\mathbf{Y}}_N$, may be uniformly improved by a nonlinear estimator under a quadratic loss function. In this chapter, we begin with the James-Stein estimator of $\boldsymbol{\theta}$ that uniformly improves over the sample mean $\overline{\mathbf{Y}}_N$ under a quadratic loss. It has inspired many researchers to travel in different paths to appreciate the impact of the Stein's theory in the parametric estimation.

We present the Bayes and the empirical Bayes methods (see Efron and Morris, 1973) of estimating $\boldsymbol{\theta}$ as well. We consider the simple linear model in a multivariate setup and propose preliminary test and Stein-type estimators of the **intercept** and the **slope**-vector parameters. Further, we discuss the related improved confidence-set problem. (e.g., see Berger, 1980b, Lu and Berger, 1989). As a particular case of the simple linear model, we obtain the results for the two-sample problem of estimating the mean vectors when they may be equal. In the course of various developments, Saleh and Sen (1978–'86) merged two diverse areas of the preliminary test and Stein-type estimation in a nonparametric setup that includes robust statistics. In this regard we considered U-statistics and the problem of estimating functionals of p-variate nonnormal distributions, and we extend the methodology to R-estimation of the **intercept** and **slope**-vectors when the error distribution is nonnormal but diagonally symmetric.

11.1 Point and Set Estimation of the Mean Vector of an MND

In this section, we consider the simplest one-sample problem of estimating the mean vector $\boldsymbol{\theta}$ of the MND, $\mathcal{N}_p(\boldsymbol{\theta}, \boldsymbol{\Sigma})$ based on the sample, $\mathbf{Y}_1, \cdots, \mathbf{Y}_N$ of size N when $\boldsymbol{\Sigma}$ is unknown.

11.1.1 Model, Estimation, and Test of Hypothesis

Consider the one-sample multivariate model:

$$\mathbf{Y}_i = \boldsymbol{\theta} + \boldsymbol{\varepsilon}_i, \quad \boldsymbol{\varepsilon}_i \sim \mathcal{N}_p(\mathbf{0}, \boldsymbol{\Sigma}), \quad i = 1, \cdots, N, \tag{11.1.1}$$

where $\mathbf{Y}_i = (Y_{i1}, \cdots, Y_{ip})'$, $\boldsymbol{\theta} = (\theta_1, \cdots, \theta_p)'$, and $\boldsymbol{\varepsilon}_i = (\epsilon_{i1}, \cdots, \epsilon_{ip})'$.

Our main objective is the estimation of $\boldsymbol{\theta}$ when it is suspected that $\boldsymbol{\theta}$ may be $\mathbf{0}$ and to study the properties of the estimator(s) under a quadratic loss function. Accordingly, we propose (1) the **unrestricted** MLE, (2) the **restricted** MLE, (3) the PTE, (4) the James-Stein estimator (JSE), and (5) the positive-rule Stein estimator (PRSE) of $\boldsymbol{\theta}$.

It is well-known that the MLE of $(\boldsymbol{\theta}, \boldsymbol{\Sigma})$ is $(\overline{\mathbf{Y}}_N, \mathbf{S})$. So we designate the unrestricted estimator (UE) of $\boldsymbol{\theta}$ to be $\tilde{\boldsymbol{\theta}}_N (= \overline{\mathbf{Y}}_N)$ and the **restricted** estimator to be $\hat{\boldsymbol{\theta}}_N (= \mathbf{0}$ in this case).

Since $\boldsymbol{\theta}$ may be $\mathbf{0}$, we test the null hypothesis $H_0 : \boldsymbol{\theta} = \mathbf{0}$ against $\boldsymbol{\theta} \neq \mathbf{0}$ based on the **liklihood ratio test** given by

$$\mathcal{L}_N = \frac{m}{np} N \tilde{\boldsymbol{\theta}}_N' \mathbf{S}^{-1} \tilde{\boldsymbol{\theta}}_N, \quad m = N - p \quad \text{and} \quad n = N - 1. \tag{11.1.2}$$

Then, from Rao (1973) and Srivastava and Khatri (1979, p. 113), we have

$$\frac{N}{n}\tilde{\boldsymbol{\theta}}'_N \mathbf{S}^{-1}\tilde{\boldsymbol{\theta}}_N = \frac{\tilde{\boldsymbol{\theta}}'_N \mathbf{S}^{-1}\tilde{\boldsymbol{\theta}}_N}{n\tilde{\boldsymbol{\theta}}'_N \boldsymbol{\Sigma}^{-1}\tilde{\boldsymbol{\theta}}_N} N\tilde{\boldsymbol{\theta}}'_N \boldsymbol{\Sigma}^{-1}\tilde{\boldsymbol{\theta}}_N. \tag{11.1.3}$$

For a given $\tilde{\boldsymbol{\theta}}_N$,

$$\frac{n\tilde{\boldsymbol{\theta}}'_N \boldsymbol{\Sigma}^{-1}\tilde{\boldsymbol{\theta}}_N}{\tilde{\boldsymbol{\theta}}'_N \mathbf{S}^{-1}\tilde{\boldsymbol{\theta}}_N} \stackrel{\mathcal{D}}{\simeq} \chi^2_{N-p}, \tag{11.1.4}$$

and we know that $\tilde{\boldsymbol{\theta}}_N \stackrel{\mathcal{D}}{\simeq} \mathcal{N}_p(\boldsymbol{\theta}, N^{-1}\boldsymbol{\Sigma})$ which leads to conclude that

$$N\tilde{\boldsymbol{\theta}}'_N \boldsymbol{\Sigma}^{-1}\tilde{\boldsymbol{\theta}}_N \stackrel{\mathcal{D}}{\simeq} \chi^2_p(\Delta^2), \quad \Delta^2 = N\boldsymbol{\theta}'\boldsymbol{\Sigma}^{-1}\boldsymbol{\theta}, \tag{11.1.5}$$

so the distributions (11.1.4) and (11.1.5) are independent. That is to say,

$$\mathcal{L}_N = \frac{m}{np} N\tilde{\boldsymbol{\theta}}'_N \mathbf{S}^{-1}\tilde{\boldsymbol{\theta}}_N \stackrel{\mathcal{D}}{\simeq} \frac{m}{p}\frac{\chi^2_p(\Delta^2)}{\chi^2_m} \stackrel{\mathcal{D}}{\simeq} F_{p,m}(\Delta^2), \tag{11.1.6}$$

where $\mathcal{L}_N$ follows a noncentral F-distribution with (p, m) d.f. and noncentrality parameter $\Delta^2/2$.

If $\boldsymbol{\theta} = \mathbf{0}$, the F-distribution is central. Thus, the PTE of $\boldsymbol{\theta}$ can be written as

$$\hat{\boldsymbol{\theta}}_N^{\text{PT}} = \tilde{\boldsymbol{\theta}}_N - \tilde{\boldsymbol{\theta}}_N I(\mathcal{L}_N < F_{p,m}(\alpha)), \tag{11.1.7}$$

where $F_{p,m}(\alpha)$ is the upper α-level critical value from the central F-distribution. Similarly, the James-Stein estimator of $\boldsymbol{\theta}$ can be written as

$$\hat{\boldsymbol{\theta}}_N^{S} = \tilde{\boldsymbol{\theta}}_N - d\tilde{\boldsymbol{\theta}}_N \mathcal{L}_N^{-1}, \quad d = \frac{(p-2)m}{p(m+2)}, \tag{11.1.8}$$

and the PRSE of $\boldsymbol{\theta}$ is given by

$$\hat{\boldsymbol{\theta}}_N^{S+} = (1 - d\mathcal{L}_N^{-1}) I(\mathcal{L}_N > d)\tilde{\boldsymbol{\theta}}_N. \tag{11.1.9}$$

We now prove that the proposed estimators, namely $\tilde{\boldsymbol{\theta}}_N$, $\hat{\boldsymbol{\theta}}_N^{\text{PT}}$, $\hat{\boldsymbol{\theta}}_N^{S}$, and $\hat{\boldsymbol{\theta}}_N^{S+}$, belong to the class of "empirical Bayes" estimators.

First, we note that $\tilde{\boldsymbol{\theta}}_N \sim \mathcal{N}_p(\boldsymbol{\theta}; N^{-1}\boldsymbol{\Sigma})$, and since $\boldsymbol{\theta}$ may be $\mathbf{0}$, we assume the prior distribution of $\boldsymbol{\theta}$ to be $\mathcal{N}_p(\mathbf{0}, \tau^2 N^{-1}\boldsymbol{\Sigma})$, where τ^2 is a scalar. Thus, a plausible estimator of $\boldsymbol{\theta}$ may be of the form $k\tilde{\boldsymbol{\theta}}_N$, where k is not known to have been chosen appropriately. Applying Bayes Theorem, we have the conditional distribution of $\boldsymbol{\theta}|\tilde{\boldsymbol{\theta}}_N$ as $\mathcal{N}_p\{k\tilde{\boldsymbol{\theta}}_N, kN^{-1}\boldsymbol{\Sigma}\}$, where $k = \tau^2/1 + \tau^2$. Hence, the Bayes estimator $\boldsymbol{\theta}$, is given by $\hat{\boldsymbol{\theta}}_N^{B} = k\tilde{\boldsymbol{\theta}}_N$, where $\tilde{\boldsymbol{\theta}}_N$ and $\mathbf{0}$ are the extreme estimators of $\boldsymbol{\theta}$ when $\tau^2 \to \infty$ and $\tau^2 \to 0$, respectively. Here k is still unknown, and equivalently τ^2 is unknown. In order to estimate k, we write it as $[1-1/(1+\tau^2)]$ so that $\hat{\boldsymbol{\theta}}_N^B = \left(1-\frac{1}{1+\tau^2}\right)\tilde{\boldsymbol{\theta}}_N$ where $1/(1+\tau^2)$ can be estimated. Thus, we consider the marginal distributions of $n\mathbf{S}$ and $\tilde{\boldsymbol{\theta}}_N$ that are independent and are given by $n\mathbf{S}\overset{\mathcal{D}}{=}W_p(\boldsymbol{\Sigma},n)$ and $\tilde{\boldsymbol{\theta}}_N \simeq \mathcal{N}_p(\mathbf{0},(1+\tau^2)N^{-1}\boldsymbol{\Sigma})$, respectively. Then $(1+\tau^2)^{-1}N\tilde{\boldsymbol{\theta}}_N'\boldsymbol{\Sigma}^{-1}\tilde{\boldsymbol{\theta}}_N\overset{\mathcal{D}}{=}\chi_p^2$ and $\mathcal{L}_N = \frac{ns}{np}N\tilde{\boldsymbol{\theta}}_N'\mathbf{S}^{-1}\tilde{\boldsymbol{\theta}}_N\overset{\mathcal{D}}{=}(1+\tau^2)^{-1}F_{p,m}$. So it is easy to see that

$$(1+\tau^2)^{-1}E(\mathcal{L}_N^{-1}) = E[F_{p,m}^{-1}] = \frac{p}{p-2} \quad \text{for } p>2. \tag{11.1.10}$$

Hence, $(1+\tau^2)^{-1}$ can be estimated by a scalar multiple of $\mathcal{L}_N^{-1}$. More generally, let $g(\mathcal{L}_N)$ a real valued function of $\mathcal{L}_N$, be an estimator of $(1+\tau^2)^{-1}$. Accordingly, we define a class of "empirical Bayes" estimators of $\boldsymbol{\theta}$ as a substitute of the Bayes estimator:

$$\hat{\boldsymbol{\theta}}_N^{EB} = (1-g(\mathcal{L}_N))\tilde{\boldsymbol{\theta}}_N. \tag{11.1.11}$$

We consider the following five choices of $g(\mathcal{L}_N)$. (1) $g(\mathcal{L}_N)=0$, $\hat{\boldsymbol{\theta}}_N^{EB}=\tilde{\boldsymbol{\theta}}_N$; (2) $g(\mathcal{L}_N)=1$, $\hat{\boldsymbol{\theta}}_N^{EB}=\mathbf{0}$; (3) $g(\mathcal{L}_N)=I(\mathcal{L}_N\le F_{p,m}(\alpha))$, $\hat{\boldsymbol{\theta}}_N^{EB}=\hat{\boldsymbol{\theta}}_N^{PT}$; (4) $g(\mathcal{L}_N)=d\mathcal{L}_N^{-1}$, $d=\frac{(p-2)m}{p(m+2)}$, $\hat{\boldsymbol{\theta}}_N^{EB}=\hat{\boldsymbol{\theta}}_N^{S}$; and (5) $g(\mathcal{L}_N)=1-(1-d\mathcal{L}_N^{-1})I(\mathcal{L}_N>d)$. In this case, $\hat{\boldsymbol{\theta}}_N^{EB}=\hat{\boldsymbol{\theta}}_N^{S+}$. Note that PSRE is a PTE, since $\hat{\boldsymbol{\theta}}_N I(\mathcal{L}_N<d)+\hat{\boldsymbol{\theta}}_N^{S}I(\mathcal{L}_N>d)=\hat{\boldsymbol{\theta}}_N^{S}I(\mathcal{L}_N>d)=\hat{\boldsymbol{\theta}}_N^{S+}$ (see also Efron and Morris, 1972-'77); and Morris, 1983).

11.1.2 Bias, QB, MSE Matrix, and Weighted Risk Expressions of the Estimators

Consider the following formulas for the bias, MSE matrices, and risks of the estimators: (1) $\mathbf{b}(\boldsymbol{\theta}_N^*)=E[\boldsymbol{\theta}_N^*-\boldsymbol{\theta}]$ and $B(\boldsymbol{\theta}_N^*)=N[\mathbf{b}(\boldsymbol{\theta}_N^*)]'\boldsymbol{\Sigma}^{-1}[\mathbf{b}(\boldsymbol{\theta}_N^*)]$, (2) $\mathbf{M}(\boldsymbol{\theta}_N^*)=E[N(\boldsymbol{\theta}_N^*-\boldsymbol{\theta})(\boldsymbol{\theta}_N^*-\boldsymbol{\theta})']$, and (3) $R(\boldsymbol{\theta}_N^*;\mathbf{Q})=\text{tr}[\mathbf{Q}\mathbf{M}(\boldsymbol{\theta}_N^*)]$, where $\boldsymbol{\theta}_N^*$ stands for any of the five estimators of $\boldsymbol{\theta}$.

For the unrestricted estimator we have $\mathbf{b}_1(\tilde{\boldsymbol{\theta}}_N)=\mathbf{0}$ and $B_1(\tilde{\boldsymbol{\theta}}_N)=0$, $\mathbf{M}_1(\tilde{\boldsymbol{\theta}}_N)=\boldsymbol{\Sigma}$ and $R_1(\tilde{\boldsymbol{\theta}}_N;\mathbf{Q})=\text{tr}(\mathbf{Q}\boldsymbol{\Sigma})$. (11.1.12)

For the PTE, we obtain $\mathbf{b}_2(\hat{\boldsymbol{\theta}}_N^{PT})=-\boldsymbol{\theta}G_{p+2,m}\big(\ell_\alpha(\alpha);\Delta^2\big)$ and $B_2(\hat{\boldsymbol{\theta}}_N^{PT})=\Delta^2\big\{G_{p+2,m}\big(\ell_\alpha(\alpha);\Delta^2\big)\big\}^2$, $\mathbf{M}_2(\hat{\boldsymbol{\theta}}_N^{PT})=\boldsymbol{\Sigma}-\boldsymbol{\Sigma}G_{p+2,m}\big(\ell_\alpha(\alpha);\Delta^2\big)+N\boldsymbol{\theta}\boldsymbol{\theta}$ $\times\big\{2G_{p+2,m}\big(\ell_\alpha(\alpha);\Delta^2\big)-G_{p+4,m}\big(\ell_\alpha^*(\alpha);\Delta^2\big)\big\}$ and $R_2(\hat{\boldsymbol{\theta}}_N^{PT};\mathbf{Q})=\text{tr}(\mathbf{Q}\boldsymbol{\Sigma})-\text{tr}(\mathbf{Q}\boldsymbol{\Sigma})G_{p+2,m}(\ell_\alpha(\alpha);\Delta^2)+N\boldsymbol{\theta}'\mathbf{Q}\boldsymbol{\theta}\big\{2G_{p+2,m}\big(\ell_\alpha(\alpha);\Delta^2\big)$ $-G_{p+4,m}\big(\ell_\alpha^*(\alpha);\Delta^2\big)\big\}$, (11.1.13)

where $\ell_\alpha=\frac{p}{p+2}F_{p,m}(\alpha)$ and $\ell_\alpha^*=\frac{p}{p+4}F_{p,m}(\alpha)$. The expressions for JSE, $\hat{\boldsymbol{\theta}}_N^S$, are given by

$$\mathbf{b}_3(\hat{\boldsymbol{\theta}}_N^S)=-dp\boldsymbol{\theta}E[\chi_{p+2}^{-2}(\Delta^2)] \text{ and } B_3(\hat{\boldsymbol{\theta}}_N^S)=d^2p^2\Delta^2\Big\{E[\chi_{p+2}^{-2}(\Delta^2)]\Big\}^2,$$

$$\mathbf{M}_3(\hat{\boldsymbol{\theta}}_N^S) = \boldsymbol{\Sigma} - dp\boldsymbol{\Sigma}\Big\{2E[\chi_{p+2}^{-2}(\Delta^2)] - (p-2)E[\chi_{p+2}^{-4}(\Delta^2)]\Big\}$$
$$+ dp(p+2)N\boldsymbol{\theta}\boldsymbol{\theta}'E[\chi_{p+4}^{-4}(\Delta^2)], \quad \text{and}$$
$$R_3(\hat{\boldsymbol{\theta}}_N^S; \mathbf{Q}) = \text{tr}(\mathbf{Q}\boldsymbol{\Sigma}) - dp\text{tr}(\mathbf{Q}\boldsymbol{\Sigma})\Big\{2E[\chi_{p+2}^{-2}(\Delta^2)] - (p-2)E[\chi_{p+2}^{-4}(\Delta^2)]\Big\}$$
$$+ dp(p+2)N\boldsymbol{\theta}'\mathbf{Q}\boldsymbol{\theta}E[\chi_{p+4}^{-4}(\Delta^2)]. \tag{11.1.14}$$

Similarly, the expressions for the PRSE, $\hat{\boldsymbol{\theta}}_N^{S+}$, are given by

$$\mathbf{b}_4(\hat{\boldsymbol{\theta}}_N^{S+}) = \mathbf{b}_3(\hat{\boldsymbol{\theta}}_N^S) - \boldsymbol{\theta}E[(1-d_1F_{p+2,m}^{-1}(\Delta^2))I(F_{p+2,m}(\Delta^2) < d_1)]$$

and

$$B_4(\hat{\boldsymbol{\theta}}_N^{S+}) = \Delta^2\{E[\chi_{p+2}^{-2}(\Delta^2)] + E[(1-d_1F_{p+2,m}^{-1}(\Delta^2))I(F_{p+2,m}(\Delta^2) < d_1)]\}^2,$$

$$\mathbf{M}_4(\hat{\boldsymbol{\theta}}_N^{S+}) = \mathbf{M}_3(\hat{\boldsymbol{\theta}}_N^S) - \boldsymbol{\Sigma}\{E[(1 - d_1F_{p+2,m}^{-1}(\Delta^2))^2I(F_{p+2,m}(\Delta^2) < d_1)]\}$$
$$+N\boldsymbol{\theta}\boldsymbol{\theta}'\{2E[(1 - d_1F_{p+2,m}^{-1}(\Delta^2))I(F_{p+2,m}(\Delta^2) < d_1)]$$
$$-E[(1 - d_2F_{p+4,m}^{-1}(\Delta^2))^2I(F_{p+4,m}(\Delta^2) < d_2)]\}$$

where $d_i = \frac{dp}{p+2i}$, $i = 1, 2$ and

$$R_4(\hat{\boldsymbol{\theta}}_N^{S+}; \mathbf{Q})$$
$$= R_3(\hat{\boldsymbol{\theta}}_N^S; \mathbf{Q}) - \text{tr}(\mathbf{Q}\boldsymbol{\Sigma})\{E[(1 - d_1F_{p+2,m}^{-1}(\Delta^2))^2I(F_{p+2,m}(\Delta^2) < d_1)]\}$$
$$+N\boldsymbol{\theta}'\mathbf{Q}\boldsymbol{\theta}\{2E[(1 - d_1F_{p+2,m}^{-1}(\Delta^2))I(F_{p+2,m}(\Delta^2) < d_1)]$$
$$-E[(1 - d_2F_{p+4,m}^{-1}(\Delta^2))^2I(F_{p+4,m}(\Delta^2) < d_2)]\}. \tag{11.1.15}$$

11.1.3 Risk and MSE Analysis of the Estimators

First, we note that

$$R_1(\tilde{\boldsymbol{\theta}}_N; \mathbf{Q}) \geq R_3(\hat{\boldsymbol{\theta}}_N^S; \mathbf{Q}) \quad \forall \Delta^2 \in \mathcal{R}_1^+, \tag{11.1.16}$$

since

$$R_3(\hat{\boldsymbol{\theta}}_N^S; \mathbf{Q}) = \text{tr}(\mathbf{Q}\boldsymbol{\Sigma}) - \frac{(p-2)m}{m+2}\text{tr}(\mathbf{Q}\boldsymbol{\Sigma})\Big\{(p-2)E[\chi_{p+2}^{-4}(\Delta^2)]$$
$$\times\Big(1 - \frac{(p+2)N\boldsymbol{\theta}'\mathbf{Q}\boldsymbol{\theta}}{2\Delta^2\text{tr}(\mathbf{Q}\boldsymbol{\Sigma})}\Big)2\Delta^2E[\chi_{p+4}^{-4}(\Delta^2)]\Big\} \geq 0 \tag{11.1.17}$$

for all Δ^2 provided that the matrix $\mathbf{Q}$ belongs to the set satisfying

$$\frac{\text{tr}(\mathbf{Q}\boldsymbol{\Sigma})}{Ch_{max}(\mathbf{Q}\boldsymbol{\Sigma})} \geq \frac{p+2}{2}. \tag{11.1.18}$$

Further,

$$R_3(\tilde{\boldsymbol{\theta}}_N^S;\mathbf{Q}) \geq R_4(\hat{\boldsymbol{\theta}}_N^{S+};\mathbf{Q}) \quad \forall\Delta^2 \in \mathcal{R}_1^+, \tag{11.1.19}$$

since

$$\begin{aligned} &R_3(\hat{\boldsymbol{\theta}}_N^S;\mathbf{Q}) - R_4(\hat{\boldsymbol{\theta}}_N^{S+};\mathbf{Q}) \\ &= \operatorname{tr}(\mathbf{Q\Sigma})E[(1-d_1F_{p+2,m}^{-1}(\Delta^2))^2 I(F_{p+2,m}(\Delta^2) < d_1)] \\ &-N\boldsymbol{\theta}'\mathbf{Q}\boldsymbol{\theta}\Big\{2E[(1-d_1F_{p+2,m}^{-1}(\Delta^2))I(F_{p+2,m}(\Delta^2) < d_1)] \\ &-E[(1-d_2F_{p+4,m}^{-1}(\Delta^2))^2 I(F_{p+4,m}(\Delta^2) < d_2)]\Big\}. \end{aligned} \tag{11.1.20}$$

The term $2E[(1-d_1F_{p+2,m}^{-1}(\Delta^2))I(F_{p+2,m}(\Delta^2) < d_1)]$ is negative, whereas $E[(1-d_2F_{p+4,m}^{-1}(\Delta^2))^2 I(F_{p+4,m}(\Delta^2) < d_2)]$ is positive. Hence, the third term of (11.2.5) is positive, and (11.2.1) holds. Thus, all together, we obtain

$$R_1(\tilde{\boldsymbol{\theta}}_N;\mathbf{Q}) \geq R_3(\hat{\boldsymbol{\theta}}_N^S;\mathbf{Q}) \geq R_4(\hat{\boldsymbol{\theta}}_N^{S+};\mathbf{Q}) \quad \forall(\Delta^2,\mathbf{Q}) \tag{11.1.21}$$

where $\mathbf{Q}$ belongs to the set (11.1.16) with $p \geq 3$.

Next, we observe that $\hat{\boldsymbol{\theta}}_N^{PT}$ is superior to $\tilde{\boldsymbol{\theta}}_N$ whenever

$$\Delta^2 \leq \frac{\operatorname{tr}(\mathbf{Q\Sigma})G_{p+2,m}(\ell_\alpha(\alpha);\Delta^2)}{Ch_{max}(\mathbf{Q\Sigma})[2G_{p+2,m}(\ell_\alpha(\alpha);\Delta^2) - G_{p+4,m}(\ell_\alpha^*(\alpha);\Delta^2)]}; \tag{11.1.22}$$

otherwise, $\tilde{\boldsymbol{\theta}}_N$ is superior whenever

$$\Delta^2 \geq \frac{\operatorname{tr}(\mathbf{Q\Sigma})G_{p+2,m}(\ell_\alpha(\alpha);\Delta^2)}{Ch_{min}(\mathbf{Q\Sigma})[2G_{p+2,m}(\ell_\alpha(\alpha);\Delta^2) - G_{p+4,m}(\ell_\alpha^*(\alpha);\Delta^2)]}. \tag{11.1.23}$$

The efficiency of $\hat{\boldsymbol{\theta}}_N^{PT}$ can be written, for example, as

$$\operatorname{Eff}[\hat{\boldsymbol{\theta}}_N^{PT} : \tilde{\boldsymbol{\theta}}_N] = R_1(\tilde{\boldsymbol{\theta}}_N;\mathbf{Q})[R_2(\hat{\boldsymbol{\theta}}_N^{PT};\mathbf{Q})]^{-1} = E[\alpha,\Delta^2], \tag{11.1.24}$$

since it depends on α and Δ^2. As we know PTE is not uniformly better than $\tilde{\boldsymbol{\theta}}_N$. Nevertheless, we can obtain a PTE with a minimum guaranteed efficiency, say, E_0, by choosing a suitable level of significance, α^*, in solving the equation

$$\max_{\alpha}\min_{\Delta^2} E(\alpha,\Delta^2) \geq E_0. \tag{11.1.25}$$

Some idea of optimum α^* may be found in the Table 4.2.2 of Chapter 4.

Next we consider an MSE analysis:

Comparison of $\tilde{\boldsymbol{\theta}}_N$ and $\hat{\boldsymbol{\theta}}_N^{\mathrm{PT}}$. Consider the MSE-matrix difference

$$\begin{aligned} &\mathbf{M}_1(\tilde{\boldsymbol{\theta}}_N) - \mathbf{M}_2(\hat{\boldsymbol{\theta}}_N^{\mathrm{PT}}) \\ &= \boldsymbol{\Sigma}G_{p+2,m}(\ell_\alpha;\Delta^2) - N\boldsymbol{\theta}\boldsymbol{\theta}'\Big\{2G_{p+2,m}(\ell_\alpha;\Delta^2) - G_{p+4,m}(\ell_\alpha^*;\Delta^2)\Big\}. \end{aligned} \tag{11.1.26}$$

The matrix difference is n.n.d. if and only if

$$\Delta^2 \leq \frac{G_{p+2,m}(\ell_\alpha;\Delta^2)}{2G_{p+2,m}(\ell_\alpha;\Delta^2) - G_{p+4,m}(\ell_\alpha;\Delta^2)}. \tag{11.1.27}$$

This means that $\hat{\boldsymbol{\theta}}_n^{\text{PT}}$ is superior to $\tilde{\boldsymbol{\theta}}_n$ whenever Δ^2 satisfies (11.2.11); otherwise, $\tilde{\boldsymbol{\theta}}_n$ is superior to $\hat{\boldsymbol{\theta}}_n^{\text{PT}}$.

Comparison of $\tilde{\boldsymbol{\theta}}_N$, $\hat{\boldsymbol{\theta}}_N^S$, and $\hat{\boldsymbol{\theta}}_N^{S+}$. First, we consider $\tilde{\boldsymbol{\theta}}_N$ vs. $\hat{\boldsymbol{\theta}}_N^S$. In this case,

$$\begin{aligned}\mathbf{M}_1(\tilde{\boldsymbol{\theta}}_N) - \mathbf{M}_3(\hat{\boldsymbol{\theta}}_n^S) &= dp\boldsymbol{\Sigma}\Big\{2E\big[\chi_{p+2}^{-2}(\Delta^2)\big] - (p-2)E\big[\chi_{p+2}^{-4}(\Delta^2)\big]\Big\}\\ &\quad - dp(p+2)N\boldsymbol{\theta\theta}'E\big[\chi_{p+4}^{-4}(\Delta^2)\big].\end{aligned} \tag{11.1.28}$$

This difference is n.n.d. if and only if

$$pE\big[\chi_{p+2}^{-2}(\Delta^2)\big] < (p-2)E\big[\chi_{p+2}^{-4}(\Delta^2)\big] \tag{11.1.29}$$

for all Δ^2, which may not hold always. Hence, $\hat{\boldsymbol{\theta}}_N^S$ does not dominate $\tilde{\boldsymbol{\theta}}_N$ uniformly based on the MSE criterian.

For the comparison of $\hat{\boldsymbol{\theta}}_n^S$ and $\hat{\boldsymbol{\theta}}_n^{S+}$, we consider the MSE-matrix difference given by

$$\begin{aligned}&\mathbf{M}_3(\hat{\boldsymbol{\theta}}_n^S) - \mathbf{M}_4(\hat{\boldsymbol{\theta}}_N^{S+})\\ &= \boldsymbol{\Sigma}E\Big[\big(1 - d_1F_{p+2,m}^{-1}(\Delta^2)\big)^2 I\big(F_{p+2,m}(\Delta^2) < d_1\big)\Big]\\ &-N\boldsymbol{\theta\theta}'\Big\{2E\big[\big(1 - d_1F_{p+2,m}^{-1}(\Delta^2)\big)I\big(F_{p+2,m}(\Delta^2) < d_1\big)\big]\\ &-E\big[\big(1 - d_2F_{p+4,m}^{-1}(\Delta^2)\big)^2 I\big(F_{p+4,m}(\Delta^2) < d_2\big)\big]\Big\}.\end{aligned} \tag{11.1.30}$$

The expression (11.1.28) is n.n.d., since $F_{p+2,m}(\Delta^2) < d_1$. Hence, $\hat{\boldsymbol{\theta}}_N^{S+}$ dominates $\hat{\boldsymbol{\theta}}_N^S$ uniformly:

$$\mathbf{M}_4(\hat{\boldsymbol{\theta}}_N^{S+}) \leq \mathbf{M}_3(\hat{\boldsymbol{\theta}}_N^S) \quad \forall\Delta^2. \tag{11.1.31}$$

We may note that $\boldsymbol{\theta}_N^{S+}$ may not dominate $\tilde{\boldsymbol{\theta}}_N$, since $\hat{\boldsymbol{\theta}}_N^S$ does not dominate $\tilde{\boldsymbol{\theta}}_N$. For the comparison of $\hat{\boldsymbol{\theta}}_N^{\text{PT}}$, $\hat{\boldsymbol{\theta}}_N^S$, and $\hat{\boldsymbol{\theta}}_N^{S+}$, see Problem 11.

Finally, we consider the confidence set estimators that are similar to those described in Chapter 4. The confidence sets are given below when $\boldsymbol{\Sigma}$ is known.

$$\text{(i)} \quad C^0(\tilde{\boldsymbol{\theta}}_N) = \{\boldsymbol{\theta} : N\|\boldsymbol{\theta} - \tilde{\boldsymbol{\theta}}_N\|^2_{\boldsymbol{\Sigma}^{-1}} \leq \chi_p^2(\gamma)\} \tag{11.1.32}$$

$$\text{(ii)} \quad C^{\text{PT}}(\hat{\boldsymbol{\theta}}_N^{\text{PT}}(\alpha)) = \{\boldsymbol{\theta} : N\|\boldsymbol{\theta} - \hat{\boldsymbol{\theta}}_N^{\text{PT}}(\alpha)\|^2_{\boldsymbol{\Sigma}^{-1}} \leq \chi_p^2(\gamma)\},$$

$$\text{(iii)} \quad C^S(\hat{\boldsymbol{\theta}}_N^S) = \{\boldsymbol{\theta} : N\|\boldsymbol{\theta} - \hat{\boldsymbol{\theta}}_N^S\|^2_{\boldsymbol{\Sigma}^{-1}} \leq \chi_p^2(\gamma)\},$$

$$\text{and (iv)} \quad C^{S+}(\hat{\boldsymbol{\theta}}_N^{S+}) = \{\boldsymbol{\theta} : N\|\boldsymbol{\theta} - \hat{\boldsymbol{\theta}}_N^{S+}(\alpha)\|^2_{\boldsymbol{\Sigma}^{-1}} \leq \chi_p^2(\gamma)\}.$$

The results hold good if $\boldsymbol{\Sigma}$ is unknown and $N \to \infty$. The analysis of the properties of the confidence sets are similar to those given in Section 4.8. For the estimation of mean vector under possible subhypothesis restriction, see Srivastava and Saleh (2005).

11.2 U-statistics Approach to Estimation

This section contains the estimation of functionals of distributions based on U-statistics and its modifications. The idea of U-statistics is due to Hoeffding (1948). Parallel to p-variate normal theory, we consider the unrestricted, PTE, SE, and PRSE of the functionals which, as a special case, contains the mean vector of distributions as target for estimation enlarging the scope of application.

Let $\{\mathbf{X}_i,\ i \geq 1\}$ be a sequence of i.i.d. random vectors with a distribution function F defined on E^q for some $q \geq 1$. Let $\mathcal{F}$ be the space of all distribution functions, and for every $F \in \mathcal{F}$, consider the vector of functionals

$$\boldsymbol{\theta} = \boldsymbol{\theta}(F) = (\theta_1(F), \dots, \theta_p(F))' \quad \text{for } p \geq 1 \tag{11.2.1}$$

whose domain is $\mathcal{F}$. If there exists a kernel $\varphi_j\ (\mathbf{x}_1, \dots, \mathbf{x}_{m_j})$, symmetric in its arguments of degree $m_j\ (\geq 1)$, such that

$$\theta_j(F) = E_F\{\varphi_j(\mathbf{X}_1, \dots, \mathbf{X}_{m_j})\}, \quad \forall\ F \in \mathcal{F}, \tag{11.2.2}$$

$j = 1, \dots, p$, then $\boldsymbol{\theta}$ is an admissible vector parameter. For $n \geq m^* = \max(m_1, \dots, m_p)$, we can define the vector of U-statistics,

$$\mathbf{U}_n = (U_{n1}, \dots, U_{np})', \tag{11.2.3}$$

by letting

$$U_{nj} = \binom{n}{m_j}^{-1} \sum_{1 \leq i_1 \leq \cdots \leq i_{m_j} \leq n} \varphi_j(\mathbf{X}_{i_1}, \dots, \mathbf{X}_{i_{m_j}}), \quad j = 1, \dots, p. \tag{11.2.4}$$

For a detailed account of U-statistics, see Rohatgi and Saleh (2001), Puri and Sen (1971), and Randles and Wolfe (1979) among others. $\mathbf{U}_n$ is symmetric and unbiased estimator of $\boldsymbol{\theta}$ with minimum variance property.

Assume that the kernels φ_j are all integrable. Then we define the following conditional expectations:

$$\varphi_{j,c} = E_F\big[\varphi_j(\mathbf{X}_1, \dots, \mathbf{X}_c;\ \mathbf{x}_{c+1}, \dots, \mathbf{x}_{m_j})\big], \quad c = 0, \dots, m_j, \tag{11.2.5}$$

and

$$\zeta_{j\ell,c} = E_F\big[\varphi_{j,c}(\mathbf{X}_1, \dots, \mathbf{X}_c)\varphi_{\ell,c}(\mathbf{X}_1, \dots, \mathbf{X}_c) - \theta_j(F)\theta_\ell(F)\big]$$

for $j, \ell = 1, \ldots, p$, and $c = 0, 1, \ldots, \min(m_j, m_\ell)$. From Hoeffding (1948) (see also Puri and Sen, 1971), we have

$$nE\big[(\mathbf{U}_n - \boldsymbol{\theta})(\mathbf{U}_n - \boldsymbol{\theta})'\big] = \boldsymbol{\Sigma} + o(n^{-1}), \tag{11.2.6}$$

where $\boldsymbol{\Sigma} = ((m_j m_\ell \zeta_{j\ell,c}))$. A convenient estimator of $\boldsymbol{\Sigma}$ is available by way of the jackknife technique. For this, let

$$\mathbf{U}_n = \mathbf{U}(\mathbf{X}_1, \ldots, \mathbf{X}_n),$$

and for every $i = 1, \ldots, n$, define one-skip U-statistics as

$$\mathbf{U}_{n-1}^{(i)} = \mathbf{U}(\mathbf{X}_1, \ldots, \mathbf{X}_{i-1}, \mathbf{X}_{i+1}, \ldots, \mathbf{X}_n)$$

and

$$\mathbf{U}_{n,i} = n\mathbf{U}_n - (n-1)U_{n-1}^{(i)}. \tag{11.2.7}$$

Then the jackknife estimate of $\boldsymbol{\Sigma}$ is given by

$$\hat{\boldsymbol{\Sigma}}_n = (n-1)^{-1} \sum_{j=1}^{n} [\mathbf{U}_{n,j} - \mathbf{U}_n][\mathbf{U}_{n,j} - \mathbf{U}_n]'. \tag{11.2.8}$$

Further, $\hat{\boldsymbol{\Sigma}}_n \xrightarrow{\text{P}} \boldsymbol{\Sigma}$ as $n \to \infty$ (see Puri and Sen, 1971).

For testing the null hypothesis $H_0 : \boldsymbol{\theta} = \mathbf{0}$, we can use the test-statistic

$$\mathcal{L}_n = n\mathbf{U}_n' \hat{\boldsymbol{\Sigma}}_n^{-1} \mathbf{U}_n = n\tilde{\boldsymbol{\theta}}_n' \hat{\boldsymbol{\Sigma}}_n^{-1} \tilde{\boldsymbol{\theta}}_n, \tag{11.2.9}$$

setting $\tilde{\boldsymbol{\theta}}_n = \mathbf{U}_n$. Then, under H_0 as $n \to \infty$, $\mathcal{L}_N$ converges to the χ_p^2-distribution.

Now, consider various estimators of $\boldsymbol{\theta}$ when it is suspected that $\boldsymbol{\theta}$ is $\mathbf{0}$. We have the *unrestricted* estimator of $\boldsymbol{\theta}$ given by $\tilde{\boldsymbol{\theta}}_n = \mathbf{U}_n$. The three estimators of $\boldsymbol{\theta}$ based on $\tilde{\boldsymbol{\theta}}_n$ and $\mathcal{L}_N$ are as follows:

$$\begin{aligned}
\hat{\boldsymbol{\theta}}_n^{\text{PT}} &= \tilde{\boldsymbol{\theta}}_n I(\mathcal{L}_n \geq \mathcal{L}_{n,\alpha}),\\
\hat{\boldsymbol{\theta}}_n^{S} &= \tilde{\boldsymbol{\theta}}_n(1 - c\mathcal{L}_n^{-1}), \qquad (c = p-2),\\
\text{and} \qquad \hat{\boldsymbol{\theta}}_n^{S+}(c) &= \tilde{\boldsymbol{\theta}}_n(1 - c\mathcal{L}_n^{-1}) I(\mathcal{L}_n > c).
\end{aligned} \tag{11.2.10}$$

We define the corresponding confidence sets as

$$\begin{aligned}
C^0(\tilde{\boldsymbol{\theta}}_n) &= \Big\{\boldsymbol{\theta} : n\|\boldsymbol{\theta} - \tilde{\boldsymbol{\theta}}_n\|^2_{\hat{\boldsymbol{\Sigma}}_n^{-1}} \leq \chi_p^2(\gamma)\Big\},\\
C^{\text{PT}}(\hat{\boldsymbol{\theta}}_n^{\text{PT}}(\alpha)) &= \Big\{\boldsymbol{\theta} : n\|\boldsymbol{\theta} - \hat{\boldsymbol{\theta}}_n^{\text{PT}}(\alpha)\|^2_{\hat{\boldsymbol{\Sigma}}_n^{-1}} \leq \chi_p^2(\gamma)\Big\},\\
C^S(\hat{\boldsymbol{\theta}}_n^{S}) &= \Big\{\boldsymbol{\theta} : n\|\boldsymbol{\theta} - \hat{\boldsymbol{\theta}}_n^{S}\|^2_{\hat{\boldsymbol{\Sigma}}_n^{-1}} \leq \chi_p^2(\gamma)\Big\},\\
\text{and} \qquad C^{S+}(\hat{\boldsymbol{\theta}}_n^{S+}) &= \Big\{\boldsymbol{\theta} : n\|\boldsymbol{\theta} - \hat{\boldsymbol{\theta}}_n^{S+}(c)\|^2_{\hat{\boldsymbol{\Sigma}}_n^{-1}} \leq \chi_p^2(\gamma)\Big\},
\end{aligned} \tag{11.2.11}$$

respectively. See also Sen (1984) and Saleh (2003).

11.2.1 Asymptotic Properties of Point and Set Estimation under Fixed Alternatives

First, we note that

$$\sqrt{n}(\tilde{\boldsymbol{\theta}}_n - \boldsymbol{\theta}) \sim \mathcal{N}_p(\mathbf{0}, \boldsymbol{\Sigma}) \quad \text{for } \boldsymbol{\theta} \neq \mathbf{0}$$

and

$$\lim_{n\to\infty} P\Big\{n\|\boldsymbol{\theta} - \tilde{\boldsymbol{\theta}}_n\|^2_{\hat{\boldsymbol{\Sigma}}_n^{-1}} \le \chi_p^2(\gamma)\Big\} = 1 - \gamma. \tag{11.2.12}$$

Next, we consider the PTE, $\hat{\boldsymbol{\theta}}_n^{\mathrm{PT}}$, and notice that

$$\begin{aligned} n\left\|\tilde{\boldsymbol{\theta}}_n - \tilde{\boldsymbol{\theta}}_n^{\mathrm{PT}}\right\|^2_{\hat{\boldsymbol{\Sigma}}_n^{-1}} &= n\tilde{\boldsymbol{\theta}}_n'\hat{\boldsymbol{\Sigma}}_n^{-1}\tilde{\boldsymbol{\theta}}_n I\big(\mathcal{L}_n \le \chi_p^2(\alpha)\big) \\ &= \mathcal{L}_n I\big(\mathcal{L}_n \le \mathcal{L}_{n,\alpha}\big) \\ &\le \mathcal{L}_{n,\alpha} I\big(\mathcal{L}_n < \mathcal{L}_{n,\alpha}\big). \end{aligned} \tag{11.2.13}$$

Similarly, for the SE, $\hat{\boldsymbol{\theta}}_n^{S}$ we obtain

$$n\left\|\tilde{\boldsymbol{\theta}}_n - \tilde{\boldsymbol{\theta}}_n^{\mathrm{PT}}\right\|^2_{\hat{\boldsymbol{\Sigma}}_n^{-1}} = c\mathcal{L}_n^{-1}, \tag{11.2.14}$$

and for the PRSE, $\hat{\boldsymbol{\theta}}_n^{S+}$ we obtain

$$n\left\|\tilde{\boldsymbol{\theta}}_n - \tilde{\boldsymbol{\theta}}_n^{S+}(c)\right\|^2_{\hat{\boldsymbol{\Sigma}}_n^{-1}} = \mathcal{L}_n I(\mathcal{L}_n < c) + c^2\mathcal{L}_n^{-1}(\mathcal{L}_n > c). \tag{11.2.15}$$

Assume that

$$E\|\varphi\|^{2r} < \infty \quad \text{for some } r \ge 1. \tag{11.2.16}$$

Now, $E_F I[\mathcal{L}_n < \mathcal{L}_{n,\alpha}] = P_{\boldsymbol{\theta}}(\mathcal{L}_n < \mathcal{L}_{n,\alpha})$, where $\mathcal{L}_{n,\alpha}$ converges to the finite limit $\chi_p^2(\alpha)$ as $n \to \infty$. Since $\hat{\boldsymbol{\Sigma}}_n \to \boldsymbol{\Sigma}$ almost surely as $n \to \infty$ due to (11.13.16) and $\tilde{\boldsymbol{\theta}}_n \overset{a.s.}{\to} \boldsymbol{\theta}$ as $n \to \infty$, then $n^{-1}\mathcal{L}_n \to \boldsymbol{\theta}'\boldsymbol{\Sigma}^{-1}\boldsymbol{\theta}$ a.s. and $\mathcal{L}_n \to \infty$ in probability as $n \to \infty$. Therefore, for every $\boldsymbol{\theta} \neq \mathbf{0}$, meaning $\Delta^2 > 0$, $\lim_{n\to\infty} P(\mathcal{L}_n < \mathcal{L}_{n,\alpha}) \to 0$. Hence, the r.h.s. of (11.2.13) converges to 0 as $n \to \infty$. Further, from the discussion above $n\mathcal{L}_n^{-1}$ converges to $(\boldsymbol{\theta}'\boldsymbol{\Sigma}^{-1}\boldsymbol{\theta})^{-1}$ almost surely, which implies that as $n \to \infty$,

$$\mathcal{L}_n^{-1} \text{ as well as } \mathcal{L}_n^{-1}I(\mathcal{L}_n > c) \to 0 \quad \text{in probability.}$$

Hence, $\lim_{n\to\infty} E_F(\mathcal{L}_n^{-1}) = 0$, and $\lim_{n\to\infty} E_F[\mathcal{L}_n^{-1}I(\mathcal{L}_n > c)] = 0$. As a result,

$$\begin{aligned} n\left\|\tilde{\boldsymbol{\theta}}_n - \hat{\boldsymbol{\theta}}_n^{\mathrm{PT}}\right\|^2_{\hat{\boldsymbol{\Sigma}}_n^{-1}} &\to 0, \\ n\left\|\tilde{\boldsymbol{\theta}}_n - \hat{\boldsymbol{\theta}}_n^{S}\right\|^2_{\hat{\boldsymbol{\Sigma}}_n^{-1}} &\to 0, \\ n\left\|\tilde{\boldsymbol{\theta}}_n - \hat{\boldsymbol{\theta}}_n^{S+}(c)\right\|^2_{\hat{\boldsymbol{\Sigma}}_n^{-1}} &\to 0, \end{aligned} \tag{11.2.17}$$

in the first mean as $n \to \infty$, and we conclude that $\tilde{\boldsymbol{\theta}}_n$, $\hat{\boldsymbol{\theta}}_n^{\mathrm{PT}}$, $\hat{\boldsymbol{\theta}}_n^{S}$, and $\hat{\boldsymbol{\theta}}_n^{S+}$ are equivalent in quadratic mean under fixed alternatives, $\boldsymbol{\theta} \neq \mathbf{0}$. Hence,

$$\sqrt{n}(\hat{\boldsymbol{\theta}}_n^{\mathrm{PT}} - \boldsymbol{\theta}) = \sqrt{n}(\hat{\boldsymbol{\theta}}_n - \boldsymbol{\theta}) + o_p(1),$$

$$\sqrt{n}(\hat{\boldsymbol{\theta}}_n^{S} - \boldsymbol{\theta}) = \sqrt{n}(\hat{\boldsymbol{\theta}}_n - \boldsymbol{\theta}) + o_p(1),$$

$$\text{(iii) } \sqrt{n}(\hat{\boldsymbol{\theta}}_n^{S+} - \boldsymbol{\theta}) = \sqrt{n}(\hat{\boldsymbol{\theta}}_n - \boldsymbol{\theta}) + o_p(1), \tag{11.2.18}$$

and

$$\lim_{n\to\infty} P\{n\|\boldsymbol{\theta} - \boldsymbol{\theta}_n^*\|^2_{\hat{\boldsymbol{\Sigma}}_n^{-1}} \le \chi_p^2(\gamma)\} = 1 - \gamma, \tag{11.2.19}$$

where

$$\boldsymbol{\theta}_n^* = \hat{\boldsymbol{\theta}}_n - \tilde{\boldsymbol{\theta}}_n g(\mathcal{L}_n) \tag{11.2.20}$$

and

$$\begin{aligned} g(\mathcal{L}_n) &= I(\mathcal{L}_n < \mathcal{L}_{n,\alpha}) \\ &= (1 - c\mathcal{L}_n^{-1}) \\ &= c\mathcal{L}_n^{-1} + (1 - c\mathcal{L}_n^{-1})I(\mathcal{L}_n < c), \end{aligned} \tag{11.2.21}$$

respectively. Also, the asymptotic distributional MSE and risks are

$$\mathbf{M}(\boldsymbol{\theta}_n^*) = \boldsymbol{\Sigma} \quad \text{and} \quad R(\boldsymbol{\theta}_n^*; \mathbf{Q}) = \operatorname{tr}(\boldsymbol{\Sigma}\mathbf{Q}). \tag{11.2.22}$$

11.2.2 Asymptotic Properties of the Point and Set Estimation under Local Alternatives

As before, we consider a class of local alternatives

$$K_{(n)} : \boldsymbol{\theta}_{(n)} = n^{-1/2}\boldsymbol{\delta}, \quad \boldsymbol{\delta} = (\delta_1, \dots, \delta_p)', \tag{11.2.23}$$

to obtain meaningful differences in the asymptotic distributions of $\sqrt{n}\hat{\boldsymbol{\Sigma}}_n^{-1/2}(\boldsymbol{\theta}_n^* - \boldsymbol{\theta}_{(n)})$, where $\boldsymbol{\theta}_n^* = \hat{\boldsymbol{\theta}}_n - \tilde{\boldsymbol{\theta}}_n g(\mathcal{L}_n)$ as $n \to \infty$. Accordingly, consider the point estimators $\hat{\boldsymbol{\theta}}_n^{\mathrm{PT}}(\alpha)$, $\hat{\boldsymbol{\theta}}_n^{S}$, and $\hat{\boldsymbol{\theta}}_n^{S+}(c)$ under $K_{(n)}$ as follows:

$$\sqrt{n}\hat{\boldsymbol{\Sigma}}_n^{-1/2}(\hat{\boldsymbol{\theta}}_n^{\mathrm{PT}}(\alpha) - \boldsymbol{\theta}_{(n)}) = \sqrt{n}\hat{\boldsymbol{\Sigma}}_n^{-1/2}(\tilde{\boldsymbol{\theta}}_n - \boldsymbol{\theta}_{(n)}) - \sqrt{n}\hat{\boldsymbol{\Sigma}}_n^{-1/2}\tilde{\boldsymbol{\theta}}_n I(\mathcal{L}_n < \mathcal{L}_{n,\alpha})$$

$$\sqrt{n}\hat{\boldsymbol{\Sigma}}_n^{-1/2}(\hat{\boldsymbol{\theta}}_n^{S}(\alpha) - \boldsymbol{\theta}_{(n)}) = \sqrt{n}\hat{\boldsymbol{\Sigma}}_n^{-1/2}(\tilde{\boldsymbol{\theta}}_n - \boldsymbol{\theta}_{(n)}) - c\sqrt{n}\hat{\boldsymbol{\Sigma}}_n^{-1/2}\tilde{\boldsymbol{\theta}}_n \mathcal{L}_n^{-1}$$

and

$$\begin{aligned} &\sqrt{n}\hat{\boldsymbol{\Sigma}}_n^{-1/2}(\hat{\boldsymbol{\theta}}_n^{S+}(c) - \boldsymbol{\theta}_{(n)}) \\ &= \sqrt{n}\hat{\boldsymbol{\Sigma}}_n^{-1/2}(\tilde{\boldsymbol{\theta}}_n - \boldsymbol{\theta}_{(n)}) - (\sqrt{n}\hat{\boldsymbol{\Sigma}}_n^{-1/2}\tilde{\boldsymbol{\theta}}_n)I(\mathcal{L}_n < c) \\ &-(\sqrt{n}\hat{\boldsymbol{\Sigma}}_n^{-1/2}\tilde{\boldsymbol{\theta}}_n)\mathcal{L}_n^{-1} + c(\sqrt{n}\hat{\boldsymbol{\Sigma}}_n^{-1/2}\tilde{\boldsymbol{\theta}}_n)\mathcal{L}_n^{-1}I(\mathcal{L}_n < c). \end{aligned} \tag{11.2.24}$$

It is known that $\hat{\mathbf{\Sigma}}_n \to \mathbf{\Sigma}$ a.s. as $n \to \infty$ and $\sqrt{n}\hat{\mathbf{\Sigma}}_n^{-1/2}(\tilde{\boldsymbol{\theta}}_n - \boldsymbol{\theta}_{(n)}) \sim \mathcal{N}_p(\mathbf{0}, \mathbf{I}_p)$ and

$$\begin{aligned}
\sqrt{n}\hat{\mathbf{\Sigma}}_n^{-1/2}\tilde{\boldsymbol{\theta}}_n &= \sqrt{n}\hat{\mathbf{\Sigma}}_n^{-1/2}(\tilde{\boldsymbol{\theta}}_n - \boldsymbol{\theta}_{(n)}) + \boldsymbol{\delta}_n^*, \ (\boldsymbol{\delta}_n^* = \hat{\mathbf{\Sigma}}_n^{-1/2}\boldsymbol{\delta}) \\
&\xrightarrow{\mathcal{D}} \mathbf{Z} + \boldsymbol{\delta}^* \text{ as } n \to \infty, \ \boldsymbol{\delta}^* = \text{Plim}\, \boldsymbol{\delta}_n^*, \\
n\tilde{\boldsymbol{\theta}}_n'\hat{\mathbf{\Sigma}}_n^{-1}\tilde{\boldsymbol{\theta}}_n &= \left[\sqrt{n}\,\hat{\mathbf{\Sigma}}_n^{-1/2}(\tilde{\boldsymbol{\theta}}_n - \boldsymbol{\theta}_{(n)}) + \boldsymbol{\delta}_n^*\right]\left[\sqrt{n}\,\hat{\mathbf{\Sigma}}_n^{-1/2}\tilde{\boldsymbol{\theta}}_n + \boldsymbol{\delta}_n^*\right]' \\
&\xrightarrow{\mathcal{D}} \|\mathbf{Z} + \boldsymbol{\delta}^*\|^2 \text{ as } n \to \infty. \qquad (11.2.25)
\end{aligned}$$

Hence, we have the following theorem:

Theorem 1. Under $\{K_{(n)}\}$ and the regularity conditions,

(i) $\lim_{n\to\infty} P_{K_{(n)}}\left\{n\|\boldsymbol{\theta}_{(n)} - \tilde{\boldsymbol{\theta}}_n\|^2_{\hat{\mathbf{\Sigma}}_n^{-1}} < \chi_p^2(\gamma)\right\} = 1 - \gamma,$

(ii) $\lim_{n\to\infty} P_{K_{(n)}}\left\{n\|\boldsymbol{\theta}_{(n)} - \hat{\boldsymbol{\theta}}_n^{\mathrm{PT}}(\alpha)\|^2_{\hat{\mathbf{\Sigma}}_n^{-1}} < \chi_p^2(\gamma)\right\}$

$= H_p(\chi_p^2(\alpha); \Delta^2)I(\Delta^2 < \chi_p^2(\gamma)) + P\{\|\mathbf{Z} + \boldsymbol{\delta}^*\|^2 < \chi_p^2(\gamma); |\mathbf{Z}|^2 \geq \chi_p^2(\alpha)\},$

(iii) $\lim_{n\to\infty} P_{K_{(n)}}\left\{n\|\boldsymbol{\theta}_{(n)} - \hat{\boldsymbol{\theta}}_n^{S}\|^2_{\hat{\mathbf{\Sigma}}_n^{-1}} \leq \chi_p^2(\gamma)\right\}$

$= P\{\|\mathbf{Z} - c(\mathbf{Z} + \boldsymbol{\delta}^*)\|\mathbf{Z} + \boldsymbol{\delta}^*\|^{-2}\| \leq \chi_p^2(\gamma)\},$

and

(iv) $\lim_{n\to\infty} P_{K_{(n)}}\left\{n\|\boldsymbol{\theta}_{(n)} - \hat{\boldsymbol{\theta}}_n^{S+}(c)\|^2_{\hat{\mathbf{\Sigma}}_n^{-1}} \leq \chi_p^2(\gamma)\right\} = H_p(c; \Delta^2)I(\Delta^2 < \chi_p^2(\gamma))$

$+P_{\delta^*}\{\|\mathbf{Z} - c(\mathbf{Z} + \boldsymbol{\delta}^*)\|\mathbf{Z} + \boldsymbol{\delta}^*\|^{-2}\|^2 < \chi_p^2(\gamma); |\mathbf{Z}|^2 > c\}. \qquad (11.2.26)$

For a proof, see Problem 10.

As for the point estimators we can easily find the ADB, ADMSE, and ADQR for $\tilde{\boldsymbol{\theta}}_n$, $\hat{\boldsymbol{\theta}}_n^{\mathrm{PT}}$, $\hat{\boldsymbol{\theta}}_n^{S}$, and $\hat{\boldsymbol{\theta}}_n^{S+}$ based on the asymptotic distribution as $n \to \infty$:

(i) $\sqrt{n}\hat{\mathbf{\Sigma}}_n^{-1/2}(\hat{\boldsymbol{\theta}}_n^{\mathrm{PT}} - \boldsymbol{\theta}_{(n)}) \xrightarrow{\mathcal{D}} \{\mathbf{Z} - (\mathbf{Z} + \boldsymbol{\delta}^*)I(\|\mathbf{Z} + \boldsymbol{\delta}^*\|^2 < \chi_p^2(\alpha)\}.$

(ii) $\sqrt{n}\hat{\mathbf{\Sigma}}_n^{-1/2}(\hat{\boldsymbol{\theta}}_n^{S} - \boldsymbol{\theta}_{(n)}) \xrightarrow{\mathcal{D}} \{\mathbf{Z} - c(\mathbf{Z} + \boldsymbol{\delta}^*)\|\mathbf{Z} + \boldsymbol{\delta}^*\|^{-2}\}.$

(iii) $\sqrt{n}\hat{\mathbf{\Sigma}}_n^{-1/2}(\hat{\boldsymbol{\theta}}_n^{S+} - \boldsymbol{\theta}_{(n)}) \xrightarrow{\mathcal{D}} \{\boldsymbol{\delta}^* - (1 - c\|\mathbf{Z} + \boldsymbol{\delta}^*\|^{-2})I(\|\mathbf{Z} + \boldsymbol{\delta}^*\|^2 > c)\}.$

(11.2.27)

Then we have the expressions for ADB, ADMSE, and ADQR of the four estimators in the theorem given below.

Theorem 2. Under $\{K_{(n)}\}$ and the assumed regularity conditions

(i) **Unrestricted estimator**

$\mathbf{b}_1(\tilde{\boldsymbol{\theta}}_n) = \mathbf{0}$ and $B_1(\tilde{\boldsymbol{\theta}}_n) = 0$,

$\mathbf{M}_1(\tilde{\boldsymbol{\theta}}_n) = \boldsymbol{\Sigma}$ and $R_1(\tilde{\boldsymbol{\theta}}_n : Q) = \operatorname{tr}(\mathbf{Q}\boldsymbol{\Sigma})$. (11.2.28a)

(ii) **PTE**

$$\mathbf{b}_2(\hat{\boldsymbol{\theta}}_n^{\mathrm{PT}}) = -\boldsymbol{\delta} H_{p+2}(\chi_p^2(\alpha);\Delta^2) \text{ and } B_2(\hat{\boldsymbol{\theta}}_n^{\mathrm{PT}}) = \Delta^2\{H_{p+2}(\chi_p^2(\alpha);\Delta^2)\}$$

$$\mathbf{M}_2(\hat{\boldsymbol{\theta}}_n^{\mathrm{PT}}) = \boldsymbol{\Sigma}\{1 - H_{p+2}(\chi_p^2(\alpha);\Delta^2)\} + \boldsymbol{\delta}\boldsymbol{\delta}'\{2H_{p+2}(\chi_p^2(\alpha);\Delta^2) - H_{p+4}(\chi_p^2(\alpha);\Delta^2)\}$$

and

$$R_2(\hat{\boldsymbol{\theta}}_n^{\mathrm{PT}} : Q) = \operatorname{tr}(\boldsymbol{\Sigma}\mathbf{Q})\{1 - H_{p+2}(\chi_p^2(\alpha);\Delta^2)\} + \boldsymbol{\delta}'\mathbf{Q}\boldsymbol{\delta}\{2H_{p+2}(\chi_p^2(\alpha);\Delta^2) - H_{p+4}(\chi_p^2(\alpha);\Delta^2)\}. \quad (11.2.28b)$$

(iii) **SE**

$$\mathbf{b}_3(\hat{\boldsymbol{\theta}}_n^S) = -c\boldsymbol{\delta} E[\chi_{p+2}^{-2}(\Delta^2)] \text{ and } B_3(\hat{\boldsymbol{\theta}}_n^S) = c^2\Delta^2\left\{E\left[\chi_{p+2}^{-2}(\Delta^2)\right]\right\}^2;$$

$$\mathbf{M}_3(\hat{\boldsymbol{\theta}}_n^S) = \boldsymbol{\Sigma} - c\boldsymbol{\Sigma}\left\{2E[\chi_{p+2}^{-2}(\Delta^2)] - (p-2)E[\chi_{p+2}^{-4}(\Delta^2)]\right\} + c(c+2)\boldsymbol{\delta}\boldsymbol{\delta}' E[\chi_{p+4}^{-4}(\Delta^2)] \text{ and}$$

$$R_3(\hat{\boldsymbol{\theta}}_n^S;\mathbf{Q}) = \operatorname{tr}(\boldsymbol{\Sigma}\mathbf{Q}) - c\operatorname{tr}(\boldsymbol{\Sigma}\mathbf{Q})\left\{2E[\chi_{p+2}^{-2}(\Delta^2)] - (p-2)E[\chi_{p+2}^{-4}(\Delta^2)]\right\} + c(c+2)(\boldsymbol{\delta}'\mathbf{Q}\boldsymbol{\delta})E[\chi_{p+4}^{-4}(\Delta^2)]. \quad (11.2.28c)$$

(iv) **PRSE**

$$\mathbf{b}_4(\hat{\boldsymbol{\theta}}_n^{S+}) = -\boldsymbol{\delta}\left\{cE[\chi_{p+2}^{-2}(\Delta^2)] + E[(1 - c\chi_{p+2}^{-2}(\Delta^2))I(\chi_{p+2}^2(\Delta^2) < c)]\right\}$$

$$B_4(\hat{\boldsymbol{\theta}}_n^{S+}) = \Delta^2\{cE[\chi_{p+2}^{-2}(\Delta^2)] + E[(1 - c\chi_{p+2}^{-2}(\Delta^2))I(\chi_{p+2}^2(\Delta^2) < c)]\}^2;$$

$$\mathbf{M}_4(\hat{\boldsymbol{\theta}}_n^{S+}) = M_3(\hat{\boldsymbol{\theta}}_n^S) - c\boldsymbol{\Sigma}\{E[(1 - c\chi_{p+2}^{-2}(\Delta^2))I(\chi_{p+2}^2(\Delta^2) < c)]\} + \boldsymbol{\delta}\boldsymbol{\delta}'\Big\{2E[(1 - c\chi_{p+2}^{-2}(\Delta^2))I(\chi_{p+2}^2(\Delta^2) < c)] - E[(1 - c\chi_{p+4}^{-2}(\Delta^2))^2 I(\chi_{p+4}^2(\Delta^2) < c)]\Big\}$$

and

$$R_4(\hat{\boldsymbol{\theta}}_n^{S+};\mathbf{Q}) = R_3(\hat{\boldsymbol{\theta}}_n^S;\mathbf{Q}) - c\operatorname{tr}(\boldsymbol{\Sigma}\mathbf{Q})\{E[(1 - c\chi_{p+2}^{-2}(\Delta^2))I(\chi_{p+2}^2(\Delta^2) < c)]\} + \boldsymbol{\delta}\boldsymbol{\delta}'\Big\{2E[(1 - c\chi_{p+2}^{-2}(\Delta^2))I(\chi_{p+2}^2(\Delta^2) < c)] - E[(1 - c\chi_{p+4}^{-2}(\Delta^2))^2 I(\chi_{p+4}^2(\Delta^2) < c)]\Big\}. \quad (11.2.28d)$$

These results may be obtained by using the random variable

$$\mathbf{Z} - (\mathbf{Z} + \boldsymbol{\delta}^*) g(\|\mathbf{Z} + \boldsymbol{\delta}^*\|^2), \qquad (11.2.29)$$

where $\mathbf{Z} \sim \mathcal{N}_p(\mathbf{0}, \mathbf{I}_p)$ and $g(\cdot)$ is defined by (11.2.21).

11.3 Nonparametric Methods: R-estimation

Consider the model

$$\mathbf{Y}_j = \boldsymbol{\theta} + \boldsymbol{\varepsilon}_j, \quad j = 1, \ldots, N, \qquad (11.3.1)$$

where $\mathbf{Y}_1, \ldots, \mathbf{Y}_N$ are i.i.d. r.v. having p-variate continuous cdf, $F_{\boldsymbol{\theta}}(\mathbf{Y})$, defined on the Euclidean space R^p that is diagonally symmetric about its location parameter, $\boldsymbol{\theta} = (\theta_1, \ldots, \theta_p)'$,

$$F_{\boldsymbol{\theta}}(\mathbf{Y}) = F(\mathbf{Y} - \boldsymbol{\theta}), \quad \mathbf{Y} \in R^p, \qquad (11.3.2)$$

where F is diagonally symmetric about $\mathbf{0}$. Let $F_{[j]}$ be the marginal cdf corresponding to F and $F_{[i,j]}$, $i \neq j$ be the bivariate marginal cdf corresponding to F. Further, let $F_{[j]}$ possess absolute continuous pdf $f_{[j]}$, and let

$$\psi_j(u) = -\frac{f'_{[j]}(F_{[j]}^{-1}(u))}{f_{[j]}(F_{[j]}^{-1}(u))}, \quad 0 < u < 1 \text{ and } j = 1, \ldots, p. \qquad (11.3.3)$$

In addition, for every N (≥ 1) and j $(= 1, \ldots, p)'$ define N scores

$$a_{Nj}^+(k) = E[\phi_j^+(U_{Nk})] \quad \text{or} \quad \phi_j^+\left(\frac{k}{N+1}\right) \text{ for } k = 1, \ldots, N, \qquad (11.3.4)$$

where U_{Nk} is the kth order statistic from sample size N from $U(0,1)$ and for every $u \in (0,1)$, consider

$$\phi_j^+(u) = \phi_j\left(\frac{1+u}{2}\right), \quad \phi_j(u) + \phi_j(1-u) = 0, \qquad (11.3.5)$$

where the functions $\{\phi_j(u)\}$ are square integrable and nondecreasing. To define the rank statistics, let $R_{ij}^+(b)$ be the rank of $|Y_{ij} - b|$ among $|Y_{1j} - b|, \ldots, |Y_{Nj} - b|$, for $i = 1, \ldots, N$ and $j = 1, \ldots, p$, and consider the vector of LRS

$$\mathbf{T}_n(\mathbf{b}) = (T_{N1}(b), \ldots, T_{Np}(b))',$$

where

$$T_{Nj}(b) = N^{-1} \sum_{i=1}^{N} a_{Nj}^+(R_{ij}^+(b)) \operatorname{sign}(Y_{ij} - b), \quad j = 1, \ldots, p. \qquad (11.3.6)$$

Notice that $T_{Nj}(b) \searrow$ in b (see Puri and Sen, 1986), and define the unrestricted estimator of θ_i as

$$\tilde{\theta}_i = \frac{1}{2}\big(\sup(b : T_{Nj}(b) > 0\} + \inf\{b; T_{Nj}(b) < 0\}\big), \quad j = 1, \ldots, p. \quad (11.3.7)$$

Then define $\tilde{\theta}_N = (\tilde{\theta}_1, \ldots, \tilde{\theta}_p)'$ as the estimator of $\boldsymbol{\theta} = (\theta_1, \ldots, \theta_p)'$. It may be verified that $\tilde{\boldsymbol{\theta}}_N$ is robust, translation-invariant and consistent estimator having (coordinatewise) median unbiasedness property (see Puri and Sen, 1986). To study various other estimators, we set

$$\lambda_{ij} = \int_{-\infty}^{\infty}\int_{-\infty}^{\infty} \phi_i(F_{[i]}(x))\phi_j(F_{[j]}(x))dF_{[ij]}(x, y) \quad \text{for } i, j = 1, 2, \ldots, p,$$

and

$$\gamma_j = \int_0^1 \phi_j(u)\psi_j(u)du, \quad j = 1, \ldots, p. \quad (11.3.8)$$

Let

$$\boldsymbol{\Lambda} = ((\lambda_{ij})), \quad \boldsymbol{\nu} = \text{Diag}(\gamma_1, \ldots, \gamma_p), \quad \text{and} \quad \boldsymbol{\Sigma} = \boldsymbol{\nu}^{-1}\boldsymbol{\Lambda}\boldsymbol{\nu}^{-1}. \quad (11.3.9)$$

It can be verified that

$$\sqrt{N}\mathbf{T}_N(0) \sim \mathcal{N}_p(\mathbf{0}, \boldsymbol{\Lambda}), \quad \mathbf{T}_N(0) = \big(T_{N1}(0), \ldots, T_{Np}(0)\big)'. \quad (11.3.10)$$

As a result, we can prove that (Puri and Sen, 1971, ch. 6)

$$\sqrt{N}(\tilde{\boldsymbol{\theta}}_N - \boldsymbol{\theta}) \sim \mathcal{N}_p(\mathbf{0}, \boldsymbol{\Sigma}). \quad (11.3.11)$$

Suppose that it is a priori suspected that $\boldsymbol{\theta}$ is $\mathbf{0}$ (null). Then, if we wish to test the hypothesis $H_0 : \boldsymbol{\theta} = \mathbf{0}$, we consider the test-statistic $\mathcal{L}_N$ defined by

$$\mathcal{L}_N = N[\mathbf{T}_N(\mathbf{0})]'\mathbf{M}_N^{\bar{*}}[\mathbf{T}_N(\mathbf{0})], \quad (11.3.12)$$

where

$$\mathbf{M}_N^{\bar{*}} = ((m_{j\ell}^*)), \quad m_{j\ell}^* = N^{-1}\sum_{i=1}^{N} a_{Nj}^+(R_{ij}^+)a_{N\ell}^+(R_{i\ell}^+)\,\text{sign}Y_{ij}\,\text{sign}Y_{i\ell}, \quad (11.3.13)$$

$j, \ell = 1, \ldots, p$, and $\mathbf{M}_N^{\bar{*}}$ is the generalized inverse of $\mathbf{M}_N^*$ (see Puri and Sen, 1967). In the sequel, we will need a **consistent** estimator of $\boldsymbol{\Sigma}$ defined by

$$\hat{\boldsymbol{\Sigma}}_N = \hat{\boldsymbol{\gamma}}_N^{-1}\mathbf{M}_N\hat{\boldsymbol{\gamma}}_N^{-1}, \quad \mathbf{M}_N = ((m_{j\ell})), \quad (11.3.14)$$

where $m_{j\ell} = N^{-1}\sum_{i=1}^{N} a_{N_j}(R_{ij})a_{N\ell}(R_{i\ell})$ and R_{ij} is the rank of Y_{ij} among $Y_{1j}, \ldots, Y_{Nj}$ $(j = 1, \ldots, p)$. Then $\mathbf{M}_N$ is robust and translation-invariant estimator of $\boldsymbol{\nu}$. Also, $\tilde{\boldsymbol{\nu}}_N = (\tilde{\gamma}_1, \ldots, \tilde{\gamma}_p)'$, where

$$\tilde{\gamma}_j = \sqrt{N}\left\{T_{Nj}\left(\tilde{\theta}_j - N^{-1/2}a\right) - T_{Nj}\left(\tilde{\theta}_j + N^{-1/2}a\right)\right\}, \quad (11.3.15)$$

$j = 1, \ldots, p$ and a is a prefixed positive number.

Let $\mathcal{L}_{N,\alpha}$ be the α-level critical value from the null distribution of $\mathcal{L}_N$. Also, as $N \to \infty$, $\mathcal{L}_{N\alpha} \to \chi_p^2(\alpha)$. This allows us to define the following three estimators of $\boldsymbol{\theta}$ as in (11.1.7) through (11.1.9):

$$\hat{\boldsymbol{\theta}}_N^{\text{PT}} = \tilde{\boldsymbol{\theta}}_N - \tilde{\boldsymbol{\theta}}_N I(\mathcal{L}_N < \mathcal{L}'_{N,\alpha})$$

$$\hat{\boldsymbol{\theta}}_N^{S} = \tilde{\boldsymbol{\theta}}_N - c\tilde{\boldsymbol{\theta}}_N \mathcal{L}_N^{-1}(\mathcal{L}_N \geq \varepsilon_N), \quad \text{where } \varepsilon_N \to 0 \text{ as } N \to \infty \ (p \geq 3), \quad \text{and}$$

$$\hat{\boldsymbol{\theta}}_N^{S+} = \tilde{\boldsymbol{\theta}}_N(1 - c\mathcal{L}_N^{-1}) I(\mathcal{L} > c), \quad c = p - 2, \tag{11.3.16}$$

respectively. Associated with the point estimators given above, we consider the corresponding confidence sets:

$$C^0(\tilde{\boldsymbol{\theta}}_N) = \left\{\boldsymbol{\theta} : N\|\theta - \tilde{\theta}_N\|^2_{\hat{\boldsymbol{\Sigma}}_N^{-1}} \leq \chi_p^2(\gamma)\right\},$$

$$C^{\text{PT}}(\hat{\boldsymbol{\theta}}_N^{\text{PT}}(\alpha)) = \left\{\boldsymbol{\theta} : N\|\boldsymbol{\theta} - \hat{\boldsymbol{\theta}}_N^{\text{PT}}\|^2_{\hat{\boldsymbol{\Sigma}}_N^{-1}} \leq \chi_p^2(\gamma)\right\}$$

$$C^S(\tilde{\boldsymbol{\theta}}_N^S) = \left\{\boldsymbol{\theta} : N\|\boldsymbol{\theta} - \hat{\boldsymbol{\theta}}_N^S\|^2_{\hat{\boldsymbol{\Sigma}}_N^{-1}} \leq \chi_p^2(\gamma)\right\}, \qquad \text{and}$$

$$C^{S+}(\tilde{\boldsymbol{\theta}}_N^{S+}) = \left\{\boldsymbol{\theta} : N\|\boldsymbol{\theta} - \hat{\boldsymbol{\theta}}_N^{S+}\|^2_{\hat{\boldsymbol{\Sigma}}_N^{-1}} \leq \chi_p^2(\gamma)\right\}, \tag{11.3.17}$$

respectively. See also Sen and Saleh (1985, 1987).

11.3.1 Asymptotic Properties of the Point Estimators

We assume that for some positive b (not necessarily ≥ 1),

$$E_F|Y_{ij}|^b < \infty \quad \text{for } j = 1.\ldots, p, \tag{11.3.18}$$

$\frac{d^r}{du^r}\phi_j(u)$, $r = 0, 1, 2$ exist for $u \in (0,1)$ almost everywhere, and there exist positive constants k and ε $\left(< \frac{1}{4}\right)$ such that

$$|\phi_j^{(r)}(u)| \leq k\{u(1-u)\}^{-\varepsilon - r} < \infty, \quad r = 0, 1, 2. \tag{11.3.19}$$

Finally, we assume that the derivative $f'_{[j]}$ is bounded almost everywhere and

$$\sup_x f_{[j]}(x)\{F_{[j]}(x)[1 - F_{[j]}(x)]\}^{-\varepsilon-\eta} < \infty, \quad 1 \leq j \leq p, \tag{11.3.20}$$

where ε and η are both positive. Then (see Sen, 1980) we conclude that for each j $(= 1, \ldots, p)$ as $n \to \infty$,

$$\sqrt{N}(\tilde{\theta}_j - \theta_j) - \gamma_j^{-1} T_{Nj}(\theta_j) = \omega_j \to 0 \quad \text{almost surely,}$$

$$E|\omega_j|^k \to 0 \quad \forall\, k < (1 - 2\varepsilon)/\varepsilon \ (> 2),$$

$$NE(\tilde{\boldsymbol{\theta}}_N - \boldsymbol{\theta})(\tilde{\boldsymbol{\theta}}_N - \boldsymbol{\theta})' \to \boldsymbol{\Sigma}. \tag{11.3.21}$$

We want to determine the asymptotic bias, MSE matrix, and quadratic risk of the estimators to compare the dominance characteristics of the estimators. First, we discuss briefly the case under fixed alternatives $\boldsymbol{\theta} \neq \mathbf{0}$. Note that

$$\|\hat{\boldsymbol{\theta}}_N^{\text{PT}} - \tilde{\boldsymbol{\theta}}_N\|^2 = \|\tilde{\boldsymbol{\theta}}_N\|^2 I(\mathcal{L}_N < \mathcal{L}_{N,\alpha}),$$
$$\|\hat{\boldsymbol{\theta}}_N^{S} - \tilde{\boldsymbol{\theta}}_N\|^2 = \|\tilde{\boldsymbol{\theta}}_N\|^2 \{I(\mathcal{L}_N < \varepsilon_n) + c^2 I(\mathcal{L}_N \geq \varepsilon_n)\mathcal{L}_N^{-2}\},$$
$$\|\hat{\boldsymbol{\theta}}_N^{S+} - \tilde{\boldsymbol{\theta}}_N\|^2 = \|\tilde{\boldsymbol{\theta}}_N\|^2 \{I(\mathcal{L}_N < c) + c^2 I(\mathcal{L}_N \geq c)\mathcal{L}_N^{-2}\}. \quad (11.3.22)$$

Now by (11.3.12),

$$P\{\mathcal{L}_N < k\} \leq \min P\{|T_{Nj}| < N^{-1/2}\sqrt{k}\} \text{ for every } k > 0. \quad (11.3.23)$$

Then it follows (see Sen 1970, Ths. 1 and 3) that under (11.3.19), whenever $\boldsymbol{\theta} \neq \mathbf{0}$, the r.h.s. of (11.3.23) is $O(N^{-2})$, while by (11.12.21), $\|\tilde{\boldsymbol{\theta}}_N\|^2$ has bounded expectation. Thus, by standard analysis, we conclude that the $I\ (\mathcal{L}_N \leq \mathcal{L}_{N\alpha})$ converges in first mean to 0 as $N \to \infty$. Next noting that $\mathcal{L}_N^{-1} \to 0$ as $N \to \infty$, $E\{I(\mathcal{L}_N > c)\mathcal{L}_N^{-2}\} \to 0$ as $N \to \infty$ as well. Thus, for fixed $\boldsymbol{\theta} \neq \mathbf{0}$, as $N \to \infty$,

$$\limsup_{N\to\infty} E\{\|\boldsymbol{\theta}_N^* - \tilde{\boldsymbol{\theta}}_N\|^2 | \boldsymbol{\theta} \neq \mathbf{0}\} = 0, \quad (11.3.24)$$

where $\boldsymbol{\theta}_N^* = \tilde{\boldsymbol{\theta}}_N - \tilde{\boldsymbol{\theta}}_N g(\mathcal{L}_N)$ and $g(\mathcal{L}_N) = I(\mathcal{L}_N < \mathcal{L}_{N\alpha})$, $(p-2)\mathcal{L}_N^{-1}$, and $(p-2)\mathcal{L}_N^{-1} + \{1-(p-2)\mathcal{L}_N^{-1}\}I(\mathcal{L}_N < (p-2))$. Thus, we have

Theorem 3. Under fixed alternatives $\boldsymbol{\theta} \neq \mathbf{0}$, the following holds as $N \to \infty$:

(i) $N^{1/2}\hat{\boldsymbol{\Sigma}}_N^{-1/2}(\hat{\boldsymbol{\theta}}_N^{\text{PT}} - \boldsymbol{\theta}) = N^{1/2}\boldsymbol{\Sigma}^{-1/2}(\tilde{\boldsymbol{\theta}}_N - \boldsymbol{\theta}) + o_P(1)$.

(ii) $N^{1/2}\hat{\boldsymbol{\Sigma}}_N^{-1/2}(\hat{\boldsymbol{\theta}}_N^{S} - \boldsymbol{\theta}) = N^{1/2}\boldsymbol{\Sigma}^{-1/2}(\tilde{\boldsymbol{\theta}}_N - \boldsymbol{\theta}) + o_P(1)$.

and

(iii) $N^{1/2}\hat{\boldsymbol{\Sigma}}_N^{-1/2}(\hat{\boldsymbol{\theta}}_N^{S+} - \boldsymbol{\theta}) = N^{1/2}\boldsymbol{\Sigma}^{-1/2}(\tilde{\boldsymbol{\theta}}_N - \boldsymbol{\theta}) + o_P(1)$. (11.3.25)

As a result, the asymptotic distributional bias, ADB, ADMSE, and ADQR of the estimators are $\mathbf{0}$, $\boldsymbol{\Sigma}$ and $\text{tr}(\mathbf{Q}\boldsymbol{\Sigma})$ respectively for each of the four estimators. Also,

$$\lim_{N\to\infty} P\Big\{N\|\boldsymbol{\theta} - \boldsymbol{\theta}_N^*\|^2_{\hat{\boldsymbol{\Sigma}}_N^{-1}} \leq \chi_p^2(\gamma)\Big\} = 1-\gamma. \quad (11.3.26)$$

Hence, under the fixed $\boldsymbol{\theta} \neq \mathbf{0}$, there is no difference among the ADB, ADMSE, and ADQR of the estimators and the coverage probabilities of the confidence sets. The situation changes when we consider local translation alternatives

$$K_{(N)} : \boldsymbol{\theta}_{(N)} = N^{-1/2}\boldsymbol{\delta}, \quad \boldsymbol{\delta} = (\delta_1, \ldots, \delta_p)'. \quad (11.3.27)$$

By virtue of the translation-invariance of $\tilde{\boldsymbol{\theta}}_N$ and the fact that $\mathbf{Y}_{N_i} = -N^{-1/2}\boldsymbol{\delta}$ has the cdf F under $K_{(N)}$ and do not depend on N, we conclude that under $\{K_{(N)}\}$,

$$\lim_{N\to\infty} E\big[N(\tilde{\boldsymbol{\theta}}_N - \boldsymbol{\theta}_{(N)})'\mathbf{Q}(\tilde{\boldsymbol{\theta}}_N - \boldsymbol{\theta}_{(N)}) = \text{tr}(\mathbf{Q}\boldsymbol{\Sigma}),$$

and

$$\lim_{N\to\infty} P\Big[N\|\tilde{\boldsymbol{\theta}}_{(N)} - \tilde{\boldsymbol{\theta}}_N\|^2_{\hat{\boldsymbol{\Sigma}}_N^{-1}} \le \chi_p^2(\gamma)\Big] = 1-\gamma. \tag{11.3.28}$$

Further, we have under $\{K_{(N)}\}$,

$$\lim_{N\to\infty} P\{\mathcal{L}_N \le x | K_{(N)}\} = H_p(x;\Delta^2), \quad \Delta^2 = \boldsymbol{\delta}'\boldsymbol{\Sigma}^{-1}\boldsymbol{\delta}. \tag{11.3.29}$$

Now, because of translation-invariance of $\tilde{\boldsymbol{\theta}}_N$, (11.3.11) holds under $\{K_{(N)}\}$ as well (replacing $\boldsymbol{\theta}$ by $\boldsymbol{\theta}_{(N)}$) so that for every k (>0) under $\{K_{(N)}\}$ with N large enough and $\mathbf{Q}$ p.d. matrix, $(N\tilde{\boldsymbol{\theta}}_N'\mathbf{Q}\tilde{\boldsymbol{\theta}}_N)^k$ and $[N(\tilde{\boldsymbol{\theta}}_N - \boldsymbol{\theta}_{(N)})'\mathbf{Q}(\tilde{\boldsymbol{\theta}}_N - \boldsymbol{\theta}_{(N)})]^k$ are integrable uniformly in $\boldsymbol{\delta}$ belonging to a compact set C^*. Hence, by (11.3.21), we obtain under $\{K_{(N)}\}$ uniformly in $\boldsymbol{\delta} \in \mathbf{C}$, as $N \to \infty$,

(1) $\mathcal{L}_N^{-1} = N\tilde{\boldsymbol{\theta}}_N'\boldsymbol{\Sigma}^{-1}\tilde{\boldsymbol{\theta}}_N + \eta_N$, $\quad \eta_N \to 0$ in probability (or a.s.)

Therefore, for $\varepsilon_n > 0$ under $\{K_{(N)}\}$ uniformly in $\boldsymbol{\delta} \in C^*$,

(2) $\mathcal{L}_N^{-1} - \big(N\tilde{\boldsymbol{\theta}}_N'\boldsymbol{\Sigma}^{-1}\tilde{\boldsymbol{\theta}}_N\big)^{-1} \xrightarrow{\text{P}} 0 \quad$ as $N \to \infty$.

Finally, under $\{K_{(N)}\}$ by (11.3.21), uniformly in $\boldsymbol{\delta} \in C^*$

(3) $N^{1/2}\boldsymbol{\Sigma}^{-1/2}\tilde{\boldsymbol{\theta}}_N \xrightarrow{\mathcal{D}} \mathcal{N}_p(\boldsymbol{\Sigma}^{-1/2}\boldsymbol{\delta}, \mathbf{I}_p), \quad$ as $N \to \infty$ (11.3.30)

(see also Huskova, 1971). As a result, we have the following theorem:

Theorem 4. Under $\{K_{(N)}\}$ and regularity conditions given above as $N \to \infty$, the following holds:

(i) **Unrestricted estimator**

$\mathbf{b}_1(\tilde{\boldsymbol{\theta}}_N) = \mathbf{0}$ and $B_1(\tilde{\boldsymbol{\theta}}_N) = 0$;

$\mathbf{M}_1(\tilde{\theta}_N) = \boldsymbol{\Sigma}$ and $R_1(\tilde{\boldsymbol{\theta}}_N;\mathbf{Q}) = \text{tr}(\mathbf{Q}\boldsymbol{\Sigma}) = p$ if $\mathbf{Q} = \boldsymbol{\Sigma}^{-1}$. (11.3.31a)

(ii) **PTE**

$$\mathbf{b}_2(\hat{\boldsymbol{\theta}}_N^{\text{PT}}) = -\boldsymbol{\delta}H_{p+2}(\chi_p^2(\alpha);\Delta^2) \text{ and } B_2(\hat{\boldsymbol{\theta}}_N^{\text{PT}}) = \Delta^2\{H_{p+2}(\chi_p^2(\alpha);\Delta^2)\}^2;$$

$$\begin{aligned}\mathbf{M}_2(\hat{\boldsymbol{\theta}}_N^{\text{PT}}) &= \boldsymbol{\Sigma}\{1 - H_{p+2}(\chi_p^2(\alpha);\Delta^2)\} \\ &\quad + \boldsymbol{\delta}\boldsymbol{\delta}'\{2H_{p+2}(\chi_p^2(\alpha);\Delta^2) - H_{p+4}(\chi_p^2(\alpha);\Delta^2)\}\end{aligned}$$

and

$$R_2(\hat{\boldsymbol{\theta}}_N^{\rm PT}; Q) = \operatorname{tr}(\mathbf{Q\Sigma})\{1 - H_{p+2}(\chi_p^2(\alpha); \Delta^2)\} + \boldsymbol{\delta}'\mathbf{Q}\boldsymbol{\delta}\{2H_{p+2}(\chi_p^2(\alpha); \Delta^2) - H_{p+4}(\chi_p^2(\alpha); \Delta^2)\}. \quad (11.3.31b)$$

(iii) **SE**

$$\mathbf{b}_3(\hat{\boldsymbol{\theta}}_N^S) = -c\boldsymbol{\delta}E[\chi_{p+2}^{-2}(\Delta^2)], \quad c = (p-2) \quad \text{and}$$

$$B_3(\hat{\boldsymbol{\theta}}_N^S) = c^2\Delta^2\{E[\chi_{p+2}^{-2}(\Delta^2)]\}^2;$$

$$\mathbf{M}_3(\hat{\boldsymbol{\theta}}_N^S) = \boldsymbol{\Sigma} - c\boldsymbol{\Sigma}\{2E[\chi_{p+2}^{-2}(\Delta^2)] - (p-2)E[\chi_{p+2}^{-4}(\Delta^2)]\} + (c^2-4)\boldsymbol{\delta}\boldsymbol{\delta}'E[\chi_{p+4}^{-4}(\Delta^2)]$$

and $$R_3(\hat{\boldsymbol{\theta}}_N^S; \mathbf{Q}) = \operatorname{tr}(\mathbf{Q\Sigma}) - c\operatorname{tr}(\mathbf{Q\Sigma})\{2E[\chi_{p+2}^{-2}(\Delta^2)] - cE[\chi_{p+4}^{-4}(\Delta^2)]\} + (c^2-4)(\boldsymbol{\delta}'\mathbf{Q}\boldsymbol{\delta})E[\chi_{p+4}^{-4}(\Delta^2)]. \quad (11.3.31c)$$

(iv) **PRSE**

$$\mathbf{b}_4(\hat{\boldsymbol{\theta}}_N^{S+}) = -\boldsymbol{\delta}\{cE[\chi_{p+2}^{-2}(\Delta^2)] + E[(1 - c\chi_{p+2}^{-2}(\Delta^2))I(\chi_{p+2}^{-2}(\Delta^2) < c)]\}$$

and

$$B_4(\hat{\boldsymbol{\theta}}_N^{S+}) = \Delta^2\{cE[\chi_{p+2}^{-2}(\Delta^2)] + E[\chi_{p+2}^{-2}(\Delta^2))I(\chi_{p+2}^{2}(\Delta^2) < c)]\}^2$$

$$\mathbf{M}_4(\hat{\boldsymbol{\theta}}_N^{S+}) = \mathbf{M}_3(\hat{\boldsymbol{\theta}}_N^S) - \boldsymbol{\Sigma}\Big\{E[(1 - \chi_{p+2}^{-2}(\Delta^2))^2 I([\chi_{p+2}^{2}(\Delta^2) < c)]\Big\} + \boldsymbol{\delta}\boldsymbol{\delta}'\Big\{2E[(1 - c\chi_{p+2}^{-2}(\Delta^2))I([\chi_{p+2}^{2}(\Delta^2) < c)] - E[1 - c\chi_{p+4}^{-2}(\Delta^2))^2 I(\chi_{p+4}^{2}(\Delta^2) < c)]\Big\}$$

and

$$R_4(\hat{\boldsymbol{\theta}}_N^{S+}; \mathbf{Q}) = R_3(\hat{\boldsymbol{\theta}}_N^S; \mathbf{Q}) - \operatorname{tr}(\mathbf{Q\Sigma})\{E[(1 - c\chi_{p+2}^{-2}(\Delta^2))^2\}I(\chi_{p+2}^{2}(\Delta^2) < c)] + (\boldsymbol{\delta}'\mathbf{Q}\boldsymbol{\delta})\Big\{2E[(1 - c\chi_{p+2}^{-2}(\Delta^2))I(\chi_{p+2}^{2}(\Delta^2) < c)] - E[(1 - c\chi_{p+4}^{-2}(\Delta^2))I(\chi_{p+4}^{2}(\Delta^2) < c)].$$

(11.3.31d)

Proof. Consider the expression

$$\sqrt{N}(\boldsymbol{\theta}_N^* - \boldsymbol{\theta}_{(N)}) = \sqrt{N}(\tilde{\boldsymbol{\theta}}_N - \theta_{(N)}) - \sqrt{N}\tilde{\boldsymbol{\theta}}_N g(\mathcal{L}_N)), \quad (11.3.32)$$

where

$$\begin{aligned} g(\mathcal{L}_N) &= I(\mathcal{L}_N < \mathcal{L}_{N,\alpha}) \quad \text{and} \quad \mathcal{L}_{N,\alpha} \to \chi_p^2(\alpha) \text{ as } N \to \infty \\ &= (p-2)\mathcal{L}_N^{-1} I(\mathcal{L}_N \geq \varepsilon_N) \quad \text{and} \quad \varepsilon_N \to 0 \text{ as } N \to \infty \\ &= (p-2)\mathcal{L}_N^{-1} + \{1-(p-2)\mathcal{L}_N^{-1}\} I(\mathcal{L}_N < p-2). \end{aligned} \tag{11.3.33}$$

Let $\mathbf{Z} = \sqrt{N}\hat{\mathbf{\Sigma}}_N^{-1/2}(\tilde{\boldsymbol{\theta}}_N - \boldsymbol{\theta}_{(N)})$. Then $\mathbf{Z} \xrightarrow{\mathcal{D}} \mathcal{N}_p(\mathbf{0}, \mathbf{I}_p)$, and $N\tilde{\boldsymbol{\theta}}_N'\hat{\mathbf{\Sigma}}_N^{-1}\tilde{\boldsymbol{\theta}}_N \approx \|\mathbf{Z}+\boldsymbol{\delta}^*\|^2$, where $\boldsymbol{\delta}^* = \mathbf{\Sigma}^{-1/2}\boldsymbol{\delta}$. Hence, under $\{K_{(N)}\}$,

$$\sqrt{N}(\boldsymbol{\theta}_N^* - \boldsymbol{\theta}_{(N)}) = \mathbf{Z} - (\mathbf{Z}+\boldsymbol{\delta}^*) g(\|\mathbf{Z}+\boldsymbol{\delta}^*\|^2) + o_P(1). \tag{11.3.34}$$

Direct calculation yields

$$\begin{aligned} \text{ADB} : \mathbf{b}(\boldsymbol{\theta}_N^*) &= -\boldsymbol{\delta} E\big[g\big(\chi_{p+2}^2(\Delta^2)\big)\big] \\ \text{ADMSE} : \mathbf{M}(\boldsymbol{\theta}_N^*) &= \mathbf{\Sigma} - \mathbf{\Sigma}\Big\{\chi_{p+2}^2(\Delta^2))\big] - E\big[g^2\big(\chi_{p+2}^2(\Delta^2)\big)\big]\Big\} \\ &\quad + \boldsymbol{\delta}\boldsymbol{\delta}'\Big\{2E\big[g\big(\chi_{p+2}^2(\Delta^2)\big)\big] - 2E\big[g\big(\chi_{p+2}^2(\Delta^2)\big)\big] \\ &\quad + E\big[g^2\big(\chi_{p+4}^2(\Delta^2)\big)\big]\Big\}, \end{aligned} \tag{11.3.35}$$

and ADQR : $R(\boldsymbol{\theta}_N^*; \mathbf{Q}) = \operatorname{tr}(\mathbf{Q}\mathbf{M}(\boldsymbol{\theta}_N^*))$. Then, specializing for $g(\mathcal{L}_N)$, we obtain the expressions in Theorem 2. For example, if $g(\mathcal{L}_N) = I(\mathcal{L}_N < \mathcal{L}_{N,\alpha})$, we have $E\big[I\big(\chi_{p+2r}^2(\Delta^2) \leq \chi_p^2(\alpha)\big)\big] = H_{p+2r}(\chi_p^2(\alpha); \Delta^2)$; $r = 1, 2$. Then we obtain

$$\begin{aligned} \mathbf{b}_3(\hat{\boldsymbol{\theta}}_N^{\mathrm{PT}}) &= -\boldsymbol{\delta} H_{p+2}(\chi_p^2(\alpha); \Delta^2), \\ \mathbf{M}_3(\hat{\boldsymbol{\theta}}_N^{\mathrm{PT}}) &= \mathbf{\Sigma}\{1 - H_{p+2}(\chi_p^2(\alpha); \Delta^2)\} \\ &\quad + \boldsymbol{\delta}\boldsymbol{\delta}'\{2H_{p+2}(\chi_p^2(\alpha); \Delta^2) - H_{p+4}(\chi_p^2(\alpha); \Delta^2)\}, \end{aligned}$$

and

$$\begin{aligned} R_3(\hat{\boldsymbol{\theta}}_N^{\mathrm{PT}}; Q) &= \operatorname{tr}(\mathbf{Q}\mathbf{\Sigma})\{1 - H_{p+2}(\chi_p^2(\alpha); \Delta^2)\} \\ &\quad + (\boldsymbol{\delta}'\mathbf{Q}\boldsymbol{\delta})\{2H_{p+2}(\chi_p^2(\alpha); \Delta^2) - H_{p+4}(\chi_p^2(\alpha); \Delta^2)\}. \end{aligned} \tag{11.3.36}$$

11.3.2 Asymptotic Properties Confidence Sets

Theorem 1 of Section 11.3.2 shows that under fixed alternatives $\boldsymbol{\theta} \neq \mathbf{0}$,

$$\lim_{N\to\infty} P\Big\{N\|\boldsymbol{\theta} - \boldsymbol{\theta}_N^*\|^2_{\hat{\mathbf{\Sigma}}_N^{-1}} \leq \chi_p^2(\gamma)\Big\} = 1 - \gamma. \tag{11.3.37}$$

Thus, we consider the local alternatives

$$K_{(N)} : \boldsymbol{\theta}_{(N)} = N^{-1/2}\boldsymbol{\delta}, \quad \boldsymbol{\delta} = (\delta_1, \ldots, \delta_p)' \tag{11.3.38}$$

as before, where $\boldsymbol{\delta} \in C^*$, a compact set in R^p. Accordingly,

$$\sqrt{N}\hat{\boldsymbol{\Sigma}}_N^{-1/2}(\boldsymbol{\theta}_N^* - \boldsymbol{\theta}_{(N)}) - \sqrt{N}\hat{\boldsymbol{\Sigma}}_N^{-1/2}(\tilde{\boldsymbol{\theta}}_N - \boldsymbol{\theta}_{(N)}) = \sqrt{N}\hat{\boldsymbol{\Sigma}}_N^{-1/2}\tilde{\boldsymbol{\theta}}_N g(\mathcal{L}_N). \tag{11.3.39}$$

Since $\hat{\boldsymbol{\Sigma}}_N \to \boldsymbol{\Sigma}$ as $N \to \infty$ and $\sqrt{N}\hat{\boldsymbol{\Sigma}}_N^{-1/2}(\tilde{\boldsymbol{\theta}}_N - \boldsymbol{\theta}_{(N)}) \sim \mathcal{N}_p(\mathbf{0}, \mathbf{I}_p)$, we obtain

$$\sqrt{N}\hat{\boldsymbol{\Sigma}}_N^{-1/2}\tilde{\boldsymbol{\theta}}_N = \sqrt{N}\hat{\boldsymbol{\Sigma}}_N^{-1/2}(\tilde{\boldsymbol{\theta}}_N - \boldsymbol{\theta}_{(N)}) + \hat{\boldsymbol{\Sigma}}_N^{-1/2}\boldsymbol{\delta} \xrightarrow{\mathcal{D}} \mathbf{Z} + \boldsymbol{\delta}^*, \quad \boldsymbol{\delta}^* = \boldsymbol{\Sigma}^{-1/2}\boldsymbol{\delta}, \tag{11.3.40a}$$

and

$$N\tilde{\boldsymbol{\theta}}_N'\hat{\boldsymbol{\Sigma}}_N^{-1}\tilde{\boldsymbol{\theta}}_n \xrightarrow{\mathcal{D}} \|\mathbf{Z} + \boldsymbol{\delta}^*\|^2 \stackrel{\mathcal{D}}{=} \chi_p^2(\Delta^2), \quad \Delta^2 = \boldsymbol{\delta}'\boldsymbol{\Sigma}^{-1}\boldsymbol{\delta}. \tag{11.3.40b}$$

Hence,

$$\sqrt{N}\hat{\boldsymbol{\Sigma}}_N^{-1/2}(\tilde{\boldsymbol{\theta}}_{(N)} - \boldsymbol{\theta}_N^*) \to \mathbf{Z} - (\mathbf{Z} + \boldsymbol{\delta}^*)g(\|\mathbf{Z} + \boldsymbol{\delta}^*\|^2), \tag{11.3.41}$$

as $N \to \infty$, and the coverage probability of the confidence set $C^*(\boldsymbol{\theta}_N^*) = \left\{\boldsymbol{\theta} : N\|\boldsymbol{\theta}_{(N)} - \boldsymbol{\theta}_N\|^2_{\hat{\boldsymbol{\Sigma}}_N^{-1}} \le \chi_p^2(\gamma)\right\}$ is given by

$$\begin{aligned}
&\lim_{N\to\infty} P_{K(N)}\left\{N\|\boldsymbol{\theta}_{(N)} - \boldsymbol{\theta}_N^*\|^2_{\hat{\boldsymbol{\Sigma}}_N^{-1}} \le \chi_p^2(\gamma)\right\} \\
&\quad = P_{\boldsymbol{\delta}^*}\left\{\left\|\mathbf{Z} - (\mathbf{Z} + \boldsymbol{\delta}^*)g(\|\mathbf{Z} + \boldsymbol{\delta}^*\|^2)\right\|^2 \le \chi_p^2(\gamma)\right\}.
\end{aligned} \tag{11.3.42}$$

If $g(\mathcal{L}_N) = 0$, we have $P_0\{\|\mathbf{Z}\|^2 < \chi_p^2(\gamma)\} = 1 - \gamma$. If $g(\mathcal{L}_N) = I(\mathcal{L}_N < \chi_p^2(\alpha))$, we obtain

$$\begin{aligned}
&P_{\boldsymbol{\delta}^*}\{\|\mathbf{Z} - (\mathbf{Z} + \boldsymbol{\delta}^*)I(\|\mathbf{Z} + \boldsymbol{\delta}^*\|^2 < \chi_p^2(\alpha))\|^2 < \chi_p^2(\gamma)\} \\
&\quad = H_p(\chi_p^2(\alpha); \Delta^2)I(\Delta^2 < \chi_p^2(\gamma)) \\
&\qquad + P\{\|\mathbf{Z}\|^2 < \chi_p^2(\gamma); \|\mathbf{Z} + \boldsymbol{\delta}^*\|^2 > \chi_p^2(\alpha)\},
\end{aligned} \tag{11.3.43}$$

which is similar to the expression (4.8.22). Theorem 4.8.5 follows where $\chi_p^2(\gamma) = k_\gamma$ and $\chi_p^2(\alpha) = k_\alpha$.

Next, consider the set $C^{S+}(\hat{\boldsymbol{\theta}}_N^{S+}(c))$. In this case, we begin with

$$\begin{aligned}
&\sqrt{N}\hat{\boldsymbol{\Sigma}}_N^{-1/2}(\boldsymbol{\theta}_{(N)} - \hat{\boldsymbol{\theta}}_N^{S+}(c)) \\
&= \sqrt{N}\hat{\boldsymbol{\Sigma}}_N^{-1/2}(\boldsymbol{\theta}_{(N)} - \tilde{\boldsymbol{\theta}}_N(c)) + \sqrt{N}\hat{\boldsymbol{\Sigma}}_N^{-1/2}\tilde{\boldsymbol{\theta}}_N I(\mathcal{L}_N < c) \\
&\quad + c\sqrt{N}\hat{\boldsymbol{\Sigma}}_N^{-1/2}\tilde{\boldsymbol{\theta}}_N\mathcal{L}_N^{-1} - c\sqrt{N}\hat{\boldsymbol{\Sigma}}_N^{-1/2}\tilde{\boldsymbol{\theta}}_N\mathcal{L}_N^{-1}I(\mathcal{L}_N < c), \quad (c = p-2) \\
&\xrightarrow{\mathcal{D}} \mathbf{Z} + (Z + \boldsymbol{\delta}^*)I(\|\mathbf{Z} + \boldsymbol{\delta}^*\|^2 < c) + c(\mathbf{Z} + \boldsymbol{\delta}^*)\|\mathbf{Z} + \boldsymbol{\delta}^*\|^{-2} \\
&\quad - c(\mathbf{Z} + \boldsymbol{\delta}^*)\|\mathbf{Z} + \boldsymbol{\delta}^*\|^{-2}I(\|\mathbf{Z} + \boldsymbol{\delta}^*\|^2 < c) \quad \text{as } N \to \infty.
\end{aligned}$$

The expression above can be rewritten as

$$= -\boldsymbol{\delta}^* + \left(1 - c\|\mathbf{Z} + \boldsymbol{\delta}^*\|^{-2}\right)I\left(\|\mathbf{Z} + \boldsymbol{\delta}^*\|^2 > c\right)(\mathbf{Z} + \boldsymbol{\delta}^*). \tag{11.3.44}$$

Hence, under $K_{(N)} : \boldsymbol{\theta}_{(N)} = N^{-1/2}\boldsymbol{\delta}$, we have

$$\begin{aligned}
&\lim_{N\to\infty} P_{K_{(N)}}\Big\{N\|\boldsymbol{\theta}_{(N)} - \boldsymbol{\theta}_N^{S+}(c)\|^2_{\hat{\boldsymbol{\Sigma}}_N^{-1}} \le \chi_p^2(\gamma)\Big\} \\
&= P_{\boldsymbol{\delta}^*}\{\|\boldsymbol{\delta}^* - (1 - c\|\mathbf{Z}+\boldsymbol{\delta}^*\|^{-2})I(\|\mathbf{Z}+\boldsymbol{\delta}^*\|^2 > c)(\mathbf{Z}+\boldsymbol{\delta}^*)\|^2 < \chi_p^2(\gamma)\} \\
&= H_p(c;\Delta^2)I(\Delta^2 < \chi_p^2(\gamma)) \\
&\qquad + P_{\boldsymbol{\delta}^*}\{\|\mathbf{Z} - c(\mathbf{Z}+\boldsymbol{\delta}^*\|^{-2}\|^2 < \chi_p^2(\gamma); \|\mathbf{Z}+\boldsymbol{\delta}^*\|^2 > c\}, \qquad (11.3.45)
\end{aligned}$$

which is the same as (4.8.32) in Theorem 4.8.7. The analysis of the confidence sets is the same as in Chapter 4 and therefore is not repeated.

11.4 Simple Multivariate Linear Regression Model

In this section, we consider a simple multivariate linear regression model that is an extension of the simple linear model of Chapter 3 in a multivariate setup. In this model, one sample $\{(x_1, Y_1), \cdots, (x_N, Y_N)\}$ of size N is drawn from the p-variate normal distributions $\{\mathcal{N}_p(\boldsymbol{\theta} + \boldsymbol{\beta}x_\alpha, \Sigma)|\alpha = 1, 2, \cdots, N\}$, where $\boldsymbol{\theta} = (\theta_1, \cdots, \theta_p)'$ and $\boldsymbol{\beta} = (\beta_1, \cdots, \beta_p)'$ are the intercept and slope vectors and $\boldsymbol{\Sigma}$ is the unknown covariance matrix. The main object of this section is the study of the properties of several estimators of $\boldsymbol{\theta}$. As a special case, we study the two-sample problem involving two p-variate normal distributions. For more details, the reader is referred to Ahmed and Saleh (1990, 1999) and Sen and Saleh (1979).

11.4.1 Model, Estimation and Tests

Model. Let

$$\mathbf{Y}_\alpha = \boldsymbol{\theta} + \boldsymbol{\beta}x_\alpha + \boldsymbol{\varepsilon}_\alpha, \quad \alpha = 1, 2, \cdots, N, \qquad (11.4.1)$$

where $\mathbf{Y}_\alpha = (Y_{\alpha 1}, \cdots, Y_{\alpha p})'$ is the response p-vector, x_α is a scalar, $\boldsymbol{\varepsilon} = (\epsilon_{\alpha 1}, \cdots, \epsilon_{\alpha p})'$ is the p-vector of errors, $\boldsymbol{\theta} = (\theta_1, \cdots, \theta_p)'$ is the intercept vector, and $\boldsymbol{\beta} = (\beta_1, \cdots, \beta_p)'$ is the slope-vector of the model. We assume $\boldsymbol{\varepsilon}_\alpha \sim \mathcal{N}_p(\mathbf{0}, \boldsymbol{\Sigma})$. Our problem is to estimate $\boldsymbol{\theta}$ when it is suspected that $\boldsymbol{\beta}$ may be null, meaning, $\mathbf{0}$ without loss of generality.

Estimation. Using LSE/MLE methods, we obtain the unrestricted estimators of $\boldsymbol{\beta}$ and $\boldsymbol{\theta}$ as

$$\tilde{\boldsymbol{\theta}}_N = \bar{\mathbf{Y}} - \tilde{\boldsymbol{\beta}}_N\bar{x}_N \text{ and } \tilde{\boldsymbol{\beta}}_N = \frac{1}{Q_N}\left\{\mathbf{x}'\begin{bmatrix}\mathbf{Y}_1\\ \vdots \\ \mathbf{Y}_N\end{bmatrix} - \frac{1}{N}(\mathbf{1}_N'\mathbf{x})\left(\mathbf{1}_N'\begin{bmatrix}\mathbf{Y}_1\\ \vdots \\ \mathbf{Y}_N\end{bmatrix}\right)\right\}, \qquad (11.4.2)$$

where $Q_N = \mathbf{x}'\mathbf{x} - \frac{1}{N}(\mathbf{1}'_N\mathbf{x})^2$ with $\mathbf{x} = (x_1, \cdots, x_N)'$ and $\bar{x}_N = \frac{1}{N}\mathbf{1}'_N\mathbf{x}$. If $\boldsymbol{\beta} = 0$, then the restricted estimator of $\boldsymbol{\theta}$ is given by

$$\hat{\boldsymbol{\theta}}_N = \bar{\mathbf{Y}}_N = \frac{1}{N}\mathbf{1}'_N \begin{bmatrix} \mathbf{Y}_1 \\ \vdots \\ \mathbf{Y}_N \end{bmatrix}. \tag{11.4.3}$$

Test of Hypothesis. For testing the null-hypothesis $H_0 : \boldsymbol{\beta} = \mathbf{0}$ against $H_A : \boldsymbol{\beta} \neq \mathbf{0}$, we use the likelihood ratio statistic given by

$$\mathcal{L}_N^* = \left\{ \frac{|\mathbf{S}_e^*|}{|\mathbf{S}_e^* + Q_N\tilde{\boldsymbol{\beta}}_N\tilde{\boldsymbol{\beta}}'_N|} \right\}^{N/2}, \tag{11.4.4}$$

where

$$\mathbf{S}_e^* = (N-2)\mathbf{S}_e = \sum_{\alpha=1}^{N}\{(\mathbf{Y}_\alpha - \overline{\mathbf{Y}}_N) - \tilde{\boldsymbol{\beta}}_N(x_\alpha - \bar{x}_N)\}\{(\mathbf{Y}_\alpha - \overline{\mathbf{Y}}_N) - \tilde{\boldsymbol{\beta}}_N(x_\alpha - \bar{x}_N)\}'. \tag{11.4.5}$$

It can be shown that the likelihood ratio statistic is equivalent to the Hotelling's T^2-statistic, given by

$$(N-2)^{-1}T_N^2 = Q_N\tilde{\boldsymbol{\beta}}'_N\mathbf{S}_e^{-1}\tilde{\boldsymbol{\beta}}_N \tag{11.4.6}$$

We define $\mathcal{L}_N = \frac{N-p-1}{(N-2)p}T_N^2$ as the test-statistic equivalent to T_N^2.

Under H_A, $\mathcal{L}_N$ follows the noncentral F-distribution with (p,m) d.f. ($m = N - p - 1$) and noncentrality parameter $\Delta^2/2$ with

$$\Delta^2 = Q_N\boldsymbol{\beta}'\boldsymbol{\Sigma}^{-1}\boldsymbol{\beta}. \tag{11.4.7}$$

11.4.2 Preliminary Test and Stein-Type Estimators

Let $F_{p,m}(\alpha)$ be the α-level critical value of the $\mathcal{L}_N$-statistic under H_0. Then we define the preliminary test and the Stein-type estimators of $\boldsymbol{\beta}$ as follows:

$$\hat{\boldsymbol{\beta}}_N^{PT} = \tilde{\boldsymbol{\beta}}_N - \tilde{\boldsymbol{\beta}}_N I(\mathcal{L}_N < F_{p,m}(\alpha)), \tag{11.4.8a}$$

$$\hat{\boldsymbol{\beta}}_N^{S} = \tilde{\boldsymbol{\beta}}_N - d\tilde{\boldsymbol{\beta}}_N\mathcal{L}_N^{-1}, \ d = \tfrac{(p-2)m}{p(m+2)} \tag{11.4.8b}$$

and $$\hat{\boldsymbol{\beta}}_N^{S+} = \hat{\boldsymbol{\beta}}_N^{S} - \tilde{\boldsymbol{\beta}}_N(1 - d\mathcal{L}_N^{-1})I(\mathcal{L}_N < d). \tag{11.4.8c}$$

Similarly, the corresponding estimators of $\boldsymbol{\theta}$ are given by

$$\begin{aligned} \hat{\boldsymbol{\theta}}_N^{PT} &= \overline{\mathbf{Y}}_N I(\mathcal{L}_N < F_{p,m}(\alpha)) + (\overline{\mathbf{Y}}_N - \tilde{\boldsymbol{\beta}}_N\bar{x}_N)I(\mathcal{L}_N > F_{p,m}(\alpha)) \\ &= \tilde{\boldsymbol{\theta}}_N + \tilde{\boldsymbol{\beta}}_N\bar{x}_N I(\mathcal{L}_N < F_{p,m}(\alpha)), \end{aligned} \tag{11.4.9a}$$

$$\hat{\boldsymbol{\theta}}_N^{S} = \tilde{\boldsymbol{\beta}}_N + d\tilde{\boldsymbol{\beta}}_N\bar{x}_N\mathcal{L}_N^{-1}, \tag{11.4.9b}$$

and $$\begin{aligned} \hat{\boldsymbol{\theta}}_N^{S+} &= \hat{\boldsymbol{\theta}}_N + (\tilde{\boldsymbol{\theta}}_N - \hat{\boldsymbol{\theta}}_N)(1 - d\mathcal{L}_N^{-1})I(\mathcal{L}_N > d) \\ &= \hat{\boldsymbol{\theta}}_N^{S} + \tilde{\boldsymbol{\beta}}_N\bar{x}_N\Big\{1 - (1 - d\mathcal{L}_N^{-1})I(\mathcal{L}_N < d)\Big\}. \end{aligned} \tag{11.4.9c}$$

11.4.3 Bias, Quadratic Bias, MSE Matrices, and Risk Expressions of the Estimators

This section contains the bias, quadratic bias, MSE matrices, and weighted risk expressions of the estimators of $\boldsymbol{\beta}$ and $\boldsymbol{\theta}$ following the definitions given in the introduction of Section (11.1.2).

Theorem 1. The bias, quadratic bias, MSE matrix, and weighted risk expressions of the estimator of $\boldsymbol{\beta}$ and $\boldsymbol{\theta}$ are given as follows:

(i) **Unrestricted estimator**
$\mathbf{b}_1(\tilde{\boldsymbol{\beta}}_N) = \mathbf{0}$ and $B_1(\tilde{\boldsymbol{\beta}}_N) = 0$.

$$\mathbf{M}_1(\tilde{\boldsymbol{\beta}}_N) = \frac{1}{Q_N}\boldsymbol{\Sigma} \text{ and } R_1(\tilde{\boldsymbol{\beta}}_N;\mathbf{Q}) = \frac{1}{Q_N}\text{tr}[\mathbf{Q}\boldsymbol{\Sigma}] = \frac{p}{Q_N} \text{ if } \mathbf{Q} = \boldsymbol{\Sigma}^{-1}. \tag{11.4.10a}$$

$\mathbf{b}_1(\tilde{\boldsymbol{\theta}}_N) = \mathbf{0}$ and $B_1(\tilde{\boldsymbol{\theta}}_N) = 0$.
$\mathbf{M}_1(\tilde{\boldsymbol{\theta}}_N) = \left(\frac{1}{N} + \frac{\bar{x}_N^2}{Q_N}\right)\boldsymbol{\Sigma}$.

$$R_1(\tilde{\boldsymbol{\theta}}_N;\mathbf{Q}) = \left(\frac{1}{N} + \frac{\bar{x}_N^2}{Q_N}\right)\text{tr}[\mathbf{Q}\boldsymbol{\Sigma}] = \left(\frac{1}{N} + \frac{\bar{x}_N^2}{Q_N}\right) p \text{ if } \mathbf{Q} = \boldsymbol{\Sigma}^{-1}. \tag{11.4.10b}$$

(ii) **Restricted estimator**

$$\begin{aligned}
&\mathbf{b}_2(\hat{\boldsymbol{\beta}}_N) = -\boldsymbol{\beta} \text{ and } B_2(\hat{\boldsymbol{\beta}}_N) = \Delta^2\\
&\mathbf{M}_2(\hat{\boldsymbol{\beta}}_N) = \boldsymbol{\beta}\boldsymbol{\beta}' \text{ and } R_2(\hat{\boldsymbol{\beta}}_N;\mathbf{Q}) = \Delta^2
\end{aligned} \tag{11.4.11a}$$

$\mathbf{b}_2(\hat{\boldsymbol{\theta}}_N) = \boldsymbol{\beta}\bar{x}_N$ and $B_2(\hat{\boldsymbol{\theta}}_N) = \bar{x}_N^2\Delta^2$.
$\mathbf{M}_2(\hat{\boldsymbol{\theta}}_N) = \frac{1}{N}\boldsymbol{\Sigma} + \bar{x}_N^2\boldsymbol{\beta}\boldsymbol{\beta}' = \left(\frac{1}{N} + \bar{x}_N^2\boldsymbol{\beta}'\boldsymbol{\Sigma}^{-1}\boldsymbol{\beta}\right)\boldsymbol{\Sigma} = \left(\frac{1}{N} + \frac{\bar{x}_N^2}{Q_N}\Delta^2\right)\boldsymbol{\Sigma}$.

$$R_2(\hat{\boldsymbol{\theta}}_N;\mathbf{Q}) = \left(\frac{1}{N} + \frac{\bar{x}_N^2}{Q_N}\Delta^2\right)\text{tr}[\mathbf{Q}\boldsymbol{\Sigma}] = \left(\frac{1}{N} + \frac{\bar{x}_N^2}{Q_N}\Delta^2\right) p \text{ if } \mathbf{Q} = \boldsymbol{\Sigma}^{-1}. \tag{11.4.11b}$$

(iii) **PTE**

$$\mathbf{b}_3(\hat{\boldsymbol{\beta}}_N^{PT}) = -\boldsymbol{\beta}G_{p+2,m}(\ell_\alpha;\Delta^2) \text{ and } B_3(\hat{\boldsymbol{\beta}}_N^{PT}) = \Delta^2\{G_{p+2,m}(\ell_\alpha;\Delta^2)\}^2.$$

$$\begin{aligned}
\mathbf{M}_3(\hat{\boldsymbol{\beta}}_N^{PT}) = \frac{1}{Q_N}\boldsymbol{\Sigma} - \frac{1}{Q_N}\boldsymbol{\Sigma}G_{p+2,m}(\ell_\alpha;\Delta^2) + \boldsymbol{\beta}\boldsymbol{\beta}'\{2G_{p+2,m}(\ell_\alpha;\Delta^2)\\
-G_{p+4,m}(\ell_\alpha^*;\Delta^2)\}
\end{aligned}$$

and

$$\begin{aligned}
R_3(\hat{\boldsymbol{\beta}}_N^{PT};\mathbf{Q}) &= \frac{1}{Q_N}\text{tr}[\mathbf{Q}\boldsymbol{\Sigma}] - \frac{1}{Q_N}\text{tr}[\mathbf{Q}\boldsymbol{\Sigma}]G_{p+2,m}(\ell_\alpha;\Delta^2)\\
&\quad + \boldsymbol{\beta}'\mathbf{Q}\boldsymbol{\beta}\{2G_{p+2,m}(\ell_\alpha;\Delta^2) - G_{p+4,m}(\ell_\alpha^*;\Delta^2)\}\\
&= \frac{p}{Q_N}[1 - G_{p+2,m}(\ell_\alpha;\Delta^2)] + \frac{\Delta^2}{Q_N}\{2G_{p+2,m}(\ell_\alpha;\Delta^2)\\
&\quad -G_{p+4,m}(\ell_\alpha^*;\Delta^2)\}, \quad \text{if } \mathbf{Q} = \boldsymbol{\Sigma}^{-1}.
\end{aligned} \tag{11.4.12a}$$

$$\mathbf{b}_3(\hat{\boldsymbol{\theta}}_N^{PT}) = \boldsymbol{\beta}\bar{x}_N G_{p+2,m}(\ell_\alpha;\Delta^2) \quad \text{and} \quad B_3(\hat{\boldsymbol{\theta}}_N^{PT}) = \bar{x}_N^2\Delta^2\{G_{p+2,m}(\ell_\alpha;\Delta^2)\}^2$$

$$\mathbf{M}_3(\hat{\boldsymbol{\theta}}_N^{\mathrm{PT}}) = \Big(\frac{1}{N} + \frac{\bar{x}_N^2}{Q_N}\Big)\boldsymbol{\Sigma} - \frac{\bar{x}_N^2}{Q_N}\boldsymbol{\Sigma}G_{p+2,m}(\ell_\alpha;\Delta^2) + \bar{x}_N^2\boldsymbol{\beta}\boldsymbol{\beta}'\{2G_{p+2,m}(\ell_\alpha;\Delta^2) - G_{p+4,m}(\ell_\alpha^*;\Delta^2)\}$$

and

$$\begin{aligned}R_3(\hat{\boldsymbol{\theta}}_N^{PT};\mathbf{Q}) &= \Big(\frac{1}{N} + \frac{\bar{x}_N^2}{Q_N}\Big)\mathrm{tr}[\mathbf{Q}\boldsymbol{\Sigma}] - \frac{\bar{x}^2}{Q_N}\mathrm{tr}[\mathbf{Q}\boldsymbol{\Sigma}]G_{p+2,m}(\ell_\alpha;\Delta^2)\\ &\quad + \bar{x}_N^2\boldsymbol{\beta}'\mathbf{Q}\boldsymbol{\beta}\{2G_{p+2,m}(\ell_\alpha;\Delta^2) - G_{p+4,m}(\ell_\alpha^*;\Delta^2)\}\\ &= \Big(\frac{1}{N} + \frac{\bar{x}_N^2}{Q_N}\Big)p - \frac{\bar{x}_N^2}{Q_N}pG_{p+2,m}(\ell_\alpha;\Delta^2)\\ &\quad + \frac{\bar{x}_N^2}{Q_N}\Delta^2\{2G_{p+2,m}(\ell_\alpha;\Delta^2) - G_{p+4,m}(\ell_\alpha^*;\Delta^2)\} \quad (11.4.12b)\end{aligned}$$

if $\mathbf{Q} = \boldsymbol{\Sigma}^{-1}$.

(iv) **Stein-type estimator (SE)**

$$\begin{aligned}\mathbf{b}_4(\hat{\boldsymbol{\beta}}_N^S) &= -dp\boldsymbol{\beta}E[\chi_{p+2}^{-2}(\Delta^2)]\\ B_4(\hat{\boldsymbol{\beta}}_N^S) &= d^2p^2\frac{1}{Q_N}\Delta^2\{E[\chi_{p+2}^{-2}(\Delta^2)]\}^2\\ \mathbf{M}_4(\hat{\boldsymbol{\beta}}_N^S) &= \frac{1}{Q_N}\boldsymbol{\Sigma} - \frac{1}{Q_N}dp\boldsymbol{\Sigma}\{2E[\chi_{p+2}^{-2}(\Delta^2)] - (p-2)E[\chi_{p+2}^{-4}(\Delta^2)]\}\\ &\quad +dp(p+2)\boldsymbol{\beta}\boldsymbol{\beta}'E[\chi_{p+4}^{-4}(\Delta^2)]\end{aligned}$$

and

$$\begin{aligned}R_4(\hat{\boldsymbol{\beta}}_N^S;\mathbf{Q}) &= \frac{1}{Q_N}\mathrm{tr}[\mathbf{Q}\boldsymbol{\Sigma}] - \frac{1}{Q_N}dp\mathrm{tr}[\mathbf{Q}\boldsymbol{\Sigma}]\{2E[\chi_{p+2}^{-2}(\Delta^2)]\\ &\quad -(p-2)E[\chi_{p+2}^{-4}(\Delta^2)]\} + dp(p+2)\boldsymbol{\beta}'\mathbf{Q}\boldsymbol{\beta}E[\chi_{p+4}^{-4}(\Delta^2)]. \quad (11.4.13a)\end{aligned}$$

$$\begin{aligned}\mathbf{b}_4(\hat{\boldsymbol{\theta}}_N^S) &= dp\boldsymbol{\beta}\bar{x}_N E[\chi_{p+2}^{-2}(\Delta^2)] \quad \text{and}\\ B_4(\hat{\boldsymbol{\theta}}_N^S) &= d^2p^2\frac{\bar{x}_N^2}{Q_N}\Delta^2\{E[\chi_{p+2}^{-2}(\Delta^2)]\}^2.\\ \mathbf{M}_4(\hat{\boldsymbol{\theta}}_N^S) &= \Big(\frac{1}{N} + \bar{x}_N^2\Big)\boldsymbol{\Sigma} - dp\frac{\bar{x}_N^2}{Q_N}\boldsymbol{\Sigma}\Big\{2E[\chi_{p+2}^{-2}(\Delta^2)]\\ &\quad -(p-2)E[\chi_{p+2}^{-4}(\Delta^2)]\Big\} + dp(p+2)\bar{x}_N^2\boldsymbol{\beta}\boldsymbol{\beta}'E[\chi_{p+4}^{-4}(\Delta^2)]\end{aligned}$$

and

$$R_4(\hat{\boldsymbol{\theta}}_N^S;\mathbf{Q}) = \left(\frac{1}{N}+\frac{\bar{x}_N^2}{Q_N}\right)\mathrm{tr}[\mathbf{Q}\boldsymbol{\Sigma}] - dp\frac{\bar{x}_N^2}{Q_N}\mathrm{tr}(\mathbf{Q}\boldsymbol{\Sigma})\Big\{2E[\chi_{p+2}^{-2}(\Delta^2)]$$
$$-(p-2)E[\chi_{p+2}^{-4}(\Delta^2)]\Big\} + dp(p+4)\bar{x}_N^2\boldsymbol{\beta}'\mathbf{Q}\boldsymbol{\beta}E[\chi_{p+4}^{-4}(\Delta^2)]$$
$$= \left(\frac{1}{N}+\frac{\bar{x}_N^2}{Q_N}\right)p - dp^2\frac{\bar{x}_N^2}{Q_N}\Big\{2E\big[\chi_{p+2}^{2}(\Delta^2)\big] - (p-2)E\big[\chi_{p+2}^{2}(\Delta^2)\big]\Big\}$$
$$+dp(p+2)\frac{\bar{x}_N^2}{Q_N}\Delta^2E\big[\chi_{p+2}^{-4}(\Delta^2)\big] \text{ if } \mathbf{Q}=\boldsymbol{\Sigma}^{-1}. \tag{11.4.13b}$$

(v) **Positive rule Stein-type estimator (PRSE)**

$$\mathbf{b}_5(\hat{\boldsymbol{\beta}}_N^{S+}) = \boldsymbol{\beta}\{d_1E[F_{p+2,m}^{-1}(\Delta^2)]$$
$$+E[(1-d_1F_{p+2,m}^{-1}(\Delta^2))I(F_{p+2,m}(\Delta^2)<d_1)]\}$$

and $$B_5(\hat{\boldsymbol{\beta}}_N^{S+}) = \Delta^2\{d_1E[F_{p+2,m}^{-1}(\Delta^2)]$$
$$+E[(1-d_1F_{p+2,m}^{-1}(\Delta^2))I(F_{p+2,m}(\Delta^2)<d_1)]\}^2.$$

$$\mathbf{M}_5(\hat{\boldsymbol{\beta}}_N^{S+}) = \mathbf{M}_4(\hat{\boldsymbol{\beta}}_N^{S})$$
$$-\frac{1}{Q_N}\boldsymbol{\Sigma}\Big\{E\big[(1-d_1F_{p+2,m}^{-1}(\Delta^2))^2I(F_{p+2,m}(\Delta^2)<d_1)\big]\Big\}$$
$$+\boldsymbol{\beta}\boldsymbol{\beta}'\Big\{2E[(1-d_1F_{p+2,m}^{-1}(\Delta^2))I(F_{p+2,m}(\Delta^2)<d_1)]$$
$$-E[(1-d_2F_{p+4,m}^{-1}(\Delta^2))^2I(F_{p+4,m}(\Delta^2)<d_2)]\Big\}$$

and $$R_5(\hat{\boldsymbol{\beta}}_N^{S+};\mathbf{Q})$$
$$= R_4(\hat{\boldsymbol{\beta}}_N^{S};\mathbf{Q}) - \frac{1}{Q_N}\mathrm{tr}[\mathbf{Q}\boldsymbol{\Sigma}]E[(1-d_1F_{p+2,m}^{-1}(\Delta^2))^2I(F_{p+2,m}(\Delta^2)<d_1)]$$
$$+\boldsymbol{\beta}'\mathbf{Q}\boldsymbol{\beta}\Big\{2E[(1-d_1F_{p+2,m}^{-1}(\Delta^2))I(F_{p+2,m}(\Delta^2)<d_1)]$$
$$-E[(1-d_2F_{p+4,m}^{-1}(\Delta^2))^2I(F_{p+4,m}(\Delta^2)<d_2)]\Big\}$$
$$= R_4(\hat{\boldsymbol{\beta}}_N^{S};\boldsymbol{\Sigma}^{-1}) - \frac{p}{Q_N}\{E[(1-d_1F_{p+2,m}^{-1}(\Delta^2))^2I(F_{p+2,m}(\Delta^2)<d_1)]\}$$
$$+\frac{\Delta^2}{Q_N}\Big\{2E[(1-d_1F_{p+2,m}^{-1}(\Delta^2))I(F_{p+2,m}(\Delta^2)<d_1)]$$
$$-E\big[1-d_2F_{p+4,m}^{-1}(\Delta^2))^2I(F_{p+4,m}(\Delta^2)<d)\big]\Big\} \text{ if } \mathbf{Q}=\boldsymbol{\Sigma}^{-1},$$

where $d_1 = \frac{p}{p+2}d$ and $d_2 = \frac{p}{p+4}d$, respectively. (11.4.14a)

$$\mathbf{b}_5(\hat{\boldsymbol{\theta}}_N^{S+}) = \mathbf{b}_4(\hat{\boldsymbol{\theta}}_N^{S}) + \boldsymbol{\beta}\bar{x}_N E[(1 - d_1 F_{p+2,m}^{-1}(\Delta^2))I(F_{p+2,m}(\Delta^2) < d_1)]$$

and

$$B_5(\hat{\boldsymbol{\theta}}_N^{S+}) = \bar{x}_N \Delta^2 \{d_1 E[F_{p+2,m}^{-1}(\Delta^2)] + E[(1 - d_1 F_{p+2,m}^{-1}(\Delta^2))I(F_{p+2,m}(\Delta^2) < d_1)]\}^2.$$

$$\begin{aligned}\mathbf{M}_5(\hat{\boldsymbol{\theta}}_N^{S+}) = \mathbf{M}_4(\hat{\boldsymbol{\theta}}_N^{S}) - \frac{\bar{x}_N^2}{Q_N}\boldsymbol{\Sigma}\Big\{E[(1 - d_1 F_{p+2,m}^{-1}(\Delta^2))^2 I(F_{p+2,m}(\Delta^2) < d_1)]\Big\} \\ + \bar{x}_N^2 \boldsymbol{\beta}\boldsymbol{\beta}'\Big\{2E[(1 - d_1 F_{p+2,m}^{-1}(\Delta^2))I(F_{p+2,m}(\Delta^2) < d_1)] \\ - E[(1 - d_2 F_{p+4,m}^{-1}(\Delta^2))^2 I(F_{p+4,m}(\Delta^2) < d_2)]\Big\}\end{aligned}$$

and

$$\begin{aligned}R_5(\hat{\boldsymbol{\theta}}_N^{S+}; \mathbf{Q}) &= R_4(\hat{\boldsymbol{\theta}}_N^{S}; \mathbf{Q}) \\ &- \frac{\bar{x}_N^2}{Q_N}\operatorname{tr}[\mathbf{Q}\boldsymbol{\Sigma}]\Big\{E[(1 - d_1 F_{p+2,m}^{-1}(\Delta^2))^2 I(F_{p+2,m}(\Delta^2) < d_1)]\Big\} \\ &+ \bar{x}_N^2 \boldsymbol{\beta}'\mathbf{Q}\boldsymbol{\beta}\Big\{2E[(1 - d_1 F_{p+2,m}^{-1}(\Delta^2))I(F_{p+2,m}(\Delta^2) < d_1)] \\ &- E[(1 - d_2 F_{p+4,m}^{-1}(\Delta^2))^2 I(F_{p+4,m}(\Delta^2) < d_2)]\Big\} \\ &= R_4(\hat{\boldsymbol{\theta}}_N; \boldsymbol{\Sigma}^{-1}) - \frac{\bar{x}_N^2}{Q_N} pE[(1 - d_1 F_{p+2,m}^{-1}(\Delta^2))^2 I(F_{p+2,m}(\Delta^2) < d_1)] \\ &+ \frac{\bar{x}_N^2}{Q_N}\Delta^2\Big\{2E[(1 - d_1 F_{p+2,m}^{-1}(\Delta^2))I(F_{p+2,m}(\Delta^2) < d_1)]\Big\}\end{aligned}$$

if $\mathbf{Q} = \boldsymbol{\Sigma}^{-1}$. (11.4.14b)

From these expressions the dominance properties of various estimators can be determined as in Chapter 7.

11.4.4 Two-Sample Problem and Estimation of the Means

Consider the model (11.4.1). Set $\mathbf{x} = (\underbrace{0, \cdots, 0}_{N_1\text{-zero's}}, \underbrace{1, \cdots, 1}_{N_2\text{-one's}})'$. Then we have

$$\bar{x}_N = \frac{N_2}{N_1 + N_2}, Q_N = \frac{N_1 N_2}{N_1 + N_2}, \quad \text{and} \quad \frac{\bar{x}_N^2}{Q_N} = \frac{N_2}{N_1(N_1 + N_2)}, \tag{11.4.15}$$

respectively. Further, let

$$\boldsymbol{\mu}_1 = \boldsymbol{\theta} \text{ and } \boldsymbol{\mu}_2 = \boldsymbol{\theta} + \boldsymbol{\beta} \tag{11.4.16}$$

so that $\boldsymbol{\mu}_2 - \boldsymbol{\mu}_1 = \boldsymbol{\beta}$. Clearly, the unrestricted estimators of $\boldsymbol{\mu}_1$ and $\boldsymbol{\mu}_2$ are given by $\tilde{\boldsymbol{\mu}}_1 = \frac{1}{N_1}\sum_{\alpha=1}^{N_1}\mathbf{Y}_\alpha = \bar{\mathbf{Y}}_1$ and $\tilde{\boldsymbol{\mu}}_2 = \frac{1}{N_2}\sum_{\alpha=N_1+1}^{N}\mathbf{Y}_\alpha = \bar{\mathbf{Y}}_2$. As a result, the estimator of the difference of the means is given by

$$\tilde{\boldsymbol{\mu}}_2 - \tilde{\boldsymbol{\mu}}_1 = \bar{\mathbf{Y}}_2 - \bar{\mathbf{Y}}_1. \tag{11.4.17}$$

The test-statistic for testing $H_0 : \boldsymbol{\mu}_1 = \boldsymbol{\mu}_2$ against $H_A : \boldsymbol{\mu}_1 \neq \boldsymbol{\mu}_2$ follows as Hotelling's T^2-test, defined by

$$T_N^2 = \frac{N_1 N_2}{N_1 + N_2}(\bar{\mathbf{Y}}_2 - \bar{\mathbf{Y}}_1)'\mathbf{S}^{-1}(\bar{\mathbf{Y}}_2 - \bar{\mathbf{Y}}_1). \tag{11.4.18}$$

Equivalently, $\mathcal{L}_N = \frac{m}{(N_1+N_2+2)p}T_N^2 \overset{\mathcal{D}}{=} F_{p,m}(\Delta^2)$, where

$$(N_1 + N_2 - 2)\mathbf{S} = \sum_{\alpha=1}^{N_1}(\mathbf{Y}_\alpha - \bar{\mathbf{Y}}_1)(\mathbf{Y}_\alpha - \bar{\mathbf{Y}}_1)' + \sum_{\alpha=N_1+1}^{N}(\mathbf{Y}_\alpha - \bar{\mathbf{Y}}_2)(\mathbf{Y}_\alpha - \bar{\mathbf{Y}}_2)'. \tag{11.4.19}$$

Now, under H_0, the pooled estimator of $\boldsymbol{\mu}_1$ is given by

$$\hat{\boldsymbol{\mu}}_1 = \frac{N_1\bar{\mathbf{Y}}_1 + N_2\bar{\mathbf{Y}}_2}{N_1 + N_2}. \tag{11.4.20}$$

The PTE, SE, and PRSE of $\boldsymbol{\mu}_1$ are given by

$$\hat{\boldsymbol{\mu}}_1^{PT} = \bar{\mathbf{Y}}_1 + \tfrac{N_2}{N_1+N_2}(\bar{\mathbf{Y}}_2 - \bar{\mathbf{Y}}_1)I(\mathcal{L}_N < F_{p,m}(\alpha)), \tag{11.4.21a}$$

$$\hat{\boldsymbol{\mu}}_1^{S} = \bar{\mathbf{Y}}_1 + d\tfrac{N_2}{N_1+N_2}(\bar{\mathbf{Y}}_2 - \bar{\mathbf{Y}}_1)\mathcal{L}_N^{-1}, \quad d = \tfrac{(p-2)m}{p(m+2)}, \tag{11.4.21b}$$

and

$$\hat{\boldsymbol{\mu}}_1^{S+} = \hat{\boldsymbol{\mu}}_1^{S} + \tfrac{N_2}{N_1+N_2}(\bar{\mathbf{Y}}_2 - \bar{\mathbf{Y}}_1)\Big\{1 - (1 - dF_{p,m}^{-1})I(\mathcal{L}_N < d)\Big\},$$

respectively.

The bias expressions of the five estimators are given by

$$\mathbf{b}_1(\tilde{\boldsymbol{\mu}}_1) = \mathbf{0} \quad \text{and} \quad B_1(\tilde{\boldsymbol{\mu}}_1) = 0, \tag{11.4.22a}$$

$$\mathbf{b}_2(\hat{\boldsymbol{\mu}}_1) = \frac{N_2}{N_1 + N_2}(\boldsymbol{\mu}_2 - \boldsymbol{\mu}_1) \quad \text{and} \quad B_2(\hat{\boldsymbol{\mu}}_1) = \frac{N_2}{N_1(N_1 + N_2)}\Delta^2, \tag{11.4.22b}$$

where $\Delta^2 = \frac{N_1 N_2}{N_1+N_2}(\boldsymbol{\mu}_2 - \boldsymbol{\mu}_1)'\boldsymbol{\Sigma}^{-1}(\boldsymbol{\mu}_2 - \boldsymbol{\mu}_1)$,

$$\mathbf{b}_3(\hat{\boldsymbol{\mu}}_1^{PT}) = \frac{N_2}{N_1 + N_2}(\boldsymbol{\mu}_2 - \boldsymbol{\mu}_1)G_{p+2,m}(\ell_\alpha; \Delta^2), \ell_\alpha = \frac{p}{p+2}F_{q,m}(\alpha), \quad \text{and}$$

$$B_3(\hat{\boldsymbol{\mu}}_1^{PT}) = \frac{N_2}{N_1(N_1 + N_2)}\Delta^2\Big\{G_{p+2,m}(\ell_\alpha; \Delta^2)\Big\}^2, \tag{11.4.22c}$$

$$\mathbf{b}_4(\hat{\boldsymbol{\mu}}_1^{S}) = \frac{N_2 dp}{N_1 + N_2}(\boldsymbol{\mu}_2 - \boldsymbol{\mu}_1)E[\chi_{p+2}^{-2}(\Delta^2)] \quad \text{and}$$

$$B_4(\hat{\boldsymbol{\mu}}_1^{S}) = \frac{N_2 d^2 p^2}{N_1(N_1 + N_2)}\Delta^2\Big(\frac{1}{N} + \frac{N_2}{N_1(N_1 + N_2)}\Big)^{-1}\{E[\chi_{p+2}^{-2}(\Delta^2)]\}^2. \tag{11.4.22d}$$

Finally,

$$\mathbf{b}_5(\hat{\boldsymbol{\mu}}_1^{S+}) = \frac{N_2 dp}{N_1+N_2}(\boldsymbol{\mu}_2-\boldsymbol{\mu}_1)E[\chi_{p+2}^{-2}(\Delta^2)] + \frac{N_2}{N_1+N_2}(\boldsymbol{\mu}_2-\boldsymbol{\mu}_1)$$
$$\times E[(1-d_1F_{p+2,m}^{-1}(\Delta^2))I(F_{p+2,m}(\Delta^2)<d_1)]$$

and $B_5(\hat{\boldsymbol{\mu}}_1^{S+}) = \frac{N_2}{N_1(N_1+N_2)}\Delta^2\Big\{d_1E[F_{p+2,m}^{-1}(\Delta^2)]$

$$+E[(1-d_1F_{p+2,m}^{-1}(\Delta^2))I(F_{p+2,m}(\Delta^2)<d_1)]\Big\}^2. \quad (11.4.22e)$$

Next, we consider MSE matrices and the weighted risk expressions of the five estimators of $\boldsymbol{\mu}_1$. They are given by

$$\mathbf{M}_1(\tilde{\boldsymbol{\mu}}_1) = \tfrac{1}{N_1}\boldsymbol{\Sigma} \text{ and } R_1(\tilde{\boldsymbol{\mu}}_1;\mathbf{Q}) = \tfrac{1}{N_1}\,\mathrm{tr}[\mathbf{Q}\boldsymbol{\Sigma}], \quad (11.4.23a)$$

$$\mathbf{M}_2(\hat{\boldsymbol{\mu}}_1) = \tfrac{1}{N}\Big(1+\tfrac{N_2}{N_1}\Delta^2\Big)\boldsymbol{\Sigma},$$

and $R_2(\hat{\boldsymbol{\mu}}_1;\mathbf{Q}) = \tfrac{1}{N}\Big(1+\tfrac{N_2}{N_1}\Delta^2\Big)\mathrm{tr}[\mathbf{Q}\boldsymbol{\Sigma}]$, (11.4.23b)

$$\mathbf{M}_3(\hat{\boldsymbol{\mu}}_1^{PT}) = \frac{1}{N}\Big(1+\frac{N_2}{N_1}\Delta^2\Big)\boldsymbol{\Sigma} - \frac{N_2}{N_1N}\boldsymbol{\Sigma}G_{p+2,m}(\ell_\alpha;\Delta^2)$$
$$+\frac{N_2}{N_1+N_2}(\boldsymbol{\mu}_2-\boldsymbol{\mu}_1)(\boldsymbol{\mu}_2-\boldsymbol{\mu}_1)'\Big\{2G_{p+2,m}(\ell_\alpha;\Delta^2)$$
$$-G_{p+4,m}\Big(\frac{p}{p+4}F_{p,m}(\alpha);\Delta^2\Big)\Big\}$$

and $R_3(\hat{\boldsymbol{\mu}}_1^{PT};\mathbf{Q}) = \tfrac{1}{N}\Big(1+\tfrac{N_2}{N_1}\Delta^2\Big)\mathrm{tr}[\mathbf{Q}\boldsymbol{\Sigma}]$

$$-\tfrac{N_2}{N_1N}\mathrm{tr}[\mathbf{Q}\boldsymbol{\Sigma}]G_{p+2,m}(\ell_\alpha;\Delta^2) + \tfrac{N_2}{N_1+N_2}\Delta^2$$
$$\times\Big\{2G_{p+2,m}(\ell_\alpha;\Delta^2) - G_{p+4,m}\Big(\tfrac{p}{p+4}F_{p,m}(\alpha);\Delta^2\Big)\Big\}, \quad (11.4.23c)$$

$$\mathbf{M}_4(\hat{\boldsymbol{\mu}}_1^{S}) = \tfrac{1}{N}\Big(1+\tfrac{N_2}{N_1}\Big)\boldsymbol{\Sigma} - \tfrac{dpN_2}{N_1N}\boldsymbol{\Sigma}\Big\{2E[\chi_{p+2}^{-2}(\Delta^2)]$$
$$-(p-2)E[\chi_{p+2}^{-4}(\Delta^2)]\Big\} + \tfrac{dp(p+4)N_2}{N_1N}(\boldsymbol{\mu}_2-\boldsymbol{\mu}_1)(\boldsymbol{\mu}_2-\boldsymbol{\mu}_1)'E[\chi_{p+4}^{-4}(\Delta^2)]$$

and

$$R_4(\hat{\boldsymbol{\mu}}_1^{S};\mathbf{Q}) = \frac{1}{N}\Big(1+\frac{N_2}{N_1}\Big)\mathrm{tr}[\mathbf{Q}\boldsymbol{\Sigma}] - \frac{dpN_2}{N_1N}\mathrm{tr}[\mathbf{Q}\boldsymbol{\Sigma}]\Big\{2E[\chi_{p+2}^{-2}(\Delta^2)]$$
$$-(p-2)2E[\chi_{p+2}^{-4}(\Delta^2)]\Big\} + \frac{dp(p+4)N_2}{N_1N}\Delta^2E[\chi_{p+4}^{-4}(\Delta^2)]. \quad (11.4.23d)$$

$$\mathbf{M}_5(\hat{\boldsymbol{\mu}}_1^{S+}) = \mathbf{M}_4(\hat{\boldsymbol{\mu}}_1^S)$$
$$-\frac{N_2}{N_1 N}\boldsymbol{\Sigma} E[(1 - d_1 F_{p+2,m}^{-1}(\Delta^2))^2 I(F_{p+2,m}(\Delta^2) < d_1)]$$
$$+\frac{N_2}{N}(\boldsymbol{\mu}_2 - \boldsymbol{\mu}_1)(\boldsymbol{\mu}_2 - \boldsymbol{\mu}_1)'\Big\{2E[(1 - d_1 F_{p+2,m}^{-1}(\Delta^2))^2 I(F_{p+2,m}(\Delta^2) < d_1)]$$
$$-E[(1 - d_2 F_{p+4,m}^{-1}(\Delta^2))^2 I(F_{p+4,m}(\Delta^2) < d_2)]\Big\},$$

and

$$R_5(\hat{\boldsymbol{\mu}}_1^{S+}; \mathbf{Q}) = R_4(\hat{\boldsymbol{\mu}}_1^S; \mathbf{Q})$$
$$-\frac{N_2}{N_1 N}\mathrm{tr}[\mathbf{Q}\boldsymbol{\Sigma}]\{E[(1 - d_1 F_{p+2,m}^{-1}(\Delta^2))^2 I(F_{p+2,m}(\Delta^2) < d_1)]\}$$
$$+\frac{N_2}{N}\Delta^2\Big\{2E[(1 - d_1 F_{p+2,m}^{-1}(\Delta^2))^2 I(F_{p+2,m}(\Delta^2) < d_1)]$$
$$-E[(1 - d_2 F_{p+4,m}^{-1}(\Delta^2))^2 I(F_{p+4,m}(\Delta^2) < d_2)]\Big\}. \quad (11.4.23e)$$

11.4.5 Confidence Sets for the Slope and Intercept Parameters

Now, consider the confidence sets for $\boldsymbol{\beta}$ based on the four estimators, $\tilde{\boldsymbol{\beta}}_N$, $\hat{\boldsymbol{\beta}}_N^{\mathrm{PT}}$, $\hat{\boldsymbol{\beta}}_N^S$, and $\hat{\boldsymbol{\beta}}_N^{S+}$

$$C^*(\boldsymbol{\beta}_N^*) = \Big\{\boldsymbol{\beta} : Q_N\|\boldsymbol{\beta} - \boldsymbol{\beta}_N^*\|_{\boldsymbol{\Sigma}^{-1}}^2 \le \chi_p^2(\gamma)\Big\},$$

where $\boldsymbol{\beta}_N^* = \tilde{\boldsymbol{\beta}}_N\,(1 - g(\mathcal{L}_N^*))$,

$g(\mathcal{L}_N^*) = 0,\ I\,(\mathcal{L}_N^* \le F_{p,m}(\alpha)),\ (p-2)\mathcal{L}_N^*$, and

$> (p-2){\mathcal{L}_N^*}^{-1} + (1 - (p-2){\mathcal{L}_N^*}^{-1} I(\mathcal{L}_N^* < (p-2))$

and $\mathcal{L}_N^* = Q_N \tilde{\boldsymbol{\beta}}_N' \boldsymbol{\Sigma}^{-1}\tilde{\boldsymbol{\beta}}_N$, where $\boldsymbol{\Sigma}$ is known.

Similarly, the confidence sets based on the five estimators of $\boldsymbol{\theta}$ can be written as

$$C^*\,(\boldsymbol{\theta}_N^*) = \left\{\boldsymbol{\theta} : \left(\frac{1}{N} + \frac{\bar{x}^2}{Q_N}\right)^{-1} \|\boldsymbol{\theta} - \boldsymbol{\theta}_N^*\|_{\Sigma^{-1}}^2 \le \chi_p^2(\gamma)\right\},$$

where

$$\boldsymbol{\theta}_N^* = \tilde{\boldsymbol{\theta}}_N + (\tilde{\boldsymbol{\theta}}_N - \hat{\boldsymbol{\theta}}_N)g(\mathcal{L}_N^*).$$

The coverage probabilities may be calculated as in Chapter 5 for the parallelism problem.

11.5 *R*-estimation and Confidence Sets for the Simple Multivariate Linear Model

11.5.1 Introduction

Let $\{\mathbf{X}_i = (X_{i1}, \dots, X_{ip})', \ i \geq 1\}$ be a sequence of independent random vectors with p (≥ 1)-variant distribution functions $\{F_i, \ i \geq 1\}$, where for every $i \geq 1$,

$$F_i(\mathbf{Y}) = F(\mathbf{Y} - \boldsymbol{\theta} - \boldsymbol{\beta} x_i), \quad \mathbf{Y} \in R^p \tag{11.5.1}$$

where $\boldsymbol{\theta} = (\theta_1, \dots, \theta_p)'$ and $\boldsymbol{\beta} = (\beta_1, \dots, \beta_p)'$ are the unknown intercept and slope parameters, respectively, and x_i ($i \geq 1$) are sequences of known constants. As in Section 11.14, we are interested in the R-estimation of $\boldsymbol{\theta}$ and $\boldsymbol{\beta}$ when it is suspected that $\boldsymbol{\beta}$ may be null vector ($\boldsymbol{\theta}$) based on the rank method.

Let $R_{ij}(a, b)$ and $R_{ij}^+(a, b)$ be the rank of $Y_{ij} - a - bx_i$ (or $|Y_{1j} - a - bx_i|$) among $Y_{1j} - a - bx_i, \dots, Y_{nj} - a - bx_n$ (or $|Y_{1j} - a - bx_1|, \dots, |Y_{nj} - a - bx_n$) for $i = 1, \dots, n$; $j = 1, \dots, p$, where a, b are real numbers. Then define the LRSs:

$$\mathbf{T}_n(\mathbf{a}, \mathbf{b}) = (T_{n1}(a_1, b_1), \dots, T_{np}(a_p, b_p))' \text{ and } \mathbf{L}_n(b) = (L_{n1}(b_1), \dots, L_{np}(b_p))', \tag{11.5.2}$$

where

$$\mathbf{T}_{nj}(a, b) = n^{-1} \sum_{i=1}^{n} a_n^+(R_{ij}^+(a, b)\text{sign}(Y_{ij} - a - bx_i)$$

and

$$\mathbf{L}_{nj}(\mathbf{b}) = n^{-1} \sum_{i=1}^{n} (x_i - \bar{x}_n) a_n(R_{ij}(\mathbf{b})), \tag{11.5.3}$$

respectively, with scores defined by

$$a_n^+(k) = E\phi_j(U_{nk}) \quad \text{or} \quad \phi_j\left(\frac{k}{n+1}\right),$$

$$a_n(k) = E\phi_j^+(U_{nk}) \quad \text{or} \quad \phi^*\left(\frac{k}{n+1}\right), \tag{11.5.4}$$

for $i = 1, \dots, n$ and U_{nk} is the order statistic with rank k among n independent samples from $U(0,1)$. The score functions $\phi_j = \{\phi_j(u); \ 0 < u < 1\}$ is absolutely continuous, nondecreasing skew-symmetric, meaning $\phi_j(u) + \phi_j(1 - u) = 0 \ \forall \ u \in (0,1)$ and are square-integrable inside $(0,1)$ whereas $\phi_j^+(u) = \phi_j\left(\frac{1+u}{2}\right)$, $u \in (0,1)$, $j = 1, \dots, p$.

Assume that

$$Q_n = \sum_{i=1}^{n} (x_i - \bar{x}_n) \text{ and } \lim_{n \to \infty} n^{-1} Q_n = Q^*$$

$$\lim_{n \to \infty} \bar{x}_n = \bar{x} \ (|\bar{x}| < \infty) \tag{11.5.5}$$

both exist. $F \in \mathcal{F}$, where $\mathcal{F}$ is the class of all p-variate continuous distribution functions that are diagonally symmetric about $\mathbf{0}$ having a finite Fisher information matrix. (11.5.6)

Note that $T_{i,j}(a,b) \searrow a$ for fixed b and $L_{nj}(a,b) \searrow$ in b independent of a for every $j = 1,\ldots,p$. Also, under $H_0 : \boldsymbol{\theta} = \boldsymbol{\beta} = \mathbf{0}$, $[\mathbf{T}_n'(\mathbf{0},\mathbf{0}), \mathbf{L}_n'(\mathbf{0})]'$ is symmetrically distributed around $(\mathbf{0},\mathbf{0})$. Hence, as in Sen and Saleh (1979), define the unrestricted estimators of $(\boldsymbol{\theta}', \boldsymbol{\beta}')'$ as follows:

$$\tilde{\boldsymbol{\theta}}_n = (\tilde{\theta}_1, \ldots, \tilde{\theta}_p)' \quad \text{and} \quad \tilde{\boldsymbol{\beta}}_n = (\tilde{\beta}_1, \ldots, \tilde{\beta}_p)',$$

$$\text{where } \tilde{\beta}_j = \frac{1}{2}\{\sup[b_j : L_{nj}(b_j) > 0] + \inf[b_j : L_{nj}(b_j) < 0]\}$$

$$\tilde{\theta}_j = \frac{1}{2}\left\{\sup[a_j : T_{nj}(a_{,j}\,\tilde{\beta}_j) > 0] + \inf[a_j : T_{nj}(a_j, \tilde{\beta}_j) < 0]\right\}.$$

Under $H_0 : \boldsymbol{\beta} = \mathbf{0}$, the restricted estimator of $\boldsymbol{\theta}$ is given by

$$\hat{\boldsymbol{\theta}}_n = (\hat{\theta}_1, \ldots, \hat{\theta}_p)', \tag{11.5.7}$$

where $\hat{\theta}_j = \frac{1}{2}\{\sup[a_j : T_{nj}(a_j;0) > 0] + \inf[a_j : T_{nj}(a_j;0) < 0]\}$ since the **restricted** estimator of $\boldsymbol{\beta}$ is $\mathbf{0}$. Then $\hat{\boldsymbol{\theta}}_n$is a translation-invariant, robust, and consistent estimator of $\boldsymbol{\theta}$ when $\boldsymbol{\beta} = \mathbf{0}$, while $\tilde{\boldsymbol{\theta}}_n$ as well as $\tilde{\boldsymbol{\beta}}_n$ are similar types of R-estimators when $\boldsymbol{\beta} \neq \mathbf{0}$.

In order to define the rest of the modified R-estimators, we consider first the test of hypothesis $H_0 : \boldsymbol{\beta} = \mathbf{0}$ based on the rank-test $\mathcal{L}_n$ given by

$$\mathcal{L}_n = [n\mathbf{L}_n'(\mathbf{0})\mathbf{M}_n^- \mathbf{L}_n(\mathbf{0})]Q_n, \tag{11.5.8}$$

where

$$\mathbf{M}_n = ((m_{j\ell})), \quad m_{j\ell} = n^{-1}\sum_{i=1}^{n}[a_{nj}(R_{ij}) - \bar{a}_j][a_{n\ell}(R_{i\ell}) - \bar{a}_\ell], \tag{11.5.9}$$

with $\bar{a}_j = n^{-1}\sum_{i=1}^{n} a_{nj}(i)$; $j,\ell = 1,\ldots,p$, and $\mathbf{M}_n^-$ is the generalized inverse of $\mathbf{M}_n$.

Now, conditionally, $\mathcal{L}_n$ is a distribution-free statistic, and for large n, $\mathcal{L}_n$ is approximately distributed as a central chi-square with p d.f. when $\mathbf{M}_n$ is of full rank. Let $\mathcal{L}_{n,\alpha}$ be a upper α-level critical value from the distribution of $\mathcal{L}_n$, and then $\mathcal{L}_{n,\alpha} \to \chi_p^2(\alpha)$ as $n \to \infty$. Based on this information, we define the three estimators of $\boldsymbol{\beta}$ given by

$$\hat{\boldsymbol{\beta}}_n^{\mathrm{PT}} = \tilde{\boldsymbol{\beta}}_n - \tilde{\boldsymbol{\beta}}_n I(\mathcal{L}_n < \mathcal{L}_{n,\alpha})$$

$$\hat{\boldsymbol{\beta}}_n^{S} = \tilde{\boldsymbol{\beta}}_n - c\tilde{\boldsymbol{\beta}}_n \mathcal{L}_n^{-1} I(\mathcal{L}_n > \varepsilon_n), \quad \varepsilon_n \to 0 \text{ as } n \to \infty; \quad c = p-2,$$

and

$$\hat{\boldsymbol{\beta}}_n^{S+} = \tilde{\boldsymbol{\beta}}_n(1 - c\mathcal{L}_n^{-1})I(\mathcal{L}_n > c). \tag{11.5.10}$$

Similarly, we have four additional estimators of $\boldsymbol{\theta}$ as follows:

$\hat{\boldsymbol{\theta}}_n$ **the restricted estimator** of $\boldsymbol{\theta}$ when $\boldsymbol{\beta} = \mathbf{0}$.

$$\hat{\boldsymbol{\theta}}_n^{\mathrm{PT}} = \tilde{\boldsymbol{\theta}}_n - (\tilde{\boldsymbol{\theta}}_n - \hat{\boldsymbol{\theta}}_n) I(\mathcal{L}_n < \mathcal{L}_{n,\alpha})$$

$$\hat{\boldsymbol{\theta}}_n^{S} = \tilde{\boldsymbol{\theta}}_n - c(\tilde{\boldsymbol{\theta}}_n - \hat{\boldsymbol{\theta}}_n)\mathcal{L}_n^{-1} I(\mathcal{L}_n \geq \varepsilon_n), \quad \varepsilon_n \to 0 \text{ as } n \to \infty \text{ and } c = (p-2),$$

and

$$\hat{\boldsymbol{\theta}}_n^{S+} = \hat{\boldsymbol{\theta}}_n + (1 - c\mathcal{L}_n^{-1}) I(\mathcal{L}_n > c)(\tilde{\boldsymbol{\theta}}_n - \hat{\boldsymbol{\theta}}_n), \tag{11.5.11}$$

respectively.

11.5.2 Asymptotic Properties of the R-estimators

Let $F_{[j]}$ and $F_{j\ell]}$ be the marginal and the joint-distribution of the jth and (j, ℓ)th variable corresponding to F in (11.5.1) as described in Section 11.3.1 and consider the matrices defined by $\boldsymbol{\Lambda}$ and $\boldsymbol{\Sigma}$ as in (11.3.9). We assume that they are of full rank and $\mathbf{M}_n$ is a consistent estimator of $\boldsymbol{\Sigma}$. Then, one may show that under fixed alternative $\boldsymbol{\beta} \neq \mathbf{0}$, the following results hold

$$\sqrt{n}\mathbf{M}_n^{-1/2}(\hat{\boldsymbol{\beta}}_n^{\mathrm{PT}} - \boldsymbol{\beta}) = \sqrt{n}\boldsymbol{\Sigma}^{-1/2}(\tilde{\boldsymbol{\beta}}_n - \boldsymbol{\beta}) + o_P(1),$$

$$\sqrt{n}\mathbf{M}_n^{-1/2}(\hat{\boldsymbol{\beta}}_n^{S} - \boldsymbol{\beta}) = \sqrt{n}\boldsymbol{\Sigma}^{-1/2}(\tilde{\boldsymbol{\beta}}_n - \boldsymbol{\beta}) + o_P(1),$$

and
$$\sqrt{n}\mathbf{M}_n^{-1/2}(\hat{\boldsymbol{\beta}}_n^{S+} - \boldsymbol{\beta}) = \sqrt{n}\boldsymbol{\Sigma}^{-1/2}(\tilde{\boldsymbol{\beta}}_n - \boldsymbol{\beta}) + o_P(1). \tag{11.5.12}$$

Thus, we consider the class of contiguous alternatives

$$K_{(n)} : \boldsymbol{\beta}_{(n)} = n^{-12}\boldsymbol{\delta}, \quad \boldsymbol{\delta} = (\delta_1, \ldots, \delta_p)'. \tag{11.5.13}$$

Now, from the basic theorems of Chapter 8 (Puri and, Sen 1985), it follows that under $\boldsymbol{\theta} = \boldsymbol{\beta} = \mathbf{0}$ and (11.5.2)

$$\begin{pmatrix} \sqrt{n}\mathbf{T}_n(\mathbf{0},\mathbf{0}) \\ \sqrt{n}\mathbf{L}_n(\mathbf{0}) \end{pmatrix} \xrightarrow{\mathcal{D}} N_{2p}\left\{ \begin{pmatrix} \mathbf{0} \\ \mathbf{0} \end{pmatrix}; \boldsymbol{\Lambda} \otimes \begin{pmatrix} 1 & 0 \\ 0 & Q^* \end{pmatrix} \right\} \quad \text{as } n \to \infty. \tag{11.5.14}$$

Further, from theorems of Chapter 6 (Puri and Sen, 1985), we have the linearity results, where k $(0 < k < \infty)$:

$$\sup_{|b_j| < k_1} \left| L_{nj}(n^{-1/2}b_j) - L_{nj}(0) + n^{1/2} b_j Q^* \gamma_j \right| \xrightarrow{P} 0 \text{ as } n \to \infty,$$

$$\sup_{\substack{|a_j| < k_2 \\ |b_j| < k_1}} \left| T_{nj}(n^{-1/2}(a_j, b_j)) - T_{nj}(0,0) + n^{1/2}(a_j + b_i \bar{x})\gamma_j \right| \xrightarrow{P} 0 \quad \text{as } n \to \infty,$$

$$\tag{11.5.15}$$

where $\gamma_j = \int_0^1 \psi_j(u)\phi_j(u)du$ and $\psi_j(u) = -\frac{f'_{[j]}(u)}{f_{[j]}(u)}$ for $j = 1, \ldots, p$. Then we have the following theorems based on (11.5.14) and (11.5.15).

Theorem 1. Under the assumed regularity conditions given in (11.5.5) through (11.5.15), as $n \to \infty$, we have

$$\begin{pmatrix} \sqrt{n}(\tilde{\boldsymbol{\theta}}_n - \boldsymbol{\theta}) \\ \sqrt{n}(\tilde{\boldsymbol{\beta}}_n - \boldsymbol{\beta}) \end{pmatrix} \xrightarrow{\mathcal{D}} N_{2p}\left\{ \begin{pmatrix} 0 \\ 0 \end{pmatrix}, \boldsymbol{\Sigma} \otimes \begin{pmatrix} 1 + \frac{\bar{x}^2}{Q^*} & -\frac{\bar{x}}{Q^*} \\ -\frac{\bar{x}}{Q^*} & \frac{1}{Q^*} \end{pmatrix} \right\}, \quad (11.5.16a)$$

and when $\boldsymbol{\beta} = \mathbf{0}$,

$$\sqrt{n}(\hat{\boldsymbol{\theta}}_n - \boldsymbol{\theta}) \xrightarrow{\mathcal{D}} N_p(\mathbf{0}, \boldsymbol{\Sigma}). \quad (11.5.16b)$$

Theorem 2. Under $\{K_{(n)}\}$ and assumed regularity conditions, as $n \to \infty$

$$\begin{pmatrix} \sqrt{n}(\tilde{\boldsymbol{\theta}}_n - \boldsymbol{\theta}) \\ \sqrt{n}\mathbf{L}_n(\mathbf{0}) \end{pmatrix} \overset{\mathcal{D}}{\simeq} N_{2p}\left\{ \begin{pmatrix} \mathbf{0} \\ \boldsymbol{\gamma}^* Q^* \end{pmatrix}; \ \boldsymbol{\Sigma} \otimes \begin{pmatrix} (1 + \frac{\bar{x}^2}{Q^*})\boldsymbol{\Sigma} & -\bar{x}\boldsymbol{\nu}^{-1}\boldsymbol{\Lambda} \\ -\bar{x}\boldsymbol{\Lambda}\boldsymbol{\nu}^{-1} & \boldsymbol{\Lambda} Q \end{pmatrix} \right\}. \quad (11.5.17a)$$

$$\begin{pmatrix} \sqrt{n}(\hat{\boldsymbol{\theta}}_n - \boldsymbol{\theta}) \\ \sqrt{n}\mathbf{L}_n(\mathbf{0}) \end{pmatrix} \overset{\mathcal{D}}{\simeq} N_{2p}\left\{ \begin{pmatrix} \boldsymbol{\delta}\bar{x} \\ \boldsymbol{\gamma}^* Q^* \end{pmatrix}; \ \boldsymbol{\Sigma} \otimes \ \text{Diag}(1, Q^*) \right\} \quad (11.5.17b)$$

where $\boldsymbol{\gamma}^* = (\gamma_1\delta_1, \ldots, \gamma_p\delta_p)'$.

$$\lim P_{K_{(n)}} \{ (n\mathbf{L}_n'(\mathbf{0})\mathbf{M}_n^- \mathbf{L}_n(\mathbf{0})Q_n) \le x \} = H_p(x; \Delta^2), \quad (11.5.17c)$$

where $\Delta^2 = Q^* \boldsymbol{\delta}' \boldsymbol{\Sigma}^{-1} \boldsymbol{\delta}$.

Proof. Under $\boldsymbol{\theta} = \boldsymbol{\beta} = \mathbf{0}$ and using componentwise linearity results, we have as $n \to \infty$,

$$\sqrt{n}\,\mathbf{L}_n(\mathbf{0}) = \boldsymbol{\nu}\sqrt{n}\,\tilde{\boldsymbol{\beta}}_n Q^* + o_P(1)$$

and

$$\sqrt{n}\mathbf{T}_n(0, 0) = \boldsymbol{\nu}\sqrt{n}(\tilde{\boldsymbol{\theta}}_n + \tilde{\boldsymbol{\beta}}_n \bar{x}) + o_P(1) \quad (11.5.18)$$

so that

$$\boldsymbol{\nu}\sqrt{n}\,\tilde{\boldsymbol{\theta}}_n = \sqrt{n}\mathbf{T}_n(\mathbf{0}, \mathbf{0}) - \frac{\bar{x}}{Q^*}\sqrt{n}\,\mathbf{L}_n(\mathbf{0}) + o_P(1). \quad (11.5.19)$$

Hence, utilizing contiguity of probability measures to those under $H_0^* : \boldsymbol{\theta} = \boldsymbol{\beta} = \mathbf{0}$, we obtain under $\{K_{(n)}\}$ as $n \to \infty$ the joint-distribution of $(\sqrt{n}\mathbf{T}_n'(0,0), \sqrt{n}\mathbf{L}_n'(0)\}'$ as

$$\begin{pmatrix} \sqrt{n}\mathbf{T}_n(\mathbf{0}, n^{-1/2}\boldsymbol{\delta}) \\ \sqrt{n}\mathbf{L}_n(n^{-1/2}\boldsymbol{\delta}) \end{pmatrix} \overset{\mathcal{D}}{\simeq} N_{2p}\left\{ \begin{pmatrix} \boldsymbol{\delta}\bar{x} \\ \boldsymbol{\gamma}^* Q^* \end{pmatrix}, \boldsymbol{\Lambda} \otimes \ \text{Diag}(1, Q^*) \right\}. \quad (11.5.20)$$

Further, since

$$\boldsymbol{\nu}\sqrt{n}\,\hat{\boldsymbol{\theta}}_n = \sqrt{n}\,\mathbf{T}_n(\mathbf{0}, \mathbf{0}) + o_P(1),$$

the proof of (11.5.16b) is complete.

Corollary 1. The marginal distributions of $\sqrt{n}(\tilde{\boldsymbol{\theta}}_n - \boldsymbol{\theta})$ and $\sqrt{n}(\hat{\boldsymbol{\theta}}_n - \boldsymbol{\theta})$ under $\{K_{(n)}\}$ as $n \to \infty$ are

$$\sqrt{n}(\tilde{\boldsymbol{\theta}}_n - \boldsymbol{\theta}) \sim N_p\left(\mathbf{0}; \left(1 + \frac{\bar{x}^2}{Q^*}\right)\boldsymbol{\Sigma}\right),$$

$$\sqrt{n}(\hat{\boldsymbol{\theta}}_n - \boldsymbol{\theta}) \sim N_p(\boldsymbol{\delta}\bar{x}; \boldsymbol{\Sigma})$$

and

$$\sqrt{n}(\tilde{\boldsymbol{\theta}}_n - \hat{\boldsymbol{\theta}}_n) \sim N_p\left(-\boldsymbol{\delta}\bar{x}; \frac{\bar{x}^2}{Q^*}\boldsymbol{\Sigma}\right). \tag{11.5.21}$$

Also, $\sqrt{n}(\hat{\boldsymbol{\theta}}_n - \boldsymbol{\theta})$ and $\sqrt{n}(\tilde{\boldsymbol{\theta}}_n - \hat{\boldsymbol{\theta}}_n)$ are mutually independent.

By the corollary we find the ADB, ADMSE, and ADQR of the estimators of $\boldsymbol{\theta}$ and $\boldsymbol{\beta}$ as given below.

(i) Unrestricted estimators

$$\mathbf{b}_1(\tilde{\boldsymbol{\beta}}_n) = 0 \quad \text{and} \quad B_1(\tilde{\boldsymbol{\beta}}_n) = 0;$$

$$\mathbf{b}_1(\tilde{\boldsymbol{\theta}}_n) = \mathbf{0} \quad \text{and} \quad B_1(\tilde{\boldsymbol{\theta}}_n) = 0;$$

$$\mathbf{M}_1(\tilde{\boldsymbol{\beta}}_n) = \frac{1}{Q^*}\boldsymbol{\Sigma} \quad \text{and} \quad M_1(\tilde{\boldsymbol{\theta}}_n) = \left(1 + \bar{x}^2\right)\boldsymbol{\Sigma};$$

$$R_1(\tilde{\boldsymbol{\beta}}_n; \mathbf{W}) = \frac{1}{Q^*}\operatorname{tr}(\mathbf{W}\boldsymbol{\Sigma}) \quad \text{and} \quad R_1(\tilde{\boldsymbol{\theta}}_n; \mathbf{W}) = \left(1 + \frac{\bar{x}^2}{Q^*}\right)\operatorname{tr}(\mathbf{W}\boldsymbol{\Sigma}). \tag{11.5.22}$$

(ii) Restricted estimators

$$\mathbf{b}_2(\hat{\boldsymbol{\beta}}_n) = -\boldsymbol{\delta} \quad \text{and} \quad B_2(\hat{\boldsymbol{\beta}}_n) = \Delta^2;$$

$$\mathbf{b}_2(\hat{\boldsymbol{\theta}}_n) = \boldsymbol{\delta}\bar{x} \quad \text{and} \quad B_2(\hat{\boldsymbol{\theta}}_n) = \bar{x}^2\Delta^2;$$

$$\mathbf{M}_2(\hat{\boldsymbol{\beta}}_n) = \boldsymbol{\delta}\boldsymbol{\delta}' \quad \text{and} \quad \mathbf{M}_2(\hat{\boldsymbol{\theta}}_n) = \boldsymbol{\Sigma} + \bar{x}^2\boldsymbol{\delta}\boldsymbol{\delta}';$$

$$R_2(\hat{\boldsymbol{\beta}}_n; \mathbf{W}) = \boldsymbol{\delta}'\mathbf{W}\boldsymbol{\delta} \quad \text{and} \quad R_2(\hat{\boldsymbol{\theta}}_n; \mathbf{W}) = \operatorname{tr}(\mathbf{W}\boldsymbol{\Sigma}) + \bar{x}^2(\boldsymbol{\delta}'\mathbf{W}\boldsymbol{\delta}). \tag{11.5.23}$$

(iii) PTE

$$\mathbf{b}_3(\hat{\boldsymbol{\beta}}_n^{\text{PT}}) = -\boldsymbol{\delta}H_{p+2}(\chi_p^2(\alpha); \Delta^2) \text{ and } B_3(\hat{\boldsymbol{\beta}}_n^{\text{PT}}) = \Delta^2\{H_{p+2}(\chi_p^2(\alpha); \Delta^2)\};$$

$$\mathbf{b}_3(\hat{\boldsymbol{\theta}}_n^{\text{PT}}) = -\boldsymbol{\delta}\bar{x}H_{p+2}(\chi_p^2(\alpha); \Delta^2) \text{ and } B_3(\hat{\boldsymbol{\theta}}_n^{\text{PT}}) = \bar{x}^2\Delta^2\{H_{p+2}(\chi_p^2(\alpha); \Delta^2)\}^2;$$

$$\mathbf{M}_3(\hat{\boldsymbol{\theta}}_n^{\text{PT}}) = \boldsymbol{\Sigma} + \frac{\bar{x}^2}{Q^*}\boldsymbol{\Sigma}\{1 - H_{p+2}(\chi_p^2(\alpha); \Delta^2)\}$$
$$+ \bar{x}^2\boldsymbol{\delta}\boldsymbol{\delta}'\{2H_{p+2}(\chi_p^2(\alpha); \Delta^2) - H_{p+4}(\chi_p^2(\alpha); \Delta^2)\}\}$$

$$R_3(\hat{\beta}_n^{\rm PT};\mathbf{W}) = {\rm tr}(\mathbf{W\Sigma})\{1 - H_{p+2}(\chi_p^2(\alpha);\Delta^2)\}$$
$$+(\boldsymbol{\delta}'\mathbf{W}\boldsymbol{\delta})\{2H_{p+2}(\chi_p^2(\alpha);\Delta^2) - H_{p+4}(\chi_p^2(\alpha);\Delta^2)\} \quad \text{and}$$

$$R_3(\tilde{\boldsymbol{\theta}}_n^{\rm PT};\mathbf{W}) = {\rm tr}(\mathbf{W\Sigma}) + \frac{\bar{x}^2}{Q^*}{\rm tr}(\mathbf{W\Sigma})\{1 - H_{p+2}(\chi_p^2(\alpha);\Delta^2)\}$$
$$+\bar{x}^2(\boldsymbol{\delta}'\mathbf{W}\boldsymbol{\delta}\{2H_{p+2}(\chi_p^2(\alpha);\Delta^2) - \{H_{p+4}(\chi_p^2(\alpha);\Delta^2)\} \quad (11.5.24)$$

(iv) SE

$$\mathbf{b}_4(\hat{\boldsymbol{\beta}}_n^S) = -c\boldsymbol{\delta}E[\chi_{p+2}^{-2}(\Delta^2)] \quad \text{and} \quad B_4(\hat{\boldsymbol{\beta}}_n^S) = c^2\Delta^2\{E[(\chi_{p+2}^{2}(\Delta^2)]\}^2$$

where $c = p - 2$;

$$\mathbf{b}_4(\hat{\boldsymbol{\theta}}_n^S) = -c\bar{x}\boldsymbol{\delta}E[\chi_{p+2}^2(\Delta^2)] \quad \text{and} \quad B_4(\hat{\boldsymbol{\theta}}_n^S) = c^2\bar{x}^2\Delta^2E[\chi_{p+2}^2(\Delta^2)]\}^2;$$

$$\mathbf{M}_4(\hat{\boldsymbol{\beta}}_n^S) = \frac{1}{Q^*}\{\boldsymbol{\Sigma} - \frac{c}{Q^*}\boldsymbol{\Sigma}\{2E[\chi_{p+2}^2(\Delta^2)] - cE[\chi_{p+2}^{-4}(\Delta^2)]\};$$
$$+ c(c+4)\boldsymbol{\delta\delta}'E[\chi_{p+4}^{-4}(\Delta^2)] \quad \text{and}$$

$$M_4(\hat{\boldsymbol{\theta}}_n^S) = \left(1 + \frac{\bar{x}^2}{Q^*}\right)\boldsymbol{\Sigma} - c\frac{\bar{x}^2}{Q^*}\boldsymbol{\Sigma}\{2E[\chi_{p+2}^{-2}(\Delta^2)] - cE[\chi_{p+2}^{-4}(\Delta^2)\}$$
$$+ \frac{\bar{x}^2}{Q^*}c(c+4)\boldsymbol{\delta\delta}'E[\chi_{p+4}^{-4}(\Delta^2)]; \quad \text{and}$$

$$R_4(\hat{\beta}_n^S;\mathbf{W}) = \frac{1}{Q^*}{\rm tr}(\mathbf{W\Sigma}) - \frac{c}{Q^*}{\rm tr}(\mathbf{W\Sigma})\{2E[\chi_{p+2}^{-2}(\Delta^2)]$$
$$- cE[\chi_{p+2}^{-4}(\Delta^2)]\} + c(c+4)(\boldsymbol{\delta}'\mathbf{W}\boldsymbol{\delta})E[\chi_{p+2}^{-4}(\Delta^2)]; \quad \text{and}$$

$$R_4(\hat{\boldsymbol{\theta}}_n^S;\mathbf{W}) = \left(1 + \frac{\bar{x}^2}{Q^*}\right){\rm tr}(\mathbf{W\Sigma}) - c\frac{\bar{x}^2}{Q^*}{\rm tr}(\mathbf{W\Sigma})\{2E[\chi_{p+2}^{-2}(\Delta^2)]$$
$$- cE[\chi_{p+2}^{-4}(\Delta^2)]\} + \frac{\bar{x}^2}{Q^*}c(c+4)(\boldsymbol{\delta}'\mathbf{W}\boldsymbol{\delta})E[\chi_{p+4}^{-4}(\Delta^2)]. \quad (11.5.25)$$

(v) PRSE

$$\mathbf{b}_5\left(\hat{\boldsymbol{\beta}}_n^{S+}\right) = -\boldsymbol{\delta}\{cE[\chi_{p+2}^{-2}(\Delta^2)] + E[(1 - c\chi_{p+2}^{-2}(\Delta^2))I(\chi_{p+2}^2(\Delta^2) < c)]\}$$

$$B_5\left(\hat{\boldsymbol{\beta}}_n^{S+}\right) = \Delta^2\{cE[\chi_{p+2}^{-2}(\Delta^2)] + E[(1 - c\chi_{p+2}^{-2}(\Delta^2))I(\chi_{p+2}^2(\Delta^2) < c)]\}^2;$$

$$\mathbf{b}_5\left(\hat{\boldsymbol{\theta}}_n^{S+}\right) = -\boldsymbol{\delta}\bar{x}\{cE[\chi_{p+2}^{-2}(\Delta^2)] + E[(1 - c\chi_{p+2}^{-2}(\Delta^2))I(\chi_{p+2}^2(\Delta^2) < c)]\}$$

$$B_5\left(\hat{\boldsymbol{\theta}}_n^{S+}\right) = \bar{x}^2\Delta^2\{cE[\chi_{p+2}^{-2}(\Delta^2)] + E[(1 - c\chi_{p+2}^{-2}(\Delta^2))I(\chi_{p+2}^2(\Delta^2) < c)]\}^2;$$

$$\mathbf{M}_5\left(\hat{\beta}_n^{S+}\right) = \mathbf{M}_4(\hat{\beta}_n^{S}) - \frac{c'}{Q^*}\Sigma E[(1 - c\chi^{2}_{p+2}(\Delta^2))^2 I(\chi^2_{p+2}(\Delta^2) < c)]$$
$$+\delta\delta'\{2E[(1 - c\chi^{-2}_{p+2}(\Delta^2))I(\chi^2_{p+2}(\Delta^2) < c)]$$
$$-E[(1 - c\chi^{-2}_{p+4}(\Delta^2))^2 I(\chi^2_{p+4}(\Delta^2) < c)]\}; \quad \text{and}$$

$$\mathbf{M}_4\left(\hat{\boldsymbol{\theta}}_n^{S+}\right) = \mathbf{M}_4(\hat{\boldsymbol{\theta}}_n^{S}) - c\frac{\bar{x}^2}{Q^*}\Sigma E[(1 - c\chi^{-2}_{p+2}(\Delta^2))^2 I(\chi^2_{p+2}(\Delta^2) < c)]$$
$$+(\boldsymbol{\delta}'\mathbf{W}\boldsymbol{\delta})\{2E[(1 - c\chi^{-2}_{p+2}(\Delta^2))I(\chi^2_{p+2}(\Delta^2) < c)]$$
$$-E[(1 - c\chi^{2}_{p+4}(\Delta^2))^2 I(\chi^2_{p+4}(\Delta^2) < c)]\}; \quad \text{and}$$

$$R_5\left(\hat{\boldsymbol{\theta}}_n^{S+}; \mathbf{W}\right) =$$

$$R_4(\hat{\boldsymbol{\theta}}_n^{S}; \mathbf{W}) - c\frac{\bar{x}^2}{Q^*}\text{tr}(\mathbf{W}\Sigma)E[(1 - c\chi^{-2}_{p+2}(\Delta^2))^2 I(\chi^2_{p+2}(\Delta^2) < c)]$$
$$+\bar{x}^2(\boldsymbol{\delta}'\mathbf{W}\boldsymbol{\delta}')\{2E[(1 - c\chi^{-2}_{p+2}(\Delta^2))I(\chi^2_{p+2}(\Delta^2) < c)]$$
$$-E[(1 - c\chi^{-2}_{p+4}(\Delta^2))^2 I(\chi^2_{p+4}(\Delta^2) < c)]\}. \tag{11.5.26}$$

Here $\mathbf{W}$ is the weight matrix of the quadratic loss function.

11.6 Conclusions

In this chapter, we first discussed the estimation of the mean vector of an MND when the covariance matrix is unknown, extending the results of Chapter 4. We covered the estimation of functionals of distribution based on U-statistics to enlarge the scope of nonnormal multivariate distributions as well as R-estimation methodology to include robustness properties of the location parameter estimators. In addition, we considered the simple multivariate linear model and discussed the various estimators of location and regression parameters with MND errors and arbitrary multivariate errors with symmetric marginals. In all cases, we included confidence set estimations with their properties.

11.7 Problems

1. Let $\mathbf{Y}_1, \ldots, \mathbf{Y}_N$ be a sample of size N from the MND, $N_{[}(\boldsymbol{\theta}, \Sigma)$. The problem is the estimation of $\boldsymbol{\theta}$ when it is suspected that $H_0 : \boldsymbol{\theta} = \theta_0 \mathbf{1}_p$ where θ_0 is a scalar and unknown.

   (a) Show that the point estimator of $\boldsymbol{\theta}$ under H_0 is the restricted estimator $\hat{\boldsymbol{\theta}}_n$ defined by

$$\hat{\boldsymbol{\theta}}_N = \mathbf{1}_p(\mathbf{1}_p'\mathbf{S}^{-1}\mathbf{1}_p)^{-1}\mathbf{1}_p'\mathbf{S}^{-1}\tilde{\boldsymbol{\theta}}_N; \quad \tilde{\boldsymbol{\theta}}_N = \bar{\mathbf{Y}}_N,$$

where $\mathbf{S}$ is the unbiased estimate of $\boldsymbol{\Sigma}$.

(b) Show that the LR-statistic for testing $H_0 : \boldsymbol{\theta} = \theta_0 \mathbf{1}_p$ against $H_A : \boldsymbol{\theta} \neq \theta_0 \mathbf{1}_p$ is given by

$$\mathcal{L}_N = \frac{m}{n(p-1)} N \tilde{\boldsymbol{\theta}}'_N (\mathbf{S}^{-1} - \mathbf{S}^{-1}\mathbf{1}_p S^{-1}\mathbf{1}_p)^{-1}\mathbf{1}'_p \mathbf{S}^{-1})\tilde{\boldsymbol{\theta}}_N,$$

where $n = (N-1)$ and $m = N - p$.

(c) Show that

(i) $\hat{\mathbf{H}} = \mathbf{SC}(\mathbf{C}'\mathbf{SC})^{-1}\mathbf{C}'$, where $\mathbf{C}'\mathbf{1}_p = \mathbf{0}$ is an *unbiased estimate* of $\mathbf{H} = \boldsymbol{\Sigma}\mathbf{C}(\mathbf{C}'\boldsymbol{\Sigma}\mathbf{C})^{-1}\mathbf{C}'$,

(ii) $E(\hat{\mathbf{H}}\boldsymbol{\Sigma}^{-1}\mathbf{H}) = \left[\frac{n-p+1}{n-p}\right] \mathbf{C}(\mathbf{C}'\boldsymbol{\Sigma}\mathbf{C})^{-1}\mathbf{C}'$.

(d) Show that a class of empirical Bayes estimators of $\boldsymbol{\theta}$ is

$$\boldsymbol{\theta}_N^{EB} = \hat{\boldsymbol{\theta}}_N + [1 - g(\mathcal{L}_N)](\tilde{\boldsymbol{\theta}}_N - \hat{\boldsymbol{\theta}}_N),$$

where $g(\mathcal{L}_N) = 0, 1, I(\mathcal{L}_N < F_{p-1,m}(\alpha)), c\mathcal{L}_N^{-1}$, and $1 - (1 - c\mathcal{L}_N^{-1})I(\mathcal{L}_N > c)$, with $c = \frac{(p-3)m}{(p-1)(m+2)}$ yielding $\tilde{\boldsymbol{\theta}}_N$, $\hat{\boldsymbol{\theta}}_N$, $\hat{\boldsymbol{\theta}}_N^{\mathrm{PT}}$, $\hat{\boldsymbol{\theta}}_N^{S}$, and $\hat{\boldsymbol{\theta}}_N^{S+}$, respectively.

2. Show that the bias vector and MSE matrix of $\boldsymbol{\theta}_N^* = \tilde{\boldsymbol{\theta}}_N - (\tilde{\boldsymbol{\theta}}_N - \hat{\boldsymbol{\theta}}_N)\times g(\mathcal{L}_N)$ are $\mathbf{b}(\boldsymbol{\theta}_N^*) = -\mathbf{H}\boldsymbol{\theta}E\left\{g\left[\frac{m}{(p-1)}\frac{\chi^2_{p+1}(\Delta^2)}{\chi^2_m}\right]\right\}$, $\mathbf{H} = \boldsymbol{\Sigma}\mathbf{C}(\mathbf{C}'\boldsymbol{\Sigma}\mathbf{C})^{-1}\mathbf{C}'$ as defined in Problem 1 (part b)

$$\begin{aligned}
\mathbf{M}(\boldsymbol{\theta}_N^*) &= \boldsymbol{\Sigma} - (\boldsymbol{\Sigma}\mathbf{C}(\mathbf{C}'\boldsymbol{\Sigma}\mathbf{C})^{-1}\mathbf{C}'\boldsymbol{\Sigma}) \\
&\times \left\{2E\left[g\left(\frac{m}{(p-1)}\frac{\chi^2_{p+1}(\Delta^2)}{\chi^2_m}\right)\right] - E\left[g^2\left(\frac{m}{(p-1)}\frac{\chi^2_{p+1}(\Delta^2)}{\chi^2_m}\right)\right]\right\} \\
&+ [\boldsymbol{\Sigma} - \boldsymbol{\Sigma}\mathbf{C}(\mathbf{C}'\boldsymbol{\Sigma}\mathbf{C})^{-1}\boldsymbol{\Sigma}]\left\{\frac{p-1}{m-2}E\left[g^2\left(\frac{m}{(p-1)}\frac{\chi^2_{p+1}(\Delta^2)}{\chi^2_m}\right)\right]\right. \\
&\left. + \frac{\Delta^2}{(m-2)}E\left[g^2\left(\frac{m}{(p-1)}\frac{\chi^2_{p+3}(\Delta^2)}{\chi^2_{m-2}}\right)\right]\right\} \\
&+ N(\mathbf{H}\boldsymbol{\theta}\boldsymbol{\theta}'\mathbf{H})\left\{2E\left[g\left(\frac{m}{p-1}\frac{\chi^2_{p+1}(\Delta^2)}{\chi^2_m}\right)\right] - 2E\left[g\left(\frac{m}{p-1}\frac{\chi_{p+3}(\Delta^2)}{\chi^2_m}\right)\right]\right. \\
&\left. + E\left[g^2\left(\frac{m}{(p-1)}\frac{\chi^2_{p+3}(\Delta^2)}{\chi^2_m}\right)\right]\right\},
\end{aligned}$$

where $\Delta^2 = N\boldsymbol{\theta}'(\boldsymbol{\Sigma}^{-1} - \boldsymbol{\Sigma}^{-1}\mathbf{1}_p(\mathbf{1}'_p\boldsymbol{\Sigma}^{-1}\mathbf{1}_p)^{-1}\mathbf{1}'_p\boldsymbol{\Sigma}^{-1})\boldsymbol{\theta}$.

3. Write down the bias vector, MSE matrix, and the quadratic risks of the five estimators explicitly for each estimator of $\boldsymbol{\theta}$.

4. Apply the U-statistic approach to the problem of estimating $\boldsymbol{\theta}$ under $\theta = \theta_0 \mathbf{1}_p$, and derive its asymptotic MSE matrix.

5. Apply the R-estimation approach to the problem of estimating θ under $\boldsymbol{\theta} = \boldsymbol{\theta}_0 \mathbf{1}_p$, and derive its asymptotic MSE matrix.

6. Prove Theorem 1.

7. Prove Theorem 2.

8. Prove Theorem 3.

9. Prove Theorem 4.

10. Prove

$$\begin{aligned}
&\lim_{n\to\infty} P_{K_{(N)}}\Big\{N||\boldsymbol{\theta}_{(N)} - \boldsymbol{\theta}^*_N||^1_{\hat{\boldsymbol{\Sigma}}_N^{-1}} \leq \chi^2_p(\gamma)\Big\} \\
&= P_{\boldsymbol{\delta}^*}\{||\mathbf{Z} - (\mathbf{Z} + \boldsymbol{\delta}^*)g(||\mathbf{Z} + \boldsymbol{\delta}^*||^2)||^2 < \chi^2_p(\gamma)\}.
\end{aligned}$$

11. Compare $\hat{\boldsymbol{\theta}}_N^{\mathrm{PT}}$, $\hat{\boldsymbol{\theta}}_N^{S}$, and $\hat{\boldsymbol{\theta}}_N^{S+}$ based on the MSE matrix criterion.

10. Prove

11. Compute

Chapter 12

Discrete Data Models

Outline

In this chapter, we consider several important discrete models that appear frequently in many applied areas, along with the related estimation strategies. We choose the following models:

Product Bernoulli models. Baseball data analysis

Product Binomial models. Pooling of odds ratios or meta-analysis

Product Multinomial Models. $r \times c$ contingency tables

We began Chapter 1 with the illustration of the Stein-type estimation methodology using the baseball data discussed by Efron and Morris (1975). In this chapter, we provide the related theoretical basis towards the Stein strategy based on the Bernoulli models. Next, as a natural extension, we consider the product binomial models to discuss the pooling of odds ratios. The pooling of odds ratios arises in many epidemiological and pharmaceutical studies. The problem occurs explicitly in combining several 2×2 tables for meta-analysis. As a further extension of our discussions, we investigate the product multinomial models to consider the estimation of **cell probabilities** in a two-way contingency table when it is plausible that the table has **independent structure** relating to the two traits.

In Section 12.1, we consider the product Bernoulli model with differing probability of success and the related problem of estimation of the probabilities when it is suspected that they are equal. These results provide the basis of baseball data analysis as discussed by Efron and Morris (1975).

In many situations, statistical data are stratified, as in meta-analysis, into several 2×2 tables in order to control for the confounding factors. Such instances occur in many case-control studies, where the primary goal is to access the disease association by means of estimating the common odds ratio. It is often uncertain that odds ratios of several tables are homogeneous across strata. Then the object is to estimate the odds ratios when homogeneity is uncertain. In Section 12.2, we consider the problem of estimating odds ratio when we suspect homogeneity of several 2×2 tables.

In many survey studies and in epidemiological surveys, data are often presented in a two-way classification leading to $r \times c$ contingency tables according to two traits. One of many statistical problems that can arise is the estimation of the cell probabilities when it is plausible that the table might have independent structure relating to the two traits. We will discuss this problem in Section 12.3 using a product of multinomial models. For some early developments on discrete data, see Kale and Bancroft (1967).

12.1 Product of Bernoulli Models, Estimation, and Test of Hypothesis

12.1.1 Model, Estimation, and Test

Let us consider the "baseball data" given in Table 1.1.1 of Chapter 1. The formulation of its analysis is based on several Bernoulli models as follows:

Consider a team consisting of p players. Let $(x_{i1}, x_{i2}, \cdots, x_{in_i})'$ be the result of hit or miss of a sequence of n_i battings of the ith player of the team. Let $\boldsymbol{\theta} = (\theta_1, \cdots, \theta_p)'$ be the vector of the batting averages for the team of p players.

Now, consider the outcome $\mathbf{x} = (x_{11}, \cdots, x_{1n_1}; \cdots; x_{p1}, \cdots, x_{pn_p})'$ of $n(= n_1 + n_2 + \cdots + n_p)$ battings of the p players in a game. The likelihood function for the parameter $\boldsymbol{\theta} = (\theta_1, \cdots, \theta_p)'$ given $\mathbf{x}$ is

$$L(\boldsymbol{\theta}|\mathbf{x}) = \prod_{i=1}^{p} \prod_{j=1}^{n_i} \theta_i^{x_{ij}} (1 - \theta_i)^{1 - x_{ij}}, \tag{12.1.1}$$

since the performances of the players are independent of each other. Then the batting averages of the p players can be written as

$$\tilde{\boldsymbol{\theta}}_n = (\tilde{\theta}_1, \cdots, \tilde{\theta}_p)', \quad \tilde{\theta}_i = \frac{1}{n_i} \sum_{j=1}^{n_i} x_{ij} = \frac{1}{n_i} y_i. \tag{12.1.2}$$

If the players are assumed to be equally good, we expect the vector of the batting averages to be $\boldsymbol{\theta} = \theta_0 \mathbf{1}_p$, $\mathbf{1}_p = (1, \cdots, 1)'$ due to strict selection of the quality of the players in the team. In this case, the likelihood function for $\boldsymbol{\theta} = \theta_0 \mathbf{1}_p$, based on $\mathbf{Y} = (y_1, \cdots, y_p)'$, is given by

$$L(\theta_0|\mathbf{x}) = \theta_0^{\sum_{i=1}^p y_i}(1-\theta_0)^{n-\sum_{i=1}^p y_i}, \tag{12.1.3}$$

and estimated common batting average of the players is

$$\hat{\boldsymbol{\theta}}_n = \hat{\boldsymbol{\theta}}_{0n}\mathbf{1}_p, \quad \mathbf{1}_p = (1, \cdots, 1)' \text{ and } \hat{\boldsymbol{\theta}}_n = \frac{\mathbf{1}_p\mathbf{1}_p'\mathbf{N}}{n}\tilde{\boldsymbol{\theta}}_n, \tag{12.1.4}$$

where $\mathbf{N} = \text{Diag}(n_1, \cdots, n_p)$. Thus, we obtain two estimates of $\boldsymbol{\theta}$, one where the players are different and the other where they are similar.

If we are not sure that $\boldsymbol{\theta} = \theta_0\mathbf{1}_p$, then it is natural to determine any deviation statistically. Thus, we test the null hypothesis

$$H_0 : \boldsymbol{\theta} = \theta_0\mathbf{1}_p \quad \text{versus} \quad H_A : \boldsymbol{\theta} \neq \theta_0\mathbf{1}_p \tag{12.1.5}$$

based on the statistics of departure from the null hypothesis:

$$D_n = n(\tilde{\boldsymbol{\theta}}_n - \hat{\boldsymbol{\theta}}_n)'\hat{\boldsymbol{\Sigma}}_n^{-1}(\tilde{\boldsymbol{\theta}}_n - \hat{\boldsymbol{\theta}}_n), \tag{12.1.6}$$

where

$$\hat{\boldsymbol{\Sigma}}_n^{-1} = [\hat{\theta}_{0n}(1-\hat{\theta}_{0n})]^{-1}\text{Diag}\Big(\frac{n_1}{n}, \cdots, \frac{n_p}{n}\Big). \tag{12.1.7}$$

It is easy to prove that for large n, D_n closely follows a central chi-square distribution with $(p-1)$ d.f. Thus, for a given level of significance, $\alpha(0 < \alpha < 1)$, the critical value $D_{n,\alpha}$ may be approximated by $\chi^2_{p-1}(\alpha)$, the upper α-level value from the central chi-square distribution with $(p-1)$ d.f. This allows us to define the PTE of $\boldsymbol{\theta}$ as

$$\hat{\boldsymbol{\theta}}_n^{\text{PT}} = \tilde{\boldsymbol{\theta}}_n - (\tilde{\boldsymbol{\theta}}_n - \hat{\boldsymbol{\theta}}_n)I(D_n < \chi^2_{p-1}(\alpha)). \tag{12.1.8}$$

Similarly, we can define the Stein-type estimator (SE) of $\boldsymbol{\theta}$ as

$$\hat{\boldsymbol{\theta}}_n^S = \tilde{\boldsymbol{\theta}}_n - d(\tilde{\boldsymbol{\theta}}_n - \hat{\boldsymbol{\theta}}_n)D_n^{-1}, \qquad d = (p-3)\ (>0). \tag{12.1.9}$$

Finally, the positive-rule Stein-type estimator (PRSE) of $\boldsymbol{\theta}$ is given by

$$\begin{aligned}\hat{\boldsymbol{\theta}}_n^{S+} &= \hat{\boldsymbol{\theta}}_n + (1 - dD_n^{-1})I(D_n > d)(\tilde{\boldsymbol{\theta}}_n - \hat{\boldsymbol{\theta}}_n)\\ &= \hat{\boldsymbol{\theta}}_n^S - (1 - dD_n^{-1})I(D_n < d)(\tilde{\boldsymbol{\theta}}_n - \hat{\boldsymbol{\theta}}_n),\end{aligned} \tag{12.1.10}$$

(see further Ali and Saleh, 1991a).

In order to assess the batting average of the players of the team, we compare the five estimators using a quadratic loss function

$$L(\boldsymbol{\theta}_n^*; \boldsymbol{\theta}) = n(\boldsymbol{\theta}_n^* - \boldsymbol{\theta})'\mathbf{W}(\boldsymbol{\theta}_n^* - \boldsymbol{\theta}), \tag{12.1.11}$$

where $\boldsymbol{\theta}_n^*$ stands for $\tilde{\boldsymbol{\theta}}_n, \hat{\boldsymbol{\theta}}_n, \hat{\boldsymbol{\theta}}_n^{\text{PT}}, \hat{\boldsymbol{\theta}}_n^S$, and $\hat{\boldsymbol{\theta}}_n^{S+}$. Before doing so, we present the Bayes and empirical Bayes estimators of $\boldsymbol{\theta}$.

12.1.2 Bayes and Empirical Bayes Estimation

In Section 12.1.1, we presented the five estimators of $\boldsymbol{\theta} = (\theta_1, \cdots, \theta_p)'$. In this section, we consider the Bayes and the empirical Bayes method of estimating $\boldsymbol{\theta}$. For the Bayes estimation of $\boldsymbol{\theta}$, we first assume that $\theta_1, \ldots, \theta_p$ are independently and identically distributed beta variables with parameter $(K\eta, K(1-\eta))$, where $0 < \eta < 1$ and $0 < K < \infty$.

Thus, the joint-distribution of $\theta_1, \cdots, \theta_p$ is given by the conjugate prior conditional on (K, η):

$$\prod_{i=1}^{p} \frac{\theta_i^{K\eta-1}(1-\theta_i)^{K(1-\eta)-1}}{B(K\eta, K(1-\eta))}, \tag{12.1.12}$$

where $B(a,b) = \frac{\Gamma(a)\Gamma(b)}{\Gamma(a+b)}$ and the **hyperparameters** (K, η) are both **known**. Hence, the joint-distribution of $(\boldsymbol{\theta}', \mathbf{Y}')'$ given (K, η) is obtained as

$$\Pi(\boldsymbol{\theta}, \mathbf{Y}|K,\eta) = [B(K\eta, K(1-\eta))]^p \prod_{i=1}^{p} \binom{n_i}{y_i} \theta_i^{y_i+K\eta-1}(1-\theta_i)^{n_i-y_i+K(1-\eta)-1}, \tag{12.1.13}$$

and the marginal distribution of $\mathbf{Y} = (y_1, \cdots, y_p)'$ is given by

$$m_1(\mathbf{Y}|K,\eta) = \prod_{i=1}^{p} \binom{n_i}{y_i} (K\eta)^{(y_i)}(K(1-\eta))^{(n_i-y_i)}K^{-(n_i)}, \tag{12.1.14}$$

where $a^{(b)} = a(a+1)\cdots(a+b-1)$, $(a^{(0)} = 1)$.

Consequently, the posterior distribution of $\boldsymbol{\theta}$ given $(\mathbf{Y}', K)'$ is

$$\Pi(\boldsymbol{\theta}|\mathbf{Y}, K) = c\int_0^1 \Pi(\boldsymbol{\theta}, \mathbf{Y}|K,\eta) m_1(\mathbf{Y}|K,\eta)[\eta(1-\eta)]^{-1} d\eta, \tag{12.1.15}$$

by the fact that $E[\eta|\mathbf{Y}, K] = Cm_1(\mathbf{Y}|K,\eta)\phi(\eta)$. Then, assigning the **noninformative** prior

$$\phi(\eta) = [\eta(1-\eta)]^{-1} \tag{12.1.16}$$

to η yields a mixture of a product of beta densities.

To obtain the Bayes estimator of $\boldsymbol{\theta} = (\theta_1, \cdots, \theta_p)'$, we compute the **posterior means**

$$E[\theta_i|\mathbf{Y}] = E\big[E(\theta_i|\mathbf{Y})\big] = \frac{y_i + KE(\eta|\mathbf{Y})}{n_i + K} \tag{12.1.17}$$

and the posterior variances as

$$\begin{aligned} V[\theta_i|\mathbf{Y}] &= E\big[\mathrm{Var}(\theta_i|\mathbf{Y},\eta)\big] + \mathrm{Var}\,\big[E(\theta_i|\mathbf{Y},\eta)\big] \\ &= E\Big[\frac{y_i + K\eta}{n_i + K}\Big(1 - \frac{y_i + K\eta}{n_i + K}\Big)(n_i + K - 1)^{-1}\Big] + \mathrm{Var}\left[\frac{y_i + K\eta}{n_i + K}\right] \\ &= \frac{1}{n_i + K - 1}E(\theta_i|\mathbf{Y})[1 - E(\theta_i|\mathbf{Y})] + \frac{K^2\,\mathrm{Var}(\eta|\mathbf{Y})}{[(n_i+K)(n_i+K+1)]}, \end{aligned} \tag{12.1.18}$$

respectively, for $i = 1, \cdots, p$.

The Bayes estimators of $\boldsymbol{\theta} = (\theta_1, \cdots, \theta_p)'$ depend on K and the conditional distribution of η given $\mathbf{Y}$. If K is known, we still have a problem of integration. Generally, K is **unknown**, then we take recourse to the empirical Bayes method due to Robbins (1956) and specifically due to Efron and Morris (1975).

To better understand the properties of the Bayes estimators, we note that $K \in (0, \infty)$ and let $K \to \infty$. Then we can show that the posterior asymptotic ($K \to \infty$) distribution of η given $\mathbf{Y}$ is a beta distribution with parameters $(n\bar{y}, n - n\bar{y})$, where $\bar{y} = \frac{1}{n}\sum_{i=1}^{p} y_i$ (see Problem 5). Hence, for example,

$$\lim_{K\to\infty} E[\theta_i|\mathbf{Y}] = \lim_{K\to\infty} E[\eta|\mathbf{Y}] = \bar{y} = \hat{\theta}_n \quad \text{and} \tag{12.1.19}$$

$$\lim_{K\to\infty} \text{Var}[\theta_i|y] = \lim_{K\to\infty} \text{Var}[\eta|y] = \frac{\bar{y}(1-\bar{y})}{n+1}. \tag{12.1.20}$$

On the other hand, if $K \to 0$, then

$$\lim_{K\to 0} E[\theta_i|\mathbf{y}] = \frac{y_i}{n_i} = \hat{\theta}_i \quad \text{and} \quad \text{Var}[\theta_i|y] = \frac{\bar{y}(1-\bar{y})}{n_i+1}. \tag{12.1.21}$$

Note that as $K \to \infty$ or 0, the posterior mean approaches $\bar{y}$ and $\hat{\theta}_i$, respectively, while the limit of variance of the estimators are similar (only n replaced by n_i) in the two cases.

Since K is unknown, we can estimate K based on the **marginal distribution** of $\mathbf{Y} = (y_1, \cdots, y_p)'$. Also, we can develop a simple approximation, as in Albert and Gupta (1981), by using a Taylor expansion to develop the **normal posterior** distribution of η. Thus, following Albert and Gupta (1981), we find the **empirical Bayes estimator** of θ_i to be

$$\hat{\theta}_i^{EB} = E[\theta_i|\mathbf{Y}] \simeq \tilde{\theta}_i\left(1 - \frac{K}{n_i+K}\right) + \hat{\theta}_n\frac{K}{n_i+K} \tag{12.1.22}$$

and

$$\text{Var}[\theta_i|\mathbf{Y}] \simeq \frac{\tilde{\theta}_i(1-\tilde{\theta}_i)}{n_i+1}\left(1 - \frac{K}{n_i+K}\right) + \frac{\hat{\theta}_n(1-\hat{\theta}_n)}{n+1}\left(\frac{K}{n_i+K}\right). \tag{12.1.23}$$

Instead of $\frac{K}{n_i+K}$, we estimate $\lambda(K) = \frac{K}{\bar{n}+K}$ $(\bar{n} = p^{-1}\sum_{i=1}^{p} n_i)$ by its **posterior mean**

$$\widehat{\lambda}(K) = E[\lambda(K)|\mathbf{Y}]$$
$$= \frac{\int_0^\infty \lambda(K)\left[\int_0^1 m_1(\mathbf{Y}|K,\eta)[\eta(1-\eta)]^{-1}d\eta\right] dK/K}{\left[\int_0^\infty \int_0^1 m_1(\mathbf{Y}|K,\eta)[\eta(1-\eta)]^{-1}d\eta dK/K\right]}. \tag{12.1.24}$$

The evaluation of these integrals are complicated except by numerical methods. However, we can show, as in Albert (1987), that the empirical Bayes estimators can be written as

$$\hat{\boldsymbol{\theta}}_n^{EB} = \tilde{\boldsymbol{\theta}}_n - (\tilde{\boldsymbol{\theta}}_n - \hat{\boldsymbol{\theta}}_n)g(D_n), \tag{12.1.25a}$$

where

$$g(D_n) = \min\left[1, (p-3)D_n^{-1}\right] \quad \text{(see Problem 8).} \tag{12.1.25b}$$

If we set $g(D_n)$, where D_n is the test-statistic, then for $g(D_n) = 0$, we have $\hat{\boldsymbol{\theta}}_n^{EB} = \tilde{\boldsymbol{\theta}}_n$; for $g(D_n) = 1$, we have $\hat{\boldsymbol{\theta}}_n^{EB} = \hat{\boldsymbol{\theta}}_n$; for $g(D_n) = I(D_n < \chi^2_{p-1}(\alpha))$, we have $\hat{\boldsymbol{\theta}}_n^{EB} = \hat{\boldsymbol{\theta}}_n^{\mathrm{PT}}$; for $g(D_n) = dD_n^{-1}$, we have $\hat{\boldsymbol{\theta}}_n^{EB} = \hat{\boldsymbol{\theta}}_n^{S}$, $d = p-3$; for $g(D_n) = 1 - (1 - dD_n^{-1})I(D_n > d)$, we have $\hat{\boldsymbol{\theta}}_n^{EB} = \hat{\boldsymbol{\theta}}_n^{S+}$. (12.1.26)

Thus, our estimators in Section 12.1.1 are approximate empirical Bayes estimators of $\boldsymbol{\theta} = (\theta_1, \cdots, \theta_p)'$.

12.1.3 Asymptotic Theory of the Estimators and the Test of Departure

Let $\boldsymbol{\delta}_n$ be an estimator of $\boldsymbol{\theta}$, and let $\mathbf{W}$ be a positive-semidefinite matrix, and consider the quadratic loss equation function

$$\begin{aligned} L_n(\boldsymbol{\delta}_n : \boldsymbol{\theta}) &= n(\boldsymbol{\delta}_n - \boldsymbol{\theta})'\mathbf{W}(\boldsymbol{\delta}_n - \boldsymbol{\theta}) \\ &= n \operatorname{tr}[\mathbf{W}(\boldsymbol{\delta}_n - \boldsymbol{\theta})(\boldsymbol{\delta}_n - \boldsymbol{\theta})']. \end{aligned} \tag{12.1.27}$$

For the p-dimensional cube Ω, let ω be the subspace for which $\boldsymbol{\theta} = \theta_0\mathbf{1}_p$. Further, the departure statistic D_n provides a consistent test as such for **fixed alternatives,** $\boldsymbol{\theta}_\delta = \theta_0\mathbf{1}_n + \boldsymbol{\delta}$, $D_n \to \infty$ as $n \to \infty$. As a consequence, under fixed alternatives the asymptotic distributions of $\sqrt{n}(\hat{\boldsymbol{\theta}}_n^{\mathrm{PT}} - \boldsymbol{\theta})$, $\sqrt{n}(\hat{\boldsymbol{\theta}}_n^{S} - \boldsymbol{\theta})$ and $\sqrt{n}(\hat{\boldsymbol{\theta}}_n^{S+} - \boldsymbol{\theta})$ are equivalent to the asymptotic distribution of $\sqrt{n}(\tilde{\boldsymbol{\theta}}_n - \boldsymbol{\theta})$, while the asymptotic distribution of $\sqrt{n}(\hat{\boldsymbol{\theta}}_n - \boldsymbol{\theta})$ is degenerate as $n \to \infty$. We show these findings in the following:

(i) **PTE,** $\hat{\boldsymbol{\theta}}_n^{\mathrm{PT}}$. Consider the fixed alternatives $\boldsymbol{\theta}_\delta$ as given above. Then we have

$$\begin{aligned} & n(\hat{\boldsymbol{\theta}}_n^{\mathrm{PT}} - \tilde{\boldsymbol{\theta}}_n)'\boldsymbol{\Sigma}_n^{-1}(\hat{\boldsymbol{\theta}}_n^{\mathrm{PT}} - \tilde{\boldsymbol{\theta}}_n) \\ &= n(\tilde{\boldsymbol{\theta}}_n - \hat{\boldsymbol{\theta}}_n)'\boldsymbol{\Sigma}_n^{-1}(\tilde{\boldsymbol{\theta}}_n - \hat{\boldsymbol{\theta}}_n)I(D_n < \chi^2_{p-1}(\alpha)) \\ &= D_n I(D_n < \chi^2_{p-1}(\alpha)) \le \chi^2_{p-1}(\alpha) I(D_n < \chi^2_{p-1}(\alpha)). \end{aligned} \tag{12.1.28}$$

By consistency of D_n, $I(D_n < \chi^2_{p-1}(\alpha)|\boldsymbol{\theta}_\delta) \to 0$ as $n \to \infty$. Hence, $\sqrt{n}(\hat{\boldsymbol{\theta}}_n^{\mathrm{PT}} - \boldsymbol{\theta}) = \sqrt{n}(\tilde{\boldsymbol{\theta}}_n - \boldsymbol{\theta}) + o_p(1)$ as $n \to \infty$ and the asymptotic distribution of $\sqrt{n}(\hat{\boldsymbol{\theta}}_n^{\mathrm{PT}} - \boldsymbol{\theta})$ is equivalent to that of $\sqrt{n}(\tilde{\boldsymbol{\theta}}_n - \boldsymbol{\theta})$.

(ii) **SE,** $\hat{\boldsymbol{\theta}}_n^{S}$. We note that the quadratic form $n(\hat{\boldsymbol{\theta}}_n^{S} - \tilde{\boldsymbol{\theta}}_n)'\boldsymbol{\Sigma}_n^{-1}(\hat{\boldsymbol{\theta}}_n^{S} - \tilde{\boldsymbol{\theta}}_n) = d^2 D_n^{-1}$. Therefore, on the set $\{D_n = 0\}$, we have $\hat{\boldsymbol{\theta}}_n^{S} = \tilde{\boldsymbol{\theta}}_n = \hat{\theta}_{0n}\mathbf{1}_p$. On the other hand, if we know that $E[D_n^{-1}] \to 0$ as $n \to \infty$, then we obtain the desired result. Lemma 1 below relates to this assertion.

Lemma 1. (Ali and Saleh 1991).

$$E[D_n I(0 < D_n < \epsilon)] \to 0 \quad \text{as} \quad n \to \infty. \tag{12.1.29}$$

Proof. Define $k_n = \max_{1\le j \le h} |\tilde{\theta}_j - \hat{\theta}_{0n}|$. Then $D_n \ge \frac{4}{n}k_n^2$. If $k_n = 0$, then $D_n = 0$ and $\hat{\boldsymbol{\theta}}_n^S = \tilde{\boldsymbol{\theta}}_n = \hat{\theta}_{0n}\mathbf{1}_p$ holds. So we consider the case where $k_n > 0$ and k_n are less than 1 with probability 1. In any case, we may set $k_n > 1$ with probability 1. Thus, on the set $\{D_n > 0\}$, we write

$$D_n^{-1}I(0 < D_n \le \epsilon) = D_n^{-1}I\left(0 < D_n < \frac{4}{n}\right) + D_n^{-1}I\left(\frac{4}{n} < D_n < \epsilon\right). \tag{12.1.30}$$

Let $k_{0n} = \frac{1}{2}\sqrt{n}\epsilon$. Then

$$E_{\boldsymbol{\theta}}\left[D_n^{-1}I(0 < D_n \le \epsilon)\right]$$

$$\frac{n}{4}\{E_{\boldsymbol{\theta}}[k_n^{-2}I(\frac{1}{n} < k_n < 1)] + E_{\boldsymbol{\theta}}[k_n^{-2}I(1 \le k_n \le k_{0n})]\}. \tag{12.1.31}$$

Replacing k_n by the largest integer $[k_n]$, we find

$$E_{\boldsymbol{\theta}}[k_n^{-2}I(1 < k_n < k_{0n})] \le E_{\boldsymbol{\theta}}[k_n^{-2}I(1 \le [k_n] \le k_{0n})] = \sum_{j=1}^{[k_{0n}]} j^{-1}P_{\boldsymbol{\theta}}\{[k_n] = j\}$$

$$\le \sum_{j=1}^{[k_{0n}]} j^{-3}P_{\boldsymbol{\theta}}\{[k_n] \le k\} + [k_{0n}]^{-2}P_{\boldsymbol{\theta}}\{[k_n] \le [k_{0n}]\}. \tag{12.1.32}$$

Now the d.f. for D_n is $p-1$. Hence, there are k^{p-1} configurations for which $k_n \le k$. For each configuration the order of probability is $O(n^{-(p-1)/2})$. Thus,

$$P_{\boldsymbol{\theta}}\{k_n \le k\} = O(n^{-(p-1)/2}) \tag{12.1.33}$$

for every $k \le k_{0n}$. Hence, by (12.1. 30) and (12.1.31),

$$\frac{n}{4}E_{\boldsymbol{\theta}}\left[k_n^{-2}I(1 < k_n < k_{0n})\right] \le \frac{n}{4}\left\{\sum_{k=1}^{[k_{0n}]} O(k^{p-4}n^{-(p-1)/2}) + O(k_{0n}^{p-3}n^{-(p-1)/2})\right\}$$

$$= O(n^{-(p-3)/2})\left\{\sum_{k=1}^{[k_{0n}]} O(k^{p-4}) + O(k_{0n}^{p-3})\right\} = O(n^{-(p-3)/2}k_{0n}^{p-3})$$

$$= O(\epsilon^{(p-3)/2}) \quad \text{for} \quad p \ge 4. \tag{12.1.34}$$

Since ϵ is small, (12.1.34) can be made small. For the first term, set $[nk] = k_{0n}$ so that on $\frac{1}{n} < k_n < 1$, we obtain $1 \le k_n^* \le n$. Thus,

$$\frac{n}{4}E_{\theta}\left\{k_n^{-2}I\left(\frac{1}{n} \le k_n \le 1\right)\right\} \le \frac{n^3}{4}E_{\boldsymbol{\theta}}\{(k_n^*)^{-2}I(1 \le k_n^* \le n-1)\}$$

$$= \frac{n^3}{4}\sum_{j=1}^{n-1} j^{-2}P_{\boldsymbol{\theta}}(k_n^* = j). \tag{12.1.35}$$

By a similar argument, we can show that (12.2.35) is $O(n^{-(p-3)/2})$, $p \geq 4$. Hence, we have proved that

$$E_{\boldsymbol{\theta}}\{D_n^{-1} I(0 < D_n < \epsilon)\} \to 0 \quad \text{as } n \to \infty. \tag{12.1.36}$$

Applying the lemma, we see that

$$d^2 E[D_n^{-1}] \to 0 \quad \text{as } n \to \infty. \tag{12.1.37}$$

Hence, we have

$$\sqrt{n}(\hat{\boldsymbol{\theta}}_n^S - \boldsymbol{\theta}) = \sqrt{n}(\tilde{\boldsymbol{\theta}}_n - \boldsymbol{\theta}) + o_p(1). \tag{12.1.38}$$

Similarly, we consider PRSE, $\hat{\boldsymbol{\theta}}_n^{S+}$:

$$\begin{aligned} n\left(\hat{\boldsymbol{\theta}}_n^{S+} - \boldsymbol{\theta}\right)' \boldsymbol{\Sigma}_n^{-1} \left(\hat{\boldsymbol{\theta}}_n^{S+} - \boldsymbol{\theta}\right) &= d^2 D_n^{-1} - (2d-1) D_n (1 - d D_n^{-1}) I(D_n < d) \\ &\leq d^2 D_n^{-1} - d(2d-1)(1 - d D_n^{-1}) I(D_n < d). \end{aligned} \tag{12.1.39}$$

By Lemma 1, we can show that as $n \to \infty$,

$$E[d^2 D_n^{-1} - d(2d-1)(1 - d D_n^{-1}) I(D_n < d)] \to 0. \tag{12.1.40}$$

Hence,

$$\sqrt{n}(\hat{\boldsymbol{\theta}}_n^{S+} - \boldsymbol{\theta}) = \sqrt{n}(\tilde{\boldsymbol{\theta}}_n - \boldsymbol{\theta}) + o_p(1). \tag{12.1.41}$$

On the other hand, as $n \to \infty$,

$$\sqrt{n}(\hat{\boldsymbol{\theta}}_n - \boldsymbol{\theta}) \overset{a.s.}{\to} (\theta_0 \mathbf{1}_p - \boldsymbol{\theta}) = \eta_0, \tag{12.1.42}$$

so

$$n(\hat{\boldsymbol{\theta}}_n - \boldsymbol{\theta})' \boldsymbol{\Sigma}_n^{-1} (\hat{\boldsymbol{\theta}}_n - \boldsymbol{\theta}) \overset{\mathrm{P}}{\to} \infty \quad \text{as} \quad n \to \infty \tag{12.1.43}$$

and also

$$n(\tilde{\boldsymbol{\theta}}_n - \boldsymbol{\theta})'(\tilde{\boldsymbol{\theta}}_n - \boldsymbol{\theta}) = \text{Diag}\left[\frac{\theta_1(1-\theta_1)}{\lambda_1}, \cdots, \frac{\theta_p(1-\theta_p)}{\lambda_p}\right], \tag{12.1.44}$$

where $\lambda_i = \lim_{n\to\infty} \frac{n_i}{n}$, $(i = 1, \cdots, p)$. This result implies that

$$n(\tilde{\boldsymbol{\theta}}_n - \boldsymbol{\theta})' \boldsymbol{\Sigma}_n^{-1} (\tilde{\boldsymbol{\theta}}_n - \boldsymbol{\theta}) (< \infty) \tag{12.1.45}$$

is bounded. It further implies the following theorem:

Theorem 1. Under fixed alternatives $\boldsymbol{\theta}_\delta = \theta_0 \mathbf{1}_p + \boldsymbol{\delta}$,

(i) $\hat{\boldsymbol{\theta}}_n$ has asymptotically unbounded risk,

(ii) $\hat{\boldsymbol{\theta}}_n^{\text{PT}}, \hat{\boldsymbol{\theta}}_n^{S}$ and $\hat{\boldsymbol{\theta}}_n^{S+}$ are all asymptotically equivalent to $\tilde{\boldsymbol{\theta}}_n$ and have bounded risk.

To compare the estimators approximately, we consider the sequence of local alternatives

$$K_{(n)} : \boldsymbol{\theta}_{(n)} = \theta_0 \mathbf{1}_p + n^{-1/2}\boldsymbol{\xi} \;\; \text{such that} \;\; \boldsymbol{\Lambda}'\boldsymbol{\xi} = \mathbf{0}, \tag{12.1.46}$$

where $\boldsymbol{\Lambda} = (\lambda_1, \cdots, \lambda_p)'$ to arrive at the next theorem on asymptotic distributions of the various statistics.

Theorem 2. (Ali and Saleh 1991). Under $\{K_{(n)}\}$ and the assumed regularity conditions, as $n \to \infty$, the following hold:

(i) $$\sqrt{n}(\tilde{\boldsymbol{\theta}}_n - \theta_0 \mathbf{1}_p) \sim \mathcal{N}_p(\boldsymbol{\xi}, \boldsymbol{\Sigma}_0), \;\; \boldsymbol{\Sigma}_0 = \theta_0(1-\theta_0)\boldsymbol{\Lambda}^{-1} \tag{12.1.47}$$

where $\boldsymbol{\Lambda} = \text{Diag}(\lambda_1, \cdots, \lambda_p)$.

(ii) $\sqrt{n}(\tilde{\boldsymbol{\theta}}_n - \hat{\boldsymbol{\theta}}_n) \sim \mathcal{N}_p(\mathbf{J}\boldsymbol{\xi}, \boldsymbol{\Sigma}_0\mathbf{J}'), \;\; \mathbf{J} = \mathbf{I}_p - \mathbf{1}_p\mathbf{1}_p'\boldsymbol{\Lambda}$.

(iii) $\sqrt{n}(\hat{\boldsymbol{\theta}}_n - \theta_0\mathbf{1}_p) \sim \mathcal{N}_p(\mathbf{0}, \mathbf{B}), \;\; \mathbf{B} = \theta_0(1-\theta_0)\mathbf{1}_p\mathbf{1}_p'$.

(iv) $$\begin{bmatrix} \sqrt{n}(\tilde{\boldsymbol{\theta}}_n - \theta_0\mathbf{1}_p) \\ \sqrt{n}(\tilde{\boldsymbol{\theta}}_n - \hat{\boldsymbol{\theta}}_n) \end{bmatrix} \sim \mathcal{N}_{2p}\left\{ \begin{bmatrix} \boldsymbol{\xi} \\ \mathbf{J}\boldsymbol{\xi} \end{bmatrix}; \begin{pmatrix} \boldsymbol{\Sigma}_0 & \boldsymbol{\Sigma}_0\mathbf{J}' \\ \mathbf{J}\boldsymbol{\Sigma}_0 & \boldsymbol{\Sigma}_0\mathbf{J}' \end{pmatrix} \right\}.$$

(v) $$\begin{bmatrix} \sqrt{n}(\hat{\boldsymbol{\theta}}_n - \theta_0\mathbf{1}_p) \\ \sqrt{n}(\tilde{\boldsymbol{\theta}}_n - \hat{\boldsymbol{\theta}}_n) \end{bmatrix} \sim \mathcal{N}_{2p}\left\{ \begin{bmatrix} \mathbf{0} \\ \mathbf{J}\boldsymbol{\xi} \end{bmatrix}; \begin{pmatrix} \mathbf{B} & \mathbf{0} \\ \mathbf{0} & \boldsymbol{\Sigma}_0\mathbf{J}' \end{pmatrix} \right\}.$$

(vi) $\lim_{n\to\infty} P\{D_n \le x | K_{(n)}\} = H_{p-1}(x : \Delta^2), \Delta^2 = \boldsymbol{\delta}'\boldsymbol{\Sigma}_0^{-1}\boldsymbol{\delta}, \boldsymbol{\delta} = \mathbf{J}\boldsymbol{\xi}$.

For a proof, see Problem 9.

12.1.4 ADB, ADQB, ADMSE, and ADQR of Estimators

We can use Theorem 3 to obtain the ADB, ADQB, ADMSE, and ADQR of the estimators as follows:

(i) **UE,** $\tilde{\boldsymbol{\theta}}_n$

$$\mathbf{b}_1(\tilde{\boldsymbol{\theta}}_n) = \mathbf{0} \quad \text{and} \quad B_1(\tilde{\boldsymbol{\theta}}_n) = 0;$$

$$\mathbf{M}_1(\tilde{\boldsymbol{\theta}}_n) = \boldsymbol{\Sigma}_0 \quad \text{and} \quad R_1(\tilde{\boldsymbol{\theta}}_n : \mathbf{Q}) = \mathrm{tr}[\mathbf{Q}\boldsymbol{\Sigma}_0].$$

(ii) **RE,** $\hat{\boldsymbol{\theta}}_n$

$$\mathbf{b}_2(\hat{\boldsymbol{\theta}}_n) = -\mathbf{J}\boldsymbol{\xi} = -\boldsymbol{\delta} \quad \text{and} \quad B_2(\hat{\boldsymbol{\theta}}_n) = \Delta^2$$

$$\mathbf{M}_2(\hat{\boldsymbol{\theta}}_n) = \mathbf{B} + \boldsymbol{\delta}\boldsymbol{\delta}' \quad \text{and} \quad R_2(\hat{\boldsymbol{\theta}}_n : \mathbf{Q}) = \mathrm{tr}[\mathbf{QB}] + \boldsymbol{\delta}'\mathbf{Q}\boldsymbol{\delta}.$$

(iii) **PTE,** $\hat{\boldsymbol{\theta}}_n^{PT}$

$$\mathbf{b}_3(\hat{\boldsymbol{\theta}}_n^{PT}) = -\boldsymbol{\delta} H_{p+1}(\chi^2_{p-1}(\alpha); \Delta^2) \quad \text{and}$$

$$B_3(\hat{\boldsymbol{\theta}}_n^{PT}) = \Delta^2\{H_{p+1}(\chi^2_{p-1}(\alpha); \Delta^2)\}^2$$

$$\begin{aligned}\mathbf{M}_3(\hat{\boldsymbol{\theta}}_n^{PT}) &= \boldsymbol{\Sigma}_0 - \boldsymbol{\Sigma}_0\mathbf{J}' H_{p+1}(\chi^2_{p-1}(\alpha); \Delta^2) \\ &\quad + \boldsymbol{\delta}\boldsymbol{\delta}'\{2H_{p+1}(\chi^2_{p-1}(\alpha); \Delta^2) - H_{p+3}(\chi^2_{p-1}(\alpha); \Delta^2)\} \quad \text{and}\end{aligned}$$

$$\begin{aligned}R_3(\hat{\boldsymbol{\theta}}_n^{PT}; \mathbf{Q}) &= \mathrm{tr}[\mathbf{Q}\boldsymbol{\Sigma}_0] - \mathrm{tr}[\mathbf{Q}\boldsymbol{\Sigma}_0\mathbf{J}'] H_{p+1}(\chi^2_{p-1}(\alpha); \Delta^2) \\ &\quad + \boldsymbol{\delta}'\mathbf{Q}\boldsymbol{\delta}\{2H_{p+1}(\chi^2_{p-1}(\alpha); \Delta^2) - H_{p+3}(\chi^2_{p-1}(\alpha); \Delta^2)\}.\end{aligned}$$

(iv) **SE,** $\hat{\boldsymbol{\theta}}_n^{S}$

$$\mathbf{b}_4(\hat{\boldsymbol{\theta}}_n^{S}) = -(p-3)\boldsymbol{\delta} E[\chi^{-2}_{p+1}(\Delta^2)] \quad \text{and}$$

$$B_4(\hat{\boldsymbol{\theta}}_n^{S}) = (p-3)^2\Delta^2\{E[\chi^{-2}_{p+1}(\Delta^2)]\}^2;$$

$$\begin{aligned}\mathbf{M}_4(\hat{\boldsymbol{\theta}}_n^{S}) &= \boldsymbol{\Sigma}_0 - (p-3)\boldsymbol{\Sigma}_0\mathbf{J}'\{2E[\chi^{-2}_{p+1}(\Delta^2)] - (p-3)E[\chi^{-4}_{p+1}(\Delta^2)]\} \\ &\quad + (p-3)(p+1)\boldsymbol{\delta}\boldsymbol{\delta}' E[\chi^{-4}_{p+3}(\Delta^2)] \quad \text{and}\end{aligned}$$

$$\begin{aligned}R_4(\hat{\boldsymbol{\theta}}_n^{S} : \mathbf{Q}) &= \mathrm{tr}[\mathbf{Q}\boldsymbol{\Sigma}_0] - (p-3)\,\mathrm{tr}[\mathbf{Q}\boldsymbol{\Sigma}_0\mathbf{J}']\{2E[\chi^{-2}_{p+1}(\Delta^2)] \\ &\quad - (p-3)E[\chi^{-4}_{p+1}(\Delta^2)]\} + (p-3)(p+1)\boldsymbol{\delta}'\mathbf{Q}\boldsymbol{\delta} E[\chi^{-4}_{p+3}(\Delta^2)].\end{aligned} \tag{12.1.48}$$

(v) **PRSE,** $\hat{\boldsymbol{\theta}}_n^{S+}$

$$\mathbf{b}_5(\hat{\boldsymbol{\theta}}_n^{S+}) = -\boldsymbol{\delta}\{(p-3)E[\chi_{p+1}^{-2}(\Delta^2)] + E[(1-(p-3)\chi_{p+1}^{-2}(\Delta^2))I(\chi_{p+1}^2(\Delta^2) < p-3)]\} \quad \text{and}$$

$$B_5(\hat{\boldsymbol{\theta}}_n^{S+}) = \Delta^2\{(p-3)E[\chi_{p+1}^{-2}(\Delta^2)] + E[(1-(p-3)\chi_{p+1}^{-2}(\Delta^2))I(\chi_{p+1}^2(\Delta^2) < p-3)]\}^2;$$

$$\begin{aligned}\mathbf{M}_5(\hat{\boldsymbol{\theta}}_n^{S+}) = \mathbf{M}_4(\hat{\boldsymbol{\theta}}_n^{S}) &- (p-3)\boldsymbol{\Sigma}_0\mathbf{J}' \\ &\times\{E[(1-(p-3)\chi_{p+1}^{-2}(\Delta^2))^2 I(\chi_{p+1}^2(\Delta^2) < p-3)]\} \\ &+ \boldsymbol{\delta}\boldsymbol{\delta}'\{2E[(1-(p-3)\chi_{p+1}^{-2}(\Delta^2))I(\chi_{p+1}^2(\Delta^2) < p-3)] \\ &- E[(1-(p-3)\chi_{p+3}^{-2}(\Delta^2))^2 I(\chi_{p+3}^2(\Delta^2) < p-3)]\} \quad \text{and}\end{aligned}$$

$$\begin{aligned}R_5(\hat{\boldsymbol{\theta}}_n^{S+};\mathbf{Q}) = R_4(\hat{\boldsymbol{\theta}}_n^{S};\mathbf{Q}) &- (p-3)\,\mathrm{tr}[\mathbf{Q}\boldsymbol{\Sigma}_0\mathbf{J}'] \\ &\times E[(1-(p-3)\chi_{p+1}^{-2}(\Delta^2))^2 I(\chi_{p+1}^2(\Delta^2) < p-3)] \\ &+ \boldsymbol{\delta}'\mathbf{Q}\boldsymbol{\delta}\{2E[(1-(p-3)\chi_{p+1}^{-2}(\Delta^2))I(\chi_{p+1}^2(\Delta^2) < p-3)] \\ &- E[(1-(p-3)\chi_{p+3}^{-2}(\Delta^2))^2 I(\chi_{p+3}^2(\Delta^2) < p-3)]\}.\end{aligned}$$

For a proof, see Problem 10.

12.1.5 Analysis of the Properties of Estimators

In this section, we consider the ADQR analysis of the five estimators based on the results of the theorems below.

Theorem 3. Under H_0, the positive-semidefinite matrix, $\mathbf{Q}$ in the loss function, ADQR of estimators, $\tilde{\boldsymbol{\theta}}_n, \hat{\boldsymbol{\theta}}_n, \hat{\boldsymbol{\theta}}_n^{\mathrm{PT}}, \hat{\boldsymbol{\theta}}_n^{S}$, and $\hat{\boldsymbol{\theta}}_n^{S+}$ can be ordered as follows:

(i) $R_2(\hat{\boldsymbol{\theta}}_n:\mathbf{Q}) \le R_5(\hat{\boldsymbol{\theta}}_n^{S+}:\mathbf{Q}) \le R_4(\hat{\boldsymbol{\theta}}_n^{S}:\mathbf{Q}) \le R_1(\tilde{\boldsymbol{\theta}}_n:\mathbf{Q})$,

(ii) $R_2(\hat{\boldsymbol{\theta}}_n:\mathbf{Q}) \le R_3(\hat{\boldsymbol{\theta}}_n^{\mathrm{PT}}:\mathbf{Q}) \le R_1(\tilde{\boldsymbol{\theta}}_n:\mathbf{Q})$,

(iii) $R_3(\hat{\boldsymbol{\theta}}_n^{\mathrm{PT}}:\mathbf{Q}) \le R_5(\hat{\boldsymbol{\theta}}_n^{S+}:\mathbf{Q}) \le R_4(\hat{\boldsymbol{\theta}}_n^{S}:\mathbf{Q}) \le R_1(\tilde{\boldsymbol{\theta}}_n:\mathbf{Q})$,

where the ordering depends on the size of the level of significance, α for the PTE, $\hat{\boldsymbol{\theta}}_n^{\mathrm{PT}}$.

Proof. Let

$$\boldsymbol{\theta}_n^* = \tilde{\boldsymbol{\theta}}_n - (\tilde{\boldsymbol{\theta}}_n - \hat{\boldsymbol{\theta}}_n)g(D_n), \tag{12.1.49}$$

where $g(D_n)$ is a decreasing function of D_n. Take the forms $g(D_n) = 0, 1, I(D_n < \chi_{p-1}^2(\alpha)), [1-(p-3)D_n^{-1}]$, and $1-[1-(p-3)D_n^{-1}]I(D_n > p-3)$

to obtain the estimators $\tilde{\boldsymbol{\theta}}_n, \hat{\boldsymbol{\theta}}_n, \hat{\boldsymbol{\theta}}_n^{\text{PT}}, \hat{\boldsymbol{\theta}}_n^{S}$ and $\hat{\boldsymbol{\theta}}_n^{S+}$, respectively. Under H_0, ADQR of the estimators are

$$\begin{aligned}
R(\boldsymbol{\theta}_n^* : \mathbf{Q}) &= \text{tr}[\mathbf{Q}\boldsymbol{\Sigma}_0] \;\text{ if }\; \boldsymbol{\theta}_n^* = \tilde{\boldsymbol{\theta}}_n \\
&= \text{tr}[\mathbf{Q}\mathbf{B}] \;\text{ if }\; \boldsymbol{\theta}_n^* = \hat{\boldsymbol{\theta}}_n \\
&= \text{tr}[\mathbf{Q}\boldsymbol{\Sigma}_0]\{1 - H_{p+1}(\chi_{p-1}^2(\alpha); 0)\} + \text{tr}[\mathbf{Q}\mathbf{B}]H_{p+1}(\chi_{p-1}^2(\alpha); 0) \\
&\qquad \text{if }\; \boldsymbol{\theta}_n^* = \hat{\boldsymbol{\theta}}_n^{\text{PT}} \\
&= \frac{2}{p-1}\text{tr}[\mathbf{Q}\boldsymbol{\Sigma}_0] + \frac{p-3}{p-1}\text{tr}[\mathbf{Q}\mathbf{B}] \quad \text{if }\; \boldsymbol{\theta}_n^* = \hat{\boldsymbol{\theta}}_n^{S} \\
&= \text{tr}[\mathbf{Q}\boldsymbol{\Sigma}_0]\Big\{\frac{2}{p-1} - (p-3)\int_0^{p-3} [1-(p-3)x^{-1}]^2 dH_{p+1}(x;0)\Big\} \\
&\quad + \text{tr}[\mathbf{Q}\mathbf{B}]\Big\{\frac{p-3}{p-1} + (p-3)\int_0^{p-3} [1-(p-3)x^{-1}]^2 dH_{p+1}(x;0)\Big\} \\
&\qquad \text{if }\; \boldsymbol{\theta}_n^* = \hat{\boldsymbol{\theta}}_n^{S+}.
\end{aligned} \tag{12.1.50}$$

The risk-difference between $\tilde{\boldsymbol{\theta}}_n$ and $\hat{\boldsymbol{\theta}}_n$ is

$$\text{tr}[\mathbf{Q}(\boldsymbol{\Sigma}_0 - \mathbf{B})] > 0. \tag{12.1.51}$$

Hence, $R_2(\hat{\boldsymbol{\theta}}_n; \mathbf{Q}) < R_1(\tilde{\boldsymbol{\theta}}_n; \mathbf{Q})$. (12.1.52a)

The risk-difference between $\tilde{\boldsymbol{\theta}}_n$ and $\hat{\boldsymbol{\theta}}_n^{\text{PT}}$ is

$$\text{tr}[\mathbf{Q}(\boldsymbol{\Sigma}_0 - \mathbf{B})]H_{p+1}(\chi_{p-1}^2(\alpha); 0) \geq 0 \;\; \forall\, \alpha \in [0,1],$$

and this implies $R_3(\hat{\boldsymbol{\theta}}_n^{\text{PT}}; \mathbf{Q}) \leq R_1(\tilde{\boldsymbol{\theta}}_n; \mathbf{Q})$. (12.1.52b)

The risk-difference between $\hat{\boldsymbol{\theta}}_n$ and $\hat{\boldsymbol{\theta}}_n^{\text{PT}}$ is

$$\text{tr}[\mathbf{Q}(\boldsymbol{\Sigma}_0 - \mathbf{B})]\{1 - H_{p+1}(\chi_{p-1}^2(\alpha); 0)\} \geq 0 \;\; \forall\, \alpha \in [0,1], \tag{12.1.52c}$$

and this implies $R_2(\hat{\boldsymbol{\theta}}_n; \mathbf{Q}) \leq R_3(\hat{\boldsymbol{\theta}}_n^{\text{PT}}; \mathbf{Q}) \;\; \forall \alpha \in [0,1]$. Hence, we have

$$R_2(\hat{\boldsymbol{\theta}}_n; \mathbf{Q}) \leq R_3(\hat{\boldsymbol{\theta}}_n^{\text{PT}}; \mathbf{Q}) \leq R_1(\tilde{\boldsymbol{\theta}}_n; \mathbf{Q}) \text{ under } H_0. \tag{12.1.52d}$$

The risk-difference between $\hat{\boldsymbol{\theta}}_n^{S}$ and $\hat{\boldsymbol{\theta}}_n^{\text{PT}}$ is

$$R_4(\hat{\boldsymbol{\theta}}_n^{S}; \mathbf{Q}) - R_3(\hat{\boldsymbol{\theta}}_n^{\text{PT}}; \mathbf{Q}) = \text{tr}[\mathbf{Q}(\boldsymbol{\Sigma}_0 - \mathbf{B})]\Big\{H_{p+1}(\chi_{p-1}^2(\alpha); 0) - \frac{p-3}{p-1}\Big\}, \; p \geq 4.$$

Hence, $\hat{\boldsymbol{\theta}}_n^{\text{PT}}$ has smaller ADQR when

$$H_{p+1}(\chi_{p-1}^2(\alpha); 0) \geq \frac{p-3}{p-1}; \tag{12.1.52e}$$

otherwise, $\hat{\boldsymbol{\theta}}_n^S$ has smaller ADQR. Also,

$$R_4(\hat{\boldsymbol{\theta}}_n^S;\mathbf{Q}) = R_2(\hat{\boldsymbol{\theta}}_n;\mathbf{Q}) + \frac{2}{p-1}\mathrm{tr}[\mathbf{Q}(\boldsymbol{\Sigma}_0 - \mathbf{B})] \geq R_2(\hat{\boldsymbol{\theta}}_n;\mathbf{Q}).$$

Hence,

$$R_2(\hat{\boldsymbol{\theta}}_n;\mathbf{Q}) \leq R_4(\hat{\boldsymbol{\theta}}_n^S;\mathbf{Q}). \tag{12.1.52f}$$

Thus, whenever $H_{p+1}(\chi^2_{p-1}(\alpha);0) > \frac{p-3}{p-1}$, the order of ADQR of $\tilde{\boldsymbol{\theta}}_n, \hat{\boldsymbol{\theta}}_n, \hat{\boldsymbol{\theta}}_n^{\mathrm{PT}}$, and $\hat{\boldsymbol{\theta}}_n^S$ under H_0 is given by

$$R_2(\hat{\boldsymbol{\theta}}_n;\mathbf{Q}) \leq R_3(\hat{\boldsymbol{\theta}}_n^{\mathrm{PT}};\mathbf{Q}) \leq R_4(\hat{\boldsymbol{\theta}}_n^S;\mathbf{Q}). \tag{12.1.52g}$$

Also note that $R(\hat{\boldsymbol{\theta}}_n^S;\mathbf{Q}) - R_1(\tilde{\boldsymbol{\theta}}_n;\mathbf{Q}) = -\frac{p-3}{p-1}\mathrm{tr}[\mathbf{Q}\boldsymbol{\Sigma}_0] + \frac{p-3}{p-1}\mathrm{tr}[\mathbf{Q}\mathbf{B}] = -\frac{p-3}{p-1}\mathrm{tr}[\mathbf{Q}(\boldsymbol{\Sigma}_0 - \mathbf{B})] < 0$. Hence, $R_4(\hat{\boldsymbol{\theta}}_n^S;\mathbf{Q}) \leq R_1(\tilde{\boldsymbol{\theta}}_n;\mathbf{Q})$, which implies that the ADQR ordering of $\tilde{\boldsymbol{\theta}}_n, \hat{\boldsymbol{\theta}}_n, \hat{\boldsymbol{\theta}}_n^{\mathrm{PT}}$, and $\hat{\boldsymbol{\theta}}_n^S$ under H_0 is

$$R_2(\hat{\boldsymbol{\theta}}_n;\mathbf{Q}) \leq R_3(\hat{\boldsymbol{\theta}}_n^{\mathrm{PT}};\mathbf{Q}) \leq R_4(\hat{\boldsymbol{\theta}}_n^S) \leq R_1(\tilde{\boldsymbol{\theta}}_n;\mathbf{Q}). \tag{12.1.53}$$

provided that $H_{p+1}(\chi^2_{p-1}(\alpha);0) > \frac{p-3}{p-1}$.

Now consider the ADQR comparison of $\hat{\boldsymbol{\theta}}_n^{S+}$ and $\hat{\boldsymbol{\theta}}_n^S, \tilde{\boldsymbol{\theta}}_n, \hat{\boldsymbol{\theta}}_n$, and $\hat{\boldsymbol{\theta}}_n^{\mathrm{PT}}$ under H_0. First, compare $\hat{\boldsymbol{\theta}}_n^S$ and $\hat{\boldsymbol{\theta}}_n^{S+}$. The risk-difference

$$\begin{aligned}
&R_4(\hat{\boldsymbol{\theta}}_n^S;\mathbf{Q}) - R_5(\hat{\boldsymbol{\theta}}_n^{S+};\mathbf{Q})\\
&= (p-3)\mathrm{tr}[\mathbf{Q}(\boldsymbol{\Sigma}_0 - \mathbf{B})]\int_0^{p-3}[1-(p-3)x^{-1}]^2 dH_{p+1}(x;0)\\
&\leq \frac{2(p-3)}{p-1}\mathrm{tr}[\mathbf{Q}(\boldsymbol{\Sigma}_0 - \mathbf{B})]\ (\geq 0).
\end{aligned}$$

Hence, under H_0,

$$R_5(\hat{\boldsymbol{\theta}}_n^{S+};\mathbf{Q}) \leq R_4(\hat{\boldsymbol{\theta}}_n^S;\mathbf{Q}) \leq R_1(\tilde{\boldsymbol{\theta}}_n;\mathbf{Q}). \tag{12.1.54}$$

Next, consider the comparison of $\hat{\boldsymbol{\theta}}_n^{\mathrm{PT}}$ and $\hat{\boldsymbol{\theta}}_n^{S+}$. The risk-difference in this case is

$$\begin{aligned}
R_5(\hat{\boldsymbol{\theta}}_n^{S+}) - R_3(\hat{\boldsymbol{\theta}}_n^{\mathrm{PT}};\mathbf{Q}) &= \mathrm{tr}[\mathbf{Q}(\boldsymbol{\Sigma}_0 - \mathbf{B})]\{H_{p+1}(\chi^2_{p-1}(\alpha);0)\\
&-\frac{p-3}{p-1} + (p-3)\int_0^{p-3}[1-(p-3)x^{-1}]^2 dH_{p+1}(x;0)\}.
\end{aligned} \tag{12.1.55}$$

We know that $\hat{\boldsymbol{\theta}}_n^{\mathrm{PT}}$ is better than $\hat{\boldsymbol{\theta}}_n^S$ whenever $H_{p+1}(\chi^2_{p-1}(\alpha);0) > \frac{p-3}{p-1}$. Hence, the risk-difference above is nonnegative whenever

$$H_{p+1}(\chi^2_{p-1}(\alpha);0) > \frac{p-3}{p-1}. \tag{12.1.56}$$

Since (12.2.53b) is equivalent to the density inequality

$$2h_{p+1}(\chi^2_{p-1}(\alpha);0) \le \frac{2}{p-1} + \alpha, \quad p \ge 4, \tag{12.1.57}$$

we note that for a range of α, $\hat{\boldsymbol{\theta}}_n^{PT}$ dominates $\hat{\boldsymbol{\theta}}_n^{S+}$ and satisfies the ordering as given below:

$$R_2(\hat{\boldsymbol{\theta}}_n;\mathbf{Q}) \le R_3(\hat{\boldsymbol{\theta}}_n^{PT};\mathbf{Q}) \le R_5(\hat{\boldsymbol{\theta}}_n^{S+};\mathbf{Q}) \le R_4(\hat{\boldsymbol{\theta}}_n^{S};\mathbf{Q}) \le R_1(\tilde{\boldsymbol{\theta}}_n;\mathbf{Q}). \tag{12.1.58}$$

The next theorem gives the dominance of $\hat{\boldsymbol{\theta}}_n^{S+}$ and $\hat{\boldsymbol{\theta}}_n^{S}$ over $\tilde{\boldsymbol{\theta}}_n$ for all Δ^2.

Theorem 4. Let $p \ge 4$ and let the matrix $\mathbf{Q}$ in the loss function, $L(\boldsymbol{\theta}_n^*;\boldsymbol{\theta}) = n(\boldsymbol{\theta}_n^* - \boldsymbol{\theta})'\mathbf{Q}(\boldsymbol{\theta}_n^* - \boldsymbol{\theta})$ be positive-semi-definite, satisfying

$$\Big\{\mathbf{Q} : \frac{\text{tr}(\mathbf{Q}\boldsymbol{\Sigma}_0)}{Ch_{max}(\mathbf{Q}\boldsymbol{\Sigma})} \ge \frac{p+1}{p}\Big\}. \tag{12.1.59}$$

Then, under $\{K_{(n)}\}$, the ADQR of $\hat{\boldsymbol{\theta}}_n^{S+}, \hat{\boldsymbol{\theta}}_n^{S}$ and $\tilde{\boldsymbol{\theta}}_n$ can be ordered as

$$R_5(\hat{\boldsymbol{\theta}}_n^{S+}) \le R_4(\hat{\boldsymbol{\theta}}_n^{S};\mathbf{Q}) \le R_1(\tilde{\boldsymbol{\theta}}_n;\mathbf{Q}) \tag{12.1.60}$$

uniformly in Δ^2.

Proof. The ADQR expression for $\hat{\boldsymbol{\theta}}_n^{S}$ can be written as

$$\begin{aligned} R_4(\hat{\boldsymbol{\theta}}_n^{S};\mathbf{Q}) &= \text{tr}[\mathbf{Q}\boldsymbol{\Sigma}_0] - (p-3)\text{tr}[\mathbf{Q}\boldsymbol{\Sigma}_0\mathbf{J}']\Big\{(p-3)E[\chi^{-4}_{p+1}(\Delta^2)] \\ &\quad + \Big[1 - \frac{(p+1)\boldsymbol{\delta}'\mathbf{Q}\boldsymbol{\delta}}{2\Delta^2\,\text{tr}[\mathbf{Q}\boldsymbol{\Sigma}_0]}\Big]2\Delta^2 E[\chi^{-4}_{p+3}(\Delta^2)]\Big\}. \end{aligned} \tag{12.1.61}$$

The risk-difference $R_1(\tilde{\boldsymbol{\theta}}_n;\mathbf{Q}) - R_4(\hat{\boldsymbol{\theta}}_n^{S};\mathbf{Q})$ is nonnegative for all $\mathbf{Q}$, satisfying (12.1.59), and for all Δ^2. Hence,

$$R_4(\hat{\boldsymbol{\theta}}_n^{S};\mathbf{Q}) \le R_1(\tilde{\boldsymbol{\theta}}_n;\mathbf{Q}) \ \ \forall\ (\Delta^2,\mathbf{Q}). \tag{12.1.62}$$

Similarly, the risk-difference

$$\begin{aligned} &R_5(\hat{\boldsymbol{\theta}}_n^{S+};\mathbf{Q}) - R_4(\hat{\boldsymbol{\theta}}_n^{S};\mathbf{Q}) \\ &= -\text{tr}[\mathbf{Q}\boldsymbol{\Sigma}_0\mathbf{J}']E[(1-(p-3)\chi^{-2}_{p+1}(\Delta^2))^2 I(\chi^2_{p+1}(\Delta^2) < p-3)] \\ &\quad + \boldsymbol{\delta}'\mathbf{Q}\boldsymbol{\delta}\{2E[(1-(p-3)\chi^{-2}_{p+1}(\Delta^2))I(\chi^2_{p+1}(\Delta^2) < p-3)]\} \\ &\quad - E[(1-(p-3)\chi^{-2}_{p+3}(\Delta^2))^2 I(\chi^2_{p+3}(\Delta^2) < p-3)]. \end{aligned}$$

The expression above is nonpositive, since

$$0 \leq \chi^2_{p+1+2j}(\Delta^2) \leq p-3 \Longrightarrow 1-(p-3)\chi^{-2}_{p-1+2j} < 0, \tag{12.1.63}$$

which implies that the risk-difference becomes

$$\begin{aligned}&-\mathrm{tr}[\mathbf{Q}\boldsymbol{\Sigma}_0\mathbf{J}']E[(1-(p-3)\chi^{-2}_{p+1}(\Delta^2))^2 I(\chi^2_{p+1}(\Delta^2) < p-3)]\\&-\boldsymbol{\delta}'\mathbf{Q}\boldsymbol{\delta}\{2E[((p-3)\chi^{-2}_{p+1}(\Delta^2)-1)I(\chi^2_{p+1}(\Delta^2) < p-3)]\}\\&+E[(1-(p-3)\chi^{-2}_{p+3}(\Delta^2))^2 I(\chi^2_{p+3}(\Delta^2) < p-3)]\} < 0 \quad \forall\Delta^2.\end{aligned}$$

Hence, the theorem holds.

Now, we consider the asymptotic distributional relative efficiency (ADRE) of the estimators compared to $\tilde{\boldsymbol{\theta}}_n$.

(i) $$\begin{aligned}ARE[\hat{\boldsymbol{\theta}}_n;\tilde{\boldsymbol{\theta}}_n] &= R_1(\tilde{\boldsymbol{\theta}}_n;\mathbf{Q})[R_2(\hat{\boldsymbol{\theta}}_n;\mathbf{Q})]^{-1}\\&= [\mathrm{tr}[\mathbf{QB}]+\boldsymbol{\delta}'\mathbf{Q}\boldsymbol{\delta}]^{-1}\mathrm{tr}[\mathbf{Q}\boldsymbol{\Sigma}_0]\\&= p[1+\Delta^2]^{-1} = p(1+\Delta^2)^{-1}, \text{ if } \mathbf{Q}=\boldsymbol{\Sigma}_0^{-1},\end{aligned} \tag{12.1.64}$$

which is a decreasing function of Δ^2. In general,

$$\begin{aligned}&[\mathrm{tr}[\mathbf{QB}]+\Delta^2 Ch_{max}(\mathbf{Q}\boldsymbol{\Sigma}_0)]^{-1}\mathrm{tr}[\mathbf{Q}\boldsymbol{\Sigma}_0] \leq ARE(\hat{\boldsymbol{\theta}}_n;\tilde{\boldsymbol{\theta}}_n)\\&\qquad\leq [\mathrm{tr}[\mathbf{QB}]+\Delta^2 Ch_{min}(\mathbf{Q}\boldsymbol{\Sigma}_0)]^{-1}\mathrm{tr}[\mathbf{Q}\boldsymbol{\Sigma}_0].\end{aligned}$$

(ii) $ARE[\hat{\boldsymbol{\theta}}_n^{\mathrm{PT}};\tilde{\boldsymbol{\theta}}_n] = [1+g_{\mathrm{PT}}(\alpha,\Delta^2)]^{-1}$,

where

$$\begin{aligned}g_{\mathrm{PT}} &= -[\mathrm{tr}(\mathbf{Q}\boldsymbol{\Sigma}_0\mathbf{J}')][\mathrm{tr}(\mathbf{Q}\boldsymbol{\Sigma}_0)]^{-1}H_{p+1}(\chi^2_{p-1}(\alpha);\Delta^2)+\boldsymbol{\delta}'\mathbf{Q}\boldsymbol{\delta}[\mathrm{tr}(\mathbf{Q}\boldsymbol{\Sigma}_0)]^{-1}\\&\qquad\times\{2H_{p+1}(\chi^2_{p-1}(\alpha);\Delta^2)-H_{p+3}(\chi^2_{p-1}(\alpha);\Delta^2)\}\\&= \left(1-\tfrac{1}{p}\right)H_{p+1}(\chi^2_{p-1}(\alpha);\Delta^2)+p^{-1}\Delta^2\{2H_{p+1}(\chi^2_{p-1}(\alpha);\Delta^2)\\&\qquad-H_{p+3}(\chi^2_{p-1}(\alpha);\Delta^2)\} \quad \text{if } \mathbf{Q}=\boldsymbol{\Sigma}_0^{-1}.\end{aligned}$$

The graph of $ARE[\hat{\boldsymbol{\theta}}_n^{\mathrm{PT}};\tilde{\boldsymbol{\theta}}_n]$ as a function of Δ^2 attains its maximum at $\Delta^2=0$ for all α $(0<\alpha<1)$ and decreases monotonically, crossing the 1-line. Table 12.1.1 gives the maximum and minimum values of the ARE together with the intersecting efficiencies of the PTE for $\alpha = 0.05(0.05)0.5$ and $p = 4(2)16$.

(iii) $ARE(\hat{\boldsymbol{\theta}}_n^{S};\tilde{\boldsymbol{\theta}}_n) = [1+g_s(\Delta^2)]^{-1}$,

where

$$g_s(\Delta^2) = -\left(1-\tfrac{3}{p}\right)[\mathrm{tr}(\mathbf{Q}\boldsymbol{\Sigma}_0\mathbf{J}')][\mathrm{tr}(\mathbf{Q}\boldsymbol{\Sigma}_0)]^{-1}\{2E[\chi^{-2}_{p+1}(\Delta^2)]$$

$$-(p-3)E[\chi_{p+1}^{-4}(\Delta^2)]\}$$
$$+\left(1-\tfrac{3}{p}\right)(p+1)\boldsymbol{\delta}'\mathbf{Q}\boldsymbol{\delta}[\mathrm{tr}(\mathbf{Q}\boldsymbol{\Sigma}_0)]^{-1}E[\chi_{p+3}^{-4}(\Delta^2)]$$
$$=-\left(1-\tfrac{3}{p}\right)\left(1-\tfrac{1}{p}\right)\{2E[\chi_{p+1}^{-2}(\Delta^2)]-(p-3)E[\chi_{p+1}^{-4}(\Delta^2)]\}$$
$$+\left(1-\tfrac{3}{p}\right)\left(1+\tfrac{1}{p}\right)\Delta^2E[\chi_{p+3}^{-4}(\Delta^2)] \text{ if } \mathbf{Q}=\boldsymbol{\Sigma}_0^{-1}.$$

Table 12.1.1 gives some ARE-values of $\hat{\boldsymbol{\theta}}_n(E_1), \hat{\boldsymbol{\theta}}_n^{\mathrm{PT}}(E_3)$ and $\hat{\boldsymbol{\theta}}_n^S(E_4)$ and the intersecting ARE-values denoted by $E_{\Delta_\alpha^2}$ for $\hat{\boldsymbol{\theta}}_n^{\mathrm{PT}}$, and $\hat{\boldsymbol{\theta}}_n^S$ and also the Δ_α^2-values where the intersection occurs. For example, for $p=6, \alpha=0.1$, and $\Delta^2 \in [0, 1.1514]$, from the table it appears that $\hat{\boldsymbol{\theta}}_n^{\mathrm{PT}}$ dominates $\hat{\boldsymbol{\theta}}_n^S$, but outside this interval, domination is reversed. For $\Delta^2=0$, $\hat{\boldsymbol{\theta}}_n$ has larger ARE ($E_1=6.000$) than $\hat{\boldsymbol{\theta}}_n^{\mathrm{PT}}$ ($E_3=4.2350$) and $\hat{\boldsymbol{\theta}}_n^S(E_4=2.5000)$ relative to $\tilde{\boldsymbol{\theta}}_n$.

12.1.6 Baseball Data Analysis

We can illustrate the theory above with baseball data as given in Efron and Morris (1975). For these data, Efron and Morris used the James-Stein (1961) estimator to predict the batting averages of 18 major league players in the remainder of the 1970 season. The data consist of the number of hits y in the first 45 bats observed for each player ($i=1,\cdots,18$). The problem is to estimate $\boldsymbol{\theta}=(\theta_1,\cdots,\theta_{18})'$, where θ_i denotes the final season batting average of the ith player. Efron and Morris (1975) used the arc-sin transformation on each y_i to obtain an approximate normal distribution with constant variance, and then used the James-Stein estimator on the transformed counts. This we denote by $\hat{\boldsymbol{\theta}}_n^{\mathrm{EM}}$. Albert (1984) proposed an **approximate empirical Bayes** estimator, $\hat{\boldsymbol{\theta}}_n^{EB}$. We present the true batting average ($\boldsymbol{\theta}^T$) together with $\tilde{\boldsymbol{\theta}}_n, \hat{\boldsymbol{\theta}}_n^{EM}, \hat{\boldsymbol{\theta}}_n^{EB}, \hat{\boldsymbol{\theta}}_n^{\mathrm{PT}}$, and $\hat{\boldsymbol{\theta}}_n^S$. The true batting average obtained from the Efron-Morris paper is 0.267 with standard deviation 0.037. To assess the performance of various estimators they used the loss defined by $(\hat{\theta}_i-\theta_i^T)^2/(0.037)^2$ for each estimator. The true values (θ_i^T) and the estimated values of θ_i based on Efron-Morris (EM), empirical Bayes (EB), and Ali and Saleh estimators, $\theta_i^{\mathrm{PT}}, \hat{\theta}_i^S$, and θ_i^{S+} are given in Table 12.1.2. The PTE is $(0.265,\cdots,0.265)'$ as a result of testing $H_0: \theta_1=\cdots=\theta_{18}$ at the 15% level of significance. In Table 12.1.3 we present the loss ($L_{\tilde{\boldsymbol{\theta}}_n}, L_{EM}, L_{EB}, L_{\mathrm{PT}}$, and L_S) of the estimators, respectively. A comparison shows that $\hat{\boldsymbol{\theta}}_n^S/\hat{\boldsymbol{\theta}}_n^{S+}$ by our method is close to $\hat{\boldsymbol{\theta}}_n^{EB}$ and $\hat{\boldsymbol{\theta}}_n^{EB}$ for all $i=1,\cdots,18$, which supports the validity of the theory.

Table 12.1.1 Maximum Relative Efficiencies of the RMLE, PTE, and SE and the Intersection Efficiencies for the PTE and SE for each α with Corresponding Δ_α-Values for p-Values for $\alpha = 0.05(0.05)0.25$ and $p = 4(2)16$

α/p		4	6	8	10	12	14	16
0.05	E_1	4.0000	6.0000	9.0000	10.0000	12.0000	14.0000	16.0000
	E_3	5.9971	7.3761	8.3363	9.0666	9.6514	10.1362	10.5481
	E_4	1.5000	2.5000	3.5000	4.5000	5.5000	6.5000	7.5000
	$E_{\Delta.05}$	1.2167	1.8128	2.4102	3.0680	3.8010	4.6266	5.5660
	$\Delta_{.05}$	2.0988	1.8974	1.7020	1.5048	1.3048	1.1025	.8983
0.10	E_1	4.0000	6.0000	9.0000	10.0000	12.0000	14.0000	16.0000
	E_3	3.5396	4.2350	4.7092	5.0648	5.3466	5.5783	5.7739
	E_4	1.5000	2.5000	3.5000	4.5000	5.5000	6.5000	7.5000
	$E_{\Delta.10}$	1.2913	1.9996	2.9013	4.0761	5.3466	5.5783	5.7739
	$\Delta_{.10}$	1.6932	1.1514	.7081	.3048	.0000	.0000	.0000
0.15	E_1	4.0000	6.0000	9.0000	10.0000	12.0000	14.0000	16.0000
	E_3	2.6415	3.1003	3.4085	3.6375	3.8177	3.9652	4.0891
	E_4	1.5000	2.5000	3.5000	4.5000	5.5000	6.5000	7.5000
	$E_{\Delta.15}$	1.3199	2.2163	3.4085	3.6375	3.8177	3.9652	4.0891
	$\Delta_{.15}$	1.3593	.5482	.0000	.0000	.0000	.0000	.0000
0.20	E_1	4.0000	6.0000	9.0000	10.0000	12.0000	14.0000	16.0000
	E_3	2.1684	2.5038	2.7274	2.8925	3.0218	3.1272	3.2155
	E_4	1.5000	2.5000	3.5000	4.5000	5.5000	6.5000	7.5000
	$E_{\Delta.20}$	1.3511	2.4973	2.7274	2.8925	3.0218	3.1272	3.2155
	$\Delta_{.20}$	1.0462	.0044	.0000	.0000	.0000	.0000	.0000
0.25	E_1	4.0000	6.0000	9.0000	10.0000	12.0000	14.0000	16.0000
	E_3	1.8729	2.1328	2.3049	2.4314	2.5300	2.6101	2.6771
	E_4	1.5000	2.5000	3.5000	4.5000	5.5000	6.5000	7.5000
	$E_{\Delta.25}$	1.3873	2.1328	2.3049	2.4314	2.5300	2.6101	2.6771
	$\Delta_{.25}$	.7340	.0000	.0000	.0000	.0000	.0000	.0000

(Table 12.1.1 – 12.1.3 are due to Ali and Saleh (1991), and reproduced with the permission of Statistica Sinica.)

Table 12.1.2 True Value (θ_1^T) and Estimated Values of θ_i Based on Efron-Morris (EM), Empirical Bayes (EB) and Ali and Saleh Estimators, $\hat{\theta}_i^{\text{PT}}, \hat{\theta}_n^{\text{S}}$, and $\hat{\theta}_i^{\text{S+}}$

i	θ_i^T	$\tilde{\theta}$	$\hat{\theta}_i^{\text{EM}}$	$\hat{\theta}_i^{\text{EB}}$	$\hat{\theta}_i^{\text{PT}}$	$\hat{\theta}_i^{\text{S}}/\hat{\theta}_i^{\text{S+}}$
1	0.346	0.400	0.290	0.279	0.265	0.294
2	0.300	0.378	0.286	0.277	0.265	0.289
3	0.279	0.356	0.281	0.274	0.265	0.242
4	0.223	0.333	0.277	0.272	0.265	0.280
5	0.276	0.311	0.273	0.270	0.265	0.275
6	0.273	0.311	0.273	0.270	0.265	0.275
7	0.266	0.289	0.268	0.268	0.265	0.270
8	0.211	0.267	0.264	0.265	0.265	0.266
9	0.271	0.244	0.259	0.263	0.265	0.261
10	0.232	0.244	0.259	0.263	0.265	0.261
11	0.266	0.222	0.254	0.261	0.265	0.256
12	0.258	0.222	0.254	0.261	0.265	0.265
13	0.306	0.222	0.254	0.261	0.265	0.265
14	0.267	0.222	0.254	0.261	0.265	0.256
15	0.228	0.222	0.254	0.261	0.265	0.265
16	0.288	0.200	0.249	0.259	0.265	0.252
17	0.318	0.178	0.233	0.257	0.265	0.247
18	0.200	0.156	0.208	0.254	0.265	0.242
mean	0.267	0.265	0.261	0.265	0.265	0.263
s.d.	0.037	0.068	0.019	0.007	0.000	0.014

Table 12.1.3 Estimated Average Loss for the Estimators

i	L_{MLE}	L_{EM}	L_{EB}	L_{PT}	$L_{S/S+}$
1	2.130	2.291	3.279	4.793	1.975
2	4.444	0.143	0.386	0.895	0.088
3	4.331	0.003	0.018	0.143	0.844
4	8.839	2.130	1.754	1.289	2.373
5	0.895	0.007	0.026	0.088	0.001
6	1.055	0.000	0.007	0.047	0.003
7	0.386	0.003	0.003	0.001	0.012
8	2.291	2.052	2.130	2.130	2.210
9	0.533	0.105	0.047	0.026	0.073
10	0.105	0.533	0.702	0.795	0.614
11	1.414	0.105	0.018	0.001	0.073
12	0.947	0.012	0.007	0.036	0.003
13	5.154	1.975	1.479	1.228	1.826
14	1.479	0.123	0.026	0.003	0.088
15	0.026	0.494	0.795	1.000	0.573
16	5.657	1.111	0.614	0.386	0.947
17	14.317	5.278	2.718	2.052	3.682
18	1.414	0.047	2.130	3.086	1.289
mean	3.079	0.912	0.897	1.000	0.926
s.e.	3.565	1.346	1.045	1.270	1.051

12.1.7 Asymptotic Properties of Confidence Sets

In this section, we consider confidence sets, which are spheres with centers $\tilde{\boldsymbol{\theta}}_n, \hat{\boldsymbol{\theta}}_n, \hat{\boldsymbol{\theta}}_n^{\text{PT}}, \hat{\boldsymbol{\theta}}_n^{S}$, and $\hat{\boldsymbol{\theta}}_n^{S+}$, respectively. Accordingly, we have the five confidence sets corresponding to the assigned estimators as follows:

$$C^0(\tilde{\boldsymbol{\theta}}_n) = \{\boldsymbol{\theta} : n\|\boldsymbol{\theta} - \tilde{\boldsymbol{\theta}}_n\|^2_{\hat{\boldsymbol{\Sigma}}_n^{-1}} \le \chi_p^2(\gamma)\},$$

$$C^0(\hat{\boldsymbol{\theta}}_n) = \{\boldsymbol{\theta} : n\|\boldsymbol{\theta} - \hat{\boldsymbol{\theta}}_n\|^2_{\hat{\boldsymbol{\Sigma}}_n^{-1}} \le \chi_p^2(\gamma)\},$$

$$C^0(\hat{\boldsymbol{\theta}}_n^{\text{PT}}) = \{\boldsymbol{\theta} : n\|\boldsymbol{\theta} - \hat{\boldsymbol{\theta}}_n^{\text{PT}}\|^2_{\hat{\boldsymbol{\Sigma}}_n^{-1}} \le \chi_p^2(\gamma)\},$$

$$C^0(\hat{\boldsymbol{\theta}}_n^{S}) = \{\boldsymbol{\theta} : n\|\boldsymbol{\theta} - \hat{\boldsymbol{\theta}}_n^{S}\|^2_{\hat{\boldsymbol{\Sigma}}_n^{-1}} \le \chi_p^2(\gamma)\},$$

$$C^0(\hat{\boldsymbol{\theta}}_n^{S+}) = \{\boldsymbol{\theta} : n\|\boldsymbol{\theta} - \hat{\boldsymbol{\theta}}_n^{S+}\|^2_{\hat{\boldsymbol{\Sigma}}_n^{-1}} \le \chi_p^2(\gamma)\}, \tag{12.1.66}$$

where $\lim_{n\to\infty} P\{n\|\tilde{\boldsymbol{\theta}}_n - \boldsymbol{\theta}\|^2_{\hat{\boldsymbol{\Sigma}}_n^{-1}} \le \chi_p^2(\gamma)\} = 1-\gamma$ and $\chi_p^2(\gamma)$ is the upper critical value from a central chi-square distribution with p d.f.

From Section 12.1.1, we know that under **fixed alternatives**, the asymptotic distribution of estimators are equivalent to that of $\sqrt{n}(\tilde{\boldsymbol{\theta}}_n - \boldsymbol{\theta}) \sim \mathcal{N}_p(\mathbf{0}, \boldsymbol{\Sigma}_0)$, except for the restricted estimator whose distribution is degenerate as $n \to \infty$. Hence, the asymptotic coverage probability of $C^{\text{PT}}(\hat{\boldsymbol{\theta}}_n^{\text{PT}}), C^S(\hat{\boldsymbol{\theta}}_n^{S})$, and $C^S(\hat{\boldsymbol{\theta}}_n^{S+})$ is $1-\gamma$ (as $n\to\infty$). Thus, to obtain meaningful asymptotic coverage probabilities of confidence sets, we consider the sequence of local alternatives, $K_{(n)} : \boldsymbol{\theta}_{(n)} = \theta_0\mathbf{1}_p + n^{-1/2}\boldsymbol{\xi}$ as given by (12.1.46). Accordingly, we have

$$P[C^0(\tilde{\boldsymbol{\theta}}_n)] = \lim_{n\to\infty} P\{n\|\boldsymbol{\theta}_{(n)} - \tilde{\boldsymbol{\theta}}_n\|^2_{\hat{\boldsymbol{\Sigma}}_0^{-1}} \le \chi_p^2(\gamma)\} = 1-\gamma \tag{12.1.67}$$

Next, we consider the sets $C^*(\boldsymbol{\theta}_n^*)$. In this case,

$$C^*(\boldsymbol{\theta}_n^*)\{\boldsymbol{\theta}_{(n)} : n\|\boldsymbol{\theta}_{(n)} - \boldsymbol{\theta}_n^*\|^2_{\hat{\boldsymbol{\Sigma}}_n^{-1}} \le \chi_p^2(\gamma)\} \tag{12.1.68}$$

We now evaluate the **asymptotic coverage probability** of $C^*(\boldsymbol{\theta}_n^*)$ (as $n\to\infty$) under $\{K_{(n)}\}$. We know that as $n\to\infty$,

$$\sqrt{n}(\tilde{\boldsymbol{\theta}}_n - \boldsymbol{\theta}_{(n)}),\ \mathcal{N}_p(\mathbf{0}, \sigma^2\Lambda^{-1}),\ \sigma^2 = \theta_0(1-\theta_0). \tag{12.1.69}$$

Now, consider the orthogonal matrix $\boldsymbol{\Gamma} = (\boldsymbol{\Gamma}_1, \boldsymbol{\Gamma}_2)$, where $\boldsymbol{\Gamma}_1$ is a $p\times(p-1)$ orthogonal matrix and $\boldsymbol{\Gamma}_2$ is a $p\times 1$ unit vector. Then $\boldsymbol{\Gamma}_2'\boldsymbol{\Gamma}_1 = \mathbf{0}$ and $\boldsymbol{\Gamma}_1\boldsymbol{\Gamma}_1' + \boldsymbol{\Gamma}_2\boldsymbol{\Gamma}_2' = \mathbf{I}_p$ so that $\boldsymbol{\Gamma}_1\boldsymbol{\Gamma}_1' = \mathbf{I}_p - \boldsymbol{\Gamma}_2\boldsymbol{\Gamma}_2'$. For our purpose, we consider $\boldsymbol{\Gamma}_2 = \boldsymbol{\Lambda}_n^{1/2}\mathbf{1}_p$ where $\boldsymbol{\Lambda} = \text{Diag}(\frac{n_1}{n}, \cdots, \frac{n_p}{n})$ with $\lambda_i = n_i/n$, $(i = 1, \cdots, p)$. Thus, $\mathbf{J}_n = \boldsymbol{\Lambda}_n^{-1/2}\boldsymbol{\Gamma}_1\boldsymbol{\Gamma}_1'\boldsymbol{\Lambda}_n^{1/2} = \mathbf{I}_p - \boldsymbol{\Lambda}_n^{-1/2}\boldsymbol{\Lambda}_n^{1/2}\mathbf{1}_p\mathbf{1}_p'\boldsymbol{\Lambda}_n^{-1/2}\boldsymbol{\Lambda}_n^{1/2} = \mathbf{I}_p - \mathbf{1}_p\mathbf{1}_p'\boldsymbol{\Lambda}_n$.

Now, consider the transformation

$$\mathbf{Z}_{(1n)} = \frac{\sqrt{n}}{\hat{\sigma}_n}\boldsymbol{\Gamma}_1'\boldsymbol{\Lambda}_n^{1/2}\tilde{\boldsymbol{\theta}}_n \text{ and, } Z_{(2n)} = \frac{\sqrt{n}}{\hat{\sigma}_n}\boldsymbol{\Gamma}_2'\boldsymbol{\Lambda}_n^{1/2}\tilde{\boldsymbol{\theta}}_n$$

$$\boldsymbol{\eta}_{(1n)} = \frac{\sqrt{n}}{\hat{\sigma}_n}\boldsymbol{\Gamma}_1'\boldsymbol{\Lambda}_n^{1/2}(\theta_0\mathbf{1}_n + n^{-1/2}\boldsymbol{\xi}) = \frac{1}{\hat{\sigma}_n}\boldsymbol{\Gamma}_1'\boldsymbol{\Lambda}_n^{1/2}\mathbf{1}_p\sqrt{n}\theta_0\frac{1}{\hat{\sigma}_n}\boldsymbol{\Gamma}_1'\boldsymbol{\Lambda}_n^{1/2}\boldsymbol{\xi},$$

and

$$\eta_{(2n)} = \frac{\sqrt{n}}{\hat{\sigma}_n}\boldsymbol{\Gamma}_2'\boldsymbol{\Lambda}_n^{1/2}(\theta_0\mathbf{1}_p + n^{-1/2}\boldsymbol{\xi}) = \frac{1}{\hat{\sigma}_n}\boldsymbol{\Gamma}_2'\boldsymbol{\Lambda}_n^{1/2}\mathbf{1}_p\sqrt{n}\theta_0 + + \frac{1}{\hat{\sigma}_n}\boldsymbol{\Gamma}_2'\boldsymbol{\Lambda}_n^{1/2}\boldsymbol{\xi}.$$

Hence,

$$\left\{ \begin{array}{c} (\mathbf{Z}_{(1n)} - \boldsymbol{\eta}_{(1n)}) \\ (Z_{(2n)} - \eta_{(2n)}) \end{array} \right\} \sim \mathcal{N}_p\left\{\begin{pmatrix} \mathbf{0} \\ 0 \end{pmatrix} : \begin{pmatrix} \mathbf{I}_{p-1} & \mathbf{0} \\ \mathbf{0} & 1 \end{pmatrix}\right\}.$$

Further, $n||\boldsymbol{\theta}_{(n)} - \boldsymbol{\theta}_n^*||^2_{\hat{\boldsymbol{\Sigma}}_n^{-1}}$ asymptotically equals

$$(\eta_2 - \omega_2)^2 + \left|\left|\boldsymbol{\eta}_1 - \boldsymbol{\omega}_1 g(||\boldsymbol{\omega}_1||^2)\right|\right|$$

where $\boldsymbol{\omega}_1 = \mathbf{Z}_1 + \boldsymbol{\delta}_1$, $\mathbf{Z}_1 \sim \mathcal{N}_{p-1}(\mathbf{0}, \mathbf{I}_{p-1})$ and $\boldsymbol{\delta} = \mathbf{J}\boldsymbol{\xi}$. Similarly, $\omega_2 = Z_2 + \eta_2$, $Z_2 \sim \mathcal{N}(0,1)$. Hence,

$$\lim_{n\to\infty} P_{K_{(n)}}\{n||\boldsymbol{\theta}_{(n)} - \boldsymbol{\theta}_n^*||^2_{\hat{\boldsymbol{\Sigma}}_n^{-1}} < \chi_p^2(\alpha)\}$$

$$= P_{\boldsymbol{\eta}}\left\{(\eta_2 - \omega_2)^2 + \left|\left|\boldsymbol{\eta}_1 - \boldsymbol{\omega}_1 g(||\boldsymbol{\omega}_1||^2)\right|\right|^2 < \chi_p^2(\gamma)\right\}$$

$$= \int_0^{\chi_p^2(\gamma)} P_{\boldsymbol{\eta}}\left\{\left|\left|\boldsymbol{\eta}_1 - \boldsymbol{\omega}_1 g(||\boldsymbol{\omega}_1||^2)\right|\right|^2 < (\chi_p^2(\gamma) - t)^+\right\} dH_1(t,0),$$

where $H_1(t,0)$ is the cdf of a central chi-square distribution with one d.f. Thus,

(i) If $g(||\boldsymbol{\omega}_1||^2) = 1$, we have

$$P_{\boldsymbol{\eta}}\{(\eta_2 - \omega_2)^2 + ||\boldsymbol{\eta}_1 - \boldsymbol{\omega}_1||^2 \le \chi_p^2(\gamma)\} = H_p(\chi_p^2(\gamma); 0) = 1 - \gamma.$$

(ii) If $g(||\boldsymbol{\omega}_1||^2) = 0$, we have

$$P_{\boldsymbol{\eta}}\{(\eta_2 - \omega_2)^2 + ||\boldsymbol{\eta}_1||^2 \le \chi_p^2(\gamma)\} = H_1(\chi_p^2(\gamma) - \Delta^2; 0).$$

(iii) If $g(||\boldsymbol{\omega}_1||^2) = I(||\boldsymbol{\omega}_1||^2 \ge \chi_{p-1}^2(\alpha))$, then we have

$$P_{\boldsymbol{\eta}}\left\{(\eta_2 - \omega_2)^2 + \left|\left|\boldsymbol{\eta}_1 - \boldsymbol{\omega}_1 I(||\boldsymbol{\omega}_1||^2 \ge \chi_{p-1}^2(\alpha))\right|\right|^2 < \chi_{p-1}^2(\gamma)\right\}$$

$$= H_1(\chi_p^2(\gamma) - \Delta^2; 0) H_{p-1}(\chi_{p-1}^2(\alpha); \Delta^2)$$

$$+ \int_0^{\chi_p^2(\gamma)} P_{\boldsymbol{\eta}_1}\{||\boldsymbol{\eta}_1 - \boldsymbol{\omega}_1||^2 \le (\chi_p^2(\gamma))^+; ||\boldsymbol{\omega}_1||^2 > \chi_{p-1}^2(\alpha)\} dH_1(t,0).$$

(iv) If $g(||\boldsymbol{\omega}_1||^2) = (1 - c||\boldsymbol{\omega}_1||^{-2})$, $0 < c < 2(p-3)$, then we have

$$P_{\boldsymbol{\eta}_1}\left\{(\eta_2 - \omega_2)^2 + \left|\left|\boldsymbol{\eta}_1 - \boldsymbol{\omega}_1(1 - c||\boldsymbol{\omega}_1||^{-2})\right|\right|^2 \le \chi_p^2(\gamma)\right\}$$

$$= \int_0^{\chi_p^2(\gamma)} P_{\eta_1}\left\{\left|\left|\eta_1 - \omega_1(1 - c||\omega||^{-2})\right|\right| \le (\chi_p^2(\gamma) - t)^+\right\} dH_1(t,0)$$

(v) If $g(||\omega_1||^2) = (1 - c||\omega_1||^{-2})I(||\omega_1||^2 > c)$, then we have

$$P_{\eta}\left\{(\eta_2 - \omega_2)^2 + \left|\left|\eta_1 - \omega_1(1 - c||\omega_1||^{-2})I(||\omega_1||^2 > c)\right|\right| \le \chi_p^2(\gamma)\right\}$$

$$= P_{\eta}\left\{(\eta_2 - \omega_2)^2 + ||\eta_1||^2 \le \chi_p^2(\gamma) : ||\omega_1||^2 < c\right\}$$

$$+ P_{\eta}\left\{(\eta_2 - \omega_2)^2 + \left|\left|\eta_1 - \omega_1(1 - c||\omega_1||^{-2})\right|\right|^2 \le \chi_p^2(\gamma); ||\omega_1||^2 < c\right\}$$

$$= H_1(\chi_p^2(\gamma) - \Delta^2; 0)H_{p-1}(c : \Delta^2)$$

$$+ \int_0^{\chi_p^2(\gamma)} P_{\eta_1}\left\{\left|\left|\eta_1 - \omega_1(1 - c||\omega_1||^{-2})\right|\right|^2 \le (\chi_p^2(\gamma) - t)^+; \right.$$

$$\left. ||\omega_1||^2 > c\right\} dH_1(t,0).$$

For the analysis of coverage probabilities, we go to Section 5.9.2 to obtain parallel conclusions on coverage probabilities of the five confidence sets.

12.2 Product Binomial Distributions, Estimation, and Test of Hypothesis: Homogeneity of Odds Ratios

12.2.1 Introduction

In many practical situations such as an epidemiological survey or a meta-analysis, statistical data are often stratified into several 2×2 tables in order to control for confounding factors. For instance, consider the case-control studies where medical records are examined and questionnaires are administered involving, say, n_{1j} experimental cases and n_{2j} control cases of the jth stratum $(j = 1, 2, \cdots, k)$ to determine the number of exposed cases x_{1j} and exposed controls x_{2j}. The primary goal is to assess the disease exposure association by examining, the **common** "odds ratio" (OR) ψ_0.

Very often it is uncertain that the ORs ψ_j, $(j = 1, 2, \cdots, k)$, of k tables are homogeneous across strata. Then the objective is to estimate the OR of the k tables, meaning $\boldsymbol{\Psi} = (\psi_1, \cdots, \psi_k)'$ when one suspects that $\boldsymbol{\Psi} = \psi_0 \mathbf{1}_k$. This section contains the study of the problem of estimating $\boldsymbol{\Psi}$ when it is suspected that $H_0 : \boldsymbol{\Psi} = \psi_0 \mathbf{1}_k$ holds.

12.2.2 Model, Estimation, and Test of Hypothesis

Consider k pairs of mutually independent binomial distributions $\{B_{1j}(n_{1j}, \theta_{1j}) \times B_{2j}(n_{2j}, \theta_{2j}) | j = 1, \cdots, k\}$, $\boldsymbol{\theta} = \{(\theta_{1j}, \theta_{2j}); j = 1, \cdots, k\}$. Consider the jth 2×2 table. Denote the number of exposed cases and exposed controls by x_{1j}

and x_{2j} from n_{1j} and n_{2j} samples, respectively. Then the likelihood function for $\boldsymbol{\theta}$ given $\mathbf{x} = \{(x_{1j}, x_{2j}); j = 1, \cdots, k\}$ is

$$L(\boldsymbol{\theta}|\mathbf{x}) = \prod_{i=1}^{2}\prod_{j=1}^{k} \theta_{ij}^{x_{ij}}(1-\theta_{ij})^{n_{ij}-x_{ij}} \tag{12.2.1}$$

We are interested in the estimation of the "odds ratio" (OR) vector, $\boldsymbol{\Psi} = (\psi_1, \cdots, \psi_k)'$, where

$$\psi_j = \frac{\theta_{1j}(1-\theta_{2j})}{\theta_{2j}\cdots(1-\theta_{1j})}, \quad j = 1, \cdots, k, \tag{12.2.2}$$

when it is suspected that $\psi_1 = \cdots = \psi_k = \psi_0$; that is, $\boldsymbol{\Psi} = \psi_0 \mathbf{1}_k$, ψ_0 is the common odds ratio of the k stratum involving the k, 2×2 tables

	1				j				k		
Exp	x_{11}	$n_{11}-x_{11}$	n_{11}	$\cdots$	x_{1j}	$n_{1j}-x_{1j}$	n_{1j}	$\cdots$	x_{1k}	$n_{1k}-x_{1k}$	n_{1k}
Control	x_{21}	$n_{21}-x_{21}$	n_{21}		x_{2j}	$n_{2j}-x_{2j}$	n_{2j}		x_{2k}	$n_{2k}-x_{2k}$	n_{2k}

(12.2.3)

Using the likelihood function (12.2.1), we can write the **unrestricted maximum likelihood estimator** (UMLE) of $\boldsymbol{\Psi}$ as

$$\tilde{\boldsymbol{\Psi}}_n = (\tilde{\psi}_1, \cdots, \tilde{\psi}_k)' \tag{12.2.4a}$$

where

$$\tilde{\psi}_j = \frac{\tilde{\theta}_{1j}(1-\tilde{\theta}_{2j})}{\tilde{\theta}_{2j}(1-\tilde{\theta}_{1j})}, \quad \tilde{\theta}_{1j} = \frac{x_{1j}}{n_{1j}}, \quad \text{and} \quad \tilde{\theta}_{2j} = \frac{x_{2j}}{n_{2j}}, \quad j = 1, \cdots, k. \tag{12.2.4b}$$

If $\psi_1 = \cdots = \psi_k = \psi_0$ holds, then maximizing $L(\boldsymbol{\theta}|\mathbf{x})$ subject to $\boldsymbol{\Psi}_0 = \psi_0 \mathbf{1}_k$ yields the **restricted MLE** (RMLE) of $\boldsymbol{\Psi}_0$, denoted $\hat{\boldsymbol{\Psi}}_{ML}$. The estimator is complicated to obtain, as it turns out to be the solution of a system of $(k-1)$ simultaneous cubic equations. To avoid this difficulty, Gart (1962) proposed a **noniterative solution.** The **restricted estimator**(RE) $\hat{\boldsymbol{\Psi}}_{0n} = \psi_{0n}\mathbf{1}_k$ is obtained by considering the **information matrix** elements for ψ_0 as follows:

$$E\Big[\frac{\partial^2 \log L}{\partial \psi_0^2}\Big] = \psi_0^2 \sum_{j=1}^{k} n_{2j}\theta_{2j}(1-\theta_{2j}), \quad E\Big[\frac{\partial^2 \log L}{\partial \theta_{1j}\partial \theta_{2j}}\Big] = 0, \quad j \neq k, \tag{12.2.5a}$$

$$E\Big[\frac{\partial^2 \log L}{\partial \Psi_0 \partial \theta_{1j}}\Big] = -\psi_0^{-1}\frac{n_{2j}\theta_{2j}(1-\theta_{2j})}{\theta_{1j}(1-\theta_{1j})}, \tag{12.2.5b}$$

$$E\Big[\frac{\partial^2 \log L}{\partial \theta_{1j}^2}\Big] = \frac{n_{1j}\theta_{1j}(1-\theta_{1j}) + n_{2j}\theta_{2j}(1-\theta_{2j})}{\theta_{1j}^2(1-\theta_{1j})}, \tag{12.2.5c}$$

Further, letting

$$\theta_{2j} = \frac{\theta_{1j}}{\theta_{1j} + \psi_0(1-\theta_{1j})}, \quad j = 1, \cdots, k, \tag{12.2.6}$$

we invert the information matrix to obtain the **asymptotic variance** of the RMLE, $\hat{\psi}_{ML}$ as

$$AVar(\hat{\psi}_{ML}) = \psi_0^2 w^{-1}, \quad w = w_1 + \cdots + w_k, \tag{12.2.7}$$

where

$$w_j^{-1} = \frac{1}{n_{1j}\theta_{1j}(1-\theta_{1j})} + \frac{1}{n_{2j}\theta_{2j}(1-\theta_{2j})}, \quad j = 1, \cdots, k. \tag{12.2.8}$$

Then there are three possible **noniterative** estimators, $\hat{\psi}_{0n}$ of ψ_0 which are given by

(i) $\hat{\psi}_{0n} = \hat{w}^{-1}\sum_{j=1}^{k}\hat{w}_j\tilde{\psi}_j, \ \hat{w} = \hat{w}_1 + \cdots + \hat{w}_k$

(ii) $\hat{\psi}_{0n} = \exp\left(\hat{w}^{-1}\sum_{j=1}^{k}\hat{w}_j ln\tilde{\psi}_j\right)$, and (iii) $\hat{\psi}_{0n} = \frac{\hat{w}}{\sum_{j=1}^{k}\hat{w}_j\tilde{\psi}_j^{-1}}$, (12.2.9)

where the estimates of $\hat{w}_j (j = 1, \cdots, k)$ are given by

$$\hat{w}_j^{-1} = \frac{1}{n_{1j}\tilde{\theta}_{1j}(1-\tilde{\theta}_{1j})} + \frac{1}{n_{2j}\tilde{\theta}_{2j}(1-\tilde{\theta}_{2j})}. \tag{12.2.10}$$

We can use any one of these estimators in our study. We prefer to use (i) in our discussion. The asymptotic variance of $\hat{\psi}_{0n}$ given by (12.2.7) is the same as that of $\hat{\Psi}_{ML}$. Thus, we can write the restricted estimator as

$$\hat{\mathbf{\Psi}}_{0n} = (\hat{\psi}_{0n}, \cdots, \hat{\psi}_{0n})' = \hat{\psi}_{0n}\mathbf{1}_k. \tag{12.2.11}$$

The next step is to test the null hypothesis $H_0 : \Psi = \psi_0\mathbf{1}_k$ against the alternatives $H_A : \mathbf{\Psi} \neq \psi_0\mathbf{1}_k$. For this, we consider the Wald-type test-statistic defined by the departure statistic due to Gupta and Saleh (1997) given by

$$D_n = n\hat{\psi}_{0n}^{-2}(\tilde{\mathbf{\Psi}}_n - \hat{\mathbf{\Psi}}_{0n})'\hat{\mathbf{\Omega}}_n(\tilde{\mathbf{\Psi}}_n - \hat{\mathbf{\Psi}}_{0n})'; \quad n = \sum_{i=l}^{k}(n_{1i} + n_{2i}), \tag{12.2.12}$$

where $\hat{\mathbf{\Omega}}_n = \text{Diag}(\hat{w}_1, \cdots, \hat{w}_k)$, which is a consistent estimator of $\mathbf{\Omega} = \text{Diag}(w_1, \cdots, w_k)$. Under the null hypothesis, D_n closely approximates the central chi-square distribution with $(k-1)$ d.f. Thus, we can define the following class of estimators of $\mathbf{\Psi}$

$$\mathbf{\Psi}_n^* = \tilde{\mathbf{\Psi}}_n - (\tilde{\mathbf{\Psi}}_n - \hat{\mathbf{\Psi}}_{0n})g(D_n), \tag{12.2.13}$$

where $g(D_n)$ is a function of the statistic, D_n:

(i) If $g(D_n) = 0$, then $\mathbf{\Psi}_n^* = \tilde{\mathbf{\Psi}}_n$. (12.2.14a)

(ii) If $g(D_n) = 1$, then $\mathbf{\Psi}_n^* = \hat{\mathbf{\Psi}}_{0n}$. (12.2.14b)

(iii) If $g(D_n) = I(D_n < \chi^2_{k-1}(\alpha))$,
then $\mathbf{\Psi}_n^* = \hat{\mathbf{\Psi}}_n^{\text{PT}} = \tilde{\mathbf{\Psi}}_n - (\tilde{\mathbf{\Psi}}_n - \hat{\mathbf{\Psi}}_n) I(D_n < \chi^2_{k-1}(\alpha))$. (12.2.14c)

(iv) If $g(D_n) = dD_n^{-1}$, then $\mathbf{\Psi}_n^* = \hat{\mathbf{\Psi}}_n^S = \tilde{\mathbf{\Psi}}_n - d(\tilde{\mathbf{\Psi}}_n - \hat{\mathbf{\Psi}}_{0n}) D_n^{-1}$, $d = k-3$.

(v) If $g(D_n) = 1 - (1 - dD_n^{-1}) I(D_n > d)$,
then $\mathbf{\Psi}_n^* = \hat{\mathbf{\Psi}}_n + (\tilde{\mathbf{\Psi}}_n - \hat{\mathbf{\Psi}}_{0n})(1 - dD_n^{-1}) I(D_n > d)$, (12.2.14d)
or $\mathbf{\Psi}_n^* = \hat{\mathbf{\Psi}}_n^S - (\tilde{\mathbf{\Psi}}_n - \hat{\mathbf{\Psi}}_{0n})(1 - dD_n^{-1}) I(D_n < d)$. (12.2.14e)

12.2.3 Asymptotic Theory of the Estimators and the Test-Statistics

Note that $\tilde{\mathbf{\Psi}}_n$ and $\hat{\mathbf{\Psi}}_{0n}$ are consistent estimators of $\mathbf{\Psi}$ and $\mathbf{\Psi}_0 = (\psi_0, \cdots, \psi_0)'$, respectively. Similarly, since D_n is a **consistent** test-statistic, under fixed alternatives,

$$\mathbf{\Psi}_{\boldsymbol{\delta}} = \psi_0 \mathbf{1}_k + \boldsymbol{\delta}, \tag{12.2.15}$$

we obtain

$$D_n = n\hat{\psi}_{0n}^{-2} (\tilde{\mathbf{\Psi}} - \mathbf{\Psi}_{0n})' \hat{\mathbf{\Omega}}_n^{-1} (\tilde{\mathbf{\Psi}} - \mathbf{\Psi}_{0n}) \to \infty \text{ as } n \to \infty. \tag{12.2.16}$$

Consequently, for fixed alternatives $\mathbf{\Psi}_{\boldsymbol{\delta}} = \psi_0 \mathbf{1}_k + \boldsymbol{\delta}$, we have

$$P_{\mathbf{\Psi}_{\boldsymbol{\delta}}}\{D_n > x\} \to 1 \quad \text{as} \quad n \to \infty \tag{12.2.17}$$

for all $x \geq 0$. Further, for fixed alternatives, $\mathbf{\Psi}_{\boldsymbol{\delta}}$.

$$n\hat{\psi}_{0n}^{-2} \| (\hat{\mathbf{\Psi}}_n^{\text{PT}} - \tilde{\mathbf{\Psi}}_n) \|^2_{\hat{\mathbf{\Omega}}_n} = D_n I(D_n < \chi^2_{k-1}(\alpha)) \leq \chi^2_{k-1}(\alpha) I(D_n < \chi^2_{k-1}(\alpha)).$$

Then, taking expectation on both sides, we have

$$P[D_n < \chi^2_{k-1}(\alpha)] \to 0 \quad \text{as} \quad n \to \infty \text{ by (12.2.17)} \tag{12.2.18a}$$

Similarly,

$$n\hat{\psi}_{0n}^{-2} \| \hat{\mathbf{\Psi}}_n^S - \tilde{\mathbf{\Psi}}_n \|^2_{\hat{\Omega}_n} = d^2 D_n^{-1} \text{ and } E[D_n^{-1}] \to 0 \text{ as } n \to \infty, \tag{12.2.18b}$$

and

$$\begin{aligned} n\tilde{\psi}_{0n}^{-2} \| \hat{\mathbf{\Psi}}_n^{S+} - \tilde{\mathbf{\Psi}}_n \|^2_{\hat{\mathbf{\Omega}}_n} &= d^2 D_n^{-1} I(D_n > d) + D_n^{-1} I(D_n > d) \\ &= d^2 D_n^{-1} + (1 - d^2) D_n^{-1} I(D_n < d). \end{aligned} \tag{12.2.18c}$$

The r.h.s. of (12.2.18a,18b) tends to 0 in the first mean as $n \to \infty$. Hence, under fixed alternatives $\mathbf{\Psi}_{\boldsymbol{\delta}}$, the asymptotic distribution of the estimators are equivalent to that of $\tilde{\mathbf{\Psi}}_n$. Further,

$$(\hat{\Psi}_{0n} - \Psi) \xrightarrow{\mathrm{P}} \eta = (\psi_0 \mathbf{1}_k - \Psi) \quad \text{as } n \to \infty \tag{12.2.19a}$$

$$\text{so that } n\hat{\psi}_{0n}^{-2} \|\hat{\Psi}_{0n} - \Psi\|^2_{\hat{\Omega}_n} \to \infty \quad \text{as } n \to \infty. \tag{12.2.19b}$$

Also,

$$E[n(\tilde{\Psi}_n - \Psi)(\tilde{\Psi}_n - \Psi)'] = \psi_0^2 \Omega^{-1}, \quad \Omega = \mathrm{Diag}(w_1, \cdots, w_k), \tag{12.2.20}$$

which is bounded,

$$n\hat{\psi}_{0n}^{-2}(\tilde{\Psi}_n - \Psi)\Omega(\tilde{\Psi}_n - \Psi) < \infty \tag{12.2.21}$$

as $n \to \infty$. Thus, we have the following theorem:

Theorem 1. Under fixed alternatives $\Psi_\delta = \psi_0 \mathbf{1}_k + \delta$,

(i) $\hat{\psi}_{0n}\mathbf{1}_k$ has **unbounded** risk, (12.2.22)

(ii) $\hat{\Psi}_n^{\mathrm{PT}}$, $\hat{\Psi}_n^{S}$ and $\hat{\Psi}_n^{S+}$ are all asymptotically equivalent to $\tilde{\Psi}_n$ in distribution and have bounded risks.

In order to compare the five estimators, we consider the asymptotic setup along with the quadratic loss function

$$L(\Psi_n^*; \Psi) = n(\hat{\Psi}_n^* - \Psi)'\mathbf{Q}(\hat{\Psi}_n^* - \Psi), \tag{12.2.23}$$

where Ψ_n^* is any of the estimators for different choices of the g-function given at (12.2.13) and $\mathbf{Q}$ is a positive-semidefinite $(k \times k)$ matrix. Then the asymptotic risk of Ψ_n^* is given by

$$R(\Psi_n^* : \mathbf{Q}) = \mathrm{tr}(\mathbf{Q}\mathbf{V}_n^*), \tag{12.2.24}$$

where $\mathbf{V}_n^*$ is the asymptotic distributional dispersion matrix of $\sqrt{n}(\Psi_n^* - \Psi)$. Further, Ψ_n^* will be called an asymptotically inadmissible estimator of Ψ if there exists an alternative estimator Ψ_n^0 such that

$$R(\Psi_n^* : \mathbf{Q}) > R(\Psi_n^0 : \mathbf{Q}) \quad \forall \Delta^2, \tag{12.2.25}$$

where Δ^2 is the departure function of the parameters of the models from the null hypothesis, with strict inequality for some Δ^2. The problem is the computation of the asymptotic distributional quadratic risks (ADQRs) of estimators, Ψ_n^*. To enable such a computation, we consider a sequence $\{K_{(n)}\}$ of local alternatives, namely

$$K_{(n)} : \Psi_{(n)} = \psi_0 \mathbf{1}_k + n^{-1/2}\xi \tag{12.2.26}$$

with a real finite vector ξ. Note that for $\xi = \mathbf{0}$, we have the null hypothesis. For an estimator Ψ_n^*, the asymptotic distribution of $\sqrt{n}(\Psi_n^* - \Psi)$ is given by

$$G_n^*(x) = \lim_{n \to \infty} P\{\sqrt{n}(\Psi_n^* - \Psi) \le x | K_{(n)}\} \tag{12.2.27}$$

whenever the limit exists. Also, let the asymptotic distributional mean square error (ADMSE) matrix be given by

$$V_n^* = \int \cdots \int \mathbf{x}\mathbf{x}' dG^*(\mathbf{x}). \tag{12.2.28}$$

Then the ADQR of $\mathbf{\Psi}_n^*$ is given by

$$\text{ADQR}(\mathbf{\Psi}_n^*; \mathbf{Q}) = \text{tr}(\mathbf{Q}\mathbf{V}_n^*). \tag{12.2.29}$$

Note that for fixed alternatives, all estimators of $\mathbf{\Psi}$ are risk-equivalent to $\tilde{\mathbf{\Psi}}_n$ (except RE, which has unbounded risk) due to the consistency of the proposed test-statistic $\mathcal{L}_n$. Thus, our results allow us to compute biases and risks of estimators under the local alternatives presented above. First, we consider the asymptotic distribution under $\{K_{(n)}\}$ of some of the basic random variables related to the estimators in the form of a theorem given below.

Theorem 2. (Gupta and Saleh 1997). Under $\{K_{(n)}\}$ and the assumed regularity conditions, as $n \to \infty$, the following results hold:

(i) $\mathbf{X}_n = \sqrt{n}(\tilde{\psi}_n - \mathbf{\Psi}_0 \mathbf{1}_k) \sim \mathcal{N}_k(\boldsymbol{\xi}, \psi_0^2 \mathbf{\Omega}^{-1})$, $\mathbf{\Omega} = \text{Diag}(\omega_1, \cdots, \omega_k)$,

where $\omega_i^{-1} = \frac{1}{\lambda_{1j}\theta_{1j}(1-\theta_{1j})} + \frac{1}{\lambda_{2j}\theta_{2j}(1-\theta_{2j})}$, $j = 1, \cdots, k$, and

$\frac{n_{ij}}{n} \to \lambda_{ij} (< \infty)$; $i = 1, 2$, $j = 1, \cdots, k$.

(ii) $\mathbf{Y}_n = \sqrt{n}(\tilde{\mathbf{\Psi}}_n - \hat{\mathbf{\Psi}}_{0n}) \sim \mathcal{N}(-\mathbf{J}\boldsymbol{\xi}, \psi_0^2 \mathbf{\Omega}^{-1}\mathbf{J}')$, $\mathbf{J} = \frac{\mathbf{1}_k \mathbf{1}_k'}{u} \mathbf{\Omega}$.

(iii) $\mathbf{Z}_n = \sqrt{n}(\hat{\mathbf{\Psi}}_{0n} - \mathbf{\Psi}_0)\mathbf{1}_k \sim \mathcal{N}_k\left(\frac{\mathbf{1}_k \mathbf{1}_k'}{u} \mathbf{\Omega}\boldsymbol{\xi}, \psi_0^2 w^{-1} \mathbf{1}_k \mathbf{1}_k'\right)$.

(iv) $\begin{pmatrix} \mathbf{X}_n \\ \mathbf{Y}_n \end{pmatrix} \sim \mathcal{N}_{2k}\left\{\begin{pmatrix} \boldsymbol{\xi} \\ -\mathbf{J}\boldsymbol{\xi} \end{pmatrix}; \psi_0^2 \begin{pmatrix} \mathbf{\Omega}^{-1} & \mathbf{\Omega}^{-1}\mathbf{J}' \\ \mathbf{J}\mathbf{\Omega}^{-1} & \mathbf{\Omega}^{-1}\mathbf{J}' \end{pmatrix}\right\}$.

(v) $\begin{pmatrix} \mathbf{Z}_n \\ \mathbf{Y}_n \end{pmatrix} \sim \mathcal{N}_{2k}\left\{\begin{pmatrix} \frac{\mathbf{1}_k \mathbf{1}_k'}{u} \mathbf{\Omega}\boldsymbol{\xi} \\ -\mathbf{J}\boldsymbol{\xi} \end{pmatrix}; \psi_0^2 \begin{pmatrix} \mathbf{\Omega}^{-1}\mathbf{1}_k\mathbf{1}_k' & \mathbf{0} \\ \mathbf{0} & \mathbf{\Omega}^{-1}\mathbf{J}' \end{pmatrix}\right\}$.

(vi) $\lim_{n\to\infty} P(D_n \le x | K_{(n)}) = H_{p-1}(x : \Delta^2)$, $\Delta^2 = \psi_0^{-2} \boldsymbol{\delta}' \mathbf{\Omega} \boldsymbol{\delta}$,
where $\boldsymbol{\delta} = \mathbf{J}\boldsymbol{\xi}$. (12.2.30)

For a proof, see Problem 10.

12.2.4 ADB, ADQB, ADMSE, and ADQR of the Estimators

With Theorem 12.2.2, we can obtain the ADB, ADQB, ADMSE, and ADQR of the five estimators given below. First, we note that the expressions for ADB, ADQB, ADMS, ADMSE, and ADQR of $\mathbf{\Psi}_n^*$ are given by

(i) $\mathbf{b}^*(\mathbf{\Psi}_n^*) = -\boldsymbol{\delta} E[g(\chi_{k+1}^2(\Delta^2))]$,

$$B^*(\Psi_n^*) = \Delta^2\{E[g(\chi^2_{k+1}(\Delta^2))]\}^2;$$

(ii) $$\mathbf{M}^* = \psi_0^2\mathbf{\Omega}^{-1} - \psi_0^2\mathbf{\Omega}^{-1}\mathbf{J}'\{2E[g(\chi^2_{k+1}(\Delta^2))] - E[g^2(\chi^2_{k+1}(\Delta^2))]\}$$
$$\boldsymbol{\delta\delta}'\{2E[g(\chi^2_{k+1}(\Delta^2))] - 2E[g(\chi^2_{k+3}(\Delta^2))] + E[g^2(\chi^2_{k+3}(\Delta^2))]\}. \quad (12.2.31)$$

The risk of Ψ_n^* is given by

$$R(\hat{\Psi}_n^*) = \psi_0^2\text{tr}[\mathbf{Q\Omega}^{-1}]$$
$$-\psi_0^2\text{tr}[\mathbf{Q\Omega}^{-1}\mathbf{J}']\{2E[g(\chi^2_{k+1}(\Delta^2))] - E[g^2(\chi^2_{k+1}(\Delta^2))]\}$$
$$\boldsymbol{\delta}'\mathbf{Q}\boldsymbol{\delta}\{2E[g(\chi^2_{k+1}(\Delta^2))] - 2E[g(\chi^2_{k+3}(\Delta^2))] + E[g^2(\chi^2_{k+3}(\Delta^2))]\}.$$

This expression yields the following risks of the five estimators.

(i) $\tilde{\Psi}_n$, **UE** (12.2.32)

$$\mathbf{b}_1(\tilde{\Psi}_n) = \mathbf{0} \text{ and } B_1(\tilde{\Psi}_n) = 0$$

$$\mathbf{M}_1(\tilde{\Psi}_n) = \psi_0^2\mathbf{\Omega}^{-1} \text{ and } R_1(\tilde{\Psi}_n; \mathbf{Q}) = \psi_0^2\text{tr}[\mathbf{Q\Omega}^{-1}]$$

(ii) $\hat{\Psi}_n$, **RE**

$$\mathbf{b}_2(\hat{\Psi}_n) = -\mathbf{J}\boldsymbol{\xi} = -\boldsymbol{\delta} \quad (\boldsymbol{\delta} = \mathbf{J}\boldsymbol{\xi})$$

and $B_1(\hat{\Psi}_n) = \Delta^2$;

$$\mathbf{M}_2(\hat{\psi}_n) = \Psi_0^2 w^{-1}\mathbf{1}_k\mathbf{1}_k' + \boldsymbol{\delta\delta}'$$

and $R_2(\hat{\psi}_n; \mathbf{Q}) = \Psi_0^2 w^{-1}\text{tr}[\mathbf{Q}\mathbf{1}_k\mathbf{1}_k'] + \boldsymbol{\delta}'\mathbf{Q}\boldsymbol{\delta}$.

(iii) $\hat{\Psi}_n^{\text{PT}}$, **PTE**

$$\mathbf{b}_3(\hat{\Psi}_n^{\text{PT}}) = -\boldsymbol{\delta}H_{k+1}(\chi^2_{k-1}(\alpha); \Delta^2)$$

and $B_2(\hat{\Psi}_n) = \Delta^2\{H_{k+1}(\chi^2_{k-1}(\alpha); \Delta^2)\}^2$;

$$\mathbf{M}_3(\hat{\Psi}_n^{\text{PT}}) = \psi_0^2\mathbf{\Omega}^{-1} - \psi_0^2\mathbf{\Omega}^{-1}\mathbf{J}'H_{k+1}(\chi^2_{k-1}(\alpha); \Delta^2)$$
$$+\boldsymbol{\delta\delta}'\{2H_{k+1}(\chi^2_{k-1}(\alpha); \Delta^2) - H_{k+3}(\chi^2_{k-1}(\alpha); \Delta^2)\}$$

and $R_3(\hat{\Psi}_n^{\text{PT}}; \mathbf{Q}) = \psi_0^2\text{tr}(\mathbf{Q\Omega}^{-1}) - \psi_0^2\text{tr}[\mathbf{\Omega}^{-1}\mathbf{J}']H_{k+1}(\chi^2_{k-1}(\alpha); \Delta^2)$
$$+\boldsymbol{\delta}'\mathbf{Q}\boldsymbol{\delta}\{2H_{k+1}(\chi^2_{k-1}(\alpha); \Delta^2) - H_{k+3}(\chi^2_{k-1}(\alpha); \Delta^2)\}.$$

(iv) $\hat{\Psi}_n^S$, **SE**

$$\mathbf{b}_4(\hat{\Psi}_n^S) = -\boldsymbol{\delta}E[\chi^{-1}_{k+1}(\Delta^2)]$$

and $B_4(\hat{\Psi}_n^S) = \Delta^2\{E[\chi^{-1}_{k+1}(\Delta^2)]\}^2$;

$$\mathbf{M}_4(\hat{\boldsymbol{\Psi}}_n^S) = \psi_0^2\boldsymbol{\Omega}^{-1} - (k-3)\psi_0^2\boldsymbol{\Omega}^{-1}\mathbf{J}'\{2E[\chi_{k+1}^{-2}(\Delta^2)]$$
$$-(k-3)E[\chi_{k+1}^{-4}(\Delta^2)]\} + (k-3)(k+1)\boldsymbol{\delta}\boldsymbol{\delta}'E[\chi_{k+3}^{-4}(\Delta^2)]$$

and $R_4(\hat{\boldsymbol{\Psi}}_n^S) = \psi_0^2\mathrm{tr}[\mathbf{Q}\boldsymbol{\Omega}^{-1}] - (k-3)\psi_0^2\mathrm{tr}[\mathbf{Q}\boldsymbol{\Omega}^{-1}\mathbf{J}']\{2E[\chi_{k+1}^{-2}(\Delta^2)]$

$$-(k-3)E[\chi_{k+1}^{-4}(\Delta^2)]\} + (k-3)(k+1)\boldsymbol{\delta}'\mathbf{Q}\boldsymbol{\delta}E[\chi_{k+3}^{-4}(\Delta^2)]. \quad (12.2.32d)$$

(v) $\hat{\boldsymbol{\Psi}}_n^{S+}$, **PRSE**

$$\mathbf{b}_5(\hat{\boldsymbol{\Psi}}_n^{S+}) = -\boldsymbol{\delta}\{(k-3)E[\chi_{k+1}^{-1}(\Delta^2)]$$
$$-E[(1-(k-3)\chi_{k+1}^{-2}(\Delta^2))I(\chi_{k+1}^{2}(\Delta^2) < k-3)]\}$$

and $B_5(\hat{\boldsymbol{\Psi}}_n^{S+}) = \Delta^2\{(k-3)E[\chi_{k+1}^{-1}(\Delta^2)]$

$$-E[(1-(k-3)\chi_{k+1}^{-2}(\Delta^2))I(\chi_{k+1}^{2}(\Delta^2) < k-3)]\}^2;$$

$$\mathbf{M}_5(\hat{\boldsymbol{\Psi}}_n^{S+}) = \mathbf{M}_4(\hat{\boldsymbol{\Psi}}_n^{S})$$
$$-\psi_0^2\boldsymbol{\Omega}^{-1}\mathbf{J}'\{E[(1-(k-3)\chi_{k+1}^{-2}(\Delta^2))^2 I(\chi_{k+1}^{2}(\Delta^2) < k-3)]\}$$
$$+\boldsymbol{\delta}\boldsymbol{\delta}'\{2E[(1-(k-3)\chi_{k+1}^{-2}(\Delta^2))I(\chi_{k+1}^{2}(\Delta^2) < k-3)]$$
$$-E[(1-(k-3)\chi_{k+3}^{-2}(\Delta^2))^2 I(\chi_{k+3}^{2}(\Delta^2) < k-3)]\}$$

and $R_5(\hat{\boldsymbol{\Psi}}_n^{S+}) = R_4(\hat{\boldsymbol{\Psi}}_n^{S})$

$$-\psi_0^2\mathrm{tr}[\mathbf{Q}\boldsymbol{\Omega}^{-1}\mathbf{J}']E[(1-(k-3)\chi_{k+1}^{-2}(\Delta^2))^2 I(\chi_{k+1}^{2}(\Delta^2) < k-3)]$$
$$+\boldsymbol{\delta}'\mathbf{Q}\boldsymbol{\delta}\{2E[(1-(k-3)\chi_{k+1}^{-2}(\Delta^2))I(\chi_{k+1}^{2}(\Delta^2) < k-3)]$$
$$-E[(1-(k-3)\chi_{k+3}^{-2}(\Delta^2))^2 I(\chi_{k+3}^{2}(\Delta^2) < k-3)]\}.$$

Based on the results above, we present the ADQR analysis of the estimators using the following theorem:

Theorem 3. Under $H_0 : \boldsymbol{\Psi} = \psi_0\mathbf{1}_k$ and loss function $L(\boldsymbol{\Psi}_n^* : \boldsymbol{\Psi}) = n(\boldsymbol{\Psi}_n^* - \boldsymbol{\Psi})'\mathbf{Q}(\boldsymbol{\Psi}_n^* - \boldsymbol{\Psi})$ with positive-semi-definite matrix, $\mathbf{Q}$, ADQR of the five estimators ($\tilde{\boldsymbol{\Psi}}_n, \hat{\boldsymbol{\Psi}}_n, \hat{\boldsymbol{\Psi}}_n^{\mathrm{PT}}, \hat{\boldsymbol{\Psi}}_n^{S}$ and $\hat{\boldsymbol{\Psi}}_n^{S+}$) can be ordered as follows:

(i) $R_2(\hat{\boldsymbol{\Psi}}_n;\mathbf{Q}) \le R_5(\hat{\boldsymbol{\Psi}}_n^{S+};\mathbf{Q}) \le R_4(\hat{\boldsymbol{\Psi}}_n^{S};\mathbf{Q}) \le R_1(\tilde{\boldsymbol{\Psi}}_n;\mathbf{Q})$.

(ii) $R_2(\hat{\boldsymbol{\Psi}}_n;\mathbf{Q}) \le R_3(\hat{\boldsymbol{\Psi}}_n^{\mathrm{PT}};\mathbf{Q}) \le R_1(\tilde{\boldsymbol{\Psi}}_n;\mathbf{Q})$.

(iii) $R_3(\hat{\boldsymbol{\Psi}}_n^{\mathrm{PT}};\mathbf{Q}) \le R_5(\hat{\boldsymbol{\Psi}}_n^{S+};\mathbf{Q}) \le R_4(\hat{\boldsymbol{\Psi}}_n^{S};\mathbf{Q}) \le R_1(\tilde{\boldsymbol{\Psi}}_n;\mathbf{Q})$. (12.2.33)

The ordering of these estimators depends on the level of significance α for the PTE.

Proof. Recall

$$\Psi_n^* = \tilde{\Psi}_n - (\tilde{\Psi}_n - \hat{\Psi}_n)g(D_n), \tag{12.2.34}$$

where $g(D_n)$ is decreasing function of D_n, the test-statistic for H_0. If

$$\begin{aligned} g(D_n) &= 0, \text{ then } \Psi_n^* = \tilde{\Psi}_n \quad \text{(UE)} \\ &= 1, \text{ then } \Psi_n^* = \hat{\Psi}_n \quad \text{(RE)} \\ &= I(D_n < \chi^2_{k-1}(\alpha)), \text{ then } \Psi_n^* = \hat{\Psi}_n^{\text{PT}} \\ &= (k-3)D_n^{-1}, \text{ then } \Psi_n^* = \hat{\Psi}_n^{S} \\ &= 1-(1-(k-3)D_n^{-1})I(D_n > k-3), \text{ then } \Psi_n^* = \hat{\Psi}_n^{S+}. \end{aligned}$$

Now, under H_0, the ADQR of Ψ_n^* is

$$\begin{aligned} &\psi_0^2 w^{-1} \operatorname{tr}[\mathbf{Q}\mathbf{1}_k\mathbf{1}_k']\{2E[g(\chi^2_{k+1}(\Delta^2))] \\ &-E[g^2(\chi^2_{k+1}(\Delta^2))]\} + \psi_0^2 \operatorname{tr}[\mathbf{Q}\boldsymbol{\Omega}^{-1}]E[1-g(\chi^2_{k+1}(\Delta^2))]^2, \end{aligned}$$

which is a increasing function of $E[1-g(\chi^2_{k+1}(\Delta^2))]^2$. Hence, we get

$$\begin{aligned} E\big[1-g(\chi^2_{k+1}(\Delta^2))\big]^2 &= 0 \text{ if } \Psi_n^* = \hat{\Psi}_n \\ &= 1 \text{ if } \Psi_n^* = \tilde{\Psi}_n \\ &= 1 - H_{k+1}(\chi^2_{k-1}(\alpha);0) \text{ if } \Psi_n^* = \hat{\Psi}_n^{\text{PT}} \\ &= \int_0^\infty \{1-(k-3)x^{-1}\}^2 dH_{k+1}(x;0) \text{ if } \Psi_n^* = \hat{\Psi}_n^{S} \\ &= \int_{k-3}^\infty \{1-(k-3)x^{-1}\}^2 dH_{k+1}(x;0) \text{ if } \Psi_n^* = \hat{\Psi}_n^{S+}, \end{aligned} \tag{12.2.35}$$

and the assertion of the theorem holds.

The next theorem deals with dominance of the Stein-type estimators.

Theorem 4. Let $k > 4$ and the matrix $\mathbf{Q}$ satisfy the condition

$$\left\{\mathbf{Q} : \frac{\operatorname{tr}[\mathbf{Q}\boldsymbol{\Omega}^{-1}]}{Ch_{max}(\mathbf{Q}\boldsymbol{\Omega}^{-1})} \geq \frac{k+1}{2}\right\}. \tag{12.2.36}$$

Then under $\{K_{(n)}\}$, the estimators, $\hat{\Psi}_n^{S}$ and $\hat{\Psi}_n^{S+}$ outperform $\tilde{\Psi}_n$ with respect to ADQR and

- $$R_5(\hat{\Psi}_n^{S+};\mathbf{Q}) \leq R_4(\hat{\Psi}_n^{S};\mathbf{Q}) \leq R_1(\tilde{\Psi}_n;\mathbf{Q}). \tag{12.2.37}$$

Proof. The ADQR of $\hat{\boldsymbol{\Psi}}_n^S$ can be written as

$$R_4(\hat{\boldsymbol{\Psi}}_n^S : \mathbf{Q}) = \psi_0^2 \operatorname{tr}[\mathbf{Q}\boldsymbol{\Omega}^{-1}] - (k-3)\psi_0^2 w^{-1} \operatorname{tr}[\mathbf{Q}\boldsymbol{\Omega}^{-1}\mathbf{J}']\Big\{(k-3)E[\chi_{k+3}^{-4}(\Delta^2)]$$

$$+\Big[1 - \frac{(k+1)\boldsymbol{\xi}'\mathbf{Q}\boldsymbol{\xi}}{2\Delta^2 \operatorname{tr}[\mathbf{Q}\boldsymbol{\Omega}^{-1}\mathbf{J}']}\Big]2\Delta^2 E[\chi_{k+3}^{-4}(\Delta^2)]\Big\} \qquad (12.2.38)$$

The risk-difference $R_1(\tilde{\boldsymbol{\Psi}}_n;\mathbf{Q}) - R_4(\hat{\boldsymbol{\Psi}}_n;\mathbf{Q}) \geq 0$ for all $(\Delta^2, \mathbf{Q})$, $\mathbf{Q}$ satisfying (12.2.36). Hence,

$$R_4(\hat{\boldsymbol{\Psi}}_n^S;\mathbf{Q}) \leq R_1(\tilde{\boldsymbol{\Psi}}_n;\mathbf{Q}) \quad \forall(\Delta^2,\mathbf{Q}). \qquad (12.2.39)$$

Similarly, the risk-difference

$$\begin{aligned}
&R_5(\hat{\boldsymbol{\Psi}}_n^{S+};\mathbf{Q}) - R_4(\hat{\boldsymbol{\Psi}}_n^S;\mathbf{Q}) \\
&\quad = -\psi_0^2 \operatorname{tr}[\mathbf{Q}\boldsymbol{\Omega}^{-1}\mathbf{J}']E[(1-(k-3)\chi_{k+1}^{-2}(\Delta^2))^2 I(\chi_{k+1}^2(\Delta^2) < k-3)] \\
&\qquad -\boldsymbol{\delta}'\mathbf{Q}\boldsymbol{\delta}\{2E[(1-(k-3)\chi_{k+1}^{-2}(\Delta^2))I(\chi_{k+1}^2(\Delta^2) < k-3)]\} \\
&\qquad -E[(1-(k-3)\chi_{k+3}^{-2}(\Delta^2))^2 I(\chi_{k+3}^2(\Delta^2) < k-3)]\} \leq 0. \qquad (12.2.40)
\end{aligned}$$

Hence,

$$R_5(\hat{\boldsymbol{\Psi}}_n^{S+};\mathbf{Q}) \leq R_4(\hat{\boldsymbol{\Psi}}_n^S;\mathbf{Q}) \leq R_1(\tilde{\boldsymbol{\Psi}}_n;\mathbf{Q}). \qquad (12.2.41)$$

The next theorem states the range of dominance of $\boldsymbol{\Psi}_n^*$ over $\tilde{\boldsymbol{\Psi}}_n$ (excluding $R_4(\hat{\boldsymbol{\Psi}}_n^S;\mathbf{Q})$ and $R_5(\hat{\boldsymbol{\Psi}}_n^{S+};\mathbf{Q})$).

Theorem 5. Under $\{K_{(n)}\}$ and positive-semi-definite, $\mathbf{Q}$, $\hat{\boldsymbol{\Psi}}_n$ and $\hat{\boldsymbol{\Psi}}_n^{\mathrm{PT}}$ have smaller ADQR than $\tilde{\boldsymbol{\Psi}}_n$ whenever

$$\Delta^2 \leq \frac{\operatorname{tr}[\mathbf{Q}\boldsymbol{\Omega}^{-1}\mathbf{J}']\{2E[g(\chi_{k+1}^2)] - E[g^2(\chi_{k+1}^2(\Delta^2))]\}}{Ch_{max}(\mathbf{Q}\boldsymbol{\Omega}^{-1})\{2E[g(\chi_{k+1}^2)] - 2E[g(\chi_{k+3}^2(\Delta^2))] + E[g^2(\chi_{k+3}^2(\Delta^2))]\}}; \qquad (12.2.42)$$

otherwise, $\tilde{\boldsymbol{\Psi}}_n$ dominates $\hat{\boldsymbol{\Psi}}_n$ and $\hat{\boldsymbol{\Psi}}_n^{\mathrm{PT}}$,

$$\Delta^2 \geq \frac{\operatorname{tr}[\mathbf{Q}\boldsymbol{\Omega}^{-1}\mathbf{J}']\{2E[g(\chi_{k+1}^2)] - E[g^2(\chi_{k+1}^2(\Delta^2))]\}}{Ch_{min}(\mathbf{Q}\boldsymbol{\Omega}^{-1})\{2E[g(\chi_{k+1}^2)] - 2E[g(\chi_{k+3}^2(\Delta^2))] + E[g^2(\chi_{k+3}^2(\Delta^2))]\}}. \qquad (12.2.43)$$

The theorem is proved by considering the risk-difference of the estimators and noting that

$$Ch_{min}(\mathbf{Q}\boldsymbol{\Omega}^{-1}) \leq \frac{\boldsymbol{\delta}'\mathbf{Q}\boldsymbol{\delta}}{\boldsymbol{\delta}'\boldsymbol{\Omega}\boldsymbol{\delta}} \leq Ch_{max}(\mathbf{Q}\boldsymbol{\Omega}^{-1}). \qquad (12.2.44)$$

From the theorem above it is clear that PRSE, $\hat{\boldsymbol{\Psi}}_n^{S+}$ is the best practical application if $k \geq 4$, while for $k < 4$, $\hat{\boldsymbol{\Psi}}_n^{\mathrm{PT}}$ is a reasonable compromise.

12.2.5 Estimation of Odds Ratio under Uncertain Zero Partial Association

In this case, we like to estimate $\boldsymbol{\Psi} = (\psi_1, \cdots, \psi_k)'$ when it is suspected that $H_0: \psi_1 = \cdots = \psi_k = 1$. This null hypothesis holds if and only if $\theta_{1j} = \theta_{2j}, j = 1, \cdots, k$, meaning that in each stratum, the probability of success is the same for the cases (treated) and control groups. This is known as θ-homogeneity. The large sample test for θ-homogeneity may be given by the asymptotic distribution of the test-statistic $\mathcal{L}_n^*$, defined by

$$\mathcal{L}_n^* = n(\tilde{\boldsymbol{\Psi}}_n - \mathbf{1}_k)'\hat{\boldsymbol{\Omega}}_n(\tilde{\boldsymbol{\Psi}}_n - \mathbf{1}_k), \mathbf{1}_k' = (1, \cdots, 1)', \tag{12.2.45}$$

where the components of $\hat{\boldsymbol{\Omega}}_n$ are given by (12.2.8) and $\tilde{\boldsymbol{\Psi}}_n$ is the **unrestricted** estimator of $\boldsymbol{\Psi}$, since the common value of the odds ratio is 1. The asymptotic distribution of $\mathcal{L}_n^*$ is a chi-square distribution under the k d.f.

The preliminary test estimator of $\boldsymbol{\Psi}$ is defined by

$$\hat{\boldsymbol{\Psi}}_n^{\text{PT}} = \mathbf{1}_k I(\mathcal{L}_n^* < \mathcal{L}_{n,\alpha}^*) + \tilde{\boldsymbol{\Psi}}_n I(\mathcal{L}_n^* > \mathcal{L}_{n,\alpha}^*), \tag{12.2.46}$$

where $\mathcal{L}_{n,\alpha}^*$ is the α-level critical value from the distribution of $\mathcal{L}_n^*$. Similarly, the shrinkage estimators (SE) of $\boldsymbol{\Psi}$ are defined by

$$\hat{\boldsymbol{\Psi}}_n^{S} = \mathbf{1}_k + \{1 - (k-2)\mathcal{L}_n^{*-1}\}(\tilde{\boldsymbol{\Psi}}_n - \mathbf{1}_k), \tag{12.2.47}$$

and

$$\hat{\boldsymbol{\Psi}}_n^{S+} = \mathbf{1}_k + \{1 - (k-2)\mathcal{L}_n^{*-1}\}I(\mathcal{L}_n^* > k-2)(\tilde{\boldsymbol{\Psi}}_n - \mathbf{1}_k) \tag{12.2.48}$$

which is the positive-rule shrinkage estimator of $\boldsymbol{\Psi}$.

For the asymptotic distributional biases and risks of the estimators, we use local alternatives of the form

$$K_{(n)}^* : \boldsymbol{\Psi}_{(n)} = \mathbf{1}_k + n^{-1/2}\boldsymbol{\xi}. \tag{12.2.49}$$

Thus, following previous developments in Section 3, we obtain the biases as

$$\begin{aligned}
\mathbf{b}(\tilde{\boldsymbol{\Psi}}_n) &= -\boldsymbol{\delta} H_{k+2}(\chi_k^2(\alpha); \Delta^2),\\
\mathbf{b}(\hat{\boldsymbol{\Psi}}_n^{S}) &= -(k-2)\boldsymbol{\delta} E[\chi_{k+2}^{-2}(\Delta^2)],\\
\mathbf{b}(\hat{\boldsymbol{\Psi}}_n^{S+}) &= -\boldsymbol{\delta}\{H_{k+1}(k-2; \Delta^2) + (k-2)E[\chi_{k+2}^{-2}(\Delta^2)],\\
&\quad -(k-2)E[\chi_{k+2}^{-2}(\Delta^2)I(\chi_{k+2}^{-2} < k-2)]\},
\end{aligned}$$

where $\Delta^2 = \boldsymbol{\delta}'\boldsymbol{\Omega}\boldsymbol{\delta}$. The asymptotic distributional quadratic risks (ADQR) are given by

$$\text{ADQR}(\tilde{\boldsymbol{\Psi}}_n) = \text{tr}[\mathbf{Q}\boldsymbol{\Omega}^{-1}]$$

$$\begin{aligned}
\text{ADQR}(\hat{\boldsymbol{\Psi}}_n^{\text{PT}}) &= \text{tr}[\mathbf{Q}\boldsymbol{\Omega}^{-1}] + \{1 - H_{k+2}(\chi_k^2(\alpha); \Delta^2)\} \\
&\quad + \boldsymbol{\delta}'\mathbf{Q}\boldsymbol{\delta}\{2H_{k+2}(\chi_k^2(\alpha); \Delta^2) - H_{k+4}(\chi_k^2(\alpha); \Delta^2)\} \\
\text{ADQR}(\hat{\boldsymbol{\Psi}}_n^{S}) &= \text{tr}[\mathbf{Q}\boldsymbol{\Omega}^{-1}] - (k-2)\text{tr}[\mathbf{Q}\boldsymbol{\Omega}^{-1}\mathbf{J}']\{2E[\chi_{k+2}^{-2}(\Delta^2)], \\
&\quad -(k-2)E[\chi_{k+4}^{-4}(\Delta^2)] + (k-2)(k+2)\boldsymbol{\delta}'\mathbf{Q}\boldsymbol{\delta}E[\chi_{k+4}^{-4}(\Delta^2)], \\
\text{ADQR}(\hat{\boldsymbol{\Psi}}_n^{S+}) &= ADQR(\hat{\boldsymbol{\Psi}}_n^{S}) \\
&\quad - \text{tr}[\mathbf{Q}\boldsymbol{\Omega}^{-1}]E[(1-(k-2)\chi_{k+2}^{-2}(\Delta^2))^2 I(\chi_{k+2}^{-2} < k-2)] \\
&\quad + \boldsymbol{\delta}'\mathbf{Q}\boldsymbol{\delta}E[2(1-(k-2)\chi_{k+2}^{-2}(\Delta^2))I(\chi_{k+2}^{-2} < k-2)]\} \\
&\quad - E[(1-(k-2)\chi_{k+4}^{-2}(\Delta^2))^2 I(\chi_{k+4}^{-2} < k-2)].
\end{aligned}$$

From these risk expressions, the asymptotic properties can be established as before with similar conclusions.

12.2.6 Odds Ratios: Application to Meta-analysis of Clinical Trials

Meta-analysis is a systematic and quantitative review of results of a set of individual studies intended to integrate their findings. Statistical methods are used as a fundamental tool in reviewing and combining the evidences from various clinical trials in medical research.

There are a number of reasons why meta-analysis is an important technique in clinical trails:

1. Narrative reviews of a set of individual clinical trials can be misleading and have the potential of misleading and distorting trial results.

2. Explosion of research evidence in the form of published trials often are not easily assimilated without formal review.

3. In assessing the benefits of a particular **medical treatment**, judgment should be based on the totality of evidences from well-conducted randomized clinical trials.

4. Since individual clinical trials may involve small samples, a collection of several similar clinical trials would increase the sample to provide reliable evidence on the medical treatment.

Meta-analysis therefore has several objectives: (1) to provide systematic qualitative as well as quantitative summaries of results from individual studies and (2) to combine these results across studies and provide overall interpretation to suggest effectiveness of the medical treatments.

In this section, we consider the principal results from nine randomized trials of **diuretics** on incidences of pre-eclampsia using a fixed effect design. The data shown in the Table 12.2.1 come from a meta-analysis of nine randomized studies investigating the use of diuretics to prevent pre-eclampsia (see Thompson and Pocock, 1991).

Table 12.2.1 Incidence of pre-eclampsia in nine randomized trials

Trials	No of Patients Diuretics	Control	odds ratio (OR)	Weights
Weseley	14/131	14/136	1.0427	6.2657
Flowers	21/385	17/134	0.3971	8.5034
Menzies	14/57	24/48	0.3256	5.6180
Fallis	6/38	18/40	0.2292	3.3454
Cuadros	12/1011	35/760	0.2488	8.7489
Landerman	138/1370	175/1336	0.7430	73.5294
Krans	15/506	20/524	0.7699	8.2850
Tervallis	6/108	2/103	2.9706	1.4569
Campbell	65/153	40/102	1.1449	14.7275

Source: Adapted from Thompson and Pocock (1991)

The object of the studies is to find whether diuretics are effective treatment for pre-eclampsia. The odds ratios and their confidence intervals are presented as initial steps nine studies. OR less than unity represents beneficial effects of diuretics.

In order to combine the nine OR values into one common value, we use the traditional log-scale and consider the modified test of homogeneity of the odds ratios by a chi-square test given by $\chi_8^2 = \sum_{i=1}^{0} w_i(\ell n\tilde{\psi}_i - \ell n\hat{\psi}_0)^2 = 27.31 > \chi_8^2(.05) = 15.51$. Thus, we have the following table for the estimators of the ORs with graphical representation in Fig. 12.2.1a.

Table 12.2.2 Various Estimators of Odd-Ratio

Samples	UE $\tilde{\psi}$	RE $\hat{\psi}_{09}$	PTE $\hat{\psi}^{PT}$	SE/PRSE $\hat{\psi}^S/\hat{\Psi}^{S+}$	95% CI for OR	95% CI using $\hat{\psi}^S/\hat{\psi}^{S+}$
Weseley	1.0427	0.6743	1.0427	0.9474	0.4766–2.2815	0.4330–2.0730
Flowers	0.3971	0.6743	0.3971	0.4461	0.2027–0.7779	0.2771–0.7181
Menzies	0.3256	0.6743	0.3256	0.3821	0.1424–0.7444	0.1671–0.8735
Fallis	0.2292	0.6743	0.2292	0.2905	0.0785–0.6692	0.0995–0.8482
Cuadros	0.2488	0.6743	0.2488	0.3097	0.1283–0.4827	0.1547–0.6007
Landerman	0.7431	0.6743	0.7431	0.7273	0.5863–0.9420	0.5738–0.9219
Krans	0.7699	0.6743	0.7699	0.7478	0.3897–1.5210	0.3785–1.4774
Tervallis	2.9705	0.6743	2.9705	2.1447	0.5857–15.0675	0.4150–10.8492
Campbell	1.1449	0.6743	1.1449	1.0192	0.6871–1.9078	0.5888–1.6349

It may be noted that 95% CI for $\hat{\psi}_0$ is (0.5680 – 0.8005).

Now, we observe that the null hypothesis $H_0 : \psi_1 = \psi_2 = \cdots = \psi_9 = \psi_0$ is rejected at 5% level of significance. Even the Stein method of pulling the 5th value of the OR is not pulling enough towards the common value (meta-analysis) 0.6743. Also, the confidence intervals do not always include the common value 0.6743. Thus, there is enough evidence to suspect heterogeneity of

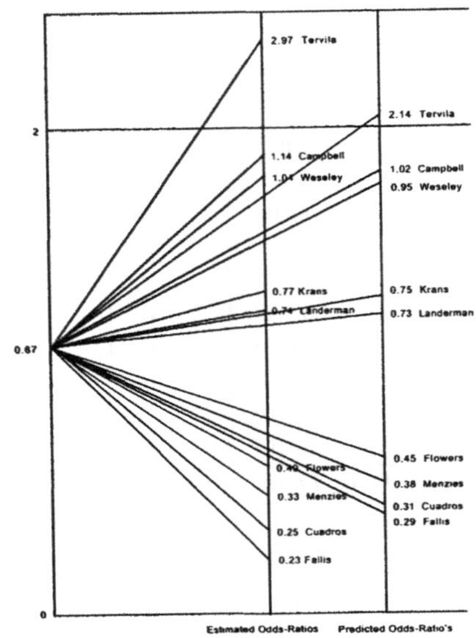

Figure 12.2.1a Predicted odds ratios

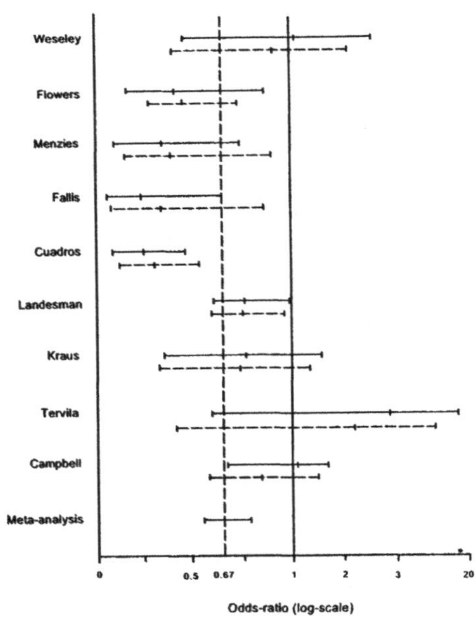

Figure 12.2.1b Confidence Intervals of odds ratios

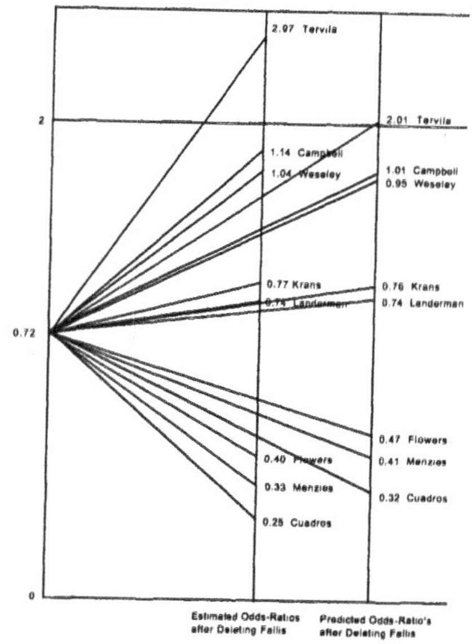

Figure 12.2.2a Predicted Odds Ratios (Deleting Fallis)

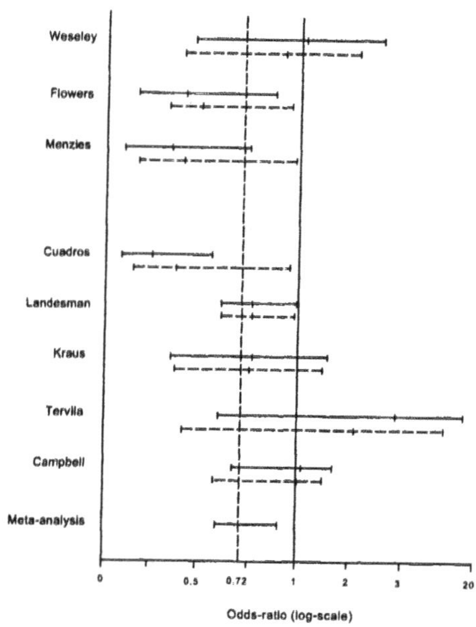

Figure 12.2.2b Confidence Intervals of odds ratios (Deleting Fallis)

nine 2×2 tables. Most of the **heterogeneity** is caused by the OR value 0.2292 (Fallis) or .2488 (Cuadros). The recomputed D_7–statistic yields the value 17.99 by deleting the "Fallis" or "Cuadros" study. But the weight for "Fallis" is less than that of "Cuadros". Hence, "Fallis" is designated as an outlier. Deleting this value, we obtain the common value as $\hat{\psi}_{08} = 0.7244$, with the corresponding chi-square value $D_7 = 17.99$ with 7 d.f. and the null hypothesis is marginally rejected. Thus, based on this D-value we obtain the following modified Table 12.2.3 of the estimators. Note that the 95% CI for $\hat{\Psi}_{08}$ is $0.6065 - 0.8652$.

Table 12.2.3 Revised Estimators of ORs after Deleting "Fallis"

Samples	UE $\tilde{\psi}$	RE $\hat{\psi}_{09}$	PTE $\hat{\psi}^{PT}$	SE/PRSE $\hat{\psi}^{S}/\hat{\psi}^{S+}$	95% CI for OR based on $\hat{\psi}^{S+}$
Weseley	1.0427	0.7244	1.0427	0.9423	0.4307 – 2.0618
Flowers	0.9371	0.7244	0.9371	0.4693	0.2396 – 0.9193
Menzies	0.3256	0.7244	0.3256	0.4066	0.1779 – 0.9296
Cuadros	0.2488	0.7244	0.2488	0.3155	0.1081 – 0.9215
Landerman	0.7431	0.7244	0.7431	0.7378	0.5870 – 0.9273
Tervillis	2.9705	0.7244	2.9705	2.0069	0.3957 – 10.1787
Krans	0.7699	0.7244	0.7699	0.7570	0.3811 – 1.4956
Campbell	1.1449	0.7244	1.1449	1.0081	0.6050 – 1.6000

The chi-square value together with the power of CI of the ORs reveal that 8 data sets are homogeneous and the common OR value is 0.7244 with 95% CI (0.6065 – 0.8652) indicating the degree of benefit of the diuretics to prevent pre-eclampsia via the odds interval (13%–39%). The reader is referred to Saleh et al. (2006) for some details.

12.3 Product of Multinomial Models, Estimation, and Test of Hypothesis

It is common practice in population census and epidemiological surveys to classify data according to two traits to obtain $r \times c$ contingency tables. The usual problem in this case is to estimate the cell probabilities when it is suspected that the two traits may be independent. For the estimation of the cell probabilities under independence structure, we consider the formulation of the model and the estimation strategies in the following subsection.

12.3.1 The Product of Multinomial Models

Let n_{ij} stand for the observed cell frequency for the (i,j)-cell in an $r \times c$ contingency table for $i = 1, \cdots, r\ (\geq 2)$ and $j = 1, \cdots, c\ (\geq 2)$. Let $n_{i+} = \sum_{j=1}^{c} n_{ij}$, and let $n_{+j} = \sum_{i=1}^{r} n_{ij}$ and $\mathbf{n} = \sum_{i=1}^{r} n_{i+} = \sum_{j=1}^{c} n_{+j}$. The probability distribution of the random vector $\mathbf{n} = (n_{11}, \cdots, n_{1c}, \cdots, n_{r1}, \cdots, n_{rc})'$ is given by the product multinomial distribution

$$n!\{\prod_{i=1}^{r}\prod_{j=1}^{c}(n_{ij}!)\}^{-1}\prod_{i=1}^{r}\prod_{j=1}^{c}(\theta_{ij})^{n_{ij}}, \tag{12.3.1}$$

where $\boldsymbol{\theta} = (\theta_{11}, \cdots, \theta_{1c}, \cdots, \theta_{r1}, \cdots, \theta_{rc})'$ with $\boldsymbol{\theta}'\mathbf{1}_{rc} = 1$ being the vector of the true cell probabilities. Define

$$\theta_{i+} = \sum_{j=1}^{c}\theta_{ij}, (1 \le i \le r), \quad \text{and} \quad \theta_{+j} = \sum_{i=1}^{r}\theta_{ij}, (1 \le j \le c). \tag{12.3.2}$$

If the independence structure of the two traits holds, then we specify the null hypothesis as

$$H_0 : \theta_{ij} = \theta_{i+}.\ \theta_{+j} \quad \forall\ (i,j). \tag{12.3.3}$$

Note that in (12.3.1) we have $(rc-1)$ independent parameters, while under H_0, we encounter $(r-1)+(c-1) = r+c-2$ parameters.

12.3.2 Estimation of the Parameters

For the estimation of θ_{ij} in the $r \times c$ contingency table, consider the **unrestricted** MLE of $\boldsymbol{\theta}$ given by

$$\tilde{\boldsymbol{\theta}}_n = (\tilde{\theta}_{11}, \cdots, \tilde{\theta}_{1c}, \cdots, \tilde{\theta}_{r1}, \cdots, \tilde{\theta}_{rc}), \text{ with } \tilde{\theta}_{ij} = \frac{n_{ij}}{n}, \tag{12.3.4}$$

while the marginal probabilities $\theta_{i+}.(1 \le i \le r)$ and $\theta_{+j}(1 \le j \le c)$ are defined by

$$\hat{\theta}_{i+} = \frac{n_{i+}}{n} \text{ and } \hat{\theta}_{+j} = \frac{n_{+j}}{n}. \tag{12.3.5}$$

Then the **restricted** MLE of θ_{ij} under H_0 is given by

$$\hat{\boldsymbol{\theta}}_n = (\hat{\theta}_{11}, \cdots, \hat{\theta}_{1c}, \cdots, \hat{\theta}_{r1}, \cdots, \hat{\theta}_{rc}), \text{ with } \hat{\theta}_{ij} = \hat{\theta}_{i+}\hat{\theta}_{+j} \quad \forall\ (i,j). \tag{12.3.6}$$

12.3.3 Test of Independence in an $r \times c$ Contingency Table

In order to test the null hypothesis

$$H_0 : \theta_{ij} = \theta_{i+}.\theta_{+j} \quad \forall\ (i,j), \tag{12.3.7}$$

we use the departure statistic D_n, defined by

$$D_n = n(\tilde{\boldsymbol{\theta}}_n - \hat{\boldsymbol{\theta}}_n)'\boldsymbol{\Sigma}_n^{-1}(\tilde{\boldsymbol{\theta}}_n - \hat{\boldsymbol{\theta}}_n), \tag{12.3.8}$$

where

$$\boldsymbol{\Sigma}_n = \text{Diag}(\hat{\theta}_{11}, \cdots, \hat{\theta}_{1c}, \cdots, \hat{\theta}_{r1}, \cdots, \hat{\theta}_{rc}). \tag{12.3.9}$$

It may be shown, following Bishop et al. (1975) and Agresti (1990), that under H_0, D_n closely follows the central chi-square distribution with m $(= (r-1)(c-1))$ d.f.

12.3.4 Preliminary Test and Stein-Type Estimators of the Cell Probabilities

Let $d_{n,\alpha}$ be the α-level critical value from the exact distribution of D_n, which may be closely approximated by $\chi^2_m(\alpha)$, and is the α-level critical value from the central chi-square distribution with m d.f. Then we define the class of estimators of $\boldsymbol{\theta}$ as

$$\boldsymbol{\theta}^*_n = \hat{\boldsymbol{\theta}}_n + (1 - g(D_n))(\tilde{\boldsymbol{\theta}}_n - \hat{\boldsymbol{\theta}}_n), \tag{12.3.10}$$

where $g(D_n)$ is a decreasing function of D_n.

(i) If we choose $g(D_n) = I(D_n < \chi^2_m(\alpha))$, we obtain the PTE of $\boldsymbol{\theta}$ as

$$\hat{\boldsymbol{\theta}}_n^{\text{PT}} = \tilde{\boldsymbol{\theta}}_n - (\tilde{\boldsymbol{\theta}}_n - \hat{\boldsymbol{\theta}}_n)I(D_n < \chi^2_m(\alpha)). \tag{12.3.11}$$

(ii) If we set $g(D_n) = dD_n^{-1}$, $d = m - 2$, then we obtain the Stein-type estimator (SE) as

$$\hat{\boldsymbol{\theta}}_n^{S} = \tilde{\boldsymbol{\theta}}_n - d(\tilde{\boldsymbol{\theta}}_n - \hat{\boldsymbol{\theta}}_n)D_n^{-1}.$$

(iii) If we choose $g(D_n) = 1 - (1 - dD_n^{-1})I(D_n > d)$, then we obtain the positive-rule Stein-type estimator (PRSE) as

$$\hat{\boldsymbol{\theta}}_n^{S+} = \hat{\boldsymbol{\theta}}_n + (1 - dD_n^{-1})I(D_n > d)(\tilde{\boldsymbol{\theta}}_n - \hat{\boldsymbol{\theta}}_n).$$

We note that for $g(D_n) = 0$ or 1, we obtain the **unrestricted** and **restricted** estimators of $\boldsymbol{\theta}$, respectively. Our estimators are restricted to the choices above of $g(D_n)$ function.

12.3.5 Bayes and Empirical Bayes Estimation of the Cell Probabilities

In order to obtain the Bayes estimator of $\boldsymbol{\theta}$, we consider a **Dirichlet distribution** as **prior** on $\boldsymbol{\theta}$ with parameter $K\boldsymbol{\eta}$ where K is a scalar with $\boldsymbol{\eta} = (\eta_{11}, \cdots, \eta_{1c}, \cdots, \eta_{r1}, \cdots, \eta_{rc})'$ having marginal row sum $\boldsymbol{\eta}_{i+}$; $i = 1, \cdots, c$, and marginal column sum $\boldsymbol{\eta}_{+j}$; $j = 1, 2, \cdots, r$. The independency configuration is then given by $n_{ij} = n_{i+} \cdot n_{+j}$ $\forall$ (i,j). If K and $\boldsymbol{\eta}$ are known, then the Dirichlet distribution with parameters $(K\boldsymbol{\eta})$ is given by

$$\pi(\boldsymbol{\theta}|K, \boldsymbol{\eta}) = \frac{\Gamma(K)}{\prod_{i=1}^r \prod_{j=1}^c \Gamma(K\eta_{ij})} \prod_{i=1}^r \prod_{j=1}^c (\theta_{ij}^{K\eta_{ij}-1}). \tag{12.3.12}$$

Hence, the joint distribution of $(\mathbf{n}', \boldsymbol{\theta}')'$ is given by

$$\frac{\Gamma(K)}{\prod_{i=1}^r \prod_{j=1}^c \Gamma(K\eta_{ij})} n!\Big\{\prod_{i=1}^r \prod_{j=1}^c n_{ij}!\Big\}^{-1} \prod_{i=1}^r \prod_{j=1}^c (\theta_{ij}^{K\eta_{ij}+n_{ij}-1}). \tag{12.3.13}$$

Accordingly, the marginal density of $\{n_{ij}\}$ is then obtained as

$$\mathcal{M}_1(\mathbf{n}|K,\boldsymbol{\eta}) = \frac{n\,!}{\prod_{i=1}^{r}\prod_{j=1}^{c}(n_{ij})\,!}\prod_{i=1}^{r}\prod_{j=1}^{c}(K\eta_{ij})^{(n_{ij})}/K^{(n)}, \tag{12.3.14}$$

where $a^{(b)} = a(a+1)\cdots(a+b-1)$ with $a^{(0)} = 1$. Consequently, the posterior distribution of $\boldsymbol{\theta}$ given $\mathbf{n}$ for fixed K is given by

$$\pi(\boldsymbol{\theta}|\mathbf{n},K) = c\int_0^1\int_0^1 \pi(\boldsymbol{\theta}|K\boldsymbol{\eta}+\mathbf{n})\mathcal{M}_1(\mathbf{n}|K,\boldsymbol{\eta})\prod_{i=1}^{c}\eta_{i+}^{-1}d\eta_{i+}\prod_{j=1}^{r}\eta_{+j}^{-1}d\eta_{+j}, \tag{12.3.15}$$

where

$$\phi(\boldsymbol{\eta}_c,\boldsymbol{\eta}_r) = \left(\prod_{i=1}^{c}\eta_{i+1}^{-1}\prod_{j=1}^{p}\eta_{+j}^{-1}\right) \tag{12.3.16}$$

is the Jeffreys' noninformative prior on the marginal totals and $\pi(\boldsymbol{\theta}|K\boldsymbol{\eta}+\mathbf{n})$ is the posterior distribution of $\boldsymbol{\theta}$ given $K\boldsymbol{\eta}+\mathbf{n}$. To obtain the Bayes estimator of $\boldsymbol{\theta}$, we compute the posterior means.

Thus, for fixed K, the posterior expectation of θ_{ij} is given by

$$E(\theta_{ij}|\mathbf{n},K) = \frac{n_{ij}+KE(\eta_{ij}|\mathbf{n},K)}{n+K} \tag{12.3.17}$$

and

$$\begin{aligned}\mathrm{Var}(\theta_{ij}|\mathbf{n},K) = {} & \frac{n_{ij}+KE(\eta_{ij}|\mathbf{n},K)}{n+K}\left\{1-\frac{n_{ij}+E(\eta_{ij}|\mathbf{n},K)}{n+K}\right\}\frac{1}{n+K+1}\\ & + K^2\frac{1}{(n+K)(n+K+1)}\mathrm{Var}(\eta_{ij}|\mathbf{n},K),\end{aligned} \tag{12.3.18}$$

respectively.

Note that the expressions for the posterior moments of θ_{ij} are complicated and intractable. Thus, we need some approximations (as in Albert and Gupta, 1982). First, note that the covariance terms of the multinomial Dirichlet distribution are given by

$$\begin{aligned}\mathrm{Cov}(n_{ij},n_{hl}) &= n\eta_{ij}(1-\eta_{ij})\frac{(K+1)}{n+K} \text{ if } (i,j)=(h,l)\\ &= -n\eta_{ij}\eta_{hl}\frac{K+1}{n+K} \text{ if } (i,j)\neq(h,l).\end{aligned} \tag{12.3.19}$$

Now, defining the random variables

$$n_{ij}^* = \tau n_{ij}, \tau = \frac{K+1}{n+K} \quad \forall\ (i,j) \tag{12.3.20}$$

we have

$$\begin{aligned}\text{Cov}(n^*_{ij}, n^*_{hl}) &= n\tau\eta_{ij}(1-\eta_{ij}) \text{ if } (i,j) = (h,l) \\ &= -n\tau\eta_{ij}\eta_{hl} \text{ if } (i,j) \neq (h,l). \qquad (12.3.21)\end{aligned}$$

We approximate the marginal density $\mathcal{M}_1(\mathbf{n}|K, \boldsymbol{\eta})$ given in (12.3.14) by a multinomial density with parameter $(n\tau, \boldsymbol{\eta}')$. Then $\mathcal{M}_1(\mathbf{n}|K, \boldsymbol{\eta})$ is approximated by

$$\begin{aligned}&\frac{\Gamma(n\tau+1)}{\prod_{i=1}^r \prod_{j=1}^c \Gamma(\tau n_{ij}+1)} \tau^{rc-1} \prod_{i=1}^r \prod_{j=1}^c (\eta_{ij})^{\tau n_{ij}} \\ &= \frac{\Gamma(n\tau+1)}{\prod_{i=1}^r \prod_{j=1}^c \Gamma(\tau n_{ij}+1)} \tau^{rc-1} \prod_{i=1}^r (\eta_{i+})^{\tau n_{i+}} \prod_{j=1}^c (\eta_{+j})^{\tau n_{+j}}. \qquad (12.3.22)\end{aligned}$$

Hence, the approximate marginal distribution $\mathbf{n}$ given K is obtained by integrating (12.3.22) with respect to the prior of $(\boldsymbol{\eta}'_r, \boldsymbol{\eta}'_c)'$ given by (12.3.16) to obtain the expression

$$M(\mathbf{n}|K) = \frac{\Gamma(n\tau+1)\tau^{rc-1}}{\prod_{i=1}^r \prod_{j=1}^c \Gamma(\tau n_{ij}+1)} \frac{\prod_{i=1}^r \Gamma(\tau n_{i+}) \prod_{j=1}^c \Gamma(\tau n_{+j})}{(\Gamma(n\tau))^2}, \qquad (12.3.23)$$

which is the approximate density of $\mathbf{n}$ given K. Now, if $\mathcal{M}_1(\mathbf{n}|K, \boldsymbol{\eta})$ at (12.3.22) is combined with the prior given at (12.3.16). Note that $\boldsymbol{\eta}_k$ and $\boldsymbol{\eta}_c$ have approximately independent posterior Dirichlet distribution with parameter $\tau\mathbf{n}_k$ and $\tau\mathbf{n}_c$, respectively, where $\mathbf{n}_r = (n_{1+}, \cdots, n_{r+})'$ and $\mathbf{n}_c = (n_{+1}, \cdots, n_{+c})'$. Then the approximation of the posterior moments (12.3.17) and (12.3.18) are obtained as

$$\begin{aligned}E[\theta_{ij}|\mathbf{n}, K] &\simeq \left(n_{ij} + \frac{K}{n^2} n_{i+}n_{+j}\right)\frac{1}{n+K} \\ &= (1-\lambda(K))\bar{\theta}_{ij} + \lambda(K)\hat{\theta}_{ij} = \bar{\theta}_{ij}, \text{ where } \lambda(K) = K(n+K)^{-1}, \qquad (12.3.24)\end{aligned}$$

and

$$\begin{aligned}\text{Var}(\theta_{ij}|n_{ij}, K) &\simeq \bar{\theta}_{ij}(1-\bar{\theta}_{ij})(n+K+1)^{-1} \\ &+K^2(n+K)^{-1}(n+K+1)^{-1}\{[\hat{\theta}_{i+}(1-\hat{\theta}_{i+})(n\tau+1)^{-1} + \hat{\theta}^2_{i+}] \\ &\times[\hat{\theta}_{+j}(1-\hat{\theta}_{+j})(n\tau+1)^{-1} + \hat{\theta}^2_{+j}]\} - [\hat{\theta}_{i+}\cdot\hat{\theta}_{+j}]^2. \qquad (12.3.25)\end{aligned}$$

Thus, the **Bayes estimator** of θ_{ij} depends on $\lambda(K)$ through K. If K is unknown we can use a noninformative prior K^{-1} and estimate K from the posterior distribution.

The **empirical Bayes** method suggests that K is to be estimated from the marginal distribution $\mathcal{M}(\mathbf{n}|K)$ given at (12.3.23). Let $\hat{\lambda}(K)$ be the estimate of $\lambda(K)$. Then the estimate of θ_{ij} is obtained as

$$\hat{\theta}_{ij}^{EB} = \widehat{\lambda(K)}\hat{\theta}_{ij} + (1 - \widehat{\lambda(K)})\tilde{\theta}_{ij}, \quad \forall(i,j)$$
$$= \tilde{\theta}_{ij} - \widehat{\lambda(K)}(\tilde{\theta}_{ij} - \hat{\theta}_{ij}), \tag{12.3.26}$$

and this can be written compactly as

$$\hat{\boldsymbol{\theta}}_n^{EB} = \tilde{\boldsymbol{\theta}}_n - \widehat{\lambda(K)}(\tilde{\boldsymbol{\theta}}_n - \hat{\boldsymbol{\theta}}_n). \tag{12.3.27a}$$

We can show (see Albert, 1987) that

$$\widehat{\lambda(K)} \simeq \min(1, dG_n^{-2}), \ G_n^2 = -2n\sum_{i=1}^{r}\sum_{j=1}^{c} ln(\hat{\theta}_{ij}|\tilde{\theta}_{ij}) \ \text{ and } d = m-2 \tag{12.3.27b}$$

(see Problem 17). Note that $(G^2 - D_n) \to 0$, as $n \to \infty$ (under H_0 and local alternatives). Hence, the empirical Bayes estimator of $\boldsymbol{\theta}$ can be written as

$$\tilde{\boldsymbol{\theta}}_n^{EB} = \hat{\boldsymbol{\theta}}_n + (1 - dD_n^{-1})I(D_n > d)(\tilde{\boldsymbol{\theta}}_n - \hat{\boldsymbol{\theta}}_n). \tag{12.3.28}$$

In the derivation of the empirical Bayes estimators of $\boldsymbol{\theta}$, we have noticed the degree of difficulty in calculating the estimator while the quasi-empirical Bayes (PTE approach) achieves the same estimators without the sequence of assumptions and derivation. Our estimators are obtainable from (12.3.28) by setting

$$\begin{aligned}\widehat{\lambda(K)} &= 0 \ \text{ to obtain } \hat{\boldsymbol{\theta}}_n^{EB} = \tilde{\boldsymbol{\theta}}_n\\ &= 1 \ \text{ to obtain } \hat{\boldsymbol{\theta}}_n^{EB} = \hat{\boldsymbol{\theta}}_n\\ &= I(D_n \le \chi_m^2(\alpha)) \ \text{ to obtain } \hat{\boldsymbol{\theta}}_n^{EB} = \hat{\boldsymbol{\theta}}_n^{\mathrm{PT}}\\ &= dD_n^{-1}, (d = m-2) \ \text{ to obtain } \hat{\boldsymbol{\theta}}_n^{EB} = \hat{\boldsymbol{\theta}}_n^{S}\\ &= 1 - (1 - dD_n^{-1})I(D_n > d) \ \text{ to obtain } \hat{\boldsymbol{\theta}}_n^{EB} = \hat{\boldsymbol{\theta}}_n^{S+},\end{aligned} \tag{12.3.29}$$

respectively.

12.3.6 Asymptotic Properties of the Estimators of Cell-Probabilities under Fixed Alternatives

For the rc-dimensional simplex $\Omega = \{\boldsymbol{\theta} : \boldsymbol{\theta}'\mathbf{1} = 1\}$, let $\omega \subset \Omega$ be the subspace for which $\boldsymbol{\theta}$ satisfies the independence hypothesis. By virtue of the consistency of the D_n-test, we note that for fixed $\boldsymbol{\theta} \in \omega$, $D_n \to \infty$ in probability as $n \to \infty$. Thus, for any fixed $\boldsymbol{\theta} \notin \omega$ and large $n(\to \infty)$, the estimators $\hat{\boldsymbol{\theta}}_n^{\mathrm{PT}}$, $\hat{\boldsymbol{\theta}}_n^{S}$, and $\hat{\boldsymbol{\theta}}_n^{S+}$ are equivalent in probability to $\tilde{\boldsymbol{\theta}}_n$, the **unrestricted estimator**, and $\hat{\boldsymbol{\theta}}_n$ has bounded risk. Thus, to obtain differing risks for the estimators, we consider Pitman's local alternatives

$$K_{(n)} : \boldsymbol{\theta}_{(n)} = \boldsymbol{\theta}_0 + n^{-1/2}\boldsymbol{\xi},\ \boldsymbol{\theta}_0 \in \omega \text{ and } \boldsymbol{\xi}'\mathbf{1} = 0, \tag{12.3.30}$$

where $\boldsymbol{\xi}$ has fixed elements.

First, we look at the **fixed** alternatives. For a positive-semidefinite matrix, $\mathbf{W}$ the loss is

$$n(\boldsymbol{\theta}_n^* - \tilde{\boldsymbol{\theta}}_n)'\mathbf{W}(\boldsymbol{\theta}_n^* - \tilde{\boldsymbol{\theta}}_n) = [g(D_n)]^2 n(\tilde{\boldsymbol{\theta}}_n - \hat{\boldsymbol{\theta}}_n)'\mathbf{W}(\tilde{\boldsymbol{\theta}}_n - \hat{\boldsymbol{\theta}}_n). \tag{12.3.31}$$

For $g(D_n) = I(D_n \leq \chi_m^2(\alpha))$ the r.h.s. of (12.3.31) reduces to

$$\begin{aligned} g(D_n) &= D_n I(D_N < \chi_m^2(\alpha)) \frac{n(\tilde{\boldsymbol{\theta}}_n - \hat{\boldsymbol{\theta}}_n)'\mathbf{W}(\tilde{\boldsymbol{\theta}}_n - \hat{\boldsymbol{\theta}}_n)}{n(\tilde{\boldsymbol{\theta}}_n - \hat{\boldsymbol{\theta}}_n)'\boldsymbol{\Sigma}_n^{-1}(\tilde{\boldsymbol{\theta}}_n - \hat{\boldsymbol{\theta}}_n)} \\ &\leq \chi_m^2(\alpha) I(D_n < \chi_m^2(\alpha)) Ch_{max}(\mathbf{W}\boldsymbol{\Sigma}_n). \end{aligned} \tag{12.3.32}$$

The elements of $\boldsymbol{\Sigma}_n$ are nonnegative and bounded by 1. Also, $I(D_n < \chi_m^2(\alpha)) \to 0$ as $n \to \infty$ by the consistency of D_n. Hence,

$$E\{D_n I(D_n < \chi_m^2(\alpha)) | \boldsymbol{\theta} \notin \omega\} \to 0 \text{ as } n \to \infty. \tag{12.3.33}$$

Thus, for $\boldsymbol{\theta} \notin \omega$, $\hat{\boldsymbol{\theta}}_n^{PT}$ and $\tilde{\boldsymbol{\theta}}_n$ are asymptotically risk equivalent. Next, for $g(D_n) = dD_n^{-1}$, $(d = m - 2)$, we have

$$d^2 D_n^{-2} n(\tilde{\boldsymbol{\theta}}_n - \hat{\boldsymbol{\theta}}_n)'\mathbf{W}(\tilde{\boldsymbol{\theta}}_n - \hat{\boldsymbol{\theta}}_n) \leq d^2 D_n^{-1} Ch_{max}(\mathbf{W}\boldsymbol{\Sigma}_n). \tag{12.3.34}$$

On the set $\{D_n = 0\}$, we have $\hat{\boldsymbol{\theta}}_n^S = \tilde{\boldsymbol{\theta}}_n = \hat{\boldsymbol{\theta}}$. Further, we prove that $E_{\boldsymbol{\theta}}\{D_n^{-1} I(D_n > 0) | \boldsymbol{\theta} \notin \omega\} \to 0$ as $n \to \infty$, which implies that $\hat{\boldsymbol{\theta}}_n^S$ and $\tilde{\boldsymbol{\theta}}_n$ are empirically risk equivalent for every $\boldsymbol{\theta} \notin \omega$.

Lemma 1. (Gupta, Saleh, and Sen, 1989).

$$\{D_n^{-1} I(D_n > 0) | \boldsymbol{\theta} \notin \omega\} \to 0 \quad \text{as } n \to \infty \tag{12.3.35}$$

Proof. We need to show that for every $\epsilon > 0$,

$$E\{D_n^{-1} I(0 < D_n < \epsilon\} \to 0 \quad \text{as } n \to \infty.$$

This we verify as follows for $\{(i,j) | i = 1, \cdots, r,\ j = 1, \cdots, c\}$,

$$u_{ij} = \frac{(n_{ij} - n\theta_{ij})}{\sqrt{n\theta_{ij}}}, \quad u_{i+} = \frac{(n_{i+} - n\theta_{i+})}{\sqrt{n\theta_{i+}}},$$

$$\text{and } u_{+j} = \frac{(n_{+j - n\theta_{+j}})}{\sqrt{n\theta_{+j}}}. \tag{12.3.36}$$

Using the Stirling approximation together with standard steps, we see that the approximate large sample distribution of $\{U_{ij}\}$ is given by

$$(2\pi n)^{-(rc-1)/2} \exp\left\{-\frac{1}{2}\sum_{i=1}^{r}\sum_{j=1}^{c} u_{ij}^2\right\}(1+O(n^{-1/2})) \tag{12.3.37}$$

for every $\mathbf{n} \in \mathcal{N}_n = \{\mathbf{n} : n_{ij} \geq 0\ \forall (i,j); \mathbf{n}'\mathbf{1} = n\}$. Similarly, the joint density function of the two marginal of $\mathbf{n}_{i+}$ and $\mathbf{n}_{+j}$ is given by

$$(2\pi n)^{-(r+c-2)/2} \exp\left\{-\frac{1}{2}\left(\sum_{i=1}^{r} u_{i+}^2 + \sum_{j=1}^{c} u_{+j}^2\right)\right\}(1+O(n^{-1/2})). \tag{12.3.38}$$

Therefore, the conditional density of $\mathbf{n}$ given marginals $(n_{i+}, n_{+j}, \boldsymbol{\theta})$ is

$$(2\pi n)^{m} \exp\left\{-\frac{1}{2}\sum_{i=1}^{r}\sum_{j=1}^{c} u_{ij}^2 + \frac{1}{2}\sum_{i=1}^{r} u_{i+}^2 + \frac{1}{2}\sum_{j=1}^{c} u_{+j}^2\right\}(1+O(n^{-1/2})), \tag{12.3.39}$$

where $m = (r-1)(c-1)(\geq 3)$.

Let

$$k_n = \max\left\{\left|n_{ij} - \frac{n_{i+}n_{+j}}{n}\right| : 1 \leq i \leq r; 1 \leq j \leq c\right\}. \tag{12.3.40}$$

Note that D_n cannot be smaller than the parallel quantity for any 2×2 table (out of an $r \times c$ table) with probability one. Thus, we have

$$D_n \geq 16n^{-1}k_n^2 \quad \text{with probability one.} \tag{12.3.41}$$

Now, if $(n_{i+}n_{+j})n^{-1}$ is an integer for every $i(1 \leq i \leq r)$ and $j(1 \leq j \leq c)$, then $k_n = 0$ with positive probability. Then $k_n = 0$ implies $D_n = 0$. Hence, $\hat{\boldsymbol{\theta}}_n^S = \tilde{\boldsymbol{\theta}}_n = \hat{\boldsymbol{\theta}}_n$. On the other hand, if $(n_{i+}n_{+j})n^{-1}$ is not an integer for at least one (i,j), then $k_n \neq 0$, though it can be less than 1 with positive probability. In any case, $k_n > n^{-1}$ with probability 1. Thus, on the set $\{D_n > 0\}$, we can write

$$D_n^{-1}I(0 < D_n < \epsilon) = D_n^{-1}\{I(0 < D_n < 16n^{-1}) + D_n^{-1}I(16n^{-1} \leq D_n \leq \epsilon)\}. \tag{12.3.42}$$

Based on (12.3.41) and (12.3.42), for every $\boldsymbol{\theta}$ (under $K_{(n)}$ as well) we obtain

$$\begin{aligned} E_{\boldsymbol{\theta}}[D_n^{-1}I(0 < D_n < \epsilon)] \leq \frac{n}{16}\{&E_{\boldsymbol{\theta}}[k_n^{-2}I(1 \leq k_n \leq k_n^0)] \\ &+ E_{\boldsymbol{\theta}}[k_n^{-2}I(n^{-1} \leq k_n \leq 1)]\}, \end{aligned} \tag{12.3.43}$$

where we choose $k_n^0 = \left[\frac{1}{4}(n\epsilon)^{1/2}\right]$ so that $(k_n^0)^2 \leq \frac{n\epsilon}{16} < (k_n^0+1)^2$. Here [ ] stands for the largest integer contained in [ ] for $k_n > 1$. Replacing k_n by $[k_n]$

we have from (12.3.39),

$$E_\theta\{k_n^{-2} I(1 < k_n < k_n^0)\} \le \{E_\theta\{[k_n]^{-2} I(1 \le k_n \le k_n^0)\}$$

$$= \sum_{k=1}^{k_n^0} k^{-2} P_\theta([k_n] = k) \le 2 \sum_{k=1}^{k_n^0 - 1} \{k^2(k+1)\}^{-1} P_\theta\{[k_n] \le k\}$$

$$+ (k_n^0)^2 P_\theta\{[k_n] = k\}. \tag{12.3.44}$$

Now for the conditional model in (12.3.39), the row or column sums $\{n_{ij} - \frac{n_{i+}+n_{+j}}{n}\}$ are zero, and essentially there are $(r-1)(c-1) = m$ linearly independent entries. Hence, given the marginals and the estimated marginal probabilities held fixed, there are $(2k)^m$ configurations for which $k_n \le k$. For each configuration, the conditional probability by (12.4.39) is $O(n^{-m/2})$. Thus, $P_\theta\{[k_n] \le k\} = O(k^m n^{-m/2})$ for every $k \le k_n^0$. In this formulation, we actually take $P_\theta(A) = E_\theta\{P_\theta(A) | n_{i+} n_{+j}\}$ and use (12.3.39). Then, by (12.3.43) and (12.3.44), we have

$$\frac{n}{16} E_\theta\{k_n^{-2} I(1 \le k_n \le k_n^0)\} \le \frac{n}{8}\left[\sum_{k=1}^{k_n^0 - 1} O(k^{-3} k^m n^{-m/2}) + O(k_n^{m-2} n^{-m/2})\right]$$

$$= O(n^{-m/2+1})\left\{\sum_{k=1}^{k_n^0 - 1} O(k^{m-3} n^{-m/2}) + O(k_n^{m-2})\right\}$$

$$= O(n^{-m/2+1}(k_n^0)^{m-2}) = O(\epsilon^{m/2-1}) = O(\epsilon^{1/2}), \quad m \ge 3. \tag{12.3.45}$$

Thus, (12.3.45) may be made arbitrarily small by suitably choosing ϵ.

For second term of (12.3.43), writing $s_n = [nk_n]$, so that $\frac{1}{n} \le k_n \le 1$ and $1 \le s_n \le n$. Thus, as in (12.3.44),

$$\frac{n}{16} E_\theta\left\{k_n^{-2} I\left(\frac{1}{n} \le k_n \le 1\right)\right\} \le \frac{n^3}{16} E_\theta\left\{s_n^{-2} I\left(1 \le s_n \le n-1\right)\right\}$$

$$\le \frac{n^3}{16} \sum_{s=1}^{n-1} s^{-2} P_\theta\{s_n = s\}. \tag{12.3.46}$$

Note that we have $k_n < 1$ so that $s_n < n$. Hence, by the definition of k_n (see 12.3.39), n_{ij} differ from $(n_{i+} n_{+j})/n$ by some number within $(-1, 1)$. Thus, for a given $\frac{(n_{i+} n_{+j})}{n}$ the number of configurations of $\mathbf{n}(s_n < n-1)$ is equal to 2^m at most. Also, the number of configuration of n_{i+} and n_{+j} for which $(n_{i+} n_{+j})/n$ differ from some number $\le k_n (= n^{-1} s_n)$ is $O(s_n^{(r-c-2)/2})$. Further, total number of the n_{i+} (or n_{+j}) is equal to n, and for every integer, a, b, $(n_{i+} + a)(n_{+j} + b)/n = (n_{i+} n_{+j})/n + a n_{+j}/n + b n_{i+}/n + ab/n$, where both n_{i+}/n and n_{+j}/n are less than 1. Thus, ab needs to be of the order $O(s_n)$, which reduces the number of configurations to the order $O(s_n^{(r-1)/2+(c-1)/2}) = O(s_n^{(r+c-2)/2})$. Finally, by (12.3.37) and (12.3.38), for each such configurations of $\mathbf{n}$, the probability is $O(n^{-m/2})$ and for the marginals , it is $O(n^{-(r+c-2)/2})$. Therefore, $P_\theta(s_n \leq s) = O\big(s^{(r+c-2)/2} n^{-m/2}\big)\big)O\big(n^{-m/2}\big)$ for $s = 1, \cdots, n-1$. Hence, (12.3.46) reduces to

$$\begin{aligned}
&\frac{n^3}{16}\sum_{s=1}^{n-1} s^{-2}\Big\{O(n^{-m/2})O(s^{-(r+c-2)/2})O(n^{-(r+c-2)/2})\Big\} \\
&\quad = O(n^{3-m/2-(r+c-2)/2})O\Big\{\sum_{s=1}^{n-1} s^{(r+c)/2-3}\Big\} \\
&\quad = O(n^{3-m/2-(r+c-2)/2})O(n^{(r+c)/2-3}) = O(n^{-m/2+1}) \\
&\quad = O(n^{-1/2}), \quad \text{as } m \geq 3. \qquad (12.3.47)
\end{aligned}$$

Hence, (12.3.46) converges to 0 as $n \to \infty$. This completes the proof that for every positive ε,

$$E_\theta\{D_n I(0 < D_n < \epsilon)|\boldsymbol{\theta} \notin \omega\} \to 0 \qquad (12.3.48)$$

as $n \to \infty$. The proof of the lemma is complete.

Next, we note that

$$n(\hat{\boldsymbol{\theta}}_n - \boldsymbol{\theta})'\mathbf{W}(\hat{\boldsymbol{\theta}}_n - \boldsymbol{\theta}) \to \infty \quad \text{in probability as } n \to \infty. \qquad (12.3.49)$$

As a result, the asymptotic risk of $\hat{\boldsymbol{\theta}}_n$ for any $\boldsymbol{\theta} \notin \omega$ goes to ∞. Also, $n(\tilde{\boldsymbol{\theta}}_n - \boldsymbol{\theta})(\tilde{\boldsymbol{\theta}}_n - \boldsymbol{\theta})' = \text{Diag}(\theta_{11}, \cdots, \theta_{rc}) - \boldsymbol{\theta}\boldsymbol{\theta}'$ so that the risk of $\tilde{\boldsymbol{\theta}}_n$ is bounded for every $\boldsymbol{\theta} \in \Omega$. Finally, we consider the asymptotic risk equivalence of $\hat{\boldsymbol{\theta}}_n^{S+}$ and $\tilde{\boldsymbol{\theta}}_n$ as follows:

$$\begin{aligned}
&n(\hat{\boldsymbol{\theta}}_n^{S+} - \boldsymbol{\theta})'\mathbf{W}(\hat{\boldsymbol{\theta}}_n^{S+} - \boldsymbol{\theta}) \\
&= n(\hat{\boldsymbol{\theta}}_n^{S} - \hat{\boldsymbol{\theta}}_n)'\mathbf{W}(\hat{\boldsymbol{\theta}}_n^{S} - \hat{\boldsymbol{\theta}}_n) + n(1 - dD_n^{-1})^2 I(D_n < d)(\tilde{\boldsymbol{\theta}}_n - \hat{\boldsymbol{\theta}}_n)'\mathbf{W}(\tilde{\boldsymbol{\theta}}_n - \hat{\boldsymbol{\theta}}_n) \\
&\quad -2n(\hat{\boldsymbol{\theta}}_n^{S} - \tilde{\boldsymbol{\theta}}_n)'\mathbf{W}(\tilde{\boldsymbol{\theta}}_n - \hat{\boldsymbol{\theta}}_n)(1 - dD_n^{-1})I(D_n < d) \\
&= n(\hat{\boldsymbol{\theta}}_n^{S} - \hat{\boldsymbol{\theta}}_n)'\mathbf{W}(\hat{\boldsymbol{\theta}}_n^{S} - \hat{\boldsymbol{\theta}}_n) \\
&\quad +(1 - dD_n^{-1})^2 I(D_n < d)D_n \frac{n(\tilde{\boldsymbol{\theta}}_n - \hat{\boldsymbol{\theta}}_n)'\mathbf{W}(\tilde{\boldsymbol{\theta}}_n - \hat{\boldsymbol{\theta}}_n)}{n(\tilde{\boldsymbol{\theta}}_n - \hat{\boldsymbol{\theta}}_n)'\boldsymbol{\Sigma}_n^{-1}(\tilde{\boldsymbol{\theta}}_n - \hat{\boldsymbol{\theta}}_n)} \\
&\quad -2n(\hat{\boldsymbol{\theta}}_n^{S} - \tilde{\boldsymbol{\theta}}_n)'\mathbf{W}(\tilde{\boldsymbol{\theta}}_n - \hat{\boldsymbol{\theta}}_n)(1 - dD_n^{-1})I(D_n < d). \qquad (12.3.50)
\end{aligned}$$

We observe that $\hat{\boldsymbol{\theta}}_n^{S}$ is asymptotically risk equivalent to $\tilde{\boldsymbol{\theta}}_n$ for fixed alternatives. Now, consider the last two terms of (12.3.49) pooled together as

$$\begin{aligned}
&[(1 - dD_n^{-1})^2 I(D_n < d)D_n + 2ndD_n^{-1}(1 - dD_n^{-1})I(D_n < d)] \\
&\qquad \times n(\tilde{\boldsymbol{\theta}}_n - \hat{\boldsymbol{\theta}}_n)'\mathbf{W}(\tilde{\boldsymbol{\theta}}_n - \hat{\boldsymbol{\theta}}_n) \\
&\leq [D_n^2(1 - dD_n^{-1})^2 I(D_n < d) + 2nd(1 - dD_n^{-1})I(D_n < d)]Ch_{min}(\mathbf{W}\boldsymbol{\Sigma}_n).
\end{aligned}$$

(12.3.51)

The item in the square bracket may be shown to tend to 0 as $n \to \infty$ using (12.3.32) and using Lemma 12.3.1 or a small modification of it. Hence, we arrive at the following theorem:

Theorem 1. Under the local alternatives $\boldsymbol{\theta}_\delta = \boldsymbol{\theta}_0 + \boldsymbol{\delta}$,

(i) the asymptotic risk of $\hat{\boldsymbol{\theta}}_n$ becomes unbounded as $n \to \infty$,

(ii) the asymptotic risk of $\tilde{\boldsymbol{\theta}}_n$ is bounded,

(iii) the asymptotic risk of $\hat{\boldsymbol{\theta}}_n^{\mathrm{PT}}, \hat{\boldsymbol{\theta}}_n^{S}$, and $\hat{\boldsymbol{\theta}}_n^{S+}$ are equivalent to that of $\tilde{\boldsymbol{\theta}}_n$ as $n \to \infty$.

12.3.7 Asymptotic Properties of the Estimators under Local Alternatives

Now, we consider the asymptotic distributions of $\tilde{\boldsymbol{\theta}}_n, \hat{\boldsymbol{\theta}}_n, \hat{\boldsymbol{\theta}}_n^{\mathrm{PT}}, \hat{\boldsymbol{\theta}}_n S$, and $\hat{\boldsymbol{\theta}}_n^{S+}$ under the local alternatives

$$K_{(n)} : \boldsymbol{\theta}_{(n)} = \boldsymbol{\theta}_0 + n^{-1/2}\boldsymbol{\xi}, \quad \boldsymbol{\xi}'\mathbf{1} = 0 \text{ and } \boldsymbol{\theta}_0 \in \omega \tag{12.3.52}$$

to obtain the required risk comparisons. First, define

$$\begin{aligned}
&\boldsymbol{\Sigma}_0 = \mathrm{Diag}(\theta_{1+}\cdot\theta_{+1}, \cdots, \theta_{1+}\cdot\theta_{+c}, \cdots, \theta_{r+}\cdot\theta_{+1}, \cdots, \theta_{r+}\cdot\theta_{+c}),\\
&\sqrt{\boldsymbol{\theta}_0} = (\sqrt{\theta_{1+}\cdot\theta_{+1}}, \cdots, \sqrt{\theta_{r+}\cdot\theta_{+c}})\\
&\mathbf{J} = \mathbf{I}_{rc} - (\sqrt{\boldsymbol{\theta}_0})(\sqrt{\boldsymbol{\theta}_0})',\\
&\mathbf{K} = \mathbf{B}(\mathbf{B}'\mathbf{B})^{-1}\mathbf{B}', \quad \mathbf{A} = \mathbf{J} - \mathbf{K},\\
&\mathbf{B} = \boldsymbol{\Sigma}_0^{1/2}\left(\frac{\partial(\theta_{i+}\theta_{+j})}{\partial\boldsymbol{\theta}^*}\right),\\
&\text{where } \boldsymbol{\theta}^* = (\theta_{1+}, \cdots, \theta_{r+}, \theta_{+1}, \cdots, \theta_{+c})'.
\end{aligned} \tag{12.3.53}$$

In this part of the discussion we shall assume the regularity conditions of Chapter 2 borrowed from Agresti (1990) and Fienberg and Holland (1975), and also Ferguson (1996). Then we arrive at the following theorem. For a proof, see Problem 17.

Theorem 2. Under $K_{(n)}$ given by (12.3.52) and assumed regularity conditions,

(i) $\mathbf{X}_{(n)} = n^{1/2}\boldsymbol{\Sigma}_0^{-1/2}(\tilde{\boldsymbol{\theta}}_n - \boldsymbol{\theta}_0) \sim \mathcal{N}_{rc}(\boldsymbol{\delta}, \mathbf{J})$, where $\boldsymbol{\delta} = \boldsymbol{\Sigma}_0^{-1/2}\boldsymbol{\xi}$;

(ii) $\mathbf{Y}_{(n)} = n^{1/2}\boldsymbol{\Sigma}_0^{-1/2}(\tilde{\boldsymbol{\theta}}_n - \hat{\boldsymbol{\theta}}_n) \sim \mathcal{N}_{rc}(\mathbf{J}^*\boldsymbol{\delta}, \mathbf{A})$, where $\mathbf{J}^* = \mathbf{I} - \mathbf{K}$;

(iii) $\mathbf{Z}_{(n)} = n^{1/2}\boldsymbol{\Sigma}_0^{-1/2}(\hat{\boldsymbol{\theta}}_n - \boldsymbol{\theta}_0) \sim \mathcal{N}_{rc}(\mathbf{K}\boldsymbol{\delta}, \mathbf{K})$;

(iv) $\begin{pmatrix} \mathbf{X}_{(n)} \\ \mathbf{Y}_{(n)} \end{pmatrix} \sim \mathcal{N}_{2rc}\left\{\begin{pmatrix} \boldsymbol{\delta} \\ \mathbf{J}^*\boldsymbol{\delta} \end{pmatrix} ; \begin{pmatrix} \mathbf{J} & \mathbf{A} \\ \mathbf{A} & \mathbf{A} \end{pmatrix}\right\}$;

(v) $\begin{pmatrix} \mathbf{Z}_{(n)} \\ \mathbf{Y}_{(n)} \end{pmatrix} \sim \mathcal{N}_{2rc}\left\{\begin{pmatrix} \mathbf{K}\boldsymbol{\delta} \\ \mathbf{J}^*\boldsymbol{\delta} \end{pmatrix} : \begin{pmatrix} \mathbf{K} & \mathbf{0} \\ \mathbf{0} & \mathbf{A} \end{pmatrix}\right\}$;

(vi) $\lim_{n\to\infty} P\{D_n \le x | K_{(n)}\} = H_m(x;\Delta^2),\ m = (r-1)(c-1)$,

where $\Delta^2 = \boldsymbol{\delta}'\mathbf{J}^*\boldsymbol{\delta}$;

(vii) $\lim_{n\to\infty} P\{\sqrt{n}\boldsymbol{\Sigma}_0^{-1/2}(\hat{\boldsymbol{\theta}}_n^{\text{PT}} - \boldsymbol{\theta}_0) \le \mathbf{x} | K_{(n)}\}$

$$= \Phi_{rc}(\mathbf{x} - \mathbf{K}\boldsymbol{\delta}; \mathbf{0}, \mathbf{K}) H_m(\chi_m^2(\alpha); \Delta^2)$$

$$= \int \cdots \int_{M(\boldsymbol{\delta})} \Phi_{rc}(\mathbf{x} - \boldsymbol{\delta} - \mathbf{z};\ \mathbf{0}, \mathbf{K}) d\Phi(\mathbf{z} : \mathbf{0}, \mathbf{A}),$$

where $M(\boldsymbol{\delta}) = \{\mathbf{z} : (\mathbf{z} + \mathbf{J}^*\boldsymbol{\delta})'(\mathbf{z} + \mathbf{J}^*\boldsymbol{\delta}) \ge \chi_m^2(\alpha)\}$.

(viii) $\sqrt{n}\boldsymbol{\Sigma}_0^{-1/2}(\hat{\boldsymbol{\theta}}_n^S - \boldsymbol{\theta}_0) = \mathbf{X}_{(n)} - (m-2)\mathbf{Y}_{(n)}(\mathbf{Y}_n'\mathbf{Y}_n)^{-1}$

$$\xrightarrow{\mathcal{D}} (\mathbf{X} + \boldsymbol{\delta}) - (m-2)(\mathbf{Y} + \mathbf{J}^*\boldsymbol{\delta})\{(\mathbf{Y} + \mathbf{J}^*\boldsymbol{\delta})'(\mathbf{Y} + \mathbf{J}^*\boldsymbol{\delta})\}^{-1},$$

where $\begin{pmatrix} \mathbf{X} \\ \mathbf{Y} \end{pmatrix} \sim \mathcal{N}_{2rc}\left\{\begin{pmatrix} \mathbf{0} \\ \mathbf{0} \end{pmatrix} : \begin{pmatrix} \mathbf{J} & \mathbf{A} \\ \mathbf{A} & \mathbf{A} \end{pmatrix}\right\}$ and $\xrightarrow{\mathcal{D}}$ converges in distribution as $n \to \infty$;

(xi) $\sqrt{n}\boldsymbol{\Sigma}_0^{-1/2}(\hat{\boldsymbol{\theta}}_n^{S+} - \boldsymbol{\theta}_0) \xrightarrow{\mathcal{D}} (\mathbf{X} + \boldsymbol{\delta}) - (m-2)(\mathbf{Y} + \mathbf{J}^*\boldsymbol{\delta})\{||(\mathbf{Y} + \mathbf{J}^*\boldsymbol{\delta})||\}^{-2}$

$$-\{1 - (m-2)||\mathbf{Y} + \mathbf{J}^*\boldsymbol{\delta}||^{-2}\} I(||\mathbf{Y} + \mathbf{J}^*\boldsymbol{\delta}||^2 \le m-2)\}$$

as $n \to \infty$. (12.3.54)

Application of Theorem 12.3.2 results in the asymptotic distributional bias expressions as

(1) $\mathbf{b}_1(\tilde{\boldsymbol{\theta}}_n) = \mathbf{0}$ and $B_1(\tilde{\boldsymbol{\theta}}_n) = 0$;

(2) $\mathbf{b}_2(\hat{\boldsymbol{\theta}}_n) = -\mathbf{J}^*\boldsymbol{\delta}$ and $B_2(\hat{\boldsymbol{\theta}}_n) = \Delta^2$;

(3) $\mathbf{b}_3(\hat{\boldsymbol{\theta}}_n^{\text{PT}}) = -\mathbf{J}^*\boldsymbol{\delta} H_{m+2}(\chi_m^2(\alpha); \Delta^2)$ and

$B_3(\hat{\boldsymbol{\theta}}_n^{\text{PT}}) = \Delta^2\{H_{m+2}(\chi_m^2(\alpha); \Delta^2)\}^2$;

(4) $\mathbf{b}_4(\hat{\boldsymbol{\theta}}_n^S) = -(m-2)\mathbf{J}^*\boldsymbol{\delta} E[\chi_{m+2}^{-2}(\Delta^2)]$ and

$B_4(\hat{\boldsymbol{\theta}}_n^S) = (m-2)^2\Delta^2\{E[\chi_{m+2}^{-2}(\Delta^2)]\}^2$;

(5) $\mathbf{b}_5(\hat{\boldsymbol{\theta}}_n^{S+}) = -(m-2)\mathbf{J}^*\boldsymbol{\delta} E[\chi_{m+2}^{-2}(\Delta^2)]$

$-\mathbf{J}^*\boldsymbol{\delta} E\{(1 - (m-2)^{-1}\chi_{m+2}^{-2}(\Delta^2)) I(\chi_{m+2}^2(\Delta^2) < m-2)\}$ and

$B_5(\hat{\boldsymbol{\theta}}_n^{S+}) = \Delta^2\{(m-2)E[\chi_{m+2}^{-2}(\Delta^2)]$

$$+E\{(1-(m-2)^{-1}\chi^{-2}_{m+2}(\Delta^2))I(\chi^2_{m+2}(\Delta^2)<m-2)\}^2. \quad (12.3.55)$$

Similarly, the asymptotic distributional MSE (ADMSE) matrices and the risks of the estimators are given by

(1) $\mathbf{M}_1(\tilde{\boldsymbol{\theta}}_n)=\mathbf{J}$

(2) $\mathbf{M}_2(\hat{\boldsymbol{\theta}}_n)=\mathbf{K}+\mathbf{J}^*\boldsymbol{\delta}\boldsymbol{\delta}'\mathbf{J}^*$

(3) $\mathbf{M}_3(\hat{\boldsymbol{\theta}}_n^{\mathrm{PT}})=\mathbf{J}\{1-H_{m+2}(\chi^2_m(\alpha);\Delta^2)+\mathbf{K}H_{m+2}(\chi^2_m(\alpha);\Delta^2)\}$
$+\mathbf{J}^*\boldsymbol{\delta}\boldsymbol{\delta}^*\mathbf{J}^*\{2H_{m+2}(\chi^2_m(\alpha);\Delta^2)-H_{m+4}(\chi^2_m(\alpha);\Delta^2);$

(4) $\mathbf{M}_4(\hat{\boldsymbol{\theta}}_n^{S})=\mathbf{J}-(m-2)\mathbf{A}\{2E[\chi^{-2}_{m+2}(\Delta^2)]-(m-2)E[\chi^{-4}_{m+2}(\Delta^2)]\}$
$+(m^2-4)(\mathbf{J}^*\boldsymbol{\delta}\boldsymbol{\delta}^*\mathbf{J}^*)E[\chi^{-4}_{m+4}(\Delta^2)];$

(5) $\mathbf{M}_5(\hat{\boldsymbol{\theta}}_n^{S+})=\mathbf{M}_4(\hat{\boldsymbol{\theta}}_n^{S})$
$-(m-2)\mathbf{A}E\big[(1-(m-2)\chi^{-2}_{m+2}(\Delta^2))I(\chi^2_{m+2}(\Delta^2)<m-2)\big]$
$+\mathbf{J}^*\boldsymbol{\delta}\boldsymbol{\delta}'\mathbf{J}^*\{2E\big[(1-(m-2)\chi^{-2}_{m+2}(\Delta^2))I(\chi^2_{m+2}(\Delta^2)<m-2)\big]$
$-E\{(1-(m-2)\chi^{-2}_{m+4}(\Delta^2))^2I(\chi^2_{m+4}(\Delta^2)<m-2)\}. \quad (12.3.55)$

Similarly,

(1) $R_1(\tilde{\boldsymbol{\theta}}_n:\mathbf{W})=\mathrm{tr}[\mathbf{WJ}];$

(2) $R_2(\hat{\boldsymbol{\theta}}_n:\mathbf{W})=\mathrm{tr}[\mathbf{WK}]+\boldsymbol{\delta}'\mathbf{J}^{*\prime}\mathbf{WJ}^*\boldsymbol{\delta};$

(3) $R_3(\hat{\boldsymbol{\theta}}_n:\mathbf{W})=\mathrm{tr}[\mathbf{WJ}]\{1-H_{m+2}(\chi^2_m(\alpha);\Delta^2)$
$+\mathrm{tr}[\mathbf{WK}]H_{m+2}(\chi^2_m(\alpha);\Delta^2)\}$
$+\boldsymbol{\delta}^*\mathbf{J}^{*\prime}\mathbf{WJ}^*\boldsymbol{\delta}\{2H_{m+2}(\chi^2_m(\alpha);\Delta^2)-H_{m+4}(\chi^2_m(\alpha);\Delta^2);$

(4) $R_4(\hat{\boldsymbol{\theta}}_n^{S}:\mathbf{W})=\mathrm{tr}[\mathbf{WJ}]-(m-2)\mathrm{tr}[\mathbf{WA}]\{(m-2)E[\chi^{-4}_{m+2}(\Delta^2)]$
$+\Big[1-\frac{(m+2)(\boldsymbol{\delta}^*\mathbf{J}^{*\prime}\mathbf{WJ}^*\boldsymbol{\delta})}{;}2(\boldsymbol{\delta}'\mathbf{J}^*\boldsymbol{\delta})\,\mathrm{tr}[\mathbf{WA}]\Big]2\Delta^2$
$\times E[\chi^{-4}_{m+4}(\Delta^2)]$

(5) $R_5(\hat{\boldsymbol{\theta}}_n^{S+}:\mathbf{W})=R_4(\hat{\boldsymbol{\theta}}_n^{S}:\mathbf{W})-(m-2)\,\mathrm{tr}[\mathbf{WA}]$
$\times E\big[(1-(m-2)\chi^{-2}_{m+2}(\Delta^2))I(\chi^2_{m+2}(\Delta^2)<m-2)\big]$
$+\ \boldsymbol{\delta}'\mathbf{J}^{*\prime}\mathbf{WJ}^*\boldsymbol{\delta}\{2E\big[(1-(m-2)\chi^{-2}_{m+2}(\Delta^2))$
$\times I(\chi^2_{m+2}(\Delta^2)<m-2)\big]-E\big[(1-(m-2)\chi^{-2}_{m+4}(\Delta^2))^2$
$\times I(\chi^2_{m+4}(\Delta^2)<m-2)\big]. \quad (12.3.56)$

12.3.8 Analysis of the Asymptotic Properties of the Estimators

We consider the asymptotic risk analysis of the five estimators.

Unrestricted and Restricted Estimators. Consider ARRE$(\hat{\boldsymbol{\theta}}_n : \tilde{\boldsymbol{\theta}}_n) = \frac{R_2(\hat{\boldsymbol{\theta}}_n:\mathbf{W})}{R_1(\tilde{\boldsymbol{\theta}}_n:\mathbf{W})} \gtrless 1$ according as

$$(\boldsymbol{\delta}'\mathbf{J}^{*\prime}\mathbf{W}\mathbf{J}^*\boldsymbol{\delta}) \gtrless \operatorname{tr}[\mathbf{WA}] \ (>0), \tag{12.3.57a}$$

which implies that $\hat{\boldsymbol{\theta}}_n$ works well near the origin. When $\boldsymbol{\delta}$ moves away from the origin, the risk of $\hat{\boldsymbol{\theta}}_n$ is unbounded while that of $\tilde{\boldsymbol{\theta}}_n$ is constant.

Unrestricted, restricted and PTE. Here we have ARRE$(\hat{\boldsymbol{\theta}}_n^{PT} : \tilde{\boldsymbol{\theta}}_n) = \frac{R_3(\hat{\boldsymbol{\theta}}_n^{PT}:\mathbf{W})}{R_1(\tilde{\boldsymbol{\theta}}_n:\mathbf{W})} \gtrless 1$ according as

$$(\boldsymbol{\delta}'\mathbf{J}^{*\prime}\mathbf{W}\mathbf{J}^*\boldsymbol{\delta}) \gtrless \frac{H_{m+2}(\chi_m^2(\alpha);\Delta^2)\operatorname{tr}[\mathbf{WA}]}{2H_{m+2}(\chi_m^2(\alpha);\Delta^2) - H_{m+4}(\chi_m^2(\alpha);\Delta^2)} \tag{12.3.57b}$$

and ARRE$(\hat{\boldsymbol{\theta}}_n^{PT} : \hat{\boldsymbol{\theta}}_n) = \frac{R_3(\hat{\boldsymbol{\theta}}_n^{PT}:\mathbf{W})}{R_2(\hat{\boldsymbol{\theta}}_n:\mathbf{W})} \gtrless 1$ according as

$$(\boldsymbol{\delta}'\mathbf{J}^{*\prime}\mathbf{W}\mathbf{J}^*\boldsymbol{\delta}) \gtrless \frac{(1 - H_{m+2}(\chi_m^2(\alpha);\Delta^2)\operatorname{tr}[\mathbf{WA}]}{1-2H_{m+2}(\chi_m^2(\alpha);\Delta^2) + H_{m+4}(\chi_m^2(\alpha);\Delta^2)} \tag{12.3.57c}$$

Here also, $\hat{\boldsymbol{\theta}}_n^{PT}$ works well near the origin. Under H_0, we have

$$R_2(\hat{\boldsymbol{\theta}}_n : \mathbf{W}) \le R_3(\hat{\boldsymbol{\theta}}_n^{PT} : \mathbf{W}) \le R_1(\tilde{\boldsymbol{\theta}}_n : \mathbf{W}). \tag{12.3.58}$$

Note that $\mathbf{J}, \mathbf{J}^*$, and $\mathbf{A}$ are all idempotent matrices when the rank of $\mathbf{A}$ is m and the rank of $\mathbf{J}^*$ is greater than m. Thus, by Courant's theorem, $\frac{\boldsymbol{\delta}'\mathbf{J}^{*\prime}\mathbf{W}\mathbf{J}^*\boldsymbol{\delta}}{\boldsymbol{\delta}'\mathbf{J}^*\boldsymbol{\delta}} \le Ch_{max}(\mathbf{W})$ for all $\boldsymbol{\delta}$. Consequently, let $\mathcal{W}$ be the class of $\mathbf{W}$ matrices such that

$$\mathcal{W} = \left\{ W_{psd} : \frac{\operatorname{tr}[\mathbf{WA}]}{Ch_{max}(\mathbf{WA})} \ge \frac{m+2}{2} \right\}. \tag{12.3.59}$$

Then we have

$$\frac{R_4(\hat{\boldsymbol{\theta}}_n^{S} : \mathbf{W})}{R_1(\tilde{\boldsymbol{\theta}}_n : \mathbf{W})} \le 1 \quad \forall\ \boldsymbol{\delta}, \text{and } \mathbf{W} \in \mathcal{W}. \tag{12.3.60}$$

We may note the choice of $\mathbf{W}$ based on (12.3.59) in the loss function for the dominance of $\hat{\boldsymbol{\theta}}_n^{S}$ over $\tilde{\boldsymbol{\theta}}_n$ for all $m > 3$. If, in particular, $\mathbf{W} = \mathbf{I}_{rc}$ or $\mathbf{J}$, then $Ch_{max}(\mathbf{W}) = 1$, and the rank of $\mathbf{I}_{rc}\mathbf{A}$ is m and the rank of $\mathbf{JA} \ge m-1$. In this case, $\mathbf{W} \in \mathcal{W}$. The dominance may not hold for arbitrary choice of $\mathbf{W}$ because the condition (12.3.59) may not hold simultaneously for all $\mathbf{A}$ (i.e., $\mathbf{K}$), and in such cases, the Stein-type estimator $\hat{\boldsymbol{\theta}}_n^S$ works with some modification due to Berger et al. (1977) as

$$\hat{\boldsymbol{\theta}}_n^{MS+} = \hat{\boldsymbol{\theta}}_n + \{1 - cd_n D_n^{-1}\mathbf{W}^{-1}\boldsymbol{\Sigma}_n^{-1}\}(\tilde{\boldsymbol{\theta}}_n - \hat{\boldsymbol{\theta}}_n), \tag{12.3.61}$$

where c is the shrinkage constant and $d_n = Ch_{min}(\mathbf{W}^{-1/2}\boldsymbol{\Sigma}_n^{-1})$.

As for the Stein-type estimator, $\hat{\boldsymbol{\theta}}_n^S$, it has smaller risk than $\tilde{\boldsymbol{\theta}}_n$ uniformly for all Δ^2 by (12.3.59). However, for the comparison of risk of the restricted estimator $\hat{\boldsymbol{\theta}}_n$ with that of $\hat{\boldsymbol{\theta}}_n^S$, we first note that

$$\begin{aligned} R_4(\hat{\boldsymbol{\theta}}_n^S : \mathbf{W}) &= R_2(\hat{\boldsymbol{\theta}}_n : \mathbf{W}) + \operatorname{tr}[\mathbf{WA}] - (\boldsymbol{\delta}'\mathbf{J}^{*\prime}\mathbf{W}\mathbf{J}^*\boldsymbol{\delta}) \\ &\quad - (m-2)\operatorname{tr}[\mathbf{WA}]\{(m-2)E[\chi_{m+2}^{-4}(\Delta^2)] \\ &\quad + \left[1 - \frac{(m+2)(\boldsymbol{\delta}^*\mathbf{J}^{*\prime}\mathbf{W}\mathbf{J}^*\boldsymbol{\delta})}{2(\boldsymbol{\delta}'\mathbf{J}^*\boldsymbol{\delta})\operatorname{tr}[\mathbf{WA}]}\right] 2\Delta^2 E[\chi_{m+4}^{-4}(\Delta^2)]. \end{aligned} \tag{12.3.62}$$

Under H_0, (12.3.62) becomes

$$R_4(\hat{\boldsymbol{\theta}}_n^S : \mathbf{W}) = R_2(\hat{\boldsymbol{\theta}}_n : \mathbf{W}) + \frac{2}{m}\operatorname{tr}[\mathbf{WA}] \geq 0. \tag{12.3.63}$$

Hence, asymptotic risk of $\hat{\boldsymbol{\theta}}_n^S$ is greater than that of $\hat{\boldsymbol{\theta}}_n$ under H_0. However, if Δ^2 diverts from 0, the risk of $\hat{\boldsymbol{\theta}}_n$ grows bigger while risk of $\hat{\boldsymbol{\theta}}_n^S$ becomes smaller and $\hat{\boldsymbol{\theta}}_n^S$ dominates $\hat{\boldsymbol{\theta}}_n$ outside an interval around the origin 0. Thus, neither $\hat{\boldsymbol{\theta}}_n^S$ nor $\hat{\boldsymbol{\theta}}_n$ dominate each other completely.

In the case of $\hat{\boldsymbol{\theta}}_n^S$ and $\hat{\boldsymbol{\theta}}_n^{\mathrm{PT}}$ we have under H_0,

$$\begin{aligned} R_4(\hat{\boldsymbol{\theta}}_n^S : \mathbf{W}) &= R_3(\hat{\boldsymbol{\theta}}_n^{\mathrm{PT}} : \mathbf{W}) + \operatorname{tr}[\mathbf{WA}]\{1 - H_{m+2}(\chi_m^2(\alpha) : 0)\} \\ &\quad + \frac{2}{m}\operatorname{tr}[\mathbf{WA}] \geq R_3(\hat{\boldsymbol{\theta}}_n^{\mathrm{PT}} : \mathbf{W}) \quad \forall\, m \geq 3. \end{aligned} \tag{12.3.64}$$

The PTE has smaller ADQR than that of $\hat{\boldsymbol{\theta}}_n^S$ at $\Delta^2 = 0$, and $\hat{\boldsymbol{\theta}}_n^S$ does not dominate $\hat{\boldsymbol{\theta}}_n^{\mathrm{PT}}$ when independence hypothesis holds. Clearly, under H_0 the risk ordering is as follows:

$$R_2(\hat{\boldsymbol{\theta}}_n : \mathbf{W}) \leq R_3(\hat{\boldsymbol{\theta}}_n^{\mathrm{PT}} : \mathbf{W}) \leq R_4(\hat{\boldsymbol{\theta}}_n^S : \mathbf{W}) \leq R_1(\tilde{\boldsymbol{\theta}}_n : \mathbf{W}). \tag{12.3.65}$$

The picture changes as soon as Δ^2 moves away from 0, namely (1) the risk of $\tilde{\boldsymbol{\theta}}_n$ remains constant, (2) the risk of $\hat{\boldsymbol{\theta}}_n$ becomes unbounded, (3) the risk of $\hat{\boldsymbol{\theta}}_n^{\mathrm{PT}}$ grows to a maximum and then drop toward the risk of $\tilde{\boldsymbol{\theta}}_n$, and (4) the risk of $\hat{\boldsymbol{\theta}}_n^S$ merges to the risk of $\tilde{\boldsymbol{\theta}}_n$ as $\Delta^2 \to \infty$ from below. It may be calculated that the risk of $\hat{\boldsymbol{\theta}}_n^S$ is greater than that of $\hat{\boldsymbol{\theta}}_n^{\mathrm{PT}}$ whenever

$$2h_{m+2}(\chi_m^2(\alpha) : 0) \leq \frac{2}{m} + \alpha, \tag{12.3.66}$$

where $h_{m+2}(\cdot; 0)$ is the pdf of $H_{m+2}(\cdot : 0)$ with $(m+2)$ d.f. and α is the level of the preliminary test of significance. As soon as (12.3.66) is violated, the ordering of the risks becomes

$$R_2(\hat{\boldsymbol{\theta}}_n : \mathbf{W}) \leq R_4(\hat{\boldsymbol{\theta}}_n^S : \mathbf{W}) \leq R_3(\hat{\boldsymbol{\theta}}_n^{\mathrm{PT}} : \mathbf{W}). \tag{12.3.67}$$

Finally, we conclude that for $m \geq 3$, $\hat{\boldsymbol{\theta}}_n^S$ is a prefered estimator than $\hat{\boldsymbol{\theta}}_n^{\mathrm{PT}}$, while for $m \leq 2$, $\hat{\boldsymbol{\theta}}_n^{\mathrm{PT}}$ is preferable.

As regards, $\hat{\boldsymbol{\theta}}_n^{S+}$, $\hat{\boldsymbol{\theta}}_n^S$ and $\tilde{\boldsymbol{\theta}}_n$, the asymptotic risks may be ordered as follows:

$$R_5(\hat{\boldsymbol{\theta}}_n^{S+} : \mathbf{W}) \leq R_4(\hat{\boldsymbol{\theta}}_n^S : \mathbf{W}) \leq R_1(\tilde{\boldsymbol{\theta}}_n : \mathbf{W}) \quad \forall\, \Delta^2. \tag{12.3.68}$$

Hence, among the five estimators of $\boldsymbol{\theta}$, $\hat{\boldsymbol{\theta}}_n^{S+}$ is the most preferred for $m \geq 3$.

12.4 Conclusions

In this chapter, we considered three discrete models related to the baseball data analysis, meta-analysis and $r \times c$ contingency tables. Asymptotic properties of the point estimators were studied, and the usual conclusions on the dominance properties of the estimators were arrived at.

12.5 Problems

1. (Refer to Section 12.1.2). Verify that the joint density of $(\boldsymbol{\theta}', \mathbf{Y}')'$ given (K, η) is

$$(B[K\eta, K(1-\eta)])^{-p} \prod_{i=1}^{p} \begin{pmatrix} \eta_j \\ y \end{pmatrix} \theta_i^{y_i + K\eta - 1}(1-\theta_i)^{n_i - y_i + K(1-\eta) - 1}.$$

2. (Continuation.) Verify that the marginal distribution of $\mathbf{Y}$ given (K, η) is

$$\prod_{i=1}^{p} \begin{pmatrix} \eta_j \\ y \end{pmatrix} (K\eta)^{(y_i)}[K(1-\eta)]^{(n_i - y_i)} K^{-(n_i)}.$$

3. (Continuation.) Verify that the posterior distribution of $\boldsymbol{\theta}$ given $(\mathbf{Y}', K)'$ is

$$C \int_0^1 \pi(\boldsymbol{\theta}, \mathbf{Y}|K, \eta) m_1(\mathbf{Y}|K, \eta)[\eta(1-\eta)]^{-1} d\eta,$$

where C is the constant of the integration.

4. Verify, based on Problem 3, that

   (a) $E[\theta_i|\mathbf{Y}] = \frac{y_i + KE(\eta|\mathbf{Y})}{n_i+K}$

   and

   (b) $\text{Var}(\theta_i|\mathbf{Y}) = E[\text{Var}(\theta_i|\mathbf{Y},\eta)] + \text{Var}[E(\theta_i|\mathbf{Y},\eta)]$
   $= E\left[\frac{y_i+K\eta}{n_i+K}\left(1-\frac{y_i+K\eta}{n_i+K}\right)(n_i+K-1)^{-1}\right] + \text{Var}\left[\frac{y_i+K\eta}{n_i+K}\right].$

5. Show that as $K \to \infty$, the posterior distribution of η is approximately a beta distribution with parameters $n\bar{y}$ and $n(1-\bar{y})$.

6. Verify that $\lim_{K\to\infty} E(\theta_i|\mathbf{Y}) = \lim_{K\to\infty} E(\eta|\mathbf{Y}) = \bar{y}$ and $\lim_{K\to\infty} \text{Var}(\theta_i|\mathbf{Y}) = \lim_{K\to\infty} \text{Var}(\eta|\mathbf{Y}) = \frac{\bar{y}(1-\bar{y})}{n+1}$.

7. Verify that $\lim_{K\to 0} E(\theta_i|\mathbf{Y}) = \lim_{K\to 0} E(\eta|\mathbf{Y}) = \bar{y}$ and $\lim_{K\to 0} \text{Var}(\theta_i|\mathbf{Y}) = \lim_{K\to 0} \text{Var}(\eta|\mathbf{Y}) = \frac{\bar{y}(1-\bar{y})}{n+1}$.

8. Show that the empirical Bayes estimators of $\boldsymbol{\theta}$ can be written as

$$\hat{\boldsymbol{\theta}}_n^{EB} = \tilde{\boldsymbol{\theta}}_n - (\tilde{\boldsymbol{\theta}}_n - \hat{\boldsymbol{\theta}}_n)g(D_n),$$

   where $g(D_n) = \min(1, (p-3)D_n^{-1})$.

9. Prove Theorem 12.1.2.

10. Refer to Section 12.1.4. Verify the expressions for ADB, ADQB, ADMSE, and ADQR from (i) through (v).

11. Prove that the asymptotic variance of the three estimators given at (12.2.8) is $\psi_0^2 w^{-1}$, $w = w_1 + \cdots + w_k$.

12. Prove Theorem 12.2.2.

13. Refer to Section 12.2.4. Verify the expressions for ADB, ADQB, ADMSE, and ADQR from (i) through (v).

14. Refer to Section 12.1.7. Verify the expressions for the asymptotic coverage probabilities of $C^*(\boldsymbol{\theta}_n^*)$.

15. Show that $M(\mathbf{n}|K) \simeq C\tau^{m/2}\exp\{-\frac{1}{2}G_n^2\}$ as $n \to \infty$, where

$$G_n = -2n\sum_{i=1}^{r}\sum_{j=1}^{c} ln(\hat{\theta}_{ij}/\tilde{\theta}_{ij}) \xrightarrow{\text{P}} D_n \quad \text{as } n \to \infty.$$

   See formula at (12.3.33).

16. Refer to (12.3.27b). Show that $\lambda(K) \approx (1 - dD_n^{-1})I(D_n > d)$ as $n \to \infty$.

17. Prove Theorem 12.3.2.

References

1. Adichie, J. N. (1967). Estimates of regression parameters based on ranks. *An. Math. Statist.* **38**:894–904.
2. Agresti, A. (1990). *Categorical Data Analysis.* Wiley, New York.
3. Ahmed, S. E., and Saleh, A. K. Md. E. (1988). Estimation strategy using a preliminary test in some normal models. *Soochow J. of Math.* **14**:135–165.
4. Ahmed, S. E., and Saleh, A. K. Md. E. (1990). Estimation strategies for intercept vector in a simple multivariate regression model. *Comput. Statist. Data Anal.*, **10**:193–206.
5. Ahmed, S. E., and Saleh, A. K. Md. E. (1993). Improved estimation for the component mean vector. *J. of the Jpn. Statist. Soc.* **43**:177–195.
6. Ahmed, S. E., and Saleh, A. K. Md. E. (1999). Improved nonparametric estimation of location vectors in multivariate regression models. *J. Nonparametric Statist.* **11**:51–78.
7. Ahsanullah, M. (1971), On the estimation of means in a bivariate normal distribution with equal marginal variances. *Biometrika* **58**:230–233.
8. Ahsanullah, M., and Saleh, A. K. Md. E. (1972), Estimation of intercept in a linear regression model with one dependent variable after a preliminary test on the regression coefficient, *Int. Statist. Rev.* **40**:139–145.
9. Akritas, M. G., and Johnson, R. A. (1982a). Efficiencies of tests and estimators in autoregression under onnormal error distribution. *An. Inst. Math. Statist*, **A34**:579–589.
10. Akritas, M. G., and Johnson, R. A. (1982b). Asyptotic inference in continuous time, diffusions and Gaussian processes with known covariance. *J. Mult. Anal.* **12**:123–135.
11. Akritas, M., Saleh, A. K. Md. E. and Sen, P. K. (1985). Nonparametric estimation of intercepts after a preliminary test on parallelism of several regression lines. *Biostatistics: Statistics in Biomedical, Public Health and Environmental Sciences* Ed. P. K. Sen, Elsevier Science, North-Holland, pp. 221–235.
12. Albert, J. H. (1984). Empirical Bayes estimation of a set of binomial probabilities. *J. Statist. Comput. Simulation* **20**:129–144.
13. Albert, J. H. (1987). Empirical Bayes estimation in contingency tables. Commun. Statist.–Theory Meth., **16**:2459–2485.
14. Albert, J. H., and Gupta, A. K. (1981). Bayesian methods for binomial data with applications to a nonresponse problem. Technical report, Department of Mathematics and Statist., Bowling Green State University.

15. Albert, J. H., and Gupta, A. K. (1982). Mixture of Dirichlet distributions and estimation in contingency tables. *An. Statist.* **10**:1260–1268.

16. Albert, J. H., and Gupta, A. K. (1983a). Bayes estimation methods for 2 × 2 contingency tables using mixtures of Dirichlet distributions. *J. Amer. Statist. Assoc.*, **78**:708–717.

17. Albert, J. H., and Gupta, A. K. (1983b). Estimation in contingency tables using prior information. *J. Roy. Statist. Soc.* B **45**:60–69.

18. Ali, M. Abdunnabi. (1990). Interface of preliminary test approach and empirical Bayes approach to shrinkage estimation. Ph.D. thesis. Carleton University, Ottawa, Canada.

19. Ali, M. Abdunnabi, and Saleh, A. K. Md. E. (1991). Estimation of means and treatment effects in a one-way ANOVA model. *Soochow J. of Math.* **17**:287–309.

20. Ali, M. Abdunnabi, and Saleh, A. K. Md. E. (1991a). Asymptotic Theory for Simultaneous Estimation of Binomial Means. *Statist. Sinica* **1**:271–294.

21. Ali, A. M., and Saleh, A. K. Md. E. (1991b). Preliminary test and empirical Bayes approach to shrinkage estimation of regression parameters. *J. Jpn. Statist. Soc.*, **21**(1):401–416.

22. Anderson, T. W. (1984). *Introduction to Multivariate Analysis.* Wiley, New York.

23. Anderson, D. R., Sweeney, D. J., and Williams, T. A. (1993). *Statistics for Business and Economica*, 5th ed., West Publishing, Boulder, CO.

24. Arnold, Steven F. (1981). *The Theory of Linear Models and Multivariate Analysis.* Wiley, New York.

25. Asano, C., and Sato, S. (1962). *A Bivariate Analogue of Pooling Data. Bull. Math. Statist.* **10**:39–59.

26. Baksalary, J. K., and Kala, R. (1983). Partial ordering between matrices one of which is of rank one. *Bull. Polish Acad. of Sci., Math.*, **31**:5–7.

27. Bancroft, T. A. (1944). On biases in estimation due to the use of preliminary tests of significance. An. Math. Statist., **15**:190–204.

28. Bancroft, T. A. (1964). Analysis and inference for incompletely specified models involving the use of preliminary test(s) of significance. *Biometrics* **20**:427–442.

29. Bancroft, T. A. (1965). Inference for incompletely specified models in the physical sciences (with discussion). Bull. ISI, Proc. 35th Session, **41**(1):497–515.

30. Bancroft, T. A., and Han, C.-P. (1977). Inference based on conditional specification: A note and a bibliography. *ISI Rev.* **45**:117–127.

31. Baranchik, A. (1970). A family of minimax estimators of the mean of a multivariate normal distribution. *An. Math. Stat.* **41**:642–645.

32. Benda, N. (1996). Pre-test estimation and design in linear model. *J. Statist. Plan and Infer.* **52**:225–240.

33. Bennett, B. M. (1952). Estimation of means on the basis of preliminary tests of significance. An. Inst. Stat. Math. **4** 31–43.

34. Bennett, B. M. (1956). On the use of preliminary tests in certain statistical procedures, An. Inst. Stat. Math. **8**:45–57.

35. Berenblutt, I. I., and Webb, G. I. (1973). A new test for autocorrelated errors in the linear regression model. *J. Roy. Statist. Soc.* B**35**:33–50.

36. Berger, J. O. (1976). Admissible minimax estimation of a multivariate normal mean with arbitrary quadratic loss. *An. Statist.* **4**:223–226.
37. Berger, J. O. (1980a). Statistical Decision Theory. Springer-Verlag, New York.
38. Berger, J. O. (1980b). A robust generlized Bayes estimator and confidence region for a multivariate normal mean. *An. Statist.* **8**:716–761.
39. Berger, J. O. (1985). *Statistical Decision Theory and Bayesian Analysis*, (2nd edition). Springer–Verlag, New York.
40. Berger, J. O., Bock, M. E., Brown, L. D., Casella, G., and Gleser, L. (1977). Minimax estimation of a normal mean vector for arbitrary quadratic loss and unknown covariance matrix. *An. Statist.* **5**:736–771.
41. Berry, J. C. (1994). Improving the James-Stein estimator using the Stein variance estimator. *Statist. Prob. Lett.* **20**:241–245.
42. Bickel, P. J., and Doksum, K. A. (2001). *Mathematical Statististics: Basic Ideas*, Vol. 1. Holden-Day, Oakland, CA
43. Bishop, Y. M. M., Fienberg, S. E., and Holland, P. W. (1975). *Discrete Multivariate Analysis.* MIT Press, Cambridge.
44. Bock, M. E. (1988). Shrinkage estimators: pseudo-Bayes rules for normal mean vectors, Vol 1. *Proc. Fourth Purdue Symp. Stat. Dec. Theo. Rel. Topics*, (eds. S. S. Gupta and J. Berger) Springer Verlag, New York.
45. Bock, M. E., Yancey, T. A., and Judge, G. G. (1973). The statistical consequence of preliminary test estimators in regression. *J. Amer. Statist. Assoc.* **68**:109–116.
46. Bolfaine, H., and Zacks, S. (1992). *Prediction Theory for Finite Population.* Springer Verlag, New York.
47. Bozivich, H., Bancroft, T. A., and Hartley, H.O. (1956). Power of analysis of variance test procedure for certain incompletely specified models. *An. Math. Statist.* **27**:1037–1043.
48. Brewster, J.F., and Zidek, J. V. (1974). Improving on equivariant estimators. *An. Statist.* **2**:21–38.
49. Brockwell, P. J., and Davis R. A. (1996). *Introduction to Times Series and Forecasting.* Spinger, New York.
50. Brown, L. D. (1966). On the admissibility of invariant estimators of one or more location parameters. *An. Math. Statist.* **37**:1087–1136.
51. Brown, L. D. (1988). The differential inequality of a statistical estimation problem. *Proc. Fourth Purdue Symp. Stat. Dec. Theo. Rel. Topics*, Vol 1 (eds. S.S. Gupta and J. Berger). Springer Verlag, New York.
52. Casella, G. (1985). An introduction to empirical Bayes data analysis. *J. Amer. Statist. Assoc.* **39** (2):83–87.
53. Casella, G., and Berger, R. L. (1990). *Statist. Inference.* Duxbury Press. Belmount, California.
54. Casella, G., and Hwang, J. T. (1983). Empirical Bayes confidence sets for the mean of a multivariate normal distribution, *J. of Amer. Statist. Assoc.*, **78**:688–698.
55. Casella, G., Hwang, J. T. (1986). Confidence sets and the Stein-effect, *Com. Stat. Theo. Meth.* **15**:2043–2063.
56. Casella, G., and Hwang, J. T. (1987). Employing vague prior information in the construction of confidence sets. *J. Mult. Anal.*, **21**:79–104.

57. Cellier, D., Fourdrinier, D., and Robert, C. (1989). Robust shrinkage estimators of the location parameter for elliptically symmetrical distributions. *J. Mult. Anal.* **29**:39–52.

58. Chen, J., and Hwang, J. T. (1988). Improved set estimators for the coefficients of a linear model when the error distribution is spherically symmetric with unknown variances. *Can. J. Statist.* **16**:293–299.

59. Chen E.J., and Saleh, A. K. Md. E. (1993). Estimation of regression parameters when the errors are autocorrelated. Proc. 3rd Pacific Area Stat. Conference Stat. Sci. and Data Analysis. VSP Utrect, The Netherlands (eds. Matrusita, Puri and Hawakawa) 61–76.

60. Chiou, P. C., and Saleh, A. K. Md. E. (2002). Preliminary test confidence sets for the mean of a multivariate normal distribution. *J. Prop. Prob. Statist.* **2**:177–189.

61. Cochrane, D., and Orcutt, G. H. (1949). *Application of least squares regression to relationships containing autocorrelated error terms. J. Amer. Statist. Assoc.*, **44**:32–61.

62. Cohen, A. (1965). Estimates of linear combinations of the parameters in the mean vector of a multivariate distribution. An. Math. Statist. **36**:78–87.

63. Cohen, A., and Strawderman, W. (1973). Admissibility implications for different criteria in confidence estimation. An. Stat. **1**:363–366.

64. Cramer, H. (1946). *Mathematical Methods of Statistics*. Princeton University Press, Princeton.

65. Deely, J., and Lindley, D. (1981). Bayes empirical Bayes. *J. Amer. Statist. Assoc.* **76**:833–841.

66. Dempster, A. P., Schatzoff, M., and Wermuth, N. (1977). *A simulation study of alternatives to ordinary least squares. J. Amer. Statist. Assoc.* **72**:77–91.

67. Doob, J.L. (1953). *Stochastic Processes*. Wiley, New York.

68. Durbin, J. (1960). Estimation of parameters in time series regression models. *J. Roy. Statist. Soc.* B**22**:139–153.

69. Durbin, J., and Watson, G. S. (1951). Testing for serial correlation in least squares regression. I. *Biometrika* **37**:409–428.

70. Durbin, J., and Watson, G.S. (1950). Testing for serial correlation in least squares regression. II. *Biometrika* **38**:159–177.

71. Dzhaparidze, K. (1986). *Parameter Estimation and Hypothesis Testing in Spectral Analysis of Stationary Time Series*. Springer Verlag, New York.

72. Efron, B. (1975). Bias versus unbiased estimation. *Adv. Math.* **16**:259–277. The Statistical Century. *RSS News* **22**(5):1–2.

73. Efron, B., and Morris, C. (1972). Limiting the risk of Bayes and empirical Bayes estimators - Part II: The empirical Bayes case. *J. Amer. Statist. Assoc.* **67**:130–139.

74. Efron, B., and Morris, C. (1973). Stein's estimation rule and its competitors - An empirical Bayes approach. *J. Amer. Statist. Assoc.*, **68**:117–130.

75. Efron, B., and Morris, C. (1975). Data analysis using Stein's estimator and its generalizations. *J. Amer. Statist. Assoc.* **70**:311–319.

76. Efron, B., and Morris, C. (1977). Stein's paradox in statistics. *Sci. Am.* **236**(5):119–127.

77. Faith, R. E. (1976). Minimax Bayes set and point estimators of a multivariate normal mean. Tech. Report 66, University of Michigan, Ann Arbor.

78. Farebrother, R. W. (1975). The minimum mean square error linear estimator and ridge regression. *Technometrics*, **17**:127–128.

79. Feller, W. (1954). *An Introduction to Probability Theory and its Applications.* Wiley, New York.

80. Ferguson, T. S. (1967). *Mathematical Statistics: A Decision Theoretical Approach.* Academic Press, New York.

81. Ferguson, T. S. (1996). *A course in Large Sample Theory.* Chapman and Hall, New York.

82. Fuller, W. (1976). Introduction to time series. Wiley, New York.

83. Gart, J. J. (1962). On the combination of relative risks. *Biometrica* **18**:471–475.

84. Gart, J. J. (1992). Pooling 2 × 2 tables: Asymptotic moments of estimators. *J. Roy. Statist. Soc.* B**54**:531–539.

85. Ghosh, M., Hwang, J., and Tsui, K. (1983). Construction of improved estimators in multiparameter estimation for discrete exponential families. Discussion by J. O. Berger, H. Malcolm Hudson and Carl Morris. *An. Statist.* **11**:351–376.

86. Ghosh, M., Saleh, A. K. Md. E., and Sen, P. K. (1989). Empirical Bayes subset estiation in regression models. *Stat. Dec.*, **7**:15–35.

87. Gibbons, D. G. (1981) A simulation study of some ridge estimators. *J. Amer. Stat. Assoc.*, **76**:131–139.

88. Graybill, F. A. (1976). *Theory and Application of the Linear Model.* Duxbury Press.

89. Griliches, Z., and Rao, P. (1969). Small sample properties of several two-stage regression methods in the context of autocorrelated errors. *J. Amer. Stat. Assoc.*, **64**:253–272.

90. Gruber, M.H.J. (1998). *Improving Efficiency by Shrinkage: The James-Stein and Ridge Regression Estimators.* Marcel Dekker, New York.

91. Gupta, A.K., Saleh, A. K. Md. E., and Sen, P. K. (1989). Improved estimation in a contingency table: Independence structure. *J. Amer. Stat. Assoc.*, **84**:525–532.

92. Gupta, A.K., and Saleh, A. K. Md. E. (1997). Estimation Odds-Ratio: Homogeneity Constraints. *J. Ital. Stat. Soc.* **6**(1):67–81.

93. Hajek, J. (1969). *Nonparametric Statistics.* Holden Day, San Francisco.

94. Hajek, J., and Sidak, Z. (1967). *Theory of Rank Tests.* Academic Press, New York.

95. Hajek, J., Sidak, Z., and Sen, P.K. (1999). *Theory of Rank Tests.* Academic Press, New York.

96. Hald, A. (1952). *Statistical Theory with Engineering Applications.* Wiley, New York.

97. Hall, P., and Heyde, C. C. (1980). *Martingale limit theory and its applications.* Academic Press, New York.

98. Hallin, M., and Puri, M. L. (1988). Optimal rank-based procedures for time-series analysis: testing an ARMA model against other ARMA models. *An. Statist.* **16**:402–432.

99. Hannan, E. J. (1970). *Multiple Time Series.* Wiley, New York.

100. Han, C. P., and Bancroft, T. A. (1968). On pooling means when variance is unknown. *J. Amer. Stat. Assoc.*, **63** 1333–1342.

101. Haq, M. S., and Kibria, B. M. G. (1996). A shrinkage estimator for the restricted linear regression model: Ridge regression approach. *J. Appl. Stat. Sci.* **3(4)**:301–316.

102. Hemmerle, W. J., and Brantle, T. F. (1978). Explicit and constraint generalized ridge estimator. *Technometrics* **2**:109–120.

103. Hocking, R. R., Speed, F. M., and Lynn, M. J. (1976). A class of biased estimators in linear regression. *Technometrics* **18**:425–438.

104. Hodges, J. L., and Lehmann, E. L. (1963). Estimates of location based on rank tests. *An. Math. Stat.* **34**:598–611.

105. Hoeffding, W. (1948). A class of statistics with asymptotically normal distribution. *An. Math. Stat.* **19**:293–325.

106. Hoerl, A. E., and Kennard, R. W. (1970). Ridge regression: biased estimation for nonorthogonal problems. *Technometrics* **12**:55–67.

107. Hoerl, A. E., Kennard, R. W., and Baldwin, K. F. (1975). Ridge regression: Some simulation. *Commun. Statist.* **4**:105–123.

108. Hoffmann, K. (1992). *Improved Estimation of Distribution Parameters: Stein-type Estimators.* Teubner Verlagsgesellschaft, Stuttgart.

109. Hogg, R. V., McKean, J. W. and Craig, A. T. (2005). *Introduction to Mathematical Statistics* (7th ed.). Prentice Hall, NJ.

110. Huntsberger, D. V. (1955). A generalization of a preliminary test procedure of pooling data. *An. Math. Stat.* **26**:734–743.

111. Hušková, M. (1971). Asymptotic distribution of rank statistics used for multivariate testing symmetry. *J. Mult. Anal.* **1**:461–484.

112. Hwang, J. T. (1985). Universal domination and stochastic domination: Estimation simultaneously under a broad class of loss functions. *An. Stat.* **13**:295–315.

113. Hwang, J. T., and Casella, G. (1982). Minimax confidence sets for the mean of a multivariate normal distribution. *An. Stat.* **10**:868–881.

114. Hwang, J. T., and Casella, G. (1984). Improved set estimators for a multivariate normal mean. *Stat. Decision, suppl.* **1**:3–16.

115. Hwang, J. T., and Chen, J. (1986). Improved confidence sets for the coefficients of a linear model with spherically symmetric errors. *An. Stat.* **14**:444–460.

116. Hwang, J. T., and Ullah, A. (1989). Confidence sets recentered at James-Stein estimators - a surprise concerning the unknown variance case. Technical Report, Mathematics Department, Cornell University.

117. Inoue, T. (2001). Improving the "HKB" ordinary type ridge estimators. *J. Jpn. Stat. Soc.* **31(1)**:67–83.

118. James, W. and Stein, C. (1961). Estimation with quadratic loss. *Proc. Fourth Berkeley Symp. on Math. Statist., and Probability,* University of California, Vol. **1** pp. 361–379.

119. Joshi, V. M. (1967). Inadmissibility of the usual confidence set for the mean of a multivariate normal population. *An. Math. Stat.* **38**:1868–1875.

120. Joshi, V. M. (1969). Admissibility of the usual confidence sets for the mean of a univariate or bivariate normal population. *An. Math. Stat.* **40**:1042–1067.

121. Johnson, N., Kotz, S., and Balakrishnan, N. S. (1994). Continuous univariate distributions - 1. Wiley, New York.

122. Judge, G. G., and Bock, M. E. (1978). *The Statistical Implications of Pre-test and Stein-rule Estimators in Econometrics.* North-Holland, Amsterdam.

123. Jurečková, J. (1969). Asymptotic linearity of a rank statistic in regression parameter. *An. Math. Statist.* **40**:1889–1900.

124. Jurečková, J., and Sen, P. K. (1996). Robust Statistical Procedures: Asymptotics and Interrelations. Wiley, New York.

125. Kakwani, T. (1968). Note on the unbiased of a mixed regression estimator. *Econometrics* **36**:610–611.

126. Kale, B. K., and Bancroft, T. A. (1967). Inference for some incompletely specified models involving normal approximations to discrete data, *Biometrics* **23**:335–348.

127. Kendall, M. G., and Stuart, A. (1963). The Advanced Theory of Statistics, 2 Vols. Hafner Publishing, New York.

128. Khan, Bashir Ul. (1997). Some contribution to positive part shrinkage estimation in various model. Ph.D. thesis. University of Regina, Regina, Canada.

129. Khan, Bashir Ul. and Saleh, A. K. Md. E. (2005). Estimation of Regression Parameters: Parallelism Restriction. *J. Statist. Theory Appl.* **4**:91–107.

130. Khan, S., and Saleh, A. K. Md. E. (1997). Shrinkage pre-test estimator of the intercept parameter for a regression model with multivariate Student's t-errors. *Biomed. J.* **2**:131–147.

131. Khan, S., and Saleh, A. K. Md. E. (2001). On the comparison of pre-test and shrinkage estimators for the univariate normal mean. *Stat. papers* **42**(4):451–473.

132. Khan, S., Hoque, Z., and Saleh, A. K. Md. E. (2002). Estimation of the slope parameter for linear regression model with uncertain prior information. *J. Stat. Res.* **36**(2):55–74.

133. Ki, F., and Tsui, K. (1985). Improved confidence set estimators of a multivariate normal mean and generalizations. *An. Inst. Statist. Math.* **37**:487–498.

134. Kibria, B. M. G. (1996a). On preliminary test ridge regression estimator for the restricted linear model with non-normal disturbances. *Commun. Stat. Theory Meth.* **25**:2349–2369.

135. Kibria, B. M. G. (1996b). On shrinkage ridge regression estimators for restricted linear models with multivariate t disturbances Students, **1 (3)** 177–188.

136. Kibria, B. M. G. (2003). Performance of some new ridge regression estimators. *Commun. Statist. Simul. Comput.* B**32**:429–435.

137. Kibria, B. M. G. and Saleh, A. K. Md. E. (1993). Performance of shrinkage preliminary test estimator in regression analysis. *Jahangirnagar Review* A**17**:133–148.

138. Kibria, B. M. G. and Saleh, A. K. Md. E. (2003). Effect of W, LR, and LM tests on the performance of preliminary test ridge regression estimators. *J. Jpn. Stat. Soc.* **33**:119–136.

139. Kibria, B. M. G. and Saleh, A. K. Md. E. (2004a). Preliminary test ridge regression estimators with Student's t errors and conflicting test-statistics, *Metrika* **59**:105–124.

140. Kibria, B. M. G. and Saleh, A. K. Md. E. (2004b). Performance of Positive Rule Ridge Regression Estimators for the Ill-Conditioned Gaussian Regression Models. *Calcutta Stat. Assoc. Bull.* **55** to appear.

141. Kim, P. T. (1987). Recentered confidence sets for the mean of a multivariate distribution when the scale parameter is unknown. Ph.D. thesis. Department of Mathematics, University of California at San Diego.

142. Kim, H. M., and Saleh, A. K. Md. E. (2003). Preliminary test estimators of the parameters of simple linear model with measurement errors. *Metrika* **57**:223–251.

143. Kim, H. M., and Saleh, A. K. Md. E. (2004). Improved estimation of regression parameters in measurement error model. *J. Mult. Anal.* **95**:273-300.

144. Kitagawa, T. (1963). Estimation after preliminary tests of significance. *UC Publ. Stat.* **3**:147–186.

145. Kmenta, J. and Gilbert, R. F. (1968). Small sample properties of alternative estimates of seemingly unrelated regressions. *J. Amer. Stat. Assoc.* **63**:1180–1200.

146. Kmenta, J. and Gilbert, R. F. (1970). Estimation of seemingly unrelated regressions with autoregressive disturbances. *J. Amer. Stat. Assoc.*, **65**:186–197.

147. Koul, H. L. (1977). Behavior of robust estimators in the regression model with dependent errors. *An. Statist.* **5**:681–699.

148. Koul, H. L., and Saleh, A. K. Md. E. (1993). *R*-Estimation of the parameters of autoregressive AR[p] models. *An. Statist.* **21**:534–551.

149. Koul, H. L., and Saleh, A. K. Md. E. (1995). Autoregression quantiles and related rank-scores processes. *An. Statist.* **23**:670-689.

150. Kubokawa, T. (1991). An approach to improving the James-Stein estimator. *J. Mult. Anal.* **36**:121–126.

151. Lambert, A., Saleh, A. K. Md. E. and Sen, P. K. (1985). On least squares estimation of intercept after a preliminary test on parallelism of regression lines. *Commun. Statist. Theory Math.* **14**(4):793–807.

152. Lawless, J.F. (1978) Ridge and related estimation procedure. *Commun. Statist.*, A**7**:139–164.

153. Lawless, J.F., and Wang, P. (1976). A simulation study of ridge and other regression estimators. *Commun. Statist.* A**5**:307-323.

154. Lindley, D. (1962). Discussion of Professor Stein's paper, "Confidence sets for the mean of a multivariate normal distribution". *J. Roy. Stat. Soc.* B**24**:265–296.

155. Lindley, D., and Smith, A. F. M. (1972). Bayes estimates for the linear model (with discussion) *J. Roy. Stat. Soc.* B**34**:1–41.

156. Lu, K., and Berger, J. (1989). Estimated confidence procedures for multidimensional normal means. J. Stat. Plann. Inf. **23**(1):1–20.

157. Maatta, J. M., and Casella, G. (1990). Variance estimation. *Statist. Sci.* **5**:107–109.

158. Malthouse, E. C. (1999). Shrinkage estimation and direct marketinmg scoring model. *J. Interactive Market.* **13(4)**:10–23.

159. Marquardt, D. W., and Snee, R. D. (1975). Ridge regression in practice. *Am. Statistician* **29**:3–20.

160. McDonald, G. C., Galarneau, D. I. (1975). A monte carlo evaluation of some Ridge-type estimators. *J. Am. Stat. Assoc.*, **70**:407–416.

161. Montgomery, D. C., Peck, E. A., and Vining, G. G. (2001). *Introduction to Linear Regression Analysis.* 3rd ed. Wiley, New York.

162. Morris, C. N. (1983). Parametric empirical Bayes inference: Theory and applications. *J. Am. Stat. Assoc.* **78**:47–65.

163. Mosteller, F. (1948). On pooling data. *J. Am. Stat. Assoc.*, **43**:231–242.

164. Nagar, A. L., and Kakwani, N. C. (1965). Note on the use of prior information in statistical estimation of economic relations. *Sankhya* A**27**:105–112.

165. Newhouse, J. P., and Oman, S. D. (1971). An Evaluation of Ridge Estimators. Rand Corporation, P-716-PR.

166. Obenchain, R. L. (1975). Ridge analysis following a preliminary test of the shrunken hypothesis. *Technometrics* **17**:431–441.

167. Prais, S. J., and Winsten, J. A. (1954). Tren estimators and series correlation. Cowles Commission Discussion Paper, **383**. Chicago.

168. Puri, M. L., and Sen, P. K. (1971). *Nonparametric Methods in Multivariate Analysis.* Wiley, New York.

169. Puri, M. L., and Sen, P. K. (1986). *Nonparametric methods in general linear models.* Wiley, New York.

170. Rao, C. R. (1973). *Linear Statistical Inference and its Applications.* Wiley, New York.

171. Rao, C. R. (1975). Simultaneous estimation of parameters in different linear models with applications to biometric problems. *Biometrics*, **31**:545–554.

172. Randles, R. H., and Wolfe, D. A. (1979). Introduction to the Theory of Nonparametric Statistics. Wiley, New York.

173. Rényi, A. (1970). *Foundations of Probability.* Holden-Day, San Francisco.

174. Robbins, H. E. (1955). An empirical Bayes approach to statistics. *Proc. 3rd Berkeley Symp. Meth. Statist.* **35**:1–20.

175. Roberg, S. J., and Stegeman, J. W. (1991). Pooling data for stability studies: testing the equality of batch degradation slopes. *Biometrics* **47**:1059–1069.

176. Robert, C. and Casella, G. (1987). Improved confidence sets in spherically symmetric distributions. *J. Mult. Anal.* **22**:300–315.

177. Robert, C. P., and Saleh, A. K. Md. E. (1991). Point estimation and confidence set estimation in a parallelism model: An empirical Bayes approach. *An. Econ. Statist.* **23**:65–89.

178. Rodrigue, J. (1987). Some shrunken predictors in finite population with a multinormal superpopulation model. *Stat. Prob.* **5**:347–351.

179. Rohatgi, V. K., and Saleh, A. K. Md. E. (2001). *Introduction to Probability and Statistics.* Wiley, New York.

180. Saleh, A. K. Md. E. (1992). On shrinkage estimation of the parameters of an autoregressive Gaussian process. *Theory Prob. Appl.* **37**:290–300.

181. Saleh, A. K. Md. E. (2003). Asymptotic properties of Stein-type confidence sets - a U-statistics Approach. *Sankhya* II, **65**:1–11.

182. Saleh, A. K. Md. E. and Han, C.-P. (1990). Shrinkage estimation in regression analysis. Estadistica **42** 40–43.

183. Saleh, A. K. Md. E. and Hassanein, K.M. (1986). On various F-estimation in a parallelism problem. *Soochow J. Math.* **12**:83–94.

184. A. K. Md. E., Hassanein, K.M., Hassein, R.S., and Kim, H.M. (2006). Quasi-empirical Bayes methodology for improving meta-analysis. *J. Biopharmaceutical Stat.* **16**:77–90.

185. Saleh, A. K. Md. E., and Kibria, B. M. G. (1993). Performances of some new preliminary test ridge regression estimators and their properties. *Commun. Stat. Theory Meth.* **22**:2747–2764.

186. Saleh, A. K. Md. E. and Sen, P. K. (1978). Non-parametric estimation of location parameter after a preliminary test on regression. *An. Statist.* **6**:154–168.

187. Saleh, A. K. Md. E., and Sen, P. K. (1983). Nonparametric tests of location after a preliminary test on regression in the multivariate case. Commun. Statist. Theory Meth. **12**(16):1855–1872.

188. Saleh, A. K. Md. E., and Sen, P. K. (1984). Least squares and rank order preliminary test estimation in general multivariate linear models. *Proc. ISI, Golden Jubilee Conf. on Statist.: Application and New Directions* (Dec. 16–19, 1981), pp. 237–253.

189. Saleh, A. K. Md. E., and Sen, P. K. (1984). Nonparametric preliminary test inference. *Handbook of Statist.* **4** (eds. P. R. Krishnaiah and P. K. Sen) North-Holland, Amsterdam, pp. 275–297.

190. Saleh, A. K. Md. E., and Sen, P. K. (1985). On shrinkage M-estimators of location parameters. Commun. Statist. Theory Meth. **14**(10):2313–2329.

191. Saleh, A. K. Md. E. and Sen, P. K. (1985a). Nonparametric shrinkage estimation in a parallelism problem. *Sankhya* **47**A:156–165.

192. Saleh, A. K. Md. E., and Sen, P. K. (1985b). Preliminary test predicted in general multivariate linear models. *Proc. Pacific Area Statistical Conf.* (ed. M. Matusita) (Dec. 15–19, 1982), North-Holland, Amsterdam, pp. 619–638,

193. Saleh, A. K. Md. E., and Sen, P. K. (1985c). Shrinkage least squares estimation in a general multivariate linear model. *Proc. 5th Pannonian Symp. on Math. Statistics* (eds. J. Mogyorodi, I. Vincze, and W. Wertz), pp. 307–325.

194. Saleh, A. K. Md. E. and Sen, P. K. (1985). On shrinkage least squares estimation in a parallelism problem. *Commun. Statist. Theory Meth.* **15**:1451–1466.

195. Saleh, A. K. Md. E. and Sen, P. K. (1986). On shrinkage R-estimation in a multiple regression model. *Commun. Statist. Theory Meth.* **15**(7):2229–2244.

196. Sarker, N. (1992). A new estimator combining the risge regression and the restricted least squares method of estimation. *Commun. Statist. Theory Meth.* **21**:1987–2000.

197. Sclove, S. L., Morris, C., and Radhakrishnan, R. (1972). Non-optimality of preliminary test estimators for the mean of a multivariate normal distribution. *An. Math. Statist.* **45**:1481–1490.

198. Sen, P. K. (1969). On a class of rank order tests for parallelism of several regression lines. *An. Math. Statist.* **40**:1668–1683.

199. Sen, P. K. (1970). On some convergence properties of one sample rank order statistics. *An. Statist.* **41**:2140–2143.

200. Sen, P. K. (1980). On almost sure linearity theorems for signed rank statistics. *An. Statist.* **8**:313–321.

201. Sen, P. K. (1981). *Sequential Nonparametrics: Invariance principle and Statistical Inference.* Wiley, New York.

202. Sen, P. K. (1984). A James-Stein type detour of U-Statistics. *Commun. Statist. Theory Meth.* **13**(22):2725–2747.

203. Sen, P. K. (1986). On the asymptotic distributional risks of shrinkage and preliminary test versions of maximum likelihood estimators. *Sankhya* A**48**:354–371.

204. Sen, P. K., and Saleh, A. K. Md. E. (1979). Nonparametric estimation of location parameter after a preliminary test on regression in multivariate case. *J. Mult. Anal.* **9**(2):322–331.

205. Sen, P. K., and Saleh, A. K. Md. E. (1985). On some shrinkage estimators of multivariate location. *An. Statist.* **13**:172–281.

206. Sen, P. K., and Saleh, A. K. Md. E. (1987). On preliminary test and shrinkage M-estimation in linear models. *An. Statist.* **15**(4):1580–1592.

207. Sen, P. K., and Singer, J.M. (1993). *Large Sample Methods in Statist.: An Introduction with Applications*, Chapman and Hall, New York.

208. Shinozaki, N. (1989). Improved confidence sets for the mean of a multivariate normal distribution. *An. Inst. Stat. Math.* **41**(2):331–346.

209. Singh, S., and Tracy, D. S. (1999). Ridge-regression using scrambled responses. *Metrika*, 147–157.

210. Snedecor, G. W. (1938). *Statistical Methods.* Collegiate Press, Iowa.

211. Srivastava, M. S., and Khatri, C. B. (1979). *An Introduction to Multivariate Statistics.* North Holland, Amsterdam.

212. Srivastava, M. S., and Saleh, A. K. Md. E. Saleh (2004). Estimation of the Mean Vector of a Multivariate Normal Distribution: Subspace Hypothesis. *J. Mult. Anal.* **96**:55-72.

213. Stein, C. (1955). Inadmissibility of the usual estimator of mean of a multivariate normal distribution. *Proc. 3rd Berkeley Symp. Math. Statist. Prob.* **1** pp. 197–206.

214. Stein, C. (1955). A necessary and sufficient condition for admissibility. *An. Math. Statist.* **26**:518–522.

215. Stein, C. (1956). Inadmissibility of the usual estimator of the mean of a multivariate normal distribution. *Proc. 3rd Berkeley Symp. Math. Statist. Probability* **1**:197–206.

216. Stein, C. (1962). Confidence sets for the mean of a multivariate normal distribution. *J. Roy. Stat. Soc.* B**24**:265–296.

217. Stein, C. (1964). Inadmissibility of the usual estimators of variance of a normal distribution with unknown mean. *An. Inst. Stat. Math.* **16**:155-160.

218. Stein, C. (1966). An approach to the recovery of inter-block information in balanced incomplete block designs. Research Papers in Statistics.

219. Stein, C. (1973). Estimation of the mean of a multivariate distribution. *Proc. Prague Symp. Asymptotic Statist.* pp. 345–381.

220. Stein, C. (1981). Estimation of the mean of multivariate normal distribution. *An. Statist.* **9**:1135–1151.

221. Strawdermann, W. E. (1971). Proper Bayes minimax estimators of the multivariate normal mean. *An. Math. Statist.* **47**:365–388.

222. Tabatabaey, S. M. (1995). Preliminary test approach estimation: regression model with spherically symmetric errors. Ph.D. thesis. Carleton University, Ottawa.

223. Tamura, R. (1967a). Small sample properties of a sometimes posted estimate in the nonparametric case. *Bull. Math. Statist.* **12**:75–83.

224. Tamura, R. (1967b). Some estimation procedures with a nonparametric preliminary test II. *Bull. Math. Statist.* **12**:47–59.

225. Tamura, R. (1967c). Some weighted estimates in the nonparametric case. *Bull. Math. Statist.* **12**:61–73.

226. Theil, H. (1963). On the use of incomplete prior information in regression analysis. *J. Am. Stat. Assoc.* **58**:401–414.

227. Theil, H. (1971). *Principle of Econometrics.* Wiley, New York.

228. Theil, H and Goldberger, A. S. (1961). On pure and mixed statistical estimation in economics. *Int. Econ. Rev.* **2**:65–78.

229. Thompson, S. G., and Pocock, S. J. (1991). Can meta-analysis be trusted? *Lancet* **338**:1127–1130.

230. Vinod, H. D., and Ullah, A. (1981). *Recent advances in regression methods.* Marcel Dekker, New York.

231. Wencheko, E. (2000). Estimation of the signal-to-noise in the linear regression model, *Statist. Papers* **41**:327–343.

232. Whittle, D. (1952). Estimation and information in time series analysis. *Skand. Akwur.* **35**:48–60.

Glossary

UE	Unrestricted estimator
UMLE	Unrestricted maximum likelihood estimator
ULSE	Unrestricted least-squares estimator
URE	Unrestricted rank estimator
UPPE	Unrestricted principal part estimator
URRE	Unrestricted ridge regression estimator
RE	Restricted estimator
RMLE	Restricted maximum likelihood estimator
RLSE	Restricted least-squares estimator
RRE	Restricted rank estimator
RPPE	Restricted principal part estimator
RUE	Restricted unbiased estimator
$R(\cdot;\cdot)$	Relative efficiency
$RRE(\cdot;\cdot)$	Risk-based relative efficiency
PTE	Preliminary test estimator
PTMLE	Preliminary test maximum likelihood estimator
PTLSE	Preliminary test least-squares estimator
PTPPE	Preliminary test principal part estimator
PTRE	Preliminary test rank estimator
PTRRE	Preliminary test ridge regression estimator
PPRSE	Positive-rule Stein-type estimator
PTPPE	Positive-rule Stein-type principal part estimator
PTRRE	Positive-rule Stein-type ridge regression estimator
JSE	James-Stein estimator
SE	Stein-type estimator
SMLE	Stein-type maximum likelihood estimator
SLSE	Stein-type least-squares estimator

SRE	Stein-type rank estimator
SRRE	Stein-type ridge regression estimator
SPPE	Stein-type principal part estimator
LSE	least-squares estimator
MLE	maximum likelihood estimator
MSE	Mean square error
MRE$(\cdot,\cdot)$	Mean square error based relative efficiency
ADB	Asymptotic distributional bias
ADQB	Asymptotic distributional quadratic bias
ADMSE	Asymptotic distributional mean square error
ADQR	Asymptotic distributional quadratic Risk
D.F.	Degrees of freedom
C.L.T.	Central limit theorem
EBE	Empirical Bayes estimation

Symbols and Meaning

$\hat{\theta}_n, \tilde{\beta}_n$	UE
$\tilde{\theta}_n, \hat{\beta}_n$	RE
$\hat{\theta}_n^{\mathrm{PT}}, \hat{\beta}_n^{\mathrm{PT}}$	PTE
$\hat{\theta}_n^{S}, \hat{\beta}_n^{S}$	SE
$\hat{\theta}_n^{S+}, \hat{\beta}_n^{S+}$	PRSE
$b(\cdot)/\mathbf{b}(\cdot)$	Bias (vector)
$B(\cdot)$	Quadratic bias
$\mathbf{M}(\cdot)$	MSE
$R(\cdot)$	Risk
$H_\nu(\cdot;\cdot)$	CDF of a noncentral chi-square distribution with ν D.F.
$G_{\nu_1,\nu_2}(\cdot;\cdot)$	CDF of a noncentral F distribution with (ν_1,ν_2) D.F.
$\mathcal{L}_\backslash$	test-statistic
Δ^2	noncentrality parameter

Authors Index

Subject Index

Estimators

Hypothesis

WILEY SERIES IN PROBABILITY AND STATISTICS
ESTABLISHED BY WALTER A. SHEWHART AND SAMUEL S. WILKS

The ***Wiley Series in Probability and Statistics*** is well established and authoritative. It covers many topics of current research interest in both pure and applied statistics and probability theory. Written by leading statisticians and institutions, the titles span both state-of-the-art developments in the field and classical methods.

Reflecting the wide range of current research in statistics, the series encompasses applied, methodological and theoretical statistics, ranging from applications and new techniques made possible by advances in computerized practice to rigorous treatment of theoretical approaches.

This series provides essential and invaluable reading for all statisticians, whether in academia, industry, government, or research.

† ABRAHAM and LEDOLTER · Statistical Methods for Forecasting
AGRESTI · Analysis of Ordinal Categorical Data
AGRESTI · An Introduction to Categorical Data Analysis
AGRESTI · Categorical Data Analysis, *Second Edition*
ALTMAN, GILL, and McDONALD · Numerical Issues in Statistical Computing for the Social Scientist
AMARATUNGA and CABRERA · Exploration and Analysis of DNA Microarray and Protein Array Data
ANDĚL · Mathematics of Chance
ANDERSON · An Introduction to Multivariate Statistical Analysis, *Third Edition*
* ANDERSON · The Statistical Analysis of Time Series
ANDERSON, AUQUIER, HAUCK, OAKES, VANDAELE, and WEISBERG · Statistical Methods for Comparative Studies
ANDERSON and LOYNES · The Teaching of Practical Statistics
ARMITAGE and DAVID (editors) · Advances in Biometry
ARNOLD, BALAKRISHNAN, and NAGARAJA · Records
* ARTHANARI and DODGE · Mathematical Programming in Statistics
* BAILEY · The Elements of Stochastic Processes with Applications to the Natural Sciences
BALAKRISHNAN and KOUTRAS · Runs and Scans with Applications
BARNETT · Comparative Statistical Inference, *Third Edition*
BARNETT and LEWIS · Outliers in Statistical Data, *Third Edition*
BARTOSZYNSKI and NIEWIADOMSKA-BUGAJ · Probability and Statistical Inference
BASILEVSKY · Statistical Factor Analysis and Related Methods: Theory and Applications
BASU and RIGDON · Statistical Methods for the Reliability of Repairable Systems
BATES and WATTS · Nonlinear Regression Analysis and Its Applications
BECHHOFER, SANTNER, and GOLDSMAN · Design and Analysis of Experiments for Statistical Selection, Screening, and Multiple Comparisons
BELSLEY · Conditioning Diagnostics: Collinearity and Weak Data in Regression

*Now available in a lower priced paperback edition in the Wiley Classics Library.
†Now available in a lower priced paperback edition in the Wiley–Interscience Paperback Series.

† BELSLEY, KUH, and WELSCH · Regression Diagnostics: Identifying Influential Data and Sources of Collinearity
BENDAT and PIERSOL · Random Data: Analysis and Measurement Procedures, *Third Edition*
BERRY, CHALONER, and GEWEKE · Bayesian Analysis in Statistics and Econometrics: Essays in Honor of Arnold Zellner
BERNARDO and SMITH · Bayesian Theory
BHAT and MILLER · Elements of Applied Stochastic Processes, *Third Edition*
BHATTACHARYA and WAYMIRE · Stochastic Processes with Applications
† BIEMER, GROVES, LYBERG, MATHIOWETZ, and SUDMAN · Measurement Errors in Surveys
BILLINGSLEY · Convergence of Probability Measures, *Second Edition*
BILLINGSLEY · Probability and Measure, *Third Edition*
BIRKES and DODGE · Alternative Methods of Regression
BLISCHKE AND MURTHY (editors) · Case Studies in Reliability and Maintenance
BLISCHKE AND MURTHY · Reliability: Modeling, Prediction, and Optimization
BLOOMFIELD · Fourier Analysis of Time Series: An Introduction, *Second Edition*
BOLLEN · Structural Equations with Latent Variables
BOLLEN and CURRAN · Latent Curve Models: A Structural Equation Perspective
BOROVKOV · Ergodicity and Stability of Stochastic Processes
BOULEAU · Numerical Methods for Stochastic Processes
BOX · Bayesian Inference in Statistical Analysis
BOX · R. A. Fisher, the Life of a Scientist
BOX and DRAPER · Empirical Model-Building and Response Surfaces
* BOX and DRAPER · Evolutionary Operation: A Statistical Method for Process Improvement
BOX, HUNTER, and HUNTER · Statistics for Experimenters: Design, Innovation, and Discovery, *Second Editon*
BOX and LUCEÑO · Statistical Control by Monitoring and Feedback Adjustment
BRANDIMARTE · Numerical Methods in Finance: A MATLAB-Based Introduction
BROWN and HOLLANDER · Statistics: A Biomedical Introduction
BRUNNER, DOMHOF, and LANGER · Nonparametric Analysis of Longitudinal Data in Factorial Experiments
BUCKLEW · Large Deviation Techniques in Decision, Simulation, and Estimation
CAIROLI and DALANG · Sequential Stochastic Optimization
CASTILLO, HADI, BALAKRISHNAN, and SARABIA · Extreme Value and Related Models with Applications in Engineering and Science
CHAN · Time Series: Applications to Finance
CHARALAMBIDES · Combinatorial Methods in Discrete Distributions
CHATTERJEE and HADI · Sensitivity Analysis in Linear Regression
CHATTERJEE and PRICE · Regression Analysis by Example, *Third Edition*
CHERNICK · Bootstrap Methods: A Practitioner's Guide
CHERNICK and FRIIS · Introductory Biostatistics for the Health Sciences
CHILÈS and DELFINER · Geostatistics: Modeling Spatial Uncertainty
CHOW and LIU · Design and Analysis of Clinical Trials: Concepts and Methodologies, *Second Edition*
CLARKE and DISNEY · Probability and Random Processes: A First Course with Applications, *Second Edition*
* COCHRAN and COX · Experimental Designs, *Second Edition*
CONGDON · Applied Bayesian Modelling
CONGDON · Bayesian Statistical Modelling
CONOVER · Practical Nonparametric Statistics, *Third Edition*
COOK · Regression Graphics

*Now available in a lower priced paperback edition in the Wiley Classics Library.
†Now available in a lower priced paperback edition in the Wiley–Interscience Paperback Series.

COOK and WEISBERG · Applied Regression Including Computing and Graphics
COOK and WEISBERG · An Introduction to Regression Graphics
CORNELL · Experiments with Mixtures, Designs, Models, and the Analysis of Mixture Data, *Third Edition*
COVER and THOMAS · Elements of Information Theory
COX · A Handbook of Introductory Statistical Methods
* COX · Planning of Experiments
CRESSIE · Statistics for Spatial Data, *Revised Edition*
CSÖRGŐ and HORVÁTH · Limit Theorems in Change Point Analysis
DANIEL · Applications of Statistics to Industrial Experimentation
DANIEL · Biostatistics: A Foundation for Analysis in the Health Sciences, *Eighth Edition*
* DANIEL · Fitting Equations to Data: Computer Analysis of Multifactor Data, *Second Edition*
DASU and JOHNSON · Exploratory Data Mining and Data Cleaning
DAVID and NAGARAJA · Order Statistics, *Third Edition*
* DEGROOT, FIENBERG, and KADANE · Statistics and the Law
DEL CASTILLO · Statistical Process Adjustment for Quality Control
DEMARIS · Regression with Social Data: Modeling Continuous and Limited Response Variables
DEMIDENKO · Mixed Models: Theory and Applications
DENISON, HOLMES, MALLICK and SMITH · Bayesian Methods for Nonlinear Classification and Regression
DETTE and STUDDEN · The Theory of Canonical Moments with Applications in Statistics, Probability, and Analysis
DEY and MUKERJEE · Fractional Factorial Plans
DILLON and GOLDSTEIN · Multivariate Analysis: Methods and Applications
DODGE · Alternative Methods of Regression
* DODGE and ROMIG · Sampling Inspection Tables, *Second Edition*
* DOOB · Stochastic Processes
DOWDY, WEARDEN, and CHILKO · Statistics for Research, *Third Edition*
DRAPER and SMITH · Applied Regression Analysis, *Third Edition*
DRYDEN and MARDIA · Statistical Shape Analysis
DUDEWICZ and MISHRA · Modern Mathematical Statistics
DUNN and CLARK · Basic Statistics: A Primer for the Biomedical Sciences, *Third Edition*
DUPUIS and ELLIS · A Weak Convergence Approach to the Theory of Large Deviations
* ELANDT-JOHNSON and JOHNSON · Survival Models and Data Analysis
ENDERS · Applied Econometric Time Series
† ETHIER and KURTZ · Markov Processes: Characterization and Convergence
EVANS, HASTINGS, and PEACOCK · Statistical Distributions, *Third Edition*
FELLER · An Introduction to Probability Theory and Its Applications, Volume I, *Third Edition,* Revised; Volume II, *Second Edition*
FISHER and VAN BELLE · Biostatistics: A Methodology for the Health Sciences
FITZMAURICE, LAIRD, and WARE · Applied Longitudinal Analysis
* FLEISS · The Design and Analysis of Clinical Experiments
FLEISS · Statistical Methods for Rates and Proportions, *Third Edition*
† FLEMING and HARRINGTON · Counting Processes and Survival Analysis
FULLER · Introduction to Statistical Time Series, *Second Edition*
FULLER · Measurement Error Models
GALLANT · Nonlinear Statistical Models
GEISSER · Modes of Parametric Statistical Inference
GEWEKE · Contemporary Bayesian Econometrics and Statistics

*Now available in a lower priced paperback edition in the Wiley Classics Library.
†Now available in a lower priced paperback edition in the Wiley–Interscience Paperback Series.

GHOSH, MUKHOPADHYAY, and SEN · Sequential Estimation
GIESBRECHT and GUMPERTZ · Planning, Construction, and Statistical Analysis of Comparative Experiments
GIFI · Nonlinear Multivariate Analysis
GIVENS and HOETING · Computational Statistics
GLASSERMAN and YAO · Monotone Structure in Discrete-Event Systems
GNANADESIKAN · Methods for Statistical Data Analysis of Multivariate Observations, *Second Edition*
GOLDSTEIN and LEWIS · Assessment: Problems, Development, and Statistical Issues
GREENWOOD and NIKULIN · A Guide to Chi-Squared Testing
GROSS and HARRIS · Fundamentals of Queueing Theory, *Third Edition*
† GROVES · Survey Errors and Survey Costs
* HAHN and SHAPIRO · Statistical Models in Engineering
HAHN and MEEKER · Statistical Intervals: A Guide for Practitioners
HALD · A History of Probability and Statistics and their Applications Before 1750
HALD · A History of Mathematical Statistics from 1750 to 1930
† HAMPEL · Robust Statistics: The Approach Based on Influence Functions
HANNAN and DEISTLER · The Statistical Theory of Linear Systems
HEIBERGER · Computation for the Analysis of Designed Experiments
HEDAYAT and SINHA · Design and Inference in Finite Population Sampling
HELLER · MACSYMA for Statisticians
HINKELMANN and KEMPTHORNE · Design and Analysis of Experiments, Volume 1: Introduction to Experimental Design
HINKELMANN and KEMPTHORNE · Design and Analysis of Experiments, Volume 2: Advanced Experimental Design
HOAGLIN, MOSTELLER, and TUKEY · Exploratory Approach to Analysis of Variance
HOAGLIN, MOSTELLER, and TUKEY · Exploring Data Tables, Trends and Shapes
* HOAGLIN, MOSTELLER, and TUKEY · Understanding Robust and Exploratory Data Analysis
HOCHBERG and TAMHANE · Multiple Comparison Procedures
HOCKING · Methods and Applications of Linear Models: Regression and the Analysis of Variance, *Second Edition*
HOEL · Introduction to Mathematical Statistics, *Fifth Edition*
HOGG and KLUGMAN · Loss Distributions
HOLLANDER and WOLFE · Nonparametric Statistical Methods, *Second Edition*
HOSMER and LEMESHOW · Applied Logistic Regression, *Second Edition*
HOSMER and LEMESHOW · Applied Survival Analysis: Regression Modeling of Time to Event Data
† HUBER · Robust Statistics
HUBERTY · Applied Discriminant Analysis
HUNT and KENNEDY · Financial Derivatives in Theory and Practice
HUSKOVA, BERAN, and DUPAC · Collected Works of Jaroslav Hajek—with Commentary
HUZURBAZAR · Flowgraph Models for Multistate Time-to-Event Data
IMAN and CONOVER · A Modern Approach to Statistics
† JACKSON · A User's Guide to Principle Components
JOHN · Statistical Methods in Engineering and Quality Assurance
JOHNSON · Multivariate Statistical Simulation
JOHNSON and BALAKRISHNAN · Advances in the Theory and Practice of Statistics: A Volume in Honor of Samuel Kotz
JOHNSON and BHATTACHARYYA · Statistics: Principles and Methods, *Fifth Edition*

*Now available in a lower priced paperback edition in the Wiley Classics Library.
†Now available in a lower priced paperback edition in the Wiley–Interscience Paperback Series.

JOHNSON and KOTZ · Distributions in Statistics
JOHNSON and KOTZ (editors) · Leading Personalities in Statistical Sciences: From the Seventeenth Century to the Present
JOHNSON, KOTZ, and BALAKRISHNAN · Continuous Univariate Distributions, Volume 1, *Second Edition*
JOHNSON, KOTZ, and BALAKRISHNAN · Continuous Univariate Distributions, Volume 2, *Second Edition*
JOHNSON, KOTZ, and BALAKRISHNAN · Discrete Multivariate Distributions
JOHNSON, KEMP, and KOTZ · Univariate Discrete Distributions, *Third Edition*
JUDGE, GRIFFITHS, HILL, LÜTKEPOHL, and LEE · The Theory and Practice of Econometrics, *Second Edition*
JUREČKOVÁ and SEN · Robust Statistical Procedures: Aymptotics and Interrelations
JUREK and MASON · Operator-Limit Distributions in Probability Theory
KADANE · Bayesian Methods and Ethics in a Clinical Trial Design
KADANE AND SCHUM · A Probabilistic Analysis of the Sacco and Vanzetti Evidence
KALBFLEISCH and PRENTICE · The Statistical Analysis of Failure Time Data, *Second Edition*
KASS and VOS · Geometrical Foundations of Asymptotic Inference
† KAUFMAN and ROUSSEEUW · Finding Groups in Data: An Introduction to Cluster Analysis
KEDEM and FOKIANOS · Regression Models for Time Series Analysis
KENDALL, BARDEN, CARNE, and LE · Shape and Shape Theory
KHURI · Advanced Calculus with Applications in Statistics, *Second Edition*
KHURI, MATHEW, and SINHA · Statistical Tests for Mixed Linear Models
* KISH · Statistical Design for Research
KLEIBER and KOTZ · Statistical Size Distributions in Economics and Actuarial Sciences
KLUGMAN, PANJER, and WILLMOT · Loss Models: From Data to Decisions, *Second Edition*
KLUGMAN, PANJER, and WILLMOT · Solutions Manual to Accompany Loss Models: From Data to Decisions, *Second Edition*
KOTZ, BALAKRISHNAN, and JOHNSON · Continuous Multivariate Distributions, Volume 1, *Second Edition*
KOTZ and JOHNSON (editors) · Encyclopedia of Statistical Sciences: Volumes 1 to 9 with Index
KOTZ and JOHNSON (editors) · Encyclopedia of Statistical Sciences: Supplement Volume
KOTZ, READ, and BANKS (editors) · Encyclopedia of Statistical Sciences: Update Volume 1
KOTZ, READ, and BANKS (editors) · Encyclopedia of Statistical Sciences: Update Volume 2
KOVALENKO, KUZNETZOV, and PEGG · Mathematical Theory of Reliability of Time-Dependent Systems with Practical Applications
LACHIN · Biostatistical Methods: The Assessment of Relative Risks
LAD · Operational Subjective Statistical Methods: A Mathematical, Philosophical, and Historical Introduction
LAMPERTI · Probability: A Survey of the Mathematical Theory, *Second Edition*
LANGE, RYAN, BILLARD, BRILLINGER, CONQUEST, and GREENHOUSE · Case Studies in Biometry
LARSON · Introduction to Probability Theory and Statistical Inference, *Third Edition*
LAWLESS · Statistical Models and Methods for Lifetime Data, *Second Edition*
LAWSON · Statistical Methods in Spatial Epidemiology
LE · Applied Categorical Data Analysis
LE · Applied Survival Analysis

*Now available in a lower priced paperback edition in the Wiley Classics Library.
†Now available in a lower priced paperback edition in the Wiley–Interscience Paperback Series.

LEE and WANG · Statistical Methods for Survival Data Analysis, *Third Edition*
LEPAGE and BILLARD · Exploring the Limits of Bootstrap
LEYLAND and GOLDSTEIN (editors) · Multilevel Modelling of Health Statistics
LIAO · Statistical Group Comparison
LINDVALL · Lectures on the Coupling Method
LINHART and ZUCCHINI · Model Selection
LITTLE and RUBIN · Statistical Analysis with Missing Data, *Second Edition*
LLOYD · The Statistical Analysis of Categorical Data
LOWEN and TEICH · Fractal-Based Point Processes
MAGNUS and NEUDECKER · Matrix Differential Calculus with Applications in Statistics and Econometrics, *Revised Edition*
MALLER and ZHOU · Survival Analysis with Long Term Survivors
MALLOWS · Design, Data, and Analysis by Some Friends of Cuthbert Daniel
MANN, SCHAFER, and SINGPURWALLA · Methods for Statistical Analysis of Reliability and Life Data
MANTON, WOODBURY, and TOLLEY · Statistical Applications Using Fuzzy Sets
MARCHETTE · Random Graphs for Statistical Pattern Recognition
MARDIA and JUPP · Directional Statistics
MASON, GUNST, and HESS · Statistical Design and Analysis of Experiments with Applications to Engineering and Science, *Second Edition*
McCULLOCH and SEARLE · Generalized, Linear, and Mixed Models
McFADDEN · Management of Data in Clinical Trials
* McLACHLAN · Discriminant Analysis and Statistical Pattern Recognition
McLACHLAN, DO, and AMBROISE · Analyzing Microarray Gene Expression Data
McLACHLAN and KRISHNAN · The EM Algorithm and Extensions
McLACHLAN and PEEL · Finite Mixture Models
McNEIL · Epidemiological Research Methods
MEEKER and ESCOBAR · Statistical Methods for Reliability Data
MEERSCHAERT and SCHEFFLER · Limit Distributions for Sums of Independent Random Vectors: Heavy Tails in Theory and Practice
MICKEY, DUNN, and CLARK · Applied Statistics: Analysis of Variance and Regression, *Third Edition*
* MILLER · Survival Analysis, *Second Edition*
MONTGOMERY, PECK, and VINING · Introduction to Linear Regression Analysis, *Third Edition*
MORGENTHALER and TUKEY · Configural Polysampling: A Route to Practical Robustness
MUIRHEAD · Aspects of Multivariate Statistical Theory
MULLER and STOYAN · Comparison Methods for Stochastic Models and Risks
MURRAY · X-STAT 2.0 Statistical Experimentation, Design Data Analysis, and Nonlinear Optimization
MURTHY, XIE, and JIANG · Weibull Models
MYERS and MONTGOMERY · Response Surface Methodology: Process and Product Optimization Using Designed Experiments, *Second Edition*
MYERS, MONTGOMERY, and VINING · Generalized Linear Models. With Applications in Engineering and the Sciences
† NELSON · Accelerated Testing, Statistical Models, Test Plans, and Data Analyses
† NELSON · Applied Life Data Analysis
NEWMAN · Biostatistical Methods in Epidemiology
OCHI · Applied Probability and Stochastic Processes in Engineering and Physical Sciences
OKABE, BOOTS, SUGIHARA, and CHIU · Spatial Tesselations: Concepts and Applications of Voronoi Diagrams, *Second Edition*
OLIVER and SMITH · Influence Diagrams, Belief Nets and Decision Analysis

*Now available in a lower priced paperback edition in the Wiley Classics Library.
†Now available in a lower priced paperback edition in the Wiley–Interscience Paperback Series.

PALTA · Quantitative Methods in Population Health: Extensions of Ordinary Regressions
PANKRATZ · Forecasting with Dynamic Regression Models
PANKRATZ · Forecasting with Univariate Box-Jenkins Models: Concepts and Cases
* PARZEN · Modern Probability Theory and Its Applications
PEÑA, TIAO, and TSAY · A Course in Time Series Analysis
PIANTADOSI · Clinical Trials: A Methodologic Perspective
PORT · Theoretical Probability for Applications
POURAHMADI · Foundations of Time Series Analysis and Prediction Theory
PRESS · Bayesian Statistics: Principles, Models, and Applications
PRESS · Subjective and Objective Bayesian Statistics, *Second Edition*
PRESS and TANUR · The Subjectivity of Scientists and the Bayesian Approach
PUKELSHEIM · Optimal Experimental Design
PURI, VILAPLANA, and WERTZ · New Perspectives in Theoretical and Applied Statistics
† PUTERMAN · Markov Decision Processes: Discrete Stochastic Dynamic Programming
QIU · Image Processing and Jump Regression Analysis
* RAO · Linear Statistical Inference and Its Applications, *Second Edition*
RAUSAND and HØYLAND · System Reliability Theory: Models, Statistical Methods, and Applications, *Second Edition*
RENCHER · Linear Models in Statistics
RENCHER · Methods of Multivariate Analysis, *Second Edition*
RENCHER · Multivariate Statistical Inference with Applications
* RIPLEY · Spatial Statistics
RIPLEY · Stochastic Simulation
ROBINSON · Practical Strategies for Experimenting
ROHATGI and SALEH · An Introduction to Probability and Statistics, *Second Edition*
ROLSKI, SCHMIDLI, SCHMIDT, and TEUGELS · Stochastic Processes for Insurance and Finance
ROSENBERGER and LACHIN · Randomization in Clinical Trials: Theory and Practice
ROSS · Introduction to Probability and Statistics for Engineers and Scientists
† ROUSSEEUW and LEROY · Robust Regression and Outlier Detection
* RUBIN · Multiple Imputation for Nonresponse in Surveys
RUBINSTEIN · Simulation and the Monte Carlo Method
RUBINSTEIN and MELAMED · Modern Simulation and Modeling
RYAN · Modern Regression Methods
RYAN · Statistical Methods for Quality Improvement, *Second Edition*
SALEH · Theory of Preliminary Test and Stein-Type Estimation with Applications
* SCHEFFE · The Analysis of Variance
SCHIMEK · Smoothing and Regression: Approaches, Computation, and Application
SCHOTT · Matrix Analysis for Statistics, *Second Edition*
SCHOUTENS · Levy Processes in Finance: Pricing Financial Derivatives
SCHUSS · Theory and Applications of Stochastic Differential Equations
SCOTT · Multivariate Density Estimation: Theory, Practice, and Visualization
SEARLE · Linear Models for Unbalanced Data
SEARLE · Matrix Algebra Useful for Statistics
SEARLE, CASELLA, and McCULLOCH · Variance Components
SEARLE and WILLETT · Matrix Algebra for Applied Economics
SEBER and LEE · Linear Regression Analysis, *Second Edition*
† SEBER · Multivariate Observations
† SEBER and WILD · Nonlinear Regression
SENNOTT · Stochastic Dynamic Programming and the Control of Queueing Systems
* SERFLING · Approximation Theorems of Mathematical Statistics
SHAFER and VOVK · Probability and Finance: It's Only a Game!

*Now available in a lower priced paperback edition in the Wiley Classics Library.
†Now available in a lower priced paperback edition in the Wiley–Interscience Paperback Series.

SILVAPULLE and SEN · Constrained Statistical Inference: Inequality, Order, and Shape Restrictions
SMALL and McLEISH · Hilbert Space Methods in Probability and Statistical Inference
SRIVASTAVA · Methods of Multivariate Statistics
STAPLETON · Linear Statistical Models
STAUDTE and SHEATHER · Robust Estimation and Testing
STOYAN, KENDALL, and MECKE · Stochastic Geometry and Its Applications, *Second Edition*
STOYAN and STOYAN · Fractals, Random Shapes and Point Fields: Methods of Geometrical Statistics
STYAN · The Collected Papers of T. W. Anderson: 1943–1985
SUTTON, ABRAMS, JONES, SHELDON, and SONG · Methods for Meta-Analysis in Medical Research
TAKEZAWA · Introduction to Nonparametric Regression
TANAKA · Time Series Analysis: Nonstationary and Noninvertible Distribution Theory
THOMPSON · Empirical Model Building
THOMPSON · Sampling, *Second Edition*
THOMPSON · Simulation: A Modeler's Approach
THOMPSON and SEBER · Adaptive Sampling
THOMPSON, WILLIAMS, and FINDLAY · Models for Investors in Real World Markets
TIAO, BISGAARD, HILL, PEÑA, and STIGLER (editors) · Box on Quality and Discovery: with Design, Control, and Robustness
TIERNEY · LISP-STAT: An Object-Oriented Environment for Statistical Computing and Dynamic Graphics
TSAY · Analysis of Financial Time Series, *Second Edition*
UPTON and FINGLETON · Spatial Data Analysis by Example, Volume II: Categorical and Directional Data
VAN BELLE · Statistical Rules of Thumb
VAN BELLE, FISHER, HEAGERTY, and LUMLEY · Biostatistics: A Methodology for the Health Sciences, *Second Edition*
VESTRUP · The Theory of Measures and Integration
VIDAKOVIC · Statistical Modeling by Wavelets
VINOD and REAGLE · Preparing for the Worst: Incorporating Downside Risk in Stock Market Investments
WALLER and GOTWAY · Applied Spatial Statistics for Public Health Data
WEERAHANDI · Generalized Inference in Repeated Measures: Exact Methods in MANOVA and Mixed Models
WEISBERG · Applied Linear Regression, *Third Edition*
WELSH · Aspects of Statistical Inference
WESTFALL and YOUNG · Resampling-Based Multiple Testing: Examples and Methods for *p*-Value Adjustment
WHITTAKER · Graphical Models in Applied Multivariate Statistics
WINKER · Optimization Heuristics in Economics: Applications of Threshold Accepting
WONNACOTT and WONNACOTT · Econometrics, *Second Edition*
WOODING · Planning Pharmaceutical Clinical Trials: Basic Statistical Principles
WOODWORTH · Biostatistics: A Bayesian Introduction
WOOLSON and CLARKE · Statistical Methods for the Analysis of Biomedical Data, *Second Edition*
WU and HAMADA · Experiments: Planning, Analysis, and Parameter Design Optimization
YANG · The Construction Theory of Denumerable Markov Processes
* ZELLNER · An Introduction to Bayesian Inference in Econometrics
ZHOU, OBUCHOWSKI, and McCLISH · Statistical Methods in Diagnostic Medicine

*Now available in a lower priced paperback edition in the Wiley Classics Library.
†Now available in a lower priced paperback edition in the Wiley–Interscience Paperback Series.

Printed and bound by CPI Group (UK) Ltd, Croydon, CR0 4YY
16/09/2026
14963442-0008